EARTH: PORTRAIT OF A PLANET

FOURTH EDITION

EARTH

Portrait of a Planet

FOURTH EDITION

Stephen Marshak

UNIVERSITY OF ILLINOIS

W. W. NORTON & COMPANY

NEW YORK   LONDON

W. W. Norton & Company has been independent since its founding in 1923, when William Warder Norton and Mary D. Herter Norton first published lectures delivered at the People's Institute, the adult education division of New York City's Cooper Union. The firm soon expanded its program beyond the Institute, publishing books by celebrated academics from America and abroad. By mid-century, the two major pillars of Norton's publishing program—trade books and college texts—were firmly established. In the 1950s, the Norton family transferred control of the company to its employees, and today—with a staff of four hundred and a comparable number of trade, college, and professional titles published each year—W. W. Norton & Company stands as the largest and oldest publishing house owned wholly by its employees.

Composition by Precision Graphics
Manufacturing by Transcontinental, Inc.
Illustrations for the Second, Third, and Fourth Editions by Precision Graphics

Editors: Jack Repcheck and Eric Svendsen
Senior project editor: Thomas Foley
Production manager: Christopher Granville
Copy editor: Chris Thillen
Managing editor, college: Marian Johnson
Media editor: Rob Bellinger
Associate editor, emedia: Matthew Freeman
Marketing manager, physical sciences: Stacy Loyal
Photography editor: Trish Marx
Photo researcher: Fay Torresyap
Editorial assistant: Hannah Bachman
See for Yourself spreads designed by Precision Graphics
Developmental editor for the First Edition: Susan Gaustad

ISBN: 978-0-393-93518-9

W. W. Norton & Company, Inc., 500 Fifth Avenue, New York, N.Y. 10110
www.wwnorton.com
W. W. Norton & Company Ltd., Castle House, 75/76 Wells Street, London W1T 3QT

5 6 7 8 9 0

Dedication

To Kathy, David, Emma, and Michelle

Brief Contents

Special Features

What a Geologist Sees

Geology at a Glance

Contents

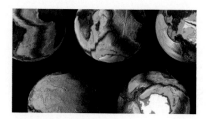

PART II

Earth Materials

CHAPTER 5

Patterns in Nature: Minerals • 104

INTERLUDE A

Rock Groups • 130

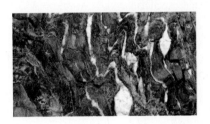

PART III

Tectonic Activity of a Dynamic Planet

CHAPTER 13

A Biography of Earth • 430

PART V
Earth Resources

PART VI

....

Processes and Problems at the Earth's Surface

INTERLUDE F

Ever-Changing Landscapes and the Hydrologic Cycle • 526

CHAPTER 16

Unsafe Ground: Landslides and Other Mass Movements • 540

CHAPTER 17

Streams and Floods: The Geology of Running Water • 568

CHAPTER 18

Restless Realm: Oceans and Coasts • 608

NARRATIVE THEMES

Why do earthquakes, volcanoes, floods, and landslides happen? What causes mountains to rise? How do beautiful landscapes develop? How have climate and life changed through time? When did the Earth form and by what process? Where do we dig to find valuable metals and where do we drill to find oil? Does sea level change? Do continents move? The study of geology addresses these important questions and many more. But from the birth of the discipline, in the late eighteenth century, until the mid-twentieth century, geologists considered each question largely in isolation, without pondering its relation to the others. This approach changed, beginning in the 1960s, in response to the formulation of two "paradigm-shifting" ideas that have unified thinking about the Earth and its features. The first idea, called the *theory of plate tectonics*, states that the Earth's outer shell, rather than being static, consists of discrete plates that slowly move relative to each other so that the map of our planet continuously changes. Plate interactions cause earthquakes and volcanoes, build mountains, provide gases that make up the atmosphere, and affect the distribution of life on Earth. The second idea emphasizes that our planet's water, land, atmosphere, and living inhabitants are dynamically interconnected as the Earth System. Materials constantly cycle among various living and nonliving reservoirs on, above, and within the planet. Thus, we have come to realize that the history of life is intimately linked to the history of the physical Earth.

Earth: Portrait of a Planet, Fourth Edition, is an introduction to the study of our planet that uses the theory of plate tectonics and the concept of the Earth System throughout to weave together a number of narrative themes, including:

1. The solid Earth, the oceans, the atmosphere, and life interact in complex ways, yielding a planet that is unique in the Solar System.

2. Most geologic processes reflect the interactions of plates, which is a constant, but very slow movement.

3. The Earth is a planet, formed like other planets from dust and gas. But, in contrast to other planets, the Earth is a dynamic place where new geologic features continue to form and old ones continue to be destroyed.

4. The Earth is very old—about 4.57 billion years have passed since its birth. During this time, the map of the planet and its surface features have changed, and life has evolved.

5. Internal processes (driven by Earth's internal heat) and external processes (driven by heat from the Sun) interact at the Earth's surface to produce complex landscapes.

6. Natural hazards—earthquakes, volcanoes, landslides, floods—can be studied and understood. In some cases, geologic knowledge can help society avoid or reduce the danger of these hazards.

7. Energy and mineral resources come from the Earth and are formed by geologic phenomena. Geologic study can help locate these resources and mitigate the consequences of their use.

8. Physical features of the Earth are linked to life processes, and vice versa.

9. Science comes from observation; people make scientific discoveries.

10. Geology utilizes ideas from physics, chemistry, and biology, so the study of geology provides an excellent means to improve science literacy overall.

These narrative themes serve as the *take-home message* of the book, a message that students hopefully will remember long after they finish their introductory geology course. In effect, they provide a mental framework on which students can organize and connect ideas, and develop a modern, coherent image of our planet.

PEDAGOGICAL APPROACH

Educational research demonstrates that students learn best when they actively engage with a combination of narrative text and narrative art. Some students respond more to the words of a textbook, which help to organize information, to provide answers to questions, to fill in the essential steps that link ideas together, and to help a student develop a personal context for understanding information. Some students respond more to narrative art—art designed to tell a story—for visual images help students comprehend and remember processes. And some respond to question-and-answer-based active learning, an approach where students can in effect "practice" their knowledge. *Earth: Portrait of a Planet*, Fourth Edition, provides all three of these learning tools. The text has been crafted to be engaging, the art has been configured to tell a story, the chapters are laid out to help students internalize key principles, and the on-line activities have been designed to both engage students and provide active feedback. For example, each chapter starts with a question—a Geopuzzle—that prompts students to think about what they already know, and prepares them to search for additional information as they read the chapter. *Take-Home Message* panels at the end of each section help students solidify key themes before proceeding to the next section. Questions at the end of each chapter not only test basic knowledge, but also stimulate critical thinking. New *SmartWork* online homework helps students prepare with automatic feedback, and visual drag and drop, labeling, and "hot spot" reviews. Finally, *See for Yourself* and *Geotour* features guide students on virtual field trips to spectacular geologic locales around the globe where they can apply their newly acquired knowledge to real-world sites and features.

ORGANIZATION

The topics covered in this book have been arranged so that students can build their knowledge of geology on a foundation of overarching principles. Thus, the book starts with cosmology and the formation of the Earth, and then introduces the architecture of our planet, from surface to center. With this basic background, students are prepared to delve into plate tectonics theory. Plate tectonics appears early in the book, so that students can relate the content of subsequent chapters to the theory. Knowledge of plate tectonics, for example, helps students understand the suite of chapters on minerals, rocks, and the rock cycle. Knowledge of plate tectonics and rocks together, in turn, provides a basis for studying volcanoes, earthquakes, and mountains. And with this background, students are prepared to see how the map of the Earth has changed through the vast expanse of geologic time, and how energy and mineral resources have developed. The book's final chapters address processes and problems occurring at or near the Earth's surface, from the unstable slopes of hills, down the course of rivers, to the shores of the sea and beyond. This section concludes with a topic of growing concern in society—global change, particularly climate change.

Although the sequence of chapters was chosen for a reason, this book is designed to be flexible enough for instructors to choose their own strategies for teaching geology. Geology is a nonlinear subject, in that individual topics are so interrelated that there is not always a single best way to order them. Thus, each chapter is self-contained, reiterating relevant material where necessary.

SPECIAL FEATURES

Earth: Portrait of a Planet, Fourth Edition contains a number of features that distinguish it from competing texts. Several of which are new to this edition.

Narrative Art and *"What a Geologist Sees"*

It's hard to understand features of the Earth System without being able to see them. To help students visualize topics, this book is lavishly illustrated, with figures that attempt to give a realistic context for a geologic feature without overwhelming students with extraneous detail. The talented artists who worked on the book have used the latest computer graphics software, resulting in the most sophisticated pedagogical art ever provided by a geoscience text. In addition to artist-produced drawings, the book provides abundant photographs from around the world, many of which were taken by the author. Where appropriate, photographs are accompanied by annotated sketches labeled *"What a Geologist Sees,"* which help students to be certain that they see the features that the photo was intended to show.

In this edition, drawings and photographs have been integrated into *narrative art* that has been laid out, labeled, and annotated to tell a story—the figures are drawn to teach! Subcaptions are positioned adjacent to the relevant parts of a figure, labels point out key features, and balloons provide important annotation. The arrangement of subparts has been designed to convey time progression, where relevant. Color schemes in drawings have been tied to those of relevant photos, so that students can easily visualize the relationships between drawings and photos.

Featured Paintings: *Geology at a Glance*

In addition to individual figures, British artist Gary Hincks has created spectacular two-page annotated paintings for each chapter, called *Geology at a Glance*. These paintings integrate key concepts introduced in the chapters and visually emphasize the relationships between components of the Earth System. They provide students a way to review a subject . . . at a glance.

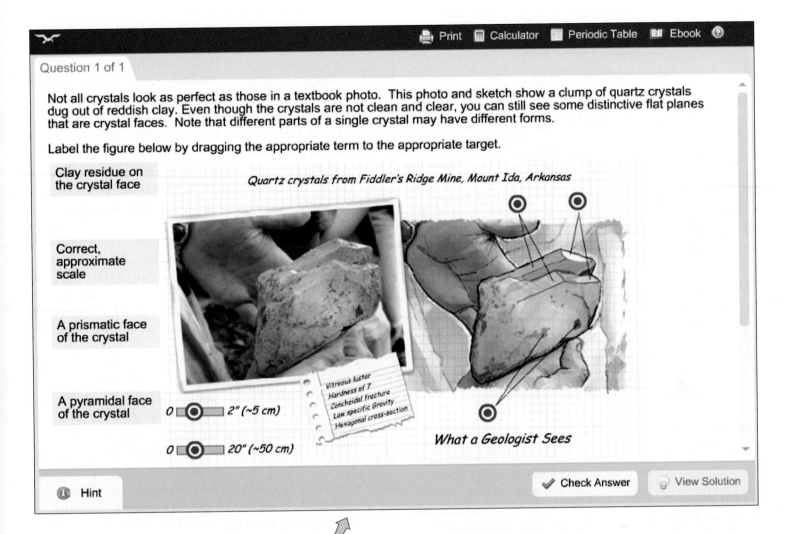

New *SmartWork* Online Homework for Geology

Designed and developed with the eye and experience of Stephen Marshak, *SmartWork* online homework features highly visual questions that provide students with just-in-time, answer-specific feedback. Students get the coaching they need to work through assignments; instructors get real-time assessment of student progress with automatic grading and item analysis. Image-based drag-and-drop and hot spot questions make use of carefully developed images, plus additional *What a Geologist Sees* figures created just for *SmartWork*. Additional types of conceptual questions challenge students to apply their understanding of the important concepts identified by the author in each chapter's *Take-Home Messages*. *SmartWork* also contains questions without feedback, including Reading Quizzes, *Geotour*-guided inquiry activities, and approximately 1,000 quiz questions from StudySpace.

Visit *See for Yourself* Sites and Take Geotours with *Google Earth*™

Through the magic of *Google Earth*™, *Earth: Portrait of a Planet*, Fourth Edition, provides a comprehensive set of active-learning Geotours that take students on virtual field trips to see outstanding examples of geology from around the world. Keyed to topics in the text, a downloadable file of *See for Yourself* sites for instructors and students automatically loads the sites at the click of a mouse. Students and instructors can also access Geotours, which include many incredible add-ons (developed by Scott and Beth Wilkerson of DePauw University), such as map overlays, outcrop photos, and geologic cross sections. A *Geotours Workbook*, prepared by Scott and Beth Wilkerson, with Stephen Marshak, provides assignable, active-learning exercises for each tour, challenging students to fly around sites and make specific observations and measurements in context. Self-grading, multiple-choice worksheets are available within *SmartWork* as well as within Norton's coursepacks.

Take-Home Messages, Geopuzzles, and On Further Thought Questions

Each chapter begins with a question, a *Geopuzzle*, that challenges students to think about what they might already know or not know about the theme of the chapter. Such prompting helps students to be on the lookout for key ideas as they read. Every chapter section ends with a *Take-Home Message*, a brief summary that helps students identify and remember the highlight of the section before moving on to the next. Take-Home messages have been used as the guiding principles for the creation of question content in *SmartWork*. Each chapter then concludes with *On Further Thought* questions that invite students to think beyond the basics.

Interludes

The book contains several interludes. These, in effect, are "mini-chapters" which focus on topics that are self-contained but are not broad enough to require an entire chapter. The feature helps keep chapters reasonable in length, and provides flexibility in sequencing topics within a course.

Societal Issues

Geology's practical applications are addressed in several chapters. Students will learn about such topics as energy resources, mineral resources, global change, and mass wasting. Further, chapters on earthquakes, volcanoes, and landscapes highlight geologic hazards. And students are encouraged to apply their geologic understanding to environmental issues, where relevant. *Science and Society* features on the free and open student website further challenge students to apply material learned in class to the interpretation of news articles and publicly available geologic data.

CHANGES IN THE FOURTH EDITION

The newest edition of *Earth: Portrait of a Planet* is not simply a cosmetic modification of the Third Edition. The text has been thoughtfully modified and thoroughly updated. Key changes include:

- *Narrative art*: Most of the art in the book has been reconfigured to produce narrative pieces that integrate drawings, photos, labels, and annotations in order to tell a story. Thus, students have two complete learning options in the book—the narrative text (written to engage), and the narrative art (drawn to teach).

- *Climate change coverage updated significantly*: Content on global warming has been updated to reflect the latest developments in the science. Material on climate change is presented alongside other kinds of natural and anthropogenic (human-caused) global change, providing students with the scientific context behind this hot-button issue and helping them to become more aware global citizens.

- *Keeping the book current*: To ensure that *Earth: Portrait of a Planet*, Fourth Edition reflects the latest research discoveries and helps students understand the geologic events that have been featured in news headlines, the author has updated many topics throughout the book. For example, this edition provides new, insightful treatment of the catastrophic earthquakes in Haiti, Japan, and New Zealand, discussion of the massive tornado outbreak of 2011, and critical sustainability issues related to our increasing reliance on rare earth elements.

- *New pedagogical tools*: This edition provides new components to help students actively retain what they learn. Each chapter begins with a *Geopuzzle*, each chapter section ends with an expanded *Take-Home Message*, and each chapter ends with critical thinking questions entitled *On Further Thought*. New "bubble questions" relate the discussion to sudent experience.

- *Coverage of planetary geology and geobiology:* This realm of study, examined by geoscientists, has expanded in recent years to include microbial life and its role in the Earth System as well as on the surfaces of other planets. Discussions of geomicrobiology and planetary geology are incorporated throughout the book so that students can understand how these topics relate to other, more traditional topics.

SUPPLEMENTARY MATERIALS

For Students and Instructors . . .

New *SmartWork*. The new *SmartWork* online assessment available for use with *Earth: Portrait of a Planet*, Fourth Edition, features highly visual questions with immediate, answer-specific feedback. Because students learn best when they receive immediate feedback, and when they can interact with art as well as with text, *SmartWork* includes drag-and-drop figure-based questions and *What a Geologist Sees* photo interpretations, all designed with the experience and perspective of Stephen Marshak. *SmartWork* further provides questions based on real field examples, using *Google Earth*™ and the Geotour Work-book, and helps students check their knowledge as they go by working with reading-based questions. Designed to be intui-tive and easy to use (for students and professors), *SmartWork* makes it a snap to assign, assess, and report on student perfor-mance, and keep the class on track.

Updated Geotours and The *Google Earth*™ Geotour Workbook. Created by Scott and Beth Wilkerson, Geo-tours are active-learning tours that take students on virtual field trips to see outstanding examples of geology around the world. Arranged by topic, questions for these Geotours have been revised for auto grading, and are available as Worksheets in print format (that come free with the book and includee complete user instructions and advanced instruction), or elec-tronically with automatic grading through *SmartWork* or your campus LMS. Request a sample copy to preview each worksheet.

See for Yourself *Google Earth*™ sample site file. Users who simply want access to sample field sites (for classroom pre-sentations or distribution to students) can download the sites from the book's *See for Yourself* appendix. This single download is available at both the Norton instructor download site, as well as from wwnorton.com/studyspace.

Two types of PowerPoints.

- *Enhanced Art*—Designed for instant classroom use, these slides optimize all photographs and line art from the book. All figures have been optimized for use in the PowerPoint™ environment. The art has been relabelled and resized for easier use in the classroom. Supplemen-tal photographs, provided by the author, are integrated into the collection and appear in context.
- *Lecture Bullets*—These slides include art and bulleted text for use in both lectures and student hand-outs.

Animations and videos. The book's accompanying Instructor Resource DVD provides a rich collection of over 60 flash animations to illustrate geologic processes. In addition, the StudySpace Student website contains four streaming, 3-D cinema-quality animations, along with 24 streaming film clips from the U.S. Geological Survey.

Zoomable art. Explore Gary Hincks's spectacular synoptic paintings in vivid detail using the zoomable art feature. Ideal for lecture use, these figures are included among the resources on the Instructor's DVD.

Ebook—nortonebooks.com. *Earth: Portrait of a Planet,* Fourth Edition, is also available in an ebook format that repli-cates actual book pages and provides links to the animations on StudySpace. Students can highlight text, use sticky notes, and work with zoomable images from the book.

***StudySpace*—wwnorton.com/studyspace.** Free and open for students, the thoroughly revised *StudySpace* offers students assignment-driven study plans for each chapter. Materials include Quiz+ diagnostic quizzes (also available in *SmartWork* and the coursepack), *Science and Society* features that challenge students to use course concepts in analyzing news articles and real-time geological data, plus animations, streaming video, reading guides, vocabulary flashcards, and a geology in the news feed.

For Instructors Only . . .

Norton Media Library Instructor's DVD-ROM. The instructor DVD offers a wealth of easy-to-use multi-media resources, all structured around the text and designed for use in lecture presentations, including:

- Fully *Editable Enhanced Art PowerPoints*™: These pre-sentations take the book's art and photographs and places them in ready-to-use *PowerPoints*™. Images are properly sized, related images are provided on the same slide or in sequenced slides, and labeling has been streamlined so it doesn't clutter the image. Each slide also includes a brief title, so it can be a stand-alone teaching or study aid. Significantly, each figure part can be easily copied and transferred to an instructor's personal *PowerPoint*™ presentation.
- *Editable PowerPoint*™ *Lecture Bullets*: These presen-tations, provided by Ron Parker of Earlham College, integrate illustrations from the book, supplemental slides, and text, into a format that can provide a basis for developing complete lectures.
- *Easy-to-use JPEGs*: All of the art and photographs from the text in JPEG format.
- *Flash animations*: The book's DVD and website pro-vide over 60 flash animations that illustrate geologic processes.
- *Cinema-quality animations*: Four animations (Forma-tion of the Earth; Plate Tectonics; Glacial Advance and Retreat; and Meandering Channel) have been developed for this book. These animations are provided as *Quicktime*™ movies on the book's DVD.

■ *Zoomable art*: The DVD and website provide zoomable versions of the synoptic paintings by Gary Hincks. These allow instructors to zoom in closely to parts of the figure, in the context of the whole figure, during class presentations.

Instructor's Manual and Test Bank. The Instructor's Manual and Test Bank prepared by John Werner of Seminole State College of Florida; Jacalyn Gorczynski of Texas A & M, Corpus Christi; Haley Wasson of Texas A & M, College Station; and Daniel Wynne of Sacramento Community College are designed to help instructors prepare lectures and exams. The Instructor's Manual contains detailed Learning Objectives, Chapters Summaries, and complete answers to the end-of-chapter Review and *On Further Thought* questions. The Test Bank has been developed using the Norton Assessment Guidelines and each chapter of the Test Bank consists of three question types classified according to Norton's taxonomy of knowledge types, as well as being classified by section and difficulty, making it easy to construct tests and quizzes that are meaningful and diagnostic. Available in print paperback, CD-ROM, and downloadable from wwnorton.com/instructors.

Instructor's Website—wwnorton.com/instructors. Online access to a rich array of resources: test bank, instructor's manual, Powerpoints, *Google Earth*™ File of sites from the text, art from the text and WebCT- and Blackboard-ready content.

Coursepacks Available at no cost to professors or students, Norton coursepacks bring high-quality Norton digital media into a new or existing online course. For *Earth: Portrait of a Planet,* Fourth Edition, content includes:

■ Test Bank
■ Reading Quizzes
■ Quiz+ questions
■ GeoTours questions
■ Guides to Reading
■ Links to the animations, streaming video, and ebook.

ACKNOWLEDGMENTS

I am very grateful for the assistance of many people in bringing this book from the concept stage to the shelf in the first place, and for helping to provide the momentum needed to make this revision take shape.

First and foremost, I wish to thank my family, who have adopted "the book" as a member of the household and have tolerated the overabundance of photo stops on family trips.

My wife, Kathy, carried out the immense task of merging changes completed for *Essentials of Geology*, Third Edition into the manuscript for *Earth: Portrait of a Planet*, Fourth Edition, and she flagged points that had been addressed by reviewers. She also edited new text, rationalized the art list, and cross-checked the many sets of proofs. Without her efforts, creation of this edition would not have been possible. My daughter, Emma, helped develop the concept of narrative art used in the book, updated the supplemental slide set, served as scale in photos, and provided feedback about how the book works from a student's perspective. My son, David, highlighted places where the writing could be improved, helped me to keep the project in perspective, and also served as scale in photographs.

I am grateful to all the staff of W. W. Norton & Company for their incredible efforts during the development of my books over the past two decades. It has been a privilege to work with a company that is willing to work so closely with its authors. In particular, I would like to thank Jack Repcheck, my editor for the past ten years. Jack has been an understanding friend and a fountain of sage advice throughout the book's evolution. He has suggested many of the innovations that have strengthened the book, and his instincts about what works have brought the book to the attention of a wider geological community than I ever thought possible. Eric Svendsen has now taken over editorial responsibility for *Earth: Portrait of a Planet*, and has already injected new enthusiasm and ideas into the project. His experience and skill have made this transition easy and fun. Thom Foley continues to do an amazing job as the production director for the book, somehow keeping track of all the drafts, the changes, and the figures for a lengthy and complicated manuscript, all while remaining incredibly calm. It's thanks to Thom that everything somehow manages to get done, and that mistakes are few and far between. I also greatly appreciate the efforts of Chris Granville for coordinating the back-and-forth between the publisher and various suppliers. Trish Marx, Fay Torresyap, and Juneoire Mitchell did a fantastic job with the Herculean task of finding, organizing, and crediting all of the photographs, right up to the last days before the book went to press. Trish's supervision has brought the management of the photo collection into the twenty-first century and has greatly streamlined the selection process. My sincere thanks also go to Rob Bellinger, for his innovative approach to ancillary development and for overseeing the development of the *SmartWork* supplement, and to Stacy Loyal, who has so ably assumed the mantle of sales manager for the book. Susan Gaustad, the outstanding developmental editor of the first edition, helped to set the style of the book and to weed out the errors that otherwise might burden a new edition. I also appreciate the work of copyeditor Chris Thillen who has kept the tradition going with her efforts on this Fourth Edition.

Production of the illustrations has involved many people over many years. I am indebted to the staff of Precision Graphics who have helped to create the style of the figures and have accommodated countless changes and tweaks without complaint. Stan Maddock and Becky Oles have been at the core of this effort, and I am forever thankful for their hard work. Jon Prince and Jeff Griffin creatively programmed the animations under the careful supervision of Andrew Troutt. For the Fourth Edition, Kristina Seymour at Precision Graphics has done a wonderful job of coordinating artwork and composition, and Stacy McDade has creatively handled page layout. Andrew Troutt, Rebecca Reid, and Amanada Bickel have creatively transformed my approach to *PowerPoint*™ presentations into reality. And, thanks once again to Ron Parker of Fronterra Geosciences for creating and revising the lecture PowerPoints that accompany this book.

It has been great fun to interact with Gary Hincks, who painted the incredible two-page spreads, in part using his own designs and geologic insights. Some of Gary's paintings originally appeared in *Earth Story* (BBC Worldwide, 1998) and were based on illustrations jointly conceived by Simon Lamb and Felicity Maxwell, working with Gary. Others were developed specifically for *Earth: Portrait of a Planet* or for *Essentials of Geology*.

Some of the chapter-opening quotes were found in *Language of the Earth*, compiled by F. T. Rhodes and R. O. Stone (Pergamon, 1981). During the initial development of the First Edition, I greatly benefited from discussions with Philip Sandberg, and during later stages in the development of the First Edition, Donald Prothero contributed text, editorial comments, and end-of-chapter material.

The four editions of this book and of its cousin, *Essentials of Geology*, have benefited greatly from input by expert reviewers for specific chapters, by new general reviewers of the entire book, and by comments from faculty and students who have used the book and were kind enough to contact me or the publisher. These include:

Jack C. Allen, Bucknell University

David W. Anderson, San Jose State University

Martin Appold, University of Missouri-Columbia

Philip Astwood, University of South Carolina

Eric Baer, Highline University

Victor Baker, University of Arizona

Julie Baldwin, University of Montana

Sandra Barr, Acadia University

Keith Bell, Carleton University

Mary Lou Bevier, University of British Columbia

Jim Black, Tarrant County College

Daniel Blake, University of Illinois

Ted Bornhorst, Michigan Technological University

Michael Bradley, Eastern Michigan University

Mike Branney, University of Leicester, UK

Sam Browning, Massachusetts Institute of Technology

Bill Buhay, University of Winnipeg

Rachel Burks, Towson University

Peter Burns, University of Notre Dame

Katherine Cashman, University of Oregon

George S. Clark, University of Manitoba

Kevin Cole, Grand Valley State University

Patrick M. Colgan, Northeastern University

Peter Copeland, University of Houston

John W. Creasy, Bates College

Norbert Cygan, Chevron Oil, retired

Michael Dalman, Blinn College

Peter DeCelles, University of Arizona

Carlos Dengo, ExxonMobil Exploration Company

John Dewey, University of California, Davis

Charles Dimmick, Central Connecticut State University

Robert T. Dodd, Stony Brook University

Missy Eppes, University of North Carolina, Charlotte

Eric Essene, University of Michigan

James E. Evans, Bowling Green State University

Susan Everett, University of Michigan, Dearborn

Dori Farthing, State University of New York, Geneseo

Grant Ferguson, St. Francis Xavier University

Eric Ferré, Southern Illinois University

Leon Follmer, Illinois Geological Survey

Nels Forman, University of North Dakota

Bruce Fouke, University of Illinois

David Furbish, Vanderbilt University

Steve Gao, University of Missouri

Grant Garvin, John Hopkins University

Christopher Geiss, Trinity College, Connecticut

Gayle Gleason, State University of New York, Cortland

Cyrena Goodrich, Kingsborough Community College

William D. Gosnold, University of North Dakota

Lisa Greer, William & Mary College

Steve Guggenheim, University of Illinois, Chicago

Henry Halls, University of Toronto, Mississuaga

Bryce M. Hand, Syracuse University

Anders Hellstrom, Stockholm University

Tom Henyey, University of South Carolina

James Hinthorne, University of Texas, Pan American

Paul Hoffman, Harvard University

Curtis Hollabaugh, University of West Georgia

Bernie Housen, Western Washington University

Mary Hubbard, Kansas State University

Paul Hudak, University of North Texas

Warren Huff, University of Cincinnati

Neal Iverson, Iowa State University

Charles Jones, University of Pittsburgh

Donna M. Jurdy, Northwestern University

Thomas Juster, University of Southern Florida

H. Karlsson, Texas Tech

Daniel Karner, Sonoma State University

Dennis Kent, Lamont Doherty/Rutgers

Charles Kerton, Iowa State University

Susan Kieffer, University of Illinois

Jeffrey Knott, California State University, Fullerton

Ulrich Kruse, University of Illinois

Robert S. Kuhlman, Montgomery County Community College

Lee Kump, Pennsylvania State University

David R. Lageson, Montana State University

Robert Lawrence, Oregon State University

Scott Lockert, Bluefield Holdings

Leland Timothy Long, Georgia Tech

Craig Lundstrom, University of Illinois

John A. Madsen, University of Delaware

Jerry Magloughlin, Colorado State University

Jennifer McGuire, Texas A&M University

Judy McIlrath, University of South Florida

Paul Meijer, Utrecht University, Netherlands

Jamie Dustin Mitchem, California University of Pennsylvania

Alan Mix, Oregon State University

Otto Muller, Alfred University

Kathy Nagy, University of Illinois, Chicago

Pamela Nelson, Glendale Community College

Robert Nowack, Purdue University

Charlie Onasch, Bowling Green State University

David Osleger, University of California, Davis

Eric Peterson, Illinois State University

Ginny Peterson, Grand Valley State University

Stephen Piercey, Laurentian University

Adrian Pittari, University of Waikato, New Zealand

Lisa M. Pratt, Indiana University

Mark Ragan, University of Iowa

Robert Rauber, University of Illinois

Bob Reynolds, Central Oregon Community College

Joshua J. Roering, University of Oregon

Eric Sandvol, University of Missouri

William E. Sanford, Colorado State University

Jeffrey Schaffer, Napa Valley Community College

Roy Schlische, Rutgers University

Sahlemedhin Sertsu, Bowie State University

Anne Sheehan, University of Colorado

Roger D. Shew, University of North Carolina, Wilmington

Doug Shakel, Pima Community College

Norma Small-Warren, Howard University

Donny Smoak, University of South Florida

David Sparks, Texas A&M University

Angela Speck, University of Missouri

Tim Stark, University of Illinois

Seth Stein, Northwestern University

David Stetty, Jacksonville State University

Kevin G. Stewart, University of North Carolina, Chapel Hill

Michael Stewart, University of Illinois

Don Stierman, University of Toledo

Gina Marie Seegers Szablewski, University of Wisconsin, Milwaukee

Barbara Tewksbury, Hamilton College

Thomas M. Tharp, Purdue University

Kathryn Thornbjarnarson, San Diego State University

Basil Tikoff, University of Wisconsin

Spencer Titley, University of Arizona

Robert T. Todd, Stony Brook University

Torbjörn Törnqvist, University of Illinois, Chicago

Jon Tso, Radford University

Stacey Verardo, George Mason University

Barry Weaver, University of Oklahoma

John Werner, Seminole State College of Florida

Alan Whittington, University of Missouri

John Wickham, University of Texas, Arlington

Lorraine Wolf, Auburn University

Christopher J. Woltemade, Shippensburg University

I apologize if I inadvertently left anyone off this list.

ABOUT THE AUTHOR

Stephen Marshak is Professor of Geology at the University of Illinois, Urbana-Champaign, where he is also the Director of the School of Earth, Society, and Environment. He holds an A.B. from Cornell University, an M.S. from the University of Arizona, and a Ph.D. from Columbia University. Steve's research interests lie in structural geology and tectonics. This work has taken him into the field on a number of continents. Steve loves teaching and has won his college's and university's highest teaching awards. In addition to research papers and *Earth: Portrait of a Planet*, Steve has authored *Essentials of Geology*; and has co-authored *Laboratory Manual for Introductory Geology; Earth Structure: An Introduction to Structural Geology and Tectonics*; and *Basic Methods of Structural Geology*.

THANKS!

I am very grateful to the faculty who selected earlier editions for their classes and to the students who engaged so energetically with the book. I particularly appreciate the comments from readers that helped to make improvements in this Fourth Edition. I continue to welcome your comments and can be reached at smarshak@illinois.edu.

Stephen Marshak

*To see the world in a grain of sand
and heaven in a wild flower.
To hold infinity in the palm of the hand
and eternity in an hour.*

—William Blake (British poet, 1757–1827)

And Just What Is Geology?

Students can see the Earth System at a glance on a sea cliff along the coast of Ireland. Here, sunlight, air, water, rock, and life all interact to produce a complex and fascinating landscape.

GEOPUZZLE

Tourists might look at this photo and see a beautiful view. What do geologists think about when they see this landscape?

Civilization exists by geological consent, subject to change without notice.

—Will Durant (1885–1981)

P.1 IN SEARCH OF IDEAS

Our C-130 Hercules transport plane rose from a smooth ice runway on the frozen sea surface at McMurdo Station, Antarctica, and headed south to spend a month studying

A plane dropping geologists on a snowfield

unusual rocks exposed on a cliff about 250 km away. We climbed past the smoking summit of Mt. Erebus, Earth's southernmost volcano, and for the next hour flew along the Transantarctic Mountains, a ridge of rock that divides the continent into two parts, East Antarctica and West Antarctica (**Fig. P.1**). Glaciers—sheets or rivers of ice that last all year—cover almost all of Antarctica. To the right of the plane we could see the continental glacier, a vast sheet of ice thousands of kilometers across and up to 4.7 km (15,000 ft) thick, that covers East Antarctica. The surface of this ice sheet forms a frigid high plain called the Polar Plateau. To the left we could see numerous valley glaciers, rivers of ice, that slowly carry ice from the Polar Plateau through gaps in the Transantarctic Mountains down to the frozen Ross Sea.

Suddenly, we heard the engines slow. As the plane descended, it lowered its ski-equipped landing gear. The loadmaster shouted an abbreviated reminder of the emergency alarm code: "If you hear three short blasts of the siren, hold on for dear life!" The plane touched the surface of our first

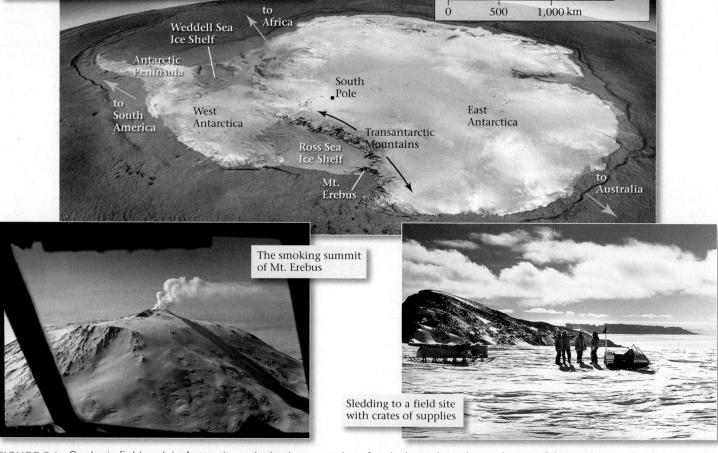

The smoking summit of Mt. Erebus

Sledding to a field site with crates of supplies

FIGURE P.1 Geologic fieldwork in Antarctica unlocks the mysteries of an icebound continent. A map of Antarctica emphasizes that the Transantarctic Mountains separate West Antarctica from East Antarctica.

FIGURE P.2 Geologists in the field, discovering Earth's history.

(a) Exploring an outcrop hidden in the woods of Ontario, Canada.

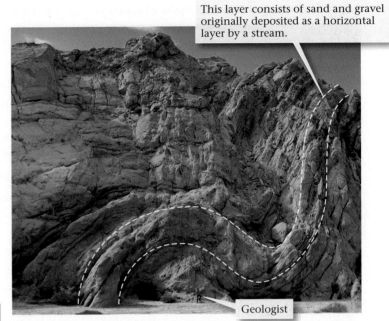

This layer consists of sand and gravel originally deposited as a horizontal layer by a stream.

Geologist

(b) Measuring contorted layers of rock near the San Andreas fault in California.

choice for a landing spot, the ice at the base of the rock cliff we wanted to study. *Wham, wham, wham, wham!!!!* As the plane's skis slammed into frozen snowdrifts on the ice surface at about 180 km an hour, it seemed as though a fairy-tale giant was shaking the plane. Seconds later, the landing aborted, we were airborne again, looking for a softer runway above the cliff. Finally, we landed in a field of deep snow, unloaded, and bade farewell to the plane. When the plane passed beyond the horizon, the silence of Antarctica hit us—no trees rustled, no dogs barked, and no traffic rumbled in this stark land of black rock and white ice. It would take us a day and a half to haul our sleds of food and equipment down to our study site (see Fig. P.1). All this to look at a few dumb rocks?

Geologists—scientists who study the Earth—explore remote regions like Antarctica almost routinely. Such efforts often strike people in other professions as a strange way to make a living, as implied by Scottish poet Walter Scott's (1771–1832) description of geologists at work: "Some rin uphill and down dale, knappin' the chucky stones to pieces like sa' many roadmakers run daft. They say it is to see how the warld was made!"

Indeed—to see how the world was made, to see how it continues to evolve, to find its valuable resources, to prevent contamination of its waters and soils, and to predict its dangerous movements. That is why geologists spend months at sea drilling holes in the ocean floor, why they scale mountains, camp in rain-drenched jungles, and trudge through scorching desert winds (**Fig. P.2a, b**). That is why geologists use electron microscopes to examine the atomic structure of minerals, use mass spectrometers to measure the composition of rock and water, and use supercomputers to model the paths of earthquake waves. For over two centuries, geologists have pored over the Earth in search of ideas to explain the processes that form and change our planet.

P.2 THE NATURE OF GEOLOGY

Geology, or geoscience, is the study of the Earth. Not only do geologists address academic questions such as the formation and composition of the Earth, the causes of earthquakes and ice ages, and the evolution of life, but they also address practical problems such as how to keep pollution out of groundwater, how to find oil and minerals, and how to avoid landslides. And in recent years, geologists have made significant contributions to the study of global climate change. When news reports begin with "Scientists say . . ." and then continue with "an earthquake occurred today off Japan," or "landslides will threaten the city," or "chemicals from the proposed toxic waste dump will ruin the town's water supply," or "there's only a limited supply of oil left," the scientists referred to are geologists. Because geologists address so many different kinds of problems, it's convenient to divide geology into many different specialities. Table P.1 lists some of the many subdisciplines of geology.

Thousands of geologists enjoy careers in oil, mining, water, engineering, and environmental companies, while a smaller number work in universities, government geological surveys, and research laboratories. Nevertheless, since most of the students reading this book will not become professional geologists, it's fair to ask the question, "Why should people, in general, study geology?"

First, geology may be one of the most practical subjects you can learn. Ask yourself the following questions, and you'll realize that geologic processes, phenomena, and materials play major roles in daily life:

TABLE P.1 **Principal Subdisciplines of Geology (Geoscience)**

Name	Subject of Study
Engineering geology	The stability of geologic materials at the Earth's surface, for such purposes as controlling landslides and building tunnels, dams, mines, roads, or foundations
Environmental geology	Interactions between the environment and geologic materials, and the contamination of geologic materials
Geochemistry	Chemical compositions of materials in the Earth and chemical reactions in the natural environment
Geochronology	The age (in years) of geologic materials, the Earth, and extraterrestrial objects
Geomorphology	Landscape formation and evolution
Geophysics	Physical characteristics of the whole Earth (such as Earth's magnetic field and gravity field) and of forces in the Earth
Hydrogeology	Groundwater, its movement, and its reaction with rock and soil
Mineralogy	The chemistry and physical properties of minerals
Paleontology	Fossils and the evolution of life as preserved in the rock record
Petrology	Rocks and their formation
Sedimentology	Sediments and their deposition
Seismology	Earthquakes and the Earth's interior as revealed by earthquake waves
Stratigraphy	The succession of sedimentary rock layers
Structural geology	Rock deformation (bending and breaking) in response to the application of force
Tectonics	Regional geologic features (such as mountain belts) and plate movements and their consequences
Volcanology	Volcanic eruptions and their products

- Do you live in a region threatened by landslides, volcanoes, earthquakes, or floods (Fig. P.3)?
- Are you worried about the price of energy or about whether there will be a war in an oil-supplying country?
- Do you ever wonder about where the copper in your home's wires comes from?
- Have you seen fields of green crops surrounded by desert and wondered where the irrigation water comes from?
- Would you like to buy a dream house on a beach or near a river?

Addressing these questions requires a basic understanding of geology. Your knowledge of geology may help you avoid building your home on a hazardous floodplain or fault zone, on an unstable slope, or along a rapidly eroding coast. With a basic understanding of groundwater, you may be able to find a good site for an irrigation well. With knowledge of the geologic controls on resource distribution, you may be able to invest more wisely in the resource industry or identify more sensible political policies.

Second, the study of geology gives you a perspective of the planet that no other field can. As you will see, the Earth is a complicated entity, where living organisms, oceans, atmosphere, and solid rock interact with one another in a great variety of ways. Geologic study reveals Earth's antiquity (it's about 4.57 *billion* years old) and demonstrates how the planet has

FIGURE P.3 Human-made cities cannot withstand the vibrations of a large earthquake. These apartment buildings collapsed during an earthquake in Turkey.

changed profoundly during its existence. What our ancestors considered to be the center of the Universe has become, with the development of geologic perspective, our "island in space" today. And what was believed to be an unchanging orb originating at the same time as humanity has become a dynamic planet that existed long before people did.

Third, the study of geology puts the accomplishments and consequences of human civilization in a broader context. View the aftermath of a large earthquake, flood, or hurricane, and it's clear that the might of natural geologic phenomena greatly exceeds the strength of human-made structures. But watch a bulldozer clear a swath of forest, a dynamite explosion remove the top of a hill, or a prairie field evolve into a housing development, and it's clear that people can change the face of the Earth at rates often exceeding those of natural geologic processes.

Finally, when you finish reading this book, your view of the world may be forever colored by geologic curiosity. If you walk in the mountains, you will think of the many forces that shape and reshape the Earth's surface. If you hear about a natural disaster, you will have insight into the processes that brought it about. And if you go on a road trip, the rock exposures along the highway will no longer be gray, faceless cliffs, but will present complex puzzles of texture and color telling a story of Earth's history.

P.3 THEMES OF THIS BOOK

A number of narrative themes appear (and reappear) throughout this text. These themes, listed below, can be viewed as this book's overall take-home message.

- *The Earth is a unique, evolving system.* Geologists increasingly recognize that the Earth is a complex system; its interior, solid surface, oceans, atmosphere, and life forms interact in many ways to yield the landscapes and environment in which we live. Within this **Earth System**, chemical elements cycle between different types of rock, between rock and sea, between sea and air, and between all of these entities and life (Fig. P.4).

- *Plate tectonics explains many Earth processes.* Earth is not a homogeneous ball, but rather consists of concentric layers—from center to surface, Earth has a core, mantle, and crust. We live on the surface of the crust, where it meets the atmosphere and the oceans. In the 1960s, geologists recognized that the crust, together with the uppermost part of the underlying mantle, forms a 100- to 150-km-thick semi-rigid shell. Large cracks separate this shell into discrete pieces, called **plates**, which move very slowly relative to one another (Fig. P.5). The theory

FIGURE P.4 In this scenic view in Switzerland, we see many aspects of the Earth System—air, water, ice, rock, life, and human activity.

FIGURE P.5 A simplified map of the Earth's plates. The arrows indicate the direction each plate is moving, and the length of the arrow indicates plate velocity (the longer the arrow, the faster the motion). We discuss the types of plate boundaries in Chapter 2.

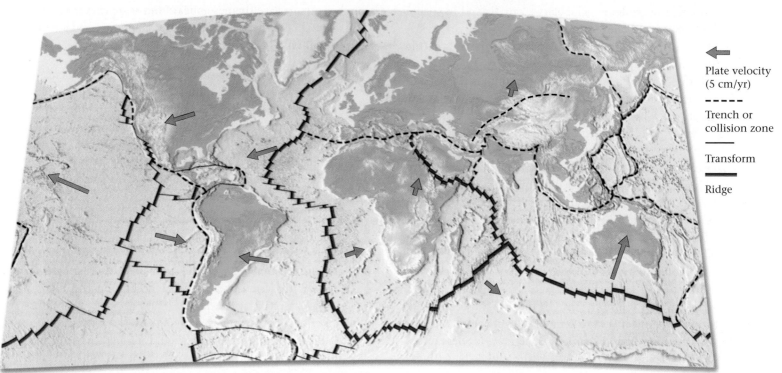

Plate velocity (5 cm/yr)

— — — —
Trench or collision zone

Transform

Ridge

that describes this movement and its consequences is called the **theory of plate tectonics**, and it is the foundation for understanding most geologic phenomena. Although plates move very slowly—generally less than 10 cm a year—their movements yield earthquakes, volcanoes, and mountain ranges, and cause the map of Earth's surface to change over time.

- *The Earth is a planet.* Despite the uniqueness of the Earth System, Earth can be viewed as a planet, formed like the other planets of the Solar System from dust and gas that encircled the newborn Sun. Though Earth resembles the other inner planets (Mercury, Venus, and Mars), it differs from them in having plate tectonics, an oxygen-rich atmosphere and liquid-water ocean, and abundant life.

- *The Earth is very old.* Geologic data indicate that the Earth formed 4.57 billion years ago—plenty of time for geologic processes to generate and destroy landscapes, for life forms to evolve, and for the map of the planet to change. Plate-movement rates of only a few centimeters per year can move a continent thousands of kilometers if those movements continue for hundreds of millions of years. There is time enough to build mountains and time enough to grind them down, many times over. To define intervals of this time, geologists developed the **geologic**

time scale. Figure P.6 depicts the major subdivisions of the geologic time scale. Geologists call the last 542 million years the Phanerozoic Eon, and all time before that the Precambrian. They further divide the Precambrian into three main intervals named, from oldest to youngest, the Hadean, the Archean, and the Proterozoic Eons. The Phanerozoic Eon is also divided into three main intervals named, from oldest to youngest, the Paleozoic, the Mesozoic, and the Cenozoic Eras. Chapter 12 will discuss these intervals in greater detail.

- *Internal and external processes drive geologic phenomena.* **Internal processes** are those phenomena driven by heat from inside the Earth. Plate movement is an example. Because plate movements cause mountain building, earthquakes, and volcanoes, we call all of these phenomena internal processes as well. **External processes** are those phenomena driven by heat supplied by radiation coming to the Earth from the Sun. This heat drives the movement of air and water, which grinds and sculpts the Earth's surface and transports the debris to new locations, where it accumulates. The interaction between internal and external processes forms the landscapes of our planet. As we'll see, gravity—the pull that one mass exerts on another—plays an important role in both internal and external processes.

FIGURE P.6 The geologic time scale.

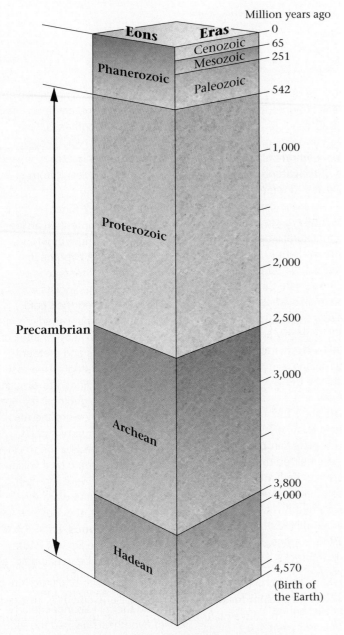

(a) The scale has been divided into eons and eras.

One thousand years ago = 1 **Ka**
(Ka stands for kilo-annum)

One million years ago = 1 **Ma**
(Ma stands for mega-annum)

One billion years ago = 1 **Ga**
(Ga stands for giga-annum)

(b) Abbreviations for time units.

- *Geologic phenomena affect our environment.* Volcanoes, earthquakes, landslides, floods, groundwater, energy sources, and mineral reserves are of vital interest to every inhabitant of this planet. Therefore, throughout this book we emphasize the linkages between geology, the environment, and society.

- *Physical aspects of the Earth System are linked to life processes.* All life on this planet depends on such physical features as the minerals in soil; the temperature, humidity, and composition of the atmosphere; and the flow of surface and subsurface water. And life in turn affects and alters physical features. For example, the oxygen in Earth's atmosphere comes primarily from plant photosynthesis, a life activity. This oxygen in turn permits complex animals to survive, and affects chemical reactions among air, water, and rock. Without the physical Earth, life could not exist; but without life, this planet's surface might have become a frozen wasteland like that of Mars, or enshrouded in clouds like that of Venus.

- *Science comes from observation, and people make scientific discoveries.* Science does not consist of subjective guesses or arbitrary dogmas, but rather of a consistent set of objective statements resulting from the application of the **scientific method** (Box P.1). Every scientific idea must be tested thoroughly, and should be used only when supported by documented observations. Further, scientific ideas do not appear out of nowhere, but are the result of human efforts. Wherever possible, this book shows where geologic ideas came from, and tries to answer the question, "How do we know that?"

- *The study of geology can increase general science literacy.* Studying geology provides an ideal opportunity to learn basic concepts of chemistry and physics, because in geology these concepts can be applied directly to understanding tangible phenomena. Thus, in this book, basic concepts of physical science are introduced in boxed features, where appropriate.

As you read this book, please keep these themes in mind. Don't view geology as a list of words to memorize, but rather as an interconnected set of concepts to digest. Most of all, enjoy yourself as you learn about what may be the most fascinating planet in the Universe.

To help illustrate the geology of our amazing world, we have created "See for Yourself" features. Using *Google Earth*™, you'll be able to see the products of geologic processes for yourself (see **See for Yourself A: How to Use Google Earth™ to see Geologic Features** on p. S-1, near the end of the book).

BOX P.1

The Scientific Method

Sometime during the past 200 million years, a large block of rock or metal, which had been orbiting the Sun, slammed into our planet. It made contact at a site in what is now the central United States, a landscape of flat cornfields. The impact of this block, a meteorite, released more energy than a nuclear bomb—a cloud of shattered rock and dust blasted skyward, and once-horizontal layers of rock from deep below the ground sprang upward and tilted on end beneath the gaping hole left by the impact. When the dust had settled, a huge crater surrounded by debris marked the surface of the Earth at the impact site. Later in Earth history, running water and blowing wind wore down this jagged scar. Some 15,000 years ago, sand, gravel, and mud carried by a vast glacier buried what remained, hiding it entirely from view (**Fig. Bx P.1a–c**). Wow! So much history beneath a cornfield. How do we know this? It takes scientific investigation.

The movies often portray science as a dangerous tool, capable of creating Frankenstein's monster, and scientists as warped or nerdy characters with thick glasses and poor taste in clothes. In reality, science is simply the use of observation, experiment, and calculation to explain how nature operates, and scientists are people who study and try to understand natural phenomena. Scientists carry out their work using the **scientific method**, a sequence of steps for systematically analyzing scientific problems in a way that leads to verifiable results. Let's see how geologists employed the steps of the scientific method to come up with the meteorite-impact story.

1. *Recognizing the problem*. Any scientific project, like any detective story, begins by identifying a mystery. The cornfield mystery came to light when water drillers discovered limestone, a rock typically made of shell fragments, just below the 15,000-year-old glacial sediment. In surrounding regions, the rock beneath the glacial sediment consists of sandstone, a rock made of cemented-together sand grains. Since limestone can be used to build roads, make cement, and produce the agricultural lime used in treating soil, workers stripped off the glacial sediment and dug a quarry to excavate the limestone. They were also amazed to find that rock layers exposed in the quarry were tilted steeply and had been shattered by large cracks. In the surrounding regions, all rock layers are horizontal like the layers in a birthday cake, the limestone layer lies underneath the sandstone, and the rocks contain relatively few cracks. Curious geologists came to investigate, and soon realized that the geologic features of the land just beneath the cornfield presented a problem to be explained: What phenomena had brought limestone up close to the Earth's surface, had tilted the layering in the rocks, and had shattered the rocks?

2. *Collecting data*. The scientific method proceeds with the collection of observations or clues that point to an answer. Geologists studied the quarry and determined the age of its rocks, measured the orientation of the rock layers, and documented (made a written or photographic record of) the fractures that broke up the rocks.

3. *Proposing hypotheses*. A scientific **hypothesis** is merely a possible explanation, involving only natural processes, that can explain a set of observations. Scientists propose hypotheses during or after their initial data collection. The geologists working in the quarry came up with two alternative hypotheses. First, the features in this region could result from a volcanic explosion; and second, they could result from a meteorite impact.

4. *Testing hypotheses*. Since a hypothesis is no more than an idea that can be either right or wrong, scientists must put hypotheses through a series of tests to see if they work. The geologists at the quarry compared their field observations with published observations made at other sites of volcanic explosions and meteorite impacts, and they studied the results of experiments designed to simulate such events. If the geologic features visible in the quarry were the result of volcanism, the quarry should contain rocks formed by the freezing of molten rock erupted by a volcano. But no such rocks were found. If, however, the features were the result of an impact, the rocks should contain **shatter cones**, small, cone-shaped cracks (see Fig. Bx P.1c). Shatter cones can be overlooked, so the geologists returned to the quarry specifically to search for them, and found them in abundance. The impact hypothesis passed the test!

Theories are scientific ideas supported by an abundance of evidence; they have passed many tests and have failed none. Scientists have much more confidence in the correctness of a theory than they do of a hypothesis. Continued study in the quarry eventually yielded so much evidence for impact, that the impact hypothesis came to be viewed as a theory. Scientists continue to test theories over a long time. Successful theories withstand these tests and are sup-

The eroded central uplift of a meteorite crater now forms the 150 m-high hills of Gosses Bluff in central Australia. Surrounding rocks have eroded away. Researchers think the impact happened about 140 Ma.

ported by so many observations that they become part of a discipline's foundation. (As you will discover in Chapter 2, geologists consider the idea that continents have moved around the surface of the Earth to be a theory, because so much evidence supports it.) However, some theories may eventually be disproven and replaced by better ones.

In a few cases, scientists have been able to devise concise statements that completely describe a specific relationship or phenomenon. Such statements, called **scientific laws**, apply without exception for a given range of conditions. Newton's law of gravitation serves as an example—it is a simple mathematical expression that always defines the invisible pull exerted by one mass upon another. Note that scientific laws do not, in themselves, explain a phenomenon, and in this regard differ from theories. For example, the *law* of gravity does not explain why gravity exists, but the *theory* of evolution does explain why evolution occurs.

FIGURE Bx P.1 An ancient meteorite impact excavates a crater and permanently changes rock beneath the surface.

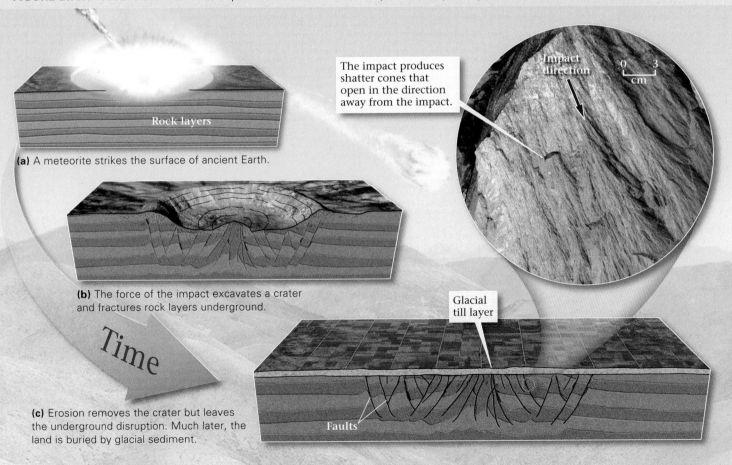

The impact produces shatter cones that open in the direction away from the impact.

Impact direction

0   3 cm

Rock layers

(a) A meteorite strikes the surface of ancient Earth.

(b) The force of the impact excavates a crater and fractures rock layers underground.

Time

Glacial till layer

(c) Erosion removes the crater but leaves the underground disruption. Much later, the land is buried by glacial sediment.

Faults

GEOPUZZLE REVISITED

Geologists looking at this landscape see clues that speak of Earth's long history. The cliff in the foreground exposes dozens of rock layers, each representing an accumulation of sand deposited under water. The fact that these now-solid layers started out as loose sand, implies that the layers must have once been buried so deeply that the sand could be compacted and slowly transformed into rock. Sand layers, when first formed, are horizontal; but the layers in this photo tilt at an angle. This tilt indicates that after becoming rock, the layers were squeezed and buckled by mountain-building forces. In fact, the landscape of low hills that we see today was probably once a towering mountain range. In examining the layers more closely, geologists find fossils of a giant fern that went extinct some 300 million years ago. Thus, when the sand was accumulating, this region hosted semi-tropical swamps, not the temperate meadows of grass we see today. Clearly, our world changes in many ways as time passes.

Guide Terms

Earth System (p. 5)
external process (p. 6)
geologic time scale (p. 6)
geologist (p. 3)
geology (p. 3)
hypothesis (p. 8)
internal process (p. 6)

plate (p. 5)
scientific laws (p. 9)
scientific method (pp. 7, 8)
shatter cones (p. 8)
theory (p. 8)
theory of plate tectonics (p. 6)

ANOTHER VIEW An image of Earth taken by an orbiting satellite (Landsat 7) emphasizes that our planet is indeed a unique and evolving system. Here, a river draining the Amazon rainforest modifies the shape of vegetation-covered islands. No other planet hosts liquid water or vegetation—no other planet looks like the Earth!

A photograph of the Earth as seen by the *Apollo 17* astronauts. This image emphasizes that our planet is a sphere with finite limits. But it's a very special sphere because Earth's air, land, water, and life all interact in ways found nowhere else in the Solar System.

PART I

Our Island in Space

When you gaze out toward the horizon from a mountaintop, the Earth seems endless, and before the modern era, many people thought it was. But to astronauts flying to the Moon, the Earth looks like a small, shining globe—they can see half the planet in a single glance. From the astronauts' perspective, we are living on an island in space.

Earth may not be endless, but it is a very special planet: its temperature and composition, unlike those of the other planets in the Solar System, make it habitable. In Part I of this book, we first learn scientific ideas about how the Earth, and the Universe around it, came to be. Then we take a quick tour of the planet to get a sense of its composition and its various layers. With this background, we're ready to encounter the twentieth-century revolution in geology that yielded the set of ideas we now call the theory of plate tectonics. We'll see that this theory, which proposes that the outer layer of the Earth consists of plates that move with respect to each other, provides a rational explanation for a great variety of geologic features—from the formation of continents to the distribution of fossils. In fact, geologists now recognize that plate interactions even led to the formation of gases from which the atmosphere and oceans formed, and without which life could not exist.

CHAPTER 1

Cosmology and the Birth of Earth

The Carina Nebula (NGC 3372) is a birthplace for stars. The nebula is about 7,500 light years away, but thanks to the Hubble Telescope, we are able to see into its clouds of dust and gas. The region in this image is approximately 40 light years wide (400 trillion km, or 235 trillion miles).

GEOPUZZLE

How did the Solar System (including the Earth) form, and what is the source of the material from which it formed?

This truth within thy mind rehearse,
That in a boundless Universe
Is boundless better, boundless worse.

—Alfred, Lord Tennyson (British poet, 1809–1892)

1.1 INTRODUCTION

Sometime in the distant past, humans developed the capacity for complex, conscious thought. This amazing ability, which distinguishes our species from all others, brought with it the gift of curiosity, an innate desire to understand and explain the workings of ourselves and of all that surrounds us—our Universe. Astronomers define the **Universe** as all of space and all the matter and energy within it. Questions that we ask about the Universe differ little from questions a child asks of a playmate: Where do you come from? How old are you? Such musings first spawned legends in which heroes, gods, and goddesses used supernatural powers to mold the planets and sculpt the landscape. Eventually, researchers began to apply scientific principles (see Box P.1) to the systematic study of the overall structure and history of the Universe, thereby establishing the modern discipline of scientific **cosmology**.

In this chapter, we begin with a brief introduction to the principles of scientific cosmology—we characterize the basic architecture of the Universe, introduce the Big Bang theory for the formation of the Universe, and discuss scientific ideas concerning the birth of the Earth.

Chapter Themes

By the end of this chapter, you should know . . .

- how people's perceptions of Earth's place in the Universe have changed over the centuries.
- modern concepts concerning the basic architecture of our Universe and its components.
- the evidence for the expanding Universe, and the Big Bang theory.
- where the elements comprising matter came from.
- the Nebula theory for forming stars and planets—i.e., a scientific model for how the Earth formed.

1.2 AN IMAGE OF OUR UNIVERSE

What Is the Structure of the Universe?

Think about the mysterious spectacle of a clear night sky. What objects are up there? How big are they? How far away are they? How do they move? How are they arranged? In addressing such questions, ancient philosophers distinguished between stars (points of light whose locations relative to each other are fixed) and planets (tiny spots of light that move relative to the backdrop of stars). Three thousand years ago (1000 B.C.E., "before the Common Era"), the Earth's human population totaled only several million, the pyramids of Egypt had already been weathering in the desert for 1,600 years, and Homer, the great Greek poet, was compiling the *Iliad* and the *Odyssey*. In Homer's day, astronomers of the Mediterranean region knew the difference between stars and planets. They had observed that, though the positions of stars remained fixed relative to each other, the whole star field slowly revolved around a fixed point (Box 1.1). They also observed that the planets moved relative to both the stars and to each other, etching seemingly complex paths across the night sky. In fact, the word *planet* comes from the Greek word *planēs*, which means wanderer. Despite their knowledge of the heavens, people of Homer's day did not realize fully that Earth itself is a planet. Some envisioned the Earth to be a flat disk, with land toward the center and water around the margins, that lay at the center of a celestial sphere, a dome to which the stars were attached. This disk supposedly lay above an underworld governed by the fearsome god Hades. Philosophers also toyed with numerous explanations for the Sun: to some, it was a burning bowl of oil, and to others, a ball of red-hot iron. Most favored the notion that movements of celestial bodies represented the activities of gods and goddesses.

Over the centuries, two schools of thought developed concerning how to explain the configuration of stars and planets, and their relationships to the Earth, Sun, and Moon. The first school advocated a **geocentric model** (Fig. 1.1a), in which the Earth sat without moving at the center of the Universe, while the Moon and the planets whirled around it within a globe of stars. The second school advocated a **heliocentric model** (Fig. 1.1b), in which the Sun lay at the center of the Universe, with the Earth and other planets orbiting around it. The geocentric image eventually gained the most followers, due to the influence of an Egyptian mathematician, Ptolemy (100–170 C.E.). Ptolemy developed equations that appeared to predict the wanderings of the planets in the context of the geocentric model. During the Middle Ages (~476–1400 C.E.), church leaders in Europe adopted Ptolemy's geocentric image as dogma, because it justified the comforting thought that humanity's home occupies the most important place in the Universe. Anyone who disagreed with this view risked charges of heresy.

Then came the Renaissance. In fifteenth-century Europe, bold thinkers spawned a new age of exploration and scientific discovery. Thanks to the efforts of Nicolaus Copernicus (1473–1543) and Galileo Galilei (1564–1642), people gradually came to realize that the Earth and planets did

BOX 1.1

How Do We Know that Earth Rotates?

As you sit reading this book, you probably aren't conscious that you are in motion. But in fact, because of the Earth's spin, you actually are moving rapidly around Earth's axis (the imaginary line that connects the North and South Poles). In fact, a person sitting on the equator is hurtling along at about 1,674 km/h (1,040 mph)—faster than the speed of sound! How do we know that the Earth spins around its axis? The answer comes from observing the *apparent* motion of the stars (**Fig. Bx 1.1a**). If you gaze at the night sky for many hours, you'll see that the stars move in a circular path around the North Star.

Curiously, it was not until the middle of the nineteenth century that Léon Foucault (1819–1868), a French physicist, proved that the Earth spins on its axis. He made this discovery by setting a heavy pendulum, attached to a long string, in motion (**Fig. Bx 1.1b**). As the pendulum continued to swing, Foucault noted that the plane in which it oscillated was perpendicular to the Earth's surface, and that this plane rotated around a vertical axis. If Newton's first law of motion—objects in motion remain in motion, objects at rest remain at rest—was correct, Foucault then deduced that this phenomenon

required the Earth to rotate under the pendulum while the pendulum continued to swing in the same plane (**Fig. Bx 1.1c**). Foucault displayed his discovery beneath the great dome of the Pantheon in Paris, to much acclaim.

We now know that, in fact, the Earth's spin axis is not perfectly fixed in orientation; rather, it wobbles. This wobble, known as precession, is like the wobble of a toy top as it spins. We'll see later in this book that the precession of the Earth's axis, which takes 23,000 years, may affect the planet's climate.

FIGURE Bx1.1 Proving the Earth rotates on its axis. The axis intersects Earth's surface at the poles.

(b) An exact replica of Foucault's original pendulum on display in the Panthéon, Paris.

(a) Time exposure of the night sky over an observatory. The stars appear to rotate around a central point near the North Star. This motion is actually due to the rotation of the Earth on its axis.

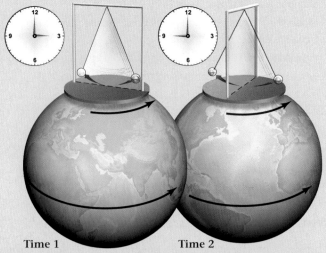

(c) Foucault's experiment. At Time 1, the plane in which the pendulum swings is the same as the plane of the frame. At Time 2, six hours later, the plane in which the pendulum swings is perpendicular to the plane of its frame.

FIGURE 1.1 Contrasting views of the Universe, as drawn by artists hundreds of years ago.

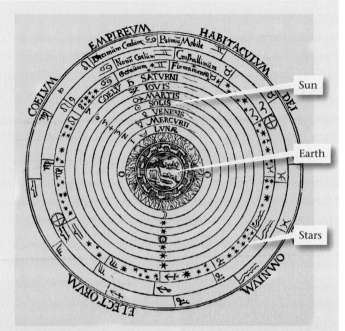

(a) The *geocentric* image of the Universe shows the Earth at the center, surrounded by air, fire, and the other planets, all contained within the globe of the stars.

(b) The *heliocentric* image of the Universe shows the Sun at the center, as envisioned by Copernicus.

indeed orbit the Sun and could not be at the center of the Universe. And when Isaac Newton (1643–1727) explained **gravity**, the attractive force that one object exerts on another (see **Box 1.2**), it finally became possible to explain the motions of these objects.

The Nature of Our Solar System

Eventually, astronomical study with powerful telescopes demonstrated that the Earth is but one of several planets orbiting the Sun, and that moons orbit most planets. Orbital motions are quite fast—Earth travels about 150 million km in the 365 days that it takes to complete an orbit. This means that our planet races along its orbit at a rate of 107,300 km/h (67,000 mph). The Sun, planets, moons, and countless other small objects held together by the "glue" of gravitational attraction comprise the **Solar System** (Fig. 1.2a, b). Let's now discuss the components of the Solar System a bit more completely.

Did you ever wonder . . . how fast you are traveling through space?

The Sun accounts for 99.8% of the mass in the Solar System. The remaining 0.2% includes a great variety of objects, the largest of which are planets. Astronomers define a **planet** as an object that orbits a star, is spherical, and has "cleared its neighborhood of other objects." The last phrase in this definition sounds a bit strange at first, but merely implies that a planet's gravity has pulled in all particles of matter in its orbit. According to this definition, which was formalized in 2005, our Solar System includes eight planets—Mercury, Venus, Earth, Mars, Jupiter, Saturn, Uranus, and Neptune. In 1930, astronomers discovered Pluto, a 2,390-km-diameter sphere of ice, whose orbit generally lies outside that of Neptune's. Until 2005, astronomers considered Pluto to be a planet. But since it does not fit the modern definition, it has been dropped from the roster (**Box 1.3**). A **moon** is a sizeable body locked in orbit around a planet. All but two planets have moons—some, such as Earth's Moon, are large and spherical, but many are small and have irregular shapes. Our Solar System is not alone in hosting planets; in recent years, astronomers have found "extra-solar planets" (over 530 by 2011) orbiting stars in many other systems.

Planets in our Solar System differ radically from one another both in size and composition. The inner planets (Mercury, Venus, Earth, and Mars), the ones closer to the Sun, are relatively small. Astronomers commonly refer to these as **terrestrial planets** because, like Earth, they consist of a shell of rock surrounding a ball of metallic iron alloy. The outer planets (Jupiter, Saturn, Uranus, and Neptune) are known as the **gas-giant planets**, or Jovian planets, because most of their mass consists of gas and "ice." (In this context, "ice" is a general term that includes the solid state of many materials besides water that could be gaseous under Earth's surface conditions.) The adjective *giant* certainly seems appropriate, for these planets are huge—Jupiter, for example, has a mass 318 times larger than that of Earth and accounts for about 71% of the non-solar mass in the Solar System.

In addition to the planets, the Solar System contains many small objects. Millions of asteroids (chunks of rock and/or metal) comprise a belt between the orbits of Mars and Jupiter. Asteroids range in size from less than a centimeter to about 930 km

BOX 1.2

The Forces of Nature

In your everyday experience, you constantly see and/or feel the effect of forces. Forces can squash objects, speed them up, or slow them down. Forces can tear, stretch, spin, and twist objects, and can make them float or sink. Sir Isaac Newton defined **force** as simply a push or pull that causes the velocity (speed) of an object to change in magnitude and/or direction.

Physicists distinguish between two general types of force. The first type, a contact force or mechanical force, results when one **mass** (a quantity of matter) moves and comes in contact with another.

You are applying a mechanical force to a boulder when you push on it (**Fig. Bx 1.2a**), and the wind applies a mechanical force to a sail when it blows. The second type, a non-contact force or field force, applies across a distance. Field forces include gravity and magnetism. **Gravity** is the force of attraction between two masses—it is what holds you to the surface of the Earth and pulls objects from higher elevations to lower ones (**Fig. Bx 1.2b**). The strength of gravity depends on the quantity of matter in the two masses and on the distance between them. For

example, you feel a much stronger gravitational pull to the huge Earth than you do to a small baseball. **Magnetism**, simplistically, is the force that is generated by electricity flowing in a wire, or by special materials called magnets. Unlike gravity, magnetic force can be attractive (pulling objects together) or repulsive (pushing them apart). Over short distances, the magnetic force of even a small magnet can be larger than the gravitational force produced by the Earth; thus, you can overcome gravity and lift objects with a magnet (**Fig. Bx 1.2c**).

FIGURE Bx1.2 Examples of the forces of nature in everyday life. Mechanical and field forces are very familiar.

(a) Pushing a boulder is a mechanical force.

(b) Gravity, a field force, pulls a person down a zip line.

(c) A magnet produces a field force sufficient to hold onto these clips.

FIGURE 1.2 The relative sizes and positions of planets in the Solar System.

(a) Relative sizes of the planets. All are much smaller than the Sun, but the gas-giant planets are much larger than the terrestrial planets. Jupiter's diameter is about 11.2 times greater than that of Earth.

BOX 1.3

Discovering Planets

Before the invention of the telescope, astronomers recognized five other planets besides Earth, for these planets (Mercury, Venus, Mars, Jupiter, and Saturn) can be seen with the naked eye. Telescopes allowed astronomers to find the next-farthest planet, Uranus, in 1781. Close examination suggested that Uranus did not follow its expected orbit exactly. The discrepancy implied that the gravity of yet another planet must be tugging on Uranus, and this prediction led to the discovery of Neptune in 1846. Discrepancies in Neptune's orbit, in turn, prompted a race to find yet another planet, leading to the discovery of Pluto in 1930. Pluto, however, proved to be a very different sort of planet—it's much smaller than the others, consists mostly of ice, and follows an orbit that does not lie in the same plane as the orbits of other planets.

In 1992, astronomers found that millions of icy objects, similar in composition to Pluto, occupy the region between the orbit of Neptune and a distance perhaps ten times the radius of Neptune's orbit. These objects together comprise the Kuiper Belt, named for the astronomer who predicted the objects' existence. As the twenty-first century dawned, astronomers learned that, though most Kuiper Belt objects are tiny, some are comparable in size to Pluto. In fact, Eris, an object found in 2003, is 20% larger than Pluto. Clearly, scientists needed to reconsider the traditional concept of a planet, or we could eventually have dozens or hundreds of planets. Thus, in August 2006, the International Astronomical Union proposed a new definition of the word *planet*. This definition states that a planet is a celestial body that orbits the Sun, has a nearly spherical shape, and has cleared its neighborhood of other objects. The last phrase means that the object has either collided with and absorbed other objects in its orbit, has captured them to make them moons, or has gravitationally disturbed their orbits sufficiently to move them elsewhere. According to

FIGURE Bx1.3 The telescope at Lowell Observatory used to discover Pluto had a 13" lens.

this definition, only the eight classical planets discovered by 1846 have the honor of being considered full-fledged planets. Pluto and Eris are sometimes called "dwarf planets."

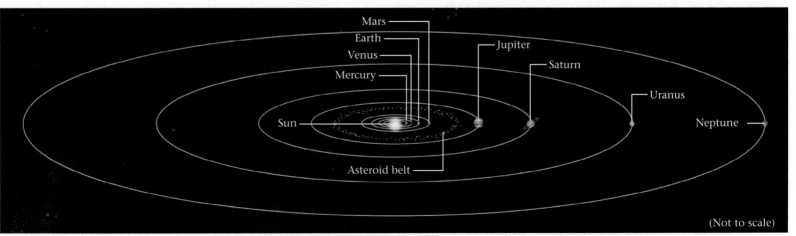

(b) Relative positions of the planets. This figure is not to scale. If the Sun in this figure was the size of a large orange, the Earth would be the size of a sesame seed 15 meters (49 feet) away. Note that all planetary orbits lie roughly in the same plane.

A view of very deep space by the Hubble Space Telescope. What looks like dark "nothingness" to the naked eye is actually filled with galaxies.

The Milky Way, as viewed from Earth.

The Milky Way, as if viewed from deep space.

FIGURE 1.3 A galaxy may contain about 300 billion stars.

in diameter. About a trillion bodies of ice lie in belts or clouds beyond the orbit of Neptune. Most of these icy fragments are tiny, but a few (including Pluto) have diameters of over 2,000 km and may be thought of as "dwarf planets." The gravitational pull of the main planets has sent some of the icy objects on paths that take them into the inner part of the Solar System, where they begin to evaporate and form long tails—we call such objects comets. We'll discuss these objects more in Chapter 2.

Stars and Galaxies

Stars look like points of light. But in fact, **stars** are immense balls of incandescent gas in which nuclear fusion reactions produce intense heat and light. Our Sun is a medium-sized star. Stars are not randomly scattered through the Universe; gravity holds them together in immense groups, called **galaxies**. The Sun and over 300 billion stars together form the Milky Way galaxy. More than 100 billion galaxies constitute the visible Universe (Fig. 1.3).

From our vantage point on Earth, the Milky Way looks like a hazy band, but if we could view the Milky Way from a great distance, it would look like a flattened spiral with great curving arms slowly swirling around a glowing, disk-like center (see Fig. 1.3). Presently, our Solar System lies near the outer edge of one of these arms and rotates around the center of the galaxy about once every 250 million years. We hurtle through space, relative to an observer standing outside the galaxy, at about 200 km per second.

Determining the Size of the Earth

The Greek astronomer Eratosthenes (c. 276–194 B.C.E.) served as chief of the library in Alexandria, Egypt, one of the great ancient centers of learning in the ancient Mediterranean region. One day, while filing papyrus scrolls, he came across a report noting that in the southern Egyptian city of Syene, the Sun lit the base of a deep vertical well precisely at noon on the first day of summer. Eratosthenes deduced that the Sun's rays at noon on this day must be exactly perpendicular to the Earth's surface at Syene, and that if the Earth was spherical, then the Sun's rays could *not* simultaneously be perpendicular to the Earth's surface at Alexandria, 800 km to the north. So on the first day of summer, Eratosthenes measured the shadow cast by a tower in Alexandria at noon. The angle between the tower and the Sun's rays, as indicated by the shadow's length, proved to be 7.2°. He then commanded a servant to pace out a straight line from Alexandria to Syene. The sore-footed servant found the distance to be 5,000 stadia (1 stadium = 0.1572 km). Knowing that a circle contains 360°, Eratosthenes then calculated the Earth's circumference using a simple equation (Fig. 1.4), and his answer came within 2% of today's accepted value of 40,008 km (24,865 miles).

The Distance from Earth to Celestial Objects

Around 200 B.C.E., Greek mathematicians, using ingenious geometric calculations, determined that the distance to the Moon was about thirty times the Earth's diameter. This number comes close to the true distance, which on average is 381,555 km (about 237,100 miles). But it wasn't until the seventeenth century that astronomers figured out that the mean distance between the Earth and the Sun is 149,600,000 km (about 93,000,000 miles). As for the stars, the ancient Greeks realized that they must be much farther away than the Sun in order for them to appear as a fixed backdrop behind the Moon and planets, but the Greeks had no way of calculating the actual distance. Our modern documentation of the vastness

FIGURE 1.4 Over 2,000 years ago, Eratosthenes calculated the circumference of the Earth, using simple geometry.

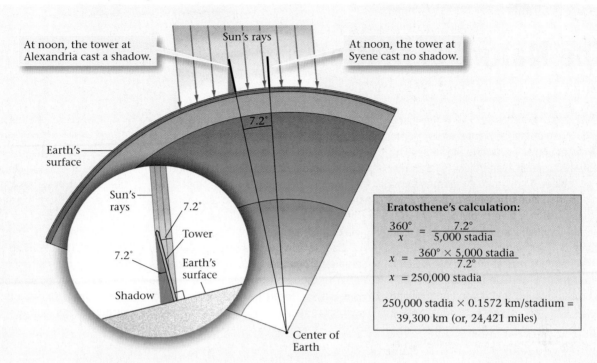

At noon, the tower at Alexandria cast a shadow.

At noon, the tower at Syene cast no shadow.

Sun's rays

7.2°

Earth's surface

Sun's rays

7.2°

Tower

7.2°

Earth's surface

Shadow

Center of Earth

Eratosthene's calculation:

$$\frac{360°}{x} = \frac{7.2°}{5,000 \text{ stadia}}$$

$$x = \frac{360° \times 5,000 \text{ stadia}}{7.2°}$$

$$x = 250,000 \text{ stadia}$$

250,000 stadia × 0.1572 km/stadium = 39,300 km (or, 24,421 miles)

Did you ever wonder...
how far away are the stars?

of the Universe began in 1838, when astronomers found that the nearest star to Earth, Alpha Centauri, lies 40.85 trillion km away.

Since it's hard to fathom the distances to planets and stars without visualizing a more reasonably sized example, imagine that the Sun is the size of an orange. At this scale, the Earth would be a sesame seed at a distance of 15 m (49 ft) from the orange. Alpha Centauri would lie 2,000 km (about 1,243 miles) from the orange.

When astronomers realized that light travels at a constant (i.e., unchanging) speed of about 300,000 km (about 186,000 miles) per second in space, they realized that they had a way to describe the huge distances between objects in space more conveniently. They defined a large distance by stating how long it takes for light to traverse that distance. For example, it takes light about 1.3 seconds to travel from the Earth to the Moon, so we can say that the Moon is about 1.3 light seconds away. Similarly, we can say that the Sun is 8.3 light minutes away. A **light year**, the distance that light travels in one Earth year, equals about 9.5 trillion km (about 6 trillion miles). When you look up at Alpha Centauri, 4.3 light years distant, you see light that started on its journey to Earth about 4.3 years ago. The Milky Way is 100,000 light years across. Other galaxies are so far away that, to the naked eye, they look like stars in the night sky. The nearest galaxy to ours, Andromeda, lies 2.2 million light years away.

Astronomers didn't develop techniques for measuring the distance to very distant stars and galaxies until the twentieth century. With these techniques (described in astronomy books), they determined that the farthest celestial objects that can be seen with the naked eye are less than 3 million light years away. Powerful telescopes allow us to see much farther. The edge of the *visible* Universe lies over 13 billion light years away, which means that light traveling to Earth from this location began its journey about 9 billion years before the Earth even existed. When such numbers became known in the middle of the twentieth century, people came to the realization that the dimensions of the Universe are truly staggering!

Take-Home Message

- Before the Renaissance, many people believed that the Earth sat at the center of the Universe.
- Modern studies indicate that the Sun is one of 300 billion stars in the Milky Way Galaxy, which is one of 100 billion galaxies in the visible Universe.
- Earth is one of 8 planets (4 terrestrial; 4 gas giant) orbiting the Sun.
- Distances in the Universe are so large that they are measured in light years.

THINK: Using just information in this Chapter, *estimate* the distance from the Sun to the center of the Milky Way.

BOX 1.4

The Nature of Matter

Matter takes up space—you can feel it and see it. We use the term **matter** to refer to any material making up the Universe. The amount of matter in an object is its **mass**, so an object with greater mass contains more matter. **Density** refers to the amount of mass within a given volume. A denser material contains more mass per unit volume.

What does matter consist of? A Greek philosopher named Democritus (ca. 460–370 B.C.E.) argued that if you kept dividing matter into progressively smaller pieces, you would eventually end up with nothing; but since it's not possible to make something out of nothing, there must be a smallest piece of matter that can't be subdivided further. He proposed the name **atom** for these smallest pieces, based on the Greek word *atomos*, which means indivisible.

Our modern understanding of matter developed in the 17th century, when chemists recognized that certain substances (such as hydrogen and oxygen) *cannot* break down into other substances, whereas others (such as water and salt) *can* break down—the former came to be known as **elements**, and the latter came to be known as **compounds**. John Dalton (1766–1844) adopted the word *atom* for the smallest piece of an element that has the

property of the element; the smallest piece of a compound that has the properties of the compound is a **molecule**. In 1869, Dmitri Mendelév (1834–1907) recognized that groups of elements share similar characteristics, and he organized the elements into a chart that we now call the periodic table of the elements (see **Appendix**).

Chemists in the 17th and 18th century identified 92 naturally occurring elements on Earth; modern physicists have created more than a dozen new ones. Each element has a name and a symbol (e.g., N = nitrogen; H = hydrogen; Fe = iron; Ag = silver).

In 1910, Ernest Rutherford, a British physicist, proved that, contrary to the view of Democritus, atoms actually can be divided into smaller pieces. Most of the mass in an atom clusters in a dense ball, called the **nucleus**, at the atom's center. The nucleus contains two types of subatomic particles: **neutrons**, which have a neutral electrical charge, and **protons**, which have a positive charge. A cloud of electrons surrounds the nucleus (**Fig. Bx 1.4a, b**); an **electron** has a negative charge and contains only 1/1,836 as much mass as a proton. ("Charge," simplistically, refers to the way in which a particle responds to a magnet or an electric current.)

Electron clouds have a complex internal structure—electrons are grouped into intervals called orbitals, energy levels, or shells. Electrons in inner shells concentrate closer to the nucleus, and those in outer shells are farther away. Roughly speaking, the diameter of an electron cloud is 10,000 times greater than that of the nucleus, yet the cloud contains only 0.05% of an atom's mass—thus, atoms are mostly empty space!

We distinguish atoms of different elements from one another by their **atomic number**, the number of protons in their nucleus. Smaller atoms have smaller atomic numbers, and larger ones have larger atomic numbers. The smallest atom, hydrogen, has an atomic number of 1, and the largest naturally occurring atom, uranium, has an atomic number of 92. Except for hydrogen nuclei, all nuclei also contain neutrons. In smaller atoms, the number of neutrons roughly equals the number of protons, but in larger atoms the number of neutrons exceed the number of protons. The **atomic mass** of an atom is roughly the sum of the number of neutrons and the number of protons. For example, an oxygen nucleus contains 8 protons and 8 neutrons, and thus has an atomic mass of 16.

Nuclear bonds serve as the "glue" that holds together subatomic particles in

1.3 FORMING THE UNIVERSE

We stand on a planet, in orbit around a star, speeding through space on the arm of a galaxy. Beyond our galaxy lie hundreds of billions of other galaxies. Where did all this "stuff"—the matter of the Universe (Box 1.4)—come from, and when did it first form? For most of human history, a scientific solution to these questions seemed intractable. But in the 1920s, unexpected observations about the nature of light from distant galaxies set astronomers on a path of discovery that ultimately led to a model of Universe formation known as the Big Bang theory. To explain these observations, we must first introduce an important phenomenon called the Doppler

effect. Thus, we begin this section by developing an understanding of how the Doppler effect modifies the light seen in telescopes. We then show how this understanding leads to the idea that the Universe expands, and then the conclusion that this expansion began during the Big Bang, 13.7 billion years ago.

Waves and the Doppler Effect

When a train whistle screams, the sound you hear moved through the air from the whistle to your ear in the form of sound waves. **Waves** are disturbances that transmit energy from one point to another in the form of periodic motions.

FIGURE Bx1.4 The nature of atoms.

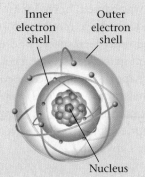

Inner electron shell

Outer electron shell

Nucleus

(a) An image of an atom with a nucleus orbited by electrons.

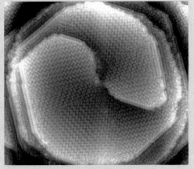

(b) An image of atoms taken with a special microscope. Each spot is an atom.

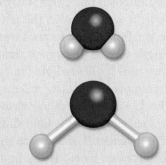

(e) Two ways of portraying a water molecule. The red ball is oxygen; the small balls are hydrogen.

Nuclear Reactions

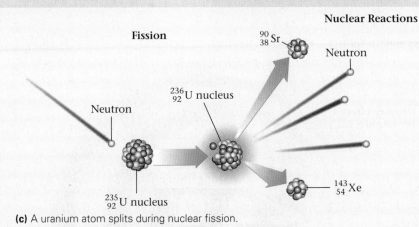

Fission

$^{90}_{38}$ Sr

Neutron

$^{236}_{92}$ U nucleus

Neutron

$^{235}_{92}$ U nucleus

$^{143}_{54}$ Xe

(c) A uranium atom splits during nuclear fission.

Fusion

Neutron

Deuterium

Helium

Tritium

(d) Two atoms (versions of hydrogen) stick together to form one atom of helium during nuclear fusion in a hydrogen bomb.

a nucleus. Atoms can change only during nuclear reactions, when nuclear bonds break or form. Physicists recognize several types of nuclear reactions. For example, during **fission**, a large nucleus breaks apart to form two smaller atoms. Fission generates energy in nuclear power plants and in the explosion of an atomic bomb. During fusion, two smaller atoms come together to form a larger atom. **Fusion** reactions power the Sun and occur during the explosion of a hydrogen bomb (**Fig. Bx 1.4c, d**).

Separate atoms are held together to form molecules by **chemical bonds**, which we discuss more fully later in the book. As an example, chemical bonds hold two hydrogen atoms to form an H_2 molecule, and two hydrogens plus an oxygen atom together form an H_2O (water) molecule (**Fig. Bx 1.4e**). During a **chemical reaction**, chemical bonds break and/or form.

As each sound wave passes, air alternately compresses, then expands. We refer to the distance between successive waves as the **wavelength**, and the number of waves that pass a point in a given time interval as the **frequency**. If the wavelength decreases, more waves pass a point in a given time interval, so the frequency increases. The "pitch" of a sound, meaning its note on the musical scale, depends on the frequency of the sound waves.

Imagine that you are standing on a station platform, and a train moves toward you. The train whistle's sound gets louder as the train approaches, but its pitch remains the same. Then, the instant the train passes, the pitch abruptly changes—it sounds like a lower note in the musical scale.

Why? When the train moves toward you, the sound has a higher frequency (the waves are closer together so the wavelength is smaller) because the sound source, the whistle, has moved slightly closer to you between the instant that it emits one wave and the instant that it emits the next (**Fig. 1.5a, b**). When the train moves away from you, the sound has a lower frequency (the waves are farther apart), because the whistle has moved slightly farther from you between the instant it emits one wave and the instant it emits the next. An Austrian physicist, C. J. Doppler (1803–1853), first interpreted this phenomenon, and thus the change in frequency that happens when a wave source moves is now known as the **Doppler effect**.

Light energy also moves in the form of waves. We can represent light waves symbolically by a periodic succession of crests and troughs. Visible light comes in many colors—the colors of the rainbow. The color you see depends on the frequency of the light waves, just as the pitch of a sound you hear depends on the frequency of sound waves. Red light has a longer wavelength (lower frequency) than does blue light. The Doppler effect also applies to light but can be noticed only if the light source moves very fast, at least a few percent of the speed of light. If a light source moves away from you, the light you see becomes redder, as the light shifts to longer wavelength or lower frequency. If the source moves toward you, the light you see becomes bluer, as the light shifts to higher frequency. We call these changes the **red shift** and the blue shift, respectively (Fig. 1.5c).

Does the Size of the Universe Change?

In the 1920s, astronomers such as Edwin Hubble, after whom the Hubble Space Telescope was named, braved many a frosty night beneath the open dome of a mountaintop observatory in order to aim telescopes into deep space. These researchers were searching for distant galaxies. At first, they documented only the location and shape of newly discovered galaxies. But then one astronomer began an additional project to study the wavelength of light produced by the distant galaxies. The results yielded a surprise that would forever change humanity's perception of the Universe.

Astronomers found, to their amazement, that the light of distant galaxies *displayed red shifts* relative to the light of nearby stars (Fig. 1.5d). Hubble pondered this mystery and, around

FIGURE 1.5 Manifestations of the Doppler effect for sound and for light.

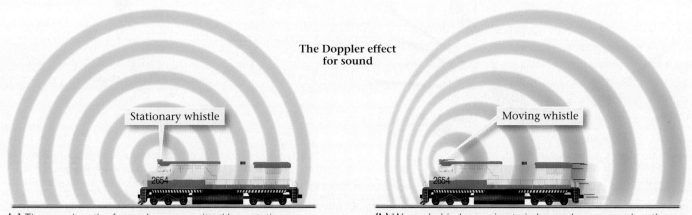

(a) The wavelength of sound waves emitted by a stationary train is the same in all directions.

(b) Waves behind a moving train have a longer wavelength than those in front.

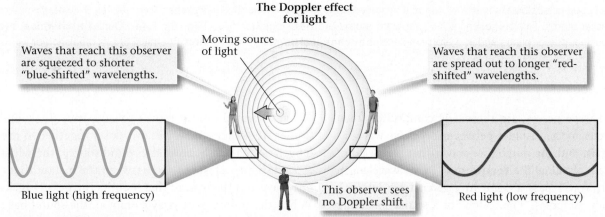

(c) The wavelength of blue light is less than that of red light. If a light source moves very fast, the Doppler effect results in a shifting of the wavelengths. The observed shift depends on the position of the observer.

Sun

Distant galaxy

(d) The atoms in a star absorb certain specific wavelengths of light. We see these wavelengths as dark lines on a light spectrum. Note that the lines from a galaxy a billion light years away are shifted toward the red end of the spectrum (i.e., to the right), in comparison to the lines from our own Sun.

1929, realized that the red shifts must be a consequence of the Doppler effect, and thus that the distant galaxies must be moving away from Earth at an immense velocity. At the time, astronomers thought the Universe had a fixed size, so Hubble initially assumed that if some galaxies were moving away from Earth, others must be moving toward Earth. But this was not the case. On further examination, Hubble concluded that the light from all distant galaxies, regardless of their direction from Earth, exhibits a red shift. In other words, *all* distant galaxies are moving rapidly away from us.

How can all galaxies be moving away from us, regardless of which direction we look? Hubble puzzled over this question and finally recognized the solution: the whole Universe must be expanding! (To picture the expanding Universe, imagine a ball of bread dough with raisins scattered throughout. As the dough bakes and expands into a loaf, each raisin moves away from its neighbors, in every direction; Fig. 1.6a.) This idea came to be known as the **expanding Universe theory**.

Hubble's ideas marked a revolution in cosmological thinking. Now we picture the Universe as an expanding bubble, in which galaxies race away from each other at incredible speeds. This image immediately triggers the key question of cosmology: did the expansion begin at some specific time in the past?

If it did, then that instant would mark the physical beginning of the Universe, the beginning of space and time.

The Big Bang

Most astronomers have concluded that expansion did indeed begin at a specific time, with a cataclysmic explosion called the Big Bang. According to the **Big Bang theory**, all matter and energy—everything that now constitutes the Universe—was initially packed into an infinitesimally small point. The point exploded and the Universe began, according to current estimates, 13.7 (±1%) billion years ago.

Take-Home Message

- The Doppler effect refers to the change in the wavelength (and, therefore, in frequency) of waves that happens if the source of the waves is moving.
- Light from all distant galaxies exhibits a red shift, meaning that the galaxies must be moving away from us at high speed. This observation requires that the Universe is expanding.
- Calculations suggest that expansion began subsequent to the Big Bang, about 13.7 Ga.

THINK: The farther a galaxy lies from the Earth, the greater the red shift it exhibits. Why?

FIGURE 1.6 The concept of the expanding Universe and the Big Bang.

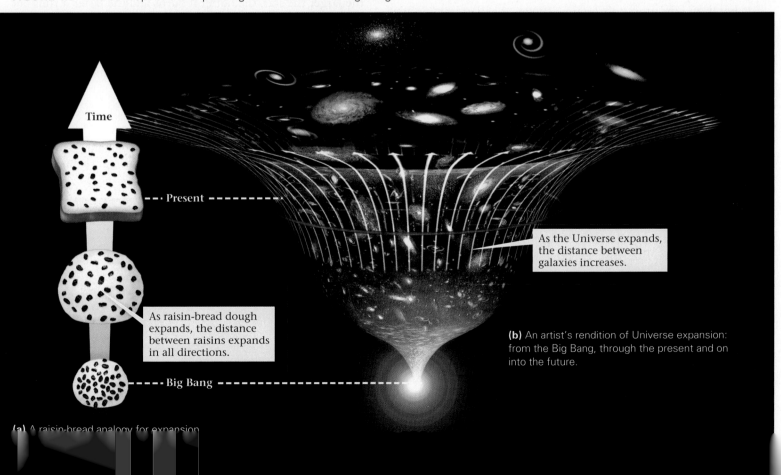

Time

Present

As the Universe expands, the distance between galaxies increases.

As raisin-bread dough expands, the distance between raisins expands in all directions.

(b) An artist's rendition of Universe expansion: from the Big Bang, through the present and on into the future.

Big Bang

(a) A raisin-bread analogy for expansion

1.4 MAKING ORDER FROM CHAOS

Aftermath of the Big Bang

Of course, no one was present at the instant of the Big Bang, so no one actually saw it happen. But by combining clever calculations with careful observations, researchers have developed a consistent model of how the Universe evolved, beginning an instant after the explosion (Fig. 1.6b). According to this model of the Big Bang, profound change happened at a fast and furious rate at the outset. During the first instant of existence, the Universe was so small, so dense, and so hot that it consisted entirely of energy—atoms, or even the smallest subatomic particles that make up atoms, could not even exist. (See Box 1.4 for an introduction to atoms.) Within a few seconds, however, hydrogen atoms could begin to form. And by the time the Universe reached an age of 3 minutes, when its temperature had fallen below 1 billion degrees, and its diameter had grown to about 53 million km (35 million miles), hydrogen atoms could fuse together to form helium atoms (see Box 1.4). Formation of new nuclei in the first few minutes of time is called Big Bang nucleosynthesis because it happened *before* any stars existed. This process could produce only small atoms, meaning ones containing a small number of protons (an atomic number less than 5), and it happened very rapidly. In fact, virtually all of the new atomic nuclei that would form by Big Bang nucleosynthesis existed by the end of the first 5 minutes.

Eventually, the Universe became cool enough for chemical bonds to bind atoms of certain elements together in molecules (see Box 1.4). Most notably, two hydrogen atoms could join to form molecules of H_2. As the Universe continued to expand and cool further, atoms and molecules slowed down and accumulated into patchy clouds called **nebulae**. The earliest nebulae of the Universe consisted almost entirely of hydrogen (74%, by volume) and helium (24%) gas.

Birth of the First Stars

When the Universe reached its 200 millionth birthday, it contained immense, slowly swirling, dark nebulae separated by vast voids of empty space. The Universe could not remain this way forever, though, because of the invisible but persistent pull of gravity (see Box 1.2). Eventually, gravity began to remold the Universe pervasively and permanently.

All matter exerts gravitational pull—a type of force—on its surroundings, and as Isaac Newton first pointed out, the amount of pull depends on the amount of mass. Somewhere in the young Universe, the gravitational pull of an initially more massive region of a nebula began to suck in surrounding gas and, in a grand example of the rich getting richer, grew in mass and, therefore, density (mass per unit volume). As this denser region attracted progressively more gas, the gas compacted into a smaller region, and the initial swirling movement of gas transformed into a rotation around an axis. As gas continued to move inward, cramming into a progressively smaller volume, the rotation rate became faster and faster. (A similar phenomenon happens when a spinning ice skater pulls her arms inward.) Because of its increased rotation, the nebula evolved into a disk shape (Geology at a Glance, pp. 30–31). As more and more matter rained down onto the disk, it continued to grow, until eventually, gravity collapsed the inner portion of the disk into a dense ball. As the gas was compressed into a smaller and smaller space, its temperature increased dramatically. Eventually, the central ball of the disk became hot enough to glow, and at this point it became a **protostar**.

A protostar continues to grow, by pulling in successively more mass, until its core becomes extremely dense and its temperature reaches about 10 million degrees. Under such conditions, hydrogen nuclei slam together so forcefully that they join, in a series of steps, to form helium nuclei. Such fusion reactions produce huge amounts of energy, and the mass becomes a fearsome furnace. When the first nuclear fusion reactions began in the first protostar, the body "ignited" and the first true star formed. When this happened, perhaps 800 million years after the Big Bang, the first starlight pierced the newborn Universe (Fig. 1.7). This process would soon happen again and again, and many first-generation stars came into existence.

First-generation stars tended to be very massive, perhaps 100 times the mass of the Sun. Astronomers have shown that the larger the star, the hotter it burns and the faster it runs out of fuel and dies. A huge star may survive only a few million years to a few tens of millions of years before it explodes to form a **supernova**. Thus, not long after the first generation of stars formed, the Universe began to be peppered with the first generation of supernova.

FIGURE 1.7 In this Hubble Space Telescope photo from 2009, we see a cloud of gas in the Carina Nebula—7,500 light years away. It is a nursery of new stars.

- Most hydrogen and helium (small atoms) formed during the first minutes after the Big Bang.
- Atoms accumulated in clouds called nebulae. Gravity pulled the gas of a nebula into a revolving disk.
- The center of the disk became a massive, hot ball called a protostar. When dense and hot enough, nuclear fusion reactions began, and the ball became a true star.
- When huge stars run out of fuel, they explode as a supernova, ejecting gas into space.

THINK: Did all *first-generation* stars form at the same time? Why or why not?

1.5 WE ARE ALL MADE OF STARDUST

Where Do Elements Come From?

Nebulae from which the first-generation stars formed consisted entirely of small atoms, because only these small atoms were generated by Big Bang nucleosynthesis. In contrast, the Universe of today contains 92 naturally occurring elements. Where do the other 87 elements come from? In other words, how did elements such as carbon, sulfur, silicon, iron, gold, and uranium form? These elements, which are common on Earth, have larger atomic numbers (see Box 1.4)—carbon has an atomic number of 6, and iron has an atomic number of 26. Physicists have shown that these elements form during the life cycle of stars, by the process of **stellar nucleosynthesis**. Because of stellar nucleosynthesis we can consider stars to be element factories, constantly fashioning larger atoms out of smaller atoms.

Did you ever wonder . . .
where the atoms in your body first formed?

What happens to the atoms formed in stars? Some escape into space during the star's lifetime, simply by moving fast enough to overcome the star's gravitational pull. The stream of

FIGURE 1.8 A photo of solar (stellar) wind streaming into space.

FIGURE 1.9 Very heavy elements form during supernova explosions. Here we see the rapidly expanding shell of gas ejected into space from an explosion whose light reached the Earth in 1054 C.E. This shell is called the Crab Nebula.

atoms emitted from a star during its lifetime is a **stellar wind** (Fig. 1.8). Some escape only when a star dies. A low-mass star (like our Sun) releases a large shell of gas as it dies, ballooning into a "red giant" during the process; whereas a high-mass star blasts matter into space during a supernova explosion (Fig. 1.9). Most very large atoms (those with atomic numbers greater than that of iron) require even more violent circumstances to form than generally occurs within a star. In fact, most very large atoms form *during* a supernova explosion. Once ejected into space, atoms from stars and supernova explosions form new nebulae or mix back into existing nebulae.

When the first generation of stars died, they left a legacy of new, heavier elements that mixed with residual gas from the Big Bang. A second generation of stars and associated planets formed out of this compositionally more diverse nebulae. Second-generation stars lived and died, and contributed heavier elements to third-generation stars. Succeeding generations contain a greater proportion of heavier elements. Because not all stars live for the same duration of time, at any given moment the Universe contains many different generations of stars. Our Sun may be a third-, fourth-, or fifth-generation star. Thus, the mix of elements we find on Earth includes relics of primordial gas from the Big Bang as well as the disgorged guts of dead stars. Think of it—the elements that make up your body once resided inside a star!

The Nebular Theory for Forming the Solar System

Earlier in this chapter, we introduced scientific concepts of how stars form from nebulae. But we delayed our discussion of how the planets and other objects in our Solar System originated until we had discussed the production of heavier atoms such as carbon, silicon, iron, and uranium, because planets consist predominantly of these elements. Now that we've discussed stars as element factories, we return to the early history of the Solar System and introduce the **nebular theory**, an explanation for the origin of planets, moons, asteroids, and comets. According to the nebular theory, the Sun and all other objects in the Solar System formed from material that had been swirling about in a nebula. This process involved several stages (see Geology at a Glance, pp. 30–31). We've already discussed the stage of forming a star like the Sun from a nebula. Now, to complete the story of Solar System formation, we consider the fate of the material in the flattened outer part of the disk, the material that did not become part of the star. This outer part is called the **protoplanetary disk**, because it is the source of planets (as well as of other moons, comets, and asteroids).

What did the protoplanetary disk consist of? The disk from which our Solar System formed contained all 92 elements, some as isolated atoms, and some bonded to others in molecules. Geologists divide the material formed from these atoms and molecules into two classes. **Volatile materials**—such as hydrogen, helium, methane, ammonia, water, and carbon dioxide—are materials that can exist as gas at the Earth's surface. In the pressure and temperature conditions of space, some volatile materials remain in gaseous form, but others freeze to form different kinds of "ice." (Note that we do not limit use of the word *ice* to water alone.) **Refractory materials** are those that melt only at high temperatures, and they condense to form solid soot-sized particles of "dust" in the coldness of space (Fig. 1.10a). Initially, the protoplanetary disk may have been fairly homogeneous, meaning that it had much the same composition throughout. But as the proto-Sun began to form, the inner part of the disk became hotter, causing volatile elements to evaporate and drift to the outer portions of the disk. Thus, the inner part of the disk ended up with higher concentrations of dust, whereas the outer portions ended up with higher concentrations

of ice. As this was happening, gravity caused the protoplanetary disk to evolve into a series of concentric rings.

How did the dusty, icy, and gassy rings transform into planets? Even before the proto-Sun ignited, the material of the surrounding rings began to clump and bind together, due to gravity. First, soot-sized particles merged to form sand-sized grains. Then, these grains stuck together to form grainy basketball-sized blocks (Fig 1.10b), which in turn collided. If the collision was slow, blocks stuck together or simply bounced apart. If the collision was fast, one or both of the blocks shattered, producing smaller fragments that recombined later. Eventually, enough blocks coalesced to form **planetesimals**, bodies whose diameter exceeded about 1 km. Because of their mass, the planetesimals exerted enough gravity to attract and pull in other objects that were nearby (see Geology at a Glance, pp. 30–31). Figuratively, planetesimals acted like vacuum cleaners, sucking in small pieces of dust and ice as well as smaller planetesimals that lay in their orbit, and in the process they grew progressively larger. Eventually, victors in the competition to attract mass grew into **protoplanets**, bodies approaching the size of today's planets. Once a protoplanet succeeded in incorporating virtually all the debris within its orbit, it became a full-fledged planet.

Early stages in the planet-forming process probably occurred very quickly—some computer models suggest that it may have taken only a few hundred thousand years to go from the dust and gas stage to the large planetesimal stage. Planets may have grown from planetesimals in 10 to 200 million years. In the inner orbits, where the protoplanetary disk consisted mostly of dust, small terrestrial planets composed of rock and metal formed. In the outer part of the Solar System, where significant amounts of ice existed, larger protoplanets grew, and these evolved into the gas-giant planets. Fragments of materials that were not incorporated in planets remain today as asteroids and comets.

FIGURE 1.10   The earliest solid components of the Solar System—dust and "grainy" planetesimals.

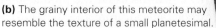

(a) This grain of interplanetary dust, collected by a satellite, is only 0.01mm across. (A pinhead is 1.0 mm across)

(b) The grainy interior of this meteorite may resemble the texture of a small planetesimal.

When did the planets form? Using techniques introduced in Chapter 12, geologists have found that special types of meteorites thought to be leftover planetesimals formed at 4.57 Ga, and thus consider that date to be the birth date of the Solar System. If this date is correct, it means that the Solar System formed about 9 billion years after the Big Bang, and thus is only about a third as old as the Universe.

The nebular theory successfully answers many questions about the Solar System. For example, why do we see such differences between the terrestrial planets and the gas-giant planets? As mentioned earlier, once the proto-Sun began to heat up, volatile materials in the inner part of the disk evaporated and migrated out, leaving refractory materials behind. When the Sun ignited and became a gigantic nuclear furnace, it emitted a strong solar wind that blew much of the remaining volatiles out of the inner Solar System. Thus, the inner rings ended up consisting mostly of dust, which coalesced to form rocky and metallic terrestrial planets. Dust existed in the outer rings as well, and thus the gas-giant planets have rocky and metallic centers. But the outer rings also contained a huge amount of volatile material, which formed the ice and gas that makes up the bulk of the gas-giant planets.

Differentiation of the Earth and Formation of the Moon

When planetesimals first formed, they had a fairly homogeneous distribution of material throughout, because the smaller pieces from which they formed all had much the same composition and collected together in no particular order. But large planetesimals did not stay homogeneous for long, because they began to heat up. The heat came primarily from three sources: the heat produced during collisions (similar to the phenomenon that happens when you bang on a nail with a hammer and they both get warm), the heat produced when matter is squeezed into a smaller volume, and the heat produced from the decay of radioactive elements. In bodies whose temperature rose sufficiently to cause internal melting, denser iron alloy separated out and sank to the center of the body, whereas lighter rocky materials remained in a shell surrounding the center. By this process, called **differentiation**, protoplanets and large planetesimals developed internal layering early in their history. As we will see later, the central ball of iron alloy constitutes the body's core and the outer shell constitutes its mantle.

In the early days of the Solar System, planets continued to be bombarded by **meteorites** (solid objects, such as fragments of planetesimals, falling from space that land on a planet) even after the Sun had ignited and differentiation had occurred. Heavy bombardment in the early days of the Solar System pulverized the surfaces of planets and eventually left huge

Did you ever wonder . . .
is the Moon as old as the Earth?

numbers of craters (See for Yourself B, p. S-4). Bombardment also contributed to heating the planets.

Based on analysis and the dating of Moon rocks, most geologists have concluded that at about 4.53 Ga, a Mars-sized protoplanet slammed into the newborn Earth. In the process, the colliding body disintegrated, along with a large part of the Earth's mantle. A ring of debris formed around the remaining, now-molten Earth, and quickly coalesced to form the Moon. Of note, not all moons in the Solar System necessarily formed in this manner. Some may have been independent protoplanets or comets that were captured by a larger planet's gravity.

Why Is the Earth Round?

Small planetesimals were jagged or irregular in shape, and asteroids today have irregular shapes. Planets, on the other hand, are more or less spherical. Why? Simply put, when a protoplanet gets big enough, gravity can change its shape. To picture how, imagine a block of cheese warming in an oven. As the cheese gets softer and softer, gravity causes it to spread out in a pancake-like blob. This model shows that gravitational force alone can cause material to change shape if the material is soft enough. Now let's apply this model to planetary growth.

The rock composing a small planetesimal is cool and strong enough so that the force of gravity is not sufficient to cause the rock to flow. But once a planetesimal grows beyond a certain critical size (about 1,000 km in diameter), its interior becomes warm and soft enough to flow in response to gravity. As a consequence, protrusions are pulled inward toward the center, and the planetesimal re-forms into a special shape that permits the force of gravity to be nearly the same at all points on its surface. This special shape is spherical because in a sphere mass is evenly distributed around the center.

Take-Home Message

- Large atoms form during nuclear fusion reactions in stars, and during the death of stars.
- Our Solar System contains atoms from the Big Bang, and from the remains of earlier generations of stars. The atoms in our bodies were manufactured in stars!
- According to the nebula theory, the Sun and planets formed from a cloud of gas and dust. The central ball became the Sun, and the matter in surrounding rings formed the planets.
- The Moon formed from debris ejected from a collision between the Earth and a protoplanet.
- When proto-planets become large and hot enough, their interiors differentiate into a core and mantle, and the body becomes spherical.

THINK: Why are the gas giant planets further from the Sun than are the terrestrial planets?

Forming the Planets and the Earth-Moon System

1. Forming the Solar System, according to the nebular theory: A nebula forms from hydrogen and helium left over from the Big Bang, as well as from heavier elements that were produced by fusion reactions in stars or during explosions of stars.

2. Gravity pulls gas and dust inward to form an accretionary disk. Eventually a glowing ball—the proto-Sun—forms at the center of the disk.

6. Gravity reshapes the proto-Earth into a sphere. The interior of the Earth differentiates into a core and mantle.

5. Forming the planets from planetesimals: Planetesimals grow by continuous collisions. Gradually, an irregularly shaped proto-Earth develops. The interior heats up and becomes soft.

8. The Moon forms from the ring of debris.

7. Soon after the Earth forms, a protoplanet collides with it, blasting debris that forms a ring around the Earth.

3. "Dust" (particles of refractory materials) concentrates in the inner rings, while "ice" (particles of volatile materials) concentrates in the outer rings. Eventually, the dense ball of gas at the center of the disk becomes hot enough for fusion reactions to begin. When it ignites, it becomes the Sun.

4. Dust and ice particles collide and stick together, forming planetesimals.

9. Eventually, the atmosphere develops from volcanic gases. When the Earth becomes cool enough, moisture condenses and rains to create the oceans. Some gases may be added by passing comets.

Chapter Summary

- The geocentric model of the Universe placed the Earth at the center of the Universe, with the planets and Sun orbiting around the Earth within a celestial sphere speckled with stars. The heliocentric model, which gained acceptance during the Renaissance, placed the Sun at the center.

- Eratosthenes was able to measure the size of the Earth in ancient times, but it was not until fairly recently that astronomers accurately determined the distances to the Sun, planets, and stars. Distances in the Universe are so large that they must be measured in light years.

- The Earth is one of eight planets orbiting the Sun, and this Solar System lies on the outer edge of a slowly revolving galaxy, the Milky Way, which is composed of about 300 billion stars. The Universe contains at least hundreds of billions of galaxies.

- The red shift of light from distant galaxies, a manifestation of the Doppler effect, indicates that all distant galaxies are moving away from the Earth. This observation leads to the expanding Universe theory. Most astronomers agree that this expansion began after the Big Bang, a cataclysmic explosion about 13.7 billion years ago.

- The first atoms (hydrogen and helium) of the Universe developed within minutes of the Big Bang. These atoms formed vast gas clouds, called nebulae.

- Gravity caused clumps of gas in the nebulae to coalesce into revolving balls. As these balls of gas collapsed inward, they evolved into flattened disks with bulbous centers. The protostars at the center of these disks eventually became dense and hot enough that fusion reactions began in them. When this happened, they became true stars, emitting heat and light.

- Heavier elements form during fusion reactions in stars; the heaviest are mostly made during supernova explosions. The Earth and the life forms on it contain elements that could only have been produced during the life cycle of stars. Thus, we are all made of stardust.

- According to the nebular theory of planet formation, planets developed from the rings of gas and dust surrounding protostars. The gas and dust condensed into planetesimals that then clumped together to form protoplanets, and finally true planets. Inner rings became the terrestrial planets. Outer rings grew into gas-giant planets.

- The Moon formed from debris ejected when a protoplanet collided with the Earth in the young Solar System.

- A planet assumes a near-spherical shape when it becomes so soft that gravity can smooth out irregularities.

GEOPUZZLE REVISITED

Geologists conclude that the Solar System formed from atoms generated by the Big Bang, and from atoms produced in stars or during the explosion of stars. Gravity pulled all this material together into a bulbous disk whose central ball became the Sun. The remainder of the disk condensed into planetesimals, which in turn coalesced to form planets.

Guide Terms

atom (p. 22)	matter (p. 22)
atomic mass (p. 22)	meteorite (p. 29)
atomic number (p. 22)	molecule (p. 22)
Big Bang theory (p. 25)	moon (p. 17)
chemical bond (p. 23)	nebula (p. 26)
chemical reaction (p. 23)	nebular theory (p. 28)
compound (p. 22)	neutron (p. 22)
cosmology (p. 15)	nucleus (p. 22)
density (p. 22)	planet (p. 17)
differentiation (p. 29)	planetesimals (p. 28)
Doppler effect (p. 23)	proton (p. 22)
electron (p. 22)	protoplanetary disk (p. 28)
element (p. 22)	protoplanets (p. 28)
expanding Universe theory (p. 25)	protostar (p. 26)
	red shift (p. 24)
fission (p. 23)	refractory materials (p. 28)
force (p. 18)	Solar System (p. 17)
frequency (p. 23)	stars (p. 20)
fusion (p. 23)	stellar nucleosynthesis (p. 27)
galaxies (p. 20)	stellar wind (p. 27)
gas-giant planets (p. 17)	supernova (p. 26)
geocentric model (p. 15)	terrestrial planets (p. 17)
gravity (pp. 17, 18)	Universe (p. 15)
heliocentric model (p. 15)	volatile materials (p. 28)
light year (p. 21)	wave (p. 22)
magnetism (p. 18)	wavelength (p. 23)
mass (pp. 18, 22)	

Review Questions

1. Why do planets appear to move with respect to stars?
2. Contrast the geocentric and heliocentric Universe concepts.

3. Describe how Foucault's Pendulum demonstrates that the Earth is rotating on its axis.

4. How did Eratosthenes calculate the Earth's circumference?

5. Imagine you hear the main character in a low-budget science-fiction movie say he will "return ten light years from now." What's wrong with his usage of the term?

6. Describe how the Doppler effect works.

7. What does the red shift of the galaxies tell us about their motion with respect to the Earth?

8. What is the Big Bang, and when did it occur?

9. When did hydrogen and helium atoms form?

10. Where did heavier elements form?

11. Describe the steps in the formation of the Solar System according to the nebular theory.

12. Why do the inner planets consist mostly of rock and metal, but the outer planets mostly of gas?

13. Describe how the Moon was formed.

14. Why is the Earth round?

On Further Thought

15. Look again at Figure Bx 1.1a. The North Star, a particularly bright star, lies just about at the center of the circles of light tracked out by other stars. (a) What does this mean about the position of the North Star relative to Earth's spin axis? Why is it called the "North Star"? (b) Consider the wobble of Earth's axis. Will the North Star be in the same position in a photograph taken from the same location as Figure 1.1 in the future? Why?

16. The horizon is the line separating sky from the Earth's surface. Consider the shape of the Earth. How does the distance from your eyes to the horizon change as your elevation above the ground increases? To answer this question, draw a semicircle to represent part of the Earth's surface, then draw a vertical tower up from the surface. With your ruler, draw a line from various elevations on the tower to where the line is tangent to the surface of the Earth. (A *tangent* is a line that touches a circle at one point and is perpendicular to a radius.)

17. Astronomers discovered that more distant galaxies move away from the Earth more rapidly than do nearer ones. Why? To answer this question, make a model of the problem by drawing three equally spaced dots along a line; the dot at one end represents the Earth, and the other two represent galaxies. "Stretch" the line by drawing the line and dots again, but this time make the line twice as long. This stretching represents Universe expansion. Notice that the dots are now farther apart. Recall that velocity = distance × time. If you pretend that it took 1 second to stretch the line (so "time" = 1 second), measurement of the distance that each galaxy moved relative to the Earth allows you to calculate velocity.

18. Consider that the deaths of stars eject quantities of heavier elements into space, and that these elements then become incorporated in nebulae from which the next generation of stars forms. Do you think that the ratio of heavier to lighter elements in, say, a sixth-generation star is larger or smaller than the ratio in a second-generation star? Why?

 For more resources, including animations, quizzes, and Norton's GeoTours, go to **wwnorton.com/studyspace**.

 If your instructor assigns exercises in SmartWork, log in at **smartwork.wwnorton.com**.

ANOTHER VIEW

(a) The Cat's Eye Nebula is about 3,300 light years away. It formed when a dying star shed a shell of gas.

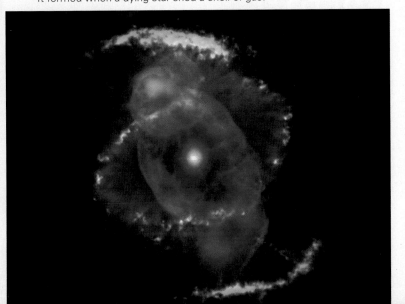

(b) The Andromeda Galaxy, 2.2 million light years away. It's about the same size as our own Milky Way.

CHAPTER 2

Journey to the Center of the Earth

The vertical elevation difference between Lake Provo and the peak of the Wasatch Mountains, in Utah, is about 2 km. Though the landscape looks dramatic, this relief represents only about 0.03% of the distance to the center of the Earth. Even the deepest canyons are mere scratches on this planet's surface.

GEOPUZZLE

If you could slice right through the Earth as if it were a hard-boiled egg, what would you see?

The Earth is not a mere fragment of dead history, stratum upon stratum like the leaves of a book . . . but living poetry like the leaves of a tree.

—Henry David Thoreau (1817–1862)

2.1 INTRODUCTION

In 1961 a Russian cosmonaut, Yuri Gagarin, became the first human to orbit the Earth, and by the end of the decade, two Americans, Neil Armstrong and Buzz Aldrin, became the first to walk on another celestial object, the Moon. These were truly amazing feats because until then, humanity had been totally earthbound. Though people have not yet traveled farther than the Moon, we have sent satellites to the planets and beyond, and with powerful telescopes, we've even seen planets orbiting other stars outside our Solar System.

Let's turn the tables and imagine that we're observers on an extrasolar planet peering at the Earth, wondering what it's like. If we were to send a satellite rocketing to this planet, what would we detect? In this chapter we outline the context of the Earth and the character of the Earth's surface by describing what an imaginary spaceship would discover as it approached and then orbited this planet. Then, we turn our attention downward, to characterize the interior of the Earth, from its surface to its center. This high-speed tour provides a foundation from which we can develop themes presented through the remainder of this book.

Chapter Themes

By the end of this chapter, you should know that . . .

- many objects besides the Sun and planets comprise the Solar System.
- a magnetic field and an atmosphere surround our planet.
- the Earth System includes many distinct, interacting realms.
- the Earth has distinct internal layers (crust, mantle, and core).
- the rigid lithosphere, Earth's outer shell, overlies a plastic asthenosphere.

2.2 WELCOME TO THE NEIGHBORHOOD

A Journey through the Solar System

For most of its journey, our spaceship travels through interstellar space, the region between stars. This region is a vacuum (an absence of matter) so profound that it contains less than one atom per liter—by comparison, air at sea level contains 27,000,000,000,000,000,000,000,000 (or, in scientific notation,

2.7×10^{22}) atoms per liter. Eventually, we reach the "edge" of the Solar System. What defines the edge? Astronomers consider it to be the invisible surface within which the Sun's gravitational pull exceeds that of other stars and can hold onto objects. Our spaceship first feels the ever-so-weak pull of the Sun's gravity at a distance of about 50,000 AU from the Sun. (An AU, or astronomical unit, is the distance between the Earth and the Sun—about 150 million kilometers, or 93 million miles.) Astronomers speculate that within this distance, the Sun's gravity holds on to specks and balls of "ice" (or, more specifically, frozen water, carbon dioxide, and methane). Together, these objects comprise the **Oort Cloud**, leftovers of the protoplanetary disk from which the Solar System formed.

Did you ever wonder . . .
what defines the edge of our Solar System?

At a distance of about 200 AU, our spaceship crosses another invisible boundary and enters the bubble-like **heliosphere**. The region within the heliosphere contains solar wind particles, and the region outside doesn't. Amazingly, two human-made objects, the Voyager satellites that launched in 1977, should reach the heliosphere within the next decade, carrying with them messages of greeting, inscribed on copper disks, from humanity to whoever—or whatever—might live on the extrasolar planets that these satellites may reach tens of thousands of years in the future.

Finally, at a distance of 30 to 55 AU, our spaceship traverses the **Kuiper Belt**, a diffuse ring of icy objects, some of which reach diameters of 1200 km, and 70,000 of which have diameters over 100 km (Fig. 2.1). Comets originate from the Kuiper Belt, and to a lesser extent, the Oort Cloud (Box 2.1).

FIGURE 2.1 An explorer coming from another planet, after passing through the Oort Cloud, would then see the heliosphere, within which lies the Kuiper Belt, the planets and moons, the asteroids, and the Sun.

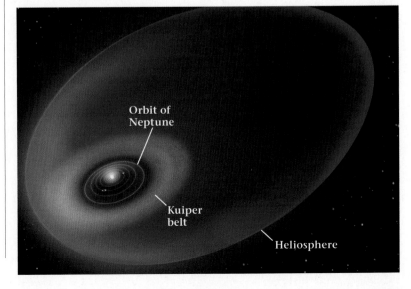

BOX 2.1

Comets and Asteroids— The Other Stuff of the Solar System

The process of planet formation, all in all, was very efficient. Much of the dust, ice, and gas that formed the accretionary disk around the proto-Sun eventually became incorporated into the planets. But some material escaped this fate and resides in two distinct classes of solid bodies—asteroids and comets. Meteorites that strike Earth provide samples of these bodies.

Asteroids are small bodies of solid rock or metal that orbit the Sun. Most reside in a belt, called the asteroid belt, between the orbits of Mars and Jupiter. Some asteroids are small, rocky planetesimals that were never incorporated into planets, whereas others are fragments of once-larger planetesimals that collided with each other and disintegrated very early in the history of the Solar System. The debris in the asteroid belt never merged to form a planet because it is constantly churned by Jupiter's gravitational pull. Asteroids are too small for their own gravity to reshape them into spheres, so they are irregular, pockmarked masses (**Fig. Bx 2.1a**). Astronomers have found 1,000 asteroids with diameters greater than 30 km and estimate that there may be 10 million more with diameters greater than 1 km. Though asteroids are numerous, their combined mass only equals that of Earth's Moon.

A **comet** is an icy planetesimal whose highly elliptical orbit brings it sufficiently close to the Sun that, during part of its journey, the comet evaporates and releases gas and dust to form a glowing tail. Comets that take less than 200 years to orbit the Sun originate from a disk-shaped region of icy fragments called the Kuiper Belt, extending from the orbit of Neptune out to a distance of about 50 times the radius of Earth's orbit. Those with longer orbits originate from a diffuse spherical region of icy fragments called the Oort Cloud, which extends much farther. All told, there could be a trillion objects in the Oort Cloud and the Kuiper Belt, with a combined mass that may exceed the mass of Jupiter. Objects from the Oort Cloud or the Kuiper Belt become comets when gravity tugs on them and sends them on a trajectory into the inner Solar System.

In recent decades, researchers have sent spacecraft to observe comets. During an approach to Halley's comet in 1986, *Giotto* photographed jets of gas and dust spurting from the comet's surface (**Fig. Bx 2.1b, c**). *Stardust* visited a comet in 2004 and returned to Earth with samples, and in 2005, *Deep Impact* dropped a copper ball on a comet and analyzed the debris ejected by the impact. Such studies confirm that comets consist of frozen water (H_2O), carbon dioxide (CO_2), methane (CH_4), ammonia (NH_3), and other volatile compounds, along with a variety

---

The orbit of Neptune, the outermost planet, defines the inner edge of the Kuiper Belt—once we've passed this orbit, we're traversing *interplanetary space*. (Pluto, once considered to be a planet, is now thought of as a large and relatively close object of the Kuiper Belt.) In interplanetary space, the concentration of atoms increases to between 5,000 and 100,000 per liter, still a profound vacuum, but much less so than interstellar space.

The orbits of the planets all lie in the same plane, called the ecliptic—as we zoom across the ecliptic, we pass the ice-giant planets (Neptune and Uranus), the gas-giant planets (Saturn and Jupiter), the asteroid belt (a ring of rocky and metallic chunks ranging from dust size to several hundred km in diameter), and then the terrestrial planets—reddish Mars, bluish Earth, greenish cloud-enshrouded Venus, and finally small, cratered Mercury (Fig. 2.2). Even from the distance of space, Earth looks special, so we pick it out to approach more closely.

Earth's Magnetic Field

As we approach the Earth, our spaceship's instruments detect the planet's magnetic field, like a signpost shouting, "Approaching Earth!" A **magnetic field**, in a general sense, is the region affected by the force emanating from a magnet. This force, which grows progressively stronger as you approach the magnet, can attract or repel another magnet and can cause charged particles to move. Earth's

FIGURE Bx2.1   Images of asteroids and comets.

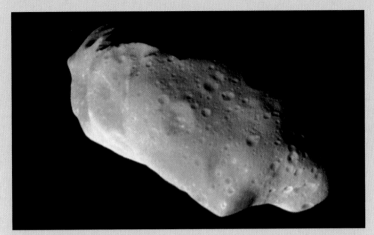

(a) Photograph of the asteroid Ida, a body that is about 56 km long.

(b) Photograph of comet Hale-Bopp, which approached the Earth in 1997. The head of this comet is about 40 km across.

of organic chemicals and dust (tiny rocky or metallic particles). Considering these components, astronomers often refer to comets as "dirty snowballs."

Some comet paths cross the orbits of planets, so collisions between comets and planets can and do occur. In 1994, astronomers observed four huge impacts between a fragmented comet and Jupiter. One of the impacts resulted in a 6-million megaton explosion. This would be equivalent to blowing up 600 times the entire nuclear arsenal on Earth all at once!

Comets have collided with the Earth during historic time. For example, one exploded in the atmosphere above Tunguska, Siberia, in 1908 and flattened trees over an area of 2,150 square kilometers. Even larger comet impacts in the geologic past may have been responsible for disrupting life on Earth, as we discuss later in this book. Researchers speculate that comets may have added significant water to the Earth over its history and perhaps even seeded the Earth with life-related chemicals.

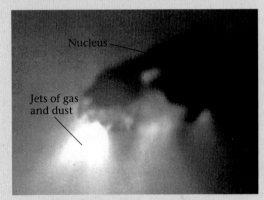

(c) A close-up of Halley's comet, in 1986. On the side facing the Sun, jets of gas and dust spew into space. The solid nucleus is 14 km long.

magnetic field, like the familiar magnetic field around a bar magnet, is largely a **dipole**, meaning that it has two poles—a north pole and a south pole (Fig. 2.3a, b). (When you bring two bar magnets close to one another, opposite poles attract and like poles repel.) By convention, we represent the orientation of a magnetic dipole by an arrow that points from the south pole to the north pole, and we represent the magnetic field of a magnet by a set of invisible magnetic field lines that curve through the space around the magnet. Arrowheads along these lines point in a direction to complete a loop from north to south. Magnetized needles, such as iron filings or compass needles, when placed in a field align with the magnetic field lines.

Simplistically, we can represent the Earth's magnetic field as emanating from an imaginary bar magnet in the planet's interior. The north pole of this bar lies near the south geographic pole of the Earth, where as the south pole of the bar lies near the north geographic pole. (The *geographic poles* are the places where the spin axis of the Earth intersects the planet's surface.) Nevertheless, geologists and geographers by convention refer to the magnetic pole closer to the north geographic pole as the *north magnetic pole*, and the magnetic pole closer to the south geographic pole as the *south magnetic pole*. This way, the *north-seeking end* of a compass points toward the north geographic pole, since opposite ends of a magnet attract.

Mercury has a crater-pocked surface and icy poles.

Dense clouds hide the surface of Venus.

The surface of Mars varies in elevation, and is host to craters and polar ice caps.

Earth's surface has both land and sea. The atmosphere is largely transparent.

FIGURE 2.2 An explorer from outer space would quickly realize that the four terrestrial planets look quite different from one another.

The solar wind, high-velocity charged particles emitted by the Sun, interacts with Earth's magnetic field, distorting it into a huge teardrop pointing away from the Sun. Fortunately, the magnetic field deflects most (but not all) of the particles, so that they do not reach Earth's surface. In this way, the magnetic field acts like a shield against the solar wind; the region inside this magnetic shield is called the magnetosphere (**Fig. 2.3c**).

Though it protects the Earth from most of the solar wind, the magnetic field does not stop our spaceship, and it continues to speed toward the planet. At distances of about 3,000 km and 10,500 km out from the Earth, it encounters the Van Allen radiation belts, named for the physicist who first recognized them in 1959. These contain solar wind particles as well as cosmic rays (nuclei of atoms emitted from supernova explosions) that were moving so fast they were able to penetrate the weaker outer part of the magnetic field, only to be trapped by the stronger magnetic field closer to the Earth. By trapping cosmic rays, the Van Allen belts protect life on Earth from dangerous radiation. Some charged particles make it past the Van Allen belts and are channeled along magnetic field lines to the polar regions of Earth. When these particles interact with gas atoms in the upper atmosphere they cause the gases to glow, like the gases in neon signs, creating spectacular aurorae (**Fig. 2.3d**).

A Plunge through the Atmosphere

As our spaceship descends further, we enter Earth's **atmosphere**, an envelope of gas consisting of 78% nitrogen (N_2) and 21% oxygen (O_2), with minor amounts (1% total) of argon, carbon dioxide (CO_2), neon, methane, ozone, carbon monoxide, and sulfur dioxide (**Fig. 2.4a, b**). Other terrestrial planets have atmospheres, but none of them are like Earth's.

The weight of overlying air squeezes on the air below, pushing gas molecules in the air below closer together. Thus, both the density (mass per unit volume) of air and the air pressure (the amount of push that the air exerts on material beneath it) increases as elevation decreases (**Fig. 2.4c**). Technically, we specify pressure in units of force, or push, per unit area. Such units include atmospheres (abbreviated atm) and bars, where 1 atm = 1.04 kilograms per square centimeter, or 14.7 pounds per square inch. An atmosphere and a bar are almost the same: 1 atm = 1.01 bars. At sea level, average air pressure is 1 atm, whereas on the peak of Mt. Everest, 8.85 km above sea level, air pressure is only 0.3 atm. In fact, *99% of atmospheric gas lies below 50 km*. Notably, humans cannot live for long at elevations greater than about 5.5 km.

The nature of the atmosphere changes with increasing distance from the Earth's surface. Because of these changes, atmospheric scientists divide the atmosphere into layers. Most winds and clouds develop only in the lowest layer, the troposphere. The layers of the atmosphere that lie above the troposphere are named, in sequence from base to top: the stratosphere, the mesosphere, and the thermosphere (**Fig. 2.4d**). Boundaries between layers are defined as elevations at which temperature stops decreasing and starts increasing, or vice versa. Boundaries are named for the underlying layer. For example, the boundary between the troposphere and the overlying stratosphere is the tropopause.

Take-Home Message

- In addition to the Sun and planets, the Solar System contains moons, asteroids, comets, and objects of the Kuiper Belt and Oort Cloud.
- A magnetic field protects the Earth from solar wind and cosmic rays.
- Comets are composed of ice, whereas asteroids consist of rock or metal.
- Most atmospheric gas (N_2 and O_2) molecules are < 50 km from the surface.

THINK: What causes the aurorae?

Did you ever wonder . . .
how thick is our atmosphere?

FIGURE 2.3 A magnetic field permeates the space around the Earth. It can be symbolized by a bar magnet.

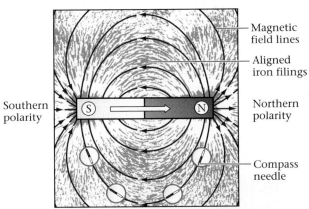

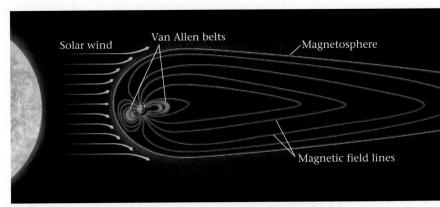

(a) A bar magnet produces a magnetic field. Magnetic field lines point into the "south pole" and out from the "north pole."

(c) Earth behaves like a magnetic dipole, but the field lines are distorted by the solar wind. The Van Allen radiation belts trap charged particles.

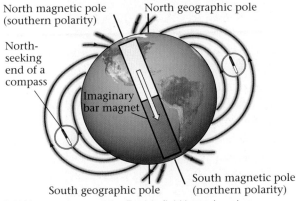

(b) We can represent the Earth's field by an imaginary bar magnet inside.

(d) Charged particles flow toward Earth's magnetic poles and cause gases in the atmosphere to glow, forming colorful aurorae in polar skies.

2.3 WHAT IS THE EARTH MADE OF?

The Earth System

As our spaceship slips into orbit around the Earth, it detects several distinct components—the atmosphere (the gaseous envelope), the hydrosphere (surface and near-surface liquid water), the cryosphere (surface and near-surface ice and snow), the biosphere (the great variety of living organisms), and the solid Earth. Each of these components interacts with others in multitudinous ways. Geologists refer to these components, and the complex interactions among them, together as the **Earth System**. Of all the planets in the Solar System, only Earth currently has life and liquid water. It lies within the habitable zone, the distance from the Sun in which liquid water can exist (Fig. 2.5).

Clearly, the Earth System is a dynamic place. Its surface and the objects on it are in motion, its atmosphere and oceans are circulating; materials from its interior spill out on the surface, and materials from the surface sink into the

interior. The forces driving all this activity ultimately come from heat inside the Earth, from gravity, and from the Sun's heat and light.

Land and Oceans

Imagine that our first task, once our spaceship has gone into orbit around the Earth, is to make a map of the planet. What features should go on this map? Starting with the most obvious, we note that land (continents and islands) forms about 30% of the surface (Fig. 2.6). Some of the land surface consists of solid rock, whereas some has a covering of **sediment** (materials such as sand and gravel, in which the grains are not stuck together). **Surface water** covers the remaining 70% of the Earth. Most surface water is salty and occupies oceans, but

FIGURE 2.4 Characteristics of the atmosphere that envelops the Earth.

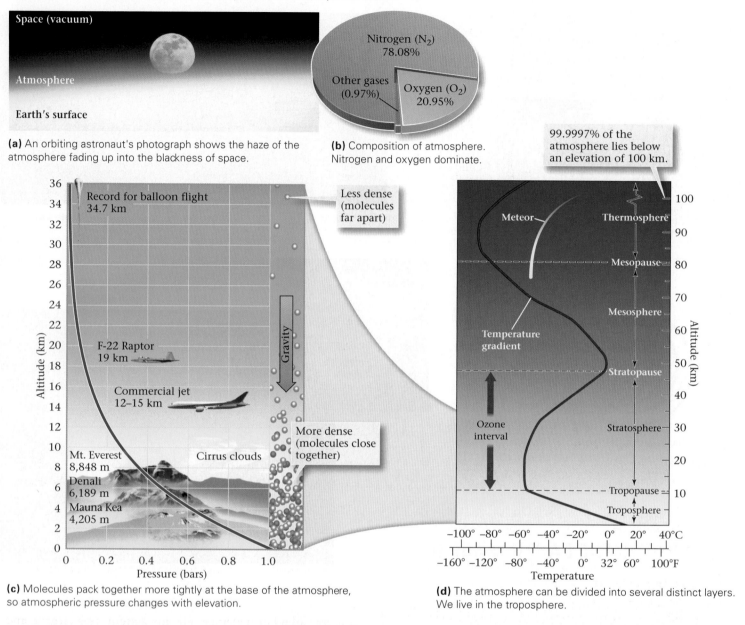

(a) An orbiting astronaut's photograph shows the haze of the atmosphere fading up into the blackness of space.

(b) Composition of atmosphere. Nitrogen and oxygen dominate.

Nitrogen (N_2) 78.08%

Other gases (0.97%)

Oxygen (O_2) 20.95%

Space (vacuum)

Atmosphere

Earth's surface

36
34 Record for balloon flight 34.7 km
32
30
28
26
24
22
20 F-22 Raptor 19 km
18
16
14 Commercial jet 12–15 km
12
10
8 Mt. Everest 8,848 m
6 Denali 6,189 m
4 Mauna Kea 4,205 m
2
0
0   0.2   0.4   0.6   0.8   1.0

Altitude (km)

Pressure (bars)

Less dense (molecules far apart)

Gravity

More dense (molecules close together)

Cirrus clouds

(c) Molecules pack together more tightly at the base of the atmosphere, so atmospheric pressure changes with elevation.

99.9997% of the atmosphere lies below an elevation of 100 km.

Meteor

Temperature gradient

Ozone interval

Thermosphere
Mesopause
Mesosphere
Stratopause
Stratosphere
Tropopause
Troposphere

100
90
80
70
60
50
40
30
20
10

Altitude (km)

−100° −80° −60° −40° −20° 0° 20° 40°C
−160° −120° −80° −40° 0° 32° 60° 100°F
Temperature

(d) The atmosphere can be divided into several distinct layers. We live in the troposphere.

FIGURE 2.5 In our Solar System, only Earth lies within the relatively narrow habitable zone.

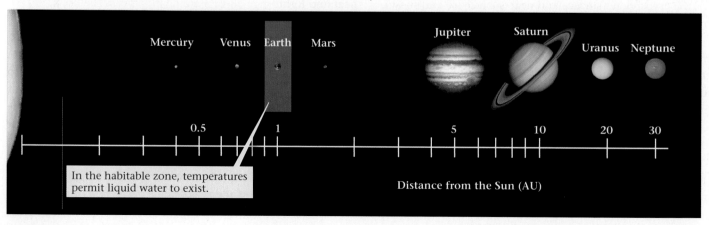

Mercury   Venus   Earth   Mars   Jupiter   Saturn   Uranus   Neptune

0.5   1   5   10   20   30

In the habitable zone, temperatures permit liquid water to exist.

Distance from the Sun (AU)

FIGURE 2.6 This map of the Earth shows variations in elevation on both the land surface and the sea floor. Darker blues are deeper water in the ocean. Greens are lower elevation on land.

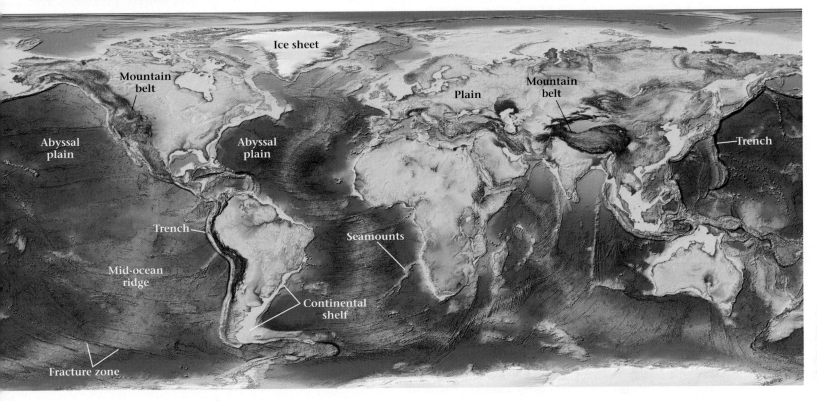

some is fresh and fills lakes and rivers. Our instruments also detect **groundwater**, which is the water that fills cracks and holes (pores) within rock and sediment under the land surface. Finally, we find that ice covers significant areas of land and sea in polar regions and at high elevations, and that living organisms populate the land, sea, air, and even the upper few kilometers of the subsurface.

To finish off our map of the Earth's surface, we note that it is not flat. **Topography**, the variation in elevation of the land surface, defines plains, mountains, and valleys (See for Yourself C, p. S-6). Similarly, **bathymetry**, the variation in elevation of the ocean floor, defines mid-ocean ridges, plains, and deep-ocean trenches (see Fig. 2.6).

A graph, called a hypsometric curve, plotting surface elevation on the vertical axis and the percentage of the Earth's surface on the horizontal axis shows that a relatively small proportion of the Earth's surface occurs at very high elevations (mountains) or at great depths (deep trenches). In fact, most of the land surface lies just within a kilometer of sea level, and most of the sea floor lies between 4 and 5 km deep (Fig. 2.7). A slight change in sea level would dramatically change the amount of dry land.

> **Did you ever wonder . . .**
> what is the average elevation of the land?

Categories of Earth Materials

At this point, we leave our fantasy space voyage and turn our attention to the materials that make up the solid Earth, because we need to be aware of these before we can discuss the architecture of the Earth's interior. Let's begin by reiterating that the Earth consists mostly of elements produced by fusion reactions in stars and supernova explosions. Only four elements (iron, oxygen, silicon, and magnesium) make up 91.2% of the Earth's mass; the remaining 8.8% consists of the other 88 elements (Fig. 2.8). The elements of the Earth comprise a great variety of materials. For reference in this chapter and the next, we introduce the basic categories of materials. All of these will be discussed in more detail later in the book.

- *Organic chemicals*. Carbon-containing compounds that either occur in living organisms, or have characteristics that resemble compounds in living organisms, are called **organic chemicals**.
- *Minerals*. A solid, natural substance in which atoms are arranged in an orderly pattern is called a **mineral**. A single coherent sample of a mineral that grew to its present shape is a crystal, whereas an irregularly shaped sample, or a fragment derived from a once-larger crystal or group of crystals, is a grain.

FIGURE 2.7 This graph shows a hypsometric curve, indicating the proportions of the Earth's solid surface at different elevations. Two principal zones—the continents and adjacent continental shelf areas (the submerged margins of continents) and the ocean floor—account for most of Earth's area. Mountains and deep trenches cover relatively little area.

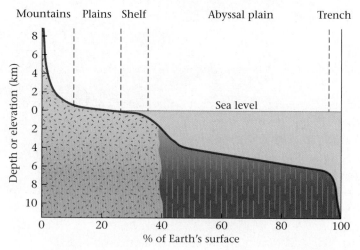

- *Glasses.* A solid in which atoms are not arranged in an orderly pattern is called **glass**.
- *Rocks.* Aggregates of mineral crystals or grains, and masses of natural glass, are called **rocks**. Geologists recognize three main groups of rocks. (1) Igneous rocks develop when hot molten (liquid) rock cools and freezes solid. (2) Sedimentary rocks form from grains that break off preexisting rock and become cemented together, or from minerals that precipitate out of a water solution. (3) Metamorphic rocks form when preexisting rocks undergo changes in response to heat and pressure.
- *Sediment.* An accumulation of loose mineral grains (grains that have not stuck together) is called sediment.
- *Metals.* Solids composed of metal atoms (such as iron, aluminum, copper, and tin) are called **metals**. In a metal, outer electrons are able to flow freely. An **alloy** is a mixture containing more than one type of metal atom.
- *Melts.* **Melts** form when solid materials become hot and transform into liquid. Molten rock is a type of melt—geologists distinguish between magma, which is mol-

ten rock beneath the Earth's surface, and lava, molten rock that has flowed out onto the Earth's surface.

- *Volatiles.* Materials that easily transform into gas at the relatively low temperatures found at the Earth's surface are called **volatiles**.

The most common minerals in the Earth contain **silica** (a compound of silicon and oxygen) mixed in varying proportions with other elements. These minerals are called silicate minerals. Not surprisingly, rocks composed of silicate minerals are **silicate rocks**. Geologists distinguish four classes of igneous silicate rocks based, in essence, on the proportion of silica to iron and magnesium. In order, from greatest to least proportion of silica to iron and magnesium, these classes are felsic (or silicic), intermediate, mafic, and ultramafic. As the proportion of silica in a rock increases, the density (mass per unit volume) decreases. Thus, felsic rocks are less dense than mafic rocks. Many different rock types occur in each class, as will be discussed in detail in Chapters 6 through 8. For now, we introduce the four rock types whose names we need to know for our discussion of the Earth's layers that follows. These are (1) **granite**, a felsic rock with large grains; (2) **gabbro**, a mafic rock with large grains; (3) **basalt**, a mafic rock with small grains; and (4) **peridotite**, an ultramafic rock with large grains.

Take-Home Message

- Geologists use the phrase "Earth System" in reference to the variety of interacting realms in, on, and around the planet.
- 70% of the surface is sea, and 30% is land; parts of both are ice covered.
- Surface elevations range from −11 km to +8.8 km. Most land elevation is < 1 km, and most sea floor lies between −4 and −5 km.
- A great variety of different materials comprise the Earth.
- Examples of common silicate rocks include granite, gabbro, basalt, and peridotite.

THINK: How do geologists distinguish among classes of silicate rocks?

2.4 HOW DO WE KNOW THAT THE EARTH HAS LAYERS?

The world's deepest mine shaft penetrates gold-bearing rock that lies about 3.5 km (2 miles) beneath South Africa. Though miners seeking this gold must begin their workday by plummeting straight down a vertical shaft for almost ten minutes aboard the world's fastest elevator, the shaft represents little more than a pinprick on Earth's surface, when compared with the planet's radius of 6,371 km. Even the deepest well ever drilled, a 12-km-deep hole in northern Russia, penetrates only the upper 0.2% of

FIGURE 2.8 The proportions of major elements making up the mass of the whole Earth. Note that iron and oxygen account for most of the mass.

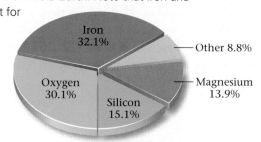

the Earth. We literally live on the thin skin of our planet, its interior forever inaccessible to our direct observation.

People have speculated about the Earth's interior since ancient times. What is the source of incandescent lavas spewed from volcanoes, of precious gems and metals, of sparkling spring water, and of the mysterious powers that shake the ground and topple buildings? Without the ability to observe the Earth's interior firsthand, pre-twentieth-century authors dreamed up fanciful images of it. For example, the English poet John Milton (1608–1674) described the underworld as a "dungeon horrible, on all sides round, as one great furnace flamed" (Fig. 2.9). Perhaps his image was inspired by volcanoes in the Mediterranean. In the eighteenth and nineteenth centuries, some European writers thought the Earth's interior resembled a sponge, containing open caverns variously filled with molten rock, water, or air. In fact, in the popular 1864 novel *Journey to the Center of the Earth*, by the French author Jules Verne, three explorers find a route through interconnected caverns to the Earth's center. Today, we picture the Earth's interior as having distinct layers, and no open spaces. This image is the end product of interpreting many clues, as we will now see.

Early Clues to Characterize the Interior

The first key to understanding the Earth's interior came from studies that provided an estimate of the planet's density (mass per unit volume). To determine Earth's density, one must first determine the amount of matter making up the Earth. In 1776, the British Royal Astronomer, Nevil Maskelyne, provided the first realistic estimate of Earth's mass. Maskelyne postulated that he could weigh the Earth by examining the deflection of a plumb bob attached to a surveying instrument. Specifically, the

angle of deflection (β) of the plumb bob caused by the gravitational attraction of a mountain indicates the magnitude of gravitational attraction exerted by the mountain's mass relative to the gravitational attraction of the Earth's mass (Fig. 2.10). Maskelyne tested his hypothesis at Schiehallion Mountain in Scotland. His results led to an estimate that the Earth's average density is 4.5 times the density of water (i.e., 4.5 g/cm^3 in the metric system). In 1778, another physicist using a different method arrived at a density estimate of 5.45 g/cm^3, fairly close to modern estimates. Significantly, typical rocks (such as granite and basalt) at the surface of the Earth have a density of only 2.5–3.0 g/cm^3, so the *average* density of Earth exceeds that of its surface rocks. Certainly, the open voids that Jules Verne described could not exist!

In 1896, a physicist named Emil Wiechert emphasized that the interior of the Earth must include metal, because only metal has a high enough density, when averaged with that of rock, to give the Earth its observed average density. He speculated that the metal occurs in a ball at the planet's center, and estimated the radius of this ball. The key to Wiechert's thinking is that the spinning Earth is only slightly flattened in response to centrifugal force. If the metal were distributed evenly, more mass would be near the surface, and centrifugal force would cause the planet to be flattened more than it is. The metallic ball in the interior came to be known as the core and the rocky shell surrounding the core as the mantle. Researchers have determined that the core has a density approaching 13 g/cm^3.

Additional insight into the nature of Earth's interior came from the study of tides, the rise and fall of the Earth's surface in response to the gravitational attraction of the Moon and Sun. If the Earth were composed of a liquid surrounded by only a thin solid crust, then the surface of the

FIGURE 2.9 John Martin (1789-1854) created an image of the Earth's interior, following a description from *Paradise Lost*, an epic poem published by John Milton in 1667.

FIGURE 2.10 A surveyor noticed that the plumb line was deflected by an angle β, owing to the gravitational attraction of the mountain. The angle represents the ratio between the mass of the mountain and the mass of the whole Earth.

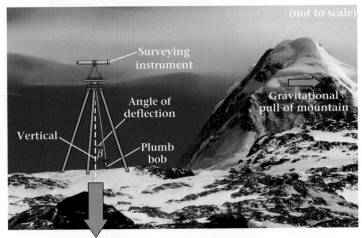

(not to scale)

Surveying instrument

Angle of deflection

Gravitational pull of mountain

Vertical

β

Plumb bob

Gravitational pull of Earth

land would rise and fall daily, like the surface of the sea. We don't observe such behavior, so the Earth's interior must be largely solid.

Taken together, basic observations about the density and shape of our planet led nineteenth-century geologists to conclude that the Earth resembled a hard-boiled egg, in that it had three principal layers (Fig. 2.11a, b): a not-so-dense **crust** (like an eggshell, composed of rocks such as granite, basalt, and gabbro), overlying a denser solid **mantle** (the "white," composed of a then-unknown material), and a very dense **core** in the middle (the "yolk," composed of a different unknown material). Clearly, many questions remained. How thick are the layers? Are the boundaries between layers sharp or gradational? And what exactly are the layers composed of?

Clues from the Study of Earthquakes: Refining the Image

When rock within the outer portion of the Earth suddenly breaks and slips along a fracture called a **fault**, it generates shock waves (abrupt vibrations) that travel through the surrounding rock outward from the break. Where these waves cause the surface of the Earth to vibrate, people feel an **earthquake**, an episode of ground shaking. You can simulate this process, at a small scale, when you break a stick between your hands and feel the snap with your hands (Fig. 2.12a, b). In 1880, a team of British researchers working in Japan, where earthquakes happen many times a year, invented an instrument called a seismograph that could detect earthquake vibrations that were too gentle to be felt directly by people. The researchers thought that their seismograph was only detecting earthquakes in Japan and nearby regions. Then, in 1889, a physicist in Germany noticed

FIGURE 2.11 An early image of Earth's internal layers.

(a) The hard-boiled egg analogy for the Earth's interior.

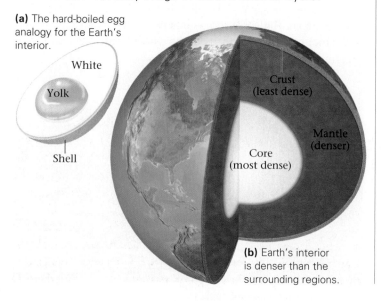

White

Yolk

Shell

Crust (least dense)

Mantle (denser)

Core (most dense)

(b) Earth's interior is denser than the surrounding regions.

FIGURE 2.12 Faulting and earthquakes.

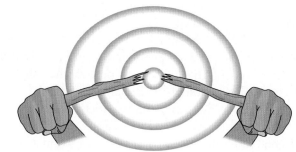

(a) Snapping a stick generates vibrations that pass through the stick to your hands.

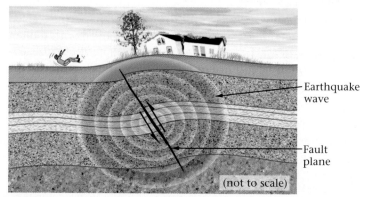

Earthquake wave

Fault plane

(not to scale)

(b) Similarly, when the rock inside the Earth suddenly breaks and slips, forming a fracture called a fault, it generates shock waves that pass through the Earth and shake the surface.

that the pendulum in his lab began to move without having been touched. He reasoned that the pendulum was actually standing still while the Earth moved under it. A few days later, he read in a newspaper that a large earthquake had taken place in Japan minutes before the movement of his pendulum began. The physicist deduced that the energy generated by the earthquake *had traveled all the way through the Earth* from Japan and had vibrated his laboratory in Germany. The energy moved in the form of waves, called either seismic waves or earthquake waves.

Geologists immediately realized that the study of seismic waves traveling through the Earth might provide a tool for exploring the Earth's insides, much as ultrasound today helps doctors study a patient's insides. Specifically, laboratory measurements demonstrated that earthquake waves travel at different velocities (speeds) through different materials. Thus, by detecting depths at which velocities suddenly change, geoscientists pinpointed the boundaries between layers and even recognized subtler boundaries within layers. For example, such studies led geoscientists to subdivide the mantle into the upper mantle and lower mantle, and subdivide the core into the inner core and outer core. (Chapter 10 provides further details about earthquakes, and Interlude D shows *how* the study of earthquake waves defines the Earth's layers.)

Pressure and Temperature Inside the Earth

In order to keep underground tunnels from collapsing under the pressure created by the weight of overlying rock, mining engineers must design sturdy support structures. It is no surprise that deeper tunnels require stronger supports: the downward push from the weight of overlying rock increases with depth, simply because the mass of the overlying rock layer increases with depth. At the Earth's center, pressure probably reaches about 3,600,000 atm.

Temperature also increases with depth in the Earth. Even on a cool winter's day, miners who chisel away at gold veins exposed in tunnels 3.5 km below the surface swelter in temperatures of about 53°C (127°F). We refer to the rate of change in temperature with depth as the **geothermal gradient**. In the upper part of the crust, the geothermal gradient averages between 20° and 30°C per km. At greater depths, the rate decreases (to 10°C per km or less). Thus, 35 km below the surface of a continent, the temperature reaches 400° to 700°C. No one has ever directly measured the temperature at the Earth's center, but calculations suggest it may exceed 4,700°C (Fig. 2.13).

> **Did you ever wonder . . .**
> how hot it gets at the center of this planet?

FIGURE 2.13 The Earth's geotherm shows how temperature increases with depth.

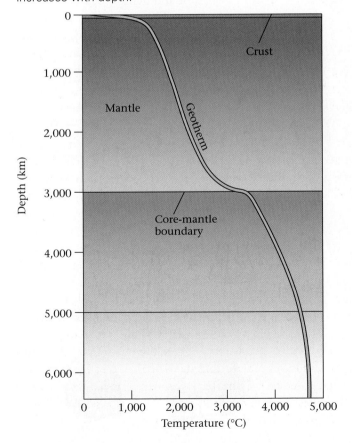

Take-Home Message

- Measurements of the Earth's mass indicate that the planet's interior is denser than rocks exposed at its surface.
- 19th century geologists proposed that the Earth has a crust, mantle, and core.
- Study of earthquake waves provides a refined image of the interior.
- Temperature and pressure increase with depth; it's about 4700°C at the center.

THINK: Is the geothermal gradient constant from Earth's surface to center?

2.5 WHAT ARE THE LAYERS MADE OF?

We saw earlier that the material composing the Earth's insides must be much denser than familiar surface rocks such as granite and basalt. To discover what this material consists of, geologists

- examined the compositions of meteorites, for some meteorites are chunks of the interiors of planetesimals and thus may resemble materials inside the Earth (Box 2.2);
- conducted laboratory experiments to determine what kinds of materials inside the Earth could be a source of magma;
- studied unusual chunks of rock that may have been carried up from the mantle in magma; and
- conducted laboratory experiments to measure densities in samples of known rock types, so that they could compare these with observed densities in the Earth.

As a result of this work, we now have a pretty clear sense of what the layers inside the Earth are made of, though this picture is constantly being adjusted as new findings become available. Let's now look at the properties of individual layers, starting with the Earth's surface (Fig. 2.14a–c).

The Crust

When you stand on the surface of the Earth, you are standing on top of its outermost layer, the crust. The crust is our home and the source of all our resources. How thick is this all-important layer? Or, in other words, what is the depth to the crust-mantle boundary? An answer came from the studies of Andrija Mohorovičić, a researcher working in Zagreb, Croatia. In 1909, he discovered that the velocity of earthquake waves suddenly increased at a depth of tens of kilometers beneath the Earth's surface, and he suggested that this increase was caused by an abrupt change in the properties of rock (see Interlude D for further detail). Later studies showed that this change can be found most everywhere around our planet, though it occurs at different depths in different locations. Specifically, it's deeper beneath continents than beneath oceans. Geologists now consider the change to be the crust-mantle boundary, and they refer to it as the **Moho** in Mohorovičić's honor. The relatively shallow depth of the Moho (7 to 70 km, depending on location) as compared to the radius of the Earth (6,371 km) emphasizes that the crust is very thin indeed. In fact, the crust is only about 0.1% to 1.0% of the Earth's radius, so if the Earth were the size of a balloon, the crust would be about the thickness of the balloon's skin.

Geologists distinguish between two fundamentally different types of crust—oceanic crust, which underlies the sea floor, and continental crust, which underlies continents. The crust is not simply cooled mantle, like the skin on chocolate pudding, but rather consists of a variety of rocks that differ in composition (chemical makeup) from mantle rock.

Oceanic crust is only 7 to 10 km thick. At highway speeds (100 km per hour), you could drive a distance equal to the thickness of the oceanic crust in about five minutes. At the top, we find a blanket of sediment, generally less than 1 km thick, composed of clay and tiny shells that settled like snow out of the sea. Beneath this blanket, the oceanic crust consists of a layer of basalt and, below that, a layer of gabbro.

Most continental crust is about 35 to 40 km thick—about four to five times the thickness of oceanic crust—but its thickness varies significantly. In some places, continental crust has been stretched and thinned so it's only 25 km from the surface to the Moho. And in some places, the crust has been crumpled and thickened so that the crust may be up to 70 km thick. In contrast to oceanic crust, continental crust contains a great variety of rock types, ranging from mafic to felsic in composition. On average, upper continental crust is less mafic than oceanic crust—it has a felsic (granite-like) to intermediate composition—so continental crust overall is less dense than oceanic crust. Notably, oxygen is the most abundant element in the crust (Fig. 2.15).

The Mantle

The mantle of the Earth forms a 2,885-km-thick layer surrounding the core. In terms of volume, it is the largest part of the Earth. In contrast to the crust, the mantle consists entirely of an ultramafic (dark and dense) rock called peridotite. This means that peridotite, though rare at the Earth's surface, is actually the most abundant rock in our planet! On the basis of the

Did you ever wonder . . .
what is the most abundant rock of the Earth?

FIGURE 2.14   A modern view of Earth's interior layers.

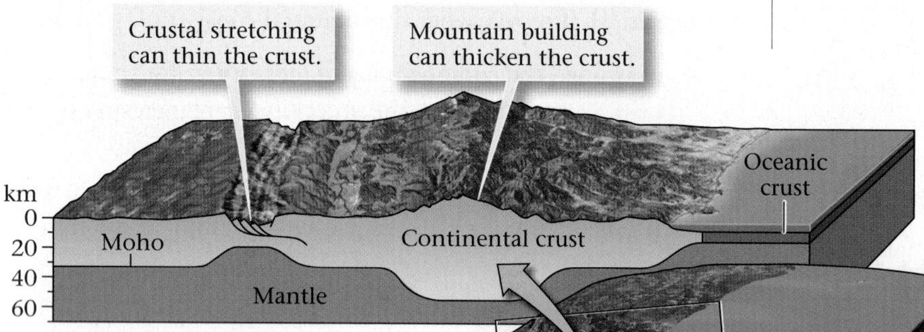

Crustal stretching can thin the crust.

Mountain building can thicken the crust.

(a) There are two basic types of crust. Oceanic crust is thinner and consists of basalt and gabbro. Continental crust varies in thickness and rock type.

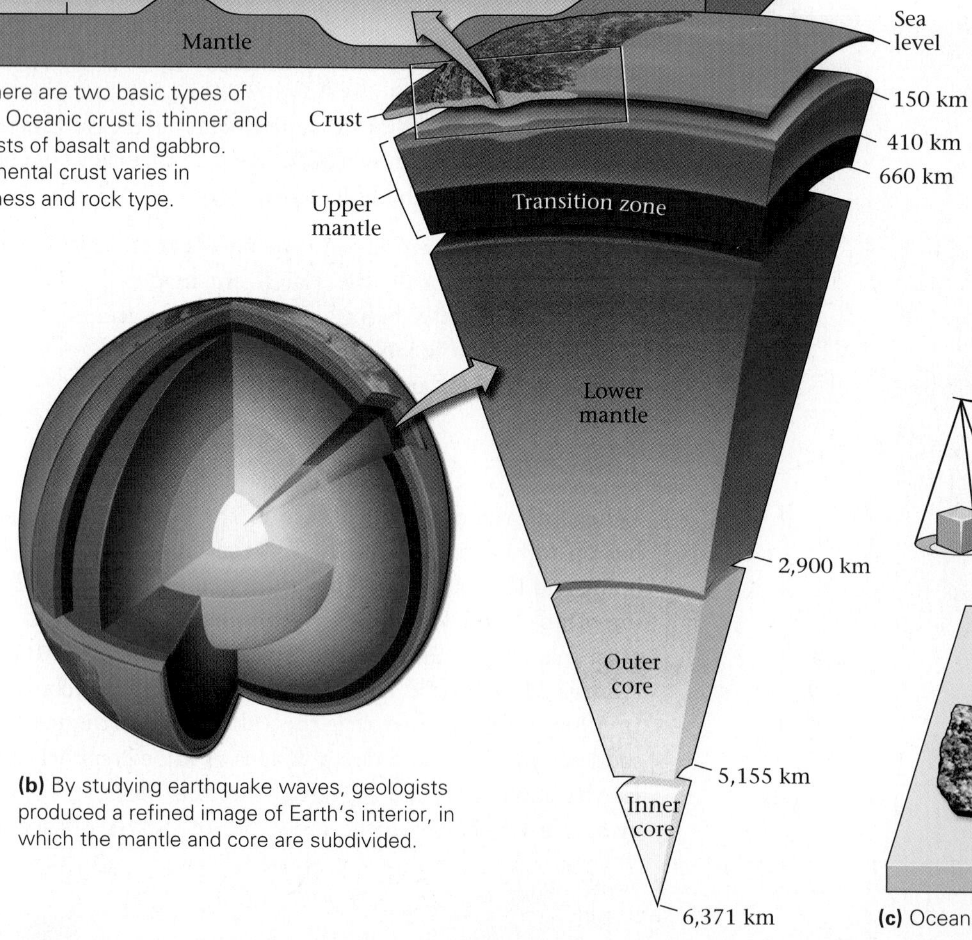

(b) By studying earthquake waves, geologists produced a refined image of Earth's interior, in which the mantle and core are subdivided.

(c) Oceanic crust is denser than continental crust.

BOX 2.2

Meteorites: Clues to What's Inside

During the early days of the Solar System, the Earth collided with and incorporated countless planetesimals and smaller fragments of solid material lying in its path. Intense bombardment ceased about 3.9 Ga, but even today collisions with space objects continue, and over 1,000 tons of material (rock, metal, dust, and ice) fall to Earth, on average, every year. The vast majority of this material consists of fragments derived from comets and asteroids sent careening into the path of the Earth after billiard-ball-like collisions with each other out in space, or because of the gravitational pull of a passing planet. Some of the material, however, consists of chips of the Moon or Mars, ejected into space when large objects collided with those bodies.

Did you ever wonder . . .
where meteorites come from?

Astronomers refer to any object from space that enters the Earth's atmosphere as a meteoroid. Meteoroids move at speeds of 20 to 75 km/s (over 45,000 mph), so fast that when they reach an altitude of about 150 km, friction with the atmosphere causes them to heat up and evaporate, leaving a streak of bright, glowing gas. The glowing streak, an atmospheric phenomenon, is a **meteor** (also known colloquially, though incorrectly, as a "falling star") (**Fig. Bx2.2a**). Most visible meteors completely evaporate by an altitude of about 30 km. But dust-sized ones may slow down sufficiently to float to Earth, and larger ones (fist-sized or bigger) can survive the heat of entry to reach the surface of the planet. In some cases, meteoroids explode in brilliant fireballs.

Objects that strike the Earth are called **meteorites**. Although almost all meteorites are small and have not caused notable damage on Earth during human history, a very few have smashed through houses, dented cars, and bruised people. During the longer term of Earth history, however, there have been some catastrophic collisions that left huge craters (**Fig. Bx 2.2b**).

Most meteorites are asteroidal or planetary fragments, for the icy material of small cometary bodies is too fragile to survive the fall. Researchers recognize three basic classes of meteorites: iron (made of iron-nickel alloy), stony (made of rock), and stony iron (rock embedded in a matrix of metal). Of all known meteorites, about 93% are stony and 6% are iron (**Fig. Bx2.2c**). From their composition, researchers have concluded that some meteors (a special subcategory of stony meteorites called carbonaceous chondrites, because they contain carbon and small spherical nodules called chondrules) are asteroids derived from planetesimals that never underwent differentiation into a core and mantle. Other stony meteorites and all iron meteorites are asteroids derived from planetesimals that had differentiated into a metallic core and a rocky mantle early in Solar System history but later shattered into fragments during collisions with other planetesimals. Most meteorites appear to be about 4.54 Ga, but carbonaceous chondrites are as old as 4.57 Ga and are the oldest solar system materials ever measured.

Since meteorites represent fragments of undifferentiated and differentiated planetesimals, geologists consider the *average* composition of meteorites to be representative of the average composition of the whole Earth. In other words, the estimates that geologists use for the proportions of different elements in the Earth are based largely on studying meteorites. Stony meteorites are probably similar in composition to the mantle, and iron meteorites are probably similar in composition to the core.

FIGURE Bx2.2 Meteors and meteorites.

(a) A shower of meteors over Hong Kong in 2001.

(b) The Barringer meteor crater in Arizona. It formed about 50,000 years ago and is 1.1 km in diameter.

The meteorite forming the crater was 50 m across.

(c) Examples of stony meteorites (left) and iron meteorites (right).

FIGURE 2.15 A table and a graph illustrating the abundance of elements in the Earth's crust.

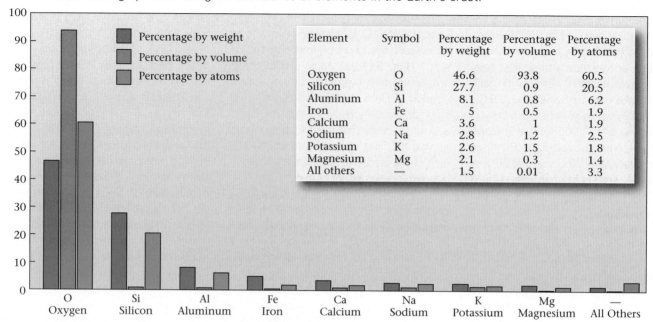

Element	Symbol	Percentage by weight	Percentage by volume	Percentage by atoms
Oxygen	O	46.6	93.8	60.5
Silicon	Si	27.7	0.9	20.5
Aluminum	Al	8.1	0.8	6.2
Iron	Fe	5	0.5	1.9
Calcium	Ca	3.6	1	1.9
Sodium	Na	2.8	1.2	2.5
Potassium	K	2.6	1.5	1.8
Magnesium	Mg	2.1	0.3	1.4
All others	—	1.5	0.01	3.3

occurrence of changes in the velocity of earthquake waves, geoscientists divide the mantle into two sublayers: the **upper mantle**, down to a depth of 660 km, and the **lower mantle**, from 660 km down to 2,900 km. The **transition zone** is the section between 400 km and 660 km deep (see Interlude D).

Almost all of the mantle is solid rock. But even though it's solid, mantle rock below a depth of 100 to 150 km is so hot that it's soft enough to flow. This flow, however, takes place extremely slowly—at a rate of less than 15 cm a year. *Soft* here does not mean liquid; it simply means that over long periods of time mantle rock can change shape, like soft wax, without breaking. We stated earlier that *almost* all of the mantle is solid. We used the word "almost" because up to a few percent of the mantle has melted. This melt occurs in films or bubbles between grains in the mantle at a depth of 100 to 200 km beneath the ocean floor.

> **Did you ever wonder . . .**
> if the Earth has the same composition throughout?

Though, overall, the temperature of the mantle increases with depth, temperature also varies significantly with location even at the same depth. The warmer regions are less dense, while the cooler regions are denser. The distribution of warmer and cooler mantle indicates that the mantle convects like water in a simmering pot (see **Box 2.3**); warm mantle gradually flows upward, while cooler, denser mantle sinks.

The Core

Early calculations suggested that the core had the same density as gold, so for many years people held the fanciful hope that vast riches lay at the heart of our planet. Alas, geologists eventually concluded that the core consists of a far less glamorous material, iron alloy (iron mixed with tiny amounts of other elements). Studies of earthquake waves led geoscientists to divide the core into two parts, the outer core (between 2,900 and 5,155 km deep) and the inner core (from a depth of 5,155 km down to the Earth's center at 6,371 km). The outer core consists of *liquid* iron alloy. It can exist as a liquid because the temperature in the outer core is so high that even the great pressures squeezing the region cannot keep atoms locked into a solid framework. The iron alloy of the outer core can flow, and this flow generates Earth's magnetic field (see **Geology at a Glance**, pp. 50–51).

The inner core, with a radius of about 1,220 km, is a *solid* iron alloy that may reach a temperature of over 4,700°C. Even though it is hotter than the outer core, the inner core is a solid because it is deeper and is subjected to even greater pressure. The pressure keeps atoms locked together tightly in very dense materials.

Take-Home Message

- Lab experiments and studies of meteorites give us clues as to the composition of the interior.
- The oceanic crust and continental crust differ in composition and thickness.
- The Moho defines the crust/mantle boundary; the mantle contains most of Earth's volume.
- Seismic studies separate the mantle and core into sublayers.
- The outer core is liquid iron alloy, and the inner core is solid iron alloy.

THINK: Which layer of the Earth generates the magnetic field?

BOX 2.3

Heat and Heat Transfer

The atoms and molecules that make up an object do not stay rigidly fixed in place, but rather jiggle and jostle with respect to one another. This vibration creates thermal energy—the faster the atoms move, the greater the thermal energy and the hotter the object. Put another way, the thermal energy in a substance represents the sum of the kinetic energy (energy of motion) of all the substance's atoms. This includes the back-and-forth displacements that an atom makes as it vibrates, as well as the movement of an atom from one place to another.

When we say that one object is hotter or colder than another, we are describing its temperature. **Temperature** is a measure of warmth relative to some standard and represents the average kinetic energy of atoms in the material. In everyday life, we generally use the freezing or boiling point of water at sea level as the standard. When using the Celsius (centigrade) scale, we arbitrarily set the freezing point of water

(at sea level) as 0°C and the boiling point as 100°C; whereas in the Fahrenheit scale, we set the freezing point as 32°F and the boiling point as 212°F.

The coldest a substance can be is the temperature at which its atoms or molecules stand still. We call this temperature absolute zero, or 0K (pronounced "zero kay"), where K stands for Kelvin (after Lord Kelvin, 1824–1907, a British physicist), another unit of temperature. You simply can't get colder than absolute zero, meaning that you can't extract any thermal energy from a substance at 0K (–273.15°C). Degrees in the Kelvin scale are the same increment as degrees in the Celsius scale.

Heat is the thermal energy transferred from one object to another. Heat can be measured in calories. One thousand calories can heat 1 kilogram of water by 1°C. There are four ways in which heat transfer takes place in the Earth System: radiation, conduction, convection, and advection.

Radiation is the process by which electromagnetic waves transmit heat into a body or out of a body (**Fig. Bx2.3a**). For example, when the Sun heats the ground during the day, radiative heating takes place. Similarly, when heat rises from the ground at night, radiative heating is occurring—in the opposite direction.

Conduction takes place when you stick the end of an iron bar in a fire (**Fig. Bx2.3b**). The iron atoms at the fire-licked end of the bar start to vibrate more energetically; they gradually incite atoms farther up the bar to start jiggling, and these atoms in turn set atoms even farther along in motion. In this way, heat slowly flows along the bar until you feel it with your hand. Conduction does not involve actual movement of atoms from one place to another.

Convection takes place when you set a pot of water on a stove (**Fig. Bx2.3c**). The heat from the stove warms the water at the base of the pot by making the molecules of water vibrate faster and move around more. As a consequence, the density of the water at the base of the pot decreases for, as you heat a liquid, the atoms move away from each other and the liquid expands. For a time, cold water remains at the top of the pot; but eventually the warm, less-dense water becomes buoyant relative to the cold, dense water. In a gravitational field, a buoyant material rises (like a styrofoam ball in a pool of water) if the material above it is weak enough to flow out of the way. Since liquid water can flow easily, hot water rises. When this happens, cold water sinks to take its place. The new volume of cold water then heats up and rises itself. Thus, during convection, the actual flow of the material itself carries heat. The trajectory of flow defines **convective cells**.

Advection, a less-familiar process, happens when heat is carried by a fluid flowing through cracks and pores within a solid material (**Fig. Bx2.3d**). The heat brought by the fluid conductively heats up the adjacent solid that the fluid passes through. Advection takes place, for example, if you pass hot water through a metal pipe and the pipe itself gets hot. In the Earth, advection occurs where molten rock rises through the crust beneath a volcano and heats up the crust in the process.

FIGURE Bx2.3 The four processes of heat transfer.

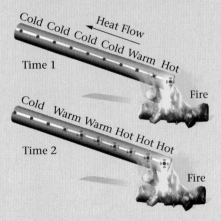

(a) Radiation from sunlight warms the Earth.

(c) Convection takes place when moving fluid carries heat with it. Hot fluid rises while cool fluid sinks, setting up a convective cell.

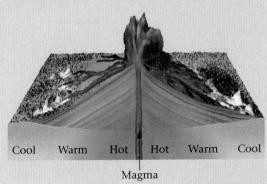

(b) Conduction occurs when you heat the end of an iron bar in a flame. Heat flows from the hot region toward the cold region, as vibrating atoms cause their neighbors to vibrate.

(d) During advection, a hot liquid (such as molten rock) rises into cooler material, and heat then conducts from the hot liquid into the cooler material.

The Earth, from Surface to Center

If we could remove all of the clouds and water, we would see that both the land areas and the seafloor have plains and mountains.

If we could break open the Earth, we would see that its interior consists of a series of concentric layers, called (in order from the surface to the center) the crust, the mantle, and the core. The crust is a relatively thin skin (7–10 km beneath oceans, 25–70 km beneath the land surface). Oceanic crust consists of basalt (mafic rock), while the average upper continental crust is intermediate to silicic. The mantle, which overall has the composition of ultramafic rock, can be divided into three layers: upper mantle, transition zone, and lower mantle. The core can be divided into an outer core of liquid iron alloy and an inner core of solid iron alloy. All terrestrial planets, as well as Earth's Moon, have differentiated to form a crust, mantle, and core. But the relative thicknesses of the different layers are not the same for all planets.

When discussing plate tectonics, it is convenient to call the outer part of the Earth, a relatively rigid shell composed of the crust and uppermost mantle, the lithosphere and the underlying warmer, more plastic portion of the mantle the asthenosphere. These are not shown in this painting.

Within the mantle and outer core, there is swirling, convective flow. Flow within the outer core generates the Earth's magnetic field.

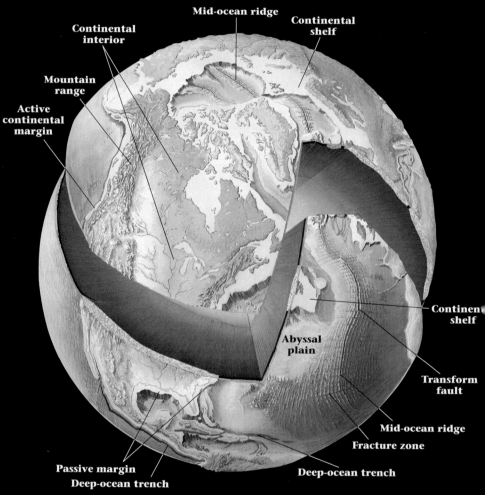

Continental interior · Mid-ocean ridge · Continental shelf · Mountain range · Active continental margin · Abyssal plain · Continental shelf · Transform fault · Mid-ocean ridge · Fracture zone · Deep-ocean trench · Passive margin · Deep-ocean trench

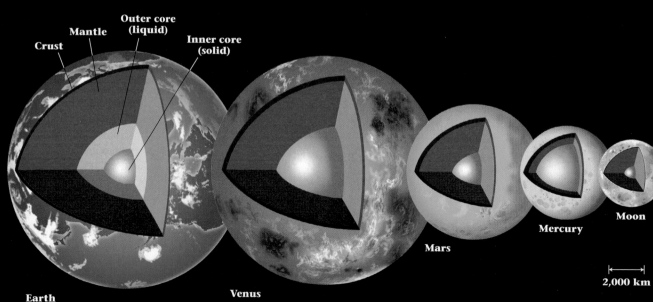

Crust · Mantle · Outer core (liquid) · Inner core (solid)

Earth · Venus · Mars · Mercury · Moon

2,000 km

sufficiently that the dipoles slow down and, like tiny compass needles, align with the Earth's magnetic field. As the rock cools still more, these tiny compass needles lock into permanent parallelism with the Earth's magnetic field at the time the cooling takes place.

Basalt is not the only rock to preserve a good record of paleomagnetism. Certain kinds of sedimentary rocks also can preserve a record of ancient magnetism. In some cases, the record forms when magnetic minerals (magnetite or another iron-bearing mineral, hematite) grow in the spaces between grains after the sediment has accumulated. These minerals form from ions that had been dissolved in groundwater passing through the sediment (Fig. 3.6c).

Apparent Polar Wander— A Proof that Continents Move

Why doesn't the paleomagnetic dipole in ancient rocks point to the present-day magnetic field? When geologists first attempted to answer this question, they assumed that continents were fixed in position, and thus concluded that the position of Earth's magnetic poles in the past were different than they are today. They introduced the term **paleopole** to refer to the *supposed position* of the Earth's magnetic north pole in the past. With this concept in mind, they set out to track what they thought was the change in position of the paleopole over time. To do this, they measured the paleomagnetism in a succession of rocks of different ages from the same general location on a continent, and they plotted the position of the associated succession of paleopole positions on a map (Fig. 3.7a, Box 3.1). The successive positions of dated paleopoles trace out a curving line that came to be known as an **apparent polar-wander path**. At first, geologists assumed that the apparent polar-wander path represented how the position of Earth's magnetic pole migrated through time. But were they in for a surprise! When they obtained polar-wander paths from many different continents, they found that *each continent has a different apparent polar-wander path*. The hypothesis that continents are fixed in position cannot explain this observation, for if the magnetic pole moved while all the continents

FIGURE 3.7 Apparent polar-wander paths and their interpretation.

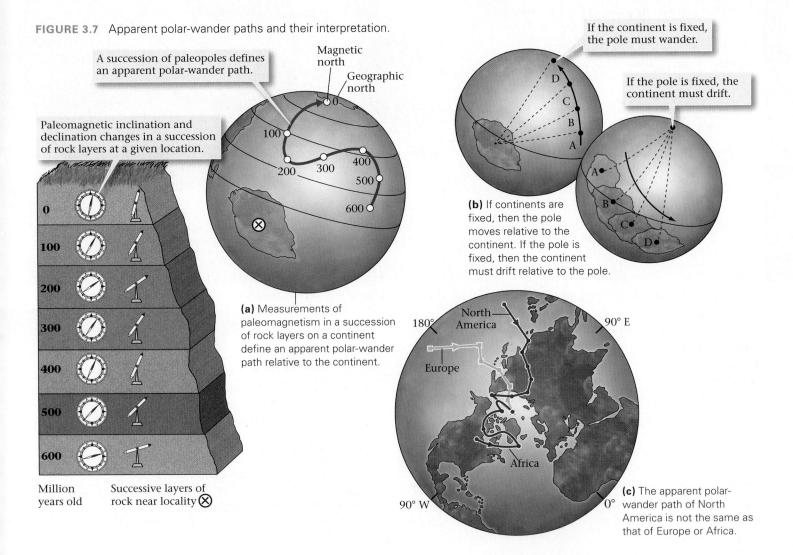

A succession of paleopoles defines an apparent polar-wander path.

Paleomagnetic inclination and declination changes in a succession of rock layers at a given location.

Magnetic north

Geographic north

(a) Measurements of paleomagnetism in a succession of rock layers on a continent define an apparent polar-wander path relative to the continent.

Million years old

Successive layers of rock near locality ⊗

If the continent is fixed, the pole must wander.

If the pole is fixed, the continent must drift.

(b) If continents are fixed, then the pole moves relative to the continent. If the pole is fixed, then the continent must drift relative to the pole.

North America

Europe

Africa

(c) The apparent polar-wander path of North America is not the same as that of Europe or Africa.

BOX 3.1

Finding Paleopoles

How do you find a paleopole position from the orientation of a paleomagnetic dipole in a rock sample? The horizontal projection of the dipole arrow on the Earth's surface (picture the *projection* as the shadow cast on the Earth's surface by the arrow if the Sun were directly overhead) is like a compass needle that points in the direction of the paleopole; that is, the projection defines an imaginary great circle around the Earth that passes through the paleopole and the sample (**Fig. Bx3.1a, b**). This great circle is like an imaginary "paleolongitude" line. Note that when drawing the circle, we assume that the declination at the time the sample was magnetized equals 0, because we assume that averaged over time, the magnetic pole coincides with the geographic pole.

To find the specific position of the paleopole on this great circle, we must look at the inclination of the paleomagnetic dipole in the rock. Recall that inclination depends on latitude (see Fig. 3.5d). Thus, the inclination of the

paleomagnetic dipole defines the *paleolatitude* of the sample with respect to the paleopole, and paleolatitude simply represents the distance (measured in degrees) from the pole along the great circle to where the sample formed.

FIGURE Bx3.1 The basic concept of how to find a paleopole, from a paleomagnetic measurement.

(a) In a sample of rock (represented by the cube), the paleomagnetic arrow has a declination, D, in the horizontal plane, and an inclination, I, in the vertical plane.

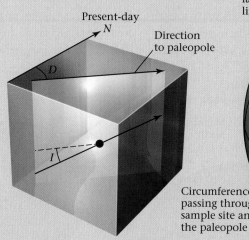

(b) The D of the sample points to the paleopole, P. Thus, the location of the rock outcrop and P lie on a circumference that represents a line of paleolongitude. The inclination indicates the paleolatitude of the sample and thus is a distance measured along the paleolongitude.

stayed fixed, measurements from all continents should produce the same apparent polar-wander paths.

Geologists suddenly realized that they were looking at apparent polar-wander paths in the wrong way. *It's not the pole that moves relative to fixed continents, but rather the continents that move relative to a fixed pole* (**Fig. 3.7b**). Since each continent has its own unique polar-wander path (**Fig. 3.7c**), the continents must move with respect to each other. The discovery proved that Wegener was essentially right all along—continents do move!

Did you ever wonder . . .
if continents actually move by large distances over time?

Take-Home Message

- Earth's magnetic field can be represented by a dipole that points from north to south and field lines that curve through space.
- A rock can preserve a paleomagnetic record that indicates the rock's position relative to the magnetic pole at the time the rock formed.
- We describe the orientation of the paleomagnetic signal in a rock by giving its inclination and declination.
- The magnetic pole position, relative to a continent, appears to change over time, a phenomenon called apparent polar wander.

THINK: How does comparison of apparent polar wander paths for different continents prove that continents move relative to each other?

3.4 SETTING THE STAGE FOR SEA-FLOOR SPREADING

New Images of Sea-Floor Bathymetry

Before World War II, we knew less about the shape of the ocean floor than we did about the shape of the Moon's surface. After all, we could at least see the surface of the Moon and could use a telescope to map its craters. But our knowledge of sea-floor **bathymetry** (the shape of the sea-floor surface) came only from scattered soundings of the sea floor. To sound the ocean depths, a surveyor let out a length of cable with a heavy weight attached. When the weight hit the sea floor, the length of the cable indicated the depth. Needless to say, it took many hours to make a single measurement, so not many could be made. Nevertheless, soundings carried out between 1872 and 1876 by the world's first oceanographic research vessel, the HMS *Challenger*, did hint at the existence of submarine mountain ranges and deep-sea troughs.

Military needs during World War II gave a boost to sea-floor exploration, for as submarine fleets grew, navies required detailed information about depth variations. The invention of echo sounding (sonar) permitted such information to be gathered quickly. Echo sounding works on the same principle that a bat uses to navigate and find insects. A sound pulse emitted from a ship travels down through the water, bounces off the sea floor, and returns up as an echo through the water to a receiver on the ship. Since sound waves travel at a known velocity, the time between the sound emission and the detection of the echo indicates the distance between the ship and the sea floor (velocity = distance/time, so distance = velocity × time). As the ship travels, observers can obtain a continuous record of the depth of the sea floor. The resulting cross section showing depth plotted against location is called a bathymetric profile (Fig. 3.8a). By cruising back and forth across the ocean many times, investigators obtained a series of bathymetric profiles and from these constructed maps of the sea floor. (Geologists can now produce such maps much more rapidly using satellite data.) Bathymetric maps reveal several important features (see Fig. 2.6).

- *Mid-ocean ridges*: The floor beneath all major oceans includes **abyssal plains**, which are broad, relatively flat regions of the ocean that lie at a depth of about 4 to 5 km below sea level; and **mid-ocean ridges**, elongate submarine mountain ranges whose peaks lie only about 2 to 2.5 km below sea level (Fig. 3.8b; Fig. 3.9a). Geologists call the crest of the mid-ocean ridge the ridge axis. All mid-ocean ridges are roughly symmetrical—bathymetry on one side of the axis is nearly a mirror image of bathymetry on the other side.

- *Deep-ocean trenches*: Along much of the perimeter of the Pacific Ocean, and in a few other localities as well, the ocean floor reaches depths of 8 to 12 km—deep enough to swallow Mt. Everest. These deep areas occur in elongate troughs that are now referred to as **trenches** (Fig. 3.9b). Trenches border **volcanic arcs**, curving chains of active volcanoes (see Fig. 3.9b inset).

- *Seamount chains*: Numerous volcanic islands poke up from the ocean floor: for example, the Hawaiian Islands lie in the middle of the Pacific. In addition to islands that

FIGURE 3.8 Bathymetry of mid-ocean ridges and abyssal plains.

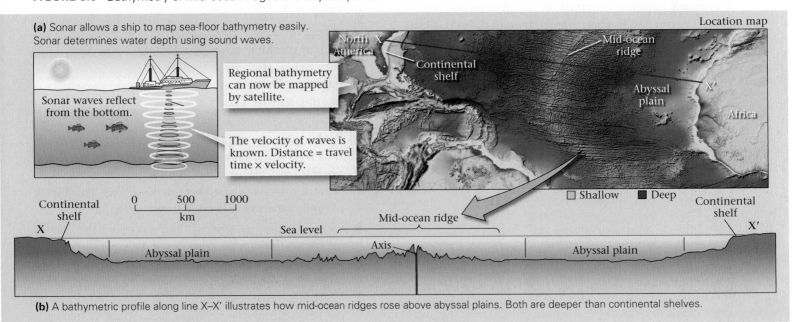

(a) Sonar allows a ship to map sea-floor bathymetry easily. Sonar determines water depth using sound waves.

Sonar waves reflect from the bottom.

Regional bathymetry can now be mapped by satellite.

The velocity of waves is known. Distance = travel time × velocity.

Location map

North America

Continental shelf

Mid-ocean ridge

Abyssal plain

Africa

☐ Shallow   ■ Deep

Continental shelf

0   500   1000
km

Mid-ocean ridge

Continental shelf

X

Sea level

Axis

Abyssal plain

Abyssal plain

X′

(b) A bathymetric profile along line X–X′ illustrates how mid-ocean ridges rose above abyssal plains. Both are deeper than continental shelves.

FIGURE 3.9   Other bathymetric features of the ocean floor.

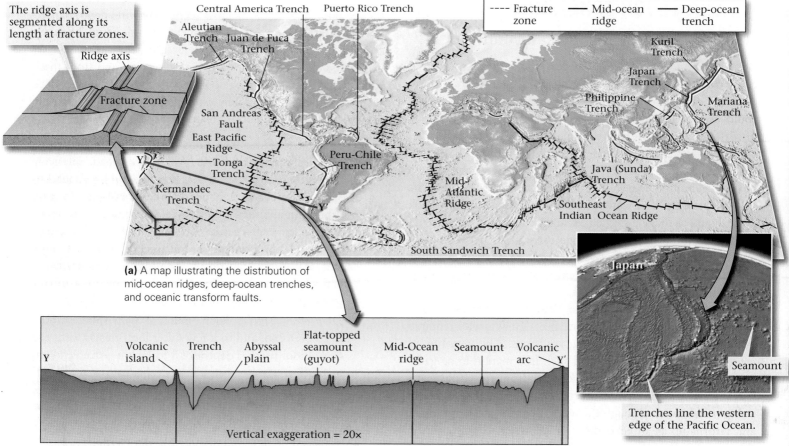

(a) A map illustrating the distribution of mid-ocean ridges, deep-ocean trenches, and oceanic transform faults.

(b) In addition to mid-ocean ridges, the sea floor displays other bathymetric features such as deep-sea trenches, oceanic islands, guyots, and seamounts, as shown in this vertically exaggerated profile.

rise above sea level, sonar has detected many **seamounts** (isolated submarine mountains), which were once volcanoes but no longer erupt. Seamounts with flat tops, due to reef growth before submergence, are called guyots. Volcanic islands and seamounts typically occur in chains, but in contrast to the volcanic arcs that border deep-ocean trenches, only one island at the end of a seamount chain remains capable of erupting volcanically today.

- *Fracture zones:* Surveys reveal that the ocean floor is diced up by narrow bands of vertical cracks and broken-up rock. These **fracture zones** lie roughly at right angles to mid-ocean ridges, effectively segmenting the ridges into small pieces (see Fig. 3.9a inset).

New Observations on the Nature of Oceanic Crust

By the mid-twentieth century, geologists had discovered many important characteristics of the sea-floor crust. These discoveries led them to realize that oceanic crust is quite different from continental crust, and further that bathymetric features of the ocean floor provide clues to the origin of the crust. Specifically:

- A layer of sediment composed of clay and the tiny shells of dead plankton covers much of the ocean floor. This layer *becomes progressively thicker* away from the mid-ocean ridge axis. But even at its thickest, the sediment layer is too thin to have been accumulating for the entirety of Earth history.

- By dredging up samples, geologists learned that oceanic crust is fundamentally different in composition from continental crust. Beneath its sediment cover, oceanic crust bedrock consists primarily of basalt—it does not display the great variety of rock types found on continents.

- Heat flow, the rate at which heat rises from the Earth's interior up through the crust, is not the same everywhere in the oceans. Rather, more heat rises beneath mid-ocean ridges than elsewhere. This observation led researchers to speculate that magma might be rising into the crust just below the mid-ocean ridge axis, for hot molten rock could bring heat into the crust.

■ When maps showing the distribution of earthquakes in oceanic regions became available in the years after World War II, it became clear that earthquakes do not occur randomly, but rather define distinct belts (Fig. 3.10). Some belts follow trenches, some follow mid-ocean ridge axes, and others lie along portions of fracture zones. Since earthquakes define locations where rocks break and move, geologists realized that these bathymetric features are places where movements of the crust take place.

Now let's see how Harry Hess used these observations to come up with the hypothesis of sea-floor spreading.

Take-Home Message

- Research about the ocean floor in the mid-20th century provided clues that ultimately led to the proposal of sea-floor spreading.
- Studies of ocean floor bathymetry reveal the existence of abyssal plains, mid-ocean ridges, deep-sea trenches, and fracture zones.
- Volcanoes occur in arcs bordering trenches and on islands at the end of seamount chains.
- Oceanic crust consists of basalt overlain by sediment.
- Heat flow increases toward the axis of a mid-ocean ridge.
- Most earthquakes occur in distinct belts.

THINK: Is the thickness of sediment the same everywhere on the sea floor? How does it vary?

FIGURE 3.10 A 1953 map showing the distribution of earthquake locations in the ocean basins. Note that earthquakes occur in belts.

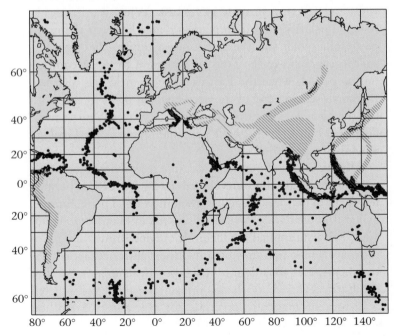

3.5 HARRY HESS AND HIS "ESSAY IN GEOPOETRY"

In the late 1950s, Harry Hess, after studying the observations described above, realized that because the sediment layer on the ocean floor was thin overall, the ocean floor might be much younger than the continents. Also, because the sediment thickened progressively away from mid-ocean ridges, the ridges themselves likely were younger than the deeper parts of the ocean floor. If this was so, then somehow *new ocean floor must be forming at the ridges*, and thus an ocean could be getting wider with time. But how? The association of earthquakes with mid-ocean ridges suggested to him that the sea floor was cracking and splitting apart at the ridge. The discovery of high heat flow along mid-ocean ridge axes provided the final piece of the puzzle, for it suggested the presence of very hot molten rock beneath the ridges. In 1960, Hess suggested that molten rock rose upward beneath mid-ocean ridges and that this material solidified to form oceanic crust (Fig. 3.11). The new sea floor then moved away from the ridge, a process we now call sea-floor spreading. Hess realized that old ocean floor must be consumed somewhere, or the Earth would have to expand. He suggested that deep-ocean trenches might be places where the sea floor sank back into the mantle, and that earthquakes at trenches were evidence of this movement, but he didn't understand how the movement took place.

Did you ever wonder... if the distance between New York and Paris changes?

Hess and his contemporaries realized that the sea-floor-spreading hypothesis instantly provided the long-sought explanation of how continental "drift" occurs. Continents passively move apart as the sea floor between them spreads at mid-ocean ridges, and they passively move together as the sea floor between them sinks back into the mantle at trenches. (Geologists now realize that it is the lithosphere that moves, not just the crust.) Thus, sea-floor spreading proved to be an important step on the route to plate tectonics—the idea seemed so good that Hess referred to his description of it as "an essay in geopoetry." But other key discoveries would have to take place before the whole theory of plate tectonics could come together.

Take-Home Message

- Hess proposed that ocean basins get wider with time by formation of new crust at mid-ocean ridges, a process called sea-floor spreading.
- Old ocean floor slides back into the mantle at trenches.
- Two continents move apart when sea-floor spreading occurs between them.

THINK: How does sea-floor spreading explain variations in ocean-floor heat flow?

FIGURE 3.11 Harry Hess's basic concept of sea-floor spreading. Hess implied, incorrectly, that only the crust moved. We will see that this sketch is an oversimplification.

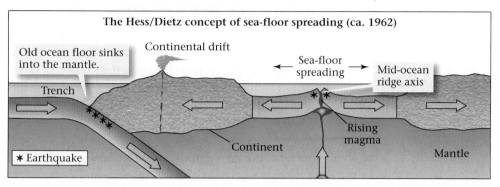

The Hess/Dietz concept of sea-floor spreading (ca. 1962)

3.6 EVIDENCE FOR SEA-FLOOR SPREADING

For a hypothesis to become a theory (see Box P.1), it must be tested—scientists must demonstrate that the idea really works. During the 1960s, geologists found that the sea-floor spreading hypothesis successfully explains several previously baffling observations. Here we discuss two: (1) the existence of orderly variations in the strength of the magnetic field over the sea floor, producing a pattern of stripes called marine magnetic anomalies; and (2) the variation in sediment thickness on the ocean crust, as measured by drilling.

Marine Magnetic Anomalies

Geologists can measure the strength of Earth's magnetic field with an instrument called a magnetometer. At any given location on the surface of the Earth, the magnetic field that you measure includes two parts: one produced by the main dipole of the Earth, due to circulation of molten iron in the outer core, and another produced by the magnetism of near-surface rock. A **magnetic anomaly** is the difference between the expected strength of the Earth's main dipole field at a certain location and the *actual* measured strength of the magnetic field at that location. Places where the field strength is stronger than expected are *positive* anomalies, and places where the field strength is weaker than expected are *negative* anomalies.

Geologists towed magnetometers back and forth across the ocean to map variations in magnetic field strength (Fig. 3.12a). As a ship cruised along its course, the magnetometer's gauge might first detect an interval of strong signal (a positive anomaly) and then an interval of weak signal (a negative anomaly). A graph of signal strength versus distance along the traverse, therefore, has a sawtooth shape (Fig. 3.12b). When geologists compiled data from many cruises on a map, these **marine magnetic anomalies** defined distinctive,

alternating bands. If we color positive anomalies dark and negative anomalies light, the pattern made by the anomalies resembles the stripes on a candy cane (Fig. 3.12c). The mystery of this marine magnetic anomaly pattern, however, remained unsolved until geologists recognized the existence of magnetic reversals.

Magnetic reversals. Recall that Earth's magnetic field can be represented by an arrow, representing the dipole, that presently points from the north magnetic pole to the south magnetic pole. When researchers measured the paleomagnetism of a succession of rock layers that had accumulated over a long period of time, they found that the polarity (which end of a magnet points north and which end points south) of the paleomagnetic field preserved in some layers was the same as that of Earth's present magnetic field, whereas in other layers it was the opposite.

At first, observations of reversed polarity were largely ignored, thought to be the result of lightning strikes or of local chemical reactions between rock and water. But when repeated measurements from around the world revealed a systematic pattern of alternating normal and reversed polarity in rock layers, geologists realized that reversals were a worldwide, not a local, phenomenon. They reached the unavoidable conclusion that, at various times during Earth history, *the polarity of Earth's magnetic field has suddenly reversed!* In other words, sometimes the Earth has normal polarity, as it does today, and sometimes it has reversed polarity (Fig. 3.13a). Times when the Earth's field flips from normal to reversed polarity, or vice versa, are called **magnetic reversals**. When the Earth has reversed polarity, the south magnetic pole lies near the north geographic pole, and the north magnetic pole lies near the south geographic pole. Thus, if you were to use a compass during periods when the Earth's magnetic field was reversed, the north-seeking end of the needle would point to the south geographic pole. Note that the Earth itself doesn't turn upside down—it is just the magnetic field that reverses (See Geology at a Glance, pp. 70–71).

In the 1950s, about the same time researchers discovered polarity reversals, they developed a technique that permitted them to define the age of a rock in years. The technique, called isotopic dating, will be discussed in detail in Chapter 12. Geologists applied the technique to determine the ages of rock layers in which they obtained their paleomagnetic measurements, and thus determined *when* the magnetic field of the Earth reversed. With this information, they constructed a history of magnetic reversals for the past 4.5 million

FIGURE 3.12 The discovery of marine magnetic anomalies.

(a) A ship towing a magnetometer detects changes in the strength of the magnetic field.

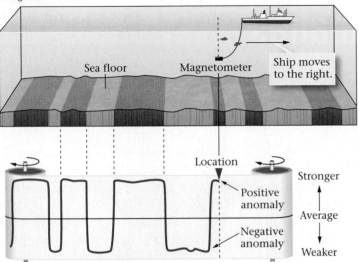

(b) On a paper record, intervals of stronger magnetism (positive anomalies) alternate with intervals of weaker magnetism (negative anomalies).

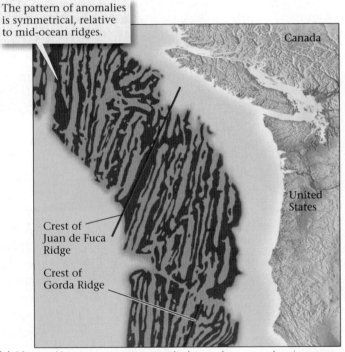

The pattern of anomalies is symmetrical, relative to mid-ocean ridges.

(c) After making many traverses, geologists make a map showing areas of positive anomalies (dark) and negative anomalies (light) off the west coast of North America. These define stripes.

years; this history is now called the **magnetic-reversal chronology**.

A diagram representing the Earth's magnetic-reversal chronology (Fig. 3.13b) shows that reversals do not occur regularly, so the lengths of different polarity "chrons," the time intervals between reversals, are different. For example, we have had a

normal-polarity chron for about the last 700,000 years. Before that, a reversed-polarity chron occurred. The youngest four polarity chrons (Brunhes, Matuyama, Gauss, and Gilbert) were named after scientists who had made important contributions to the study of rock magnetism. As more measurements became available, investigators realized that some short-duration reversals (less than 200,000 years long) took place within the chrons, and they called these shorter durations polarity subchrons.

The question of *why* reversals take place still puzzles geologists, but researchers using supercomputer models are getting close to an answer. Their models show that changes in the fluid motion of the outer core can trigger reversals, and that during reversals, the magnetic field weakens and becomes complicated, before reconfiguring with a different polarity (Fig. 3.13c).

The interpretation of marine anomalies. Why do marine magnetic anomalies exist? A graduate student in England, Fred Vine (working with his adviser, Drummond Matthews), and a Canadian geologist, Lawrence Morley (working independently), discovered a solution to this riddle. Simply put, a positive anomaly occurs over areas of sea floor where basalt has normal polarity. In these areas, the magnetic force produced by the basalt *adds* to the force produced by Earth's dipole and creates a stronger magnetic signal than expected, as measured by the magnetometer (Fig. 3.14a). A negative anomaly occurs over regions of sea floor where basalt has reversed polarity. Here, the magnetic force of the basalt *subtracts* from the force produced by the Earth's dipole and results in a weaker magnetic signal.

Vine, Matthews, and Morley pointed out that marine magnetic anomalies would be an expected consequence of sea-floor spreading—sea floor yielding positive anomalies developed at times when the Earth had normal polarity, whereas sea floor yielding negative anomalies formed when the Earth had reversed polarity. If this was so, then the sea-floor-spreading hypothesis implies that the pattern of anomalies should be symmetric with respect to the axis of a mid-ocean ridge. With this idea in mind, geologists set sail to measure magnetic anomalies near mid-ocean ridges (Fig. 3.14b). By 1966, the story was complete—in the examples studied, the magnetic anomaly pattern on one side of a ridge was indeed a mirror image of the anomaly pattern on the other. Let's see why.

At Time 1 (sometime in the past), a time of normal polarity, the normal polarity (dark) stripe of the sea floor forms (Fig. 3.14c). The tiny dipoles of magnetite grains in basalt making up this stripe align with the Earth's field. As it forms, the rock in this stripe migrates away from the ridge axis, half to the right and half to the left. Later, at Time 2, the field has reversed, and the reversed polarity (light-gray) stripe forms. As it forms, it too moves away from the axis, and still younger crust begins to develop along the axis. As the process continues over millions of years, many stripes form. A positive anomaly

Magnetic Reversals and Marine Magnetic Anomalies

Normal polarity

The Earth behaves like a giant magnet, and thus is surrounded by a magnetic field. The magnetism is due to the flow of liquid iron alloy in the outer core.

Reversed polarity

The age of oceanic crust varies with location. The youngest crust lies along a mid-ocean ridge, and the oldest along the coasts of continents. Here, the different color stripes correspond to different ages of oceanic crust. Red is youngest, purple is oldest.

The rock of oceanic crust preserves a record of the Earth's magnetic polarity at the time the crust formed. Eventually, a symmetric pattern of polarity stripes develops.

Marine magnetic anomalies are stripes representing alternating bands of oceanic crust that differ in the measured strength of the magnetic field above them. Stronger fields are measured over crust with normal polarity, whereas weaker fields are measured over crust with reversed polarity.

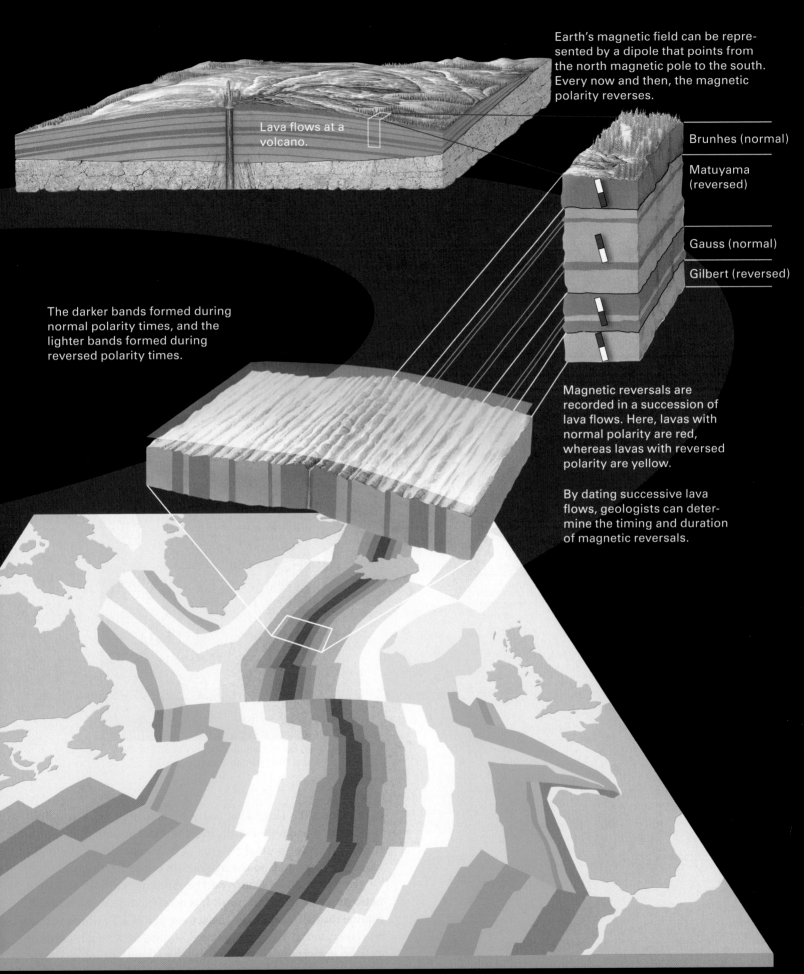

Earth's magnetic field can be represented by a dipole that points from the north magnetic pole to the south. Every now and then, the magnetic polarity reverses.

Lava flows at a volcano.

Brunhes (normal)

Matuyama (reversed)

Gauss (normal)

Gilbert (reversed)

The darker bands formed during normal polarity times, and the lighter bands formed during reversed polarity times.

Magnetic reversals are recorded in a succession of lava flows. Here, lavas with normal polarity are red, whereas lavas with reversed polarity are yellow.

By dating successive lava flows, geologists can determine the timing and duration of magnetic reversals.

FIGURE 3.13 Magnetic polarity reversals, and the chronology of reversals.

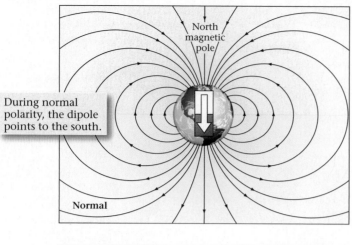

During normal polarity, the dipole points to the south.

North magnetic pole

Normal

During reversed polarity, the dipole points to the north.

North magnetic pole

Reversed

(a) Geologists proposed that the Earth's magnetic field reverses polarity every now and then.

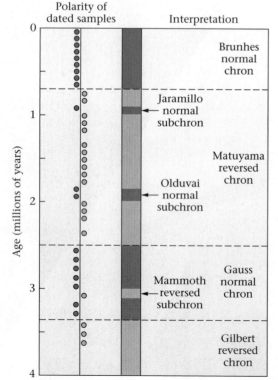

Polarity of dated samples

Interpretation

Age (millions of years)

Brunhes normal chron

Jaramillo normal subchron

Matuyama reversed chron

Olduvai normal subchron

Gauss normal chron

Mammoth reversed subchron

Gilbert reversed chron

(b) Observations led to the production of a reversal chronology, with named polarity intervals.

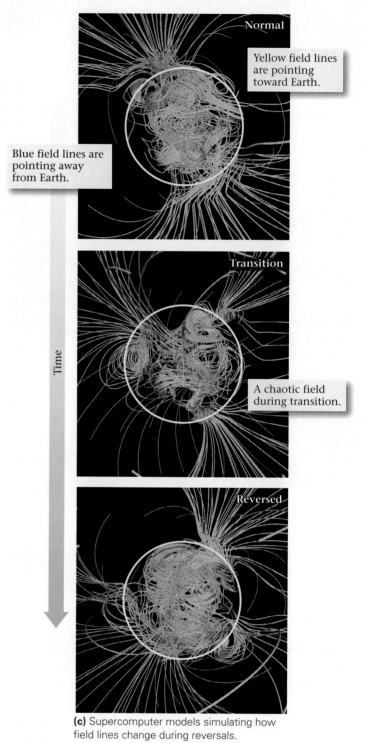

Normal

Yellow field lines are pointing toward Earth.

Blue field lines are pointing away from Earth.

Time

Transition

A chaotic field during transition.

Reversed

(c) Supercomputer models simulating how field lines change during reversals.

exists along the ridge axis today because this represents sea floor that has developed during the most recent interval of time, a chron of normal polarity. The magnetism of the rock along the ridge adds to the magnetism of the Earth's field.

By relating the stripes on the Atlantic sea floor to magnetic reversals found in basalt layers on land, geologists dated marine magnetic anomalies back to an age of 4.5 million years, and they found that the relative widths of anomaly stripes exactly corresponded to the relative durations of polarity chrons in the magnetic-reversal chronology (Fig. 3.14d). The relationship between anomaly-stripe width and polarity-chron duration indicates that the rate of spreading has been constant in the Atlantic for at least the last 4.5 million years.

To determine the rate of spreading, researchers used simple arithmetic. Remember that velocity = distance/time. In the North Atlantic Ocean, 4.5-million-year-old sea floor lies 45 km away from the ridge axis. Therefore, the velocity (v) at which the sea floor moves away from the ridge axis can be calculated as follows:

$$v = \frac{45 \text{ km}}{4,500,000 \text{ years}} = \frac{4,500,000 \text{ cm}}{4,500,000 \text{ years}} = 1 \text{ cm/y}$$

This means that a point on one side of the ridge moves away from a point on the other side by 2 cm per year. We call this number the **spreading rate**. In the Pacific Ocean, sea-floor spreading occurs at the East Pacific Rise. (Geographers named this a "rise" because it is not as rough and jagged as the Mid-Atlantic Ridge.) The anomaly stripes bordering the East Pacific Rise are much wider, and 4.5-million-year-old sea floor lies about 225 km from the rise axis. This requires the sea floor to move away from the rise at a rate of about 5 cm per year, so the spreading rate for the East Pacific Rise is about 10 cm per year.

Using this concept, researchers were able to estimate the ages of all the stripes that were eventually identified on the sea floor (Fig. 3.15a, b). The oldest are less than 200 Ma.

Evidence from Deep-Sea Drilling

In the late 1960s, a drilling ship called the *Glomar Challenger* set out to sail around the ocean drilling holes into the sea floor. This amazing ship could lower enough drill pipe to drill in 5-km-deep water and could continue to drill until the hole reached a depth of about 1.7 km (1.1 miles) below the sea floor. Drillers brought up cores of rock and sediment that geoscientists then studied on board.

On one of its early cruises, the *Glomar Challenger* drilled a series of holes through sea-floor sediment to the basalt layer. These holes were spaced at progressively greater distances from the axis of the Mid-Atlantic Ridge. If the model of sea-floor spreading was correct, then not only should the sediment layer be progressively thicker away from the axis, but the age of the oldest sediment just above the basalt should be progressively older away from the axis. When the drilling and the analyses were complete, the prediction was confirmed (Fig 3.15c).

So by the early 1960s, it had become clear that Wegener had been right all along—continents do drift. But, though the case for drift had been greatly strengthened by the

FIGURE 3.14 The progressive development of magnetic anomalies, and the long-term reversal chronology.

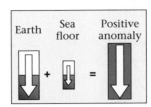
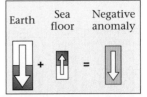

(a) Positive anomalies form when sea-floor rock has the same polarity as the present magnetic field. Negative anomalies form when sea-floor rock has polarity that is opposite to the present field.

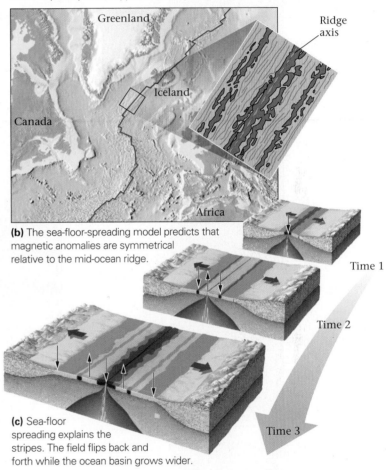

(b) The sea-floor-spreading model predicts that magnetic anomalies are symmetrical relative to the mid-ocean ridge.

(c) Sea-floor spreading explains the stripes. The field flips back and forth while the ocean basin grows wider.

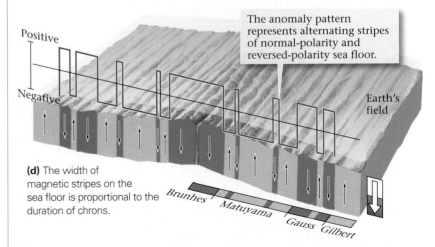

The anomaly pattern represents alternating stripes of normal-polarity and reversed-polarity sea floor.

(d) The width of magnetic stripes on the sea floor is proportional to the duration of chrons.

FIGURE 3.15 Magnetic anomalies of the eastern Pacific.

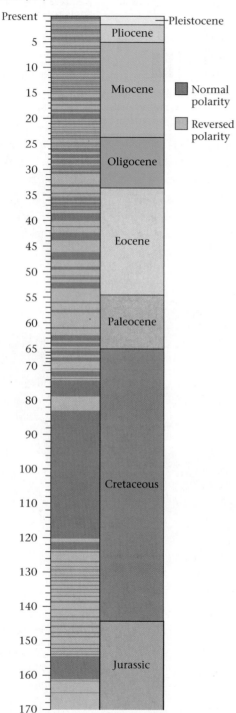

(a) The reversal chronology for the past 170 million years.

Million years (Ma)

Legend:
- Normal polarity
- Reversed polarity

Geologic periods shown: Pleistocene, Pliocene, Miocene, Oligocene, Eocene, Paleocene, Cretaceous, Jurassic

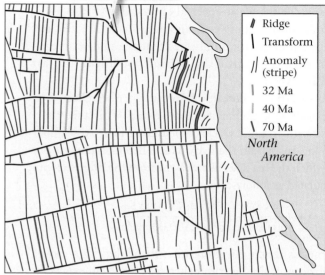

Anomalies are offset at transforms.

Legend:
- Ridge
- Transform
- Anomaly (stripe)
- 32 Ma
- 40 Ma
- 70 Ma

North America

(b) Anomalies get progressively older to the west, off the coast of North America, except in the region off of Oregon and Washington. Here, anomalies get younger toward the ridge.

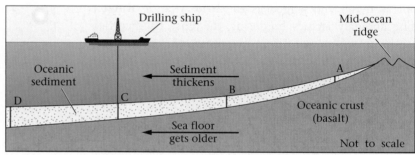

Drilling ship · Mid-ocean ridge · Oceanic sediment · Sediment thickens · Sea floor gets older · Oceanic crust (basalt) · Not to scale

(c) Drilling of the sea floor confirmed that sediment thickness at D is greater than at A, and that the age of the oldest sediment increases away from the ridge. Thus, the sea floor gets progressively older away from the ridge.

Take-Home Message

- Earth's magnetic field reverses polarity every now and then.
- Measurements show that there are alternating stripes of positive and negative magnetic anomalies on the sea floor.
- Positive anomalies exist over sea floor formed when Earth's magnetic field has normal polarity, whereas negative anomalies exist over sea floor formed during chrons of reversed polarity.
- Magnetic anomaly stripes on one side of a mid-ocean ridge are a mirror image of those on the other side, an observation compatible with the sea-floor spreading hypothesis.
- Drilling shows that the age of the oldest sediment increases progressively away from the ridge, as predicted by the sea-floor spreading hypothesis.

THINK: What determines the widths of marine magnetic anomalies?

discovery of apparent polar-wander paths, it really took the proposal and proof of sea-floor spreading to make believers of most geologists. Very quickly, as we will see in the next chapter, these ideas became the basis of the theory of plate tectonics.

Chapter Summary

- Alfred Wegener proposed that continents had once been joined together to form a single huge supercontinent (Pangaea) and had subsequently drifted apart. This idea is the continental drift hypothesis.

- Wegener drew from several different sources of data to support his hypothesis: (1) coastlines on opposite sides of the ocean match up; (2) the distribution of late Paleozoic glaciers can be explained if the glaciers made up a polar ice cap over the southern end of Pangaea; (3) the distribution of late Paleozoic equatorial climatic belts is compatible with the concept of Pangaea; (4) the distribution of fossil species suggests the existence of a supercontinent; (5) distinctive rock assemblages that are now on opposite sides of the ocean were adjacent on Pangaea.

- Despite all the observations that supported continental drift, most geologists did not initially accept the idea because no one could explain *how* continents could move. It took decades of new data collection before the idea could be reconsidered.

- Rocks retain a record of the Earth's magnetic field that existed at the time the rocks formed. This record is called paleomagnetism. By measuring paleomagnetism in successively older rocks, geologists found that the apparent position of the Earth's magnetic pole relative to the rocks changes through time. Successive positions of the pole define an apparent polar-wander path.

- Apparent polar-wander paths are different for different continents. This observation can be explained if continents move with respect to each other, while the Earth's magnetic poles remain roughly fixed.

- The invention of echo sounding permitted explorers to make detailed maps of the sea floor. These maps revealed the existence of mid-ocean ridges, deep-ocean trenches, seamount chains, and fracture zones. Heat flow is generally greater near the axis of a mid-ocean ridge.

- Around 1960, Harry Hess proposed the hypothesis of sea-floor spreading. According to this hypothesis, new sea floor forms at mid-ocean ridges, above a band of upwelling mantle, then spreads symmetrically away from the ridge axis. As a consequence, an ocean basin gets progressively wider with time, and the continents on either side of the ocean basin drift apart. Eventually, the ocean floor sinks back into the mantle at deep-ocean trenches.

- Magnetometer surveys of the sea floor revealed marine magnetic anomalies. Positive anomalies, where the magnetic field strength is greater than expected, and negative anomalies, where the magnetic field strength is less than expected, are arranged in alternating stripes.

- During the 1950s, geologists documented that the Earth's magnetic field reverses polarity every now and then. The record of reversals, dated by isotopic techniques, is called the magnetic-reversal chronology.

- A proof of sea-floor spreading came from the interpretation of marine magnetic anomalies. Sea floor that forms when the Earth has normal polarity results in positive anomalies, whereas sea floor that forms when the Earth has reversed polarity results in negative anomalies. Anomalies are symmetric with respect to a mid-ocean ridge axis, and their widths are proportional to the duration of polarity chrons. Study of anomalies allows us to calculate the rate of spreading.

- Drilling of the sea floor confirmed that its age increases away from the mid-ocean ridge axis and served as another proof of sea-floor spreading.

GEOPUZZLE REVISITED

Many lines of evidence indicate that the position of continents indeed changes over time, and thus that the map of the Earth's surface is not fixed. Not only do coastlines match, but the observed distribution of rock units, fossils, and climate belts, and evidence of past glaciations all point to the occurrence of continental drift. Movement occurs because ocean basins grow wider by the process of sea-floor spreading, or get narrower by the process of subduction. The documentation and interpretation of marine magnetic anomalies proved that sea-floor spreading does happen.

Guide Terms

abyssal plains (p. 65)

apparent polar-wander path (p. 63)

bathymetry (p. 65)

continental drift (p. 56)

fracture zones (p. 66)

magnetic anomaly (p. 68)

magnetic declination (p. 61)

magnetic dipole (p. 60)

magnetic inclination (p. 61)

magnetic poles (p. 60)

magnetic-reversal chronology (p. 68)

magnetic reversals (p. 68)

marine magnetic anomalies (p. 68)

mid-ocean ridges (p. 65)

paleomagnetism (p. 62)

paleopole (p. 63)

Pangaea (p. 56)

plate tectonics (p. 57)

sea-floor spreading (p. 56)

seamounts (p. 66)

spreading rate (p. 73)

subduction (p. 56)

trenches (p. 65)

volcanic arcs (p. 65)

Review Questions

1. What was Wegener's continental drift hypothesis?
2. How does the fit of the coastlines around the Atlantic support continental drift?
3. Explain the distribution of late Paleozoic glaciation.
4. How does the distribution of climatic belts support continental drift?
5. Was it possible for a dinosaur to walk from New York to Paris when Pangaea existed? Explain your answer.
6. Why were geologists initially skeptical of Wegener's continental drift hypothesis?
7. Describe how the angle of inclination of the Earth's magnetic field varies with latitude. How can paleomagnetic inclination be used to determine the ancient latitude of a continent?
8. Describe the basic characteristics of mid-ocean ridges, deep-ocean trenches, and seamount chains.
9. Describe the hypothesis of sea-floor spreading.
10. How did the observations of heat flow and seismicity support the hypothesis of sea-floor spreading?
11. What is a magnetic reversal?
12. What is a marine magnetic anomaly? How is it detected?
13. Describe the pattern of marine magnetic anomalies across a mid-ocean ridge. How do geologists explain the pattern?
14. How do geologists calculate rates of sea-floor spreading?
15. Did drilling into the sea floor contribute further proof of sea-floor spreading? If so, how?

ANOTHER VIEW This image was produced by Christoph Hormann, using computer rendering techniques. It shows the Caucasus Mountains between the Black Sea and the Caspian Sea. This range is forming due to the collision between two moving continental masses.

On Further Thought

The following questions will be answered, in large part, by Chapter 4. But by thinking about them now, you can get a feel for the excitement of discovery that geologists enjoyed in the wake of the proposal of sea-floor spreading.

16. Alfred Wegener's writings implied that all continents had been linked to form Pangaea from the formation of the Earth until Pangaea's breakup in the Mesozoic. Modern geologists do not agree. Geologic evidence suggests that Pangaea itself was formed by the late Paleozoic collision of continents that had been separate during most of the Paleozoic, and that other supercontinents had formed and broken up prior to the Paleozoic. What geologic evidence led geologists to this conclusion? (*Hint*: Keep in mind that modern geologists, unlike Wegener, understand that mountain belts such as the Appalachians form when two continents collide, and that modern geologists, unlike Wegener, are able to determine the age of rocks using isotopic dating.)
17. Dating methods indicate that the oldest rocks on continents are almost 4 billion years old, whereas the oldest ocean floor is only 200 million years old. Why?
18. The geologic record suggests that when supercontinents break up, a pulse of rapid evolution, with many new species appearing and many existing species becoming extinct, takes place. Why might this be? (*Hint*: Consider how the environment, both global and local, might change as a result of breakup, and keep in mind the widely held idea that competition for resources drives evolution.)
19. Why are the marine magnetic anomalies bordering the East Pacific Rise in the southeastern Pacific Ocean wider than those bordering the Mid-Atlantic Ridge in the South Atlantic Ocean?

 For more resources, including animations, quizzes, and Norton's GeoTours, go to **wwnorton.com/studyspace**.

 If your instructor assigns exercises in SmartWork, log in at **smartwork.wwnorton.com**.

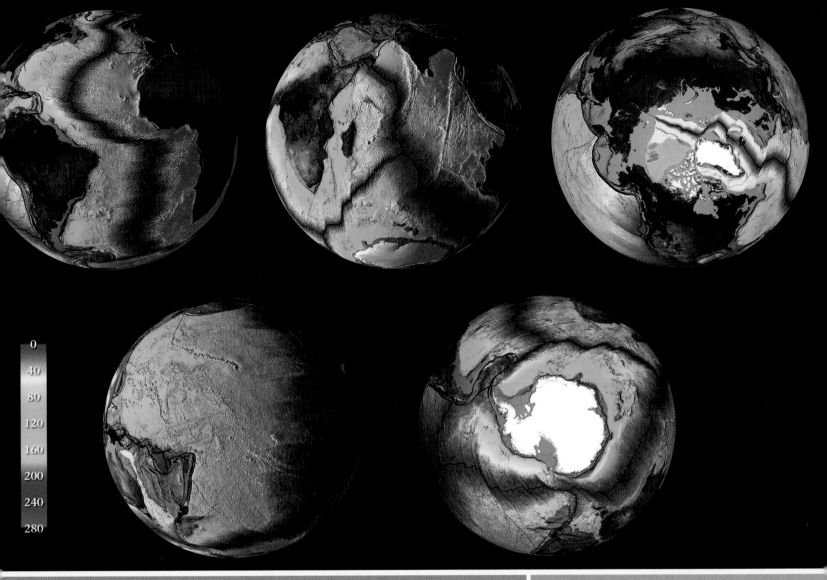

0
40
80
120
160
200
240
280

CHAPTER 4

The Way the Earth Works: Plate Tectonics

Colors on these globes depict the age of the ocean floor (red is youngest and blue is oldest). The pattern of colors indicates that ocean-floor age increases systematically away from the axis of mid-ocean ridges. Plate tectonics successfully explains this observation, and many more!

GEOPUZZLE

Why do earthquakes, volcanoes, and mountain belts occur where they do? And why does Earth's surface differ so much from the surfaces of other planets?

4.1 INTRODUCTION

Thomas Kuhn, an influential historian of science working in the 1960s, argued that scientific thought evolves in fits and starts. According to Kuhn, scientists base their interpretation of the natural world on an established line of reasoning, a scientific paradigm. But once in a while, a revolutionary thinker or group of thinkers proposes a radically new point of view that invalidates an old paradigm. Almost immediately, the scientific community scraps old hypotheses and formulates others consistent with the new paradigm. Kuhn called such abrupt changes in thought a "scientific revolution."

In physics, Isaac Newton's mathematical description of moving objects established the paradigm that natural phenomena obey physical laws; older paradigms suggesting that natural phenomena followed the dictates of Greek philosophers had to be scrapped. In biology, Charles Darwin's proposal that species evolve by natural selection required biologists to rethink hypotheses based on the older paradigm that all species originated at the same time. And in geology, a scientific revolution in the 1960s yielded the new paradigm that the outer layer of the Earth, the lithosphere, consists of separate pieces, or plates, that move with respect to each other. This idea, which we now call the theory of plate tectonics, or simply **plate tectonics**, required geologists to cast aside hypotheses rooted in the paradigm of fixed continents, and thus led to a complete restructuring of how geologists think about Earth history. Compare this book with a geology textbook from the 1950s, and you will instantly see the difference.

Alfred Wegener planted the seed of plate tectonics theory with his proposal of continental drift in 1915. But until 1960 this seed lay dormant while geoscientists focused on collecting new data about the Earth. Discoveries about the ocean floor and about apparent polar wander led to the germination of the seed in 1960, with Harry Hess and Robert Dietz's proposal of sea-floor spreading. The roots took hold three years later when marine magnetic anomalies supplied proof of sea-floor spreading. During the next five years, the study of geoscience turned into a feeding frenzy as many investigators dropped what they'd been doing and turned their attention to examining the broader implications of sea-floor spreading. By 1968, thanks primarily to the work of at least two dozen different investigators, the sea-floor-spreading hypothesis had bloomed into plate tectonics theory. Geologists clarified the concept of a plate, described the types of plate boundaries, calculated plate motions, related plate tectonics to earthquakes and volcanoes, showed how plate motions generate mountain belts and seamount chains, and defined the history of past plate motions. After these scientific revolutionaries had presented their new ideas to standing-room-only audiences at many conferences between 1968 and 1970, the geoscience community, with few exceptions, embraced plate tectonics theory and has built on it ever since.

To begin our explanation of the key elements of plate tectonics theory, we first learn about lithosphere plates, distinguishing among the three types of plate boundaries. Next, we characterize the nature of geologic activity that occurs at each boundary. We then look at hot spots and other special locations on plates. Finally, we examine how continents break apart and how they collide, and we learn about what makes plates move. Because plate tectonics theory is geology's grand unifying theory, it is now an essential foundation for the discussion of all geology.

Chapter Themes

By the end of this chapter, you should know that . . .

- Earth's lithosphere is divided into about 20 moving plates.
- plate movement takes place at plate boundaries; this movement causes earthquakes, so the distribution of earthquake belts defines plate boundaries.
- geologists distinguish three types of plate boundaries based on the relative movement across the boundary. Distinct geologic features characterize each type.
- we can directly measure plate motion by using GPS.

4.2 WHAT DO WE MEAN BY PLATE TECTONICS?

The Concept of a Lithosphere Plate

We learned earlier that geoscientists divide the outer part of the Earth into two layers. The **lithosphere** consists of the crust plus the top (cooler) part of the upper mantle. It behaves relatively rigidly, meaning that when a force pushes or pulls on it, it does not flow but rather bends or breaks (**Fig. 4.1a**). The lithosphere floats on a relatively soft, or "plastic," layer called the **asthenosphere**, composed of warmer (> 1,280°C) mantle that *can flow* slowly when acted on by force. As a result, the asthenosphere convects, like water in a pot, though much more slowly (see Box 2.3).

Continental lithosphere and oceanic lithosphere differ markedly in their thickness. On average, continental lithosphere has a thickness of 150 km, whereas *old* oceanic lithosphere has a thickness of about 100 km (**Fig. 4.1b**). (For reasons discussed later in this chapter, new oceanic lithosphere at a mid-ocean ridge is much thinner.) Recall that the crustal part of continental lithosphere ranges from 25 to 70 km thick and consists largely of low-density felsic and intermediate rock (see Chapter 2). In contrast, the crustal part of oceanic lithosphere is only 7 to 10 km thick and consists largely of relatively high-density mafic rock. The mantle part of both continental and oceanic lithosphere consists of very high-density ultramafic rock.

FIGURE 4.1 Nature of the lithosphere and its behavior.

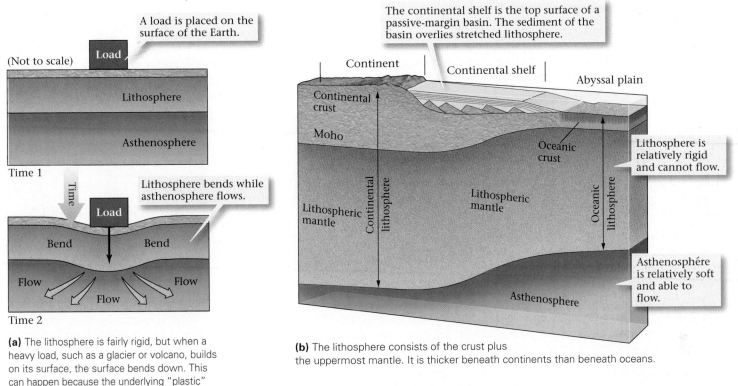

(a) The lithosphere is fairly rigid, but when a heavy load, such as a glacier or volcano, builds on its surface, the surface bends down. This can happen because the underlying "plastic" asthenosphere can flow out of the way.

(b) The lithosphere consists of the crust plus the uppermost mantle. It is thicker beneath continents than beneath oceans.

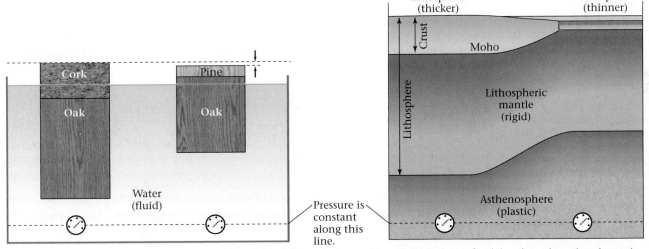

(c) We can picture continental lithosphere as a thick oak block (lithospheric mantle) overlaid by a layer of cork (continental crust), and oceanic lithosphere as a thinner block of oak overlaid by a layer of pine (oceanic crust). The pine layer is thinner than the cork layer. If both blocks float in a tub of water, the surface of the thick cork/oak block lies at a higher elevation than that of the pine/oak block. Similarly, the ocean floor lies 4–5 km below the surface of the continents, on average, because lithosphere, like the wood blocks, floats on the asthenosphere.

Because of the differences in density and thickness that we've just noted, the surface of continental lithosphere lies at a higher elevation than the surface of oceanic lithosphere. To picture why, let's consider an analogy. Imagine that we have two blocks of oak (a high-density wood), one 15 cm thick and one 10 cm thick. On top of the thicker block, we glue a 4-cm-thick layer of cork (a low-density wood), and on top of the thinner block, we glue a 1-cm-thick layer of pine (a medium-density

wood). Now we place the two composite blocks in water (Fig. 4.1c). The total mass of the cork-covered block exceeds the total mass of the pine-covered block, so the base of the cork-covered block sinks deeper into the water. But because the cork-covered block is thicker and has a lower overall density, it floats higher. In our analogy, the cork-covered block represents continental lithosphere, with its thick crust of low-density rock, whereas the pine-covered block represents oceanic lithosphere, with its

BOX 4.1

Archimedes' Principle of Buoyancy

Archimedes (c. 287–212 B.C.E.), a Greek mathematician and inventor, left an amazing legacy of discoveries. He described the geometry of spheres, cylinders, and spirals; introduced the concept of a center of gravity; and was the first to understand buoyancy. **Buoyancy** is the upward force acting on an object immersed or floating in a fluid. According to legend, Archimedes recognized this concept suddenly, while floating in a public bath, and was so inspired that he jumped out of the bath and ran home naked, shouting "Eureka!"

Archimedes realized that when you place a solid object in water, the object displaces a volume of water equal in mass to the object (**Fig. Bx 4.1**). An object denser than water, such as a stone, sinks through the water because even when completely submerged, the stone's mass exceeds the mass of the water it displaces. When submerged, however,

FIGURE Bx4.1 According to Archimedes' principle of buoyancy, an iceberg sinks until the total mass of the water displaced equals the total mass of the whole iceberg. Since water is denser, the volume of the water displaced is less than the volume of the iceberg, so only 20% of the iceberg protrudes above the water.

the stone weighs less than it does in air. (For this reason, a scuba diver can lift a heavy object underwater.) An object less dense than water, such as an iceberg, sinks only until the mass of the water displaced equals the total mass of the iceberg. This condition happens while part of the iceberg still protrudes up into the air. Put another way, an object placed in a fluid feels a "buoyancy force" that tends to push it up. If the object's weight is less than the buoyancy force, the object floats; but if its weight is greater than the buoyancy force, the object sinks.

---

thin crust of dense rock. The oak represents the very high-density rock constituting the mantle part of the lithosphere, which is thicker for the continent than for the ocean. Our analogy reflects Archimedes' principle (Box 4.1).

The lithosphere forms the Earth's relatively rigid shell. But unlike the shell of a hen's egg, the lithospheric shell contains a number of major breaks, which separate it into distinct pieces. As noted earlier, we call the pieces **lithosphere plates**, or simply plates.[1] The breaks between plates are known as **plate boundaries** (Fig. 4.2a, b). Geoscientists distinguish 12 major plates and several microplates. Note that some plates consist entirely of oceanic lithosphere whereas some plates consist of both oceanic *and* continental lithosphere. Some plates have familiar names (the North American Plate, the African Plate), whereas some do not (the Cocos Plate, the Juan de Fuca Plate).

Some plate boundaries follow continental margins, the boundary between a continent and an ocean, but others do not. For this reason, we distinguish between **active margins**, which are plate boundaries, and **passive margins**, which are not plate boundaries. Along passive margins, continental crust is thinner

than normal. (As we'll discuss later, this thinning occurs when continents break apart. During this process, the upper part breaks into wedge-shaped slices.) Thick (10–15 km) accumulations of sediment cover this thinned crust; this accumulation is a "passive-margin basin" (see Fig. 4.1b). The surface of this sediment layer is a broad, shallow (less than 500 m deep) region called the **continental shelf**, home to the major fisheries of the world. Some plates consist entirely of oceanic lithosphere or entirely of continental lithosphere, whereas some plates consist of both.

The Basic Principles of Plate Tectonics

With the background provided above, we can restate plate tectonics theory concisely as follows. The Earth's lithosphere is divided into plates that move relative to each other. Plate movement occurs at rates of about 1 to 15 cm per year. As a plate moves, its internal area remains mostly, but not perfectly, rigid and intact. But rock along plate boundaries undergoes intense deformation (cracking, sliding, bending, stretching, and squashing) as the plate grinds or scrapes against its

---

1. Note that the above definition equates the base of a plate with the base of the lithosphere. This definition works well for oceanic plates, because the asthenosphere directly beneath oceanic lithosphere is particularly weak. But some geologists now think that the upper 100 to 150 km of the asthenosphere beneath continents actually moves with the continental lithosphere. Thus, the base of continental plates—the base of the layer that moves during plate motion—may actually lie within the asthenosphere. To simplify our discussion, we ignore this complication.

FIGURE 4.2 The locations of plate boundaries and the distribution of earthquakes.

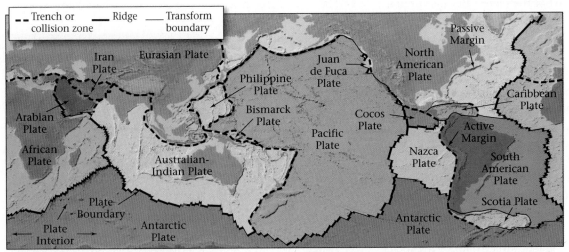

(a) A map of major plates shows that some consist entirely of oceanic lithosphere, whereas some consist of both continental and oceanic lithosphere. Active continental margins lie along plate boundaries; passive margins do not.

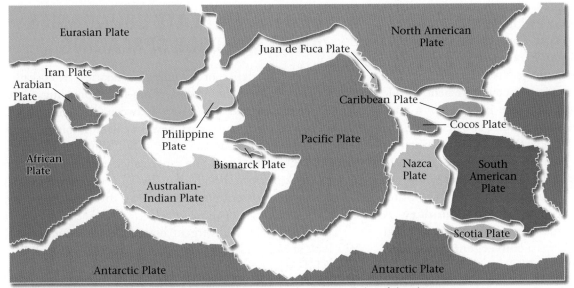

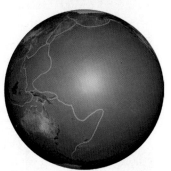

(b) An exploded view of the plates emphasizes the variation in shape and size of the plates.

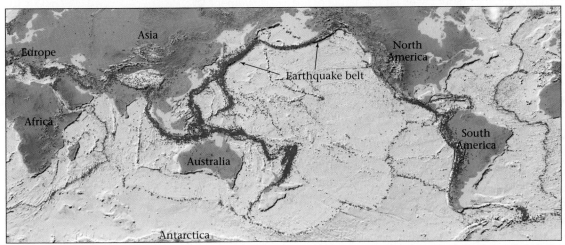

(c) The locations of earthquakes (red dots) mostly fall in distinct bands that correspond to plate boundaries. Relatively few earthquakes occur in the stabler plate interiors.

(d) You can get a sense of the scale of the plates by looking at their distribution on a globe.

neighbors or pulls away from its neighbors. As plates move, so do the continents that form part of the plates. Because of plate tectonics, the map of Earth's surface constantly changes.

Identifying Plate Boundaries

How do we recognize the location of a plate boundary? The answer becomes clear from looking at a map showing the locations of earthquakes (Fig. 4.2c). Recall from Chapter 2 that earthquakes are vibrations caused by shock waves that are generated where rock breaks and suddenly shears (slides) along a fault (a fracture on which sliding occurs). The focus of the earthquake is the spot where the fault slips, and the epicenter marks the point on the surface of the Earth directly above the focus. Earthquake epicenters do not speckle the Earth's surface randomly, like buckshot on a target. Rather, the majority occur in relatively narrow, distinct belts. These earthquake belts define the position of plate boundaries (see Fig. 4.2b–d), because the fracturing and slipping that occur along plate boundaries as plates move generate earthquakes. Plate interiors, regions away from the plate boundaries, remain relatively earthquake-free because they do not accommodate as much movement. While earthquakes serve as the most definitive indicator of a plate boundary, other prominent geologic features also develop along plate boundaries, as you will learn by the end of this chapter.

> **Did you ever wonder . . .**
> why earthquakes don't occur everywhere?

Geologists define three types of plate boundaries, based simply on the *relative motions* of the plates on either side of the boundary (Fig. 4.3a–c ⏵▮). A boundary at which two plates move apart from each other is a **divergent boundary**. A boundary at which two plates move toward each other so that one plate sinks beneath the other is a **convergent boundary**. And a boundary at which two plates slide sideways past each

other is a **transform boundary**. Each type of boundary looks and behaves differently from the others, as we will now see.

4.3 DIVERGENT PLATE BOUNDARIES AND SEA-FLOOR SPREADING

At divergent boundaries, or spreading boundaries, two oceanic plates move apart by the process of sea-floor spreading. Note that an *open space does not develop* between diverging plates. Rather, as the plates move apart, new oceanic lithosphere forms along the divergent boundary (Fig. 4.4a). This process takes place at a submarine mountain range called a mid-ocean ridge that rises 2 km above the adjacent abyssal plains of the ocean. Thus, geologists also commonly refer to a divergent boundary as a mid-ocean ridge, or simply a ridge.

To characterize a divergent boundary more completely, let's look at one mid-ocean ridge in more detail (Fig. 4.4b). The Mid-Atlantic Ridge extends from the waters between

FIGURE 4.3 The three types of plate boundaries differ based on the nature of relative movement. ⏵▮

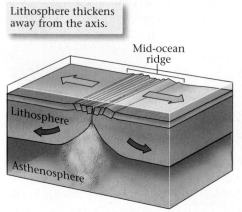

(a) At a divergent boundary, two plates move away from the axis of a mid-ocean ridge. New oceanic lithosphere forms.

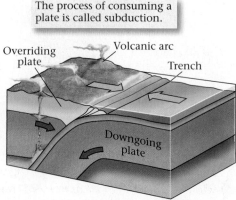

(b) At a convergent boundary, two plates move toward each other; the downgoing plate sinks beneath the overriding plate.

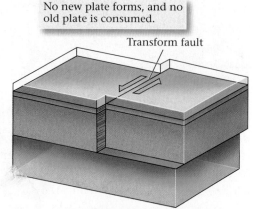

(c) At a transform boundary, two plates slide past each other on a vertical fault surface.

FIGURE 4.4 The process of sea-floor spreading. ▶️

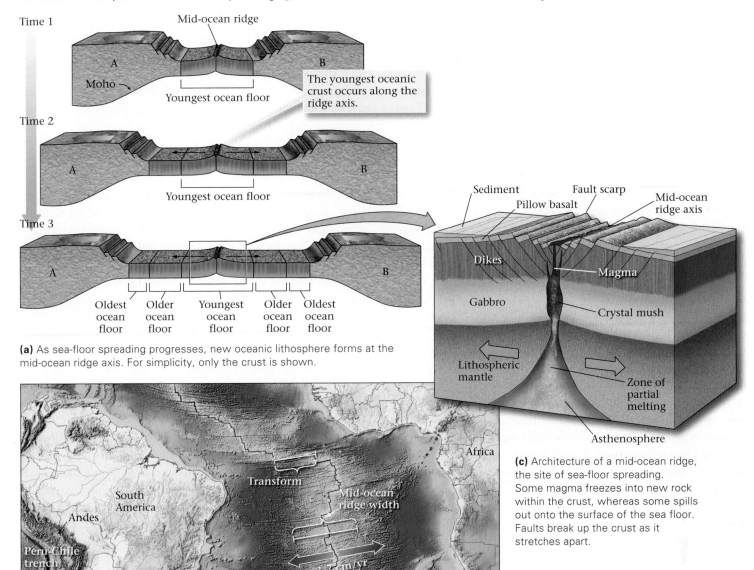

Time 1

Mid-ocean ridge

A B

Moho

Youngest ocean floor

The youngest oceanic crust occurs along the ridge axis.

Time 2

A B

Youngest ocean floor

Time 3

A B

| Oldest ocean floor | Older ocean floor | Youngest ocean floor | Older ocean floor | Oldest ocean floor |

(a) As sea-floor spreading progresses, new oceanic lithosphere forms at the mid-ocean ridge axis. For simplicity, only the crust is shown.

Sediment Fault scarp Mid-ocean ridge axis

Pillow basalt

Dikes Magma

Gabbro Crystal mush

Lithospheric mantle Zone of partial melting

Asthenosphere

(c) Architecture of a mid-ocean ridge, the site of sea-floor spreading. Some magma freezes into new rock within the crust, whereas some spills out onto the surface of the sea floor. Faults break up the crust as it stretches apart.

Africa

South America

Andes

Peru-Chile trench

Transform

Mid-ocean ridge width

Passive continental margin

Fracture zone

1.7 cm/yr

Hot-spot track

Abyssal plain

Transform

Continental shelf

Seamount

Triple junction

(b) The bathymetry of the Mid-Atlantic Ridge in the Southern Atlantic Ocean. The lighter shades of blue are shallower depths.

500 to 800 km from the ridge axis (see Fig. 3.8). Roughly speaking, the Mid-Atlantic Ridge is symmetrical—its eastern half looks like a mirror image of its western half. The ridge consists, along its length, of short segments (tens to hundreds of km long) that step over at breaks called transform faults, which we will discuss later.

How Does Oceanic Crust Form at a Mid-Ocean Ridge?

northern Greenland and northern Scandinavia southward across the equator to the latitude of the southern tip of South America. Geologists have found that the formation of new sea floor takes place *only* along the axis (centerline) of the ridge. The axis lies at water depths of about 2 to 2.5 km.

The sea floor slopes away from the ridge axis, reaching the depth of the abyssal plain (4 to 5 km) at a distance of about

As sea-floor spreading takes place, hot asthenosphere rises beneath the ridge (Fig. 4.4c ▶️). The rising asthenosphere begins to melt (for reasons described in Chapter 6), producing molten rock, or magma. Magma has a lower density than solid rock, so it behaves buoyantly and rises, like oil rises above vinegar in salad dressing. Molten rock eventually accumulates in the crust below the ridge axis, filling a region called a magma chamber. As the magma cools, it turns into a mush of crystals. Some of the magma solidifies completely along the side of the chamber to make the coarse-grained, mafic igneous rock called gabbro. The rest rises still higher to fill vertical cracks, where it solidifies and forms wall-like sheets, or dikes, of basalt. Some magma rises all the way to the surface of the sea floor at the ridge axis and spills out of small submarine volcanoes. The resulting lava cools to form a layer of basalt blobs or "pillows" (Fig. 4.5a). We can't easily see the submarine volcanoes because they occur at depths of more than 2 km beneath sea level, but they have been observed by the research submarine *Alvin*. Observers in the submarine have also detected chimneys spewing hot, mineralized water rising from cracks in the sea floor after being heated by magma below the surface. These chimneys are called **black smokers** because the water they emit looks like a cloud of dark smoke; the color comes from a suspension of tiny mineral grains that precipitate in the water the instant that the water cools (Fig. 4.5b).

> **Did you ever wonder . . .**
> whether the width of oceans remains constant?

As soon as it forms, new oceanic crust moves away from the ridge axis, and when this happens, more magma rises from below, so still more crust forms. In other words, like a vast, continuously moving conveyor belt, magma from the mantle rises to the Earth's surface at the ridge, solidifies to form oceanic crust, and then moves laterally away from the ridge. Because all sea floor forms at mid-ocean ridges, the youngest sea floor occurs on either side of the ridge axis, and sea floor becomes progressively older away from the ridge. In the Atlantic Ocean, the oldest sea floor, therefore, lies adjacent to the passive continental margins on either side of the ocean (Fig. 4.6). The oldest ocean floor on our planet underlies the western Pacific Ocean; this crust formed 200 million years ago.

The tension (stretching force) applied to newly formed solid crust as spreading takes place breaks this new crust, resulting in the formation of faults. Movement (slip) on the faults causes divergent-boundary earthquakes and produces numerous cliffs, or scarps, that lie parallel to the ridge axis.

How Does the Lithospheric Mantle Form at a Mid-Ocean Ridge?

So far, we've seen how oceanic crust forms at mid-ocean ridges. What about the formation of the mantle part of the oceanic lithosphere? This part consists of the cooler uppermost layer of the mantle, in which temperatures are less than about 1,280°C. At the ridge axis, such temperatures occur almost at the base of the crust, because of the presence of rising hot asthenosphere and hot magma, so lithospheric mantle beneath the ridge axis effectively doesn't exist. But as the newly formed oceanic crust moves away from the ridge axis, the crust and the uppermost mantle directly beneath it gradually cool by losing heat to the ocean above. As soon as mantle rock cools below 1,280°C, it becomes, by definition, part of the lithosphere because at this temperature, it begins to behave rigidly.

As oceanic lithosphere continues to move away from the ridge axis, it continues to cool, so the lithospheric mantle, and therefore the oceanic lithosphere as a whole, grows progressively thicker (Fig. 4.7a, b). Note that this process doesn't change the thickness of the oceanic crust, for the crust formed entirely at the ridge axis. The rate at which cooling and lithospheric thickening occur decreases with distance from the ridge axis. In fact, by the time the lithosphere is about 80 million years old, it has just about reached its maximum thickness. As lithosphere

FIGURE 4.5 Eruptions on mid-ocean ridges.

(a) Recently erupted pillow basalt from the Juan de Fuca Ridge.

(b) A column of superhot water gushing from a vent (known as a "black smoker") along the mid-ocean ridge. The cloud of "smoke" actually consists of tiny mineral grains. A local ecosystem of bacteria, shrimp, and worms lives around the vent.

FIGURE 4.6 This map of the world shows the age of the sea floor. Note how the sea floor grows older with increasing distance from the ridge axis. (Ma = million years ago.)

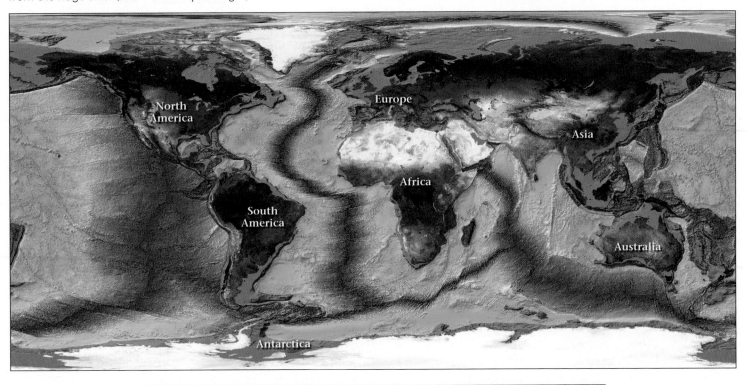

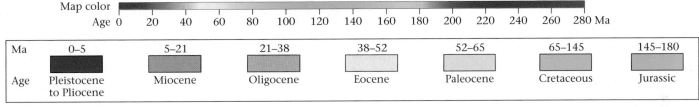

Ma	0–5	5–21	21–38	38–52	52–65	65–145	145–180
Age	Pleistocene to Pliocene	Miocene	Oligocene	Eocene	Paleocene	Cretaceous	Jurassic

thickens and gets cooler and denser, it sinks down into the asthenosphere, like a ship taking on ballast. Thus, the ocean is deeper over older ocean floor than over younger ocean floor.

Take-Home Message

- At a divergent plate boundary, plates move apart due to sea-floor spreading. As the plates move, new ocean lithosphere forms in between.
- The youngest plate at a divergent boundary is along the ridge axis; the sea floor gets progressively older away from the axis.
- Ocean crust forms from magma that rises along the mid-ocean ridge axis.
- As plates move away from the axis, they cool and the lithospheric mantle forms and thickens.

THINK: What is a black smoker, and why do they form?

4.4 CONVERGENT PLATE BOUNDARIES AND SUBDUCTION

At convergent plate boundaries, two plates, at least one of which is oceanic, move toward one another. But rather than butting each other like angry rams, one oceanic plate bends and sinks down into the asthenosphere beneath the other plate. Geologists refer to the sinking process as **subduction**, so convergent boundaries are also known as subduction zones. Because subduction at a convergent boundary consumes old ocean lithosphere and thus closes (or "consumes") oceanic basins, geologists also refer to convergent boundaries as con-suming boundaries, and because they are delineated by deep-ocean trenches, they are sometimes simply called **trenches** (see Fig. 3.9). The amount of oceanic plate consumption worldwide, averaged over time, equals the amount of sea-floor spreading worldwide, so the surface area of the Earth remains constant through time.

FIGURE 4.7 Changes accompanying the aging of lithosphere.

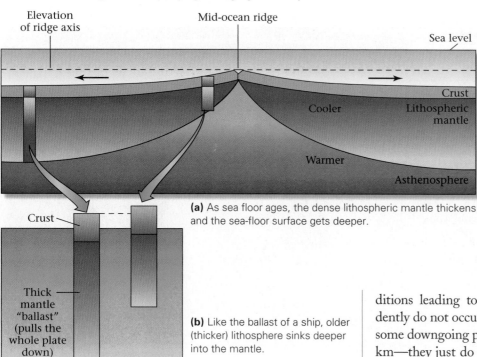

(a) As sea floor ages, the dense lithospheric mantle thickens and the sea-floor surface gets deeper.

(b) Like the ballast of a ship, older (thicker) lithosphere sinks deeper into the mantle.

Earthquakes and the Fate of Subducted Plates

At convergent plate boundaries, the downgoing plate grinds along the base of the overriding plate, a process that generates large earthquakes. These earthquakes occur fairly close to the Earth's surface, so some of them cause massive destruction in coastal cities. But earthquakes also happen in downgoing plates at greater depths. In fact, geologists have detected earthquakes within downgoing plates to a depth of 660 km. The band of earthquakes in a downgoing plate is called a **Wadati-Benioff zone**, after its two discoverers (Fig. 4.8b).

At depths greater than 660 km, conditions leading to earthquakes in subducted lithosphere evidently do not occur. Recent observations, however, indicate that some downgoing plates do continue to sink *below* a depth of 660 km—they just do so without generating earthquakes. In fact, the lower mantle may be a graveyard for old subducted plates.

Geologic Features of a Convergent Boundary

To become familiar with the various geologic features that occur along a convergent plate boundary, let's look at an example, the boundary between the western coast of the South American Plate and the eastern edge of the Nazca Plate (a portion of the Pacific Ocean floor). A deep-ocean trench, the Peru-Chile Trench, delineates this boundary (see Fig. 4.4b). Such trenches form where the plate bends as it starts to sink into the asthenosphere.

In the Peru-Chile Trench, as the downgoing plate slides under the overriding plate, sediment (clay and plankton) that had settled on the surface of the downgoing plate, as well as sand that fell into the trench from the shores of South America, gets scraped up and incorporated in a wedge-shaped mass known as an **accretionary prism** (Fig. 4.8c ⊙). An accretionary prism forms in basically the same way as a pile of snow in front of a plow, and like the snow, the sediment tends to be squashed and contorted.

A chain of volcanoes known as a **volcanic arc** develops behind the accretionary prism (see **See for Yourself** C on p. S-6 to see a volcanic arc and other features of plate boundaries). As we will see in Chapter 6, the magma that feeds these volcanoes forms just above the surface of the downgoing plate where the plate reaches a depth of about 150 km below the Earth's surface. If the volcanic arc forms where an oceanic plate subducts beneath continental lithosphere, the resulting chain of volcanoes grows on the continent and forms a continental volcanic arc. (In some cases, the plates compress together, causing a belt of faults to form behind the arc.) If, however, the volcanic arc forms where one

Subduction occurs for a simple reason: oceanic lithosphere, once it has aged at least 10 million years, is denser than asthenosphere and thus can sink through the asthenosphere. When it lies flat on the surface of the asthenosphere, oceanic lithosphere doesn't sink, because the resistance of the asthenosphere to flow is too great. However, once the end of the convergent plate bends down and slips into the mantle, it begins to sink like an anchor falling to the bottom of a lake (Fig. 4.8a). As the lithosphere sinks, asthenosphere flows out of its way, just as water flows out of the way of an anchor. But unlike water, the asthenosphere can flow only very slowly, so oceanic lithosphere can sink only very slowly, at a rate of less than about 15 cm per year. To visualize the difference, imagine how much faster a coin can sink through water than it can through honey.

Note that the downgoing plate, the plate that has been subducted, *must* be composed of oceanic lithosphere. The overriding plate, which does not sink, can consist of either oceanic or continental lithosphere. Continental crust cannot be subducted because it is too buoyant; the low-density rocks of continental crust act like a life preserver keeping the continent afloat. If continental crust moves into a convergent margin, subduction eventually stops. Because of subduction, all ocean floor on the planet is less than about 200 million years old. Because continental crust cannot subduct, some continental crust has persisted at the surface of the Earth for over 3.8 billion years.

FIGURE 4.8 During the process of subduction, oceanic lithosphere sinks back into the deeper mantle. ▶❚❚

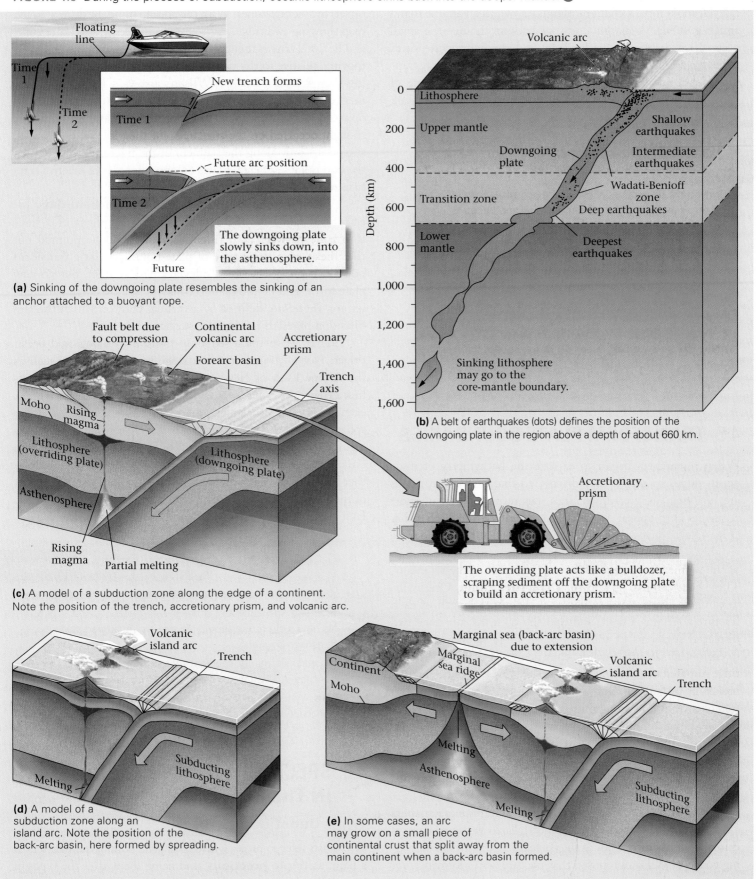

(a) Sinking of the downgoing plate resembles the sinking of an anchor attached to a buoyant rope.

(b) A belt of earthquakes (dots) defines the position of the downgoing plate in the region above a depth of about 660 km.

(c) A model of a subduction zone along the edge of a continent. Note the position of the trench, accretionary prism, and volcanic arc.

The overriding plate acts like a bulldozer, scraping sediment off the downgoing plate to build an accretionary prism.

(d) A model of a subduction zone along an island arc. Note the position of the back-arc basin, here formed by spreading.

(e) In some cases, an arc may grow on a small piece of continental crust that split away from the main continent when a back-arc basin formed.

oceanic plate subducts beneath another oceanic plate, the result-ing volcanoes form a chain of islands known as a volcanic island arc (**Fig. 4.8d**). A marginal sea, or back-arc basin—the small ocean basin between an island arc and the continent—forms either in cases where subduction happens to begin offshore, trap-ping ocean lithosphere behind the arc, or where stretching of the lithosphere behind the arc leads to the formation of a small, spreading ridge between the arc and the continent (**Fig. 4.8e**).

Take-Home Message

- Convergent plate boundaries, where oceanic lithosphere subducts (sinks into the mantle) beneath another plate, are delineated by trenches bordered by volcanic arcs.
- A band of earthquakes, the Wadati-Benioff zone, delineates the subducted plate down to a depth of around 660 km.
- During subduction, ocean-floor sediment gets scraped up to build an accretionary prism.
- Continental arcs form on continental lithosphere; island arcs occur in the ocean.

THINK: Why is the oldest oceanic lithosphere less than about 200 Ma?

4.5 TRANSFORM PLATE BOUNDARIES

When researchers began to explore the bathymetry of mid-ocean ridges in detail, they discovered that mid-ocean ridges are not long, uninterrupted lines, but rather consist of short segments that appear to be offset laterally from each other (**Fig. 4.9a**). Narrow belts of broken and irregular sea floor lie roughly at right angles to the ridge segments, intersect the ends of the segments, and extend beyond the ends of the segments. These belts came to be known as **fracture zones**. Originally, researchers *incorrectly* assumed that the entire length of each fracture zone was a fault, and that slip on a fracture zone had displaced segments of the mid-ocean ridge sideways, relative to each other. In other words, they imagined that a mid-ocean ridge initiated as a continuous, fence-like line that only later was broken up by faulting. But when information about the distribu-tion of earthquakes along mid-ocean ridges became available, it was clear that this model could not be correct. Earthquakes, and therefore active fault slip, occur *only* on the segment of a frac-ture zone that lies between two ridge segments. The portions of fracture zones that extend beyond the edges of ridge segments, out into the abyssal plain, are not active.

The distribution of movement along fracture zones remained a mystery until a Canadian researcher, J. Tuzo Wilson, began to think about fracture zones in the context of the sea-floor spread-ing concept. Wilson proposed that fracture zones formed *at the same time* as the ridge axis itself, and thus the ridge consisted of

separate segments to start with. These segments were linked (not offset) by fracture zones. With this idea in mind, he drew a sketch map showing two ridge-axis segments linked by a fracture zone, and he drew arrows to indicate the direction that ocean crust was moving, relative to the ridge axis, as a result of sea-floor spread-ing (**Fig. 4.9b** ●). Look at Wilson's arrows. Clearly, the move-ment direction on the active portion of the fracture zone must be opposite to the movement direction that researchers originally thought occurred on the structure. Further, in Wilson's model, slip occurs only along the segment of the fracture zone between the two ridge segments. Plates on opposite sides of inactive frac-ture zones move together, as one plate.

Wilson introduced the term **transform boundary** (or transform fault) for the actively slipping segment of a frac-ture zone between two ridge segments, and he pointed out that these are a third type of plate boundary. At a transform boundary, one plate slides sideways past another, but no new plate forms and no old plate is consumed. Transform bound-aries are, therefore, defined by a vertical fault on which the slip direction parallels the Earth's surface.

So far we've discussed only transforms along mid-ocean ridges. Not all transforms link ridge segments. Some, such as the Alpine Fault of New Zealand, link trenches, while others link a trench to a ridge segment. Further, not all transform faults occur in oceanic lithosphere; a few cut across continental litho-sphere. The San Andreas Fault, for example, which cuts across California, defines part of the plate boundary between the North American Plate and the Pacific Plate—the portion of California that lies to the west of the fault (including Los Angeles) is part of the Pacific Plate, while the portion that lies to the east of the fault is part of the North American Plate (**Fig. 4.9c, d**).

Take-Home Message

- At a transform boundary, one plate slides in a direction parallel to the plate boundary.
- No new plate forms and no old plate is consumed along a transform boundary.
- Most transforms link segments of mid-ocean ridges, but some link trenches. A few transforms cut through continental lithosphere.

THINK: What is the difference between an active transform fault and an inactive fracture zone?

4.6 SPECIAL LOCATIONS IN THE PLATE MOSAIC

Triple Junctions

Geologists refer to places where *three* plate boundaries intersect at a point as **triple junctions**, and name them after the types of boundaries that intersect. For example, the triple junction formed

FIGURE 4.9 The concept of transform faulting. ▶️

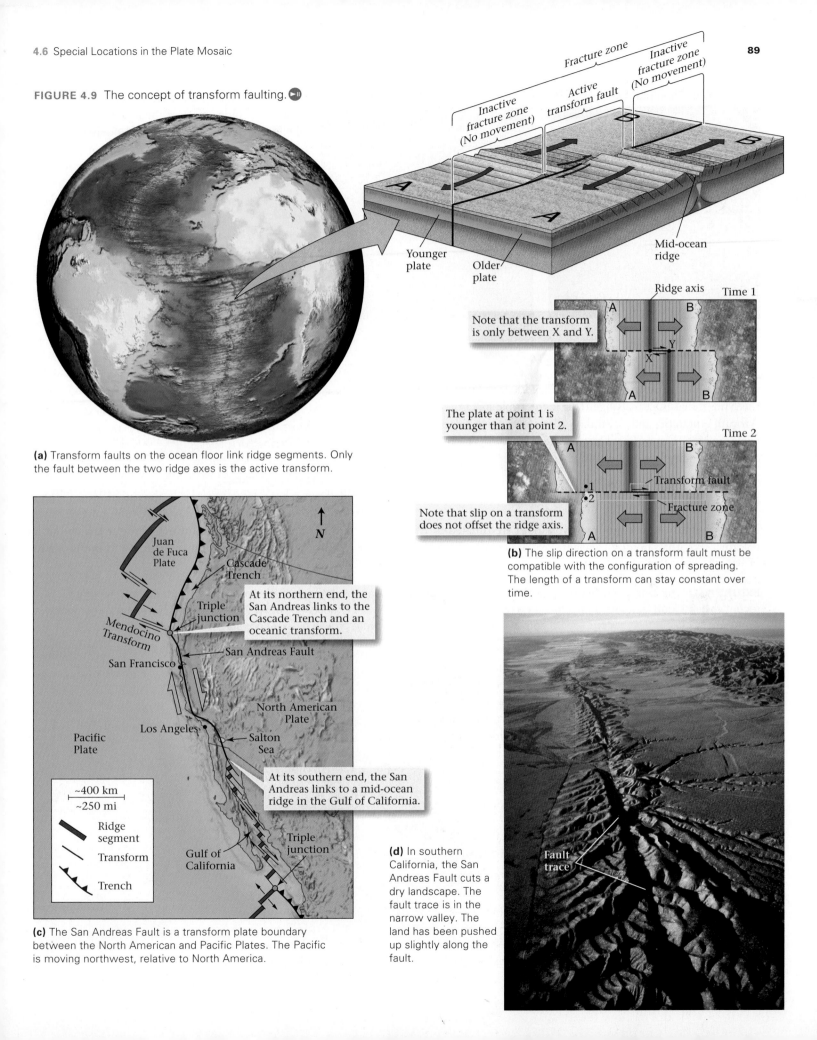

(a) Transform faults on the ocean floor link ridge segments. Only the fault between the two ridge axes is the active transform.

Note that the transform is only between X and Y.

The plate at point 1 is younger than at point 2.

Note that slip on a transform does not offset the ridge axis.

(b) The slip direction on a transform fault must be compatible with the configuration of spreading. The length of a transform can stay constant over time.

At its northern end, the San Andreas links to the Cascade Trench and an oceanic transform.

At its southern end, the San Andreas links to a mid-ocean ridge in the Gulf of California.

(c) The San Andreas Fault is a transform plate boundary between the North American and Pacific Plates. The Pacific is moving northwest, relative to North America.

(d) In southern California, the San Andreas Fault cuts a dry landscape. The fault trace is in the narrow valley. The land has been pushed up slightly along the fault.

FIGURE 4.10 Examples of triple junctions. The triple junctions are marked by dots.

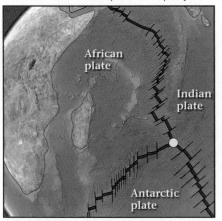

(a) A ridge-ridge-ridge triple junction occurs in the Indian Ocean.

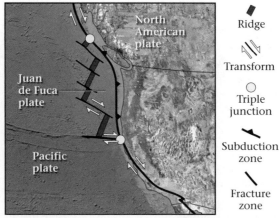

(b) A trench-transform-transform triple junction occurs at the north end of the San Andreas Fault.

mid-ocean ridges, but ocean water hides most of them. The volcanoes of volcanic arcs and mid-ocean ridges are *plate-boundary volcanoes*, in that they formed as a consequence of movement along the boundary. Not all volcanoes on Earth are plate-boundary volcanoes, however. For example, the big island of Hawaii, a large volcano, lies in the middle of the Pacific Plate, and Yellowstone National Park, site of recent volcanic activity, lies in the interior of the North American Plate. Worldwide, geoscientists have identified about 100 volcanoes that exist as isolated points and are not a consequence of movement at a plate boundary. These are called hot-spot volcanoes, or simply **hot spots** (Fig. 4.11). Most hot spots are located in the interiors of plates, away from the boundaries, but a few grew on mid-ocean ridges.

where the Southwest Indian Ocean Ridge intersects two arms of the Mid–Indian Ocean Ridge (this is the triple junction of the African, Antarctic, and Australian Plates) is a ridge-ridge-ridge triple junction (Fig. 4.10a). The triple junction north of San Francisco is a trench-transform-transform triple junction (Fig. 4.10b).

Hot Spots

Most "subaerial" (above sea level) volcanoes are situated in the volcanic arcs that border trenches. Volcanoes also lie along

> **Did you ever wonder...**
> why Hawaii rises above the middle of the ocean?

What causes hot-spot volcanoes? In the early 1960s, J. Tuzo Wilson noted that *active* hot-spot volcanoes (examples that are erupting or may erupt in the future) occur at the end of a chain of *dead* volcanic islands and seamounts (formerly active

FIGURE 4.11 The dots represent the locations of selected hot-spot volcanoes. The red lines represent hot-spot tracks. The most recent volcano (dot) is at one end of this track. Some of these volcanoes are extinct, indicating that the mantle plume no longer exists. Some hot spots are fairly recent and do not have tracks. Dashed tracks were broken by sea-floor spreading.

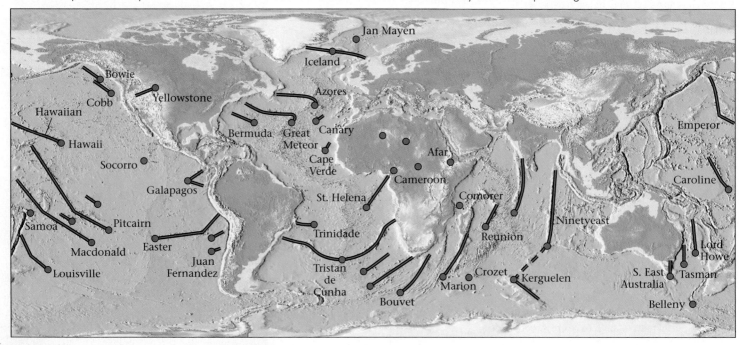

volcanoes that will never erupt again). This configuration is different from that of volcanic arcs along convergent plate boundaries—at volcanic arcs, all of the volcanoes are active. With this image in mind, Wilson suggested that the position of the heat source causing a hot-spot volcano is fixed, relative to the moving plate. In Wilson's model, the active volcano represents the present-day location of the heat source, whereas the chain of dead volcanic islands represents locations on the plate that were once over the heat source but progressively moved off.

A few years later, Jason Morgan suggested that the heat source for hot spots is a **mantle plume**, a column of very hot rock rising up through the mantle to the base of the lithosphere (Fig. 4.12a–d 🔊). In Morgan's model, plumes originate deep in the mantle. Rock in the plume, though solid, is soft enough to flow, and rises buoyantly because it is less dense than surrounding cooler rock. When the hot rock of the plume reaches the base of the lithosphere, it partially melts (for reasons discussed in Chapter 6) and produces magma that seeps up through the lithosphere to the Earth's surface. The chain of extinct volcanoes, or **hot-spot track**, forms when the overlying plate moves over a fixed plume. This movement slowly carries the volcano off the top of the plume, so that it becomes extinct. A new, younger volcano grows over the plume.

The Hawaiian chain provides a clear example of the volcanism associated with a hot-spot track. Volcanic eruptions occur today only on the big island of Hawaii. Other islands to the northwest are remnants of dead volcanoes, the oldest of which is Kauai. To the northwest of Kauai, still older volcanic remnants are found. About 1750 km northwest of Midway Island, the track bends in a more northerly direction, and the volcanic remnants no longer poke above sea level; we refer to this northerly-trending segment as the Emperor seamount chain. The bend is due to a change in the direction of Pacific Plate motion at about 40 Ma.

The deep-mantle plume model of hot-spot volcanism has been widely, but not universally, accepted by geologists. Some researchers suggest alternative models in which the plumes either originate at shallow depths in the upper mantle, or don't exist at all and instead represent places where the lithosphere cracked open above a region of special asthenosphere whose melting can produce particularly large quantities of magma.

Some hot spots lie within continents. For example, several have been active in the interior of Africa, and one now underlies Yellowstone National Park. The famous geysers (natural steam and hot-water fountains) of Yellowstone exist because hot magma, formed above the Yellowstone hot spot, lies not far below the surface of the park. A few hot spots lie on mid-ocean ridges. Where this happens, a volcanic island may protrude above sea level, because the hot spot produces far more magma than does a normal mid-ocean ridge. Iceland, for example, formed where a hot spot underlies the Mid-Atlantic Ridge.

The extra volcanism of the hot spot built up the island of Iceland so that it rises almost 3 km above other places on the Mid-Atlantic Ridge.

> ### Take-Home Message
>
> - Triple junctions exist where three plate boundaries intersect at a point.
> - Hot spots are places where volcanism happens independently of plate motion.
> - Most hot spots occur in the interior of plates, away from plate boundaries, but a few lie at mid-ocean ridges.
> - As plates move over a hot spot, a hot-spot track of inactive volcanoes forms.
> - Hot spots appear to be associated with mantle plumes, but this idea remains controversial.
>
> **THINK:** Why is the volcano at the end of a hot-spot track the only one to be active?

4.7 HOW DO PLATE BOUNDARIES FORM AND DIE?

The configuration of plates and plate boundaries visible on our planet today has not existed for all of geologic history, and will not exist indefinitely into the future. Because of plate motion, oceanic plates form and are later consumed, while continents merge and later split apart. How does a new divergent boundary come into existence, and how does an existing convergent boundary eventually cease to exist? Most new divergent boundaries form when a continent splits and separates into two continents. We call this process **rifting**. A convergent boundary ceases to exist when a piece of buoyant lithosphere, such as a continent or an island arc, moves into the subduction zone and, in effect, jams up the system. We call this process **collision**.

Continental Rifting

A **continental rift** is a linear belt in which continental lithosphere undergoes rifting, or pulls apart (Fig. 4.13a 🔊). The lithosphere stretches horizontally, so it thins vertically, much like a piece of taffy you pull between your fingers. Near the surface of the continent, where the crust is cold and brittle, stretching causes rock to break and faults to develop. As a consequence of faulting, blocks of rock slide down, leading to the formation of a low area that gradually becomes buried by sediment. Deeper

Did you ever wonder . . .
can continents really split apart, and if so, how?

FIGURE 4.12 The deep mantle plume hypothesis for the formation of hot-spot tracks.

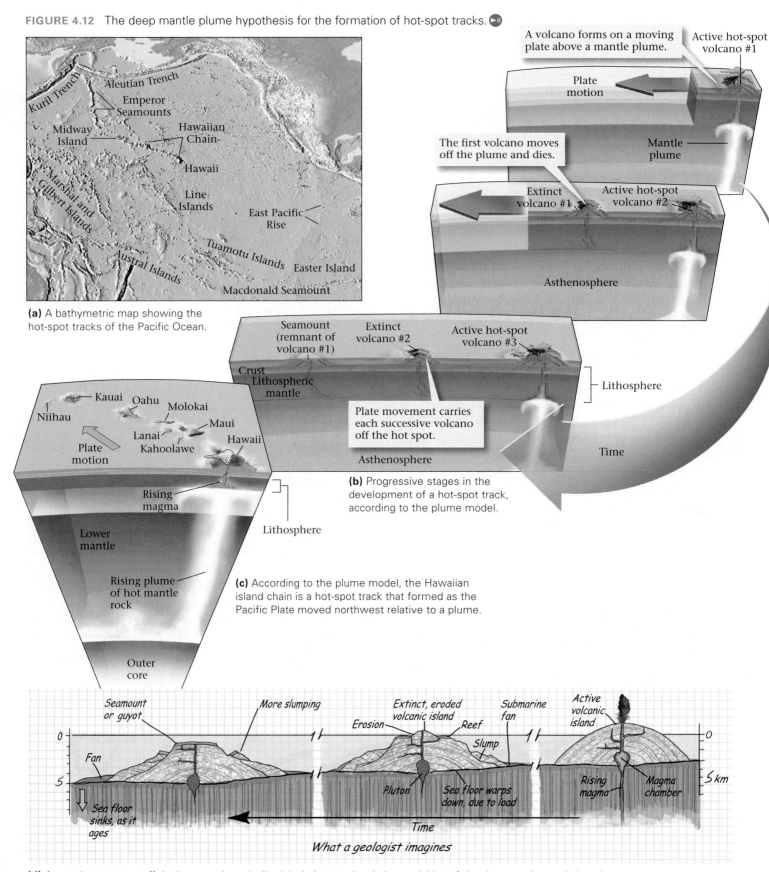

(a) A bathymetric map showing the hot-spot tracks of the Pacific Ocean.

(b) Progressive stages in the development of a hot-spot track, according to the plume model.

(c) According to the plume model, the Hawaiian island chain is a hot-spot track that formed as the Pacific Plate moved northwest relative to a plume.

(d) As a volcano moves off the hot spot, it gradually sinks below sea level, due to sinking of the plate, erosion, and slumping.

FIGURE 4.13 During the process of rifting, lithosphere stretches. ▶❙❙

Time 1

Moho

Time 2

New rift

Time 3

Range   Wide rift

Basin

Time 4

New sediment

New mid-ocean ridge

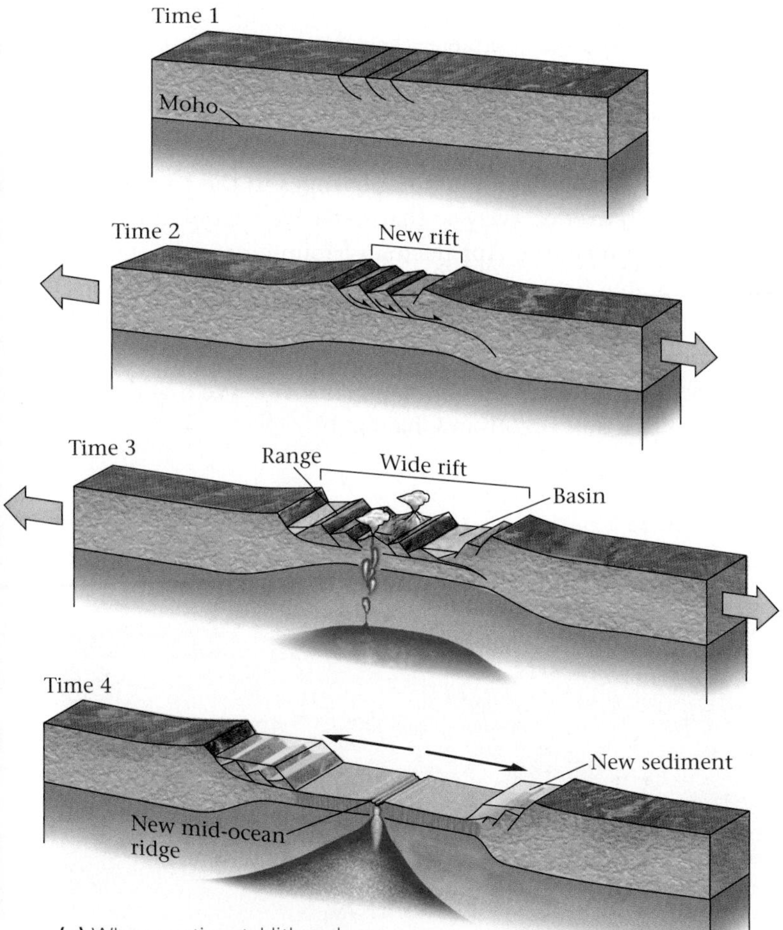

(a) When continental lithosphere stretches and thins, faulting takes place, and volcanoes erupt. Eventually, the continent splits in two and a new ocean basin forms.

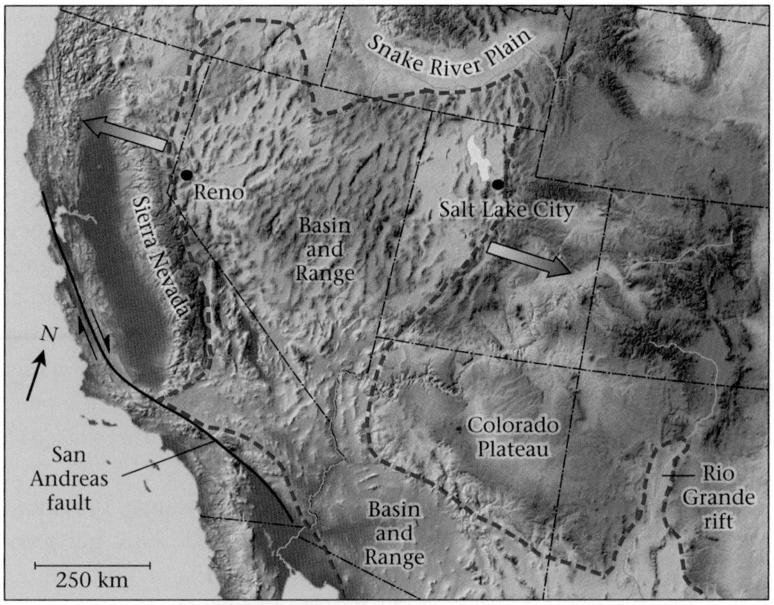

(b) The Basin and Range Province is a rift. Faulting bounds the narrow north-south-trending mountains, separated by basins. The arrows indicate the direction of stretching.

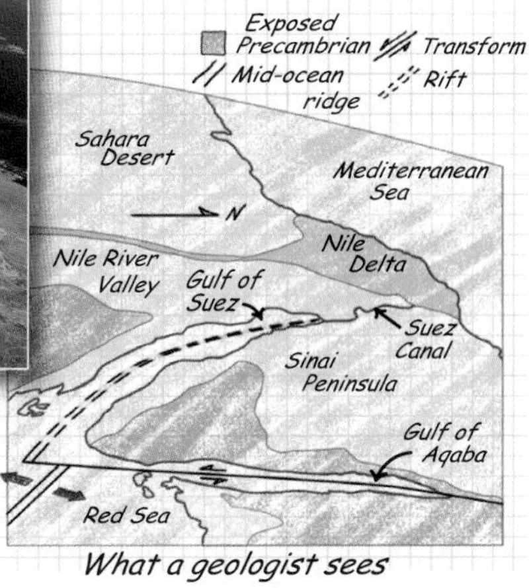

(d) Astronauts can see how rifting has opened up gulfs on either side of the Sinai Peninsula.

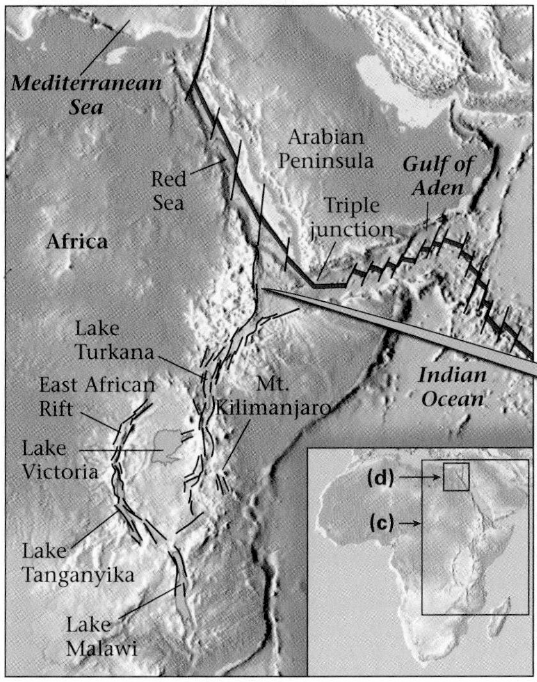

What a geologist sees

(c) The East African Rift is growing today. The Red Sea started as a rift. The inset shows map locations.

Photo taken at the north end of the East African Rift

Fault   Volcano

in the crust, and in the underlying lithospheric mantle, rock is warmer and softer, so stretching takes place in a plastic manner without breaking the rock. The whole region that stretches is the rift.

As continental lithosphere thins, hot asthenosphere rises beneath the rift and starts to melt. Molten rock erupts at volcanoes along the rift. If rifting continues for a long enough time, the continent breaks in two, a new mid-ocean ridge forms, and sea-floor spreading begins. The relict of the rift evolves into a passive margin (see Fig. 4.1b). In some cases, however, rifting stops before the continent splits in two; it becomes a low-lying trough that fills with sediment. Then, the rift remains as a permanent scar in the crust, defined by a belt of faults, volcanic rocks, and a thick layer of sediment.

A major rift, known as the Basin and Range Province, breaks up the landscape of the western United States between Salt Lake City, Utah, and Reno, Nevada (Fig. 4.13b). Here, movement on numerous faults tilted blocks of crust to form narrow mountain ranges, while sediment that eroded from the blocks filled the adjacent basins (the low areas between the ranges). Perhaps the most spectacular example of an active rift is located in eastern Africa; geoscientists aptly refer to it as the East African Rift (Fig. 4.13c, d ●⫫). To astronauts in orbit, the rift looks like a giant gash in the crust. On the ground, it consists of a deep trough bordered on both sides by high cliffs formed by faulting. Along the length of the rift, several major volcanoes smoke and fume; these include the snow-crested Mt. Kilimanjaro, towering over 6 km above the savannah. At its north end, the rift joins the Red Sea Ridge and the Gulf of Aden Ridge at a triple junction.

Collision

India was once a small, separate continent that lay far to the south of Asia. But subduction consumed the ocean between India and Asia, and India moved northward, finally slamming into the southern margin of Asia about 40 to 50 million years ago. Continental crust, unlike oceanic crust, is too buoyant to subduct. So when India collided with Asia, the attached oceanic plate broke off and sank down into the deep mantle while India pushed hard into Asia, squeezing the rocks and sediment that once lay between the two continents into the 8-km-high welt that we now know as the Himalayan Mountains. During this process, not only did the surface of the Earth rise, but the crust became thicker. The crust beneath a collisional mountain range can be up to 60 to 70 km thick, about twice the thickness of normal continental crust. The boundary between what was once two separate continents is called a suture; a sliver of trapped ocean crust may remain along a suture.

Geoscientists refer to the process during which two buoyant pieces of lithosphere converge and squeeze together

as collision (Fig. 4.14). Some collisions involve two continents, whereas some involve continents and an island arc. When a collision is complete, the convergent plate boundary that once existed between the two colliding pieces ceases to exist. Collisions yield some of the most spectacular mountains on the planet, such as the Himalayas and the Alps. They also yielded major mountain ranges in the past, which subsequently eroded away so that today we see only their relicts. For example, the Appalachian Mountains in the eastern United States formed as a consequence of three collisions. After the last one, a collision between Africa and North America around 280 million years ago, North America became part of the Pangaea supercontinent. We'll add more detail to this description in Chapter 11.

FIGURE 4.14 Continental collision (not to scale).

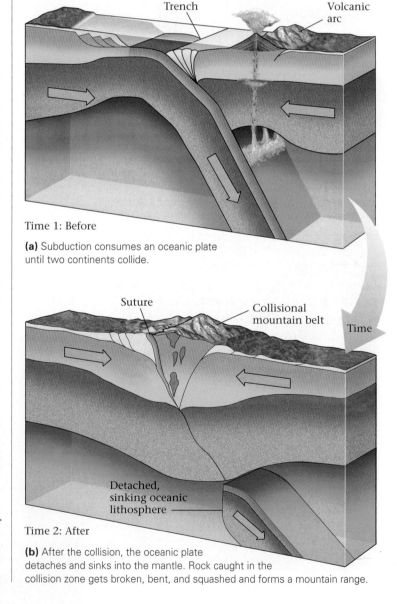

Time 1: Before

(a) Subduction consumes an oceanic plate until two continents collide.

Time 2: After

(b) After the collision, the oceanic plate detaches and sinks into the mantle. Rock caught in the collision zone gets broken, bent, and squashed and forms a mountain range.

Take-Home Message

- At a rift, continental lithosphere stretches and starts to pull apart.
- Rifts are defined by a set of faults bounding sediment- and/or volcanic-filled troughs.
- If rifting successively splits a continent in two, a new mid-ocean ridge forms.
- If one continent moves into a continental convergent plate boundary, it collides with the continent on the overriding plate and builds a large mountain range.
- During collision, two continents suture together and subduction ceases.

THINK: Why can't continental crust subduct?

4.8 WHAT DRIVES PLATE MOTION, AND HOW FAST DO PLATES MOVE?

Forces Acting on Plates

We've now discussed the many facets of plate tectonics theory (see **Geology at a Glance**, pp. 96–97). But to complete the story, we need to address the major question of what drives plate motion? When geoscientists first proposed plate tectonics, they thought the process occurred simply because convective flow in the asthenosphere actively dragged plates along, as if the plates were rafts on a flowing river. Thus, early images depicting plate motion showed simple convection cells—elliptical flow paths—in the asthenosphere. Conveniently, these cells were positioned so that asthenosphere rose beneath mid-ocean ridges and sank at subduction zones. At first glance, this hypothesis looked pretty good, but on closer examination, *it failed*. Among other reasons, it is impossible to draw a global arrangement of convection cells that can really explain the complex geometry of plate boundaries on Earth. Gradually, geoscientists came to the conclusion that, while convective flow within the asthenosphere does occur, it does not *directly* drive motion. In other words, hot asthenosphere does rise in some places and sink in others, probably because of temperature contrasts, and this convection does influence plate motion. But the local directions of this convective flow do not necessarily define the local directions of plate motion. Today, geoscientists favor the hypothesis that two forces—ridge-push force and slab-pull force—strongly influence the motion of individual plates.

Ridge-push force develops because the lithosphere of mid-ocean ridges lies at a higher elevation than that of the adjacent abyssal plains (Fig. 4.15a). To understand ridge-push force, imagine you have a glass containing a layer of water over a layer of honey. By tilting the glass momentarily

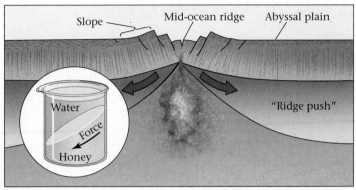

(a) Ridge push develops because the region of a rift is elevated. Like a wedge of honey with a sloping surface, the mass of the ridge pushes sideways.

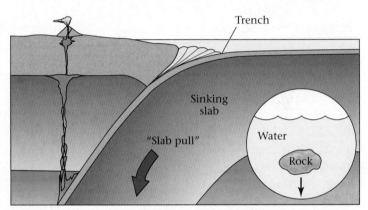

(b) Slab pull develops because lithosphere is denser than the underlying asthenosphere, and sinks like a stone in water (though much more slowly).

and then returning it to its upright position, you can create a temporary slope in the boundary between these substances. While the boundary has this slope, gravity causes the elevated honey to push against the glass adjacent to the side where the honey surface lies at lower elevation. The geometry of a mid-ocean ridge resembles this situation, for the surface of the sea floor is higher along a mid-ocean ridge axis than in adjacent abyssal plains. Gravity causes the elevated lithosphere at the ridge axis to push on the lithosphere that lies farther from the axis, making it move away. As lithosphere moves away from the ridge axis, new hot asthenosphere rises to fill the gap. Note that the local upward movement of asthenosphere beneath a mid-ocean ridge is a *consequence* of sea-floor spreading, not the cause.

Slab-pull force, the force that subducting, downgoing plates apply to oceanic lithosphere at a convergent margin, arises simply because lithosphere that was formed more than 10 million years ago is denser than asthenosphere, so it can sink into the asthenosphere (Fig. 4.15b). Thus, once an oceanic plate starts to sink, it gradually pulls the rest of the plate along behind

The Theory of Plate Tectonics

Labels in illustration:
Hot-spot volcano
Transform plate boundary
Volcanic arc
Trench
Continental rift
Convergent plate boundary
Subducting oceanic lithosphere
Collisional mountain belt
Continental crust
Continental lithosphere
Lithospheric mantle
Asthenosphere

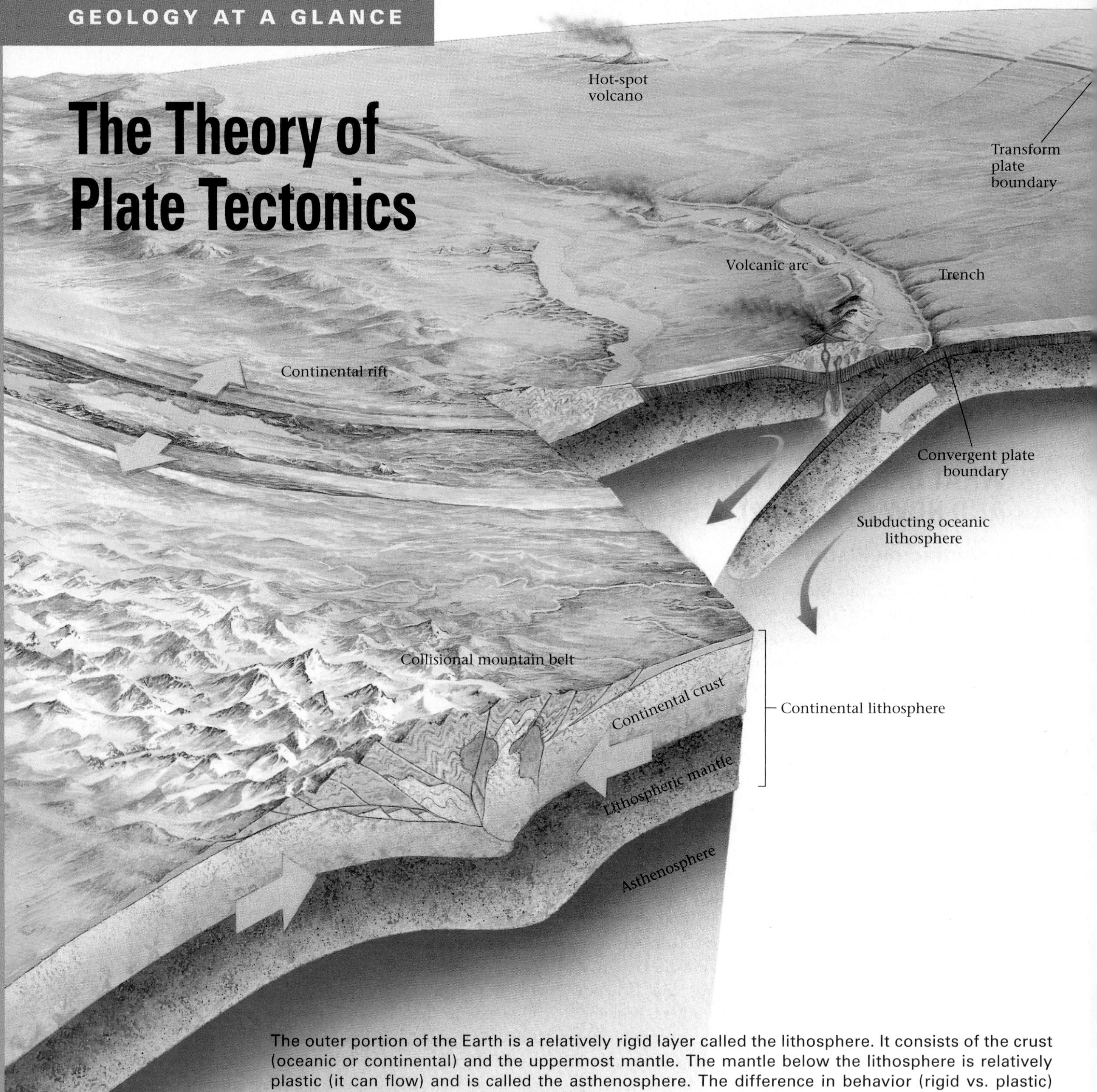

The outer portion of the Earth is a relatively rigid layer called the lithosphere. It consists of the crust (oceanic or continental) and the uppermost mantle. The mantle below the lithosphere is relatively plastic (it can flow) and is called the asthenosphere. The difference in behavior (rigid vs. plastic) between lithospheric mantle and asthenospheric mantle is a consequence of temperature—the former is cooler than the latter. Continental lithosphere is typically about 150 km thick, while oceanic lithosphere is about 100 km thick. (*Note:* They are not drawn to scale in this image.)

According to the theory of plate tectonics, the lithosphere is broken into about 20 plates that move relative to each other. Most of the motion takes place by sliding along plate boundaries (the edges of plates); plate interiors stay relatively unaffected by this motion. There are three kinds of plate boundaries.

1. Divergent boundaries: Here, two plates move apart by a process called sea-floor spreading. A mid-ocean ridge delineates a divergent boundary. Asthenospheric mantle rises beneath a mid-ocean

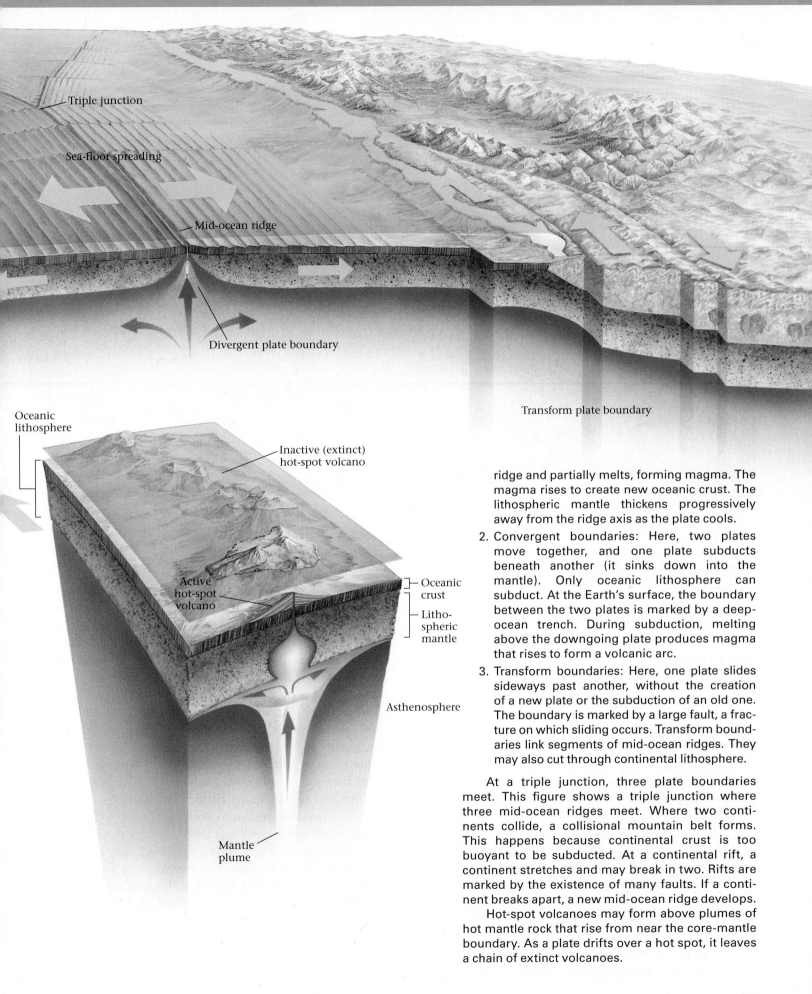

Triple junction

Sea-floor spreading

Mid-ocean ridge

Divergent plate boundary

Transform plate boundary

Oceanic
lithosphere

Inactive (extinct)
hot-spot volcano

Active
hot-spot
volcano

Oceanic
crust

Litho-
spheric
mantle

Asthenosphere

Mantle
plume

ridge and partially melts, forming magma. The magma rises to create new oceanic crust. The lithospheric mantle thickens progressively away from the ridge axis as the plate cools.

2. Convergent boundaries: Here, two plates move together, and one plate subducts beneath another (it sinks down into the mantle). Only oceanic lithosphere can subduct. At the Earth's surface, the boundary between the two plates is marked by a deep-ocean trench. During subduction, melting above the downgoing plate produces magma that rises to form a volcanic arc.

3. Transform boundaries: Here, one plate slides sideways past another, without the creation of a new plate or the subduction of an old one. The boundary is marked by a large fault, a fracture on which sliding occurs. Transform boundaries link segments of mid-ocean ridges. They may also cut through continental lithosphere.

At a triple junction, three plate boundaries meet. This figure shows a triple junction where three mid-ocean ridges meet. Where two continents collide, a collisional mountain belt forms. This happens because continental crust is too buoyant to be subducted. At a continental rift, a continent stretches and may break in two. Rifts are marked by the existence of many faults. If a continent breaks apart, a new mid-ocean ridge develops.

Hot-spot volcanoes may form above plumes of hot mantle rock that rise from near the core-mantle boundary. As a plate drifts over a hot spot, it leaves a chain of extinct volcanoes.

FIGURE 4.16 *Relative plate velocities*: The blue arrows show the rate and direction at which the plate on one side of the boundary is moving with respect to the plate on the other side. Outward-pointing arrows indicate spreading (divergent boundaries), inward-pointing arrows indicate subduction (convergent boundaries), and parallel arrows show transform motion. The length of an arrow represents the velocity. *Absolute plate velocities*: The red arrows show the velocity of the plates with respect to a fixed point in the mantle.

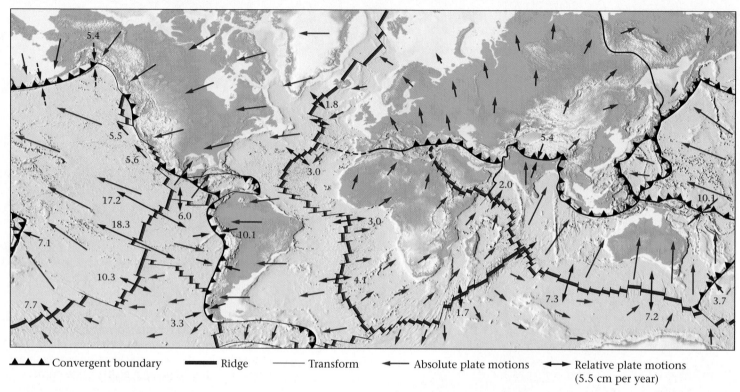

▲▲▲ Convergent boundary   ▬▬ Ridge   ─── Transform   ←── Absolute plate motions   ←──→ Relative plate motions (5.5 cm per year)

it, like an anchor pulling down the anchor line. This "pull" is the slab-pull force.

The convecting asthenosphere probably applies a force (a shear) to the base of the plate. In some places, this force helps move the plate along, but it's also possible that the force may slow the plates down.

The Velocity of Plate Motions

How fast do plates move? *It depends on your frame of reference.* To illustrate this concept, imagine two cars speeding in the same direction down the highway. From the viewpoint of a tree along the side of the road, car A zips by at 100 km an hour, while car B moves at 80 km an hour. But relative to car B, car A moves at only 20 km an hour. Geologists use two different frames of reference for describing plate velocity (remember: velocity = distance/time). If we describe the movement of plate A with respect to plate B, then we are speaking about **relative plate velocity**. But if we describe the movement of both plates relative to a fixed location in the mantle below the plates, then we are speaking of **absolute plate velocity** (Fig. 4.16).

To determine relative plate motions, geoscientists measure the distance of a known magnetic anomaly from the axis of a mid-ocean ridge and then calculate the velocity of a plate relative to the ridge axis by applying this equation: plate velocity = distance from the anomaly to the ridge axis divided by the age of the anomaly. The velocity of the plate on one side of the ridge relative to the plate on the other is twice this value.

To estimate absolute plate motions, we can *assume* that the position of a mantle plume does not change much for a long time. If this is so, then the track of hot-spot volcanoes on the plate moving over the plume provides a record of the plate's absolute velocity and indicates the direction of movement. (In reality, plumes are not completely fixed; geologists must use other, more complex methods to calculate absolute plate motions.) The Hawaiian-Emperor seamount chain, for example, defines the absolute velocity of the Pacific Plate (see Fig. 4.12a). Note that the Hawaiian chain runs northwest, whereas the Emperor chain trends north-northwest. Isotopic dates of volcanic rocks from the seamount at the bend indicate that they formed about 43 million years ago. Thus, the direction in which the Pacific Plate moved changed significantly at this time.

Working from the calculations described above, geologists have determined that relative plate motions on Earth today occur at rates of about 1 to 15 cm per year. But these rates, though small, can yield large displacements given the immensity of geologic time. At a rate of 10 cm per year, a plate can move 100 km in a million years! Can we detect such slow rates? Until the last decade, the answer was no. Now the answer is yes, because of satellites orbiting the Earth with **global positioning system (GPS)** technology (**Fig. 4.17**). Automobile drivers use GPS receivers to find their destinations, and geologists use them to monitor plate motions. If we calculate carefully enough, we can detect displacements of millimeters per year. In other words, we can now see the plates move—this serves as the ultimate proof of plate tectonics.

Taking into account many data sources that define the motion of plates, geologists have greatly refined the image of continental drift that Wegener tried so hard to prove nearly a century ago. We can now see how the map of our planet's surface has evolved radically during the past 400 million years (**Fig. 4.18**), and even before.

> **Did you ever wonder . . .**
> whether we can really "see" continents drift?

FIGURE 4.17 The Global Positioning System (GPS) is used to measure plate motions at many locations on Earth.

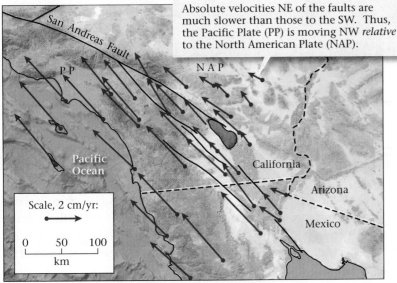

Absolute velocities NE of the faults are much slower than those to the SW. Thus, the Pacific Plate (PP) is moving NW *relative* to the North American Plate (NAP).

(a) GPS measurements in southern California show the region west of the San Andreas Fault system, a plate boundary, is moving northwest up to 6 cm/y. The length of the arrows indicates the magnitude of velocity.

> ## Take-Home Message
>
> - Convection in the mantle applies force to the base of a plate, but the plate motion is probably controlled primarily by ridge push and slab pull.
> - Ridge push occurs because elevated lithosphere at mid-ocean ridges creates a sideways force. Slab pull occurs because older oceanic lithosphere is denser than asthenosphere and sinks.
> - Geologists distinguish between relative plate motion (velocity of one plate with respect to its neighbor) and absolute plate motion (velocity of a plate relative to a fixed point).
> - Most plates move at 1 to 15 cm/year. Modern GPS measurements can detect this motion.
>
> **THINK:** The North Atlantic Ocean is about 3600 km wide. Sea-floor spreading along the Mid-Atlantic Ridge occurs at about 2 cm per year. When did the ocean start to form, assuming that the spreading rate has remained constant?

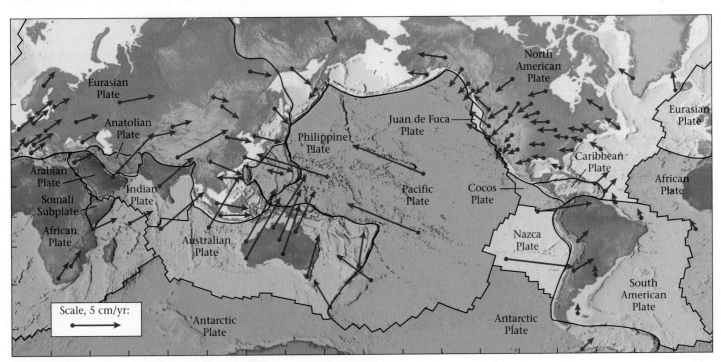

(b) A global map of plate velocities, determined by GPS.

FIGURE 4.18 Due to plate tectonics, the map of Earth's surface slowly changes. Here we see the assembly, and later the breakup, of Pangaea during the past 400 million years.

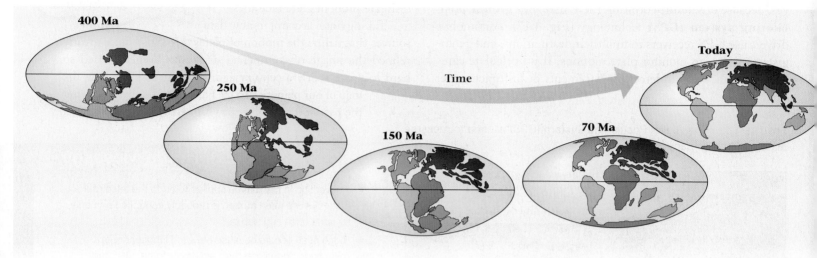

Chapter Summary

- The lithosphere, the rigid outer layer of the Earth, is broken into discrete plates that move relative to each other. Plates consist of the crust and the uppermost (cooler) mantle. Lithosphere plates effectively float on the underlying soft asthenosphere. Continental drift and sea-floor spreading are manifestations of plate movement.

- Most earthquakes and volcanoes occur along plate boundaries; the interiors of plates remain relatively rigid and intact.

- There are three types of plate boundaries—divergent, convergent, and transform—distinguished from each other by the movement the plate on one side of the boundary makes relative to the plate on the other side.

- Divergent boundaries are marked by mid-ocean ridges. At divergent boundaries, sea-floor spreading takes place, a process that forms new oceanic lithosphere.

- Convergent boundaries are marked by deep-ocean trenches and volcanic arcs. At convergent boundaries, oceanic lithosphere of the downgoing plate subducts beneath an overriding plate.

- Transform boundaries are marked by large faults at which one plate slides sideways past another. No new plate forms and no old plate is consumed at a transform boundary.

- Triple junctions are points where three plate boundaries intersect.

- Hot spots are places where volcanism occurs at an isolated volcano. As a plate moves over the hot spot, the volcano moves off and dies, and a new volcano forms over the hot spot; the chain of volcanoes defines a hot-spot track. Hot spots may be caused by mantle plumes.

- A large continent can split into two smaller ones by the process of rifting. During rifting, continental lithosphere stretches and thins. If it finally breaks apart, a new mid-ocean ridge forms and sea-floor spreading begins.

- Convergent plate boundaries cease to exist when a buoyant piece of crust (a continent or an island arc) moves into the subduction zone. When that happens, collision occurs. Collision can thicken crust and build large mountain belts.

- Ridge-push force and slab-pull force contribute to driving plate motions. Plates move at rates of about 1 to 15 cm per year. We can describe plate motions relative to each other, or relative to a fixed point. Modern satellite measurements can detect these motions.

GEOPUZZLE REVISITED

The outer shell of our planet consists of plates that move relative to each other. Interactions at plate boundaries generate most major geologic features—earthquakes, volcanoes, and mountain belts—of the Earth. Plate tectonics does not occur on other planets, so their surfaces do not look like Earth's.

Guide Terms

absolute plate velocity (p. 98)

accretionary prism (p. 86)

active margin (p. 80)

asthenosphere (p. 78)

black smoker (p. 84)

buoyancy (p. 80)

collision (p. 91)

continental rift (p. 91)

continental shelf (p. 80)

convergent boundary (p. 82)

divergent boundary (p. 82)

fracture zone (p. 88)

global positioning system
  (GPS) (p. 99)

hot spot (p. 90)

hot-spot track (p. 91)

lithosphere (p. 78)

lithosphere plate (p. 80)

mantle plume (p. 91)

passive margin (p. 80)

plate boundary (p. 80)

plate tectonics (p. 78)

relative plate velocity (p. 98)

ridge-push force (p. 95)

rifting (p. 91)

slab-pull force (p. 95)

subduction (p. 85)

transform boundary
  (pp. 82, 88)

trench (p. 85)

triple junction (p. 88)

volcanic arc (p. 86)

Wadati-Benioff zone (p. 86)

Review Questions

1. What are the characteristics of a lithosphere plate?

2. How does oceanic lithosphere differ from continental lithosphere in thickness, composition, and density?

3. What are the basic premises of plate tectonics?

4. How do we identify a plate boundary?

5. Describe the three types of plate boundaries.

6. How does crust form along a mid-ocean ridge?

7. Why is subduction necessary on a nonexpanding Earth with spreading ridges?

8. Describe the major features of a convergent boundary.

9. Describe the motion that takes place on a transform boundary.

10. What is a triple junction?

11. How is a hot-spot track produced, and how can hot-spot tracks be used to track the past motions of a plate?

12. Describe the characteristics of a contintental rift, and give examples of where this process is occurring today.

13. Describe the process of continental collision, and give examples of where this process has occurred.

14. Discuss the major forces that move lithosphere plates.

15. Explain the difference between relative plate velocity and absolute plate velocity.

On Further Thought

16. Why are the marine magnetic anomalies bordering the East Pacific Rise in the southeastern Pacific Ocean wider than those bordering the Mid-Atlantic Ridge in the South Atlantic Ocean?

17. The Pacific Plate moves north relative to the North American Plate at a rate of 6 cm per year. How long will it take Los Angeles (a city on the Pacific Plate) to move northward by 480 km, the present distance between Los Angeles and San Francisco?

18. Look at a map of the western Pacific Ocean, and examine the position of Japan with respect to mainland Asia. Japan's older crust contains rocks similar to those of eastern Asia. Presently, there are many active volcanoes along the length of Japan. With these facts in mind, explain how the Japan Sea (the region between Japan and the mainland) formed.

 For more resources, including animations, quizzes, and Norton's GeoTours, go to **wwnorton.com/studyspace**.

 If your instructor assigns exercises in SmartWork, log in at **smartwork.wwnorton.com**.

These towering cliffs of red rock that rise from the desert of New Mexico consist of sand-sized grains of a mineral called quartz. As time passes, blocks break off and tumble to the ground below, forming a pile of debris or sediment, some of which becomes soil. Rock, sediment, and soil—these are the materials that underlie the Earth's surface.

Earth Materials

What is the Earth made of? There are four basic components: the solid Earth (the crust, mantle, and core), the biosphere (living organisms), the atmosphere (the envelope of gas surrounding the planet), and the hydrosphere (the liquid and solid water at or near the ground surface), all of which perpetually interact, as the "Earth System." In this part of the book, we focus on the materials that make up the crust and mantle of the solid Earth. We will find that these consist primarily of rock. Most rock, in turn, contains minerals, so minerals are, in effect, the building blocks of our planet.

We therefore begin in Chapter 5 by learning about minerals and how they grow. Then we see, in Interlude A, how geologists distinguish three categories of rock—igneous, sedimentary, and metamorphic—based on how the rocks form. In each of the following three chapters (6, 7, and 8), we look at one of these rock categories. Interlude B helps explain the origin of the sediments that develop into sedimentary rocks. Finally, Interlude C shows us how materials in the Earth System pass through a rock cycle, for atoms constituting one rock type may end up being incorporated into a succession of other rock types.

A cluster of quartz crystals from Brazil displays nature's own ability to produce abstract sculpture.

CHAPTER 5

Patterns in Nature: Minerals

GEOPUZZLE

In the game of "Twenty Questions," you try to guess the identity of an object that your friend is thinking about and start by asking, "Is it animal, vegetable, or mineral?" Do geologists consider everything on Earth that is not "animal" or "vegetable" to be "mineral?"

I died a mineral, and became a plant. I died as plant and rose to animal, I died as animal and I was Man. Why should I fear?

—Jalal-Uddin Rumi
(Persian mystic and poet, 1207–1273)

5.1 INTRODUCTION

Zabargad Island rises barren and brown above the Red Sea, about 70 km off the coast of southern Egypt. Nothing grows on Zabargad, except for scruffy grass and a few shrubs, so no one lives there now. But in ancient times many workers toiled on this 5-square-km patch of desert, gradually chipping their way into the side of its highest hill. They were searching for glassy green, pea-sized pieces of peridot, a prized gem. Carefully polished peridots were worn as jewelry by ancient Egyptians. Eventually, some of the gems appeared in Europe, set into crowns and scepters (Fig. 5.1). These peridots now glitter behind glass cases in museums, millennia after first being pried free from the Earth, and perhaps 10 million years after first being formed by the bonding together of still more ancient atoms.

Peridot is one of about 4,000 minerals that have been identified on Earth so far. **Mineralogists**, people who specialize in the study of minerals, discover 50 to 100 new minerals every year. Each different mineral has a name. Some names come from Latin, Greek, German, or English words describing a certain characteristic (for instance, *albite* comes from the Latin word for white, *orthoclase* comes from the German words meaning splits at right angles, and *olivine* is olive-colored); some honor a person (*sillimanite* was named for Benjamin Silliman, a famous nineteenth-century mineralogist); some indicate the place where the mineral was first recognized (illite was first identified in rocks from Illinois); and some reflect a particular

FIGURE 5.1 A royal crown containing a variety of valuable jewels.

element in the mineral (chromite contains chromium). Several minerals have more than one name—for example, peridot is the gem-quality version of olivine, a common mineral. Although the vast majority of mineral types are rare, forming only under special conditions, many are quite common and occur in a variety of rock types at Earth's surface.

Though ancient Greek philosophers pondered minerals and medieval alchemists puttered with minerals, true scientific study of minerals did not begin until 1556, when Georgius Agricola, a German physician, published *De Re Metallica* ("On the Nature of Metals"), in which he gave basic descriptions of minerals. In 1669, more than a century after Agricola's work, Nicholas Steno, a Danish physician and scientist who later became a priest, discovered important geometric characteristics of minerals. Steno's work became the basis for systematic descriptions of minerals, a task that occupied many researchers during the next two centuries.

The study of minerals with an optical microscope began in 1828, but though such studies helped in mineral identification, they could not reveal the arrangement of atoms inside minerals. That understanding had to wait until 1912, when Max von Laue of Germany proposed that X-rays, electromagnetic radiation whose wavelength is comparable to the tiny distance between atoms in a mineral, could be used to study the internal structure of minerals. A father-and-son team, W. H. and W. L. Bragg of England, published the first X-ray study of a mineral, work for which they shared the 1915 Nobel Prize in physics. In subsequent decades, researchers developed progressively more complex instruments to aid their study of minerals. For example, in the 1960s, mineralogists began to use electron microscopes to obtain actual images of the internal structure of minerals, and electron microprobes to analyze the chemical composition of grains that are almost too small to see.

Why study minerals? Without exaggeration, we can say that *minerals are the building blocks of our planet.* To a geologist, almost any study of Earth materials depends on an understanding of minerals, for minerals make up most of the rocks and sediments comprising the Earth and its landscapes. Minerals are also important from a practical standpoint (see Chapter 15). *Industrial minerals* serve as the raw materials for manufacturing chemicals, concrete, and wallboard. *Ore minerals* are the source of valuable metals like copper and gold and provide energy resources like uranium (Fig. 5.2a, b). And certain forms of minerals—gems—delight the eye as jewelry. Unfortunately, though, some minerals pose environmental hazards. No wonder **mineralogy**, the study of minerals, fascinates professionals and amateurs alike.

In this chapter, we begin by presenting the geologic definition of a mineral, then look at how minerals form and at the main characteristics that enable us to identify specific samples. Finally, we describe the basic scheme that mineralogists use to classify minerals. This chapter assumes that you understand

FIGURE 5.2 Copper ore is a useful mineral that serves as a source of copper metal.

Malachite grows by precipitation, in a succession of layers.

(a) Malachite is a type of copper ore [$Cu_2(CO_3)(OH)_2$]; it contains copper plus other chemicals.

(b) The copper for pots is produced by processing ore minerals.

the fundamental concepts of matter and energy, especially the nature of atoms, molecules, and chemical bonds. If you are rusty on these topics, please study Box 5.1.

Chapter Themes

By the end of this chapter, you should know that . . .

- the term "mineral" has a very special meaning in geologic contexts.
- there are thousands of minerals that can be divided into relatively few classes based on chemical composition.
- silicates are the most common minerals on Earth; they are the building blocks of this planet.
- you can identify mineral specimens based on diagnostic physical properties.
- gems are particularly rare and beautiful minerals.

5.2 WHAT IS A MINERAL?

To a geologist, a **mineral** is a naturally occurring solid, formed by geologic processes, that has a crystalline structure and a definable chemical composition. Almost all minerals are inorganic. Let's pull apart this mouthful of a definition and examine its meaning in detail.

- *Naturally occurring*: True minerals are formed in nature, not in factories. We need to emphasize this point because in recent decades, industrial chemists have learned how to synthesize materials that have characteristics virtually identical to those of real minerals. These materials are not minerals in a geologic sense, though they are referred to in the commercial world as synthetic minerals.

- *Solid*: A solid is a state of matter that can maintain its shape indefinitely, and thus will not conform to the shape of its container (Fig. 5.3a–c). Liquids (such as oil or water) and gases (such as air) are not minerals (Fig. 5.3d, e).

- *Formed by geologic processes*: Minerals, as we see later in this chapter, can form by the freezing of molten rock, by precipitation out of a water solution, or by chemical reactions within or on the surface of preexisting rocks. All of these processes are considered to be "geologic" since they occur naturally on or in the Earth. This component of the definition has a caveat, however. Some substances are identical in character to minerals produced by geologic processes, but are a by-product of living organisms—the calcite in a clamshell is an example. Such materials are minerals, but they are called **biogenic minerals** to emphasize their origin. Significantly, biogenic substances that cannot also be formed by geologic processes are not considered to be minerals.

- *Crystalline structure*: The atoms that make up a mineral are not distributed randomly and cannot move around easily. Rather, they are fixed in a specific, orderly pattern. A material in which atoms are fixed in an orderly pattern is called a crystalline solid. Mineralogists refer to the pattern itself (the imaginary framework representing the arrangement of atoms) as a **crystal lattice** (Fig. 5.4a). A *glass*, though solid, is not crystalline, because atoms in a glass are not arranged in a regular pattern.

  Did you ever wonder . . . how window glass differs from clear quartz?

- *Definable chemical composition*: This simply means that it is possible to write a chemical formula for a mineral (Box 5.1). Some minerals contain only one element, but most are compounds of two or more elements. For example, diamond and graphite have the formula C, because they consist entirely of carbon. Quartz has the formula SiO_2—

it contains the elements silicon and oxygen in the proportion of one silicon atom for every two oxygen atoms. Some formulas are more complicated: for example, the formula for biotite is $K(Mg,Fe)_3(AlSi_3O_{10})(OH)_2$.

- *Inorganic*: Organic chemicals are molecules containing carbon-hydrogen bonds. Although some organic chemicals contain only carbon and hydrogen, others also contain oxygen, nitrogen, and other elements in varying quantities. Sugar ($C_{12}H_{22}O_{11}$), for example, is an organic chemical. *Almost all minerals are inorganic.* Thus, sugar and protein are not minerals. But, we have to add the qualifier "almost all" because mineralogists do consider about 30 organic substances formed by "the action of geologic processes on organic materials" to be minerals. Examples include the crystals that grow in ancient deposits of bat guano.

With these definitions in mind, we can make an important distinction between minerals and glasses. Both minerals and glasses are solids, in that they can retain their shape indefinitely. But a mineral is crystalline, and glass is not. Whereas atoms, ions, or molecules in a mineral are ordered into a crystal lattice, like soldiers standing in formation, those in a glass are arranged in a semi-chaotic way, like people at a party, in small clusters or chains that are neither oriented in the same way nor spaced at regular intervals (**Fig. 5.4b**).

If you ever need to figure out whether a substance is a mineral or not, just check it against the criteria listed above. Is motor oil a mineral? No—it's an organic liquid. Is table salt a mineral? Yes—it's a solid crystalline compound with the formula NaCl.

FIGURE 5.3 The four states of matter—solid, liquid, gas, or plasma. Whether a substance is a solid, liquid, or gas depends on how tightly atoms or molecules are held in place.

Solid

(a) A solid retains its shape regardless of the size of the container. In a solid, atoms or molecules are locked together tightly.

Liquid

(b) A liquid conforms to the shape of the container, so its density does not change when the shape of the container changes. In a liquid, chains or clumps of atoms or molecules can move.

Gas

(c) A gas expands to fill whatever volume it occupies, so its density changes if the volume changes. In a gas, atoms and molecules are not held together and can drift.

The gas heated by an electrical spark becomes a plasma.

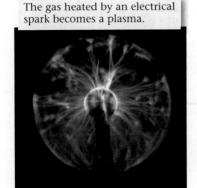

(d) Plasma is a gas-like substance in which atoms have been stripped of electrons. It exists only at very high temperatures. The glowing gas around an electric spark is a plasma.

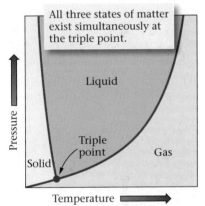

All three states of matter exist simultaneously at the triple point.

Pressure

Liquid

Triple point

Solid

Gas

Temperature

(e) The state of matter depends on pressure and temperature, as depicted on this graph, called a phase diagram.

5.3 BEAUTY IN PATTERNS: CRYSTALS AND THEIR STRUCTURE

What Is a Crystal?

The word *crystal* brings to mind sparkling chandeliers, elegant wine goblets, and shiny jewels. But, as is the case with the word *mineral*, geologists have a more precise definition. A **crystal** is a single, continuous (that is, uninterrupted) piece of a crystalline solid, typically bounded by flat surfaces called **crystal faces** that grow naturally as the mineral forms. The word comes from

FIGURE 5.4 The nature of crystalline and noncrystalline materials.

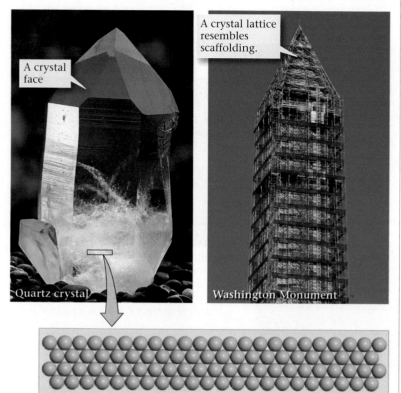

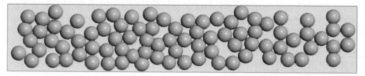

(a) This quartz crystal contains an orderly arrangement of atoms. The arrangement resembles scaffolding.

(b) Atoms in noncrystalline solids, such as glass, are not orderly.

the Greek *krystallos*, meaning ice. Many crystals have beautiful shapes that look like they belong in the pages of a geometry book. The angle between two adjacent crystal faces of one specimen is identical to the angle between the corresponding faces of another specimen. For example, a perfectly formed quartz crystal looks like an obelisk (Fig. 5.5a). The angle between the faces of the columnar part of a quartz crystal is exactly 120°. This rule, discovered by Steno, holds regardless of whether the whole crystal is big or small and regardless of whether all of the faces are the same size. Crystals come in a great variety of shapes, including cubes, trapezoids, pyramids, octahedrons, hexagonal columns, blades, needles, columns, and obelisks (Fig. 5.5b).

Because crystals have a regular geometric form, people have always considered them to be special, perhaps even a source of magical powers. For example, shamans of some cultures relied on talismans or amulets made of crystals, which supposedly brought power to their wearer or warded off evil spirits.

FIGURE 5.5 Some characteristics of crystals.

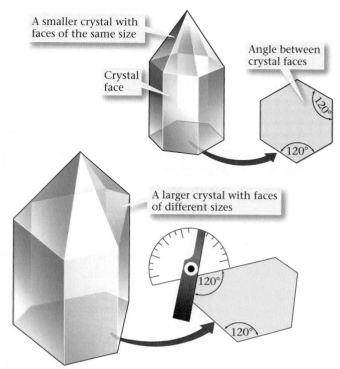

(a) Regardless of specimen size, the angle between two adjacent crystal faces is consistent in a particular mineral.

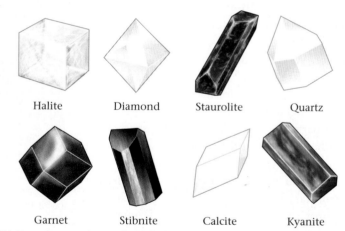

(b) Crystals come in a variety of shapes, including cubes, prisms, blades, and pyramids. Some terminate at a point and some terminate with flat surfaces.

Scientists have concluded, however, that crystals have no effect on health or mood. For millennia, crystals have inspired awe because of the way they sparkle; but such behavior is simply a consequence of how crystal structures interact with light.

What's Inside a Crystal?

What do the insides of a mineral actually look like? We can picture atoms in minerals as tiny balls packed together tightly and held in place by chemical bonds. The way in which atoms are packed defines the **crystal structure** of the mineral. The physical properties of a mineral (for example, the shape of its crystals, how hard it is, how it reacts chemically with other substances) depend both on the identity of the elements making up the mineral and on the way these elements are arranged and bonded in a crystal structure.

Not all minerals have the same kind of bonding, and in some minerals more than one type of bonding occurs. The type of bonding, the ease with which bonds form or are broken, and the geometric arrangement of bonds play an important role in determining the characteristics of minerals. As your intuition might suggest, bonds are stronger in harder minerals and in minerals with higher melting temperatures. In some minerals, the nature and strength of bonding vary with direction in the mineral. If bonds form more easily in one direction than another, a crystal will grow faster in one direction than another. And if a mineral has weak bonds in one direction and strong bonds in another direction, it will break more easily in one direction than in the other.

To illustrate crystal structures, we look at a few examples. Halite (rock salt) consists of oppositely charged ions that stick together because opposite charges attract. In halite, six chloride (Cl^-) ions surround each sodium (Na^+) ion, producing an overall arrangement of atoms that defines the shape of a cube (Fig. 5.6a, b). Diamond, by contrast, is a mineral made entirely of carbon. In diamond, each atom bonds to four neighbors arranged in the form of a tetrahedron; some naturally formed diamond crystals have the shape of a double tetrahedron (Fig. 5.6c). Graphite, another mineral composed entirely of carbon, behaves very differently from diamond. In contrast to diamond, graphite is so soft that we use it as the "lead" in a pencil; when a pencil moves across paper, tiny flakes of graphite peel off the pencil point and adhere to the paper. This behavior occurs because the carbon atoms in graphite are not arranged in tetrahedra, but rather occur in sheets (Fig. 5.6d). The sheets are bonded to each other by weak bonds and thus can separate from each other easily. Two different minerals (such as diamond and graphite) that have the same composition but different crystal structures are **polymorphs**.

What determines how atoms pack together in a crystal? The size of an ion depends on the number of electrons orbiting the nucleus (Fig. 5.7a); so, since anions have extra electrons, they tend to be bigger than cations. Thus, cations nestle snugly in the spaces between anions in many crystal structures, and as many anions will try to fit around a cation as there is room for. Depending on the identity of an ion, different geometries of packing can occur (Fig. 5.7b).

Note that in halite, as described above, each ion is a single atom. In many ionically bonded minerals, the ions building the minerals consist of more than one atom. For example, the

FIGURE 5.6 The nature of crystalline structure in minerals.

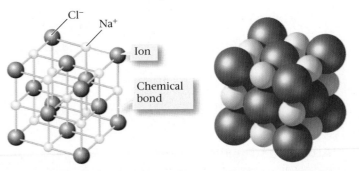

(a) In a ball-and-stick model of halite, the balls are ions, and the sticks are chemical bonds.

(b) This ball model gives a better sense of how ions pack together in crystal.

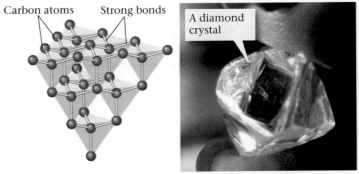

(c) In a diamond, carbon atoms are arranged in tetrahedra. All of the bonds are strong.

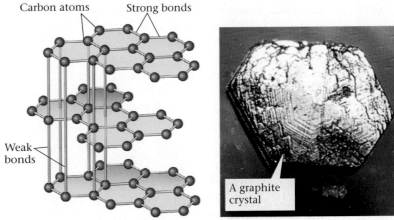

(d) Graphite consists of carbon atoms arranged in hexagonal sheets. The sheets are connected by weak bonds.

mineral calcite ($CaCO_3$) consists of calcium (Ca^{2+}) cations and carbonate (CO_3^{2-}) anions—each carbonate anion (or anionic group) consists of four atoms and thus is quite large.

The orderly arrangement of atoms inside a crystal—its crystal structure—provides one of nature's most spectacular examples of a pattern. The pattern on wallpaper may be defined by the regular spacing of clumps of flowers. Similarly, the pattern in a crystal is defined by the regular spacing of

BOX 5.1

The Basics of Chemistry

Minerals are chemicals or chemical compounds, so to understand how they form and behave requires us to utilize the vocabulary and concepts of chemistry. In this box, we briefly review basics about atoms, ions, chemical bonding, and solutions.

Atoms and Their Structure

As noted in Chapter 1, an **element** is a pure substance that cannot be separated into other elements. There are 92 naturally occurring elements on Earth, and physicists have produced 26 more. Each element has a name (e.g., hydrogen; carbon; chlorine; silicon; oxygen; sulfur; iron; uranium) and a symbol (e.g., H; C; Cl; Si; O; S; Fe; U). Dmitri Mendeleev, in 1869, arranged elements in a chart, called the periodic table, based on their behavior (see **Appendix**).

The smallest piece of an element that retains the characteristics of the element is an **atom**. Atoms are so small that over 5 trillion (5,000,000,000,000) can fit on the head of a pin. An atom can be subdivided into smaller pieces with different electrical charges. (The "charge" of a particle refers to the behavior of the particle when exposed to an electrical current or a magnetic field.) The central ball, or **nucleus**, of all atoms except hydrogen contains two types of subatomic particles: **protons**, with a positive charge, and **neutrons** that are electrically neutral. Most hydrogen nuclei consist of just a single proton. Negatively charged **electrons** zip around the nucleus in clouds with distinct geometries; the clouds are called orbitals or **shells** (**Fig. Bx 5.1a**). A specific number of electrons can fit into a shell—if there are fewer than that number, the shell is "incomplete." Electrons are even tinier than protons or neutrons. In fact, about 1,839 electrons could fit inside a proton or neutron. The number of protons in an atom is called the **atomic number** of the atom—all atoms of a given element have the same atomic number. The **atomic weight** of an atom is a measure of its mass. Strictly speaking, atomic weight is the ratio of an atom's mass to 1/12 the mass of a carbon atom; simplistically, it is approximately equal to the number of protons plus the number of neutrons in the atom.

The Nature of Ions

If the number of protons equals the number of electrons in an atom, the atom is electrically neutral. Not all atoms are neutral, because electrons may be added to or removed from the outer shells. A non-neutral atom is called an **ion**; an ion with more electrons than protons has a net negative charge and is called an **anion**, whereas an ion with fewer electrons than protons has a net positive charge and is called a **cation**. We indicate the charge of an ion with a superscript. For example, when chlorine gains an electron, it forms an anion (Cl^-), and when iron loses two electrons, it forms a cation (Fe^{2+}). Some ions have unique names that differ from the neutral atom from which they were derived—for example, the anion of chlorine is called chloride.

FIGURE Bx5.1 Atoms and elements.

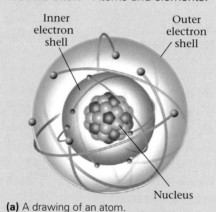

Inner electron shell

Outer electron shell

Nucleus

(a) A drawing of an atom.

Salt crystal

Sodium metal

Chlorine gas

(b) A compound, salt, can be subdivided to form two elements, sodium metal and chlorine gas. Neither sodium nor chlorine can be divided further.

Chemical Bonds and Compounds

Some atoms and ions exist in isolation, but some attract each other and stick (or bond) together. The "glue" that holds two or more atoms (or ions) together is called a **chemical bond**, and a piece of matter containing two or more atoms (or ions) bonded together is called a **molecule**. Some molecules contain only atoms of the same element, but others consist of two or more different elements. Chemists use the term **compound** for substances that can be subdivided into two or more elements—the smallest piece of a compound that has the properties of the compound is a molecule. Many compounds occur in daily life. For example, common table salt (a mineral called halite) is a compound whose molecules contain chloride anions (Cl^-) bonded to sodium cations (Na^+). A compound does not resemble the elements within it—salt is a grayish solid, whereas pure chlorine is a noxious gas and pure sodium is a shiny metal (**Fig. Bx 5.1b**). Note that we can use the term **chemical** for any pure substance, whether it is a single element or a compound. A **chemical formula** is a shorthand recipe that itemizes the elements in a compound and indicates their relative proportions. For example, the chemical formula for water, H_2O, indicates that water contains molecules in which two hydrogen atoms have bonded to one oxygen atom.

Why do chemical bonds form? Bonds form because of the interaction among the electron clouds of nearby atoms. Chemists recognize four distinct types based on the nature of the interaction:

- *Ionic bonds*: As an inviolate rule of nature, "like" electrical charges repel (e.g., two positive charges push each other away), while "unlike" electrical charges attract (a negative charge sticks to a positive charge). Bonds that form between ions with opposite charges are called **ionic bonds (Fig. Bx 5.1c)**. For example, in a molecule of salt, positively charged sodium ions (Na^+) attract negatively charged chloride (Cl^-) ions.

- *Covalent bonds*: The atoms of carbon making up a diamond do not transfer electrons to one another, but rather share electrons. Bonding that involves the *sharing* of electrons is called **covalent bonding (Fig. Bx 5.1d)**. Because of the sharing, the electron shells of all the carbon atoms in a diamond are complete, and all the carbon atoms have a neutral charge. Water molecules exist because of covalent bonding: in a water molecule, two hydrogen atoms are covalently bonded to one oxygen atom.

- *Metallic bonds*: In metals, electrons of the outer shells move easily from atom to atom and bind the atoms to each other. We call this type of bonding **metallic bonding (Fig. Bx 5.1e)**. Because outer-shell electrons move so freely, metals conduct electricity easily—when you connect a metal wire to an electrical circuit, a current of electrons flows through the metal.

- *Bonds resulting from the polarity of atoms or molecules*: Chemists long recognized that some materials break or split so easily that they must be held together by particularly weak chemical bonds. Eventually, they realized that these bonds are due to the permanent or temporary polarity of molecules. **Polarity** means that the molecule has a positive charge on one side and a negative charge on the other. Polarity-related bonds form because the negative side of one molecule attracts the positive side of another.

Bonds resulting from the polarity of molecules are important in many geologic materials. For example, the hydrogen atoms in water both lie on the same side of the oxygen atom. The oxygen nucleus, which contains eight protons, exerts more attraction to the shared electrons than do the hydrogen nuclei, each of which contains only one proton. As a consequence, the shared electrons concentrate closer to the oxygen side of the molecule, making the oxygen side more negative than the hydrogen side. Thus, the water molecules are polar, and they attract each other. This attraction is called a **hydrogen bond (Fig. Bx 5.1f)**.

Johannes van der Waals (1837–1923), a Dutch physicist, discovered another type of weak chemical bonding that depends on polarity. This type, now known as **van der Waals bonding**, links one covalently bonded molecule to

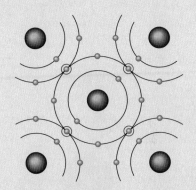

● Unshared electron   ◎ Shared electron   ⬤ Nucleus

(d) Covalent bonding. Carbon has only four electrons in its outer shell, which has a capacity of eight. Thus, the carbon atom shares electrons with four other carbon atoms.

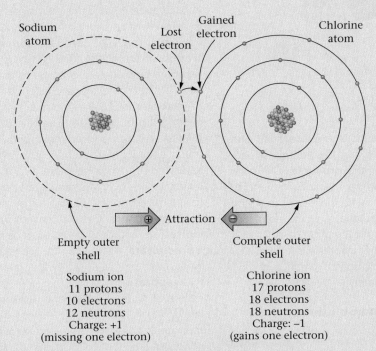

Sodium atom   Lost electron   Gained electron   Chlorine atom

⬅➡ Attraction ⊕ ⊖

Empty outer shell   Complete outer shell

Sodium ion
11 protons
10 electrons
12 neutrons
Charge: +1
(missing one electron)

Chlorine ion
17 protons
18 electrons
18 neutrons
Charge: –1
(gains one electron)

(c) An ionic bond between sodium and chlorine. The sodium atom gives up its outer electron, and thus has one more proton than electron, whereas the chlorine gains an electron and thus has one more electron than proton. The two ions attract each other.

(e) In a metallically bonded material, nuclei and their inner shells of electrons float in a "sea" of free electrons. The electrons can stream through the metal.

111

BOX 5.1

The Basics of Chemistry *(continued)*

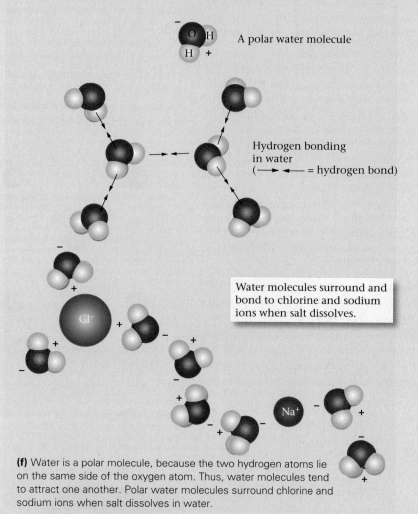

A polar water molecule

Hydrogen bonding
in water
(———▶ ◀——— = hydrogen bond)

Water molecules surround and
bond to chlorine and sodium
ions when salt dissolves.

(f) Water is a polar molecule, because the two hydrogen atoms lie
on the same side of the oxygen atom. Thus, water molecules tend
to attract one another. Polar water molecules surround chlorine and
sodium ions when salt dissolves in water.

another. The bonds exist because electrons temporarily cluster on one side of each molecule, giving it a polarity.

Do chemical bonds last forever? Not necessarily. Existing bonds can be broken and new ones can form during a **chemical reaction**. Chemical reactions, in effect, break molecules apart, or produce new molecules and/or isolated atoms. Typically, chemical reactions occur when two or more chemicals (the **reactants**) are placed together—the reaction may happen spontaneously, or may take place only if the reactants are heated. Some reactions absorb heat

and some reactions produce heat. The new chemicals that form during a reaction are called the **products**. Chemists represent chemical reactions in the form of an equation, with the reactants on the left and the products on the right. For example, in the reaction $C_{10}H_8 + 12\ O_2 \rightarrow 10\ CO_2 + 4\ H_2O$, napthalene (mothballs) reacts with oxygen to produce carbon dioxide and water.

Mixtures and Solutions

In addition to compounds, in which atoms bond together to form molecules, chemists recognize two other

kinds of materials in which substances have combined. In a **mixture**, two elements or compounds have been "stirred together" so that one is dispersed in the other. But unlike a compound, the components of a mixture can be separated *without* a chemical reaction. For example, succotash is a mixture of corn and beans—you can separate the two components with a fork. In a **solution**, one chemical (the solute) dissolves or becomes completely incorporated in another (the solvent). During the dissolving process, the solute may separate into ions. For example, when salt (NaCl) dissolves in water, it separates into sodium (Na^+) and chloride (Cl^-) ions. In a solution, molecules of the solvent surround the ions of the solute. The polarity of water molecules makes water a good solvent (meaning, it can dissolve substances), because the polar water molecules attract and surround ions of soluble materials and pull them apart.

The amount of solute per unit volume of solvent is called the **concentration** of the solute, and can be represented by a percent or by ppm (parts per million). The average concentration of seawater is 3.5%—this means that every 100g of seawater contains 3.5g of salt. A solution can hold only so much solute before it becomes oversaturated (or overconcentrated). When solutions are oversaturated, new solid grains may form and separate from the solution—this process is called **precipitation**, and the new solid is called a **precipitate**. As an example, consider what happens when saltwater starts to **evaporate** (meaning that the molecules of water, the solvent, leave the liquid and turn into gas)—eventually, the remaining saltwater becomes overconcentrated and solid grains of salt precipitate and settle out.

FIGURE 5.7 The various sizes of ions and the ways they pack together in minerals.

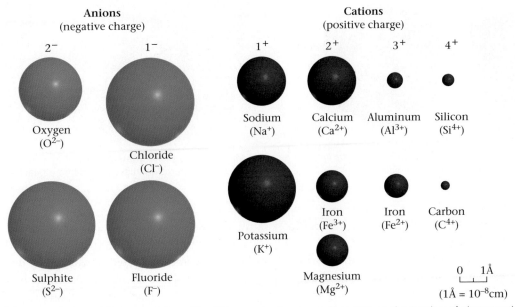

(a) Ions come in a wide range of sizes. The difference in size depends, in part, on the number of electrons they contain.

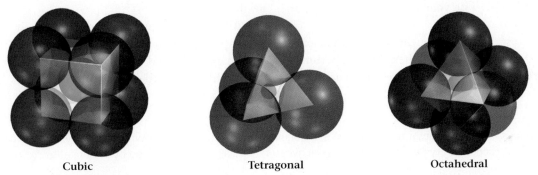

(b) Ions can pack together in different ways. Each configuration can be described by a geometric shape.

atoms (Fig. 5.8a, b; Box 5.2). If the crystal contains more than one type of atom, the atoms alternate in a regular way. The orderly arrangement controls the outward shape of crystals. For example, if the atoms are packed into the shape of a cube, a crystal of the mineral will have faces that intersect at 90° angles.

The pattern of atoms or ions in a mineral displays **symmetry**, meaning that the shape of one part of a mineral is a mirror image of the shape of another part. For example, if you were to cut a halite crystal in half and place one half against a mirror, the crystal would look whole again (Fig. 5.8c).

The Formation and Destruction of Minerals

New mineral crystals can form in five ways. First, they can form by the solidification of a melt, meaning the freezing of a liquid. For example, ice crystals, a type of mineral, are made by freezing water. Second, they can form by precipitation from a solution, meaning that atoms, molecules, or ions dissolved in water bond together and separate out of the water (Fig. 5.9). Salt crystals, for example, develop when you evaporate salt water (see Box 5.1). Third, they can form by solid-state diffusion, the movement of atoms or ions through a solid to arrange into a new crystal structure, a process that takes place very slowly. For example, garnets grow by diffusion in solid rock. Fourth, minerals can form at interfaces between the physical and biological components of the Earth System by a process called biomineralization. This occurs when living organisms cause minerals to precipitate either within or on their bodies, or immediately adjacent to their bodies. For example, clams and other shelled organisms extract ions from water to produce mineral shells. Fifth, minerals can precipitate directly from a gas. This process typically occurs around volcanic vents or around geysers, for at such locations volcanic gases or steam enter the atmosphere and cool

114

FIGURE 5.8 Patterns and symmetry in minerals.

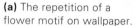

(a) The repetition of a flower motif on wallpaper.

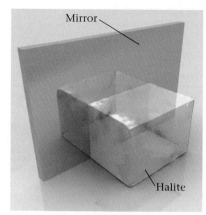

(b) The repetition of alternating sulfur and lead atoms in the mineral galena (PbS).

(c) Minerals display symmetry. One-half of a halite crystal is a mirror image of the other. Snowflakes are symmetrical hexagons.

abruptly. Some of the bright yellow sulfur deposits found in volcanic regions form in this way.

The first step in forming a crystal is the chance formation of a seed, or an extremely small crystal (**Fig. 5.10a**). Once the seed exists, other atoms in the surrounding material attach themselves to the face of the seed. As the crystal grows, crystal faces move outward but maintain the same orientation (**Fig. 5.10b** ⏺). The youngest part of the crystal is at its outer edge.

In the case of crystals formed by the solidification of a melt, atoms begin to attach to the seed when the melt becomes so cool that thermal vibrations can no longer break apart the

attraction between the seed and the atoms in the melt. Crystals formed by precipitation from a solution develop when the solution becomes saturated, meaning the number of dissolved ions per unit volume of solution becomes so great that they can get close enough to each other to bond together. (If a solution is not saturated, dissolved ions are surrounded by solvent molecules, which shield the ions from the attractive forces of their neighbors.)

As crystals grow, they develop their particular crystal shape, based on the geometry of their internal structure. The shape is defined by the relative dimensions of the crystal (needle-like, sheet-like, etc.) and the angles between crystal

FIGURE 5.9 This photo is real, not a computer collage! We're seeing the world's largest-known mineral crystals jutting from the walls of a cave in Chihuahua, Mexico. The crystals are of the mineral gypsum ($CaSO_4 \cdot 2H_2O$), formed by precipitation from a water solution.

faces. Typically, the growth of minerals is restricted in one or more directions, because existing crystals act as obstacles. In such cases, minerals grow to fill the space that is available, and their shape is controlled by the shape of their surroundings. Minerals without well-formed crystal faces are *anhedral* grains (Fig. 5.10c). If a mineral's growth is uninhibited so that it displays well-formed crystal faces, then it is a *euhedral* crystal. The surface crystals of a **geode**, a mineral-lined cavity in rock, may be euhedral (Fig. 5.10d).

A mineral can be destroyed by melting, dissolving, or some other chemical reaction. Melting involves heating a mineral to a temperature at which thermal vibration of the atoms or ions in the lattice break the chemical bonds holding them to the lattice. The atoms or ions then separate, either individually or in small groups, to move around again freely. Dissolution occurs when you immerse a mineral in a solvent, such as water. Atoms or ions then separate from the crystal face and are surrounded by solvent molecules. Chemical reactions can destroy a mineral when it comes in contact with reactive materials. For example, iron-bearing minerals react with air and water to form rust. The action of microbes in the environment can also destroy

minerals. In effect, some microbes can "eat" certain minerals; the microbes use the energy stored in the chemical bonds that hold the atoms of the mineral together as their source of energy for metabolism.

FIGURE 5.10 The growth of crystals.

Ions attach to the crystal face.

(a) New crystals nucleate and begin to precipitate out of a water solution. As time progresses, they grow into the open space.

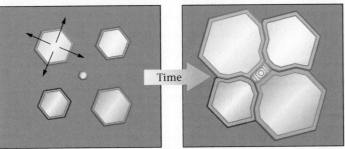

Time

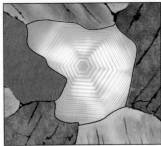

(b) New crystals grow outward from the central seed. As time passes, they maintain their shape until they interfere with each other.

(c) A crystal growing in a confined space will be anhedral.

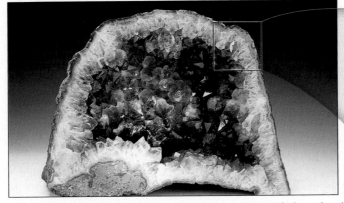

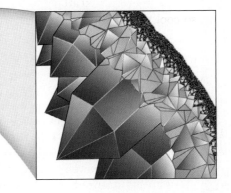

(d) A geode from Brazil consists of purple quartz crystals (amethyst) that grew from the wall into the center. The enlargement sketch indicates that the crystals are euhedral.

Take-Home Message

- A mineral contains an orderly arrangement of atoms, its crystal structure; glass does not.
- If a mineral crystal grows without obstruction, its surface has smooth faces; specific angles occur between adjacent faces.

- Geologists recognize many different shapes of crystals; crystal structures display symmetry.
- Minerals can grow by solidification from a melt, by diffusion, or by precipitation from a solution or a gas.

THINK: Why do some mineral grains have an irregular shape?

CONSIDER THIS . . .

BOX 5.2

How Do We "See" the Arrangement of Atoms in a Crystal?

We can see crystals and hold them in our hands. But atoms are so small that the human eye cannot possibly distinguish them, even with the strongest optical microscope (a microscope using light that passes through lenses). So how do mineralogists come up with the models that depict atoms arranged in crystals? To explain, we must first provide some background about waves.

The energy of light, radio signals, and X-rays moves from one place to another in the form of waves. For simplicity, we picture such waves as having the same form as familiar water waves, with crests (high parts) and troughs (low parts). A set of waves moving in a given direction is called a wave train, and the distance between adjacent crests (or troughs) is the wavelength. When a train of straight waves strikes a wall that contains a small opening whose width is comparable to the wavelength, the opening acts as a new point source of waves (just as a pebble striking a pond surface is a point source of water waves), and a set of curving waves propagates outward from the opening. This phenomenon is

called **diffraction**. If the obstacle contains many equally spaced openings, the new waves emitting from each opening interfere with each other (**Fig. Bx 5.2a**). This means that in some places, crests from one point opening overlap crests from another, producing larger crests (manifested as a stronger signal). In other places, crests from one point opening overlap troughs from another and cancel each other out (manifested as a weaker signal).

In 1912, Max von Laue proposed that an X-ray passing through a crystal would undergo diffraction if the atoms were arranged in orderly columns and rows and if the spacing between columns was comparable to the wavelength of an X-ray. He surmised that the space between two adjacent columns would act as a point source of waves. Thus, the curving waves emitting from spaces between adjacent columns would interfere. Because of this interference, the diffracted beams would produce a distinctive pattern of strong and weak signals, which would appear as spots on a photographic plate (**Fig. Bx 5.2b**). (This phenomenon doesn't happen

with light waves because the wavelengths of light are too big.) When von Laue's students tested this proposal, they indeed produced diffraction patterns. The existence of such X-ray diffraction (XRD) patterns requires that the atoms in crystals have an orderly arrangement. From the patterns, researchers can now deduce the geometric arrangement of atoms in a crystal.

More recently, investigators have been able to use transmission electron microscopes (TEM) to "see" the arrangement of atoms in a mineral directly. A TEM shoots a beam of electrons at a thin slice of a crystal. Since electrons are much smaller than the spaces between atoms, the electrons can pass through spaces between atoms and can strike a detector, forming a light spot. Electrons that interact with atoms bounce off and scatter in all directions, and therefore don't reach the detector (**Fig. Bx 5.2c**). As a result, a dark spot (somewhat like a shadow) remains under a column of atoms. The overall pattern of dark and light spots on the detector represents the distribution of atoms.

5.4 HOW CAN YOU TELL ONE MINERAL FROM ANOTHER?

Amateur and professional mineralogists get a kick out of recognizing minerals. They might hover around a display case in a museum and name specimens without bothering to look at the labels. How do they do it? The trick lies in learning to recognize the basic physical properties (visual and material characteristics) that distinguish one mineral from another. Some physical properties, such as shape and color, can be seen from a distance.

Others, such as hardness and magnetization, can be determined only by handling the specimen or by performing an identification test on it. Identification tests include scratching the mineral against another object, placing it near a magnet, weighing it, tasting it, or placing a drop of acid on it. Let's examine some of the physical properties most commonly used in mineral identification. (The Appendix provides charts for identifying minerals on the basis of their physical properties.)

- *Color*: **Color** results from the way a mineral interacts with light. Sunlight contains the whole spectrum of colors;

FIGURE Bx5.2 Using X-rays and electron beams to characterize the internal structure of minerals.

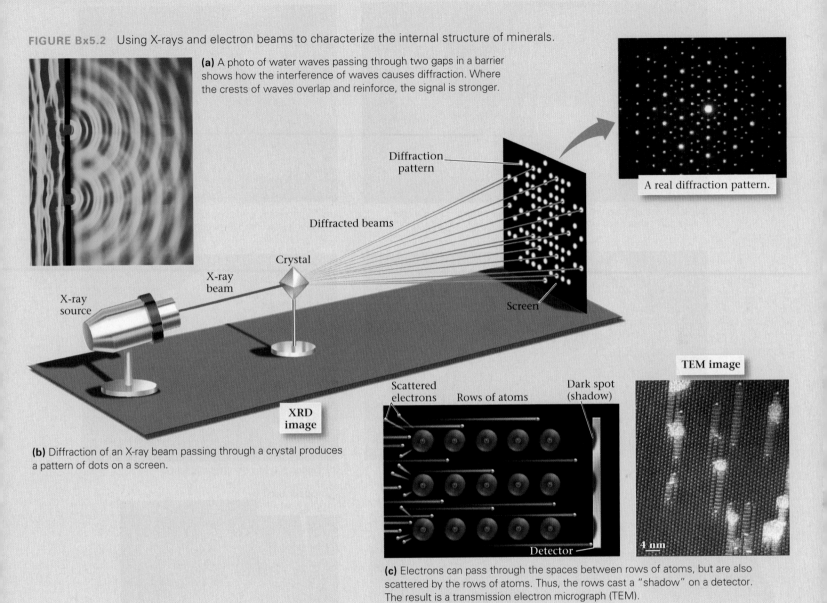

(a) A photo of water waves passing through two gaps in a barrier shows how the interference of waves causes diffraction. Where the crests of waves overlap and reinforce, the signal is stronger.

A real diffraction pattern.

Diffraction pattern

Diffracted beams

Crystal

X-ray beam

X-ray source

Screen

XRD image

(b) Diffraction of an X-ray beam passing through a crystal produces a pattern of dots on a screen.

TEM image

Scattered electrons   Rows of atoms   Dark spot (shadow)

4 nm

Detector

(c) Electrons can pass through the spaces between rows of atoms, but are also scattered by the rows of atoms. Thus, the rows cast a "shadow" on a detector. The result is a transmission electron micrograph (TEM).

each color has a different wavelength. A mineral absorbs certain wavelengths, so the color you see when looking at a specimen represents the wavelengths the mineral does not absorb. Certain minerals always have the same color, but many show a range of colors (Fig. 5.11a). Color variations in a mineral reflect the presence of impurities.

■ *Streak*: The **streak** of a mineral refers to the color of a powder produced by pulverizing the mineral. You can obtain a streak by scraping the mineral against an unglazed ceramic plate (Fig. 5.11b). The color of a mineral powder

tends to be less variable than the color of a whole crystal, and thus provides a fairly reliable clue to a mineral's identity. Calcite, for example, always yields a white streak even though pieces of calcite may be white, pink, or clear.

■ *Luster*: **Luster** refers to the way a mineral surface scatters light. Geoscientists describe luster simply by comparing the appearance of the mineral with the appearance of a familiar substance. For example, minerals that look like metal have a metallic luster, whereas those that do not have a nonmetallic luster—the adjectives are

FIGURE 5.11 Physical characteristics of minerals.

(a) Color is diagnostic of some minerals, but not all. For example, quartz can come in many colors.

(b) To obtain the streak of a mineral, rub it against a porcelain plate. The streak consists of mineral powder.

(c) Pyrite has a metallic luster because it gleams like metal.

(d) Feldspar has a nonmetallic luster.

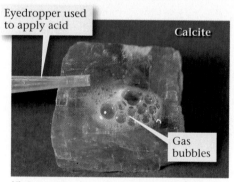

(f) Calcite reacts with hydrochloric acid to produce carbon dioxide gas.

(e) Crystal habit refers to the shape or character of the crystal. The blue kyanite crystals on the left are bladed, and the chrysotile on the right is fibrous.

(g) Magnetite is magnetic.

self-explanatory (**Fig. 5.11c, d**). Terms used for types of nonmetallic luster include *silky, glassy, satiny, resinous, pearly,* or *earthy.*

- *Hardness*: **Hardness** is a measure of the relative ability of a mineral to resist scratching, and it therefore represents the resistance of bonds in the crystal structure to being broken. The atoms or ions in crystals of a hard mineral are more strongly bonded than those in a soft mineral. Hard minerals can scratch soft minerals, but soft minerals cannot scratch hard ones. Diamond, the hardest mineral known, can scratch anything, which is why it is used to cut glass. In the early 1800s, a mineralogist named Friedrich Mohs listed some minerals in sequence of relative hardness; a mineral with a hardness of 5 can scratch all minerals with a hardness of 5 or less. This list, the **Mohs hardness scale**, helps in mineral identification. When you use the scale (**Table 5.1**), it helps to compare the hardness of a mineral with a common item such as your fingernail, a penny, or a glass plate. Note that not all of the minerals in Table 5.1 are common or familiar. Also, it's important to realize that the numbers on the Mohs hardness scale do not specify the true relative differences in hardness of minerals. For example, on the Mohs scale, talc has a hardness of 1 and quartz has a hardness of 7. But this does not mean that quartz is 7 times harder than talc. Careful tests show that quartz is actually about 100 times harder than talc, as indicated by how difficult it is to make an indentation in the mineral (see Table 5.1).

- *Specific gravity*: **Specific gravity** represents the density of a mineral, as represented by the ratio between the weight of a volume of the mineral and the weight of an equal volume of water at 4°C. For example, one cubic centimeter of quartz has a weight of 2.65 grams, whereas one cubic centimeter of water has a weight of 1.00 gram. Thus, the specific gravity of quartz is 2.65. In practice, you can develop a feel for specific gravity by hefting minerals in your hands. A piece of galena (lead ore) "feels" heavier than a similar-sized piece of quartz.

- *Crystal habit*: The **crystal habit** of a mineral refers to the shape of a single crystal with well-formed crystal faces, or to the character of an aggregate of many well-formed crystals that grew together as a group (**Fig. 5.11e**). The habit depends on the internal arrangement of atoms in the crystal. The arrangement, in turn, controls the geometry of crystal faces (e.g., triangular, square, rectangular, parallelogram) and the angular relationships among the faces. Mineralogists use a great many adjectives when describing habit. For example, a crystal may be compared with a geometric shape by using adjectives such as *cubic* or *prismatic*. A description

of habit generally includes adjectives that define relative dimensions of the crystal and the geometric shape of the crystal. For example, crystals that are roughly the same length in all directions are called equant or blocky, those that are much longer in one dimension than in others are columnar or needle-like, those shaped like sheets of paper are platy, and those shaped like knives are bladed. The relative dimensions depend on relative rates of crystal growth in different directions. A crystal that grows equally fast in three directions will be blocky, one that grows rapidly in two directions and slowly in the third direction will be platy, and one that grows rapidly in one direction but slowly in the other two directions will be needle-like. If the crystal grows as part of an aggregate, other adjectives can be used. For example, a group of needle-like crystals is a fibrous array.

- *Special properties*: Some minerals have distinctive properties that readily distinguish them from other minerals. For example, calcite ($CaCO_3$) reacts with dilute hydrochloric acid (HCl) to produce carbon dioxide (CO_2) gas (**Fig. 5.11f**). Dolomite ($CaMg[CO_3]_2$) also reacts with acid, but not as strongly. Graphite makes a gray mark on

TABLE 5.1 Mohs hardness scale. Mohs' numbers are relative—in reality, diamond is 3.5 times harder than corundum, as the graph shows.

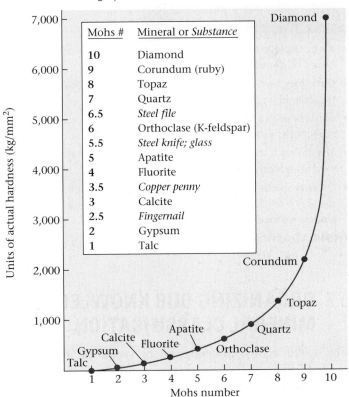

Mohs #	Mineral or *Substance*
10	Diamond
9	Corundum (ruby)
8	Topaz
7	Quartz
6.5	*Steel file*
6	Orthoclase (K-feldspar)
5.5	*Steel knife; glass*
5	Apatite
4	Fluorite
3.5	*Copper penny*
3	Calcite
2.5	*Fingernail*
2	Gypsum
1	Talc

paper, magnetite attracts a magnet (Fig. 5.11g), halite tastes salty, and plagioclase has striations (thin parallel corrugations or stripes) on its surface.

- *Fracture and cleavage*: Different minerals fracture (break) in different ways, depending on the internal arrangement of atoms. If a mineral breaks to form distinct planar surfaces that have a specific orientation in relation to the crystal structure, then we say that the mineral has **cleavage** and we refer to each surface as a cleavage plane. Cleavage forms in directions where the bonds holding atoms together in the crystal are the weakest (Fig. 5.12a–e). Some minerals have one direction of cleavage. For example, mica has very weak bonds in one direction but strong bonds in the other two directions. Thus, it easily splits into parallel sheets; the surface of each sheet is a cleavage plane. Other minerals have two or three directions of cleavage that intersect at a specific angle. For example, halite has three sets of cleavage planes that intersect at right angles, so halite crystals break into little cubes. Materials that have no cleavage at all (because bonding is equally strong in all directions) break either by forming irregular fractures or by forming conchoidal fractures (Fig. 5.12f). **Conchoidal fractures** are smoothly curving, clamshell-shaped surfaces; they typically form in glass. Cleavage planes are sometimes hard to distinguish from crystal faces (Fig. 5.12g, h).

Take-Home Message

- It's possible to distinguish one type of mineral from another based on physical properties.
- Color may be diagnostic, but it is not as reliable as the streak, the color of a powder.
- Minerals with a metallic luster look like metals; those with nonmetallic luster do not.
- The strength of chemical bonds in a mineral determines its hardness; Mohs scale indicates relative hardness.
- Other properties include specific gravity, crystal habit, and the way a crystal fractures. Some minerals break on specific smooth surfaces called cleavage planes.

THINK: Which minerals react with acid to produce CO_2 bubbles?

5.5 ORGANIZING OUR KNOWLEDGE: MINERAL CLASSIFICATION

The 4,000 known minerals can be separated into a small number of groups, or mineral classes. You may think, "Why bother?" Classification schemes are useful because they help organize information and streamline discussion. Biologists,

for example, classify animals into groups based on how they feed their young and on the architecture of their skeletons, and botanists classify plants according to the way they reproduce and by the shape of their leaves. In the case of minerals, a good means of classification eluded researchers until it became possible to determine the chemical makeup of minerals. A Swedish chemist, Baron Jöns Jacob Berzelius (1779–1848), analyzed minerals and noted chemical similarities among many of them. Berzelius, along with his students, established that most minerals can be *classified by specifying the principal anion* (negative ion) *or anionic group* (negative molecule) within the mineral. (Note again that some anions consist of single atoms, while others consist of a group of atoms that act as a unit.) We now take a look at principal mineral classes, focusing especially on silicates, the class that constitutes most of the rock in the Earth.

The Mineral Classes

Mineralogists distinguish several principal classes of minerals. Here are some of the major ones.

- *Silicates*: The fundamental component of most silicates in the Earth's crust is the SiO_4^{4-} anionic group. A well-known example, quartz (Fig. 5.11a), has the formula SiO_2. We will learn more about silicates in the next section.
- *Oxides*: Oxides consist of metal cations bonded to oxygen anions. Typical oxide minerals include hematite (Fe_2O_3; Fig. 5.11b) and magnetite (Fe_3O_4; Fig. 5.11g).
- *Sulfides*: Sulfides consist of a metal cation bonded to a sulfide anion (S^{2-}). Examples include galena (PbS) and pyrite (FeS_2; Fig. 5.11c). As with oxides, the metal forms a high proportion of the mineral, so many sulfides are considered ore minerals. Many sulfide minerals have a metallic luster.
- *Sulfates*: Sulfates consist of a metal cation bonded to the SO_4^{2-} anionic group. Many sulfates form by precipitation out of water at or near the Earth's surface. An example is gypsum ($CaSO_4 \cdot 2H_2O$), in which water molecules bond to the calcium-sulfate molecules. Pulverized gypsum mixed with water can be spread out in thin sheets that harden when they dry. Contractors use these sheets as wallboard ("sheetrock") in houses.
- *Halides*: The anion in a halide is a halogen ion (such as chloride [Cl^-] or fluoride [F^-]), an element from the second column from the right in the periodic table (see Appendix). Halite, or rock salt (NaCl; Fig. 5.12d), and fluorite (CaF_2), a source of fluoride, are common examples.

FIGURE 5.12 The nature of mineral cleavage and fracture.

(a) Mica has one strong plane of cleavage and splits into sheets.

(b) Pyroxene has two planes of cleavage that intersect at 90°.

(c) Amphibole has two planes that intersect at 60°.

Halite breaks into cubes.

(d) Halite has three mutually perpendicular planes of cleavage.

Calcite breaks into rhombs.

(e) Calcite has three planes of cleavage, one of which is inclined.

Irregular fracture

Crystal face

Garnet

Quartz

Conchoidal fracture

(f) Minerals without cleavage can develop irregular or conchoidal fractures.

Crystal face

Cleavage steps

(h) Cleavage and crystal faces in a sample of fluorite (CaF_2).

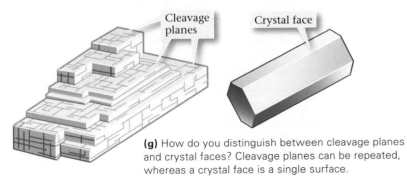

Cleavage planes

Crystal face

(g) How do you distinguish between cleavage planes and crystal faces? Cleavage planes can be repeated, whereas a crystal face is a single surface.

- *Carbonates*: In **carbonates**, the molecule CO_3^{2-} serves as the anionic group. Elements such as calcium or magnesium bond to this group. The two most common carbonates are calcite ($CaCO_3$; Fig. 5.12e) and dolomite ($CaMg[CO_3]_2$).
- *Native metals*: Native metals consist of pure masses of a single metal. The metal atoms are bonded by metallic bonds. Copper and gold, for example, may occur as native metals. A gold nugget is a mass of native gold that has been broken out of a rock.

Silicates: The Major Rock-Forming Minerals

Silicate minerals, or **silicates**, make up over 95% of the continental crust. Rocks of the oceanic crust and of the Earth's mantle consist almost entirely of silicates. Thus, silicates are the most common minerals on Earth.

As noted earlier, most silicates in the Earth's crust and upper mantle contain the SiO_4^{4-} anionic group. In this group, four oxygen atoms surround a single silicon atom, thereby defining the corners of a tetrahedron, a pyramid-like shape with four triangular faces (Fig. 5.13a). We refer to this anionic group as the **silicon-oxygen tetrahedron** (or, informally, as the "silica tetrahedron"), and it acts, in effect, as the building block of silicate minerals. (Of note, at the very high pressures found deep in the mantle, silicon and oxygen arrange in a different, more compact configuration.)

Mineralogists distinguish among several groups of silicate minerals based on the way in which silica tetrahedra are arranged (Fig. 5.13b). The arrangement, in turn, determines the degree to which tetrahedra share oxygen atoms. Below, we describe five of the major silicate groups in sequence from the group in which *no* oxygens are shared, to the group in which *all* oxygens are shared. Note that the number of shared oxygens determines the ratio of silicon (Si) to oxygen (O) in the mineral. Each group includes several distinct minerals that differ from each other by the cations that they contain.

- *Independent tetrahedra*: In this group, the tetrahedra are independent and do not share any oxygen atoms. The attraction between the tetrahedra and positive ions holds such minerals together. This group includes olivine, a glassy green mineral, and garnet (Fig. 5.12f).
- *Single chains*: In a single-chain silicate, the tetrahedra link to form a chain by sharing two oxygen atoms. The most common of the many different types of single-chain silicates are pyroxenes, a group of black or dark-green minerals that occur in elongate crystals with two cleavage directions at 90° to one another (Fig. 5.12b).

- *Double chains*: In a double-chain silicate, the tetrahedra link to form a double chain by sharing two or three oxygen atoms. Amphiboles are the most common type. These typically are black or dark-brown elongate crystals, with two cleavage directions (Fig. 5.12c). You can distinguish amphiboles from pyroxenes because the cleavage planes of amphiboles lie at about 60° to one another.
- *Sheet silicates*: The tetrahedra in this group share three oxygen atoms and therefore link to form two-dimensional sheets. Other ions and, in some cases, water molecules fit between the sheets in some sheet silicates. Because of their structure, sheet silicates have a single strong cleavage in one direction, and they occur in books of very thin sheets. In this group we find micas (Fig. 5.12a), a type of sheet silicate including muscovite (light-brown or clear mica) and biotite (black mica). Clay minerals are also sheet silicates; they have a crystal structure similar to that of mica, but clay occurs only in extremely tiny flakes.
- *Framework silicates*: In a framework silicate, each tetrahedron shares all four oxygen atoms with its neighbors, forming a three-dimensional structure. Examples include feldspar and quartz. The two most common feldspars are plagioclase, which tends to be white, gray, or blue; and orthoclase (also called potassium feldspar, or K-feldspar), which tends to be pink (Fig. 5.11d). Feldspars contain aluminum (which substitutes for silicon in the tetrahedra), as well as varying proportions of other elements, such as calcium, sodium, and potassium. Quartz, in contrast, contains only silicon and oxygen. As the ratio of silicon to oxygen in quartz is 1:2, the mineral has the familiar formula SiO_2.

Take-Home Message

- There are over 4,000 individual minerals; they are classified into relatively few groups.
- Mineralogists distinguish a group based on the principal anion or anionic group it contains.
- Examples include silicates, oxides, sulfides, sulfates, halides, and carbonates.
- Most minerals on Earth are silicates, containing silicon-oxygen tetrahedra. Mineralogists distinguish groups of silicates based on the arrangement of tetrahedra.

THINK: What is the principal anionic group in carbonate minerals?

FIGURE 5.13 The structure of silicate minerals.

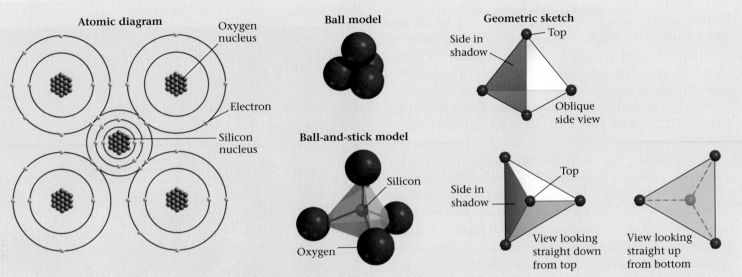

(a) The fundamental building block of a silicate mineral is the silicon-oxygen tetrahedron. Oxygens occupy the corners of the tetrahedron, and silicon lies at the center. Geologists portray the tetrahedron in a number of different ways.

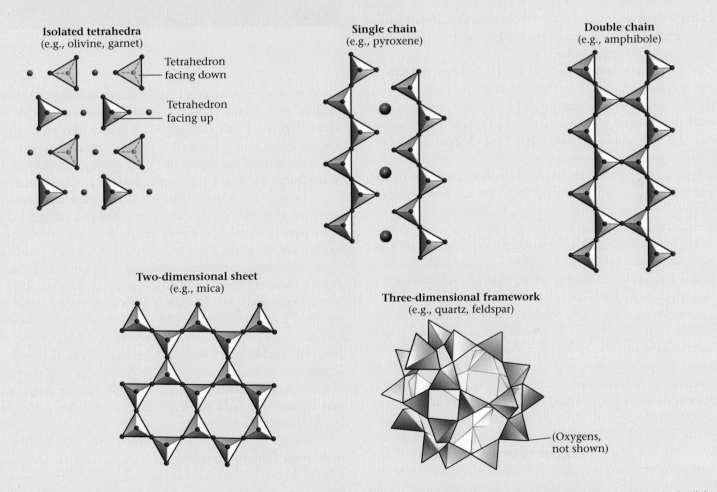

(b) The classes of silicate minerals differ from one another by the way in which the silicon-oxygen tetrahedra are linked. Where the tetrahedra link, they share an oxygen atom. Oxygen atoms are shown in red. Positive ions (not shown) occupy spaces between tetrahedra.

5.6 **SOMETHING PRECIOUS—GEMS!**

Mystery and romance follow famous gems. Consider the stone now known as the Hope Diamond, recognized by name the world over. No one knows who first dug it out of the ground (Box 5.3). Was it mined in the 1600s, or was it stolen off an ancient religious monument? What we do know is that in the 1600s, a French trader named Jean Baptiste Tavernier obtained a large (112.5 carats, where 1 carat = 200 milligrams), rare

> **Did you ever wonder . . .**
> where diamonds come from and how they form?

blue diamond in India, perhaps from a Hindu statue, and carried it back to France. King Louis XIV bought the diamond and had it fashioned into a jewel of 68 carats. This jewel vanished in 1762 during a burglary. Perhaps it was lost forever— perhaps not. In 1830, a 44.5-carat blue diamond mysteriously appeared on the jewel market for sale. Henry Hope, a British banker, purchased the stone, which then became known as the Hope Diamond (Fig. 5.14). It changed hands several times until 1958, when a famous New York jeweler named Harry Winston donated it to the Smithsonian Institution in Washington, D.C., where it now sits behind bulletproof glass in a heavily guarded display.

What makes stones such as the Hope Diamond so special that people risk life and fortune to obtain them? What is the difference between a gemstone, a gem, and any other mineral? A gemstone is a mineral that has special value because it is rare and people consider it beautiful. A **gem** is a cut and finished stone ready to be set in jewelry. Jewelers distinguish between precious stones (such as diamond, ruby, sapphire, and emerald), which are particularly rare and expensive, and semiprecious stones (such as topaz, tourmaline, aquamarine, and garnet), which are less rare and less expensive. All the stones mentioned so far are transparent crystals, though most have some color (see Table 5.2). The category of semiprecious stones also includes opaque or translucent minerals such as lapis, malachite (Fig. 5.2a), and opal.

In everyday language, pearls and amber may also be considered gemstones. Unlike diamonds and garnets, which form inorganically in rocks, pearls form in living oysters when the oyster extracts calcium and carbonate ions from water and precipitates them around an impurity, such as a sand grain, embedded in its body. Thus, pearls are a result of biomineralization. Most pearls used in jewelry today are "cultured" pearls, made by artificially introducing round sand grains into oysters in order to stimulate pearl production. Amber is also formed by organic processes—it consists of fossilized tree sap. But because amber consists of organic compounds that are not arranged in a crystal structure, it does not meet the definition of a mineral.

FIGURE 5.14 The Hope Diamond, now on display at the Smithsonian Institution in Washington, D.C.

Rare means hard to find, and some gemstones are indeed hard to find. Many diamond localities, for example, occur in isolated regions of Congo, South Africa, Brazil, Canada, Russia, India, and Borneo (Box 5.3). In some cases, it is not the mineral itself but rather the "gem-quality" versions of the mineral that are rare. For example, garnets are found in many rocks in such abundance that people use them as industrial abrasives. But most garnets are quite small and contain inclusions (specks of other minerals and/or bubbles) or fractures, so they are not particularly beautiful. Gem-quality garnets— clean, clear, large, unfractured crystals—are unusual. In some cases, gemstones are merely pretty and rare versions of more common minerals. For example, ruby is a special version of the common mineral corundum (Al_2O_3), and emerald is a special version of the common mineral beryl (Fig. 5.15a). As for the beauty of a gemstone, this quality lies basically in its color and, in the case of transparent gems, its "fire"—the way the mineral bends and internally reflects the light passing through it, and disperses the light into a spectrum. Fire makes a diamond sparkle more than a similarly cut piece of glass.

Gemstones form in many ways. Some solidify from a melt, some form by diffusion, some precipitate out of a water solution in cracks, and some are a consequence of the chemical interaction of rock with water near the Earth's surface. Many gems come from pegmatites, particularly coarse-grained rocks formed by the solidification of steamy melt.

BOX 5.3

CONSIDER THIS . . .

Where Do Diamonds Come From?

Diamonds consist of carbon, which typically accumulates only at or near Earth's surface. Experiments demonstrate that the temperatures and pressures needed to form diamond are so extreme that, in nature, they generally occur only at depths of around 150 km below the Earth's surface. Under these conditions, the carbon atoms that were arranged in hexagonal sheets in graphite rearrange to form the much stronger and more compact structure of diamond. (Of note, engineers can duplicate these conditions in the laboratory; corporations manufacture several tons of synthetic diamonds a year.)

How does carbon get down into the mantle, where it transforms into diamond? Geologists speculate that subduction or collision provides the means of carrying carbon-containing rocks and sediments from the Earth's surface down to depth. This carbon transforms into diamond, some of which becomes trapped beneath continents.

FIGURE Bx5.3 Diamond occurrences.

A diamond mine pit.

But if diamonds form at great depth, then how do they return to the surface? One possibility is that the process of rifting cracks the continental crust and causes a small part of the underlying lithospheric mantle to melt. Magma generated during this process rises to the surface, bringing the diamonds with it. Near the surface, the magma cools and solidifies to form a special kind of igneous rock called kimberlite (named for Kimberley, South Africa, where it was first found). Diamonds brought up with the magma are embedded in the kimberlite (**Fig. Bx 5.3**). Kimberlite magma contains a lot of dissolved gas and thus froths to the surface very rapidly. Kimberlite rock commonly occurs in carrot-shaped bodies called kimberlite pipes that are 50 to 200 m across and at least 1 km deep.

Controversial measurements suggest that many of the diamonds that sparkle on engagement rings today were formed 3.2 billion years ago. The diamonds sat at depths of 150 km in the Earth until two rifting events, one of which took place in the late Precambrian and the other during the late Mesozoic, released them to the surface, like genies out of a bottle. The Mesozoic rifting event led to the breakup of Pangaea.

In places where diamonds occur in solid kimberlite, they can be obtained only by digging up the kimberlite and crushing it, to separate out the diamonds. But nature can also break diamonds free from the Earth. In places where kimberlite has been exposed at the ground surface for a

A diamond embedded in solid kimberlite.

long time, the rock chemically reacts with water and air (a process called weathering; see Interlude B). These reactions cause most minerals in kimberlite to disintegrate, creating sediment that washes away in rivers. Diamonds are so hard that they remain as solid grains in river gravel. Thus, many diamonds have been obtained simply by separating them from recent or ancient river gravel.

Diamond-bearing kimberlite pipes occur in many places around the world, particularly where very old continental lithosphere exists. Southern and central Africa, Siberia, northwestern Canada, India, Brazil, Borneo, Australia, and the U.S. Rocky Mountains all have pipes (**See for Yourself D**, p. S-8). Rivers and glaciers, however, have transported diamond-bearing sediments great distances from their original sources. In fact, diamonds have even been found in farm fields of the midwestern United States. Not all natural diamonds are valuable; value depends on color and clarity. Diamonds that contain imperfections (cracks, or specks of other material), or are dark gray in color, are not used for jewelry. These stones, called industrial diamonds, are used instead as abrasives, for diamond powder is so hard (10 on the Mohs hardness scale) that it can be used to grind away any other substance.

Gem-quality diamonds come in a range of sizes. Jewelers measure diamond size in carats, where one carat equals 200 milligrams (0.2 grams). In English units of measurement, one ounce equals 142 carats. (Note that a *carat* measures gemstone weight, whereas a *karat* specifies the purity of gold.) The largest diamond ever found, a stone called the Cullinan Diamond, was discovered in South Africa in 1905. It weighed 3,106 carats (621 grams) before being cut. By comparison, the diamond on a typical engagement ring weighs less than one carat. Diamonds are rare, but not as rare as their price perhaps suggests. A worldwide consortium of diamond producers stockpile the stones so as not to flood the market and drive the price down.

TABLE 5.2 Precious and semiprecious materials

Gem Name	Material/Formula	Comments
Amber	Fossilized tree sap	Composed of organic chemicals; amber is not strictly a mineral.
Amethyst	Quartz/SiO_2	The best examples precipitate from water in openings in igneous rocks; a deep-purple version of quartz.
Aquamarine	Beryl/$Be_3Al_2Si_6O_{18}$	A bluish version of emerald.
Diamond	Diamond/C	Brought to the surface from the mantle in igneous bodies called diamond pipes; may later be mixed in deposits of sediment.
Emerald	Beryl/$Be_3Al_2Si_6O_{18}$	Occurs in coarse igneous rocks (pegmatites; see Chapter 6).
Garnet	Garnet/(e.g., $Mg_3Al_2[SiO_4]_3$)	A variety of types differ in composition (Ca, Fe, Mg, and Mn versions); occurs in metamorphic rocks (see Chapter 8).
Jade	Jadeite/$NaAlSi_2O_6$ Nephrite/$Ca_2(Mg,Fe)_5Si_8O_{22}(OH)_2$	Jade can be one of two minerals, jadeite (a pyroxene) or nephrite (an amphibole); both occur in metamorphic rocks.
Opal	Composed of microscopic spheres of hydrated silica packed together	Most opal comes from a single mining district in central Australia; occurs in bedrock that has reacted with water near the surface.
Pearl	Aragonite/$CaCO_3$	Formed by oysters, which secrete coatings around sand grains that are accidentally embedded in the soft parts of the organism. Cultured pearls are formed the same way, but the impurity is a spherical bead that is intentionally introduced.
Ruby	Corundum/Al_2O_3	The red color is due to chromium impurities; found in coarse igneous rocks called pegmatites and as a result of contact metamorphism (see Chapters 6 and 8).
Sapphire	Corundum/Al_2O_3	A blue version of ruby.
Topaz	$Al_2SiO_4(F,OH)_2$	Found in igneous rocks, and as a result of the reaction of rock with hot water.
Tourmaline	$Na(Mg,Fe)_3Al_6(BO_3)_3(Si_6O_{18})(OH,F)_4$	Forms in igneous and metamorphic rocks.
Turquoise	$CuAl_6(PO_4)_4(OH)_8 \cdot 4H_2O$	Found in copper-bearing rocks; a popular jewelry gem in the American Southwest.

Most gems used in jewelry are "cut" stones. The smooth **facets** on a gem are ground and polished surfaces made with a faceting machine (Fig. 5.15b). Facets are not the natural crystal faces of the mineral, nor are they cleavage planes, though gem cutters sometimes make the facets parallel to cleavage directions and will try to break a large gemstone into smaller pieces by splitting it on a cleavage plane. A faceting machine consists of a doping arm, a device that holds a stone in a specific orientation; and a lap, a rotating disk covered with a wet paste of grinding powder and water. The gem cutter fixes a gemstone to the end of the doping arm and positions the arm so that it holds the stone against the moving lap. The movement of the lap grinds a facet. When the facet is complete, the gem cutter rotates the arm by a specific angle, lowers the stone, and grinds another facet. The geometry of the facets defines the cut of the stone. Different cuts have names, such as "brilliant," "French," "star," and "pear." Grinding facets is a lot of work—a typical engagement-ring diamond with a brilliant cut has 57 facets (Fig. 5.15c)!

Some mineral specimens have special value simply because their geometry and color before cutting are beautiful. Prize specimens exhibit shapes and colors reminiscent of fine art and may sell for tens of thousands of dollars (Fig. 5.16). It's no wonder that mineral "hounds" risk their necks looking for a cluster of crystals protruding from the dripping roof of a collapsing mine or hidden in a crack near the smoking summit of a volcano.

Did you ever wonder . . .

how jewelers make the facets on a jewel?

FIGURE 5.15 Cutting gemstones.

Non-gem-quality beryl

Rough emerald

Cut emerald

(a) Emerald is a green, transparent variety of the mineral beryl.

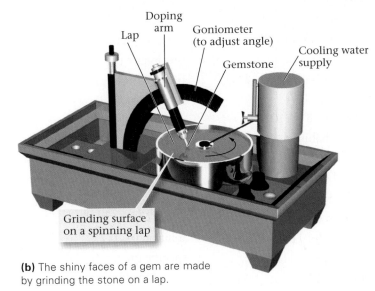

Doping arm

Lap

Goniometer (to adjust angle)

Gemstone

Cooling water supply

Grinding surface on a spinning lap

(b) The shiny faces of a gem are made by grinding the stone on a lap.

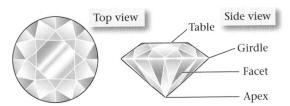

Top view

Side view

Table

Girdle

Facet

Apex

(c) There are many different "cuts" for a gem. Here we see the top and side views of a brilliant-cut diamond.

FIGURE 5.16 A spectacular museum specimen of a mineral cluster. The arrangement of colors and shapes is like abstract art. The multicolored minerals in this specimen are called watermelon tourmaline.

Chapter Summary

- Minerals are naturally occurring, solid substances, formed by geologic processes, with a definable chemical composition and an internal structure characterized by an orderly arrangement of atoms, ions, or molecules in a crystalline lattice. Most minerals are inorganic.
- In the crystalline lattice of minerals, atoms occur in a specific pattern—one of nature's finest examples of ordering.
- Minerals can form by the solidification of a melt, precipitation from a water solution, diffusion through a solid, the metabolism of organisms, and precipitation from a gas.
- About 4,000 different types of minerals are known, each with a name and distinctive physical properties (such as color, streak, luster, hardness, specific gravity, crystal habit, and cleavage).
- The unique physical properties of a mineral reflect its chemical composition and crystal structure. By observing these physical properties, you can identify minerals.
- The most convenient way for classifying minerals is to group them according to their chemical composition. Mineral classes include silicates, oxides, sulfides, sulfates, halides, carbonates, and native metals.
- The silicate minerals are the most common on Earth. The silicon-oxygen tetrahedron, a silicon atom surrounded by four oxygen atoms, is the fundamental building block of silicate minerals.
- Groups of silicate minerals are distinguished from each other by the ways in which the silicon-oxygen tetrahedra that constitute them are linked.
- Gemstones are minerals known for their beauty and rarity. The facets on cut gems used in jewelry are made by grinding and polishing the stones with a faceting machine.

GEOPUZZLE REVISITED

The geologic definition of a mineral is much narrower than the definition used in everyday conversation. Just because something is not or was not alive doesn't mean that it's a mineral. Minerals have an orderly internal crystalline structure and must have formed by geologic processes.

Guide Terms

anion (p. 110)
atom (p. 110)
atomic number (p. 110)
atomic weight (p. 110)
biogenic minerals (p. 106)
carbonates (p. 122)
cation (p. 110)
chemical (p. 110)
chemical bond (p. 110)
chemical formula (p. 110)
chemical reaction (p. 112)
cleavage (p. 120)
color (p. 116)
compound (p. 110)
concentration (p. 112)
conchoidal fracture (p. 120)
covalent bonding (p. 111)
crystal (p. 107)
crystal face (p. 107)
crystal habit (p. 119)
crystal lattice (p. 106)
crystal structure (p. 109)
diffraction (p. 116)
electron (p. 110)
element (p. 110)
evaporate (p. 112)
facet (p. 126)
gem (p. 124)
geode (p. 115)
hardness (p. 119)

hydrogen bond (p. 111)
ion (p. 110)
ionic bond (p. 111)
luster (p. 117)
metallic bonding (p. 111)
mineral (p. 106)
mineralogist (p. 105)
mineralogy (p. 105)
mixture (p. 112)
Mohs hardness scale (p. 119)
molecule (p. 110)
neutron (p. 110)
nucleus (p. 110)
polarity (p. 111)
polymorph (p. 109)
precipitate (p. 112)
precipitation (p. 112)
product (p. 112)
proton (p. 110)
reactant (p. 112)
shell (p. 110)
silicates (p. 122)
silicon-oxygen tetrahedron (p. 122)
solution (p. 112)
specific gravity (p. 119)
streak (p. 117)
symmetry (p. 113)
van der Waals bonding (p. 111)

Review Questions

1. What is a mineral, as geologists understand the term? How is this definition different from the everyday usage of the word?
2. Why is glass not a mineral?

3. Salt is a mineral, but the plastic making up an inexpensive pen is not. Why not?

4. Describe the several ways that mineral crystals can form.

5. Why do some minerals occur as euhedral crystals, whereas others occur as anhedral grains?

6. List and define the principal physical properties used to identify a mineral.

7. How can you determine the hardness of a mineral? What is the Mohs hardness scale?

8. How do you distinguish cleavage surfaces from crystal faces on a mineral? How does each type of surface form?

9. What is the prime characteristic that geologists use to separate minerals into classes?

10. On what basis do mineralogists organize silicate minerals into distinct groups?

11. What is the relationship between the way in which silicon-oxygen tetrahedra bond in micas and the characteristic cleavage of micas?

12. Why are some minerals considered gemstones? How do you make the facets on a gem?

On Further Thought

13. Compare the chemical formula of magnetite with that of biotite. Why is magnetite mined as iron ore, but biotite is not?

14. Imagine that you are given two milky white crystals, each about 2 cm across. You are told that one of the crystals is composed of plagioclase and the other of quartz. How can you determine which is which?

15. Could you use crushed calcite to grind and form facets on a diamond? Why or why not?

 For more resources, including animations, quizzes, and Norton's GeoTours, go to **wwnorton.com/studyspace**.

 If your instructor assigns exercises in SmartWork, log in at **smartwork.wwnorton.com**.

ANOTHER VIEW Corundum comes in many colors. Gem-quality versions, including ruby and sapphire, can be cut into many shapes.

Rock Groups

It took years of back-breaking labor for nineteenth-century workers to chisel and chip ledges and tunnels through hard rock to run a rail line across a mountain range. In the process, the workers became very familiar with the nature of rock.

A.1 INTRODUCTION

During the 1849 gold rush in the Sierra Nevada of California, only a few lucky individuals actually became rich. The rest of the "forty-niners" either slunk home in debt or took up less glamorous jobs in new towns such as San Francisco. These towns grew rapidly, and soon the American west coast was demanding large quantities of manufactured goods from east-coast factories. Making the goods was no problem, but getting them to California meant either a stormy voyage around the southern tip of South America or a trek with stubborn mule teams through the deserts of Nevada and Utah. The time was ripe to build a railroad linking the east and west coasts of North America, and, with much fanfare, the Central Pacific line decided to punch one right across the peaks of the Sierras. In 1863, while the Civil War raged elsewhere in the United States, the company transported six thousand Chinese laborers across the Pacific in the squalor of unventilated cargo holds and set them to work chipping ledges and blasting tunnels. Along the way, untold numbers of laborers died of frostbite, exhaustion, mistimed blasts, landslides, or avalanches.

Through their efforts, the railroad laborers certainly gained an intimate knowledge of how rock feels and behaves—it's solid, heavy, and hard! They also found that some rocks break easily into layers but others do not, and some rocks are dark-colored while others are light-colored. They realized, like anyone who looks closely at rock exposures, that rocks are not just gray, featureless masses, but rather come in a great variety of colors and textures.

Why are there so many distinct types of rocks? The answer is simple: rocks can form in many different ways and from many different materials. Because of the relationship between rock type and the process of formation, *rocks provide a historical record of geologic events and give insight into interactions among components of the Earth System.*

The next few chapters are devoted to a discussion of rocks and a description of how rocks form. This interlude serves as a general introduction to these chapters. We learn what the term *rock* means to geologists, what rocks are made of, and how to distinguish among the three principal groups of rocks. We also look at how geologists study rocks.

A.2 WHAT IS ROCK?

To geologists, **rock** is a coherent, naturally occurring solid, consisting of an aggregate of minerals or, less commonly, a mass of glass. Now let's take this definition apart.

- *Coherent*: A rock holds together, and thus must be broken to be separated into pieces. As a result of its coherence, rock can form cliffs or can be carved into sculptures. A pile of unattached mineral grains does not constitute a rock.
- *Naturally occurring*: Geologists consider only naturally occurring materials to be rocks, so manufactured materials, such as concrete and brick, do not qualify.
- *An aggregate of minerals or a mass of glass*: The vast majority of rocks consist of an aggregate (a collection) of many mineral grains, and/or crystals, stuck or grown together. (*Note*: A grain is any fragment or piece of mineral, rock, or glass.) Technically, a single mineral crystal is a "mineral specimen," not a rock, even if it is meters long. Some rocks contain only one kind of mineral, whereas others contain several different kinds. A few of the rock types that form at volcanoes consist of glass, which may occur either as a homogeneous mass or as an accumulation of tiny glass shards.

What holds rock together? Grains in rock stick together to form a coherent mass either because they are bonded by natural **cement**, mineral material that precipitates from water and fills the space between grains (**Fig. A.1a**), or because they interlock with one another like pieces in a jigsaw puzzle (**Fig. A.1b**). Rocks whose grains are stuck together by cement are called **clastic**, whereas rocks whose crystals interlock with one another are called **crystalline**. Glassy rocks hold together either because they originate as a continuous mass (that is, they have no separate grains) or because separate glassy grains were welded together while still hot or cemented together at a later time.

All rocks, in the most basic sense, are just masses of chemicals bonded in molecules of varying size and complexity. But not all rocks contain the same chemicals. For example, granite—a rock commonly used for gravestones, building facades, and kitchen counters—contains oxygen, silicon, aluminum, calcium, iron, magnesium, and potassium. Marble—a rock favored by fine sculptors—contains oxygen, carbon, and calcium. Note, as we pointed out in Chapter 2, that the elements oxygen and silicon are the most common elements in the Earth's crust; indeed, oxygen constitutes 93.8% of the volume of the crust. Most of the rock in the crust as a whole consists of silicate minerals (minerals containing the silicon-oxygen tetrahedron). Very close to the Earth's surface, however, the activity of life plays a role in rock formation, so a significant proportion of bedrock exposed at the surface of the Earth consists of carbonate minerals (minerals containing the CO_3^{2-} ion) extracted from water to form shells. Other minerals (such as oxides, sulfides, sulfates) are important as resources for metals and industrial materials, but they constitute only a small percentage of rocks in the crust.

FIGURE A.1 Rocks, aggregates of mineral grains and/or crystals, can be clastic or crystalline.

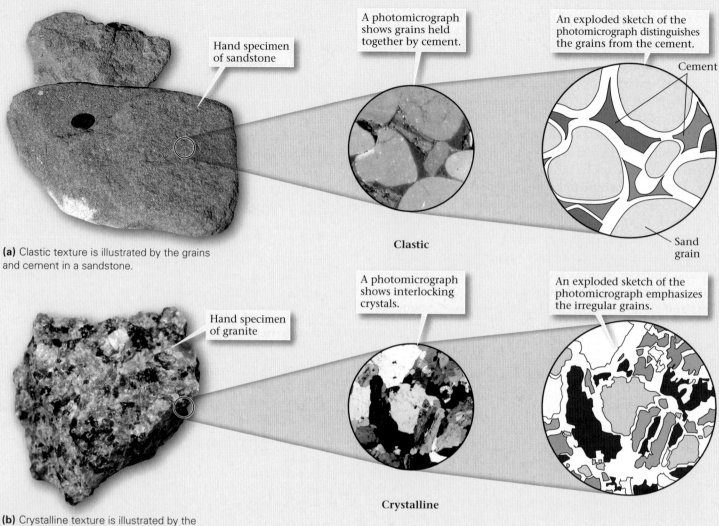

(a) Clastic texture is illustrated by the grains and cement in a sandstone.

(b) Crystalline texture is illustrated by the interlocking crystals in a granite.

A.3 ROCK OCCURRENCES

At the surface of the Earth, rock occurs either as broken chunks (pebbles, cobbles, or boulders) that have moved by falling down a slope or by being transported in ice, water, or wind, or as **bedrock** that is still attached to the Earth's crust. Geologists refer to an exposure of bedrock as an **outcrop**. An outcrop may appear as a rounded knob out in a field, as a ledge forming a cliff or ridge, on the face of a stream cut (where running water dug down into bedrock), or along human-made road cuts and excavations (Fig. A.2a–d).

To people who live in cities or forests or on farmland, outcrops of bedrock may be unfamiliar, since bedrock may be completely covered by vegetation, sand, mud, gravel, soil, water, asphalt, concrete, or buildings. Outcrops are particularly rare in regions such as the midwestern United States, where,

during the past million years, ice-age glaciers melted and left behind thick deposits of debris, which completely buried pre-existing valleys and hills. The depth of bedrock plays a key role in urban planning, because architects prefer to set foundations of large buildings on bedrock rather than on loose sand or mud. Because of this preference, the skyscrapers of New York City rise in two clusters on the island of Manhattan, one at the south end and the other in the center, locations where bedrock lies close to the surface.

A.4 THE BASIS OF ROCK CLASSIFICATION

Beginning in the eighteenth century, geologists struggled to develop a sensible way to classify rocks, for they realized, as did miners from centuries past, that not all rocks are the

FIGURE A.2 Types of rock exposures.

(a) Outcrops are natural rock exposures. These outcrops rise as cliffs above the forest of the Rocky Mountains in Colorado.

(c) By blasting into the ground to produce a more level grade for roads, highway engineers produce roadcuts.

(b) In arid (dry) climates, a lack of vegetation leaves outcrops unobscured.

(d) Stream cuts form where flowing water grinds into the land and strips away soil and vegetation.

same. Classification schemes help us organize information and remember significant details about materials or objects, and they help us recognize similarities and differences among them. One of the earliest classification schemes divided rocks into three groups—so-called primary, secondary, and tertiary—based on the incorrect perception that the groups had formed in a time succession. In light of this concept, a German mineralogist named Abraham Werner proposed that a "universal ocean" containing dissolved and suspended minerals once had covered the Earth. According to Werner, the earliest rocks formed by precipitation from this solution; later, as sea level dropped, the action of rivers, waves, and wind wore down exposed rocks and produced debris that consolidated to form younger rocks. Werner was an influential teacher, and his followers came to be known as the *Neptunists*, after the Roman god of the sea.

At about the same time that Werner was developing his ideas, a Scottish gentleman farmer and doctor named James Hutton began exploring the outcrops of his native land. Hutton was a keen thinker who lived in Edinburgh, a hotbed of intellectual argument during the Age of Enlightenment where everything from political institutions to scientific paradigms became fodder for debate. He associated with prominent philosophers and scientists in Edinburgh and, like them, was open to new ideas. Hutton began to ponder the issue of how rocks formed, and rather than force his perceptions to fit established dogma, he developed alternative ideas based on his own observations. For example, he watched sand settle on a beach and realized that some rocks could have formed from cementation of sand grains. He examined exposures in which bodies of certain crystalline rocks appeared to have pushed into other rocks, heating the other rocks in the process, and concluded that some rocks could have formed by solidification from a melt. He also noticed that rocks adjacent to bodies of now-solid melts had somehow been altered, and he proposed that these rocks formed by change of preexisting rocks as a result of what he referred to as "subterranean heat." Hutton, like Werner, attracted followers—Hutton's group came to be known as the *Plutonists*, after the Roman god of the underworld, because they favored the idea that the formation of certain rocks involved melts that had risen from deeper within the Earth.

FIGURE A.3 Examples of the three major rock groups.

Igneous

(a) Lava (molten rock) freezes to form igneous rock. Here, the molten tip of a brand-new flow still glows red. Older flows are already solid.

Sedimentary

(b) Sand, formed from grains eroded off the rock cliffs, collects on the beach. If buried and turned to rock, it becomes layers of sandstone, such as those making up the cliffs.

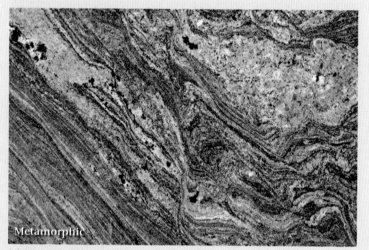

Metamorphic

(c) Metamorphic rock forms when preexisting rocks endure changes in temperature and pressure and/or are subjected to shearing, stretching, or shortening. New minerals and textures form, as seen in this decorative floor tile.

In the last decades of the eighteenth century, as the armed rebellions that led to the formation of the United States and the Republic of France raged, a battle of ideas concerning the origin of rocks rattled the infant science of geology. This battle, pitting the Neptunists against the Plutonists, lasted for years. In the end the Plutonists won, for they demonstrated beyond a shadow of a doubt that certain crystalline rocks must have been in molten form when emplaced. As a consequence, it became clear that different rocks formed in different ways, and that, as a starting point, rocks can best be *classified on the basis of how they formed*. This principle became the foundation of the modern system of classification of rocks. Because Hutton fostered this idea, as well as many others that we will describe later in the book, modern geologists revere Hutton as the "father of geology."

By the end of the eighteenth century, most geologists had accepted the *genetic scheme* for classifying rocks—a scheme that focuses on the origin (genesis) of rocks—and this is the approach that we continue to use today. Using this approach,

FIGURE A.4 A cross section illustrating various geologic settings in which rocks form.

Deep ocean   Passive Margin   Rift   Mountain belt

Metamorphic rock formation

Erosion

Erosion

Sedimentary rock formation

Sediment deposition

Sedimentary rock formation

Sedimentary rock formation

geologists recognize three basic groups: (1) **igneous rocks**, which form by the freezing (solidification) of molten rock (Fig. A.3a); (2) **sedimentary rocks**, which form either by the cementing together of fragments (grains) broken off preexisting rocks or by the precipitation of mineral crystals out of water solutions at or near the Earth's surface (Fig. A.3b); and (3) **metamorphic rocks**, which form when preexisting rocks change character in response to a change in pressure and temperature conditions, and/or as a result of squashing, stretching, or shear (Fig. A.3c). Metamorphic change occurs in the solid state, which means that it does not require melting. In the context of modern plate tectonics theory, different rock types form in tectonic settings (Fig. A.4). In succeeding chapters, we will explore these settings in more detail.

Each of the three groups contains many different individual rock types, distinguished from one another by physical characteristics.

- *Grain size*: The dimensions of individual grains in a rock may be measured in millimeters or centimeters. Some grains are so small that they can't be seen without a microscope, whereas others are as big as a fist or larger. Some grains are **equant**, meaning that they have the same dimensions in all directions; some are **inequant**, meaning that the dimensions are not the same in all directions (Fig. A.5a, b). In some rocks, all the grains are the same size, whereas other rocks contain a variety of grain sizes.

- *Composition*: A rock is a mass of chemicals. These chemicals may be ordered into mineral grains or, less commonly, may be disordered and constitute glass. The term *rock composition* refers to the proportions of differ-

ent chemicals making up the rock. The proportion of chemicals, in turn, affects the proportion of different minerals constituting the rock. As you will see, however, chemical composition does not completely control the minerals present in a rock. For example, two rocks with exactly the same chemical composition can have totally different assemblages of minerals, if each rock formed under different pressure and temperature conditions. That's because the process of mineral formation is affected by environmental factors such as pressure and temperature.

FIGURE A.5 Describing grains in rock.

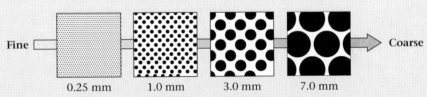

Equant   Inequant

Magnification reveals a variety of grains.

This rock is an aggregate of mineral grains.

1 millimeter

Inequant grains align to form foliation.

1 meter

(a) Grains in rock come in a variety of shapes. Some are equant, whereas some are inequant. In this example of metamorphic rock, inequant grains align to define a foliation.

Fine   Coarse

0.25 mm   1.0 mm   3.0 mm   7.0 mm

(b) Geologists define grain size by using this comparison chart.

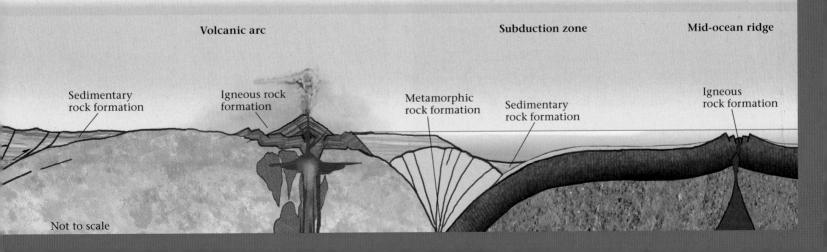

Volcanic arc   Subduction zone   Mid-ocean ridge

Sedimentary rock formation

Igneous rock formation

Metamorphic rock formation

Sedimentary rock formation

Igneous rock formation

Not to scale

■ *Texture*: This term refers to the arrangement of grains in a rock, that is, the way grains connect to one another and whether or not inequant grains are aligned parallel to each other. The concept of rock texture will become easier to grasp as we look at different examples of rocks in the following chapters.

■ *Layering*: Some rock bodies appear to contain distinct layering, defined either by bands of different compositions or textures, or by the alignment of inequant grains so that they trend parallel to each other. Different types of layering occur in different kinds of rocks. For example, the layering in sedimentary rocks is called **bedding**, whereas the layering in metamorphic rocks is called **metamorphic foliation** (Fig. A.6a, b).

Each distinct rock type has a name. Names come from a variety of sources. Some come from the dominant component making up the rock, some from the region where the rock was first discovered or is particularly abundant, some from a root word of Latin origin, and some from a traditional name used by people in an area where the rock is found. All told, there are hundreds of different rock names, though in this book we will introduce only about 30.

A.5 STUDYING ROCK

Outcrop Observations

The study of rocks begins by examining a rock in an outcrop. If the outcrop is big enough, such an examination will reveal relationships between the rock you're interested in and the rocks around it, and will allow you to detect layering. Geologists carefully record observations about an outcrop, then break off a **hand specimen**, a fist-sized piece, that they can examine more closely with a hand lens (magnifying glass). Observation with a hand lens enables geologists to identify sand-sized or larger mineral grains, and may enable them to describe the texture of the rock.

Thin-Section Study

Geologists often must examine rock composition and texture in minute detail in order to identify a rock and develop a hypothesis for how it formed. To do this, they take a specimen back to the lab, make a very thin slice (about 0.03 mm thick, the thickness of a human hair) and mount it on a glass slide (Fig. A.7a–c). They study the resulting **thin section** with a petrographic microscope (*petro* comes from the Greek word for rock). A petrographic microscope differs from an ordinary microscope in that it illuminates the thin section with transmitted polarized light. This means that the illuminating light beam first passes through a special filter that makes all the light waves in the beam vibrate in the same plane, and then the light passes up through the thin section. An observer looks through the thin section as if it were a stained glass window. When illuminated with transmitted polarized light, each type of mineral grain displays a unique suite of colors (Fig. A.7d). The specific color the observer sees depends on both the identity of the grain and its orientation with respect to the waves of polarized light, for a crystal interferes with polarized light and allows only certain wavelengths to pass through.

FIGURE A.6 Layering in rock.

(a) Bedding in a sedimentary rock, here defined by alternating layers of coarser and finer grains, as exposed on a cliff along an Oregon beach. Older beds were tilted before younger ones were deposited.

(b) Foliation in metamorphic rock, here defined by alternating layers of light minerals and dark minerals.

FIGURE A.7 Studying rocks in thin section.

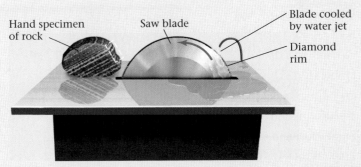

(a) Using a special saw, a geologist cuts a thin chip of a rock specimen.

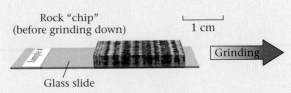

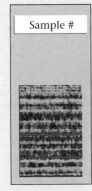

(b) The geologist glues the chip to a glass slide and grinds it down until it is so thin that light can pass through it.

(c) With a petrographic microscope, it's possible to view thin sections with light that shines through the sample from below.

(d) If the light is polarized, different minerals display different colors when viewed through the microscope.

The brilliant colors and strange shapes in a thin section rival the beauty of an abstract painting. By examining a thin section with a petrographic microscope, geologists can identify most of the minerals constituting the rock and can describe the way in which the grains connect to each other. A photograph taken through a petrographic microscope is called a **photomicrograph**.

Note that to make a thin section, a geologist first uses a special rock saw with a spinning, diamond-studded blade to cut a small rectangular block, or chip, of rock (see Fig. A.7). Rock saws work by grinding their way through rock—the hard diamonds embedded in the saw blade scratch and pulverize minerals as the saw blade rubs against the rock. The geologist then cements the chip to a glass microscope slide with an epoxy adhesive (glue). By using a lap, a spinning plate coated with abrasive, the geologist can grind the chip down until only a thin slice, still cemented to the glass slide, remains, ready for examination with a petrographic microscope.

High-Tech Analytical Equipment

Beginning in the 1950s, high-tech electronic instruments became available that enabled geologists to examine rocks on an even finer scale than is possible with a petrographic microscope. Modern research laboratories typically boast instruments such as electron microprobes, which can focus a beam of electrons on a small part of a grain to create a signal that defines the chemical composition of the mineral (Fig. A.8); mass spectrometers, which analyze the proportions of atoms with different atomic weights contained in a rock; and X-ray diffractometers, which identify minerals by looking at the way X-ray beams pass through crystals in a rock. Such instruments, in conjunction with optical examination, can provide geologists with highly detailed characterizations of rocks, which in turn help them understand how the rocks formed and where the rocks came from. This information enables geologists to use the study of rocks as a basis for deciphering Earth history.

FIGURE A.8 An electron microprobe uses a beam of electrons to analyze the chemical composition of minerals.

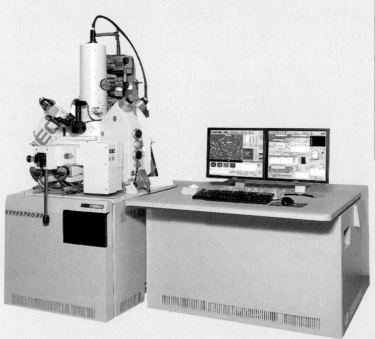

Guide Terms

bedding (p. 136)	inequant (p. 135)
bedrock (p. 132)	metamorphic foliation (p. 136)
cement (p. 131)	metamorphic rock (p. 135)
clastic (p. 131)	outcrop (p. 132)
crystalline (p. 131)	photomicrograph (p. 137)
equant (p. 135)	rock (p. 131)
hand specimen (p. 136)	sedimentary rock (p. 135)
igneous rock (p. 135)	thin section (p. 136)

ANOTHER VIEW This quarry, in northwestern Italy, provides blocks of pure white marble, some of which were carved into beautiful sculptures. It also provides a view into the bedrock that lies beneath rugged peaks.

CHAPTER 6

Up from the Inferno: Magma and Igneous Rocks

Massive blocks of granite crop out in Joshua Tree National Monument, California. The rock formed by the slow solidification of magma 15 km underground, about 140 million years ago. Erosion of overlying crust has exposed the granite.

GEOPUZZLE

Where does the red-hot molten rock that spills and/or blasts out of a volcano come from, and what does it turn into?

6.1 INTRODUCTION

Every now and then, an incandescent liquid—hot molten rock, or melt—fountains from a crater or crack on the big island of Hawaii. Hawaii is a **volcano**, a vent at which melt from inside the Earth spews onto the planet's surface. The transfer of melt from inside the Earth onto its surface is a volcanic eruption. Geologists refer to melt *beneath* the ground surface as **magma**,

and melt that has emerged at the surface as **lava**. Some lava pools around the vent, while some runs down the mountainside as a syrupy red-yellow stream called a **lava flow**. Near its source, lava on Hawaii has a temperature of 1,100 to 1,200°C and moves swiftly, cascading over escarpments at speeds of up to 60 km per hour (**Fig. 6.1a**). At the base of the mountain, the lava flow slows but advances nonetheless, engulfing roads, houses, or vegetation in its path (**Fig. 6.1b**). As the flow cools, its surface darkens and crusts over, occasionally breaking to reveal the hot, sticky mass that continues to ooze within. Finally, the flow stops moving entirely, and within days or weeks the once red-hot melt has become a hard, black solid

FIGURE 6.1 Formation and evolution of lava flows.

Lava fountain

Active lava flow

As the lava cools, it darkens.

(a) Lava erupts as a fountain from a volcanic vent on Hawaii. A fast-moving river of lava then flows downslope.

Smoke comes from burning vegetation.

(b) At a distance from the vent, the lava has completely crusted over with new rock, but the interior of the flow remains molten.

Time

The reddish color comes from weathering.

(d) Over time, many lava flows can accumulate one on top of another to build a large volcano. A canyon cut into the volcano exposes dozens of ancient flows.

(c) Eventually, the flow cools completely and becomes a layer of new rock. This flow engulfed a road on Hawaii.

through and through (**Fig. 6.1c, d**). New **igneous rock**, rock made by the freezing of a melt, has formed. Considering the fiery heat of the melt from which igneous rocks solidify, the name *igneous*—from the Latin *ignis*, meaning fire—makes sense. Igneous rocks are very common on Earth. They make up all of the oceanic crust and much of the continental crust.

It may seem strange to speak of "freezing" in the context of forming rock, for most people think of freezing as the transformation of liquid water to solid ice when the temperature drops below 0°C (32°F). Nevertheless, the freezing of liquid melt to form solid igneous rock represents the same phenomenon, solidification of a liquid, except that igneous rocks freeze at high temperatures—between 650°C and 1,100°C. To put such temperatures in perspective, keep in mind that home ovens attain a maximum temperature of only 260°C (500°F).

Rock that forms by the freezing of lava above ground, after it spills out (extrudes) onto the surface of the Earth and comes into contact with the atmosphere or ocean, is **extrusive igneous rock** (**Fig. 6.2a**). Extrusive igneous rock includes both solid lava flows, formed when streams or mounds of lava solidify on the surface of the Earth, and deposits of pyroclastic debris (from the Greek word *pyro*, meaning fire) (**Fig. 6.2b, c**). Such debris includes **volcanic ash**, consisting of fine particles of glass that form when a spray of lava erupts into the air and freezes instantly.

FIGURE 6.2 The intrusive and extrusive realms.

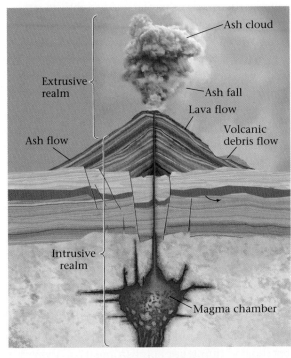

(a) The intrusive realm lies underground and the extrusive realm lies above ground. Lava flows, as well as various types of ash eruptions, all produce extrusive rocks.

(b) Extrusive rocks include lava flows and pyroclastic layers.

(c) Broken blocks of lava were blasted out of this Alaskan volcano, and now litter its slopes.

(d) An intrusion of basalt (dark rock) cuts across a mass of granite (light rock).

Some ash billows up several kilometers into the sky above a volcano and eventually drifts down in a snow-like ash fall. But some ash rushes down the side of the volcano in a scalding avalanche called an ash flow. **Pyroclastic debris** also includes larger fragments formed when clots of lava freeze in the air, or when the force of the eruption blasts apart preexisting rock of a volcano. We'll provide further detail about pyroclastic debris in Chapter 9.

Although some igneous rocks solidify at the surface during volcanic eruptions, a vastly greater volume results from solidification of magma underground, out of sight. Rock made by the freezing of magma underground, after it has pushed its way (intruded) into preexisting rock of the crust, is **intrusive igneous rock**. Some intrusive rock freezes within a relatively large, irregular volume called a magma chamber, whereas some freezes in tabular cracks or chimney-like columns (Fig. 6.2d).

Note that a significant amount of the material making up a volcano actually consists of pyroclastic debris accumulations that moved downslope *after* being initially deposited. Volcanic debris may slip in a landslide; it may be carried away by streams and redeposited as stream sediment; or it may mix with water to form a slurry of mud and debris, called a volcanic debris flow, that oozes down the mountain like wet concrete. If the volcano is an island or seamount, the mud and debris move underwater.

A great variety of igneous rocks exist on Earth (See for Yourself F, p. S-10); they make up all of the oceanic crust, and most of the continental crust. To understand why and how these rocks form, and why there are so many different kinds, we first discuss why magma forms, why it rises, how it flows, and how it freezes in intrusive and extrusive environments. We then look at the scheme that geologists use to classify igneous rocks.

Chapter Themes

By the end of this chapter, you should know . . .

- why melting produces molten rock at special places in the Earth.
- why magma moves into intrusive or extrusive settings, where it solidifies.
- how lava can erupt at the Earth's surface.
- why there are different types of igneous rock, and how to classify these rocks.
- where igneous activity takes place, in the context of plate tectonics.

6.2 WHY DOES MAGMA FORM?

The Earth is very hot inside. But despite the high temperatures that occur at depth, the popular image that the solid crust of the Earth floats on a sea of molten rock is *not* correct. Magma forms only in special places where preexisting solid rock melts. Below, we describe

Did you ever wonder . . .
whether Earth's crust floats on a magma sea?

conditions that lead to melting. We'll relate these conditions to plate tectonics theory later in the chapter.

Why Is It Hot Inside the Earth?

Clearly, if the Earth were not hot inside, igneous processes would not take place. Where does our planet's internal heat come from? Much of the heat is left over from the Earth's early days. According to the nebula theory, this planet formed from the collision and merging of countless planetesimals. Every time a collision occurred, the kinetic energy (energy of motion) of the colliding planetesimals transformed into heat energy. (You can simulate this phenomenon by banging a hammer repeatedly on a nail—the head of the nail becomes quite warm.) As the Earth grew, gravity pulled matter inward until eventually the weight of overlying material squeezed the matter inside tightly together. Such compression made the Earth's insides even hotter, just as compressing a gas with a piston makes it hotter. Even after the Earth had grown to become a planet, intense bombardment continued to add heat energy. Eventually, the Earth became hot enough for iron to melt. The iron sank to the center to form the core. Friction between the sinking iron and its surroundings generated still more heat, just as rubbing your hands together generates heat. (Note that this process transformed gravitational potential energy into heat.) Soon after Earth's formation, but probably after differentiation (see Chapter 1), a Mars-sized object collided with the Earth. This collision generated vast amounts of heat. Taken together, collisions and differentiation made the early Earth so hot that it was at least partially molten throughout, and its surface may have been an ocean of lava.

Ever since the heat-producing catastrophes of its early days, the Earth has radiated heat into space and thus has slowly cooled (see Box 2.3 for a discusssion of heat tranfer processes). Eventually, the sea of lava solidified and formed igneous rock, the first rock on Earth. If no heat had been added to the Earth after the end of intense bombardment (at about 3.9 Ga), the Earth might be too cold by now for igneous activity to take place. This hasn't happened because of the presence of radioactive elements, primarily in the crust. Decay of a single radioactive atom produces only a tiny amount of heat, but the cumulative effect of radioactive decay throughout the Earth has been sufficient to slow the cooling of this planet. Thus, Earth remains very hot today, with temperatures at the base of the lithosphere reaching almost 1,300°C, and temperatures at the planet's center exceeding 4,700°C. (By comparison, the surface of our Sun is about 5,700°C.)

Causes of Melting

Magma forms both in the upper part of the asthenosphere and in the lower crust. Here we discuss the physical conditions that lead to this melting.

Melting due to a decrease in pressure (decompression).
We can portray the variation in temperature with depth in the Earth on a graph by a curving line called the **geotherm**. Beneath typical oceanic crust, temperatures comparable to those of lava occur in the upper mantle (**Fig. 6.3a**). But even though the upper mantle is very hot, its rock stays solid because it is also under great pressure from the weight of overlying rock. Basically, pressure at great depth prevents atoms from breaking free of solid mineral crystals. Because pressure prevents melting, *a decrease in pressure can permit melting*. Specifically, if the pressure affecting hot mantle rock decreases while the temperature remains unchanged, magma forms. This kind of melting, called **decompression melting**, occurs where hot mantle rock rises to shallower depths in the Earth. Such movement occurs in mantle plumes, beneath rifts, and beneath mid-ocean ridges (**Fig. 6.3b**).

Melting as a result of the addition of volatiles. Magma also forms at locations where chemicals called volatiles mix with hot mantle rock. Volatiles are substances such as water

(H_2O) and carbon dioxide (CO_2) that evaporate easily and can exist in gaseous forms at the Earth's surface. When volatiles mix with hot rock, they help break chemical bonds. So if you add volatiles to a solid, hot, dry rock, the rock begins to melt. In effect, adding volatiles decreases a rock's melting temperature. Melting due to addition of volatiles is sometimes called **flux melting**. Of the common volatiles, water plays the most important role in triggering melting.

Flux melting happens in the mantle above subducting oceanic crust. As the downgoing plate heats up, water bonded to oceanic crustal minerals separates and rises into the overlying asthenosphere, causing the asthenosphere to melt (**Fig. 6.4a**).

Melting as a result of heat transfer from rising magma.
When magma from the mantle rises up into the crust, it brings heat with it. This heat raises the temperature of the surrounding crustal rock and, in some cases, the rise in temperature may

FIGURE 6.3 Decompression melting.

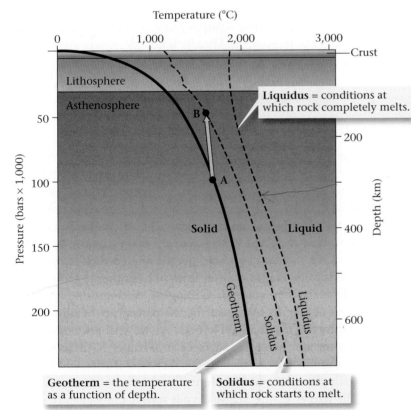

(a) Decompression melting takes place when the pressure acting on hot rock decreases. As this graph of pressure and temperature conditions in the Earth shows, when rock rises from point A to point B, the pressure decreases a lot, but the rock cools only a little, so the rock begins to melt.

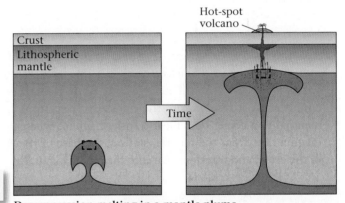

Decompression melting in a mantle plume

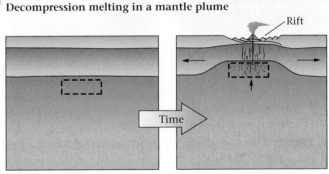

Decompression melting beneath a rift

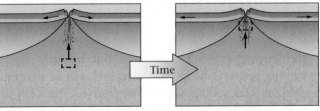

Decompression melting beneath a mid-ocean ridge

(b) The conditions leading to decompression melting occur in several different geologic environments. In each case, a volume of hot asthenosphere (outlined by dashed lines) rises to a shallower depth, and magma (red dots) forms.

FIGURE 6.4   Flux melting and heat-transfer melting.

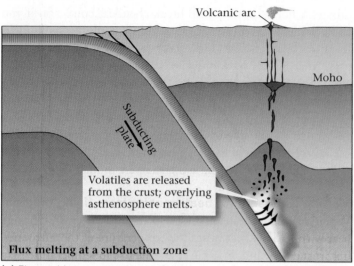

(a) Flux melting occurs where volatiles enter hot mantle; this happens at subduction zones.

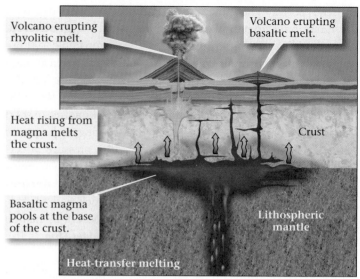

(b) Heat-transfer melting occurs when rising magma brings heat up with it and melts overlying or surrounding rock. (Not to scale.)

be sufficient for the crustal rock to begin melting. To picture the process, imagine injecting hot fudge into ice cream; the fudge transfers heat to the ice cream, raises its temperature, and causes it to melt. We call such melting **heat-transfer melting**, because it results from the transfer of heat from a hotter material to a cooler one. Heat transfer happens when *very* hot magma rises from the mantle, stalls at the base of the crust or in the crust, and transfers heat into the crust (Fig. 6.4b).

Take-Home Message

- The Earth is hot inside because of two factors: the heat left over from its formation and because of radioactive decay.
- But even though its hot, pressure keeps most of the mantle and crust in the solid state.
- Melting takes place due to decompression, the addition of volatiles, or because of heat transfer.

THINK: What does the geotherm tell us about the Earth's interior?

6.3 WHAT IS MAGMA MADE OF?

All magmas contain silicon and oxygen, which bond to form the silicon-oxygen tetrahedron. But magmas also contain varying proportions of other elements such as aluminum (Al), calcium (Ca), sodium (Na), potassium (K), iron (Fe), and magnesium (Mg); each of these ions also bonds to oxygen. Because magma is a liquid, its molecules do not lie in an orderly crystal-line lattice but are grouped instead in clusters or short chains, relatively free to move with respect to one another.

Geologists distinguish between "dry" magmas, which contain no volatiles, and "wet" magmas, which do. In fact, wet magmas include up to 15% dissolved volatiles such as water, carbon dioxide, nitrogen (N_2), hydrogen (H_2), and sulfur dioxide (SO_2). These volatiles come out of the Earth at volca-noes in the form of gas. Usually water constitutes about half of the gas erupting at a volcano. Thus, magma contains not only the molecules that constitute solid minerals in rocks but also the molecules that become water or air.

The Major Types of Magma

Imagine four pots of molten chocolate simmering on a stove. Each pot contains a different type of chocolate. One pot con-tains white chocolate, one milk chocolate, one semisweet chocolate, and one baker's chocolate. It's no surprise that dif-ferent kinds of molten chocolate yield different kinds of solid chocolate; each type differs from the others in taste and color.

Like molten chocolate, not all molten rock is the same. Specifically, magmas differ from one another in terms of the chemicals they contain. Geologists distinguish four major types of magma by measuring the percentage of silica (SiO_2) within the magma (Table 6.1). As we will see later in this chapter and we've already seen in Chapter 5, these varieties of magma differ in their physical characteristics, such as how easily they flow and freeze to form different kinds of rocks. The names of the magma types reflect the elements within the magma. For example, **mafic magma** gets its name because it contains a relatively small amount of silica and a relatively high proportion of iron oxide (FeO, or Fe_2O_3) and magnesium oxide (MgO)—"ma" stands for magnesium and "–fic" stands for iron, from the Latin *ferric*. **Ultramafic**

TABLE 6.1 **The Four Categories of Magma**

Felsic (or silicic) magma	66–76% silica*
Intermediate magma	52–66% silica
Mafic magma	45–52% silica
Ultramafic magma	38–45% silica

*The numbers provided are "weight percent," meaning the proportion of the magma's weight that consists of silica.

magma, as the name suggests, contains even less silica than does mafic magma. **Felsic magma** contains a high proportion of the elements in feldspar and silica. (At an elementary level, we can use the terms *felsic* and *silicic* interchangeably.) **Intermediate magma** gets its name simply because this type has a composition between felsic and mafic.

Why are there so many kinds of magma? Several factors control magma composition, including those described below.

Source rock composition. When you melt ice, you get water, and when you melt wax, you get liquid wax. There is no way to make water by melting wax. Clearly, the composition of a melt reflects the composition of the solid from which it was derived. Not all magmas form from the same source rock, so not all magmas have the same composition; magmas formed from crustal sources don't have the same composition as magmas formed from mantle sources, because the crust and mantle have different compositions to start with (see Chapter 2).

Partial melting. The melting of rock differs markedly from the melting of pure water ice, in that ice contains only one kind of mineral (crystals of H_2O), whereas most rocks contain a variety of minerals. Because not all minerals melt by the same amount under given conditions, and because chemical reactions take place during melting, the magma that forms as a rock begins to melt does *not* have the same composition as the original rock from which it formed. Typically, the magma contains more silica than did the original rock. Because magma can flow, it moves out of the original rock long before the rock completely melts. Therefore, the melt carries away silica. The process during which only part of a rock melts to form a magma that can then move away is called **partial melting**. Note that the solid rock left behind as partial melting occurs will be more mafic than the original rock.

Geologists find that, in general, only about 2% to 30% of a rock melts to produce a magma. The melt forms films on grain surfaces, and collects in little pockets between grains, until it eventually migrates away (Fig. 6.5a). "Crystal mush" forms where there is enough melt to keep the remaining crystals suspended (not in contact with each other) in the melt. Because partial melting removes silica, magma formed during later stages of melting contains less silica than magma formed during earlier stages. Because melts formed by partial melting tend to be richer in silica than the rock they were derived from, partial melting of ultramafic rock can produce mafic magma. Similarly, partial melting of intermediate rock can produce a felsic magma.

Assimilation. As magma sits in a magma chamber before completely solidifying, it may incorporate chemicals derived from the wall rocks of the chamber. This process of **assimilation** (Fig. 6.5b) takes place when rocks fall into the magma and then partially melt, when heat from the magma partially melts the walls of the chamber, or when the magma chemically reacts with the wall rock. In some cases, selected elements migrate out

FIGURE 6.5 Phenomena that can affect the composition of magma.

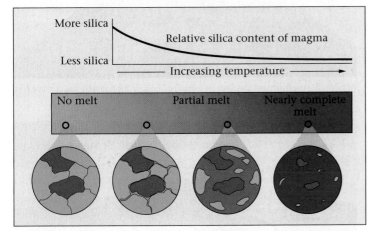

(a) Partial melting: The first-formed melt will be richer in silica than the original rock. As melting continues, magma becomes more mafic.

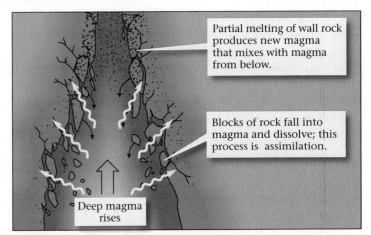

(b) Mixing and assimilation: Heat provided by deep magma partially melts wall rock; the new magma may then mix with deep magma. Also, blocks of wall rock can dissolve (assimilate) in the deep magma, and the wall rock may chemically react with the magma.

of the wall and into the magma without the wall melting. This process may be accelerated if hot water circulates through the wall rock, for the water may dissolve elements and carry them into the magma. Geologists do not agree about how much assimilation takes place during magma formation.

Magma mixing. Different magmas formed in different locations from different sources may come in contact within a magma chamber prior to freezing. When this happens, the originally distinct magmas may mix to create a new, different magma. For example, thoroughly mixing a felsic magma with a mafic magma in equal proportions produces an intermediate magma. Sometimes different magmas that come in contact do not mix together completely; when this occurs, blobs of rock formed from one magma remain suspended within a mass formed from the other magma after solidification occurs.

Take-Home Message

- Most magma consists of silicon, oxygen, and various cations.
- We represent magma composition fundamentally by the proportion of silica to the sum of iron and magnesium oxide.
- The major compositional groups, in order from high silica to low silica, are: felsic, intermediate, mafic, and ultramafic.
- Magma composition reflects the original source, the degree of partial melting, the amount of assimilation, and/or the amount of magma mixing.

THINK: Why does partial melting of a rock produce magma that is more felsic than the source rock?

6.4 MOVEMENT OF MOLTEN ROCK

If magma stayed put once it formed, new igneous rocks would not develop in or on the crust. But it doesn't stay put; magma tends to move upward, away from where it formed. In some cases, it reaches the Earth's surface and erupts at a volcano. This movement is a key component of the Earth System, because it transfers material from deeper parts of the Earth upward and provides the raw material from which new rocks and the atmosphere and ocean form.

Why Does Magma Rise?

Magma rises for two reasons. First, magma rises because it is less dense than surrounding rock; buoyancy drives magma upward just as it drives a wooden block up through water. Second, magma rises because the weight of overlying rock creates pressure at depth that literally squeezes magma upward. The same process happens when you step into a puddle barefoot and mud squeezes up between your toes.

What Controls the Speed of Flow?

Viscosity, or resistance to flow, affects the speed with which magmas or lavas move. Magmas with low viscosity flow more easily than those with high viscosity, just as water flows more easily than molasses. Viscosity depends on temperature, volatile content, and silica content. Hotter magma is less viscous than cooler magma, just as hot tar is less viscous than cool tar, because thermal energy breaks bonds and allows atoms to move more easily. Similarly, magmas or lavas containing more volatiles are less viscous than dry (volatile-free) magmas, because the volatiles also tend to break apart silicate molecules and may collect into bubbles. Mafic magmas are less viscous than felsic magmas, because silicon-oxygen tetrahedra tend to link together in the magma to create long chains that can't move past each other easily, and there are more chains in a felsic magma. Thus, hotter mafic lavas have relatively low viscosity and flow in thin sheets over wide regions, but cooler felsic lavas are highly viscous and clump at the volcanic vent (**Fig. 6.6a, b**).

Take-Home Message

- Magma rises from its source to shallower depths in the crust, or on out to the surface.
- The rise occurs due to buoyancy, and because of pressure due to the weight of overlying rock.
- The velocity of magma or lava flow depends on its viscosity, which in turn is influenced by temperature and composition.

THINK: Why is a felsic lava more viscous than a mafic lava?

FIGURE 6.6 Viscosity affects lava behavior.

(a) Felsic to intermediate lava is very viscous. When it erupts, it may form a mound-like lava dome around the volcano's vent.

(b) Mafic lava has relatively low viscosity. It can erupt in fountains, move long distances, and form thin lava flows.

6.5 TRANSFORMING MAGMA AND LAVA INTO ROCK

What makes magma freeze? In some cases, magma freezes when volatiles bubble away, for removal of volatiles makes freezing temperature higher. In most cases, magma freezes simply when it cools below its freezing temperature and crystals start to grow. Because temperatures decrease toward the Earth's surface, cooling automatically happens when magma rises. The cooler environment may be cool **wall rock** (meaning the rock surrounding the intrusion) if the magma intrudes underground, or it may be the atmosphere or the ocean if the magma extrudes as lava at the Earth's surface. Let's now consider factors controlling the *rate of cooling*, and consider how magma changes as it cools.

FIGURE 6.7 Factors that affect the freezing of molten rock.

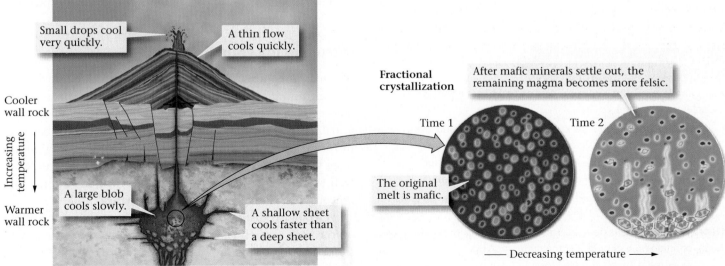

(a) The larger the ratio of its surface area to volume, the faster an intrusion cools. This ratio depends on both size and shape.

(b) The cooling rate of molten rock depends on the size and shape of the magma or lava body, and on its depth.

How Fast Does Magma Cool?

The time it takes for a magma to cool depends on how fast it is able to transfer heat into its surroundings. To see why, think about the process of cooling coffee. If you pour hot coffee into a thermos bottle and seal it, the coffee stays hot for hours; because it's insulated, the coffee in the thermos loses heat to the air outside only very slowly. Like the thermos bottle, surrounding rock acts as an insulator in that it transports heat away from a magma only very slowly, so magma underground (in an intrusive environment) cools slowly. In contrast, if you spill coffee on a table, it cools quickly because it loses heat to the cold air. Similarly, lava that erupts at the ground surface cools quickly because the air or water surrounding it can conduct heat away quickly.

Three factors control the cooling time of magma that freezes below the surface in the intrusive realm.

- *The depth of intrusion*: Magma intruded deep in the crust, where it is surrounded by warm wall rock, cools more slowly than does magma intruded into cold wall rock near the ground surface.
- *The shape and size of a magma body*: Heat escapes from magma at an intrusion's surface, so the greater the surface area for a given volume of intrusion, the faster it cools. Thus, a body of magma roughly with the shape of a pancake cools faster than one with the shape of a melon. And since the ratio of surface area to volume increases as size decreases, a body of magma the size of a car cools faster than one the size of a ship (Fig. 6.7a, b).
- *The presence of circulating groundwater*: Water passing through magma absorbs and carries away heat, much like the coolant that flows around an automobile engine.

(c) The process of fractional crystallization results in a progressive change in magma composition during freezing.

The same factors also control cooling of lava. Specifically, a thin flow cools faster than a thick one, and lava immersed in water cools faster than lava immersed in air. Tiny droplets of lava freeze extremely fast.

Changes in Magma During Cooling: Fractional Crystallization

Most people are familiar with the process of forming ice out of liquid water—cool the water to a temperature of 0°C and crystals of ice start to form. Keep the temperature cold enough for long enough and all the water becomes solid, composed entirely of one type of mineral—water ice. The process of freezing magma or lava is much more complex, because molten rock contains many different compounds, not just water, so *during freezing of molten rock, many different minerals form*. Further, not all of these minerals form at the same time. In the early twentieth century, an American researcher, N. L. Bowen, worked out the sequence in which minerals form, as described in Box 6.1. To get a sense of this complexity, let's look at an example.

When a mafic magma starts to freeze, mafic (iron- and magnesium-rich) minerals such as olivine and pyroxene start to crystallize first. These solid crystals are denser than the remaining magma, so they start to sink (Fig. 6.7c). Some react chemically with the remaining magma as they sink, but some reach the floor of the magma chamber and become isolated from the magma. This process of sequential crystal formation and settling is called **fractional crystallization**—it progressively extracts iron and magnesium from the magma, so the remaining magma becomes more felsic. If a magma freezes completely before much fractional crystallization has occurred, the magma becomes mafic igneous rock. But freezing of a magma that has been left over after lots of fractional crystallization has occurred produces felsic igneous rock.

Take-Home Message

- Magma freezes when it loses volatiles or when it enters a cooler environment.
- Intrusive rocks freeze as they rise into cooler crust, and lava freezes in flows or in ash clouds.
- The rate of cooling is influenced by the temperature of the surroundings, and/or by the shape of the molten rock body.
- During fractional crystallization, magma composition evolves as crystals settle out.

THINK: Which cools faster—a large blob of magma intruded at depth, or a thin flow extruded at the surface? Why?

6.6 HOW DO EXTRUSIVE AND INTRUSIVE ENVIRONMENTS DIFFER?

With a background on how melts form and freeze, we can now introduce key features of the two settings—intrusive and extrusive—in which igneous rocks form.

Extrusive Igneous Settings

Not all volcanic eruptions are the same, so not all extrusive rocks are the same. Some volcanoes erupt streams of low-viscosity lava that flood down the flanks of the volcano and then cover swaths of the countryside. When this lava freezes, it forms a relatively thin lava flow. Such flows may cool in days to months. In contrast, some volcanoes erupt viscous masses of lava that pile into domes, and still others erupt explosively. During some types of explosions, volcanic ash and debris billow several kilometers into the sky as an eruption plume (Fig. 6.8a). Larger framents fall like hail on the volcano. Finer ash may be carried by the wind for great distances, but eventually, this material falls from the sky like snow. Ash and debris may also avalanche down the sides of the volcano forming a very fast-moving density current. (A density current is a layer of denser fluid flowing down a slope beneath a layer of less-dense fluid—in this case the denser fluid is a mixture of ash and air, whereas the less-dense fluid is clear air.) When the density current slows, it deposits ash and debris on the ground surface. Such density currents are also known as ash flows or **pyroclastic flows** (from the Greek *pyro*, meaning fire, and the Greek *klastos*, meaning broken).

Which type of eruption occurs depends largely on a magma's composition and volatile content. As noted earlier, volatile-rich felsic lavas tend to erupt explosively and form thick ash and debris deposits (Fig. 6.8b, c). Mafic lavas tend to have low viscosity and spread in broad, thin flows (Fig. 6.8d, e). We discuss the products of extrusive settings in more detail in Chapter 9.

Intrusive Igneous Settings

Magma rises and intrudes into preexisting rock by slowly percolating upward between grains and/or by forcing open cracks. The magma that doesn't make it to the surface freezes solid underground in contact with preexisting rock and becomes intrusive igneous rock. As we noted, geologists commonly refer to the preexisting rock into which magma intrudes as wall rock. The boundary between wall rock and an intrusive igneous rock is called an **intrusive contact**.

Geologists distinguish among different types of intrusions on the basis of their shape. Tabular intrusions, or sheet intrusions, are planar and are of roughly uniform thickness. Most are in the range of centimeters to tens of meters thick, and

FIGURE 6.8 Examples of eruptions and extrusive volcanic materials.

Ash cloud

Pyroclastic flow

(a) This volcanic explosion produced two styles of ash eruption.

(b) Thick layers of tuff deposited by explosive eruptions in New Mexico, about 1.14 Ma. Note the highway, for scale.

Some of the lava freezes in mid-air.

Lava flow

(d) This volcano is producing lava flows and fountains.

A golf-ball-sized fragment of pumice

(c) A close-up of tuff, showing ash and fragments of pumice.

A group of people

(e) A stack of over 50 lava flows, capped by debris, visible from inside Mt. Vesuvius, Italy.

tens of meters to tens of kilometers long. A **dike** is a tabular intrusion that cuts across preexisting layering (bedding or foliation), whereas a **sill** is a tabular intrusion that injects parallel to layering (Fig. 6.9a–d ⏵). In places where tabular intrusions cut across rock that does not have layering, a nearly vertical, wall-like tabular intrusion is a dike, whereas a nearly horizontal, tabletop-shaped tabular intrusion is a sill. Some intrusions start to inject between layers but then dome upward, creating a blister-shaped intrusion known as a **laccolith**.

Plutons are irregular or blob-shaped intrusions that range in size from tens of meters across to tens of kilometers across (Fig. 6.10a–d). The intrusion of numerous plutons in a region

FIGURE 6.9   Igneous sills and dikes, examples of tabular intrusions. ▶❚❚

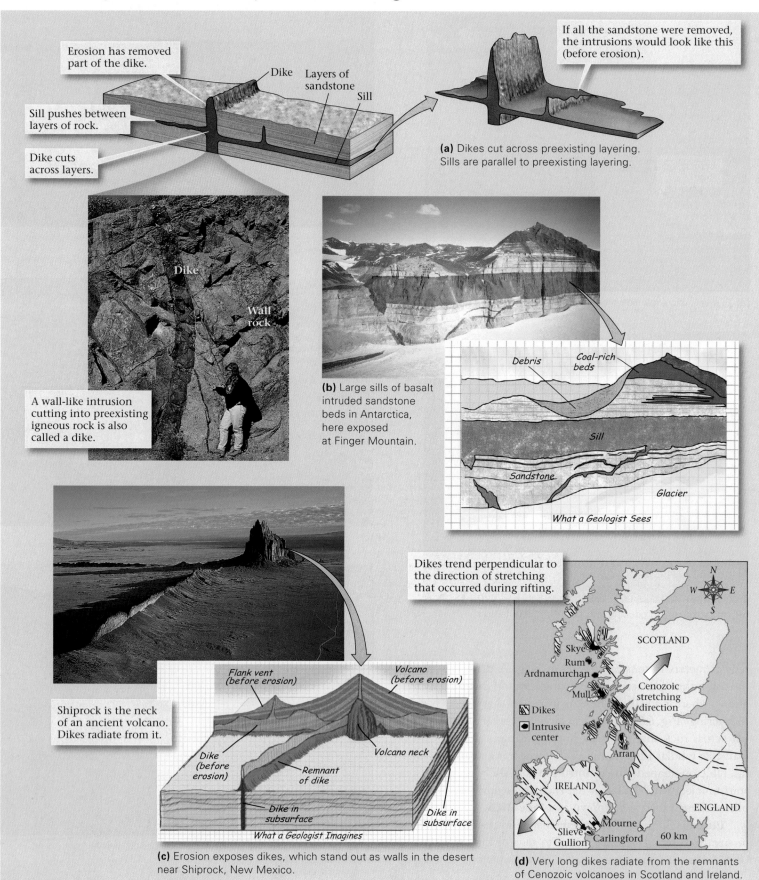

Erosion has removed part of the dike.

Sill pushes between layers of rock.

Dike cuts across layers.

Dike

Layers of sandstone

Sill

If all the sandstone were removed, the intrusions would look like this (before erosion).

(a) Dikes cut across preexisting layering. Sills are parallel to preexisting layering.

Dike

Wall rock

A wall-like intrusion cutting into preexisting igneous rock is also called a dike.

(b) Large sills of basalt intruded sandstone beds in Antarctica, here exposed at Finger Mountain.

Debris

Coal-rich beds

Sill

Sandstone

Glacier

What a Geologist Sees

Dikes trend perpendicular to the direction of stretching that occurred during rifting.

Shiprock is the neck of an ancient volcano. Dikes radiate from it.

Flank vent (before erosion)

Volcano (before erosion)

Dike (before erosion)

Remnant of dike

Volcano neck

Dike in subsurface

Dike in subsurface

What a Geologist Imagines

(c) Erosion exposes dikes, which stand out as walls in the desert near Shiprock, New Mexico.

SCOTLAND

Skye
Rum
Ardnamurchan
Mull

Cenozoic stretching direction

Arran

⊠ Dikes
◨ Intrusive center

IRELAND

ENGLAND

Slieve Gullion
Mourne
Carlingford

60 km

(d) Very long dikes radiate from the remnants of Cenozoic volcanoes in Scotland and Ireland.

FIGURE 6.10 Igneous plutons, "blob-shaped" intrusions.

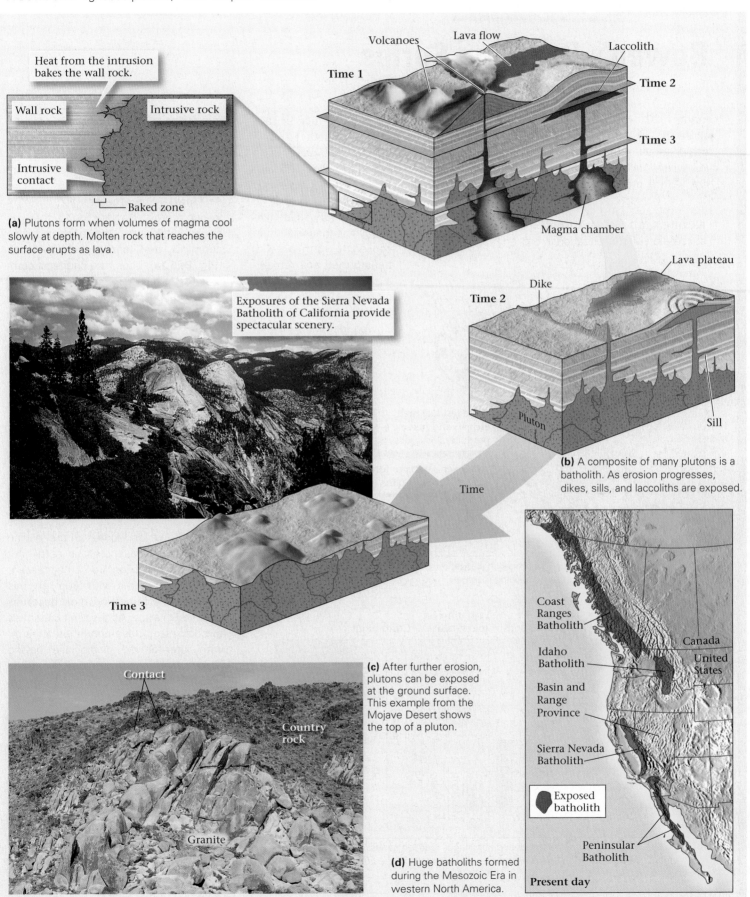

Heat from the intrusion bakes the wall rock.

Wall rock

Intrusive rock

Intrusive contact

Baked zone

(a) Plutons form when volumes of magma cool slowly at depth. Molten rock that reaches the surface erupts as lava.

Volcanoes

Lava flow

Laccolith

Time 1

Time 2

Time 3

Magma chamber

Exposures of the Sierra Nevada Batholith of California provide spectacular scenery.

Time 2

Dike

Lava plateau

Pluton

Sill

(b) A composite of many plutons is a batholith. As erosion progresses, dikes, sills, and laccoliths are exposed.

Time

Time 3

(c) After further erosion, plutons can be exposed at the ground surface. This example from the Mojave Desert shows the top of a pluton.

Contact

Country rock

Granite

Coast Ranges Batholith

Canada

Idaho Batholith

United States

Basin and Range Province

Sierra Nevada Batholith

Exposed batholith

Peninsular Batholith

Present day

(d) Huge batholiths formed during the Mesozoic Era in western North America.

BOX 6.1

Bowen's Reaction Series

In the 1920s, Norman L. Bowen began a series of laboratory experiments designed to determine the sequence in which silicate minerals crystallize from a melt. First, Bowen melted powdered mafic igneous rock by raising its temperature to about 1,280°C. Then he cooled the melt just enough to cause part of it to solidify. Finally, he "quenched" the remaining melt by submerging it quickly in cold mercury. Quenching, which means sudden cooling to form a solid, transformed any remaining liquid into glass. The glass trapped the earlier-formed crystals within it. Bowen identified mineral crystals formed before quenching with a microscope, and he analyzed the chemical composition of the remaining glass. The glass composition represented the composition of the magma.

After experiments at different temperatures, Bowen found that, as new crystals form, they extract certain chemicals preferentially from the melt (**Fig. Bx6.1a**). Thus, the chemical composition of the remaining melt progressively changes as the melt cools. Bowen described the specific sequence of mineral-producing reactions that take place in a cooling, initially mafic, magma. This sequence is now called **Bowen's reaction series** in his honor.

Let's examine the sequence more closely. In a cooling melt, olivine and calcium-rich plagioclase form first. The Ca-plagioclase reacts with the melt to form more plagioclase, which contains more sodium (Na). Meanwhile, some olivine crystals react with the remaining melt to produce pyroxene, which may encase olivine crystals or even replace them. However, some of the olivine and Ca-plagioclase crystals settle out of the melt, taking iron, magnesium, and calcium atoms with them. By this process, the remaining melt becomes enriched in silica. As the melt continues to cool, plagioclase continues to form, with later-formed plagioclase having progressively more sodium (Na) than earlier-formed plagioclase. Pyroxene crystals react with melt to form amphibole, and then amphibole reacts with the remaining melt to form biotite. All the while, crystals continue to settle out, so the remaining melt continues to become more felsic. At temperatures of between 650°C and 850°C, only about 10% melt remains, and this melt has a high silica content. At this stage, the final melt freezes, yielding quartz, K-feldspar, and muscovite.

FIGURE Bx6.1 Bowen's reaction series indicates the succession of crystallization in cooling magma.

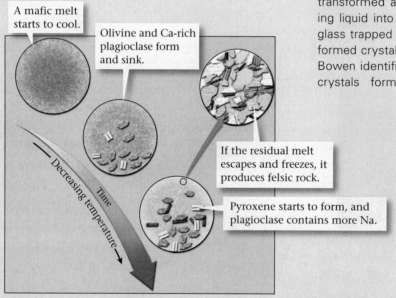

(a) With decreasing temperature, fractional crystallization begins and the composition of the remaining magma becomes more felsic.

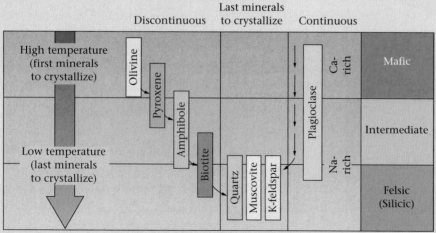

(b) This chart displays the discontinuous and continuous reaction series. Rocks formed from minerals at the top of the series are mafic, whereas rocks formed from the bottom of the series are felsic.

On the basis of his observations, Bowen realized that there are two tracks to the reaction series. The "discontinuous reaction series" refers to the sequence olivine, pyroxene, amphibole, biotite, K-feldspar/muscovite/quartz: each step yields a different class of silicate mineral. The "continuous reaction series" refers to the progressive change from calcium-rich to sodium-rich plagioclase: the steps yield different versions of the same mineral (**Fig. Bx6.1b**). It's important to note that not all minerals listed in the series appear in all igneous rock. For example, a mafic magma may completely crystallize before felsic minerals such as quartz or K-feldspar have a chance to form.

Also note that the succession of minerals in the discontinuous series is not random—it begins with minerals having isolated tetrahedra (olivine), progresses to those having single chains of tetrahedra (pyroxene), then double chains (amphibole), and finally sheets (mica) or 3-D networks (quartz). Put another way, the minerals that crystallize later in the discontinuous series have more Si-O-Si bonds (i.e., are more "polymerized") than those that crystallize earlier; minerals at the end of the series have smaller O/Si ratios than those at the beginning. This happens because, as crystallization progresses, the oxygen atoms in the melt are used up more quickly than the silicon atoms in the melt.

creates a vast composite body that may be several hundred kilometers long and over 100 km wide; such immense masses of igneous rock are called **batholiths**. The rock making up the Sierra Nevada of California is a batholith formed from plutons that intruded between 145 and 80 million years ago.

Where does the space for intrusions come from? Dikes form in regions where the crust is being stretched horizontally, such as in a rift. Thus, as the magma that makes a dike forces its way up into a crack, the crust opens up sideways (Fig. 6.11a). Intrusion of sills occurs near the surface of the Earth, so the pressure of the magma effectively pushes up the rock above the sill, leading to uplift of the Earth's surface (Fig. 6.11b).

How does the space for a pluton develop? Some geologists propose that a pluton is a frozen "diapir," meaning a light-bulb-shaped blob of magma that pierced overlying rock and pushed it aside as it rose (Fig. 6.11c). Alternatively, plutons form by injection of several superimposed dikes or sills, which coalesce and recrystallize to become a single, massive body (Fig. 6.11d). Another explanation involves **stoping**, a process during which magma assimilates wall rock, and blocks of wall rock break off and sink into the magma (Fig. 6.11e). If a stoped block does not melt entirely, but rather becomes surrounded by new igneous rock, it is a **xenolith**, after the Greek word *xeno*, meaning foreign (Fig. 6.11f).

If intrusive igneous rocks form beneath the Earth's surface, why can we see them exposed today? Over long periods of geologic time, mountain building slowly uplifts belts of crust. Moving water, wind, and ice eventually strip away great thicknesses of overlying rock and expose the intrusive rock that has formed below.

> **Take-Home Message**
>
> - Molten rock can extrude either as a lava flow, or as pyroclastic debris.
> - Intrusion underground juxtaposes magma with wall rock along an intrusive contact.
> - Tabular intrusions include dikes, that cut across layers, and sills, that inject between layers.
> - Blob-shaped intrusions are plutons; composites of many plutons produce a batholith.
>
> **THINK:** Intrusions take up space—where does the room come from?

6.7 HOW DO YOU DESCRIBE AN IGNEOUS ROCK?

Wander around the downtown of a big city and you'll find that the facades of many modern office buildings consist of slices of polished rock. In recent years, similar polished rock has also become a desirable material for kitchen countertops. Architects refer to this rock in a generic way as "granite," and like to use it because it's both beautiful and durable. Its durability comes from the minerals it contains—most architectural "granite" contains feldspar, quartz, and garnet, which have high numbers on the Mohs hardness scale (see Chapter 5). But, if you look at the rock more closely, you'll see that it doesn't all look the same. In fact, by the end of this section you'll see that much of it is not really "granite," as a geologist would use the term. Now imagine that you wanted to describe a sample of this rock to a friend—what adjectives would you use?

> **Did you ever wonder . . .**
> why "granite" used in floors and countertops is so hard?

FIGURE 6.11 Making room for an igneous intrusion.

Dike

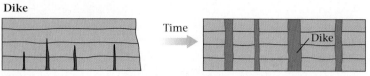

(a) Dikes fill space formed when crust undergoes horizontal stretching.

Sill

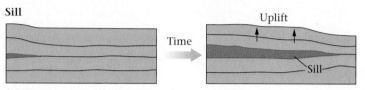

(b) Sills intrude between layers and may cause the uplift of the land surface.

Pluton

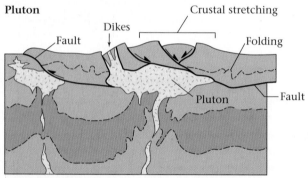

(c) The magma comprising a pluton may rise as a buoyant blob, filling space produced by crustal stretching.

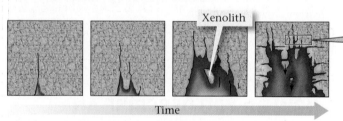

(e) Plutons may intrude by stoping. When blocks of wall rock fall into the magma, some dissolve but others may remain as xenoliths.

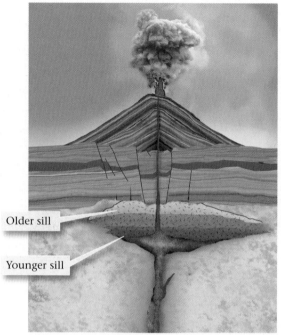

(d) Plutons may intrude as a succession of sill-like sheets.

The white rock (granite) is intruding into the dark wall rock.

(f) A xenolith surrounded by light-colored granite.

You would probably start by noting the rock's color. Overall, is the rock dark or light? More specifically, is it gray, pink, white, or black? Describing color may not be easy, because some igneous rocks contain many visible mineral grains, each with a different color; but even so, you'll probably be able to characterize the overall hue of the rock. What physical factors control the color of igneous rock? Generally, the color reflects the rock's composition, but it isn't always so simple, because color may also be influenced by grain size and by the presence of trace amounts of impurities. (For example, the presence of a small amount of iron oxide gives rock a reddish tint.)

Next, you would probably characterize the rock's texture. A description of igneous texture indicates whether the rock consists of glass, crystals, or fragments. If the rock consists of crystals or fragments, a description of texture also specifies the grain size. Here are the common terms for defining texture:

- *Interlocking texture*: Rocks that consist of mineral crystals that intergrow when the melt solidifies, and thus fit together like pieces of a jigsaw puzzle, have an interlocking texture (Fig. 6.12a). Rocks with an interlocking texture are **crystalline igneous rocks**. The interlocking of crystals in these rocks occurs because once some grains have developed, they interfere with the growth of later-formed grains. The last grains to form end up filling irregular spaces between already existing grains.

   Geologists distinguish subcategories of crystalline igneous rocks according to the size of the crystals. Coarse-grained (phaneritic) rocks have crystals large enough to be identified with the naked eye. Fine-grained (aphanitic) rocks have crystals too small to be identified with the naked eye. Porphyritic rocks have larger crystals surrounded by a mass of fine crystals. In a porphyritic rock, the larger crystals are called phenocrysts, while the mass of finer crystals is called groundmass.

- *Fragmental texture*: Rocks consisting of igneous chunks and/or shards that are packed together, welded together, or cemented together after having solidified are **fragmental igneous rocks** (Fig. 6.12a).

- *Glassy texture*: Rocks made of a solid mass of glass, or of tiny crystals surrounded by glass, are **glassy igneous rocks** (Fig. 6.12a). Glassy rocks fracture conchoidally (Fig. 6.12b).

What factors control the texture of igneous rocks? In the case of nonfragmental rocks, *texture largely reflects cooling rate*. The presence of glass indicates that cooling happened so quickly that the atoms within a lava didn't have time to arrange into crystal lattices. Crystalline rocks form when a melt cools more slowly. In crystalline rocks, grain size depends on cooling time. A melt that cools rapidly, but not rapidly enough to make glass, forms fine-grained rock, because many crystal seeds form but none has time to grow large (Fig. 6.12c). A melt that cools very slowly forms a coarse-grained rock, because relatively few seeds form and each crystal has time to grow large.

Because of the relationship between cooling rate and texture, lava flows, dikes, and sills tend to be composed of fine-grained igneous rock. In contrast, plutons tend to be composed of coarse-grained rock. Plutons that intrude into hot wall rock at great depth cool very slowly and thus tend to have larger crystals than plutons that intrude into cool country rock at shallow depth, where they cool relatively rapidly. Porphyritic rocks form when a melt cools in two stages. First, the melt cools slowly at depth, so that phenocrysts form. Then, the melt erupts and the remainder cools quickly, so that groundmass forms around the phenocrysts.

There is, however, an exception to the standard cooling rate and grain size relationship. A very coarse-grained igneous rock called **pegmatite** doesn't necessarily cool slowly. Pegmatite contains crystals up to tens of centimeters across and occurs in dikes. Because pegmatite occurs in dikes, which generally cool quickly, the coarseness of the rock may seem surprising. Researchers have shown that pegmatites are coarse because they form from water-rich melts in which atoms can move around so rapidly that large crystals can grow very quickly.

Take-Home Message

- In a crystalline igneous rock, there are interlocking crystals, whereas in a glassy rock there are no crystals.
- Fragmental igneous rocks consist of chunks and shards held together.

THINK: Why is the grain size of a pegmatite puzzling?

6.8 CLASSIFYING IGNEOUS ROCKS

Because melts can have a variety of compositions and can freeze to form igneous rocks in many different environments above and below the surface of the Earth, we observe a wide spectrum of igneous rock types. We classify these according to their texture and composition. Studying a rock's texture tells us about the rate at which it cooled, as we've seen, and therefore the environment in which it formed (see Geology at a Glance, p. 159). Studying its composition tells us about the original source of the magma and the way in which the magma evolved before finally solidifying. Below, we introduce some of the most important igneous rock types.

Crystalline Igneous Rocks

The scheme for classifying the principal types of crystalline igneous rocks is quite simple. The different compositional classes are distinguished on the basis of silica content—ultramafic, mafic, intermediate, or felsic—whereas the different textural classes are distinguished according to whether the grains are coarse or fine. Figure 6.13a gives the texture and composition of the most commonly used crystalline rock names. As a rough guide, the color of an igneous rock reflects its composition: mafic rocks tend to be black or dark gray, intermediate rocks tend to be lighter gray or greenish gray, and felsic rocks tend to be light tan to pink or maroon. Figure 6.12 provides images of some of these rocks.

FIGURE 6.12 Textures and types of igneous rocks.

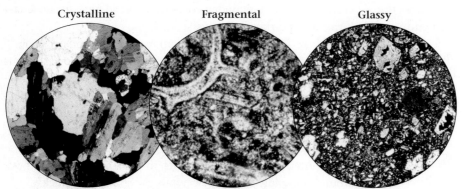

Crystalline   Fragmental   Glassy

(a) Photomicrographs of thin sections reveal the different textures of igneous rocks.

Obsidian

(b) Obsidian fractures conchoidally.

Fine grained   Coarse grained

Felsic

Increasing silica content

Mafic

Rhyolite

Granite, cut by pegmatite

Andesite

Diorite

Basalt

Gabbro

(c) Examples of igneous rocks, arranged by grain size and composition. Different environments yield different rock types.

FIGURE 6.13 Igneous rocks are classified based on composition and texture.

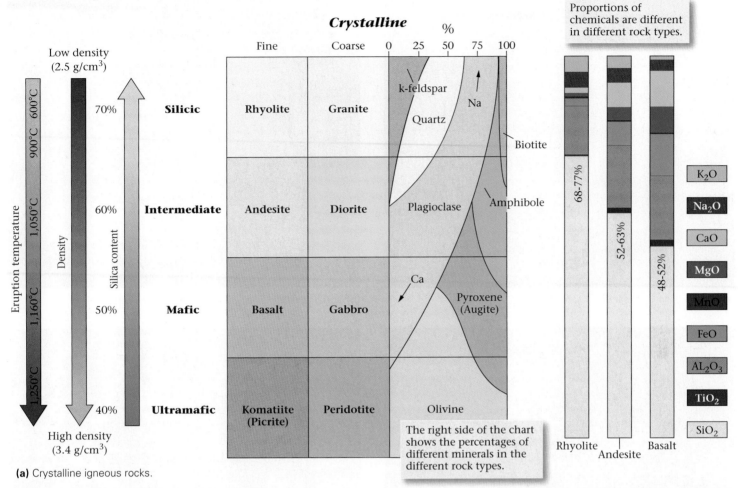

(a) Crystalline igneous rocks.

The right side of the chart shows the percentages of different minerals in the different rock types.

Proportions of chemicals are different in different rock types.

	Nonfragmental	
Nonvesicular		
Obsidian	felsic solid glass	
Tachylite	mafic solid glass	
Vesicular		
Scoria	mafic glassy rock with abundant vesicles	
Pumice	felsic glassy rock with abundant vesicles	

Fragmental	
Tuff	ash ± tiny pumice fragments
Volcanic breccia	angular fragments of lava broken by flow
Hyaloclastite	angular fragments formed by shattering under water or ice

(b) Non-crystalline igneous rocks.

Note that, according to Figure 6.13, rhyolite and granite have the same chemical composition but differ in grain size. Which of these two rocks develops from a melt of felsic composition depends on the cooling rate. A felsic lava that solidifies quickly at the Earth's surface or in a thin dike or sill turns into fine-grained rhyolite; but the same magma, if solidifying slowly at depth in a pluton, turns into coarse-grained granite. A similar situation holds for mafic lavas—a mafic lava that cools quickly in a lava flow forms basalt, but a mafic magma that cools slowly forms gabbro.

Glassy Igneous Rocks

Glassy texture develops more commonly in felsic igneous rocks because the high concentration of silica inhibits the easy growth of crystals. But basaltic and intermediate lavas can form glass if they cool rapidly enough. In some cases, a rapidly cooling lava freezes while it still contains a high concentration of gas bubbles—these bubbles remain as open holes known as **vesicles**. Geologists distinguish among several different kinds of glassy rocks (Fig. 6.13b).

FIGURE 6.14 Vesicle-filled volcanic rocks.

(a) Pumice is so light that paper can hold it up.

(b) Scoria looks like a dark sponge, though very hard.

(c) Volcanic tuff composed of glass shards (too small to see) and small lapilli.

(d) Volcanic breccia, composed of angular basalt chunks.

- **Obsidian** is a mass of solid, felsic glass. It tends to be black or brown (Fig. 6.12b). Because it breaks conchoidally, sharp-edged pieces split off its surface when you hit a sample with a hammer. Pre-industrial people worldwide used such pieces for arrowheads, scrapers, and knife blades.

> **Did you ever wonder . . .**
> how black glass once used
> for arrowheads formed?

- **Tachylite** is a bubble-free mass consisting of more than 80% mafic glass. This rock is relatively rare, in comparison with obsidian.

- **Pumice** is a felsic volcanic rock that contains abundant vesicles, giving it the appearance of a sponge. Pumice forms by the quick cooling of frothy lava that resembles the head of foam in a glass of beer. In some cases, pumice contains so many air-filled pores that it can actually float on water, like styrofoam (Fig. 6.14a). Ground-up pumice makes the grainy abrasive that blue-jean manufacturers use to "stonewash" jeans. Pumice tends to be light gray to tan in color.

Formation of Igneous Rocks

Igneous rock forms by the cooling of magma underground, or of lava at the surface. Igneous rocks that solidify underground are intrusive, whereas those that solidify at the surface are extrusive. The type of igneous rock that forms depends on the composition of the melt and the environment of cooling.

In the extrusive environment, melt may cool quickly and have a glassy texture. Melt that explodes into the air forms ash and other debris with fragmental texture.

Stratified volcanic tuff

Increasing silica content

Fast cooling

Slow cooling

MAFIC	FELSIC
Scoria (glassy)	Obsidian (glassy)
Basalt (fine grained)	Rhyolite (fine grained)
Gabbro (coarse grained)	Granite (coarse grained)

In the intrusive environment, crystals grow together to form an interlocking texture. Slower cooling makes coarser grains.

Cooler

Minerals in an igneous rock crystallize in succession as the melt cools.

Hotter

EXTRUSIVE ENVIRONMENT

Pyroclastic flow

Lava flow

Dike swarm

Sills

Lava dome

Laccolith

Ring dikes

Volcanic neck

Irregular stock

INTRUSIVE ENVIRONMENT

Pluton

Magma chamber

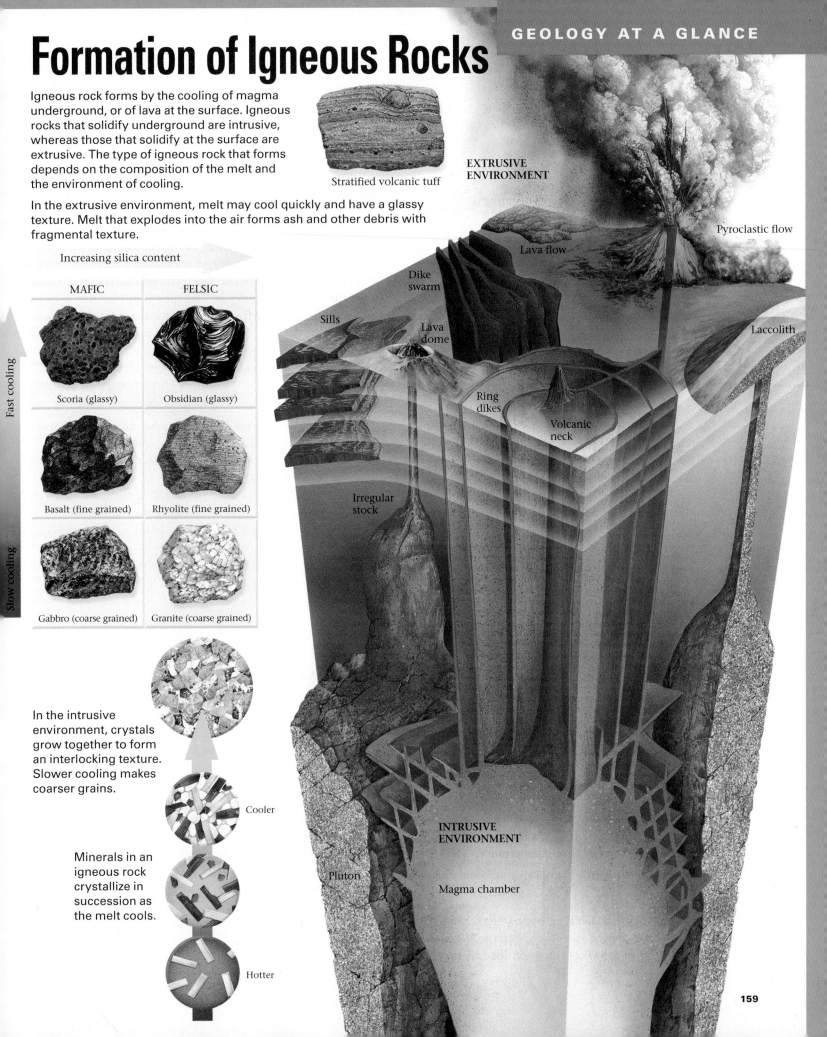

- **Scoria** is a mafic volcanic rock that contains abundant vesicles (more than about 30%). Generally, the bubbles in scoria are bigger than those in pumice, and the rock, overall, looks darker (Fig. 6.14b).

Pyroclastic Igneous Rocks

As we have noted, when volcanoes erupt explosively, they spew out fragments of lava that had started to solidify inside the volcano, and in some cases droplets and clots of still-molten lava that freezes solid in the air as it flies. Powerful explosions may also blast apart preexisting volcanic rocks from the walls of the volcano and send them flying too. Geologists refer to all such fragments as pyroclasts. Pyroclasts come in a great range of sizes—dust-sized specks or flakes of glass or pulverized rock make up ash; pea- to marble-sized pellets are called lapilli; and still larger chunks form bombs (if streamlined) or blocks if angular. (Chapter 9 provides further details.)

Accumulations of fragmental volcanic debris are called pyroclastic deposits, and when the material in these deposits consolidates into a solid mass, due either to welding together of still-hot clasts or to cementation by minerals precipitating from water passing through, it becomes a **pyroclastic rock**. Geologists distinguish among several types of pyroclastic rocks based on grain size. We introduce just a few below.

- **Tuff** is a fine-grained pyroclastic rock composed of volcanic ash, or of ash mixed with pumice fragments (Fig. 6.14c). Tuff can form from debris that settled out of the air like snow, accumulating in beds that blanket the landscape. Or it may form from debris that rushed down the volcano's side in an avalanche-like pyroclastic flow. Of note, ash in the interior of a pyroclastic flow may be so hot that it welds together, when the flow stops moving, to produce "welded tuff."

- **Volcanic breccia** consists of pea-sized or larger angular fragments of volcanic debris that have been cemented together (Fig. 6.14d). A special type of breccia, called **hyaloclastite**, forms when lava erupts under water or ice and violently shatters into glassy fragments that react chemically with the surrounding water. But not all volcanic breccias form from debris falling from the sky or tumbling down a volcano. Some forms from the surface crust of a lava flow, if the crust breaks up as the interior continues to move.

- **Volcanic agglomerate** consists of accumulations of lapilli or bombs.

When we investigate volcanic eruptions further in Chapter 9, we will see that pyroclastic debris, because it generally does not necessarily become consolidated (bound together)

when it initially accumulates, may move down the slope of a volcano before eventually getting buried and turned to rock. In some cases, this movement occurs in landslides; but commonly, the movement takes place when water from rain or from melting ice or snow saturates and weakens the debris. The debris then becomes a muddy slurry that flows downslope like wet concrete. In some cases, the slurry, containing a mixture of fragments of many sizes suspended in a muddy matrix, solidifies to rock. But sometimes running water carries the debris away from the volcano, and sorts it to produce layers of sand or gravel composed entirely of volcanic material. Geologists use the term **volcaniclastic rock** in a general sense to refer to any rock that contains a large proportion of volcanic fragments; this category includes both pyroclastic rocks and rocks formed from water-transported volcanic debris.

Take-Home Message

- Crystalline igneous rocks are distinguished based on grain size (coarse vs. fine) and composition (e.g., mafic, felsic, etc.).
- The composition of an igneous rock reflects the assemblage of minerals comprising the rock.
- Each rock type has a distinct name (e.g., basalt, granite, etc.).
- Separate names are used for glassy and fragmental rocks. Some glassy rocks are solid masses, whereas others are full of holes, relicts of gas bubbles.
- Pyroclastic rocks are formed from ash particles and other debris.

THINK: What kind or rock will form from felsic magma cooled in a large pluton at depth?

6.9 WHERE DOES IGNEOUS ACTIVITY OCCUR, AND WHY?

If you look at a map showing the distribution of igneous activity—the formation, movement, and in some cases eruption of molten rock—around the world (Fig. 6.15 ◉), you'll see that igneous activity occurs in four settings: (1) in volcanic arcs bordering deep-ocean trenches, (2) at isolated hot spots, (3) within continental rifts, and (4) along mid-ocean ridges. As evident from this list, most igneous activity takes place at established or newly forming plate boundaries. (Hot-spot volcanoes, which erupt in the interiors of plates, violate this rule.) Most volcanic activity along mid-ocean ridges happens underwater, at submarine volcanoes. Most volcanic activity in rifts, along convergent margins, and at hot spots takes place under the air, at subaerial volcanoes.

FIGURE 6.15 The tectonic setting of igneous rocks.

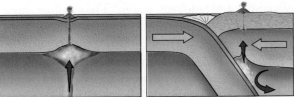
Mantle plume and
a hot-spot volcano

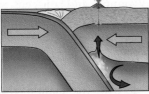

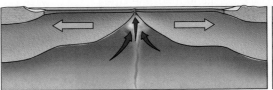
Subduction yields
a volcanic arc.

Melting occurs beneath
a mid-ocean ridge.

Melting occurs beneath
a continental rift.

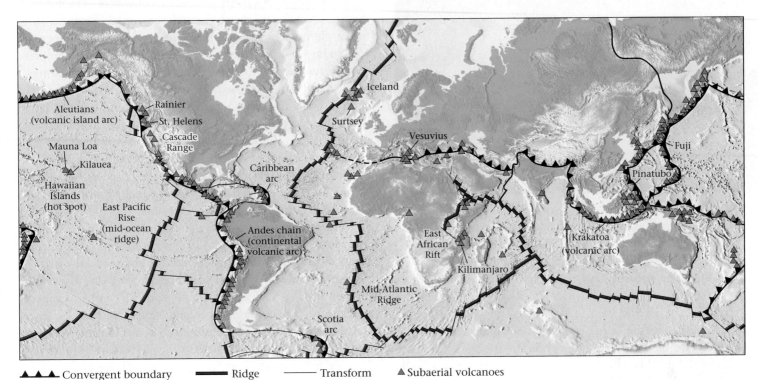

▲▲▲ Convergent boundary   ▬▬ Ridge   ── Transform   ▲ Subaerial volcanoes

The Formation of Igneous Rocks at Volcanic Arcs—the Product of Subduction

Most subaerial volcanoes on Earth occur in long, curving chains, called volcanic arcs (or just arcs), adjacent to the deep-ocean trenches that mark convergent plate boundaries. Some of these arcs, such as the Andes arc in South America, fringe the edge of a continent and are called continental arcs. Others, such as the volcanoes of the Aleutian Islands in Alaska, form oceanic islands and are called island arcs. They are called "arcs" because in many locations the volcanic chain defines a curve on a map.

Recall that at convergent plate boundaries, where volcanic arcs form, oceanic lithosphere subducts and sinks into the mantle. How does this process trigger volcanic activity? The oceanic lithosphere that subducts at convergent boundaries is made up of oceanic crust and the underlying lithospheric mantle. Some of the minerals constituting the oceanic crust contain volatiles; these minerals formed by reaction of the crust with sea water. When the subducting crust sinks into the

mantle to a depth of about 150 km, it is heated by the surrounding mantle to such a degree that the volatile compounds separate and enter the adjacent hot asthenosphere. The addition of volatiles, mainly water, causes the very hot ultramafic rock of the asthenosphere to partially melt and produce a basaltic magma. This magma either rises directly, to erupt as basaltic lava, or undergoes fractional crystallization before erupting and evolves into intermediate or felsic lava. Whether the crust on the downgoing slab partially melts and contributes to the volcanic system remains a subject of debate. If crustal melting makes a contribution, it is small.

In continental volcanic arcs, not all the mantle-derived basaltic magma rises directly to the surface; some gets trapped at the base of the continental crust, and some in magma chambers deep in the crust. When this happens, heat transfers into the continental crust and causes partial melting of this crust. Because much of the continental crust is mafic to intermediate in composition to start with, the resulting magmas are intermediate to felsic in composition. This magma rises, leaving the

FIGURE 6.16 The light-colored rock of the Torres del Paines in Chile is an intrusion of granite. The granite formed when melt froze deep in the crust. It has been exposed due to mountain building and erosion.

basalt behind, and either cools higher in the crust to form plutons (Fig. 6.16) or rises to the surface and erupts. For this reason, granitic plutons and andesite lavas form at continental arcs.

Subduction presently occurs along 60% of the margin of the present Pacific Plate, so volcanic arcs border 60% of the margin of the Pacific Ocean; geologists refer to the Pacific rim, therefore, as the "Ring of Fire." The volcanic arcs of the Ring of Fire include the Andes of South America, the Cascades of the northwestern United States, the Aleutian Islands of Alaska, the Kuril Islands off the eastern coast of Russia, Japan, and several arcs in the southwestern Pacific.

The Formation of Igneous Rocks at Hot Spots—a Surprise of Nature

Hawaii and some other South Pacific island volcanoes are **hot-spot volcanoes**, isolated volcanoes that are not a consequence of plate-boundary interactions. There are about 50 to 100 currently active hot-spot volcanoes scattered around the world (see Figs. 4.11 and 4.12). Most oceanic hot-spot volcanoes erupt in the interior of oceanic plates, away from convergent or divergent boundaries. Some, such as Iceland, sit astride a divergent boundary. (Geoscientists associate Iceland with a hot spot because its volcanoes generate far more lava than normal mid-ocean ridge volcanoes do.) The recent eruption of Eyjafjallajokull in Iceland, and the disruption it caused airline traffic worldwide, emphasizes the volume and impact of this volcanism.

Did you ever wonder . . .
why volcanic islands like Iceland and Hawaii exist?

Some hot-spot volcanoes erupt in the interior of continents. For example, a continental hot-spot volcano produced the stunning landscape of Yellowstone National Park in northwestern Wyoming and adjacent states. The "yellow stone" exposed in the park consists of sulfur- and iron-stained layers of volcanic ash.

As we learned in Chapter 4, most researchers think that many hot-spot volcanoes form above plumes of hot mantle rock from deep in the mantle, though some studies suggest that some hot spots may originate at shallower depths. According to the plume hypothesis, a plume consists of very hot rock that rises like soft plastic up through the overlying mantle. Note that a plume does *not* consist of magma. When the hot rock of a plume reaches the base of the lithosphere, decompression causes the rock (peridotite) to undergo partial melting, a process that generates mafic magma. The mafic magma then rises through the lithosphere, pools in a magma chamber in the crust, and eventually erupts at the surface, forming a volcano. In the case of oceanic hot spots, mostly mafic magma erupts. In the case of continental hot spots, some of the mafic magma erupts to form basalt; but some transfers heat to the continental crust, which then partially melts itself, producing felsic magmas that erupt to form rhyolite.

The plume hypothesis suggests that the position of a hot-spot plume remains relatively fixed with respect to the moving plates. So, in the case of plate-interior hot spots, the drift of the plates causes a volcano that grew during one interval of time eventually to move off the hot spot. When this happens, the volcano dies and a new volcano forms. As a consequence, active hot-spot volcanoes commonly occur at the end of a chain of dead volcanoes, and this chain is called a **hot-spot track**. Hawaii, for example, is at the end of a track consisting of the Emperor seamount chain and the Hawaiian Islands (see Fig. 4.12a). Similarly, Yellowstone Park lies at the end of a track manifested by a chain of now-dead volcanoes along the Snake River Plain in Idaho.

Large Igneous Provinces (LIPs) and Their Effects on the Earth System

In many places on Earth, *particularly voluminous* quantities of mafic magma have erupted and/or intruded (Fig. 6.17). Some of these regions occur along the margins of continents, some in the interior of oceanic plates, and some in the interior of continents. The largest of these, the Ontong Java Oceanic Plateau of the western Pacific, covers an area of about 5,000,000 km^2 of the sea floor and has a volume of about 50,000,000 km^3. Such provinces also occur on land. It's no surprise that these huge volumes of igneous rock are called **large igneous provinces (LIPs)**. More recently, this term LIP has been applied to huge eruptions of felsic rocks too. Using this concept, Yellowstone and the Snake River Plane also constitute a LIP.

Mafic LIPs may form when the bulbous head of a mantle plume first reaches the base of the lithosphere. More partial melting can occur in a plume head than in normal asthenosphere, because temperatures are higher in a plume head. Thus, an unusually large quantity of unusually hot basaltic magma forms in the plume head; when volcanic eruptions begin, huge quantities of basaltic lava spew out of the ground. The particularly hot basaltic lava that erupts at such localities has such low viscosity that it can flow tens to hundreds of kilometers across the landscape. Geoscientists refer to such flows as **flood basalts**. Flood basalts make up the bedrock of the Columbia River Plateau in Oregon and Washington (Fig. 6.18a, b), the Paraná Plateau in southeastern Brazil (Fig. 6.18c), the Karoo region of southern Africa, and the Deccan region of southwestern India.

The volume of material that erupted at the biggest LIPs may have exceeded the amount that erupted along the Earth's entire mid-ocean ridge system during the same time. Thus, it seems that eruption of such LIPs are special events in Earth history. Geologists attribute them to **superplumes** in the mantle—plumes that bring up vastly more hot asthenosphere than do normal plumes. Formation of immense LIPs associated with superplumes might have affected sea level and climate, and may have caused the extinction of some species.

The Formation of Igneous Rocks at Rifts— a Record of Continental Breakup

According to the theory of plate tectonics (Chapter 4), rifts are places where continental lithosphere is being stretched and thinned. Successful rifting splits a continent in two and gives birth to a new mid-ocean ridge. As the continental lithosphere thins (before the continent splits), the weight of rock overlying the asthenosphere decreases, so pressure in the asthenosphere decreases and decompression melting

FIGURE 6.18 Flood basalts form when vast quantities of low-viscosity mafic lava "floods" over the landscape and freezes into a thin sheet. Accumulation of successive flows builds a flat-topped plateau.

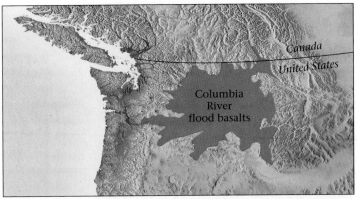

(a) Flood basalts underlie the Columbia River Plateau in Washington and Oregon, the dark area on this map.

(b) Flood basalts form the layers exposed in Palouse Canyon, Washington.

(c) Iguazu Falls, on the Brazil-Argentina border. The falls flow over the huge flood basalt sheet (the black rock) of the Paraná Plateau. Flood basalt underlies all of the region in view.

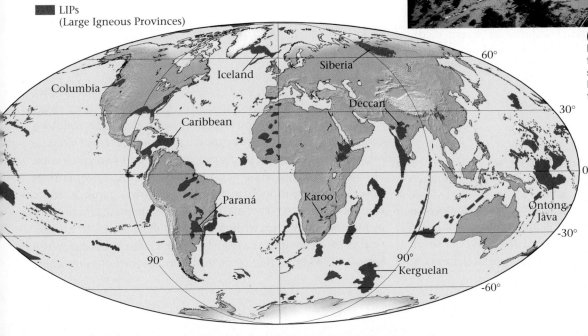

FIGURE 6.17 A map showing the distribution of large igneous provinces (LIPs) on Earth. The red areas are or once were underlain by immense volumes of basalt; not all of this basalt is exposed.

FIGURE 6.19 Pillow basalt forms when lava erupts under water and cools very quickly.

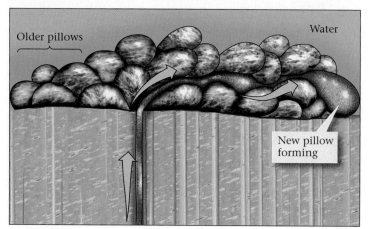

(a) The formation of pillow basalt. Lava squeezes through a conduit and emerges at the surface like toothpaste.

(b) This pillow basalt forms part of an ophiolite, a slice of sea floor that was pushed up onto the surface of a continent during mountain building.

(c) A cross section through a single pillow shows the glassy rind, with a more crystalline center.

takes place, producing basaltic magma, which rises into the crust. Some of this magma makes it to the surface, perhaps flowing along the cracks that appear in the crust as a consequence of the stretching and breaking that accompany rifting, and erupts as basalt. However, some of the magma gets trapped in the crust and transfers heat to the crust. The resulting partial melting of the crust yields felsic (silicic) magmas that erupt as rhyolite. Thus, a sequence of volcanic rocks in a rift generally includes basaltic flows and sheets of rhyolitic lava or ash. Locally, the felsic and mafic magmas mix to form intermediate magma.

The most famous active rift, the East African Rift, presently forming a 4,000-km-long gash in the crust of Africa, has produced numerous volcanoes, including Mount Kilimanjaro. Recent rifting in North America has yielded the Basin and Range Province of Utah, Nevada, Arizona, and southeastern California. Though no currently erupting volcanoes exist in this region, the abundance of recent volcanic deposits suggests that igneous activity could occur again. Geoscientists are now monitoring the Mono Lakes volcanic area of California along the western edge of the Basin and Range, because of the possibility that volcanoes in this area may erupt in the near future.

The Formation of Igneous Rocks at Mid-Ocean Ridges—Hidden Plate Formation

Most igneous rocks at the Earth's surface form at mid-ocean ridges, that is, along divergent plate boundaries. Think about it—the entire oceanic crust, a 7- to 10-km-thick layer of basalt and gabbro that covers 70% of the Earth's surface, forms at

mid-ocean ridges. And this entire volume gets subducted and replaced by new crust, over a period of about 200 million years.

Igneous magmas form at mid-ocean ridges for much the same reason they do at hot spots and rifts. As sea-floor spreading occurs and oceanic lithosphere plates drift away from the ridge, hot asthenosphere rises to keep the resulting space filled. As this asthenosphere rises, it undergoes decompression, which leads to partial melting and the generation of basaltic magma. As noted in Chapter 4, this magma rises into the crust and pools in a shallow magma chamber. Some cools slowly along the margins of the magma chamber to form massive gabbro, while some intrudes upward to fill vertical cracks that appear as newly formed crust splits apart (see Fig. 4.4c). Magma that cools in the cracks forms basalt dikes, and magma that makes it to the sea floor and extrudes as lava forms basalt flows. The basalt flows of the sea floor don't look like those that erupt on land, because the seawater cools the lava so rapidly that it can't flow very far before solidifying into a blob with a glassy rind. The blobs are commonly shaped like pillows, though some are elongate and roughly tube-shaped. Eventually, the pressure of the lava inside a pillow breaks the glassy rind, and another pillow extrudes. Thus, sea-floor basalt is made up of a pile of pillows, known by geologists as **pillow basalt** (Fig. 6.19a–c).

In this chapter, we've focused on the diversity of igneous rocks, and why and where they form. We see that extrusive rocks develop at volcanoes. There's a lot more to say about volcanoes. Proceed directly to Chapter 9 if you want to consider volcanic eruptions in detail at this point in your course.

Take-Home Message

- Volcanic arcs form because of flux melting in the asthenosphere accompanying the release of volatiles from the subducting plate.
- Melting at hot spots is probably due to decompression melting at the top of a mantle plume.
- Particularly voluminous amounts of melting yield large igneous provinces (LIPs), some of which are now manifested by extensive plateaus of flood basalts.
- Melting at rifts is due to decompression in the asthenosphere as the overlying lithosphere thins.
- Melting at mid-ocean ridges occurs due to decompression melting as asthenosphere rises to fill the space formed as plates drift apart.

THINK: Why do intermediate and felsic rocks, in addition to basalt, also erupt at continental volcanic arcs, hot spots, and rifts?

Chapter Summary

- Magma is liquid rock (melt) under the Earth's surface. Lava is melt that has erupted from a volcano at the Earth's surface.

- Magma forms when hot rock in the Earth partially melts. This process occurs only under certain circumstances—when the pressure decreases (decompression), when volatiles (such as water or carbon dioxide) are added to hot rock, and when heat is transferred to adjacent rock by magma rising from below.

- Magma occurs in a range of compositions: felsic (silicic), intermediate, mafic, and ultramafic. The composition of magma is determined in part by the original composition of the rock from which the magma formed and in part by the way the magma evolves by such processes as assimilation and fractional crystallization.

- During partial melting, only part of the source rock melts to form magma. Magma tends to be more felsic than the rock from which it was extracted.

- Magma rises from the depth because of its buoyancy and because the pressure caused by the weight of overlying rock squeezes magma upward.

- Magma viscosity (its resistance to flow) depends on its composition. Felsic magma is more viscous than mafic magma.

- Geologists distinguish between two types of igneous rocks. Extrusive igneous rocks form from lava that erupts out of a volcano and freezes in contact with air or water. Intrusive igneous rocks develop from magma that freezes inside the Earth.

- Lava may solidify to form flows or domes, or it may explode into the air to form ash, lapilli, and blocks.

- Intrusive igneous rocks form when magma intrudes into preexisting rock (country rock) below Earth's surface. Blob-shaped intrusions are called plutons. Sheet-like intrusions that cut across layering in country rock are dikes, and sheet-like intrusions that form parallel to layering in country rock are sills. Huge intrusions, made up of many plutons, are known as batholiths.

- The rate at which intrusive magma cools depends on the depth at which it intrudes, on the size and shape of the magma body, and on whether circulating groundwater is present. The cooling time controls the texture of an igneous rock.

- Crystalline (nonglassy) igneous rocks are classified according to texture and composition. Glassy igneous rocks are classified according to texture (a solid mass is obsidian; ash that has cemented or welded together is tuff).

- The origin of igneous rocks can readily be understood in the context of plate tectonics. Magma forms at continental or island volcanic arcs along convergent margins, mostly because of the addition of volatiles to the asthenosphere above the subducting slab. Igneous rocks form at hot spots, owing to the decompression melting of a rising mantle plume. Igneous rocks form at rifts as a result of decompression melting of the asthenosphere below the thinning lithosphere or heat transfer from mantle melts into crustal rocks. Igneous rocks form along mid-ocean ridges because of decompression melting of the rising asthenosphere.

GEOPUZZLE REVISITED

Molten rock, or magma, forms in the upper mantle and lower crust, but only at special localities—along convergent and divergent plate boundaries, at hot spots, and in rifts. Most magma freezes underground, but some erupts as lava or ash at volcanoes. When melt cools and solidifies, it becomes igneous rock.

Guide Terms

assimilation (p. 145)

batholith (p. 153)

Bowen's reaction series (p. 152)

crystalline igneous rock (p. 155)

decompression melting (p. 143)

dike (p. 149)

extrusive igneous rock (p. 141)

felsic magma (p. 145)

flood basalt (p. 163)

flux melting (p. 143)

fractional crystallization (p. 148)

fragmental igneous rock (p. 155)

geotherm (p. 143)

glassy igneous rock (p. 155)

heat-transfer melting (p. 144)

hot-spot track (p. 162)

hot-spot volcanoes (p. 162)

hyaloclasite (p. 160)

igneous rock (p. 141)

intermediate magma (p. 145)

intrusive contact (p. 148)

intrusive igneous rock (p. 142)

laccolith (p. 149)

large igneous province (LIP) (p. 162)

lava (p. 140)

lava flow (p. 140)

liquidus (p. 143)

mafic magma (p. 144)

magma (p. 140)

obsidian (p. 158)

partial melting (p. 145)

pegmatite (p. 155)

pillow basalt (p. 165)

plutons (p. 149)

pumice (p. 158)

pyroclastic debris (p. 142)

pyroclastic flow (p. 148)

pyroclastic rock (p. 160)

scoria (p. 160)

sill (p. 149)

solidus (p. 143)

stoping (p. 153)

superplume (p. 163)

tachylite (p. 158)

tuff (p. 160)

ultramafic magma (p. 144)

vesicle (p. 157)

viscosity (p. 146)

volcanic agglomerate (p. 160)

volcanic ash (p. 141)

volcanic breccia (p. 160)

volcaniclastic rock (p. 160)

volcano (p. 140)

wall rock (p. 147)

xenolith (p. 153)

Review Questions

1. How is the process of freezing magma similar to that of freezing water? How is it different?

2. What is the source of heat in the Earth? How did the first igneous rocks on the planet form?

3. Describe the three processes that are responsible for the formation of magmas.

4. Why are there so many different types of magmas?

5. Why do magmas rise from depth to the surface of the Earth?

6. What factors control the viscosity of a melt?

7. What factors control the cooling time of a magma within the crust?

8. How does grain size reflect the cooling time of a magma?

9. What does the mixture of grain sizes in a porphyritic igneous rock indicate about its cooling history?

10. Describe the way magmas are produced in subduction zones.

11. What process in the mantle may be responsible for causing hot-spot volcanoes to form?

12. Describe how magmas are produced at continental rifts.

13. What is a large igneous province (LIP)? How might the formation of LIPs have affected the Earth System?

14. Why does melting take place beneath the axis of a mid-ocean ridge?

On Further Thought

15. If you look at the Moon, even without a telescope, you see broad areas where its surface appears relatively darker and smoother. These areas are individually called *mare* (plural: *maria*), from the Latin word for sea. The term is misleading, for they are not bodies of water but rather plains of igneous rock formed after huge meteors struck the Moon and formed very deep craters. These impacts occurred early in the history of the Moon, when its interior was warmer. With this background information in mind, propose a cause for the igneous activity, and suggest the type of igneous rock that fills the mare. (*Hint:* Think about how the presence of a deep crater affects pressure in the region below the crater, and think about the viscosity of a magma that could spread over such a broad area.)

16. The Cascade volcanic chain of the northwestern United States is only about 800 km long (from the southernmost volcano in California to the northernmost one in Washington State). The volcanic chain of the Andes is several thousand kilometers long. Look at a map showing the Earth's plate boundaries, and explain why the Andes volcanic chain is so much longer than the Cascade volcanic chain.

17. A 250-m-high cliff, known as the Palisades, forms the western shore of the Hudson River at the latitude of New York City (Fig. 6.20a, b). This cliff exposes a sill of dark igneous rock that intruded into cool sedimentary rock between 186 and 192 Ma, when what is now the east coast of North America was an active rift. The rock in the sill is not homogeneous. At its top and bottom, the sill consists of several meters of basalt. The interior of the sill contains layers of different minerals. The bottom layer consists mostly of olivine, the next layer mostly of pyroxene, and the top layer mostly of plagioclase (Fig. 6.20b). It is significant that the composition of the plagioclase lower in the sill contains more Ca, and the plagioclase at the top contains more Na. Explain the internal character of the Palisades sill.

FIGURE 6.20 Explaining mineral distribution in the Palisades.

(a) The Palisades sill in New Jersey rises above the Hudson River.

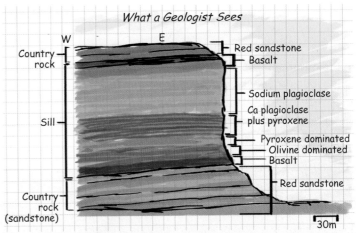

(b) In this cross-sectional sketch, we see that the mineral makeup of the sill changes with depth.

 For more resources, including animations, quizzes, and Norton's GeoTours, go to **wwnorton.com/studyspace**.

 If your instructor assigns exercises in SmartWork, log in at **smartwork.wwnorton.com**.

ANOTHER VIEW The lighter-colored elliptical areas in this satellite image of a region called the Pilbara block, in northwestern Australia, are complex batholiths consisting of Archean granite. The darker rock surrounding the batholiths consists of basalt, komatiite, and sedimentary rock. The Pilbara block is about 300 km measured side to side.

A Surface Veneer: Sediments and Soils

In the mountains east of Los Angeles, California, earthquake- and storm-triggered landslides dump vast amounts of coarse rock debris into valleys. Over time, weathering and transport will break the boulders and cobbles into progressively finer sediment that can be washed further downstream. Where sediment remains in place long enough, further alteration by rain and organisms can transform it into soil.

B.1 **INTRODUCTION**

In the 1950s, the government of Egypt decided to build the Aswan High Dam to trap water of the Nile River in a huge reservoir before the water could reach the Mediterranean Sea. In the process of identifying a good site for the dam's foundation, geologists discovered that the present-day Nile River flows on the surface of a 1.5-km-thick layer of gravel, sand, and mud that fills what was once a canyon as large as the Grand Canyon (Fig. B.1). How could the river once have carved a canyon 1.5 km deep, and why did this canyon later fill with debris?

The origin of the pre-Nile canyon remained a mystery until the summer of 1970, when geologists began to study the material that forms the floor of the Mediterranean Sea. They expected to find layers consisting of the shells of plankton (tiny floating organisms) that had settled out of the water, or of clay that rivers had carried to the sea. To their surprise, however, they also found a 2-km-thick layer of halite and gypsum. Such minerals, types of salts, form when seawater dries up. The researchers proposed that to yield a salt layer that is 2 km thick, the entire Mediterranean would have had to dry up completely several times, with the sea refilling after each drying event. They realized that this discovery solved the mystery of the pre-Nile canyon. When the Mediterranean Sea dried up, the Nile River could dig down and cut a canyon.

And when the sea refilled with water, this canyon flooded and filled with sand and gravel.

Why did the Mediterranean Sea dry up? Only 10% of its water comes from rivers, so for the Mediterranean Sea to remain full, water must flow in from the Atlantic Ocean through the Strait of Gibraltar. Put another way, if water flow in from the Atlantic were to be blocked, water would exit the Mediterranean by evaporation at a rate 10 times faster than it would enter the Mediterranean by river drainage. About 6 million years ago, the northward-drifting African Plate collided with the European Plate, forming a natural dam separating the Mediterranean from the Atlantic. When global sea level dropped, water stopped flowing over this dam from the Atlantic and the Mediterranean evaporated. All the salt that had been dissolved in its water accumulated as a solid deposit of halite and gypsum on the floor of the resulting basin, and the pre-Nile canyon formed. When sea level rose, water flooded in from the Atlantic into the Mediterranean, filling the basin again. This process was repeated many times. About 5.5 million years ago, the Mediterranean rose to its present level, and gravel, sand, and mud carried by the Nile River filled the Nile canyon to its present level.

Geologists refer to the kinds of deposits just described—sand, mud, gravel, halite and gypsum layers, and shell fragments—as sediment. **Sediment**, broadly defined, consists of loose fragments of rocks or minerals broken off bedrock, mineral crystals that precipitate directly out of water, and shells or shell fragments.

Sediments form by the weathering (physical and chemical breakdown) of preexisting rock. They form a surface veneer, or "cover," on bedrock (Fig. B.2). This veneer ranges in thickness from nonexistent, in places where bedrock crops out at the Earth's surface, to a few kilometers. Some sediments transform into soil, essential for life. Let's now look at how

FIGURE B.1 The present Nile River flows on sediment that filled a deep canyon, cut when the Mediterranean basin was dry.

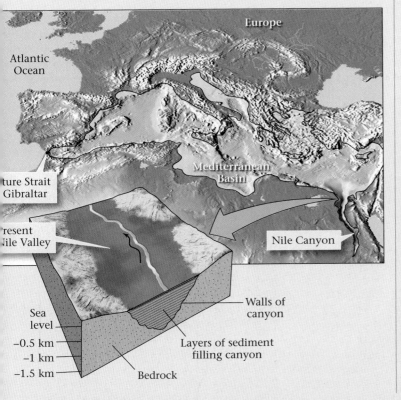

FIGURE B.2 A layer of unconsolidated sediment (sand, clay, and cobbles), topped by dark soil, overlies bedrock in this outcrop along the coast of western Ireland.

weathering produces sediment, and how soils form and evolve from sediment or from rock.

B.2 WEATHERING: FORMING SEDIMENT

Weathering refers to the combination of processes that break up and corrode solid rock, eventually transforming

FIGURE B.3   Examples of rock outcrops undergoing weathering.

(a) The texture of granite changes as it weathers.

Weathered granite breaks into loose grains.

Unweathered granite is solid.

it into sediment. We can say that rock that has not undergone weathering is unweathered or "fresh" (Fig. B.3a). Rock exposed at the Earth's surface sooner or later crumbles away because of weathering (Fig. B.3b). Just as a plumber can unclog a drain by using physical force (with a plumber's snake) or by causing a chemical reaction (with a dose of liquid drain opener), nature can attack rocks in two ways. So geologists distinguish between two types of weathering: physical and chemical.

Physical Weathering

Physical weathering, sometimes also referred to as mechanical weathering, breaks intact rock into unconnected *clasts* (grains or chunks), collectively called debris or detritus. Each size range of clasts has a name (Table B.1). Many different phenomena contribute to physical weathering, as described below.

Jointing.   Rocks buried deep in the Earth's crust endure enormous pressure due to the weight of overlying rock or overburden. Rocks at depth are also warmer than rocks nearer the surface because of the Earth's geothermal gradient. Over long periods of time, moving water, air, and ice at the Earth's surface grind away and remove overburden, so rock formerly at depth rises closer to the Earth's surface. As a result, the pressure squeezing this rock decreases, and the rock becomes cooler. A change in pressure and temperature causes rock to change shape slightly. Such changes cause hard rock to break. Natural cracks that form in rocks due to removal of overburden or due to cooling (and for other reasons as well; see Chapter 11) are known as **joints**.

(b) These slopes have a smooth, rubbly surface, because the rocks they are made of have weathered.

TABLE B.1 Clasts Are Classified by Grain Diameter

Boulders	More than 256 millimeters (mm)
Cobbles	Between 64 mm and 256 mm
Pebbles	Between 2 mm and 64 mm
Sand	Between 1/16 mm and 2 mm
Silt	Between 1/256 mm and 1/16 mm
Mud	Less than 1/256 mm

Almost all rock outcrops contain joints. Some joints are fairly planar, some curving, and some irregular. Large granite plutons may split into onion-like sheets along joints that lie parallel to the mountain face; this process is called exfoliation. Sedimentary rock beds, however, tend to break into rectangular blocks (**Fig. B.4a**). Regardless of their orientation, the formation of joints turns formerly intact bedrock into separate blocks. Eventually, these blocks fall from the outcrop at which they formed. After a while, they may collect in an apron of talus, the rock rubble at the base of a slope (**Fig. B.4b**).

Frost wedging. Freezing water bursts pipes and shatters bottles because water expands when it freezes and pushes the walls of the container apart. The same phenomenon happens in rock. When the water trapped in a joint freezes, it forces the joint open and may cause the joint to grow. Such frost wedging helps break blocks free from intact bedrock (**Fig. B.5a**).

FIGURE B.4 Joints (natural cracks) break bedrock into blocks and sheets, which can tumble down a slope.

(a) When buried deeply, rocks are subjected to a large downward pressure. Later, after erosion removes overburden, rocks expand and crack.

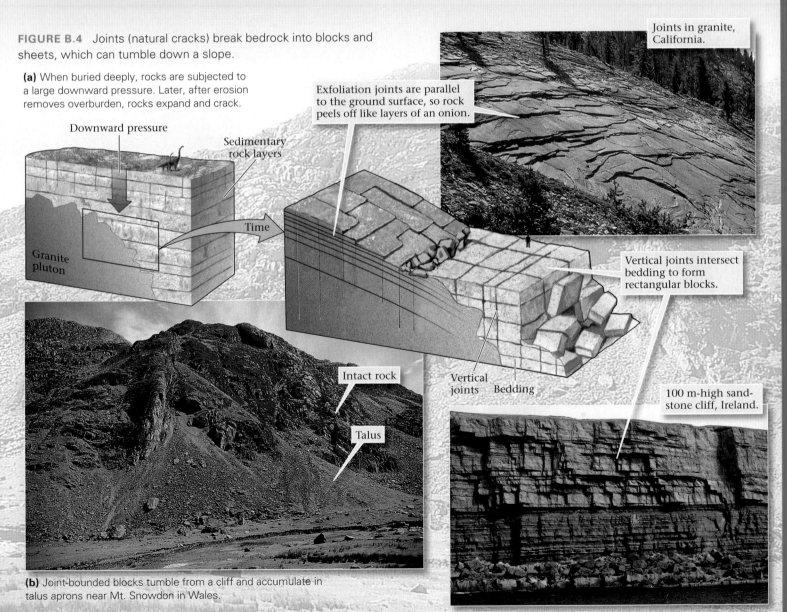

(b) Joint-bounded blocks tumble from a cliff and accumulate in talus aprons near Mt. Snowdon in Wales.

FIGURE B.5 Wedging is one type of physical (mechanical) weathering.

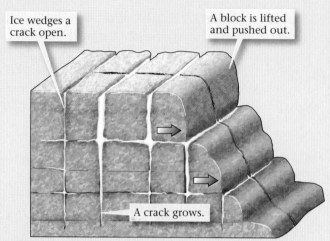

Ice wedges a crack open.

A block is lifted and pushed out.

A crack grows.

(a) When the water that fills cracks freezes, it expands and wedges the cracks open.

(c) Salt wedging led to disintegratIon of gravestones in Whitby, England.

Tree growing in a joint.

(b) Root wedging pushes open a joint, slowly separating a block from the cliff.

Eventually, the blocks tumble to the base of the cliff.

Root wedging. Have you ever noticed how the roots of an old tree can break up a sidewalk? Even though the wood of roots doesn't seem very strong, as roots expand they apply pressure to their surroundings. Tree roots that grow into joints can push joints open in a process known as root wedging (Fig. B.5b).

Salt wedging. In arid climates, dissolved salt in groundwater precipitates and grows as crystals in open pore spaces in rocks. This process, called salt wedging, pushes apart the surrounding grains and so weakens the rock that when exposed to wind and rain, the rock disintegrates into separate grains. The same phenomenon happens in coastal areas, where salt spray percolates into rock and then dries (Fig. B.5c).

Thermal expansion. When the heat of an intense forest fire bakes a rock, the outer layer of the rock expands. On cooling, the layer contracts. This change creates forces in the rock sufficient to make the outer part of the rock break off in sheet-like pieces.

Animal attack. Animal life also contributes to physical weathering: burrowing creatures, from earthworms to gophers, push open cracks and move rock fragments. And in the past century, humans have become perhaps the most energetic agent of physical weathering on the planet. When we excavate quarries, foundations, mines, or roadbeds by digging and blasting, we shatter and displace rock that might otherwise have remained intact for millions of years more.

Chemical Weathering

Up to now we've taken the plumber's-snake approach to breaking up rock. Now let's look at the liquid-drain-opener approach. **Chemical weathering** refers to the chemical reactions that alter or destroy minerals when rock comes in contact with water solutions or air. Common reactions involved in chemical weathering include the following:

■ *Dissolution.* Chemical weathering during which minerals dissolve into water is called dissolution. Dissolution primarily affects salts and carbonate minerals (Fig. B.6a), but even quartz dissolves slightly. Some miner-

als, such as halite, can dissolve rapidly in pure rainwater. But some, such as calcite, dissolve rapidly only when the water is acidic, meaning that it contains an excess of hydrogen ions (H^+). Acidic water reacts with calcite to form a solution and bubbles of CO_2 gas (see Chapter 5).

How does the water in rock near the surface of the Earth become acidic? As rainwater falls, it dissolves carbon dioxide gas in the atmosphere, and as the water sinks down through soil containing organic debris, it reacts with the debris. Both processes yield carbonic acid. Because of the solubility of calcite, limestone and marble (two types of rock composed of

FIGURE B.6 Examples of chemical weathering.

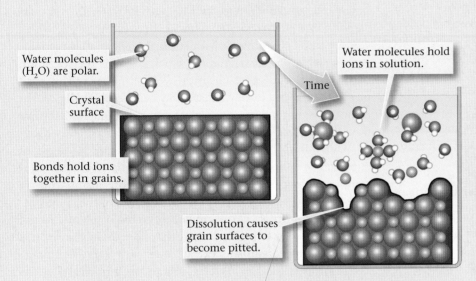

Water molecules (H_2O) are polar.

Crystal surface

Bonds hold ions together in grains.

Time

Water molecules hold ions in solution.

Dissolution causes grain surfaces to become pitted.

(a) Dissolution occurs when water molecules pluck ions off of grain surfaces.

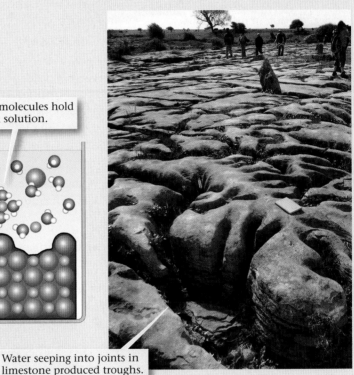

Water seeping into joints in limestone produced troughs.

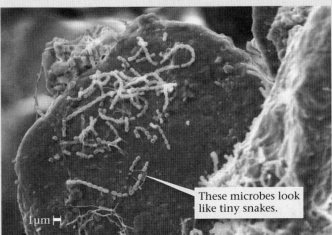

Weathered pyrite crystals

These microbes look like tiny snakes.

1 μm

(b) Pyrite (FeS_2) is a sulfide mineral that reacts with air to form iron oxide.

(c) Certain types of microbes obtain their life energy from the chemical bonds in minerals.

calcite) dissolve, widening joints and leading to the formation of caverns.

- *Hydrolysis.* During hydrolysis, water chemically reacts with minerals and breaks them down (*lysis* means loosen in Greek) to form other minerals. For example, potassium feldspar, a common mineral in granite, reacts with acidic water to produce kaolinite (a type of clay) and other dissolved ions.

    Hydrolysis reactions break down not only feldspars, but many other silicate minerals as well—amphiboles, pyroxenes, micas, and olivines all react slowly and transform into various types of clay. Quartz also undergoes hydrolysis but does so at such a slow rate that it survives weathering in most climates.

- *Oxidation.* Chemists refer to a reaction during which an element loses electrons as an oxidation reaction, because such a loss commonly takes place when elements combine with oxygen. The oxidation, or rusting, of iron serves as an example. Oxidation reactions in rocks transform iron-bearing minerals (such as biotite and pyrite) into a rusty-brown mixture of various iron-oxide and iron-hydroxide minerals (Fig. B.6b).

- *Hydration.* Hydration, the absorption of water into the crystal structure of minerals, causes some minerals,

such as certain types of clay, to expand. Such expansion weakens rock.

Not all minerals undergo chemical weathering at the same rates (Table B.2). Some weather in a matter of months or years, whereas others remain unweathered for millions of years. In humid climates, for example, halite and calcite weather faster than most silicate minerals. Of the silicate minerals, those that crystallize at the highest temperatures are generally less stable under the cool temperatures of the Earth's surface than those that crystallize at lower temperatures (Table B.2). The difference depends partly on crystal structure and partly on chemical composition. Specifically, minerals with fewer linkages between silicon-oxygen tetrahedra tend to have weaker structures and thus weather faster than do minerals with stronger structures. And minerals containing iron, magnesium, sodium, potassium, and aluminum tend to weather faster than minerals without these elements. Thus, quartz (pure SiO_2), which consists of a 3-D network with strong bonds in all directions, is very stable. When a granite (which contains quartz, mica, and feldspar) undergoes chemical weathering, everything but quartz transforms to clay. That's why beaches typically consist of quartz sand; quartz is the most common mineral left after the other minerals have turned to clay and washed away.

Until fairly recently, geoscientists tended to think of chemical weathering as a strictly inorganic chemical reaction, occurring entirely independently of life forms. But we now realize that organisms play a major role in the chemical-weathering process. For example, the roots of plants, fungi, and lichens secrete organic acids that help dissolve minerals in rocks; these organisms extract nutrients from the minerals. Microbes, such as bacteria, are amazing in that they literally eat minerals for lunch (Fig. B.6c). Bacteria pluck off molecules from minerals and use the energy from the molecules' chemical bonds to supply their own life force.

Physical and Chemical Weathering Working Together

So far we've looked at the processes of chemical and physical weathering separately, but in the real world they happen together, aiding one another in disintegrating rock to form sediment (see Geology at a Glance, pp. 176–177).

Physical weathering speeds up chemical weathering. To understand why, keep in mind that chemical-weathering reactions take place at the surface of a material. Thus, the overall rate at which chemical weathering occurs depends on the ratio of surface area to volume—the greater the surface area, the faster the volume as a whole can chemically weather. When jointing (physical weathering) breaks a large block of rock into smaller pieces, the surface area increases, so chemical weathering happens faster (Fig. B.7a).

TABLE B.2 **Relative Stability of Minerals at the Earth's Surface**

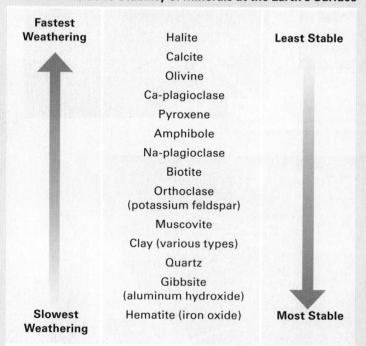

Fastest Weathering		Least Stable
	Halite	
	Calcite	
	Olivine	
	Ca-plagioclase	
	Pyroxene	
	Amphibole	
	Na-plagioclase	
	Biotite	
	Orthoclase (potassium feldspar)	
	Muscovite	
	Clay (various types)	
	Quartz	
	Gibbsite (aluminum hydroxide)	
Slowest Weathering	Hematite (iron oxide)	Most Stable

Note that minerals that form early in Bowen's reaction series (see Box 6.1) are among the least stable minerals at the Earth's surface. Minerals that are the products of weathering reactions (e.g., hematite) are among the most stable minerals at the Earth's surface. Mafic minerals weather by oxidation, felsic minerals by hydrolysis, carbonates and salts by dissolution, and oxide minerals don't weather at all.

Similarly, chemical weathering speeds up physical weathering by dissolving away grains or cements that hold a rock together, transforming hard minerals (like feldspar) into soft minerals (like clay) and causing minerals to absorb water and expand. These phenomena make rock weaker, so it can disintegrate more easily (Fig. B.7b). If you drop a block of fresh granite on the ground, it will most likely stay intact, but if you drop a block of chemically weathered granite on the ground, it will crumble into a pile of sand and clay.

Note that weathering happens faster at edges, and even faster at the corners of broken blocks. This is because weathering attacks a flat face from only one direction, an edge from two directions, and a corner from three directions. Thus, with time, edges of blocks become blunt and corners become rounded (Fig. B.8a). In rocks such as granite, which do not contain layering that can affect weathering rates, rectangular blocks transform into a spheroidal shape (Fig. B.8b).

When different rocks in an outcrop undergo weathering at different rates, we say that the outcrop has undergone "differential weathering." As a result of differential weathering, cliffs composed of a variety of rock layers take on a stair-step or sawtooth shape (Fig. B.8c). Weak layers may weather away beneath a more resistant layer, creating an overhang. Similarly, the rate at which the land surface weathers depends on the rock type, so valleys tend to develop over weak rocks, while strong rocks hold up hills.

You can easily see the consequences of differential weathering if you walk through a graveyard. The inscriptions on some headstones are sharp and clear, whereas those on other stones have become blunted or have even disappeared (Fig. B.8d). That's because the minerals in these different stones have different resistances to weathering. Granite, an igneous rock with a high quartz content, retains inscriptions the longest. But marble, a metamorphic rock composed of calcite, dissolves away relatively rapidly in acidic rain.

B.3 SOIL

If you've ever had the chance to dig in a garden, you've seen firsthand that the material in which flowers grow looks and feels different from beach sand or potter's clay. We call the material in a garden dirt or, more technically, soil. **Soil** consists of rock or sediment that has been modified by physical and chemical interaction with organic material and rainwater, over time, to produce a substrate that can support the growth of plants. Soil is one of our planet's most valuable resources, for without it there could be no agriculture, forestry, ranching, or home gardening.

FIGURE B.7 Physical and chemical processes work together during the weathering process.

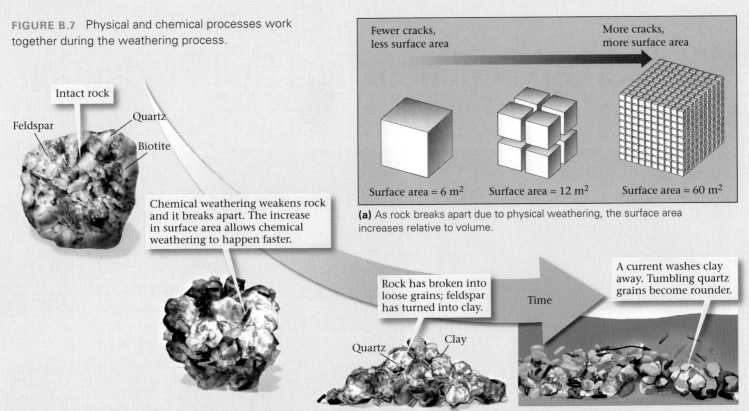

(a) As rock breaks apart due to physical weathering, the surface area increases relative to volume.

(b) Chemical weathering weakens rock so it breaks apart. As this happens, the surface area increases, so chemical weathering happens still faster. Eventually, the rock completely disaggregates to form sediment. Weathering of granite produces quartz sand and clay.

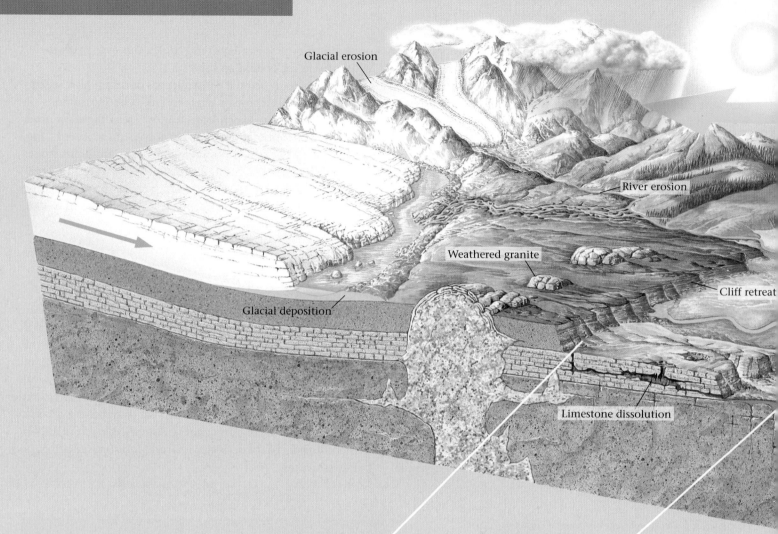

Glacial erosion

River erosion

Weathered granite

Cliff retreat

Glacial deposition

Limestone dissolution

Weathering, Sediment, and Soil Production

New boulders tumble from a cliff in New Mexico.

Silt collects along a stream in Indiana.

Wind

Tectonic processes uplift the land surface above sea level. Once exposed, rock interacts with air and water and undergoes chemical and physical weathering, ultimately breaking down to produce sediment. Convection in the atmosphere generates wind, rain, and snow. Flowing water, ice, and air erode and transport sediment to sites of deposition. Leaching by downward-percolating rainwater, along with the addition of organic material, produces soil.

Coastal erosion

River deposition

Soil formation

Coastal deposition

Soil forms on bedrock of chalk in southern England.

Waves move sand on a beach in Brazil.

Erosion carves the coastal cliffs of Ireland.

FIGURE B.8 Differential weathering.

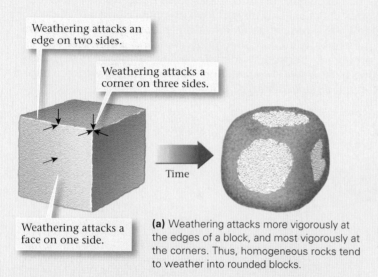

Weathering attacks an edge on two sides.

Weathering attacks a corner on three sides.

Weathering attacks a face on one side.

Time

(a) Weathering attacks more vigorously at the edges of a block, and most vigorously at the corners. Thus, homogeneous rocks tend to weather into rounded blocks.

(b) Weathering along cracks in granite of the Mojave Desert led to the formation of rounded blocks. This style is called spheroidal weathering.

Weak shale

Strong sandstone

(c) Sawtooth weathering profiles develop in sequences of alternating strong and weak layers on this exposure in New Mexico. Weak layers are indented, whereas strong layers protrude.

(d) Inscriptions on a granite headstone (left) last for centuries, but those on a marble headstone (right) may weather away in decades. These gravestones are in the same cemetery and are about the same age.

How Does Soil Form?

Three processes taking place at or just below the surface of the Earth contribute to soil formation. First, chemical and physical weathering produces loose debris, new mineral grains (such as clay), and ions in solution. Second, rainwater percolates through the debris and carries dissolved ions and clay flakes downward. The region in which this downward transport occurs is called the **zone of leaching**, because leaching means extracting, absorbing, and removal. Farther down, new mineral crystals precipitate directly out of the water or form by reaction of the water with debris. Also, the water leaves behind its load of fine clay. The region in which new minerals and

clay collect is the **zone of accumulation** (Fig. B.9a). Third, microbes, fungi, plants, and animals interact with sediment by producing acids that weather grains, by absorbing nutrient atoms, and by leaving behind organic waste and remains. Plant roots and burrowing animals (insects, worms, and gophers) churn and break up the soil, and microbes metabolize mineral grains and release chemicals.

As a consequence of the above processes, regolith and rock evolve into soil—the soil's character (its texture and composition) becomes very different from that of the starting material. Note that the biologic and physical components of the Earth System interact profoundly in the soil. Indeed, soils serve as home for a remarkable number of organisms—a single cubic

FIGURE B.9 Formation of soil horizons.

Rain enters ground.

Plant debris accumulates.

Worms churn.

Microbes and fungi metabolize.

Roots weather minerals.

Downward-percolating water transports ions and clay.

Ions and clay accumulate.

Zone of leaching

Zone of accumulation

Time

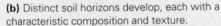

O

A

E

B

C

Topsoil

Transition

Subsoil

Weathered bedrock

Solid bedrock

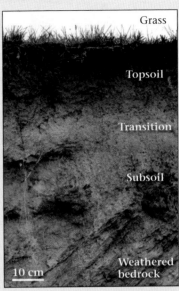

Grass

Topsoil

Transition

Subsoil

10 cm

Weathered bedrock

(a) Soil character depends on climate, for climate controls rainfall and vegetation.

(b) Distinct soil horizons develop, each with a characteristic composition and texture.

(c) Soil horizons exposed on the wall of a gully in eastern Brazil.

centimeter of moist soil in a warm region hosts over 1 billion bacteria and 1 million protozoans. Over 1.5 million earthworms wriggle through each acre of such soil.

Because different soil-forming processes operate at different depths, soils typically develop distinct zones, known as **soil horizons**, arranged in a vertical sequence called a **soil profile** (Fig. B.9b, c). Let's look at an idealized soil profile, from top to bottom, using a soil formed in a temperate forest as our example. The highest horizon is the O-horizon (the prefix stands for organic), so called because it consists almost entirely of organic matter and contains barely any mineral matter. Below the O-horizon we find the A-horizon, in which humus (organic debris) has decayed further and has mixed with mineral grains (clay, silt, and sand). Water percolating through the A-horizon causes chemical weathering reactions to occur and produces ions in solution and new clay minerals. The downward-moving water eventually carries soluble chemicals and fine clay deeper into the subsurface. The O- and A-horizons constitute dark-gray to blackish-brown *topsoil*, the fertile portion of soil that farmers till for planting crops. In some places, the A-horizon grades downward into the E-horizon, a soil level that has undergone substantial leaching but has not yet mixed with organic material. Ions and clay accumulate in the B-horizon, or subsoil. Note from our description that the O-, A-, and E-horizons make up the zone of leaching, whereas the B-horizon makes up the zone of accumulation.

Finally, at the base of a soil profile we find the C-horizon, which consists of material derived from the substrate that's been chemically weathered and broken apart, but has not yet undergone leaching or accumulation. The C-horizon grades downward into unweathered bedrock, or into unweathered sediment.

As farmers, foresters, and ranchers well know, the soil in one locality can differ greatly from the soil in another, in terms of composition, thickness, and texture. And crops that grow well in one type of soil may wither and die in another. Such diversity exists because the makeup of a soil depends on several soil-forming factors (Fig. B.10a, b).

- *Climate*: The total rainfall, the distribution of rainfall during the year, and the range and average of temperature during the year determine the rate and amount of chemical weathering and leaching that take place at a given location. Large amounts of rainfall and warm temperatures accelerate chemical weathering and cause most of the soluble elements to be leached. In regions with small amounts of rainfall and cooler temperatures, soils take a long time to develop and can retain unweathered minerals and soluble components. Climate seems to be the single most important factor in determining the nature of soils that develop.

- *Substrate composition*: Some soils form on basalt, some on granite, some on volcanic ash, and some on recently deposited quartz silt. These different substrates consist of different materials, so the soils formed on them end up with different chemical compositions. For example, a soil formed on basalt tends to be richer in iron than a soil formed on granite. Also, soils tend to develop faster

FIGURE B.10 Factors that control the character of soil.

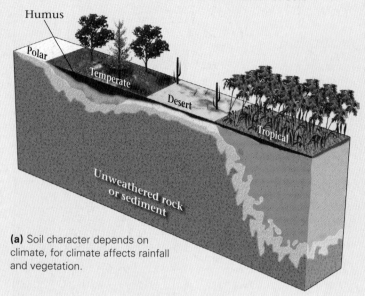

(a) Soil character depends on climate, for climate affects rainfall and vegetation.

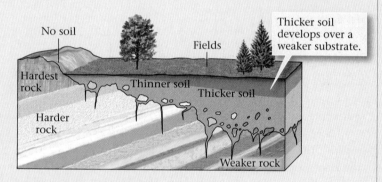

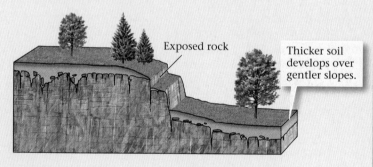

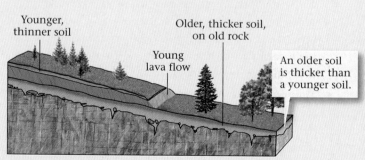

(b) Soil character also depends on the strength of the substrate, the steepness of the slope, and the length of time soil has been forming.

on unconsolidated material (ash or sediment) than on hard bedrock.

- *Slope steepness*: A thick soil can accumulate under land that lies flat. But on a steep slope, regolith may wash away before it can evolve into a soil. Thus, all other factors being equal, soil thickness increases as the slope angle decreases.

- *Wetness*: Depending on the details of local topography and on the depth below the surface at which ground-water occurs, some soil is wetter than other soil in the same region. Wet soils tend to contain more organic material than do dry soils.

- *Time*: Because soil formation is an evolutionary process, a young soil tends to be thinner and less developed than an old soil. The rate of soil formation varies greatly with environment. In a protected, moist, warm region, soils may develop over the course of a few years to a few decades. But in an exposed, cold, dry region, soils may take thousands of years or more to develop. In temperate regions, soil forms at a rate of 0.02 to 0.20 mm per year, thereby producing 1 meter of soil in about 10,000 years.

- *Vegetation type*: Different kinds of plants extract or add different nutrients and quantities of organic matter to a soil. Also, some plants have deeper root systems than others and help prevent soil from washing away.

Soil Classification

Depending on their evolution and composition, soils come in a variety of textures, structures, and colors. Soil texture reflects the relative proportions of sand, silt, and clay-sized grains in the soil. For many crops, farmers prefer to sow in **loam**, a type of soil consisting of about 10 to 30% clay and the rest silt and sand. In loam, pores (open spaces) can remain between grains so that water and air can pass through and roots can easily penetrate. In soils with too much clay, the clay packs together and prevents water movement. Soil structure refers to the degree to which soil grains clump together to form lumps or clods, which soil scientists refer to as "peds" (from the Latin *pedo*, meaning soil). The structure changes as a soil develops, because structure depends on clay content and organic content, both of which change with time. These materials give soil its stickiness. Soil color reflects its composition: organic-rich soil is gray or black, organic-poor and calcite-rich soil is whitish, and iron-rich soil is red or yellow.

Soil scientists worldwide have struggled mightily to develop a rational scheme for classifying soils. Not all schemes utilize the same criteria, and even today there is not world-wide agreement on which works best. In the United States, a country that includes many climates at midlatitudes, many soil scientists use the U.S. Comprehensive Soil Classification System, which distinguishes among 12 orders of soil based on the physical characteristics and environment of soil formation (see Table B.3 and Fig. B.11). Canadians use a different scheme focusing only on soils that develop north of the 40th parallel. This scheme works well for cooler, high-latitude climates.

Rainfall and vegetation play a key role in determining the type of soil that forms. For example, in deserts, where there is very little rainfall and sparse vegetation, an aridisol forms (Fig. B.12a). Aridisols have no O-horizon (because there is so little organic material), and the A-horizon is thin. Soluble minerals, specifically calcite, that would be washed away entirely if there were more rainfall, instead accumulate in the B-horizon. In fact, capillary action may bring calcite *up* from deeper down as water evaporates at the ground surface. Calcite cements clasts together in the B-horizon to form a rock-like mass called caliche or calcrete. In temperate environments, an alfisol forms—this soil has an O-horizon, and because of moderate amounts of rainfall, materials leached from the A-horizon accumulate in the B-horizon (Fig. B.12b). In a tropical climate, oxisols develop. Here, so much rainfall percolates down into the ground that all reactive minerals in the soil undergo chemical weathering, producing ions and clay that flush downward. This process leaves an A-horizon that contains substantial amounts of stable iron-oxide, aluminum-oxide,

TABLE B.3 Soil Orders: U.S. Comprehensive Soil Classification System (see also the map below)

Alfisol	Gray/brown, has subsurface clay accumulation and abundant plant nutrients. Forms in humid forests.
Andisol	Forms in volcanic ash.
Aridisol	Low in organic matter, has carbonate horizons. Forms in arid environments.
Entisol	Has no horizons. Formed very recently.
Gelisol	Underlain with permanently frozen ground.
Histosol	Very rich in organic debris. Forms in swamps and marshes.
Inceptisol	Moist, has poorly developed horizons. Formed recently.
Mollisol	Soft, black, and rich in nutrients. Forms in subhumid to subarid grasslands.
Oxisol	Very weathered, rich in aluminum oxide and iron oxide, low in plant nutrients. Forms in tropical regions.
Spodosol	Acidic, low in plant nutrients, ashy, has accumulations of iron and aluminum. Forms in humid forests.
Ultisol	Very mature, strongly weathered soils, low in plant nutrients.
Vertisol	Clay-rich soils capable of swelling when wet, and shrinking and cracking when dry.

FIGURE B.11 U.S. Department of Agriculture map of soil types around the world.

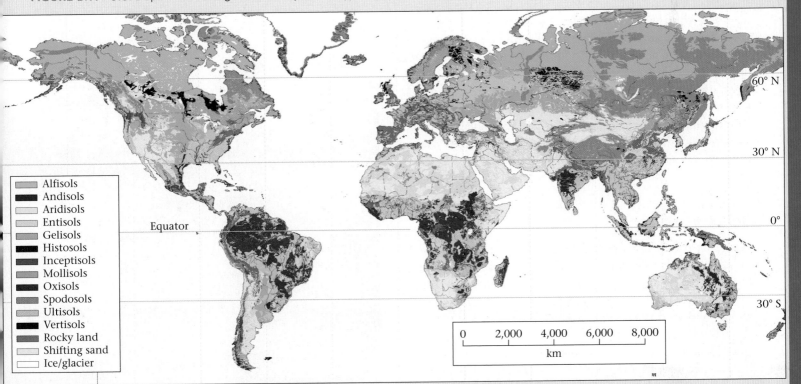

Alfisols
Andisols
Aridisols
Entisols
Gelisols
Histosols
Inceptisols
Mollisols
Oxisols
Spodosols
Ultisols
Vertisols
Rocky land
Shifting sand
Ice/glacier

Equator

60° N

30° N

0°

30° S

0   2,000   4,000   6,000   8,000

km

FIGURE B.12 Examples of soil classification.

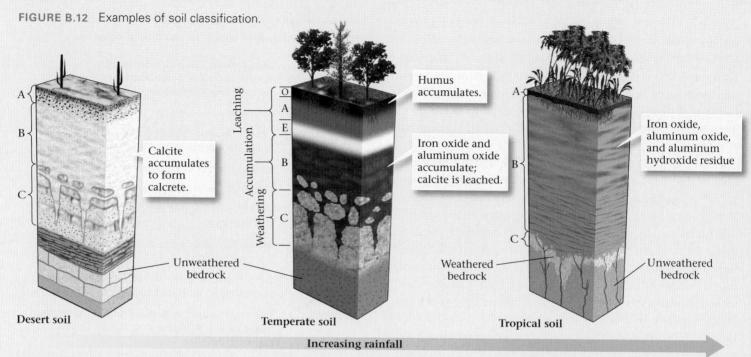

Increasing rainfall

(a) Aridisol forms in deserts. Rainfall is so low that no O-horizon forms, and soluble minerals accumulate in the B-horizon.

(b) Alfisol forms in temperate climates. An O-horizon forms, and less-soluble materials accumulate in the B-horizon.

(c) Oxisol forms in tropical climates where percolating rainwater leaches all soluble minerals, leaving only iron- and aluminum-rich residues.

and aluminum-hydroxide residues (Fig. B.12c). The resulting soil tends to be brick-red (Fig. B.13). So much water flushes down through some oxisols that a B horizon cannot develop.

Laterite is a type of oxisol that tends to be very rich in iron oxides and can be hard enough to be broken into blocks that can be used for construction—in fact the name comes from the Latin word, *later*, which means brick. Laterite typically forms in areas that have a distinct dry season, during which capillary action brings additional oxide minerals back up into the A horizon, where they accumulate. Some laterite contains so much iron that it can be used as an ore from which iron metal can be extracted. An oxisol that develops from an aluminum-rich source (e.g., granite) may accumulate so much aluminum hydroxide that it can be quarried to provide ore from which aluminum metal can be extracted (see Chapter 15). Such aluminum ore is called bauxite.

Soil Erosion

As we have seen, soils take time to form, so soils capable of supporting crops or forests should be considered a natural resource worthy of protection. However, agriculture, overgrazing, and clear-cutting have led to the destruction of soil. Crops rapidly remove nutrients from soil, so if they are not replaced, the soil will not contain sufficient nutrients to maintain plant life. When the natural plant cover disappears, the surface of the soil becomes exposed to wind and water. Actions such as

the impact of falling raindrops or the rasping of a plow break up the soil at the surface, with the result that it can wash away in water or blow away as dust. When this happens, **soil erosion**, the removal of soil by running water or by wind, takes place (Fig. B.14). In some cases, almost six tons of soil may be lost from an acre of land per year. Human activities increase

FIGURE B.13 Laterite in Brazil.

rates of soil erosion by 10 to 100 times, so that it far exceeds the rate of soil formation. Droughts exacerbate the situation. For example, during the 1930s a succession of droughts killed off so much vegetation in the American plains that wind stripped the land of soil and caused devastating dust storms. Large numbers of people were forced to migrate away from the Dust Bowl of Oklahoma and adjacent areas.

The consequences of rain-forest destruction on soil are particularly profound. In an established rain forest, lush growth provides sufficient organic debris so that trees can grow. But if the forest is logged, or cleared for agriculture, the humus rapidly disappears, leaving laterite that contains few nutrients. Crop plants consume whatever nutrients there are so rapidly that the soil becomes infertile after only a year or two, useless for agriculture and unsuitable for regrowth of rain-forest trees.

Soil erosion is but one of several problems that face society. The overuse of fertilizers, pesticides, and herbicides, as well as spills of a great variety of toxic chemicals, have contaminated soils. Too much irrigation in arid climates can make soils too saline for plant growth, for irrigation water contains trace amounts of salts. Fortunately, people have begun to realize the fragility of soil and have been working on ways to conserve soil. Most countries now have soil-conservation agencies.

Guide Terms

chemical weathering (p. 173)

joints (p. 170)

laterite (p. 182)

loam (p. 180)

physical weathering (p. 170)

sediment (p. 169)

soil (p. 175)

soil erosion (p. 182)

soil horizon (p. 179)

soil profile (p. 179)

weathering (p. 170)

zone of accumulation (p. 178)

zone of leaching (p. 178)

FIGURE B.14 In this image, we can see that the lack of natural plant coverage has led to severe soil erosion by wind. Similar conditions produced the Dust Bowl of the 1930s.

Pages of Earth's Past: Sedimentary Rocks

The rugged landscape of the Grand Canyon exposes a blanket of sedimentary rock layers. These layers formed by the hardening (lithification) of mud, shell accumulations, and sand. The succession of changes in sediment type that we can observe indicate that the environment in the region of this view has changed radically over geologic time—beaches, reefs, mud flats, river floodplains, and desert sand dunes all existed here in the past.

GEOPUZZLE

The rim of the Grand Canyon now lies at an elevation of over 2 km above sea level. How did layers representing the sands of beaches and the shells of reefs accumulate in widespread layers here? Why do some layers form vertical cliffs while others do not?

7.1 INTRODUCTION

On an isolated, windswept drilling platform in the North Sea off the coast of Scotland, a group of roughnecks ready a multi-million-dollar drill—their plan is to penetrate the sea floor and see what lies beneath. The North Sea formed as a consequence of rifting that began tens of millions of years ago. During rifting, what was once dry land between Great Britain and continental Europe slowly sank or, in geologic parlance, "subsided." Rivers carried sediments from the surrounding land into the newborn North Sea, and these sediments collected in layers. At certain stages in the process, salts precipitated from seawater, and the shells of sea creatures settled and collected on the sea floor. As the drilling begins, a geologist stationed on the deck of the platform examines the material flushed out of the lengthening drill hole by high-pressure fluids. At first, drilling brings up soft mud and loose sand, silt, pebbles, and shell fragments. But as the hole goes deeper, the material coming up holds together in soft but coherent clumps. Eventually, when the hole has entered layers that now lie almost a kilometer below the sea floor, the drilling fluid flushes out chips and chunks of solid rock. When the geologist studies these fragments, she finds that the composition of fragments changes with depth. At some depths, fragments consist of grains of sand or silt cemented together, tightly packed clay that is harder than pottery. At other depths, they consist of broken shell aggregates or of crystalline salt masses. The geologist has observed the transition of loose sediment into solid layers of sedimentary rock as burial depth progressively increases.

Formally defined, **sedimentary rock** is rock that forms at or near the surface of the Earth in one of several ways: by the cementing together of loose **clasts** (fragments or grains) that had been produced by physical or chemical weathering of preexisting rock; by the growth of shell masses or the cementing together of shells and shell fragments; by the accumulation and subsequent alteration of organic matter from living organisms; or by the precipitation of minerals from water solutions. Layers of sedimentary rock are like the pages of a book, recording tales of ancient events and ancient environments on the ever-changing face of the Earth. They occur only in the upper part of the crust, and form a "cover" that buries the underlying "basement" of igneous and/or metamorphic rock (Fig. 7.1 ◑).

In Interlude B, we introduced the concept of weathering, and showed how it attacks bedrock, breaking it down into ions and loose sediment grains. What can happen next? Some of the sediment may become incorporated in soil, as we have seen. But some becomes buried and transformed into sedimentary rock. In this chapter, we discuss the various kinds of sedimentary rock and the ways in which they form, and we consider what sedimentary rocks can tell us about the history of the Earth System.

Chapter Themes

By the end of this chapter, you should know . . .

- what distinguishes various classes of sedimentary rocks from one another.
- how clastic sedimentary rocks form, and how to recognize and name major types.
- the role of life in the production of rocks such as limestone and coal.
- how the layering (bedding) in sedimentary rock forms.
- how shapes and textures preserved in sedimentary rocks reflect depositional environments.
- why thick accumulations of sedimentary rock can be found only in certain locations.

FIGURE 7.1 In this close-up of the inner gorge of the Grand Canyon, we see a sedimentary-rock blanket, that forms a "cover" over older "basement" of metamorphic and igneous rock. ◑

(a) Sedimentary cover here consists of horizontal layers. Stronger rock types form steep cliffs. Metamorphic basement does not exhibit layering.

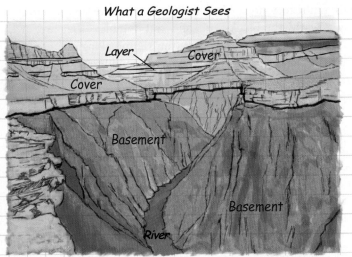

(b) A geologist's sketch emphasizes the contact between the sedimentary cover and basement.

Mesa Verde
cliff dwellings

Sandstone
layer

7.2 CLASSES OF SEDIMENTARY ROCKS

Geologists divide sedimentary rocks into four major classes, based on their mode of origin. (1) **Clastic sedimentary rock** consists of cemented-together clasts, solid fragments and grains broken off of preexisting rocks (*Clastic* comes from the Greek *klastos*, meaning broken); (2) **Biochemical sedimentary rock** consists of shells; (3) **Organic sedimentary rock** consists of carbon-rich relicts of plants or other organisms; and (4) **chemical sedimentary rock** is made up of minerals that precipitated directly from water solutions. Geologists also distinguish among different kinds of sedimentary rocks on the basis of their mineral composition—siliceous rocks contain quartz, argillaceous rocks contain clay, and carbonate rocks contain calcite or dolomite. It's hard to determine relative proportions of different types of sedimentary rocks, but by some estimates 70% to 85% of all the sedimentary rocks on Earth are siliceous or argillaceous clastic rocks, whereas 15% to 25% are carbonate biochemical or chemical rocks. Other kinds of sedimentary rocks occur only in minor quantities. Let's now look at these major classes in more detail.

Clastic Sedimentary Rocks

Formation. Nine hundred years ago, a thriving community of Native Americans inhabited the high plateau of Mesa Verde, Colorado. In hollows beneath huge overhanging ledges, they built multistory stone-block buildings that have survived to this day. Clearly, the blocks are solid and durable—they are, after all, rock. But if you were to rub your thumb along one, it would feel gritty, and small grains of quartz would break free and roll under your thumb, for the block consists of quartz sand grains cemented together. Geologists call such rock a **sandstone**.

Sandstone is an example of detrital, or clastic sedimentary rock. It consists of detritus (loose clasts) that has been stuck together to form a solid mass. The grains can consist of individual minerals (such as grains of quartz or flakes of clay) or of fragments of rock (such as pebbles of granite). The loose grains of sediment transform into clastic sedimentary rock by the following five steps (Fig. 7.2a–c).

- *Weathering*: Detritus forms by disintegration of bedrock into separate grains due to physical and chemical weathering.

- *Erosion*: **Erosion** refers to the combination of processes that separate rock or regolith (surface debris) from its substrate. Erosion involves abrasion, plucking, scouring, and dissolution, and is caused by moving air, water, or ice.

- *Transportation*: Gravity, wind, water, or ice carry sediment. The ability of a medium to carry sediment depends on its viscosity and velocity. Solid ice can transport sediment of any size, regardless of how slowly the ice moves. Very fast-moving, turbulent water can transport coarse fragments (cobbles and boulders), moderately fast-moving water can carry only sand and gravel, and slowly moving water carries only silt and mud. Strong winds can move sand and dust, but gentle breezes carry only dust.

- *Deposition*: **Deposition** is the process by which sediment settles out of the transporting medium. Sediment settles out of wind or moving water when these fluids slow, because as the velocity decreases, the fluid no longer has the ability to carry sediment. Sediment is deposited by ice when the ice melts.

- *Lithification*: Geologists refer to the transformation of loose sediment into solid rock as **lithification**. The lithification of clastic sediment involves two steps. First, when the sediment has been buried, pressure caused by the weight of overlying material squeezes out water and air that had been trapped between clasts, and clasts press together tightly, a process called **compaction**. Compacted sediment may then be bound in place to make coherent sedimentary rock by the process of **cementation**. Cement consists of minerals (commonly quartz or calcite) that precipitate from groundwater and fill the spaces between clasts. Cement acts like glue and holds grains together.

Classifying clastic sedimentary rocks. Say that you pick up a clastic sedimentary rock and want to describe it sufficiently so that, from your words alone, another person can picture the rock. What characteristics should you mention? Geologists find the following characteristics most useful.

- *Clast size*. Size refers to the diameter of fragments or grains making up a rock. Names used for clast size,

FIGURE 7.2 The five steps in clastic sedimentary rock formation.

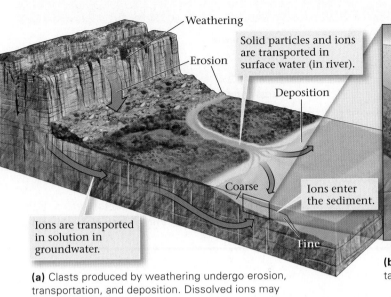

Weathering

Erosion

Solid particles and ions are transported in surface water (in river).

Deposition

Coarse

Ions enter the sediment.

Fine

Ions are transported in solution in groundwater.

(a) Clasts produced by weathering undergo erosion, transportation, and deposition. Dissolved ions may eventually become cement.

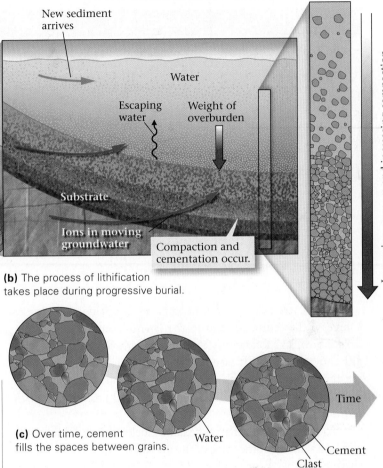

New sediment arrives

Water

Escaping water

Weight of overburden

Substrate

Ions in moving groundwater

Compaction and cementation occur.

Increasing pressure and increasing compaction

(b) The process of lithification takes place during progressive burial.

(c) Over time, cement fills the spaces between grains.

Water

Clast

Cement

Time

listed in order from coarsest to finest, are boulder, cobble, pebble, sand, silt, and clay (see Table B.1 in Interlude B). Geologists informally use the term "gravel" for an accumulation of pebbles and cobbles, and "mud" for wet clay. In this context, clay refers to extremely small clasts—grains of this size typically consist of clay minerals (varieties of sheet silicate that occur in tiny flakes; see Chapter 5), but specks of quartz may be included.

- *Clast composition.* Composition refers to the makeup of clasts in sedimentary rock. Larger clasts (pebbles or larger) typically consist of rock fragments, meaning the clasts themselves are an aggregate of many mineral grains, whereas smaller clasts (sand or smaller) typically consist of individual mineral grains. In some cases, chips of fine-grained rock may be mixed in with sand grains. Such chips are called lithic clasts. Some sedimentary rocks contain only clasts of one composition, but others contain several different kinds of clasts.

- *Angularity and sphericity.* Angularity indicates the degree to which clasts have smooth or angular corners and edges (Fig. 7.3a, b). Sphericity, in contrast, refers to the degree to which a clast is equidimensional, or resembles a sphere.

- *Sorting.* **Sorting** of clasts indicates the degree to which the clasts in a rock are all the same size or include a variety of sizes (Fig. 7.3c, d). Well-sorted sediment consists entirely of sediment of the same size, whereas poorly sorted sediment contains a mixture of more than one

grain size. If a sedimentary rock contains larger clasts surrounded by much smaller clasts (for example, cobbles surrounded by sand), then the mass of smaller grains constitutes the "matrix" of the rock.

- *Character of cement.* Not all clastic sedimentary rocks have the same kind of cement. In some, the cement consists predominantly of quartz, whereas in others, it consists predominantly of calcite. Other kinds of mineral cements do occur, but they are rare.

With these characteristics in mind, we can distinguish among several common types of clastic sedimentary rocks, listed in Table 7.1. This table provides *common* rock names—specialists sometimes use other, more precise names based on more complex classification schemes. Though no single characteristic serves as a complete basis for classifying clastic rocks, grain size is the most important one. Geologists further distinguish among different kinds of sandstone (quartz sandstone, arkose, wacke) on the basis of clast composition and/or sorting, and they distinguish between shale and mudstone on the basis of the way in which the rock breaks. (Shale splits into thin sheets, whereas mudstone does not.)

TABLE 7.1 Classification of Clastic Sedimentary Rocks

Clast Size*	Clast Character	Rock Name (Alternate Name)
Coarse to very coarse	Rounded pebbles and cobbles Angular clasts Large clasts in muddy matrix	Conglomerate Breccia Diamictite
Medium to coarse	Sand-sized grains • quartz grains only • quartz and feldspar sand • sand-sized rock fragments • sand and rock fragments in a clay-rich matrix	Sandstone • quartz sandstone (quartz arenite) • arkose • lithic sandstone • wacke (informally called graywacke)
Fine	Silt-sized clasts	Siltstone
Very fine	Clay and/or very fine silt	Shale (if it breaks into platy sheets) Mudstone (if it doesn't break into platy sheets)

*For precise diameters, see Table B.1 on p. 171.

Characteristics of a sedimentary rock provide clues to the source of the sediment, and to the environment of deposition. To see how, let's follow the fate of rock fragments as they gradually move from a cliff face in the mountains via a river to the seashore. Different kinds of sediment develop along the route. Each kind, if buried and lithified, would yield a different type of sedimentary rock.

To start, imagine that some large blocks of rock tumble off a cliff and slam into other blocks already at the bottom. The impact shatters the blocks, producing a pile of angular fragments with sharp edges. If these fragments were to be cemented together, the resulting rock would be **breccia** (Fig. 7.4a). Later, a storm causes the fragments (clasts) to slide downslope into a turbulent river. In the water, clasts bang into each other and into the riverbed, a process that shatters them into still smaller pieces and breaks off their sharp edges. Angular clasts gradually become rounded clasts. When the river water slows, pebbles and cobbles stop moving and form a mound or bar of gravel. Burial and lithification of these rounded clasts produces **conglomerate** (Fig. 7.4b).

> **Did you ever wonder . . .**
> where beach sand comes from?

If the gravel stays put for a long time, it undergoes chemical weathering. As a consequence, cobbles and pebbles break apart into individual mineral grains, eventually producing a mixture of quartz, feldspar, and clay. Clay is so fine that it can be carried far downstream, leaving sand containing a mixture of quartz and some feldspar grains—this sediment, if buried and lithified, becomes **arkose** (Fig. 7.4c). Over time, feldspar grains in sand continue to weather into clay so that gradually, during successive events that wash the sediment downstream, the sand loses feldspar and ends up being composed almost entirely of durable quartz grains. Some of the sand may make it to the sea, where waves carry it to beaches, and some may end up in desert dunes. This sediment, when buried and lithified,

becomes quartz sandstone (Fig. 7.4d). Meanwhile, silt and clay may accumulate in the flat areas bordering streams, regions called floodplains (see Chapter 17) that become inundated only during floods. And some silt and mud settles in a wedge, called a delta, at the mouth of the river, or in lagoons or mudflats along the shore. The silt, when lithified, becomes **siltstone**, and the mud, when lithified, becomes **shale** or **mudstone** (Fig. 7.4e).

Two of the rock names in Table 7.1 do not appear in the above narrative, because they don't form in the depositional settings just described. Diamictite forms either from debris flows (viscous slurries consisting of mud mixed with larger clasts) both on land and underwater, or in glacial settings where ice deposits clasts of all sizes. Wacke typically forms from the deposits of submarine avalanches. (Most wacke has a grayish color, and thus geologists informally refer to it as graywacke). Note that diamictites and wackes are poorly sorted.

You may have sensed from this narrative that as sediment moves downstream, grains overall become smaller and rounder; grains composed of minerals that are susceptible to chemical weathering progressively break down and disappear; and sediment becomes better sorted. Geologists use the term *sediment maturity* to refer to the degree to which a sediment has evolved from being just a crushed-up version of the original rock to a sediment that has lost its easily weathered components and has become well sorted and rounded. Thus, we can say that an immature sandstone is one that contains angular clasts, both durable and easily weathered minerals, and is poorly sorted. In contrast, a mature sandstone is one that contains only well-sorted grains of resistant minerals (Fig. 7.3).

Biochemical Sedimentary Rocks

The Earth System involves interactions between living organisms and the physical planet. Numerous organisms

FIGURE 7.3 Grain characteristics, and their evolution with increasing transport and weathering.

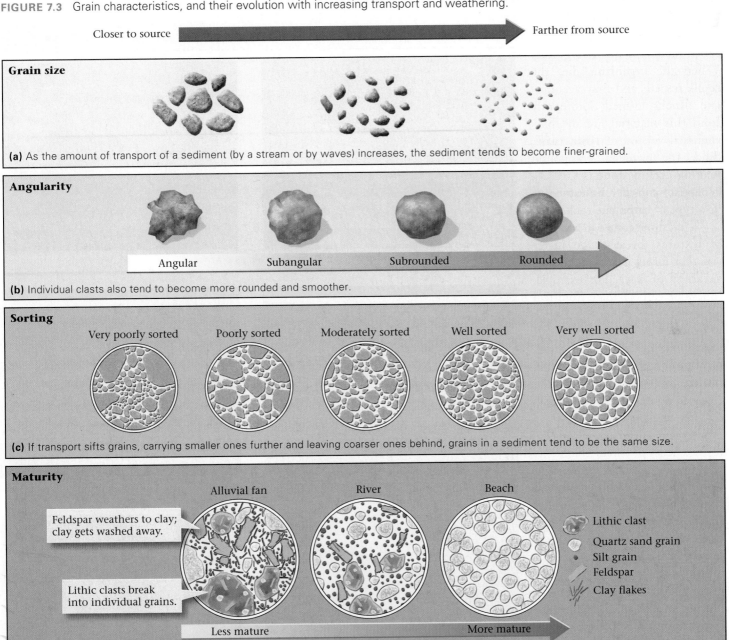

Closer to source → Farther from source

Grain size

(a) As the amount of transport of a sediment (by a stream or by waves) increases, the sediment tends to become finer-grained.

Angularity

Angular   Subangular   Subrounded   Rounded

(b) Individual clasts also tend to become more rounded and smoother.

Sorting

Very poorly sorted   Poorly sorted   Moderately sorted   Well sorted   Very well sorted

(c) If transport sifts grains, carrying smaller ones further and leaving coarser ones behind, grains in a sediment tend to be the same size.

Maturity

Alluvial fan   River   Beach

Feldspar weathers to clay; clay gets washed away.

Lithic clasts break into individual grains.

Lithic clast
Quartz sand grain
Silt grain
Feldspar
Clay flakes

Less mature → More mature

(d) A less mature sediment consists of fragments of the original rock, and contains both resistant and non resistant minerals. A mature sediment contains only well-sorted resistant minerals.

have evolved the ability to extract dissolved ions from seawater to make solid shells. Some organisms construct their shells out of calcium (Ca^{2+}) and carbonate (CO_3^{2-}) ions, which they merge to make the mineral calcite ($CaCO_3$) or its polymorph, aragonite, whereas other organisms make their shells out of dissolved silica (SiO_2). When the organisms die, the solid material in their shells survives and settles to the floor of lakes, seas, or rivers. This material, when lithified, comprises biochemical sedimentary rock. Geologists recognize several different types of biochemical sedimentary rocks, which we now describe.

Limestone (biochemical). A snorkeler gliding above a reef sees an incredibly diverse community of coral and algae, around which creatures such as clams, oysters, snails (gastropods), and lampshells (brachiopods) live, and above which plankton float (Fig. 7.5a). Though they look so different from each other, many of these organisms share an important characteristic: they make solid shells of calcium carbonate (CaCO₃), which occurs either as the minerals calcite or aragonite. When the organisms die, the shells remain and may accumulate. Rocks formed dominantly from this material are the biochemical version of **limestone**. Since the principal compound making up limestone is $CaCO_3$, geologists consider limestone to be a type of carbonate rock.

Limestone comes in a variety of textures, because the material that forms it accumulates in a variety of ways. For example, limestone can originate from reef builders (such as coral) that grew in place, from shell debris that was broken up and transported, or from carbonate mud that settled like snow out of water. Because of this variety, geologists distinguish among fossiliferous limestone, consisting of visible fossil shells or shell fragments (Fig. 7.5b); micrite, consisting of very fine carbonate mud; and chalk, consisting of plankton shells. Experts recognize many other types as well.

Typically, limestone is a massive light-gray to dark-bluish-gray rock that breaks into chunky blocks—it doesn't look much like a pile of shell fragments (Fig. 7.5c, d). That's because several processes take place that change the texture of the rock over time. Prior to lithification, organisms may burrow into recently formed or deposited shells and break them up, and may convert some to lime mud. Later, water passing through the rock not only precipitates cement but also dissolves some carbonate grains and causes new ones to grow. Thus, original crystals may be replaced by new ones; typically, all aragonite originally in shells transforms into calcite, a more stable mineral, and smaller crystals of calcite are replaced by larger ones.

FIGURE 7.4 Different kinds of clasts lithify into different kinds of sedimentary rocks.

Sediment ⟶ Lithification ⟶ Sedimentary rock

(a) Lithification of an accumulation of angular clasts yields breccia.

(b) Layers of river gravel lithify into conglomerate.

(c) Sediment deposited in an alluvial fan, close to its source, can be feldspar rich. Lithification of this sediment yields arkose.

Alluvial fan

Chert (biochemical).   If you walk beneath the northern end of the Golden Gate Bridge in California, you will find outcrops of reddish, almost porcelain-like rock occurring in 3- to 15-cm-thick layers (Fig. 7.6a). Hit it with a hammer, and the rock cracks, almost like glass, creating smooth, spoon-shaped (conchoidal) fractures. Geologists call this rock biochemical chert; it's made from cryptocrystalline quartz (*crypto* is Greek

(d) Layers of beach or dune sand lithify into sandstone.

Sandstone

Shale

(e) Layers of mud, exposed beneath marsh grass, lithify to form shale. Here, the thin-bedded shale is interbedded with sandstone.

for *hidden*), meaning quartz grains that are too small to be seen without the extreme magnification of an electron microscope. The chert beneath the Golden Gate Bridge formed from the shells of silica-secreting plankton that accumulated on the sea floor. Gradually, after burial, the shells dissolved, forming a silica-rich gel. Chert then precipitated from this gel.

Organic Sedimentary Rocks

We've seen how the mineral shells of organisms ($CaCO_3$ or SiO_2) can accumulate and lithify to become *biochemical* sedimentary rocks. What happens to the "guts" of the organisms—the cellulose, fat, carbohydrate, protein, and other organic compounds that make up living matter? Commonly, this organic debris gets eaten by other organisms or decays at the Earth's surface. But in some environments, such as oxygen-poor water in swamps, lagoons, or lakes, the debris settles along with other sediment and eventually gets buried. At the elevated temperatures and pressures that exist at depth below the surface, organic matter undergoes chemical reactions that transform it into organic sedimentary rock, distinctive from other sedimentary rock in that it contains abundant organic chemicals. Since the dawn of the industrial revolution (early 19th century), organic sedimentary rock has provided

the fuel of modern industry and transportation (see Chapter 14), for the organic chemicals within burn to produce energy. We will briefly mention two types, coal and oil shale.

Coal is a black, combustible rock consisting of over 50 to 90% carbon. The remainder consists of oxygen, nitrogen, hydrogen, sulfur, silica, and minor amounts of other elements. Typically, the carbon in coal occurs in large, complex organic molecules made of many rings—note that it does *not* occur in $CaCO_3$. As discussed further in Chapter 14, coal forms when plant remains have been buried deeply enough and long enough for the material to become compacted and to lose significant amounts of volatiles (hydrogen, water, CO_2, and ammonia); as the volatiles seep away, a concentration of carbon remains (Fig. 7.6b).

Not all organic rocks come from plant material. Organic material, in the form of chemicals derived from fats and proteins that made up the flesh of plankton or algae, can mix with mud and be incorporated in shale. The organic material, which tends to color shale black, may gradually transform into oil consisting of chain-like organic molecules. A fine-grained clastic rock containing a large amount of the organic precursor of oil is called *oil shale*.

Chemical Sedimentary Rocks

The colorful terraces, or mounds, that grow around the vents of hot-water springs; the immense layers of salt that underlie the floor of the Mediterranean Sea; the smooth, sharp point of an ancient arrowhead—these materials all have something in common. They all consist of rock formed primarily by the precipitation of minerals directly out of water solutions. We call such rocks chemical sedimentary rocks. They typically have a crystalline texture, partly formed during their original precipitation and partly when, at a later time, new crystals grow at the expense of old ones through a process called recrystallization.

Evaporites: The products of saltwater evaporation. In 1965, two daredevil drivers in jet-powered cars battled to be the first to set the land speed record of 600 mph. On November

FIGURE 7.5 The formation of carbonate rocks (limestone).

(a) In this modern coral reef, corals produce shells. If buried and preserved, these become limestone.

Relict of a small reef

(b) A Vermont quarry shows the gray color of 400-Ma limestone. The white mounds are relicts of small reefs.

(c) Wave energy breaks up shells, forming carbonate clasts that later become cemented together.

(d) Beds of bioclastic limestone in a Pennsylvania roadcut.

7, Art Arfons, in the *Green Monster*, peaked at 576.127 mph; but eight days later Craig Breedlove, driving the *Spirit of America*, reached 600.601 mph. Traveling at such speeds, a driver must maintain an absolutely straight line; any turn will catapult the vehicle out of control. Thus, high-speed trials take place on extremely long and flat racecourses. Not many places can provide such conditions—the Bonneville Salt Flats of Utah do. The salt flats formed by the evaporation of an ancient salt lake. Under the heat of the Sun, the water turned to vapor and drifted up into the atmosphere, but the salt that had been dissolved in the water stayed behind. Such salt precipitation occurs wherever saturated saltwater develops—along desert lakes with no outlet and along margins of restricted seas (Fig. 7.7a–c). For thick deposits of salt to form, large volumes of water must evaporate. This may happen when plate tectonic movements temporarily cut off arms of the sea (as we saw in the case of the Mediterranean Sea; see Interlude B) or during continental rifting, when seawater first begins to spill into the rift valley.

Because salt deposits form as a consequence of evaporation, geologists refer to them as **evaporites**. The specific type of salt minerals comprising an evaporite depends on the amount of evaporation. When 80% of the water evaporates, gypsum

FIGURE 7.6 Examples of biochemical and organic sedimentary rocks.

(a) This bedded chert developed on the deep-sea floor by the deposition of plankton that secrete silica shells.

Sandstone and shale

Coal layer

(b) Coal is deposited in layers (beds), just like other kinds of sedimentary rocks.

FIGURE 7.7 The formation of evaporite deposits.

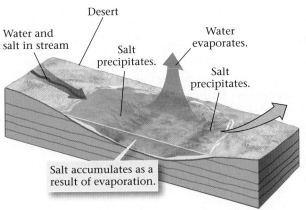

Desert

Water and salt in stream

Salt precipitates.

Water evaporates.

Salt precipitates.

Salt accumulates as a result of evaporation.

Salt on the floor of Death Valley, California

(a) In lakes with no outlet, tiny amounts of salt brought in by streams stay behind as the water evaporates. When the water evaporates entirely, a white crust of salt remains.

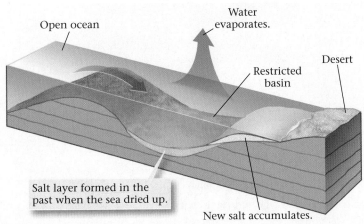

Open ocean

Water evaporates.

Restricted basin

Desert

Salt layer formed in the past when the sea dried up.

New salt accumulates.

(b) Salt precipitation can also occur along the margins of a restricted marine basin, if saltwater evaporates faster than it can be resupplied.

(c) Thick layers of salt may be buried deeply. Here, salt is being mined deep underground.

forms; and when 90% of the water evaporates, halite precipitates. If seawater were to evaporate entirely, the resulting evaporite would consist of 80% halite, 13% gypsum, and the remainder of other salts and carbonates.

Travertine (chemical limestone). **Travertine** is a rock composed of crystalline calcium carbonate ($CaCO_3$) formed by chemical precipitation from groundwater that has seeped out at the ground surface either in hot- or cold-water springs, or on the walls of caves. What causes this precipitation? It happens, in part, when the groundwater degasses, meaning that some of the carbon dioxide that had been dissolved in the groundwater bubbles out of solution, for removal of carbon dioxide decreases the ability of the water to hold dissolved carbonate. Precipitation also occurs when water evaporates, thereby increasing the concentration of carbonate. Various kinds of microbes live in the environments in which travertine accumulates, so biologic activity may contribute to the precipitation process. Travertine produced at springs forms terraces and mounds that are meters or even hundreds of meters thick (Fig. 7.8a). Spectacular terraces of travertine grew at Mammoth Hot Springs in Yellowstone National Park. Amazing column-like mounds of travertine grew up from the floor of Mono Lake, California, where hot springs seeped into the cold water of the lake. Travertine also grows on the walls of caves where groundwater seeps out (Fig. 7.8b). In cave settings, travertine builds up beautiful and complex growth forms called speleothems (see Chapter 19).

Travertine has been quarried for millennia to make building stones and decorative stones. The rock's beauty comes in part because in thin slices it is translucent, and in part because it typically displays colored growth bands. Bands develop in response to changes in the composition of groundwater, or in the environment into which the water drains. Some travertines (a type called tufa) contain abundant large pores (open spaces).

Dolostone. Another carbonate rock, **dolostone**, differs from limestone in that it contains the mineral dolomite ($CaMg[CO_3]_2$). Where does the magnesium come from? Most dolostone forms by a chemical reaction between solid calcite and magnesium-bearing groundwater. Much of the dolostone you may find in an outcrop actually originated as limestone but later changed as dolomite crystals replaced calcite. This change may take place beneath lagoons along a shore soon after the limestone formed, or a long time later, after the limestone has been buried deeply.

Chert (replacement). A tribe of Native Americans, the Onondaga, once lived off the land in eastern New York State. Here, outcrops of limestone contain layers or nodules (lenses or lumps) of a black chert (Fig. 7.9a). Because of the way it breaks, the tribe's artisans could fashion sharp-edged tools (arrowheads and scrapers) from this chert, so the Onondaga collected it for their own toolmaking industry and for use in trade with other people. Unlike the deep-sea (biochemical) chert described earlier, the chert collected by the Onondaga formed when cryptocrystalline quartz gradually *replaced* calcite crystals within a body of limestone long after the limestone was deposited; geologists call such material replacement chert.

> **Did you ever wonder...**
> how flint used for arrowheads first formed?

FIGURE 7.8 Examples of travertine (chemical limestone) deposits.

(a) Travertine accumulates in terraces at Mammoth Hot Springs in Yellowstone Park, Wyoming.

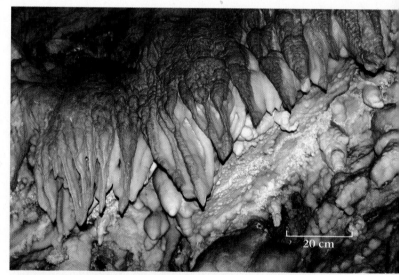

(b) Travertine speleothems form as calcite-rich water drips from the ceiling of Timpanogos Cave in Utah.

FIGURE 7.9 Examples of chert that precipitated in place.

(a) Replacement chert forms as layers of nodules between tilted limestone beds in New York.

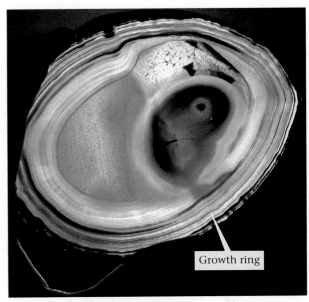

(b) A thin slice of Brazilian agate, lit from the back, shows growth rings.

Chert comes in many colors (black, white, red, brown, green, gray), depending on the impurities it contains. Petrified wood is chert that forms when silica-rich sediment, such as ash from a volcanic eruption, buries a forest. Dissolved silica precipitates as cryptocrystalline quartz within wood, gradually replacing the wood's cellulose. The chert deposit retains the shape of the wood and the growth rings within it. Some chert, known as agate, precipitates in concentric rings inside hollows in a rock and ends up with a striped appearance, caused by variations in the content of impurities while precipitation took place (Fig. 7.9b).

Take-Home Message

- The grains that comprise clastic sedimentary rocks (such as sandstone) come from weathering and erosion of preexisting rock.
- Clasts are transported by water, wind, or ice. When fluids slow, or ice melts, deposition takes place. After burial, compaction and cementation cause lithification.
- Geologists distinguish among different types of clastic sedimentary rocks based on clast size.
- Biochemical sedimentary rocks, such as limestone, consist of the shells of organisms.
- Organic sedimentary rocks, such as coal, form from the organic remains of organisms.
- Chemical sedimentary rocks precipitate from water solutions.

THINK: Do all chemical sedimentary rocks have the same composition? Why or why not?

7.3 SEDIMENTARY STRUCTURES

Geologists use the term **sedimentary structure** for the layering of sedimentary rocks, for surface features on layers formed during deposition, and for the arrangement of grains within layers. Here, we examine some of the more important types.

Bedding and Stratification

Let's start by introducing the jargon for discussing sedimentary layers. A single layer of sediment or sedimentary rock with a recognizable top and bottom is called a **bed**; the boundary between two beds is a bedding plane; several beds together constitute **strata** (singular stratum, from the Latin *stratum*, meaning pavement); and the overall arrangement of sediment into a sequence of beds is bedding, or stratification. From the word *strata*, we derive other words, such as *stratigrapher* (a geologist who specializes in studying strata) and *stratigraphy* (the study of the record of Earth history preserved in strata).

When you examine strata in a region with good exposure, the bedding generally stands out clearly. Beds appear as bands or stripes across a cliff face (Fig. 7.10). Typically, a contrast in rock type distinguishes one bed from adjacent beds. For example, a sequence of strata may contain a bed of sandstone, overlain by a bed of shale, overlain by a bed of limestone. Each bed has a definable thickness (from a couple of centimeters to tens of meters) and may display a contrast in composition, color, and/or grain size, that distinguishes it

FIGURE 7.10　Bedding in sedimentary rocks.

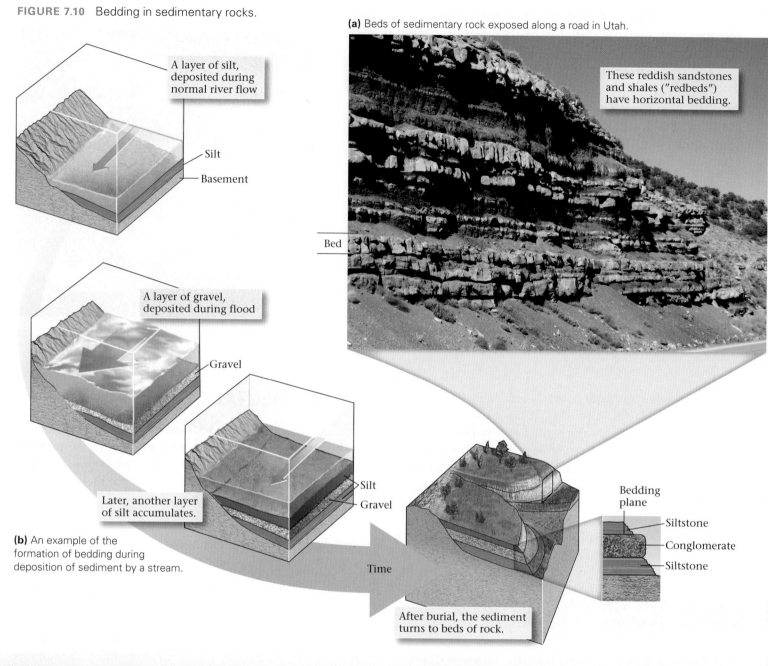

(a) Beds of sedimentary rock exposed along a road in Utah.

A layer of silt, deposited during normal river flow

Silt

Basement

These reddish sandstones and shales ("redbeds") have horizontal bedding.

Bed

A layer of gravel, deposited during flood

Gravel

Later, another layer of silt accumulates.

Silt

Gravel

(b) An example of the formation of bedding during deposition of sediment by a stream.

Time

After burial, the sediment turns to beds of rock.

Bedding plane

Siltstone

Conglomerate

Siltstone

(c) Differential weathering makes stratification stand out. It's clear—even from a distance—that these hills near Las Vegas, Nevada, consist of sedimentary rocks.

FIGURE 7.11 The concept of a stratigraphic formation. 🔊

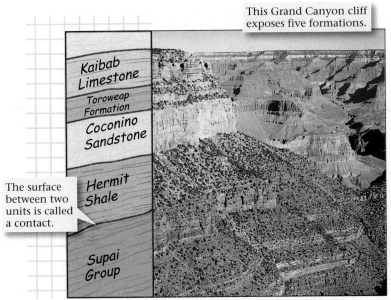

This Grand Canyon cliff exposes five formations.

Kaibab Limestone

Toroweap Formation

Coconino Sandstone

The surface between two units is called a contact.

Hermit Shale

Supai Group

(a) The names of formations consisting of one rock type may indicate the rock type (e.g., Kaibab Limestone). The name of a formation including more than one rock type includes the word *formation* (Toroweap Formation). Several related formations comprise a group (Supai Group).

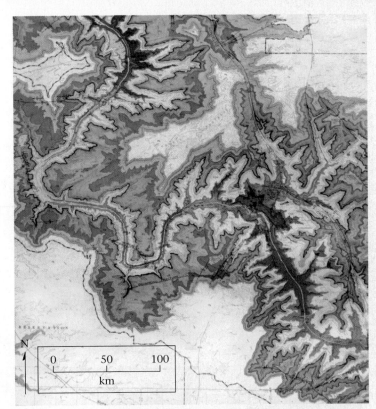

(b) A geologic map portrays the distribution of formations in a portion of the Grand Canyon. Each color band is a specific formation.

from its neighbors. But in many examples, adjacent beds all appear to have the same composition. In such cases, bedding may be defined by subtle changes in grain size, by surfaces that represent interruptions in deposition, or by cracks that formed parallel to bed surfaces.

Why does bedding form? To find the answer, we need to think about how sediment accumulates. Changes in the climate, water depth, current velocity, or the sediment source control the type of sediment deposited at a location at a given time. For example, on a normal day a slow-moving river may carry only silt, which collects on the riverbed (Fig. 7.10b). During a flood, the river flows faster and carries sand and pebbles, so a layer of sandy gravel forms over the silt layer. Then, when the flooding stops, more silt buries the gravel. If this succession of sediments become lithified and exposed for you to see, they appear as alternating beds of siltstone and sandy conglomerate. Bedding is not always well preserved. In some environments, burrowing organisms disrupt the layering. Worms, clams, and other creatures churn sediment and may leave behind burrows; this process is called bioturbation.

During geologic time, long-term changes in a depositional environment can take place. Thus, a given sequence of strata may differ markedly from sequences of strata above or below. A sequence of strata that is distinctive enough to be traced as a package across a fairly large region is called a **stratigraphic formation**, or simply a "formation" (Fig. 7.11a 🔊). For example, a region may contain a succession of alternating sandstone and shale beds deposited by rivers, overlain by beds

of marine limestone deposited later when the region was submerged by the sea. A stratigrapher might identify the sequence of sandstone and shale beds as one formation and the sequence of limestone beds as another. Formations are often named after the locality where they were first found and studied. A map that portrays the distribution of stratigraphic formations is called a **geologic map** (Fig. 7.11b 🔊).

Ripple Marks, Dunes, and Cross Bedding: Consequences of Deposition in a Current

Many clastic sediments accumulate in moving fluids (wind, rivers, or waves). Fascinating sedimentary structures develop at the interface between the sediment and the fluid. These structures are called bedforms. Bedforms that develop at a given location reflect factors such as the velocity of the flow and the size of the clasts. Though there are many types of bedforms, we'll focus on only two—ripple marks and dunes. The growth of both produces cross bedding, a special type of lamination within beds.

Ripple marks are relatively small (generally no more than a few centimeters high), elongated ridges that form on a bed surface at right angles to the direction of current flow. If the current always flows in the same direction, the ripple marks are asymmetric, with a steeper slope on the downstream (lee) side (Fig. 7.12a). Along the shore, where water flows back and forth due to wave action, ripples tend to be symmetric. The crest (the high ridge) of a symmetric ripple is a sharp ridge, whereas the trough between adjacent ridges is a

FIGURE 7.12 Ripple marks, a type of sedimentary structure, are visible on the surface of modern and ancient beds.

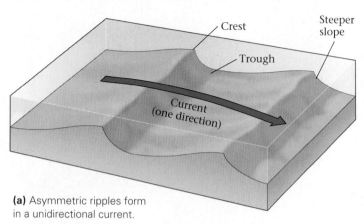

(a) Asymmetric ripples form in a unidirectional current.

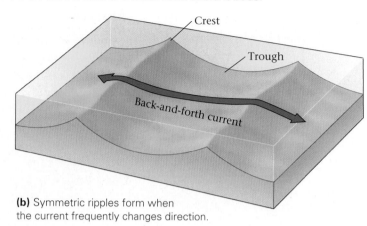

(b) Symmetric ripples form when the current frequently changes direction.

(c) Modern ripples exposed at low tide along a sandy beach on the shore of Cape Cod, Massachusetts.

(d) These 145-million-year-old ripples are preserved on a tilted bed of solid sandstone at Dinosaur Ridge, Colorado.

smooth, concave-up curve (Fig. 7.12b). You can find ripples on modern beaches and preserved on bedding planes of ancient rocks (Fig. 7.12c, d).

Dunes look like ripples, only they are larger. For example, dunes on the bed of a stream may be tens of centimeters high, and wind-formed dunes formed in deserts may be tens to over 100 meters high.

If you examine a vertical slice cut into a ripple or dune, you will find distinct internal laminations that are inclined at an angle to the boundary of the main sedimentary layer. Such laminations are called **cross beds**. To see how cross beds develop, imagine a current of air or water moving uniformly in one direction (Fig. 7.13a, b ▶). The current erodes and picks up clasts from the upstream part of the bedform and deposits them on the downstream or leeward part. Sediment builds up on the leeward side until gravity causes it to slip down. With time, the leeward side of the bedform builds in the downstream direction. The surface of the slip face establishes the shape of the cross beds. Eventually, a new cross-bedded layer builds out over a preexisting one. The boundary between two successive layers is called the main bedding, and the internal curving surfaces within the layer constitute the cross bedding (Fig. 7.13c, d ▶).

Turbidity Currents and Graded Beds

Sediment deposited on a submarine slope might not stay in place forever. For example, an earthquake or storm might disturb this sediment and cause it to slip downslope. If the sediment is loose enough, it mixes with water to create a murky, turbulent cloud. This cloud is denser than clear water and thus flows downslope like an underwater avalanche (Fig. 7.14a–c). We call this moving submarine suspension of sediment a **turbidity current**. Downslope, the turbidity current slows, and the sediment that it has carried starts to settle out. Larger grains sink faster through a fluid than do finer grains, so the coarsest sediment settles out first. Progressively finer grains accumulate on top, with the finest

FIGURE 7.13 Cross bedding, a type of sedimentary structure within a bed. ⏸

(a) Dunes in Death Valley, Califorinia.

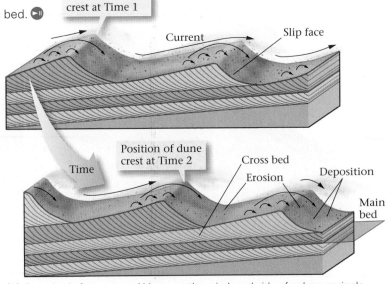

(b) Cross beds form as sand blows up the windward side of a dune or ripple and then accumulates on the slip face. With time, the dune crest moves.

(c) Slip face of a small sand dune, Death Valley. The edges of the slip face are highlighted.

(d) A cliff face in Zion National Park, Utah, displays large cross beds formed between 200 and 180 million years ago, when the region was a desert with large sand dunes.

sediment (clay) settling out last. This process forms a **graded bed**—that is, a layer of sediment in which grain size varies from coarse at the bottom to fine at the top. Geologists refer to a deposit from a turbidity current as a turbidite.

Bed-Surface Markings

A number of features appear on the surface of a bed as a consequence of events that happen during deposition or soon after, while the sediment layer remains soft. These bed-surface markings include the following.

- *Mud cracks:* If a mud layer dries up after deposition, it cracks into roughly hexagonal plates that typically curl up at their edges. We refer to the openings between the plates as mud cracks. Later, these fill with sediment and can be preserved (Fig. 7.15a, b).

- *Scour marks:* As currents flow over a sediment surface, they may erode small troughs, called scour marks, parallel to the current flow. These indentations can be buried and preserved.

- *Fossils:* Fossils are relics of past life. Some fossils are shell imprints or footprints on a bedding surface (see Interlude E).

FIGURE 7.14 The development of graded bedding from turbidity currents.

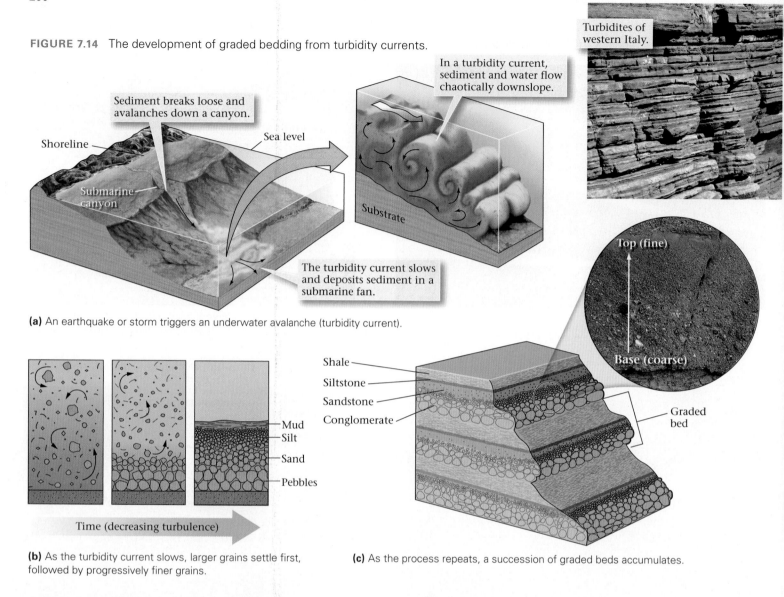

Sediment breaks loose and avalanches down a canyon.

Shoreline

Submarine canyon

Sea level

In a turbidity current, sediment and water flow chaotically downslope.

Substrate

The turbidity current slows and deposits sediment in a submarine fan.

Turbidites of western Italy.

(a) An earthquake or storm triggers an underwater avalanche (turbidity current).

Top (fine)

Base (coarse)

Mud
Silt
Sand
Pebbles

Time (decreasing turbulence)

Shale
Siltstone
Sandstone
Conglomerate

Graded bed

(b) As the turbidity current slows, larger grains settle first, followed by progressively finer grains.

(c) As the process repeats, a succession of graded beds accumulates.

FIGURE 7.15 Mud cracks, a sedimentary structure formed by drying out of mud.

20 cm

(a) Mud cracks in red mud at Bryce Canyon, Utah. Note how the edges of the mud plates curl up.

(b) Mud cracks preserved in a 410-million-year-old bed exposed on the base of a cliff in New York.

Why Study Sedimentary Structures?

Sedimentary structures are not just a curiosity, but are important clues that help geologists understand the environment in which clastic sedimentary beds were deposited. For example, the presence of ripple marks and cross bedding indicates that layers were deposited in a current; the presence of mud cracks indicates that the sediment layer was exposed to the air and dried out on occasion; and graded beds indicate deposition by turbidity currents. Also, fossil types can tell us whether sediment was deposited along a river or in the deep sea, for different species of organisms live in different environments. In the next section of this chapter, we examine these environments in greater detail.

Take-Home Message

- Sediments are deposited in layers or beds.
- A distinct succession of beds that can be traced over a region is a stratigraphic formation.
- Bedforms develop because of the interaction between sediment and currents. Examples include ripples and dunes, within which there are cross beds.
- The surfaces of beds may display distinctive structures that formed during or after deposition.

THINK: Which sedimentary structures tell you that sediment was deposited in a current?

7.4 HOW DO WE RECOGNIZE DEPOSITIONAL ENVIRONMENTS?

Geologists refer to the conditions in which sediment was deposited as the **depositional environment**. Examples include beach, glacial, and river environments. To identify these environments, geologists, like detectives, look for such clues in the rocks. For example, grain size, composition, sorting, and roundness of clasts, tell us how far the sediment has traveled from its source and whether it was deposited by the wind, by a fast-moving current, or in a stagnant body of water. Fossil content and sedimentary structures can help us decide whether the sediments were deposited on land, just off the coast, or in the deep sea. Let's now look at some examples of different depositional environments and the sediments deposited in them, by imagining that we are taking a journey from the mountains to the sea, examining sediments as we go (See Geology at a Glance on pp. 202–203 and See for Yourself H on p. S-14). We will see that geologists distinguish among three basic categories of depositional environment: terrestrial, coastal, and marine.

Terrestrial (Nonmarine) Sedimentary Environments

Terrestrial depositional environments are those which develop inland, far enough away from the shoreline that they are not affected by ocean tides and waves. The sediments of terrestrial deposits do *not* accumulate under seawater—rather, they settle on dry land, or under and adjacent to freshwater streams, glaciers, and lakes. In some settings, oxygen in surface water or groundwater reacts with the iron in terrestrial sediments to produce rust-like iron-oxide minerals, which give the sediment an overall reddish hue. Strata which have this hue are called, informally, "redbeds."

Glacial environments. We begin high in the mountains, where it's so cold that more snow collects in the winter than melts away, so glaciers—rivers or sheets of ice—develop and slowly flow. Because ice is a solid, it can move sediment of any size. So as a glacier moves down a valley in the mountains, it carries along *all* the sediment that falls on its surface from adjacent cliffs or gets plucked from the ground at its base or sides. At the end of the glacier, where the ice finally melts away, it drops its sedimentary load and makes a pile of "glacial till" (Fig. 7.16a). Till is unsorted and unstratified—it contains clasts ranging from clay size to boulder size all mixed together—and thus becomes a type of diamicton (see Table 7.1). We'll provide further details on glacial sediments in Chapter 22.

Mountain stream environments. As we walk down beyond the end of the glacier, we enter a realm where turbulent streams rush downslope in steep-sided valleys. This fast-moving water has the power to carry large clasts; in fact, during floods, boulders and cobbles can tumble down the stream bed. Between floods, when water flow slows, the largest clasts settle out to form gravel and boulder beds, while the stream carries finer sediments like sand and mud farther downstream (Fig. 7.16b). Sedimentary deposits of a mountain stream would, therefore, include breccia and conglomerate (depending on the degree of rounding).

Alluvial-fan environments. Our journey now takes us to the mountain front, where the fast-moving stream empties onto a plain. In arid regions, where there is not enough water for the stream to flow continuously, the stream deposits its load of sediment near the mountain front, producing a wedge-shaped apron of gravel and sand called an alluvial fan (Fig. 7.16c). Deposition takes place here because when the stream pours from a canyon mouth and spreads out over a broader region, friction with the ground causes the water to slow down, and slow-moving water does not have the power to move coarse sediment. The sand here still contains feldspar grains, for these have not yet weathered into clay. Alluvial-fan sediments become arkose and conglomerate.

The Formation of Sedimentary Rocks

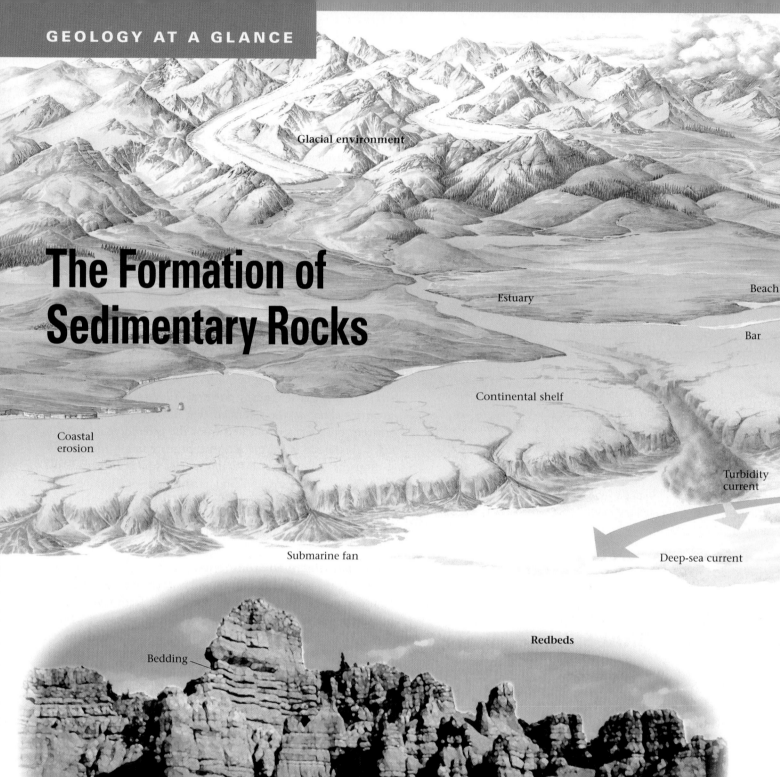

Glacial environment

Estuary

Beach

Bar

Continental shelf

Coastal erosion

Turbidity current

Submarine fan

Deep-sea current

Bedding

Redbeds

Categories of sedimentary rocks include clastic sedimentary rocks, chemical sedimentary rocks (formed from the precipitation of minerals out of water), and biochemical sedimentary rocks (formed from the shells of organisms). Clastic sedimen- tary rocks develop when grains (clasts) break off preexisting rock by weathering and erosion and are transported to a new location by wind, water, or ice; the grains are deposited to create sediment layers, which are then cemented together.

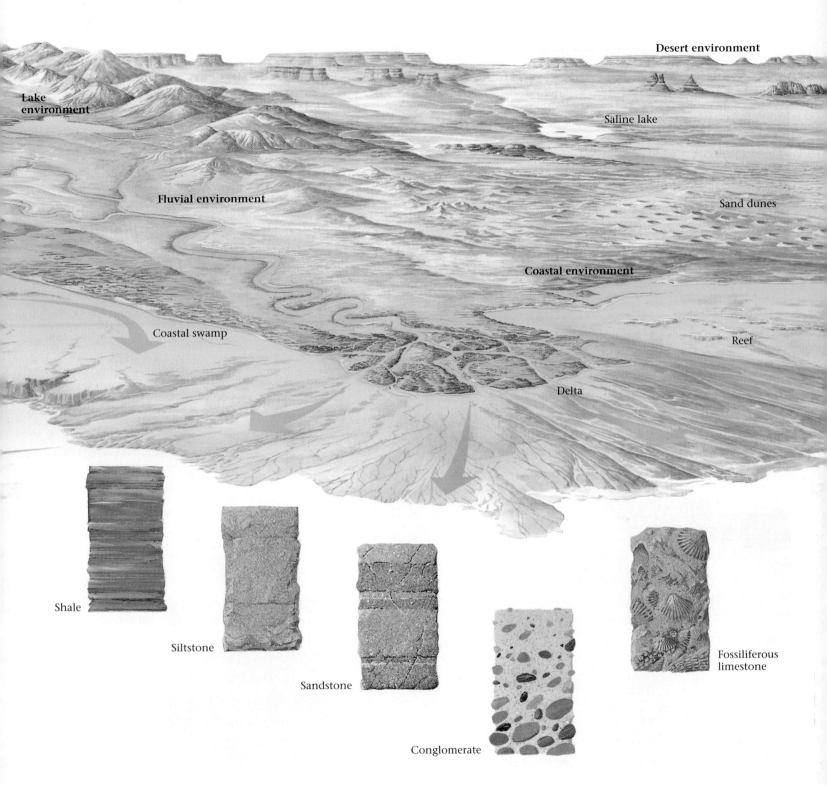

Lake environment

Desert environment

Saline lake

Fluvial environment

Sand dunes

Coastal environment

Coastal swamp

Reef

Delta

Shale

Siltstone

Sandstone

Conglomerate

Fossiliferous limestone

We distinguish among types of clastic sedimentary rocks on the basis of grain size.

The character of a sedimentary rock depends on the composition of the sediment and on the environment in which it accumulated. For example, glaciers carry sediment of all sizes, so they leave deposits of poorly sorted (different-sized) till; streams deposit coarser grains in their channels and finer ones on floodplains; a river slows down at its mouth and deposits an immense pile of silt in a delta. Fossiliferous limestone develops on coral reefs. In desert environments, sand accumulates into dunes and evaporites precipitate in saline lakes. Offshore, submarine canyons channel avalanches of sediment, or turbidity currents, out to the deep-sea floor.

Sedimentary rocks tell the history of the Earth. For example, the layering, or bedding, of sedimentary rocks is initially horizontal. So where we see layers bent or folded, we can conclude that the layers were deformed during mountain building. Where horizontal layers overlie folded layers, we have an unconformity: for a time, sediment was not deposited, and/or older rocks were eroded away.

FIGURE 7.16   Examples of nonmarine depositional environments. ▶❙❙

(a) Glacial till at the end of a glacier in France.

(b) Boulders and cobbles deposited by a mountain stream in Colorado.

(c) An alluvial fan in Death Valley, California.

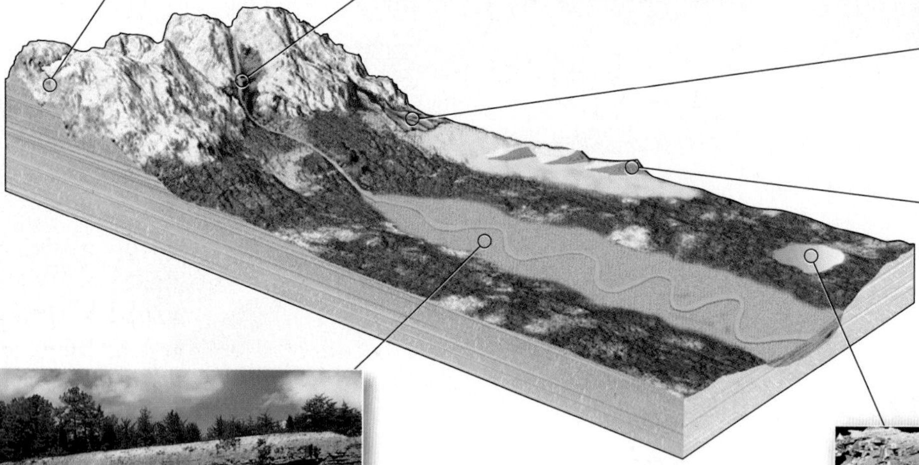

(d) Sand dunes in Brazil.

(e) Deposits of an ancient river channel in Indiana. Note how the floor of the channel cuts across older strata. The geologist's sketch emphasizes the relationship.

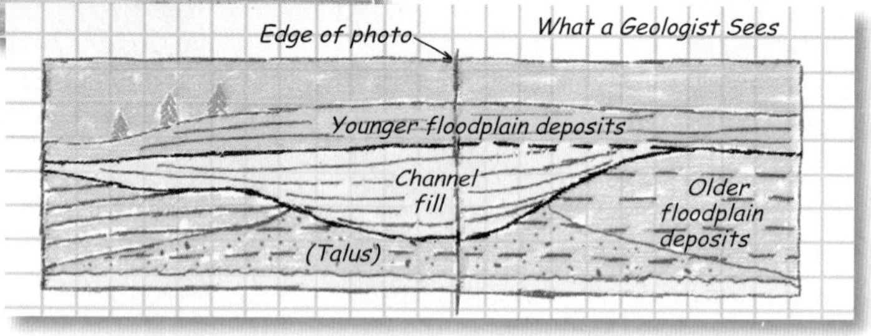

(f) Laminated mud from a lake bed.

What a Geologist Sees

Edge of photo

Younger floodplain deposits

Channel fill

Older floodplain deposits

(Talus)

Sand-dune environments.   If the climate is very dry, few plants can grow and the ground surface lies exposed. Strong winds can move dust and sand. The dust gets carried away, and the resulting well-sorted sand can accumulate in dunes. Thus, thick layers of well-sorted sandstone, in which we see large cross beds, are relics of desert sand-dune environments (Fig. 7.16d).

River environments.   In climates where streams flow, we find several distinctive depositional environments. Rivers transport gravel, sand, silt, and mud. The coarser sediments tumble along the bed in the river's channel and collect in cross-bedded, rippled layers while the finer sediments drift along, suspended in the water. This fine sediment settles out along the banks of the river, or on the floodplain, the flat land on either side of the river that is covered with water only during floods. On the floodplain, mud layers dry out between floods, leading to the formation of mud cracks. River sediments lithify to form sandstone, siltstone, and shale. Typically, channels of coarser

FIGURE 7.17 A simple "Gilbert-type" delta formed where a stream enters a lake.

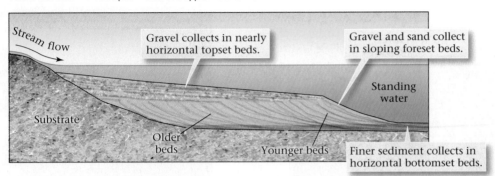

sediment are surrounded by layers of fine-grained floodplain deposits; in cross section, the channel has a lens-like shape (Fig. 7.16e ⏵). Geologists commonly refer to river deposits as fluvial sediments, from the Latin word *fluvius*, for river.

Lake environments. In temperate climates, where water remains at the surface throughout the year, lakes form. In lakes, the relatively quiet water can't move coarse sediment; any coarse sediment brought into the lake by a stream settles out at the stream's outlet. Only fine clay makes it out into the center of the lake, where it settles to form mud on the lake bed. Thus, lake sediments typically consist of finely laminated shale (Fig. 7.16f). At the mouths of streams that empty into lakes, small deltas may form. A **delta** is a wedge of sediment that accumulates where moving water enters standing water. Deltas were so named because the map shape of some deltas resembles the Greek letter *delta* (Δ), as we discuss further in Chapter 17. In 1885, an American geologist named G. K. Gilbert showed that such deltas contain three components (Fig. 7.17): topset beds composed of gravel, foreset beds of gravel and sand, and silty bottomset beds.

Coastal and Marine Sedimentary Environments

Along the sea shore, a variety of distinct coastal environments occur; the character of each reflects the nature of the sediment supply and the climate. Marine environments start at the high tide line and extend offshore, to include the deep ocean floor. The type of sediment deposited at a location depends on the environment, water depth, and whether or not clastic grains are available.

Marine delta deposits. After following the river downstream for a long distance, we reach its mouth, where it empties into the sea. Here, the river builds a delta of sediment out into the sea. River water stops flowing when it enters the sea, so sediment settles out.

Large deltas are much more complex than the lake examples that Gilbert studied, for they include many different sedimentary environments including swamps, channels, floodplains, and submarine slopes. Sea-level changes may cause the position of the different environments to move with time. Thus,

deposits of an ocean-margin delta produce a great variety of sedimentary rock types (Fig. 7.18a).

Coastal beach sands. Now we leave the delta and wander along the coast. Oceanic currents transport sand along the coastline. The sand washes back and forth in the surf, so it becomes well sorted (waves winnow out mud and silt) and well rounded, and because of the back-and-forth movement of ocean water over the sand, the sand surface may become rippled (Fig. 7.18b). Thus, if you find well-sorted, medium-grained sandstone, perhaps with ripple marks, you may be looking at the remnants of a beach environment.

Shallow-marine clastic deposits. From the beach, we proceed offshore. In deeper water, where wave energy does not stir the sea floor, finer sediment can accumulate. Because the water here may be only meters to a few tens of meters deep, geologists refer to this depositional setting as a shallow-marine environment. Clastic sediments that accumulate in this environment tend to be fine-grained, well-sorted, well-rounded silt, and they are inhabited by a great variety of organisms such as mollusks and worms. Thus, if you see beds of siltstone and mudstone containing marine fossils, you may be looking at shallow-marine clastic deposits.

Shallow-water carbonate environments. In shallow-marine settings in places where relatively little clastic sediment (sand and mud) enters the water, warm, clear, nutrient-rich water hosts an abundance of organisms. Their shells, which consist of carbonate minerals, make up most of the sediment that accumulates (Fig. 7.19a, b). The nature of carbonate sediment depends on the water depth. Beaches collect sand composed of shell fragments; lagoons (protected bodies of quiet water) are sites where carbonate mud accumulates; and reefs consist of coral and coral debris. Farther offshore of a reef, we can find a sloping apron of reef fragments. Shallow-water carbonate environments transform into various kinds of limestone.

Deep-marine deposits. We conclude our journey by sailing far offshore. Along the transition between coastal regions and the deep ocean, turbidity currents deposit graded beds. In the deep-ocean realm, only fine clay and plankton provide a source for sediment. The clay eventually settles out onto the deep-sea floor, forming deposits of finely laminated mudstones, and plankton shells settle to form chalk (from calcite shells; Fig. 7.20a, b) or chert (from siliceous shells). Thus, deposits of mudstone, chalk, or bedded chert indicate a deep-marine origin.

FIGURE 7.18 Examples of coastal depositional environments.

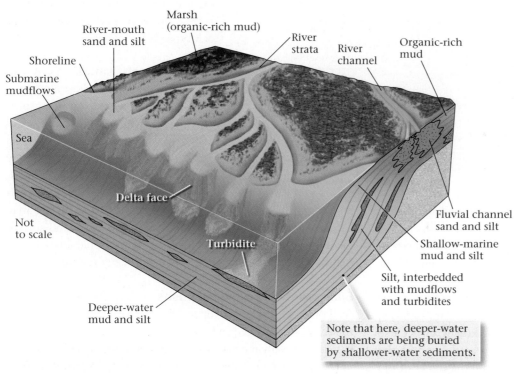

(a) A major river delta along an ocean coast is a complex depositional environment. Sea-level changes affect locations of depositional settings.

(b) Waves on this California beach wash and sort the sand.

FIGURE 7.19 Reef environments for the deposition of carbonate rocks.

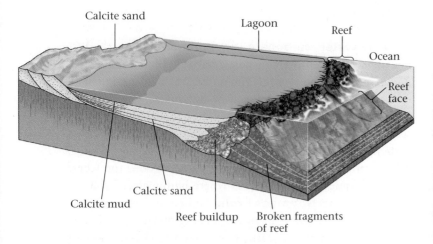

(a) Carbonate reefs form along shorelines in warm-water environments. In detail, reefs include many distinct depositional environments.

(b) A dramatic reef surrounds an island in the tropical Pacific. Deeper water is darker. Note the surf along the edge of the reef.

Take-Home Message

- A depositional environment is a setting in which sediment accumulates. There are three main categories: terrestrial, coastal, and marine.
- Terrestrial environments include streams, glaciers, alluvial fans, and lakes.
- Coastal environments include beaches and the near-shore parts of deltas.
- Marine environments include reefs and deeper ocean water.

THINK: How can you distinguish sediment deposited in an alluvial fan from sediment deposited in a shallow marine environment?

FIGURE 7.20 Examples of deep-marine sediment.

(a) These plankton shells, which make up some kinds of deep-marine sediment, are so small that they could pass through the eye of a needle.

(b) The chalk cliffs of southeastern England consist of plankton shells deposited on the sea floor tens of millions of years ago.

7.5 SEDIMENTARY BASINS

The sedimentary veneer on the Earth's surface varies greatly in thickness. If you stand in central Siberia or south-central Canada, you will find yourself on igneous and metamorphic basement rocks that are over a billion years old—sedimentary rocks are nowhere in sight. Yet if you stand along the southern coast of Texas, you would have to drill through over 15 km of sedimentary beds before reaching igneous and metamorphic basement. *Thick accumulations of sediment form only in special regions* where the surface of the Earth's lithosphere sinks, providing space in which sediment collects. Geologists use the term **subsidence** to refer to the process by which the surface of the lithosphere sinks, and the term **sedimentary basin** for the sediment-filled depression. In what geologic settings do sedimentary basins form? An understanding of plate tectonics theory provides the answers.

Categories of Basins in the Context of Plate Tectonics Theory

Geologists distinguish among different kinds of sedimentary basins on the basis of the region of a lithosphere plate in which the basin formed as defined in the context of plate tectonics theory. Let's consider a few examples.

- *Rift basins*: These form in continental rifts, regions where the lithosphere has been stretched. During the early stages of rifting, the surface of the Earth subsides simply because crust becomes thinner as it stretches. (To picture this process, imagine pulling on either end of a block of clay with your hands—as the clay stretches, the central region of the block thins and sinks lower than the ends.) As the rift grows, slip on faults drops blocks of crust down, producing low areas bordered by narrow mountain ridges. Alluvial-fan deposits form along the base of the mountains, and salt flats or lakes develop in the low areas between the mountains.

  Thinning is not the only reason that rifted lithosphere subsides. During rifting, warm asthenosphere rises beneath the rift and heats up the thin lithosphere. When rifting ceases, the rifted lithosphere then cools, thickens, and becomes denser. This heavier lithosphere sinks down, causing more subsidence, just as the deck of a tanker ship drops to a lower elevation when the ship is filled with ballast. Sinking due to cooling of the lithosphere is called thermal subsidence.

- *Passive-margin basins*: These form along the edges of continents that are not plate boundaries. They are underlain by stretched lithosphere, the remnants of a rift whose evolution successfully led to the formation of a mid-ocean ridge. Passive-margin basins form because thermal subsidence of stretched lithosphere continues long after rifting ceases and sea-floor spreading begins. They fill with sediment carried to the sea by rivers and with carbonate rocks formed in coastal reefs. Sediment in a passive-margin basin can reach an astounding thickness of 15 to 20 km.

- *Intracontinental basins*: These develop in the interiors of continents, initially because of subsidence over a rift. They may continue to subside in pulses even hundreds of millions of years after they first formed, for reasons that are not yet well understood. Illinois and Michigan are each underlain with an intracontinental basin (the Illinois basin and the Michigan basin, respectively) in which up to 7 km of sediment has accumulated. Most of this sediment is fluvial, deltaic, or shallow marine. At times, extensive swamps formed along the shoreline in these basins. The plant matter of these swamps was buried to form coal.

■ *Foreland basins*: These form on the continent side of a mountain belt because the forces produced during convergence or collision push large slices of rock up faults and onto the surface of the continent. The weight of these slices pushes down on the surface of the lithosphere, producing a wedge-shaped depression adjacent to the mountain range that fills with sediment eroded from the range. Fluvial and deltaic strata accumulate in foreland basins.

As indicated by the above descriptions, different assemblages and thicknesses of sedimentary rocks form in different sedimentary basins. So geologists may be able to determine the nature of the basin in which ancient sedimentary deposits accumulated by looking at the character and thickness of the deposits.

Transgression and Regression

Sea-level changes, relative to the land surface, control the succession of sediments that we see in a sedimentary basin. At times during Earth history, sea level has risen by as much as a couple of hundred meters, creating shallow seas that submerge the interiors of continents. At other times sea level has fallen by a couple of hundred meters, exposing the continental shelves to air. Global sea-level changes may be due to a number of factors, including climate changes, which control the amount of ice stored in polar ice caps and cause changes in the volume of ocean basins. Sea level at a location may also be due to the local uplift or sinking of the land surface.

When relative sea level rises, the shoreline migrates inland. We call this process **transgression**. During this process, terrestrial sediments are buried by coastal sediments, and coastal sediments are buried by deeper-water sediment. Thus, as transgression occurs, an extensive layer of beach sand eventually forms. This layer may look like a blanket of sand that was deposited all at once, but in fact the sand deposited at one location differs in age from the sand deposited at another location. When sea level falls, the coast migrates seaward. We call this process **regression** (Fig. 7.21 ◑). Typically, the

record of a regression will not be well preserved, because as sea level drops, areas that had been sites of deposition become exposed to erosion. A succession of strata deposited during a cycle of transgression and regression is called a "depositional sequence."

Diagenesis

Earlier in this chapter we discussed the process of lithification, by which sediment hardens into rock. Lithification is an aspect of a broader phenomenon called diagenesis. Geologists use the term **diagenesis** for all the physical, chemical, and biological processes that transform sediment into sedimentary rock and that alter characteristics of sedimentary rock after the rock has formed.

In sedimentary basins, sedimentary rocks may become buried very deeply. As a result, the rocks endure higher pressures and temperatures and come in contact with warm groundwater. Diagenesis, under such conditions, can cause chemical reactions in the rock that produce new minerals and can also cause cement to dissolve or precipitate.

As temperature and pressure increase still deeper in the subsurface, the changes that take place in rocks become more profound. At sufficiently high temperature and pressure, metamorphism begins, in that a new assemblage of minerals forms, and/or mineral grains become aligned parallel to each other. The transition between diagenesis and metamorphism in sedimentary rocks is gradational and occurs between temperatures of 150°C and 300°C. In the next chapter, we enter the realm of true metamorphism.

Take-Home Message

- A sedimentary basin is a region where the surface of the lithosphere sinks, or subsides, creating a depression in which sediment accumulates.
- Sedimentary basins form in a variety of tectonic settings, including rifts, passive margins, mountain-belt forelands, and within the interior of continents.
- During a transgression, sea level rises and the coastline migrates inland; during a regression, sea level falls and the coastline migrates seaward.
- Diagenesis changes sediment over time. In deeply buried beds, reaction with groundwater may produce or destroy cement.

THINK: How can sandstone formed in a beach environment, comprise a formation that blankets a broad region?

FIGURE 7.21 The concept of transgression and regression, during deposition of sedimentary sequence. ▶⏸

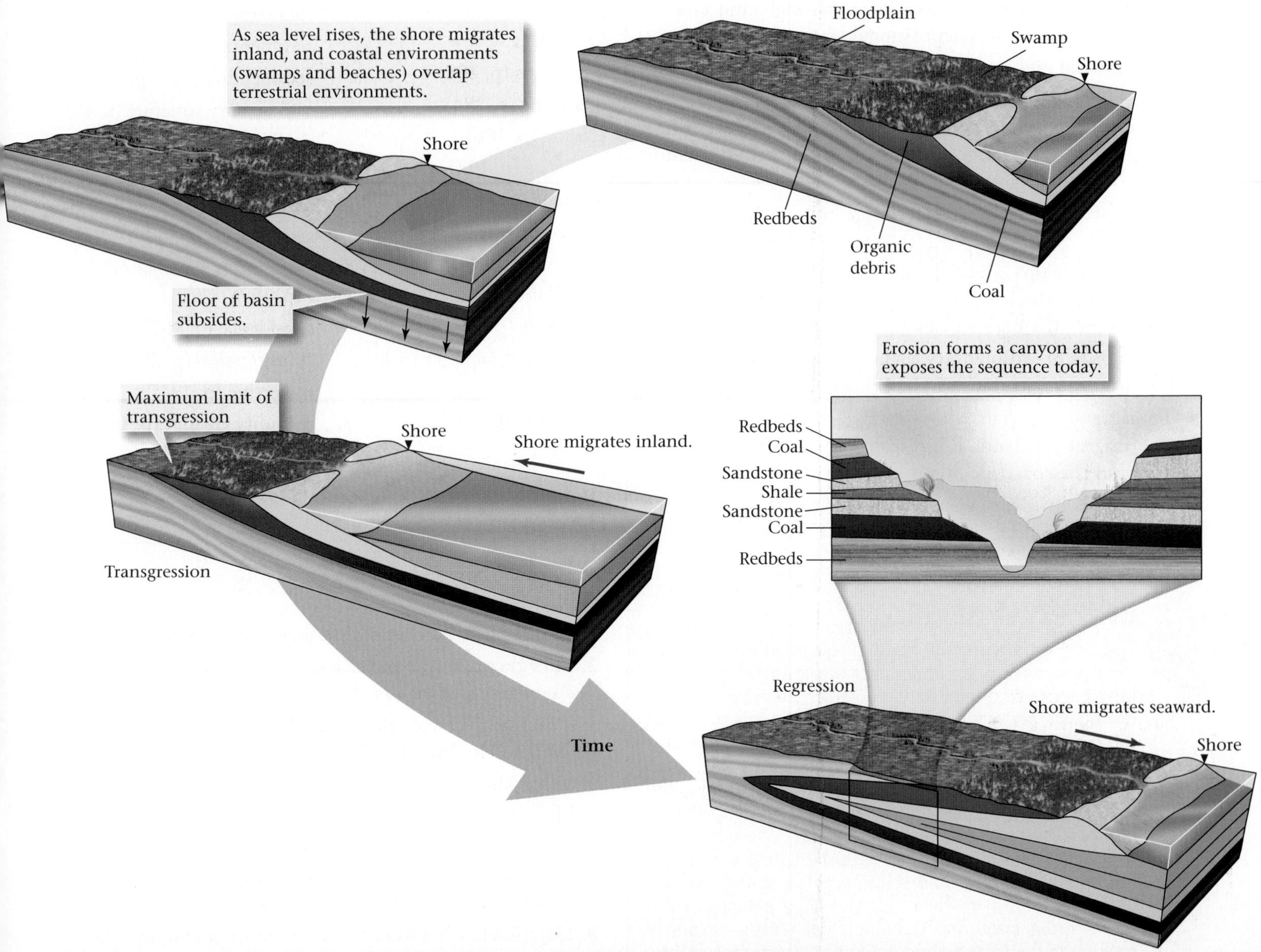

As sea level rises, the shore migrates inland, and coastal environments (swamps and beaches) overlap terrestrial environments.

Shore

Floodplain

Swamp

Shore

Redbeds

Organic debris

Coal

Floor of basin subsides.

Maximum limit of transgression

Shore

Shore migrates inland.

Transgression

Time

Erosion forms a canyon and exposes the sequence today.

Redbeds
Coal
Sandstone
Shale
Sandstone
Coal
Redbeds

Regression

Shore migrates seaward.

Shore

Chapter Summary

■ Geologists recognize four major classes of sedimentary rocks. Clastic rocks form from cemented-together grains that were first produced by weathering, then were transported, deposited, and lithified. Biochemical rocks develop from the shells of organisms. Organic rocks consist of plant debris or of altered plankton remains. Chemical rocks precipitate directly from water.

■ Formation of clastic rocks begins when grains erode from preexisting rock. Moving water, air, or ice transport these grains to a site of deposition, where they accumulate. Lithification, involving compaction and cementation, converts loose sediment into rock.

■ Clastic rocks are classified based primarily on grain size. Characteristics (roundness; sorting; composition) help to distinguish sediment source and depositional setting.

- Sedimentary structures include bedding, cross bedding, graded bedding, ripple marks, dunes, and mud cracks. They serve as clues to depositional settings.
- Biochemical and organic rocks form from materials produced by living organisms.
- Limestone consists dominantly of calcite, chert forms from silica, coal from carbon, shale from clay, and sandstone from quartz grains.
- Evaporites consist of minerals precipitated from saline water.
- Glaciers, streams, alluvial fans, deserts, rivers, lakes, deltas, beaches, shallow seas, and deep seas each accumulate a different, distinctive assemblage of sedimentary strata.
- Thick piles of sedimentary rocks accumulate in sedimentary basins, regions where the lithosphere sinks.
- Transgressions occur when sea level rises and the coastline migrates inland. Regressions occur when sea level falls and the coastline migrates seaward.
- Diagenesis involves processes leading to lithification and processes that alter sedimentary rock once it has formed.

GEOPUZZLE REVISITED

The Grand Canyon cuts down through an over 1.5-km-thick succession of sedimentary strata, recording a long history of deposition in a variety of environments during successive transgressions and regressions of the sea. At the time the layers were deposited, the surface of the crust lay at an elevation close to sea level. The present high elevation is due to uplift during the past 10 million years. Contrasting layers consist of contrasting rock types—each rock type formed in a different depositional environment. Not only are different layers different colors (in part due to the amount of oxidized iron in the rock), but they also have different grain sizes and bedding thicknesses. Stronger rock units (sandstone and limestone) form steep cliffs, whereas weaker units (shale) form gentler slopes.

Guide Terms

arkose (p. 188)
bed (p. 195)
biochemical sedimentary
  rock (p. 186)
breccia (p. 188)
cementation (p. 186)
chemical sedimentary rock
  (p. 186)

clastic sedimentary rock
  (p. 186)
clasts (p. 185)
coal (p. 191)
compaction (p. 186)
conglomerate (p. 188)
cross bed (p. 198)
delta (p. 205)

deposition (p. 186)
depositional environment
  (p. 201)
diagenesis (p. 208)
dolostone (p. 194)
erosion (p. 186)
evaporite (p. 192)
geologic map (p. 197)
graded bed (p. 199)
limestone (p. 190)
lithification (p. 186)
mudstone (p. 188)
organic sedimentary rock
  (p. 186)
regression (p. 208)

ripple marks (p. 197)
sandstone (p. 186)
sedimentary basin (p. 207)
sedimentary rock (p. 185)
sedimentary structure (p. 195)
shale (p. 188)
siltstone (p. 188)
sorting (p. 187)
strata (p. 195)
stratigraphic formation
  (p. 197)
subsidence (p. 207)
transgression (p. 208)
travertine (p. 194)
turbidity current (p. 198)

Review Questions

1. Describe how a clastic sedimentary rock forms from its unweathered parent rock.
2. Explain how biochemical sedimentary rocks form.
3. How do grain size and shape, sorting, sphericity, and angularity change as sediments move downstream?
4. Describe the two different kinds of chert. How are they similar? How are they different?
5. What kinds of conditions produce evaporites?
6. How does dolostone differ from limestone, and how does dolostone form?
7. What are cross beds, and how do they form? How can you read the current direction from cross beds?
8. Describe how a turbidity current forms and moves. How does it produce graded bedding?
9. Compare deposits of an alluvial fan with those of a deep-marine deposit.
10. What kinds of sediments accumulate in river and delta systems?
11. Why don't sediments accumulate everywhere? What types of tectonic conditions are required to create basins?
12. What kinds of changes may take place during diagenesis?

On Further Thought

13. Recent exploration of Mars by robotic vehicles suggests that layers of sedimentary rock cover portions of the

planet's surface. On the basis of examining images of these layers, some researchers claim that the layers contain cross bedding and relics of gypsum crystals. At face value, what do these features suggest about depositional environments on Mars in the past? (*Note*: Interpretation of the images remains controversial.)

14. The Gulf Coast of the United States is a passive-margin basin that contains a very thick accumulation of sediment. Drilling reveals that the base of the sedimentary succession in this basin consists of redbeds. These are overlain by a thick layer of evaporite. The evaporite, in turn, is overlain by deposits composed dominantly of sandstone and shale. In some intervals, sandstone occurs in channels and contains ripple marks, and the shale contains mud cracks. In other intervals, the sandstone and shale contain fossils of marine organisms. The sequence contains hardly any conglomerate or arkose. Be a sedimentary detective, and explain the succession of sediment in the basin.

15. Examine the Bahamas with *Google Earth*™ or NASA World Wind. (You can find a high-resolution image at Lat 23°58'40.98"N Long 77°30'20.37"W.) Note that broad expanses of very shallow water surround the islands, that white-sand beaches occur along the coast of the islands, and that small reefs occur offshore. What does the sand consist of, and what rock will it become if it eventually becomes buried and lithified? Compare the area of shallow water in the Bahamas area with the area of Florida. The bedrock of Florida consists mostly of shallow-marine limestone. What does this observation suggest about the nature of the Florida peninsula in the past? Keep in mind that sea level on Earth changes over time. Presently, most of the land surface of Florida lies at less than 50 m (164 feet) above sea level.

 For more resources, including animations, quizzes, and Norton's GeoTours, go to **wwnorton.com/studyspace**.

 If your instructor assigns exercises in SmartWork, log in at **smartwork.wwnorton.com**.

ANOTHER VIEW These cliffs, near Bryce Canyon (Utah), expose beds of sedimentary rock deposited in lakes, and by streams, over 40 million years ago. Present-day erosion has produced aprons of debris at the base of the cliffs.

CHAPTER 8

Metamorphism: A Process of Change

A geologist examines an outcrop of 2.7-billion-year-old metamorphic rock in the Canadian Shield of Ontario. The curved layering in this rock gives the impression that the rock flowed like plastic. The layering is metamorphic foliation.

GEOPUZZLE

Marble, the rock from which Michelangelo carved his sculptures, contains the same chemicals as limestone, a sedimentary rock. But grains in marble interlock and do *not* resemble fossil fragments, cements, or water precipitates. Further, in some examples, features of marble suggest it flowed like soft plastic. The rock can't be igneous because its composition is unlike that of any known magma. So, how do rocks like marble form?

Nothing in the world lasts, save eternal change.

—Honorat de Bueil (1589–1650)

8.1 INTRODUCTION

Cool winds sweep across Scotland for much of the year. In this blustery climate, vegetation has a hard time taking hold, so the landscape provides countless outcrops of barren rock. During the latter half of the eighteenth century, James Hutton became fascinated with the Earth and examined these outcrops, hoping to learn how rock formed. Hutton found that many features in the outcrops resembled the products of present-day sediment deposition and volcanic activity, and he came to an understanding of how sedimentary and igneous rock form. But Hutton also found rock that contained minerals and textures quite different from those in sedimentary and igneous samples. He described this puzzling rock as "a mass of matter which had evidently formed originally in the ordinary manner . . . but which is now extremely distorted in its structure . . . and variously changed in its composition."

The rock that so puzzled Hutton is now known as metamorphic rock, from the Greek words *meta*, meaning change, and *morphe*, meaning form. In modern terms, a **metamorphic rock** is one that forms when a preexisting rock, or **protolith**, undergoes a solid-state change in response to the modification of its environment. This process of change is called **metamorphism**. Let's look at the components of our definition more closely. By *solid-state*, we mean that a metamorphic rock does *not* form by solidification of magma—remember that geologists consider rocks that solidified from magma to be igneous. By *change*, we mean that metamorphism produces new minerals that did not occur in the protolith, and/or produces a new texture (arrangement of mineral grains) that is distinct from that of the protolith. And by *modification of environment*, we mean that metamorphism takes place when a protolith endures a rise or fall in temperature and/or pressure, undergoes compression and shear, or reacts with "hydrothermal fluids" (very hot-water solutions).

Hutton did more than just note the existence of metamorphic rock—he also tried to understand why metamorphism takes place. Because he found metamorphic rocks adjacent to igneous intrusions, he concluded that metamorphism can take place when heat from an intrusion "cooks" the rock into which it intrudes. And because he found that metamorphic rocks can occur over broad regions in the absence of intrusions, he speculated that metamorphism can also take place when rock becomes deeply buried, as occurs during mountain building.

From Hutton's day to the present, geologists have undertaken field studies, laboratory experiments, and theoretical calculations to better characterize metamorphism. In this chapter, we introduce the results of their work. We begin by explaining the causes of metamorphism and the basis for classifying metamorphic rocks. We conclude by discussing the geologic settings in which these rocks form. As you will see, Hutton's speculations on the origin of metamorphic rock were basically correct, but they represented only part of the story—the rest of the story could not take shape until the theory of plate tectonics was proposed.

Did you ever wonder . . .
once formed, do rocks ever change?

Chapter Themes

By the end of this chapter, you should know that . . .

- In metamorphic rock, the texture and/or mineral assemblage changed in the solid state.
- Metamorphism happens due to changes in temperature and/or pressure as hydrothermal fluids pass through the rock, or in response to squeezing, shear, and shock.
- Different metamorphic textures and mineral assemblages form in different environments.
- Metamorphic rocks can form during mountain building or in response to heating by magma.
- Some metamorphic rocks develop foliation, manifested either by parallel alignment of mineral grains or by compositional layering.
- We distinguish among different kinds of metamorphic rocks based on the presence or absence of foliation, character of foliation (if present), and on mineral content.

8.2 CONSEQUENCES AND CAUSES OF METAMORPHISM

What Is a Metamorphic Rock?

If someone were to put a rock on a table in front of you, how would you know that it is metamorphic? First, metamorphic rocks can possess **metamorphic minerals**, new minerals that grow in place within the solid rock only under metamorphic temperatures and pressures. In fact, metamorphism can produce a group of minerals which together make up a metamorphic mineral assemblage. And second, metamorphic rocks can have **metamorphic texture** defined by the arrangement of mineral grains. Commonly, the texture results in **metamorphic foliation**, defined by the parallel alignment of platy minerals (such as mica) and/or the presence of alternating light-colored and dark-colored layers. When these characteristics develop, a metamorphic rock becomes as different from its protolith as a butterfly is from a caterpillar. For example, metamorphism of red shale can yield a metamorphic rock consisting of aligned mica flakes and brilliant garnet crystals (Fig. 8.1a). Metamorphism of limestone composed of cemented-together fossil fragments can yield a metamorphic rock consisting of large interlocking crystals of calcite (Fig. 8.1b). And metamorphism of granite, a rock with randomly oriented crystals, can produce a rock with crystals that align parallel to one another (Fig. 8.1c).

The formation of metamorphic minerals and textures takes place very slowly—it may take thousands to millions of years—and it involves several processes, which sometimes occur alone and sometimes together. The most common processes are as follows:

- *Recrystallization*, which changes the shape and size of grains without changing the identity of the mineral making up the grains (Fig. 8.2a). For example, during recrystallization of sandstone, a tightly fitting mosaic of irregularly shaped, large quartz grains may replace a cluster of cemented-together, small, round quartz grains (Fig. 8.2a).

- *Phase change*, which transforms one mineral into another mineral with the same composition but a different crystal structure. For example, the transformation of quartz into a denser mineral called coesite represents a phase change, for these minerals have the same formula (SiO_2) but different crystal structures. On an atomic scale, phase change involves the rearrangement of atoms.

- *Metamorphic reaction*, or *neocrystallization* (from the Greek *neos*, for new), which results in the growth of new mineral crystals that differ from those of the protolith (Fig. 8.2b). During neocrystallization, chemical reactions digest minerals of the protolith to produce new minerals of the metamorphic rock. For this to take place, atoms migrate (diffuse) through solid crystals, a very slow process, and/or dissolve and reprecipitate at grain boundaries.

FIGURE 8.1 Metamorphism causes changes in mineral makeup and texture.

(a) A specimen of red shale (left) contains clay, quartz, and iron oxide. When this rock undergoes intense metamorphism, it may change into a gneiss containing different minerals. This gneiss sample (right) contains biotite, quartz, feldspar, and purple garnet.

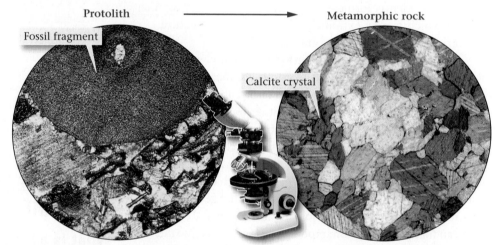

(b) A photomicrograph of a limestone in thin section (left) shows tiny fossil fragments and lime mud. After metamorphism, the texture of the rock changes completely and we see large, interlocking crystals of calcite (right). Note that the color seen here depends on the orientation of the crystals.

(c) Metamorphism can transform a rock in which crystals are randomly oriented (left) into one in which they are aligned (right).

- *Pressure solution*, which happens when a wet rock is squeezed more strongly in one direction than in others. Mineral grains dissolve where their surfaces are pressed against other grains, producing ions that migrate through the water to precipitate elsewhere (Fig. 8.2c). Precipitation may take place on faces where the grains are squeezed together less strongly. Thus, pressure solution can cause grains to become shorter in one direction and new growth to occur in another (Fig. 8.2c). Pressure solution takes place under both nonmetamorphic and metamorphic conditions.

- *Plastic deformation*, which happens when a rock is squeezed or sheared at elevated temperatures and pressures. Under these conditions, minerals behave like soft plastic and change shape without breaking (Fig. 8.2d). Such deformation can take place without changing either the composition or the crystal structure of the mineral. But commonly it takes place while metamorphic reactions are also happening. The atomic-scale processes causing plastic deformation are complex, so we must defer an explanation of them to more advanced books.

Caterpillars undergo metamorph*osis* because of hormonal changes in their bodies. Rocks undergo metamorph*ism* when they are subjected to heat, pressure, compression and shear, and/or very hot water. Let's now consider the details of how these *agents of metamorphism* operate.

Metamorphism Due to Heating

When you heat cake batter, the batter transforms into a new material—cake. Similarly, when you heat a rock, its ingredients transform into a new material—metamorphic rock. Why? Think about what happens to atoms in a mineral grain as the grain warms. Heat causes the atoms to vibrate rapidly, stretching and bending chemical bonds that lock atoms to their neighbors. If bonds stretch too far and break, atoms detach from their original neighbors, move slightly, and form new bonds with other atoms. Repetition of this process leads to rearrangement of atoms within grains, or to migration of atoms into and out of grains. As a consequence, recrystallization and/or neocrystallization take place, enabling a metamorphic mineral assemblage to grow in solid rock.

Metamorphism takes place at temperatures between those at which diagenesis occurs and those that cause melting. Roughly speaking, this means that most metamorphic rocks you find in outcrops on continents formed at temperatures of between 250°C and 850°C. However, melting temperature depends on composition and water content (see Chapter 6), so the upper limit of the metamorphic realm actually ranges between 650°C and 1,200°C, depending on rock composition and water content.

FIGURE 8.2 Metamorphic processes, as seen through a microscope.

Protolith ⟶ Metamorphic rock

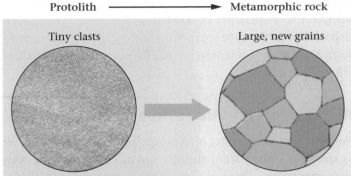

(a) Mineral grains recrystallize to form new, interlocking grains of the same mineral. Typically, grains get larger.

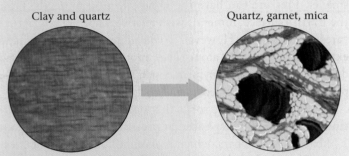

(b) Chemical reactions change the original assemblage of minerals into a new, metamorphic assemblage of minerals.

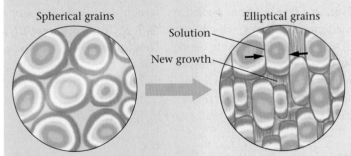

(c) Pressure solution dissolves grains on the sides undergoing more pressure, and precipitates new mineral material where the pressure is lower. Arrows indicate the squeezing direction.

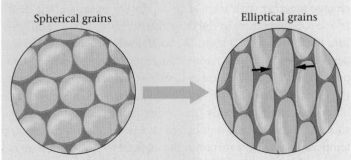

(d) Plastic deformation changes the shape of grains—without breaking them—as a result of squeezing or shear at high temperatures.

The depth in the Earth at which metamorphic temperatures exist depends on the geothermal gradient, which, in turn, reflects the geologic setting. For example, near a hot, igneous intrusion, a temperature of 500°C can develop at very shallow depths. But in the upper part of *average* continental crust, away from intrusions, a temperature of 500°C occurs at a depth of about 20 to 25 km.

Metamorphism Due to Pressure

As you swim underwater in a swimming pool, water squeezes against you equally from all sides—in other words, your body feels **pressure**. Pressure can cause a material to collapse inward. For example, if you pull an air-filled balloon down to a depth of 10 m in a lake, the balloon becomes significantly smaller. Pressure can have the same effect on minerals. Near the Earth's surface, minerals with relatively open crystal structures can be stable. However, if you subject these minerals to extreme pressure, the atoms pack more closely together and denser minerals tend to form. Such transformations involve phase changes and/or neocrystallization.

Most metamorphic rocks that occur in outcrops on continents were metamorphosed at pressures of less than about 12 kbar (= 12,000 bars, which is about 12,000 times the pressure at Earth's surface for 1 bar ≈ 1 atm). Pressure increases at about 270 to 300 bars/km due to the weight of overlying rock, so a pressure of 12 kbar occurs at depths of about 40 km. However, in a few locations geologists have found *ultrahigh-pressure* metamorphic rocks, which appear to have formed at pressures of up to 29 kbar, meaning depths of about 80 to 100 km. Rocks subjected to ultrahigh pressure contain grains of coesite, a phase of SiO_2 that is much denser than familiar quartz. In fact, some of these rocks contain tiny grains of diamond, a phase of carbon that forms only under very high pressure.

Changing Pressure and Temperature Together

So far, we've considered changes in pressure and temperature as separate phenomena. But in the Earth, pressure and temperature change together with increasing depth. For example, at a depth of 8 km, temperature in the crust reaches about 200°C and pressure reaches about 2.3 kbar. If a rock slowly becomes buried to a depth of 20 km, as can happen during mountain building, temperature in the rock increases to more than 500°C, and pressure to 5.5 kbar. Experiments and calculations show that the "stability" of certain minerals (the ability of a mineral to form and survive) and mineral assemblages depends on *both* pressure and temperature. Thus, a metamorphic rock formed at 8 km does not contain the same minerals as one formed at 20 km.

We can illustrate the relationship of mineral stability to pressure and temperature by studying the behavior of Al_2SiO_5

(aluminum silicate) as portrayed on a phase diagram, a graph with temperature indicated by one axis and pressure indicated by the other (Fig. 8.3). Al_2SiO_5 can exist as three different minerals (or "phases"): kyanite, andalusite, and sillimanite. Each of these minerals exists only under a specific range of temperatures and pressures, indicated by an area called a stability field, on the phase diagram. If a protolith containing the elements necessary to produce Al_2SiO_5 is taken to a depth in the Earth where the pressure is 2 kbar and the temperature is 450°C (point X), then andalusite grows. If the temperature stays at 450° but pressure on the rock increases to 5 kbar (point Y), then andalusite becomes unstable and kyanite grows. And if the pressure stays at 5 kbar but the temperature increases to 650°C (point Z), then sillimanite grows.

Compression, Shear, and Development of Preferred Orientation

Imagine that you have just built a house of cards and, being in a destructive mood, you step on it. The structure collapses because the downward push you apply with your foot exceeds the push provided by air in other directions. If a material is squeezed (or stretched) *unequally* from different sides, we say that it is subjected to **differential stress**. (Note that differential stress differs from pressure—in the latter, the push is the *same* in all directions.) In other words, under conditions of differential stress, the push or pull in one direction differs in magnitude from the push or pull in another direction (Fig. 8.4a). We distinguish two kinds of differential stress:

- *Normal stress*: Normal stress pushes or pulls perpendicular to a surface. We call a push *compression* and a pull *tension*. Compression flattens a material (Fig. 8.4b), whereas tension stretches a material.

FIGURE 8.3 The stability fields for three metamorphic minerals (kyanite, andalusite, and sillimanite) that are polymorphs of Al_2SiO_5 (aluminum silicate) can be depicted on a phase diagram.

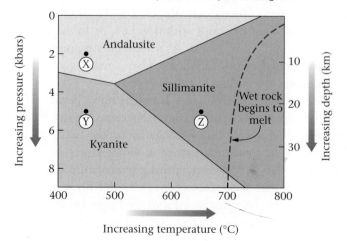

■ *Shear stress*: Shear stress, or shear, moves one part of a material sideways, relative to another. If, for example, you place a deck of cards on a table, then set your hand on top of the deck and move your hand parallel to the table, you shear the deck (Fig. 8.4c).

Compression and shear at elevated temperatures and pressures can cause a rock to change shape without breaking. As it changes shape, the internal texture of a rock also changes. For example, platy (pancake-shaped) grains become parallel to one another, and elongate (cigar-shaped) grains align in the same direction. Both platy and elongate grains are "inequant grains," meaning that the dimension of a grain is not the same in all directions; in contrast, equant grains have roughly the same dimensions in all directions (Fig. 8.4d). The alignment of inequant minerals in a rock results in a **preferred orientation** (Fig. 8.4e).

How does preferred orientation form? In wet rocks at relatively low temperatures, pressure solution dissolves on the faces perpendicular to the direction of compression. Commonly, precipitation of the dissolved minerals takes place on faces where compression is less. So as a result of pressure solution, grains become shorter in one direction. At relatively high temperatures, existing grains flatten in response to differential stress by means of plastic deformation (Fig. 8.5a). As a rock undergoes flattening, relatively rigid, inequant grains distributed throughout a soft matrix may rotate into parallelism as the overall rock changes shape, much like logs scattered in a flowing river align with the current. Shear can have the same effect (Fig. 8.5b ◗). Finally, grain growth (by neocrystallization) may produce preferred orientation, because certain minerals grow faster in the direction in which a rock is stretching than in other directions (Fig. 8.4e).

FIGURE 8.4   Compression and shear changes rock grains during metamorphism.

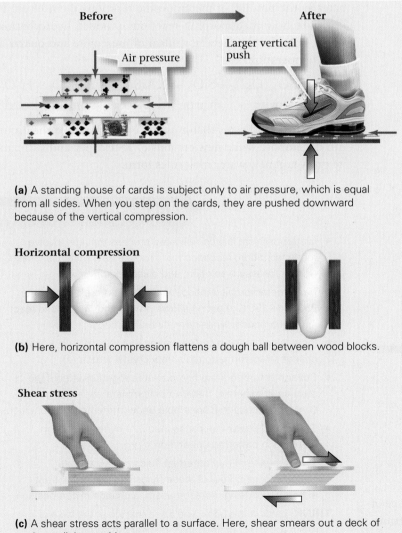

(a) A standing house of cards is subject only to air pressure, which is equal from all sides. When you step on the cards, they are pushed downward because of the vertical compression.

(b) Here, horizontal compression flattens a dough ball between wood blocks.

(c) A shear stress acts parallel to a surface. Here, shear smears out a deck of cards parallel to a table top.

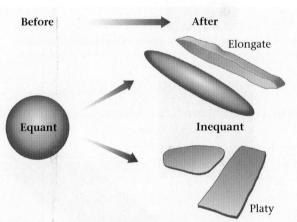

(d) Compression and shear can transform equant grains into inequant grains. Inequant grains can be elongate (cigar-shaped) or platy (pancake-shaped).

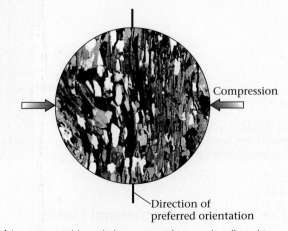

(e) In metamorphic rock, inequant grains may be aligned to form a preferred orientation. As seen through a microscope, the flat planes of grains are perpendicular to the compression direction.

FIGURE 8.5 Changes in grain shape and orientation, and the development of preferred orientation. ▶II

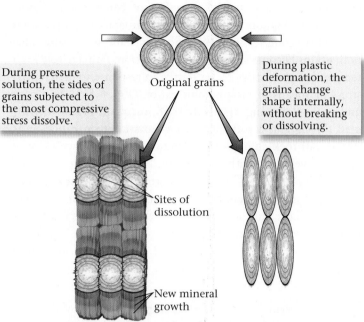

Original grains

During pressure solution, the sides of grains subjected to the most compressive stress dissolve.

During plastic deformation, the grains change shape internally, without breaking or dissolving.

Sites of dissolution

New mineral growth

(a) Application of differential stress (directed pressure) during metamorphism can change the shape of grains in two ways.

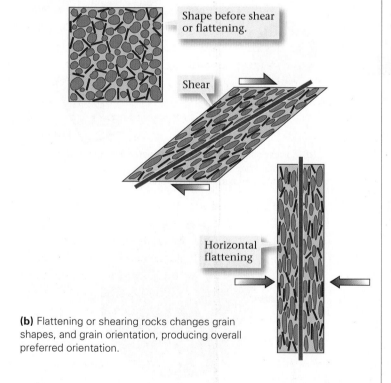

Shape before shear or flattening.

Shear

Horizontal flattening

(b) Flattening or shearing rocks changes grain shapes, and grain orientation, producing overall preferred orientation.

The Role of Hydrothermal Fluids

Metamorphic reactions commonly take place in the presence of very hot water, or "hydrothermal fluids," because water permeates the crust. We initially defined hydrothermal fluids simply as "very hot-water solutions." In fact, they actually can include hot water, steam, and so-called supercritical fluid as well as dissolved ions. (A supercritical fluid is a substance that forms under high temperatures and pressures and has characteristics of both liquid and gas; it can permeate rock, seeping into every conceivable opening.) Hydrothermal fluids chemically react with rock. For example, hydrothermal fluids accelerate metamorphic reactions, because atoms involved in the reactions can migrate faster through a fluid than they can through a solid, and hydrothermal fluids provide water that can be absorbed by minerals during metamorphic reactions. Finally, fluids passing through a rock may pick up some dissolved ions and drop off others, resulting in the overall chemical composition of a rock during metamorphism, a process called **metasomatism**. In some cases, the dissolved ions carried by hydrothermal fluids precipitate to form mineral-filled cracks known as veins (Fig. 8.6).

The water constituting hydrothermal fluids comes from several sources. Some originates as groundwater that entered the crust at the Earth's surface and then sank down, some was released from magma when the magma rose, and some originates as the product of metamorphic reactions themselves. To illustrate how metamorphic reactions produce hydrothermal fluids, consider the metamorphism of muscovite and quartz at high temperature:

$$KAl_3Si_3O_{10}(OH)_2 + SiO_2 \rightarrow KAlSi_3O_8 + Al_2SiO_5 + H_2O$$
  muscovite   quartz   K-feldspar sillimanite water

This formula indicates that muscovite and quartz of the protolith decompose while new crystals of K-feldspar and sillimanite grow and new water molecules form.

Take-Home Message

- Metamorphism is a complicated process that can involve recrystallization, reactions that produce new minerals, phase change, pressure solution, and plastic deformation.
- Most metamorphism happens at temperatures of between 250° and 850°C; heating allows preexisting minerals to react and decompose, while new minerals grow.
- Increasing pressure can cause minerals to undergo a phase change (transformation to a new crystal structure).
- Change in temperature and pressure together affects the stability of minerals; metamorphic mineral assemblages contain minerals that are stable under metamorphic conditions.
- Differential stress can cause inequant minerals to align, resulting in preferred orientation.
- The presence of hydrothermal fluids can accelerate metamorphism; flow of these fluids can add or subtract elements, causing metasomatism, a change in composition.

THINK: Later in this book, you'll see that ice at the base of a glacier is coarser-grained than ice near the surface and has undergone slow flow without breaking. Is glacial ice metamorphic?

FIGURE 8.6 Veins in metamorphic rock on the coast of Wales. Note that the veins themselves have been plastically deformed into complex shapes.

8.3 TYPES OF METAMORPHIC ROCKS

Coming up with a way to classify and name the great variety of metamorphic rocks on Earth hasn't been easy. After decades of debate, geologists have found it most convenient to divide metamorphic rocks into two fundamental classes: foliated rocks and nonfoliated rocks. Each class contains several rock types. We distinguish nonfoliated rocks from each other primarily by their component minerals, whereas we distinguish foliated rocks from each other partly by their component minerals and partly by the nature of their foliation.

Foliated Metamorphic Rocks

To understand this class of rocks, we first need to discuss the nature of **foliation** in more detail. The word comes from the Latin *folium*, for leaf. Geologists use *foliation* to refer to an assemblage of parallel planar surfaces and/or layers in a metamorphic rock (Fig. 8.7a). Foliation can give metamorphic rocks a striped or streaked appearance in an outcrop, and/or give them the ability to split into thin sheets. A foliated metamorphic rock has foliation either because it contains inequant mineral crystals that are aligned parallel to one another, defining preferred mineral orientation, and/or because the rock has alternating dark-colored and light-colored layers.

Foliated metamorphic rocks can be distinguished from one another according to their composition, their grain size, and the nature of their foliation. The most common types include

- *Slate*: The finest-grained foliated metamorphic rock, **slate**, forms by metamorphism of shale or mudstone—rocks composed of clay—under relatively low pressures and temperatures. Slate contains a type of foliation called slaty cleavage, which allows it to split into thin sheets that make excellent roofing shingles (see Fig. 8.7a). The slaty cleavage develops when compression causes clay flakes to reorient and regrow into an orientation perpendicular to

the direction of compression. For example, end-on compression of a sequence of horizontal shale beds produces vertical slaty cleavage (Fig. 8.7b). Commonly, such compression also causes the layers to bend and become folded. The development of aligned clay is primarily a consequence of pressure solution and recrystallization—grains lying at an angle to the cleavage plane dissolve, whereas grains parallel to the cleavage plane grow. In addition, during this process, less-soluble grains may passively rotate into the plane of cleavage, and new grains may grow.

- *Phyllite*: **Phyllite** is a fine-grained metamorphic rock with a foliation caused by the preferred orientation of very fine-grained white mica. The word comes from the Greek word *phyllon*, meaning leaf, as does the word *phyllo*, the flaky dough in Greek pastry. The parallelism of translucent fine-grained mica gives phyllite a silky sheen known as phyllitic luster (Fig. 8.8a). Phyllite forms by the metamorphism of slate at a temperature high enough to cause neocrystallization of white mica; metamorphic reactions produce a new assemblage of minerals (fine-grained mica and chlorite) out of clay. The formation of foliation in phyllite is due to differential stress during metamorphism.

- *Metaconglomerate*: Under the metamorphic conditions that produce slate or phyllite, a protolith of conglomerate becomes **metaconglomerate**. Specifically, pressure solution and plastic deformation flatten pebbles and cobbles into pancake-like shapes. The alignment of inequant clasts defines a foliation (Fig. 8.8b).

- *Schist*: **Schist** is a medium- to coarse-grained metamorphic rock that possesses a type of foliation, called schistosity, defined by the preferred orientation of *large* mica flakes (muscovite and/or biotite; Fig. 8.8c). Schist forms at a higher temperature than does phyllite, and it differs from phyllite in that the mica grains are larger. Typically, schists also contain other minerals such as quartz, feldspar, kyanite, garnet, and amphibole—the specific minerals that grow depend on the chemical composition of the protolith. Schist can form from a shale but also from a great variety of other protoliths, as long as the protolith contains the appropriate elements to make mica.

In many cases, certain mineral grains in schists grow to be much larger than surrounding minerals. For example, garnet crystals in schist may become many times larger than those of other minerals (see Fig. 8.2b). Especially large crystals that grow in a metamorphic rock are called porphyroblasts. Smaller grains surrounding porphyroblasts constitute a rock's matrix.

- *Gneiss*: **Gneiss** is a compositionally layered metamorphic rock, typically composed of alternating dark-colored

FIGURE 8.7   Slate is a foliated metamorphic rock that forms at relatively low temperature and pressure.

(a) A block of slate splits easily along cleavage planes, which may be at a high angle to the bedding planes. Workers split slate to produce roof shingles that, when overlapped, make a watertight surface (see insert).

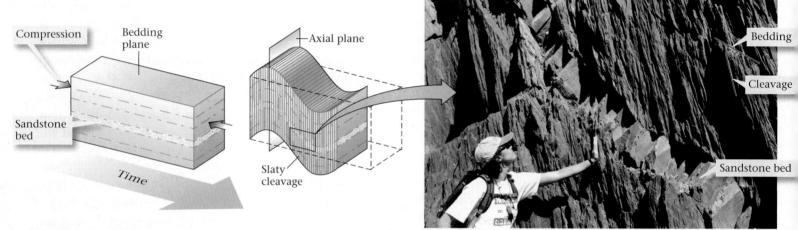

(b) Slaty cleavage forms in response to compressive stress. In this example, layers also bend to form folds, as slaty cleavage develops. The cleavage tends to be oriented parallel to the axial plane, an imaginary surface that, simplistically, divides the fold in half.

and light-colored layers that range in thickness from millimeters to meters. This compositional layering, or gneissic banding, gives gneiss a striped appearance (Fig. 8.9a ◉). The contrasting colors represent contrasting compositions. Light-colored layers contain predominantly felsic minerals such as quartz and feldspar, whereas the dark-colored layers contain predominantly mafic minerals such as amphibole, pyroxene, and biotite. If gneiss contains mica, the mica-rich layers may have schistosity. Gneiss that formed at very high temperatures, however, does not contain mica, because at high temperatures, mica reacts to form other minerals.

How does the banding in gneiss form? Banding in some examples of gneiss evolved directly from the original bedding in a rock. For example, metamorphism of a protolith consisting of alternating beds of sandstone and shale produces a gneiss consisting of alternating beds of

quartzite and mica. Gneissic banding can also form when the protolith undergoes an extreme amount of shearing under conditions in which the rock can flow like soft plastic (Fig. 8.9b). Such flow stretches, folds, and smears out preexisting compositional contrasts in the rock and transforms them into aligned sheets. To picture this process, imagine slowly stirring vanilla batter in which there are blobs of chocolate batter—eventually you will see thin, alternating layers of dark and light batter. Similarly, imagine what happens if you take a rolling pin and flatten a ball consisting of two different colors of dough, then fold it in half, and flatten it again—you end up with thin parallel layers of contrasting colors. Finally, banding in some gneisses can develop by an incompletely understood process called metamorphic differentiation. Differentiation may involve dissolution of minerals in some layers, migration of the chemical components of those

FIGURE 8.8 Examples of foliated metamorphic rocks formed at higher temperatures and pressures.

(a) During formation of phyllite, clay recrystallizes to form tiny mica flakes that reflect light, giving the rock a sheen.

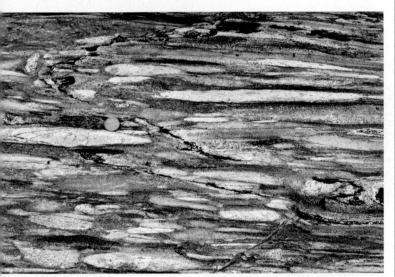

(b) In metaconglomerate, pebbles and cobbles flatten into a pancake shape without cracking.

(c) A schist contains coarse mica flakes, along with other metamorphic minerals.

minerals to other layers, where new minerals then grow. In effect, chemical reactions segregate different minerals into different layers (Fig. 8.9c).

- *Migmatite*: Under certain conditions, gneiss may *begin* to melt, producing felsic magma and residual, still solid, mafic rock. If the melt freezes again before flowing out of the source area, a mixture of igneous rock and relict metamorphic rock forms. This mixture is called **migmatite** (Fig. 8.10). Plastic flow during the process can contort the light-colored felsic rock and the dark-colored mafic rock into complex shapes.

Nonfoliated Metamorphic Rocks

Nonfoliated metamorphic rocks contain minerals that recrystallized or grew during metamorphism, but have no foliation. The lack of foliation means either that metamorphism occurred in the absence of compression and shear, or that most of the new crystals are equant. We list below some of the rock types that can occur without foliation.

- *Hornfels*: **Hornfels** is a fine-grained nonfoliated rock that contains a variety of metamorphic minerals. The specific mineral assemblage in a hornfels depends on the composition of the protolith and on the temperature and pressure of metamorphism.

- *Granofels*: A granofels is a coarse-grained nonfoliated rock consisting primarily of quartz and feldspar.

- *Amphibolite*: Metamorphism of mafic rocks (basalt or gabbro) can't produce quartz and muscovite when metamorphosed, for these rocks don't contain the right mix of chemicals to yield such minerals. Rather, they transform into amphibolite, a dark-colored metamorphic rock containing predominantly hornblende (a type of amphibole) and plagioclase (a type of feldspar) and, in some cases, a tiny bit of biotite. Where subjected to differential stress, amphibolites can develop a foliation; but the foliation tends to be poorly defined because the rock contains very little mica (Fig. 8.11).

- *Quartzite*: **Quartzite** forms by the metamorphism of pure quartz sandstone. During metamorphism, preexisting quartz grains recrystallize, creating new, larger grains. In the process, the distinction between cement and grains disappears, open pore space disappears, and the grains become interlocking. When quartzite cracks, the fracture cuts across grain boundaries—in contrast, fractures in sandstone curve *around* grains. Quartzite looks glassier than sandstone and does not have the grainy, sandpaper-like surface characteristic of sandstone (Fig. 8.12a). Depending on the impurities it contains, quartzite can vary in color from white to gray, purple, or green. Most quartzite is nonfoliated because

FIGURE 8.9 The formation of gneiss, which takes place at very high temperatures and pressures. ▶◉

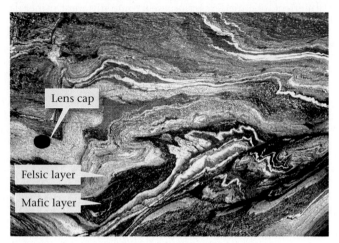

(a) Gneiss displays compositional banding (alternating felsic and mafic layers). The layers can become contorted.

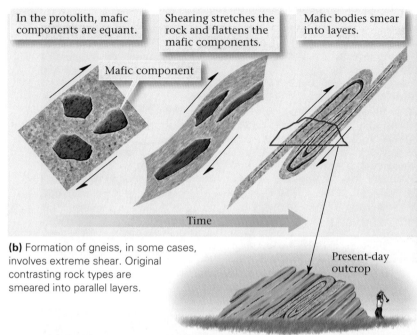

(b) Formation of gneiss, in some cases, involves extreme shear. Original contrasting rock types are smeared into parallel layers.

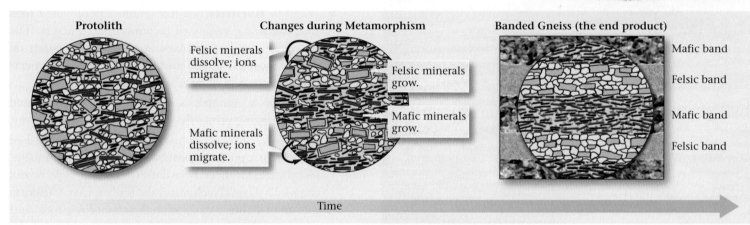

(c) Gneiss may also form by metamorphic differentiation, during which chemical reactions cause felsic and mafic minerals to grow in distinct, separate layers.

it does not contain aligned mica or compositional layering. In some cases, however, quartz grains deform plastically and become pancake shaped. The alignment of "pancakes" yields a foliation. (To avoid confusion, quartzite with this texture should be called foliated quartzite.)

- *Marble*: The metamorphism of limestone yields **marble**. During the formation of marble, calcite composing the protolith recrystallizes, so fossil shells, pore space, and the distinction between grains and cement disappear. Thus, marble typically consists of a fairly uniform mass of interlocking calcite crystals. It may also contain other, less-familiar minerals formed from the reaction of calcite with quartz, clay, and iron oxide, if these minerals had existed in the protolith.

Sculptors love to work with marble because the rock is relatively soft and has a uniform texture that gives it the cohesiveness and homogeneity needed to fashion large, smooth, highly detailed sculptures. Marble comes in a variety of colors—white, pink, green, and black—depending on the impurities it contains. Michelangelo, one of the great Italian Renaissance artists, sought large, unbroken blocks of creamy white marble from quarries in the Italian Alps for his masterpieces (**Fig. 8.12b**). If the original protolith contained layers with different impurities, the resulting marble has color banding that makes it a prized decorative stone (**Fig. 8.12c**). Because marble is a relatively weak rock, it flows like soft plastic under metamorphic conditions, and this flow can smear out different-colored portions of marble into beautiful contorted, curving bands.

Taking Chemical Composition into Account in Classifying Metamorphic Rocks

Up to this point, we've emphasized the importance of temperature and pressure on determining the mineral assemblage in a metamorphic rock. Let's not forget that composition plays a key role in determining which minerals form as well. For example, it's impossible to form a biotite-rich schist from a pure quartz sandstone, because biotite contains elements, such as iron, that do not occur in quartz.

To distinguish among different compositions of metamorphic rock, geologists use the following terms. (1) *Pelitic metamorphic rocks* form from sedimentary protoliths such as shale, which contain a relatively high proportion of aluminum. Metamorphism of these rocks produces aluminum-rich metamorphic minerals such as muscovite. (2) *Basic* (or *mafic*) *metamorphic rocks* contain relatively little silica and an abundance of iron and magnesium. During metamorphism, minerals such as biotite and hornblende grow. (3) *Calcareous metamorphic rocks* form from calcium-rich sedimentary rocks (i.e., limestone) and contain calcite ($CaCO_3$). (4) *Quartzo-feldspathic metamorphic rocks* form from protoliths (e.g., granite) that contained mostly quartz and feldspar. These metamorphic rocks contain mostly the same minerals as the protolith, because quartz and feldspar remain stable under metamorphic conditions, but during metamorphism the quartz tends to flow and transform into thin ribbons that define a foliation (see Fig. 8.2d).

Take-Home Message

- Geologists separate metamorphic rocks into two classes, foliated and nonfoliated, based on whether the rock contains a preferred mineral orientation and/or compositional layering.
- There are several distinct types of foliated metamorphic rocks (slate, phyllite, schist, and gneiss) based on the nature of foliation.
- The progression from slate to gneiss reflects metamorphism under progressively higher pressures and temperatures.
- Nonfoliated rocks (hornfels, marble, and quartzite) are distinguished by composition; marble consists largely of calcite, and quartzite consists largely of quartz.

THINK: Why don't builders use gneiss to make roofing shingles?

8.4 HOW DO WE DEFINE THE INTENSITY OF METAMORPHISM?

Not all metamorphism takes place under the same physical conditions. For example, rocks carried to a great depth beneath a mountain range undergo more intense metamorphism than do rocks closer to the surface. Geologists use the term

Light (felsic) bands formed from melt.

Dark (mafic) bands remained solid.

FIGURE 8.10 An outcrop of migmatite in northern Michigan contains both light-colored (felsic) igneous rock and dark-colored (mafic) metamorphic rock. The mixture of the two rock types makes migmatite resemble marble cake.

metamorphic grade in a somewhat informal way to indicate the intensity of metamorphism, meaning the amount or degree of metamorphic change. (To provide a more complete indication of the intensity of metamorphism, geologists use the concept of metamorphic facies; see **Box 8.1**.) Classification of metamorphic grade depends primarily on temperature, because temperature plays the dominant role in determining the extent of recrystallization and neocrystallization during metamorphism. Metamorphic rocks that form at relatively low temperatures (between about 250°C and 400°C) are low-grade rocks, and metamorphic rocks that form at relatively high temperatures (over 600°C) are

FIGURE 8.11 A black amphibolite from the island of Zabargad, Egypt. The fold defined by the streaks of light-colored gneiss indicates the rock, overall, was intensely sheared. But the pure amphibolite away from the light streaks (e.g., in the region above the jackknife) does not have an obvious foliation.

FIGURE 8.12 Examples of quartzite and marble—typically, but not always, these are nonfoliated metamorphic rocks.

Quartz sandstone—the protolith of quartzite.

(a) This unfoliated maroon quartzite from Wisconsin breaks on smooth fractures that cut across grains.

Italian marble quarry sliced into a mountain.

(b) Michelangelo used nonfoliated white marble for his spectacular sculptures.

Limestone—the protolith of marble.

(c) The marble in this floor tile has color banding inherited from original bedding. The layers became contorted during metamorphism.

high-grade rocks. Intermediate-grade rocks form at temperatures between these two extremes (Fig. 8.13a).

Different grades of metamorphism yield different metamorphic mineral assemblages (Fig. 8.13b). As grade increases, recrystallization and neocrystallization tend to produce coarser grains and new mineral assemblages that are stable at higher temperatures and pressures (Fig. 8.13c).

Metamorphism that occurs while temperature and pressure progressively *increase* is called prograde metamorphism. As grade increases, metamorphic reactions release water, so high-grade rocks tend to be "drier" than low-grade rocks. This means that minerals in high-grade rocks do not contain minerals with −OH in their chemical formula, whereas minerals in lower-grade rocks can.

To understand prograde metamorphism, consider the changes that a shale undergoes when it starts near the Earth's surface and ends up at great depth beneath a mountain range (Fig. 8.13b). The clay flakes in shale lie more or less parallel to the bedding. Under low-grade metamorphic conditions and differential stress, shale transforms into slate. In slate, the clay flakes are a bit larger, are better formed, and align parallel to cleavage. As metamorphic grade increases a bit more, the clay flakes decompose and new crystals of fine-grained white mica as well as new crystals of chlorite and quartz form, transforming

FIGURE 8.13   Intensity of metamorphism is indicated by metamorphic grade.

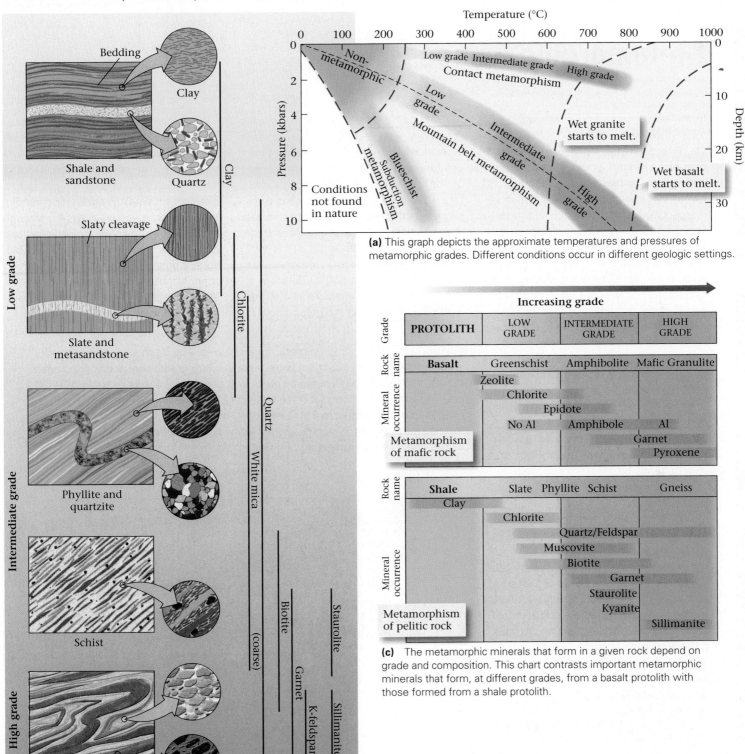

(a) This graph depicts the approximate temperatures and pressures of metamorphic grades. Different conditions occur in different geologic settings.

(c) The metamorphic minerals that form in a given rock depend on grade and composition. This chart contrasts important metamorphic minerals that form, at different grades, from a basalt protolith with those formed from a shale protolith.

(b) Here, we see the consequences of the progressive metamorphism of shale and sandstone from low grade to high grade during mountain building. Lines on the right indicate the range of grade in which key metamorphic minerals form.

BOX 8.1

Metamorphic Facies

In the early years of the twentieth century, geologists working in Scandinavia, where erosion by glaciers has left beautiful, nearly unweathered exposures, came to realize that metamorphic rocks, in general, do not consist of a hodgepodge of minerals formed at different times and in different places, but rather consist of a distinct assemblage of minerals that grew in association with each other at a certain pressure and temperature. It seemed that the mineral assemblage present in a rock more or less represented a condition of chemical equilibrium, meaning that the chemicals making up the rock had organized into a group of mineral grains that were—to anthropomorphize a bit—comfortable with each other and their surroundings, and thus did not feel the need to change further. These geologists also realized that the specific mineral assemblage in a rock depends on pressure and temperature conditions, *and* on the composition of the protolith.

This discovery led the geologists to propose the concept of metamorphic facies. A **metamorphic facies** is a set of metamorphic mineral assemblages indicative of a certain range of pressure and temperature. Each specific assemblage in a facies reflects a specific protolith composition. According to this definition, a given metamorphic facies includes several different kinds of rocks that differ from each other in terms of chemical composition and, therefore, mineral content—but all the rocks of a given facies formed under roughly the same temperature and pressure conditions. Geologists recognize several facies, of which the major ones are zeolite, hornfels, greenschist, amphibolite, blueschist, eclogite, and granulite. The names of the different facies are based on a distinctive feature or mineral found in some of the rocks of the facies.

We can represent the approximate conditions under which metamorphic facies formed by using a pressure-temperature graph (**Fig. Bx8.1**). Each area on the graph, labeled with a facies name, represents the approximate range of temperatures and pressures in which mineral assemblages characteristic of that particular facies form. For example, a rock subjected to the pressure and temperature at Point A (4.5 kbar and 400°C) develops a mineral assemblage characteristic of the greenschist facies. As the graph implies,

the boundaries between facies cannot be precisely defined, and there are likely broad transitions between facies. We can also portray the geothermal gradients of different crustal regions on the graph. Beneath mountain ranges, for example, the geothermal gradient passes through the zeolite, greenschist, amphibolite, and granulite facies. In contrast, in the accretionary prism that forms at a subduction zone, temperature increases slowly with increasing depth, so blueschist assemblages can form.

FIGURE Bx8.1 The common metamorphic facies. The boundaries between the facies are depicted as wide bands because they are gradational and approximate. Note that some amphibolite-facies rocks and all granulite-facies rocks form at pressure-temperature conditions to the right of the melting curve for *wet* granite. Thus, such metamorphic rocks develop only if the protolith is dry. One of the facies depicted on the graph is not mentioned in the text: specifically, P-P ("prehnite-pumpellyite") facies, named for two metamorphic minerals.

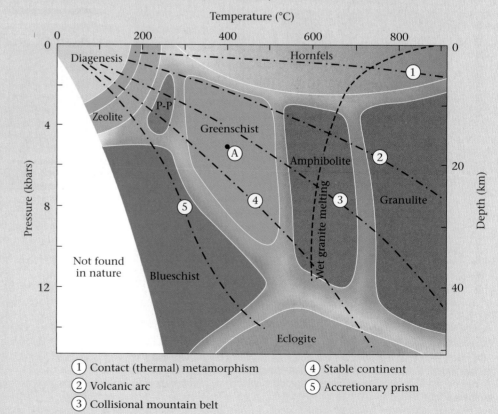

1. Contact (thermal) metamorphism
2. Volcanic arc
3. Collisional mountain belt
4. Stable continent
5. Accretionary prism

FIGURE 8.14 The metamorphic history of a rock can be portrayed on a graph showing variations in temperature and pressure. This graph shows one of many possible paths. As the temperature and pressure decrease, and if water is present, the rock undergoes retrograde metamorphism. P_m is the peak pressure, and T_m is the peak temperature.

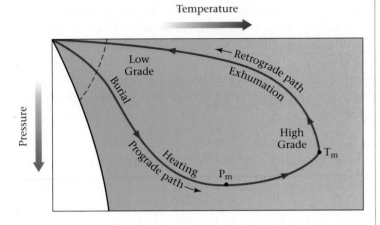

the rock into phyllite. Under intermediate-grade conditions, the minerals in phyllite react and decompose, yielding atoms that combine to produce large crystals of mica (such as muscovite and biotite) as well as other minerals such as garnet. The reactions also release water, which escapes. In our example, this metamorphism is taking place under differential stress, so the micas grow with a preferred orientation and the rock becomes a schist. During metamorphic reactions under high-grade conditions, yet another assemblage of minerals forms. High-grade rocks do not contain much mica, if any, because mica contains –OH and tends to decompose and release water at high temperatures. In fact, high-grade rock typically includes water-free minerals, such as feldspar, quartz, pyroxene, and garnet. As mica disappears, the rock loses its schistosity but can develop gneissic layering.

Metamorphism that takes place when temperatures and pressures progressively *decrease* is known as retrograde metamorphism. For retrograde metamorphism to occur, hydrothermal fluids must enter the rock to add back water. In fact, under cold and dry conditions, retrograde metamorphic reactions cannot proceed or take place too slowly to cause much change—atoms cannot diffuse easily at low temperature. It is for this reason that high-grade rocks formed early during Earth history have survived and can be exposed at the surface of the Earth today.

We can represent the concept of prograde and retrograde metamorphism as a path plotted on a pressure-temperature graph (Fig. 8.14). When a rock gets buried progressively deeper, temperature and pressure increase; the rock follows a prograde path on the graph. (In this particular example the rock was buried quickly, so pressure increased faster than temperature; rocks are good insulators and thus heat up very slowly.) Later, when the rock moves back toward the Earth's surface, because uplift raises the rock and erosion strips away overlying rock, it follows the retrograde path. Geologists have developed techniques to determine the times at which a rock reached particular locations along the pressure-temperature path. This information defines a "P-T-t path" (pressure-temperature-time path) for the rock. Knowledge of these factors helps geologists interpret the geologic history of the rock.

The presence of certain minerals known as **index minerals** in a rock indicates the approximate metamorphic grade of the rock. The line on a map along which an index mineral first appears is called an isograd (from the Greek *iso*, meaning equal). All points along an isograd have approximately the same metamorphic grade. **Metamorphic zones** are regions between two isograds; zones are named after an index mineral that was not present in the previous, lower-grade zone. To compare rocks of different grades, you could take a hike from central New York State eastward into central Massachusetts in the eastern United States. Your path starts in a region where rocks were not metamorphosed, and it takes you into the internal part of the Appalachian Mountain belt, where rocks were intensely metamorphosed. As a consequence, you cross several metamorphic zones (Fig. 8.15).

FIGURE 8.15 Metamorphic zones and isograds.

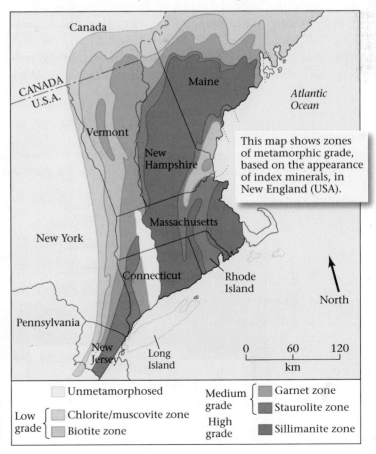

This map shows zones of metamorphic grade, based on the appearance of index minerals, in New England (USA).

☐ Unmetamorphosed	Medium { ☐ Garnet zone
Low { ☐ Chlorite/muscovite zone	grade { ☐ Staurolite zone
grade { ☐ Biotite zone	High grade { ☐ Sillimanite zone

- "Metamorphic grade" is an informal specification of the intensity of metamorphism; high-grade rocks form at higher temperatures, and low-grade rocks at lower temperatures.
- The grade of a rock is reflected by the mineral assemblage it contains.
- A region in which a specific grade of rock occurs is a metamorphic zone, and the boundary between zones is an isograd.
- The mineral assemblage in a metamorphic rock depends on temperature and pressure, and on protolith composition.
- A metamorphic facies is a set of mineral assemblages indicative of certain pressure and temperature conditions.

THINK: The geothermal gradient on Mars (~ 8°C/km) is much less than that on Earth, and in places the crust of Mars is only 30 km thick. Do high-grade metamorphic rocks exist in this crust?

8.5 WHERE DOES METAMORPHISM OCCUR?

So far, we've discussed the nature of changes that occur during metamorphism, the agents of metamorphism (heat, pressure, differential stress, and hydrothermal fluids), the rock types that form as a result of metamorphism, and the concepts of metamorphic grade and metamorphic facies. With

this background, let's now examine the geologic settings on Earth where metamorphism takes place, as viewed from the perspective of plate tectonics theory (see **Geology at a Glance**, pp. 230–231). Because of the wide range of possible metamorphic environments, metamorphism occurs at a wide range of conditions in the Earth (**See for Yourself I**, p. S-16). You will see that the conditions under which metamorphism occurs are not the same in all geologic settings. That's because the geothermal gradient (the relation between temperature and depth; **Fig. 8.16**), the extent to which rocks endure differential stress during metamorphism, and the extent to which rocks interact with hydrothermal fluids all depend on the geologic environment.

Thermal or Contact Metamorphism

Imagine a hot magma that rises from great depth beneath the Earth's surface and intrudes into cooler rock at a shallow depth. Heat flows from the magma into the wall rock, for heat always flows from hotter to colder materials. As a consequence, the magma cools and solidifies while the wall rock heats up. In addition, hydrothermal fluids circulate through both the intrusion and the wall rock. As a consequence of the heat and hydrothermal fluids, the wall rock undergoes metamorphism, with the highest-grade rocks forming immediately adjacent to the pluton, where the temperatures were highest, and progressively lower-grade rocks forming farther away. The distinct belt of metamorphic rock that forms around an igneous intrusion is called a **metamorphic aureole** or contact aureole (**Fig. 8.17**). The width of an aureole

FIGURE 8.16 The change in temperature with depth at different locations in a continent. The solid lines are isotherms (in °C) —they connect locations where the temperature has the same value.

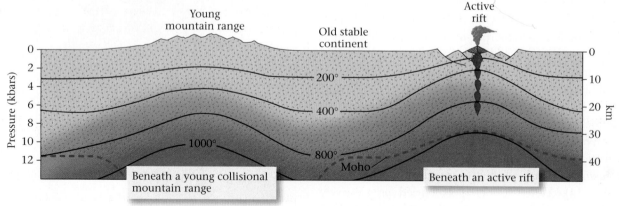

BOX 8.2

Pottery Making—An Analog for Thermal Metamorphism

A brick for the wall of an adobe house, an earthenware pot, a stoneware bowl, or a translucent porcelain teacup may all be formed from the same lump of soft clay, scooped from the surface of the Earth and shaped by human hands. This pliable and slimy muck is a mixture of very fine clay minerals and quartz grains formed during the chemical weathering of rock and water. Fine potter's clay for making white china contains a particular clay mineral called kaolinite, named after the locality in China (called "Kauling," meaning high ridge) where it was originally discovered.

Did you ever wonder . . . why porcelain is harder than the clay it's made from

People in arid climates make adobe bricks by forming damp clay into blocks, which they then dry in the sun. Such bricks can be used for construction *only* in arid climates, because if it rains heavily, the bricks will rehydrate and turn back into sticky muck—drying clay in the sun does not change the structure of the clay minerals.

To make a more durable material, brick makers place clay blocks in a kiln and bake ("fire") them at high temperatures. This process makes the bricks hard and impervious to water. Potters use the same process to make jugs. In fact, fired clay jugs that were used for storing wine and olive oil have been found intact in sunken Greek and Phoenician ships that have rested on the floor of the Mediterranean Sea for thousands of years! Clearly, the firing of a clay pot fundamentally and permanently changes clay in a way that makes it physically different (see Fig. 8.17). In other words, firing causes a thermal metamorphic change in the mineral assemblage that composes pottery. The extent of the transformation depends on the kiln temperature, just as the grade of metamorphic rock depends on temperature. Potters usually fire earthenware at about 1,100°C and stoneware (which is harder than a knife or fork) at about 1,250°C. To produce porcelain—fine china—the clay must partially melt at even higher temperatures. Just as it begins to melt, the potter cools it quickly. Such quenching of the melt creates glass, which gives porcelain its translucent, vitreous (glassy) appearance.

---

depends on the size and shape of the intrusion, and on the amount of hydrothermal circulation—larger intrusions produce wider aureoles.

The local metamorphism caused by igneous intrusion can be called either **thermal metamorphism** (see Box 8.2), to emphasize that it develops in response to heat without a change in pressure and without differential stress, or **contact metamorphism**, to emphasize that it develops adjacent to the contact of an intrusion with its wall rock. Because this metamorphism takes place without application of compression or shear, aureoles contain hornfels, a nonfoliated metamorphic rock.

You can see a classic example of contact metamorphism by traveling to the state of Maine in the northeastern United States. Here you will find a 14-km-long by

FIGURE 8.17 Heat radiated from a large pluton can produce a metamorphic aureole, in which hornfels develops. Grade decreases progressively away from the pluton contact.

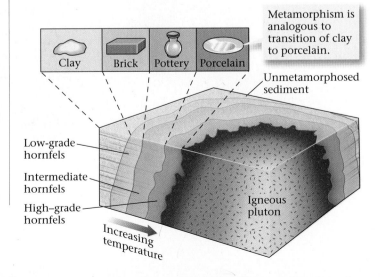

Clay | Brick | Pottery | Porcelain

Metamorphism is analogous to transition of clay to porcelain.

Unmetamorphosed sediment

Low-grade hornfels

Intermediate hornfels

High-grade hornfels

Igneous pluton

Increasing temperature

Environments of Metamorphism

Regional Metamorphism
in an Orogenic Belt

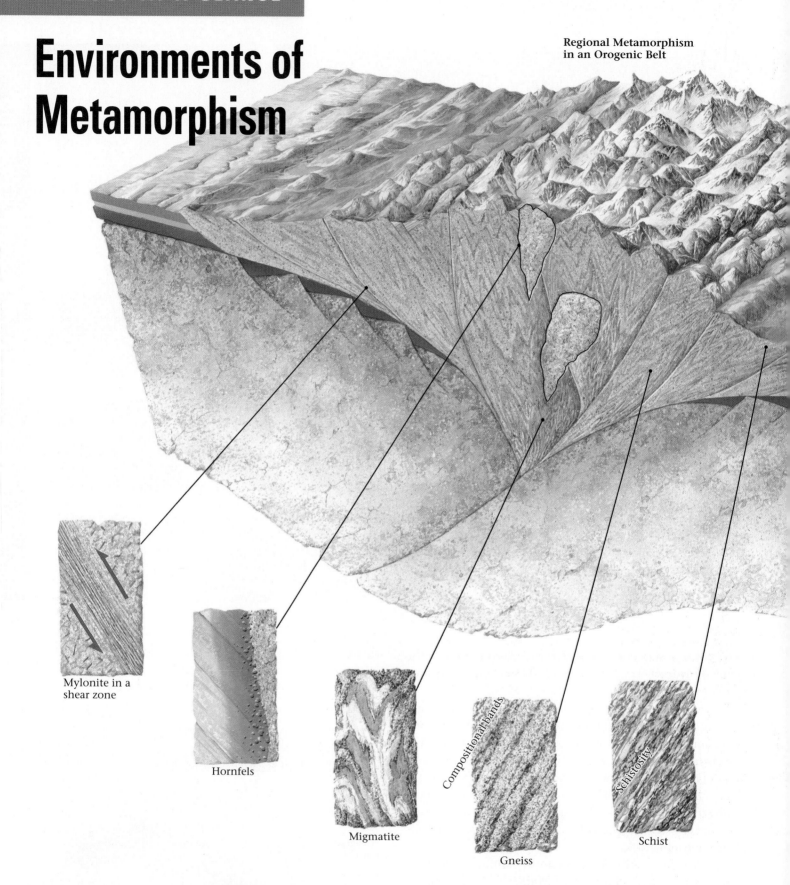

Mylonite in a
shear zone

Hornfels

Migmatite

Compositional bands

Gneiss

Schistosity

Schist

Slate

Unmetamorphosed shale

Metamorphic rocks form when a preexisting rock (a protolith) undergoes changes in texture and/or mineral content in the solid state, in response to changes in temperature, pressure, and/or differential stress. Metamorphism may also reflect interaction with hydrothermal fluids. Some metamorphic rocks are nonfoliated (they do not have metamorphic layering), whereas others are foliated (they do have metamorphic layering). Foliation results when rock is compressed or sheared during metamorphism, causing minerals to grow or rotate into parallelism with each other. Dynamothermal (regional) metamorphism occurs during mountain building. Contact metamorphism takes place around an igneous intrusion, or pluton, caused by the heat released by the pluton.

Geologists distinguish among metamorphic rocks according to the type of foliation and the mineral assemblage a rock contains. Hornfels is unfoliated and forms as a result of contact metamorphism. Mylonite develops when shearing creates a foliation but not necessarily a change in types of minerals. Slate, which forms from shale, contains slaty cleavage; clay flakes are typically aligned at an angle to bedding. Schist contains coarse grains of mica (muscovite and/or biotite) aligned parallel to each other. Gneiss has compositional banding. (Migmatite forms when part of the rock melts, and thus it is a mixture of metamorphic and igneous rock.) Quartzite is composed predominantly of quartz (it is metamorphosed sandstone), whereas marble is composed predominantly of calcite or dolomite (it is metamorphosed limestone or dolostone). Quartzite and marble are usually unfoliated.

The types of minerals and foliation in a metamorphic rock indicate the rock's grade. High-grade rocks, such as gneiss, form at higher temperatures and pressures, whereas low-grade rocks, such as schist, form at lower pressures and temperatures. Blueschist is an unusual metamorphic rock that develops under relatively high pressures but relatively low temperatures—the environment of an accretionary prism.

Metamorphism at a Convergent Margin

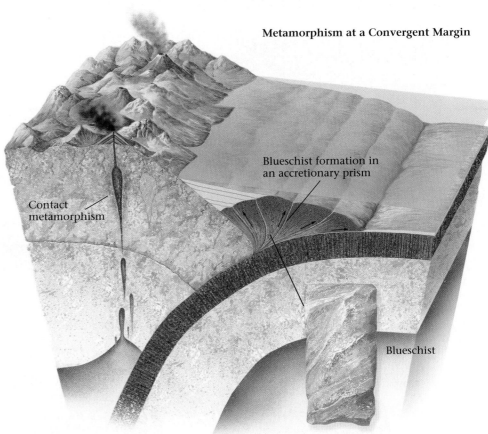

Contact metamorphism

Blueschist formation in an accretionary prism

Blueschist

Foliation resulting from deformation

No foliation

Foliation due to squashing

Foliation due to shearing

Increasing temperature

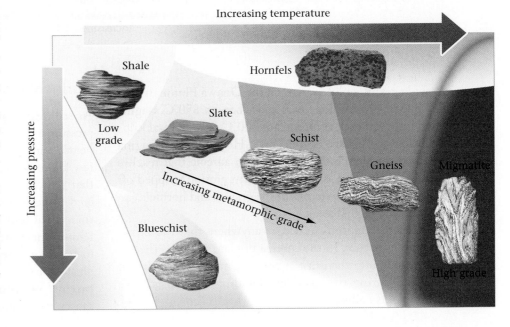

Increasing pressure

Increasing metamorphic grade

Shale

Low grade

Slate

Schist

Hornfels

Gneiss

Migmatite

Blueschist

High grade

FIGURE 8.18 Contact ("thermal") metamorphism occurs in an aureole adjacent to an igneous intrusion. This type of metamorphism produces hornfels. The grade of hornfels decreases away from the contact.

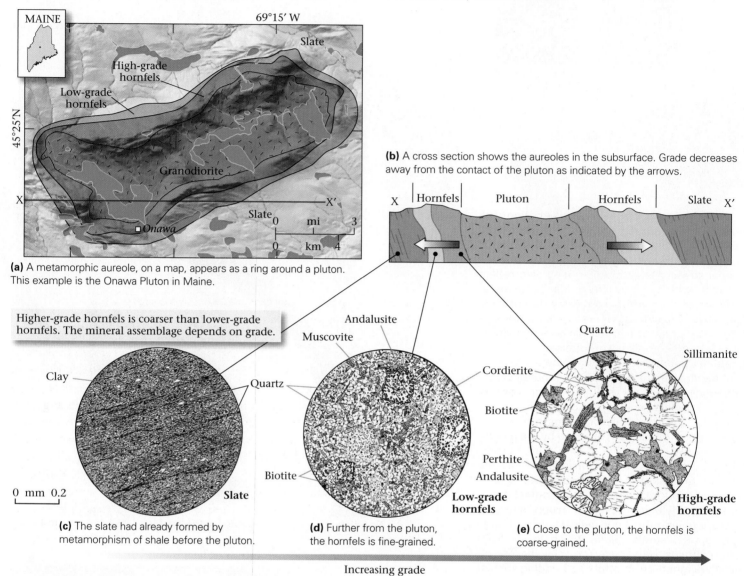

(a) A metamorphic aureole, on a map, appears as a ring around a pluton. This example is the Onawa Pluton in Maine.

(b) A cross section shows the aureoles in the subsurface. Grade decreases away from the contact of the pluton as indicated by the arrows.

Higher-grade hornfels is coarser than lower-grade hornfels. The mineral assemblage depends on grade.

(c) The slate had already formed by metamorphism of shale before the pluton.

(d) Further from the pluton, the hornfels is fine-grained.

(e) Close to the pluton, the hornfels is coarse-grained.

Increasing grade

4-km-wide granitic pluton, named the Onawa Pluton, which formed about 400 million years ago when an 850°C magma intruded into wall rock composed of 300°C slate, several kilometers below the surface of the Earth. Heat from the magma transformed the slate into hornfels in an aureole that reaches a maximum width of 2 km. Subsequently, erosion stripped off overlying rock, so that outcrops of the granite and hornfels can be seen today (Fig. 8.18a–e).

Contact metamorphism occurs anywhere that the intrusion of plutons occurs. In the context of plate tectonics theory, plutons intrude into the crust at convergent plate boundaries, in rifts, and during the mountain building that takes place when continents collide.

Burial Metamorphism

As sediment gets buried in a subsiding sedimentary basin, the pressure increases due to the weight of overburden, and the temperature increases due to the geothermal gradient. In the upper few kilometers, the temperatures and pressures are low enough that the changes taking place in the sediment can be considered to be manifestations of diagenesis. But at depths greater than about 8 to 15 km, depending on the geothermal gradient, temperatures may be great enough for metamorphic reactions to begin, and low-grade metamorphic rocks form. Metamorphism due only to the consequences of very deep burial is called **burial metamorphism**. Of note, burial metamorphism causes the organic molecules of oil to break up; for this reason,

FIGURE 8.19 The formation of mylonite during dynamic metamorphism in a shear zone.

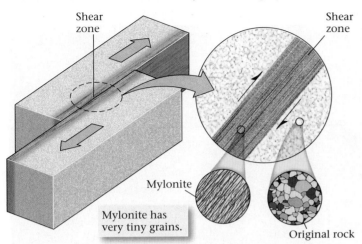

(a) Shearing of a rock under plastic conditions causes original crystals to divide into tiny crystals without breaking to form a mylonite.

(b) Typically, a mylonite has a strong, very fine grain and a strong foliation, as in this example, cutting granite, from Ontario.

oil drillers stop drilling when the bottom of the hole reaches depths at which burial metamorphism has begun.

Dynamic Metamorphism

Faults are surfaces on which one piece of crust slides, or shears, past another. Near the Earth's surface (in the upper 10 to 15 km) this movement can fracture rock, breaking it into angular fragments or even crushing it to a powder. But at greater depths, rock is so warm that it behaves like soft plastic as shear along the fault takes place. During this process, the minerals in the rock recrystallize. We call this process **dynamic metamorphism**, because it occurs as a *consequence of shearing alone* under metamorphic conditions, without requiring a change in temperature or pressure. The resulting rock, a **mylonite**, has a foliation that roughly parallels the fault (Fig. 8.19). Mylonites are very fine-grained, due to processes during dynamic metamorphism, that replace larger crystals with a mass of very tiny ones. Dynamic metamorphism takes place anywhere that faulting occurs at depth in the crust. Thus, mylonites can be found at all plate boundaries, in rifts, and in collision zones.

Dynamothermal or Regional Metamorphism

During the development of large mountain ranges, in response to either convergent-margin tectonics or continental collision, regions of crust are squeezed and large slices of continental crust slip up and over other portions of the crust along faults. As a consequence, rock that was once near the Earth's surface along the margin of a continent ends up at great depth beneath the mountain range (Fig. 8.20). In this environment, three changes happen: (1) the protolith heats up because of

the geothermal gradient and because of igneous activity; (2) the protolith endures greater pressure because of the weight of overburden; and (3) the protolith undergoes compression and shearing. As a result of these changes, the protolith transforms into foliated metamorphic rock. The type of foliated rock that forms depends on the grade of metamorphism—slate forms at shallower depths, whereas schist and gneiss form at greater depths. Since the metamorphism we've just described involves not only heat but also compression and shearing, we can call it **dynamothermal metamorphism**. Typically, such metamorphism affects a large region, so geologists also call it **regional metamorphism**. Erosion eventually removes the mountains, exposing a belt of metamorphic rock that once lay at depth. Such belts may be hundreds of kilometers wide and thousands of kilometers long.

Hydrothermal Metamorphism at Mid-Ocean Ridges

Hot magma rises beneath the axis of mid-ocean ridges, so when cold seawater sinks through cracks down into the oceanic crust along ridges, it heats up and transforms into hydrothermal fluid. This fluid then rises through the crust, near the ridge, causing **hydrothermal metamorphism** of ocean-floor basalt (Fig. 8.21). Eventually, the fluid escapes through vents back into the sea; these vents are called black smokers (see Chapter 2).

Metamorphism in Subduction Zones

Blueschist is a relatively rare rock that contains an unusual blue-colored amphibole. Laboratory experiments indicate that

FIGURE 8.20 The formation of dynamothermal metamorphic rocks during the development of mountain belts. Since this type of metamorphism affects broad regions, it is also called regional metamorphism.

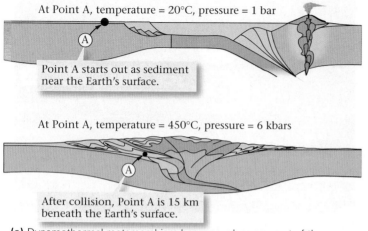

(a) Dynamothermal metamorphism happens when one part of the crust shoves over another part, so that rocks once near the surface end up at great depth.

(b) This roadcut of gneiss in Ontario exposes metamorphic rock that underwent dynamothermal metamorphism over 20 km beneath the Earth's surface.

FIGURE 8.21 Metamorphism due to hydrothermal circulation along mid-ocean ridges.

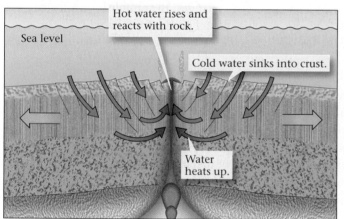

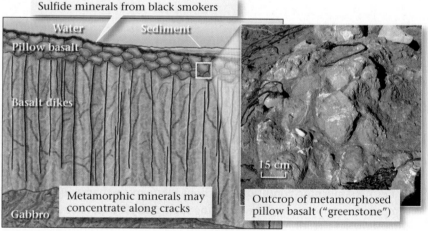

(a) The heat of rising magma at a ridge axis heats water, which then convects. The hot water reacts with the crust and forms metamorphic minerals.

(b) Hydrothermal metamorphism in oceanic crust produces greenish minerals (chlorite and epidote). Sulfide minerals typically precipitate from water mostly at the surface.

formation of this mineral requires very high pressure but relatively low temperature. Such conditions do not develop in continental crust—usually, at the high pressure needed to produce blue amphibole, temperature in continental crust is also high (see Box 8.1). So to figure out where blueschist forms, we must determine where high pressure can develop at relatively low temperature.

Plate tectonics theory provides the answer to this puzzle. Researchers found that blueschist occurs only in the accretionary prisms that form at subduction zones (see Geology at a Glance, pp. 230–231). They realized that because prisms grow to be over 20 km thick, rock at the base of the prism feels high pressure (due to the weight of overburden). But because the

subducted oceanic lithosphere beneath the prism is cool, temperatures at the base of the prism remain relatively low. Under these conditions, blue amphibole can form. Because of shear between the subducting plate and the overriding plate, blueschist develops a foliation.

Shock Metamorphism

When large meteorites slam into the Earth, a vast amount of kinetic energy instantly transforms into heat, and a pulse of extreme compression (a shock wave) propagates into the Earth. The heat may be sufficient to melt or even vaporize rock at the impact site, and the extreme compression of the shock wave causes the crystal structure of quartz grains in rocks below the impact site to suddenly undergo a phase change to a more compact mineral called coesite. The changes in rock due to the passage of a shock wave are called **shock metamorphism**. The discovery of shock metamorphism at Meteor Crater, Arizona, proved that the structure indeed resulted from impact, and when astronauts sampled the Moon, they discovered that the regolith covering the lunar surface contains the products of shock metamorphism produced by countless impacts.

Where Do You Find Metamorphic Rocks?

When you stand on an outcrop of metamorphic rock, you are standing on material that once lay many kilometers beneath the surface of the Earth. How does metamorphic rock return to the Earth's surface? Geologists refer to the overall process by which deeply buried rocks end up back at the surface as **exhumation**.

Exhumation results from several processes in the Earth System that happen simultaneously. Let's look at the specific processes that contribute to bringing high-grade metamorphic rocks from below a collisional mountain range back to the surface (Fig. 8.22). First, as two continents progressively push together, the rock caught between them squeezes upward, much like dough pressed in a vise; the upward movement takes place by slip on faults and by plastic-like flow of rock. Second, as the mountain range grows, the crust at depth beneath it warms up and becomes softer and weaker. Eventually, the range starts to collapse under its own weight, much like a block of soft cheese placed in the hot sun. As a result of this collapse, the upper crust spreads out laterally. Stretching of the upper part of the crust in the horizontal direction causes it to become thinner in the vertical direction, and as the upper part of the crust becomes thinner, the deeper crust ends up closer to the surface. Third, erosion takes place at the surface; weathering, landslides, river flow, and glacial flow together play the role of a giant rasp, stripping away rock at the surface and exposing rock that was once below the surface.

Keeping in mind the processes that form metamorphic rock and cause exhumation, let's ask the question, "Where are metamorphic rocks presently exposed?" You can start your quest to find metamorphic rock outcrops by hiking into a mountain range. As we've seen, the process of mountain building produces and eventually exhumes metamorphic rocks. The towering cliffs in the interior of a mountain range typically reveal schist, gneiss, and quartzite (Fig. 8.23a). Even after the peaks have eroded away, the record of mountain building remains in the form of a belt of metamorphic rock at the ground surface.

Vast expanses of metamorphic rock crop out in continental shields. A **shield** is a broad region of long-lived, stable continental crust where Phanerozoic sedimentary cover either was not deposited or has been eroded away so that Precambrian rocks are exposed (Fig. 8.23b, c). These rocks were metamorphosed during a succession of Precambrian mountain-building events that led to the original growth of continents.

FIGURE 8.22 Processes that bring metamorphic rock back to Earth's surface. Three phenomena contribute to exhumation of rocks at depth. Here, the red dot (representing metamorphic rocks formed at the base of a mountain range) gets progressively closer to the surface over time.

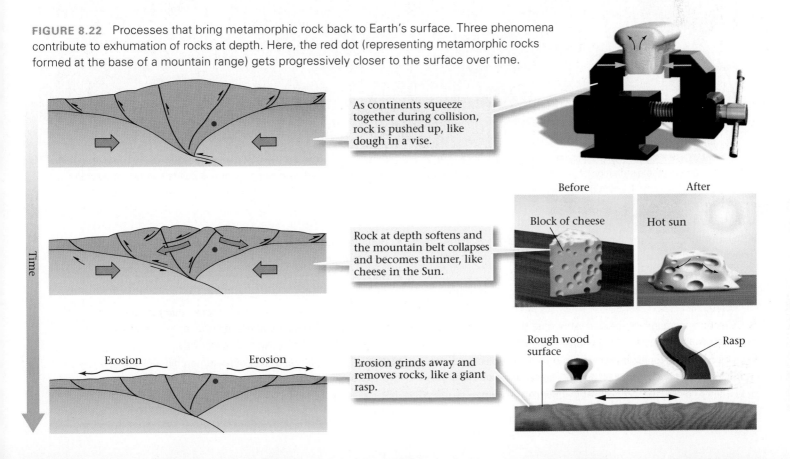

FIGURE 8.23 Examples of rock exposures consisting of Precambrian metamorphic rocks.

(a) The Gunnison River carved a deep canyon into the Precambrian basement rock in Colorado.

(b) A photograph from an airplane window of the flat landscape of the eastern Canadian Shield.

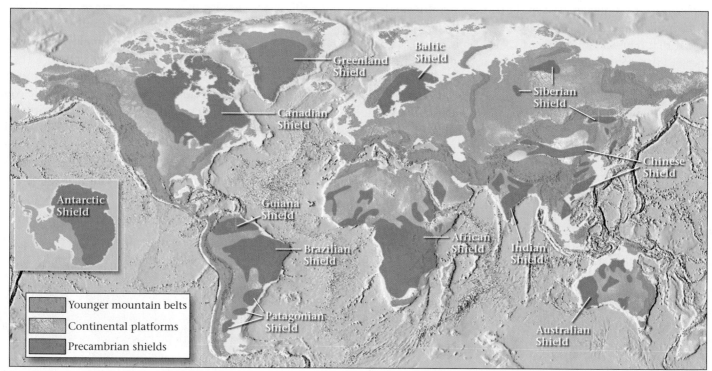

(c) A map showing the distribution of shields, areas where broad expanses of Precambrian crust, including Precambrian metamorphic rocks, crop out.

Take-Home Message

- Thermal (or contact) metamorphism occurs in response to the heat emitted from an igneous intrusion; higher-grade rocks form closer to the intrusion.
- Burial metamorphism occurs at the base of deep sedimentary basins.
- Dynamic metamorphism produces mylonite and develops where rock has been intensely sheared under metamorphic conditions.
- Dynamothermal (or regional) metamorphism takes place at depth in the crust beneath mountain ranges; this type of metamorphism produces foliated rocks because of differential stress.

- Hydrothermal metamorphism develops when hot-water solutions pass through rock.
- Shock metamorphism occurs during the impact of a meteorite; examples are relatively rare on Earth.
- Exhumation causes metamorphic rocks that formed at depth in the crust to be exposed at the surface; examples occur in mountain belts and on continental shields.

THINK: Could you find a layer of metamorphic rock in between the layers of sedimentary rock in a sedimentary basin? Why?

Chapter Summary

- Metamorphism refers to changes in a rock that result in the formation of a metamorphic mineral assemblage, and/or metamorphic texture, in response to change in temperature and/or pressure, to the application of differential stress, and to interaction with hydrothermal fluids (hot-water solutions).

- Metamorphism involves recrystallization, phase changes, metamorphic reactions (neocrystallization), pressure solution, and/or plastic deformation. If hydrothermal fluids bring in or remove elements, we say that metasomatism has occurred.

- Metamorphic foliation can be defined either by preferred mineral orientation (aligned inequant crystals) or by compositional banding. Preferred mineral orientation develops where differential stress causes the compression and shearing of a rock so that its inequant grains align parallel with each other.

- Geologists separate metamorphic rocks into two classes, foliated rocks and nonfoliated rocks, depending on whether the rocks contain foliation.

- The class of foliated rocks includes slate, phyllite, metaconglomerate, schist, and gneiss. The class of nonfoliated rocks includes hornfels, quartzite, and marble. Migmatite, a mixture of igneous and metamorphic rock, forms under conditions where melting begins.

- Rocks formed under relatively low temperatures are known as low-grade rocks, whereas those formed under high temperatures are known as high-grade rocks. Intermediate-grade rocks develop between these two extremes. Different metamorphic mineral assemblages form at different grades.

- Geologists track the distribution of different grades of rock by looking for index minerals. Isograds indicate the location at which index minerals first appear. A metamorphic zone is the region between two isograds.

- A metamorphic facies is a group of metamorphic mineral assemblages that develop under a specified range of temperature and pressure conditions.

- Thermal metamorphism (also called contact metamorphism) occurs in an aureole surrounding an igneous intrusion. Burial metamorphism occurs at depth in a sedimentary basin. Dynamically metamorphosed rocks form along faults, where rocks undergo plastic shearing. Dynamothermal metamorphism (also called regional metamorphism) results when rocks undergo heating and shearing during mountain building. Hydrothermal metamorphism takes place due to the circulation of hot water in oceanic crust at mid-ocean ridges. Shock metamorphism happens during the impact of a meteorite.

- We find belts of metamorphic rocks in mountain ranges. Blueschist forms in accretionary prisms. Shields expose broad areas of Precambrian metamorphic rocks.

GEOPUZZLE REVISITED

Marble is one of many different types of metamorphic rocks, formed due to changes in mineral content and/or rock texture, which happen when a rock is subjected to changes in temperature, pressure, and/or differential stress. The marble in Michelangelo's sculptures started as fossiliferous limestone (in the Jurassic). It was buried deeply, heated, and sheared during the formation of the Apennine mountains. Exhumation, due partly to extensional collapse and partly to erosion, has returned the rock to the surface.

Guide Terms

burial metamorphism (p. 232)

contact metamorphism (p. 229)

differential stress (p. 216)

dynamic metamorphism (p. 233)

dynamothermal metamorphism (p. 233)

exhumation (p. 235)

foliation (p. 219)

gneiss (p. 219)

hornfels (p. 221)

hydrothermal metamorphism (p. 233)

index mineral (p. 227)

marble (p. 222)

metaconglomerate (p. 219)

metamorphic aureole (p. 228)

metamorphic facies (p. 226)

metamorphic foliation (p. 213)

metamorphic grade (p. 223)

metamorphic mineral (p. 213)

metamorphic rock (p. 213)

metamorphic texture (p. 213)

metamorphic zone (p. 227)

metamorphism (p. 213)

metasomatism (p. 218)

migmatite (p. 221)

mylonite (p. 233)

phyllite (p. 219)

preferred orientation (p. 217)

pressure (p. 216)

protolith (p. 213)

quartzite (p. 221)

regional metamorphism (p. 233)

schist (p. 219)

shield (p. 235)

shock metamorphism (p. 235)

slate (p. 219)

thermal metamorphism (p. 229)

Review Questions

1. How are metamorphic rocks different from igneous and sedimentary rocks?

2. What two features characterize most metamorphic rocks?

3. What phenomena can cause metamorphism?

4. What is metamorphic foliation, and how does it form?

5. How does slate differ from a phyllite? How does phyllite differ from a schist? How does schist differ from a gneiss?

6. Why is hornfels nonfoliated?

7. What is a metamorphic grade, and how can it be determined? How does grade differ from facies?

8. Describe the geologic settings where thermal, dynamic, and dynamothermal metamorphism take place.

9. Why does metamorphism happen at the site of meteor impacts and along mid-ocean ridges?

10. How does plate tectonics explain the combination of low-temperature but high-pressure minerals found in a blueschist?

11. Where would you go if you wanted to find exposed metamorphic rocks, and how did such rocks return to the surface of the Earth after being at depth in the crust?

On Further Thought

12. Do you think that you would be likely to find a broad region (hundreds of km across and hundreds of km long) in which the outcrop consists of high-grade hornfels? Why or why not? (*Hint*: Think about the causes of metamorphism and the conditions under which a hornfels forms.)

13. Would we likely find broad regions of gneiss and schist on the Moon? Why or why not?

 For more resources, including animations, quizzes, and Norton's GeoTours, go to **wwnorton.com/studyspace**.

 If your instructor assigns exercises in SmartWork, log in at **smartwork.wwnorton.com**.

ANOTHER VIEW   This map shows a geologic map of a portion of the Adirondack Mountains in northern New York State. The geologist who made this map distinguished different metamorphic rock types, and it depicts the distribution of these various types with different colors. Various symbols depict the orientation of foliation and other structures. But even without understanding the details, it is clear from the map that the geology of metamorphic rocks can be very complex.

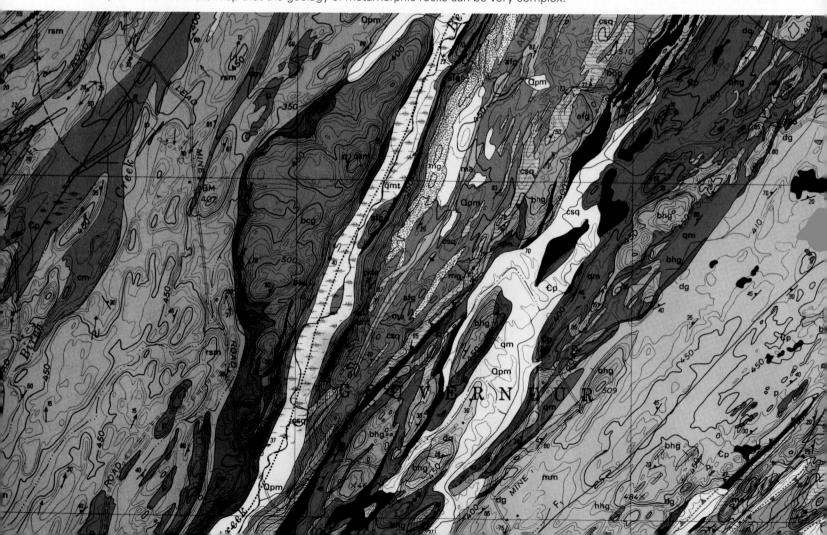

The Rock Cycle

The photos below show three major rock groups of the Earth System. Over time, the atoms in one rock type may be incorporated in another, and then another. This transfer is called the rock cycle.

Beds of red sandstone and siltstone, Red Rocks, Colorado.

Contorted layering in metamorphic rock, Italy.

Molten lava extruding from a volcano in Hawaii, in the process of freezing to form igneous rock.

C.1 INTRODUCTION

"Stable as a rock." This familiar expression implies that a rock is permanent, unchanging over time. But it isn't. In the time frame of Earth history, a span of over 4.57 billion years, atoms making up one rock type may be rearranged or moved elsewhere, eventually becoming part of another rock type. Later, the atoms may move again to form a third rock type, and so on. Geologists refer to the progressive transformation of Earth materials from one rock type to another as the **rock cycle** (Fig. C.1), one of many examples of cycles acting in or on the Earth. A discussion of the rock cycle illustrates the relationships among the rock types described in the previous three chapters.

A **cycle** (from the Greek word for circle or wheel) is a series of interrelated events or steps that occur in succession and can be repeated, perhaps indefinitely. During temporal cycles, such as the phases of the Moon or the seasons of the year, events happen according to a timetable, but the materials involved do not necessarily change. The rock cycle, in contrast, is an example of a geologic **mass-transfer cycle**, one that involves the transfer or movement of materials (mass) to different parts of the Earth System. (The hydrologic cycle, which we will learn about in Interlude F, is another mass-transfer cycle.)

By following the arrows in Figure C.1, you can see that there are many paths around or through the rock cycle. For example, igneous rock may weather and erode to produce

FIGURE C.1 The stages of the rock cycle, showing various alternative pathways.

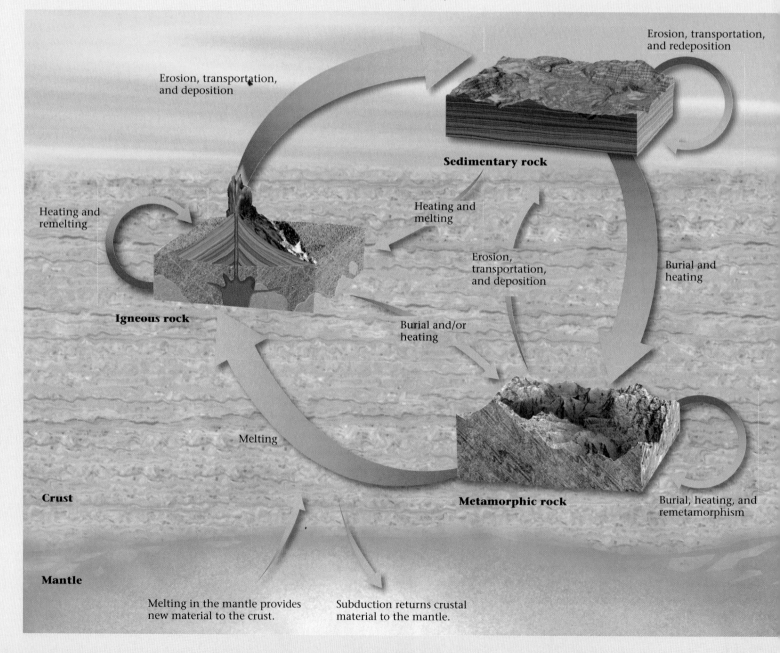

sediment, which lithifies to form sedimentary rock. The new sedimentary rock may become buried so deeply that it transforms into metamorphic rock, which then could partially melt and produce magma. This magma later solidifies to form new igneous rock. We can symbolize this path as

igneous → sedimentary → metamorphic → igneous.

But alternatively, the metamorphic rock could be uplifted and eroded to form new sediment and, later, new sedimentary rock, without melting. This path takes a shortcut through the cycle that we can symbolize as

igneous → sedimentary → metamorphic → sedimentary.

Likewise, the igneous rock could be metamorphosed directly, without first turning to sediment. This metamorphic rock could then be eroded to produce sediment that becomes sedimentary rock, defining another shortcut path:

igneous → metamorphic → sedimentary.

To get a clearer sense of how the rock cycle works, let's look at one example in the context of plate tectonics.

C.2 A CASE STUDY OF THE ROCK CYCLE

Material can enter the rock cycle when magma rises from the mantle. Suppose the magma erupts and forms basalt (an igneous rock) at a continental hot-spot volcano (Fig. C.2a). Interaction with wind, rain, and vegetation gradually weathers the basalt, physically breaking it into smaller fragments and chemically weathering it to yield clay. Water washes the newly formed clay away and transports it downstream—if you've ever seen a brown-colored river, you've seen clay traveling to a site of deposition. Eventually the river reaches the sea, where the water slows down and the clay settles out.

Let's imagine, for this example, that the clay settles out along the margin of Continent X and forms a deposit of mud. Gradually, through time, the mud becomes buried and the clay flakes pack tightly together, resulting in a new sedimentary rock, shale. The shale resides 6 km below the continental shelf for millions of years, until the adjacent oceanic plate subducts and another Continent, Y, collides with X. The edge of the encroaching continent pushes over the shale and buries it very deeply. As the mountains grow, the shale that had once been 6 km below the surface ends up 20 km below the surface, and under the pressure and temperature conditions present at this depth, it metamorphoses into schist (Fig. C.2b).

The story's not over. Once mountain building stops, erosion grinds away the mountain range, and exhumation brings some of the schist to the ground surface. Some of this schist erodes to form sediment, which is carried off and deposited elsewhere to form new sedimentary rock. But other schist remains preserved below the surface (Fig. C.2c). Eventually, continental rifting takes place at the site of the former mountain range, and the crust containing the schist begins to split apart. When this happens, some of the schist partially melts and a new felsic magma forms. This felsic magma rises to the surface of the crust and freezes into rhyolite, a new igneous rock (Fig. C.2d). In terms of the rock cycle, we're back at the beginning, having once again made igneous rock.

Note that atoms, as they pass through the rock cycle, do not always stay within the same mineral. In our example, a silicon atom in a pyroxene crystal of the basalt may become part of a clay crystal in the shale, part of a muscovite crystal in the schist, and part of a feldspar crystal in the rhyolite.

C.3 RATES OF MOVEMENT THROUGH THE ROCK CYCLE

We have seen that not all atoms pass through the rock cycle in the same way. Similarly, not all atoms pass through the rock cycle at the same rate, and for that reason we find rocks of many different ages at the surface of the Earth. Some rocks remain in one form for less than a few million years, while others stay unchanged for most of Earth history. A rock in the Appalachian Mountains has passed through stages of the rock cycle many times during the past few hundred million years, because the eastern margin of North America has been subjected to multiple events of basin formation, mountain building, and rifting since the end of the Precambrian. In contrast, some Precambrian rocks in the interior of continents have remained unchanged for over a billion years.

Most atoms that comprise continental rocks never return to the mantle, because continental crust is buoyant and does not subduct. However, a small amount of sediment that erodes off a continent ends up in deep-ocean trenches, and some of this gets carried back into the mantle by subduction. In fact, recent research suggests that metamorphic and igneous rocks at the base of the continental crust may be scraped off and transported down into the mantle by subduction.

Our tour of the rock cycle has focused on continental rocks. What about the oceans? Oceanic crust consists of igneous rock (basalt and gabbro) overlain by sediment. Because a layer of water blankets the crust, erosion does not affect it, so oceanic crustal rock generally does not follow the path into the sedimentary loop of the rock cycle. But sooner or later, oceanic crust subducts. When this happens, the rock of the crust undergoes metamorphism, for as it sinks, it is subjected to progressively higher temperatures and pressures. Most oceanic crust eventually gets carried back into

FIGURE C.2   An example of the rock cycle in the context of plate tectonics.

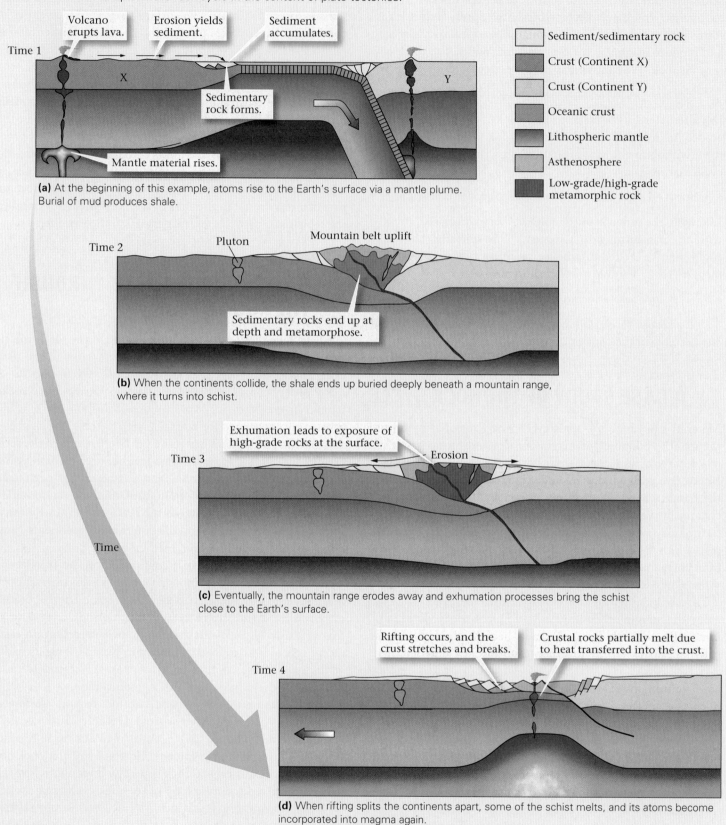

Volcano erupts lava.

Erosion yields sediment.

Sediment accumulates.

Time 1

Sedimentary rock forms.

Mantle material rises.

Sediment/sedimentary rock

Crust (Continent X)

Crust (Continent Y)

Oceanic crust

Lithospheric mantle

Asthenosphere

Low-grade/high-grade metamorphic rock

(a) At the beginning of this example, atoms rise to the Earth's surface via a mantle plume. Burial of mud produces shale.

Pluton

Mountain belt uplift

Time 2

Sedimentary rocks end up at depth and metamorphose.

(b) When the continents collide, the shale ends up buried deeply beneath a mountain range, where it turns into schist.

Exhumation leads to exposure of high-grade rocks at the surface.

Time 3

Erosion

Time

(c) Eventually, the mountain range erodes away and exhumation processes bring the schist close to the Earth's surface.

Rifting occurs, and the crust stretches and breaks.

Crustal rocks partially melt due to heat transferred into the crust.

Time 4

(d) When rifting splits the continents apart, some of the schist melts, and its atoms become incorporated into magma again.

the mantle. A small part of this subducted crust may melt and be incorporated in magma that rises and becomes new igneous rock.

C.4 WHAT DRIVES THE ROCK CYCLE IN THE EARTH SYSTEM?

The rock cycle occurs because the Earth is a dynamic planet and there are many rock-forming environments (see Geology at a Glance, pp. 244–245). The planet's internal heat and gravitational field drive plate movements and plume-associated hot spots. Plate interactions cause the uplift of mountain ranges, a process that exposes rock to weathering, erosion, and sediment production. Plate interactions also generate the geologic settings in which metamorphism occurs, where rock melts to provide magma, and where sedimentary basins develop. Plumes may also lead to magma generation.

At the surface of the Earth, the gases initially released by volcanism collect to form the ocean and atmosphere. Heat (from the Sun) and gravity drive convection in the atmosphere and oceans, leading to wind, rain, ice, and currents—the agents of weathering and erosion. Weathering and erosion grind away at the surface of the Earth and send material into the sedimentary loop of the cycle. In the Earth System, life also plays a key role by adding corrosive oxygen to the atmosphere and by directly contributing to weathering. In sum, external energy (solar heat), internal energy (Earth's internal heat), gravity, and life all play a role in driving the rock cycle by keeping the mantle, crust, atmosphere, and oceans in motion.

Guide Terms

cycle (p. 240)
mass-transfer cycle (p. 240)
rock cycle (p. 240)

ANOTHER VIEW This scenic view on the coast of Brittany provides a snapshot of one stage in the rock cycle. The cobbles, a sediment, were derived from the granite bedrock, an igneous rock. They are in the process of being broken up and rounded by waves. If the cobbles were buried and lithified as is, they'd become part of a conglomerate, and a step in the rock cycle would have come to completion.

Rock-Forming Environments and the Rock Cycle

Drainage networks collect surface water that can transport sediment to the ocean.

In a desert environment, rock weathers and fragments. Debris falls in landslides.

Sand dunes form from grains carried by the wind.

Flash floods carry sediment out of canyons to form an alluvial fan.

km

0

Volcanic eruptions emit lava and ash, which form new igneous rock at Earth's surface.

10

Sedimentary rocks make a cover on the surface of continents.

The crust and lithospheric mantle stretch and thin in a rift.

20

Magma rises from the mantle. Heat from this magma causes contact metamorphism.

30

Deep levels of continents consist of ancient metamorphic and igneous rocks. This is the basement of the continents.

40

Continental margins slowly sink and are buried by new sediment.

50

60

70

80

90

100

Partial melting occurs in the asthenosphere to produce new magma.

Rocks form in many different environments. Igneous rocks develop where melt rises from depth and cools. Intrusive igneous rocks form where magma cools underground; extrusive igneous rocks form where lava and ash erupt at the surface.

Weathering and erosion break up existing rock and produce sediment. Different kinds of sediments develop in different places, reflecting both the composition of the source and the setting in which the sediment accumulates. We distinguish among sediment that collects in alluvial fans, desert dunes, river channels and floodplains, deltas, coral reefs, coastlines, the continental shelf, the deep sea, and the toe of a glacier. When this sediment eventually gets buried and undergoes lithification, new sedimentary rocks form.

Under certain conditions, preexisting rocks can undergo change in the solid state—metamorphism—which produces

Glaciers erode rock and can transport sediment of all sizes.

In a region of continental collision, rocks that were near the surface are deeply buried and metamorphosed.

humid climates, ick soils develop.

Magma that cools and solidifies underground forms igneous intrusions.

Along coastal plains, rivers meander. Sediment collects in the channel and floodplain.

Reefs grow from calcite-secreting organisms. These will eventually turn into limestone.

Where a river enters the sea, sediment settles out to form a delta.

Many different kinds of sediment accumulate along coastlines, building out a continental shelf.

Underwater avalanches carry a cloud of sediment that settles to form a submarine fan.

Fine clay and plankton shells settle on the oceanic crust.

The oceanic crust consists of igneous rocks formed at a mid-ocean ridge.

metamorphic rocks. Contact metamorphism is due to heat released by an intrusion of magma. Regional metamorphism occurs where tectonic processes cause rocks from the surface to be buried very deeply.

Because the Earth is dynamic, environments change through time. Tectonic processes cause new igneous rocks to form. When exposed at the surface, these rocks weather to make sediment. The slow sinking of some regions creates sedimentary basins in which sediment accumulates and new sedimentary rocks form. Later, these rocks may be buried deeply and metamorphosed. Uplift as a result of mountain building exposes the rocks to the surface, where they may once again be transformed into sediment. This progressive transformation is called the rock cycle.

In mountain belts, we see the complex texture of the Earth's surface. Such textures reflect the complex interplay of plate-tectonic forces pushing up the Earth's surface, and erosive forces tearing them down. The eruption of volcanoes, the shaking of earthquakes, and the uplift of mountains provide dramatic proof that the Earth remains a dynamic planet.

PART III

Tectonic Activity of a Dynamic Planet

Earlier in this book, we learned that the map of the Earth constantly, but ever so slowly, changes in response to plate movements and interactions. We now turn our attention to the dramatic consequences of such tectonic activity in the Earth System: volcanoes (Chapter 9), earthquakes (Chapter 10), and mountains (Chapter 11).

Why does molten rock rise like a fountain out of the ground, or explode into the sky at a volcano? Why does the ground shake and heave, in some cases so violently that whole cities topple, during an earthquake? How can the energy released by an earthquake tell us about the insides of the Earth, thousands of kilometers below the surface? What processes cause the land surface to rise several kilometers above sea level to form mountain belts? How do rocks bend, squash, stretch, and break in response to forces caused by plate interactions? Read on, and you will be able not only to answer these questions but also to see how the answers help people deal with some of the deadliest natural hazards that threaten society.

The Wrath of Vulcan:
Volcanic Eruptions

On the coast of Hawaii, glowing red streams of lava spill into the sea, instantly vaporizing the water. Each flow adds a bit more land to the island.

GEOPUZZLE

Why do volcanoes exist? Why do they occur where they do? Are all eruptions the same?

Glowing waves rise and flow, burning all life on their way, and freeze into black, crusty rock which adds to the height of the mountain and builds the land, thereby adding another day to the geologic past. . . . I became a geologist forever, by seeing with my own eyes: the Earth is alive!

—Hans Cloos (1886–1951),
on seeing the eruption of Mt. Vesuvius (Italy)

9.1 INTRODUCTION

Every few hundred years, one of the hills on Vulcano, an island in the Mediterranean Sea off the western coast of Italy, rumbles and spews out molten rock, glassy cinders, and dense "smoke" (actually a mixture of various gases, fine ash, and very tiny liquid droplets). Ancient Romans thought that such eruptions happened when Vulcan, the god of fire, fueled his forges beneath the island to manufacture weapons for the other gods. Geologic study suggests, instead, that eruptions take place when hot magma, formed by melting inside the Earth, rises through the crust and emerges at the surface. No one believes the Roman myth anymore, but the island's name evolved into the English word **volcano**, which geologists use to designate either an erupting vent through which molten rock reaches the Earth's surface or a mountain built from the products of eruption.

On the main peninsula of Italy, not far from Vulcano, another volcano, Mt. Vesuvius, towers over the Bay of Naples. Two thousand years ago, a prosperous Roman resort and trading town named Pompeii sprawled at the foot of Vesuvius. One morning in 79 C.E., earthquakes signaled the mountain's awakening. At 1:00 P.M. on August 24, a dark mottled cloud boiled up above Mt. Vesuvius's summit to a height of 27 km. As lightning sparked in its crown, the cloud drifted over Pompeii, turning day into night. Blocks and pellets of rock fell like hail, while fine ash and choking fumes enveloped the town (**Fig. 9.1a, b** 🔊). Frantic people rushed to escape, but for many it was too late. As the growing weight of volcanic debris began to crush buildings, a scalding, turbulent current of ash mixed with pumice fragments surged down the flank of the volcano and swept over Pompeii. By the next day the town and its neighbor, Herculaneum, had vanished beneath a 6-m-thick gray-black blanket (**Fig. 9.1c**). This covering protected the ruins of Pompeii and Herculaneum so well that when archaeologists excavated the towns 1,800 years later, they found an amazingly complete record of Roman daily life (**Fig. 9.1d**). In addition, they discovered open spaces in the debris covering Pompeii. Out of curiosity, they filled the spaces with plaster and then dug away the surrounding ash. The spaces turned out to be fossil casts of Pompeii's unfortunate inhabitants, their bodies forever twisted in agony or huddled in despair (**Fig. 9.1e**).

Clearly, volcanoes are unpredictable and dangerous. Volcanic activity can build a towering, snow-crested mountain or can blast one apart. It can provide the fertile soil and mineral deposits that enable a civilization to thrive, or a rain of destruction that can snuff one out. The gas emitted from volcanoes can change the composition of the atmosphere, and the heat within can provide geothermal power. Because of the diversity of volcanic activity and its consequences, this chapter sets out ambitious goals. We first look more closely at the products of volcanic eruptions and the basic characteristics of volcanoes. Then we consider the different kinds of volcanic eruptions on Earth and why they occur where they do. Finally, we consider the hazards posed by volcanoes, efforts by geoscientists to predict eruptions and help minimize the damage they cause, and the possible influence of eruptions on climate and civilization.

Chapter Themes

By the end of this chapter, you should know that . . .

- volcanic eruptions produce a great variety of materials, including lava and pyroclastic debris.
- some eruptions yield rivers of lava, whereas others produce catastrophic explosions.
- the type of eruption depends on the character of the magma being erupted, which may reflect the geologic setting of the volcano.
- volcanoes are hazards, for lava flows, clouds and cascades of debris, slurries of wet ash, and invisible gases can cost lives and destroy property.
- in some cases, impending eruptions can be predicted, allowing people to take precautions.
- volcanoes can affect climate, evolution, and perhaps the future of civilizations.

9.2 THE PRODUCTS OF VOLCANIC ERUPTIONS

The drama of a volcanic eruption transfers materials from inside the Earth to our planet's surface. Products of an eruption come in three forms—lava flows, pyroclastic debris, and gas. Note that we use the name *flow* for both a molten, moving layer of lava and for the solid layer of rock that forms when the lava freezes.

Lava Flows

Sometimes it races down the side of a volcano like a fast-moving, incandescent stream, sometimes it builds into a rubble-covered mound at a volcano's summit, and sometimes it oozes

FIGURE 9.1   The eruption of Vesuvius devastated and buried Pompeii and nearby Herculaneum in 79 C.E.

(a) In this 1817 painting, the British artist J.M.W. Turner depicted the cataclysmic explosion.

(b) Excavations exposed the ruins of Pompeii. Vesuvius towers behind; dashed lines give its profile before eruption. ▶️

A modern suburb of Naples has been built on top of the ash.

The ash layer here is about 25 m thick.

Excavation exposes Roman columns.

(c) Ash still covers much of Herculaneum.

You can see ruts carved into streets by chariots.

(d) Streets and buildings of Pompeii are well preserved.

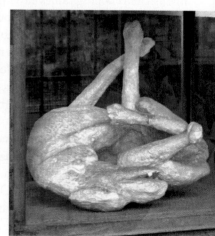

(e) Casts of people and animals convey the terror of that awful day.

FIGURE 9.2 The character of a lava flow depends on its viscosity.

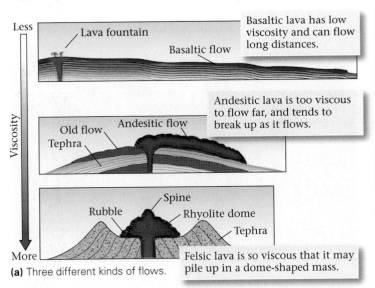

Less

Viscosity

More

Lava fountain
Basaltic flow

Basaltic lava has low viscosity and can flow long distances.

Andesitic flow
Old flow
Tephra

Andesitic lava is too viscous to flow far, and tends to break up as it flows.

Spine
Rubble
Rhyolite dome
Tephra

Felsic lava is so viscous that it may pile up in a dome-shaped mass.

(a) Three different kinds of flows.

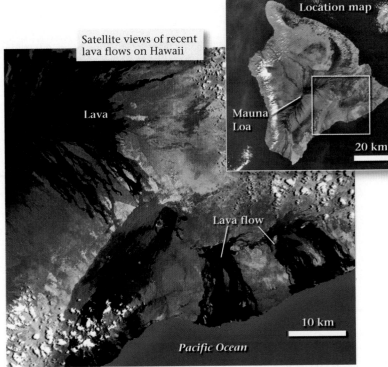

Location map

Satellite views of recent lava flows on Hawaii

Lava

Mauna Loa

20 km

Lava flow

10 km

Pacific Ocean

(b) Basaltic lava flows travel for a long distance, on the south shore of Hawaii.

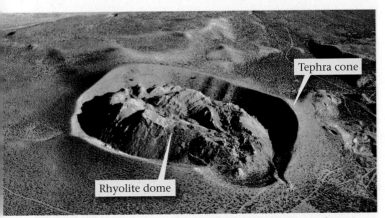

Tephra cone

Rhyolite dome

(c) This rhyolite dome formed about 650 years ago, in Panum Crater, California. Tephra (cinders) accumulated around the vent.

like a sticky but scalding paste. Clearly, not all lava behaves in the same way when it rises out of a volcano. Therefore, not all lava flows look the same. Why? The character of a lava primarily reflects its *viscosity* (resistance to flow), and not all lavas have the same viscosity. Differences in viscosity depend, in turn, on a variety of factors including chemical composition, temperature, gas content, and crystal content. Silica content, the proportion of SiO_2, one of many chemicals making up lava, plays a particularly key role in controlling viscosity. Specifically, silica-poor (basaltic) lava is less viscous, and thus flows farther than does silica-rich (rhyolitic) lava (Fig. 9.2a–c). That's because when silica is abundant, silicon-oxygen tetrahedra can link in strongly bonded chains or networks. If there is less silica, the tetrahedra form weaker bonds with relatively abundant metal ions. To illustrate the different ways in which lava behaves, we now examine flows of different compositions.

Basaltic lava flows. Basaltic (mafic) lava has very low viscosity when it first emerges from a volcano because it contains relatively little silica and is very hot. Thus, on the steep slopes near the summit of a volcano, it can flow very quickly, sometimes at speeds of over 30 km per hour (Fig. 9.3a). The lava slows down to less-than-walking pace after it has traveled several kilometers and has started to cool (Fig. 9.3b). Most flows measure less than 10 km long, but some flows reach as far as 600 km from the source.

How can lava travel such distances? Although all the lava in a flow moves when it first emerges, rapid cooling causes the surface of the flow to crust over after the flow has moved a short distance from the source. The solid crust serves as insulation, allowing the hot interior of the flow to remain liquid and continue to move. On gentle slopes, new molten lava injects between the original ground surface and the new solid crust. The addition of this lava effectively inflates the flow, jacking up the hardened crust and making the overall flow thicker. As time progresses, part of the flow's interior solidifies, so eventually, molten lava moves only through a tunnel-like passageway, or **lava tube**, within the flow—the largest of these may be tens of meters in diameter. In some cases, lava tubes drain and eventually become empty tunnels (Fig. 9.3c).

The surface texture of a basaltic lava flow when it finally freezes reflects the timing of freezing relative to its movement. Basalt flows with warm, pasty surfaces wrinkle into smooth, glassy, rope-like ridges; geologists have adopted the Hawaiian word **pahoehoe** (pronounced "pa-hoy-hoy") for such flows (Fig. 9.3d). If the surface layer of the lava freezes and then

FIGURE 9.3 Features of basaltic lava flows. They have low viscosity and thus can flow long distances. Their surface and interior can be complex.

(a) A fast-moving flow coming from Mt. Etna, Sicily.

(b) A basaltic lava flow covers a highway in Hawaii.

(c) In a lava tube, still-molten lava flows beneath a crust of already-solid basalt.

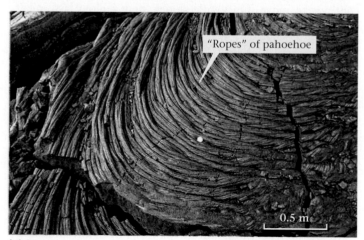

(d) Pahoehoe from a recent lava flow in Hawaii. Note the coin for scale.

(e) The rubbly surface of an a'a' flow, Sunset Crater, Arizona.

breaks up due to the continued movement of lava underneath, it becomes a jumble of sharp, angular fragments, creating a rubbly flow also called by its Hawaiian name, **a'a'** (pronounced "ah-ah") (Fig. 9.3e). Footpaths made by people living in basaltic volcanic regions follow the smooth surface of pahoehoe flows rather than the rough, foot-slashing surface of a'a' flows.

During the final stages of cooling, lava flows contract, because rock shrinks as it loses heat, and may fracture into polygonal columns. This type of fracturing is called **columnar jointing** (Fig. 9.4a).

Basaltic flows that erupt underwater look different from those that erupt on land because the lava cools so much more quickly in water. Because of rapid cooling, submarine basaltic lava can travel only a short distance before its surface freezes, producing a glass-encrusted blob, or "pillow" (Fig. 9.4b). The rind of a pillow momentarily stops the flow's advance, but within minutes the pressure of the lava squeezing into the pillow breaks the rind, and a new blob of lava squirts out, freezes, and produces another pillow. In some cases, successive pillows add to the end of previous ones, forming worm-like chains.

Andesitic and rhyolitic lava flows. Because of its higher silica content and thus its greater viscosity, andesitic lava cannot flow as easily as basaltic lava. When erupted, andesitic lava first forms a large mound above the vent. This mound then advances slowly down the volcano's flank at only about 1 to 5 m a day, in a lumpy flow with a bulbous snout. Typically, andesitic flows are only a few km long; unusually hot flows may travel further. Because the lava moves so slowly, the outside of the flow has time to solidify; so as it moves, the surface breaks up into angular blocks, and the whole flow looks like a jumble of rubble called **blocky lava**. On steep slopes, the blocks may tumble downhill. In the interior of the flow, the lava may contain subtle banding, for the lava is not perfectly uniform in texture, and variations smear out into layers during flow.

Rhyolitic lava is the most viscous of all lavas because it is the most silicic and the coolest. Therefore, it tends to accumulate either above the vent in a dome-like mass, called a lava dome (see Fig. 9.2c), or in short and bulbous flows rarely more than 1 to 2 km long. Sometimes rhyolitic lava freezes while still in the vent and then pushes upward as a column-like spire or spine up to 100 m above the vent. Rhyolitic flows, where they do form, have broken and blocky surfaces.

Volcaniclastic Deposits

On a mild day in February 1943, as Dionisio Pulido prepared to sow the fertile soil of his field 330 km (200 miles) west of Mexico City, an earthquake jolted the ground again, as it had dozens of times in the previous days. But this time, to Dionisio's amazement, the surface of his field visibly bulged upward by a few meters and then cracked. Ash and sulfurous fumes filled the air, and Dionisio fled. When he returned the following morning, his rich soil lay buried beneath a 40-m-high mound of gray cinders—Dionisio had witnessed the birth of Paricutín, a new volcano.

> **Did you ever wonder . . .**
> has anyone ever seen a brand-new volcano form?

FIGURE 9.4 Examples of structures within lava flows.

(a) Columnar jointing develops when the interior of a flow cools and cracks. Such jointing develops in dikes and sills. This example is Devils Postpile in California.

(b) Pillow basalt develops when lava erupts underwater. Later uplift may expose pillows above sea level, as in this Oregon outcrop.

During the next several months, Paricutín erupted continuously, at times blasting clots of lava into the sky like fireworks. By the following year, it had become a steep-sided cone 330 m high. Nine years later, when the volcano ceased all activity, its lava and debris covered 25 square km, and Dionisio's farm and those of his neighbors were gone.

This description of Paricutín's eruption, and that of Vesuvius at the beginning of this chapter, emphasizes that volcanoes can erupt large quantities of fragmental igneous material. Geologists use the general term **volcaniclastic deposits** for accumulations of this material. Volcaniclastic deposits include **pyroclastic debris** (from the Greek *pyro*, meaning fire), which forms from lava that flies into the air and freezes. It also includes the debris formed when an eruption blasts apart preexisting volcanic rock that surrounds the volcano's vent, the debris that accumulates after tumbling down the volcano in landslides or after being transported in water-rich slurries, and the debris formed as lava flows break up or shatter. Let's look at these components in more detail—you'll see that different types form in association with different kinds of eruptions.

Pyroclastic debris from basaltic eruptions.   Basaltic magma rising in a volcano may contain dissolved volatiles (such as water). As such magma approaches the surface, the volatiles form bubbles. When the bubbles reach the surface, they burst and eject clots and drops of molten magma upward to form dramatic fountains (**Fig. 9.5a**). To picture this process, think of the droplets that spray from a newly opened bottle of soda. Pea-sized fragments of glassy lava and scoria are a type of **lapilli**, from the Latin word for little stones; they are informally known as cinders. Rarely, flying droplets may trail thin strands of lava, which freeze into filaments of glass known as Pelé's hair, after the Hawaiian goddess of volcanoes, and the droplets themselves freeze into tiny streamlined glassy beads known as Pelé's tears. Apple- to refrigerator-sized fragments are called **blocks** (**Fig. 9.5b**). Blocks may consist of already-solid volcanic rock, broken up during the eruption—such blocks tend to be angular and chunky. In some cases, however, blocks form from soft lava squirting out of the vent—such blocks, also known as **bombs**, have streaked, polished surfaces.

Pyroclastic debris from andesitic or rhyolitic eruptions. Andesitic or rhyolitic lava is more viscous than basalt, and may be more gas-rich. The lava flows tend to be blocky to start with, and blocks of flows may tumble down the volcano. As we discuss later, eruptions of these lavas also tend to be explosive. Volcanic explosions can produce immense quantities of pyroclastic debris, much more than can come from a basaltic volcano (**Fig. 9.6a**). Debris ejected from explosive eruptions includes fragments of pumice and ash (**Fig. 9.6b**). **Ash** consists of particles less than 2 mm in diameter, made from both glass shards formed when frothy lava explosively fragments

FIGURE 9.5 Pyroclastic debris from basaltic eruptions.

(a) Fountains of lava may erupt from basaltic volcanoes.

Blocks and bombs litter the slope of a Hawaiian volcano.

Bombs have smooth, streaked surfaces.

(b) Apple-sized or larger chunks of rock blasted out of a volcano are called blocks. If the lava is soft when it squirts out, it forms streamlined bombs.

FIGURE 9.6 The components of a large explosive volcanic eruption.

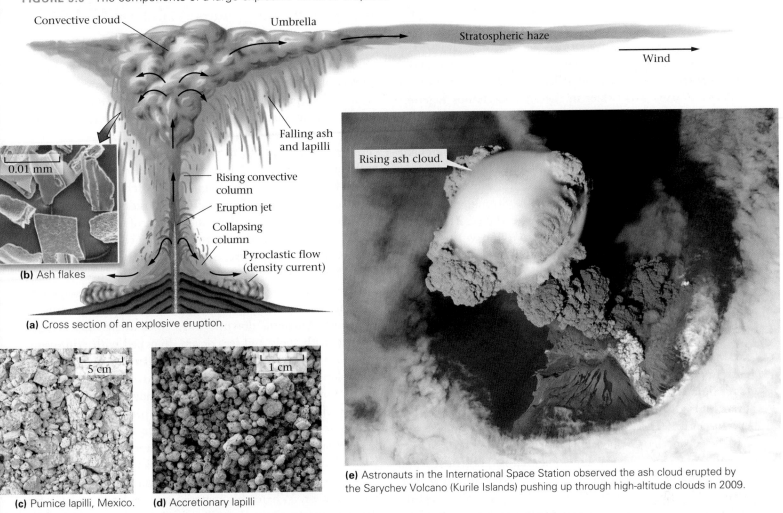

(a) Cross section of an explosive eruption.

(b) Ash flakes

0.01 mm

Convective cloud

Umbrella

Stratospheric haze

Wind

Falling ash and lapilli

Rising convective column

Eruption jet

Collapsing column

Pyroclastic flow (density current)

Rising ash cloud.

(c) Pumice lapilli, Mexico.

5 cm

(d) Accretionary lapilli

1 cm

(e) Astronauts in the International Space Station observed the ash cloud erupted by the Sarychev Volcano (Kurile Islands) pushing up through high-altitude clouds in 2009.

(f) Ash erupts from Mt. Merapi, Indonesia, in 2006. Some of the ash rises into the sky, whereas some rushes down the volcano's flank as a pyroclastic flow.

The ash destroyed the forest.

The ash built a fan offshore.

(g) The aftermath of a pyroclastic flow down the flank of the Soufriere Hills Volcano, Montserrat (Caribbean).

during an eruption, and from pulverized preexisting volcanic rock. Geologists distinguish between two types of lapilli produced by explosive eruptions: **pumice lapilli** consists of angular pumice fragments formed when frothy lava in the vent breaks up during an explosion (Fig. 9.6c); **accretionary lapilli** consists of snowball-like lumps of ash formed when ash mixes with water in the air and clumps together (Fig. 9.6d).

Much of the pyroclastic debris erupted from an exploding volcano rushes upward in a turbulent, convecting cloud that may reach stratospheric heights (Fig. 9.6e). Some of this debris falls from the air, like hail and snow, and blankets the countryside near the volcano. Strong winds may waft substantial amounts of ash great distances (hundreds or thousands of kilometers) away from the volcano. But gravity can cause part of the column of pyroclastic debris to collapse, forming a mixture of air, hot ash, and pumice lapilli that rushes down the side of a volcano in a scalding current. This current or "surge" stays relatively close to the ground because the mixture of gas and debris is denser than the clear air above (Fig. 9.6f, g). Such avalanche-like clouds are known as **pyroclastic flows**, or as pyroclastic density currents (by analogy to turbidity currents underwater). In older literature, a pyroclastic flow was called a *nuée ardente*, French for glowing cloud.

In April 1902, a devastating pyroclastic flow erupted from Mt. Pelée on the Caribbean island of Martinique. On the morning of May 8, the rhyolite dome that had obstructed the throat of the volcano broke. The immense pressure that had been building beneath the obstruction was suddenly released, and in the same way champagne bursts out of a bottle when the cork is pulled, a cloud of ash and pumice lapilli spewed out of Mt. Pelée. Collapse of the ash column formed a pyroclastic flow—at a temperature of 200°C to 450°C—that swept down Pelée's flank. This flow rode on a cushion of air and reached speeds of up to 300 km per hour before slamming into the busy port town of St. Pierre. Within moments, the town's buildings were flattened and its 28,000 inhabitants were dead of incineration or asphyxiation. Only two people survived—one was a prisoner who was protected by the stout walls of his underground cell. Similar eruptions have happened more recently on the nearby island of Montserrat, but with a much smaller death toll because of timely evacuation.

Pyroclastic deposits. Unconsolidated deposits of pyroclastic grains, regardless of size, constitute **tephra**. Ash, or ash mixed with lapilli, becomes **tuff** when buried and transformed into coherent rock. Tuff that formed from ash and/or pumice lapilli that fell like snow from the sky is called air-fall tuff, whereas a sheet of tuff that formed from a pyroclastic flow is an **ignimbrite**. Ash and pumice lapilli in an ignimbrite is sometimes so hot that it welds together to form a hard mass.

Other volcaniclastic deposits.   The nature and origin of debris produced during and after eruptions continues to be the subject of active research; some of the deposits prove to be difficult to interpret. Geologists use the term *volcaniclastic deposit* to refer to any material that consists of volcanic igneous fragments; they distinguish among three categories:

Pyroclastic deposits, as we have just seen, consist of fragments that accumulated directly from the clouds of debris produced by an eruption. The fragments in such deposits have not moved, subsequent to their original deposition.

Volcani-sedimentary deposits consist of volcanic material (lava and pyroclastic debris) that later moved downslope and was redeposited elsewhere. Some of this material tumbles as a landslide, breaking up to varying degrees as it moves. Notably, large volcanoes are fundamentally unstable edifices, so gravity eventually causes large parts of them to collapse, resulting in transport of material downslope. In cases where volcanoes are covered with snow and ice, or are drenched with rain, water mixes with debris to form a **volcanic debris flow** that moves downslope like wet concrete (Fig. 9.7a–d). Very wet, ash-rich debris flows down in a relatively fast-moving slurry called a **lahar**, which can reach speeds of 50 km per hour. Lahars tend to follow preexisting river channels and may travel for tens of kilometers. When debris flows and lahars stop moving, they yield a layer consisting of volcanic blocks suspended in ashy mud. Rivers may eventually sort and transport some volcanic sediment. Where this material accumulates, perhaps far downstream, it forms deposits of volcanic sandstone and/or volcanic conglomerate.

Fragmented lava deposits consist of debris produced when lava breaks up into angular clasts while flowing, without ever being ejected into the air. Fragmentation (or brecciation) happens when the inside of a lava flow continues to move after its surface has frozen—the crust breaks up due to the movement. Fragmentation may also happen when lava freezes very quickly and shatters upon erupting into water or ice or when a submarine volcano erupts gassy lava that spatters up into the water. The resulting material, consisting of glassy fragments embedded in ash that has reacted with hot water, is called **hyaloclastite**.

Volcanic Gas

Most magma contains dissolved gases, including water, carbon dioxide, sulfur dioxide, and hydrogen sulfide (H_2O, CO_2, SO_2, and H_2S) which, over time, contribute atoms to the atmosphere. In fact, up to 9% of a magma may consist of

FIGURE 9.7 Examples of volcaniclastic deposits (debris flows and lahars).

(a) Water-soaked volcanic debris slid down the side of this volcano in Nicaragua.

(b) Deposits of a debris flow that accumulated about 35 Ma in Utah. Note that the debris flow filled a channel cut into finer, stratified ash.

(c) Lahars-choked stream valleys near Mt. Saint Helens in the days following the 1980 eruption.

(d) Deposits of a lahar from Mt. Saint Helens, 20 years later, include logs ripped off hill slopes.

gaseous components. Generally, lavas with more silica contain a greater proportion of gas. Volcanic gases come out of solution when the magma approaches the Earth's surface and pressure decreases, just as bubbles come out of solution in a soda when you pop the bottle top off. Because of the sulfur in volcanic gas, the cloud above a volcano typically smells like rotten eggs. Sulfurous gases react with water in the air to create an aerosol of corrosive sulfuric acid. **Aerosols** are very tiny liquid droplets or solid particles that can remain suspended in air.

In low-viscosity (mafic) magma, gas bubbles can rise faster than the magma moves, and thus most reach the surface of the magma and enter the atmosphere before the lava does. Thus some volcanoes may, for a while, produce large quantities of steam, without much lava (Fig. 9.8a). The last bubbles to form, however, freeze into the lava and become holes called **vesicles** (Fig. 9.8b). A mafic rock in which more than 50% of the rock's volume consists of bubbles is called scoria. In high-viscosity (silicic) magmas, the gas has trouble escaping because bubbles can't push through the sticky lava. The buoyancy caused by these bubbles pulls the magma up. At shallower depths, even more bubbles grow, and the magma becomes a sticky foam. As such foam-like magmas approach the Earth's surface, and the weight of overlying lava decreases, the gas expands so that in some cases bubbles may account for as much as 50 to 75% of the volume of the magma. Pumice forms when this material freezes. Pumice has such low density that it can float.

FIGURE 9.8 The gas component of volcanic eruptions.

(a) A volcano in Alaska erupting large quantities of steam.

(b) Gas bubbles frozen in lava produce vesicles, as in this block from Sunset Crater, Arizona.

Take-Home Message

- Lava viscosity depends on composition and temperature. Basaltic lava has low viscosity and can flow far, but rhyolite lava is so sticky that it forms a dome-like mass above the vent.
- Volcaniclastic deposits (aggregates of volcanic debris) include pyroclastic debris erupted directly from a volcano, and volcani-sedimentary material that moved since initial deposition, commonly in the presence of water.
- Ash is very fine, lapilli is marble-sized, and blocks and bombs are bigger. Such material may fall from eruptive clouds or may rush down the flank of a volcano in a pyroclastic flow.
- Lahars consists of a slurry-like mixture of volcanic debris and water, which can travel far.
- Volcanoes emit gas, some of which is trapped as bubbles within lava.

THINK: Were the inhabitants of Pompeii killed by lava, by pyroclastic debris, or by another kind of volcaniclastic sediment?

9.3 THE STRUCTURE AND ERUPTIVE STYLE OF VOLCANOES

Volcanic Architecture

As we saw in Chapter 6, melting in the upper mantle and lower crust produces magma, which rises into the upper crust. Typically, this magma accumulates underground in a **magma chamber**, a zone of open spaces and/or fractured rock that can contain a large quantity of magma. Some of the magma freezes in the magma chamber and transforms into intrusive igneous rock, but some rises through an opening, or conduit, to the Earth's surface and erupts to build a volcanic "edifice" (a hill or mountain more commonly known as a volcano). In some volcanoes, the conduit has the shape of a vertical pipe, while in others the conduit is a crack called a **fissure** (Fig. 9.9a, b). Initially, a fissure may erupt a curtain of lava.

With time, the solid products of eruption (lava and/or pyroclastic debris) accumulate around a vent. At the top of the resulting edifice, a circular depression called a **crater** (shaped like a bowl, up to 500 m across and 200 m deep) may develop, either during eruption as material accumulates around the summit vent or just after eruption as the summit collapses into the drained conduit. Eruptions that happen in the summit crater are summit eruptions. In some volcanoes, a secondary

FIGURE 9.9 Crater eruptions and fissure eruptions come from conduits of different shapes.

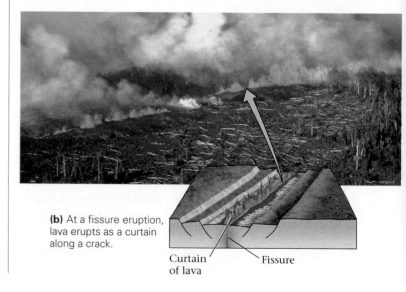

(a) At a crater eruption, lava spouts from a chimney-shaped conduit.

(b) At a fissure eruption, lava erupts as a curtain along a crack.

conduit or fissure breaks through along the sides, or flanks, of the volcano, causing a flank eruption.

During major eruptions, the center of the volcanic edifice may collapse into the large, drained magma chamber below, producing a **caldera**, a big circular depression up to thousands of meters across and up to several hundred meters deep. Typically, a caldera has steep walls and a fairly flat floor and may be partially filled with ignimbrite (Fig. 9.10a–d). Note that calderas differ from craters in terms of size, shape, and mode of formation.

Geologists distinguish among several different shapes of subaerial (above sea level) volcanic edifices. **Shield volcanoes** are broad, gentle domes that are so named because they resemble a soldier's shield lying on the ground. They form when the products of eruption have low viscosity and thus are weak, so they cannot pile up around the vent but rather spread out over large areas. Volcanoes such as those of Hawaii, which produce layer upon layer of low-viscosity basaltic lava, are shield volcanoes (Fig. 9.11a), as are some volcanoes that erupt successive ignimbrites. **Scoria cones** (informally called **cinder cones**) consist of cone-shaped piles of basaltic lapilli and blocks, generally from a single eruption (Fig. 9.11b). **Stratovolcanoes**, also known as composite volcanoes, are large and cone-shaped, generally with steeper slopes near the summit, and consist of interleaved layers of lava, tephra, and volcaniclastic debris (Fig. 9.11c ▶️). Their shape, exemplified by Japan's Mt. Fuji, supplies the classic image that most people have of a volcano; the prefix *strato-* emphasizes that they can grow to be kilometers high.

FIGURE 9.10 The formation of volcanic calderas.

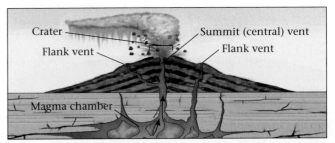

(a) As an eruption begins, the magma chamber inflates with magma. There can be a central vent and one or more flank vents.

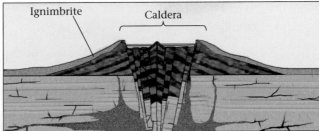

(b) During an eruption, the magma chamber drains, and the central portion of the volcano collapses downward.

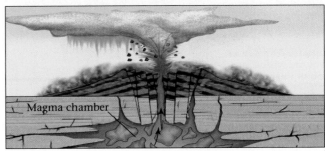

(c) The collapsed area becomes a caldera. Later, a new volcano may begin to grow within the caldera.

Time

(d) This caldera in Oregon formed about 7,700 years ago. Afterward, it filled with water to become Crater Lake. Wizard Island, protruding from the lake, is a cinder cone that grew on top of the caldera floor.

FIGURE 9.11 Different shapes of volcanoes.

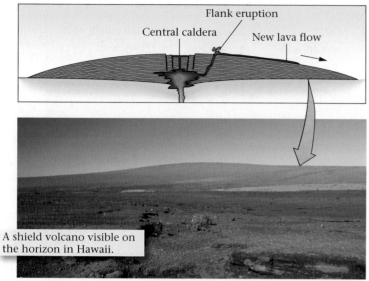

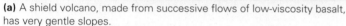

(a) A shield volcano, made from successive flows of low-viscosity basalt, has very gentle slopes.

(b) A cinder cone on the flank of a larger volcano in Arizona. The pile of cinders has assumed the angle of repose. A lava flow covers the land surface in the distance.

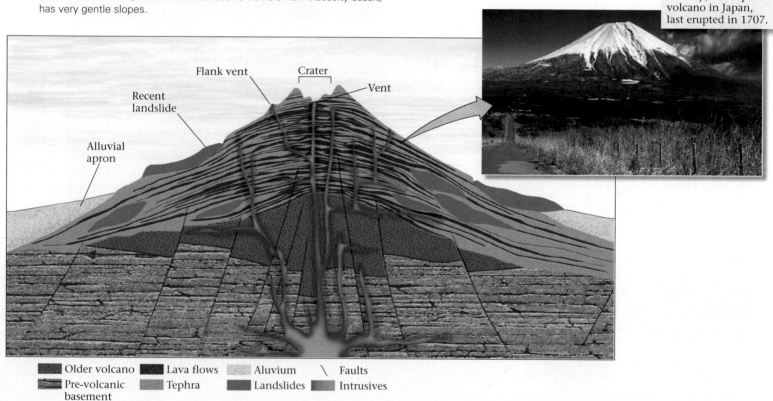

(c) A composite volcano consists of layers of tephra and lava. Volcanic debris flows and ash avalanches modify slopes and contribute to the development of a classic cone-like shape. ▶❚❚

Let's look at the nature of stratovolcanoes a little more closely. Their edifices build from the products of many eruptions over an extended period of time; not all the eruptions produce the same kind of material. Specifically, eruptions producing pyroclastic debris produce layers of tephra, whereas eruptions of low-viscosity lava produce flows that cascade down the flanks of the volcano. The lava flows armor the underlying tephra and prevent it from eroding away. The Fuji-like symmetrical shape of a stratovolcano does not last indefinitely, for many reasons. Large masses of pyroclastic debris and solid lava can detach on failure surfaces and slump down the volcano's flanks. Heavy rains can trigger floods and debris flows, which transport volcaniclastic sediment out to alluvial fans surrounding the volcano.

FIGURE 9.12 These profiles emphasize that volcanoes come in different sizes. Large shield volcanoes, like those on Hawaii, are many times larger than cinder cones.

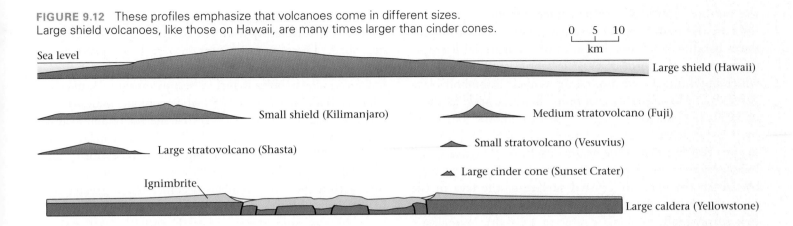

Finally, explosive eruptions can blast away a large portion of the volcano's edifice.

The hills or mountains resulting from volcanic eruptions come in a great range of sizes (Fig. 9.12). Shield volcanoes tend to be the largest, followed by stratovolcanoes. Cinder cones tend to be relatively small and are often found on the surface of larger volcanoes. Submarine volcanoes don't fit any of these categories because they usually grow as irregularly shaped mounds, modified by huge submarine landslides along their margins.

Concept of Eruptive Style: Will It Flow, or Will It Blow?

Kilauea, a volcano on Hawaii, produces rivers of lava that cascade down the volcano's flanks. Mt. Saint Helens, a volcano near the Washington–Oregon border, exploded catastrophically in 1980 and blanketed the surrounding countryside with tephra. Clearly, different volcanoes erupt differently and, as we've noted, successive eruptions from the same stratovolcano may differ markedly in character from one another. Geologists refer to the character of an eruption as **eruptive style**. Below, we describe several distinct eruptive styles and explore why the differences occur (see Geology at a Glance, pp. 268–269).

Effusive Eruptions

The term *effusive* comes from the Latin word for pour out, and indeed that's what happens during an **effusive eruption**—lava pours out a summit vent or fissure, filling a lava lake around the crater and/or flowing in molten rivers for great distances. Geologists refer to small- to moderate-sized effusive eruptions as Hawaiian eruptions, because they are common on Hawaii (Fig. 9.13). Really huge effusive eruptions producing flood basalts, as we discuss later, have not happened in historic time.

Effusive eruptions occur where the magma feeding the volcano is hot and mafic and, therefore, has low viscosity. Pressure, applied to the magma chamber by the weight of overlying rock, squeezes magma upward and out of the vent; in some cases, the

pressure is great enough to drive the magma up into a fountain over the vent. As we have seen, when the magma rises, gas comes out of solution and forms bubbles; the presence of bubbles decreases the overall density and viscosity of the magma, allowing it to rise faster. If the conduit through which the magma rises narrows in the throat of the volcano, the velocity of the rising magma increases even more, much like the velocity of water coming out of a hose increases if you pinch the end of the hose. Such pinching of conduits may play a role in producing very high magma fountains spurting up to 500 m above the vent.

Explosive Eruptions

Volcanoes that forcefully emit significant quantities of pyroclastic debris are called **explosive eruptions**. The 1980 eruption of Mt. Saint Helens clearly fits in this category. But not all explosive eruptions are like that of Mt. Saint Helens, as we now see.

Ancient Romans referred to the island of Stromboli as "the lighthouse of the Mediterranean" because it has erupted about every 10 to 20 minutes through recorded history, and

FIGURE 9.13 A 1986 effusive eruption on Hawaii.

the red-hot clots that it ejects trace out glowing arcs of light in the night sky. Occasionally, Stromboli also produces basaltic lava flows, but most of the material it erupts comes out in the form of basaltic scoria lapilli and blocks which build into a cone around the vent. If Stromboli's lava is basaltic, why aren't its eruptions the same style as those of Hawaii? Geologists suggest that Stromboli's "burps" of pyroclastic debris happen because gas bubbles within the magma rise faster than the magma itself. As they rise, small bubbles collect into larger ones, and when these large bubbles reach the partially crusted surface of the lava in the vent, they burst, shooting up fragments of red-hot scoria. Not surprisingly, geologists refer to a basaltic pyroclastic eruption as a Strombolian eruption, regardless of where it occurs (Fig. 9.14a).

Andesitic and rhyolitic magmas rising in a volcano behave very differently from the basaltic magmas we've just described. These magmas contain very large quantities of gas to start with, and when this gas comes out of solution, it forms so many bubbles that they occupy most of the magma's volume. Because of their high silica content, andesitic and rhyolitic magmas are so viscous that the gas bubbles cannot rise to the vent and escape. Rather, as the gassy magma rises in the conduit, bubbles grow and new bubbles form, until the magma becomes a froth as we noted earlier. The pressure within the rising bubbles becomes much greater than in the air above the volcano—only the thin bubble walls keep the gas from bursting free. Eventually, as the rising froth shears against the walls of the conduit, the walls of the bubbles stretch and break. Many of the bubble walls

FIGURE 9.14 Examples of explosive eruptive styles. No two eruptions are exactly alike.

(a) A large Strombolian eruption on Mt. Etna, Sicily.

(b) A 2008 Vulcanian eruption of the Chaiten Volcano, Chile.

(c) The Plinean eruption of Mt. Pinatubo, in the Philippines.

(d) A Surtseyan eruption of a subsea volcano near Tonga.

shatter into dust-sized pieces of ash, and these surround the pumice lapilli (larger chunks that still contain both bubbles and bubble walls). When the very hot gas that had been held in by bubble walls is released, it expands violently, producing immense pressure that pushes surrounding debris upward. If the pressure cracks the lava dome capping the vent, the mixture of fragments and gas blast out of the volcano's vent at very high velocity (over 300 km/h). The sudden release of this material decreases the pressure on magma deeper in the conduit, allowing bubbles in the deeper magma to expand and shatter, so more and more pyroclastic debris erupts until the magma chamber drains.

Simplistically, such explosive eruptions of andesitic or rhyolitic volcanoes resemble the blast of a shotgun. When a hunter pulls the trigger, detonation of the gunpowder causes sudden gas expansion that propels the pellets out of the gun barrel. (Note, however, that volcanic explosions differ from shotgun blasts in that the driving force in a volcano is the expansion of gas released when existing bubbles break, not a sudden chemical reaction as happens when gunpowder explodes.) In some cases, the force of a volcanic explosion shatters the volcanic edifice, just like a shotgun blast may occasionally rupture a gun barrel.

Moderate-sized explosive eruptions that emit pyroclastic debris are known as Vulcanian eruptions, named for the island of Vulcano, where they occur on occasion (**Fig. 9.14b**). Really huge explosive eruptions of stratovolcanoes are known as Plinean eruptions, named for the Roman scholar, Pliny the Younger, who observed and described an example of one, the eruption of Vesuvius. Plinean eruptions (**Fig. 9.14c**) can eject many cubic kilometers of material into the atmosphere and may destroy a substantial part of the stratovolcano's edifice itself. In fact, the explosion can completely change the profile of the volcano, so that it no longer has a cone-like shape (**Box 9.1**). Particularly large explosions may happen when a volcano erupts in the sea, because large quantities of seawater come in contact with the magma as the edifice cracks. When this happens, the water abruptly turns to steam and adds to the explosive pressure.

What happens during the climatic stage of a Plinian eruption? The pressure of the eruption pushes a jet of pyroclastic debris upward in a column that rises several hundred meters above the summit. At the top of this jet, heavier debris slows and collapses downward, to cascade down the flank of the volcano in pyroclastic flows (see Fig. 9.6a). The cloud of smaller, lighter hot ash, along with air heated by the eruption, is buoyant relative to surrounding, cool air, and thus wafts (convects) upward above the eruption jet in a column of billowing clouds that can rise to elevations of 10 to 40 km above the Earth; this part of the eruptive cloud is called the convective cloud. Eventually, the convective cloud cools sufficiently and is no longer buoyant, so it spreads laterally to form a vast, high-altitude umbrella from which ash gradually sifts down like snow. Strong winds may carry some of this ash for great distances, so Plinean eruptions can blanket large regions with tephra.

Some explosive eruptions are distinctive because they involve significant quantities of water. For example, Surtseyan eruptions, named for the island of Surtsey off the coast of Iceland, are ones at which the vent lies in relatively shallow seawater—heating the water produces prodigious amounts of steam that billow out of the sea, along with fountains of wet ash (**Fig. 9.14d**). In Phreatic eruptions, magma interacts with groundwater beneath the volcano—these eruptions emit lots of steam, which rises so forcefully that it can pluck off chunks of preexisting rock from the walls of the conduit and eject them as blocks; but such eruptions do not produce much magma.

In recent years, geologists have realized that some volcanic explosions of the geologic past dwarf any that have been observed during human history (see Box 9.1). Such incomprehensibly huge volcanoes have come to be known informally as **supervolcanoes**, and they can produce hundreds to a few thousand cubic kilometers of pyroclastic debris. After the eruption, a huge caldera forms as the land collapses into the drained magma chamber (see Fig. 9.10). Eruptions that occurred hundreds of thousands of years ago in the area that is now Yellowstone Park serve as an example—in the aftermath of one of these explosions, a caldera 72 km across formed! (To explore Yellowstone and other past and current volcanoes, go to **See for Yourself G** on p. S-12.)

Take-Home Message

- A magma chamber lies in or below a volcano. Molten rock rising from this chamber erupts at a vent. Some vents are chimney-like pipes, and others are long fissures.
- Accumulation of lava and debris around a vent builds a volcanic edifice. Some are broad domes (shield volcanoes), some are piles of scoria (scoria cones), and some contain alternating debris and lava and can grow quite tall (stratovolcanoes).
- Effusive eruptions emit fountains and rivers of lava; explosive eruptions emit pyroclastic debris.
- There are many types of explosive eruptions—basaltic ones spatter out scoria; but felsic ones can explode, producing pyroclastic flows and immense clouds of ash. Volcanoes erupting in water produce billows of steam and clots of wet ash.
- A crater may form at the top of a volcano, when magma drains. Large-scale drainage, or an explosion, can produce a large depression called a caldera. Eruption of supervolcanoes produced calderas that are tens of kilometers across.

THINK: How can you distinguish between a volcanic caldera and a meteorite crater?

BOX 9.1

Volcanic Explosions to Remember

Explosions of volcanoes generate enduring images of destruction. An explosion can rip a volcano apart (**Fig. Bx9.1a**). The historical record shows a vast range in the volume of debris erupted, even though the largest *observed* eruption (Tambora in 1815) was small compared to a super-explosion that took place over 600,000 years ago in what is now Yellowstone National Park, Wyoming (**Fig. Bx9.1b**). Let's look at two notable cases.

Mt. St. Helens, a snow-crested strato-volcano in the Cascades of the northwestern United States, had not erupted since 1857. However, geologic evidence suggested that the mountain had a violent past, punctuated by many explosive eruptions. On March 20, 1980, an earthquake announced that the volcano was awakening once again. A week later, a crater 80 m in diameter burst open at the summit and began emitting gas and pyroclastic debris. Geologists who set up monitoring stations to observe the volcano noted that its north side was beginning to

bulge markedly, suggesting that the volcano was filling with magma and that the magma was making the volcano expand like a balloon. Their concern that an eruption was imminent led local authorities to evacuate people in the area.

The climactic eruption came suddenly. At 8:32 A.M. on May 18, a geologist, David Johnston, monitoring the volcano from a distance of 10 km, shouted over his two-way radio, "Vancouver, Vancouver, this is it!" An earthquake had triggered a huge

FIGURE Bx9.1 Examples of explosive eruptions.

The more recent edifice of Vesuvius

Estimated profile, before explosion

The eroded surface of the older volcano

(a) The recently active part of Mt. Vesuvius is growing within the blasted remnant of an older volcano.

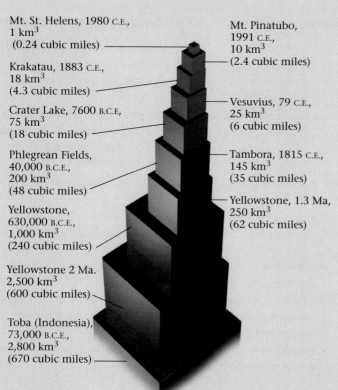

Mt. St. Helens, 1980 C.E.,
1 km^3
(0.24 cubic miles)

Krakatau, 1883 C.E.,
18 km^3
(4.3 cubic miles)

Crater Lake, 7600 B.C.E,
75 km^3
(18 cubic miles)

Phlegrean Fields,
40,000 B.C.E.,
200 km^3
(48 cubic miles)

Yellowstone,
630,000 B.C.E.,
1,000 km^3
(240 cubic miles)

Yellowstone 2 Ma.
2,500 km^3
(600 cubic miles)

Toba (Indonesia),
73,000 B.C.E.,
2,800 km^3
(670 cubic miles)

Mt. Pinatubo,
1991 C.E.,
10 km^3
(2.4 cubic miles)

Vesuvius, 79 C.E.,
25 km^3
(6 cubic miles)

Tambora, 1815 C.E.,
145 km^3
(35 cubic miles)

Yellowstone, 1.3 Ma,
250 km^3
(62 cubic miles)

(b) The relative amounts of pyroclastic debris (in cubic km) ejected during major explosive eruptions.

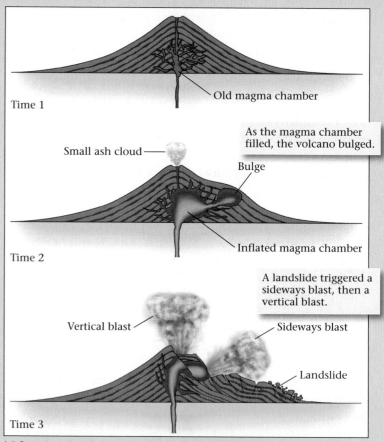

Old magma chamber

Time 1

Small ash cloud

As the magma chamber filled, the volcano bulged.

Bulge

Inflated magma chamber

Time 2

A landslide triggered a sideways blast, then a vertical blast.

Vertical blast

Sideways blast

Landslide

Time 3

(c) Stages during the eruption of Mt. St. Helens, 1980.

The blast knocked trees down as if they were toothpicks.

30 years later, the downed trees remain.

Legend:
- Mud and debris flow
- Pyroclastic flows
- Eruptive dome
- Trees blown down (lateral blast); arrows indicate direction
- Scoured area/mud flow deposits
- Less affected area above tree line
- Less affected forest
- Lake

Johnston Ridge Observatory

Spirit Lake

Windy Ridge Viewpoint

Mt. St. Helens 8,363 ft 2,549 m

N

0 mi 2
0 km 2

Products of Mt. St. Helens 1980 Eruption

(d) A map shows the dimensions of the region destroyed by the eruption of Mt. St. Helens. The arrows indicate the blast direction. The neighboring forest was flattened by a blast of rock, steam, and ash.

landslide that caused 3 cubic km of the volcano's weakened north side to slide away. The sudden landslide released pressure on the magma in the volcano, causing a sudden and violent expansion of gases that blasted through the side of the volcano (**Fig. Bx9.1c**). Rock, steam, and ash screamed north at the speed of sound and flattened a forest and everything in it over an area of 600 square km (**Fig. Bx9.1d**). Tragically, Johnston, along with 60 others, vanished forever.

Seconds after the sideways blast, a vertical column carried about 540 million tons of ash (about 1 cubic km) 25 km into the sky, where the jet stream carried it away so that it was able to circle the globe. In towns near the volcano, a blizzard of ash choked roads and buried fields. Water-saturated ash formed viscous slurries, or lahars, that flooded river valleys, carrying away everything in their path.

When the eruption was finally over, the once cone-like peak of Mt. St. Helens had disappeared—the summit now lay 440 m lower, and the once snow-covered

mountain was a gray mound with a large gouge in one side. The volcano came alive again in 2004, but did not explode.

An even greater explosion happened in 1883. Krakatau, a volcano in the sea between Indonesia and Sumatra, where the Indian Ocean floor subducts beneath Southeast Asia, had grown to become a 9-km-long island rising 800 m (2,600 feet) above the sea. On May 20, the island began to erupt with a series of large explosions, yielding ash that settled as far as 500 km away. Smaller explosions continued through June and July, and steam and ash rose from the island, forming a huge black cloud that rained ash into the surrounding straits. Ships sailing by couldn't see where they were going, and their crews had to shovel ash off the decks.

Krakatau's demise came at 10 A.M. on August 27, perhaps when the volcano cracked and the magma chamber suddenly flooded with seawater. The resulting blast, five thousand times greater than the Hiroshima atomic bomb explosion, could be heard as far as 4,800 km away, and subaudi-

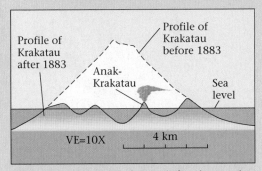

Profile of Krakatau after 1883

Profile of Krakatau before 1883

Anak-Krakatau

Sea level

VE=10X

4 km

(e) Profile of Krakatau, before and after the eruption. Note that a new volcano (Anak-Krakatau) has formed.

ble sound waves traveled around the globe seven times. Giant waves pushed out by the explosion slammed into coastal towns, killing over 36,000 people. Near the volcano, a layer of ash up to 40 m thick accumulated. When the air finally cleared, Krakatau was gone, replaced by a submarine caldera some 300 m deep (**Fig. Bx9.1e**). All told, the eruption shot 20 cubic km of rock into the sky. Some ash reached elevations of 27 km. Because of this ash, people around the world could view spectacular sunsets during the next several years.

9.4 **GEOLOGIC SETTINGS OF VOLCANISM**

Different styles of volcanism occur at different locations on earth (See for Yourself G, p. S-12). Most eruptions occur along plate boundaries, but major eruptions also occur at hot spots (Fig. 9.15). We'll now look at the settings in which eruptions occur, in the context of plate tectonics theory.

Mid-Ocean Ridges

Products of mid-ocean ridge volcanism cover 70% of our planet's surface. We don't generally see this volcanic activity, however, because the ocean hides most of it beneath a blanket of water. Mid-ocean ridge volcanoes, which develop along fissures parallel to the ridge axis, are not all continuously active.

Each one turns on and off in a time scale measured in tens to hundreds of years. They erupt basalt, formed when hot mantle rock rises from great depth to shallow depths beneath the ridge and undergoes decompression melting (see Chapter 6). This basalt, because it cools so quickly underwater, forms pillow-lava mounds. The pillow basalts commonly occur in association with hyaloclastites. Water that heats up as it circulates through the crust near the magma chamber bursts out of hydrothermal (hot-water) vents along these mounds. As mentioned in Chapter 2, these vents are called black smokers.

Convergent Boundaries

Most of the subaerial volcanoes on Earth lie along convergent plate boundaries (subduction zones). The volcanoes form

FIGURE 9.15 A map showing the distribution of volcanoes around the world and the basic geologic settings in which volcanoes form, in the context of plate tectonics theory.

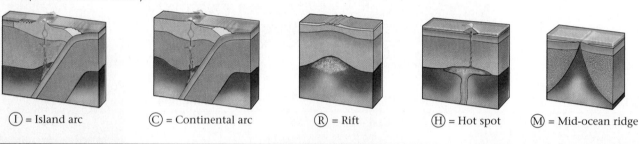

(I) = Island arc   (C) = Continental arc   (R) = Rift   (H) = Hot spot   (M) = Mid-ocean ridge

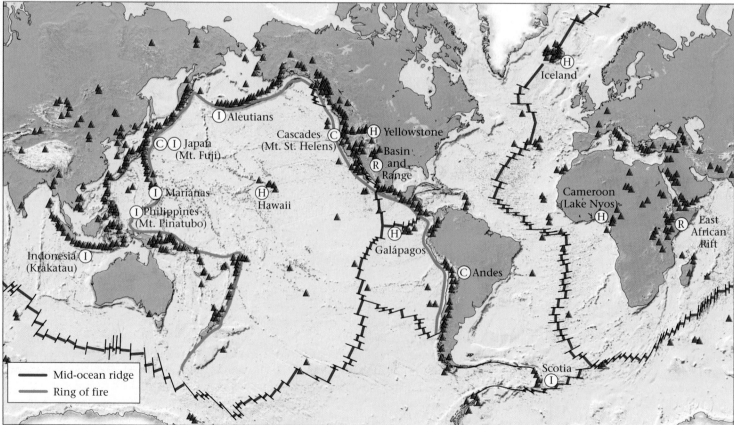

FIGURE 9.16 The interior of an oceanic hot-spot volcano is complicated. Initially, eruption produces pillow basalts. When the volcano emerges above sea level, it becomes a shield volcano. The margins of the island frequently undergo slumping, and the weight of the volcano pushes down the surface of the lithosphere. The Hawaiian islands exemplify this architecture.

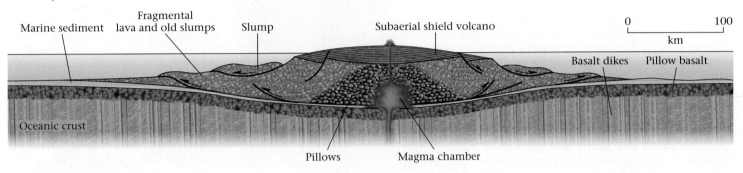

when volatile compounds such as water and carbon dioxide are released from the rock of the subducting plate and rise into the overlying hot mantle, causing melting and producing magma, which then rises through the lithosphere and erupts. Some of these volcanoes grow on oceanic crust and become volcanic island arcs, such as the Marianas of the western Pacific. Others grow on continental crust, building continental volcanic arcs such as the Cascade volcanic chain of Washington and Oregon or the Andes chain of South America. Typically, individual volcanoes in volcanic arcs lie about 50 to 100 km apart. Subduction zones border over 60% of the Pacific Ocean, creating a 20,000-km-long chain of volcanoes known as the Ring of Fire.

In island arcs, where magma rises from the mantle through oceanic crust, volcanoes initially primarily produce basalt, formed by partial melting of the mantle. This lava builds piles of pillows and hyaloclastites until the edifice rises from the sea, and then the lava can build shields. Eventually, processes such as fractional crystalization and assimilation (see Chapter 6) lead to the production of andesites and rhyolites. In continental arcs, some basalt rises to the surface, but andesitic and rhyolitic eruptions are more common because more of the magma undergoes fractional crystallization and assimilation within the crust, and heat transfer partially melts some of the continental crust (see Chapter 6). Because many different kinds of magma form at volcanic arcs, these volcanoes sometimes have effusive eruptions and sometimes pyroclastic eruptions—and occasionally they explode. Such eruptions yield stratovolcanoes, such as the elegant symmetrical cone of Mt. Fuji (see Fig. 9.11c) and the blasted-apart hulk of Mt. St. Helens (see Box 9.1).

Continental Rifts

The igneous activity of rifts happens because thinning of the continental lithosphere allows the underlying asthenosphere to rise to shallower depths, where it undergoes decompression and partially melts to produce basaltic magma. Some of this magma rises straight to the surface and erupts as basalt, but

some gets trapped at the base or within the continental crust, and undergoes fractional crystallization and/or assimilates surrounding crust to produce intermediate or felsic magmas. Trapped magma can also transfer enough heat to cause surrounding mafic and intermediate rocks to partially melt, a process that also produces rhyolitic magma.

Because of the diversity of magmas that can form beneath rifts, rifts can host both basaltic fissure eruptions, in which curtains of lava fountain up or linear chains of cinder cones develop, and explosive rhyolitic volcanoes. In some places, they even host stratovolcanoes such as Mt. Kilamanjaro in Africa.

Oceanic Hot-Spot Volcanoes (Hawaii)

Oceanic hot-spot volcanoes form where asthenosphere undergoes decompression melting and produces voluminous amounts of basaltic magma. Most oceanic hot-spot volcanoes, such as the ones that produced Hawaii, occur in the interior of plates, away from plate boundaries. A few, such as the ones that produced Iceland, sit astride a mid-ocean ridge. Most geologists conclude that hot-spot volcanoes lie above mantle plumes, localized upwellings from the deep mantle, but some argue for alternative interpretations.

When a hot-spot volcano first forms on oceanic lithosphere, basaltic magma erupts at the surface of the sea floor. At first, such submarine eruptions yield an irregular mound of pillow lava. With time, the volcano grows up above the sea surface and becomes an island. When the volcano emerges from the sea, the basalt lava that erupts no longer freezes so quickly, and thus flows as a thin sheet over a great distance. Thousands of thin basalt flows pile up, layer upon layer, to build a broad, dome-shaped shield volcano with gentle slopes (Fig. 9.16). As the volcano grows, portions of it can't resist the pull of gravity and slip seaward, creating large submarine slumps.

The big island of Hawaii, the largest oceanic hot-spot volcano on Earth today, currently consists of five shield volcanoes,

Volcanoes

Beneath a volcano, magma rises to fill a pervasively cracked region of crust and forms a magma chamber. Buoyancy of the magma, along with pressure applied by the weight of overlying crust, causes the magma to migrate upward, either through chimney-like conduits or fissures, until it eventually erupts at a surface vent. An eruption is a sight to behold, and often a hazard to fear.

Once molten rock has erupted at the surface, it is called lava. Some lava spills down the side of the volcano in a river of molten rock called lava flows. Lava may fountain out of a vent to form scoria fragments that pile up in

a cone around the vent. Eruptions may eject larger chunks as ballistic blocks or streamlined bombs. The nature of eruptions depends on the viscosity of the lava. Basaltic lava is less viscous and tends to flow easily. Andesitic and rhyolitic lava is very viscous, so viscous that bubbles forming within it cannot escape. As a result, the lava erupts explosively, blasting up a cloud of ash and lapilli—the explosion may pulverize preexisting volcanic rock and send it skyward too. Some of the debris from an explosive eruption rises

Cinder cone

A caldera

Vulcanian eruptions occur when a buildup of gas and magma explodes.

Strombolian crater explosions frequently burst through thinly crusted lava.

Hawaiian fountain explosions are caused by escaping gas.

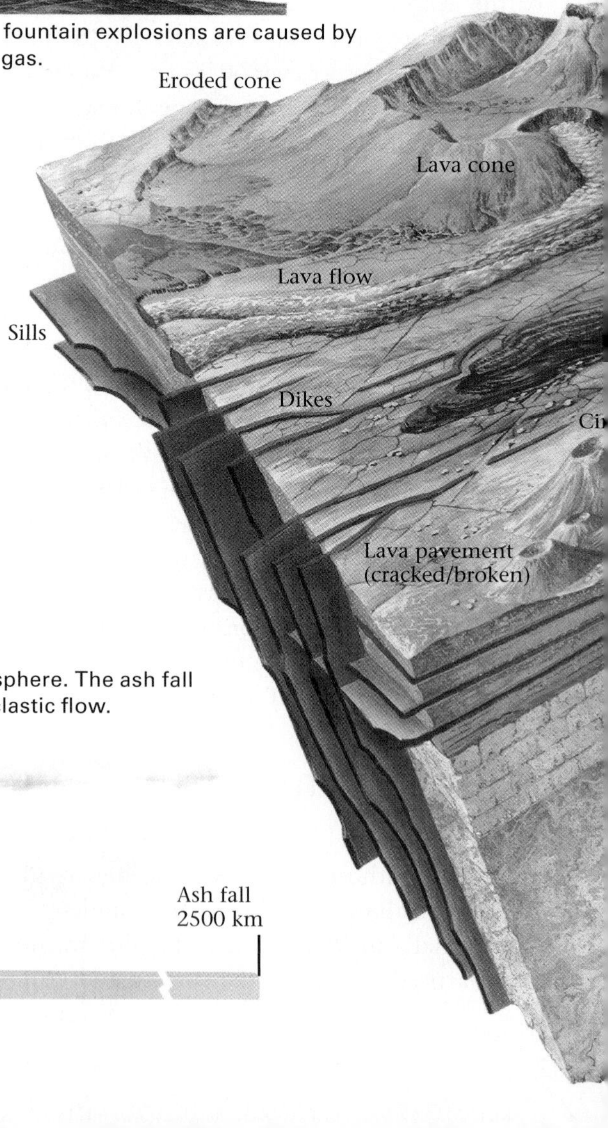

Eroded cone

Side

Lava cone

Lava flow

Sills

Dikes

Ci

Lava pavement (cracked/broken)

Plinean explosions shoot a huge column of pumice fragments up to 50 km into the atmosphere. The ash fall rains down and the column collapses back around the vent, traveling overland as a pyroclastic flow.

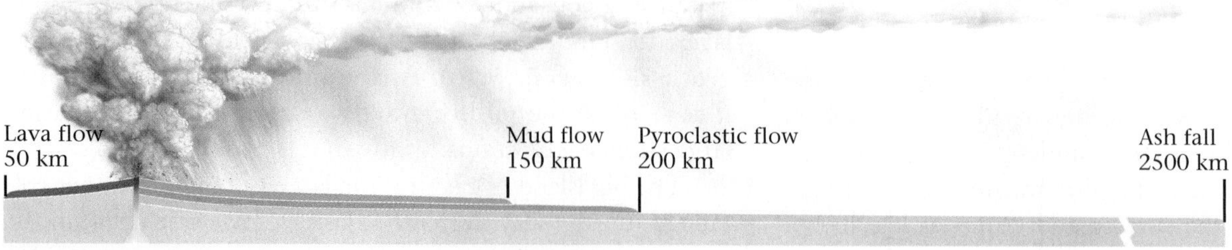

| Lava flow 50 km | Mud flow 150 km | Pyroclastic flow 200 km | Ash fall 2500 km |

The distance volcanic hazards can travel from an eruption.

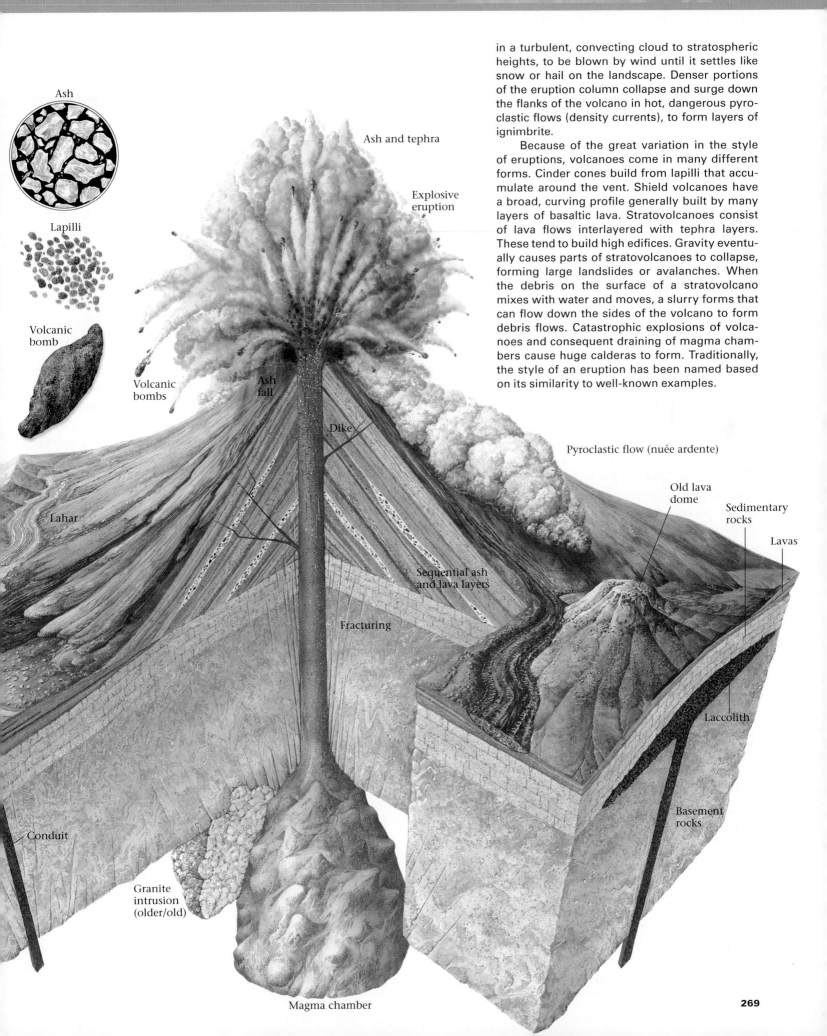

Ash

Lapilli

Volcanic
bomb

Ash and tephra

Explosive
eruption

Volcanic
bombs

Ash
fall

Dike

Lahar

Sequential ash
and lava layers

Fracturing

Conduit

Granite
intrusion
(older/old)

Magma chamber

Pyroclastic flow (nuée ardente)

Old lava
dome

Sedimentary
rocks

Lavas

Laccolith

Basement
rocks

in a turbulent, convecting cloud to stratospheric heights, to be blown by wind until it settles like snow or hail on the landscape. Denser portions of the eruption column collapse and surge down the flanks of the volcano in hot, dangerous pyroclastic flows (density currents), to form layers of ignimbrite.

Because of the great variation in the style of eruptions, volcanoes come in many different forms. Cinder cones build from lapilli that accumulate around the vent. Shield volcanoes have a broad, curving profile generally built by many layers of basaltic lava. Stratovolcanoes consist of lava flows interlayered with tephra layers. These tend to build high edifices. Gravity eventually causes parts of stratovolcanoes to collapse, forming large landslides or avalanches. When the debris on the surface of a stratovolcano mixes with water and moves, a slurry forms that can flow down the sides of the volcano to form debris flows. Catastrophic explosions of volcanoes and consequent draining of magma chambers cause huge calderas to form. Traditionally, the style of an eruption has been named based on its similarity to well-known examples.

each built around a different vent. The island now towers over 9 km above the adjacent ocean floor (about 4.2 km above sea level), the greatest relief from base to top of any mountain on Earth; by comparison, Mt. Everest rises 8.85 km above the plains of India. Calderas up to 3 km wide have formed at the summit, and basaltic lava has extruded from both conduits and fissures (see Fig. 9.2b).

Continental Hot-Spot Volcanoes (Yellowstone National Park)

Yellowstone National Park lies at the northeast end of a string of calderas whose remnants crop out in the Snake River Plain of Idaho. This string is the Yellowstone hot-spot track. The oldest of these calderas, at the southwest end of the track, erupted 16 million years ago (Fig. 9.17a, b). Recent studies have found evidence of a mantle plume beneath Yellowstone, adding support to the hypothesis that the Yellowstone hot-spot track formed as the North American plate moved over the plume.

Ongoing activity beneath Yellowstone has yielded fascinating landforms, volcanic rock deposits, and geysers. Eruptions at the Yellowstone hot spot differ from those in Hawaii in an important way: unlike Hawaii, the Yellowstone hot spot erupts both basaltic lava and rhyolitic pyroclastic debris. This happens because basaltic magma rising from the asthenosphere heats up and partially melts the continental crust.

About 630,000 years ago, immense pyroclastic flows, as well as convective clouds of ash and pumice lapilli, blasted out of the Yellowstone region. Close to the eruption, ignimbrites up to tens of meters thick formed, and ash and lapilli from the giant cloud sifted down over the United States as far east as the Mississippi River (Fig. 9.17c). The 0.63 Ma eruption produced an immense caldera, up to 72 km across, that overlaps earlier calderas (Fig. 9.17b,d). When the debris settled, it blanketed an area of 2,500 square km with tuffs that, in the park, reached a thickness of 400 m. The park's name reflects the brilliant color of volcaniclastic debris exposures in the park's canyons (Fig. 9.17d). Eruptive activity, producing basalt and rhyolite lavas and more pyroclastic debris, continued until about 70,000 years ago. Magma remains in the crust beneath the park today; energy radiating from this magma heats groundwater that rises to fill hot springs and spurt out of steaming geysers. Eruptions will likely happen in the future.

Flood-Basalt Eruptions

In several locations around the world, *huge* sheets of low-viscosity lava erupted and spread out in vast sheets. Geologists refer to the lava of these sheets as **flood basalt** (Fig. 9.18a). Over time, many successive eruptions of flood basalt can build up a broad plateau. The aggregate volume of rock in such a plateau may be so great (over 175,000 km^3), that geologists also refer to the region as a **large igneous province** (**LIP**). (The term has also been used for regions of immense rhyolite eruption.)

What causes flood-basalt eruptions? A popular hypothesis suggests that flood basalts form when a mantle plume starts to rise beneath a region that is undergoing rifting (Fig. 9.18b). As the plume reaches the base of the lithosphere, it has a bulbous head containing a large amount of partially molten rock. Stretching and thinning of the overlying lithosphere results in further decompression of the plume head, and causes even more melt to form. The melt intrudes along fissures that form in the rift, and erupts spectacularly at the surface. Once the plume head no longer exists, the volume of eruption decreases, and "normal" hot-spot volcanism (with less magma production) takes place.

An example of a LIP, the Columbia River Plateau, occurs in Washington and Oregon (see Fig. 9.17a). The basalt here, which erupted around 15 million years ago, reaches a thickness of 3.5 km. Geologists have identified about 300 individual flows in the Columbia River Plateau. Lava in some of these flows traveled great distances—up to 600 km—from its source. Eventually, basalt covered an area of 220,000 km. Even larger flood-basalt provinces occur in eastern Siberia (an occurrence known as the Siberian Traps; Fig. 9.18c), the Deccan Plateau of India, the Paraná region of Brazil, and the Karroo Plateau of south Africa.

Iceland—a Hot Spot on a Ridge

Iceland is one of the few places on Earth where mid-ocean ridge volcanism has built a mound of basalt that protrudes above the sea. The island formed where a hot spot lies beneath the Mid-Atlantic Ridge—the presence of this hot spot (probably due to an underlying mantle plume) means that far more magma erupted here than beneath other places along the ridge. Because Iceland straddles a divergent plate boundary, it is being stretched apart, with faults forming as a consequence. Indeed, the central part of the island is a narrow rift, in which the youngest volcanic rocks of the island have erupted (Fig. 9.19a–c). This rift *is* the trace of the Mid-Atlantic Ridge. Faulting cracks the crust and so provides a conduit to a magma chamber. Thus, some eruptions on Iceland tend to be fissure eruptions, yielding either curtains of lava that are many kilometers long or linear chains of small cinder cones.

Not all volcanic activity on Iceland occurs subaerially. Some eruptions take place under glaciers. During 1996, for example, an eruption at the base of a 600-m-thick glacier melted the ice and produced a column of steam that rose several kilometers into the air. Meltwater accumulated under the ice for six days, until it burst through the edge of the glacier

FIGURE 9.17 Hot-spot volcanic activity in Yellowstone National Park.

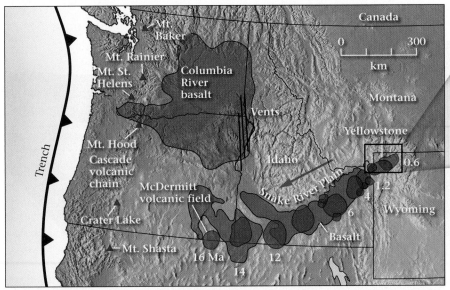

(a) Yellowstone lies at the end of a continental hot-spot track. Progressively older calderas follow the Snake River Plain. The blue arrow indicates plate motion.

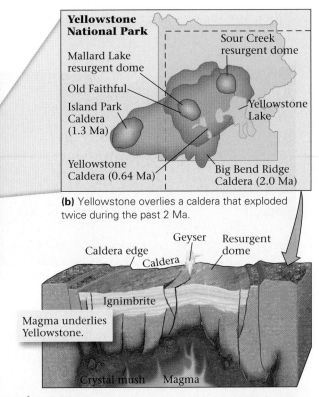

(b) Yellowstone overlies a caldera that exploded twice during the past 2 Ma.

Magma underlies Yellowstone.

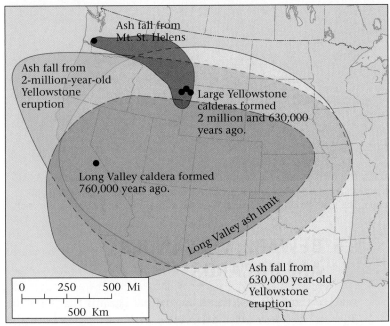

(c) The ash produced by explosions of the Yellowstone calderas covered vast areas—much more than Mt. St. Helens.

(d) Felsic tuffs form the colorful walls of Yellowstone Canyon.

and became a flood (called a *jokulhlaup* in Icelandic) that lasted two days and destroyed roads, bridges, and telephone lines. The 2010 eruption of the Eyjafjallajökull volcano caused similar problems, but was also disruptive in other ways (see. p. 275).

Some of Iceland's volcanic activity occurs off the coast; such activity produced the island of Surtsey. The birth of Surtsey was heralded by huge quantities of steam bubbling up from the ocean. Eventually, steam pressure explosively ejected ash as high as 5 km into the atmosphere. Surtsey finally emerged from the sea on November 14, 1963, building up a cone of ash

and lapilli that rose almost 200 m above sea level in just three months. Waves could easily have eroded the cinder cone away, but the island has survived because lava erupted from the vent and flowed over the cinders, effectively encasing them in an armor-like blanket of solid rock.

FIGURE 9.18   According to one hypothesis, flood basalts erupt when the head of a plume reaches the base of rifting lithosphere.

(a) Flood-basalt layers exposed on the wall of a canyon in Idaho.

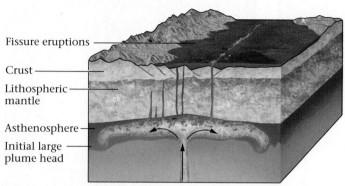

(b) The plume model for forming flood basalts.

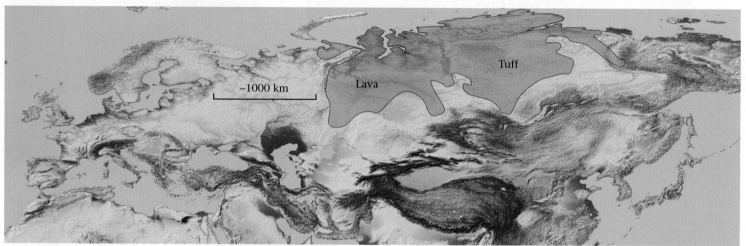

(c) An area the size of Europe was covered by flood basalts (the "Siberian traps") and associated tuffs in Siberia. The time of the eruption, about 250 Ma, coincides with the end of the Paleozoic, when there was mass extinction. The map shows the maximum extent of the volcanic rock before erosion.

Take-Home Message

- Volcanoes form in many geologic settings; causes of volcanism can be understood in the context of plate tectonics theory.
- Mid-ocean ridge volcanism, which takes place beneath the sea, produced all ocean floor.
- Island arc volcanoes tend to be basaltic; continental arcs host large stratovolcanoes and commonly erupt andesite.
- Rifts are home to both effusive basaltic volcanoes and explosive felsic volcanoes. Locally, stratovolcanoes form along them.
- Hot-spot volcanoes occur both on oceanic and continental lithosphere. Oceanic hot-spot eruptions are primarily basaltic (examples include Hawaii and Iceland). Continental hot spots, like Yellowstone, produce both basalt eruptions and hugely explosive felsic eruptions.
- At times during geologic history, immense eruptions of flood basalts occurred, producing large igneous provinces.

THINK: Why don't volcanoes occur along passive continental margins?

9.5 BEWARE: VOLCANOES ARE HAZARDS!

Like earthquakes, volcanoes are natural hazards that have the potential to cause great destruction to humanity, in both the short term and the long term. According to one estimate, volcanic eruptions in the last two thousand years have caused about a quarter of a million deaths—much fewer than those caused by earthquakes, but nevertheless a sizable number. Considering the rapid expansion of cities, far more people live in dangerous proximity to volcanoes today than ever before, so if anything, the hazard posed by volcanoes has gotten worse—imagine if a large explosion were to occur next to a major city today. Let's now look at the different kinds of threats posed by volcanic eruptions.

Hazards Due to Eruptive Materials

Threat of flows.   When you think of an eruption, perhaps the first threat that comes to mind is the lava that flows from a

FIGURE 9.19 Iceland, a hot spot on the Mid-Atlantic Ridge.

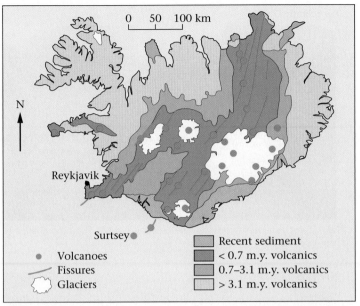

0 50 100 km

N

Reykjavik

Surtsey

- Volcanoes
- Fissures
- Glaciers

Recent sediment
< 0.7 m.y. volcanics
0.7–3.1 m.y. volcanics
> 3.1 m.y. volcanics

(a) A geologic map of Iceland shows how the youngest volcanics occur in the central rift, effectively the on-land portion of the Mid-Atlantic Ridge.

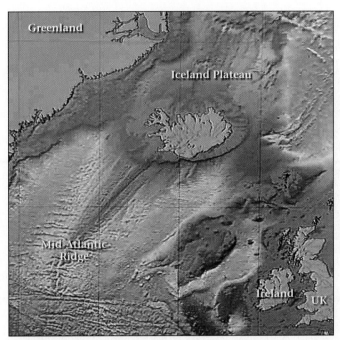

Greenland

Iceland Plateau

Mid-Atlantic Ridge

Iceland   UK

(b) A bathymetric map shows that Iceland sits atop a huge plateau straddling the Mid-Atlantic Ridge. Red is shallower water, blue is deeper.

(c) A fissure and volcanic edifice on Iceland.

volcano. Indeed, lava is a threat to real estate; on many occasions lava has overwhelmed towns (Fig. 9.20a–d). Basaltic lava from effusive eruptions is the greatest threat, because it can spread over a broad area. In Hawaii, recent lava flows have buried roads, housing developments, and vehicles. Although people usually have time to get out of the way of such flows, they might have to watch helplessly from a distance as an advancing flow engulfs their homes. Before the lava even touches it, the building bursts into flames from the intense heat. The most disastrous lava flow in recent times came from the 2002 eruption of Mt. Nyiragongo in the Congo. Lava flows traveled almost 50 km and flooded

the streets in the city of Goma, encasing them with a 2-m-thick layer of basalt. The flows destroyed almost half the city.

Pyroclastic flows can move extremely fast (100 to 300 km/h) and are so hot (500° to 1,000°C) that they represent a profound hazard to humans and the environment (Fig. 9.20e). Relatively small examples, such as the flow that struck St. Pierre on Martinique, can flatten towns and devastate fields even though they leave only a few centimeters of ash and lapilli behind. People caught in the direct path of such flows may be incinerated, and even those protected from the ash may die from inhaling toxic, superhot gases.

FIGURE 9.20 Hazards due to lava and ash from volcanic eruptions.

Lava Flows

(a) A lava flow reaches a house in Hawaii and sets it on fire.

(b) Lava from Mt. Etna threatens a town and olive grove in Sicily.

(c) Residents rescue household goods after a lava flow filled the streets of Goma, along the East African Rift.

(d) This empty school bus was engulfed by lava in Hawaii.

Pyroclastic Debris

(e) A pyroclastic flow from the 1991 eruption of Mt. Pinatubo chases a fleeing vehicle.

(f) This 1989–1990 eruption of Redoubt Volcano, Alaska, sent immense clouds of ash to high elevations.

(g) A blizzard of ash fell from the cloud erupted by Mt. Pinatubo in the Philippines.

Threat of falling ash and lapilli. During a pyroclastic eruption, large quantities of ash and lapilli erupt into the air, later to fall back to Earth (Fig. 9.20f). Close to the volcano, pumice and lapilli tumble out of the sky; these materials can accumulate into a blanket up to several meters thick. When rain saturates the ash with water, the sodden mass can cause roofs and power lines to collapse. Winds can carry fine ash over a broad region. In the Philippines, for example, a typhoon spread heavy air-fall ash from the 1991 eruption of Mt. Pinatubo so that it covered a 4,000-square-km area (Fig. 9.20g). Ash buries crops, may spread toxic chemicals that poison the soil, and insidiously infiltrates machinery, causing moving parts to wear out.

Threat to aircraft. Fine ash from an eruption can also present a hazard to airplanes. Like a sandblaster, the sharp, angular shards of ash abrade turbine blades, greatly reducing engine efficiency. The ash, along with sulfuric acid formed from the volcanic gas, scores windows and damages the fuselage. Also, when heated inside a jet engine, the ash melts, creating a liquid that coats interior parts of the engine and freezes, restricting the airflow and cause the engine to flame out. In 1982, a British Airways 747 flew through the ash cloud over a volcano on Java. Corrosion turned the windshield opaque, and ingested ash caused all four engines to fail. For 13 minutes, the plane glided earthward, dropping from 11.5 km (37,000 feet) to 3.7 km (12,000 feet) above the black ocean below. As passengers assumed a brace position for ditching at sea, the pilots tried repeatedly to restart the engines. Suddenly, in the oxygen-rich air of lower elevations, the engines roared back to life. The plane swooped back into the sky and headed for an emergency landing in Jakarta, where, without functioning instruments and with an opaque windshield, the pilot brought the 263 passengers and crew back to the ground safely. To land, he had to squint out an open side window, with only his toes touching the controls. A similar near-disaster occurred in 1989 when a jumbo jet encountered ash billowing from Redoubt Volcano, near Anchorage. Officials estimate that more than 80 jets have encountered ash; of these, seven experienced engine failure.

The 2010 eruption of the Eyjafjallajökull volcano in Iceland had a profound impact on air traffic. All told, the eruption sent about 0.25 cubic km of pyroclastic material up into the air, which was not particularly large in comparison with other known explosive eruptions (see Box 9.1). But because part of Eyjafjallajökull's eruption took place under a glacier, abundant meltwater interacted with the lava and triggered energetic steam explosions. This process produced convecting clouds of very fine ash that were particularly glassy and, therefore, very abrasive. The jet stream, a high-altitude current of rapidly moving air, was passing over Iceland at the time of the eruption and thus dispersed the ash throughout European air space. Because of concern that this ash could damage planes, officials shut down almost all air traffic across Europe for 6 days. The closure directly cost airlines $200 million/day, disrupted travel plans for countless passengers, and halted shipment of everything from electronic goods to flowers, thus impacting economies worldwide.

Other Hazards Related to Eruptions

Threat of the blast. Most exploding volcanoes direct their fury upward. But some, like Mt. St. Helens, explode sideways. The forcefully ejected gas and ash, like the blast of a bomb, flattens everything in its path. In the case of Mt. St. Helens, the region around the volcano had been a beautiful pine forest; but after the eruption, the once-towering trees were stripped of bark and needles and lay scattered over the hill slopes like matchsticks (see Box 9.1).

Threat of landslides. Eruptions commonly trigger large landslides along a volcano's flanks. The debris, composed of ash and solidified lava that erupted earlier, can move quite fast (250 km per hour) and far. During the eruption of Mt. St. Helens, 8 billion tons of debris took off down the mountainside, careened over a 360-m-high ridge, and tumbled down a river valley, until the last of it finally came to rest over 20 km from the volcano.

Threat of lahars. When volcanic ash and other debris mix with water, the result is a lahar that resembles wet concrete. A lahar can move downslope at speeds of over 50 km per hour. Because lahars are denser and more viscous than clear water, they pack more force than clear water and can literally carry away everything in their path. The lahars of Mt. St. Helens traveled along existing drainages for more than 40 km from the volcano. When they had passed, they left a gray and barren wake of mud, boulders, broken bridges, and crumpled houses, as if a giant knife had scraped across the landscape. The lahars generated during the 1991 eruption of Pinatubo were even more devastating, for the eruption was bigger, and coincided with the drenching rains of a typhoon.

Lahars may develop in regions where snow and ice cover an erupting volcano, for the eruption melts the snow and ice, thereby creating a supply of water. Perhaps the most destructive lahar of recent times accompanied the eruption of the snow-crested Nevado del Ruiz in Colombia on the night of November 13, 1985. The lahar surged down a valley like a 40-m-high wave, hitting the sleeping town of Armero, 60 km from the volcano. Ninety percent of the buildings in the town vanished, replaced by a 5-m-thick layer of mud, which now entombs the bodies of 25,000 people (Fig. 9.21a).

Threat of earthquakes. Earthquakes accompany almost all major volcanic eruptions, for the movement of magma breaks rocks underground. Such earthquakes may trigger landslides on the volcano's flanks and can cause buildings to collapse and dams to rupture, even before the eruption itself begins.

FIGURE 9.21 Hazards due to lahars and gas.

(a) A lahar completely buried Armero, Columbia.

(b) Cattled died where they stood, due to CO_2 from Lake Nyos, Cameroon.

Threat of tsunamis (giant waves). Where explosive eruptions occur in the sea or along a coast, the blast and the underwater collapse of a caldera, pyroclastic density currents, or giant landslides from ocean island volcanoes can generate huge sea waves, or tsunamis, tens of meters high. Most of the 36,000 deaths attributed to the 1883 eruption of Krakatau were not due to ash or lava, but rather to tsunamis that slammed into nearby coastal towns (see Box 9.1).

Threat of gas and aerosols. We have already seen that volcanoes erupt not only solid material but also large quantities of gases such as water vapor, carbon dioxide, sulfur dioxide, and hydrogen sulfide. Usually the gas eruption accompanies the lava and ash eruption, with the gas contributing only a minor part of the calamity. For example, sulfuric acid aerosols can cause respiratory problems in people who live downwind. But occasionally the gas alone snuffs out life in its path without causing any other damage. Such an event occurred in 1986 near Lake Nyos in western Africa.

Lake Nyos is a small but deep lake filling the crater of an active volcano in Cameroon. Though only 1 km across, the lake reaches a depth of over 200 m. Because of its depth, the cool bottom water of the lake does not mix with warm surface water. Carbon dioxide gas slowly bubbles out of cracks in the floor of the crater and dissolves in the cool bottom water. Apparently, by August 21, 1986, the bottom water had become supersaturated in carbon dioxide. On that day, perhaps because a landslide or wind disturbed the water, the lake "burped" and expelled a forceful froth of CO_2 bubbles. Because it is denser than air, this invisible gas flowed down the flank of the volcano and spread out over the countryside for a distance of about 23 km before dispersing. Although not toxic, carbon dioxide cannot provide oxygen for metabolism or oxidation. When the gas cloud engulfed the village of Nyos, it quietly put out cooking fires and suffocated the sleeping inhabitants. The next morning, the landscape looked exactly as it had the day before, except for the lifeless bodies of 1,742 people and about 6,000 head of cattle (see **Fig. 9.21b**).

Take-Home Message

- Basaltic lava flows can flow over farms and towns, but generally the flows are slow enough that people can evacuate before the flow arrives. Pyroclastic flows, however, can move so fast that they can engulf fleeing victims.
- A fall of ash and lapilli can crush buildings, especially if the fall is wet.
- Volcanic explosions can be catastrophic. The explosion, plus accompanying pyroclastic flows and tsunamis can eliminate cities.
- Downslope movement of landslides and lahars can bring disaster even to people some distance from the volcano.

THINK: Why can a lahar do so much more damage than an equivalent-sized flood of clear water?

9.6 PROTECTION FROM VULCAN'S WRATH

Volcanic eruptions are a natural hazard of extreme danger. Can anything be done to protect lives and property from this danger? The answer is, yes. Below, we first examine the evidence that geologists use to determine if a volcano has the potential to erupt and then we consider the suite of observations that may allow geologists to predict the timing of an eruption.

Active, Dormant, and Extinct Volcanoes

Geologists refer to volcanoes that are erupting, have erupted recently, or are likely to erupt soon as **active volcanoes** and distinguish them from **dormant volcanoes**, which have not erupted for hundreds to thousands of years but may erupt again in the future. Volcanoes that were active in the geologic past but have shut off entirely and will not erupt ever again are called **extinct volcanoes**. As examples, geologists consider Hawaii's Kilauea to be active, for it currently erupts and has erupted frequently during recorded history. In contrast, Mt. Rainier in the Cascades last erupted centuries to millennia ago, but since subduction continues along the western edge of Oregon and Washington, the volcano could erupt in the future, and so it is considered dormant. Devils Tower, in eastern Wyoming, is the remnant of a shallow igneous intrusion, formed beneath a volcano and subsequently exposed by erosion (Fig. 9.22). Eastern Wyoming was a volcanic region at the time the magma of Devils Tower intruded, tens of millions of years ago. But volcanism could not happen again in this region because the geologic cause for volcanism no longer exists there. Thus, we can say volcanism in the Devils Tower area is extinct.

How do you determine if a volcano is active, dormant, or extinct? One way is to examine the historical record. Another is to determine the age of erupted rocks, and to search for evidence that the volcano still lies within a tectonically active area. Finally, you can examine the landscape character of the volcano. Specifically, the shape (shield, stratovolcano, or cinder cone) of an erupting volcano depends primarily on the eruptive style, because at an erupting volcano, the process of construction happens faster than the process of erosion. Once a volcano stops erupting, erosion attacks. The rate at which erosion destroys a volcano depends on whether it's composed of pyroclastic debris or lava. Cinder cones and ash piles can wash away quickly. In contrast, composite or shield volcanoes, which have been armor-plated by lava flows, can withstand the attack of water and ice for quite some time. In the end, however, erosion wins out, and you can distinguish a dormant volcano from an active volcano by the extent to which river or glacial valleys have been carved into its flanks (Fig. 9.23). In some cases, the softer exterior of a volcano completely erodes away, leaving behind the plug of harder frozen magma that once lay within or beneath the volcano, as well as the network of dikes that radiate from this plug. You can see good examples of such landforms at Shiprock, New Mexico (Fig. 6.9c).

Predicting Eruptions

Predicting the time of an eruption decades or even years in advance is impossible. The only way to constrain a long-term prediction is to determine the recurrence interval (the average time between eruptions). Geologists determine recurrence intervals by determining the age of erupted layers comprising the volcano's edifice. For example, Mt. Fuji has erupted about

FIGURE 9.22 Devils Tower, Wyoming, is not an eroded volcano. It is probably an intrusion into sedimentary rocks; it solidified hundreds of meters below the surface of the Earth and has been exposed by erosion. Cooling produced spectacular columnar joints.

65 times in the last 10,000 years. So, its eruptive recurrence interval is about 150 years. A recurrence interval does not indicate periodicity—some of Fuji's eruptions were decades apart while others were centuries apart.

Short-term (days to weeks to months) predictions of impending volcanic activity, unlike short-term predictions of earthquakes, *are* actually feasible. Some volcanoes send out distinct warning signals announcing that an eruption may take place very soon, for as magma squeezes into the magma chamber, it causes a number of changes that geologists can measure.

> **Did you ever wonder . . .**
> could a volcano erupt beneath London?

- *Earthquake activity*: Movement of magma generates vibrations in the Earth. When magma flows into a volcano, rocks surrounding the magma chamber crack, and

FIGURE 9.23 The shape of a volcano changes as it is eroded.

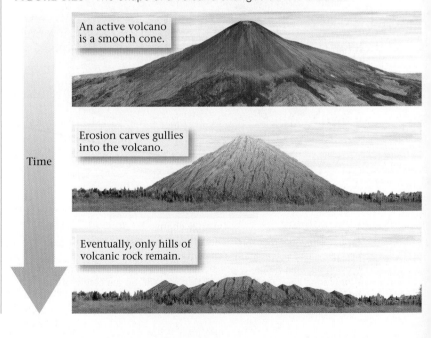

An active volcano is a smooth cone.

Erosion carves gullies into the volcano.

Eventually, only hills of volcanic rock remain.

Time

blocks slip with respect to each other. Such cracking and shifting also causes earthquakes. Thus, in the days or weeks preceding an eruption, the region between 1 and 7 km beneath a volcano becomes seismically active.

■ *Changes in heat flow*: The presence of hot magma increases the local heat flow, the amount of heat passing up through rock. In some cases, the increase in the heat flow melts snow or ice on the volcano, triggering floods and lahars even before an eruption occurs.

■ *Changes in shape*: As magma fills the magma chamber inside a volcano, it pushes outward and can cause the surface of the volcano to bulge; the same effect happens when you blow into a balloon. Geologists now use laser sighting, accurate tiltmeters, surveys using global positioning systems (GPS), and a technique called satellite interferometry (InSAR), which uses radar beams emitted by satellites to measure elevation changes, to detect modification of a volcano's shape due to the rise of magma (Fig. 9.24a).

■ *Increases in gas and steam emission*: Even though magma remains below the surface, gases bubbling out of the magma, or steam formed by the heating of groundwa-ter by the volcano, percolate upward through cracks in the Earth and rise from the volcanic vent. So an increase in the volume of gas emission, or of new hot springs, may indicate that magma has entered the ground below.

Because geologists can determine when magma has moved into the magma chamber of a volcano, government agencies now send monitoring teams to a volcano at the first sign of activity. These teams set up instruments to record earthquakes, measure the heat flow, determine changes in the volcano's shape, and analyze emissions.

Mitigating Volcanic Hazards

Danger assessment maps. Let's say that a given active volcano has the potential to erupt in the near future. What

FIGURE 9.24 Studies designed to mitigate volcanic disasters in the Cascade volcanic chain, western USA.

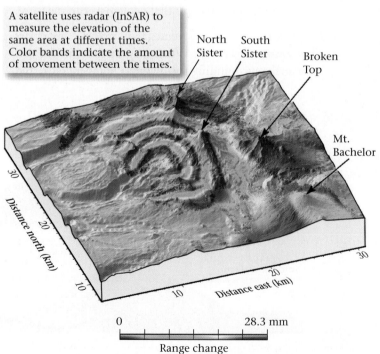

A satellite uses radar (InSAR) to measure the elevation of the same area at different times. Color bands indicate the amount of movement between the times.

(a) Satellite radar measurements of the Three Sisters Wilderness in Oregon, indicate that the region centered over the concentric bands is moving up rapidly, due to intrusion of magma below. Each rainbow band (from red to blue) represents ~28 mm of uplift. So the center of the bulls-eye has gone up over 8 cm relative to the edge.

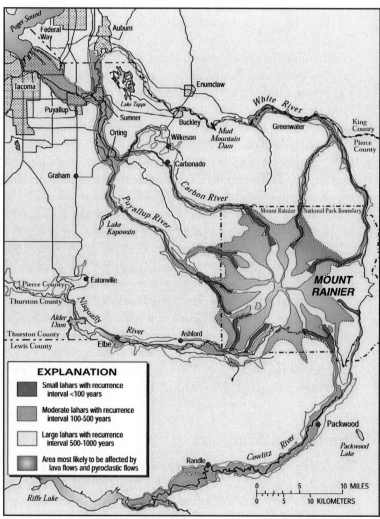

b) A volcanic hazard assessment map for Mt. Rainier, Washington, showing the regions that might be affected by flows and lahars.

FIGURE 9.25 Efforts to divert lava flows away from inhabited locations.

(a) Workers spray the lava flow to solidify it and use a bulldozer to build an embankment to divert it on the flanks of Mt. Etna.

(b) Firefighters pumping 6 million cubic meters of water on a lava flow, in an effort to freeze it and stop it.

can we do to prevent the loss of life and property? Since we can't prevent the eruption, the first and most effective precaution is to define the regions that can be directly affected by the eruption—to compile a volcanic-hazard assessment map (Fig. 9.24b). These maps delineate areas that lie in the path of potential lava flows, lahars, or pyroclastic flows.

River valleys originating on the flanks of a volcano are particularly dangerous places to be because lahars may flow down them. Before the 1991 eruption of Mt. Pinatubo in the Philippines, geologists had defined areas potentially in the path of pyroclastic flows, and had predicted which river valleys were likely hosts for lahars. Although the predicted pyroclastic-flow paths proved to be accurate, the region actually affected by lahars was much greater. Nevertheless, many lives were saved by evacuating people in areas thought to be under threat.

Evacuation. Unfortunately, because of the uncertainty of prediction, the decision about whether or not to evacuate is a hard one. In the case of Mt. St. Helens, hundreds of lives were saved in 1980 by timely evacuation, but in the case of Mt. Pelée in 1902, thousands of lives were lost because warning signs were ignored. In the Philippines, evacuation of people from the danger area around Mt. Pinatubo, over many years, succeeded in saving thousands of lives.

In 1976, debate over the need for an evacuation around a volcano on Guadeloupe, in the French West Indies, was fierce. Eventually, the population of a threatened town was evacuated, but as months passed, the volcano did not erupt. Instead, tempers did, and anger at the cost of the evacuation translated into lawsuits.

Diverting flows. In traditional cultures, people believed that gods or goddesses controlled volcanic eruptions, so when a volcano rumbled, they provided offerings to appease the deity. Sometimes, people have used direct force to change the direction of a flow or even to stop it. For example, during a 1669 eruption of Mt. Etna, an active volcano on the Italian island of Sicily, basaltic lava formed a glowing orange river that began to spill down the side of the mountain. When the flow approached the town of Catania, 16 km from the summit, 50 townspeople protected by wet cowhides boldly hacked through the chilled side of the flow to create an opening through which the lava could exit. They hoped thereby to cut off the supply of lava feeding the end of the flow, near their homes. Their strategy worked, and the flow began to ooze through the new hole in its side. But unfortunately, the diverted flow began to move toward the neighboring town of Paterno. Five hundred men of Paterno then chased away the Catanians so that the hole would not be kept open, and eventually the flow swallowed part of Catania.

More recently, people use high explosives to blast breaches in the flanks of flows, and use bulldozers to build dams and channels to divert flows. Major efforts to divert flows from a 1983 eruption of Mt. Etna, and again in 1992, were successful (Fig. 9.25a). Inhabitants of Iceland used a particularly creative approach in 1973 to stop a flow before it overran a town—they sprayed cold seawater onto the flow to freeze it in its tracks (Fig. 9.25b). The flow did stop short of the town, but whether this was a consequence of the cold shower it received remains unknown.

Did you ever wonder . . .
can you redirect a lava flow?

FIGURE 9.26 The ash cloud from the 2011 eruption of a volcano in the Puyehue-Cordón Caulle chain of Chile. The ash circled the globe within two weeks, disrupting air traffic throughout the southern hemisphere.

Take-Home Message

- Geologists distinguish among active, dormant, and extinct volcanoes based on whether they are erupting, have erupted in the past and are likely to do so in the future, or erupted a long time ago and can never erupt again.
- Eruptions can be predicted based on observations that earthquake activity, heat flow, shape, and quantities of gas emission are changing.
- Geologists can produce maps showing where lava flows, pyroclastic flows, and lahars are likely to go. These areas should be evacuated if an eruption is impending.
- In a few cases, people have worked to divert lava flows to protect property.

THINK: Using the Web, determine how many times Vesuvius has erupted since 79 C.E. and calculate its recurrence interval. Is it wise to continue building housing on its flanks?

9.7 EFFECT OF VOLCANOES ON CLIMATE AND CIVILIZATION

Can Eruptions Affect Climate?

In 1783, Benjamin Franklin was living in Europe, serving as the American ambassador to France. The summer of that year seemed to be unusually cool and hazy. Franklin, who was an accomplished scientist as well as a statesman, couldn't resist seeking an explanation for this phenomenon, and learned that in June of 1783, a huge volcanic eruption had taken place in Iceland. He wondered if the "smoke" from the eruption had prevented sunlight from reaching the Earth, thus causing the cooler temperatures. Franklin reported this idea at a meeting, and by doing so, may well have been the first scientist ever to suggest a link between eruptions and climate.

Franklin's idea seemed to be confirmed in 1815, when Mt. Tambora in Indonesia exploded. Tambora's explosion ejected over 100 cubic km of ash and pumice into the air (compared with 1 cubic km from Mt. St. Helens). Ten thousand people were killed by the eruption and the associated tsunami. Another 82,000 died of starvation. The sky became so hazy that stars dimmed by a full magnitude. Temperatures dipped so low in the northern hemisphere that 1816 became known as "the year without a summer." The unusual weather of that year inspired artists and writers. For example, memories of fabulous sunsets and the hazy glow of the sky inspired the luminous and atmospheric quality that made the landscape paintings of the English artist J. M. W. Turner so famous (see Fig. 9.1a), and Byron's 1816 poem "Darkness" contains the gloomy lines "The bright Sun was extinguish'd, and the stars / Did wander

darkling in the eternal space . . . Morn came and went—and came, and brought no day." Two years later, Mary Shelley, trapped in her house by bad weather, wrote *Frankenstein*, with its numerous scenes of gloom and doom.

Geoscientists have witnessed other examples of eruption-triggered coolness more recently. In the months following the 1883 eruption of Krakatau and the 1991 eruption of Pinatubo, global temperatures noticeably dipped. Classical literature provides more evidence of the volcanic impact on climate. For example, Plutarch wrote around 100 C.E., "Among events of divine ordering there was . . . after Caesar's murder . . . the obscuration of the Sun's rays. For during all the year its orb rose pale and without radiance . . . and the fruits, imperfect and half ripe, withered away." Similar conditions appear to have occurred in China the same year, as described in records from the Han dynasty, and may have been a consequence of volcanic eruption.

To study the effect of volcanic activity on climate even further in the past, geologists have studied ice from the glaciers of Greenland and Antarctica. Glacial ice has layers, each of which represents the snow that fell in a single year. Some layers contain concentrations of sulfuric acid, formed when sulfur dioxide from volcanic gas dissolves in the water from which snow forms. These layers indicate years in which major eruptions occurred. Years in which ice contains acid correspond to years during which the thinness of tree rings elsewhere in the world indicates a cool growing season.

How can a volcanic eruption create these cooling effects? When a large explosive eruption takes place, fine ash and aerosols (tiny liquid droplets) enter the stratosphere (Fig. 9.26). The aerosols form when volcanic SO_2 dissolves in water to form sulfuric acid. It takes only about two weeks for the ash and aerosols to circle the planet. They stay suspended in the stratosphere for many months to years, because they are above the weather and do not get washed away by rainfall. The resulting haze causes cooler average temperatures because it reflects and/or scatters incoming visible solar radiation during the day but not the infrared radiation that rises from the Earth's surface at night. Thus, it keeps energy from reaching the Earth, but unlike greenhouse gases (such as CO_2), it does not prevent heat from escaping.

A Krakatau-scale eruption can lead to a drop in global average temperature of about 0.3° to 1°C. According to some calculations, a series of large eruptions over a short period of time could cause a global average temperature drop of 6°C. The temperature dip seems to last up to a few years. For example, the eruption of Toba (Indonesia) about 73,000 years ago, the largest eruption known (see Box 9.1), may have cooled the climate for 6 years. Researchers have suggested that this was long enough to cause most humans to die, leaving only a small group of individuals to repopulate the planet.

Some researchers are currently exploring the possibility that the eruption of LIPs may have longer-lasting effects, perhaps changing climate so much that species go extinct. The apparent coincidence of the eruption of the Siberian traps with the extinctions that define the end of the Paleozoic may be an example.

The observed effect of volcanic eruptions on the climate provides a model with which to predict the consequences of a nuclear war. Researchers have speculated that so much dust and gas would be blown into the sky in the mushroom clouds of nuclear explosions that a "nuclear winter" would ensue.

Volcanoes and Civilization

Not all volcanic activity is bad. Over time, volcanic activity has played a major role in making the Earth a habitable planet. Eruptions and underlying igneous intrusions produced the rock making up the Earth's crust, and gases emitted by volcanoes provided the raw materials from which the atmosphere and oceans formed. The black smokers surrounding vents along mid-ocean ridges may have served as a birthplace for life, and volcanic islands in the oceans have hosted populations whose evolution adds to the diversity of life on the planet. Volcanic activity continues to bring nutrients (potassium, sulfur, calcium, and phosphorus) from Earth's interior to the surface, and to provide fertile soils that nurture plant growth. And in more recent times, people have exploited the mineral and energy resources generated by volcanic eruptions. Thus, volcanoes and people have lived in close association since the first human-like ancestors walked the Earth 3 million years ago. In fact, one of the earliest relics of human ancestors consists of footprints fossilized in a volcanic ash layer in East Africa.

But as we have seen, volcanic eruptions also pose a hazard. Eruptions may even lead to the demise of civilizations. The history of the Minoan people, who inhabited several islands in the eastern Mediterranean during the Bronze Age, illustrates this possibility. Beginning around 3000 B.C.E., the Minoans built elaborate cities and prospered. Then their civilization waned and disappeared (**Fig. 9.27a**). Geologists have discovered that the disappearance of the Minoans came within 150 years of a series of explosive eruptions of the Santorini Volcano in 1623 B.C.E. Remnants of the volcano now constitute Thera, one of the islands of Greece (**Fig. 9.27b**). After a huge eruption, the center of the volcano collapsed into the sea, leaving only a steep-walled caldera (**Fig. 9.27c**). Archaeologists speculate that pyroclastic debris from the eruptions periodically darkened the sky, burying Minoan settlements and destroying crops. In addition, related earthquakes crumbled homes, and tsunamis generated by the eruptions damaged Minoan seaports. Perhaps the Minoans took these calamities as a sign of the gods' displeasure, became demoralized, and left the region. Or perhaps trade was disrupted, and bad times led to political unrest. Eventually the Mycenaeans moved in, bringing the culture that evolved into that of classical Greece. The Minoans, though, were not completely forgotten. Plato, in his dialogues, refers to a lost city, home of an advanced civilization that bore many similarities to that of the Minoans. According to Plato, this city, which he named Atlantis, disappeared beneath the waves of the sea. Perhaps this legend evolved from the true history of the Minoans, as modified by Egyptian scholars who passed it on to Plato.

Numerous cultures living along the Pacific Ring of Fire have evolved religious practices that are based on volcanic activity—no surprise, considering the awesome might of a volcanic eruption in comparison with the power of humans. In some cultures, this reverence took the form of sacrifice in hopes of preventing an eruption that could destroy villages and bury food supplies. In traditional Hawaiian culture, Pelé, goddess of the volcano, created all the major landforms of the Hawaiian Islands. She gouged out the craters that top the volcanoes, her fits and moods bring about the eruptions, and her tears are the smooth, glassy lapilli ejected from the lava fountains.

Take-Home Message

- The aerosols and ash produced by a large eruption rise into the stratosphere, where winds carry them around the globe. The resulting haze reflects or scatters sunlight back into space.
- Huge LIP eruptions may have changed the climate enough to trigger mass extinction events.
- Eruptions in the Mediterranean may have played a role in the demise of ancient civilizations.

THINK: Not all eruptions of equivalent size (defined by the volume of material erupted) trigger equivalent amounts of global cooling. Why? (*Hint:* Think about eruptive style.)

FIGURE 9.27   The eruption of Santorini, 1645 B.C.E., may have led to the demise of the Minoan culture in the eastern Mediterranean.

(a) The ruins of palaces left by the Minoans.

(c) The inner wall of the caldera reveals the typical geology of a stratovolcano; we see both lava and tuff layers.

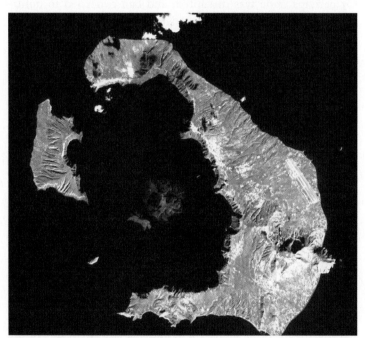

(b) From space, we can see the 11-km-wide caldera, all that is left of Santorini.

9.8 VOLCANOES ON OTHER PLANETS

We conclude this chapter by looking beyond the Earth, for our planet is not the only one in the Solar System to have hosted volcanic eruptions. We can see the effects of volcanic activity on our nearest neighbor, the Moon, just by looking up on a clear night. The broad darker areas of the Moon, the maria (singular *mare*, after the Latin word for sea), consist of flood basalts that erupted more than 3 billion years ago (Fig. 9.28a). Geologists propose that the flood basalts formed when huge meteors collided with the Moon, blasting out giant craters. Crater formation decreased the pressure in the Moon's mantle so that it underwent partial melting, producing basaltic magma that rose to the surface and filled the craters.

On Venus, about 22,000 volcanic edifices have been identified. Some of these even have caldera structures at their crests (Fig. 9.28b). Though no volcanoes currently erupt on Mars, the planet's surface displays a record of a spectacular volcanic past. The largest known mountain in the Solar System, Olympus Mons (Fig. 9.28c), is an extinct shield volcano on Mars. The base of Olympus Mons is 600 km across, and its peak rises 27 km above the surrounding plains.

Active volcanism currently occurs on Io, one of the many moons of Jupiter. Cameras in the *Galileo* spacecraft have recorded these volcanoes in the act of spraying plumes of sulfur gas into space (Fig. 9.28d) and have tracked immense, moving lava flows. Different colors of erupted material make the surface of this moon resemble a pizza. Researchers have proposed that the volcanic activity is due to tides: the gravitational pull exerted by Jupiter and by other moons alternately stretches and then squeezes Io, generating sufficient friction to keep Io's mantle hot. Geologists have also detected eruptions from moons of Saturn (Fig. 9.28e).

> **Did you ever wonder . . .**
> is the Earth the only planet with volcanoes?

Take-Home Message

- Volcanic eruptions occurred on other terrestrial planets and on our Moon in the past, but no terrestrial planet besides Earth appears to have active volcanism today.
- Active volcanism does happen today on some of the moons orbiting gas-giant planets. The character of this volcanism is very different from that on Earth.

THINK: Since most volcanic activity on Earth is a result of plate tectonics, what does the lack of volcanic activity on the Moon tell us about whether or not plate tectonics happens on the Moon?

FIGURE 9.28 Volcanism on other planets and moons in the Solar System.

(a) Maria of the Moon were seas of basaltic lava.

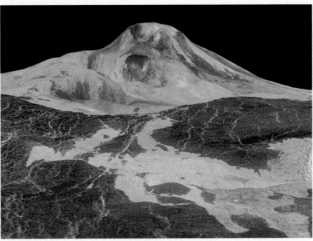

(b) A volcano rises above the plains of Venus.

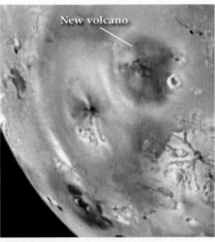

(d) An active volcano erupts sulfur on Io, a moon of Jupiter.

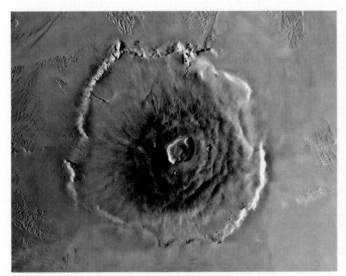

(c) Olympus Mons rises 27 km above the surface of Mars. It's the largest volcano in our Solar System.

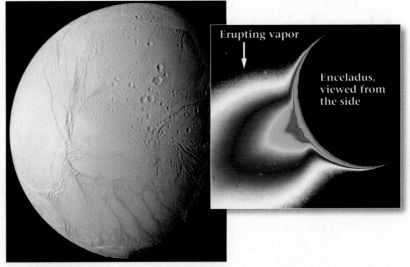

(e) Large cracks at the southern end of Enceladus, a moon of Saturn, erupt water vapor.

Chapter Summary

- Volcanoes are vents at which molten rock (lava), pyroclastic debris, gas, and aerosols erupt at the Earth's surface. A hill, mountain, crater, or caldera depression created by volcanism is also called a volcano.

- The characteristics of a lava flow depend on its viscosity, which in turn depends on its temperature and composition.

- Basaltic lava can flow great distances. Pahoehoe flows have smooth, ropy surfaces, whereas a'a' flows have rough, rubbly surfaces. Andesitic and rhyolitic lava flows tend to pile into mounds at the vent.

- Pyroclastic debris includes powder-sized ash, marble-sized lapilli, and apple- to refrigerator-sized blocks and bombs. Some falls from the air, whereas some forms incandescent pyroclastic flows (density currents) that rush away from the volcano.

- Eruptions may occur at a volcano's summit or from fissures on its flanks. The summit of an erupting volcano may collapse to form a bowl-shaped depression called a caldera.

- A volcano's shape depends on the type of eruption. Shield volcanoes are broad, gentle rises. Cinder cones are steep-sided, symmetrical hills composed of tephra. Composite volcanoes (stratovolcanoes) can become quite large and consist of alternating layers of pyroclastic debris and lava.

- The type of eruption depends on several factors, including the lava's viscosity and gas content. Effusive eruptions produce only flows of lava, whereas explosive eruptions produce clouds and flows of pyroclastic debris.
- Different kinds of volcanoes form in different geologic settings as defined by plate tectonics theory.
- Volcanic eruptions pose many hazards: lava flows overrun roads and towns, ash falls blanket the landscape, pyroclastic flows incinerate towns and fields, landslides and lahars bury the land surface, earthquakes topple structures and rupture dams, tsunamis wash away coastal towns, and invisible gases suffocate people and animals.
- Eruptions can be predicted by earthquake activity, changes in heat flow, changes in shape of the volcano, and the emission of gas and steam.
- We can minimize the consequences of an eruption by avoiding construction in danger zones and by drawing up evacuation plans.
- Immense flood basalts cover portions of the Moon. The largest known volcano, Olympus Mons, towers over the surface of Mars. Satellites have documented evidence for eruptions on moons of Jupiter and Saturn.

GEOPUZZLE REVISITED

Earth's volcanoes develop because there are places in the upper mantle and crust where partial melting takes place and magma forms. Though some magma freezes underground, some rises through conduits to the surface and erupts as lava or pyroclastic debris. The special places where melting takes place can be understood in the context of plate tectonics theory. Not all eruptions are the same, in part because not all lava has the same composition. Some eruptions spew out lava that flows in fast-moving streams, whereas others end in a cataclysmic explosion that blankets the countryside in ash.

Guide Terms

a'a' (p. 253)
accretionary lapilli (p. 256)
aerosols (p. 257)
active volcano (p. 277)
ash (p. 254)
blocks (p. 254)
blocky lava (p. 253)
bombs (p. 254)
caldera (p. 259)

cinder cone (p. 259)
columnar jointing (p. 253)
crater (p. 258)
dormant volcano (p. 277)
effusive eruption (p. 261)
eruptive style (p. 261)
explosive eruption (p. 261)
extinct volcano (p. 277)
fissure (p. 258)

flood basalt (p. 270)
hyaloclastite (p. 256)
ignimbrite (p. 256)
lahar (p. 256)
lapilli (p. 254)
large igneous province (LIP) (p. 270)
lava tube (p. 251)
magma chamber (p. 258)
pahoehoe (p. 251)
pumice lapilli (p. 256)
pyroclastic debris (p. 254)

pyroclastic flow (p. 256)
scoria cone (p. 259)
shield volcano (p. 259)
stratovolcano (p. 259)
supervolcano (p. 263)
tephra (p. 256)
tuff (p. 256)
vesicles (p. 257)
volcanic debris flow (p. 256)
volcaniclastic deposits (p. 254)
volcano (p. 249)

Review Questions

1. Describe the three different kinds of material that can erupt from a volcano.
2. Describe the differences between a pyroclastic flow and a lahar.
3. Describe the differences among shield volcanoes, stratovolcanoes, and cinder cones. How are these differences explained by the composition of their lavas and other factors?
4. Why do some volcanic eruptions consist mostly of lava flows, while others are explosive and do not produce flows?
5. Describe the activity in the mantle that leads to hot-spot eruptions.
6. How do continental-rift eruptions form flood basalts?
7. Contrast an island volcanic arc with a continental volcanic arc. What is a hot-spot volcano?
8. Identify some of the major volcanic hazards, and explain how they develop.
9. How do geologists predict volcanic eruptions?
10. Explain how steps can be taken to protect people from the effects of eruptions.

On Further Thought

11. The Long Valley Caldera, near the Sierra Nevada mountains, exploded about 700,000 years ago and produced a huge ignimbrite (a lapilli tuff deposited from vast pyroclastic density currents) called the Bishop Tuff. About 30 km to the northwest lies Mono Lake (see photo), with an island in the

middle and a string of craters extending south from its south shore. Hot springs and tufa deposits can be found along the lake. You can see the lake on *Google Earth*™ at Lat 37° 59' 56.58" N Long 119° 2' 18.20" W. Explain the origin of Mono Lake. Do you think that it represents a volcanic hazard?

12. Mt. Fuji is a 3.6-km-high stratovolcano in Japan formed as a consequence of subduction (see photo below). With *Google Earth*™ you can reach the volcano at Lat 35° 21' 46.72" N Long 138° 43' 49.38" E. It contains volcanic rocks with a range of compositions, including some andesitic rocks. Why do andesites erupt at Mt. Fuji? Very little andesite occurs on the Marianas Islands, which are also subduction-related volcanoes. Why?

13. The city of Albuquerque lies along the Rio Grande River in New Mexico. Within the valley, numerous volcanic features crop out. Using *Google Earth*™ you can fly to Albuquerque and then along the river to find many examples. Many of the volcanoes are basaltic, but in places you will see huge caldera remnants. In fact, the city of Los Alamos lies atop thick ignimbrites. What causes the volcanism in the Rio Grande Valley, and why are there different kinds of volcanism? Look at the photo of the volcanic cluster (see photo below). These occur north of Santa Fe, at Lat 36° 45' 27.51" N Long 105° 47' 24.85" W—use *Google Earth*™ to get a closer view. Judging from the character of these volcanoes, would you say they are active? Why? How would you evaluate the volcanic hazard of this region? (*Hint*: Use the Web to find a map of seismicity for the Rio Grande Valley region, and think about its implications.)

For more resources, including animations, quizzes, and Norton's GeoTours, go to **wwnorton.com/studyspace**.

If your instructor assigns exercises in SmartWork, log in at **smartwork.wwnorton.com**.

ANOTHER VIEW Mt. Etna, Sicily, erupts fairly frequently and produces voluminous quantities of lava as well as clouds of ash.

Close up, you can see a lava fountain in the plume.

View of the ash from the International Space Station.

Sicily

Ash

Ash from a 2002 eruption of Mt. Etna drifts to Africa.

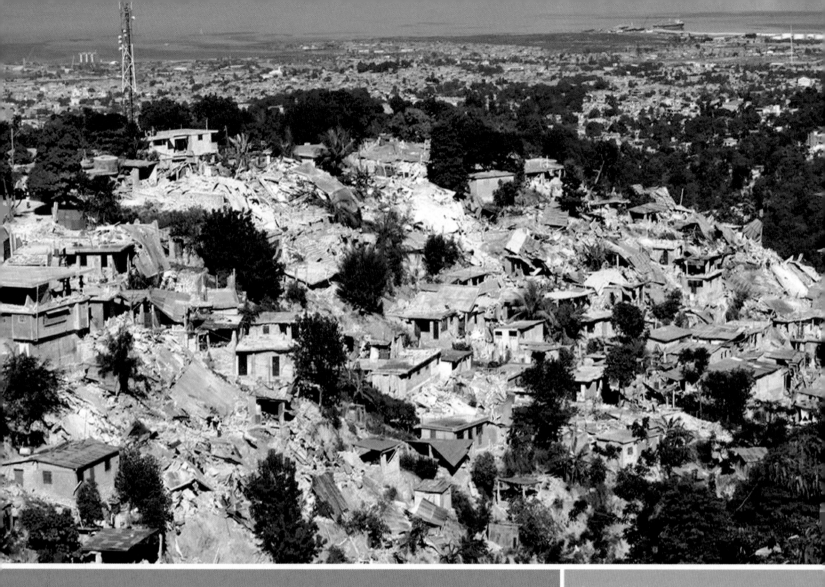

CHAPTER **10**

A Violent Pulse: Earthquakes

Ground shaking during the 2010 Haiti earthquake caused most of the houses in this residential neighborhood to collapse. Earthquakes are a natural feature of our dynamic planet, but they can be catastrophes to society.

GEOPUZZLE

Why do earthquakes happen where they do? Can people predict when an earthquake will occur?

We learn geology the morning after the earthquake.

—Ralph Waldo Emerson (American poet, 1803–1882)

10.1 INTRODUCTION

As the afternoon of March 11, 2011, began, fishing fleets were unloading their catch, factories were churning out goods, office workers were at their desks, shoppers filled store aisles, and children studied in school in numerous towns along the eastern coast of Honshu, the largest island of Japan. All this changed at 2:46 P.M., local time. Japan lies along a convergent plate boundary, at which the floor of the Pacific Ocean slips underneath the edge of the Eurasian Plate and sinks back into the mantle. The boundary itself is a **fault**, a large fracture on which sliding occurs, that slopes (dips) gently to the west. Motion along a fault doesn't happen continuously. Rather, for many years, rocks along the fault slowly bend. But like a stick that you flex with your hands, rock can bend only so far before it snaps (**Fig. 10.1a, b**). The "snap" on March 11 happened about 130 km (80 miles) east of Japan's coast, at a depth of about 24 km (15 miles) below the sea floor. In the course of a few seconds, Eurasia lurched eastward by a few meters, relative to the Pacific Plate. During the event, rocks cracked and ruptured over an area of the fault that measured 600 km long by 200 km wide. This movement abruptly uplifted the overlying sea floor by as much as tens of centimeters. A triple disaster was about to strike Japan.

Shock waves generated by the shearing and fracturing of rock along the fault raced through the crust at an average speed of 11,000 km (7,000 miles) per hour, ten times the speed of sound in air. When the vibrations reached Japan, the land surface lurched back and forth and bounced up and down for over a minute. A major **earthquake**—an episode of ground shaking had occurred. People lost their balance and crouched or fell, bottles and plates flew off shelves and crashed to the floor, buildings twisted and swayed, causing their ceilings and facades to collapse in a shower of debris, and power lines stretched and sparked. In addition, natural gas tanks and pipes broke, sending flammable vapors into the air—in places, the gas ignited in billows of flame that set fire to damaged buildings. This earthquake is now known as the Great Tohoku-oki earthquake—*Tohoku* is the eastern province of Honshu, and *oki* means open sea.

In most cases, the societal calamity due to an earthquake is a direct consequence of ground shaking, because many buildings collapse and crush their inhabitants, and debris tumbles down slopes in landslides. For example, during the May 12, 2008, earthquake of Sichuan, in central China, shaking leveled cities—70,000 people died, and millions were left homeless. And in February, 2011, sudden rupture

FIGURE 10.1 What happens during an earthquake?

(a) Before an earthquake, rock bends elastically, like a stick that you arch between your hands. The drawing exaggerates the amount of bending.

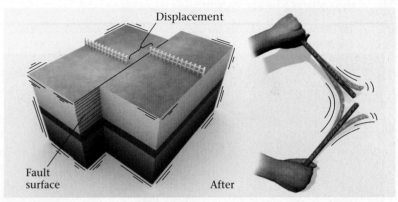

(b) Eventually, the rock breaks, and sliding suddenly occurs on a fault. This break generates vibrations. You feel such vibrations when you break a stick.

on a fault near Christchurch, New Zealand, toppled buildings and a stalwart stone cathedral (**Fig. 10.2a**). But in Japan, a country struck by earthquakes fairly frequently, officials have established stringent building codes that successfully prevented widespread building collapse. Unfortunately, even strong buildings could not resist what happened minutes after the earthquake shock. When the sea floor off Japan's coast suddenly uplifted during the Tohoku-oki earthquake, a massive amount of water was displaced. This water formed immense waves, known as tsunamis, which inundated low-lying coastal areas. Each tsunami was like a plateau of water that completely overran the beach and kept moving inland at about 30 km per hour, picking up people, boats, houses, cars, planes, trains, and trees as it moved. The roaring jumble of water and debris submerged everything in its path, as far as 10 km in from the shore line (**Fig. 10.2b**). To make matters worse, the water destroyed power sources used to run the pumps that cooled reactors in a coastal nuclear power plant, causing partial meltdown and release of radioactivity. The immensity of the devastation has been labeled the worst to impact Japan since World War II.

Earthquakes have affected the Earth since the solid lithosphere first formed. Most are a consequence of plate movement; they punctuate each step in the growth of mountains, the drift of continents, and the opening and closing of ocean basins. And, perhaps of more relevance to us, earthquakes have afflicted human civilization since the construction of the first village; they have directly caused the deaths of over 3.5 million people during the past two millennia (Table 10.1). Ground shaking, giant waves, landslides, and fires during earthquakes turn cities to rubble and ash. The destruction caused by some earthquakes may even have changed the course of civilization.

What does an earthquake feel like? When you're in one, time seems to stand still, so even though most earthquakes last

FIGURE 10.2 Recent earthquake devastation.

(a) Damage caused during the Christchurch, New Zealand earthquake in February 2011.

(b) A tsunami washes over coastal Japan (March 11, 2011).

TABLE 10.1 Some Notable Earthquakes

Year	Location	Number of Deaths
2011	Tohoku, Japan (tsunami)	20,000
2011	Christchurch, New Zealand	180
2010	Haiti	230,000
2010	Concepcion, Chile	1,000
2008	Sichuan, China	70,000
2005	Pakistan	80,000
2004	Sumatra (tsunami)	230,000
2003	Bam, Iran	41,000
2001	Bhuj, India	20,000
1999	Calaraca/Armenia, Colombia	2,000
1999	Izmit, Turkey	17,000
1995	Kobe, Japan	5,500
1994	Northridge, California	51
1990	Western Iran	50,000
1989	Loma Prieta, California	65
1988	Spitak, Armenia	24,000
1985	Mexico City	9,500
1983	Turkey	1,300
1978	Iran	15,000
1976	T'ang-shan, China	255,000
1976	Caldiran, Turkey	8,000
1976	Guatemala	23,000
1972	Nicaragua	12,000
1971	San Fernando, California	65
1970	Peru	66,000
1968	Iran	12,000
1964	Anchorage, Alaska	131
1963	Skopje, Yugoslavia	1,000
1962	Iran	12,000
1960	Agadir, Morocco	12,000
1960	Southern Chile	6,000
1948	Turkmenistan, USSR	110,000
1939	Erzincan, Turkey	40,000
1939	Chillan, Chile	30,000
1935	Quetta, Pakistan	60,000
1932	Gansu, China	70,000
1927	Tsinghai, China	200,000
1923	Tokyo, Japan	143,000
1920	Gansu, China	180,000
1915	Avezzano, Italy	30,000
1908	Messina, Italy	160,000
1906	San Francisco	500
1896	Japan	22,000
1886	Charleston, South Carolina	60
1866	Peru and Ecuador	25,000
1811–12	New Madrid, Missouri (3 events)	few
1783	Calabria, Italy	50,000
1755	Lisbon, Portugal	70,000
1556	Shen-shu, China	830,000

seconds to, at most, a few minutes, they seem much longer. Because of the lurching, bouncing, and swaying of the ground and buildings, people become disoriented, panicked, and even seasick. Some people recall hearing a dull rumbling or a series of dull thumps, as well as crashing and clanging. Earthquakes may even shake dust into the air, creating a fine, fog-like mist.

Earthquakes are a fact of life on planet Earth—almost 1 million detectable earthquakes happen every year. Fortunately, most cause no damage or casualties, because they are too small or they occur in unpopulated areas. But a few hundred earthquakes per year rattle the ground sufficiently to damage buildings and injure their occupants, and every 5 to 20 years, on average, a great earthquake triggers a horrific calamity. What geologic phenomena generate earthquakes? Why do earthquakes take place where they do? How do they cause damage? Can we predict when earthquakes will happen, or even prevent them from happening? These questions have puzzled seismologists (from the Greek word *seismos*, for shock or earthquake), the geoscientists who study earthquakes, for decades. In this chapter, we seek some of the answers.

Chapter Themes

By the end of this chapter, you should understand . . .

- the causes of earthquakes, and why earthquakes occur where they do.
- how to determine the size and location of an earthquake.
- the many ways in which earthquakes cause damage.
- the limitations on people's ability to predict earthquakes.
- the steps that people can take to prevent earthquake damage.

10.2 **WHAT CAUSES EARTHQUAKES?**

Ancient cultures offered a variety of explanations for **seismicity** (earthquake activity), most of which involved the action or mood of a giant animal or god. For example, in Japanese folklore, a giant catfish, Namazu, is said to have lived in the mud below the surface of the ground. If the gods did not restrain him, he would thrash about and shake the ground. In Indian folklore, earthquakes happened when one of eight elephants holding up the Earth shook its head, and in Siberia, earthquakes were thought to take place when a dog hauling the Earth in a sled stopped to scratch. Native American cultures of the west coast thought earthquakes were caused by arguments among the turtles holding up the Earth. Scientific study suggests that seismicity instead occurs for several reasons, including

- the sudden formation of a new fault (a fracture or rupture on which sliding occurs),

- sudden slip on an existing fault,
- a sudden change in the arrangement of atoms in the minerals of rock,
- movement of magma in a volcano,
- the explosion of a volcano,
- a giant landslide,
- a meteorite impact, or
- an underground nuclear-bomb test.

Of these various reasons, faulting related to plate movements is by far the most significant. In other words, *most earthquakes are due to slip on faults.*

As we learned in Chapter 2, the place within the Earth where rock ruptures and slips, or the place where an explosion occurs, is the **hypocenter** (or focus) of the earthquake. Energy radiates from the hypocenter. The point on the surface of the Earth that lies directly above the hypocenter is the **epicenter**, so maps can portray the position of epicenters (Fig. 10.3a, b).

FIGURE 10.3 Earthquake hypocenters and epicenters.

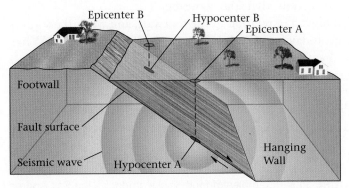

(a) The hypocenter, or focus, is the place underground where rock breaks and generates earthquake waves. The epicenter is the point on the surface of the Earth directly above the hypocenter.

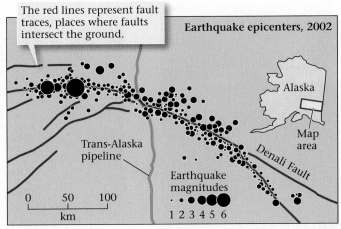

(b) Geologists represent epicenters as dots on a map; dot size indicates the size (magnitude) of the earthquake. Epicenters typically lie on or near a fault trace.

Typically the hypocenter of an earthquake coincides with a fault surface. Since slip on faults causes most earthquakes, we focus the discussion that follows on faults.

Faults in the Crust

Faults can be pictured, simplistically, as fractures that cut the crust. Nineteenth-century miners who encountered faults in mine tunnels referred to the rock mass above a sloping fault plane as the hanging wall, because it hung over their heads, and the rock mass below the fault plane as the foot-wall, because it lay beneath their feet. The miners described the direction in which rock masses slipped on a fault by specifying the direction that the hanging

FIGURE 10.4 The basic types of faults. Fault types are distinguished from one another by the direction of slip relative to the fault surface.

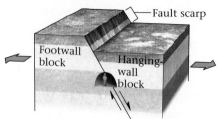
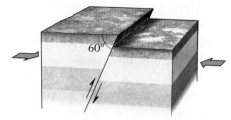

(a) Normal faults form during extension of the crust. The hanging wall moves down.

(b) Reverse faults form during shortening of the crust. The hanging wall moves up and the fault is steep.

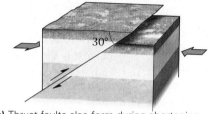

(c) Thrust faults also form during shortening. The fault's slope is gentle (less than 30°).

(d) On a strike-slip fault, one block slides laterally past another, so no vertical displacement takes place.

wall moved in relation to the footwall, and we still use these terms today (see **Geology at a Glance**, pp. 292–293). When the hanging wall slips down the slope of the fault, it's a normal fault. When the hanging wall slips up the slope, it's a reverse fault if steep and a thrust fault if shallowly sloping (**Fig. 10.4a–c**). Strike-slip faults are near-vertical planes on which slip occurs parallel to an imaginary horizontal line, called a strike line, on the fault plane—no up or down motion takes place here (**Fig. 10.4d**). In Chapter 11, we will consider other types of faults.

Normal faults form in response to stretching or extension of the crust. Reverse or thrust faults develop in response to squeezing (compression) and shortening of the crust, and strike-slip faults form where one block of crust slides

FIGURE 10.5 Examples of fault displacement on the San Andreas Fault in California.

(a) A wooden fence built across the fault was offset during the 1906 San Francisco earthquake. The displacement indicates that the motion was strike-slip.

(b) An aerial photograph shows offset of a dirt road by slip on the fault in 1999.

past another laterally. By measuring the distance between the two ends of a fence, a distinctive sedimentary bed, or an igneous dike that's been offset by a fault, geologists can define the **displacement**, the amount of slip, on the fault (Fig. 10.5a 🔊).

Faults are found in many locations—but don't panic! Not all of them are likely to be the source of earthquakes. Faults that have moved recently or are likely to move in the near future are called *active* faults (and if they generate earthquakes, news media sometimes refer to them as "earthquake faults"). Faults that last moved in the distant past and probably won't move again in the near future are called *inactive* faults. Some faults have been inactive for billions of years. The intersection between a fault and the ground surface is the fault trace, or fault line.

In places where an active normal or reverse fault intersects the ground, movement on the fault displaces the ground surface and generates a step called a **fault scarp** (see Fig. 10.4a). Slip on strike-slip faults tends to form narrow bands of low ridges and narrow depressions because faults are not perfect planes, and at bends along the fault the ground pulls apart or squeezes together (Fig. 10.5b). Not all active faults, though, intersect the ground surface. Those that don't are called blind faults or hidden faults. The trace of the fault that we now see on the ground may once have been far below the surface of the Earth, and it became visible only because overlying material has been eroded away.

Generating Earthquake Energy: Stick-Slip

What is the relationship between faulting and earthquakes? Earthquakes can happen either when rock breaks and a new fault forms, or when a preexisting fault suddenly slips again. Let's look more closely at these two causes.

- *Earthquakes due to fault formation*: Imagine that you grip each side of a brick-shaped block of rock with a clamp. Apply an upward push on one of the clamps and a downward push on the other. By doing so, you have applied a "stress" to the rock. (**Stress** refers to a push, pull, or shear.) At first, the rock bends slightly but doesn't break (Fig. 10.6a). In fact, if you were to stop applying stress at this stage, the rock would return to its original shape. Geologists refer to such a phenomenon as elastic behavior—the same phenomenon happens when a rubber band returns to its original shape or a bent stick straightens out after you let go. Now repeat the experiment, but bend the rock even more. If you bend the rock far enough, a number of small cracks or breaks start to form. Eventually the cracks connect to one another to form a fracture that cuts across the entire block of rock (Fig. 10.6b). The instant that such a throughgoing fracture forms, the block breaks in two. The rock on one side suddenly slides past the rock on the other side, and any elastic bending that had built up is released so the rock straightens out (Fig. 10.6c). Because sliding occurs,

FIGURE 10.6 A model representing the development of a new fault. Rupturing can generate earthquake-like vibrations.

Time

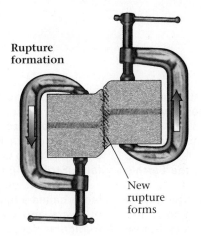

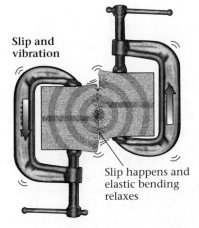

(a) Imagine a block of rock gripped by two clamps. Move one clamp up, and the rock starts to bend. Small cracks develop along the bend.

(b) Eventually, the cracks link. When this happens, a throughgoing rupture forms.

(c) The instant that the rupture forms, the rock breaks into two pieces that slide past each other. The energy that is released generates vibrations (earthquake energy).

Faulting in the Crust

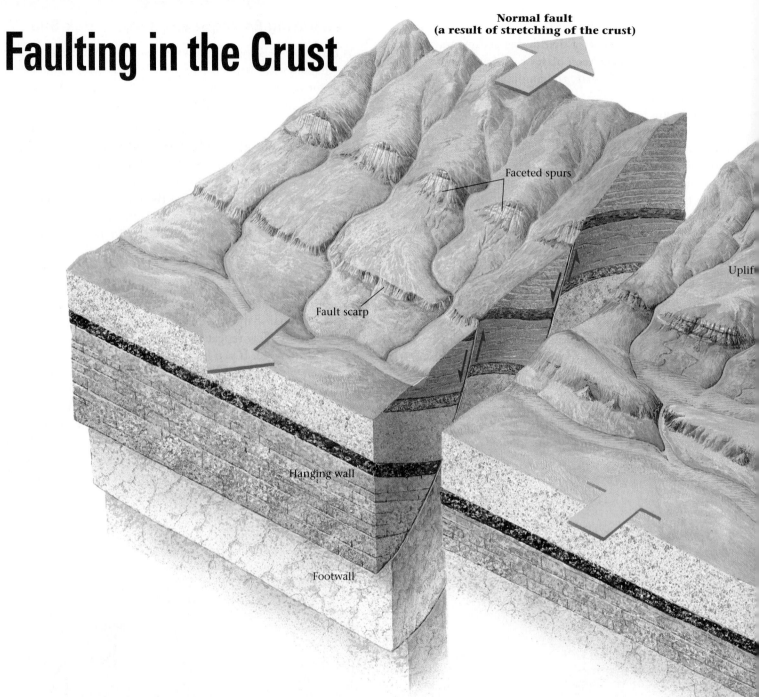

**Normal fault
(a result of stretching of the crust)**

Faceted spurs

Fault scarp

Uplif

Hanging wall

Footwall

Faults are fractures along which one block of crust slides past another block. Sometimes movement takes place slowly and smoothly, without earthquakes, but other times the movement is sudden, and rocks break as a consequence. The sudden breaking of rock sends shock waves, called seismic waves, through the crust, creating vibrations at the Earth's surface—an earthquake.

Geologists recognize three types of faults. If the hanging-wall block (the rock above a fault plane) slides down the fault's slope relative to the footwall block (the rock below the fault plane), the fault is a normal fault. (Normal faults form where the crust is being stretched apart, as in a continental rift.) If the hanging-wall block is being pushed up the

slope of the fault relative to the footwall block, then the fault is a reverse fault. (Reverse faults develop where the crust is being compressed or squashed, as in a collisional mountain belt.) If one block of rock slides past another and there is no up or down motion, the fault is a strike-slip fault. Strike-slip fault planes tend to be nearly vertical.

If a fault displaces the ground surface, it creates a ledge called a fault scarp. Where fault scarps cut a system of rivers and valleys, the ridges are truncated to make triangular facets. Strike-slip faults may offset ridges, streams, and orchards sideways. If there is a slight extension along the fault, the land surface sinks, and a sag pond develops.

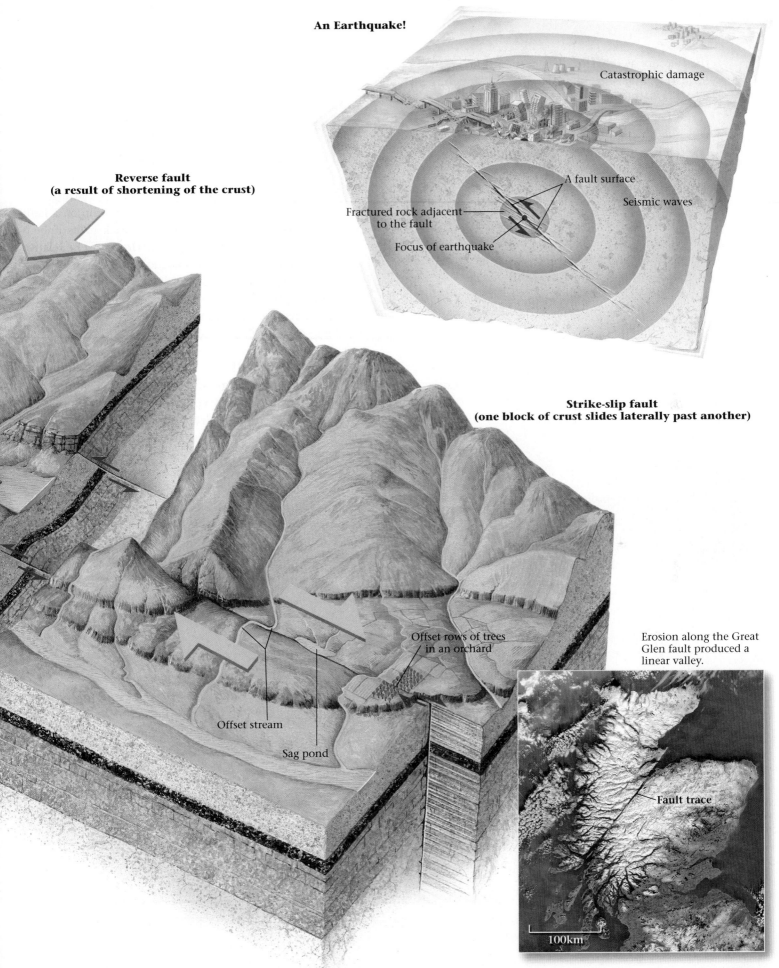

An Earthquake!

Catastrophic damage

A fault surface

Seismic waves

Fractured rock adjacent
to the fault

Focus of earthquake

Reverse fault
(a result of shortening of the crust)

Strike-slip fault
(one block of crust slides laterally past another)

Offset rows of trees
in an orchard

Offset stream

Sag pond

Erosion along the Great
Glen fault produced a
linear valley.

Fault trace

100km

293

the fracture at this stage has become a fault. Such fault formation and slip, like the snap of a stick, generate vibrations—an earthquake.

A fault can't slip forever, for friction eventually slows and stops the movement, just as friction slows and stops a book sliding across a table. **Friction**, defined as the force that resists sliding on a surface, is caused by the existence of bumps or protrusions on fault surfaces that act like tiny anchors (Fig. 10.7a, b).

■ *Earthquakes due to slip on a preexisting fault*: Once a fault comes into being, it is a scar in the Earth's crust that can remain weaker than surrounding, intact crust. Most earthquakes happen on preexisting faults. When stress builds sufficiently, it overcomes friction and the fault slips. This movement takes place before stress becomes

great enough to cause fracturing of surrounding intact rock. Note that after each slip event, stress builds again (Fig. 10.7c). Geologists refer to such alternation between stress buildup and slip events (earthquakes) as **stick-slip behavior**.

The breaking of rock that occurs when a fault slips, generates the energy of an earthquake. The concept that earthquakes happen because stresses build up, causing rock adjacent to the fault to bend elastically until slip on a fault occurs, allowing the bending to twang back to its unbent shape, is called the **elastic-rebound theory**. The relative movement of one plate with respect to another, along a plate boundary, can generate the stress that drives this process.

Recently, researchers have developed a new technique, called interferometric synthetic aperture radar (abbreviated InSAR), to help detect subtle ground-surface distortion associated with the elastic bending that preceeds earthquakes. To create an InSAR image, a satellite uses radar to make a precise map of ground elevation of a region at two different times (weeks to years apart). A computer compares the two images and detects differences in elevation as small as the wavelength of radar energy. A printout of the result portrays these differences as a rainbow of color bands (Fig. 10.8a, b). Each complete rainbow represents a certain amount of change in elevation.

Of note, a major earthquake along a fault may be preceded by smaller ones, called **foreshocks**, which possibly result from the development of the smaller cracks in the vicinity of what will be the major rupture. Smaller earthquakes, called **aftershocks**, follow a major earthquake.

FIGURE 10.7 The concept of friction and stick-slip behavior. On a microscopic scale, fault surfaces are rough and friction temporarily inhibits sliding.

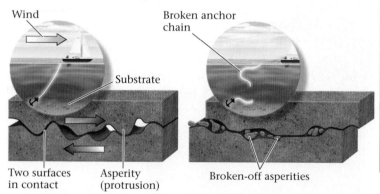

(a) Before movement, protrusions lock together, causing friction that prevents sliding. Similarly, an anchor keeps a boat in place.

(b) When a fault slips, the protrusions break off, and like a ship moving when its anchor chain breaks, the block slides.

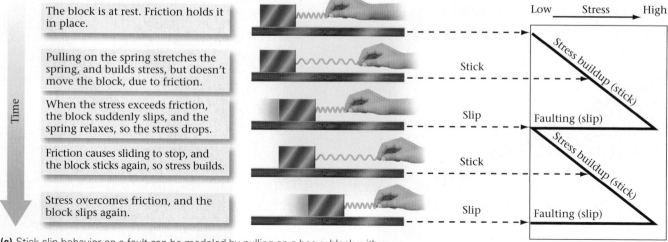

(c) Stick-slip behavior on a fault can be modeled by pulling on a heavy block with a spring. As the spring stretches, it applies stress to the block.

FIGURE 10.8 Detecting distortion of the Earth's surface using InSAR.

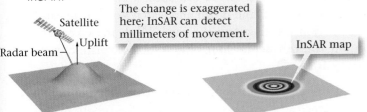

The change is exaggerated here; InSAR can detect millimeters of movement.

Satellite
Uplift
Radar beam
InSAR map

(a) Measurements made using InSAR, a type of satellite-based radar, portray warping of the Earth's surface by color bands on a map. The span of each rainbow indicates a specified amount of uplift.

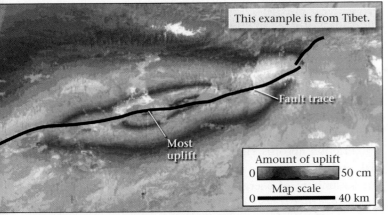

This example is from Tibet.

Fault trace

Most uplift

Amount of uplift
0   50 cm

Map scale
0   40 km

(b) InSAR studies show that buildup of stress before an earthquake locally warps the Earth's surface adjacent to a fault. Here, warping dies out toward the end of the fault.

Aftershocks can occur for weeks or even years; their distribution indicates the area of the fault that slipped during the initial large earthquake. The largest aftershock tends to be ten times smaller than the main shock; most are much smaller. Aftershocks happen because the movement of rock during the main earthquake generates new stresses, which may be large enough to reactivate small portions of the main fault or to activate small, nearby faults.

The Amount of Slip on Faults

How much of a fault surface slips during an earthquake? The answer depends on the size of the earthquake: the larger the earthquake, the larger the slipped area and the greater the amount of slip. For example, the major earthquake that hit San Francisco, California, in 1906 ruptured a 430-km-long (measured parallel to the Earth's surface) by 15-km-deep (measured perpendicular to the Earth's surface) segment of the San Andreas Fault. The amount of slip varies along the length of a fault (Fig. 10.9)—the maximum observed displacement during the 1906 earthquake was 7 m, in a strike-slip sense. Slip on a

thrust fault that caused the 1964 Good Friday earthquake in southern Alaska reached a maximum of 12 m. Smaller earthquakes, such as the one that hit Northridge, California, in 1994, resulted in only about 0.5-m slip—even so, this earthquake toppled homes, ruptured pipelines, and killed 51 people. The smallest-felt earthquakes result from displacements measured in millimeters to centimeters.

Although the cumulative movement on a fault during a human life span may not amount to much, over geologic time the cumulative movement becomes significant. For example, if earthquakes occurring on a strike-slip fault cause 1 cm of displacement per year, on average, the fault's movement will yield 10 km of displacement after 1 million years. Thus, earthquakes mark the incremental movements that create mountains. To take another example, movement on the San Andreas Fault in California averages around 6 cm per year. At this rate, Los Angeles, which is to the west of the fault, could move northward by 6,000 km in 100 million years.

Can Faults Slip without Earthquakes?

When a material breaks along fractures, we say that it has undergone brittle deformation. For example, a glass plate shattering on the floor is a type of brittle deformation. Faulting that generates earthquakes represents brittle deformation. In some cases, a rock bends without breaking. We call such behavior ductile deformation. If the glass plate were heated to a high temperature, it could be bent in a ductile manner, like chewing gum. Ductile deformation does *not* cause earthquakes.

FIGURE 10.9 During an earthquake on a preexisting fault, not all of the fault slips. The slipped area starts at a point and then grows outward. Red represents the most slip; yellow; the least slip.

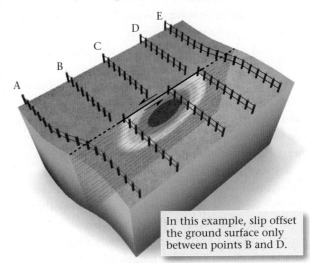

E
D
C
B
A

In this example, slip offset the ground surface only between points B and D.

Because the temperature of the Earth increases with depth, most brittle deformation and, therefore, earthquake-generating faulting in continental crust generally occurs only in the upper 15 to 20 km of the crust. At greater depths, shear and movement can take place by ductile deformation without earthquakes. In oceanic plates that have been subducted, earthquakes can happen much deeper as we see later in this chapter.

In some cases, movement on faults in the upper 15 to 20 km of the continental crust takes place slowly and steadily, without generating earthquakes. When movement on a fault happens without generating earthquakes, we call the movement **fault creep**. Seismologists do not completely understand fault creep, but speculate that it occurs in particularly weak rock, which can change shape without breaking or can slip smoothly without creating shock waves.

Take-Home Message

- In the vast majority of cases, earthquakes are due to faulting.
- The hypocenter is the point where the earthquake occurs; the epicenter is the point on the Earth's surface above the hypocenter.
- Displacement during the largest earthquakes may be 15 m.
- On faults that display stick-slip behavior, stress builds until a sudden movement takes place.

THINK: Why do foreshocks and aftershocks take place?

10.3 SEISMIC WAVES

How does the energy produced at the hypocenter of an earthquake travel to the surface or even pass through the entire Earth? Earthquake energy travels through rock and sediment in the form of waves. We call these waves **seismic waves** (or earthquake waves). You feel such waves when you touch one end of a brick and strike the other end with a hammer.

Seismologists distinguish among different types of seismic waves on the basis of where and how the waves move. **Body waves** pass through the interior of the Earth, whereas **surface waves** travel along the Earth's surface. Waves that cause particles of material to move back and forth *parallel* to the direction in which the wave itself moves are called **compressional waves**. As a compressional wave passes, the material first compresses (or squeezes together), then dilates (or expands). To see this kind of motion in action, push on the end of a spring and watch as the little pulse of compression moves along the length of the spring. Waves that cause particles of material to move back and forth *per-*

pendicular to the direction in which the wave itself moves are called **shear waves**. To see shear-wave motion, jerk the end of a rope up and down and watch how the up-and-down motion travels along the rope. With these concepts in mind, we can define four basic types of seismic waves (Fig. 10.10a–c ●)):

- P-waves (*P* stands for primary) are compressional body waves.
- S-waves (*S* stands for secondary) are shear body waves.
- L-waves (*L* stands for Love, the name of a seismologist) are surface waves that cause the ground to ripple back and forth, producing a snake-like movement.
- R-waves (*R* stands for Rayleigh, the name of a physicist) are surface waves that cause the ground to ripple up and down.

The different types of seismic waves travel at different velocities. P-waves travel the fastest and thus arrive first. S-waves travel more slowly, at about 60% of the speed of P-waves, so they arrive later. Surface waves are the slowest of all.

Friction absorbs energy as waves pass through a material, and waves bounce off layers and obstacles in the Earth, so the amount of energy carried by seismic waves decreases the farther they travel. Thus, people near the epicenter may be thrown off their feet by a large earthquake, but those 100 km away barely feel it. Similarly, an earthquake caused by slip on a fault deep in the crust causes less damage than one caused by slip on a fault near the surface.

Take-Home Message

- Earthquake energy travels through rock in the form of seismic waves; body waves travel through the Earth, and surface waves along the surface.
- Body waves include P-waves, which are compressional, and S-waves, which are shear. P-waves travel the fastest.

THINK: How does the amount of energy carried by earthquake waves change with increasing distance from the hypocenter?

10.4 HOW DO WE MEASURE AND LOCATE EARTHQUAKES?

Most news reports about earthquakes provide information on the size and location of an earthquake. What does this information mean, and how do we obtain it? What's the difference between a large earthquake and a minor one? How do seismologists locate an epicenter? To answer these questions we must first understand how a seismograph works and how to read the information it provides.

FIGURE 10.10 Different types of earthquake waves.

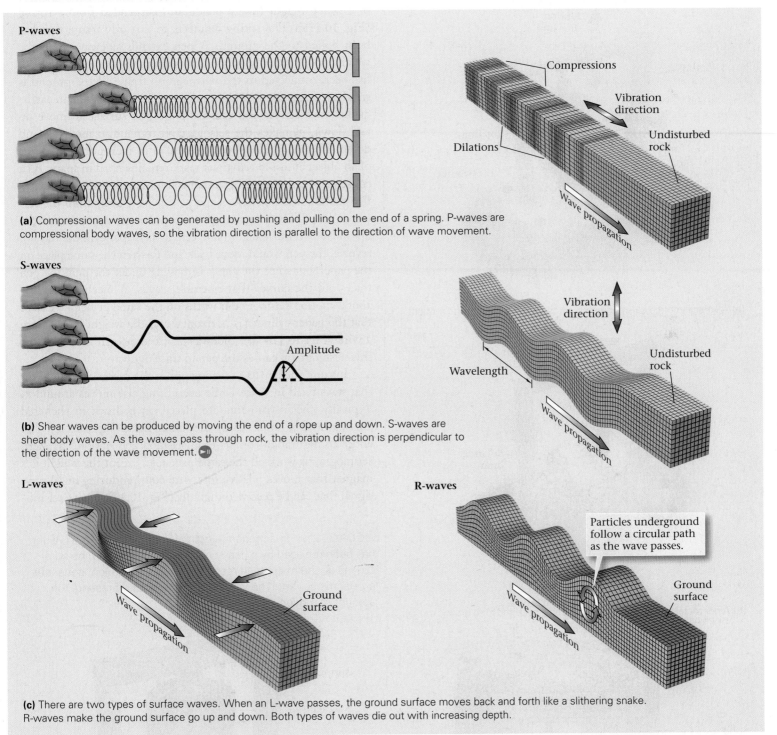

P-waves

(a) Compressional waves can be generated by pushing and pulling on the end of a spring. P-waves are compressional body waves, so the vibration direction is parallel to the direction of wave movement.

S-waves

(b) Shear waves can be produced by moving the end of a rope up and down. S-waves are shear body waves. As the waves pass through rock, the vibration direction is perpendicular to the direction of the wave movement.

L-waves

R-waves

(c) There are two types of surface waves. When an L-wave passes, the ground surface moves back and forth like a slithering snake. R-waves make the ground surface go up and down. Both types of waves die out with increasing depth.

Seismographs and the Record of an Earthquake

In 1889, a German physicist noted that a pendulum in his lab moved some time after a deadly earthquake had happened in Japan. This observation confirmed hypotheses that earthquake energy in the form of sesimic waves can pass through the planet. On reading of this discovery, other researchers saw a way to construct an instrument, called a **seismograph**, that can systematically record the ground motion from an earthquake happening anywhere on Earth. Seismologists now use two basic configurations of seismographs, one for measuring vertical (up-and-down) ground motion and the other for measuring horizontal (back-and-forth) ground motion.

FIGURE 10.11 The basic operation of a seismograph.

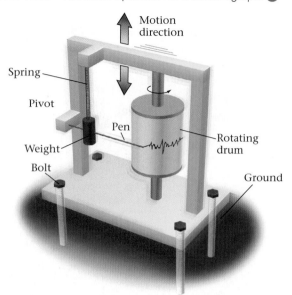

(a) A vertical-motion seismograph records up-and-down ground motion.

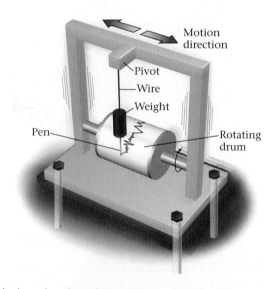

(b) A horizontal-motion seismograph records back-and-forth ground motion.

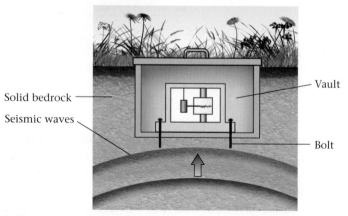

(c) Seismographs are bolted to bedrock in a protected shelter or vault.

A classic mechanical vertical-motion seismograph consists of a heavy weight (like a pendulum) suspended from a spring (Fig. 10.11a). The spring connects to a sturdy frame that has been bolted to the ground. A pen extends sideways from the weight and touches a vertical revolving cylinder of paper that has been connected to the seismograph frame. If the ground is steady, it traces out a straight reference line. But when an earthquake wave arrives and causes the ground surface to move up and down, it makes the seismograph frame also move up and down. The weight, however, because of its inertia (the tendency of an object at rest to remain at rest), remains fixed in space. As a consequence, the revolving paper roll moves up and down under the pen and the position of the pen deflects away from the reference line. The deflection of the pen, therefore, represents the up-and-down movement. Note that if the paper cylinder did not revolve, the pen would move back and forth in the same place on the paper, but since the paper is moving under the pen, the pen traces out the curves that resemble waves. A mechanical horizontal-motion seismograph works on the same principle, except that the paper cylinder is horizontal and the weight hangs from a wire (Fig. 10.11b). Sideways back-and-forth movement of this seismograph causes the pen to trace out waves.

In sum, the key to a seismograph is the presence of a weight that stays fixed in space while everything else moves around it. Typically, the instruments are placed on bedrock in sheltered areas, away from traffic and other urban noise (Fig. 10.11c). The entire configuration comprises a "seismograph station." Modern seismographs work on the same principle, except the weight is a magnet that moves relative to a wire coil, producing an electric signal that can be recorded digitally (Fig. 10.12). These seismo-

FIGURE 10.12 In a modern electronic seismograph, the weight has been replaced by a heavy magnet. Movement of the wire coil inside the magnet, in response to ground motion, generates an electric current. The voltage of the current represents the amount of motion.

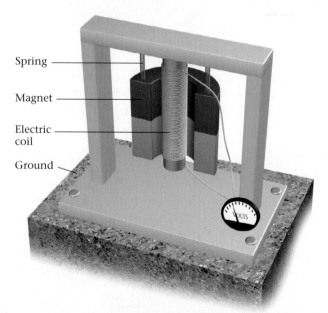

FIGURE 10.13 The nature of seismograms.

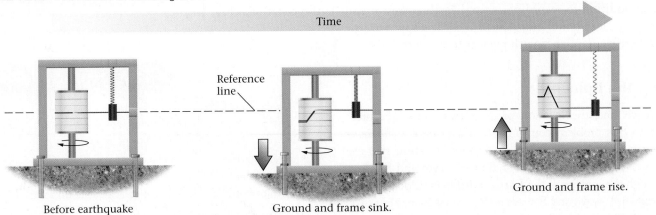

(a) Before an earthquake, the pen traces a straight line. During an earthquake, the paper roll moves up and down while the pen stays in place.

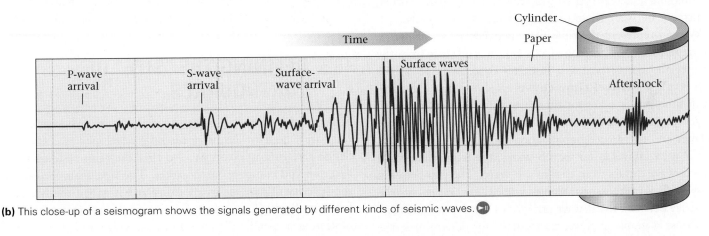

(b) This close-up of a seismogram shows the signals generated by different kinds of seismic waves.

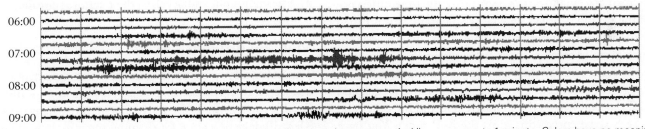

(c) A digital seismic record from a seismograph station in Arkansas. The space between vertical lines represents 1 minute. Colors have no meaning but make the figure more readable. Each color band represents the record of 15 minutes.

graphs are so sensitive that they can record ground movements of a millionth of a millimeter (only ten times the diameter of an atom)—movements that people can't feel.

The deflections traced by a seismograph provide a record of the earthquake called a **seismogram** (Fig. 10.13a–c). In order for seismologists to determine when a particular earthquake wave arrives, the record also displays lines representing time. At first glance, a typical seismogram looks like a messy squiggle of lines, but to a seismologist it contains a wealth of information. The horizontal axis represents time, and the vertical axis represents the amplitude (the size) of the seismic waves. The instant at which an earthquake wave appears at a

seismograph station is the "arrival time" of the wave. The first squiggles on the record represent P-waves, because P-waves travel the fastest. Next come the S-waves, and finally the surface waves (R-waves and L-waves). Typically, the surface waves have the largest amplitude and arrive over a relatively long interval of time.

Seismologists the world over have agreed to certain standards for measuring earthquakes, and they calibrate their seismographs using GPS (global positioning satellite) signals so that they can compare seismograms from different parts of the world. Today, seismologists can obtain data from thousands of stations constituting the world-wide seismic network. Governments have supported this network because in addition to recording natural earthquakes, it can also detect nuclear-bomb tests. Such records allow governments to verify compliance with nuclear test-ban treaties.

Finding the Epicenter

How do we find the location of an earthquake's epicenter? The key to this problem comes from measuring the *difference* between the time that the P-wave arrives and the time that the S-wave arrives at a seismograph station. P-waves and S-waves pass through the interior of the Earth at different velocities. The delay between P-wave and S-wave arrival times increases as the distance from the epicenter increases (Fig. 10.14a). To picture why, imagine a car race. If one car travels faster than another, the distance between them increases as the race proceeds.

We can represent the time delay between P- and S-waves on a graph where the horizontal axis indicates distance from the epicenter and the vertical axis indicates time. A curve on the graph is called a **travel-time curve**—it shows how the duration of time that it takes for the wave to move from its origin to a seismograph station increases as the distance between the epicenter and the seismograph station increases (Fig. 10.14b). Since P-waves travel faster than S-waves, the travel-time curve for an S-wave differs from that of a P-wave. The gap between the two curves (as measured along a vertical line) represents the time delay between the P-wave arrival and the S-wave arrival.

To use a graph of travel-time curves for determining the distance to an epicenter, start by measuring the time difference between the P- and S-waves on your seismogram; this is called the S–P (pronounced "S minus P") time. Then draw a line segment on a piece of tracing paper to represent this amount of time, at the scale used for the vertical axis of the graph. Orient the line segment parallel to the time axis and move it back and forth until one end lies on the P-wave curve and the other end lies on the S-wave curve (this gives the S–P time). You have now identified the gap between the two travel-time curves for the seismic station. Extend the line down to the horizontal axis, and simply read off the distance to the epicenter.

The analysis of *one* seismogram tells you only the distance between the epicenter and the seismograph station; it does not tell you in which direction from the station the epicenter lies. To determine the map position of the epicenter, we use a method called triangulation, by plotting the distance from the epicenter to *three* stations. For example, say you know that the epicenter lies 2,000 km from Station 1, 4,000 km from Station 2, and 6,000 km from Station 3. On a map, draw a circle around each station, such that the radius of the circle is the distance between the station and the epicenter, at the scale of the map. The epicenter lies at the intersection of the three circles, for this is the only point at which

the epicenter has the appropriate measured distance from all three stations (Fig. 10.14c).

Take-Home Message

- A seismograph is an instrument that records earthquakes; in a seismograph, a heavy weight stays fixed in position, due to inertia, relative to the crust moving around it.
- The record of an earthquake is a seismogram. On a seismogram, P-waves arrive first.
- Using three seismograms we can calculate epicenter location.

THINK: Will the duration of ground shaking be longer or shorter as the distance from the epicenter increases?

10.5 DEFINING THE "SIZE" OF EARTHQUAKES

Some earthquakes shake the ground violently whereas others can barely be felt. Seismologists have developed two scales to define size in a uniform way, so that they can systematically describe and compare earthquakes. The first scale focuses on the severity of damage at a locality and is called the Mercalli Intensity Scale. The second focuses on the amount of ground motion at a specific distance from the epicenter, as measured by a seismograph, and is called the magnitude scale.

Mercalli Intensity Scale

The **intensity** of an earthquake refers to the effect or consequence of an earthquake's ground shaking at a locality on the Earth's surface. In 1902, an Italian scientist named Giuseppe Mercalli devised a scale for defining intensity by systematically assessing the damage that the earthquake caused. A version of this scale, called the **Modified Mercalli Scale**, continues to be used today (Table 10.2). The Modified Mercalli Scale uses Roman numerals from I (low intensity) to XII (extremely high intensity). Assignment of lower numbers on the scale comes from recording people's perceptions of the earthquake's shaking, whereas assignment of higher numbers comes from documenting the damage caused by the earthquake. To use the scale, seismologists visit the affected region to interview residents and observe the damage. For example, if an earthquake at a location causes people to feel shaking, but does not damage buildings, then that earthquake had an intensity of II at that location. In contrast, if ground shaking topples most chimneys, destroys poorly constructed buildings, and causes some damage in substantial buildings, then the earthquake had an intensity of VIII.

FIGURE 10.14 The method for locating an earthquake epicenter.

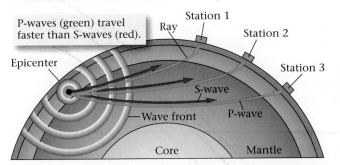

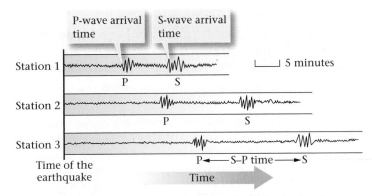

(a) Seismic waves travel at different velocities. The greater the distance between the epicenter and the seismograph station, the longer it takes for earthquake waves to arrive, and the greater the delay between the P-wave and S-wave arrival times.

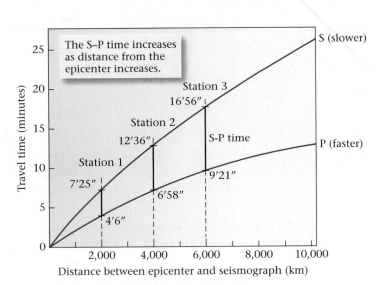

(b) We can represent the different arrival times of P-waves and S-waves on a graph of travel-time curves.

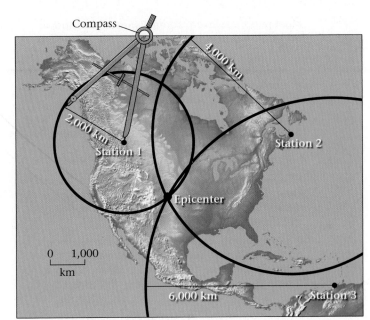

(c) If an earthquake epicenter lies 2,000 km from Station 1, we draw a circle with a radius of 2,000 km around the station at the scale of the map. We repeat for the other two stations. The intersection is the epicenter.

Note that the specification of earthquake intensity depends on a *subjective* assessment of perception and damage, not on a direct measurement with an instrument. Also, the Mercalli intensity value *varies with location* for a given seismic event—we cannot assign a single Mercalli number to a given earthquake. Typically, the intensity is greater near the epicenter, and decreases progressively away from the epicenter. To illustrate how intensity varies over a region for a given earthquake, seismologists draw lines, called contours, which delimit zones where the earthquake had a given intensity (Fig. 10.15). Of note, the maximum intensity of a "large" earthquake is greater than that of a "small" earthquake, and the distance from the epicenter to a given contour is greater for a "large" earthquake than for a "small" one. In detail, the spacing between intensity contours for a given earthquake depends on the strength of the crust in the region where the earthquake occurred.

Where the crust is strong, it can transmit earthquake energy more efficiently, so the contours are farther apart. In contrast, in regions where the crust is weak (because it contains lots of fractures and/or consists of soft rock), the contours are close together, and the area affected by the earthquake is smaller.

Earthquake Magnitude Scales

When you read a report of an earthquake disaster in the news, you will likely come across a phrase that reads something like, "An earthquake with a magnitude of 7.2 struck the city yesterday at noon." What does this mean? The **magnitude** of an earthquake is a number that indicates its size at a specified distance from the epicenter, as determined by measuring the maximum amplitude of ground motion recorded by a seismograph. By *amplitude of ground motion*, we mean the amount of

TABLE 10.2 Modified Mercalli Intensity Scale

MMI	Destructiveness (Perceptions of the Extent of Shaking and Damage)
I	Detected only by seismic instruments; causes no damage.
II	Felt by a few stationary people, especially in upper floors of buildings; suspended objects, such as lamps, may swing.
III	Felt indoors; standing automobiles sway on their suspensions; it seems as though a heavy truck is passing.
IV	Shaking awakens some sleepers; dishes and windows rattle.
V	Most people awaken; some dishes and windows break, unstable objects tip over; trees and poles sway.
VI	Shaking frightens some people; plaster walls crack, heavy furniture moves slightly, and a few chimneys crack, but overall little damage occurs.
VII	Most people are frightened and run outside; a lot of plaster cracks, windows break, some chimneys topple, and unstable furniture overturns; poorly built buildings sustain considerable damage.
VIII	Many chimneys and factory smokestacks topple; heavy furniture overturns; substantial buildings sustain some damage, and poorly built buildings suffer severe damage.
IX	Frame buildings separate from their foundations; most buildings sustain damage, and some buildings collapse; the ground cracks, underground pipes break, and rails bend; some landslides occur.
X	Most masonry structures and some well-built wooden structures are destroyed; the ground severely cracks in places; many landslides occur along steep slopes; some bridges collapse; some sediment liquifies; concrete dams may crack; facades on many buildings collapse; railways and roads suffer severe damage.
XI	Few masonry buildings remain standing; many bridges collapse; broad fissures form in the ground; most pipelines break; severe liquefaction of sediment occurs; some dams collapse; facades on most buildings collapse or are severely damaged.
XII	Earthquake waves cause visible undulations of the ground surface; objects are thrown up off the ground; there is complete destruction of buildings and bridges of all types.

up-and-down or back-and-forth motion of the ground. The larger the ground motion, the greater the deflection of a seismograph pen or needle as it traces out a seismogram. Since seismologists take into account the distance between the epicenter and the seismograph when calculating a magnitude, a magnitude value does *not* depend on this distance. Use of the

FIGURE 10.15 This map shows modified Mercalli intensity contours for the 1886 Charleston, South Carolina, earthquake. Note that near the epicenter, ground shaking reached MMI = X, and in New York City, ground shaking reached MMI = II to III.

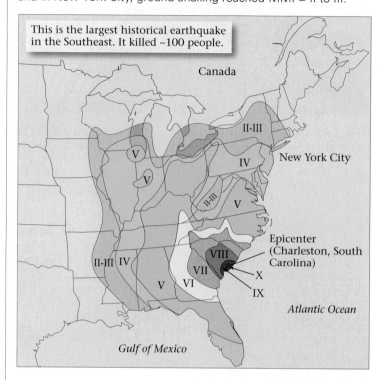

magnitude scale, therefore, allows seismologists to define the size of an eathquake in a uniform way.

The American seismologist Charles Richter developed a method for defining and measuring earthquake magnitude in 1935. The scale he proposed came to be known as the **Richter scale**. To ensure that a magnitude on this scale has the same meaning, regardless of who measures it, Richter prescribed detailed guidelines for determining magnitude. Specifically, seismologists determine a Richter magnitude by measuring the amplitude of the largest deflection on a seismograph as it would appear at a seismic station 100 km from the epicenter. (For this measurement, seismologists use waves with a "period" of 1 second. The period for a set of earthquake waves is the time interval between the arrival of successive waves; period = 1/frequency.) Because the amount of deflection depends on the distance between the seismograph and the epicenter, and since most seismograph stations do not happen to lie exactly 100 km from the epicenter, seismologists use a chart to adjust for distance of the station from the epicenter (Fig. 10.16a, b).

Richter's scale became so widely used that news reports often include wording such as, "The earthquake registered a 7.2 on the Richter scale." These days, however, seismologists actually use several different magnitude scales, not just the Richter scale, because the original Richter scale works well

FIGURE 10.16 Using the Richter magnitude scale.

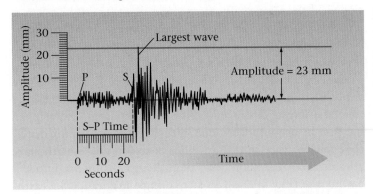

(a) To calculate the Richter magnitude from a seismograph, first measure the S–P (S minus P) time, to determine the distance to the epicenter. Then measure the amplitude of the largest wave.

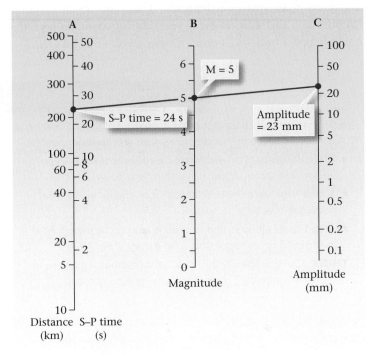

(b) Draw a line from the point on Column A representing the S–P time or the distance to the epicenter to the point on Column C representing the wave amplitude. Read the Richter magnitude off Column B.

only for shallow earthquakes that are close to the seismograph. Because of the distance limitation, a number on the original Richter scale is now also called a local magnitude (M_L). To apply Richter's concept to the description of distant earthquakes, seismologists developed a new scale based on measuring the amplitudes of certain R-waves. A number on this scale is called a surface-wave magnitude (M_S). The surface-wave magnitude scale, however, is not suitable for an earthquake whose hypocenter is more than 50 km below the surface, because such earthquakes do not create large surface waves. So to describe the size of deeper earthquakes, seismologists deter-

mine a body-wave magnitude (m_b), which is based on measurement of P-waves. Note that by an unfortunately confusing convention, some magnitudes use uppercase M and some use lowercase m.

The M_L, m_b, and M_S scales have limitations—they cannot accurately define the sizes of very large earthquakes, because for an earthquake above a given size, the scales give roughly the same magnitude regardless of how large the earthquake actually is. For example, an earthquake with an M_L, m_b, or M_S of 8.3 might be much larger. Because of this problem, seismologists developed the **moment magnitude scale** (M_W).

The moment magnitude scale provides the most accurate representation of an earthquake's size. To calculate the moment magnitude (M_W), seismologists measure the amplitude of several different seismic waves, determine the dimensions of the slipped area on the fault, and determine the displacement that occurred. The M_W scale does a better job of distinguishing among very large earthquakes. For example, the largest recorded earthquake in history, the great 1960 Chilean quake, registered as an 8.5 on the M_L scale and as a 9.5 on the M_W scale. The larger number makes more sense, because this earthquake was indeed much larger than other known events for which M_L = 8.5. Of note, the catastrophic 2011 Tohoku-oki earthquake off Honshu, Japan had a magnitude of M_W = 9.0.

What magnitude is given in modern reports of earthquakes in the newspaper? For early reports, seismologists report a preliminary magnitude, such as an M_L, which can be calculated quickly. Later on, after they have had the chance to collect the necessary data, seismologists report an M_W, which becomes the number generally used for archival records.

All magnitude scales are logarithmic, meaning that an increase of one unit of magnitude represents a tenfold increase in the maximum ground motion. Thus, a magnitude 8 earthquake results in ground motion that is 10 times greater than that of a magnitude 7 earthquake, and 1,000 times greater than that of a magnitude 5 earthquake. To make discussion easier, seismologists use familiar adjectives to describe the size of an earthquake, as listed in **Table 10.3**.

Note that, while there is an approximate relationship between the maxiumum intensity of an earthquake, as implied by Table 10.3, earthquake *magnitude* and earthquake *intensity* are very different concepts. As we have seen, the former is based on an instrument measurement and has a single value for an earthquake, while the latter is based on observations in the field and varies with location for a single earthquake.

Energy Release by Earthquakes

As we pointed out earlier, earthquakes release energy. Seismologists can calculate the energy release from equations that relate

TABLE 10.3 Adjectives for Describing Earthquakes

Adjective	Magnitude	Approximate maximum intensity at epicenter	Effects
Great	> 8.0	X to XII	Major to total destruction
Major	7.0 to 7.9	IX to X	Great damage
Strong	6.0 to 6.9	VII to VIII	Moderate to serious damage
Moderate	5.0 to 5.9	VI to VII	Slight to moderate damage
Light	4.0 to 4.9	IV to V	Felt by most; slight damage
Minor	< 3.9	III or smaller	Felt by some; hardly any damage

moment magnitude to energy. Not all versions of this calculation yield the same result, so energy estimates must be taken as an approximation. According to some researchers, a magnitude 6 earthquake releases about as much energy as the atomic bomb that was dropped on Hiroshima in 1945. The 1964 Alaska Good Friday earthquake, during which up to 15 m of slip occurred on a thrust fault, near Anchorage, released significantly more energy than the largest hydrogen bomb ever detonated. Notably, an increase in magnitude by one integer represents approximately a 32-fold increase in energy. Thus, a magnitude 8 earthquake releases about 1 million times more energy than a magnitude 4 earthquake (Fig. 10.17a). In fact, a single magnitude 8.9 earthquake releases as much energy as the entire *average* global annual release of seismic energy coming from all other earthquakes combined! Fortunately, such large earthquakes occur much less frequently than small earthquakes. There are about 100,000 magnitude 3 earthquakes every year, but a magnitude 8 earthquake happens only about once or twice a year (Fig. 10.17b).

Take-Home Message

- The intensity of an earthquake is a measure of ground shaking at a location, based on human perceptions and damage severity.
- The magnitude of an earthquake represents the size of an earthquake by a number, calculated from seismograms.
- A very large earthquake releases more energy than a giant hydrogen bomb.
- Magnitudes are logarithmic; ground shaking by a magnitude 8 event is 10 times larger than from a magnitude 7 event.

THINK: Can you specify the "size" of an earthquake by giving one intensity number? How about a magnitude number?

FIGURE 10.17 Measuring the energy released by earthquakes.

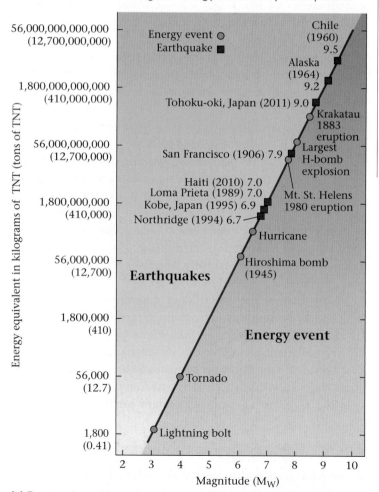

(a) Energy released by earthquakes increases dramatically with magnitude. Great earthquakes release vastly more energy than the largest bombs.

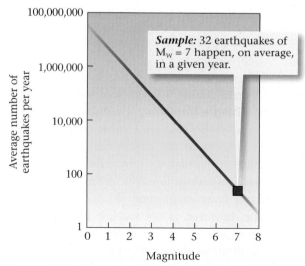

(b) The number of earthquakes per year of a given magnitude decreases with increasing magnitude.

10.6 WHERE AND WHY DO EARTHQUAKES OCCUR?

Earthquakes do not take place everywhere on the globe. By plotting the distribution of earthquake epicenters on a map, seismologists find that most, but not all, earthquakes occur in fairly narrow **seismic belts**, or seismic zones. Most seismic belts correspond with plate boundaries and earthquakes within these belts are called plate-boundary earthquakes. Earthquakes that occur away from plate boundaries are called intraplate earthquakes (the prefix *intra* means within) (Fig. 10.18). Eighty percent of the earthquake energy released on Earth comes from the plate-boundary earthquakes in the belts surrounding the Pacific Ocean.

Earthquakes do not occur at random depths in the Earth. Seismologists distinguish three classes of earthquakes based on hypocenter (focus) depth: shallow earthquakes occur in the top 20 km of the Earth, intermediate earthquakes take place between 20 and 300 km, and deep earthquakes occur down to a depth of about 660 km. Earthquakes cannot happen still deeper in the Earth, because rock at great depth cannot rupture or change in a manner that produces shock waves. In this section, we look at the characteristics of earthquakes in various geologic settings and learn why earthquakes take place where they do.

Earthquakes at Plate Boundaries

As we've noted, the majority of earthquakes happen at faults along plate boundaries, for the relative motion between plates causes slip on faults. We find different kinds of faulting at different types of plate boundaries.

Divergent-plate-boundary seismicity. At divergent plate boundaries (mid-ocean ridges), two oceanic plates form and move apart. Divergent boundaries are broken into spreading segments linked by transform faults. Therefore, two kinds of faults develop at divergent boundaries. Along spreading segments, stretching generates normal faults, whereas along transform faults strike-slip displacement occurs (Fig. 10.19). Seismicity along mid-ocean ridges takes place at shallow depths (less than 10 km). Since most ridges lie out in the ocean, far away from settled areas, mid-ocean ridge earthquakes don't cause damage. Only a few populated localities (such as Iceland) lie astride divergent-plate-boundaries.

FIGURE 10.18 A map of epicenters emphasizes that most earthquakes occur in distinct belts along plate boundaries.

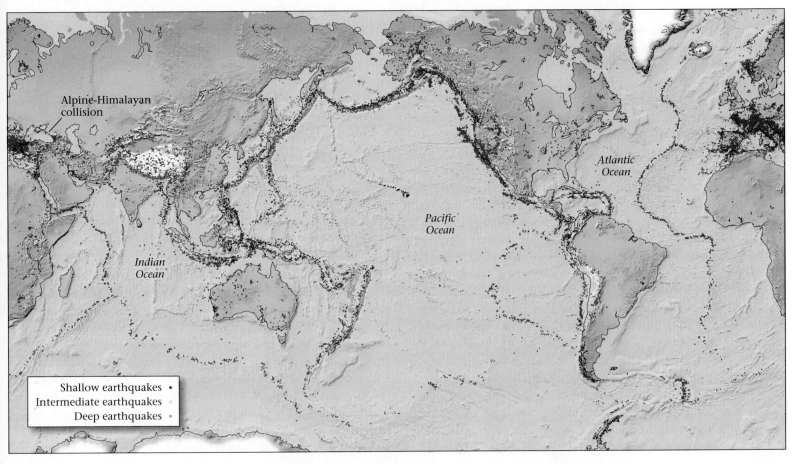

Alpine-Himalayan collision

Atlantic Ocean

Pacific Ocean

Indian Ocean

Shallow earthquakes •
Intermediate earthquakes •
Deep earthquakes •

FIGURE 10.19 The distribution of earthquakes at a mid-ocean ridge. Note that normal faults occur along the ridge axis, and strike-slip faults occur along active transform faults. Earthquakes don't take place along inactive fracture zones.

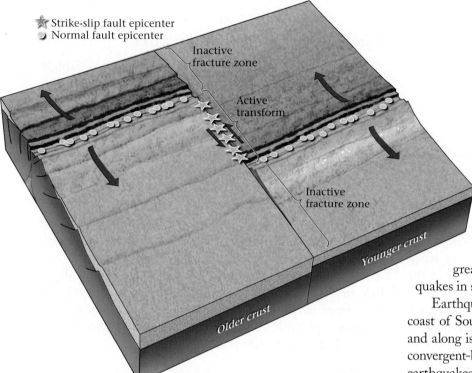

★ Strike-slip fault epicenter
● Normal fault epicenter

Inactive fracture zone

Active transform

Inactive fracture zone

Younger crust

Older crust

Convergent-plate-boundary seismicity. Convergent-plate boundaries are complicated regions at which several different kinds of earthquake take place. Shallow-focus earthquakes occur in both the subducting plate and the overriding plate. Specifically, as the downgoing plate begins to subduct, it bends and scrapes along the base of the overriding plate. Bending causes normal faults to develop in the downgoing plate, seaward of the trench. Large thrust faults develop along the contact between the downgoing and overriding plates, and shear on these faults can produce disastrous, shallow earthquakes. In some cases, the push applied by the downgoing plate compresses the overriding plate and triggers shallow faulting in the overriding plate.

In contrast to other types of plate boundaries, convergent-plate boundaries also host intermediate and deep earthquakes. These occur in the downgoing slab as it sinks into the mantle, defining the sloping band of seismicity called a **Wadati-Benioff zone**, after the seismologists who first recognized it (Fig. 10.20a). Intermediate and deep earthquakes happen partly in response to stresses caused by shear between the downgoing plate and the mantle, and partly by the pull of the sinking deeper part of the plate on the shallower part.

Why can intermediate and deep earthquakes of a Wadati-Benioff zone take place? Shouldn't the rock of a subducted plate at these depths be too warm and soft to break brittlely? To answer these questions, seismologists began by studying

the rate at which a subducting slab warms up as it sinks down through hot asthenosphere. They determined that rock is such a good insulator that the interior of a plate actually remains cool enough to fracture seismically, even down to a depth of about 300 km. To explain deeper earthquakes, seismologists studied the stability of minerals comprising the rock in the lithosphere. They found that at the extreme pressures developed in deeply subducted lithosphere, certain minerals collapse to form new, denser minerals. As such sudden "phase change" (see Chapter 8) takes place, the minerals abruptly decrease in volume, a process that could generate an earthquake. The fate of subducted lithosphere below a depth of 660 km remains uncertain. Some of the subducted plate may accumulate at 660 km, whereas some sinks still deeper. But at depths greater than 660 km, processes that generate earthquakes in subducted plate can no longer take place.

Earthquakes in southern Alaska, eastern Japan, the western coast of South America, the coast of Oregon and Washington, and along island arcs in the western Pacific serve as examples of convergent-boundary earthquakes (Fig. 10.20a). Some of these earthquakes are large, and occur near populated areas, so they can be devasting. Notable examples include the 1960 $M_W = 9.5$ earthquake of Chile, the largest earthquake on record; the 1964 $M_W = 9.2$ Good Friday earthquake near Anchorage, Alaska; the 1995 $M_W = 6.9$ earthquake of Kobe, Japan, which devastated the city (Fig. 10.20b, c); the 2004 $M_W = 9.3$ Sumatra earthquake, which triggered the giant Indian Ocean waves that killed 230,000 people; the 2010 $M_W = 8.8$ Chilean earthquake; and the 2011 Tohoku-oki earthquake, which also triggered a tsunami.

Recently, geologists have recognized that, in 1700, a huge earthquake, with a magnitude (M_w) of 8.7–9.2, accompanied slip on the Cascadia subduction zone off the now-densely populated coast of Oregon and Washington. An earthquake of similar size today would be disastrous to the region. GPS measurments indicate that the crust of the region is moving, and thus that stresses are likely building (Fig. 10.21a, b).

Transform-plate-boundary seismicity. At transform-plate boundaries, where one plate slides past another without the production or consumption of oceanic lithosphere, most faulting results in strike-slip motion. The majority of transform faults in the world link segments of oceanic ridges, but a few, such as the San Andreas Fault of California, the Alpine Fault of New Zealand, and the Anatolian Faults in Turkey, cut through continental lithosphere or volcanic arcs. All transform-fault earthquakes have a shallow focus, so the larger ones on land can cause disaster. The 2010 earthquake of Haiti is a tragic example of such an earthquake (Box 10.1).

FIGURE 10.20 An example of a convergent-boundary earthquake.

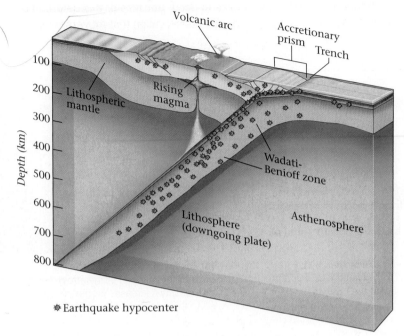

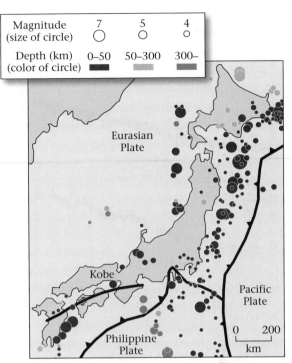

(a) At a convergent-plate boundary, earthquakes occur along the contact between the two plates, as well as in the downgoing plate and overriding plate.

(b) Map of earthquake epicenters and subduction zones in and near Japan.

(c) A collapsed building after the Kobe earthquake.

As another example of a transform-fault earthquake, consider the slip of the San Andreas Fault near San Francisco in 1906 (Fig. 10.22a). In the wake of the gold rush, San Francisco was a booming city with broad streets and numerous large buildings. But it was built on the plate-boundary along which the Pacific Plate moves northwest an average rate of 6 cm per year, relative to North America. Because of the stick-slip behavior of the fault, this movement doesn't occur smoothly but happens in jerks, each of which causes an earthquake. At 5:12 A.M. on April 18, the fault near San Francisco slipped by

as much as 7 m, and earthquake waves slammed into the city. Witnesses watched in horror as the street undulated like ocean waves. Buildings swayed and banged together, laundry lines stretched and snapped, church bells rang, and then towers, facades, and houses toppled. Judging from the damage, seismologists estimate that the largest shock would have registered a seismic moment magnitude of 7.9. Most building collapse took place downtown. Fire followed soon after, consuming huge areas of the city, for most buildings were made of wood (Fig. 10.22b). In the end, about five hundred people died, and a quarter of a million were left homeless.

The San Francisco earthquake has not been the only one to strike along the San Andreas and nearby related faults. Over a dozen major earthquakes have happened on these faults during the past two centuries, including the 1857 magnitude 7.7 earthquake just east of Los Angeles, the 1989 magnitude 7.1 Loma Prieta earthquake, which occurred 100 km south of San Francisco but nevertheless shut down a World Series game and caused the collapse of a double-decker freeway, and the Northridge quake that struck near Los Angeles in 1994 (Fig. 10.22c).

Earthquakes at Continental Rifts and Collision Zones

Continental rifts. The stretching of continental crust at continental rifts generates normal faults (Fig. 10.23). Active rifts today include the East African Rift, the Basin and Range Province (mostly in Nevada, Utah, and Arizona), and the Rio Grande

FIGURE 10.21 GPS (global positioning satellite) measurements made as part of the *EarthScope* program show that the crust is actively shortening in western Oregon and Washington, due to subduction. The last earthquake to occur because of this deformation was in 1700; it was an M_W = 8.7 to 9.2 (i.e., a *great* earthquake), and caused a major tsunami. The observation that movement is occurring leads to the prediction that another earthquake may happen in the future.

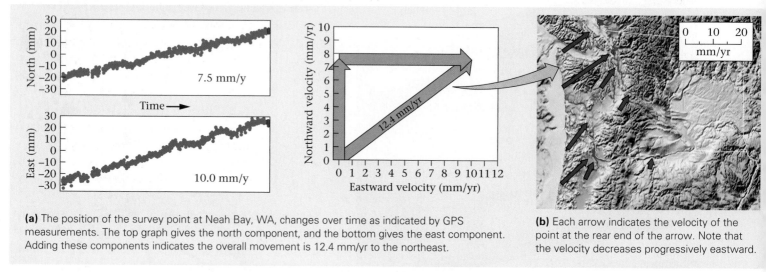

(a) The position of the survey point at Neah Bay, WA, changes over time as indicated by GPS measurements. The top graph gives the north component, and the bottom gives the east component. Adding these components indicates the overall movement is 12.4 mm/yr to the northeast.

(b) Each arrow indicates the velocity of the point at the rear end of the arrow. Note that the velocity decreases progressively eastward.

FIGURE 10.22 The San Andreas fault system, a continental transform.

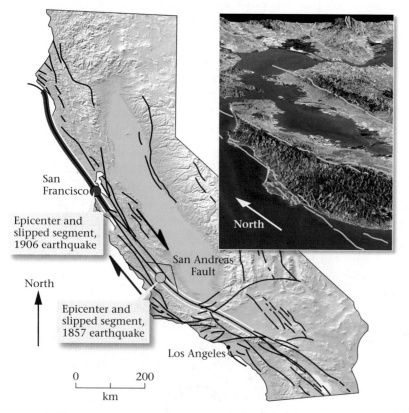

(a) The San Andreas Fault system in California. Note that the system includes many faults in a 100-km-wide band. A satellite image shows fault traces around San Francisco Bay.

(b) A street in San Francisco after the 1906 earthquake. Huge fires swept through the city.

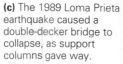

(c) The 1989 Loma Prieta earthquake caused a double-decker bridge to collapse, as support columns gave way.

Rift (in New Mexico). In all these places, shallow earthquakes occur, similar in nature to the earthquakes at mid-ocean ridges. But in contrast to mid-ocean ridges, these seismic zones can be located under populated areas, and thus cause major damage.

Collision zones. Two continents collide when the oceanic lithosphere that once separated them has been completely subducted. Such collisions produce great mountain ranges such as the Alpine-Himalayan chain. Though a variety of earthquakes happen in collision zones, the most common result from movement on thrust faults (see Fig. 10.23).

The magnitude 9.0 earthquake that occurred on All Saint's Day, November 1, 1755 serves as an example of a collision-related thrust event. The event happened when stresses arising from the northward push of Africa against Europe caused a thrust fault beneath the Atlantic Ocean floor west of Lisbon, Portugal to slip. The resulting ground shaking toppled 85% of the city's buildings. Fires set by overturned stoves then consumed much of the wreckage. Uplift of the sea floor by the thrust movement also produced a tsunami that inundated the coast and washed away Lisbon's harbor. Not only did the catastrophe destroy irreplaceable structures (including the library that housed all the records of Portuguese exploration) and countless Renaissance artworks, but it led the intelligentsia of that time to question long-held beliefs about philosophy. Influential works by Voltaire (1694-1778) and Kant (1724-1804) address some of the philosophical implications of that event.

Another example happened in October, 2005, when collision between the Indian subcontinent and Asia caused a magnitude 7.6 earthquake in the Kashmir region along the border of India and Pakistan. In this region of poorly constructed homes, ground shaking resulted in the collapse of whole towns. Associated landslides ripped out roads, denying access to potential rescuers. At least 86,000 people perished, and another 4 million were left homeless.

Intraplate Earthquakes

Some earthquakes occur in the interiors of plates and are not associated with plate boundaries, active rifts, or collision zones (see Fig. 10.23). These **intraplate earthquakes** account for only about 5% of the earthquake energy released in a year. Almost all have a shallow focus. Seismologists are still trying to understand the causes of intraplate earthquakes. Most favor the idea that force applied to the boundary of a plate can cause the interior of the plate to break suddenly at weak, preexisting fault zones, some of which may date back to the Precambrian. Alternatively, the activity may be due to forces resulting from shear between the lithosphere and the underlying asthenosphere, or when the upper crust readjusts to loading caused by growth of glaciers or accumulation of sediment, or unloading due to melting of glaciers or erosion of sediment.

In Europe, a number of intraplate earthquakes have been recorded. For example, an earthquake with a magnitude of 4.8 hit central England, near Birmingham, in 2002. In North America, intraplate earthquakes occur in the vicinities of New Madrid, Missouri; Charleston, South Carolina; eastern Tennessee; Montréal, Quebec; and the Adirondack Mountains, New York. A magnitude 7.3 earthquake occurred near Charleston in 1886, ringing church bells up and down the coast and vibrating buildings as far away as Chicago. In Charleston itself, over 90% of the buildings were damaged, and sixty people died. More recently, a magnitude 5.9 earthquake struck central Virginia, abruptly reminding residents of the eastern United States that the region is not immune to seismicity. The tremor was felt from the Carolinas to New England, and people evacuated buildings in Washington D.C. and in New York City. Some buildings in Washington were damaged slightly by the shaking.

The largest intraplate earthquakes to affect the United States happened in the early 19th century, near New Madrid, which lies near the Mississippi River in southernmost Mis-

FIGURE 10.23 The tectonic settings in which earthquakes occur in continental lithosphere. Subduction-related earthquakes in continental crust are not shown.

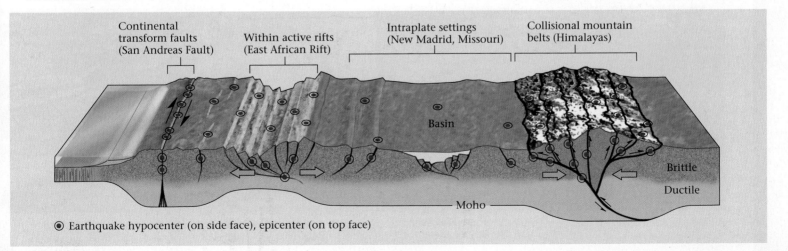

• Earthquake hypocenter (on side face), epicenter (on top face)

BOX 10.1

CONSIDER THIS . . . .

The Haiti Catastrophe of 2010

The history of Haiti changed forever on the sunny afternoon of January 12, 2010. At 4:53 P.M., a 70-km-long segment of the Enriquillo-Plantain Garden (EPG) fault suddenly slipped by an average amount of 1.8 m and, locally, by as much as 4 m. The motion began at a hypocenter only 25 km west-southwest of Port-au-Prince, and 13 km beneath the ground surface, so the shock waves of the resulting magnitude M_w-7 earthquake reached the capital city in a matter of seconds, causing the ground to lurch violently back and forth, side to side, and up and down, over a duration of 35 seconds. The event released as much energy as a 32-megaton nuclear weapon, 1,000× the energy released by the Nagasaki atomic bomb.

Ground shaking during the quake wrenched, bounced, and twisted buildings, causing insufficiently reinforced structures to crack and crumble (**Fig. Bx10.1a**). Unable to hold up the shifting weight of the building above, columns gave way, bringing floors down into gruesome pancake-like stacks. Similarly, brick or block walls broke apart, roads buckled, and hill slopes slumped. And beneath the harbor, sediment liquified, causing the wharfs to sink into the sea and the giant crane used to unload ships to topple sideways. When the shaking, which reached an intensity of IX on the Mercalli scale (**Fig. Bx10.1b**), finally stopped, most of Port-au-Prince, a city of 2 million in a country of 9 million, had collapsed. As a dense cloud of white dust slowly rose over the rubble, survivors began the frantic scramble to dig out victims, a task made more hazardous by aftershocks, of which there were over 50 with magnitudes between 4.5 and 6.1 (**Fig. Bx10.1c**)—the renewed shaking caused still-standing, but weakened walls to collapse on rescuers. Sadly, only about 150 people were eventually pulled from the rubble. No one will ever know exactly how many people died during the earthquake or of injuries in the weeks afterward, but some estimates place the death toll at 230,000, about 2.5% of the country's population. More than a million people lost their homes.

Was the earthquake a surprise? The answer depends on what is meant by surprise. Seismologists cannot predict *exactly* when and where an earthquake will occur, nor can they predict how strong one will be. But they can identify locations where large earthquakes are more likely, and plate boundaries, by definition, are such places. Haiti sits astride the transform plate boundary along which the North American plate moves westward at about 2 cm per year, relative to the Caribbean plate (**Fig. Bx10.1d–f**). So earthquakes in Haiti are inevitable. Slip-accommodating plate motion in Haiti is divided among a few large faults. The EPG fault is the southernmost of these and accommodates about half the slip. The last major earthquakes on this plate boundary happened about 240 years ago, so stress has been building on locked faults for quite some time, and the likelihood of an earthquake increased every year. But which year, exactly? No one could have known.

FIGURE Bx10.1 The disastrous January 2010 earthquake in Haiti, and its geologic setting.

(a) Survivors salvage what they can and try to carry on in Port-au-Prince, the capital of Haiti, after the devastating earthquake of January 2010.

FIGURE Bx10.1 *(continued)*

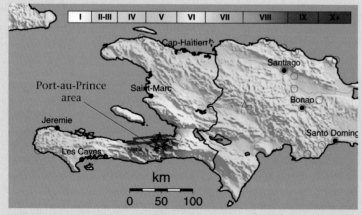

(b) The Mercalli intensity of shaking in Haiti.

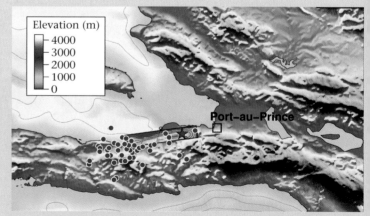

(c) The distribution of aftershocks, and the length of the slipped area.

A magnitude 7 earthquake is a "major" earthquake, not a "great" earthquake; recall that the largest measured earthquake (Chile, 1960) had a magnitude of 9.5. But the impact of an earthquake on society depends not only on its size but also on the nature of the substrate, on the steepness of the slopes, on construction practices, and on the quality of emergency services in the affected area. Much of Port-au-Prince was built on a basin of weak sediment—as earthquake waves passed into this basin from regions of harder bedrock, they were amplified, and therefore caused particularly large ground movements. In addition, many of the city's neighborhoods perch on steep slopes, which slumped downhill during the quake. And sadly, buildings in Haiti were not designed to withstand ground vibration, and local emergency services were overwhelmed. The country did not have enough rescue workers, doctors, or supplies to cope with the catastrophe, and debris made streets impassable. To make things worse, the country has only a small airport, slowing down delivery of supplies by air, and the destruction of the port prevented ships from bringing relief.

What does the future hold? One possibility is that the earthquake (a "slip" event) may have released enough stress buildup (an interval of "sticking") that another large event will not happen for another century or two. (If about 1 cm/year of motion should be accommodated on the EPG fault, and the fault has been stuck for 240 years, the sudden mean slip of 1.8 m during the earthquake could have relaxed a significant amount of the local stress buildup.) But many geologists worry that the 2010 earthquake might be the beginning of a chain of earthquakes along the plate boundary. In fact, the slip that then occurred along the EPG fault on January 23 may have caused stress to build on other segments, so that they might slip in the near future. Because nothing can stop the movement of plates, the safety of the region's inhabitants will depend on the strength of the new buildings to rise from the rubble, and on efforts to reinforce buildings in other regions along the fault.

FIGURE Bx10.1 *(continued)*

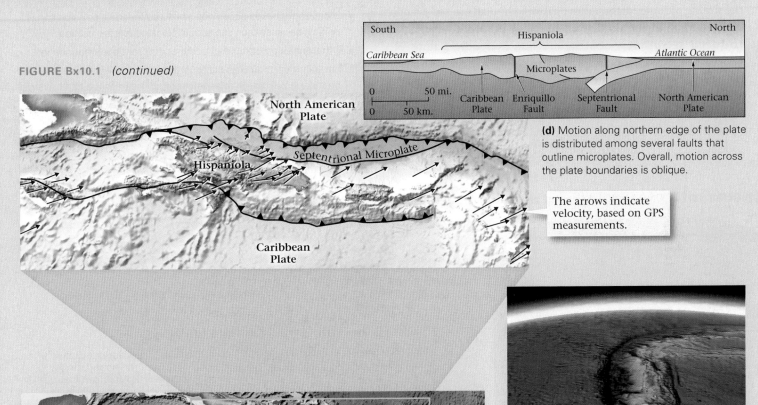

(d) Motion along northern edge of the plate is distributed among several faults that outline microplates. Overall, motion across the plate boundaries is oblique.

The arrows indicate velocity, based on GPS measurements.

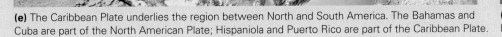

(e) The Caribbean Plate underlies the region between North and South America. The Bahamas and Cuba are part of the North American Plate; Hispaniola and Puerto Rico are part of the Caribbean Plate.

(f) Oblique satellite view looking due east at Hispaniola and Puerto Rico. The Puerto Rico trench stops where the Bahamas collided with Hispaniola.

souri. At the time, the region was inhabited by a small population of Native Americans and an even smaller population of European descent. During the winter of 1811–1812, three magnitude 7 to 7.4 earthquakes struck the region. The ground motion temporarily reversed the flow of the Mississippi River and toppled cabins (Fig. 10.24a). The earthquakes resulted from slip on thrust and strike-slip faults that underlie the Mississippi Valley (Fig. 10.24b). These faults may represent renewed slip faults that originated during rifting in the Precambrian. Both St. Louis, Missouri, and Memphis, Tennessee, lie close to the epicenter, so earthquakes in the area could be disastrous if they occurred today.

Induced Seismicity

Most earthquakes happen independent of human activity. But the timing of some earthquakes relative to human-caused events suggests that, in certain cases, people *can* influence seismicity. **Induced seismicity**, or seismic events caused by actions of people, generally occurs in response to changes in groundwater pressure. In the Earth, the pressure of water can slightly push apart the opposing surfaces of faults and thereby decrease the friction that resists motion on them. So when people increase groundwater pressure in a region containing a fault, a fault may suddenly slip. Seismologists observed such a relationship near Denver, Colorado, when engineers pumped wastewater from a military installation down a well; as soon as the pumping began, earthquakes started.

Induced seismicity is a particular danger when people build dams and create large reservoirs in valleys overlying active faults. Faulting generally breaks up rock, making it more erodible by rivers, so it is no surprise that deep river valleys commonly form over large faults. When a reservoir fills over a fault, water seeps down into the fault and, under the pressure caused by the water column above, triggers earthquakes. A big earthquake could destroy the dam, causing calamity downstream.

> **Did you ever wonder . . .**
> can people trigger earthquakes?

Take-Home Message

- Most earthquakes occur along plate boundaries, collision zones, or rifts, because fault slip accommodates relative motion.
- Different kinds of faulting occur at different plate boundaries.
- Intraplate earthquakes occur within plates, away from plate boundaries. Though relatively rare, some of these can be large.

THINK: On a global basis, why is more death and damage associated with convergent-margin earthquakes than with mid-ocean ridge earthquakes?

FIGURE 10.24 New Madrid, Missouri, is an example of intraplate seismic activity.

(a) The earthquakes of 1811–1812 destroyed cabins and disrupted the Mississippi River.

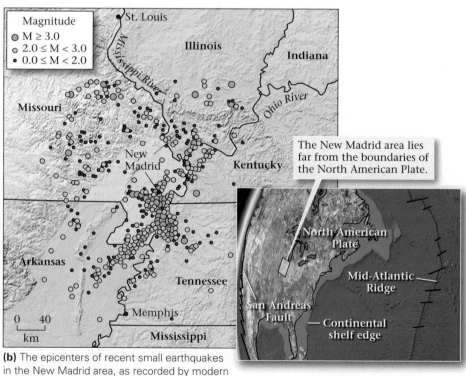

Magnitude
- M ≥ 3.0
- 2.0 ≤ M < 3.0
- 0.0 ≤ M < 2.0

The New Madrid area lies far from the boundaries of the North American Plate.

(b) The epicenters of recent small earthquakes in the New Madrid area, as recorded by modern seismic instruments. The region remains active.

10.7 HOW DO EARTHQUAKES CAUSE DAMAGE?

An area ravaged by a major earthquake is a heartbreaking sight. The terror and sorrow etched on the faces of survivors mirror the inconceivable destruction. This destruction comes as a result of many processes.

Ground Shaking and Displacement

Did you ever wonder . . . how long does an earthquake last?

An earthquake starts suddenly and may last from a few seconds to a couple of minutes. This statement applies only to the "felt area," meaning the area in which shaking is noticeable and has an effect. Since different kinds of earthquake waves travel at different speeds, their arrival at a seismic station on the other side of the Earth may happen over a much longer period (up to an hour). Different kinds of earthquake waves cause different kinds of ground motion (Fig. 10.25).

Generally, P-waves are almost perpendicular to the ground surface when they arrive and cause the ground to buck up and down. Next come the S-waves, which also reach the surface at a steep angle. These waves are more complicated but tend to cause back-and-forth motion, parallel to the ground surface. Almost immediately afterward L-waves, the first surface waves, arrive and cause snake-like side-to-side motion. Finally, the R-waves arrive and cause a rolling motion as particles near the surface of the ground follow elliptical movements. Rayleigh waves last longer than other waves and may cause the most damage. Interference among the different kinds of waves causes motion to be anything but regular, like choppiness on the surface of a pond. In great earthquakes, the ground's movement can have an amplitude of as much as 1 m, but in moderate earthquakes, motions fall in the range of a few centimeters or less.

Feeling the ground move is a terrifying and disorienting experience. Ground accelerations caused by moderate earthquakes are in the range of 10 to 20% of g (where g is the acceleration resulting from gravity), and during a great earthquake accelerations may approach 1 g. If the motions are large enough, they may knock you down or even toss you in the air.

The nature and severity of the shaking at a given location depend on four factors: (1) the magnitude of the earthquake, because larger-magnitude events release more energy; (2) the distance from the hypocenter, because earthquake energy decreases as waves pass through the Earth; (3) the nature of the substrate at the location (that is, the character and thickness of different materials beneath the ground surface) because earthquake waves tend to be amplified in weaker substrate (Box 10.2); and (4) the "frequency" of the earthquake waves (where frequency equals the number of oscillations that pass a point in a specified interval of time).

If you're out in an open field during an earthquake, ground motion alone won't kill you—you may be knocked off your feet and bounced around a bit, but your body is too flexible to break. Buildings and bridges aren't so lucky (Fig. 10.26a–d). When earthquake waves pass, they sway, twist back and forth, or lurch up and down, depending on the type of wave motion. As a result, connectors between the frame and facade of a building may separate, so the facade crashes to the ground. The flexing of walls shatters windows and makes roofs collapse. Floors or

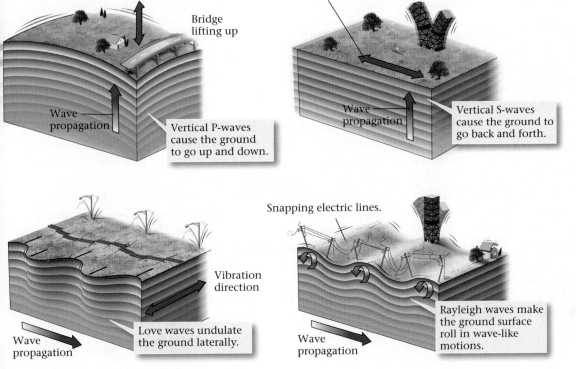

FIGURE 10.25 Types of ground motion during earthquakes. The ground can shake in many ways at once, causing surface structures to move.

CONSIDER THIS...

BOX 10.2

When Earthquake Waves Resonate—Beware!

When you shine a flashlight, you produce white light. If you aim that beam at a prism, the light spreads into a spectrum of different colors with each color representing light waves of a different frequency. Thus, the original beam of white light contained all these frequencies. Similarly, when an earthquake occurs, it produces seismic waves with a variety of frequencies. As the waves travel away from the hypocenter, however, the Earth acts like a filter in that high-frequency waves (waves with short wavelengths) lose energy more rapidly than low-frequency waves (waves with long wavelengths). You've experienced this phenomenon if you've ever heard a car stereo playing loud rap music—when you stand near the car, you hear all the sound frequencies and can make out soprano voices and high-frequency guitar notes, but if you are far away, all you can hear are the low-frequency thump-thump-thumps of the bass guitar and bass drum. Because of this effect, the

frequency content (the variety of wave frequencies) of the earthquake changes with distance from the hypocenter.

Why is frequency content important? Different frequencies of waves cause different amounts of ground acceleration. The frequency of waves is also important because of a phenomenon called resonance. **Resonance** happens when each new wave arrives at just the right time to add more energy to a system. To understand resonance, picture a boy on a swing. If the boy pumps his legs at just the right time, he swings higher; but if not, he slows down. The same phenomenon occurs if you slide a block of Jell-O that is resting on a plate back and forth on a table. If you move the plate too fast, the Jell-O merely trembles, but if your motion is at just the right frequency, resonance begins and the block sways wildly.

Resonance played a major role in accentuating the damage during an earthquake in Mexico City. On September 19, 1985, a magnitude 8.1 earthquake

occurred 350 km away, on the convergent plate boundary along which the Cocos Plate grinds beneath North America. Though the epicenter was far away, ground movements in Mexico City were very intense. That's because Mexico City sits on a thick sequence of unconsolidated lake-bed sediments, exposed when the Spanish conquistadors drained Lake Texcoco. The sedimentary basin, somewhat like a lens, focused earthquake energy on the city. Because of the distance from the epicenter, high-frequency waves had weakened, but because of the nature of the sedimentary basin, low-frequency waves were amplified. These waves had just the right frequency to make buildings between 8 and 18 stories high begin to resonate. In some cases, neighboring buildings slammed together like clapping hands. Engineers had not designed the buildings to accommodate such motion, so many buildings collapsed. Between 8,000 and 30,000 people died, and another 250,000 were left homeless.

---

bridge decks may rise up and slam down on the columns that support them, thereby crushing the columns. Some buildings collapse with their floors piling on top of one another like pancakes in a stack, whereas others simply tip over. The majority of earthquake-related deaths and injuries happen when people are hit by debris or are crushed beneath falling walls or roofs. Aftershocks worsen the problem, because they may topple already weakened buildings, trapping rescuers. During earthquakes, roads, rail lines, and pipelines may also buckle and rupture. If a building, fence, road, pipeline, or rail line straddles a fault rupture, slippage on the fault can crack the structure and separate it into two pieces.

Ground motion can cause water in lakes, bays, reservoirs, and pools to slosh back and forth, in some cases thousands of kilometers from the epicenter. The water's rhythmic movement, known as a seiche, can build up waves almost 10 m high and can last for hours. Seiches capsize small boats and flood

shoreline homes. And if they occur in reservoirs, seiches may wash over and weaken retaining dams.

Landslides

The shaking of an earthquake can cause ground on steep slopes or ground underlain by weak sediment to give way. This movement results in a landslide, the tumbling and flow of soil and rock downslope (see Chapter 16). Landslides occur commonly along the coast of California, for movement on faults has rapidly uplifted this coastline in the past few million years, resulting in the development of steep cliffs. When earthquakes take place, the cliffs collapse, often carrying expensive homes down to the beach below (Fig. 10.27a, b). Such events lead to the misperception that "California will someday fall into the sea." Although small portions of the coastline do collapse, the state as

Did you ever wonder...
will California fall into the sea?

FIGURE 10.26 Examples of earthquake damage due to vibration.

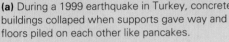

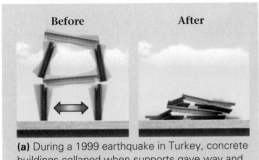

(a) During a 1999 earthquake in Turkey, concrete buildings collaped when supports gave way and floors piled on each other like pancakes.

(b) An elevated bridge tipped over during the 1995 Kobe, Japan, earthquake.

(c) Concrete bridge supports were crushed during the 1994 Northridge earthquake when the overlying bridge bounced up and slammed back down.

(d) A neighborhood of masonry buildings in Armenia collaped during a 1999 earthquake because the walls broke apart.

Turkey

Japan

California

Armenia

a whole remains firmly attached to the continent, despite what Hollywood scriptwriters say.

Landslides into water may cause huge waves. For example, in 1958, a magnitude 8.3 earthquake in southeastern Alaska triggered a landslide at the head of Lituya Bay. The splash from the displaced water washed the forest off the opposite wall of the bay up to an elevation of 516 m.

Sediment Liquefaction

Places where the substrate contains wet sediment can be particularly hazardous during an earthquake. In beds of wet sand or silt, ground shaking causes the sediment grains to try and settle together. But because the spaces (pores) between grains are filled with water, water pressure in the pores increases and pushes the grains apart, and the wet silt or sand becomes a fluid-like slurry—it becomes quicksand. The abrupt loss of strength of a wet, sandy sediment in response to ground shaking is a phenomenon called **liquefaction**, and it can cause major damage during an earthquake. Buildings whose foundations lie in liquefied material may sink or even tip over (**Fig. 10.27c**).

Similarly, in certain types of damp clay, the clay flakes stick together only because of weak bonds between their wet surfaces. When still, the clay behaves like a solid gel, but when it is shaken, the bonds break and the clay transforms into a viscous liquid. Clay that displays this behavior is called thixotropic clay, or quickclay. The behavior of thixotropic clay resembles that of ketchup—if

FIGURE 10.27 Examples of earthquake damage due to slumping and liquefaction.

(b) During the 1964 Alaska earthquake, slumping caused the land to give way beneath parts of Anchorage.

(a) Shaking triggers landslides, which caused a steep slope along the coast of California to collapse, carrying part of a home with it.

The dark area in this aerial photograph is the slump that was Turnagain Heights.

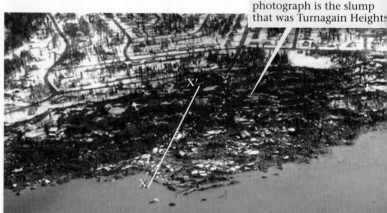

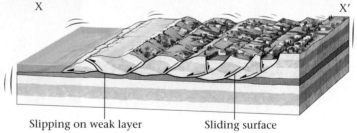

(c) Liquefaction under their foundations caused these apartment buildings in Niigata, Japan to tip over during a 1964 earthquake.

Slipping on weak layer   Sliding surface

(d) During the 1964 Turnagain Heights disaster, a landslide carried a neighborhood out to sea. Slip occured on a weak layer.

you slowly turn a bottle of ketchup over, the ketchup doesn't move, but if you shake the bottle first, the ketchup flows easily.

If shaking weakens sediment, the ground above may give way. Such an event happened during the 1964 Alaska earthquake, when the sediment beneath a housing development in the coastal suburb of Turnagain Heights liquefied (Fig. 10.27d). The land broke into a series of blocks that slid seaward, carrying with it dozens of houses, which were transformed into a jumble of splintered wood and shattered windows.

In some cases, the liquefaction of sand layers below the ground surface makes the sand erupt through holes or cracks in overlying sediment, producing small mounds of sand called sand volcanoes, sand boils, or sand blows (Fig. 10.28a–c). Liquefaction may also cause bedding in unconsolidated sequences of sediment to break up, and ground settling due to underlying liquefaction may cause large fissures to develop in the overlying sediments (Fig. 10.28d, e).

FIGURE 10.28 The ground shaking of an earthquake can disrupt unlithified beds of sediment. Liquefaction of sand can cause ground cracking and sand volcanoes.

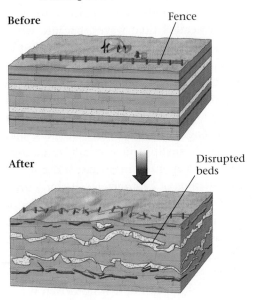

Before Fence

After Disrupted beds

(a) Ground shaking can break up and contort beds of sediment.

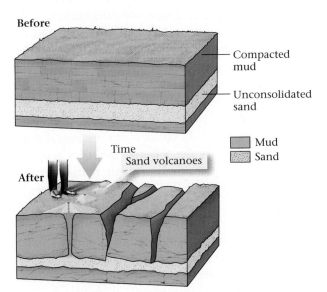

Before — Compacted mud
— Unconsolidated sand

Time
Sand volcanoes

After

Mud
Sand

(b) Liquefaction of a sand layer causes the ground to crack and sand volcanoes to erupt.

(c) A sand volcano (sand blow) formed in a farm field during the 1979 El Centro, California, earthquake.

(d) Liquefaction of a sand layer beneath the dry soil of this field caused the soil to crack and fissures to develop.

(e) During the 2011 Christchurch earthquake, liquefied sand spurted out and spread over the pavement. The process produced open space underground, so the pavement collapsed to form a sinkhole.

Fire

The shaking during an earthquake can make lamps, stoves, or candles with open flames tip over, and it may break wires or topple power lines, generating sparks. As a consequence, areas already turned to rubble, and even areas not so badly damaged, may be consumed by fire. Ruptured gas pipelines and oil tanks feed the flames, sending columns of fire erupting skyward (Fig. 10.29). Firefighters might not even be able to reach the fires, because the doors to the firehouse won't open or rubble blocks the streets. Moreover, firefighters may find themselves without water, for ground shaking and landslides damage water lines.

Once a fire starts to spread, it can become an unstoppable inferno. Most of the destruction of the 1906 San Francisco earthquake, in fact, resulted from fire. For three days,

FIGURE 10.29 Broken gas tanks erupt in fountains of flame after the 2011 Tohoku-oki, Japan, earthquake.

the blaze spread through the city until firefighters contained it by blasting a firebreak. By then, 500 blocks of structures had turned to ash, causing 20 times as much financial loss as the shaking itself. When a large earthquake hit Tokyo in September 1923, fires set by cooking stoves spread quickly through the wood-and-paper buildings, creating an inferno that heated the air above the city. The hot air rose like a balloon, and when cool air rushed in, creating wind gusts of over 100 mph, the wind stoked the blaze and incinerated 120,000 people.

Tsunamis

The azure waters and palm-fringed islands of the Indian Ocean's east coast hide one of the most seismically active plate boundaries on Earth—the Sunda Trench. Along this convergent boundary, the Indian Ocean floor subducts at about 4 cm per year, leading to slip on large thrust faults. Just before 8:00 A.M. on December 26, 2004, the crust above a 1,300-km-long by 100-km-wide portion of one of these faults lurched westward by as much as 15 m. The rupture started at the hypocenter and then propagated north at 2.8 km/s; thus, the rupturing process took 9 minutes. This slip triggered a great earthquake (M_W = 9.3) and pushed the sea floor up by tens of centimeters. The rise of the sea floor, in turn, shoved up the overlying water. Because the area that rose was so broad, the volume of displaced water was immense. As a consequence, tragedy of an unimaginable extent was about to unfold. Water from above the upthrust sea floor began moving outward from above the fault zone, a process that generated a series of giant waves, or tsunamis, traveling at speeds of about 800 km per hour (500 mph)—almost the speed of a jet plane (Fig. 10.30a). *Tsunami* is a Japanese word that translates literally as "harbor wave," an apt name because tsunamis can be particularly damaging to harbor towns. Specifically, geologists now use the term **tsu-**

nami for a large wave produced by displacement of the sea floor; the displacement can be due to an earthquake, submarine landslide (see Chapter 16), or volcanic explosion. In older literature such waves were called tidal waves, because when one arrives, water rises as if a huge tide were coming in, but in fact the waves have nothing to do with daily tidal cycles.

Regardless of cause, tsunamis are very different from familiar, wind-driven storm waves. Large wind-driven waves can reach heights of 10 to 30 meters in the open ocean. But even such monsters have wavelengths of only tens of meters, and thus only contain a relatively small volume of water. In contrast, although a tsunami in deep water may cause a rise in sea level of at most only a few tens of centimeters—a ship crossing one wouldn't even notice—tsunamis have wavelengths of tens to hundreds of kilometers and an individual wave can be several kilometers wide, as measured perpendicular to the wave front. Thus, the wave involves a *huge volume of water.* In simpler terms, we can think of the width of a tsunami, in map view, as being more than 100 times the width of a wind-driven wave. Because of this difference, a storm wave and a tsunami have very different effects when they strike the shore.

When a water wave approaches the shore, friction between the base of the wave and the sea floor slows the bottom of the wave, so the back of the wave catches up to the front, and the added volume of water builds the wave higher (Fig. 10.30b). The top of the wave may fall over the front of the wave and cause a breaker. In the case of a wind-driven wave, the breaker may be tall when it washes onto the beach, but because the wave doesn't contain much water, the wave makes it only partway up the beach before it runs out of water, friction slows it to a stop, and gravity causes the water to spill back seaward. In the case of a tsunami, the wave is so wide that, as friction slows the wave, it builds into a "plateau" of water that can be tens of meters high, many kilometers wide, and hundreds of kilometers long. Thus, when a tsunami reaches shore, it contains so much water that it crosses the beach and just keeps on going, eventually covering a huge area (Fig. 10.30c, d). A single earthquake may generate several tsunamis that arrive on distant shores as much as an hour apart (Fig. 10.30e).

Tsunami damage can be catastrophic. The December 2004 wave struck Banda Aceh, a city at the north end of the island of Sumatra, as it was waking to a beautiful, cloudless day (Fig. 10.31a). First, the sea receded much farther than anyone had ever seen, exposing large areas of reefs that normally remained submerged even at low tide. People walked out onto the exposed reefs in wonder. But then, with a rumble that grew to a roar, a wall of frothing water began to build in the distance and approach land (Fig. 10.31b). Puzzled bathers first watched, then ran inland in panic when the threat became clear. As the tsunami approached shore, friction with the sea floor had slowed it to less than 30 km an hour, but it still

FIGURE 10.30 Tsunamis can be generated when faulting displaces the sea floor over a broad area.

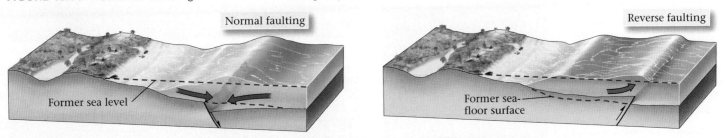

Normal faulting

Reverse faulting

Former sea level

Former sea-floor surface

(a) Tsunamis are generated by sudden displacement of the sea floor during either normal or reverse faulting.

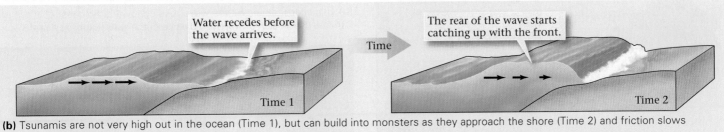

Water recedes before the wave arrives.

The rear of the wave starts catching up with the front.

Time

Time 1

Time 2

(b) Tsunamis are not very high out in the ocean (Time 1), but can build into monsters as they approach the shore (Time 2) and friction slows the base of the wave.

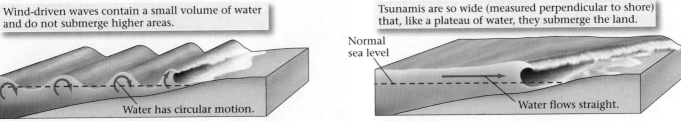

Wind-driven waves contain a small volume of water and do not submerge higher areas.

Tsunamis are so wide (measured perpendicular to shore) that, like a plateau of water, they submerge the land.

Normal sea level

Water has circular motion.

Water flows straight.

(c) Tsunamis are very different from storm-generated waves in that they contain much greater volumes of water.

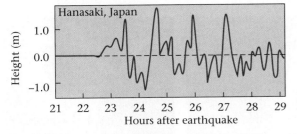

Hanasaki, Japan

(e) A single earthquake can generate several tsunamis, as recorded here by a tidal gauge in Japan, after the great 1960 earthquake in Chile.

(d) A tsunami inundated the shore of Valdez, Alaska, in 1964, and washed away the snow cover. The town's port was destroyed.

moved faster than people could run. In places, the wave front reached heights of 15 to 30 m (45 to 100 feet) as it slammed into Banda Aceh (Fig. 10.31c).

The impact of the water ripped boats from their moorings, snapped trees, battered buildings into rubble, and tossed cars and trucks like toys. And the water just kept coming, eventually flooding low-lying land as far as about 7 km inland (Fig. 10.31d). It drenched forests and fields with salt water (deadly to plants) and buried fields and streets with up to a meter of sand and mud. Eventually the water slowed and then began to rush back to the shore, but at Banda Aceh, at least two more tsunamis struck before the first one had entirely receded, so the water remained high for some time. When the water level finally returned to normal, a jumble of flotsam, as well as the bodies of unfortunate victims, floated out to sea and drifted away.

FIGURE 10.31 The great Indian Ocean tsunami of 2004.

(a) A devastating tsunami was triggered by an earthquake off Sumatra. Three hours later, the leading wave struck the coasts of Sri Lanka and India. A computer model shows the wave.

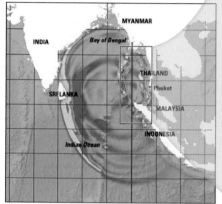

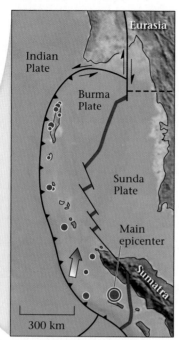

Colors represent wave height—yellow is highest. There were several waves.

The earthquake was caused by subduction at a trench. Red dots are epicenters.

(b) This snapshot shows the wave rushing toward the coast of Sumatra. Recession of water in advance of the wave exposed a reef.

(c) The wave blasts through a grove of palm trees as it strikes the coast of Thailand.

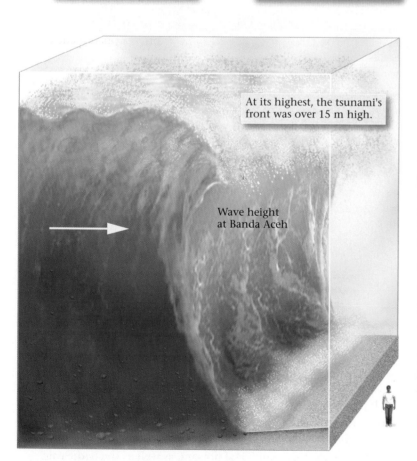

At its highest, the tsunami's front was over 15 m high.

Wave height at Banda Aceh

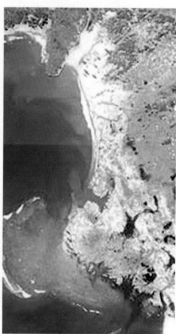

(d) Satellite photos of the Indonesian province of Aceh before and after the tsunami struck. Note that the city was washed away and the beach vanished.

Geologists refer to the tsunami that struck Banda Aceh as a near-field (or local) tsunami, because of its proximity to the earthquake. But the horror of Banda Aceh was merely a preamble to the devastation that would soon visit other stretches of Indian Ocean coast. Far-field (or distant) tsunamis crossed the ocean and struck Sri Lanka 2.5 hours after the earthquake, the coast of India half an hour after that, and the coast of Africa, on the west side of the Indian Ocean, 5.5 hours after the earthquake. Coastal towns vanished, fishing fleets sank, and beach resorts collapsed into rubble. In the end, perhaps more than 230,000 people died that day.

FIGURE 10.32 Damage due to the 2011 Tohoku-oki tsunami.

(a) The complete destruction of a Japanese coastal town by a tsunami, which followed the Tohoku-oki, Japan earthquake of March 2011.

After

Before

The power plant was built next to the shore.

Reactor building 4

(b) Each cubic building houses a reactor of the Fukushima nuclear power plant. The tsunami washed over the sea walls and inundated the plant to a depth of 14 m (46 ft), destroying power to the cooling pumps. Hydrogen explosions destroyed the reactor buildings.

The 2004 Indian Ocean event remains etched in people's minds because of the immense death toll and the nonstop news coverage. But it is not unique. Tsunamis generated by the great Chilean earthquake of 1960 ($M_W = 9.5$) destroyed coastal towns of South America and crossed the Pacific, causing a 10.7-m-high wall of water to strike Hawaii 15 hours later. When, 21 hours after the earthquake, the tsunami reached Japan, it flattened coastal villages and left 50,000 people homeless. The tsunami following the 1964 Good Friday earthquake in Alaska destroyed ports at Valdez and Kodiak.

A tsunami struck Japan again, soon after the 2011 magnitude 9 Tohoku-oki earthquake. Its rush over the shoreline and its dramatic inundation of coastal areas was vividly captured in high-definition video that was seen throughout the world, generating a new level of international awareness. The huge earthquake's epicenter lay about 130 km (80 miles) offshore, so groundshaking was not extremely intense on land. But, since tsunamis travel so fast, the first waves reached the shore only about 10 minutes after the earthquake struck.

Even though warning sirens were blaring, there was not enough time for all residents of coastal towns to escape. The earthquake displaced an area of sea floor that measured 600 km (parallel to the coast) by 200 km (perpendicular to the coast). This produced a 10 m-high wave (locally, where focused, the water rose as high as 30 meters). Though much of the coast was fringed by seawalls, they proved to be a minor impediment to the advance of the wave. Water picked up boats and ships in the harbor and flung them over walls. On land the wave smashed through houses and picked up debris. It tumbled cars, trucks, and trains as if they were pebbles in a mountain stream. Racing inland at 30 km/h, the wave erased whole towns, submerging airports and fields (see Fig. 10.2b). As the churning wave picked up dirt and debris, it became a viscous slurry, resembling a volcanic lahar, moving with such force that nothing could withstand its impact. When the wave finally ran out of water, it receded to the sea, carrying debris and victims with it. When equilibrium returned, the coast of Japan for several kilometers out to sea was covered with floating wood, overturned boats, plastic tubs, and countless items of everyday life. The devastation of coastal towns was so complete that they looked as though they had been struck by nuclear bombs (Fig. 10.32a).

But the catastrophe was not over. The wave had also hit a nuclear power plant. Though the plant had withstood ground shaking and had automatically shut down, its radioactive core still needed to be cooled by water in order to remain safe. The tsunami not only destroyed power lines, cutting the plant off from the electrical grid, but it also eliminated backup diesel generators and cut water lines. Thus, cooling pumps stopped functioning. Eventually, water surrounding the heat-producing radioactive core of the reactors, as well as the water cooling spent fuel, boiled away. Some of the water separated into hydrogen and oxygen gas. These gases exploded, and ultimately, the integrity of the nuclear plant was breached (Fig. 10.32b). Radioactivity entered the environment.

Because of the danger of tsunamis, predicting their arrival can save thousands of lives. A tsunami warning center in Hawaii keeps track of earthquakes around the Pacific and uses data relayed from tide gauges and sea-floor pressure gauges to determine whether a particular earthquake has generated a tsunami. If observers detect a tsunami, they flash warnings to authorities around the Pacific. In recent years, to help in the effort, researchers have used computer models that predict how tsunamis propagate. Unfortunately, no warning system existed in the Indian Ocean in 2004, and even though Hawaiian observers detected the earthquake and realized that a tsunami was likely, they did not have the means to contact many local authorities. Even if they had, affected towns had no evacuation plans in place. A multinational team has worked to remedy this situation.

Disease

Once the ground shaking and fires have stopped, disease may still threaten lives in an earthquake-damaged region. Earthquakes cut water and sewer lines, destroying clean water supplies and exposing the public to bacteria, and they cut transportation lines, preventing food and medicine from reaching the area. The severity of such problems depends on the ability of emergency services to cope.

Take-Home Message

- Earthquakes cause damage in many ways. Ground shaking, landslides, sediment liquefaction, and fire lead to the destruction of buildings and loss of life.
- Tsunamis are waves that can be generated when the sea floor suddenly moves up or down; the movement displaces immense amounts of water.
- A tsunami can inundate coastal areas.

THINK: Is ground shaking the major cause of loss of life in all earthquakes?

10.8 CAN WE PREDICT THE "BIG ONE"?

We have seen that large earthquakes occurring near population centers cause catastrophe (See for Yourself J, p. S-18). Needless to say, many lives could be saved if only it were possible to know exactly when and where an earthquake will happen, so people could evacuate dangerous buildings, turn off gas and electricity, and, with more warning, build stronger structures.

Can seismologists predict earthquakes? The answer depends on the time frame of the prediction. With our present understanding of the distribution of seismic zones and the frequency at which earthquakes occur, we can make long-term predictions (on the time scale of decades to centuries). For example, with some certainty, we can say that a major earthquake will probably rattle California during the next 100 years, and that a major earthquake probably won't strike central Canada during the next 10 years. But despite extensive research, seismologists *cannot* make accurate short-term predictions (on the time scale of hours to weeks or even years). Thus we cannot say, for example, that an earthquake will happen in Montreal at 2:43 P.M. on January 17.

In this section, we look at the scientific basis of both long- and short-term predictions and consider the consequences of a prediction. Seismologists refer to studies leading to predictions as seismic-risk, or seismic-hazard assessment. On the basis of this work, they produce maps that assign levels of seismic risk to different regions.

Long-Term Predictions

When making a prediction, we use the word *probability*, because a prediction only gives the likelihood of an event. For example, a seismologist may say, "The probability of a major earthquake occurring in the next 20 years in this state is 20%." This sentence implies that there's a 1-in-5 chance that the earthquake will happen during a 20-year period. Urban planners and civil engineers can use long-term predictions to help create building codes for a region—codes requiring stronger, more expensive buildings make sense for regions with greater seismic risk. They may also use predictions to determine whether to build vulnerable structures such as nuclear power plants, hospitals, or dams in potentially seismic areas. Seismologists base long-term earthquake predictions on two pieces of information: the identification of seismic zones and the **recurrence interval** (the average time between successive events).

To identify a seismic zone, seismologists produce a map showing the epicenters of earthquakes that have happened during a set period of time (say, 30 years). Clusters or belts of epicenters define the seismic zone. The basic premise of long-term earthquake prediction can be stated as follows: a region in which there have been many earthquakes in the past

FIGURE 10.33 Identifying recent fault movement and recurrence interval.

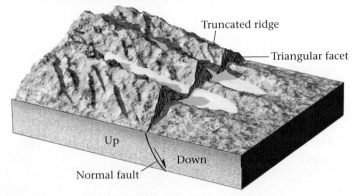

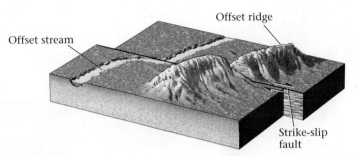

(a) Displacement on a normal fault truncates ridges, creating triangular facets.

(b) A strike-slip fault may offset ridges and streams.

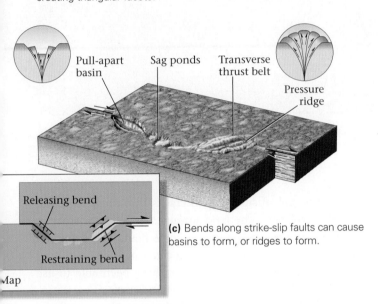

(c) Bends along strike-slip faults can cause basins to form, or ridges to form.

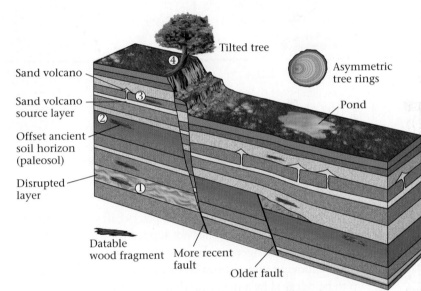

(d) Earthquake events are represented by a layer of disrupted bedding, an offset ancient soil horizon (or paleosol), a layer of sand volcanoes, and a bent tree. Dating several events helps define the recurrence interval.

will likely experience more earthquakes in the future. Seismic zones, therefore, are regions of greater seismic risk. This doesn't mean that a disastrous earthquake can't happen far from a seismic zone—they can and do—but the risk that an event will happen in a given time window is less.

Epicenter maps can be produced with data from only the past fifty years or so, because before that time seismologists did not have enough seismographs to locate epicenters accurately. Fortunately, geologists can also gain insight into seismic risk by examining landforms for evidence of recent faulting. For example, the presence of a distinct fault scarp in a landscape indicates that faulting has happened so recently that erosion has not yet had time to grind away the evidence (Fig. 10.33a–c).

To determine the recurrence interval for large earthquakes in a seismic zone, seismologists must determine when large earthquakes happened there in the past. Since the historical record does not provide information far enough back in time,

they study geologic evidence for great earthquakes. For example, a trench cut into sedimentary strata near a fault may reveal layers of sand volcanoes and disrupted bedding in the stratigraphic record. Each layer, whose age can be determined by using radiocarbon dating of plant fragments, records the time of an earthquake (Fig. 10.33d). By calculating the number of years between successive events and taking the average, seismologists obtain the recurrence interval. As an example, imagine that disrupted layers formed 260, 820, 1,200, 2,100, and 2,300 years ago. We can say that the recurrence interval between events is about 510 years. Note that a recurrence interval does *not* specify the exact number of years between events, only the average number. Since stress builds up over time on a fault, the probability that an earthquake will happen in any given year probably increases as time passes. Information on a recurrence interval allows seismologists to refine regional maps illustrating seismic risk (Fig. 10.34a, b).

FIGURE 10.34 Examples of seismic-hazard maps.

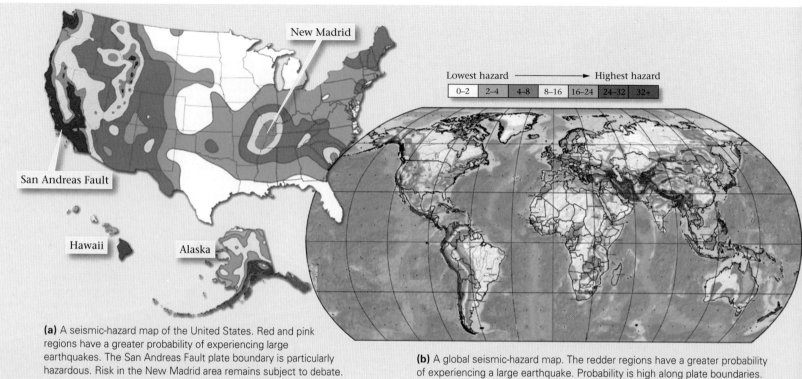

(a) A seismic-hazard map of the United States. Red and pink regions have a greater probability of experiencing large earthquakes. The San Andreas Fault plate boundary is particularly hazardous. Risk in the New Madrid area remains subject to debate.

(b) A global seismic-hazard map. The redder regions have a greater probability of experiencing a large earthquake. Probability is high along plate boundaries.

Some seismologists suspect that places called seismic gaps, where a known active fault has not slipped for a long time, may be particularly dangerous. In a seismic gap, either the fault must be moving nonseismically, or the stress must be building up to be released by a major earthquake in the future.

Short-Term Predictions

Short-term predictions, which could lead to such precautions as evacuating dangerous buildings, shutting off gas and electricity, and readying emergency services, are not reliable and may never be. Nevertheless, there are clues to imminent earthquakes, and seismologists have been working hard to understand them. The first clue comes from the detection of foreshocks. A swarm (a cluster of events during a short period of time) of foreshocks may indicate the cracking that precedes a major rupture along a fault zone. In this regard, foreshocks are analogous to the cracking noises you hear just before a tree limb breaks off and falls down. But foreshocks do not always occur, and even if they do, they usually can be identified only in hindsight, because they may be indistinguishable from other small earthquakes.

Another possible data source for short-term predictions comes from precise laser surveying of the ground. Before an earthquake, a region of crust may undergo distortion, either in response to the buildup of elastic strain in the rock or because of the development of small, open cracks that cause the crust to increase in volume. The land surface may bulge or sink, or straight lines on the ground may become bent. The detection of such movements by laser surveying can hint at an upcoming earthquake. More recently, researchers have been able to use satellite data to detect distortion of the land surface.

Other changes that have been explored but have not been confirmed as precursors of earthquakes include the following: changes in the water level in wells; appearance of gases, such as radon or helium, in wells; changes in the electrical conductivity of rock underground; and unusual animal behavior. Believers in these proposed clues suggest that they all reflect the occurrence of cracking in the crust prior to an earthquake, but most investigators remain skeptical.

Geologists have also begun to use computer models of stress to predict where stress buildups may lead to earthquakes. Use of these "stress-triggering models" has provided substantial insight into seismic activity along the North Anatolian Fault, in Turkey (Fig. 10.35a). This fault is a large strike-slip fault along which Turkey slips westward. (Turkey is essentially being squeezed like a watermelon seed out of the way of Arabia, which is moving northward and is colliding with Asia.) Earthquakes happen again and again along the fault—in fact, ruins of several ancient cities lie along the trace of the fault. Since 1939, 11 major earthquakes have

FIGURE 10.35 Major earthquakes of the last century along the Anatolian fault of Turkey have occurred roughly in sequence from east to west.

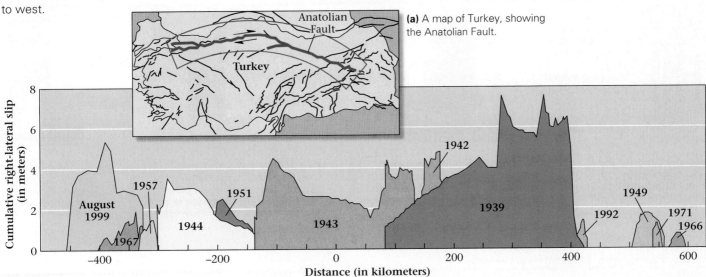

(a) A map of Turkey, showing the Anatolian Fault.

(b) A graph representing regions of the fault; the vertical axis represents the amount of slip.

occurred along the fault—each ruptured a different portion of the fault (Fig. 10.35b). Overall, there has been a westward progression of faulting. By modeling the stress changes resulting from each earthquake, geologists identified places where stress accumulates, and predicted that the next big earthquake would happen toward the western end of the fault. It did. In August 1999, a magnitude (M_W) 7.4 earthquake devastated Izmit, a city just south of Istanbul, killing more than 17,000 people and leaving more than 600,000 homeless. Recently, seismologists have begun to explore the possibility that shock waves from a large earthquake may trigger other earthquakes far away.

As long as short-term predictions remain questionable, emergency service planners must ask, What if a prediction is wrong? Should schools and offices be shut because of a prediction? Should millions of dollars be spent to evacuate people? Should a city be deserted, allowing for the possibility of looters? Should the public be notified, or should only officials be notified, creating a potential for rumor? If the prediction proves wrong, can seismologists be sued? No one really knows the answers to these questions.

Even though seismologists *cannot* provide weeks to years of advance notice that an earthquake will happen, they have successfully developed an **earthquake warning system**, meaning an automatic communications network that broadcasts information about a large earthquake taking place seconds to tens of seconds *before* damaging vibrations arrive. Though the warning time seems very short, it can be sufficient for automatic controls to start braking high-speed trains,

to cut off flow on gas lines, and to reroute electrical power on the grid. The system could also trigger sirens to provide people with just enough time to take cover. At the heart of an earthquake warning system is a computer-linked network of seismographs which can almost instantly calculate the location and potential size of an earthquake from P-waves—in some cases, the calculation can be complete before the fault has even stopped rupturing. If the calculation implies that the earthquake will be large enough to be dangerous, the computer sends a signal—which travels at the speed of light—to automatic control devices. Some earthquake-prone regions (e.g., Japan and California) have already installed commercial earthquake warning systems and have them operating.

> ## Take-Home Message
>
> - Seismologists identify seismic belts based on the distribution of epicenters. The risk of an earthquake is greater in a seismic belt.
> - The recurrence interval, the average time between earthquakes, can be determined by dating past earthquakes.
> - Long-term prediction, based on recurrence interval, gives the probability that an earthquake will happen in a given year.
> - Short-term prediction is pretty much impossible. However, methods to provide early warning of earthquakes, before the main vibrations arrive, are promising.
>
> **THINK:** What is the relation between recurrence interval and the likelihood that an earthquake will happen in a given location in a given year?

FIGURE 10.36 Preventing damage and injury during an earthquake.

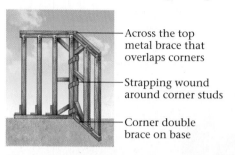
Across the top metal brace that overlaps corners
Strapping wound around corner studs
Corner double brace on base

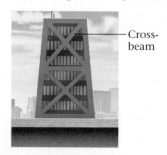
Cross-beam

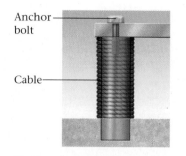
Anchor bolt
Cable

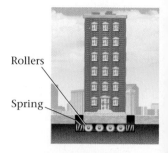
Rollers
Spring

Adding corner struts, braces, and connectors can substantially strengthen a wood-frame house.

Buildings are less likely to collapse if they are wider at the base and if crossbeams are added for strength.

Wrapping a bridge's support columns in cable and bolting the span to the columns will prevent the bridge from collapsing so easily.

Placing buildings on rollers or shock absorber lessens the severity of the vibrations.

(a) Damage can be prevented if buildings are designed to withstand vibration.

Unreinforced building: insufficient shear strength
Reinforced building: Sufficient shear strength

(b) An unreinforced building will shear side-to-side in a way that causes floor to shift out of alignment.

(c) Buoys can detect a tsunami in the open ocean, so people on land can be warned.

(d) If an earthquake strikes, take cover under a sturdy table near a wall.

10.9 EARTHQUAKE ENGINEERING AND ZONING

The loss of human life from an earthquake of a given size varies widely and depends on a number of factors. The most important factors include the proximity of an epicenter to a population center, the depth of the hypocenter, the style of construction in the epicentral region, whether the earthquake occurred in a region of steep slopes or along the coast, whether building foundations are on solid bedrock or on weak substrate, whether the earthquake happened when people were outside or inside, and whether the government was able to provide emergency services promptly.

For example, a 1988 earthquake in Armenia was not much bigger than the 1971 San Fernando earthquake in southern California, but it caused almost 500 times as many deaths (24,000 versus 65). The difference in death toll reflects differences in the style and quality of construction and the characteristics of the substrate. The unreinforced concrete-slab buildings and masonry houses of Armenia collapsed, whereas the structures in California had, by and large, been erected according to building codes that take into account stresses caused by earthquakes. Most flexed and twisted but did not fall down and crush people. The terrible 1976 earthquake in T'ang-shan, China, killed over a quarter of a million people because the ground beneath the epicenter had been weakened by coal mining and collapsed, and because buildings were poorly constructed. During the M_W-7.9 2008 Sichuan earthquake in China, 70,000 people died and almost 5 million were left homeless because of building failures. Mexico City's 1985 earthquake proved disastrous because the city lies over a sedimentary basin whose composition and bowl-like shape focused seismic energy (see Box 10.2). During the 1989 Loma Prieta quake in California, portions of Route 880 in Oakland that were built on a weak substrate collapsed, whereas portions built on bedrock or gravel remained standing.

We can mitigate or diminish the consequences of earthquakes by taking sensible precautions. Clearly, earthquake engineering (the designing of buildings that can withstand shaking) and earthquake zoning (the determination of where land is stable and where it might collapse) can help save lives and property. In regions prone to large earthquakes, buildings and bridges should be constructed so they are able to withstand vibrations without collapsing (Fig. 10.36a, b). They should be somewhat flexible so that ground motions can't crack them, but they should have sufficient bracing so movements don't become too severe. Also, supports should be strong enough to maintain loads far in excess of the loads caused by their static (nonmoving) weight. Wrapping steel cables around bridge support columns makes them many times stronger. Bolting the bridge spans to the top of a column prevents the spans from bouncing off. Bolting buildings to foundations keeps them in place, and adding diagonal braces to frames keeps them from twisting and shearing too much.

Certain kinds of construction should be avoided. For example, concrete-block, unreinforced concrete, and unreinforced brick buildings crack and tumble under conditions in which wood-frame, steel girder, or reinforced concrete buildings remain standing. Traditional heavy, brittle tile roofs shatter and bury the inhabitants inside, whereas sheet-metal or asphalt-shingle roofs do not. Loose decorative stone and huge open-span roofs also do not fare well when vibrated. Of note, inadequate structures can be made safer by **seismic retrofitting**, the process of strengthening existing buildings in potentially seismically hazardous areas. Examples of retrofitting include adding vibration dampness to foundations, adding braces, jacketing support columns, and coating masonry with resins. In some cases, substrates can be strengthened by draining water.

Similarly, developers should avoid construction on land underlain by weak mud that could liquefy. They should not build on top of, on, or at the base of steep escarpments because the escarpments could fail and produce landslides, and they should avoid locating large population centers downstream of dams (which could crack and collapse, causing a flood). And they should avoid constructing vulnerable buildings directly over active faults, because fault movement could crack and destroy the buildings. Cities in seismic zones need to draw up emergency plans to deal with disaster. Communication centers should be situated in safe localities, and strategies need to be implemented for providing supplies under circumstances where roads may be impassable. And in coastal areas, tsunami warning systems need to be implemented (Fig. 10.36c).

Finally, communities and individuals should learn to protect themselves during an earthquake. In your home, keep emergency supplies, bolt bookshelves to walls, strap the water heater in place, install locking latches on cabinets, know how to shut off the gas and electricity, know how to find the exit, have a fire extinguisher handy, and know where to go to find family members. Schools and offices should have earthquake-preparedness drills. When an earthquake strikes, stay away from buildings. If you are trapped inside, a heavy table or solid door frame may provide some protection (Fig. 10.36d). As long as lithosphere plates continue to move, earthquakes will continue to shake. But we can learn to live with them.

> ## Take-Home Message
>
> - People can reduce the death and damage due to earthquakes by taking common sense precautions.
> - Builders should avoid construction on weak substrates and should reinforce buildings to withstand earthquake vibrations.
> - Existing buildings can be strengthened by retrofitting.
>
> **THINK:** What factors influence the degree of devastation during an earthquake?

Chapter Summary

- Earthquakes are episodes of ground shaking. Earthquake activity is called seismicity.

- Most earthquakes happen when rock slips during faulting. A fault is a fracture on which sliding occurs. The place where rock breaks and earthquake energy is released is called the hypocenter (focus), and the point on the ground directly above the hypocenter is the epicenter.

- Active faults are faults on which movement is likely. Inactive faults ceased being active long ago, but can still be recognized because of the displacement across them. Dis-

placement on active faults that intersect the ground surface may yield a fault scarp.

- During fault formation, rock elastically bends, then cracks. Eventually, cracks link to form a throughgoing rupture on which sliding occurs. When this happens, the rock breaks and vibrates, and this generates an earthquake.

- Faults exhibit stick-slip behavior, in that they move in sudden increments. Most earthquakes happen when stress overcomes friction on a preexisting fault, and the fault slips again.

- Earthquake energy travels in the form of seismic waves. Body waves, which pass through the interior of the Earth, include P-waves (compressional waves) and S-waves (shear waves).

Surface waves, which pass along the surface of the Earth, include R-waves (Rayleigh waves) and L-waves (Love waves).

- We can detect earthquake waves by using a seismograph.
- Seismograms demonstrate that different earthquake waves arrive at different times, because they travel at different velocities. Using the difference between P-wave and S-wave arrival times, seismologists can pinpoint the epicenter location.
- The Mercalli Intensity Scale is based on documenting the damage caused by an earthquake.
- Magnitude scales, such as the Richter scale, are based on measuring the amount of ground motion, as indicated by traces of waves on a seismogram. The moment-magnitude scale takes into account the amount of slip, the length and depth of the rupture, and the strength of the ruptured rock.
- A magnitude 8 earthquake yields about 10 times as much ground motion as a magnitude 7 earthquake and releases about 32 times as much energy.
- Most earthquakes occur in seismic belts, or zones, of which the majority lie along plate boundaries. Intraplate earthquakes happen in the interior of plates. Different kinds of earthquakes happen at different kinds of plate boundaries.
- Earthquake damage results from ground shaking (which can topple buildings), landslides (set loose by vibration), sediment liquefaction (the transformation of compacted clay into a muddy slurry), fire, and tsunamis (giant waves).
- Seismologists predict that earthquakes are more likely in seismic zones than elsewhere, and can determine the recurrence interval (the average time between successive events) for great earthquakes. But it may never be possible to pinpoint the exact time and place at which an earthquake will happen.
- Earthquake hazards can be reduced with better construction practices and zoning, and by educating people about what to do during an earthquake.

GEOPUZZLE REVISITED

Most earthquakes happen when rock abruptly breaks during the formation of a new fault, or during renewed slip on an existing fault. Stress drives the process, building up slowly until it exceeds the rock's strength. Major seismic belts (regions of frequent earthquake activity) coincide with plate boundaries, collision zones, and rifts, and thus delineate regions where relative plate motion is being accommodated. A few seismic belts, however, do occur along ancient, weak faults within plates. Seismic risk is clearly greater in seismic belts, for stress builds up more rapidly in these regions. But it is not possible to predict exactly where or when an earthquake will occur within a zone.

Guide Terms

aftershocks (p. 294)
body waves (p. 296)
compressional waves (p. 296)
displacement (p. 291)
earthquake (p. 287)
earthquake warning system (p. 325)
elastic-rebound theory (p. 294)
epicenter (p. 289)
fault (p. 287)
fault creep (p. 296)
fault scarp (p. 291)
foreshocks (p. 294)
friction (p. 294)
hypocenter (p. 289)
induced seismicity (p. 312)
intensity (p. 300)
intraplate earthquake (p. 309)
liquefaction (p. 315)
magnitude (p. 301)

Modified Mercalli Scale (p. 300)
moment magnitude scale (p. 303)
recurrence interval (p. 322)
resonance (p. 314)
Richter scale (p. 302)
seismic belt (p. 305)
seismicity (p. 289)
seismic retrofitting (p. 327)
seismic waves (p. 296)
seismogram (p. 299)
seismograph (p. 297)
shear waves (p. 296)
stick-slip behavior (p. 294)
stress (p. 291)
surface waves (p. 296)
travel-time curve (p. 300)
tsunami (p. 318)
Wadati-Benioff zone (p. 306)

Review Questions

1. Compare normal, reverse, and strike-slip faults.
2. Describe elastic-rebound theory and the concept of stick-slip behavior.
3. Describe the motions of the four types of seismic waves. Which are body waves, and which are surface waves?
4. Explain how the vertical and horizontal components of an earthquake are detected on a seismograph.
5. Explain the differences among the scales used to describe the size of an earthquake.
6. How does seismicity on mid-ocean ridges compare with seismicity at convergent or transform boundaries? Do all earthquakes occur at plate boundaries?
7. What is a Wadati-Benioff zone, and why was it important in understanding plate tectonics?
8. Describe the types of damage caused by earthquakes.
9. What is a tsunami, and why does it form?
10. Explain how liquefaction occurs in an earthquake and how it can cause damage.
11. How are long-term and short-term earthquake predictions made? What is the basis for determining a recurrence interval, and what does a recurrence interval mean?
12. What types of structure are most prone to collapse in an earthquake? What types are most resistant to collapse?

On Further Thought

13. Is seismic risk greater in a town on the west coast of South America or in one on the east coast? Explain your answer.

14. The northeast-trending Ramapo Fault crops out north of New York City near the east coast of the United States. Precambrian gneiss forms the hills to the northwest of the fault, and Mesozoic sedimentary rock underlies the lowlands to the southeast. (You can see the fault on *Google Earth*™ by going to Lat 41° 10' 21.12" N Long 74° 5' 12.36" W. Once you're there, tilt the image and fly northeast along the fault.) Where the fault crosses the Hudson River, there is an abrupt bend in the river. A nuclear power plant was built near this bend. There are numerous bogs, containing sediments deposited over the past several thousand years, on the surface of the basin. Geologic studies suggest that the Ramapo Fault first formed during the Precambrian, was reactivated during the Paleozoic, and was the site of major displacement during the Mesozoic rifting that separated North America from Africa. Imagine that you are a geologist with the task of determining the seismic risk of the fault. What evidence of present-day or past seismic activity could you look for? What does the long-term geologic history of the fault imply about its strength?

15. On the seismogram of an earthquake recorded at a seismic station in Paris, France, the S-wave arrives six minutes after the P-wave. On the seismogram obtained by a station in Mumbai, India, for the same earthquake, the difference between the P-wave and S-wave arrival times is 4 minutes. Which station is closer to the epicenter? From the information provided, can you pinpoint the location of the epicenter? Explain.

 For more resources, including animations, quizzes, and Norton's GeoTours, go to **wwnorton.com/studyspace**.

 If your instructor assigns exercises in SmartWork, log in at **smartwork.wwnorton.com**.

ANOTHER VIEW A model showing the predicted height of the tsunami in the Pacific Ocean generated by the 2011 Tohoku-oki earthquake. Height depends on both water depth, and on distance from the source. Note that the height generally decreases away from the source off eastern Japan, but increases near coasts.

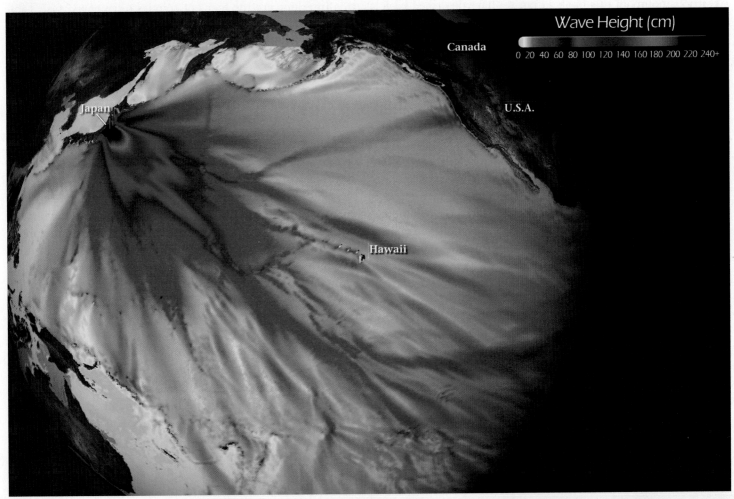

The Earth's Interior— Revisited: Seismic Layering, the Magnetic Field, and Gravity

D.1 INTRODUCTION

At various points in this book so far, we have described basic features of the Earth's interior and have noted how characteristics of the interior affect the Earth's surface and the space around our planet. We needed to provide a rudimentary understanding of the interior in order to be able to discuss several topics, such as plate tectonics, that were introduced in early chapters. But to provide the *rest of the story* about our planet's interior, and its influence on its surroundings, we had to wait until now, after Chapter 10's discussion of earthquakes and seismic waves, for constraints on important details about the interior can come only from the study of how seismic waves pass through the Earth.

The purpose of this Interlude is to help you complete a modern image of the Earth's interior, now that you have the background to understand how this image came to be. We begin by examining how seismic waves interact with layer boundaries, in a general sense, and then discuss the discoveries of specific layer boundaries within the Earth and of three-dimensional variations within each layer. With knowledge of the seismically defined character of the Earth's interior at hand, we can turn our attention to the gravity field of the Earth and examine why gravitational pull varies with location. We conclude by reconsidering the Earth's magnetic field, this time focusing on the question of why the magnetic field exists. Interlude D ties together material covered in Chapters 2, 3, and 10, so we recommend that you

read (or reread) relevant parts of these chapters before proceeding. At a more advanced level, the subjects covered in this Interlude are the focus of courses in *geophysics* (see Table P.1).

D.2 SETTING THE STAGE FOR SEISMIC STUDY OF THE INTERIOR

Strange as it may seem, researchers had developed an understanding of Solar System structure long before they had any clear concept of the Earth's *internal* structure. Why? We can see light years into space just by looking up; but since rock is opaque, our eyes cannot see below the surface of our planet's skin. Tunnels and drill holes don't help much, for they literally only scratch the surface—the deepest mine, a gold mine in South Africa, reaches a depth of 3.9 km below the surface (only 0.06% of the Earth's radius), and the deepest drill hole penetrates to a depth of 12.3 km beneath northwest Russia (only 0.19% of the Earth's radius). Thus, to study the Earth's interior, geologists have had to utilize several types of indirect measurements.

As we discussed in Chapter 2, the first clues to what's inside this planet came from measurements of the Earth's overall mass and shape. These measurements led geologists to conclude that the Earth consists of three concentric layers that differ from each other in terms of relative density (Fig. D.1a).

From the surface down, these are (1) the crust (of low density); (2) the mantle (of intermediate density); and (3) the core (of very high density).

To gain further insight into the nature of the crust, geologists examined samples of rocks exposed on land or samples dredged and drilled from the sea floor. To characterize the mantle and core, geologists studied meteorites, because meteorites come from the interiors of planetesimals and protoplanets similar to those from which the Earth formed. They also analyzed igneous rocks solidified from mantle melts, because the composition of magma provides information about the composition of the magma's source. Overall, such work has led geologists to conclude that, beneath a veneer of sediment, the oceanic crust consists dominantly of mafic rock (basalt), whereas the continental crust consists of a variety of igneous and metamorphic rocks ranging from mafic to felsic in composition. The deeper interior differs from both kinds of crust in that the mantle consists of ultramafic rock (peridotite), and the core consists of iron alloy (metal) (Fig. D.1b).

To go beyond this basic understanding, researchers searched for a tool that could provide an actual image of the interior. Study of seismic waves provides that tool. By measuring how fast seismic waves travel through the Earth, and how the waves bend and/or reflect as they travel, researchers can determine the exact depth to the crust-mantle boundary and to the mantle-core boundary, and they can identify sublayers within the crust, mantle, and core. To understand these discoveries, we first examine how seismologists describe the direction in which seismic waves move and then consider how seismic waves interact with layer boundaries.

D.3 THE MOVEMENT OF SEISMIC WAVES THROUGH THE EARTH

Wave Fronts and Travel Times

Energy travels from one location to another in the form of waves. For example, the impact of a pebble on the surface of a pond produces water waves that eventually can cause a stick meters away to bob up and down, and the takeoff of a jet produces sound waves that rattle the windows of nearby houses. A sudden rupture of intact rock or the sudden frictional slip of rock on a fault produces seismic waves. These waves transmit energy outward from the point of rupture—the earthquake's hypocenter or focus—in all directions at once, eventually reaching distant seismographs.

A single earthquake produces many kinds of waves, distinguished from each other by where and how they move (see Chapter 10). Surface waves (R-waves and L-waves) propagate along the planet's surface, whereas body waves (P-waves and S-waves) pass through the interior. P-waves are compressional, and resemble the waves generated when you push a spring back and forth in a direction parallel to the length of the spring. S-waves are shear, and resemble the waves generated when you wiggle a rope back and forth perpendicular to the length of the rope (see Fig. 10.10b).

The boundary between the rock through which a wave has passed and the rock through which it has not yet passed is called a **wave front**. In 3-D, a wave front expands outward from the earthquake focus like a growing bubble. We can represent a succession of waves in a drawing by a series of concentric wave fronts. The changing position of an imaginary point on a wave front as the front moves through rock is called a **seismic ray**. Seismic rays are lines drawn perpendicular to wave fronts; each point on a curving wave front follows a slightly different ray (Fig. D.2a). The time it takes for a wave to travel from the focus to a seismograph station along a given ray is the **travel time** along that ray.

The ability of a seismic wave to travel through a certain material, and the velocity at which it travels, depend on the character of the material. Factors such as *density* (mass per unit volume), *rigidity* (how stiff or resistant to bending a material is), and *compressibility* (how easily a material's volume changes in response to

FIGURE D.1 Simplified images of the Earth's interior.

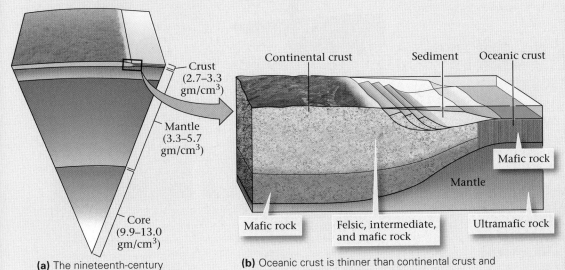

Crust (2.7–3.3 gm/cm^3)

Mantle (3.3–5.7 gm/cm^3)

Core (9.9–13.0 gm/cm^3)

(a) The nineteenth-century three-layer image of the Earth.

Continental crust   Sediment   Oceanic crust

Mafic rock

Mantle

Mafic rock   Felsic, intermediate, and mafic rock   Ultramafic rock

(b) Oceanic crust is thinner than continental crust and has a different composition.

squashing) all affect seismic-wave movement. Studies of seismic waves reveal the following:

■ Seismic waves travel at different velocities in different rock types (Fig. D.2b). For example, P-waves travel at 8 km per second in peridotite (an ultramafic igneous rock), but at only 3.5 km per second in sandstone (a porous sedimentary rock). Therefore, waves accelerate or slow down if they pass from one rock type into another. P-waves in rock travel about 10 to 25 times faster than sound waves in air. But even at this rate, they take about 20 minutes to pass entirely through the Earth along a diameter.

FIGURE D.2 The propagation of earthquake waves.

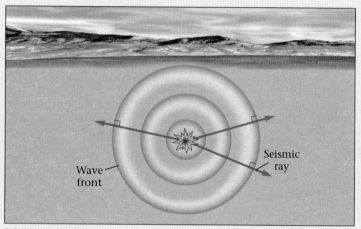

(a) An earthquake sends out waves in all directions.

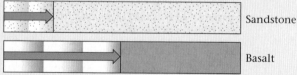

(b) Seismic waves travel at different velocities in different rock types. After a given time, the wave will have traveled farther in basalt than in sandstone.

(c) P-waves travel faster in solid iron alloy than in liquid, such as molten iron alloy.

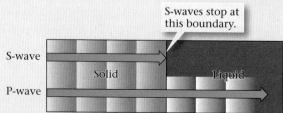

(d) Both P-waves and S-waves can travel through a solid, but only P-waves can travel through a liquid.

■ In general, seismic waves travel faster in a solid than in a liquid. Specifically, seismic waves travel more slowly in magma than in solid rock, and more slowly in molten iron alloy than in solid iron alloy (Fig. D.2c).

■ Both P-waves and S-waves can travel through a solid, but only P-waves can travel through a liquid (Fig. D.2d). To see why, picture what happens if you push down on the water surface in a pool—you send a pulse of compression (a P-wave) to the bottom of the pool. Now move your hand sideways through the water (shear). The water in front of your hand simply slides or flows past the water deeper down—your shearing motion has no effect on the water at the bottom of the pool (Fig. D.3a, b).

Reflection and Refraction of Wave Energy

Shine a flashlight or laser into a container of water so that the light ray hits the boundary (or interface) between water and air at an angle. Some of the light bounces off the water surface and heads back up into the air, while some enters the water (Fig. D.4a). The light ray that enters the water bends at the air-water boundary, so that the angle between the ray and the boundary in the air is different from the angle between the ray and the boundary in the water (Fig. D.4b). Physicists refer to the light ray that bounces off the air-water boundary and heads back into the air as the *reflected* ray, and the ray that bends at the boundary as the *refracted* ray. The phenomenon of bouncing off is **reflection**, and the phenomenon of bending

FIGURE D.3 Shear waves don't pass through a liquid.

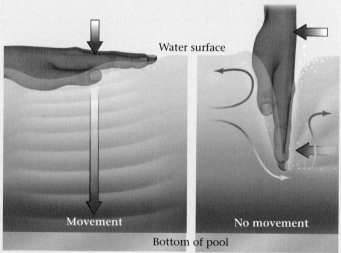

(a) Pushing down on a liquid produces a compressive pulse (P-wave) that can travel far through a liquid.

(b) Moving your hand sideways does *not* generate a shear wave; the moving water simply flows past deeper water.

FIGURE D.4 Refraction and reflection of waves.

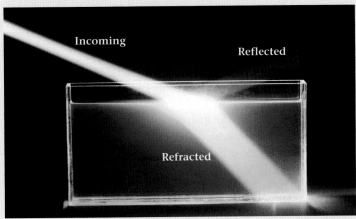

(a) A lab experiment showing the refraction and reflection of a light beam at the air-water interface.

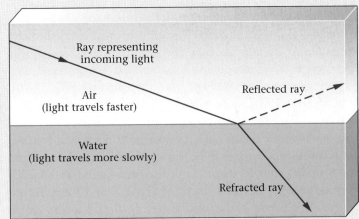

(b) A ray of light reflects and partly refracts when it crosses the boundary between two different materials.

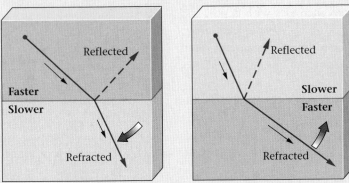

(c) A ray that enters a material through which it travels more slowly bends away from the boundary. A ray that enters a faster medium bends toward the boundary.

is **refraction**. Wave reflection and refraction take place at the interface between two materials, if the wave travels at different velocities in the two materials.

Seismic energy travels in the form of waves, so seismic waves, like rays of light in water, reflect and/or refract when reaching the interface between two rock layers if the waves travel at different velocities in the two layers. For example, imagine a layer of sandstone overlying a layer of basalt. Seismic velocities in sandstone are slower than in basalt, so as seismic waves reach the boundary, some reflect, and some refract.

The amount and direction of refraction at a boundary depend on the contrast in wave velocity across the boundary and on the angle at which a wave hits the interface. As a rule, if waves enter a material through which they will travel more slowly, the rays representing the waves bend down and away from the interface. For example, the light ray in Figure D.4b bends down when hitting the air-water boundary because light travels more slowly in water. This relation makes sense if you picture a car driving from a paved surface diagonally onto a sandy beach—the wheel that

rolls onto the sand first slows down relative to the wheel still on the pavement, causing the car to turn toward the sand. Alternatively, if the ray were to pass from a layer in which it travels slowly into one in which it travels more rapidly, the rays representing the waves would bend up and toward the interface (Fig. D.4c).

D.4 SEISMIC STUDY OF EARTH'S INTERIOR

Let's now utilize your knowledge of seismic velocity, refraction, and reflection to see how each of the major layer boundaries inside the Earth was discovered.

Discovering the Crust-Mantle Boundary

The concept that seismic waves refract at boundaries between different layers led to the first documentation of the core-mantle boundary. In 1909, Andrija Mohorovičić, a Croatian seismologist, noted that P-waves arriving at seismograph stations less than 200 km from the epicenter traveled at an *average* speed of 6 km per second, whereas P-waves arriving at seismographs more than 200 km from the epicenter traveled at an *average* speed of 8 km per second. To explain this observation, he suggested that P-waves reaching nearby seismographs followed a shallow path through the crust, in which they traveled relatively slowly, whereas P-waves reaching distant seismographs followed a deeper path through the mantle, in which they traveled relatively rapidly.

To understand Mohorovičić's proposal, examine Figure D.5a, which shows P-waves, depicted as rays, generated by an earthquake in the crust. Ray C, the shallower wave, travels through the crust directly to a seismograph. Ray M, the deeper wave, heads downward, refracts at the crust-mantle boundary, curves through the mantle, refracts again at the crust-mantle boundary, and then proceeds through the crust up to the seismograph. At stations less than 200 km from the epicenter, Ray C arrives first, because it has a shorter distance to travel. But at stations more than 200 km from the epicenter (Fig. D.5b), Ray M arrives first, even though it has farther to go, because it travels faster for much of its length. Calculations based on this observation require the crust-mantle boundary beneath most continental regions to be at a depth of about 35 to 40 km. Later studies showed that the crust-mantle boundary beneath oceanic regions lies at a depth of about 7 to 10 km. As we learned in Chapter 2, the crust-mantle boundary is now called the **Moho**, in honor of Mohorovičić.

Subsequent studies indicate that seismic velocities in the lower continental crust are faster than those in the upper continental crust. To interpret these observations, geologists have measured velocities in rock samples, of various compositions, under laboratory conditions. Such experiments suggest that the upper crust has, on average, felsic to intermediate composition (i.e., it's comparable to granite or diorite), whereas the lower continental crust has, on average, mafic composition (i.e., it has the composition of basalt or gabbro), as indicated in Figure D.1.

Defining the Structure of the Mantle

After studying materials erupted from volcanoes and the material comprising meteorites, geologists have concluded that the entire mantle has roughly the chemical composition of peridotite, an ultramafic rock. If the density, rigidity, and compressibility of peridotite were exactly the same at all depths, seismic velocities would be the same everywhere in the mantle, and seismic rays would be straight lines. But by studying travel times, seismologists have determined that seismic waves travel at different velocities at different depths. Let's now look at variations in seismic velocity that depend on mantle depth, and consider how these variations affect the shape of seismic-ray paths.

Between about 100 and 200 km deep in the mantle beneath oceanic lithosphere, seismic velocities are slower than in the overlying lithospheric mantle (Fig. D.6a). In this **low-velocity zone** (**LVZ**), the prevailing temperature and pressure conditions cause peridotite to melt by up to 2% to 6%. The melt, a liquid, coats solid grains and fills voids between grains. Because seismic waves travel more slowly through liquids than through solids, the coatings of melt slow seismic waves down. In the context of plate tectonics theory, the low-velocity zone is the weak layer on which oceanic lithosphere plates move. Below the low-velocity zone, the mantle does *not* contain melt. Of note, geologists do not find a well-developed low-velocity zone beneath continents.

Below about 200 km, seismic-wave velocities in the mantle increase with depth (Fig. D.6d). Seismologists interpret this increase to mean that mantle peridotite becomes progressively less compressible, more rigid, and denser with depth. This proposal makes sense, considering that the weight of overlying rock increases with depth, and as pressure increases, the atoms making up rock squeeze together more tightly and are not so free to move. Because of refraction, the progressive increase in seismic velocity with depth causes seismic rays to curve in the mantle. To understand the shape of a curved ray, look at Figure D.6b, c, which represents a portion of the mantle by a series of imaginary layers, each of which has a slightly greater seismic-wave velocity than the layer above. Every time a seismic ray crosses the boundary between adjacent layers, it refracts a little toward the boundary. After the ray has crossed several layers, it has bent so much that it begins to head back up toward the top of the stack. Now if we replace the stack of distinct layers with a single layer in which velocity increases with depth at a constant rate, the wave follows a smoothly curving path (Fig. D.6b, c).

At depths between 410 km and 660 km (see Fig. D.6a), seismic velocity increases in a series of abrupt steps, so the stack of layers in Figure D.6b is actually a somewhat realistic image. Experiments suggest that these **seismic-velocity discontinuities** occur at depths where pressure causes atoms in minerals to rearrange and pack together more tightly, thereby abruptly changing the rock's compressibility and rigidity—in other words, each seismic-velocity discontinuity corresponds to a *phase change*

FIGURE D.5 Discovery of the Moho.

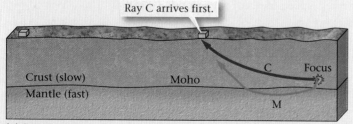

(a) Seismic waves traveling only in the crust reach a nearby seismograph first.

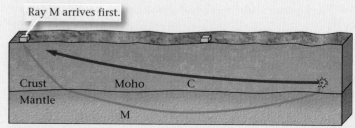

(b) Seismic waves traveling for most of their path in the mantle reach a distant seismograph first.

FIGURE D.6 The velocity of P-waves in the mantle changes because the physical properties of the mantle change with depth.

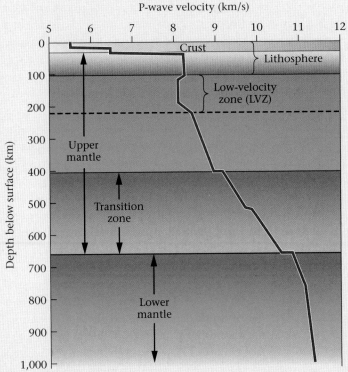

(a) The velocity of P-waves changes with depth in the mantle.

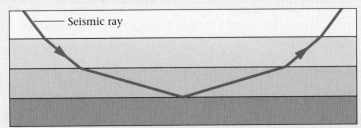

(b) In a stack of discrete layers, rays bend at each boundary. If the velocity is progressively faster in each lower layer, the ray eventually bends back and returns to the surface.

(c) In a material in which the velocity increases gradually with depth, rays curve smoothly and eventually return to the surface.

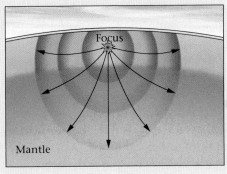

(d) The velocity of seismic waves increases with depth in the mantle, so rays curve and wave fronts are oblong.

(see Chapter 8). For example, at 410 km, the mineral olivine, a major constituent of peridotite, becomes unstable and collapses to form a different mineral called magnesium spinel. Magnesium spinel has the same chemical composition as olivine but a different internal arrangement of atoms and stays stable to a depth of about 660 km. At this depth, atoms rearrange to form an even denser crystal structure known as the perovskite structure. Because of these seismic-velocity discontinuities, as we learned in Chapter 2 but now can see more clearly, seismologists subdivide the mantle into the **upper mantle** (above 660 km), the **transition zone** (between 410 and 660 km), and the **lower mantle** (below 660 km; see Fig. 2.14b).

At this point, you may be wondering: Is there any way to test ideas about the identity of minerals that occur deep in the mantle? The answer is yes. Researchers can use a device called a diamond anvil to simulate deep-mantle pressure and temperature conditions. In a diamond-anvil experiment, tiny samples of minerals thought to occur in the mantle are squeezed between two diamonds. Simultaneously, the samples are subjected to a beam of intense laser light (Fig. D.7). Diamond is the hardest mineral known, and the pressure between the two diamonds can reach values comparable to those found in the deep mantle. Laser light passes through clear diamonds, and thus can heat the sample to mantle temperatures. Using

sophisticated instruments, researchers measure seismic velocities in the sample while it remains in the diamond anvil at high pressures and temperatures. By comparing the velocities observed in the laboratory with those observed for the real mantle, researchers gain insight into whether or not the sample could be a mantle mineral.

Discovering the Core-Mantle Boundary

During the first decade of the twentieth century, seismologists installed seismographs at many stations around the world, expecting to be able to record waves produced by a large earthquake anywhere on Earth. In 1914, one of these seismologists, Beno Gutenberg, discovered that P-waves from an earthquake do not arrive at seismographs lying in a band between 103° and 143°, as measured along the surface of the Earth from the earthquake epicenter. This band is now called

FIGURE D.7 A laboratory apparatus for studying the characteristics of minerals under very high pressures and temperatures. The green laser beam is heating up a microscopic sample being squeezed between two diamonds hidden at the center of the metal cylinder.

the **P-wave shadow zone** (Fig. D.8a). If the density of the Earth increased *gradually* with depth all the way to the center, the shadow zone would not exist because rays passing into the interior would curve up and reach every point on the surface. Thus, the presence of a shadow zone means that deep in the Earth a major interface exists where seismic waves *abruptly* refract down (implying that the velocity of seismic waves suddenly decreases). This interface, now called the **core-mantle boundary**, lies at a depth of about 2,900 km. Notably, the density contrast across this boundary is greater than the density contrast between the crust and water.

To see why the P-wave shadow zone exists, follow the two seismic rays labeled A and B in Figure D.8a. Ray A curves smoothly in the mantle (we are ignoring seismic-velocity discontinuities in the mantle) and passes just above the core-mantle boundary before returning to the surface. It reaches the surface 103° from the epicenter. In contrast, Ray B penetrates the boundary and refracts down into the core. Ray B then curves through the core and refracts again when it crosses back into the mantle. As a consequence, Ray B intersects the surface at more than 143° from the epicenter.

Discovering the Nature of the Core

The downward bending of seismic waves when they pass from the mantle down into the core indicates that seismic velocities in, at least, the outer core are slower than in the mantle. Thus, even though the core is deeper and denser than the mantle, at least the outer part of the core must be less rigid than the mantle. How can this be? What does the core consist of?

Based on density calculations and on the study of meteorites thought to be fragments of a large planetesimal's interior, seismologists concluded that the core consists of iron alloy. This alloy contains about 85% iron, 5% nickel, and 10% of a lighter element (probably oxygen, silicon, and/or sulfur). On further study, seismologists found that S-waves do not arrive at stations located between 103° and 180° from the epicenter (a band called the **S-wave shadow zone**). This means that S-waves *cannot* pass through the core at all—otherwise, an S-wave headed straight down through the Earth would appear on the other side. Remember that S-waves are shear waves, which by their nature can travel only through solids. Thus, the fact that S-waves do not pass through the core means that the core, or at least part of it, *consists of liquid* (Fig. D.8b).

At first, seismologists thought that the entire core might be liquid iron alloy. But in 1936, a Danish seismologist, Inge Lehmann, discovered that P-waves passing through the core reflected off a boundary within the core. She then proposed that the core is made up of two parts: an **outer core** consisting of liquid iron alloy and an **inner core** consisting of solid iron alloy. Lehmann's work defined the existence of the inner core but could not locate the depth at which the inner core–outer core interface occurs. This depth was eventually located by measuring the exact time it took for seismic waves generated by nuclear explosions to penetrate the Earth, bounce off the inner core–outer core boundary, and return to the surface (Fig. D.8c). The measurements showed that the inner core–outer core boundary occurs at a depth of about 5,155 km.

Why does the core have two layers—a liquid outer layer and a solid inner one? An examination of Figure D.8d provides some insight. This graph shows two curves: (1) the geotherm, which indicates how temperature changes with increasing depth in the Earth; and (2) the melting curve, which indicates that the temperature at which materials melt changes with increasing depth in the Earth (at temperatures to the right of the melting curve, Earth materials are molten). As the graph shows, the geotherm lies to the left of the melting curve through most of the mantle and in the inner core. This means that temperatures in most of the mantle and in the inner core are not high enough to cause melting, under the high pressures found in these regions, so these regions are solid. But the geotherm lies to the right of the melting curve in the low-velocity zone of the mantle and in the outer core, so these regions contain molten material.

Before we leave the core, let's look again at one important characteristic of the liquid outer core, its flow. As noted earlier in the book, the liquid outer core undergoes convection, and this convection generates Earth's magnetic field. Because of the Earth's rotation, convective cells in the outer core may take the form of spirals that rise in a direction parallel to the Earth's rotation axis, and thus lead to a dipolar field that roughly parallels the rotation axis. Geologists suggest that

FIGURE D.8 Shadow zones and the discovery of the Earth's core.

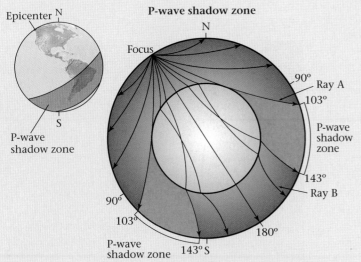

P-wave shadow zone

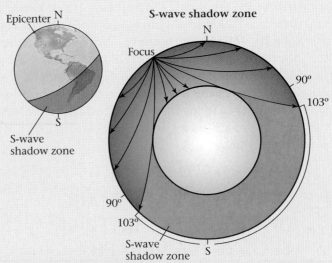

S-wave shadow zone

(a) P-waves do not arrive in the P-wave shadow zone because they refract at the core-mantle boundary.

(b) S-waves do not arrive in the S-wave shadow zone because they cannot pass through the liquid outer core.

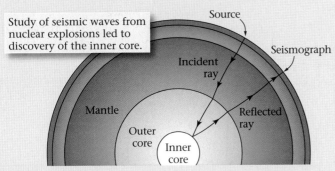

Study of seismic waves from nuclear explosions led to discovery of the inner core.

(c) Seismic waves reflect off the inner core–outer core boundary.

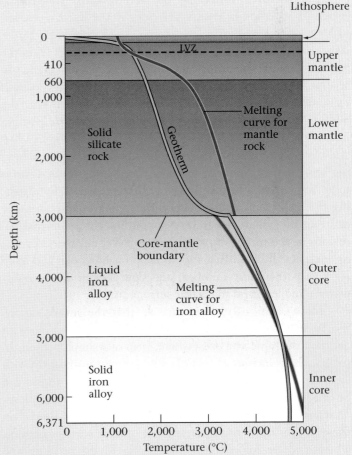

(d) A graph of the geotherm and melting curve for the Earth. Note that the melting temperature is less than the Earth's temperature in the outer core, so the outer core is molten.

convection in the outer core cannot be explained by thermal contrasts between the top and bottom of the core, for the outer core consists of metal, which is a very good heat conductor. Recent research suggests that the density differences leading to convection are largely due to differences in chemical composition between the top and bottom of the outer core. These compositional differences arise because the solid *inner core is slowly growing* as the Earth, overall, cools. The new crystals of solid iron that form along the surface of the inner core do not have room in their crystal structure for low-density elements such as silicon, sulfur, or oxygen. So these elements are expelled into the base of the molten outer core. Thus, at any given time, the base of the outer core has a lower density than the upper parts, and it starts to rise convectively.

The Velocity-versus-Depth Profile

By the time of World War II, seismologists had identified the crust-mantle boundary (the Moho) and the core-mantle boundary, and had recognized that the core is divided into two parts. In other words, the prevailing image of the Earth's inte-

rior being an onion-like sequence of concentric zones had been established. After World War II, seismologists set to the task of refining this image, a task made easier by the Cold War: because of the need to detect nuclear explosions, the Western nuclear powers built a vast array of precise seismograph stations scattered around the world. Through painstaking effort, seismologists used data from this array to develop a graph known as a **velocity-versus-depth curve**. We've already presented the upper part of the curve (Fig. D.6); Figure D.9 presents the rest of it. This curve shows the depths at which seismic velocity suddenly changes. These changes define the principal layers and sublayers in the Earth. Note that the graph does not show a velocity for S-waves in the outer core because S-waves cannot travel through a liquid.

D.5 FINE-TUNING OUR IMAGE OF THE EARTH'S INTERIOR: SEISMIC TOMOGRAPHY

In recent years, seismologists have developed a technique, called **seismic tomography**, to produce three-dimensional images of variation in seismic velocities in the Earth's interior. (This technique resembles the method used to produce three-dimensional CAT or CT scans of the human body. *CAT*, in the medical context, stands for computerized axial tomography.) In seismic tomography studies, researchers compare the *observed* travel time of seismic waves following a specific ray path with the *predicted* travel time that waves following the same path would have if the average velocity-versus-depth model of Figures D.6 and D.9 were completely correct. They found that waves following some paths take more time than predicted, whereas waves following other paths take less time than predicted. By repeating the measurements for many different wave paths in many different directions, researchers can outline three-dimensional regions of the mantle in which waves travel unexpectedly fast or unexpectedly slow. Fast regions are probably cooler and more rigid than their surroundings, and slow regions are probably warmer and less rigid than their surroundings.

Tomographic studies emphasize that the simple onion-like layered image of the Earth, with velocities increasing with depth at the same rate everywhere, is an oversimplification. In reality, the velocities of seismic waves vary significantly with location at a given depth. Results of tomographic studies can be displayed by three-dimensional models, cross sections, or maps (Fig. D.10a–c). Generally, warmer colors (reds) on these images indicate slower and presumably warmer regions, whereas cooler colors (blues and purples) indicate faster and presumably cooler regions.

Seismic tomography studies provide new insight into the Earth's interior. For example, the studies show that a region of fast (cool) mantle lies in the asthenosphere beneath North America (see Fig D.10b). This region may represent the remnants of oceanic lithosphere that was subducted during the

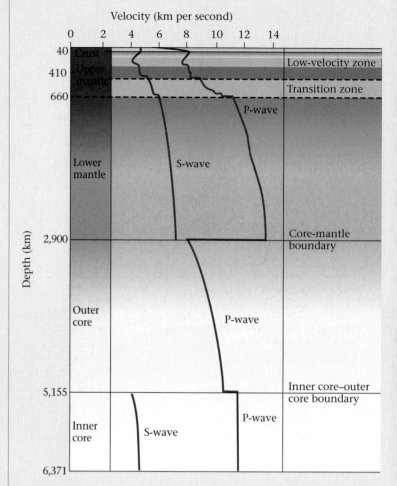

FIGURE D.9 The velocity-versus-depth profile of the whole Earth.

Mesozoic. The fact that the fast region extends into the lower mantle suggests that some subducted plates may sink into the lower mantle. In fact, some researchers suggest that there is a "plate graveyard" at the base of the mantle, where denser portions of subducted plates sink and accumulate. But others maintain that most subducted plate material remains in the upper mantle. Tomographic images of the uppermost mantle (70-km depth) show a region of slow (warm) mantle that occurs in the region beneath mid-ocean ridges (Fig. D.10d). This result supports the concept that warm asthenosphere rises beneath ridges.

More recently, tomography studies have defined a 200-km-thick layer of slow (warm) mantle just above the core-mantle boundary. This layer, called the D" layer (pronounced "dee double prime") may represent the region in which the mantle has absorbed heat radiating from the core. Some researchers speculate that it may be the source of mantle plumes, but this proposal remains a subject for debate.

Tomographic studies of the inner core have shown that the inner core also has a pattern of fast and slow regions.

FIGURE D.10 Tomographic images of the Earth's interior, and their interpretation.

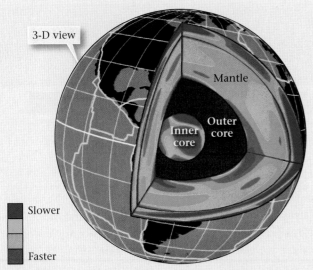

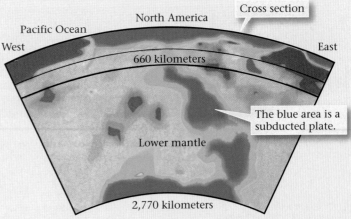

(b) Close-up tomographic image of the mantle beneath North America and the Pacific Ocean.

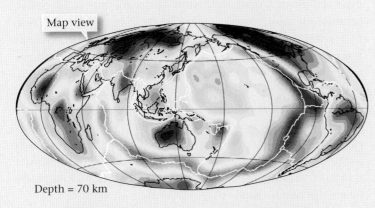

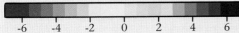

(a) Tomographic image of the Earth. Slower areas may be relatively cooler than their surroundings, and faster areas may be relatively warmer.

(c) A tomographic image of the Earth showing relative seismic velocities at 70-km depth. The red areas have slower velocities, and the blue areas have faster velocities.

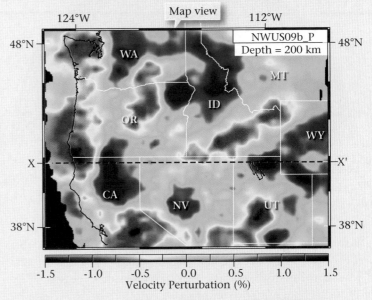

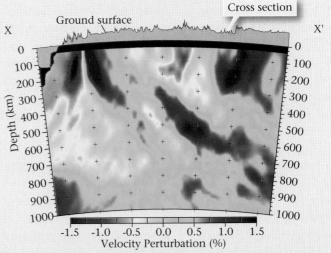

(d) A map (of a surface at a depth of 200 km) and cross section of the Earth beneath the western United States, constructed using tomographic data from the EarthScope array. Red areas have faster than expected velocities, and probably are warmer, whereas blue areas are cooler.

Significantly, the orientation of this pattern changes over time, relative to the orientation of this pattern in the mantle. Researchers interpret this to mean that *the inner core rotates slightly faster* than the rest of the Earth—it makes an extra rotation about once every 20 to 25 years.

Taken together, tomographic studies have begun to give us an image of mantle convection in which warmer, less dense regions rise (or undergo "upwelling"), while cooler, denser areas sink (or undergo "downwelling"). This image has inspired researchers to develop supercomputer models of what this convection looks like (**Fig. D.11a**). Even though many important questions remain, tomography has led geologists to picture the Earth's insides as a dynamic place, not

FIGURE D.11 Images of convection in the Earth's interior.

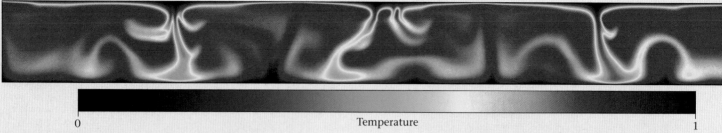

0   Temperature   1

(a) A computer model that simulates convective cells formed in part of the mantle. Blue areas are cooler, downwelling regions, whereas dark red areas are warmer, upwelling regions. Computer models provide insight into how the real Earth might work.

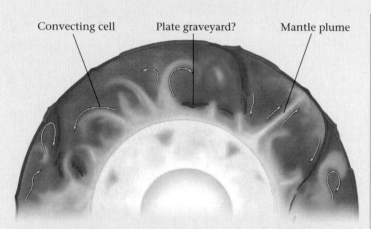

Convecting cell   Plate graveyard?   Mantle plume

(b) An artist's rendition of a modern view of the complex and dynamic Earth interior. Note the convecting cells, the mantle plumes, and the graveyards of subducted plates.

just a region of static, concentric shells (Fig. D.11a, b). This image should become even clearer, for a major new research initiative, called **EarthScope**, has begun (Fig. D.10d). This initiative involves placing hundreds of seismographs in an array across the United States. Just as digital photographs have higher resolution when taken by a camera with a 10-megapixel sensor than one with a 2-megapixel sensor, the greater number of seismographs in the EarthScope array will provide a higher-resolution image of the Earth's interior. Researchers are also using other measurements, such as variation in the pull of Earth's gravity, to characterize the interior.

D.6 SEISMIC-REFLECTION PROFILING

Seismic techniques are also letting us fine-tune our image of the crust. During the past half century, geologists have found that by exploding dynamite, by banging large weights against the Earth's surface, or by releasing bursts of compressed air into the water, they can create artificial seismic waves that propagate down into the Earth and reflect off the boundaries between different layers of rock in the crust. By recording

the time at which these reflected waves return to the surface, geologists can determine the depth to these boundaries. With this information, they can produce a cross-sectional view of the crust called a **seismic-reflection profile** (Fig. D.12a–d). This image can define subsurface bedding and stratigraphic formation contacts, and can reveal the presence of subsurface folds (bends in layers) and faults. Oil companies must obtain seismic-reflection profiles, despite their high cost, because they allow geologists to identify likely locations for oil and gas reserves underground. Research geologists at universities have used the technique to obtain images of the Moho and even the upper mantle.

In the past decade, computers have become so sophisticated that geologists can now produce three-dimensional seismic-reflection images of the crust. These provide so much detail that geologists can even trace out a ribbon of sand, representing the channel of an ancient stream, buried kilometers below the Earth's surface (Fig. D.12e).

D.7 EARTH'S GRAVITY

Now that we have an image of the complexity of the Earth's interior, from our study of seismic waves, we can develop additional understanding of gravity and magnetism, the two field forces that emanate from our planet. Recall that a **field force** is a push or pull that applies across a distance—a field force exerted by one object can cause another object to *accelerate* (change speed and/or direction) without ever touching it (Box D.1). In this section we discuss gravity, and in the next, we turn our attention to magnetism.

The Geoid

Gravity is an *attractive* field force that one mass exerts on another. As Newton showed, the magnitude of gravitational pull depends on the size of the masses and on the distance between them—larger masses exert stronger pulls than do smaller ones, and closer masses produce stronger pulls than do

FIGURE D.12 Seismic-reflection profiling.

(a) Trucks thumping on the ground to generate the signal needed for making a seismic-reflection profile.

(b) Geologists at Shell Oil Company process seismic data.

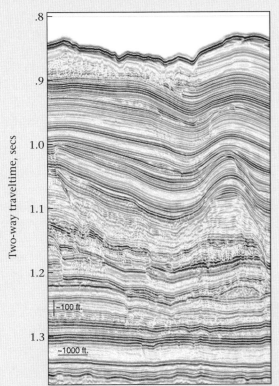

(c) After computer analysis, the data yield a seismic-reflection profile. Color bands represent horizons in the stratigraphic sequence.

(d) A ship collects seismic data by towing "air guns" that send a pulse of energy through the water into the strata below.

(e) The data can be used to produce a 3-D image of the subsurface. Geologists can study both map surfaces and cross sections through the block.

farther ones. When gravity acts on an object but can't make the object move, then the object has **gravitational potential energy**. For example, a boulder sitting at rest on a hill slope has gravitational potential energy (Fig. D.13a). If the boulder starts to tumble, the potential energy transforms into kinetic energy, the energy of motion; and when the boulder comes to rest at a lower elevation, it has less potential energy. Note that two boulders at the same elevation have the same gravitational potential energy; a boulder higher on a hill has more potential energy than one lower on a hill. Similarly, every point on the surface of standing water in a still pond has the same gravitational potential energy. If you throw a pebble in the water, the

CONSIDER THIS . . .

BOX D.1

The Meaning of Acceleration

In scientific discussion, velocity is a measure of distance traveled per unit time ($v = D/t$), and is a quantity that has both a magnitude and a direction. An **acceleration** is a *change* in magnitude and/or direction of velocity. For example, we can describe the velocity of a car by saying that it is traveling west at 100 km/h. In this description, the number "100 km/h" is its speed (the magnitude of its velocity), and the heading "west" is its direction. If the car increases its speed to 110 km per hour, it has undergone an acceleration. If it stays at 100

km per hour, but turns and starts heading northwest, it has also undergone an acceleration, even though its speed has not changed. A quantity that has both magnitude and direction is a **vector**, which we can represent by an arrow whose length gives the magnitude and whose orientation gives direction.

What units do we use to specify an acceleration? Since acceleration is a change in velocity per unit time, we use units of "distance per second per second" (written as d/s^2). For exam-

ple, the acceleration due to gravity in a vacuum at the surface of the Earth is approximately 9.8 m/s^2. Thus, if the Earth did not have an atmosphere, a bungee jumper leaping from a tower would travel 9.8 m/s after the first second, 19.6 m/s after the second, and so on. Of course, the real Earth does have an atmosphere; air resistance pushes back as the jumper falls, so the jumper's acceleration is less than 9.8 m/s^2, and after a while, the jumper reaches terminal velocity and can't travel any faster.

---

energy of impact temporarily forms waves (Fig. D.13b). The waves eventually disappear because water is so weak that gravity eventually evens out highs and lows of the surface (Fig. D.13c). A surface on which all points have the same potential energy, such as the surface of standing water in a pond, is an **equipotential surface**.

What does the gravitational field produced by a planet look like? In the jargon of physics, we can restate this question by asking, what does the gravitational equipotential surface of the Earth look like? We can picture this surface by imagining what sea level *would look like* (not what it really is) if an ocean completely surrounded the world and was not subjected to winds or currents. If the Earth were a stationary, perfect sphere (meaning there are no hills, valleys, or basins), the equipotential surface (the surface of an imaginary global ocean) representing its gravity would itself be a sphere. In reality, however, the equipotential surface is *not* a perfect sphere, in part because the Earth spins on an axis. Rotation produces centrifugal force, which flattens the Earth

into a **spheroid** (Fig. D.14a). The imaginary spheroid that most closely has the shape of the Earth—it's not exactly the real Earth, because in reality our planet has continents, ocean basins, mountains, and valleys—has a radius at the equator of 6,378 km (3,963 miles) and a radius at the pole of 6,357 km (3,950 miles). Put another way, we can say that the equipotential surface has a slight "equatorial bulge." Geologists refer to this imaginary equipotential surface as the **reference geoid**.

While the reference geoid is a starting point for describing the Earth's gravity, it doesn't provide a completely accurate

FIGURE D.13 The concept of gravitational potential energy.

(a) Potential energy increases as you go up the hill. Boulder 1 has the most potential energy. Boulders 2 and 3 have the same potential energy, boulder 4 has less.

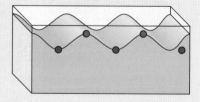

(b) The crests of waves on a pond have higher potential energy than the troughs.

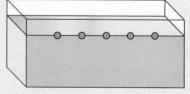

(c) When there are no waves, the surface of the water is an equipotential surface.

FIGURE D.14 Representations of the geoid, the shape of an equipotential surface representing Earth's gravity. Colors represent *elevations* of the geoid relative to the reference geoid.

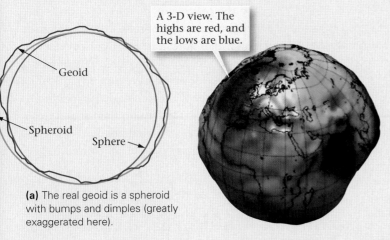

A 3-D view. The highs are red, and the lows are blue.

Geoid

Spheroid

Sphere

(a) The real geoid is a spheroid with bumps and dimples (greatly exaggerated here).

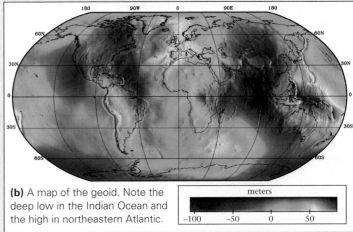

(b) A map of the geoid. Note the deep low in the Indian Ocean and the high in northeastern Atlantic.

meters

−100   −50   0   50

picture for many reasons, including (1) the density of material within the Earth's layers is not uniform throughout the layer, as shown by tomographic studies; and (2) the outermost layer of the Earth, the lithosphere, is strong enough to hold up heavy loads such as volcanoes or glaciers, or to allow depressions such as trenches to persist. Because of these factors, and others, there can be more mass or less mass at a location than the reference geoid predicts, so the real equipotential surface for the Earth has lumps and dimples. Geologists refer to this irregular surface, that provides the best representation of the Earth's gravity, simply as the **geoid** (Fig. D.14b). The surface of the Earth's real geoid differs from that of the reference geoid only slightly—the highest point on the real geoid lies 85 m above the reference geoid, and the lowest point lies 107 m below the reference geoid. But even these small differences can noticeably influence the orbits of satellites or the accuracy of surveys.

Gravity Anomalies

Geologists can measure the pull of gravity at a point on the Earth by using an instrument called a **gravimeter**, which consists of a weight hanging from a delicate spring—stronger gravity pulls the weight down more; weaker gravity pulls it down less. In recent years, satellites have been able to determine the variation in the Earth's gravitational pull at the surface over broad areas very rapidly. Researchers have found that the actual surface of the Earth is *not* an equipotential surface, so the pull of gravity at the surface varies with location. Recall that gravity is a force, so it causes an unrestrained object to accelerate. Researchers describe the **acceleration** caused by the Earth's gravity using a unit called the Gal (named for Galileo, 1564–1642, the Italian astronomer), where 1 Gal = 1 cm/s^2. On average, the acceleration due to Earth's gravity is 981 Gals ($\approx$ 9.8 m/s^2; see Box D.1), but it ranges from 976 Gals to 983 Gals.

Differences in the gravitational pull between one location and an adjacent one may be so tiny that geologists must specify them in milliGals (mGal), where 1 mGal = 1/1,000 Gal.

Geologists refer to a deviation of the observed (real) geoid from the reference geoid as a **gravity anomaly**. Over a positive anomaly, the pull is stronger, whereas over a negative anomaly, the pull is weaker. A positive anomaly indicates that there is extra mass below the site, perhaps due to a body of particularly dense rock underground, whereas a negative anomaly means that there is a deficit of mass, perhaps due to the presence of less-dense rock (Fig. D.15). Gravity anomalies provide important information about the interior of the Earth. As a practical point, study of them helps geologists to find reserves of valuable metal ores underground; and from an academic perspective, they may allow geologists to detect upwelling and downwelling zones in the mantle.

FIGURE D.15 A gravity anomaly map of Illinois, Indiana, and Ohio. On this map, colors represent variations in the magnitude of gravitational pull.

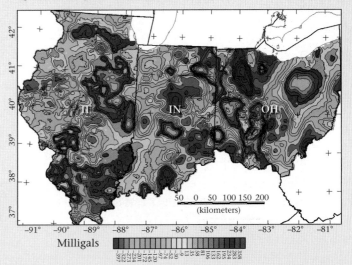

50 0 50 100 150 200
(kilometers)

Milligals

Isostasy and Gravity Anomalies

According to Archimedes' principle, a cargo ship anchored in a harbor floats at just the right level so that the mass of the water displaced by the ship equals the mass of the whole ship. Even though the ship consists of heavy steel, the inside of the ship is filled with air, so the ship, in effect, is a steel-sheathed bubble—it floats because its overall *average* density is less than that of water. When the ship remains empty, the distance that its deck lies above the water, its so-called freeboard, is large (Fig. D.16). Addition of heavy cargo causes the ship's keel to sink deeper into the water, the deck to move down, and the freeboard, therefore, to decrease. This movement can take place because the water beneath the ship can flow out of the way as the ship settles downward—if the ship were sitting on solid ground, addition of cargo would not change the freeboard (relative to the ground) because the ground is strong enough to hold up the ship. When the freeboard of the ship is just right for a given cargo, so that ship's freeboard does not "want" to rise or sink, we say that the ship is in **isostatic equilibrium**, or that a condition of **isostasy** exists.

As we learned in Chapter 4, the outer layer of the Earth, the lithosphere, can be thought of as a relatively rigid shell that "floats" on the underlying, relatively soft asthenosphere. Asthenosphere can flow out of the way of sinking lithosphere or flow in under rising lithosphere, just like water does beneath a ship, but much more slowly. Thus the lithosphere approaches a condition of isostatic equilibrium in many places, so the elevation of the lithosphere's surface reflects the density and thickness of the lithosphere.

The lithosphere behaves somewhat like the ship—addition of a load, such as growth of an ice sheet or the building of a volcano, causes the surface of the lithosphere to sink, whereas removal of a load, say by melting of an ice sheet, causes the surface of the lithosphere to rise as it tries to attain isostatic equilibrium. But since asthenosphere flows so slowly, there are places on Earth where isostasy has not been achieved at a given time. We observe gravity anomalies over such places. Examples include deep-sea trenches, where the lithosphere has been bent and held down by the process of subduction—since the density of water is much less than that of rock, there is a deficit of mass over a trench and, therefore, a negative gravity anomaly. In places where a volcano or small mountain range builds so quickly that the lithosphere does not have time to sink, or if the lithosphere's strength prevents a region of dense rock from sinking, or a region of light rock from rising, then again, isostasy is not achieved and there is a gravity anomaly. We discuss isostasy again in Chapter 11, to see how it is involved in the uplift of mountain ranges.

D.8 EARTH'S MAGNETIC FIELD, REVISITED

As we discussed in Chapters 2 and 3, the Earth behaves like a weak magnet and thus produces a magnetic field. Information about this magnetic field can be locked into rock and preserved as paleomagnetism. Here, in Interlude D, we look more closely at the magnetic field, focusing on why the field exists and on why there are variations in the field strength on land and on sea. To set the stage, we briefly review the fundamentals of magnetism.

A Brief Review of Magnetism

If you hold a magnet over a pile of steel paper clips, it lifts the clips, for a magnet can exert an attractive force whose pull on the clips, at close range, exceeds the force of gravity. A magnet can also produce a repulsive force capable of pushing an object away—when oriented appropriately, one magnet can levitate another. The push or pull exerted by a magnet is its **magnetic force**. (Note that a magnetic force can be attractive *or* repulsive, whereas gravitational force is only attractive.) The region

FIGURE D.16 A ship analog for the concept of isostasy. At isostatic equilibrium, the freeboard reflects Archimedes' principle, so the freeboard decreases as the mass of the ship increases.

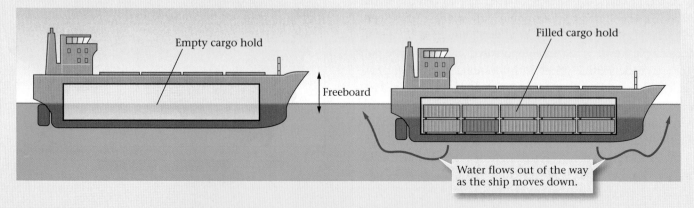

Empty cargo hold

Filled cargo hold

Freeboard

Water flows out of the way as the ship moves down.

around a magnet in which we can detect a magnetic force is a **magnetic field**. Magnetic fields have a characteristic geometry that we can represent by curving lines, called **magnetic field lines** (see Fig. 2.3a).

Magnetic fields always surround a **permanent magnet**, a special material that can remain magnetic for a very long time all by itself. Magnetic forces can also be produced temporarily by an electrical current passing through a wire—an electrical device that produces a magnetic force is called an **electromagnet**. The stronger the magnet, meaning the stronger the magnetic force it produces, the greater its **magnetization**. The strength of the magnetic force that an object feels depends on the magnetization and on the distance of the object from the magnet.

All magnets have two **magnetic poles**, a north pole at one end and a south pole at the other. "Opposite" poles attract, and "like" poles repel—thus the north end of one magnet attracts the south end of another, whereas it repels the north end of another. Physicists represent the direction from the north pole to the south pole by a vector called a dipole moment or, more simply, a **dipole**. (As noted in Box D.1, a vector is a quantity with both magnitude and direction; we can represent a vector by an arrow.) Magnetic field lines form a continuous loop through the magnet's dipole (see Fig. 2.3a).

Where does the magnetization of materials come from? To answer this question, recall that all atoms contain electrons, negatively charged particles that orbit a nucleus. Each electron also spins on its axis. The electron's motion (both its spin and its orbit) produces a tiny electrical current that, in turn, produces a tiny magnetic field. Each atom, therefore, can be pictured as a tiny dipole (Fig. D.17a). But even though all materials consist of atoms, not all materials behave like permanent magnets. In fact, most materials (wood, plastic, glass, gold) are essentially nonmagnetic, simplistically because the atomic dipoles in the materials are randomly oriented, and the magnetic field of one atom cancels out the field of another that is oriented in the opposite direction (Fig. D.17b). In a permanent magnet, the shape of the electron orbits and the spacing between atoms is such that the dipoles of adjacent atoms lock into alignment with one another. When this happens, the magnetization of each atom *adds* to that of its neighbor, so the material, as a whole, becomes measurably magnetic (Fig. D.17c).

In Chapter 2, we learned that Earth has a magnetic field that deflects the solar wind and traps deadly cosmic rays. We also learned that the angle between the dipole representing most of the Earth's field and the Earth's spin axis is currently about 8°, equivalent to a distance of about 855 km. (This angle is changing because the magnetic pole is currently moving at about 50 km/year.) Geologists, by convention, refer to the tail end of the dipole, which lies close to the north geographic pole

(the point where the Earth's spin axis intersects the Earth's surface), as the **north magnetic pole**, and the arrowhead end of the dipole, which lies near the south geographic pole, as the **south magnetic pole**. This convention can be a bit confusing because, to a physicist, the north end of a magnet corresponds to the arrowhead end, and the arrowhead representing the Earth's dipole currently lies in the south. But because of the geologic convention, the *north-seeking* end of a compass (i.e., the tail end of the compass needle's dipole) points to the north magnetic pole. Magnetic field lines are parallel to the Earth's surface at the equator, and they are vertical at the pole. Over thousands of years, Earth's magnetic poles follow looping paths that don't stray more than about 15° from the geographic pole.

In Chapter 3, we learned that the "polarity" of the Earth's magnetic dipole reverses every now and then. During times of normal polarity, the dipole points in the same direction that it does now, whereas during times of reversed polarity, the dipole points in the opposite direction and the north magnetic pole would lie near the south geographic pole. The duration of a polarity interval, or chron, can last from tens of thousands of years to millions of years; the last reversal happened about 730,000 years ago. Recent studies suggest that reversals begin by a gradual weakening of the field until it nearly disappears. Eventually, when the magnetic field reappears, its dipole may point in the opposite direction. The whole reversal process takes only 1,000 to 6,000 years.

Origin of the Earth's Magnetic Field

Why does the Earth have a magnetic field? We take the existence of the field for granted because we are so accustomed to using compasses. But not all planets have fields—in fact, neither Mars

FIGURE D.17 The origin of permanent magnetization in a material.

(a) The spin and orbital motion of electrons produces a tiny magnetic dipole.

(b) In a nonmagnetic material, the dipoles are randomly oriented and cancel each other out.

(c) In a magnetic material, the dipoles lock into alignment.

nor Venus, the two planets most like the Earth, have measurable fields. The path toward discovering the origin of the Earth's field started in 1926, when researchers learned that our planet's outer core consists of liquid iron alloy. Since flow of metal can produce an electric current, it seemed that the flow of the outer core must be responsible for the magnetic field. But how?

Clues to understanding the generation of the Earth's magnetic field come from understanding how an electric power plant works. In a power plant, water or wind power spins a wire coil (an electrical conductor) around an iron bar (a permanent magnet); the motion of the wire in the bar's magnetic field generates an electric current in the wire. Such an apparatus is called a **dynamo**. In the case of the Earth, flow in the outer core serves the role of the spinning wire coil; but the inner core probably can't serve the role of the permanent magnet, because it is so hot that its atoms tumble chaotically and their dipoles cannot align. Researchers suggest instead that the Earth is a **self-exciting dynamo**. Evidently, during the Earth's early history, flow in the outer core took place in the presence of a magnetic field. This flow generated an electric current, and once the current existed it generated a magnetic field. Continued flow in the presence of the generated magnetic field produced more electric current that, in turn, maintained the magnetic field. Once started, the system perpetuates itself, as long as there's an input of energy to keep the outer core in motion.

So the question now becomes, what causes the outer core to keep moving? As we've noted, recent calculations suggest that as the Earth cools overall, the diameter of the solid inner core grows (by about 0.1 to 1 mm in diameter per year) at the expense of the outer core. The process of crystallization releases both heat and lower-density elements (such as silicon, sulfur, hydrogen, carbon, or oxygen) into the base of the outer core. The expulsion of "light" elements occurs because the solid iron crystals comprising the inner core do not have room in their crystal structure for atoms of these elements. The higher temperatures and higher concentration of lighter elements makes the base of the outer core *less dense* than the top, so the base of the outer core becomes buoyant. Since the molten outer core is liquid, and therefore weak, the buoyancy force can cause the less-dense material at the base of the outer core to rise. As this happens, denser material at the top of the outer core starts to sink. This process, as described in Chapter 2, is called convection. During convection, material follows a curving path known as a convection cell.

Researchers suspect that the key to understanding the *polarity* of the Earth's magnetic field comes from understanding the *geometry* of the convection cells in the outer core. According to some researchers, the spin of the Earth produces a force (called the Coriolis force, which we introduce in Chapter 18) that causes the convective cells to form into somewhat cylindrical spirals that are roughly parallel to the Earth's axis of rotation. The magnetic dipole produced by the spirals, there-fore, aligns with the spin axis of the Earth. In other words, the spin of the Earth influences the geometry of the outer core's convection, making the magnetic axis and the spin axis attain roughly the same orientation. Convection spirals in the outer core may wobble, or their shape may change over time, so at any given time, Earth's magnetic dipole does not necessarily align exactly with the Earth's spin axis. Notably, the convective spirals apply a force to the surface of the inner core, like a conveyor belt applies a force to the roller bearing, and this force could cause the inner core to rotate faster, as observed in the seismic tomography studies that we noted earlier.

Why does the polarity of the Earth's field reverse every now and then? Perhaps this behavior reflects the instability of convective cells in the outer core. Adjacent cells interact, so over time the geometry of their flow will change. This may decrease the field strength for a while, and as new spirals become established, the field may regrow with different polarity. The details of this process remain a subject of active research.

Magnetic Anomalies, Revisited

Close examination of the Earth's magnetic field indicates that it *is not a perfect dipole*. If it were, physicists could predict what the field strength would be at every point on the planet with a high degree of confidence. That isn't the case, because two other factors contribute to the field strength measured at a given location, and cause the field to be either stronger than or weaker than expected at the location. As mentioned in Chapter 3, the difference between the expected field and the measured field at a locality is called a **magnetic anomaly**. A positive anomaly occurs where the field is stronger than expected, and a negative anomaly occurs where the field is weaker than expected.

The first factor affecting the magnetic field at a locality is the non-dipolar field, which may be a consequence of complexities (turbulence) in the flow of the outer core near the core-mantle boundary—the non-dipolar field may account for as much as 10% of the Earth's field strength measured at a location. The second factor is the magnetization of rock in the crust—some rocks contain small amounts of magnetic minerals such as magnetite and hematite, enough to make the rock, overall, behave as a weak magnet. The presence of magnetic rocks in the crust will produce local magnetic anomalies that are independent of the Earth's internal magnetic field (Fig. D.18).

We've seen in Chapter 3 that magnetic anomalies due to the magnetization of rocks take the form of parallel stripes on the sea floor, because the basalt comprising the upper layer of oceanic crust was produced at the mid-ocean ridge axis by sea-floor spreading during successive polarity reversals. A positive anomaly stripe occurs over basalt with normal polarity because the magnetization of the basalt adds to the Earth's field,

FIGURE D.18 Earth's magnetic field may be due to spiral-like convection cells in the outer core. Interaction between the magnetic field and the inner core may cause the inner core to rotate a little faster than the rest of the Earth.

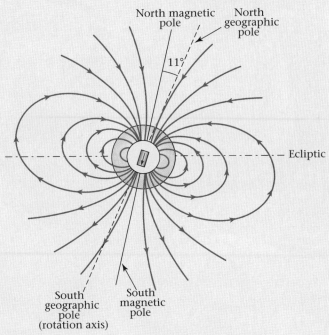

(a) The geometry of Earth's dipole field.

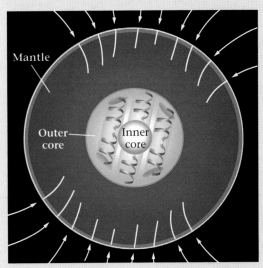

(b) An artist's image of spiral convection cells in the outer core.

FIGURE D.19 Gravity anomaly map of Iowa. Red areas have anomalously high gravity, and blue areas have anomalously low gravity.

MCR is the Midcontinent Rift, containing up to 15 km of basalt.

whereas a negative anomaly stripe occurs over basalt with reversed polarity because the magnetization of this basalt subtracts from that of the Earth's field.

On continents, in contrast, the pattern of magnetic anomalies is much more complex, for the distribution of rock types is more complex. We see anomalies due to igneous intrusions, due to lava flows, due to concentrations of iron-rich sediments, and due to variations in the depth to basement (Fig. D.19). The shapes of the anomalies may reflect the shape of intrusions or extrusions, or the shape of iron-rich sedimentary layers.

Guide Terms

acceleration (pp. 342, 343)
core-mantle boundary (p. 336)
dipole (p. 345)
dynamo (p. 346)
EarthScope (p. 340)
electromagnet (p. 345)
equipotential surface (p. 342)
field force (p. 340)
geoid (p. 343)
gravimeter (p. 343)

gravitational potential energy (p. 341)
gravity anomaly (p. 343)
inner core (p. 336)
isostasy (p. 344)
isostatic equilibrium (p. 344)
lower mantle (p. 335)
low-velocity zone (LVZ) (p. 334)
magnetic anomaly (p. 346)

magnetic field (p. 345)
magnetic field line (p. 345)
magnetic force (p. 344)
magnetic pole (p. 345)
magnetization (p. 345)
Moho (p. 334)
north magnetic pole (p. 345)
outer core (p. 336)
permanent magnet (p. 345)
P-wave shadow zone (p. 335)
reference geoid (p. 342)
reflection (p. 332)
refraction (p. 333)
seismic ray (p. 331)
seismic-reflection profile (p. 340)

seismic tomography (p. 338)
seismic-velocity discontinuity (p. 334)
self-exciting dynamo (p. 346)
south magnetic pole (p. 345)
spheroid (p. 342)
S-wave shadow zone (p. 336)
transition zone (p. 335)
travel time (p. 331)
upper mantle (p. 335)
velocity-versus-depth curve (p. 338)
wave front (p. 331)
vector (p. 342)

CHAPTER 11

Crags, Cracks, and Crumples: Crustal Deformation and Mountain Building

The complex, swirling curves, visible in this photo of a coastal outcrop in Wales, were once simple horizontal planes between layers of siltstone and shale. Squeezing and shearing, a consequence of convergence between two plates hundreds of millions of years ago, contorted the layers into these geologic structures. Mountain-building processes not only uplift the land, but also bend and/or break rocks, and produce new fabrics. Note the purple flowers for scale.

GEOPUZZLE

Mountain belts include some of the most spectacular scenery on Earth. Why do mountain ranges form? Why are the rocks within these ranges bent and broken? And how long can a mountain range survive?

Innumerable peaks, black and sharp, rose grandly into the dark blue sky, their bases set in solid white, their sides streaked and splashed with snow, like ocean rocks with foam. . . . [Mountains] are nature's poems carved on tables of stone. . . . How quickly these old monuments excite and hold the imagination!

—John Muir (1838–1914), American naturalist

11.1 INTRODUCTION

Geographers call the peak of Mt. Everest "the top of the world," for this mountain, which lies in the Himalayas of south Asia, rises higher than any other on Earth. The cluster of flags on Mt. Everest's summit flap at 8.85 km (29,029 feet) above sea level—almost the cruising height of modern jets. No one can survive very long at the top, for the air there is too thin to breathe. In 1953, Sir Edmund Hillary, from New Zealand, and Tenzing Norgay, a Nepalese guide, became the first to reach the summit. By 2011, more than 3,000 other people had also succeeded—but more than 200 people have died trying. Success depends not just on the skill of the climber, but also on the path of the jet stream, a 200-km-per-hour current of air that flows at high elevations (see Chapter 20). If the jet stream crosses the summit, it engulfs climbers in heat-robbing winds that can freeze a person's face, hands, and feet even if they're swaddled in high-tech clothing.

Mountains draw nonclimbers as well, for everyone loves a vista of snow-crested peaks. The stark cliffs, clear air, meadows, forests, streams, and glaciers of mountains provide a refuge from the mundane. For millennia, mountain beauty has inspired the work of artists and poets. Geologists feel a special fascination with mountains, for they provide one of the most obvious indications of dynamic activity on Earth. To make a mountain, cubic kilometers of rock rise skyward against the pull of gravity. With the exception of the large volcanoes formed over hot spots, mountains do not occur in isolation, but rather as part of linear ranges variously called mountain belts or **orogens** (from the Greek words *oros*, meaning mountain, and *genesis*, meaning formation). Geographers define about a dozen major orogens and numerous smaller ones worldwide (Fig. 11.1).

The process of forming a mountain belt not only raises the surface of the crust, a process called **uplift**, but also causes rocks to undergo **deformation**, a process by which rocks bend, break, or flow in response to compression, tension, or shearing. Deformation produces geologic structures, including **joints**

FIGURE 11.1 Digital map of world topography, showing the locations of major mountain ranges.

(cracks), **faults** (fractures on which one body of rock slides past another), **folds** (bends or wrinkles), and **foliation** (layering resulting from the alignment of mineral grains or the development of compositional bands). Mountain building may also involve metamorphism and igneous activity.

A mountain-building event, or orogeny, may last for tens of millions of years. As the land rises, erosion starts to grind it away, producing sediment and sculpting awesome, jagged topography (**Fig. 11.2**). When the process of uplift ceases, erosion can eventually bevel a range down to near sea level in tens of millions of years. But even after the high peaks are gone, a low-lying belt of fractured, contorted, and metamorphosed rock remains. Such crustal scars serve as a permanent monument to what had once been a region of high peaks.

> **Did you ever wonder . . .**
> do mountain belts last forever?

In this chapter, we learn about deformation, uplift, and other phenomena that happen during mountain building (**See for Yourself K**, p. S-20). We will also discover how to describe and interpret geologic structures. Finally, we consider why mountains form, in the context of plate tectonics theory.

By the end of this chapter, you should understand . . .

- that rocks deform (crack; bend; flow) during mountain building in response to stresses.
- the characteristics of basic geologic structures (joints; faults; folds; foliations) and how to describe them.
- where and why mountain belts form, what processes cause uplift, and their relation to plate tectonics.

11.2 ROCK DEFORMATION IN THE EARTH'S CRUST

Deformation and Strain

As noted above, deformation (bending, breaking, shortening, stretching, or shearing) produces a variety of geologic structures. To get a visual sense of deformation, let's compare a road cut along a highway in a region that has not undergone orog-

FIGURE 11.2 A panorama of the Alps, a mountain range in Europe. The tall peak is the Matterhorn, first climbed in 1865.

eny, the central Great Plains of North America, with a cliff face in a region that has, the Alps of Europe (**Fig. 11.3a, b**).

The road cut, which lies at an elevation of only about 100 m above sea level, exposes nearly horizontal beds of sandstone, shale, and limestone—these beds have the same orientation that they had when first deposited. Sand grains in sandstone beds of this outcrop have a nearly spherical shape (the same shape they had when deposited), and clay flakes in the shale lie roughly parallel to the bedding, because of compaction. Rock of this outcrop is undeformed, meaning that it contains no geologic structures other than a few joints (**Fig. 11.3c**).

In an Alpine cliff, exposed at an elevation of 3 km, rocks look very different. Here, we find layers of quartzite, slate, and marble (the metamorphic equivalent of sandstone, shale, and limestone) in contorted beds whose surfaces are curved. These wrinkles are folds (**Fig. 11.3d**). Grains in the quartzite may resemble flattened eggs, and the clay flakes in the

slate are aligned at a steep angle to the bedding, forming slaty cleavage (see Chapter 8). Finally, if we try tracing the quartzite and slate layers along the outcrop face, we might find that they abruptly terminate at a sloping surface marked by broken-up rock. This surface is a fault. In the example of Figure 11.3b, thick layers of marble lie below this surface, so the fault juxtaposes two different rock units—the quartzite and slate must have moved along the fault from where they first formed to get to their present location.

Clearly, the beds in the Alpine cliff have been deformed, and as a result the cliff exposes a variety of geologic structures (faults, folds, foliation). Beds no longer have the same shape and position that they had when first formed, and the shape and orientation of grains has changed. In sum, deformation includes one or more of the following (**Fig. 11.4a–c**): (1) a change in location (displacement); (2) a change in orientation (rotation); and (3) a change in shape (distortion).

FIGURE 11.3 Deformation changes the character and configuration of rocks.

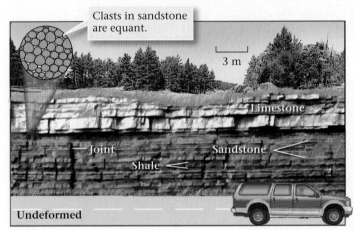

(a) Flat-lying beds of strata along a highway in the Great Plains of North America are essentially undeformed. A few joints, formed when overlying rock eroded away, are visible.

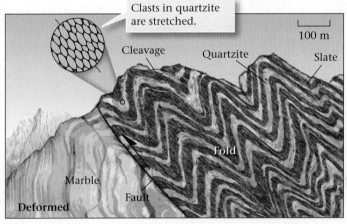

(b) In a mountain belt, deformation may cause layers to undergo folding and faulting. In addition, foliation (such as slaty cleavage) and stretched clasts may develop.

(c) Another example of undeformed, horizontal beds. These form a 100-m-high sea cliff in western Ireland. The vertical lines are joints.

(d) Another example of deformed beds. These have arched into a fold, along the coast of Wales. Note how the beds look "wrinkled."

FIGURE 11.4 The components of deformation include displacement, rotation, and distortion.

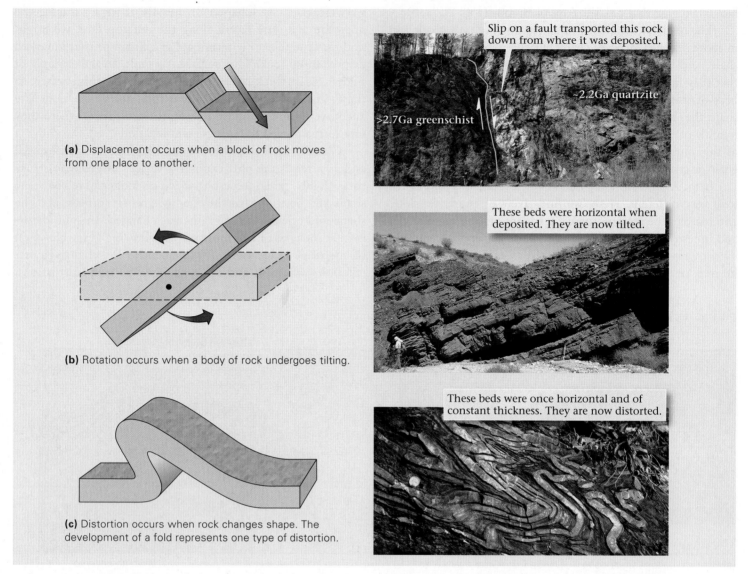

(a) Displacement occurs when a block of rock moves from one place to another.

(b) Rotation occurs when a body of rock undergoes tilting.

(c) Distortion occurs when rock changes shape. The development of a fold represents one type of distortion.

Slip on a fault transported this rock down from where it was deposited.

>2.7Ga greenschist

~2.2Ga quartzite

These beds were horizontal when deposited. They are now tilted.

These beds were once horizontal and of constant thickness. They are now distorted.

Deformation can be fairly obvious when observed in an outcrop (See Fig. 11.3c, d).

Geologists refer to a measure of the *change in shape* or distortion that deformation causes as **strain**. We distinguish among different kinds of strain according to the nature of the shape change. If a layer of rock becomes longer, it has undergone stretching, but if the layer becomes shorter, it has undergone shortening (Fig. 11.5a–c). If a change in shape involves the movement of one part of a rock body past another so that angles between features in the rock change, the result is called shear strain (Fig. 11.5d, e ▶).

Brittle versus Ductile Deformation

Imagine that a plate tumbles off a table and lands on a hard floor—the plate breaks and smashes into pieces. Similarly, if you strike a glass window with a ball, the window cracks and

may even shatter. Such phenomena serve as familiar examples of **brittle deformation** (Fig. 11.6a, b). Now, imagine that you squeeze a ball of soft dough between a book and a tabletop—the dough flattens into a pancake. Similarly, if you bend a stick of chewing gum, it changes from a plane into a curve. During such **ductile deformation**, objects change shape *without* visibly breaking (Fig. 11.6c, d).

What actually happens within mineral grains during these two different kinds of deformation? Recall that the atoms that make up mineral grains are connected by chemical bonds. During brittle deformation, many of the bonds break and stay broken, leading to the formation of a permanent crack across which material no longer connects. During ductile deformation, simplistically, some bonds break but new ones quickly form. In this way, the atoms within grains rearrange, and the grains change shape without permanent cracks forming.

FIGURE 11.5 Different kinds of strain in rock. Strain is a measure of the distortion, or change in shape, that takes place in rock during deformation.

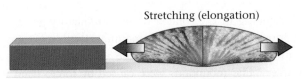

(a) An unstrained cube and an unstrained fossil shell (brachiopod).

(b) Horizontal stretching changes the cube into a horizontal brick and elongates the shell.

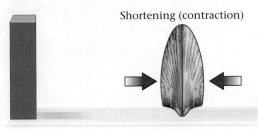

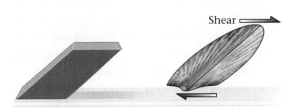

(c) Horizontal shortening changes the cube into a vertical brick and makes the shell narrower.

(d) Shear strain tilts the cube and transforms it into a parallelogram, and changes angular relationships in the shell.

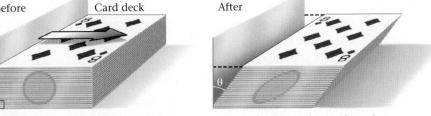
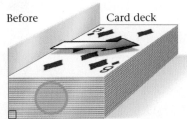
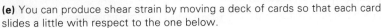

(e) You can produce shear strain by moving a deck of cards so that each card slides a little with respect to the one below.

Why do rocks inside the Earth sometimes deform brittlely and sometimes ductilely? The behavior of a rock depends on:

- *Temperature*: Warmer rocks tend to deform ductilely, whereas colder rocks tend to deform brittlely. To see this contrast, try an experiment with a candle. Chill a candle in a freezer, then press its middle against the edge of a table—the candle will brittlely snap in two. But if you first warm the candle in an oven, it will ductilely bend without breaking when pressed against the table. Heat makes materials softer.

- *Pressure*: Under great pressures deep in the Earth, rock behaves more ductilely than it does under low pressures near the surface. Pressure effectively prevents rock from separating into fragments.

- *Deformation rate*: A sudden change in shape causes brittle deformation, whereas a slow change in shape causes ductile deformation. For example, if you hit a marble bench with a hammer, it shatters, but if you leave the bench alone for a century, it gradually sags without breaking.

- *Composition*: Some rock types are softer than others; for example, halite (rock salt) deforms ductilely under conditions in which granite deforms brittlely.

Considering that pressure and temperature both increase with depth in the Earth, geologists find that, in typical continental crust, rocks generally behave brittlely above about 10 to 15 km, and ductilely below this depth. The depth at which this change in behavior takes place is called the brittle-ductile transition. Earthquakes in continental crust happen only above this depth because these earthquakes involve brittle breaking.

In some cases, you can see both brittle and ductile structures in the same outcrop. For example, our Alpine cliff (Fig. 11.3b) displays both faulting (brittle deformation) and folding (ductile deformation). Such an occurrence may seem like a paradox at first. But a juxtaposition of different styles can happen because of changes in the deformation rate during orogeny. Slow deformation yielded the folds, whereas a pulse of rapid deformation caused the fault to form.

Force, Stress, and the Causes of Deformation

Up to this point, we've focused on picturing the *consequences* of deformation. Describing the *causes* of deformation is a bit more challenging in the context of an introductory geology book. In captions for displays about mountain building, museums and national parks typically dispense with the issue by using the phrase "The mountains were caused by forces deep within the Earth." But what does this mean? Isaac Newton stated that a force can cause an object to speed up, slow down, or change direction. In the context of geology, plate interactions and continent-continent collisions apply forces to rock and thus cause rock to change location, orientation, or shape. In other words, the application of forces in the Earth indeed causes deformation.

Geologists, however, use the word *stress* instead of *force* when talking about the cause of deformation. We define the **stress** acting on a plane as the force applied *per unit area* of the plane. The need to distinguish between stress and force arises because the actual consequences of applying a force depend not just on the amount of force but also on the area over which the force acts. A pair of simple experiments shows why (Fig. 11.7a). Experiment 1: Stand on a single, empty aluminum can. All of your weight—a force—focuses entirely on the can, and the can crushes. Experiment 2: Place a board atop 100 cans and stand on the board. In this case, your weight is distributed across 100 cans, and the cans don't crush. In both experiments, the force caused by the weight of your body was the same, but in Experiment 1 the force was applied over a small area so a large stress developed, whereas in Experiment 2 the same force was applied over a large area so only a small stress developed; a large stress could crush a can, but a small one could not. How does this concept apply to geology? During mountain building, the force of one plate interacting with another is distributed across the area of contact between the two plates, so the deformation resulting at any specific location actually depends on the stress developed at that location, not on the total force produced by the plate interaction.

Different kinds of stress occur in rock bodies (Fig. 11.7b–e). **Compression** takes place when a rock is squeezed, **tension** occurs when a rock is pulled apart, and **shear stress** develops when one part of a rock body moves sideways past another. **Pressure** refers to a special stress condition that happens when the same push acts on all sides of an object.

Note that *stress* and *strain* have very different meanings to geologists, even though we tend to use them interchangeably in everyday English: stress refers to the amount of force applied per unit area of a rock, whereas strain refers to a change in shape of a rock. Thus, stress *causes* strain. Specifically, compression causes shortening, tension leads to stretching, and shear stress produces shear strain. Pressure can cause an object to become smaller, but will not cause it to change shape. With our knowledge of stress and strain, we can now look at the nature and origin of various classes of geologic structures.

FIGURE 11.6   Brittle versus ductile deformation.

(a) Brittle deformation occurs when you drop a plate and it shatters.

(b) Cracks in an Australian outcrop are due to brittle deformation.

(c) Ductile deformation occurs when you squash a ball of dough.

(d) Swirls in the marble of this Brazilian cliff formed ductilely.

Take-Home Message

- Strain refers to the change in shape (e.g., shortening or stretching) of rock that develops during deformation.
- During brittle deformation, rock fractures; during ductile deformation, rock flows without breaking.
- Temperature, pressure, composition, and strain rate determine whether a rock behaves brittlely or ductilely
- Stress refers to compression, tension, and shear.

THINK: What is the difference between stress and strain to a geologist?

11.3 JOINTS AND VEINS: NATURAL CRACKS IN ROCKS

If you look at the photographs of rock outcrops in this book, you'll notice thin dark lines that cross the rock faces. These lines represent traces of natural cracks along which the rock broke and separated into two pieces during brittle deformation. Geologists refer to such natural cracks as **joints** (Fig. 11.8a, b). Rock bodies do *not* slide past each other on joints. Since joints are roughly planar structures, we define their orientation by their strike and dip, as described in **Box 11.1**.

> **Did you ever wonder...**
> do rocks ever "break" in nature?

Joints develop in response to tensile stress in brittle rock: a rock splits open because it has been pulled slightly apart. Joints may form for a variety of geologic reasons. For example, some joints form when a rock cools and contracts, because contraction makes one part of a rock pull away from the adjacent part. Others develop when rock formerly at depth undergoes a decrease in pressure as overlying rock erodes away, and thus changes shape slightly. Still others form when rock layers bend.

Rock bodies may contain two categories of joints. *Systematic* joints are long planar cracks that occur fairly regularly through a rock body, whereas *nonsystematic* joints are short cracks that occur in a range of orientations and are randomly spaced. A group of systematic joints constitutes a joint set, a spectacular example of which can be seen in sandstone beds of Arches National Park, in Utah (see Fig. 11.8a). Erosion along the joints produced narrow gullies. In sedimentary rocks, systematic joints typically are vertical planes (see Fig. 11.8b).

If groundwater seeps through joints for a long period of time, minerals such as quartz or calcite may precipitate out of the groundwater and fill the joint. Such mineral-filled joints are called **veins** and look like white stripes cutting across a body of rock (Fig. 11.8c). Some veins contain small quantities of valuable metals, such as gold.

FIGURE 11.7 There are several kinds of stress.

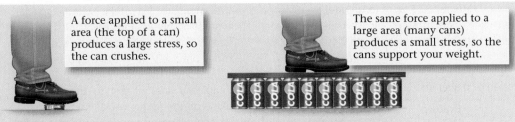

A force applied to a small area (the top of a can) produces a large stress, so the can crushes.

The same force applied to a large area (many cans) produces a small stress, so the cans support your weight.

(a) The difference between stress and force. Stress is the force per unit area.

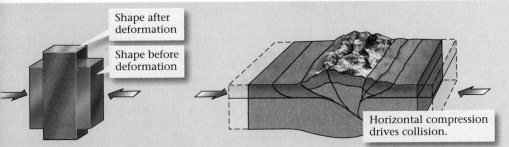

Shape after deformation

Shape before deformation

Horizontal compression drives collision.

(b) Compression takes place when an object is squeezed.

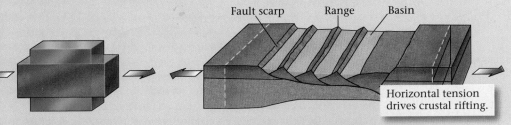

Fault scarp   Range   Basin

Horizontal tension drives crustal rifting.

(c) Tension occurs when the opposite ends of an object are pulled in opposite directions.

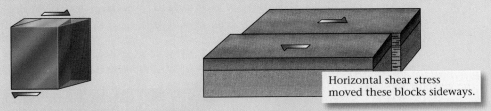

Horizontal shear stress moved these blocks sideways.

(d) Shear stress develops when one surface of an object slides relative to the other surface.

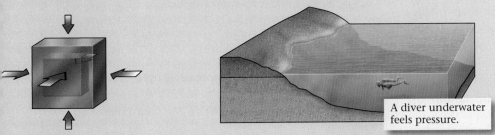

A diver underwater feels pressure.

(e) Pressure occurs when an object feels the same stress on all sides.

BOX 11.1

Describing the Orientation of Geologic Structures

When discussing geologic structures, it's important to be able to communicate information about their orientation. For example, does a fault exposed in an outcrop at the edge of town continue beneath the nuclear power plant 3 km to the north, or does it go beneath the hospital 2 km to the east? If we knew the fault's orientation, we might be able to answer this question. To describe the orientation of a geologic structure, geologists picture the structure as a simple geometric shape, then specify the angles that the shape makes with respect to a horizontal plane (a flat surface parallel to sea level), a vertical plane (a flat surface perpendicular to sea level), and the north direction (a line of longitude).

Let's start by considering planar structures such as faults, beds, and joints.

We call these structures *planar* because they resemble a geometric plane. A planar structure's orientation can be specified by its strike and dip. The **strike** is the angle between an imaginary horizontal line (the strike line) on the structure and the direction to true north (**Fig. Bx11.1a, b**). We measure the strike with a special type of compass (**Fig. Bx11.1c**). The **dip** is the angle of the structure's slope—more precisely, it is the angle between a horizontal plane and the dip line (an imaginary line parallel to the steepest slope on the structure), as measured in a vertical plane perpendicular to the strike (**Fig. Bx11.1d**). We measure the dip angle with a clinometer, a type of protractor. A horizontal plane has a dip of 0°, and a vertical plane has a dip of 90°. We represent

FIGURE Bx11.1 Specifying the orientation of planar and linear structures.

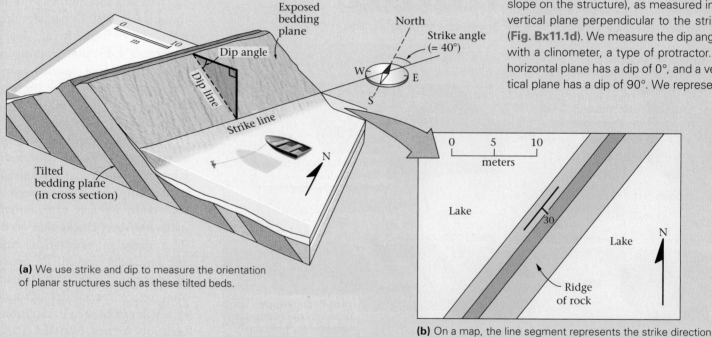

(a) We use strike and dip to measure the orientation of planar structures such as these tilted beds.

(b) On a map, the line segment represents the strike direction and the tick on the segment represents the dip direction. The number indicates the dip angle as measured in degrees.

Geotechnical engineers, people who study the geologic setting of construction sites, pay close attention to jointing when recommending where to put roads, dams, and buildings. Water flows much more easily through joints than it does through solid rock, so it would be a bad investment to situate a water reservoir over rock containing lots of joints—the water would leak down into the joints. Also, building a road on a steep cliff composed of jointed rock could be risky, for joint-bounded blocks separate easily from bedrock, and the cliff might collapse.

Take-Home Message

- Joints are natural cracks in rocks, formed in response to cooling, removal of overlying rock, or bending of rock layers.
- Systematic joints are relatively long, and they are spaced roughly evenly.
- Veins form when minerals precipitate out of water solutions in joints.

THINK: How can you distinguish a joint from a vein on an outcrop?

strike and dip on a geologic map using the symbol shown in Figure Bx11.1b.

A *linear* structure resembles a geometric line rather than a plane; examples of linear structures include scratches or grooves on a rock surface. Geologists specify the orientation of linear structures by giving their plunge and bearing (**Fig. Bx11.1e**). The plunge is the angle between a line and horizontal in the vertical plane that contains the line. A horizontal line has a plunge of 0°, and a vertical line has a plunge of 90°. The bearing is the compass heading of the line—more precisely, the angle between the projection of the line on the horizontal plane and the direction to true north.

The edge of the compass is parallel to the strike.

The intersection of the water and the rock is horizontal.

(c) Geologists use a Brunton compass to measure strike and dip.

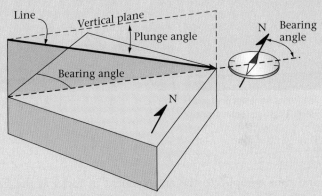

Line

Vertical plane

Plunge angle

Bearing angle

Bearing angle

N

N

Bearing angle

(e) To specify the orientation of a line, we use plunge and bearing.

Dipping beds intersect a calm sea.

Bedding

(d) The intersection of dipping beds with the horizontal water surface is a strike line. The slope of the bed surface is the dip.

11.4 FAULTS: SURFACES OF SLIP

After the San Francisco earthquake of 1906, geologists found a rupture that had torn the land surface near the city. Where this rupture crossed orchards, it offset rows of trees, and where it crossed a fence, it broke the fence in two; the western side of the fence moved northward by about 2 m (see Fig. 10.5a). The rupture represents the trace of the San Andreas Fault. As we have seen, a **fault** is a fracture on which sliding occurs, and slip events, or faulting, can generate earthquakes. Faults, like joints, are planar structures, so we represent their orientation by strike and dip.

Faults have formed throughout Earth history. Some are currently *active* in that sliding has been occurring on them in recent geologic time, but most are *inactive*, meaning that sliding on them ceased long ago. Some faults, such as the San Andreas, intersect the ground surface and thus displace the ground when they move. Others accommodate the sliding of rocks in the crust at depth and remain invisible at the surface unless they are later exposed by erosion.

Geologists study faults not only because the movement on some faults causes earthquakes but also because they juxtapose bodies of rock that did not originally lie adjacent to each other

FIGURE 11.8   Examples of joints and veins.

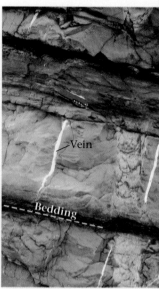

(a) Prominent vertical joints cut red sandstone beds in Arches National Park, Utah, as seen from the air.

(b) Vertical joints on a cliff face in shale near Ithaca, New York.

(c) Milky white quartz veins cut across gray limestone beds.

and thus complicate the arrangement of rocks at the Earth's surface. For example, in our Alpine cliff, Fig. 11.3b, movement on a fault placed quartzite and slate beds against marble beds. Geologists must understand these rearrangements in order to interpret orogenic events, and to predict where resources lie underground.

Fault Classification

Geologists have developed terminology to classify faults and describe movement on them. The fault plane dip (see Box 11.1) can be vertical, horizontal, or at some angle in between. We define the hanging-wall block as the rock above a non-vertical fault plane, and the footwall block as the rock below the fault plane (Fig. 11.9a). If you stand in a tunnel along a fault plane, the hanging-wall block looms over your head, and the footwall block lies under your feet. With this terminology, we can distinguish several types of faults (Fig. 11.9b–d).

- *Dip-slip versus strike-slip versus oblique-slip faults*: On dip-slip faults, sliding occurs up or down the slope of the fault (therefore, up or down the dip); on **strike-slip faults**, one block slides past another horizontally (therefore, parallel to the strike line); and on oblique-slip faults, sliding occurs diagonally on the fault plane.
- *Types of dip-slip faults*: We subdivide dip-slip faults into two kinds, depending on which way the hanging-wall block moves relative to the footwall block. On **thrust faults** and **reverse faults**, the hanging-wall block

moves up the slope of the fault. Thrust faults differ from reverse faults only in terms of the fault plane's slope (or dip). Generally, the term thrust fault refers to surfaces with a dip of less than about 30°. On **normal faults**, the hanging-wall block moves down the slope of the fault. "Normal" and "reverse" are relics of nineteenth-century miners' jargon. Normal faults were simply more common in the mines where faults were first recognized. But globally, normal faults aren't any more common or typical than reverse faults.

- *Types of strike-slip faults*: Geologists distinguish between two types of strike-slip faults, based on the relative movement of one side of the fault with respect to the other. If you stand facing the fault, you can say it is a *left-lateral* strike-slip fault if the block on the far side slipped to your left, and that it is a *right-lateral* strike-slip fault if the block on the far side slipped to your right. Note that strike-slip faults commonly have a vertical dip, so we generally cannot define the hanging-wall or footwall block on such faults.

Recognizing Faults

How do you recognize a fault when you see one? The most obvious criterion is the occurrence of **displacement**, or offset, meaning the amount of movement across a fault plane. Displacement disrupts the layers in rocks, so that layers on one side of a fault are not continuous with layers on the other

FIGURE 11.9 The different categories of faults.

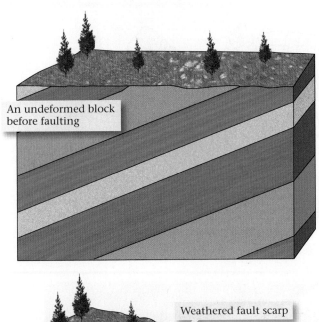

An undeformed block before faulting

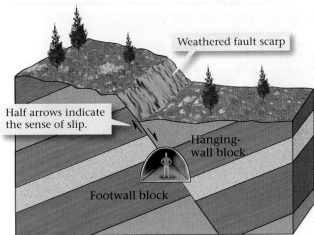

Weathered fault scarp

Half arrows indicate the sense of slip.

Hanging-wall block

Footwall block

(a) The block of crust above a non-vertical fault is the hanging wall, whereas the block below is the footwall.

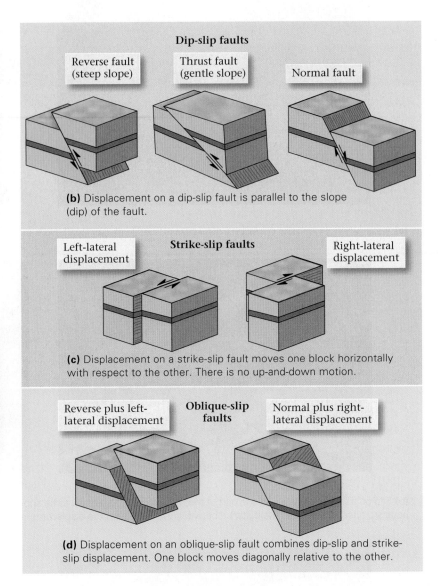

Dip-slip faults

Reverse fault (steep slope)

Thrust fault (gentle slope)

Normal fault

(b) Displacement on a dip-slip fault is parallel to the slope (dip) of the fault.

Strike-slip faults

Left-lateral displacement

Right-lateral displacement

(c) Displacement on a strike-slip fault moves one block horizontally with respect to the other. There is no up-and-down motion.

Oblique-slip faults

Reverse plus left-lateral displacement

Normal plus right-lateral displacement

(d) Displacement on an oblique-slip fault combines dip-slip and strike-slip displacement. One block moves diagonally relative to the other.

side (Fig. 11.10a, b ⏺). The displacement on the fault shown in Figure 11.10a is about 2 m. In some cases, faults juxtapose two different rock units (for example, Fig. 11.3b). Note that in the example of Figure 11.3, we can't see distinctive markers, so without additional information we can't measure the displacement directly. Typically, thrust or reverse faults cutting sedimentary beds place older beds on younger ones, whereas normal faults place younger beds on older. In some cases, layers of rock cut by a fault undergo folding during or just before slip; the resulting folds are informally called drag folds (Fig. 11.4c).

Faults may also leave their mark on the landscape. Those that intersect the ground surface while they are active can displace natural landscape features (such as stream valleys or glacial moraines; **Fig. 11.10c** ⏺) and human-made features

(such as highways, fences, or rows of trees in orchards). Displacement on a dip-slip or oblique-slip fault will make a small step on the ground surface; this step is called a **fault scarp** (Fig. 11.11a). And because faults tend to break up and weaken rock, the fault may be preferentially eroded. If this happens, the fault trace (the line of intersection between the fault and the ground surface) will be marked by a linear valley (see inset photo on p. 293).

Fault surfaces and their borders typically look different from bedding planes. For example, faulting under brittle conditions may crush or break adjacent rock. If this shattered rock consists of visible angular fragments, then it is called fault breccia (**Fig. 11.11b**), but if it consists of a fine powder, then it is called fault gouge. Some fault surfaces are polished and

FIGURE 11.10 Recognizing fault displacement in the field. ▶❚

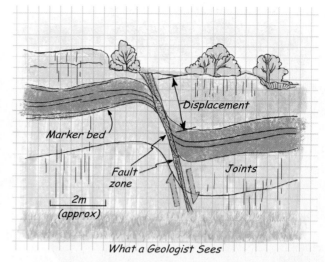

What a Geologist Sees

(a) A steep normal fault has displaced a distinctive red bed (a marker layer). Note that the displacement formed a 0.5-m-wide fault zone of broken rock.

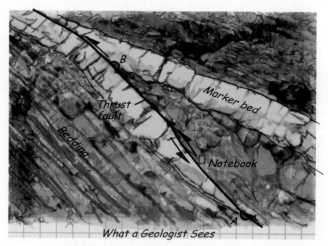

What a Geologist Sees

(b) Slip on a thrust fault caused one part of the light-colored marker bed to be shoved over another part, as emphasized in the drawing. Note that the beds are tilted to the right. The distance between the red dots is the displacement.

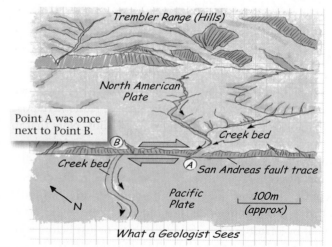

What a Geologist Sees

(c) An aerial photograph of a portion of the San Andreas Fault. As emphasized by the sketch, the fault offsets a stream channel in a right-lateral sense.

FIGURE 11.11 Features of exposed fault surfaces.

A scarp due to slip on a normal fault in Nevada.

(a) A fault scarp develops where faulting displaces the land surface.

(b) This fault breccia consists of irregular fragments of light-colored rock.

Striation orientation

(c) Striations on the surface of a strike-slip fault look like grooves.

grooved by the movement of the hanging wall past the foot-wall. Polished fault surfaces are called slickensides, and linear grooves on fault surfaces are slip lineations (Fig. 11.11c). We specify the orientation of a slip lineation by giving its plunge and bearing (see Box 11.1).

The shear on some faults takes place under ductile conditions at depth in the crust. Where this happens, rock does not break up into breccia or gouge along the fault zone, because the rock is too soft; rather it shears ductilely to form a band of fine-grained foliated rock called mylonite. The fine grain size of mylonite results not from brittle fracturing during shear, but rather from a type of metamorphic recrystallization that subdivides large grains into small ones. Geologists refer to faults in which movement occurred ductilely as **shear zones** (Fig. 11.12).

Fault Systems and the Resulting Strain

Numerous related faults are often found in groups called fault systems. Individual faults in a system may merge at depth with a nearly horizontal **detachment fault**. The faulting in a thrust-fault system shortens the crust; slip makes slices of the crust overlap like shingles (Fig. 11.13a). Faulting in normal-fault systems, on the other hand, stretches the crust (Fig. 11.13b). Many normal fault systems consist of parallel faults that curve to shallower dips at depth, where they merge with the detachment. As slip progresses on these curved faults, the hanging-wall block rotates, and a wedge-shaped space between the fault surface and the tilted top of

the hanging-wall block develops. This space, called a half graben, fills with sediments to form a basin. Locally, two normal faults of a thrust system may dip toward each other—the keystone-shaped block that drops down between them is a graben, whereas the high block between adjacent grabens is a horst (Fig. 11.14 ⏺).

FIGURE 11.12 An outcrop of a small shear zone in granite, from Michigan. Mylonite in the shear zone is finer-grained and has foliation.

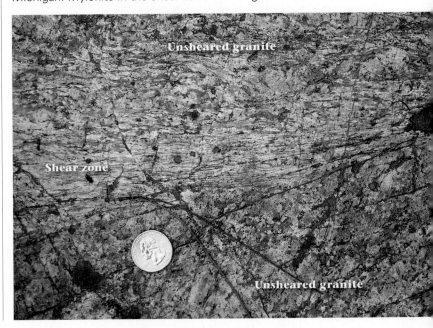

Unsheared granite

Shear zone

Unsheared granite

FIGURE 11.13 Examples of fault systems, arrays of related faults formed in sequence during a deformation event.

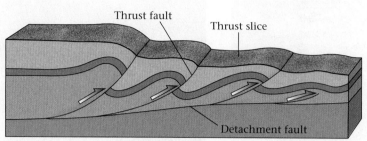

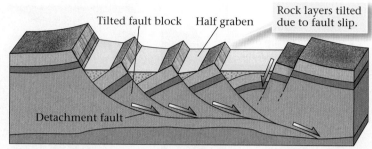

(a) In this thrust-fault system, several related thrust faults merge at depth with a detachment fault. Note that displacement on the thrusts shortens the layers of rock above the detachment.

(b) A normal-fault system consists of several related normal faults. In this example, several faults dip in the same direction and merge with a detachment fault at depth.

FIGURE 11.14 An example of horsts and grabens exposed in the wall of a marble quarry in Brazil.

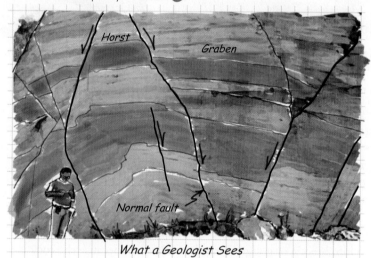

What a Geologist Sees

Take-Home Message

- Faults are fractures on which sliding occurs. On reverse or thrust faults, the hanging wall moves up; on normal faults, the hanging wall moves down.
- Strike-slip faults tend to be vertical; displacement on them is parallel to strike (i.e., horizontal).
- Movement along a fault can crush rock to form breccia and gouge.
- Normal-fault systems develop during crustal stretching, whereas thrust-fault systems develop during crustal shortening.

THINK: How can you recognize a fault in the field?

11.5 FOLDS AND FOLIATIONS

Geometry of Folds

Imagine a carpet lying flat on the floor. Push on one end of the carpet, and it will wrinkle or contort into a series of wave-like curves. Stresses developed during mountain building or during other tectonic processes can similarly warp or bend bedding and foliation (or other planar features) in rock. The result—a curve in the shape of a rock layer—is called a **fold**.

Not all folds look the same—some look like arches, some like troughs, and some have other shapes. To describe these shapes, we must first label the parts of a fold (Fig. 11.15a). The **hinge** refers to a line along which the curvature is greatest, and the **limbs** are the sides of the fold that display less curvature. The **axial surface** is an imaginary plane that contains the hinges of successive layers. With these terms in hand, we can now describe types of folds.

- *Anticlines, synclines, and monoclines*: Folds that have an arch-like shape in which the limbs dip away from the hinge are called **anticlines** (Fig. 11.15a), whereas folds with a trough-like shape in which the limbs dip toward the hinge are called **synclines** (Fig. 11.15b). On the ground surface, the oldest beds crop out near the hinge, and the youngest away from the hinge, in an anticline. The opposite is true in a syncline. A **mono-**

FIGURE 11.15 Geometric characteristics of folds.

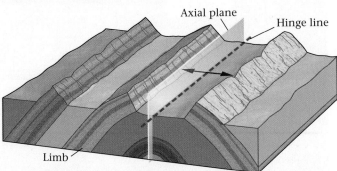

(a) An anticline looks like an arch. The beds dip away from the hinge.

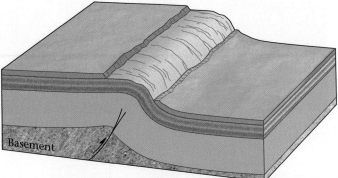

(c) A monocline looks like a stair step, and is commonly draped over a fault block.

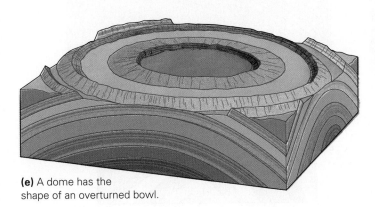

(e) A dome has the shape of an overturned bowl.

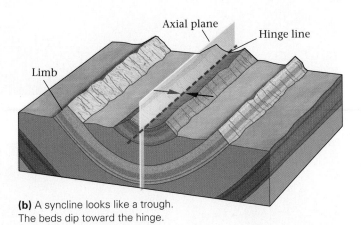

(b) A syncline looks like a trough. The beds dip toward the hinge.

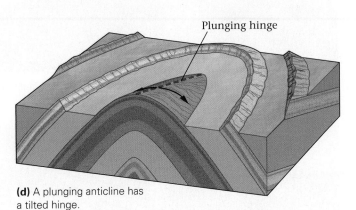

(d) A plunging anticline has a tilted hinge.

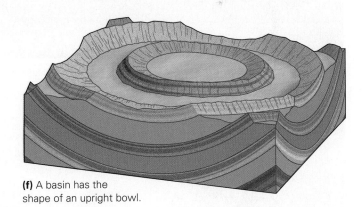

(f) A basin has the shape of an upright bowl.

cline has the shape of a carpet draped over a stair step (Fig. 11.15c).

- *Open and tight folds:* If the angle between the limbs is large, then the fold is an open fold, but if the angle between the limbs is small, then the fold is a tight fold.

- *Nonplunging and plunging folds:* If the hinge is horizontal, the fold is called a nonplunging fold, but if the hinge is tilted, the fold is called a plunging fold (Fig. 11.15d).

- *Domes and basins:* A fold with the shape of an overturned bowl is called a **dome**, whereas a fold shaped like an upright bowl is called a **basin** (Fig. 11.15e, f). Domes and basins both display circular outcrop patterns that look like bull's-eyes—the oldest layer occurs in the center of a dome, whereas the youngest layer is located in the center of a basin. Domes and basins occur in a range of sizes. The largest measure hundreds of kilometers across.

FIGURE 11.16 Characteristics of folds on outcrops and in the landscape.

(a) This anticline, exposed in a road cut near Kingston, New York, involves beds of Paleozoic limestone.

(b) This syncline, exposed in a road cut in Maryland, involves beds of Paleozoic sandstone and shale.

Fold hinge

Fold limb

(c) This fold, exposed along the coast of Brazil, occurs in Precambrian gneiss.

(d) A train of folds exposed in sea cliffs in eastern Ireland includes anticlines and synclines. The folds affect beds of Paleozoic sandstone and shale. ⏸

What a Geologist Sees

Unconformity

Younger sediment

Anticline

Syncline

Axial surface trace

(e) The plunging anticline of Sheep Mountain, Wyoming, is easy to see because of the lack of vegetation. Resistant rock layers (sandstone) stand out as ridges, whereas weak rock layers (shale) erode away.

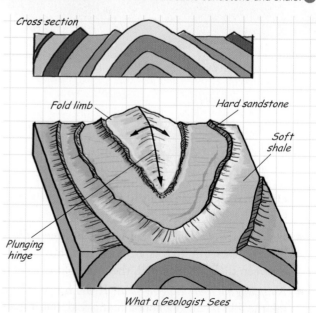

Cross section

Fold limb

Hard sandstone

Soft shale

Plunging hinge

What a Geologist Sees

Now see if you can identify the various folds shown in Figure 11.16a–e 🔊.

You can recognize folds not only in cross-sectional view (a vertical slice through the crust), but also by the pattern of rock layers in map view (a horizontal plane representing the ground surface). For example, a nonplunging anticline involving sedimentary layers appears as a series of parallel stripes, with the oldest layer in the center and progressively younger layers away from the center; the stripes are symmetrically positioned around the hinge. In a nonplunging syncline involving sedimentary layers, the youngest layers crop out in the center and the oldest at the margins. Layers in a plunging fold have a U-shape on the ground surface (Fig. 11.16d, e). We can represent the hinge of the fold by a heavy line bordered by outward-pointing arrows for an anticline and inward-pointing arrows for a syncline. Note that because some layers erode more easily than others, the shapes of folds may be indicated by the pattern of ridges and valleys.

Formation of Folds

Folds develop in two principal ways (Fig. 11.17a, b 🔊). During formation of *flexural-slip* folds, a stack of layers bends, and slip occurs between the layers. The same phenomenon happens when you bend a deck of cards—to accommodate the change in shape, the cards slide with respect to each other. *Passive-flow* folds form when the rock, overall, is so soft that it behaves like weak plastic and slowly flows; these folds develop simply because different parts of the rock body move at different rates.

Why do folds form? Some layers wrinkle up, or buckle, in response to end-on compression (Fig. 11.18a–d). Others form where shear stress gradually shifts one part of a layer up and over another part. Still others develop where rock layers move up and over step-like bends in a fault and must curve to conform with the fault's shape. Finally, some folds form when new slip on a fault causes a block of basement to move up so that the overlying sedimentary layers must warp.

FIGURE 11.17 Fold development in flexural-slip and passive-flow folding. 🔊

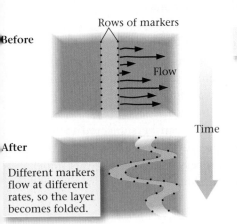
Before

After

Time

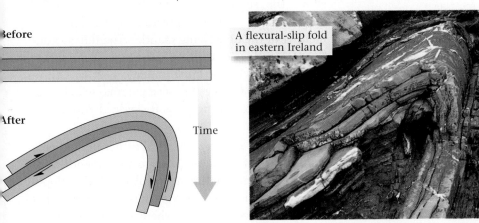
A flexural-slip fold in eastern Ireland

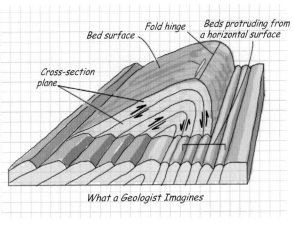
Bed surface — Fold hinge — Beds protruding from a horizontal surface

Cross-section plane

What a Geologist Imagines

(a) During the formation of flexural-slip folds, layers maintain constant thickness, so for the fold to form, layers must bend. To accommodate the bending, each bed slips relative to its neighbor.

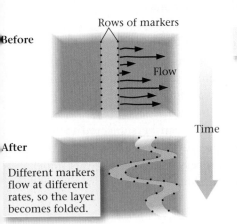
Rows of markers

Before

Flow

Time

After

Different markers flow at different rates, so the layer becomes folded.

A passive-flow fold in northern Scotland

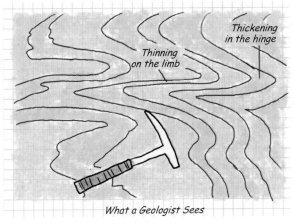
Thickening in the hinge

Thinning on the limb

What a Geologist Sees

(b) During the formation of passive-flow folds, the rock slowly flows. Different points along a marker line flow at different rates, causing the layer to become folded. Note that the layer's thickness doesn't stay constant during folding.

FIGURE 11.18 Folding is caused by several different processes as illustrated by the following cross sections.

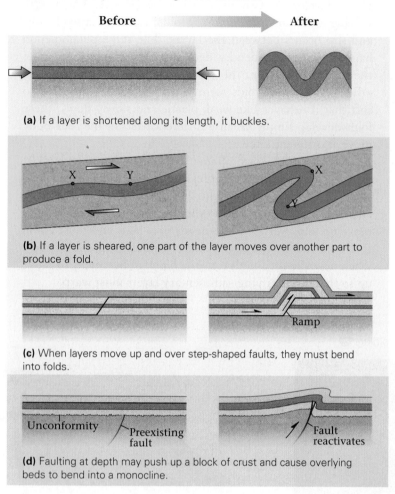

(a) If a layer is shortened along its length, it buckles.

(b) If a layer is sheared, one part of the layer moves over another part to produce a fold.

(c) When layers move up and over step-shaped faults, they must bend into folds.

(d) Faulting at depth may push up a block of crust and cause overlying beds to bend into a monocline.

Tectonic Foliation in Rocks

In an undeformed sandstone, the grains of quartz are roughly spherical, and in an undeformed shale, clay flakes press together into the plane of bedding so that shales tend to split parallel to the bedding. During ductile deformation, however, internal changes take place in a rock that gradually modify the original shape and arrangement of grains. For example, quartz grains may transform into cigar shapes, elongate ribbons, or tiny pancakes, and clay flakes may recrystallize or reorient so that they lie at an angle to the bedding. Overall, deformation can produce inequant grains and can cause them to align parallel to each other. We refer to layering developed by the alignment of grains in response to deformation as **tectonic foliation**.

We introduced foliation, such as slaty cleavage, schistosity, and gneissic layering, in Chapter 8 while discussing the effects of metamorphism. Here we add to the story by noting that such foliation forms in response to flattening and shearing in ductilely deforming rocks—in other words, foliation indicates that the rock has developed a strain (Fig. 11.19a, b ◖). For example, in rocks with slaty cleavage, the cleavage planes lie perpendicular to the direction of the shortening strain, so the cleavage may be parallel to the axial plane of folds. In schists and gneisses, the foliation commonly lies parallel to or at a slight angle to planes on which shear takes place, because shear smears grains out (Fig. 11.20 ◖).

Forming Rocks in and Near Mountains

The process of orogeny establishes geologic conditions appropriate for the formation of a great variety of rocks. We'll consider examples from all three rock categories (Fig. 11.21):

- *Igneous activity during orogeny*: In convergent plate boundaries, melting takes place in the mantle above the subducting plate. In rifts, stretching and thinning of lithosphere causes decompression melting of the underlying mantle. And during continental collision, melting may take place where deep portions of the crust undergo heating. All of these melting regimes produce magma, which rises and freezes to form igneous rocks in the overlying mountains.

- *Sedimentation during orogeny*: Weathering and erosion in mountain belts generate vast quantities of sediment. This sediment tumbles down slopes and gets carried away by glaciers or streams that transport it to low areas where it accumulates in large fans or deltas. In some locations the weight of mountain belts pushes down the surface of the lithosphere, thereby producing a deep sedimentary basin at the border of the range.

- *Metamorphism during orogeny*: Contact metamorphic aureoles (see Chapter 8) form adjacent to igneous intrusions in orogens. And regional metamorphism occurs where mountain building thrusts one part of the crust over another; for when this happens, rock of the footwall ends up at great depth and thus can be subjected to high temperature and pressure. Because deformation accompanies this process, the resulting metamorphic rocks contain tectonic foliation.

FIGURE 11.19 The development of tectonic foliation in rock. ⏸

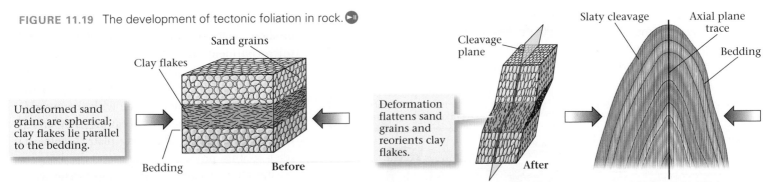

(a) Compression shortens beds, flattens sand grains, and reorients clay flakes. Clay flakes were originally parallel to bedding, but they become parallel to slaty cleavage during deformation. Folding may accompany cleavage formation.

What a Geologist Sees

(b) An example of slaty cleavage developed in Paleozoic strata exposed in a stream cut in New York. Relict bedding is still visible, but note that the rock breaks more easily on the cleavage. This cleavage formed in association with folding.

FIGURE 11.20 Development of a foliation due to shearing. In this rock, quartz and mica-rich layers have become very fine-grained and strongly foliated. The foliation is parallel to the shearing direction. ⏸

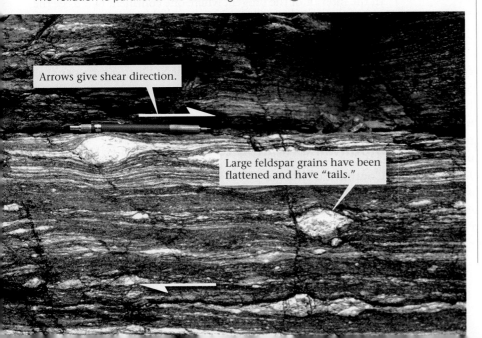

> Arrows give shear direction.

> Large feldspar grains have been flattened and have "tails."

Take-Home Message

- Folds are bends or curves defined by the shape of rock layering; anticlines are arch-shaped, synclines trough-shaped, and monoclines step-shaped.
- Basins are bowl-like, whereas domes are shaped like upside-down bowls.
- Folds can develop due to end-on compression, shear, movement of layers over fault bends, or a push from below.
- Foliation, defined by aligned flattened minerals, forms in response to compression and shear.
- Distinctive igneous and metamorphic rocks form in mountain belts; basins adjacent to mountain belts accumulate sediment.

THINK: What mechanisms allow layers to undergo a change in shape during folding?

FIGURE 11.21 Various rocks form during orogeny. Sediment eroded from the orogen fills a sedimentary basin next to the mountain; rocks buried deeply in the orogen become squeezed, sheared, and heated to form metamorphic rocks; and igneous rocks intrude from below.

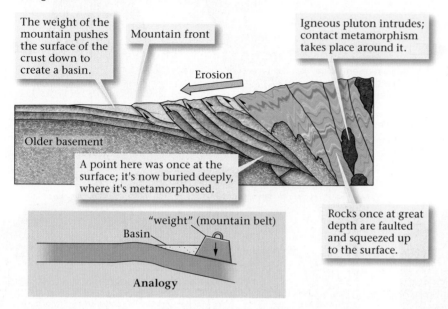

The weight of the mountain pushes the surface of the crust down to create a basin.

Mountain front

Igneous pluton intrudes; contact metamorphism takes place around it.

Erosion

Older basement

A point here was once at the surface; it's now buried deeply, where it's metamorphosed.

"weight" (mountain belt)

Basin

Rocks once at great depth are faulted and squeezed up to the surface.

Analogy

11.6 CAUSES OF MOUNTAIN BUILDING

Did you ever wonder . . .
why do mountains occur in distinct belts?

Before plate tectonics theory became established, geologists were just plain confused about how mountains formed. In the context of the new theory, however, the many processes driving mountain building became clear: mountains form in response to convergent-boundary deformation, continental collisions, and rifting. Since collison zones, rifts, and plate boundaries are linear, mountain belts are linear. Below, we look at these different settings and the types of mountains and geologic structures that develop in each one.

Mountains Related to Subduction

As we have seen in Chapter 9, subduction at a convergent plate boundary produces a volcanic arc. But that's not the only feature to develop in response to plate interactions at such boundaries. At some convergent plate boundaries, compressional stresses develop, and these cause crustal shortening and uplift in the overriding plate. Such shortening may produce a **fold-thrust belt**, in which a thrust system develops (Fig. 11.22a). As a consequence of this faulting, thrust slices (sheets of crust) push up and over their neighbors, and rocks within thrust slices bend and become folded (see Fig. 11.13a). The Andes orogen of western South America displays the rugged topography that can develop in compressional convergent-margin orogens (Fig. 11.22b).

If subduction continues over a long time, offshore volcanic arcs, oceanic plateaus, and microcontinents may drift into the convergent margin (see Fig. 11.22a). Such crustal blocks are too buoyant to subduct and sink back into the mantle, so instead they collide with the overriding plate and "suture" to (meaning they attach to) the edge of the overriding plate. Geologists refer to this process as accretion; the buoyant crustal block is called an **exotic terrane** when it is offshore, and an "accreted terrane" once it has attached to the overriding plate. Once accretion occurs, the convergent plate boundary may jump to the seaward side of the accreted terrane, so that subduction can continue. The process of accretion can add substantial new crust to a convergent-margin orogen. For example, the western half of the North American Cordillera, a region that is up to 500 km wide, consists of accreted terranes (Fig. 11.22c). Orogens that grew laterally by the attachment of exotic terranes have come to be known as **accretionary orogens**.

Mountains Related to Continental Collision

Once the oceanic lithosphere between two continents completely subducts, the continents themselves collide with each other. Continental collision results in the formation of large mountain ranges such as the present-day Himalayas or the Alps and the Paleozoic Appalachian Mountains (See Geology at a Glance, pp. 372–373). The final stage in the growth of the Appalachians happened when Africa and North America collided.

During collision, intense compression generates fold-thrust belts on the margins of the orogen (Fig. 11.23a–c ▶). In the interior of the orogen, where one continent overrides the edge of the other, high-grade metamorphism occurs, accompanied by formation of passive-flow folds and tectonic foliation. During this process, the crust below the orogen thickens to as much as twice its normal thickness. During such **crustal thickening**, rocks squeeze upward in the hanging walls of large thrust faults.

Mountains Related to Continental Rifting

Continental rifts are places where continents are splitting in two. During rifting, stretching causes normal faulting in the brittle crust (Fig. 11.24a). Movement on the normal faults drops down blocks of crust, producing deep, sediment-filled basins separated by narrow, elongate mountain ranges that contain tilted rocks. These ranges are sometimes called fault-

FIGURE 11.22 Characteristics of convergent-margin orogens.

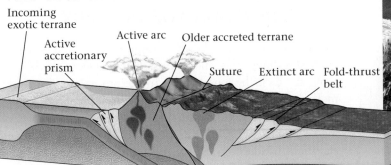

(a) At some convergent margins, compression between the downgoing and overriding plates uplifts a mountain range in which volcanism occurs. Subduction may bring in exotic crustal blocks that collide and become incorporated in the orogen.

(b) The Andes in Chile formed along a convergent margin.

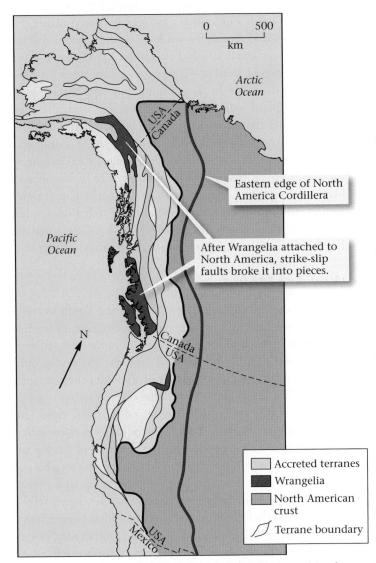

(c) The western portion of the North American Cordillera consists of accreted terranes that attached to the continent during the Mesozoic. During docking, a distinct terrane called Wrangelia (highlighted in red) was sliced into pieces that were displaced by strike-slip faults.

block mountains. Stretching thins the lithosphere, allowing hot asthenosphere to rise and undergo decompression melting (see Chapter 6). This process produces magmas that rise to form volcanoes within the rift. Today, the East African Rift clearly shows the configuration of rift-related mountains and volcanoes. And in North America, rifting yielded the broad Basin and Range Province of Utah, Nevada, and Arizona (Fig. 11.24b).

Take-Home Message

- Stresses that generate mountain belts develop mainly due to plate interactions.
- Compression, due to collision and convergence, shortens the crust horizontally and thickens it vertically.
- Convergent tectonics can attach exotic crustal blocks to continental margins.
- Rifting breaks the crust apart and produces tilted "fault-block" mountains.

THINK: Could fold-thrust belts develop in rifts? Why, or why not?

FIGURE 11.23 Characteristics of collisional orogens. 🔊

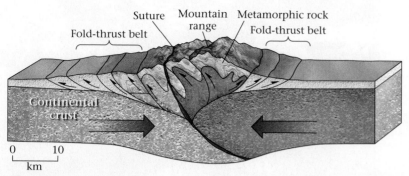

(a) During collision, continents squeeze together and deform. Thrusting brings metamorphic rock up to shallower levels.

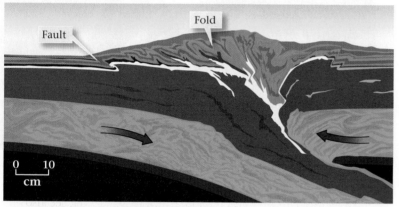

(b) Geologists simulate collision in the laboratory using layers of colored sand. Dragging the left side of the model under the right produces structures and uplift as shown in this sketch of a model.

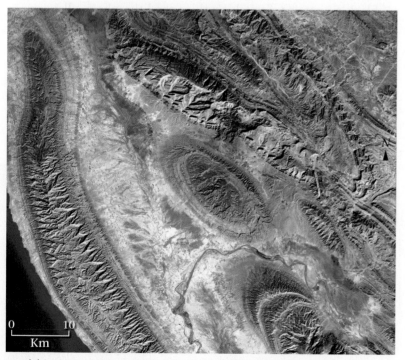

(c) In this satellite photo; the ridges and valleys of the Zagros Mountains along the coast of Iran represent eroded folds of a collisional orogen. Ridges consist of durable sandstone beds.

11.7 UPLIFT AND THE FORMATION OF MOUNTAIN TOPOGRAPHY

Leonardo da Vinci, the Renaissance artist and scientist, enjoyed walking in the mountains, sketching ledges and examining the rocks he found there. In the process, he discovered marine shells (fossils) in limestone beds cropping out a kilometer above sea level, and he suggested that the rock containing the fossils had risen from below sea level up to its present elevation. Modern geologists agree with Leonardo, and now refer to the process by which the surface of the Earth moves vertically from a lower to a higher elevation as **uplift**. Mountain building requires substantial uplift of the Earth's surface (Fig. 11.25). Think about it—to form a mountain range, on the order of 1 million km^3 of rock moves up.

What kinds of distances are we talking about when referring to uplift in mountain ranges? As noted earlier, Mt. Everest rises 8.85 km above sea level. Although this distance may seem monumental—that's equal to 5,000 people standing one on top of another—it represents only about 0.06% of the Earth's diameter. In fact, if the Earth were shrunk to the size of a billiard ball, its surface (mountains and all) would feel smoother than that of an actual billiard ball. Note that, in general, the individual peaks that you see in a mountain range represent only a fraction of the range's total height, for the plains at the base of the mountains may be significantly higher than sea level. Nevertheless, mountain heights are spectacular, and in this section we look at why uplift occurs, how erosion carves rugged landscapes out of uplifted crust, and why Earth's mountains can't get much higher than Mt. Everest.

Why Are Mountains High?

What processes can cause the surface of the Earth to rise to mountainous heights? There are many because, as we have seen, mountain building happens in numerous different geologic settings. To understand how the variety of uplift processes work, we must begin by introducing the concept of isostasy. (See Interlude D for further discussion.)

The lithosphere, which consists of relatively rigid crust and lithospheric mantle, "floats" on the softer asthenospheric mantle below. As a consequence, the elevation of the top surface of the lithosphere, *over a broad region*, represents a balance between buoyancy force pushing lithosphere up, and gravitational force pulling the lithosphere down. Geologists refer to the condition that exists when this balance has been achieved as **isostasy**, or isostatic equilibrium. Put another way, isostasy exists where the elevation of the Earth's surface reflects the level at which the lithosphere naturally floats.

To picture isostasy, imagine placing a block of wood into a bathtub full of water. If the block is less dense than water, it floats, with part of the block remaining above the water surface,

FIGURE 11.24 Rift-related orogens.

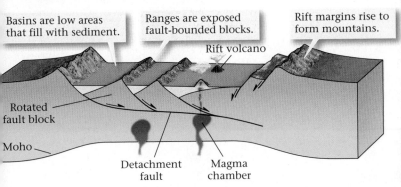

Basins are low areas that fill with sediment.

Ranges are exposed fault-bounded blocks.

Rift margins rise to form mountains.

Rift volcano

Rotated fault block

Moho

Detachment fault

Magma chamber

(a) Rifting leads to the development of numerous narrow mountain ranges. Rift-margin mountains also may form.

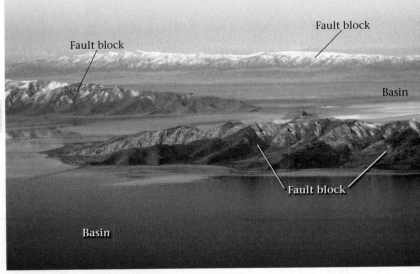

Fault block

Fault block

Basin

Fault block

Basin

(b) The Basin and Range Province of the western United States formed during Cenozoic rifting.

and most of the block submerged below. Recall that the relationship between the amount above and the amount below was first recognized by the Greek scientist Archimedes (287–212 B.C.E.), who pointed out that a floating object sinks, or subsides, only until it has displaced an amount of water whose mass equals the mass of the whole object—this relationship is known as Archimedes' principle (see Box 4.1). Because of Archimedes' principle, the top of denser block sits lower than that of a less-dense block of the same thickness. Similarly, the top surface of a thicker block sits higher than the top surface of a thinner block of the same density. If you were to add another block of wood on top of one that is already floating, the lower block would subside to adjust for the addition, so as to main isostatic equilibrium. Since the asthenosphere, unlike water, can flow at only a few centimeters per year, such isostatic compensation takes a long time.

From our bathtub experiment, we can deduce that any phenomenon that changes the thickness and/or density of layers in the lithosphere will affect the elevation of the land. So to answer the question of why mountain belts rise, we must identify geo-

logic processes that can change the thickness and/or density of layers in the lithosphere. Here are some of the major ones.

Crustal Shortening and Thickening During collisional orogeny or during certain types of convergent-margin orogeny, horizontal compression causes the crust to shorten horizontally and thicken vertically. In fact, the folding, faulting, and plastic flow that take place during such events can almost double the crust's thickness. For example, the crust beneath the tallest range, the Himalayas, is 70 km thick, whereas crust beneath the plains of the central United States is 35 km thick (Fig. 11.26a; see also Geology at a Glance, pp. 372–373). To isostatically compensate for the thickening of the crust (the geologic equivalent to adding another block of low-density wood to the top of a floating block), the base of the crust and underlying lithospheric mantle subside (Fig. 11.26b). Indeed, since the Himalayas are ~8 km high, most of the thickened crust extends downward beneath the range. This downward protrusion of crust is called a **crustal root**. Simplistically, the higher the mountains, the deeper the

FIGURE 11.25 There are over 2 km of vertical relief between the base of Wyoming's Grand Teton Mountains and the peak. And the base already lies at a high elevation. The exposed rocks once lay 12 km or more beneath the surface of the Earth. Clearly, mountain building results in large vertical displacements of the surface of the crust.

The Collision of India with Asia

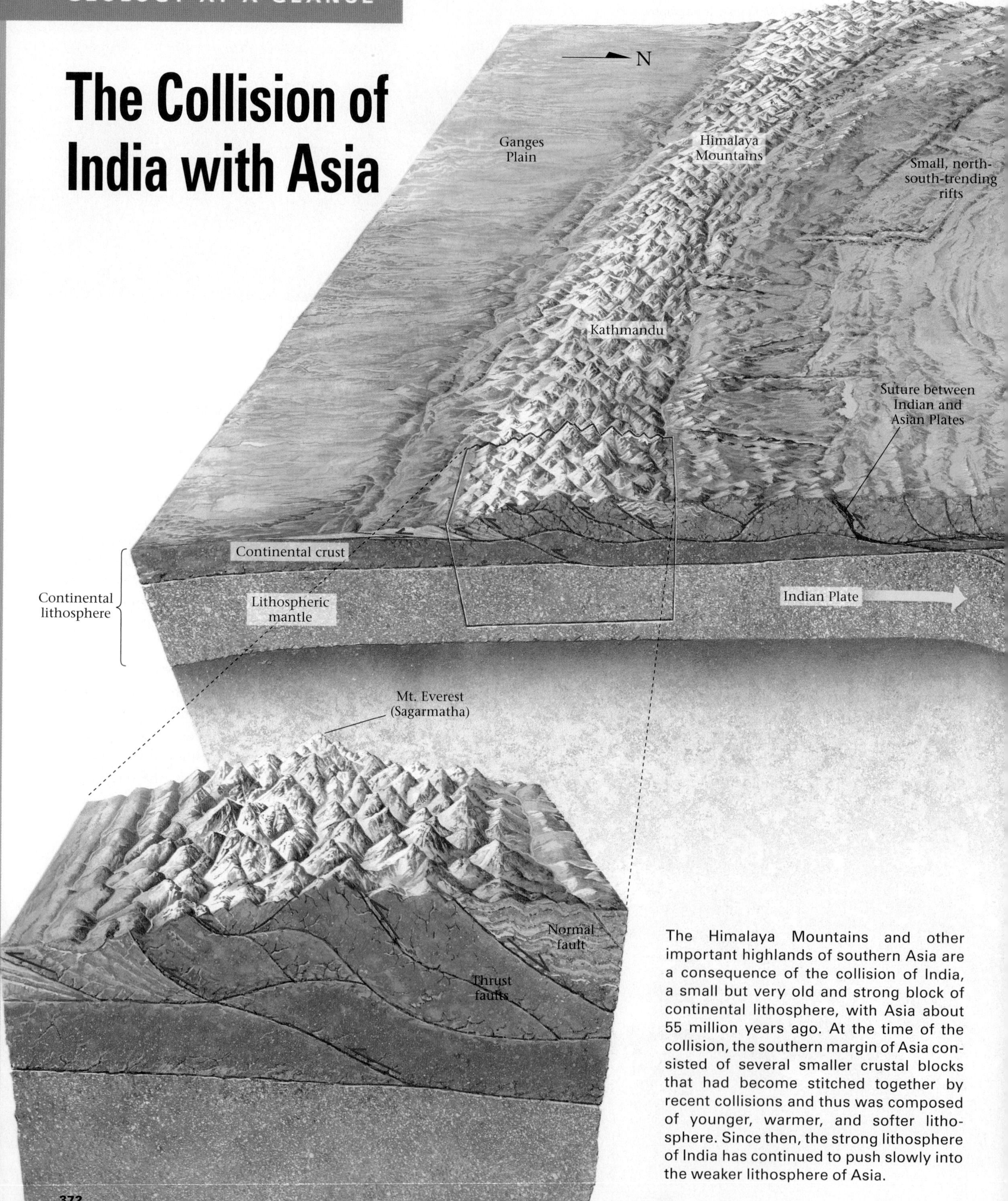

→N

Ganges Plain

Himalaya Mountains

Small, north-south-trending rifts

Kathmandu

Suture between Indian and Asian Plates

Continental crust

Continental lithosphere

Lithospheric mantle

Indian Plate →

Mt. Everest (Sagarmatha)

Normal fault

Thrust faults

The Himalaya Mountains and other important highlands of southern Asia are a consequence of the collision of India, a small but very old and strong block of continental lithosphere, with Asia about 55 million years ago. At the time of the collision, the southern margin of Asia consisted of several smaller crustal blocks that had become stitched together by recent collisions and thus was composed of younger, warmer, and softer lithosphere. Since then, the strong lithosphere of India has continued to push slowly into the weaker lithosphere of Asia.

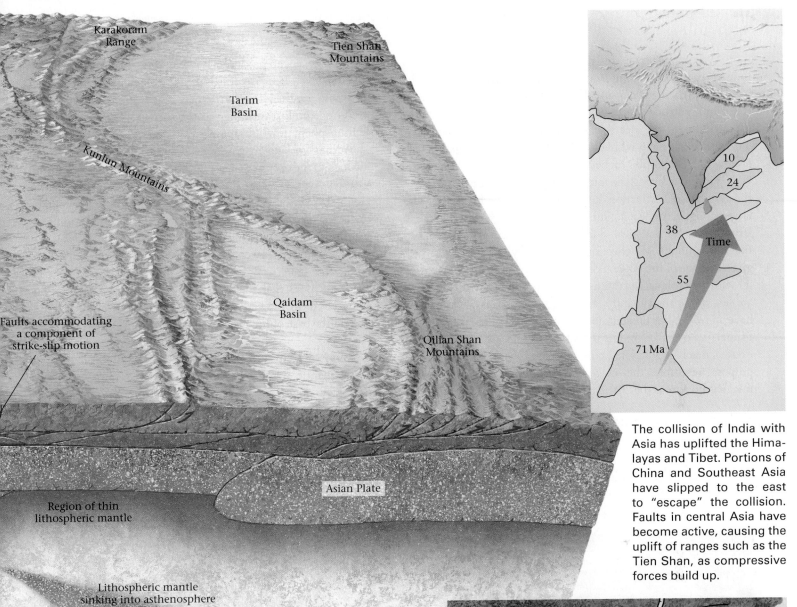

Karakoram Range

Tien Shan Mountains

Tarim Basin

Kunlun Mountains

Faults accommodating a component of strike-slip motion

Qaidam Basin

Qilian Shan Mountains

Asian Plate

Region of thin lithospheric mantle

Lithospheric mantle sinking into asthenosphere

The collision of India with Asia has uplifted the Himalayas and Tibet. Portions of China and Southeast Asia have slipped to the east to "escape" the collision. Faults in central Asia have become active, causing the uplift of ranges such as the Tien Shan, as compressive forces build up.

The development of large thrust faults has uplifted the curving Himalayan chain where Asia begins to thrust over India. Why the broad plateau of Tibet has risen remains something of a mystery. In part, the uplift may be a consequence of the thickening of the crust as it is squashed horizontally; continental crust is relatively weak, and so may spread laterally (like soft cheese in the sun), leading to the formation of normal faults and small rifts in the upper crust and a plastic-like flow in the deep crust. The uplift may also be due to the heating of the region when slabs of the underlying lithospheric mantle drop off and sink, to be replaced by hot asthenosphere.

As India has pushed into Asia, it may have squeezed blocks of China and Southeast Asia sideways, toward the east; this motion is accommodated by slip on strike-slip faults. The collision may also have caused reverse faults in the interior of Asia to become active, uplifting a succession of small mountain ranges, such as the Tien Shan.

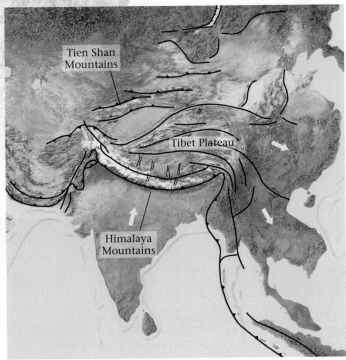

Tien Shan Mountains

Tibet Plateau

Himalaya Mountains

root. We can illustrate this relationship in a bathtub model by lining up a row of floating blocks of different thickness—if all the blocks have the same density, the thicker blocks rise farther above water and protrude deeper below the surface (Fig. 11.26c).

Where did the idea that mountains have roots come from? The discovery began with an observation by Sir George Everest, after whom the world's highest mountain was named. When surveying in India during the mid-1800s, Sir Everest discovered that the gravitational attraction due to the mass of the Himalayas was large enough to deflect a plumb bob (a lead weight at the end of a string) away from vertical. Subsequent calculations demonstrated that the amount of deflection was actually *less* than expected, given the size of the range. A British scientist, George Airy, came up with an explanation. Airy, keeping in mind that crustal rocks are less dense than mantle rocks, suggested that the Himalayas have a low-density crustal root, and that because of this root, the Himalayas overall have less mass than expected and thus do not exert as much gravitational pull on a plumb bob. Note that Airy's proposal resembles the image shown in Figure 11.26c, so this image is known as the "Airy model" of isostasy.

Adding Igneous Rock to the Crust   When a volcano erupts, a thick pile of pyroclastic debris and/or lava may build up on the surface of the crust. Successive eruptions over millions of years, and granitic intrusions within and below the vol-

canic sequence, can produce a range whose peaks rise kilometers above sea level. Because of isostasy, the elevation of the range is not simply the thickness of the new igneous rock accumulation, for adding the load of range to the surface of the Earth bends the lithosphere down, to maintain isostasy.

Adding volcanic material to the surface of the crust, and granitic rock to the upper crust, is not the only way that igneous activity causes uplift. Geologists suspect that in some locations, basaltic magma formed by partial melting of the mantle gets trapped at or near the base of the crust and accumulates in sills. Addition of this subsurface, the hidden basalt, thickens the crust *relative* to the lithospheric mantle. Even though basalt (a mafic rock) is denser than granite (a felsic rock), it is less dense than the peridotite (an ultramafic rock) that makes up the lithospheric mantle. Thus, adding basalt to the base of the crust causes the land surface to rise, and the base of the crust to sink, to maintain isostasy.

Removal of Lithospheric Mantle   The weight of the lithospheric mantle (composed of very dense peridotite) pulls the lithosphere down, just like heavy ballast makes a ship settle deeper into the water. Removal of some or all of the lithospheric mantle from the base of a plate, therefore, causes the surface of the remaining lithosphere to rise to maintain isostasy, even if the thickness of the crustal component remains unchanged (Fig. 11.27a, b). Such removal, a process known as **delamination**,

FIGURE 11.26   The concept of isostasy as applied to collisional mountain ranges.

(a) The Himalayas are the world's highest mountain range. The crust beneath them is almost twice the normal thickness.

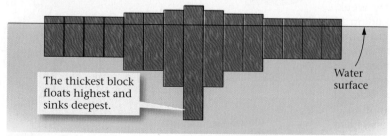

(c) Because of isostasy, blocks of wood floating in water sink to a depth such that the mass of the water displaced is the same as the mass of the block.

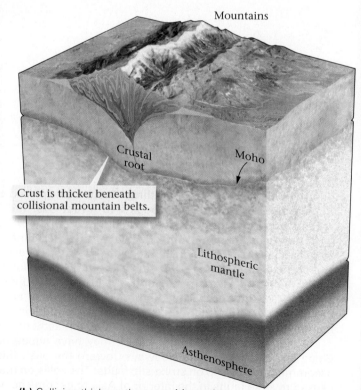

(b) Collision thickens the crust. Mountains form where the low-density crust is thicker.

resembles removal of ballast from the hold of a ship—as the weight of the ballast disappears, the deck of the ship rises. A variety of data sources suggest that the region of the Tibet Plateau in Asia underwent an episode of uplift when some of the underlying lithospheric mantle dripped or peeled off the base of the plate and sank down into the asthenosphere (see Geology at a Glance, pp. 372–373).

Thinning and Heating the Lithosphere In rifts, the lithosphere undergoes stretching and thinning. As a result, relatively less-dense asthenosphere rises beneath the rift, and the remaining lithosphere heats up. Replacing dense lithospheric mantle with less-dense hot asthenosphere, and heating the remaining, overlying lithosphere (thereby causing rocks to expand so their density increases) results in uplift of the rift and its borders in order to maintain isostasy. As the rifted lithosphere ages, and therefore thickens and becomes denser, the surface of the lithosphere subsides.

Emplacement or Local-Scale Loads We mentioned earlier that isostasy is a regional-scale phenomenon. By this we mean the regions that move up or down to establish isostatic equilibrium are on the order of hundreds of kilometers across. A local-scale load (only tens of kilometers across, or smaller), when placed on the surface of the lithosphere, can be held up by the strength of the lithosphere, just like a weight placed on a stiff wooden board can be supported by the board. Thus, a localized igneous eruption can build a small volcanic mountain peak without being isostatically compensated.

What Goes Up Must Come Down

When the land surface rises significantly, for whatever reason, it doesn't remain a smooth welt or bulge on the Earth's surface. As soon as a difference in elevation between one location and an adjacent one develops, gravity begins to drive a variety of erosive processes. For example, as slopes steepen, landslides of various types cause rock and debris to tumble from higher to lower elevations (see Chapter 16); when rain falls, streams form and sculpt valleys and canyons (see Chapter 17); and if it remains cold enough, glaciers grow and flow, carving pointed peaks at their origin, and widening and deepening valleys downslope (see Chapter 22). The net result of all these processes is to grind away elevated areas and produce the jagged landscapes that we associate with mountain terrains (Fig. 11.28a, b).

It's important to keep in mind that uplift and erosion happen simultaneously in active mountain belts, so for the elevation of a range to increase over time, the rate of uplift must *exceed* the rate of erosion. If instead the erosion rate exceeds the uplift rate, then the range's elevation will decrease over time, even if the tectonic processes driving uplift continue to operate; and if the rate of erosion equals the rate of uplift, the elevation of the range stays the same. Notably, the balance between erosion rate and uplift rate evolves during the history of an orogen, either due to a change in tectonic processes (e.g., a change in relative plate motion) or due to a change in climate. For example, if rainfall or snowfall increases, so rivers and glaciers grow larger and flow faster, the erosion rate increases.

When tectonic processes driving uplift eventually stop, erosion continues and can bevel a mountain range back to near sea level. But the process takes a long time, perhaps 30 to 50 million years. It takes so long, in part, because of isostasy. Specifically, removal of rock from the top of a mountain range is like taking cargo off the deck of a ship. Removing cargo causes the ship to float higher, and erosion of rock from the range causes isostatic rebound (uplift). Roughly speaking, for every 3 cm eroded, the range rises by 1 cm.

FIGURE 11.27 Uplift, due to delamination of the lithosphere root, may happen after collision.

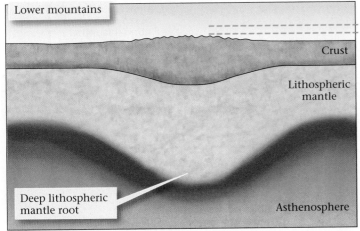

(a) Soon after collision, the crust and lithospheric mantle have thickened. The lithospheric mantle root is like ballast, holding the surface of the crust down.

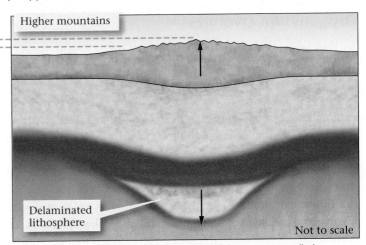

(b) If the lithospheric mantle root detaches and sinks, a process called delamination, the surface of the lithosphere may rise, like a balloon dropping ballast.

FIGURE 11.28 Manifestations of erosion in mountain ranges. When land rises, water and ice start cutting into it.

(a) Glaciers carved these rugged peaks in Switzerland.

(b) Streams cut valleys into weathered bedrock in Brazil.

The highest point on Earth, the peak of Mt. Everest, lies 8.85 km above sea level—can our planet's mountain ranges get significantly higher? Probably not. Mountains as high as Olympus Mons on Mars, which rises 27 km above the plain at its base, couldn't form on Earth because of the relatively high geothermal gradient (the rate of increase in temperature with depth) in Earth's crust. Due to the gradient, quartz-rich crustal rocks at mid-crustal depths (15–30 km) become so warm and weak that they can flow ductilely. When this flow begins, overlying mountains above begin to collapse under their own weight, and spread laterally like soft cheese that has been left out in the summer sun (Fig. 11.29a, b). Geologists call this process **orogenic collapse**. During orogenic collapse, the upper crust breaks and a system of normal faults develop, to accommodate the horizontal stretching.

Orogeny: An Overview

In sum, several processes affecting mountain height happen simultaneously as an orogen evolves. To emphasize this point, consider what happens over time in a collisional orogen. As collision progresses, the crust shortens horizontally and thickens vertically by development of folds, faults, and foliation. During shortening, thrust faulting brings rocks from deeper crustal levels up and over rocks that were closer to the surface. Thickening leads to development of a crustal root, and because of isostasy, the added weight of the mountains causes the base of the lithosphere to sink downward. Eventually, crust at depth warms and weakens so orogenic collapse takes place. And while all these processes are taking place underground, erosion grinds away rock from the uplifted land surface, providing sediment that gets carried away to be deposited elsewhere. Because

of isostasy, rebound accompanies erosion. As a result of this complex combination of phenomena, rock metamorphosed at depth eventually becomes exposed at the surface—this process of revealing deeper rocks by removal of the overlying crust is called unroofing or **exhumation**. In large mountain ranges,

FIGURE 11.29 The concept of orogenic collapse.

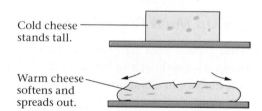

Cold cheese stands tall.

Warm cheese softens and spreads out.

(a) Cheese spreads out sideways as it warms up and softens.

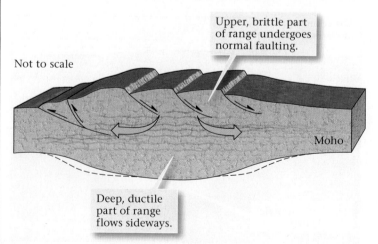

Upper, brittle part of range undergoes normal faulting.

Not to scale

Moho

Deep, ductile part of range flows sideways.

(b) Similarly, mountain belts spread out sideways once they reach a certain thickness. The ductile crust at depth flows, whereas the upper (brittle) crust is broken by normal faults.

rocks that were metamorphosed at depths of over 20 km may be exhumed, eventually to be exposed in outcrops at the surface.

Take-Home Message

- Mountain elevation depends, in part, on isostasy, the condition that exists when elevation reflects the level at which the lithosphere floats.
- Changes in thickness and/or density of layers in the lithosphere influence uplift.
- Crustal thickening due to collision produces the highest ranges. Mountains also form where igneous activity thickens crust, and rifting heats lithosphere.
- As soon as elevation exists, erosion starts to tear it down. In regions where crust thickens substantially, it may "collapse" under its own weight. Exhumation exposes rocks from depth.

THINK: How do changes in climate influence mountain building?

11.8 THE BASINS AND DOMES OF CRATONS

A **craton** consists of crust that has not been affected by orogeny for at least about the last 1 billion years and, as a result, consists of crust that has become quite cool, and therefore relatively strong and stable. Geologists divide cratons into two provinces: **shields**, in which Precambrian metamorphic and igneous rocks crop out at the ground surface, and **cratonic platforms**, where a relatively thin layer of Phanerozoic sediment covers the Precambrian rocks (Fig. 11.30).

In shield areas, we find widespread exposures of intensively deformed metamorphic rocks with abundant examples of flow folds and tectonic foliation. That's because the crust making the cratons was deformed during a succession of orogenies in the Precambrian. Recent studies of the Canadian Shield, which occupies much of the eastern two-thirds of Canada, for example, reveal the traces of Himalaya-like collision zones, Andean-like

FIGURE 11.30 North America's craton consists of a shield, where Precambrian rock is exposed, and a platform, where Paleozoic sedimentary rock covers the Precambrian.

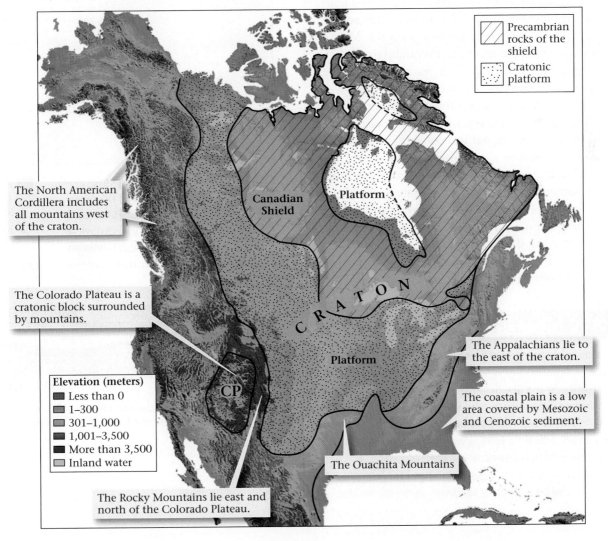

The North American Cordillera includes all mountains west of the craton.

The Colorado Plateau is a cratonic block surrounded by mountains.

The Appalachians lie to the east of the craton.

The coastal plain is a low area covered by Mesozoic and Cenozoic sediment.

The Ouachita Mountains

The Rocky Mountains lie east and north of the Colorado Plateau.

Elevation (meters)
- Less than 0
- 1–300
- 301–1,000
- 1,001–3,500
- More than 3,500
- Inland water

Canadian Shield

Platform

CRATON

Platform

CP

convergent boundaries, and East African–like rifts, all formed more than 1 billion years ago (some over 3 billion years ago). These orogens are so old that erosion has worn away the original topography, in the process exhuming deep crustal rocks.

In the cratonic platform, the pattern of contacts between stratigraphic formations defines regional domes and basins. These are broad areas that gradually sank or rose, respectively, over geologic time (Fig. 11.31a, b). For example, in Missouri, strata arch across a broad dome, the Ozark Dome, whose diameter is 300 km. Individual sedimentary layers thin toward the top of the dome, because less sediment accumulated on the dome than in adjacent basins. Erosion during more recent geologic history has produced the characteristic bull's-eye pattern of a dome, with the oldest rocks (Precambrian granite) exposed near the center. In the Illinois Basin, strata warp downward into a huge bowl that is also about 300 km across. Strata get thicker toward the basin center, indicating that the floor of the basin was subsiding *as* sediment was accumulating. There was more room for sediment to accumulate where the basin subsided the most. The basin also has a bull's-eye shape, but here the youngest strata are exposed in the center. Geologists refer to the broad vertical movements that generate huge, but gentle, mid-continent domes and basins as **epeirogeny**.

Faults occur locally within the cratonic platform. Most of these structures originated during the Precambrian, perhaps in response to episodes of rifting. They reactivated in pulses during the Paleozoic, generally at times when orogenies were active along the margins of the continent. Evidently, stress generated during the orogenies was sufficient to cause slip on preexisting faults in the craton, but was not sufficient to generate the pervasive folds and tectonic foliation found within orogens. Faults within the cratonic platform typically root in the Precambrian basement, cut across the Precambrian/Paleozoic unconformity, and displace contacts in Paleozoic strata. Most intracratonic faults die out up-dip, so they do not reach the ground surface. But the displacement on them, like the rise of a trapdoor beneath a carpet, produced monoclines (step-shaped folds) in the overlying strata.

Take-Home Message

- Cratons, relatively stable parts of continents, include shields and platforms.
- Epeirogeny (upwarps and downwarps) produces gentle basins, domes, and arches. Strata are thicker in basins than on domes because deposition occurs while basins subside.
- Locally, Precambrian faults in cratons have been reactivated; these may be overlain by monoclines.

THINK: Imagine coal occurs in a particular stratigraphic unit. Will mines to reach the coal be deeper, or shallower, in the center of a basin?

FIGURE 11.31 Domes and basins of the North American cratonic platform.

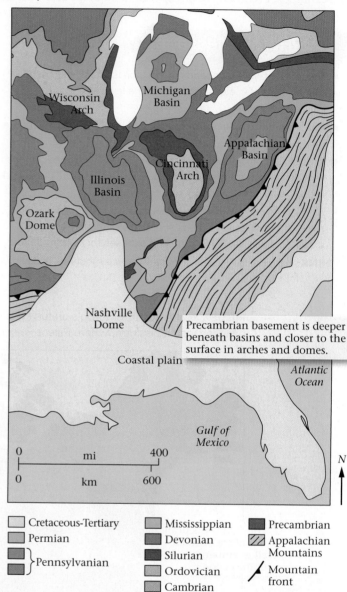

(a) A geologic map of the U.S. mid-continent platform region showing the locations of basins and domes.

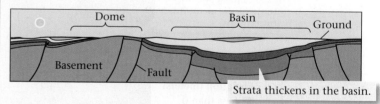

(b) A cross section showing how strata thin toward the crest of a dome and thicken toward the center of a basin. The cross section is vertically exaggerated.

11.9 LIFE STORY OF A MOUNTAIN RANGE: A CASE STUDY

Perhaps the easiest way to bring together all the information in this chapter is to look at the life story of one particular mountain range—let's take the Appalachian Mountains of eastern North America as our example. (We've simplified the story a bit, for ease of reading.) Geologists have constructed the range's life story by studying structures, by determining the ages of igneous and metamorphic rocks, and by searching for strata formed from sediment eroded from the range.

At about 1.1 Ga, the eastern margin of North America was involved in a massive collision with another continent (**Fig. 11.32**). This event, called the Grenville orogeny, yielded a belt of deformed and metamorphosed rocks that underlie the eastern fifth of the continent. For a while after the Grenville event, eastern North America lay in the middle of a supercontinent. But this supercontinent rifted apart around 600 million years ago. Eventually, new ocean formed to the east, and the former rifted margin of eastern North America cooled, subsided, and evolved into a passive-margin sedimentary basin. From 600 to about 420 million years ago, this basin gradually filled with a thick sequence of sediment.

Between around 420 and 370 million years ago, two collisions took place between North America and exotic terranes. During the first convergent event, called the Taconic orogeny, a volcanic arc collided with eastern North America, and during the second convergent event, the Acadian orogeny, continental crustal slivers accreted to the continent. The accretion of these terranes deformed the sediment that had accumulated in the passive-margin basin, and made the continent grow eastward. Significant strike-slip displacement occurred during these events; thus, slivers of crust were transported along the margin of the continent. Then, 270 million years ago, Africa collided with North America. This event, the Alleghenian orogeny, yielded a huge mountain range resembling the present-day Himalayas and generated a wide fold-thrust belt along the mountains' western margin. Eroded folds of this belt make up the topog-

raphy of the present Valley and Ridge Province in Pennsylvania (**Fig. 11.33a, b**).

When the Alleghenian orogeny ceased, the Appalachian region once again lay in the interior of a supercontinent (Pangaea), where it remained until about 180 million years ago. At that time, rifting split the region open again, creating the Atlantic Ocean. As you can see from this example, major ranges such as the Appalachians incorporate the products of multiple orogenies and reflect the opening and closing of ocean basins (a sequence of events called the Wilson cycle after J. Tuzo Wilson).

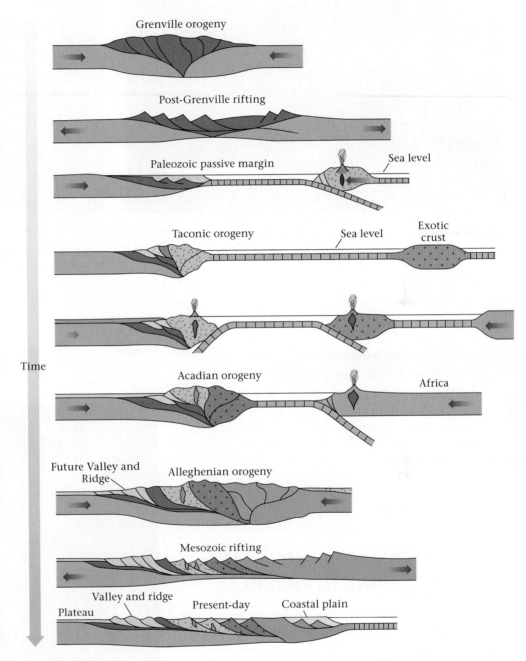

FIGURE 11.32 These idealized stages show the tectonic evolution of the Appalachian Mountains. Note that although mountains do form during rifting events, geologists traditionally assign names only to the collisional or convergent events.

Grenville orogeny

Post-Grenville rifting

Paleozoic passive margin — Sea level

Taconic orogeny — Sea level — Exotic crust

Acadian orogeny — Africa

Future Valley and Ridge — Alleghenian orogeny

Mesozoic rifting

Plateau — Valley and ridge — Present-day — Coastal plain

Time

FIGURE 11.33 The relationship between deformation and topography.

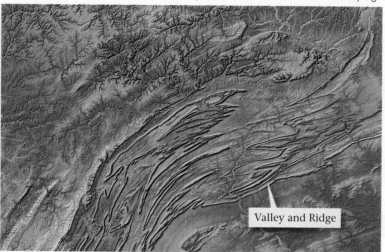

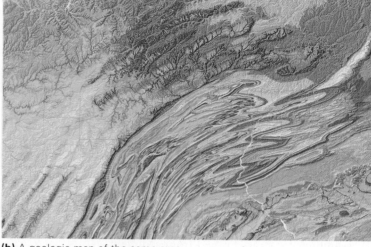

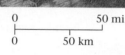

(a) The topography of eastern Pennsylvania, here shown in shaded relief, shows how erosional patterns reveal the folds of the Appalachians, producing a region called the Valley and Ridge.

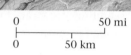

(b) A geologic map of the same area emphasizes the relation between the distribution of stratigraphic formations (indicated by different colors) and the topographic features. A resistant sandstone formation forms the ridges.

Take-Home Message

- The Appalachian Mountains expose remnants of structures and rocks formed during three distinct orogenic events, each associated with a collision.
- The last event, at the end of the Paleozoic, produced a huge fold-thrust belt.

THINK: How did geologists determine the numerical age of the three events?

11.10 MEASURING MOUNTAIN BUILDING IN PROGRESS

Not all mountains are just "old monuments," as John Muir mused. The rumblings of earthquakes and the eruptions of volcanoes attest to present-day, continuing movements in some ranges. Geologists can measure the rates of these movements through field studies and satellite technology. For example, geologists can determine where coastal areas have been rising relative to the sea level by locating ancient beaches that now lie high above the water. And they can tell where the land surface has risen relative to a river by identifying places where a river has recently carved a new valley down into sediments that it had previously deposited. In addition, geologists now use the satellite **global posi-**

FIGURE 11.34 GPS measurements of shortening in the Andes. The lines indicate the velocity of the red dots relative to the interior of South America. The line at the yellow dot indicates relative plate motion.

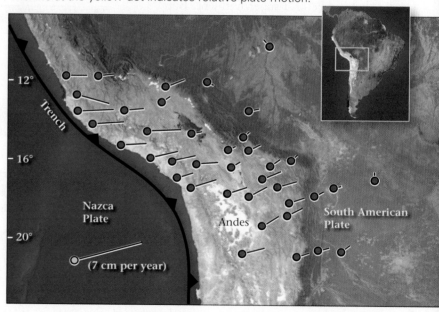

tioning system (GPS) to measure rates of uplift and horizontal shortening in orogens. With present technology, we can "see" the Andes shorten horizontally at a rate of a couple of centimeters per year (**Fig. 11.34**), and we can "watch" as mountains along this convergent boundary rise by a couple of millimeters per year.

Chapter Summary

- Mountains occur in linear ranges called mountain belts, orogenic belts, or orogens. An orogen forms during an orogeny, or mountain-building event.

- Mountain building causes rocks to bend, break, shorten, stretch, and shear. Because of such deformation, rocks can change their location, orientation, and shape.

- During brittle deformation, rocks crack and break into two or more pieces. During ductile deformation, rocks change shape without breaking.

- Rocks undergo three kinds of stress: compression, tension, and shear. Strain refers to the way rocks change shape when subjected to a stress.

- Deformation results in the development of geologic structures.

- Joints are natural cracks in rock, formed in response to tension under brittle conditions. Veins develop when minerals precipitate out of water passing through joints.

- Faults are fractures on which there has been shearing. Geologists distinguish among normal, reverse, thrust, strike-slip, and oblique faults.

- Folds are curved layers of rock. Anticlines are arch-like, synclines are trough-like, monoclines resemble the shape of a carpet draped over a stair step, basins are shaped like a bowl, and domes are shaped like an overturned bowl.

- Tectonic foliation forms when grains flatten, rotate, or grow so that they align parallel with one another.

- Uplift in mountains, over broad regions, is controlled by the isostasy, meaning that the elevation of the Earth's surface reflects the level at which lithosphere naturally floats.

- Large collisional mountain ranges are underlain by buoyant roots. Within these ranges, folds, faults, and foliations develop.

- Fold-thrust belts form on the continental edge of collisional and convergent-margin orogens.

- Mountain areas may also rise if part of the underlying lithospheric mantle detaches and sinks into the asthenosphere.

- Once uplifted, mountains are sculpted by erosion. When crust thickens during mountain building, the deep crust eventually becomes warm and weak, leading to orogenic collapse.

- Mountain belts formed by convergent-margin tectonism may incorporate accreted terranes.

- Rifting produces tilted blocks of crust that become narrow, elongate mountain ranges. Heating of the lithosphere causes uplift of rifts.

- Cratons are the old, relatively stable parts of continental crust. They include shields and platforms. Broad regional domes and basins form in platform areas.

- With modern GPS technology, it is now possible to measure the slow shortening and uplift of mountains.

GEOPUZZLE REVISITED

Mountain ranges form in response to the interaction of lithosphere plates. Some ranges form along convergent plate boundaries, some in association with rifting, and some where two continents collide after the sea floor between them has been subducted. Stress (compression, tension, or shear) that develops during mountain building causes deformation, producing such geologic structures as folds, faults, and foliations. Mountain building involves uplift, and in many cases igneous activity and metamorphism. As soon as a range has risen above sea level, erosion attacks it. In fact, if the geologic processes causing uplift slow, erosion can eventually bevel a range nearly back to sea level. Thus, mountain ranges do not last forever.

Guide Terms

accretionary orogen (p. 368)

anticline (p. 362)

axial surface (p. 362)

basin (p. 363)

brittle deformation (p. 352)

compression (p. 354)

craton (p. 377)

cratonic platform (p. 377)

crustal root (p. 371)

crustal thickening (p. 368)

deformation (p. 349)

delamination (p. 374)

detachment fault (p. 361)

dip (p. 356)

displacement (p. 358)

dome (p. 363)

ductile deformation (p. 352)

epeirogeny (p. 378)

exhumation (p. 376)

exotic terrane (p. 368)

fault (p. 350, 357)

fault scarp (p. 359)

fold (p. 350, 362)

fold-thrust belt (p. 368)

foliation (p. 350)

global positioning system (GPS) (p. 380)

hinge (p. 362)

isostasy (p. 370)

joints (pp. 349, 355)

limb (of fold) (p. 362)

monocline (p. 362–363)

normal fault (p. 358)

orogen (p. 349)

orogenic collapse (p. 376)

pressure (p. 354)

reverse fault (p. 358)

shear stress (p. 354)

shear zone (p. 361)

shield (p. 377)

strain (p. 352)

stress (p. 354)

strike (p. 356)

strike-slip fault (p. 358)

syncline (p. 362)

tectonic foliation (p. 366)

tension (p. 354)

thrust fault (p. 358)

uplift (p. 349, 370)

vein (p. 355)

Review Questions

1. What changes do rocks undergo during formation of an orogenic belt such as the Alps?

2. What is the difference between brittle and ductile deformation?

3. What factors determine whether a rock will behave in brittle or ductile fashion?

4. How are stress and strain different?

5. How is a fault different from a joint?

6. Compare the motion of normal, reverse, and strike-slip faults.

7. How do you recognize faults in the field?

8. Describe the differences among an anticline, a syncline, and a monocline.

9. Discuss the relationship between foliation and deformation.

10. Describe the principle of isostasy.

11. Discuss the processes by which mountain belts form in convergent margins, in continental collisions, and in continental rifts.

12. How are the structures of a craton different from those of an orogenic belt?

On Further Thought

13. Imagine that a geologist sees two outcrops of resistant sandstone, as depicted in the cross-section sketch. The region between the outcrops is covered by soil. A distinctive bed of cross-bedded sandstone occurs in both outcrops, so the geologist correlated the western outcrop (on the left) with the eastern outcrop. The curving lines in the bed indicate the shape of the cross beds.

   (a) Keeping in mind how cross beds form (see Chapter 7), sketch how the cross-bedded bed connected from one outcrop to the other, before erosion. What geologic structure have you drawn?

   (b) Is the bedrock directly beneath the geologist older than or younger than the sandstone bed of the outcrop?

 For more resources, including animations, quizzes, and Norton's GeoTours, go to **wwnorton.com/studyspace**.

 If your instructor assigns exercises in SmartWork, log in at **smartwork.wwnorton.com**.

Every outcrop tells a story of our planet's past. Here, in an exposure from northern Michigan, we see banded iron formation (BIF) in which grey hematite alternates with red jasper. Geologists speculate that such rock represents a time in Earth history when the atmosphere began to accumulate significant quantities of oxygen. Folding of the layers indicates that, after it was deposited and lithified, this BIF became incorporated in a mountain belt.

History before History

Perhaps the most important contribution that the science of geology has made to humanity's understanding of the Earth is the demonstration that our planet existed long, long before humans took their first steps. In Part IV, we peer back into this history. First we look at fossils, remnants of ancient life that allow geologists to correlate life's evolution with that of Earth. Then, in Chapter 12, we learn how geologists gaze into "deep time"—geologic time, or the time since Earth formed—first by determining the relative ages of geologic features (whether one feature is older or younger than another) and then by learning how to calculate numerical ages (ages in years) on the basis of the ratios of radioactive elements to their daughter products in minerals. With the background provided in Chapter 12, we're ready for Chapter 13's brief synopsis of Earth history, from the birth of the planet to the present. We see how plate tectonics has redistributed continents and built mountains, how sea level has risen and fallen, and how Earth's climate has changed over time.

Memories of Past Life: Fossils and Evolution

E.1 THE DISCOVERY OF FOSSILS

If you look at bedding surfaces of sedimentary rock, you may find shapes that resemble shells, bones, leaves, or footprints (Fig. E.1a). The origin of these shapes mystified early thinkers. Some thought that the shapes had simply grown underground, in solid rock. Today, geologists consider such **fossils** (from the Latin word *fossilis*, which means dug up) to be remnants or traces of ancient living organisms now preserved in rock, and that fossils formed when organisms were buried by sediment. This viewpoint, though first proposed by the Greek historian Herodotus in 450 B.C.E. and revived by Leonardo da Vinci in 1500 C.E., did not become widely accepted until the publication in 1669 of a book on the subject by a Danish physician named Nicholaus Steno (1638–1686). Steno argued that once buried, shells, teeth, and bones could transform into rock without losing their distinctive shape. The understanding of fossils increased greatly thanks to the efforts of a British scientist, Robert Hooke (1635–1703), who realized that most fossils represent *extinct* species, meaning species that lived in the past but no longer exist today. During the following two centuries, **paleontologists**, scientists who study fossils, described thousands of specimens and established comprehensive museum collections (Fig. E.1b, c).

The nineteenth century saw **paleontology**, the study of fossils, ripen into a science. Work with fossils went beyond description alone when William Smith, a British engineer who supervised canal construction in England during the 1830s, noted that different fossils occur in different layers of strata within a sequence of sedimentary rocks. In fact, Smith realized that strata contain a predictable succession of fossils, and that a given species can be found only in a specific interval of strata. This discovery made it possible for geologists to use fossils as a basis for determining the age of one sedimentary rock layer relative to another. Fossils, therefore, have become an indispensable tool for studying geologic history and the evolution of life. In this interlude, we introduce fossils, an understanding of which serves as essential background for the next two chapters.

E.2 FOSSILIZATION

What Kinds of Rocks Contain Fossils?

Most fossils are found in sediments or sedimentary rocks. Fossils form when organisms die and become buried by sediment, or when organisms travel over or through sediment and leave imprints or debris. Rocks formed from sediments deposited under anoxic (oxygen-free) conditions in quiet water (such as lake beds or lagoons) can preserve particularly fine specimens. Rocks made from sediments deposited in high-energy environments, on the other hand—where strong currents tumble shells and bones and break them up—contain at best only small fragments of fossils mixed with other clastic grains.

Fossils can survive low grades of metamorphism, but not the recrystallization and new mineral growth, and in some cases shearing, that occur during intermediate- and high-grade metamorphism. Similarly, fossils generally do not occur in igneous rocks that crystallize directly from melt, for organisms can't live in molten rock, and if engulfed by molten rock, they will be incinerated. Occasionally, however, lava flows preserve the shapes of tree trunks, because lava may surround a tree and freeze before the tree completely burns up—the resulting hole in the lava is, strictly speaking, a fossil. Also, fossils can occur in deposits formed from air-fall ash, for the ash settles just like sediment and can bury an organism or a footprint. In fact, ash preserved the footprints of 3.6-million-year-old human ancestors in deposits now exposed in Olduvai Gorge, in the East African Rift (Fig. E.2).

FIGURE E.1 Paleontologists collect fossils for study.

(a) This bedding plane contains the imprint of species that lived hundreds of millions of years ago and are now extinct.

(b) Paleontologists painstakingly chip and dig away at rock layers in search of fossils.

(c) A drawer of labeled fossils in a museum. Paleontologists study such collections to help identify new specimens.

FIGURE E.2 The famous fossil footprints at a site called Laetoli, in Olduvai Gorge, Tanzania. They were left when an adult and child of a human ancestor, *Australopithecus,* walked on two feet over ash that had recently been erupted by a nearby volcano. The ash had been dampened by rain when a second ash eruption buried it and thereby preserved the footprints.

Forming a Fossil

Paleontologists refer to the process of forming a fossil as **fossilization**. To see how a typical fossil develops in sedimentary rock, let's follow the fate of an old dinosaur as it searches for food along a riverbank (Fig. E.3). On a scalding summer day, the hungry dinosaur, plodding through the muddy ground, succumbs to the heat and collapses dead into the mud. Over the coming days, scavengers strip the skeleton of meat and scatter the bones. But before the bones have had time to weather away, the river floods and buries the bones, along with the dinosaur's footprints, under a layer of silt. More silt from succeeding floods buries the bones and prints still deeper, so that the bones cannot be reworked by currents or disrupted by burrowing organisms. Later, sea level rises and a thick sequence of marine sediment buries the fluvial sediment.

Eventually, the sediment containing the bones and footprints turns to rock (siltstone and shale). The footprints remain outlined by the contact between the siltstone and shale, while the bones reside within the siltstone. Minerals precipitating from groundwater passing through the siltstone gradually replace some of the chemicals constituting the bones, until the bones themselves have become rock-like. The buried bones and footprints are now fossils. One hundred million years later, uplift and erosion expose the dinosaur's grave. Part of a fossil bone protrudes from a rock outcrop. A lucky paleontologist observes the fragment and starts excavating, gradually uncovering enough of the bones to permit reconstruction of the beast's skeleton. Further digging uncovers the footprints. The dinosaur rises again, but this time in a museum. In recent years, bidding wars have made some fossil finds extremely valuable. For example, a skeleton of a *Tyrannosaurus rex*, a 67-million-year-old dinosaur, sold at auction in 1997 for $7.6

million. The specimen, named Sue after its discoverer, now stands in the Field Museum of Chicago.

Similar tales can be told for fossil seashells buried by sediment settling in the sea, for insects trapped in hardened tree sap (amber), and for mammoths drowned in the muck of a tar pit. In all cases, fossilization involves the burial and preservation of an organism or the trace (a footprint or burrow) of an organism.

The Many Different Kinds of Fossils

Perhaps when you think of a fossil, you picture either a dinosaur bone or the imprint of a seashell in rock. In fact, paleontologists distinguish many different kinds of fossils, according to the specific way in which the organism was fossilized. General categories include **body fossils**, which are whole bodies or pieces of bodies; **trace fossils**, which are features left by an organism as it passed by; and **chemical fossils**, which are chemicals first formed by organisms and now preserved in rock. Let's look at examples of these categories.

FIGURE E.3 How a dinosaur eventually becomes a fossil. This takes many steps, over a long period of time.

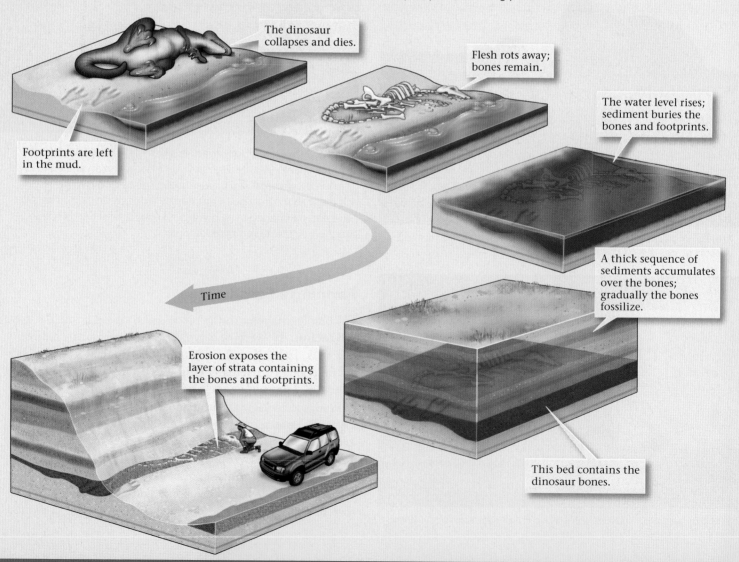

The dinosaur collapses and dies.

Flesh rots away; bones remain.

The water level rises; sediment buries the bones and footprints.

Footprints are left in the mud.

A thick sequence of sediments accumulates over the bones; gradually the bones fossilize.

Time

Erosion exposes the layer of strata containing the bones and footprints.

This bed contains the dinosaur bones.

- *Frozen or dried body fossils*: In a few environments, whole bodies of organisms may be preserved. Most of these fossils are fairly young, by geologic standards—their ages can be measured in thousands, not millions, of years. Examples include woolly mammoths that became incorporated in the permafrost (permanently frozen ground) of Siberia (Fig. E.4a). In desert climates, organisms can become desiccated (dried out). Such "mummified" corpses can survive for millennia in caves.

- *Body fossils preserved in amber or tar*: Insects landing on the bark of trees may become trapped in the sticky sap or resin the trees produce. This golden syrup envelops the insects and over time hardens into **amber**, the semiprecious "stone" used for jewelry. Amber can preserve insects, as well as other delicate organic material such as feathers, for 40 million years or more (Fig. E.4b). Tar similarly acts as a preservative. In isolated regions where oil has seeped to the surface, the more volatile components of the oil evaporate away and bacteria degrade what remains, leaving behind a sticky residue of tar. At one such locality, the La Brea Tar Pits in Los Angeles, tar accumulated in a swampy area. While grazing, drinking, or hunting at the swamp, animals became mired in the tar and sank into it. Their bones have been remarkably well preserved for over 40,000 years (Fig. E.4c).

- *Preserved or replaced bones, teeth, and shells*: Bones (the internal skeletons of vertebrate animals) and shells (the external skeletons of invertebrate animals) consist of durable minerals, which may survive in rock (Fig. E.4d). Some bone or shell minerals are not stable, and they recrystallize. But even when this happens, the shape of the bone or shell can be preserved in the rock.

- *Permineralized organisms*: **Permineralization** refers to the process by which minerals precipitate in porous material, such as wood or bone, from groundwater solutions that have seeped into the pores. **Petrified wood**, for example, forms by permineralization of wood, so that cell interiors are replaced with silica, causing the wood to become chert. In fact, the word *petrified* literally means turned to stone. The more resistant cellulose of the wood transforms into an organic film that remains after permineralization, so that the fine detail of the wood's cell structure can be seen in a petrified log (Fig. E.4e). The colorful bands in a petrified log come from impurities such as iron or carbon. Petrified wood typically forms when a volcanic eruption rapidly buries a forest in silica-containing ash.

- *Molds and casts of bodies*: As sediment compacts around a shell, it conforms to the shape of the shell or body. If the shell or body later disappears, because of weathering and dissolution, a cavity called a **mold** remains (Fig. E.4f). (Sculptors use the same term to refer to the receptacle into which they pour bronze or plaster.) A mold preserves the delicate shape of the organism's surface; it looks like an indentation on a rock bed. The sediment that had filled the mold also preserves the organism's shape; this **cast** protrudes from the surface of the adjacent bed.

- *Carbonized impressions of bodies*: Impressions are simply flattened molds created when soft or semisoft organisms (leaves, insects, shell-less invertebrates, sponges, feathers, jellyfish) get pressed between layers of sediment. Chemical reactions eventually remove the organic material, leaving only a thin film of carbon on the surface of the impression (Fig. E.4g).

- *Trace fossils*: These include footprints, feeding traces, burrows, and dung (coprolites) that organisms leave behind in sediment (Fig. E.4h, i).

- *Chemical fossils*: Living things consist of complex organic chemicals. When buried with sediment and subjected to diagenesis, some of these chemicals are destroyed, but some either remain intact or break down to form different, but still distinctive, chemicals. A distinctive chemical derived from an organism and preserved in rock is called a chemical fossil. (Such chemicals may also be called molecular fossils or **biomarkers**.)

Paleontologists also find it useful to distinguish among different fossils on the basis of their size. Macrofossils are large enough to be seen with the naked eye. But some rocks and sediments also contain abundant **microfossils**, which can be seen only with a microscope or even an electron microscope. Microfossils include remnants of plankton, algae, bacteria, and pollen (Fig. E.5).

Fossil Preservation

Not all living organisms become fossils when they die. In fact, only a small percentage do, for it takes special circumstances—namely, one or more of the following three—to produce a fossil and allow it to survive.

- *Death in an anoxic (oxygen-poor) environment*: A dead squirrel by the side of the road won't become a fossil. As time passes, birds, dogs, or other scavengers may come along and eat the carcass. And if that doesn't happen, maggots, bacteria, and fungi infest the carcass and gradually digest it. Flesh that has not been eaten, or does not rot, can react with oxygen in the atmosphere and transform into carbon dioxide gas. The remaining skeleton weathers in air and turns to dust. Thus, before road kill can become incorporated in sediment, it has

FIGURE E.4 Examples of different kinds of fossils.

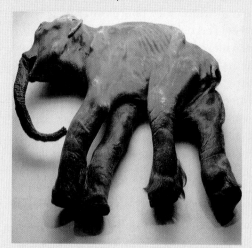

(a) A 100,000-year-old frozen mammoth found in Siberia still has flesh and fur.

(b) A piece of amber containing two fossil insects.

(c) A fossil skeleton of a 2-m-high giant ground sloth from the La Brea Tar Pits in California.

(d) Fossil dinosaur bones on a tilted bed of sandstone in Utah.

(e) Petrified wood from Arizona. It is so hard that it remains after the rock that surrounded it has eroded away.

(f) Molds and casts of a shell. Even fine detail is preserved.

(g) The carbonized impressions of fern fronds in a shale.

(h) Dinosaur footprints in a mudstone in Arizona.

(i) Worm burrows on siltstone in Ireland (lens cap for scale).

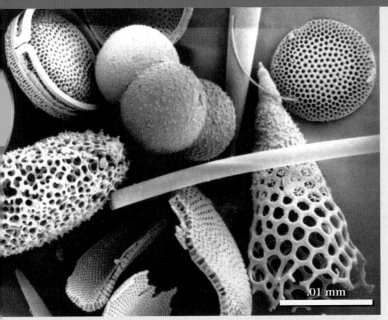

FIGURE E.5 Examples of fossil plankton shells. Because of their size, geologists refer to these as microfossils or nannofossils.

.01 mm

vanished. If, however, a carcass settles into an oxygen-poor environment, oxidation reactions happen slowly, scavenging organisms aren't abundant, and bacterial metabolism takes place very slowly. In such environments the organism won't rot away before it has a chance to be buried and preserved, so the likelihood that the organism becomes fossilized increases.

■ *Rapid burial*: If an organism dies in a depositional environment where sediment accumulates rapidly, it may be buried before it has time to rot, oxidize, be eaten, be consumed by burrowing organisms, or undergo complete weathering. For example, if a storm suddenly buries an oyster bed with a thick layer of silt, the oysters die and become part of the sedimentary rock derived from the sediment.

■ *The presence of hard parts*: Organisms without durable shells or skeletons, or other "hard parts," usually won't be fossilized, for soft flesh decays long before hard parts do under most depositional conditions. For this reason, paleontologists have learned more about the fossil record of bivalves (a class of organisms, including clams and oysters, with strong shells) than they do about the fossil record of jellyfish (which have no shells) or spiders (which have very fragile shells).

By carefully studying modern organisms, paleontologists have been able to provide rough estimates of the **preservation potential** of organisms, meaning the likelihood that an organism will be buried and eventually transformed into a fossil. For example, in a typical modern-day shallow-marine environment, such as the mud-and-sand sea floor close to a beach, about 30% of the organisms have sturdy shells and thus a high preservation potential, 40% have fragile shells and a low preservation potential, and the remaining 30% have no hard parts at all and are not likely to be fossilized except in special circumstances. Of the 30% with sturdy shells, though, few happen to die in a depositional setting where they actually *do* become fossilized. Thus, fossilization is the exception rather than the rule.

Extraordinary Fossils: A Special Window to the Past

Though only hard parts survive in most fossilization environments, paleontologists have discovered a few special locations where rock contains relics of soft parts as well; such fossils are known as **extraordinary fossils** (Fig. E.6a–d). We've already seen how extraordinary fossils such as insects and even feathers can be preserved in amber, and how complete skeletons have been found in tar pits. Extraordinary fossils may also be preserved in sediment that accumulates on the anoxic floor of lakes or lagoons or the deep ocean. Here, oxidation cannot occur, and flesh does not rot before burial. Carcasses of animals that settle into the mud gradually become fossils, but because they were buried before the destruction of their soft parts, fossil impressions of their soft parts surround the fossils of their bones.

A small quarry near Messel, in western Germany, for example, has provided extraordinary fossils of 49-million-year-old mammals, birds, fish, and amphibians that died in a shallow-water lake (Fig. E.6a). Bird fossils from the quarry include the delicate imprints of feathers; bat fossils come complete with impressions of ears and wings, and other mammal fossils have an aura of carbonized fur. In southern Germany, exposures of the Solenhofen Limestone, an approximately 150-million-year-old rock derived from $CaCO_3$ mud deposited in a stagnant lagoon, contain extraordinary fossils of about 600 species, including *Archaeopteryx*, one of the earliest birds (Fig. E.6b). And exposures of the Burgess Shale in the Canadian Rockies of British Columbia have yielded a plentitude of fossils showing what shell-less invertebrates that inhabited the deep-sea floor about 510 million years ago looked like (Fig. E.6c). The Burgess Shale fauna (animal life) is so strange—for example, it includes organisms with circular jaws—that it has been hard to determine how these organisms are related to present-day ones. Indeed, many of the forms of life represented by Burgess Shale fossils became extinct *without* leaving any descendants that evolved into modern species.

In a few cases, extraordinary fossils include actual tissue, a discovery that has led to a research race to find the oldest preserved DNA. (DNA, short for deoxyribonucleic acid, is the complex molecule, shaped like a double helix, that contains the code that guides the growth and development of

FIGURE E.6 Extraordinary fossils. These fossils are particularly well preserved.

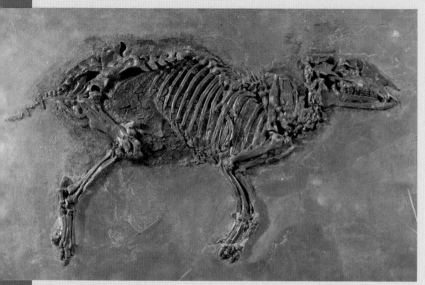

(a) A 50-million-year-old mammal fossil was chiseled from oil shale near Messel, Germany. It still contains the remains of skin.

(b) *Archaeopteryx* from the 150-million-year-old Solenhofen Limestone of Germany. The imprints of feathers are clearly visible.

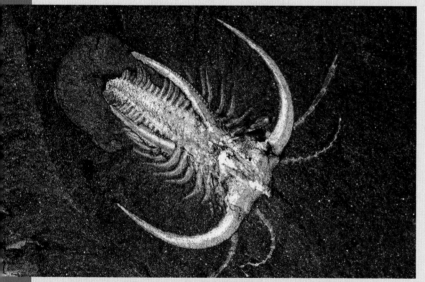

(c) The Burgess Shale of the Canadian Rockies contains unique Cambrian arthropods, such as Marella, shown here.

(d) An artist's reconstruction of a Cambrian ecosystem. The paintings are based on Burgess Shale fossils.

an organism. Individual components of this code are called genes.) Paleontologists have isolated small segments of DNA from amber-encased insects that are over 40 million years old. The amounts are not enough, however, to clone extinct species, as suggested in the popular 1993 film, *Jurassic Park*.

E.3 CLASSIFYING LIFE

The classification of fossils follows the same principles used for the classification of living organisms. These principles were first proposed in the eighteenth century by Carolus Linnaeus, a Swedish biologist. The study of how to classify organisms is now referred to as **taxonomy** (Fig. E.7). Linnaeus's scheme has a hierarchy of divisions.

First, all life is divided into three "domains," named archaea, bacteria, and eukaryote. The domains differ from one another based on fundamental characteristics of their genes. Archaea include a vast array of tiny single-celled microorganisms that occur not only in the mild environments of oceans, soils, and wetlands but also in the harsh environments of hot springs, black smokers, salt lakes, very

FIGURE E.7 The taxonomic subdivisions. The center column shows how the names apply to human beings.

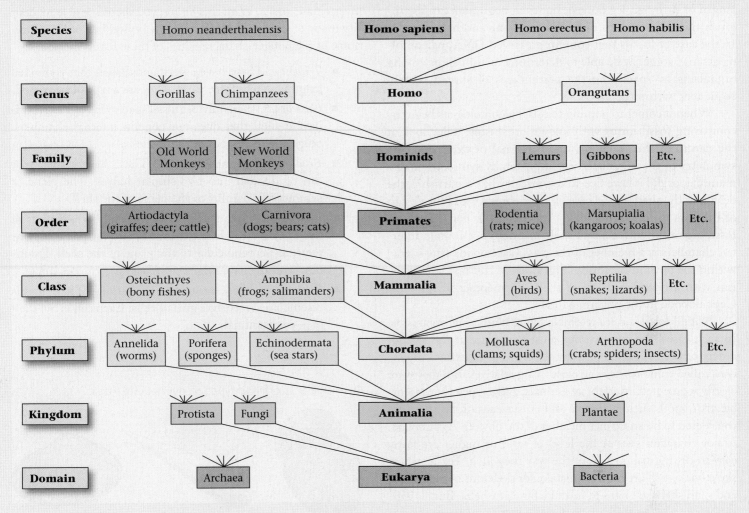

acidic sediments, and deep subsurface water. Organisms that can survive in harsh environments are known as extremophiles; they do not need light, but rather live off the energy stored in the chemical bonds of minerals. Bacteria are also tiny single-celled organisms, species of which inhabit almost all livable environments on Earth. Visually, it may be difficult to distinguish bacteria from archaea, but on a genetic level they are profoundly different. Nevertheless, both archaea and bacteria are **prokaryotes**, meaning that their cells *do not* contain nuclei or other organelles surrounded by membranes. In this regard, archaea and bacteria differ from **eukaryotes**, for cells of organisms of the latter domain do contain nuclei and other organelles. Taxonomists divide the eukaryote domain into several kingdoms. Examples of these, with some representative organisms indicated, include:

- *Protista*: various unicellular and simple multicellular organisms such as diatoms and forams, two of the major plankton types in the oceans;

- *Fungi*: mushrooms and yeast;
- *Plantae*: trees, grasses, and ferns; and
- *Animalia*: sponges, corals, snails, dinosaurs, ants, and people.

Each kingdom consists of one or more phyla. A phylum, in turn, consists of several classes; a class, of several orders; an order, of several families; a family, of several genera; and a genus, of one or more species. Kingdoms, therefore, are the broadest category of eukaryota and species are the narrowest.

E.4 CLASSIFYING FOSSILS

Paleontologists traditionally distinguish among different fossil species based on **morphology** (the form or shape) of specimens. A fossil clam is a fossil clam because it looks like one! In some cases, the characteristics that distinguish one fossil species from another may be quite subtle (such as the number of ridges on the surface of a shell, or the relative length of different leg bones), but even beginners can distinguish the major groups of fossils from one another on sight. In the case of fossils that contain preserved DNA, paleontologists may someday be able to determine relationships among organisms by specifying the percentages of shared protein sequences within the DNA.

When it comes to naming fossils, paleontologists begin by comparing fossil forms with known organisms. They look at the nature of the skeleton (was it internal or external?), the symmetry of the organism (was it bilaterally symmetric like a mammal, or did it have five-fold symmetry like a starfish?), the design of the shell (in the case of invertebrates), and the design of the jaws or feet (in the case of vertebrates). For example, a fossil organism with a spiral shell that does not contain internal chambers is a member of the class Gastropoda (the snails) within the phylum Mollusca (Fig. E.8). In contrast, an organism with a chambered, spiral shell is a member of the class Cephalopoda in the phylum Mollusca.

Not all fossil species resemble living families of organisms. In such cases, the comparisons must take place at the level of orders or even higher. For example, a group of extinct organisms called trilobites have no close living relatives. But they were clearly segmented invertebrate animals, and as such they resemble arthropods such as insects and crustaceans. Thus, they are considered to be an extinct member of the phylum Arthropoda. Major characteristics at the level of class or higher are fairly easy to distinguish just from the way they look. For instance, skeletons of fish are hard to mistake for skeletons of mammals, and snail shells are hard to mistake for clamshells. But classification can also be difficult, especially if specimens are incomplete, and in many cases classification remains controversial.

Figure E.9 shows examples of some of the major types of invertebrate fossils. With this figure, you should be able to identify many of the fossils you'll find in a typical bed of limestone. Some of the notable characterisitics of these fossils include:

- *Trilobites*: These have a segmented shell that is divided lengthwise into three parts. They are a type of arthropod.
- *Gastropods* (snails): Most fossil specimens of gastropods have a shell that does not contain internal chambers. Slugs are gastropods without shells.
- *Bivalves* (clams and oysters): These have a shell that can be divided into two similar halves. The plane of symmetry is parallel to the plane of the shell.
- *Brachiopods* (lamp shells): The top and bottom parts of these shells have different shapes, and the plane of symmetry is perpendicular to the plane of the shell. Examples typically have ridges radiating out from the hinge.
- *Bryozoans*: These are colonial animals. Their fossils resemble a screen-like grid of cells. Each cell is the shell of a single animal.

FIGURE E.9 Common types of invertebrate fossils.

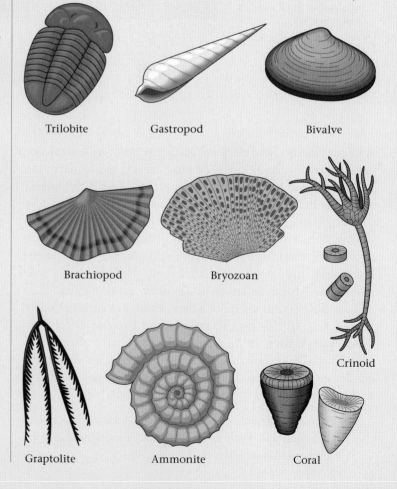

Trilobite   Gastropod   Bivalve

Brachiopod   Bryozoan   Crinoid

Graptolite   Ammonite   Coral

FIGURE E.8 Examples of the diversity of gastropods (snails). Note that although all these shells have a spiral shape, they differ in detail from each other. Some gastropods have no shells at all.

- *Crinoids* (sea lilies): These organisms look like a flower but actually are animals. Their shells have a stalk consisting of numerous circular plates stacked one on top of the other.
- *Graptolites*: These look like tiny carbon-saw blades in a rock. They are remnants of colonial animals that floated in the sea.
- *Cephalopods*: These include ammonites, with a spiral shell, and nautiloids, with a straight shell. Their shells contain internal chambers and have ridged surfaces. These organisms were squid-like.
- *Corals*: These include colonial organisms that form distinctive mounds or columns. Paleozoic examples are solitary.

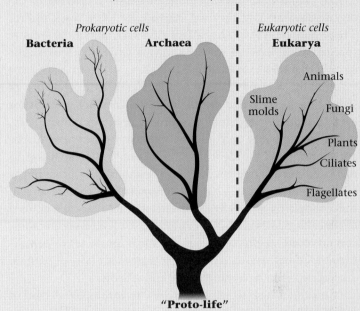

FIGURE E.10 The "tree of life" symbolically illustrates relations among different domains of organisms on Earth. Archaea and bacteria are both prokaryotes and don't have nuclei. All other life-forms are eukaryotes because they have cells with a nucleus.

E.5 THE FOSSIL RECORD

A Brief History of Life

Archaea and bacteria fossils appear in rocks as old as about 3.7 billion years. Paleontological evidence has not determined when or where the first living cell metabolized and reproduced. Based on laboratory experiments conducted in the 1950s, researchers speculated that reactions in concentrated "soups" of chemicals that formed when seawater evaporated in shallow, coastal pools led to the formation of the earliest protein-like organic chemicals ("proto-life"). More recent studies suggest, instead, that such reactions took place in warm groundwater beneath the Earth's surface or at hydrothermal vents on the sea floor.

While the nature of proto-life remains a mystery, an image of early life has begun to take shape, based on detailed analysis of the oldest sedimentary rocks. The fossil record defines the subsequent long-term record of life's evolution on planet Earth. And, of course, that record is more complete in younger strata. For the first billion years or so of life history, archaea and bacteria were the *only* types of life on Earth. Then, about 2.5 Ga, organisms of the Protista kingdom first appeared. Early multicellular organisms, shell-less invertebrates of the animal kingdom, and fungi came into existence at perhaps 1.0 to 1.5 Ga. Within each kingdom, life radiated (divided) into different phyla, and within each phylum, life radiated into different classes. The great variety of shelly invertebrate classes appeared a little over 530 million years ago, during an event called the **Cambrian explosion**, after the geologic era in which it occurred. Many organisms that persist today appeared at different times since then: first came invertebrates, and then in sequence: fish, land plants, amphibians, reptiles, and finally birds and mammals.

Researchers have been working hard to understand the phylogeny (the evolutionary relationships) among organisms, using both the morphology of organisms and, more recently, the study of genetic material. Ideas about which groups radiated from which ancestors is shown in a chart called the tree of life, or, more formally, the **phylogenetic tree** (Fig. E.10). Study of the DNA of different organisms is beginning to enable researchers to understand the relationship between molecular processes and evolutionary change.

Is the Fossil Record Complete?

By some estimates, more than 250,000 species of fossils have been collected and identified to date, by thousands of geologists working on all continents during the past two centuries. These fossils define the framework of life evolution on planet Earth. But the record is not complete—known fossils cannot account for every intermediate step in the evolution of every organism. Considering that as many as 5 million prokaryote species may be living on Earth today (not counting bacteria), over the billions of years that life has existed there may have been 5 billion to 50 billion species. Clearly, known fossils represent at most a tiny percentage of these species. Why is the record so incomplete?

First, despite all the fossil-collecting efforts of the past two centuries, paleontologists have not even come close to sampling every cubic centimeter of sedimentary rock exposed on Earth. Just as biologists have not yet identified every living species of insect, paleontologists have not yet identified every species of fossil. New species and even genera of fossils continue to be discovered every year.

Second, not all organisms are represented in the rock record, because not all organisms have a high preservation potential. As noted earlier, fossilization occurs only under special conditions, and thus only a minuscule fraction of the organisms that have lived on Earth have left a fossil record. There may be few, if any, fossils of a vast number of extinct species, so we have no way of knowing what they looked like or even that they ever existed.

Finally, as we will learn in Chapter 12, the sequence of sedimentary strata that exists on Earth does not account for every minute of time since the formation of our planet. Sediments accumulate only in environments where conditions are appropriate for deposition and not for erosion—sediments do not accumulate, for example, on the dry great plains or on mountain peaks, but do accumulate in the sea and in the floodplains and deltas of rivers. Because Earth's climate changes through time and because the sea level rises and falls, certain locations on continents are sometimes sites of deposition and sometimes aren't, and on occasion are sites of erosion. Therefore, strata accumulate only episodically.

In sum, a rock sequence provides an incomplete record of Earth history, organisms have a low probability of being preserved, and paleontologists have found only a small percentage of the fossils preserved in rock. So the incompleteness of the fossil record comes as no surprise.

E.6 EVOLUTION AND EXTINCTION

Darwin's Grand Idea

As a young man in England in the early nineteenth century, Charles Darwin had been unable to settle on a career but had developed a strong interest in natural history. Therefore, he jumped at the opportunity to serve as a naturalist aboard HMS *Beagle* on an around-the-world surveying cruise. During the five years of the cruise, from 1831 to 1836, Darwin made detailed observations of plants, animals, and geology in the field and amassed an immense specimen collection from South America, Australia, and Africa. Just before Darwin departed on the voyage, a friend gave him a copy of Charles Lyell's 1830 textbook, *Principles of Geology*, which argued in favor of James Hutton's proposal that the Earth had a long history and that geologic time extended much further into the past than did human civilization. A visit to the Galápagos Islands, off the coast of Peru, led to a turning point in Darwin's thinking during the voyage. The naturalist was most impressed with the variability of Galápagos finches. He marveled not only at the fact that different varieties of the bird occurred on different islands, but at how each variety had adapted to utilize a particular food supply. With Lyell's writings in mind, Darwin developed a hypothesis

that the finches had begun as a single species but had branched into several different species when populations of the birds became isolated on different islands. This proposal implied that a species could change, or undergo **evolution**, throughout long periods of time and that new species could appear.

The crux of Darwin's argument is simply this: populations of organisms cannot increase in numbers forever, because they are limited by competition for scarce resources in the environment. In nature, only organisms capable of survival can pass on their characteristics to the next generation. In each new generation, some individuals have characteristics that make them more fit, whereas some have characteristics that make them less fit. The fitter organisms are more likely to survive and produce offspring. Thus, beneficial characteristics that they possess get passed on to the next generation. Darwin called this process **natural selection**, because it occurs on its own in nature. According to Darwin, when natural selection takes place over long periods of time—geologic time—it eventually produces new organisms that differ so significantly from their distant ancestors that the new organisms can be considered to constitute a new species. If environmental conditions change, or if competitors enter the environment, species that do not evolve and become better adapted to survive eventually die off.

Darwin's view of evolution has been successfully supported by many observations, and so far has not been definitively disproved by any observation or experiment. Also, it can be used to make testable predictions. Thus, scientists now refer to Darwin's idea as the **theory of evolution** by natural selection (see Box P.1 in the Prelude for the definition of a theory).

In the century and a half since Darwin published his work, genetics (the study of genes) has developed and has provided insight into *how* evolution works. Progress began in the late nineteenth century when an Austrian monk, Gregor Mendel, studied peas in the garden of his monastery and showed that genetic mutations led to new traits that could be passed on to offspring. Traits that make an organism less likely to survive are not passed on, either because the organism dies before it has offspring or because the offspring themselves cannot survive, but traits that make an organism better suited to survival are passed on to succeeding generations. With the discovery of DNA in 1953, biologists began to understand the molecular nature of genes and mutations, and thus of evolution. And with the genome projects of the twenty-first century, which define the detailed architecture of DNA molecules for a given species, it is now possible to pinpoint the exact arrangement of genes responsible for specific traits.

The theory of evolution provides a conceptual framework in which to understand paleontology. By studying fossils in sequences of strata, paleontologists are able to observe progressive changes in species through time, and can determine when some species die out and other species appear. But because of the incompleteness of the fossil record, many questions remain as to the *rates* at which evolution takes place during the course of geologic time. Originally, it was assumed that evolution happened at a constant, slow rate—this concept is called **gradualism**. More recently, however, researchers have suggested that evolution takes place in fits and starts: evolution occurs very slowly for quite a while (the species are in equilibrium) and then, during a relatively short period, it takes place very rapidly. This concept is called **punctuated equilibrium**. Factors that could cause sudden pulses of evolution include (1) a sudden mass extinction event during which many organisms disappear, leaving ecological niches open for new species to colonize; (2) a sudden change in the Earth's climate that puts stress on organisms—organisms that evolve to survive the new stress survive, whereas others become extinct; (3) the sudden formation of new environments, as may happen when rifting splits apart a continent and generates a new ocean with new coastlines; and (4) the isolation of a breeding population.

Extinction: When Species Vanish

Extinction occurs when the last members of a species die, so there are no parents to pass on their genetic traits to offspring. Some species become extinct as a population evolves into new species, whereas other species just vanish, leaving no hereditary offspring. These days, we take for granted that species become extinct, because a great number have, unfortunately, vanished from the Earth during human history. Before the 1770s, however, few geologists thought that extinction occurred; they thought fossils that didn't resemble known species must have living relatives somewhere on the planet. Considering that large parts of the Earth remained unexplored, this idea wasn't so far-fetched. But by the end of the eighteenth century, it became clear that numerous fossil organisms did not have modern-day counterparts. The bones of mastodons and woolly mammoths, for example, were too different from those of elephants to be of the same species, but the animals were too big to hide.

Twentieth-century studies concluded that many different phenomena can contribute to extinction. Some extinctions may happen suddenly, when all members of a species die off in a short time, whereas others may occur over longer periods, when the replacement rate of a population simply becomes lower than the mortality rate. By examining the number of species on Earth through time (in other words, by studying variations in the diversity of life, or in **biodiversity**), paleontologists have found that the rate of extinction varies through time. Generally, the rate is fairly slow; but on occasion a **mass extinction event** occurs, during which a large number of species worldwide disappear. At least five major mass extinc-

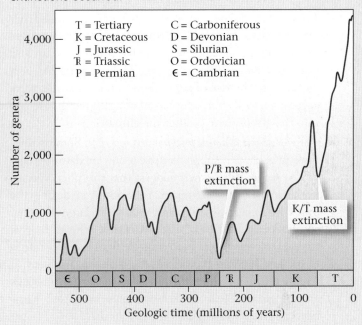

FIGURE E.11 This graph shows how the diversity of life has changed with time. Sudden drops indicate periods when mass extinctions occurred.

tion events have happened during the past half billion years (Fig. E.11). These events define the boundaries between some of the major intervals into which geologists divide time. For example, a major extinction event marks the end of the Cretaceous Period, 65 million years ago. During this event, all dinosaur species (with the exception of their modified descendants, the birds) vanished, along with many marine invertebrate species. A huge extinction event also brought the Permian to a close. Some researchers have suggested that extinction events are periodic, but this idea remains controversial.

Following are some of the geologic factors that may cause extinction.

- *Global climate change*: At times, the Earth's mean temperature has been significantly colder than today's, whereas at other times it has been much warmer. Because of a change in climate, an individual species may lose its habitat, and if it cannot adapt to the new habitat or migrate to stay with its old one, the species will disappear.

- *Tectonic activity*: Tectonic activity causes vertical movement of the crust over broad regions, changes in sea-floor spreading rates, and changes in the amount of volcanism. These phenomena can modify the distribution and area of habitats. Species that cannot adapt die off.

- *Asteroid or comet impact*: Many geologists have concluded that impacts of large meteorites with the Earth have been catastrophic for life. A large impact would send dust and debris into the atmosphere that could blot out the Sun and plunge the Earth into darkness and cold (see Chapter 23). Such a change, though relatively short lived, could interrupt the food chain.

- *Voluminous volcanic eruption*: Several times during Earth history, incredible quantities of lava have spilled out on the surface and/or incredible volumes of ash and gas have spewed into the air. These eruptions, perhaps due to the rise of superplumes in the mantle, were accompanied by the release of enough greenhouse gas into the atmosphere to alter the climate.

- *The appearance of a new predator or competitor*: Some extinctions may happen simply because a new predator appears on the scene. Researchers suggest that this phenomenon explains the mass extinction that occurred during the past 20,000 years, when a vast number of large mammal species vanished from North America. The timing of these extinctions appears to coincide with the appearance of the first humans (fierce predators) on the continent. If a more efficient competitor appears, the competitor steals an ecological niche from the weaker species, whose members can't obtain enough food and thus die.

Guide Terms

amber (p. 389)

biodiversity (p. 397)

biomarker (p. 389)

body fossil (p. 388)

Cambrian explosion (p. 395)

cast (p. 389)

chemical fossil (p. 388)

eukaryote (p. 393)

evolution (p. 396)

extinction (p. 397)

extraordinary fossil (p. 391)

fossil (p. 386)

fossilization (p. 387)

gradualism (p. 397)

mass extinction event (p. 397)

microfossil (p. 389)

mold (p. 389)

morphology (p. 393)

natural selection (p. 396)

paleontologist (p. 386)

paleontology (p. 386)

permineralization (p. 389)

petrified wood (p. 389)

phylogenetic tree (p. 395)

preservation potential (p. 391)

prokaryote (p. 393)

punctuated equilibrium (p. 397)

taxonomy (p. 392)

theory of evolution (p. 396)

trace fossil (p. 388)

ANOTHER VIEW This Cretaceous fish fossil was found in a sandstone bed in Brazil.

This road cut of sedimentary rocks in eastern New York contains a wealth of information about events in Earth's past. The tilting of the beds and the nature of the contacts between them tells us about mountain-building events that happened hundreds of millions of years ago.

CHAPTER 12

Deep Time: How Old Is Old?

GEOPUZZLE

The outcrop in the photo is not particularly dramatic—but holds the narrative of at least two mountain-building events. How can we determine when these events took place?

If the Eiffel Tower were now representing the world's age, the skin of paint on the pinnacle-knob at its summit would represent man's share of that age; and anybody would perceive that that skin was what the tower was built for. I reckon they would, I dunno.

—Mark Twain (1835–1910)

12.1 INTRODUCTION

In May of 1869, a one-armed Civil War veteran named John Wesley Powell set out with a team of nine geologists and scouts to explore the previously unmapped expanse of the Grand Canyon, the greatest gorge on Earth. Though Powell and his companions battled fearsome rapids and the pangs of starvation, most managed to emerge from the mouth of the canyon three months later (Fig. 12.1). During their voyage, seemingly insurmountable walls of rock both imprisoned and amazed the explorers, and led them to pose important questions about the Earth and its history, questions that even casual tourists to the canyon ponder today: Did the Colorado River sculpt this marvel, and if so, how long did it take? When did the rocks making up the walls of the canyon form?

FIGURE 12.1 Woodcut illustration of the "noonday rest in Marble Canyon," from J. W. Powell's *The Exploration of the Colorado River and Its Canyons* (1895). "We pass many side canyons today that are dark, gloomy passages back into the heart of the rocks."

Was there a time *before* the colorful layers accumulated? Such questions pertain to **geologic time**, the span of time since Earth's formation.

In this chapter, we first learn the geologic principles that allowed geologists to develop the concept of geologic time and thus to develop a frame of reference for describing the relative ages of rocks, fossils, structures, and landscapes. This information sets the stage for introducing the geologic column, the way that geologists divide time into intervals. Then we look at the tools geologists use to determine the numerical age of the Earth and its features in years; specifically, we introduce isotopic (radiometric) dating and the geologic time scale. With the concept of geologic time in hand, a hike down a trail into the Grand Canyon becomes a trip into what authors call *deep time*. The geological discovery that our planet's history extends billions of years into the past changed humanity's perception of time and the Universe as profoundly as did the astronomical discovery that the limit of space extends billions of light years beyond the edge of our Solar System (Box 12.1).

Chapter Theme

By the end of this chapter, you should understand . . .

- the meaning of geologic time, and the difference between relative and numerical ages.
- geologic principles (uniformitarianism; superposition; fossil succession) and their implications.
- how unconformities form and what they represent.
- the basis for correlating stratigraphic formations, and how correlation led to development of the geologic column.
- how geologists determine the numerical age of rocks by using isotopic dating.
- the basis for determining dates on the geologic time scale, and for determining the age of the Earth.

12.2 THE CONCEPT OF GEOLOGIC TIME

Setting the Stage for Studying the Past

Until relatively recently, people in most cultures have believed that geologic time began about the same time that human history began, and that our planet has been virtually unchanged since its birth. With this viewpoint in mind, an Irish archbishop named James Ussher (1581–1656) tallied successions of lineages and reigns described in the Old and New Testaments to determine the age of the Earth and, in 1654, stated his conclusion: the birth of the Earth took place on October 23, 4004 B.C.E.

BOX 12.1

CONSIDER THIS . . .

Time: A Human Obsession

When you plan your daily schedule, you have to know not only where you need to be, but when you need to be there. Because time assumes such significance in human consciousness today, we have developed elaborate tools to measure it and formal scales to record it. We use a second as the basic unit of time measurement. What exactly is a second? From 1900 to 1968, we defined the second as 1/31,556,925.9747 of the year 1900, but now we define it as the duration of time that it takes for a cesium atom to change back and forth between two

energy states 9,192,631,770 times. This change is measured with a device called an atomic clock, which is accurate to about 1 second per 30 million years. We sum 60 seconds into 1 minute, 60 minutes into 1 hour, and 24 hours into 1 day, about the time it takes for Earth to spin once on its axis.

In the pre-industrial era, each locality kept its own time, setting noon as the moment when the Sun reached the highest point in the sky. But with the advent of train travel and telegraphs, people needed to calibrate schedules from place

to place. So in 1883, countries around the globe agreed to divide the world into 15°-wide bands of longitude called time zones—in each time zone, all clocks keep the same standard time. The times in each zone are set in relation to Greenwich mean time (GMT), the time at the astronomical observatory in Greenwich, England. Today, the world standard for time is determined by a group of about 200 atomic clocks that together define Coordinated Universal Time (UTC). UTC is the basis for the global positioning system (GPS), used for precise navigation.

Not long after Ussher had implied that, in effect, the Earth has existed for only about 250 human generations, Nicholas Steno (1638–1686) proposed an idea that established the foundation for a very different approach to thinking about geologic time. Steno, born Niels Stenson in Denmark, was serving as a physician in a nobleman's court in Florence, Italy, when he realized that unusual triangular rocks which locals considered to be petrified tongues of dragons closely resembled the teeth of sharks (Fig. 12.2). Steno became convinced that "tongue stones" were actually **fossils** (remnants of once-living organisms that were buried in sediment that had later turned into rock) of shark teeth. He soon came to the realization that the seashell-like shapes which he found in strata of the nearby Apennine mountains were also fossils, so he speculated that the rock had formed from sediment that had accumulated on the sea floor and then was later incorporated in mountains (see Interlude E). In 1669, Steno published his ideas in a book entitled *Forerunner to a Dissertation on a Solid Naturally Occurring Within a Solid*. This title sounds strange until you think about the puzzle that he wanted to address: How does a solid object (a fossil) get inside another solid (a sedimentary rock)? The implications of Steno's ideas—that fossils were remnants of ancient life buried in loose sediment, that loose sediment might later transform into rock, and that rock may be uplifted from the sea floor to mountain heights—were explored in greater depth by James Hutton (1726–1797), a Scottish gentleman farmer and doctor, about a century later.

Hutton lived during the Age of Enlightenment when, sparked by the discovery of physical laws by Sir Isaac Newton, many people were led to seek natural, rather than supernatural, explanations for features of the world around them. While wandering in the highlands of Scotland, a region where rocks are well exposed, Hutton noted that many features (such as ripple marks and cross beds) found in sedimentary rock types resembled features he could see forming today in modern depositional environments. Based on such observations, Hutton

FIGURE 12.2 A fossilized shark's tooth. Before Steno explained the origin of such fossils, they were known as dragons' tongues.

developed an idea, which later came to be known as the principle of uniformitarianism, and published it in a 1785 book called *The Theory of the Earth*.

The principle of **uniformitarianism** states that physical processes we observe today also operated in the past at roughly the same rates, and that these processes were responsible for the formation of geologic features that we now see in outcrops. More concisely, the priniciple can be stated as: *the present is the key to the past*. Hutton deduced that the development of individual geologic features took a long time, and that not all features formed at the same time, so the Earth must have a history that includes a succession of slow events. Since no one in recorded history has seen the entire process of sediment first turning into rock and then later rising into mountains, Hutton also deduced that there must have been a long time *before* human history began. In fact, he speculated that we could see "no vestige of a beginning, nor prospect of an end."

Hutton was not a particularly clear writer, and it took the efforts of subsequent geologists to clarify the implications of the principle of uniformitarianism and to publicize them. Once this had been accomplished, geologists around the world began to apply their growing understanding of geologic processes to define and interpret geologic events of the Earth's past.

Relative versus Numerical Age

Like historians, geologists strive to establish both the sequence of events that created an array of geologic features (such as rocks, structures, and landscapes) and, when possible, the date on which each event happened. We specify the age of one feature with respect to another in a sequence as its **relative age**, and the age of a feature given in years as its **numerical age** (or, in older literature, its "absolute age") (Fig. 12.3). Geologists learned how to determine relative age long before they could determine numerical age, so we will look at the principles leading to relative-age determination next.

Take-Home Message

- The idea that fossils are relicts of past life proved to be a key to unraveling Earth's history.
- James Hutton first articulated the principle of uniformitarianism, which states that geologic processes observed today also happened in the past, at similar rates.
- Relative age refers to the time of one geologic feature with respect to another; numerical age indicates the date, in years, before present that a geologic feature formed.

THINK: What observations led Hutton to propose uniformitarianism?

FIGURE 12.3 The difference between relative and numerical age.

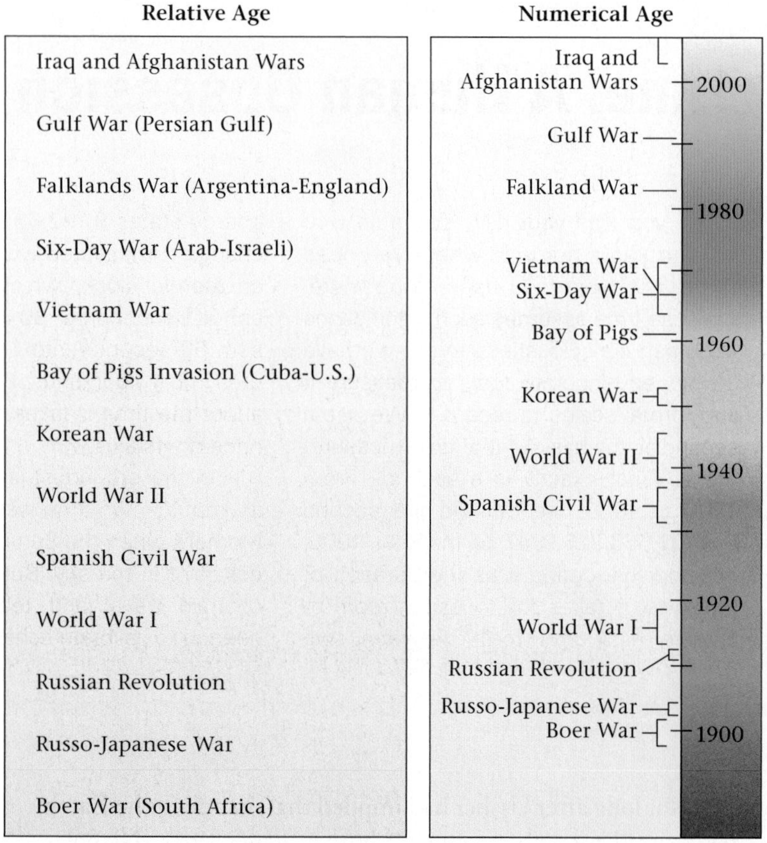

(a) The relative ages of selected wars in the last 100 or so years.

(b) The numerical ages of these same wars. Clearly, this chart provides more information, for it displays the duration of events and indicates the amount of time between events.

12.3 PHYSICAL PRINCIPLES FOR DEFINING RELATIVE AGE

Building from the work of Steno, Hutton, and others, the British geologist Charles Lyell (1797–1875) laid out a set of formal, usable geologic principles in the first modern textbook of geology (*Principles of Geology*, published in 1830–1833). These principles, defined below, continue to provide the basic framework within which geologists read the record of Earth history and determine relative ages.

- *The principle of uniformitarianism*: As noted earlier, physical processes we observe operating today also operated in the past, at roughly comparable rates (Fig. 12.4a, b). Again, the present is the key to the past.

- *The principle of original horizontality*: Sediments on Earth settle out of a fluid in a gravitational field. Typically, the surfaces on which sediments accumulate (such as a floodplain or the bed of a lake or sea) are fairly horizontal.

FIGURE 12.4 The major geologic principles.

Present-day mudcracks form in clay-rich sediment.

Ancient mudcracks in solid rock.

(a) Uniformitarianism: The processes that formed cracks in the dried-up mud puddle on the left also formed the mudcracks preserved in the ancient, solid rock on the right. We can see these ancient mudcracks because erosion removed the adjacent bed.

Present-day volcanism produces molten lava.

Layers of basalt formed during volcanic activity.

(b) Uniformitarianism (cont.): We can observe lava flows forming today, so we can infer that solid lava flows represent the products of volcanic eruptions in the past.

Sediment exposed at low tide, Mont St. Michel, France.

(c) Original horizontality: Gravity causes sediment to accumulate in fairly horizontal sheets on a flat plain.

Youngest bed

Bedding plane

Cross beds

Oldest bed

What a Geologist Sees

(d) Horizontal bedding in an outcrop of sandstone in Wisconsin.

FIGURE 12.4 *(continued)*

Time 1   Time 2   Time 3

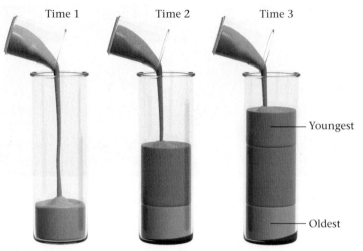

— Youngest

— Oldest

(e) Superposition: In a sequence of strata, the oldest bed is on the bottom, and the youngest on top. Pouring sand into a glass illustrates this point.

Therefore, layers of sediment when originally deposited are also fairly horizontal (Fig. 12.4c, d). If sediments were deposited on a steep slope, they would likely slide downslope before lithification, and so would not be preserved as sedimentary rocks. With this principle in mind, we realize that when we see folds and tilted beds, we are seeing the consequences of deformation that postdates deposition.

- *The principle of superposition*: In a sequence of sedimentary rock layers, each layer must be younger than the one below, for a layer of sediment cannot accumulate unless there is already a substrate on which it can collect. Thus, the layer at the bottom of a sequence is the oldest, and the layer at the top is the youngest (Fig. 12.4e).

- *The principle of lateral continuity*: Sediments generally accumulate in continuous sheets within a given region. If today you find a sedimentary layer cut by a canyon, then you can assume that the layer once spanned the area that was later eroded by the river that formed the canyon (Fig. 12.4f).

At time of deposition

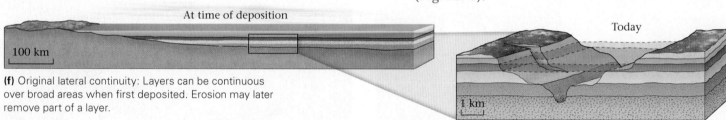

100 km

Today

1 km

(f) Original lateral continuity: Layers can be continuous over broad areas when first deposited. Erosion may later remove part of a layer.

Dike

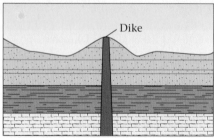

(g) Cross-cutting relations: The dike cuts across the sedimentary beds, so the dike is younger.

Baked contact

Pluton

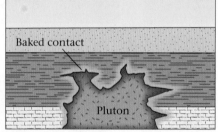

(h) Baked contact: The pluton baked the adjacent rock, so the adjacent rock is older.

The pebbles of basalt in a conglomerate must be older than the conglomerate.

Flow

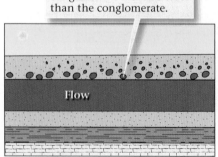

Xenoliths of sandstone must be older than the basalt containing them.

Sill

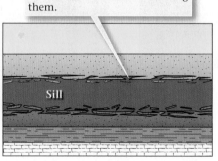

(i) Inclusions: Rock that occurs as an inclusion in another rock must be the older of the two.

Chilled Margin

Younger   Older

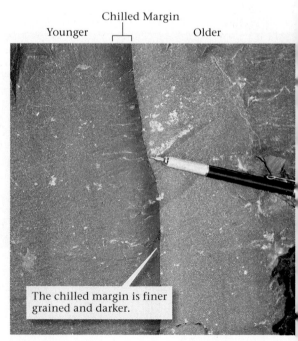

The chilled margin is finer grained and darker.

(j) The edge of the younger dike is finer grained, for it cools quickly when it contacts the older, cooler rock.

■ *The principle of cross-cutting relations*: If one geologic feature cuts across another, the feature that has been cut is older. For example, if an igneous dike cuts across a sequence of sedimentary beds, the beds must be older than the dike (**Fig. 12.4g**). If a fault cuts across and displaces layers of sedimentary rock, then the fault must be younger than the layers. But if a layer of sediment buries a fault, the sediment must be younger than the fault.

■ *The principle of baked contacts*: An igneous intrusion "bakes" (metamorphoses) surrounding rocks. The rock that has been baked must be older than the intrusion (**Fig. 12.4h**). Similarly, since an intrusion injects into cooler rocks, the margin of the intrusion cools rapidly and is finer grained than the interior. Such chilled margins are visible in outcrops (**Fig. 12.4j**).

■ *The principle of inclusions*: If a layer of sediment deposited on an igneous layer includes pebbles of the igneous rock, then the sedimentary layer must be younger (**Fig. 12.4i, j**). If an igneous intrusion contains fragments of another rock, the fragments must be older than the intrusion. The fragments (xenoliths) in an igneous body and the pebbles in the sedimentary layer are inclusions, or pieces of one material incorporated in another. The rock containing the inclusion must be younger than the inclusion. We can use these principles to determine the relative ages of geologic features. In so doing, we *develop a geologic history* of the region, defining the relative ages of events that took place there.

Let's consider an example of how to develop a geologic history by interpreting features portrayed in a drawing (**Fig. 12.5a** ▶). First, note what kinds of rocks and structures the

FIGURE 12.5 Interpreting the geologic history of a region, using geologic principles as a guide. ▶

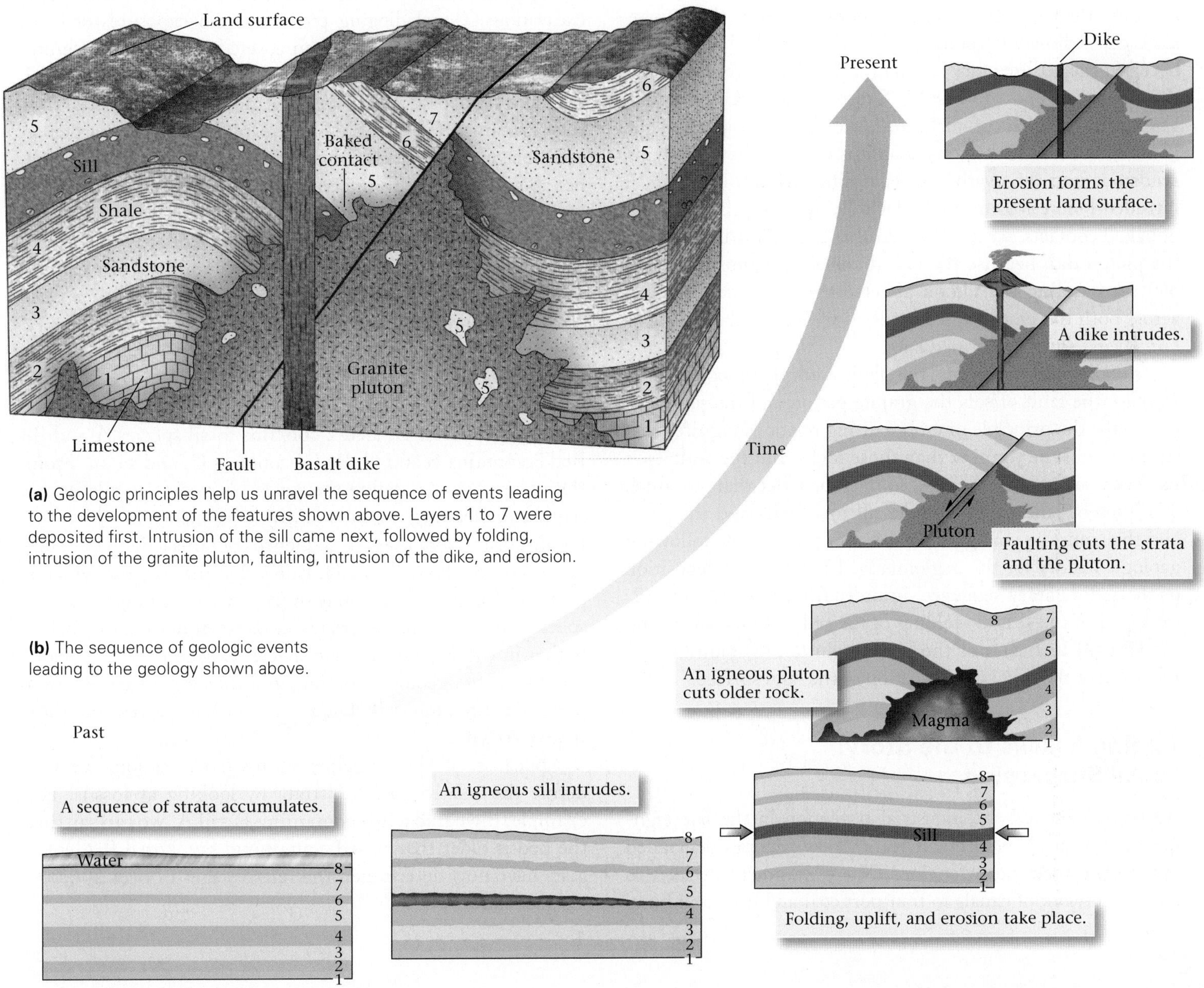

(a) Geologic principles help us unravel the sequence of events leading to the development of the features shown above. Layers 1 to 7 were deposited first. Intrusion of the sill came next, followed by folding, intrusion of the granite pluton, faulting, intrusion of the dike, and erosion.

(b) The sequence of geologic events leading to the geology shown above.

figure contains. Most of the rocks constitute a sequence of folded sedimentary beds, but we also see a granite pluton, a basalt dike, and a fault. Let's start our analysis by looking at the sedimentary sequence. By the principle of superposition, we conclude that the oldest layer is the limestone labeled 1, for it occurs at the bottom. Progressively younger beds lie above the limestone. We can confirm that layer 1 predates layer 2 by applying the principle of inclusions, for layer 2 contains pebbles (inclusions) of layer 1. Thus, the sedimentary beds from oldest to youngest are 1, 2, 3, 4, 5, 6, 7. (There are other rocks below layer 1, but we do not see them.) Considering the principle of original horizontality, we conclude that the layers were folded sometime after deposition.

Now let's look at the relationships between the igneous rocks and the sedimentary rocks. The granite pluton cuts across the folded sedimentary rocks, so the intrusion of the pluton occurred after the deposition of the sedimentary beds and after they were folded. The layer of igneous rock that parallels the sedimentary beds could be either a sill that intruded between the sedimentary layers or a flow that spread out over the sandstone and solidified before the shale was deposited. From the principle of inclusions, we deduce that the layer is a sill, because it contains xenoliths of both the underlying sandstone and the overlying shale. Since the sill is folded, it intruded before folding took place. By applying the principle of baked contacts, we also can tell that the sill intruded before the pluton did, because the baked zones (metamorphic aureole) surrounding the pluton affected the sill. The dike cuts across both the pluton and the sill, as well as all the sedimentary layers, and thus formed later.

Finally, let's consider the fault and the land surface. Because the fault offsets the granite pluton and the sedimentary beds, the principle of cross-cutting relations requires that the fault must be younger than those rocks. But the fault itself has been cut by the dike, and so must be older than the dike. The present land surface erodes all rock units and the fault, and thus must be younger. We can now propose the following geologic history for this region (Fig. 12.5b ▶❙❙): (1) deposition of the sedimentary sequence, in order from layers 1 to 8; (2) intrusion of the sill; (3) folding of the sedimentary layers and the sill; (4) intrusion of the granite pluton; (5) faulting; (6) intrusion of the dike; (7) formation of the land surface.

Adding Fossils to the Story: Fossil Succession

As Britain entered the industrial revolution in the late eighteenth and early nineteenth centuries, new factories demanded coal to fire their steam engines. The government decided to build a network of canals to transport coal and iron, and hired

FIGURE 12.6 A bedding surface containing many fossils.

an engineer named William Smith (1769–1839) to survey the excavations. Canal digging provided fresh exposures of bedrock, which previously had been covered by vegetation. Smith learned to recognize distinctive layers of sedimentary rock and to identify the fossil assemblage (the group of fossil species) that they contained (Fig. 12.6). He also realized that a particular assemblage can be found only in a limited interval of strata, and not above or below this interval. Thus, once a fossil species disappears at a horizon in a sequence of strata, it never reappears higher in the sequence. In other words, extinction is forever. Smith's observation has been repeated at millions of locations around the world, and has been codified as the *principle of fossil succession*. It provides the geologic underpinning for the theory of evolution (see Interlude E).

To see how this principle works, examine Figure 12.7, which depicts a sequence of strata. Bed 1 at the base contains fossil species A, Bed 2 contains fossil species A and B, Bed 3 contains B and C, Bed 4 contains C, and so on. From these data, we can define the range of specific fossils in the sequence, meaning the interval in the sequence in which the fossils occur. Note that the sequence contains a definable succession of fossils (A, B, C, D, E, F), that the range in which a particular species occurs may overlap with the range of other species, and that once a species vanishes, it does not reappear higher in the sequence. Some species existed only for a short interval of the geologic column, and thus are diagnostic of a particular time interval. The fossils of such species are called **index fossils**.

Because of the principle of fossil succession, we can define the relative ages of strata by looking at fossils. For example, if we find a bed containing Fossil A, we can say that the bed is older than a bed containing, say, Fossil F. Geologists have now determined the relative ages of over 200,000 fossil species.

FIGURE 12.7 The principle of fossil succession.

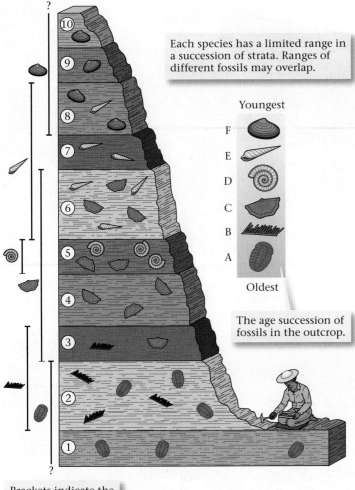

Each species has a limited range in a succession of strata. Ranges of different fossils may overlap.

Youngest

Oldest

The age succession of fossils in the outcrop.

Brackets indicate the range of a species.

- The determination of relative age utilizes a number of fundamental geologic principles in addition to uniformitarianism.
- Strata originally accumulate in fairly horizontal layers that are laterally continuous over a region. Strata at the base of a succession are older than those at the top.
- If one geologic feature cuts across another, it is the younger of the two features.
- Fragments of rock embedded in another rock must be older than the rock containing them.
- The succession of fossil species in strata is not haphazard; younger fossil species occur above older ones, and once a species goes extinct, it doesn't reappear higher in the succession.

THINK: If the fault in Figure 12.5 had formed a fault scarp, how would the geologic history of the region shown by the figure be different than the history described in the text?

12.4 UNCONFORMITIES: GAPS IN THE RECORD

James Hutton boated along the coast of Scotland because the shore cliffs provided good exposures of rock, stripped of soil and shrubbery. He was particularly puzzled by an outcrop along the shore at Siccar Point. One sequence of rock exposed there consisted of alternating beds of gray sandstone and shale, but another consisted of red sandstone and conglomerate (Fig. 12.8a, b ◖▶). The beds of gray sandstone and shale were nearly vertical, whereas the beds of red sandstone and conglomerate had a dip of about 20° (see Box 11.1). Further, the gently dipping layers seemed to lie across the truncated ends of the vertical layers, like a handkerchief lying across a row of books. We can imagine that as Hutton examined this odd geometric relationship, the tide came in and deposited a new layer of sand on top of the rocky shore. With the principle of uniformitarianism in mind, Hutton suddenly realized the significance of what he saw. The gray sandstone–shale sequence had been deposited, turned into rock, tilted, and truncated by erosion *before* the red sandstone–conglomerate beds had been deposited.

Hutton deduced that the surface between the gray and red rock sequences represented a time interval during which new strata had not been deposited at Siccar Point and the older strata had been eroded away. We now call such a surface, representing a period of nondeposition and possibly erosion, an **unconformity**. The gap in the geologic record that an unconformity represents is called a hiatus. Geologists recognize three kinds of unconformities:

- *Angular unconformity*: Rocks below an angular unconformity were tilted or folded before the unconformity developed (Fig. 12.9a). Thus, an angular unconformity cuts across the underlying layers, and the orientation of layers below an unconformity differs from that of the layers above. (The outcrop at Siccar Point exposes an angular unconformity.) Angular unconformities form where rocks were tilted by either folding or faulting, before being exposed at the Earth's surface.

- *Nonconformity*: A nonconformity is a type of unconformity at which sedimentary rocks overlie generally much older intrusive igneous rocks and/or metamorphic rocks (Fig. 12.9b). The igneous or metamorphic rocks underwent cooling, uplift, and erosion prior to becoming the substrate or basement on which new sediments accumulated.

- *Disconformity*: Imagine that a sequence of sedimentary beds has been deposited beneath a shallow sea. Then sea level drops, exposing the beds for some time. During this time, no new sediment accumulates, and some of the preexisting sediment gets eroded away. Later,

FIGURE 12.8 Examples of unconformities, as visible in outcrops. ▶ⅠⅠ

What a Geologist Sees

(a) James Hutton found this unconformity along the east coast of Scotland. He deduced that the layers above were deposited long after the beds below had been tilted. Geologists have since determined that strata above are about 50 million years younger than strata below.

What a Geologist Sees

(b) This unconformity, exposed along a reservoir in Missouri, shows sedimentary beds deposited on top of much older rhyolite. The rhyolite mass was a hill at the time of deposition, so the beds pinch out near the contact. Geologists have determined that the rhyolite is almost a billion years older than the beds.

sea level rises, and a new sequence of sediment accumulates over the old. The boundary between the two sequences is a disconformity (Fig. 12.9c, d ▶ⅠⅠ). Even though the beds above and below the disconformity are parallel, the contact between them represents an interruption in deposition.

The succession of strata at a particular location provides a record of Earth history there. But because of unconformities, *the record preserved in the rock layers is incomplete*. It's as if geologic history is being chronicled by a tape recorder that turns on only intermittently—when it's on (times of deposition), the rock

record accumulates, but when it's off (times of nondeposition and possibly erosion), an unconformity develops. Because of unconformities, no single location on Earth contains a complete record of Earth history.

How do you recognize an disconformity in the field? At angular unconformities, the difference in dip is an easy giveaway, and at nonconformities, the juxtaposition of cover over basement serves as a clue. Across a disconformity, there is a gap in the fossil succession, but if you are unfamiliar with fossils, they are hard to spot. If erosion and weathering took place

Did you ever wonder . . .
do Grand Canyon strata represent all Earth's history?

FIGURE 12.9 The three kinds of unconformities and their formation.

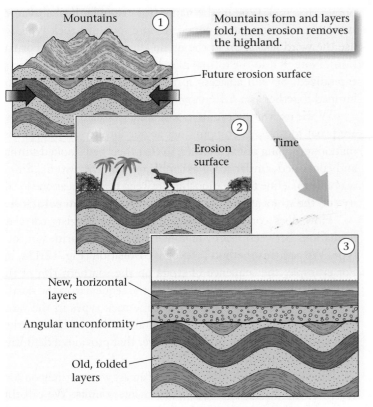

(a) An angular unconformity: (1) layers undergo folding; (2) erosion produces a flat surface; (3) sea level rises and new layers of sediment accumulate.

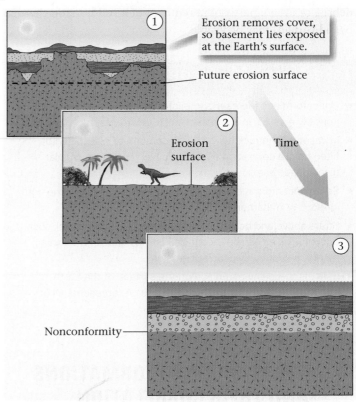

(b) A nonconformity: (1) a pluton intrudes; (2) erosion cuts down into the crystalline rock; (3) new sedimentary layers accumulate above the erosion surface.

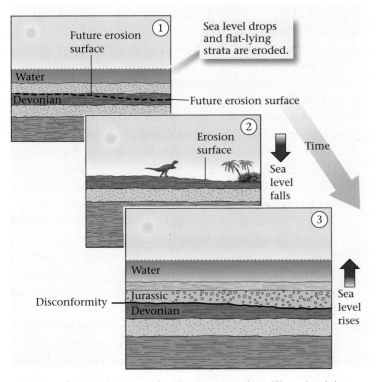

(c) A disconformity: (1) layers of sediment accumulate; (2) sea level drops and an erosion surface forms; (3) sea level rises and new sedimentary layers accumulate.

(d) This roadcut in Utah shows a sand-filled channel cut down into floodplain mud. The mud was exposed between floods, and a soil formed on it. When later buried, all the sediment turned into rock; the channel floor is an unconformity, and the ancient soil is a "paleosol." Note that the channel cut across the paleosol. The paleosol represents a disconformity.

prior to deposition of the younger sequence, a pebbly layer of debris, or even a paleosol (a remnant of a soil horizon that has been lithified) may be visible.

Take-Home Message

- Unconformities represent time intervals of nondeposition, and possibly erosion, in a region.
- In cases where rock below the unconformity was folded or tilted before deposition of strata above, there is an angular discordance across the unconformity.
- Some unconformities juxtapose sedimentary strata above with igneous or metamorphic basement below.
- Strata above and below an unconformity may be parallel; such unconformities can be recognized only by gaps in the fossil record, and/or by evidence of erosion.

THINK: Is there any one place on the surface of the Earth where the exposed stratigraphic succession represents all of geologic time?

12.5 STRATIGRAPHIC FORMATIONS AND THEIR CORRELATION

Geologists summarize information about the sequence of sedimentary strata at a location by drawing a **stratigraphic column**. Typically, we draw columns to scale, so that the relative thicknesses of layers portrayed on the column reflect the thicknesses of layers in the outcrop. Then, geologists divide the sequence of strata represented on a column into **stratigraphic formations** (formations, for short), a sequence of beds of a specific rock type or group of rock types that can be traced over a fairly broad region. The boundary surface between two formations is a type of geologic **contact**. (Fault surfaces and the boundary between an igneous intrusion and its wallrock are also types of contacts.) Typically, a formation has a specific geologic age.

Let's see how the concept of a stratigraphic formation applies to the Grand Canyon. The walls of the canyon look striped, because they expose a variety of rock types that differ in color and in resistance to erosion. Geologists identify major contrasts distinguishing one interval of strata from another, and use them as a basis for dividing the strata into formations, each of which may consist of many beds (Fig. 12.10a–c ▶). Note that some formations include a single rock type, whereas others include interlayered beds of two or more rock types. Also, note that not all formations have the same thickness, and the thickness of a single formation can vary with location. Commonly, geologists name a formation after a locality where it was first identified or first studied. For example, the Schoharie Formation was first defined based on exposures in Schoharie Creek, of eastern New York.

If a formation consists of only one rock type, we may incorporate that rock type in the name (for example, Kaibab Limestone), but if a formation contains more than one rock type, we use the word *formation* in the name (such as Toroweap Formation). Note that in the formal name of a formation, all words are capitalized. Several adjacent formations in a succession may be lumped together as a *stratigraphic group*.

While excavating canals in England, William Smith discovered that formations cropping out at one locality resembled formations cropping out at another, in that their beds looked similar and contained similar fossil assemblages. In other words, Smith was able to define the relationship between the strata at one locality and the strata at another, a process now called **correlation**.

How does correlation work? Typically, geologists correlate formations between *nearby* regions based on similarities in rock type. We call this method lithologic correlation (Fig. 12.11a, b). For example, the sequence of strata on the southern rim of the Grand Canyon clearly correlates with the sequence on the northern rim, because they contain the same rock types in the same order. In some cases, a sequence contains a key bed, or marker bed, which is a particularly unique layer that provides a definitive basis for correlation.

To correlate rock units over *broad* areas, we must rely on fossils to define the relative ages of sedimentary units. We call this method fossil correlation. Geologists use fossil correlation for studies of broad areas because sources of sediments and depositional environments may change from one location to another. The beds deposited at one location during a given time interval may look quite different from the beds deposited at another location during the same time interval. But if *fossils of the same relative age* occur at both locations, we can say that the strata at the two locations correlate. (Note that the fossils are not necessarily the same species—they won't be if the depositional environments are different.) Fossil correlation may also come in handy when rock types are not distinctive enough to allow correlation. For example, imagine that the Santuit Sandstone and Oswaldo Sandstone of Figure 12.11a look the same. In Column C, only fossils may distinguish one layer from the other, if the intervening Milo Limestone is absent. The contact between the two formations is an unconformity.

Now let's look at an example of fossil correlation by tracing the individual formations exposed in the Grand Canyon into the mountains just north of Las Vegas, 150 km to the west (Fig. 12.12a, b). Near Las Vegas, we find a sequence of sedimentary rocks that includes a limestone formation called the Monte Cristo Limestone. The Monte Cristo Limestone contains fossils of the same relative age as occur in the Redwall Limestone of the Grand Canyon, but it is much thicker. Because the formations contain fossils of the same relative age, we conclude that they were deposited during the same time interval, and thus we say that they correlate with one another. Note also that not only are the units thicker in the Las Vegas area than in the Grand Canyon area, but there are more of them. This discovery indicates that during part of the time

FIGURE 12.10 The stratigraphic column and stratigraphic formation: examples from the Grand Canyon in Arizona.

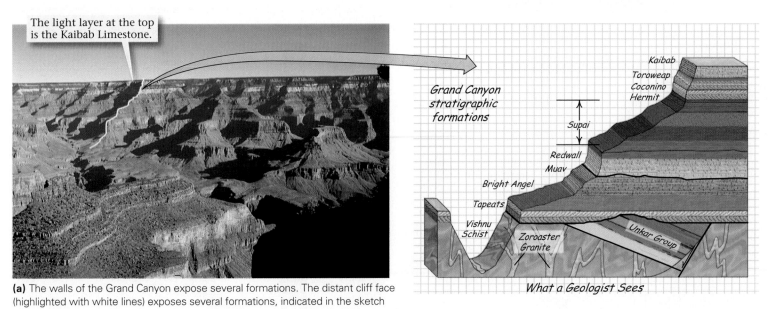

The light layer at the top is the Kaibab Limestone.

Grand Canyon stratigraphic formations

Kaibab
Toroweap
Coconino
Hermit
Supai
Redwall
Muav
Bright Angel
Tapeats
Vishnu Schist
Zoroaster Granite
Unkar Group

What a Geologist Sees

(a) The walls of the Grand Canyon expose several formations. The distant cliff face (highlighted with white lines) exposes several formations, indicated in the sketch on the right.

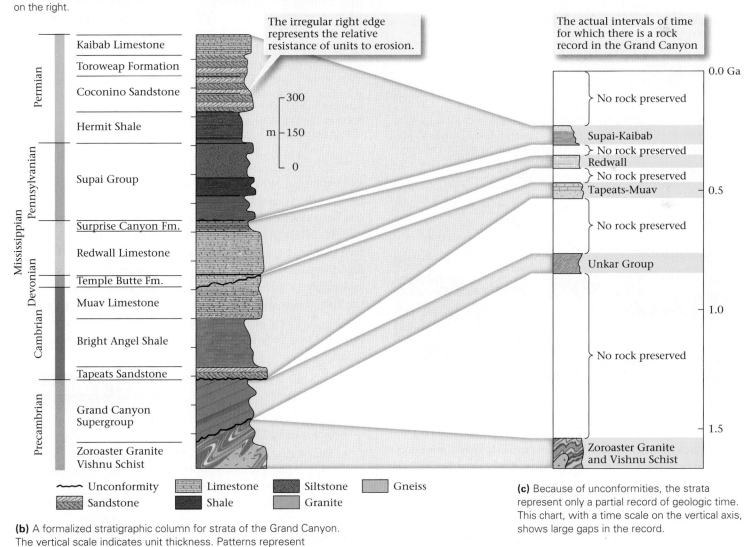

The irregular right edge represents the relative resistance of units to erosion.

The actual intervals of time for which there is a rock record in the Grand Canyon

Permian	Kaibab Limestone
Permian	Toroweap Formation
Permian	Coconino Sandstone
	Hermit Shale
Pennsylvanian	Supai Group
Mississippian	Surprise Canyon Fm.
	Redwall Limestone
Devonian	Temple Butte Fm.
	Muav Limestone
Cambrian	Bright Angel Shale
	Tapeats Sandstone
Precambrian	Grand Canyon Supergroup
Precambrian	Zoroaster Granite Vishnu Schist

m 300 — 150 — 0

0.0 Ga

No rock preserved

Supai-Kaibab
No rock preserved
Redwall
No rock preserved
Tapeats-Muav — 0.5

No rock preserved

Unkar Group

— 1.0

No rock preserved

— 1.5

Zoroaster Granite and Vishnu Schist

~~~ Unconformity

Limestone   Siltstone   Gneiss

Sandstone   Shale   Granite

(b) A formalized stratigraphic column for strata of the Grand Canyon. The vertical scale indicates unit thickness. Patterns represent different rock types.

(c) Because of unconformities, the strata represent only a partial record of geologic time. This chart, with a time scale on the vertical axis, shows large gaps in the record.

FIGURE 12.11 The principles of correlation.

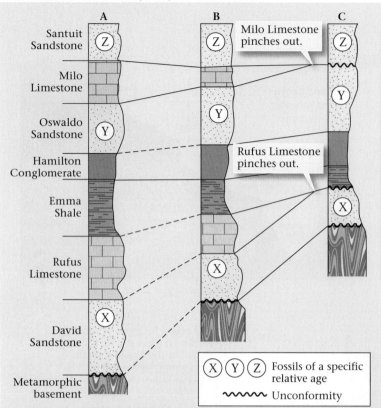

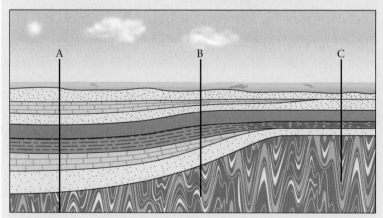

(a) Stratigraphic columns can be correlated by matching rock types (lithologic correlation). The Hamilton Conglomerate is a marker horizon. Because some strata pinch out, Column C contains unconformities. Fossil correlation indicates that the youngest beds in C are Santuit Sandstone.

(b) At the time of deposition, locations A, B, and C were in different parts of a basin. The basin floor was subsiding fastest at A.

FIGURE 12.12 An application of correlation to relate strata near Las Vegas to strata in the Grand Canyon.

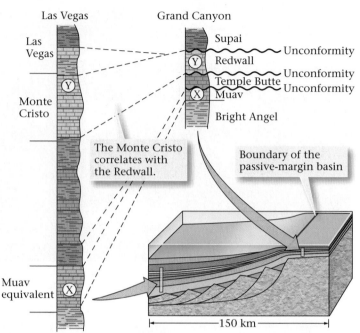

(a) The section between fossils of age X and fossils of age Y is much thicker near Las Vegas than in the Grand Canyon. The edge of a passive-margin basin lay between the two localities at the time of deposition.

(b) A passive-margin basin forms over crust that has stretched and thinned.

when thick sediments were deposited near Las Vegas, none accumulated near the Grand Canyon. Thus, the contact beneath the Grand Canyon's Redwall Limestone is an unconformity.

The reason that there is such a contrast between the stratigraphic columns of the Las Vegas and the Grand Canyon areas is because at a given time, the two regions hosted different depositional settings. The Las Vegas column accumulated on the thinner crust of a passive-margin basin that sank (subsided) rapidly

and remained submerged below the sea almost continuously, whereas the Grand Canyon column accumulated on the thicker crust of a craton that episodically emerged above sea level.

By correlating strata at many locations, William Smith realized that he could trace individual formations of strata over fairly broad regions. In 1815, he plotted the distribution of formations and created the first modern **geologic map**, which portrays the spatial distribution of rock units at the Earth's surface. As Figure 12.13a–c shows, we can use stratigraphic information provided by a geologic map to identify geologic structures.

Take-Home Message

- A stratigraphic formation is a definable succession of strata that can be traced over a region.
- Since exposures (outcrops) of strata are limited, geologists must examine fossils and other features to correlate strata at one locality with strata at another.
- The spatial distribution of stratigraphic formations and other geologic features can be portrayed on a geologic map.

THINK: Would correlation be possible if the principle of fossil succession had not been established?

FIGURE 12.13 A geologic map depicts the distribution of rock units and structures.

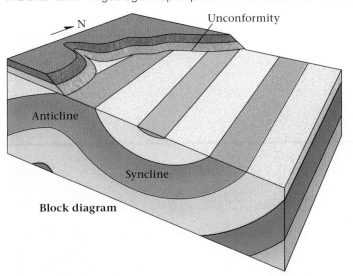

Block diagram

(a) A block diagram provides a three-dimensional representation. Here, we see an angular unconformity over folds.

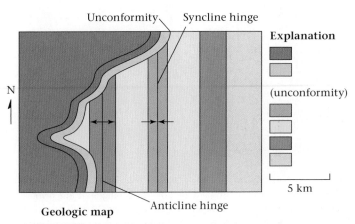

Geologic map

(b) A geologic map shows what the distribution of units would look like if viewed from above. Contacts occur between units.

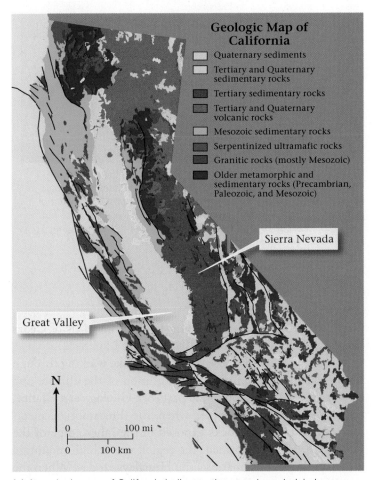

Geologic Map of California

☐ Quaternary sediments

☐ Tertiary and Quaternary sedimentary rocks

■ Tertiary sedimentary rocks

■ Tertiary and Quaternary volcanic rocks

☐ Mesozoic sedimentary rocks

■ Serpentinized ultramafic rocks

■ Granitic rocks (mostly Mesozoic)

■ Older metamorphic and sedimentary rocks (Precambrian, Paleozoic, and Mesozoic)

(c) A geologic map of California indicates the state is underlain by many different rock units. Granite underlies the Sierras, and Quaternary sediments underlie the Great Valley. The black lines are fault traces.

12.6 THE GEOLOGIC COLUMN

As stated earlier, no one locality on Earth provides a complete record of our planet's history, because stratigraphic columns can contain unconformities. But by correlating rocks from locality to locality at millions of places around the world, geologists have pieced together a *composite* stratigraphic column, called the **geologic column**, that *represents* the entirety of Earth history (Fig. 12.14a, b). The column is divided into segments, each of which represents a specific interval of time. The largest subdivisions break Earth history into the Hadean, Archean, Proterozoic, and Phanerozoic **Eons**. (The first three together constitute the **Precambrian**.) The suffix –*zoic* means life, so Phanerozoic means visible life, and Proterozoic means first life. (After the eons had been named, geologists discovered that the earliest life, bacteria and archaea, actually appeared during the Archean Eon.)

The Phanerozoic Eon is subdivided into **eras**. In order from oldest to youngest, they are the Paleozoic (ancient life), Mesozoic (middle life), and Cenozoic (recent life) Eras. We further divide each era into **periods** and each period into **epochs**.

Where do the names of the periods come from? They refer either to localities where a fairly complete stratigraphic column representing that time interval was first identified (for example, rocks representing the Devonian Period crop out near Devon, England) or to a characteristic of the time (rocks from the Carboniferous Period contain a lot of coal). The terminology was not set up in a planned fashion that would make it easy to learn. Instead, it grew haphazardly in the years between 1760 and 1845, as geologists began to refine their understanding of geologic history and fossil succession.

FIGURE 12.14 Global correlation of strata led to the development of the geologic column. (Not to scale)

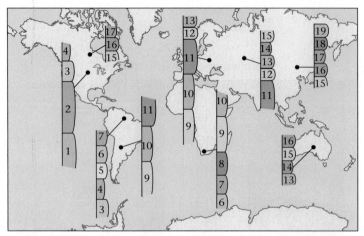

(a) Each of these small columns represents the stratigraphy at a given location. By correlating these columns, geologists determined their relative ages, filled in the gaps in the record, and produced the geologic column.

	Eon	Era	Period	Epoch
19	Phanerozoic	Cenozoic	Quaternary	Holocene
18				Pleistocene
17			Neogene	Pliocene
16			Tertiary	Miocene
15				Oligocene
14			Paleogene	Eocene
13				Paleocene
12				
11		Mesozoic	Cretaceous	
10			Jurassic	
9			Triassic	
8		Paleozoic	Permian / Pennsylvanian	
7			Carboniferous / Mississippian	
6			Devonian	
5			Silurian	
4			Ordovician	
3			Cambrian	
2	Precambrian	Proterozoic		
1		Archean		

(b) By correlation, the strata from localities around the world were stacked in a chart representing geologic time to create the geologic column. Geologists assigned names to time intervals, but since the column was built without knowledge of numerical ages, it does not depict the duration of these intervals. Subdivisions of eons in the Precambrian are not shown.

The succession of fossils preserved in strata of the geologic column defines the course of life's evolution throughout Earth history (**Fig. 12.15**). Simple bacteria and archaea appeared during the Archean Eon, but complex shell-less invertebrates did not evolve until the late Proterozoic. The appearance of invertebrates with shells defines the Precambrian-Cambrian boundary. At this time there was a sudden diversification in life, with many new genera appearing over a relatively short interval—this event is called the **Cambrian explosion** (**Fig. 12.16**).

The first vertebrates, fish, appeared during the Ordovician Period. Before the Silurian Period, the land surface was barren of multicellular life, but in the Silurian Period, land plants spread over the continents. Amphibians appeared during the Devonian. Though reptiles appeared during the Pennsylvanian Period, the first dinosaurs did not pound across the land until the Triassic. Dinosaurs continued to inhabit the Earth until their sudden extinction at the end of the Cretaceous Period. For this reason, geologists refer to the Mesozoic Era as the Age of Dinosaurs. Small mammals appeared during the Triassic Period, but the diversification (development of many different species) of mammals to fill a wide range of ecological niches did not happen until the beginning of the Cenozoic Era, so geologists call the Cenozoic the Age of Mammals. Birds also appeared during the Mesozoic (specifically, at the beginning of the Cretaceous Period), but underwent great diversification in the Cenozoic Era.

To conclude our discussion, let's see how the geologic column comes into play when correlating strata across a region. We return to the Colorado Plateau of Arizona and Utah, in the southwestern United States (**Fig. 12.17a, b**). Because of the lack of vegetation in this region, you can easily see bedrock exposures on the walls of cliffs and canyons; some of these exposures are so beautiful that they have become national parks. The oldest sedimentary rock of the region crops out at the base of the

Grand Canyon, whereas the youngest form the cliffs of Cedar Breaks and Bryce Canyon (**See for Yourself L**, p. S-22).

Walking through these parks is thus like walking through time—each rock layer gives an indication of the climate and topography of the region in the past (See **Geology at a Glance**, pp. 418–419). For example, when the Precambrian metamorphic and igneous rocks exposed in the inner gorge of the Grand Canyon first formed, the region was a high mountain range, perhaps as dramatic as the Himalayas today. When the fossiliferous beds of the Kaibab Limestone at the rim of the canyon first developed, the region was a Bahama-like carbonate reef and platform, bathed in a warm, shallow sea. And when the rocks making up the towering red cliffs of sandstone in Zion Canyon were deposited, the region was a Sahara-like desert, blanketed with huge sand dunes.

FIGURE 12.15 Life evolution in the context of the geologic column. The Earth formed at the beginning of the Hadean Eon.

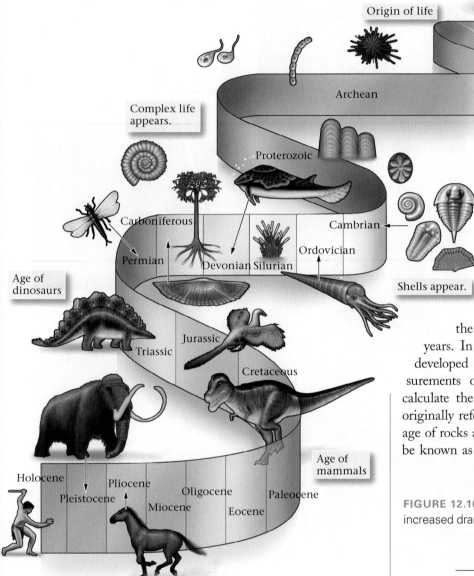

12.7 HOW DO WE DETERMINE NUMERICAL AGE?

Geologists since the days of Hutton could determine the relative ages of geologic events, but they had no way to specify numerical ages (called "absolute ages" in older literature). Thus, they could not define a timeline for Earth history or determine the duration of events. This situation changed with the discovery of radioactivity. Simply put, radioactive elements decay at a constant rate that can be measured in the lab and can be specified in years. In the 1950s, geologists first developed techniques for using measurements of radioactive elements to calculate the ages of rocks. Geologists originally referred to this process of determining the numerical age of rocks as radiometric dating; more recently, it has come to be known as **isotopic dating**. The overall study of numerical

Did you ever wonder...
how geologists can specify the age of some rocks in years?

Take-Home Message

- By correlating successions of strata from around the world, geologists established a "geologic column" that represents all of geologic time.
- The geologic column, which was established prior to the availability of numerical ages, is divided into eras, periods, and epochs.
- Specific fossil species lived for specific intervals of the geologic column. Thus, fossils can be used as a basis for specifying the interval in which the host rock was deposited.

THINK: What feature of certain living organisms appeared at the Precambrian/Cambrian boundary?

FIGURE 12.16 The "Cambrian explosion." The diversity of genera increased dramatically at about 530 Ma.

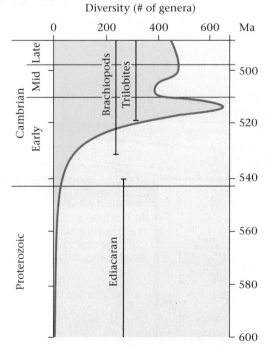

FIGURE 12.17 Correlation of strata among the national parks of Arizona and Utah.

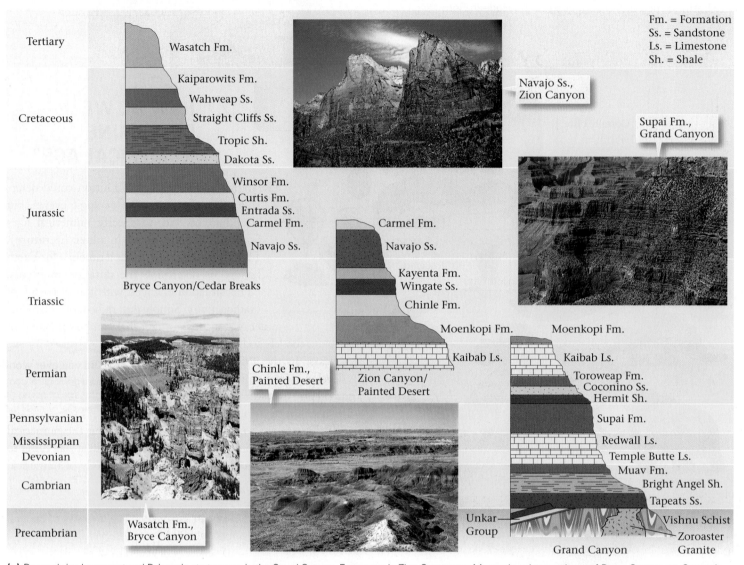

Fm. = Formation
Ss. = Sandstone
Ls. = Limestone
Sh. = Shale

Tertiary — Wasatch Fm.

Cretaceous — Kaiparowits Fm. / Wahweap Ss. / Straight Cliffs Ss. / Tropic Sh. / Dakota Ss.

Jurassic — Winsor Fm. / Curtis Fm. / Entrada Ss. / Carmel Fm. / Navajo Ss.

Bryce Canyon/Cedar Breaks

Navajo Ss., Zion Canyon

Supai Fm., Grand Canyon

Carmel Fm.
Navajo Ss.
Kayenta Fm.
Wingate Ss.
Chinle Fm.
Moenkopi Fm.
Kaibab Ls.

Zion Canyon/Painted Desert

Moenkopi Fm.
Kaibab Ls.
Toroweap Fm.
Coconino Ss.
Hermit Sh.
Supai Fm.
Redwall Ls.
Temple Butte Ls.
Muav Fm.
Bright Angel Sh.
Tapeats Ss.
Unkar Group
Vishnu Schist
Zoroaster Granite

Grand Canyon

Triassic

Permian

Pennsylvanian

Mississippian
Devonian

Cambrian

Precambrian

Chinle Fm., Painted Desert

Wasatch Fm., Bryce Canyon

(a) Precambrian basement and Paleozoic strata occur in the Grand Canyon. Exposures in Zion Canyon are Mesozoic, whereas those of Bryce Canyon are Cenozoic.

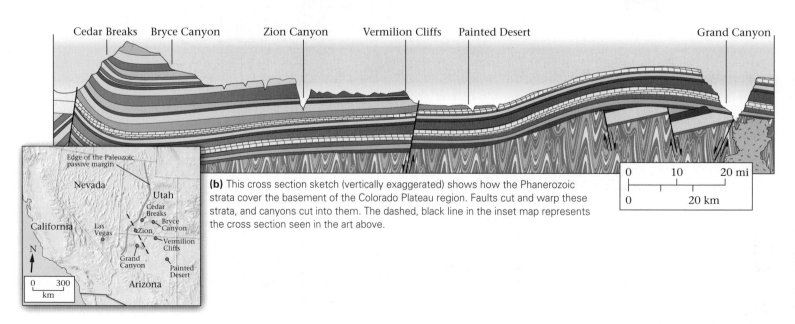

Cedar Breaks   Bryce Canyon   Zion Canyon   Vermilion Cliffs   Painted Desert   Grand Canyon

Edge of the Paleozoic passive margin
Nevada
Utah
Cedar Breaks
California
Las Vegas
Zion
Bryce Canyon
Vermilion Cliffs
Grand Canyon
Painted Desert
Arizona
N
0   300
km

0   10   20 mi
0   20 km

(b) This cross section sketch (vertically exaggerated) shows how the Phanerozoic strata cover the basement of the Colorado Plateau region. Faults cut and warp these strata, and canyons cut into them. The dashed, black line in the inset map represents the cross section seen in the art above.

ages is **geochronology**. Techniques of isotopic dating have been vastly improved over the years. We begin our discussion of this technique by first learning more about radioactive decay.

Isotopes, Radioactive Decay, and the Concept of a Half-Life

All atoms of a given element have the same number of protons in their nucleus—we call this number the atomic number (see Chapter 1). However, not all atoms have the same number of neutrons in their nucleus. Therefore, not all atoms of a given element have the same atomic weight—roughly the number of protons plus neutrons. Different versions of an element, called **isotopes** of the element, have the same atomic number but a different atomic weight (see Box 1.4). For example, all uranium atoms have 92 protons, but the uranium-238 isotope (abbreviated ^{238}U) has an atomic weight of 238 and thus has 146 neutrons, whereas the ^{235}U isotope has an atomic weight of 235 and thus has 143 neutrons.

Some isotopes of an element are stable, meaning that they last essentially forever. Radioactive isotopes are unstable in that eventually, they undergo a change called **radioactive decay**, which converts them into a different element. Radioactive decay can take place by a variety of reactions that change the atomic number of the nucleus and thus form a different element. In these reactions, the isotope that undergoes decay is the **parent isotope**, while the decay product is the **daughter isotope**. For example, rubidium-87 (^{87}Rb) decays to strontium-87 (^{87}Sr), potassium-40 (^{40}K) decays to argon-40 (^{40}Ar), and uranium-238 (^{238}U) decays to lead-206 (^{206}Pb).

Physicists cannot specify how long an individual radioactive isotope will survive before it decays, but they can measure how long it takes for half of a group of parent isotopes to decay. This time is called the **half-life** of the isotope. Figure 12.18a–c can help you visualize the concept of a half-life. Imagine a crystal containing 16 radioactive parent isotopes. (In real crystals, the number of atoms would be much larger.) After one half-life, 8 isotopes have decayed, so the crystal now contains 8 parent and 8 daughter isotopes. After a second half-life, 4 of the remaining parent isotopes have decayed, so the crystal contains 4 parent and 12 daughter isotopes. And after a third half-life, 2 more parent isotopes have decayed, so the crystal contains 2 parent and 14 daughter isotopes. For a given decay reaction, the half-life is a constant.

Isotopic Dating Techniques

Since radioactive decay proceeds at a known rate, like the tick-tock of a clock, it provides a basis for telling time. In other words, because an element's half-life is a constant, we can cal-

FIGURE 12.18 The concept of a half-life, in the context of radioactive decay.

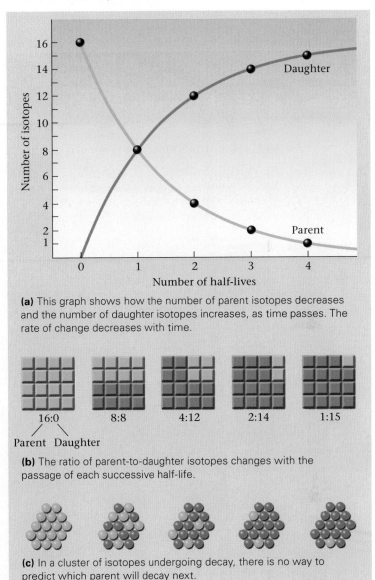

(a) This graph shows how the number of parent isotopes decreases and the number of daughter isotopes increases, as time passes. The rate of change decreases with time.

(b) The ratio of parent-to-daughter isotopes changes with the passage of each successive half-life.

(c) In a cluster of isotopes undergoing decay, there is no way to predict which parent will decay next.

culate the age of a mineral by measuring the ratio of parent to daughter isotopes in the mineral.

How do geologists actually obtain an isotopic date? First, we must find the right kind of elements to work with. Although there are many different pairs of parent and daughter isotopes among the known radioactive elements, only a few have long enough half-lives, and occur in sufficient abundance in minerals, to be useful for isotopic dating. Particularly useful elements are listed in Table 12.1. Each radioactive element has its own half-life. Note that carbon dating is *not* used for dating rocks because appropriate carbon isotopes only occur in organisms and have a very short half-life (Box 12.2).

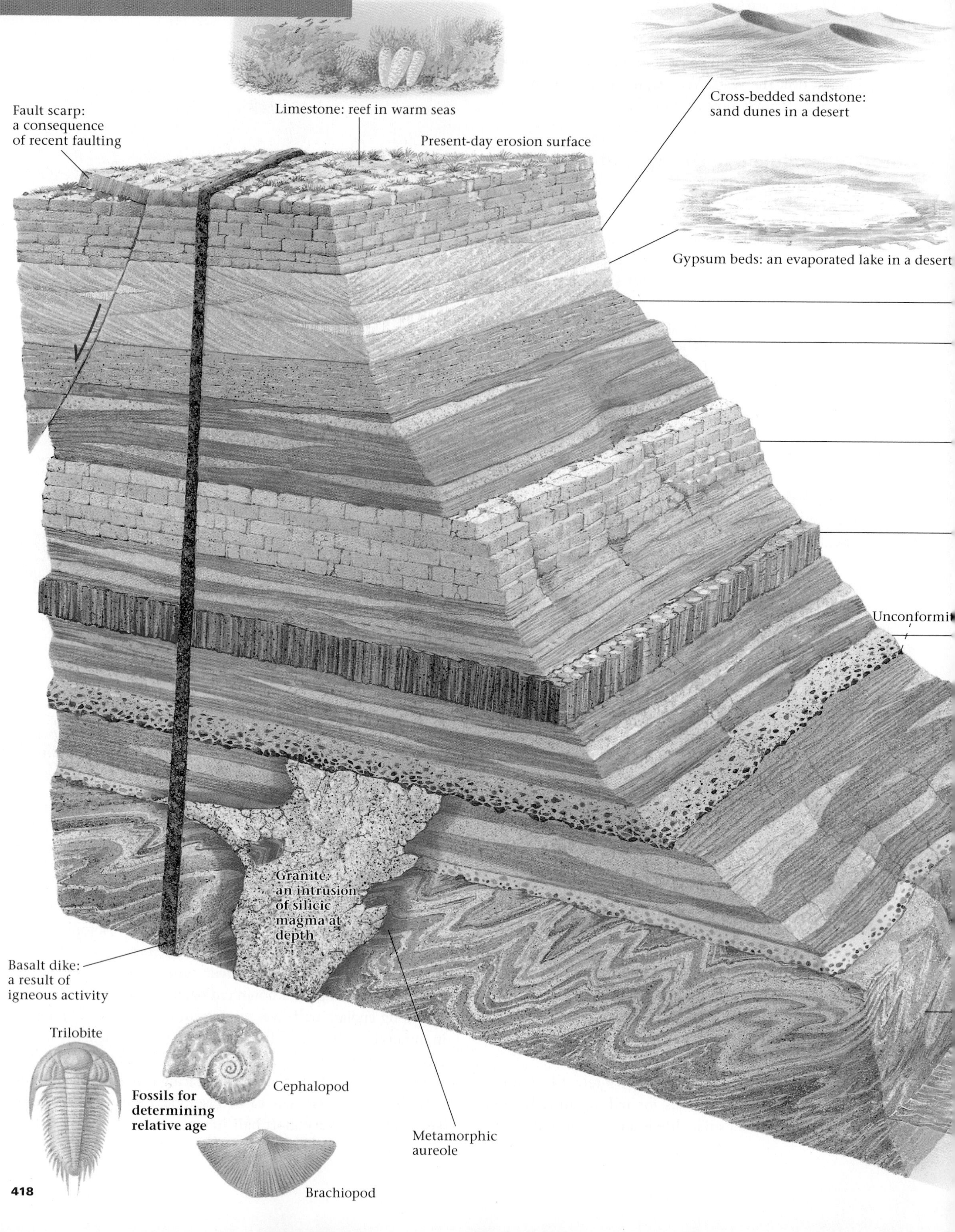

Limestone: reef in warm seas

Fault scarp:
a consequence
of recent faulting

Present-day erosion surface

Cross-bedded sandstone:
sand dunes in a desert

Gypsum beds: an evaporated lake in a desert

Unconformi

Granite:
an intrusion
of silicic
magma at
depth

Basalt dike:
a result of
igneous activity

Trilobite

Cephalopod

Fossils for
determining
relative age

Metamorphic
aureole

Brachiopod

The Record in Rocks: Reconstructing Geologic History

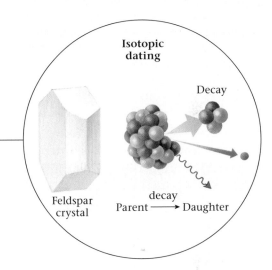

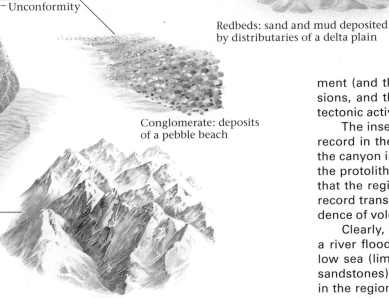

Ignimbrite (welded tuff): an explosive volcanic eruption

Limestone: reef in warm seas

Redbeds: sand and mud deposited in a river channel and bordering floodplain

Basalt lava: flows from a volcano

Isotopic dating

Decay

Feldspar crystal

decay
Parent ⟶ Daughter

Conglomerate: debris eroded from a cliff

– – Unconformity

Redbeds: sand and mud deposited by distributaries of a delta plain

Conglomerate: deposits of a pebble beach

Gneiss: metamorphism at depth beneath a mountain belt

When geologists examine a sequence of rocks exposed on a cliff, they see a record of Earth history that can be interpreted by applying the basic principles of geology, by searching for fossils, and by using isotopic dating. On this cliff, we see evidence for many geologic events. The layers of sediment (and the sedimentary structures they contain), the igneous intrusions, and the geologic structures tell us about past climates and past tectonic activity.

The insets show the way the region looked in the past, based on the record in the rocks. For example, the presence of gneiss at the base of the canyon indicates that at one time the region was a mountain belt, for the protoliths of the gneiss were buried deeply. Unconformities indicate that the region underwent uplift and erosion. Sedimentary successions record transgressions and regressions of the sea, igneous rocks are evidence of volcanic and intrusive activity, and faults indicate deformation.

Clearly, the land surface portrayed in this painting was sometimes a river floodplain or a delta (indicated by redbeds), sometimes a shallow sea (limestone), and sometimes a desert dune field (cross-bedded sandstones). And at several times in the past, volcanic activity occurred in the region. We can gain insight into the age of the sedimentary rocks by studying the fossils they contain, and into the age of the igneous and metamorphic rocks by using isotopic dating methods.

BOX 12.2

Carbon-14 Dating

Many people who have heard of **carbon-14 (^{14}C)** dating assume that it can be used to define the numerical age of rocks. But this is not the case. Rather, ^{14}C dating tells us the ages of organic materials—such as wood, cotton fibers, charcoal, flesh, bones, and shells—that contain carbon originally extracted from the atmosphere by photosynthesis in plants. ^{14}C, a radioactive isotope of carbon, forms naturally in the atmosphere when cosmic rays (charged particles from space) bombard atmospheric nitrogen-14 (^{14}N) atoms.

When plants consume carbon dioxide during photosynthesis, or when animals consume plants, they ingest a tiny amount of ^{14}C along with ^{12}C, the more common isotope of carbon. After an organism dies and can no longer exchange carbon with the atmosphere, the ^{14}C in its body begins to decay back to ^{14}N. Thus, the ratio of ^{14}C to ^{12}C changes at a rate determined by the half-life of ^{14}C.

We can use ^{14}C dating to determine the age of prehistoric fire pits or of organic debris in sediment. ^{14}C has a short half-life—only 5,730 years. Thus, the method cannot be used to date anything older than about 70,000 years, for after that time essentially no ^{14}C remains in the material. But this range makes it a useful tool for geologists studying sediments of the last ice age and for archaeologists studying ancient cultures or prehistoric peoples. Again, since rocks do not contain organic carbon, and may be significantly older than 70,000 years, we cannot determine the age of rocks by using the ^{14}C dating method.

TABLE 12.1 **Isotopes Used in the Isotopic Dating of Rocks**

Parent → Daughter	Half-Life (years)	Minerals Containing the Isotopes
^{147}Sm → ^{143}Nd	106 billion	Garnets, micas
^{87}Rb → ^{87}Sr	48.8 billion	Potassium-bearing minerals (mica, feldspar, hornblende)
^{238}U → ^{206}Pb	4.5 billion	Uranium-bearing minerals (zircon, uraninite)
^{40}K → ^{40}Ar	1.3 billion	Potassium-bearing minerals (mica, feldspar, hornblende)
^{235}U → ^{207}Pb	713 million	Uranium-bearing minerals (zircon, uraninite)

Sm = samarium, Nd = neodymium, Rb = rubidium, Sr = strontium, U = uranium, Pb = lead, K = potassium, Ar = argon

Second, we must identify the right kind of minerals to work with. Not all minerals contain radioactive elements, but fortunately some common minerals do. Now we can set to work using the following steps.

- *Collecting the rocks*: Geologists collect unweathered rocks for dating, for the chemical reactions that happen during weathering may lead to the loss of some isotopes.
- *Separating the minerals*: The rocks are crushed, and the appropriate minerals are separated from the debris.
- *Extracting parent and daughter isotopes*: To separate out the parent and daughter isotopes from minerals, geologists use several techniques, including dissolving the minerals in acid or evaporating portions of them with a laser.

- *Analyzing the parent-daughter ratio*: Geologists pass the atoms through a mass spectrometer, an instrument that uses a strong magnet to separate isotopes from one another according to their respective weights (**Fig. 12.19**). The instrument can count the number of atoms of specific isotopes separately.

At the end of the laboratory process, geologists can define the ratio of parent to daughter isotopes in a mineral, and from this ratio calculate the age of the mineral. Needless to say, the description of the procedure here has been simplified—in reality, obtaining an isotopic date is time-consuming and expensive and requires complex calculations.

What Does an Isotopic Date Mean?

At high temperatures, atoms in a crystal lattice vibrate so rapidly that chemical bonds can break and reattach relatively easily. As a consequence, isotopes escape from or move into crystals, so parent-daughter ratios are meaningless. Because isotopic dating is based on the parent-daughter ratio, the "isotopic clock" starts only when crystals become cool enough for isotopes to be locked into the lattice. The temperature below which isotopes are no longer free to move is called the **closure temperature** of a mineral. When we specify an isoto-

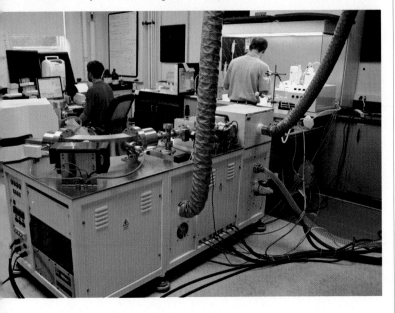

FIGURE 12.19 In an isotopic dating laboratory, samples are analyzed using a mass spectrometer. This instrument measures the ratio of parent to daughter isotopes.

pic date for a rock, we are defining the time at which a specific mineral in the rock cooled below its closure temperature.

With the concept of closure temperature in mind, we can interpret the meaning of isotopic dates. In the case of igneous rocks, isotopic dating tells you when a magma or lava cooled to form a solid, cool igneous rock. In the case of metamorphic rocks, an isotopic date tells you when a rock cooled from a metamorphic temperature above the closure temperature to a temperature below.

Can we isotopically date a clastic sedimentary rock directly? No. If we date minerals in a sedimentary rock, we determine only when these minerals first crystallized as part of an igneous or metamorphic rock, not the time when the minerals were deposited as sediment nor the time when the sediment lithified to form a sedimentary rock. For example, if we date the feldspar grains contained within a granite pebble in a conglomerate, we're dating the time the granite cooled below feldspar's closure temperature, not the time the pebble was deposited by a stream. The age of mineral grains in sediment, however, can be useful. In recent years, geologists have used the ages of detrital (clastic) grains to learn the age of the rocks in the sediment's source region.

Other Methods of Determining Numerical Age

Counting rings in trees or layers in sediment. The changes in seasons affect a wide variety of phenomena, including the following.

- *The growth rate of trees*: Trees grow seasonally, with rapid growth during the spring and no growth during the winter.

- *The organic productivity of lakes and seas*: During the winter, less light reaches the Earth, water temperatures decrease, and the supply of nutrients decreases.

- *The growth rate of chemically precipitated sedimentary rocks (such as travertine)*: Factors that control precipitation rates, such as the rate of groundwater flow and temperature, change seasonally in some locations.

- *The growth rate of shell-secreting organisms*: Organisms grow and produce new shell material on a seasonal basis.

- *The layering in glaciers*: During the snowy season, more snow falls relative to dust than during the dry season; annual snowfall includes a dusty layer and a clean layer. These contrasts are preserved as layers in sediment.

As a consequence of these seasonal changes, **growth rings** develop in trees, travertine deposits, and shelly organisms, and **rhythmic layering** develops in sedimentary accumulations and glacier ice (Fig. 12.20a–c). By counting rings or layers, geologists can determine how long an organism survived, or how long a sedimentary accumulation took to form. If rings or layers have developed right up to the present, we can count backward and determine how long ago they began to form.

Dendrochronologists, scientists who study and date tree rings, have found that not only do tree rings count time, they also preserve a record of the climate: in warm, rainy years, trees grow much faster than they do in drought years. Thus, by studying tree rings, dendrochronologists can provide a dated record of the climate in prehistoric times. The oldest living trees, bristlecone pines, are almost 4,000 years old. By correlating the older rings of these trees with rings in logs preserved in sediment accumulations, dendrochronologists have extended the tree-ring record back for many thousands of years (Fig. 12.20d).

Similarly, geologists have found that glacier ice preserves a valuable record of past climate, for the ratio of different oxygen isotopes in the water molecules making up the ice reflects the global temperature at the time the snow fell to create the ice. Ice cores drilled through the Greenland ice cap contain a continuous record of climate back through 750,000 years.

Magnetostratigraphy. As was discussed in Chapter 3, the polarity of Earth's magnetic field flips every now and then through geological time. Geologists have determined when the reversals took place, and have constructed a reference column showing the succession of reversals through time (Fig. 12.21). This reference column resembles the bar code on a supermarket product. By comparing the pattern of the reversals in a sequence of strata with the pattern of

FIGURE 12.20 Using layering and rings as a basis for dating.

(a) Dust settling during the summer highlights boundaries between snow layers, now turned to ice in this Oregon glacier.

(b) The ridges in this giant clamshell from the Red Sea form during successive seasons of growth.

(c) Each ring in this slice of wood represents the growth of one year. The width of each ring depends on the temperature and rainfall during growth. Patterns of rings are like bar codes—the pattern in a living tree can be correlated with the pattern in a dead tree, which can be correlated with the pattern in an ancient, buried tree. Such correlation extends the tree-ring record back in time.

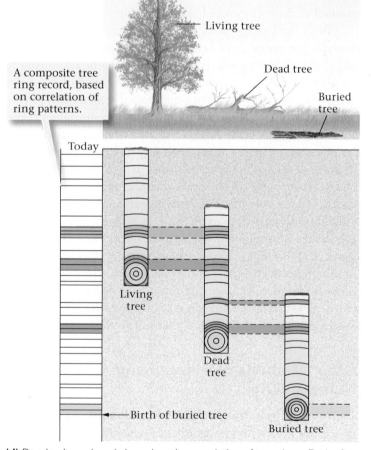

(d) Dendrochronology is based on the correlation of tree rings. Each of the columns in the diagram represents a core drilled out of a tree. Distinctive clusters of closely spaced rings indicate dry seasons. By correlation, researchers extend the climate record back in time before the oldest living tree started to grow.

reversals in a reference column, a study known as **magnetostratigraphy**, geologists can determine the age of the sequence.

Fission tracks. In certain minerals, the ejection of an atomic particle during the decay of a radioactive isotope damages the nearby crystal, creating a line called a **fission track**. This

FIGURE 12.21 Magnetostratigraphy involves comparing the sequence of polarity reversals in strata with the sequence of polarity reversals in a global reference column to determine the age of the strata.

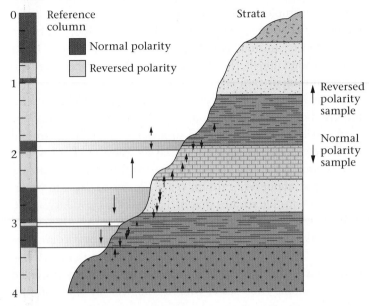

FIGURE 12.22 The method of fission-track dating.

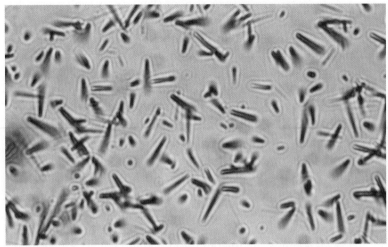

(a) This high-magnification photomicrograph shows fission tracks in a crystal. When formed, each track is only a few atoms wide. Treating the sample with acid enlarges the tracks so they are visible.

(b) A fission track forms when a radioactive atom decays and blasts out particles which disrupt the crystal structure in a narrow band.

track resembles the line of crushed grass left behind when you roll a tire across a lawn. As time passes, more atoms undergo fission, so the number of fission tracks in the crystal increases (Fig. 12.22a, b). Therefore, the number of fission tracks in a given volume of a crystal represents the age of the crystal. Geologists have been able to measure the rate at which fission tracks are produced, and thus can determine the age of a mineral grain by counting the fission tracks within it.

Take-Home Message

- Certain minerals contain radioactive parent atoms that decay to daughter atoms.
- Radioactive decay of a population of atoms occurs at a rate defined by the half-life.
- The ratio of parent to daughter in a mineral reflects the age of the mineral. Measuring this ratio is called isotopic dating; the result provides a numerical age.
- For younger materials, growth rings and ice layers can give numerical ages.
- The isotopic clock in a rock starts when crystals cool below a certain temperature.
- Isotopic ages can give the age that an igneous rock forms or a metamorphic rock cools; they cannot directly provide the age of deposition of a sedimentary rock.

THINK: What other tools, besides isotopic dating, can be used to constrain numerical ages?

12.8 HOW DO WE ADD NUMERICAL AGES TO THE GEOLOGIC COLUMN?

We have seen that isotopic (radiometric) dating can be used to date the time when igneous rocks formed and when metamorphic rocks metamorphosed, but not when sedimentary rocks were deposited. So how do we determine the numerical age of a sedimentary rock? We must answer this question if we want to add numerical ages to the geologic column—remember, the column was originally constructed by studying only the *relative* ages of fossil-bearing sedimentary rocks.

Geologists obtain dates for sedimentary rocks by studying cross-cutting relationships between sedimentary rocks and datable igneous or metamorphic rocks. For example, if we find a sedimentary rock layer deposited unconformably on a datable igneous rock, we know that the sedimentary rock must be younger than the igneous rock. If we find a datable igneous dike that cuts across beds of sedimentary rock, then the datable rock must be younger. And if a datable ash layer spread out over a layer of sediment and then was buried by another layer of sediment, the datable rock or ash must be younger than the underlying sediment and older than the overlying sediment (Fig. 12.23).

Let's look at an example of dating sedimentary rocks (see Fig. 12.23). Imagine that a layer of sandstone contains fossilized dinosaur bones. Geologists assign the fossils to the Cretaceous Period by correlating them with fossils in known Cretaceous strata from elsewhere. The sandstone was deposited unconformably over an eroded granite pluton that has a numerical age, determined by uranium-lead dating, of 125 million years. This same layer of sandstone has been cut by a basalt dike whose numerical age, based on potassium-argon dating, is 80 million years. These measurements mean that this Cretaceous sandstone bed was deposited sometime between 125 million and 80 million years ago. Note that the data provide an age range, not an exact age. Thus, from this observation we can only conclude that the Cretaceous Period includes the time interval of *at least* 125 to 80 million years ago.

Geologists have searched the world for localities where they can recognize cross-cutting relations between datable igneous rocks and sedimentary rocks or for layers of datable volcanic rocks interbedded with sedimentary rocks. By isotopically dating the igneous rocks, they have been able to provide numerical ages for the boundaries between all geologic periods. For example, work from around the world shows that the Cretaceous Period actually began about 145 million years ago and ended 65 million years ago. So the bed in Figure 12.23 was deposited during the middle part of the Cretaceous, not at the beginning or end.

The discovery of new data may cause the numbers defining the boundaries of periods to change, which is why the term *numerical age* is preferred to *absolute age*. In fact, around 1995, new dates on rhyolite ash layers above and below the Cambrian/Precambrian boundary showed that this boundary occurred at about 542 million years ago, in contrast to previous, less definitive studies that had placed the boundary at 570 million years ago. Figure 12.24 shows the currently favored numerical ages of periods and eras in the geologic column as of 2009. This dated column is commonly called the **geologic time scale**. Because of the numerical constraints provided by the geologic time scale, when geologists say that the first dinosaurs appeared during the Triassic Period, they mean that dinosaurs appeared later than 251 million years ago.

Take-Home Message

- To add numerical ages to the geologic column, geologists first determine the relative age of sedimentary strata relative to datable igneous or metamorphic rocks.
- Numerical ages on datable rocks that cross-cut, underlie, overlie, or are interlayered with strata provide numerical constraints on the age of the strata.
- By adding numerical ages to boundaries between intervals on the geologic column, geologists have produced a geologic time scale.

THINK: Why don't all periods on the geologic time column have the same duration in years?

12.9 WHAT IS THE AGE OF THE EARTH?

The mind grows giddy gazing so far back into the abyss of time.

—John Playfair (1747–1819),
British geologist who popularized the works of Hutton

Isotopic Dating of the Earth

During the eighteenth and nineteenth centuries, before the discovery of isotopic dating, scientists came up with a great variety of clever solutions to the question, "How old is the

FIGURE 12.23 The Cretaceous sandstone bed was deposited on the granite, so it must be younger than 125 Ma. The dike cuts the bed, so the bed must be older than 80 Ma. Thus, the Cretaceous bed was deposited between 125 and 80 Ma. The Paleocene sandstone was unconformably deposited over the dike and lies beneath a 50-million-year-old layer of ash. Therefore, it must have been deposited between 80 and 50 Ma.

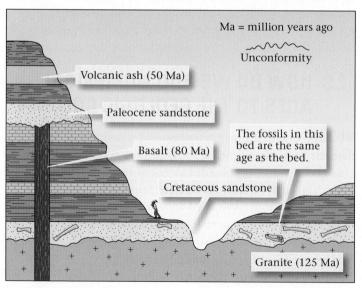

Ma = million years ago

〜〜〜
Unconformity

Volcanic ash (50 Ma)

Paleocene sandstone

Basalt (80 Ma)

The fossils in this bed are the same age as the bed.

Cretaceous sandstone

Granite (125 Ma)

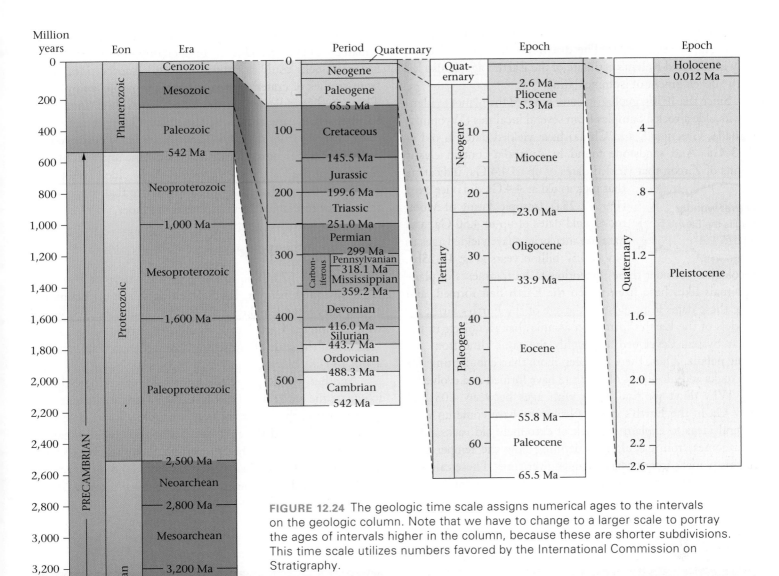

FIGURE 12.24 The geologic time scale assigns numerical ages to the intervals on the geologic column. Note that we have to change to a larger scale to portray the ages of intervals higher in the column, because these are shorter subdivisions. This time scale utilizes numbers favored by the International Commission on Stratigraphy.

Earth?"—all of which have since been proven wrong. Lord William Kelvin, a nineteenth-century physicist renowned for his discoveries in thermodynamics, made the most influential scientific estimate of the Earth's age of his time. Kelvin calcu-lated how long it would take for the Earth to cool down from a temperature as hot as the Sun's, and concluded that this planet is about 20 million years old.

Kelvin's estimate contrasted with those being promoted by followers of Hutton, Lyell, and Darwin, who argued that if the concepts of uniformitarianism and evolution were correct, the Earth must be much older. They argued that physical pro-cesses that shape the Earth and form its rocks, as well as the process of natural selection that yields the diversity of species, all take a very long time. Geologists and physicists continued to debate the age issue for many years. The route to a solution appeared in 1896, when Henri Becquerel announced the dis-covery of radioactivity. Geologists immediately realized that the Earth's interior was producing some heat from the decay of radioactive material. This realization uncovered one of the flaws in Kelvin's argument: Kelvin had assumed that no new heat was produced after the Earth first formed. Because radio-activity constantly generates new heat in the Earth, the planet has cooled down much more slowly than Kelvin had calculated and could be much older. The discovery of radioactivity not only invalidated Kelvin's estimate of the Earth's age, it also led to the development of isotopic dating.

Since the 1950s, geologists have scoured the planet to identify its oldest rocks. Samples from several localities (Wyoming, Canada, Greenland, and China) have yielded dates as old as 4.03 Ga. And sandstone found in Australia contain clastic grains of Zircon that yielded dates of up to 4.4 Ga, indicating that rock as old as 4.4 Ga did once exist (Fig. 12.25a). Isotopic dating of Moon rocks yield dates of up to 4.50 Ga, and dates on meteorites have yielded ages as old as 4.57 billion years (Fig. 12.25b).

> **Did you ever wonder . . .**
> how old is the Earth's oldest rock?

Geologists consider these meteorites to be fragments of planetesimals like those from which the Earth first formed, and take these dates to be close to the age of the Earth's birth, for models of the Earth's formation assume that all objects in the Solar System developed at roughly the same time from the same nebula. Thus, there has been more than enough time for the rocks and life-forms of Earth to have formed and evolved.

Why don't we find rocks with ages between 4.03 and 4.57 Ga in the Earth's crust? Geologists have come up with several ideas to explain the lack of extremely old rocks. One idea comes from calculations defining how the temperature of this planet's interior has changed over time. These calculations indicate that during the first half-billion years of its exis-tence, the Earth might have been so hot that rocks in the crust remained above the closure temperature for minerals, and isotopic clocks could not start "ticking." Another idea comes from studies of cratering events on other moons and planets. These studies indicate that the inner planets were bombarded so intensely by meteorites at about 4.0 Ga, that almost all crust formed earlier than 4.0 Ga was completely destroyed. As noted earlier, geologists have named the time interval between the birth of the Earth and the origin of the oldest isotopically dated rock as the Hadean Eon, to emphasize that conditions at the surface resembled literary images of Hades.

Picturing Geologic Time

The number 4.57 billion is so staggeringly large that we can't begin to comprehend it. If you lined up this many pennies in a row, they would make an 87,400-km-long line that would wrap around the Earth's equator more than twice (Fig. 12.26). Notably, at the scale of our penny chain, human history is only about 100 city blocks long.

Another way to grasp the immensity of geologic time is to equate the entire 4.57 billion years to a single calendar year. On this scale, the oldest rocks preserved on Earth date from early February, and the first bacteria appear in the ocean on February 21. The first shelly invertebrates appear on October 25, and the first amphibians crawl out onto land on November 20. On December 7, the continents coalesce into the super-

FIGURE 12.25 Examples of really, really old rocks of the Earth, and even older ones on the Moon.

(a) This 3.96-Ga gneiss from the Northwest Territories of Canada is one of the oldest isotopically dated rocks on Earth.

(b) Isotopic dates on Moon rocks, such as this boulder being examined by an Apollo astronaut, are as old as 4.5 Ga.

FIGURE 12.26 We can use the analogy of distance to represent the duration of geologic time.

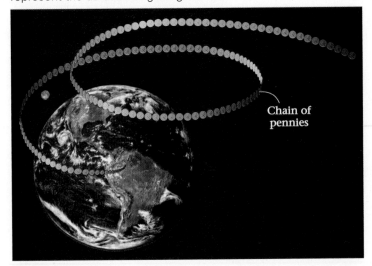

Chain of pennies

continent of Pangaea. The first mammals and birds appear about December 15, along with the dinosaurs, and the Age of Dinosaurs ends on December 25. The last week of December represents the last 65 million years of Earth history, including the entire Age of Mammals. The first human-like ancestor appears on December 31 at 3 P.M., and our species, *Homo sapiens*, shows up an hour before midnight. The last ice age ends a minute before midnight, and all of recorded human history takes place in the last 30 seconds. To put it another way, human history occupies the last 0.000001% of Earth history.

> ## Take-Home Message
>
> - To determine the numerical age of the Earth, geologists date meteorites that are thought to have formed at the same time as the Earth. The oldest meteorites yield an age of 4.57 Ga.
> - The oldest rocks found on continents are 4.03 Ga, and the oldest crystals of zircon are 4.40 Ga.
> - No rock yet found on Earth is as old as the whole planet, because tectonic processes heat and, in some cases, melt rocks of the crust and reset the isotopic clock.
> - Geologic time is so immense that it's hard to picture. Humans have been around for only 0.000001% of geologic time.
>
> **THINK:** Why couldn't Kelvin provide a correct age of the Earth when he tried in the nineteenth century?

Chapter Summary

- Geologic time refers to the time span since the Earth's formation.

- Relative age specifies whether one geologic feature is older or younger than another; numerical age provides the age of a geologic feature in years.

- Using such principles as uniformitarianism, original horizontality, superposition, and cross-cutting relations, we can construct the geologic history of a region.

- The principle of fossil succession states that the assemblage of fossils in strata changes from base to top of a sequence. Once a species becomes extinct, it never reappears.

- Strata are not necessarily deposited continuously at a location. An interval of nondeposition and/or erosion is called an unconformity. Geologists recognize three kinds: angular unconformity, nonconformity, and disconformity.

- A stratigraphic column shows the succession of strata in a region. A given succession of strata that can be traced over a fairly broad region is called a stratigraphic formation. The process of determining the relationship between strata at one location and strata at another is called correlation. A geologic map shows the distribution of formations.

- A composite chart that represents the entirety of geologic time is called the geologic column. The column's largest subdivisions, each of which represent a specific interval of time, are eons. Eons are subdivided into eras, eras into periods, and periods into epochs.

- The numerical age of rocks can be determined by isotopic (radiometric) dating. This is because radioactive elements decay at a rate characterized by a known half-life.

- The isotopic date of a mineral specifies the time at which the mineral cooled below a certain temperature. We can use isotopic dating to determine when an igneous rock solidified and when a metamorphic rock cooled from high temperatures. To date sedimentary strata, we must examine cross-cutting relations with dated igneous or metamorphic rock.

- Other methods for dating materials include counting growth rings in trees and seasonal layers in glaciers.

- From the isotopic dating of meteors and Moon rocks, geologists conclude that the Earth formed about 4.57 billion years ago. Our species, *Homo sapiens*, has been around for only 0.004% of geologic time.

GEOPUZZLE REVISITED

The geologist is standing on a sandstone and shale formed from sediment deposited in a trench along a convergent margin. These strata were folded and therefore tilted during the collision of an island arc with North America. Her hammer points to an unconformity indicating that the mountains resulting from the collision were eroded back to sea level before the overlying units were deposited. The overlying units consist of dolostone and limestone deposited in a shallow sea.

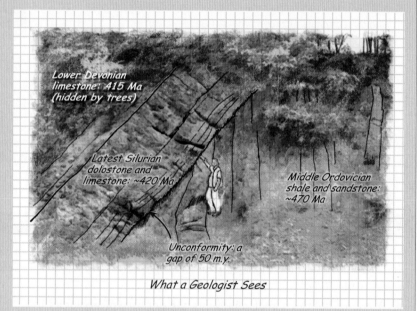

Lower Devonian limestone: 415 Ma (hidden by trees)

Latest Silurian dolostone and limestone: ~420 Ma

Middle Ordovician shale and sandstone: ~470 Ma

Unconformity: a gap of 50 m.y.

What a Geologist Sees

To determine ages of these events, geologists must search for fossils. Graptolites from the turbidites turn out to be Middle Ordovician, and brachiopods from the limestone and dolostone turn out to be Late Silurian to Early Devonian. The numerical age of these events can be ascertained by looking for these periods on the Geologic Time Scale. The sandstone and shale sequence is about 465 Ma, whereas the limestone and dolostone sequence is about 415 Ma. Thus, the unconformity represents a 50-million-year-long gap in the stratigraphic record.

The fact that the limestone and dolostone beds, as well as the unconformity, are themselves tilted means that the region was subjected to a second mountain-building event, subsequent to Lower Devonian time.

Guide Terms

Cambrian explosion (p. 414)

carbon-14 (^{14}C) (p. 420)

closure temperature (p. 420)

contact (p. 410)

correlation (p. 410)

cross-cutting relations (p. 405)

daughter isotope (p. 417)

eon (p. 413)

epoch (p. 413)

era (p. 413)

fission track (p. 422)

fossil (p. 401)

fossil succession (p. 406)

geochronology (p. 417)

geologic column (p. 413)

geologic map (p. 412)

geologic time (p. 400)

geologic time scale (p. 424)

growth rings (p. 421)

half-life (p. 417)

index fossil (p. 406)

isotope (p. 417)

isotopic dating (p. 415)

magnetostratigraphy (p. 422)

numerical age (p. 402)

original horizontality (p. 402)

parent isotope (p. 417)

period (p. 413)

Precambrian (p. 413)

radioactive decay (p. 417)

relative age (p. 402)

rhythmic layering (p. 421)

stratigraphic column (p. 410)

stratigraphic formation (p. 410)

superposition (p. 404)

unconformity (p. 407)

uniformitarianism (p. 402)

Review Questions

1. Compare numerical age and relative age.
2. Describe the principles that allow us to determine the relative ages of geologic events.
3. How does the principle of fossil succession allow us to determine the relative ages of strata?
4. How does an unconformity develop? Describe the differences among the three kinds of unconformities.
5. Describe two different methods of correlating rock units. How was correlation used to develop the geologic column? What is a stratigraphic formation?
6. What does the process of radioactive decay entail?
7. How do geologists obtain an isotopic date? What are some of the pitfalls in obtaining a reliable one?
8. Why can't we date sedimentary rocks directly?
9. How are growth rings and ice cores useful in determining the ages of geologic events?
10. What is the age of the oldest rocks on Earth? What is the age of the oldest rocks known? Why is there a difference?

On Further Thought

11. Imagine an outcrop exposing a succession of alternating sandstone and conglomerate beds. A geologist studying the outcrop notes the following:

- The sandstone beds contain fragments of land plants, but the fragments are too small to permit identification of species.
- A layer of volcanic ash overlies the sandstone bed. Isotopic dating indicates that this ash is 300 Ma.
- A paleosol occurs at the base of the ash layer.

- A basalt dike, dated at 100 Ma, cuts both the ash and the sandstone-conglomerate sequence.
- Pebbles of granite in the conglomerate yield radiometric dates of 400 Ma.

On the basis of these observations, how old is the sandstone and conglomerate? (Specify both the numerical age range and the period or periods of the geologic column during which it formed.) If the igneous rocks were not present, could you still specify the maximum or minimum ages of the sedimentary beds? Explain.

For more resources, including animations, quizzes, and Norton's GeoTours, go to **wwnorton.com/studyspace**.

If your instructor assigns exercises in SmartWork, log in at **smartwork.wwnorton.com**.

ANOTHER VIEW The pages of Earth history stand on end in Namibia, southwestern Africa. Here, in a false-color image, the Ugab River cuts across layer upon layer of strata that were tilted to near-vertical by a Precambrian mountain-building event. Subsequent erosion exposed the strata, and the desert climate keeps it clear of vegetation. The field of view is 45 km.

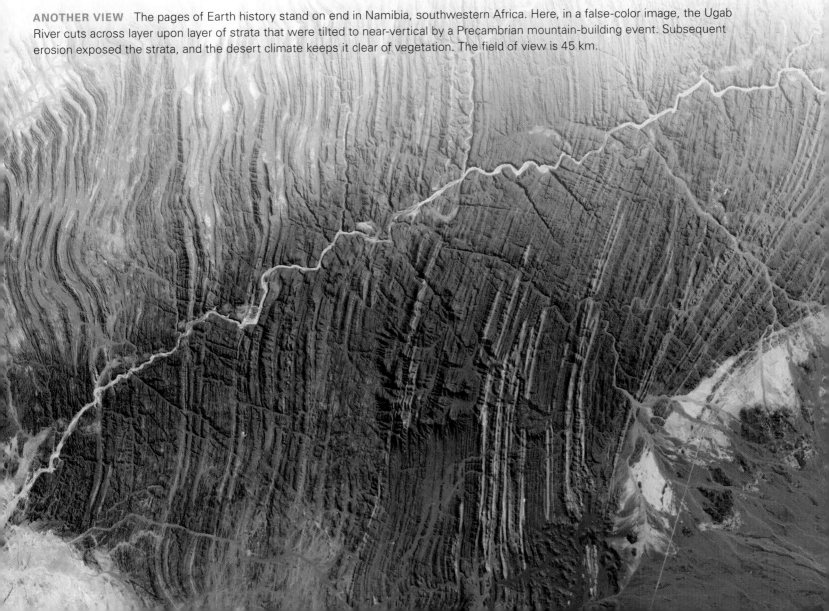

CHAPTER 13

A Biography of Earth

GEOPUZZLE

Is life on Earth as old as the Earth? Has land always been land? When did the mountain belts that we see today first form?

I weigh my words well when I assert that the man who should know the true history of the bit of chalk which every carpenter carries about in his breeches pocket, though ignorant of all other history, is likely, if he will think his knowledge out to its ultimate results, to have a truer and therefore a better conception of this wonderful universe and of man's relation to it than the most learned student who [has] deep-read the records of humanity [but is] ignorant of those of nature.

—Thomas Henry Huxley, from *On a Piece of Chalk* (1868)

13.1 INTRODUCTION

In 1868, a well-known British scientist, Thomas Henry Huxley, presented a public lecture on geology to an audience in Norwich, England. Seeking a way to convey his fascination with Earth history to people with no previous geologic knowledge, he focused his audience's attention on the piece of chalk he had been writing with (see the epigraph above). And what a tale the chalk has to tell! Chalk, a type of limestone, consists of microscopic marine algae shells and shrimp feces. The specific chalk that Huxley held came from beds deposited in Cretaceous time (the name *Cretaceous*, in fact, derives from the Latin word for chalk). These beds now form the white cliffs bordering the shore of England (Fig. 13.1). Geologists in Huxley's day knew of similar chalk beds in outcrops throughout much of Europe, and had discovered that the chalk contains not only plankton shells but also fossils of bizarre swimming reptiles, fish, and invertebrates—species absent in the seas of today. Clearly, when the chalk was deposited, warm seas holding unfamiliar creatures covered some of what is dry land today.

Clues in his humble piece of chalk allowed Huxley to demonstrate to his audience that the geography and inhabitants of the Earth in the past differed markedly from those today, and thus that *the Earth has a history*. We now see a complex, evolving Earth System in which physical and biological components interact pervasively in ways that transformed a formerly barren, crater-pocked surface into countless environments supporting a diversity of life. In this chapter, we offer a concise geologic biography of our planet, from its birth 4.57 billion years ago to the present. We illustrate how continents came into existence and have waltzed across the globe ever since. We also describe mountain-building events, changes in Earth's climate and sea level through time, and the evolution of life. To simplify the discussion, remember that we use the following abbreviations: Ga (for "billion years ago"), Ma (for "million years ago"), and Ka (for "thousand years ago").

FIGURE 13.1 Horizontal chalk beds exposed along the coast of southern England, formed from layers of deep-sea sediment. The thin dark beds, between white chalk layers, consist of chert. The chalk erodes easily, so the pebbles on the beach all consist of chert.

Bedding

Person

Chapter Themes

By the end of this chapter, you should understand that . . .

- many geologic clues indicate that Earth changes over time.

- oceans and life could be sustained only after ~ 3.8 Ga; it took about another billion years for the first continents to take form.

- continental blocks came together to form supercontinents, only to rift apart to form new, smaller continents, several times during Earth history.

- the fossil record indicates that early life consisted of archaea and bacteria; after oxygen levels increased, complex multicellular life became possible; after shells evolved, organisms diversified.

- due to sea-level rise and fall, continental interiors sometimes hosted shallow seas.

- mountain belts formed at several times in the past; eventually, the belts eroded away.

- the most recent ice age occurred during the past ~ 2 million years; fossils indicate that humans evolved during this time.

13.2 METHODS FOR STUDYING THE PAST

When historians outline human history, they describe daily life, wars, economics, governments, leaders, inventions, and explorations. When geologists outline Earth history, we describe the distribution of depositional environments, mountain-building events (orogenies), past climates, life evolution, the changing positions of continents, the past configuration of plate boundaries, and changes in the composition of the atmosphere and oceans. Historians collect data by reading written accounts, examining relics and monuments, and, for more recent events, listening to

recordings or watching videos. Geologists collect data by examining rocks, geologic structures, and fossils and, for more recent events, by studying sediments, ice cores, and tree rings.

Figuring out Earth's past hasn't been an easy task for geologists. The record that we can see isn't complete, because the materials that hold the record of the past don't form continuously through time and may have been eroded away or covered by younger rocks. Also, it can be very challenging to obtain and interpret accurate isotopic ages of old rocks. Nevertheless, we can find enough of the record to outline major geologic events of the past. Following are a few examples of how we use observational data to study Earth history.

- *Identifying ancient orogens*: We identify present-day orogens (mountain belts) by finding regions of high, rugged peaks. However, since it takes as little as 50 million years to erode a mountain range entirely away, we cannot identify orogens of the past simply by studying topography. Rather, we must look for the rock record that orogeny leaves behind. Orogeny causes igneous activity, deformation (folds, faults, and foliation), and metamorphism. Thus a belt of crust containing these features represents an ancient orogen (Fig. 13.2). We can determine the age of an orogeny by isotopically dating the metamorphic and igneous rocks that crop out in the orogen. Orogeny also leads to the development of unconformities, for uplift exposes rocks to erosion. Finally, orogeny can lead to the formation of sedimentary basins within which a several-kilometer-thick wedge of detritus, produced by erosion of mountains, accumulates. These basins, which border the orogen, develop because the weight of the orogen pushes the lithosphere's surface down, forming a depression that traps sediment.

FIGURE 13.2 Evidence of mountain building in the past. Even after the topography of a mountain belt has eroded away, a record of mountain building remains; deformed and metamorphosed rock defines a distinct belt.

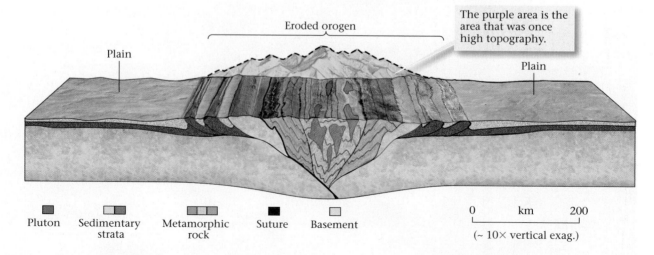

The purple area is the area that was once high topography.

Eroded orogen

Plain   Plain

Pluton   Sedimentary   Metamorphic   Suture   Basement   0   km   200
          strata   rock   (~ 10× vertical exag.)

- *Recognizing the growth of continents*: Not all continental crust formed at the same time. In order to determine how a continent grew, geologists find the ages of different regions of the crust by using modern isotopic dating techniques. They can figure out not only when rocks originally formed from magmas rising out of the mantle but also when the rocks were metamorphosed during a subsequent orogeny. The identities of the rock types making up the crust indicate the tectonic environment in which the crust formed.

- *Recognizing past depositional environments*: The environment at a particular location changes through time. To learn about these changes, we study successions of sedimentary rocks, for the depositional environment controls both the type of sediment accumulating at a location and the type of organisms that live there.

- *Recognizing past changes in relative sea level*: We can determine when sea level has gone up or down by looking for changes in the depositional environment. For example, occurrence of a marine limestone above an alluvial-fan conglomerate indicates a rise in sea level at that site.

- *Recognizing positions of continents in the past*: To help us find out where a continent was located in the past, we have three sources of information. First, the study of apparent polar-wander paths, using the techniques of paleomagnetism can help constrain the relative positions of the continents in the past (see Chapter 3; Fig.

FIGURE 13.3 Paleomagnetic tools that geologists used to determine the past positions of the continents.

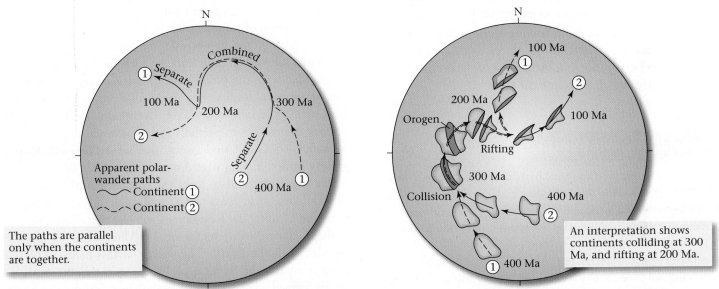

(a) Comparison of the apparent polar-wander paths for two continents indicates when they moved together and when they moved separately.

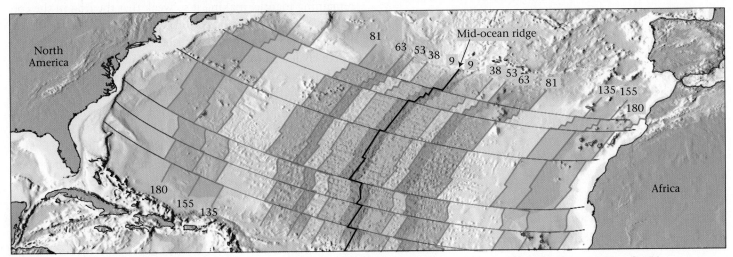

(b) Marine magnetic anomalies indicate how the distance between continents separated by a mid-ocean ridge changes over time. On this map, the age of anomaly boundaries is indicated in Ma.

13.3a). Second, we can study marine magnetic anomalies to reconstruct changes in the width of an ocean basin between continents over time (Fig. 13.3b). Third, we can compare rocks and/or fossils from different continents to see if there are correlations that could indicate that the continents were adjacent.

- *Recognizing past climates*: We can gain insight into past climates by looking at fossils and rock types that formed at given latitudes. For example, if organisms requiring semitropical conditions lived near the poles during a given time period, then the atmosphere overall must have been warmer. Geologists have also learned how to use the ratios of certain isotopes in fossil shells as a measure of past temperatures.

- *Recognizing life evolution*: Progressive changes in the assemblage of fossils in a sequence of strata represent changes in the assemblage of organisms inhabiting Earth through time, and thus characterize life's evolution.

As you read the following sections describing specific event in Earth's history, try to think about how geologists used the tools that we've just described to identify and characterize the event.

Take-Home Message

- Outcrops of deformed and metamorphosed rocks indicate where mountain belts once existed.
- The character and distribution of sedimentary strata provide clues to the timing of sea-level rise and fall, and to the nature of past environments.
- Study of paleomagnetism and correlation of rock units allows researchers to decipher the past positions of continents relative to one another.
- The fossil record provides a window into the evolution of life.

THINK: How can you determine the numerical age of an ancient orogen?

13.3 THE HADEAN EON: HELL ON EARTH?

A viscid pitch boiled in the fosse below
and coated all the bank with gluey mire.
I saw the pitch; but I saw nothing in it
except the enormous bubbles of its boiling
which swelled and sank, like breathing, through all the pit.

—Dante, *The Inferno*, Canto XXI

James Hutton, the eighteenth-century Scottish geologist who was the first to provide convincing evidence that the Earth was very old, could not measure Earth's age directly, and indeed speculated that there may be "no vestige of a beginning." But isotopic dating studies conducted in recent decades have shown that it is possible, in fact, to assign a numerical age to our planet's formation. Specifically, dates obtained for a class of meteorites thought to be remnants of the planetesimal cloud out of which the Earth formed consistently yield an age of 4.57 Ga. Geologists currently take this age to be the Earth's birth date. But the oldest whole rock yet found is only 4.03 Ga, and a clear record of Earth history, as recorded in continental crustal rocks, does not begin until about 3.85 Ga. Geologists refer to the mysterious time interval between the birth of Earth and 3.85 Ga as the **Hadean Eon** (from Hades, the Greek god of the underworld) because during this interval the Earth's surface was, perhaps at times, like an inferno.

The Hadean Eon began with the formation of the Earth by the accretion of planetesimals (see Chapter 1). As the planet grew, collection and compression of matter into a dense ball generated substantial heat. Further, each time another meteorite collided with the proto-Earth, kinetic energy transformed into more heat. Radioactive decay also produced substantial heat within the newborn planet. Eventually, the proto-Earth became hot enough to partially melt, and when this happened, by about 4.5 Ga, it underwent internal **differentiation**—gravity pulled molten iron down to the center of the Earth, where it accumulated to form the core. A mantle, composed of ultramafic rock, remained as a thick shell surrounding the core (see Chapter 2). Researchers suggest that soon after—or perhaps during—differentiation, a Mars-sized protoplanet collided with the Earth. The energy of this collision blasted away a significant fraction of Earth's mantle; the resulting debris formed a ring of silicate-rock debris orbiting the Earth. Heat generated by the collision substantially melted Earth's remaining mantle, making it so weak that the iron in the core of the colliding body sank through it and merged with the Earth's core. Meanwhile, the ring of debris surrounding the Earth coalesced to form the Moon, which, when first formed, was less than 20,000 km away. (By comparison, the Moon is 384,000 km from Earth today.)

In the wake of differentiation and Moon formation, the Earth was so hot that much of its surface was an ocean of seething magma (Fig. 13.4). Rafts of solid rock formed temporarily on the surface of the magma ocean, but these eventually sank and remelted. This stage lasted at least until about 4.4 Ga. After that time, the amount of radioactive heat generation decreased (because elements with short half-lives had decayed), so the Earth *might* have become cool enough for solid rocks to form at its surface. The evidence for this statement comes from western Australia, where geologists

FIGURE 13.4 This speculative painting depicts the Hadean Earth's surface as a magma ocean pummeled by meteorites. The Moon was much closer then, but might not have been visible through the dense atmosphere.

have found 4.4-Ga grains of a durable mineral called zircon mixed with the quartz of much younger sandstone beds. The zircons presumably formed originally in an igneous rock.

During the Hadean Eon, outgassing of the Earth's mantle began to take place. This means that volatile (gassy) elements or compounds originally incorporated in mantle minerals were released and erupted at the Earth's surface, along with lava. The gases accumulated to constitute a toxic atmosphere of water (H_2O), methane (CH_4), ammonia (NH_3), hydrogen (H_2), nitrogen (N_2), carbon dioxide (CO_2), sulfur dioxide (SO_2), and other gases. Some researchers speculate that gases from comets colliding with Earth may have contributed additional gases to the early atmosphere. If the Hadean Earth's surface was sufficiently cool for an extensive solid crustal rock to form beginning at 4.4 Ga, then the first oceans may have accumulated soon thereafter, when water in the atmosphere condensed and fell as rain. In fact, detailed analyses of oxygen isotopes in the 4.4-Ga zircons of Australia suggest that the grains formed at a time when cool, liquid water existed at the Earth's surface. But the question of whether very early oceans existed remains a subject of intense debate; there are alternative explanations for the zircon data.

Though mineral grains as old as 4.4 Ga exist, the oldest *whole rock* yet found on Earth has an age of only 4.03 Ga, as we noted earlier. What destroyed most, if not all, of the pre-4.03-Ga crust (and oceans, if they existed) of the Earth? The answer comes from studies of cratering on the Moon. These studies suggest that the Moon—and, therefore, all inner planets of the Solar System—underwent intense meteor bombardment between 4.0 and 3.9 Ga. Researchers speculate that this bombardment would have pulverized and/or melted almost all crust that had existed on Earth at the time, and would have destroyed the existing atmosphere and ocean. Only after the bombardment ceased could long-lasting crust, atmosphere,

and oceans begin to form. The discovery of 3.85-Ga marine sedimentary rocks in Greenland suggests that the appearance of land and sea happened quite soon after bombardment ceased. What did the Earth's surface look like at this time? An observer probably would have found small, barren landmasses, spotted with volcanoes, poking up above an acidic sea. But both land and sea would have been obscured by murky, dense (CO_2- and SO_2-rich) air.

Take-Home Message

- The Hadean is the earliest interval of Earth history (4.57 to 3.8 Ga).
- Early in the Hadean, the Earth differentiated to form a core and mantle.
- Either the Earth remained too hot for isotopic clocks to start, or meteorite bombardment destroyed the early surface of the planet; either way, hardly any Hadean rock remains.
- During the Hadean, the atmosphere was an unbreathable mix of volcanic gases.

THINK: How old are the oldest Earth rocks and the oldest minerals yet found? Why can we find mineral grains that are older than the oldest rocks?

13.4 THE ARCHEAN EON: THE BIRTH OF THE CONTINENTS AND THE APPEARANCE OF LIFE

The boundary between the **Archean** (from the Greek word for beginning) **Eon** and the Hadean Eon occurs at about 3.85 Ga. Effectively, this date marks the time at which substantial quantities of crustal rocks, including rocks that originated as marine sediments, formed. With the advent of the Archean, crust was locally cool and stable enough for isotopic clocks to start ticking.

Geologists still argue about whether plate tectonics in the form that occurs today operated in the early part of the Archean Eon. Most researchers picture an early Archean Earth with rapidly moving small plates, numerous volcanic island arcs, and abundant hot-spot volcanoes. Others propose that early Archean lithosphere was too warm and buoyant to subduct, and that plate tectonics could not have operated until the later part of the Archean; these authors argue that plume-related volcanism was the main source of new crust until the late Archean. Regardless of which model ultimately proves more accurate, it is clear that the Archean was a time of significant change in the map of the Earth. During that time, the volume of continental crust increased significantly.

FIGURE 13.5 A model for crust formation during the Archean Eon.

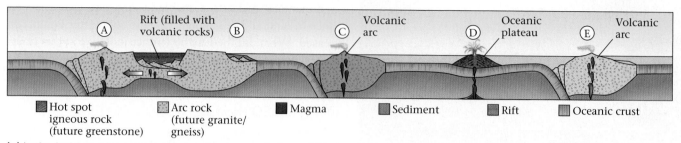

(a) In the Archean, island arcs and hot-spot volcanoes built small blocks of buoyant crust. Rifting of these blocks may have produced flood basalts, and erosion of the blocks produced sediment. (Not to scale.)

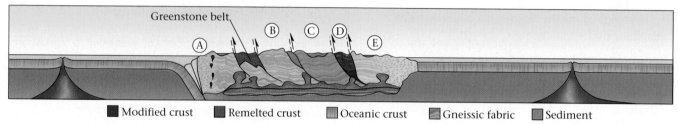

(b) Buoyant blocks collided and sutured together, forming protocontinents. Melting at depth produced granite. Eventually, regions of crust cooled, stabilized, and became cratons. (Not to scale.)

How did the continental crust come to be? According to one model, relatively buoyant (felsic and intermediate) crustal rocks formed both at subduction zones and at hot-spot volcanoes. Frequent collisions sutured volcanic arcs and hot-spot volcanoes together, creating progressively larger blocks called protocontinents (Fig. 13.5a, b). Some of these protocontinents underwent stretching, forming rifts that filled with basalt. As time passed, protocontinents became cooler and stronger; and between 3.2 and 2.7 Ga, the first cratons, long-lived blocks of durable continental crust, had developed. By the end of the Archean Eon, about 80% of continental crust had formed (Fig. 13.6).

Archean cratons contain five principal rock types: *gneiss* (relicts of Archean metamorphism in collisional zones), *greenstone* (metamorphosed relicts of ocean crust trapped between colliding blocks of continental crust; basalts that had filled early continental rifts; flood basalts produced at hot spots; or basaltic volcanoes of island arcs), *granite* (formed from magmas generated by the partial melting of the crust in continental volcanic arcs or above hot spots), *greywacke* (a mixture of sand and clay eroded from the volcanic areas and dumped into the ocean), and *chert* (formed by the precipitation of silica in the deep sea). Archean shallow-water sediments are rare, either because continents were so small that depositional environments in which such sediments could accumulate didn't exist, or because any that were present have since eroded away.

Once land areas had formed, rivers flowed over their stark, unvegetated surfaces. Geologists reached this conclusion because sedimentary beds from this time contain clastic grains that were clearly rounded by transport in liquid water. Salts

FIGURE 13.6 As time progressed, the area of the Earth covered by continental crust increased. Most crust had formed by the beginning of the Proterozoic.

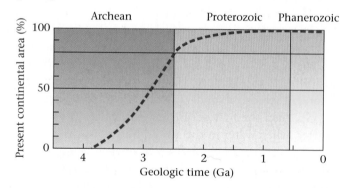

that weathered out of rock and were transported to the sea in rivers made the oceans salty.

Clearly, the Archean Eon saw many firsts in Earth history. Not only did the first continents appear during the Archean, but probably also the first life. Geologists use three sources of evidence to identify early life:

- *Chemical (molecular) fossils, or biomarkers*: These are durable chemicals that are produced only by the metabolism of living organisms.

- *Isotopic signatures*: By analyzing the ratio of ^{12}C to ^{13}C in carbon-rich sediment, geologists can determine if the sediment once contained the bodies of organisms, because organisms preferentially incorporate ^{12}C.

FIGURE 13.7 Archean life-forms.

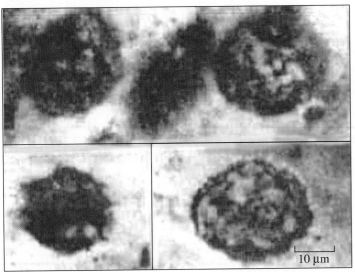

10 μm

(a) These shapes in 3.2-Ga chert from South America are thought to be fossil bacteria or archaea.

(b) This weathered outcrop of 1.85-Ga dolostone near Marquette, Michigan, reveals the layer-like structure of stromatolites. The delicate ridges represent the fossilized remnants of bacterial mats. Similar stromatolites also occur in exposures of Archean rocks.

- *Fossil forms*: Given appropriate depositional conditions, fossils of bacteria or archaea cells can be preserved in rock. However, identification of such fossil forms can be controversial—similar shapes can result from inorganic crystal growth.

The search for the earliest evidence of life continues to make headlines in the popular media. Most geologists currently conclude that life has existed on Earth since at least 3.5 Ga, and perhaps since 3.8 Ga, for rocks of this age contain clear isotopic signatures of organisms. The oldest undisputed fossil forms of bacteria and archaea occur in 3.2-Ga rocks (Fig. 13.7a). Shapes resembling such organisms occur in rocks as old as 3.4 to 3.5 Ga, but their identity remains less certain. Some rocks of this age contain **stromatolites**, distinctive mounds of sediment produced by mats of cyanobacteria. Stromatolites form because cyanobacteria secrete a mucus-like substance to which sediment settling from water sticks. As the mat gets buried, new cyanobacteria colonize the top of the sediment, building a mound upward (Fig. 13.7b); modern examples locally occur in shallow, tropical waters (Fig. 13.7c). Biomarkers in Archean sediments indicate that photosynthetic organisms appeared by 2.7 Ga.

What specific environment on the Archean Earth served as the cradle of life? Laboratory experiments conducted in the 1950s led many researchers to think that life began in warm pools of surface water, beneath a methane- and ammonia-rich atmosphere streaked by bolts of lightning. The only problem with this hypothesis is that more recent evidence suggests that the early atmosphere consisted mostly of CO_2 and N_2, with relatively little methane and ammonia. Thus, some researchers suggest instead that submarine hot-water vents, so-called black smokers, served as the hosts of the first organisms. These vents

(c) Modern stromatolites in Sharks Bay, Western Australia.

emit clouds of ion-charged solutions from which sulfide minerals precipitate and build chimneys. The earliest life in the Archean Eon may well have been thermophilic (heat-loving) bacteria or archaea that dined on pyrite at dark depths in the ocean alongside these vents. Wet clays underground may have provided the substrate for the earliest life.

As the Archean Eon came to a close, the first continents had formed, and life had colonized not only the depths of the sea but also the shallow marine realm. Plate tectonics had commenced, continental drift was taking place, collisional mountain belts were forming, and erosion was occurring. The atmosphere was gradually accumulating oxygen, although probably this gas still accounted for only a very small percentage of the air; Archean air was unbreathable. The stage was set for another major change in the Earth System.

Take-Home Message

- The Archean lasted from 3.85 Ga to 2.5 Ga. The stratigraphic record indicates that oceans existed by the beginning of the Archean.

- Formation of the oceans removed H_2O from the atmosphere. CO_2 then dissolved in the oceans, thereby leaving N_2 as the dominant gas of the atmosphere.

- During the Archean, the Earth was hotter, and plate tectonic activity may have been different; also, there may have been huge mantle plumes.

- Rocks incorporated continental crust, formed initially by igneous properties; by 2.7 Ga, the first substantial continents existed; some of these cooled and stabilized to become cratons.

- Only archaea and bacteria lived during the Archean.

THINK: How did the first continents grow?

13.5 THE PROTEROZOIC EON: TRANSITION TO THE MODERN WORLD

The **Proterozoic** (from the Greek words meaning earlier life) **Eon** spans roughly 2 billion years, from about 2.5 Ga to the beginning of the Cambrian Period at 542 Ma—thus, it encompasses almost half of Earth's history. During Proterozoic time, Earth's surface environment changed from being an unfamiliar world of fast-moving plates, small continents, and an oxygen-poor atmosphere, to the more familiar world of mostly large plates, large continents, and an oxygen-rich atmosphere.

First, let's look at changes to the continents. New continental crust continued to form during the Proterozoic Eon, but at progressively slower rates—in fact, by the middle of the eon, over 90% of the Earth's continental crust had formed. Collisions between the Archean continents, as well as with volcanic island arcs and hot-spot volcanoes, resulted in the assembly of relatively large continents whose interiors could remain isolated from orogeny. These continents cooled and strengthened to become cratons. A **craton** is a region of old, relatively stable continental crust. The large cratons that exist today came into existence by the end of the Proterozoic (Fig. 13.8).

FIGURE 13.8 Major crustal provinces of Earth. The black lines indicate the borders of regions underlain by Precambrian crust. Shields are regions where broad areas of Precambrian rocks are exposed.

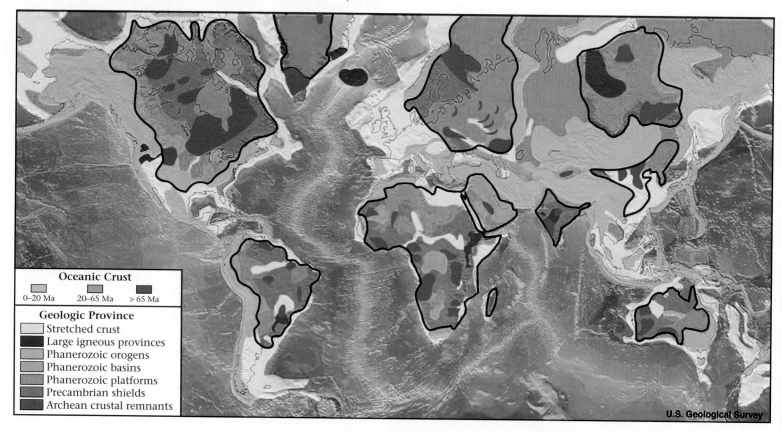

Oceanic Crust
0–20 Ma   20–65 Ma   > 65 Ma

Geologic Province
- Stretched crust
- Large igneous provinces
- Phanerozoic orogens
- Phanerozoic basins
- Phanerozoic platforms
- Precambrian shields
- Archean crustal remnants

U.S. Geological Survey

FIGURE 13.9 On this map of North America, we see four different geologic provinces: shield areas, where Precambrian rocks of the craton crop out; platform areas, where Phanerozoic sedimentary rocks have buried Precambrian rocks (in the eastern platform, the strata exposed are Paleozoic, whereas in the western platform, Mesozoic and Cenozoic strata overlie the Paleozoic strata); Phanerozoic orogenic belts composed of rocks deformed during the past half-billion years, and the coastal plain which is underlain by Cretaceous and Cenozoic strata.

Let's consider the geology of a large craton a little more closely, by examining the interior of North America (Fig. 13.9). North America's craton consists of two parts, the Canadian Shield and the cratonic platform (see Fig. 11.30). A **shield** is a broad, low-lying area in which basement older than about 1 Ga forms the bedrock and is exposed in abundant surface outcrops. In the **cratonic platform**, which occupies the Midwest and High Plains of the United States, as well as the region between the Rocky Mountains and the Canadian Shield of Canada, Precambrian basement lies hidden beneath Phanerozoic sedimentary cover (see Chapter 11). Crust of the North American craton includes several Archean blocks that were sutured together along collisional belts. Each of these blocks and belts has been given a name (Fig. 13.10). The southern portion of the craton grew when a series of volcanic island arcs and continental slivers accreted (attached) to the margin of the Canadian Shield between 1.8 and 1.6 Ga. In the Midwest, granite plutons intruded much of this accreted region, and rhyolite covered it, between 1.5 and 1.3 Ga.

Successive collisions ultimately brought together most continental crust on Earth into a single supercontinent, named **Rodinia**, by around 1 Ga. The last major collision during the formation of Rodinia produced a large orogen called the Grenville orogen. If you look at a popular (though not totally accepted) reconstruction of Rodinia, you can identify the crustal provinces that would eventually become the familiar continents of today (Fig. 13.11a). In this reconstruction,

FIGURE 13.10 The North American craton consists of a collage of different belts and blocks stitched together during collisional and accretionary orogenies of Precambrian time.

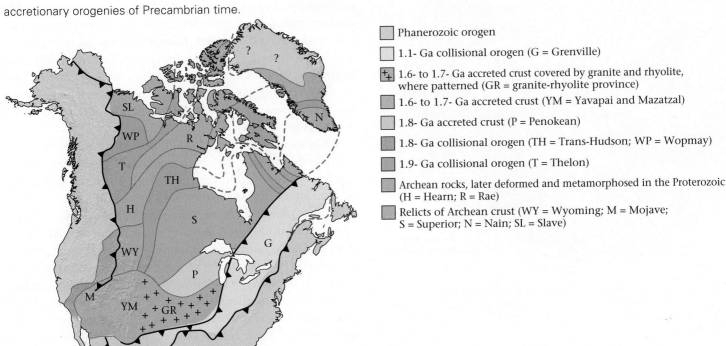

- Phanerozoic orogen
- 1.1- Ga collisional orogen (G = Grenville)
- 1.6- to 1.7- Ga accreted crust covered by granite and rhyolite, where patterned (GR = granite-rhyolite province)
- 1.6- to 1.7- Ga accreted crust (YM = Yavapai and Mazatzal)
- 1.8- Ga accreted crust (P = Penokean)
- 1.8- Ga collisional orogen (TH = Trans-Hudson; WP = Wopmay)
- 1.9- Ga collisional orogen (T = Thelon)
- Archean rocks, later deformed and metamorphosed in the Proterozoic (H = Hearn; R = Rae)
- Relics of Archean crust (WY = Wyoming; M = Mojave; S = Superior; N = Nain; SL = Slave)

1,000 km

FIGURE 13.11 Supercontinents in the late Precambrian.

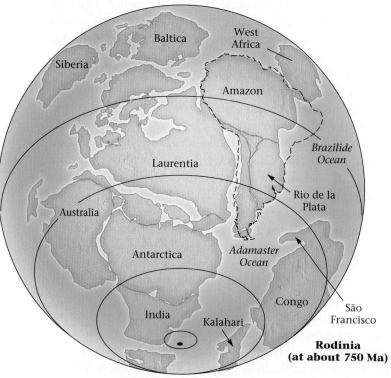

**Rodinia
(at about 750 Ma)**

(a) Rodinia formed around 1-Ga and lasted until about 700 Ma. North America and Greenland together comprise Laurentia.

**Pannotia
(at about 570 Ma)**

(b) According to one model, by 570 Ma Rodinia had broken apart; continents that once lay to the west of Laurentia ended up to the east of Africa. The resulting supercontinent, Pannotia, broke up soon after it formed.

the crust of what later became Antarctica and Australia lay somewhere along the western coast of North America.

Several studies suggest that sometime between 800 and 600 Ma, Rodinia "turned inside out," in that Antarctica, India, and Australia broke away from western North America and swung around and collided with the future South America, possibly forming a short-lived supercontinent that some geologists refer to as Pannotia (Fig. 13.11b). The breakup of Rodinia resulted in the formation of a passive-margin basin (a thick accumulation of sediment along a tectonically inactive coast) along the western edge of North America.

The map of the Earth clearly changed radically during the Proterozoic. But that's not all that changed—fossil evidence suggests that this eon saw important steps in the evolution of life as well. When the Proterozoic began, most life was *prokaryotic*, meaning that it consisted of single-celled organisms (archaea and bacteria) without a nucleus. Though studies of chemical fossils ("biomarkers"; see Interlude E) hint that *eukaryotic* life, consisting of cells that have nuclei, originated as early as 2.7 Ga, the first possible body fossil of a *eukaryotic* organism occurs in 2.1-Ga rocks, and abundant body fossils of eukaryotic organisms can be found only in rocks younger than about 1.2 Ga. Thus the proliferation of eukaryotic life, the foundation from which complex organisms eventually evolved, took place during the Proterozoic.

The last half-billion years of the Proterozoic Eon saw the remarkable transition from simple organisms into complex ones. Ciliate protozoans (single-celled organisms coated with fibers that give them mobility) appear at about 750 Ma. A great leap forward in complexity of organisms occurred during the next 150 million years of the eon, for sediments deposited perhaps as early as 620 Ma and certainly by 565 Ma contain several types of multicellular organisms that together constitute the **Ediacaran fauna**, named for a region in southern Australia. Ediacaran species survived into the beginning of the Cambrian before becoming extinct. Their fossil forms suggest that some of these invertebrate (shell-less) organisms resembled jellyfish, while others resembled worms (Fig. 13.12a).

The evolution of life played a key role in the evolution of Earth's atmosphere. Before life appeared, there was hardly any free oxygen (O_2) in the atmosphere. With the appearance of photosynthetic organisms (which take in CO_2 and produce O_2), probably during the early Archean, oxygen began to enter the atmosphere. But almost all of this oxygen reacted with minerals and became trapped in mineral structures, so the amount of free oxygen in the atmosphere remained miniscule. At about 2.4 Ga, however, photosynthetic organisms became abundant enough—either because of the evolution of new organisms or the development of new, more hospitable environments—that more oxygen entered the atmosphere

FIGURE 13.12 Major changes in the Earth System during the Proterozoic Eon.

(a) *Dicksonia*, a fossil of the Ediacaran fauna. These complex, soft-bodied marine organisms appeared in the late Proterozoic.

(b) An outcrop of BIF in the Iron Ranges of Michigan's Upper Peninsula. The red stripes are jasper (red chert) and the gray stripes are hematite. The rock was folded during a mountain-building event long after deposition. The hammer indicates scale.

Till contains boulders and cobbles surrounded by mudstone, for glaciers can carry clasts of all sizes.

Bedding

(c) Layers of Proterozoic glacial till crop out in Africa, indicating that low-latitude landmasses were glaciated during the Proterozoic.

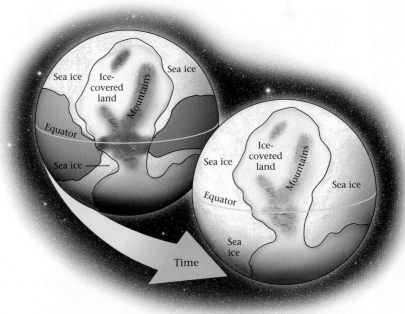

Sea ice   Ice-covered land   Sea ice

Equator

Sea ice   Mountains

Ice-covered land   Mountains

Sea ice

Equator

Sea ice

Time

(d) This planet may have frozen over completely to form "snowball Earth." Glaciers first grew on land, and eventually the sea surface froze over.

than could be absorbed by minerals, and oxygen began to collect in the atmosphere. This **great oxygenation event** changed the world radically. For example, up until 2.4 Ga, iron existed primarily in a chemical form that could dissolve in seawater—addition of oxygen made the water "oxidizing" and transformed iron into a chemical form that could not dissolve. As a result, immense volumes of iron precipitated out of the ocean and collected as sediment on the sea floor. When buried and lithified, this sediment became a rock type now known as banded iron formation (BIF); the "banding" refers to the alternation of layers of iron oxide minerals (hematite and/or magnetite) and chert (Fig. 13.12b). Though BIF did form in the Archean, most of the world's BIF, humanity's primary source of iron, accumulated between 2.4 and 1.8 Ga, implying that the transition to an oxygenated atmosphere was complete by about 1.8 Ga. (Other evidence reinforces this idea, as described in Box 13.1).

The great oxygenation event then profoundly influenced the evolution of life, for it permitted larger, multicelled organisms to prosper. Life could become more complex because

CONSIDER THIS . . . .

BOX 13.1

The Evolution of Atmospheric Oxygen

Without oxygen, the great variety of life that exists on Earth could not survive. Presently, the atmosphere contains 21% oxygen, but this has not always been the case. Throughout most of the Archean Eon, and into the beginning of the Proterozoic Eon, the atmosphere contained less than 1% oxygen. Several lines of evidence lead geologists to conclude that a transformation from an oxygen-poor to an oxygen-rich atmosphere occurred during the period from about 2.4 to 1.8 Ga, the early part of the Proterozoic.

As noted earlier, one line of evidence comes from studying the deposition of banded iron formations (BIFs), a process that could only have happened in an oxygenated environment. Another line of evidence comes from examining clastic grains in sandstones. In sediments deposited *before* 1.8 Ga, well-formed pyrite (iron

sulfide) occurs as clasts in sediment. This could only be possible if the atmosphere before 1.8 Ga contained very little oxygen, for in an oxygen-rich atmosphere, pyrite undergoes chemical weathering (oxidation) or rusting and doesn't survive long enough at the Earth's surface to become a sedimentary clast.

A third line of evidence comes from studying the age of redbeds, clastic sedimentary rocks colored by the presence of bright red hematite (iron oxide). Redbeds form when oxygen-rich groundwater flows through sediment during lithification, and such rocks appear in the geologic record only *after* 1.8 Ga.

The concentration of oxygen in the atmosphere has changed considerably over Earth's history. As we have noted, it constituted less than 1% of the atmosphere during the Archean. By 1.8 Ga,

after the great oxygenation event, it had risen to about 3%, and stayed at roughly that level until about 0.6 Ga (**Fig. Bx13.1**). Then, because all the "sinks" on land and in the sea that could absorb oxygen had become saturated, atmospheric concentrations gradually grew, reaching about 12% at the end of the Proterozoic. The proportion grew substantially when photosynthetic organisms began to prosper on land, and it has oscillated between about 35% and 15% during the past half-billion years. Oxygen concentration has remained at 21% since about 25 Ma.

Keep in mind that oxygen in the atmosphere comes from living, photosynthetic organisms. Thus, geologists suggest that the increases reflect proliferation of photosynthetic life (including land-based plants) and that the decreases reflect mass-extinction events (see Interlude E).

FIGURE Bx.13.1 The change in the proportion of oxygen content in the atmosphere over time. Oxygen level remained low until the end of the Protoerozoic.

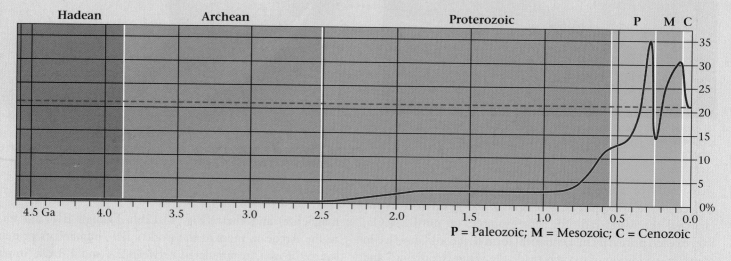

P = Paleozoic; M = Mesozoic; C = Cenozoic

oxygen-dependent (aerobic) metabolism can produce energy much more efficiently than can oxygen-free (anaerobic) metabolism—eating sulfide minerals may sustain an archaea cell, but it can't keep a multicellular organism alive. Addition of oxygen to the atmosphere also made the surface realms

habitable, because some oxygen reacted to produce ozone (O_3) molecules, which absorb harmful ultraviolet radiation from the Sun and prevent it from reaching the surface.

Radical climate shifts occurred on Earth at the end of the Proterozoic Eon. Specifically, accumulations of glacial sediments occur worldwide in the latest Proterozoic stratigraphic sequences. What's strange about the occurrence of these sediments is that they can be found even in regions that were located at the equator during these times (**Fig. 13.12c**). This observation implies that the *entire* planet was cold enough for glaciers to form at the end of the Proterozoic. Geologists still are debating the history of these global ice ages (see Chapter 22); but in one model, glaciers covered the land and perhaps the entire ocean surface froze, resulting in **snowball Earth** (**Fig. 13.12d**). The shell of ice cut off the oceans from the atmosphere, causing oxygen levels in the sea to drop drastically, so many life-forms died off.

> **Did you ever wonder . . .**
> have the oceans ever frozen over entirely?

What brought an end to snowball Earth conditions? According to one model, the icy sheath also prevented atmospheric CO_2 from dissolving in seawater, but it did not prevent volcanic activity from continuing to add CO_2 to the atmosphere. Earth would have remained a snowball forever, were it not for volcanic CO_2. CO_2 is a greenhouse gas, meaning that it traps heat in the atmosphere much as glass panes trap heat in a greenhouse (see Chapter 23), so as the CO_2 concentration increased, Earth warmed up and eventually the glaciers melted. Life may have survived snowball Earth conditions only near submarine black smokers and near hot springs. When the ice vanished, life rapidly expanded into new environments, and new species, such as the Ediacaran fauna, could evolve.

Take-Home Message

- The Proterozoic lasted from 2.5 Ga to 0.542 Ga, almost half of Earth's history.
- During the Proterozoic, supercontinents formed and broke apart.
- Oxygen levels in the atmosphere increased, as photosynthetic organisms evolved, eventually permitting larger and more complex organisms to evolve.
- During an ice age near the end of the Proterozoic, glaciers and sea ice may have formed worldwide to produce the "snowball Earth."

THINK: Has new continental crust formed at the same rate throughout geologic time?

13.6 THE PHANEROZOIC EON: LIFE DIVERSIFIES, AND TODAY'S CONTINENTS FORM

The **Phanerozoic** (Greek for visible life) **Eon** encompasses the last 542 million years of Earth history. Its name reflects the appearance of diverse organisms with hard shells or skeletons that became the well-preserved fossils you can find easily in sedimentary rock outcrops. Geologists divide the Phanerozoic into three eras—the Paleozoic (Greek for ancient life), the Mesozoic (middle life), and the Cenozoic (recent life)—to emphasize the changes in Earth's living population throughout the eon. The solid Earth also changed during this time. Continental blocks rearranged, and new supercontinents formed, only to break apart again. Numerous orogenies built mountain ranges, the most recent of which persist today (**See for Yourself M**, p. S-24). Shell-secreting organisms, burrowing organisms, and plants invaded the sedimentary environment and changed the style of deposition. Below, we look at the changes in the planet's map (its paleogeography), and then in its life-forms, during the three eras of the Phanerozoic Eon.

Take-Home Message

- The Phanerozoic refers to the interval of geologic time between the end of the Precambrian and the present-day. It includes the Paleozoic, Mesozoic, and Cenozoic.
- The beginning of the Phanerozoic is the time when abundant multicellular organisms with shells appear in the fossil record.

THINK: What does the suffix *-zoic* mean?

13.7 THE PALEOZOIC ERA: FROM RODINIA TO PANGAEA

The Early Paleozoic Era (Cambrian–Ordovician Periods, 542–444 Ma)

Paleogeography. As the Proterozoic Eon drew to a close, most continental crust had accumulated in a supercontinent called Pannotia, formed when Rodinia turned inside out. Recall that during this reorganization, Antarctica, India, and Australia rifted away from the future North America, and a passive-margin basin formed along North America's western margin. At the beginning of the Paleozoic Era, Pannotia broke up, yielding smaller continents including **Laurentia** (composed of North America and Greenland), **Gondwana** (South America, Africa, Antarctica, India, and Australia), Baltica (Europe), and Siberia (**Fig. 13.13a**). Passive-margin basins formed along the edges of these new continents.

In addition, sea level rose, so that vast areas of continental interiors were flooded with shallow seas called **epicontinental seas** (**Fig. 13.13b**). (These regions are now cratonic platforms.) In many places, water depths in epicontinental seas reached only a few meters, creating a well-lit marine environment in which life abounded. Deposition in these seas, therefore, yielded layers of fossiliferous sediment. Sea level,

FIGURE 13.13 Land and sea in the early Paleozoic Era.

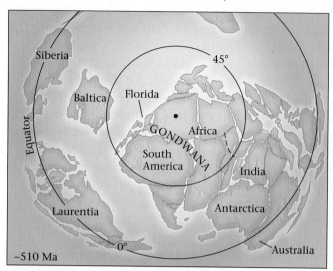

(a) The distribution of continents in the Cambrian Period (510 Ma), as viewed looking down on the South Pole.

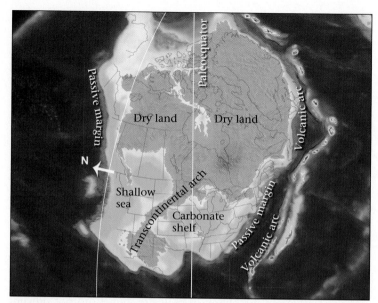

(b) A paleogeographic map of North America shows the regions of dry land and shallow sea in the Late Cambrian Period.

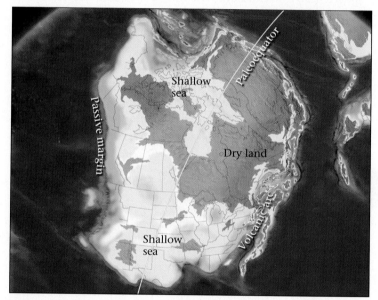

(c) During the Middle Ordovician Period, shallow seas covered much of North America. A volcanic arc formed off the east coast.

however, did not stay high for the entire early Paleozoic Era; regressions and transgressions took place, the former marked by unconformities and the latter by accumulations of sediment that you can see today in the Grand Canyon (Box 13.2).

The geologically peaceful world of the early Paleozoic Era in Laurentia abruptly came to a close in the Middle Ordovician Period, for at this time its eastern margin rammed into a volcanic island arc and other crustal fragments. The resulting collision, called the Taconic orogeny, deformed and metamorphosed strata of the continent's margin and produced a mountain range in what is now the eastern part of the Appalachians (Fig. 13.13c).

Biologic evolution. Beginning with the Cambrian Period, life on Earth left a clear record of evolution because, as we have noted, so many organisms developed shells of durable minerals. The fossil record indicates that soon after the Cambrian began, life underwent remarkable diversification. This event, which paleontologists refer to as the **Cambrian explosion** of life, took about 20 million years. What caused this event? No one can say for sure, but considering that it occurred roughly at the time a supercontinent broke up, it may have had something to do with the production of new ecological niches and the isolation of populations that resulted when small continents formed and drifted apart.

The first animals to appear in the Cambrian Period had simple tube- or cone-shaped shells, but soon thereafter, the shells became more complex. Small, eel-like organisms called conodonts are also found in Cambrian strata. Conodonts contained hard elements resembling tiny teeth.

Shells on other organisms may have evolved as a means of protection against predation by organisms such as conodonts. By the end of the Cambrian, trilobites were grazing the sea floor. Trilobites shared the environment with mollusks, brachiopods, nautiloids, gastropods, graptolites, and echinoderms (Fig. 13.14; see Interlude E). Thus, a complex food chain arose, which included plankton, bottom feeders, and at the top, giant predators. Many of the organisms crawled over or swam around reefs composed of mounds of sponges with mineral skeletons.

FIGURE 13.14 A museum diorama illustrates what early Paleozoic marine organisms may have looked like.

Nautiloid

Brachiopod Trilobite Coral

The Ordovician Period saw the first crinoids and the first vertebrate animals, jawless fish. At the end of the Ordovician, mass extinction took place, perhaps because of the brief glaciation and associated sea-level lowering of the time.

Although the sea teemed with organisms during the early Paleozoic Era, there were no land organisms for most of this time, so the land surface was a stark landscape of rock and sediment, subjected to rapid erosion rates. Our earliest record of primitive land plants and green algae comes from the Late Ordovician Period, but these plants were very small and occurred only along bodies of water. As we mentioned earlier, the invasion of the land could begin only when there was enough ozone in the atmosphere to protect the land surface from UV radiation

The Middle Paleozoic Era (Silurian–Devonian Periods, 444–359 Ma)

Paleogeography. As the world entered the Silurian Period, global climate warmed (leading to so-called greenhouse conditions), sea level rose, and the continents flooded once again. In some places, where water in the epicontinental seas was clear and could exchange with water from the oceans, huge reef complexes grew, forming a layer of fossiliferous limestone on the continents.

More orogeny took place during the middle Paleozoic Era. A succession of collisions on the eastern side of Laurentia during Silurian and Devonian time produced the Caledonian orogen (affecting eastern Greenland, western Scandinavia, and Scotland) and the Acadian orogen in the region that is now the Appalachians (Fig. 13.15a).

Throughout much of the middle Paleozoic, the western margin of North America continued to be a passive-margin basin. Finally, in the Late Devonian, the quiet environment of the west-coast passive margin ceased, possibly because of a collision with an island arc. During this event, known as the Antler orogeny, deep-marine strata were shoved eastward on thrust faults, sliding up and over shallow-water strata. It was the first of many orogenies to affect the western margin of the continent. The Caledonian, Acadian, and Antler orogenies all shed deltas of sediment onto the continents; these deposits formed thick successions of red beds, such as those visible today in the Catskill Mountains (Fig. 13.15b, d).

Biologic evolution. Life on Earth underwent radical changes in the middle Paleozoic Era. In the sea, new species of trilobites, eurypterids, gastropods, crinoids, and bivalves replaced species that had disappeared during the mass extinction at the end of the Ordovician Period. On land, vascular plants with woody tissues, seeds, and veins (for transporting water and food) rooted for the first time. With the evolution of veins and wood, plants could grow much larger, and by the Late Devonian Period the land surface hosted swampy forests with tree-sized relatives of club mosses and ferns. Also at this time, spiders, scorpions, insects, and crustaceans began to exploit both dry-land and freshwater habitats, and jawed fish such as sharks and bony fish began to cruise the oceans. Finally, at the very end of the Devonian Period, the first amphibians crawled out onto land and inhaled air with lungs (Fig. 13.15c).

The Late Paleozoic Era (Carboniferous–Permian Periods, 359–251 Ma)

Paleogeography. After the peak of the greenhouse conditions and high sea levels in the middle Paleozoic Era, the climate cooled significantly (leading to so-called icehouse conditions) in the late Paleozoic. Seas gradually retreated from the continents, so that during the Carboniferous Period, regions that had hosted the limestone-forming reefs of epicontinental seas now became coastal areas and river deltas in which sand, shale, and organic debris accumulated. In fact, during the Carboniferous Period, Laurentia lay near the equator, so it enjoyed tropical and semitropical conditions that favored lush growth in swamps. This growth left thick piles of woody debris that transformed into coal after burial. Much of Gondwana and Siberia, in contrast, lay at high latitudes, and by the Permian Period became covered by ice sheets.

The late Paleozoic Era also saw a succession of continental collisions, culminating in the formation of a single

BOX 13.2

Stratigraphic Sequences and Sea-Level Change

As we have seen in this chapter, there have been intervals in geologic history when the interior of North America was submerged beneath a shallow sea, and times when it was high and dry. This observation implies that sea level, relative to the surface of the continent, rises and falls through geologic time. When the continental surface was dry and exposed to the atmosphere, weathering and erosion ground away at previously deposited rock and, as a result, created a continent-wide unconformity. In the early 1960s, an American stratigrapher named Larry Sloss introduced the term **stratigraphic sequence** to refer to the strata deposited on the continent during periods when continents were submerged. Such sequences are bounded above and below by regional unconformities.

We can picture the deposition of an *idealized* sequence as follows. As the starting condition, much of the surface of a continent lies above sea level, and an unconformity develops. Then, as a transgression takes place, the shoreline at the time floods first, then the shoreline migrates inland and the continent's interior progressively floods. Thus, the preexisting unconformity gets progressively buried, and the bottom layers of the newly deposited sequence tend to be older near the margin of the continent than toward the interior.

During transgression, the depositional environment at a location changes as the sea deepens. Thus, the base of a sequence consists of terrestrial strata (river alluvium). This sediment is buried by nearshore sediment, then by deeper-water sediment (**Fig. Bx13.2a**). Eventually, nearly the entire width of the continent, with the exception of highlands and mountain belts, lies below sea level. As regression proceeds, the interior of the continent is exposed first and then the margins. So a new unconformity forms first in the interior and then later along the margins. Many shorter-duration transgressions and regressions may happen during a single long-duration rise or fall.

Sloss recognized six major stratigraphic sequences in North America, and he named each after a nation of Native Americans (**Fig. Bx13.2b**). Each

---

supercontinent, **Pangaea** (Fig. 13.16a). The largest collision occurred during Carboniferous and Permian time, when Gondwana rammed into Laurentia and Baltica, causing the Alleghanian orogeny of North America (Fig. 13.16b). During this event, the final stage in the development of the Appalachians, eastern North America rammed against northwestern Africa, and what is now the Gulf Coast region of North America squashed against the northern margin of South America. The collision was oblique, not head-on, so in addition to the development of thrust faults and folds, significant strike-slip faulting took place. A vast mountain belt grew, in which deformation generated huge faults and folds. We now see the eroded remnants of rocks deformed during this event in the Appalachian and Ouachita mountains.

Along the continental side of the Alleghanian orogeny, a wide band of deformation called the Appalachian fold-thrust belt formed. In this province, a distinctive style of deformation, called *thin-skinned* deformation, developed. In a thin-skinned fold-thrust belt, an array of thrust faults cuts across what was originally a layer cake of sedimentary strata (Fig. 13.17a–c). Movement on the faults displaces the strata and results in the formation of large folds. At depth, the thrust faults merge with a near-horizontal sliding surface, called a detachment, that lies just above the Precambrian basement. The adjective *thin-skinned* highlights the fact that the thrust faults do *not* cut through the basement, but occur only in the overlying "skin" of sedimentary strata.

Stresses generated during the Alleghanian orogeny were so strong that preexisting faults in the continental crust clear across North America became active again. The movement produced uplifts (local high areas) and sediment-filled basins in the Midwest and in the region of the present-day Rocky Mountains (Colorado, New Mexico, and Wyoming). Geologists refer to the late Paleozoic uplifts of the Rocky Mountain region as the **Ancestral Rockies**.

The assembly of Pangaea involved a number of other collisions around the world as well. Notably, Africa collided with southern Europe to form the Hercynian orogeny. Also, a rift or small ocean in Russia closed, leading to the uplift of the Ural Mountains, and parts of China along with other fragments of Asia attached to southern Siberia.

Biologic evolution. The fossil record indicates that during the late Paleozoic Era, plants and animals continued to evolve

FIGURE Bx13.2 The rise and fall of sea level, and its manifestation in the stratigraphic record.

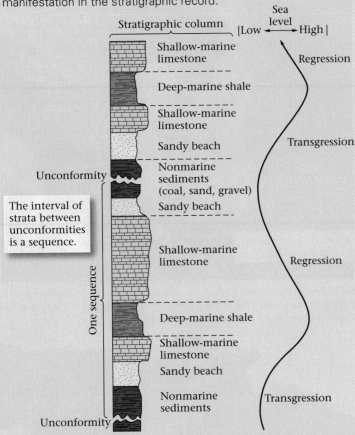

The interval of strata between unconformities is a sequence.

(a) Different types of strata are deposited as sea level rises and falls. Unconformities develop when sea level is low.

sequence represents deposition during an interval of time lasting tens of millions of years. The transgressions and regressions could have been caused by global ("eustatic") sea-level change, or they could reflect mountain-building processes, or they could reflect the subsidence or uplift of continents.

In more recent decades, studies of strata along continental passive margins have provided an even more detailed record of sequences, because there is less erosion in these regions.

Taken together, this information may provide insight into the history of the global rise and fall of sea level, though this interpretation remains controversial (**Fig. Bx13.2c**). Some sea-level changes may reflect changes in sea-floor-spreading rates and thus in the volume of mid-ocean ridges; some may reflect hot-spot activity; some may reflect variations in Earth's climate that caused the formation or melting of ice sheets; others may represent changes in areas of continents accompanying continental collisions; and still others may reflect the warping of continental surfaces as a result of mountain building.

Tan areas were being buried by sediment during a given time. Grey areas were exposed and undergoing erosion.

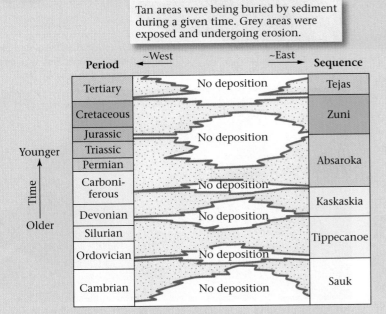

(b) Major stratigraphic sequences of North America. Sloss (1962) named them for Native American tribes. The graph represents a cross section of the continent.

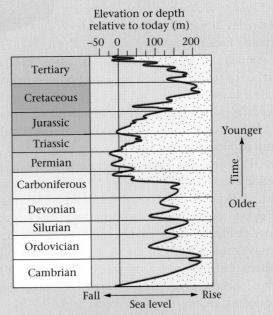

(c) An interpretation of the rise and fall of sea level during the Paleozoic.

FIGURE 13.15 Paleogeography and fossils of Silurian and Devonian time.

The Taconic orogen was a relict of an Ordovician collision.

(a) During the Silurian and Devonian Periods, Laurentia collided with Baltica, Avalonia, and South America in succession, as oceans in between were consumed. The Antler arc formed off the west coast.

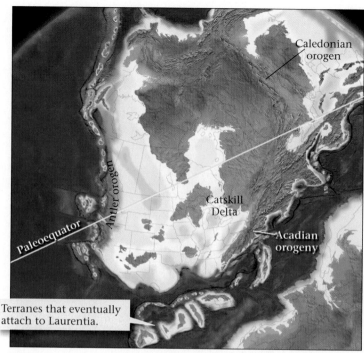

Terranes that eventually attach to Laurentia.

(b) During the Devonian Period, the Acadian orogeny shed sediments into a shallow sea to form the Catskill Delta on the east coast of Laurentia. The Antler orogeny shed sediments in the west.

~ 20 cm

(c) A Late Devonian fossil skeleton of *Tiktaalik*; this lobe-finned fish was one of the first animals to walk on land.

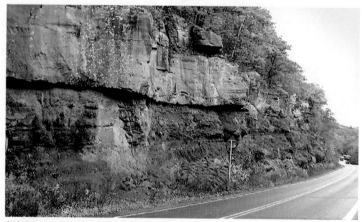

(d) A roadcut exposing Devonian redbeds (sandstone and shale) in New York.

toward more familiar forms. In coal swamps, fixed-wing insects including huge dragonflies flew through a tangle of ferns, club mosses, and scouring rushes, and by the end of the Carboniferous Period insects such as the cockroach, with foldable wings, appeared (Fig. 13.18). Forests containing gymnosperms ("naked seed" plants such as conifers) and cycads (trees with a palm-like stalk peaked by a fan of fern-like fronds) became widespread in the Permian Period. Amphibians and, later, reptiles populated the land. The appearance of reptiles marked the evolution of a radically new component in animal reproduction: eggs with a protective covering. Such eggs permitted reptiles to reproduce without returning to the water, and thus allowed the group to populate previously uninhabitable environments.

The late Paleozoic Era came to a close with two major mass-extinction events, during which over 95% of marine species disappeared. Why these particular events occurred remains a subject of debate. According to one hypothesis, the terminal Permian mass extinction occurred as a result of an episode of extraordinary volcanic activity in the region that is now Siberia; basalt sheets extruded during the event are known as the "Siberian traps." Eruptions could have clouded the atmosphere, acidified the oceans, and disrupted the food chain.

FIGURE 13.16 Paleogeography at the end of the Paleozoic Era.

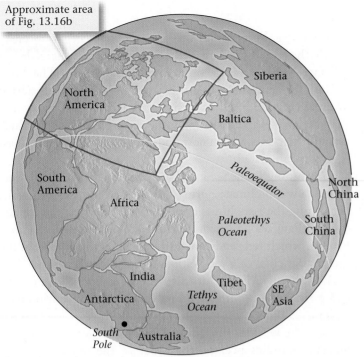

Approximate area of Fig. 13.16b

(a) At the end of the Paleozoic, almost all land had combined into a single supercontinent called Pangaea.

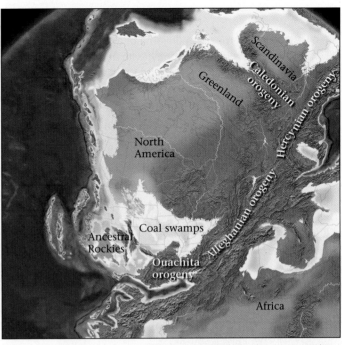

(b) During the Alleghanian and Hercynian orgenies, a huge mountain belt formed. The Caledonian orogeny had formed earlier. Coal swamps bordered interior seas, and the Ancestral Rockies rose.

FIGURE 13.17 Features of the Appalachian Mountains in the eastern United States.

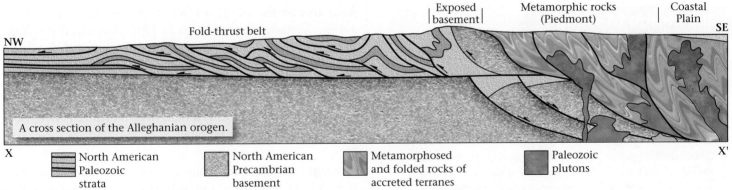

A cross section of the Alleghanian orogen.

| North American Paleozoic strata | North American Precambrian basement | Metamorphosed and folded rocks of accreted terranes | Paleozoic plutons |

(a) In the fold-thrust belt, strata have been pushed westward, and were folded and faulted. The Blue Ridge exposes a slice of Precambrian basement. Metamorphic and plutonic rocks underlie the Piedmont. Cross section location is shown in (b).

(b) The eroded remnants of the Appalachian orogeny stand out in the eastern United States. The white line shows the approximate cross-section position.

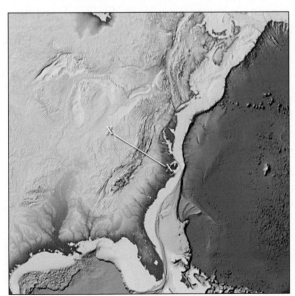

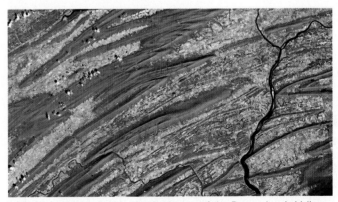

(c) Folds stand out in this satellite image of the Pennsylvania Valley and Ridge. Resistant sandstone layers form the ridges. The field of view is 80 km wide.

FIGURE 13.18 A museum diorama of a Carboniferous coal swamp includes a giant dragonfly, with a wingspan of about 1 m. The inset photo gives a sense of its size relative to a human.

Take-Home Message

- As the Paleozoic began, the late Precambrian supercontinent broke up.
- Sea level rose and fell over geologic time. When sea level was high, the interior of continents flooded with water.
- The Appalachian orogen began to form in the middle Paleozoic, and finished forming at the end.
- Life diversified during the Cambrian explosion. In the middle Paleozoic, the land surface began to host plants, insects, and amphibians, and by the end of the Paleozoic, reptiles had appeared.
- By the end of the Paleozoic, all land had assembled into the Pangaea supercontinent.

THINK: What event, indicated by the fossil record, marks the end of the Paleozoic?

13.8 THE MESOZOIC ERA: WHEN DINOSAURS RULED

The Early and Middle Mesozoic Era (Triassic–Jurassic Periods, 251–145 Ma)

Paleogeography. Pangaea, the supercontinent formed at the end of the Paleozoic Era, existed for about 100 million years, until rifting commenced during the Late Triassic and Early Jurassic Periods and the supercontinent began to break up. By the end of the Jurassic Period, rifting had succeeded in splitting North America from Europe and Africa. The Mid-Atlantic Ridge formed, and the North Atlantic Ocean started to grow (Fig. 13.19a). At first, the Atlantic was narrow and shallow, and evaporation made its water so salty that thick evaporite deposits, which now underlie much of the Gulf Coast region, accumulated.

According to the record of sedimentary rocks, Earth overall had a warm climate during the Triassic and Early Jurassic. But during the Late Jurassic and Early Cretaceous, the climate cooled. Pangaea's interior remained a nonmarine environment in which red sandstones and shales, now exposed in the spectacular cliffs of Zion National Park, were deposited (Fig. 13.19b). By the Middle Jurassic Period, sea level began to rise, and a shallow sea submerged much of the Rocky Mountain region.

On the western margin of North America, convergent-margin tectonics became the order of the day. Beginning with Late Permian and continuing through Mesozoic time, subduction generated volcanic island arcs and caused them, along with microcontinents and hot-spot volcanoes, to collide with North America. Thus, North America grew in land area by the addition ("accretion") of crustal fragments on its western margin (Fig. 13.19c). Because these fragments consist of crust that formed elsewhere, not originally on or adjacent to the continent, geologists call them **exotic terranes**. From the end of the Jurassic through the Cretaceous Period, a major continental volcanic arc, the Sierran arc, grew along the western margin of North America; we'll learn more about this arc later.

Biologic evolution. During the early Mesozoic Era, a variety of new plant and animal species appeared, filling the ecological niches left vacant by the Late Permian mass extinction. Swimming reptiles plied the oceans, and corals became the predominant reef builders. On land, gymnosperms and reptiles diversified, and the Earth saw its first turtles and flying reptiles. And at the end of the Triassic Period, the first true dinosaurs evolved.

> **Did you ever wonder...**
> when did the dinosaurs live?

Dinosaurs differed from other reptiles in that their legs were positioned under their bodies rather than off to the sides, and they were possibly warm-blooded. By the end of the Jurassic Period, gigantic sauropod dinosaurs (weighing up to 100 tons), along with other familiar examples such as *stegosaurus*, thundered across the landscape, and the first feathered birds, such as Archaepteryx, took to the skies (Fig. 13.19d). The earliest ancestors of mammals appeared at the end of the Triassic Period, in the form of small, rat-like creatures.

FIGURE 13.19 Aspects of Early and Middle Mesozoic paleogeography and paleobiology.

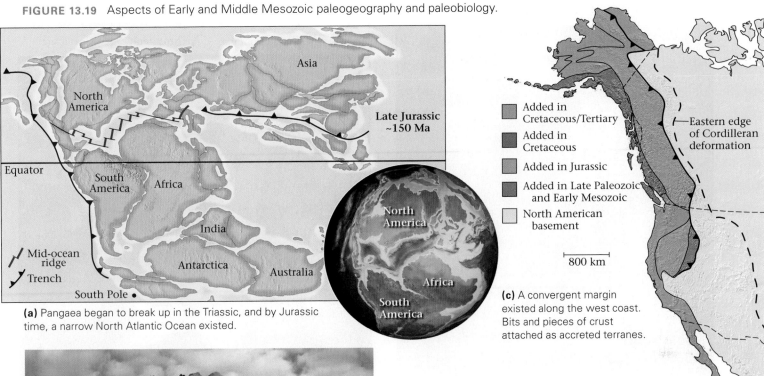

(a) Pangaea began to break up in the Triassic, and by Jurassic time, a narrow North Atlantic Ocean existed.

(c) A convergent margin existed along the west coast. Bits and pieces of crust attached as accreted terranes.

(b) During the Jurassic, immense sand dunes blanketed the southwestern United States. Sandstone beds in Zion Park are the relicts of these dunes.

(d) During the Jurassic, giant dinosaurs roamed the land. This painting shows several species.

The Late Mesozoic Era (Cretaceous Period, 145–65 Ma)

Paleogeography. During the Cretaceous Period, the Earth's climate continued to shift to warmer, greenhouse conditions, and sea level rose significantly, reaching levels that had not been attained for the previous 200 million years. Great seaways flooded most of the continents (Fig. 13.20a, b). In the latter part of the Cretaceous Period, a shark could have swum from the Gulf of Mexico to the Arctic Ocean, or across much of western Europe.

The breakup of Pangaea continued through the Cretaceous Period, with the opening of the South Atlantic Ocean and the separation of South America and Africa from Antarctica and Australia. India broke away from Gondwana and headed rapidly northward toward Asia (Fig. 13.21a, b). Along the continental margins of the newly formed Mesozoic oceans, passive-margin basins developed that filled with great thicknesses of sediments. For example, along the Gulf Coast, a wedge of sediment over 15 km thick has accumulated.

In western North America, the **Sierran arc**, a large continental volcanic arc that was initiated at the end of the Jurassic Period, continued to be active. This arc resembled the present-day Andean arc of western South America. Though the volcanoes of the Sierran arc have long since eroded away, we can see their roots in the form of the plutons that now constitute the granitic batholith of the Sierra Nevada mountains. A thick accretionary prism, formed from sediments and debris scraped off the subducting oceanic plate, piled up to the west of the Sierran arc and now crops out in the Coast Ranges of

FIGURE 13.20 Cretaceous paleogeography

(a) In Early Cretaceous, several island arcs collided with western North America.

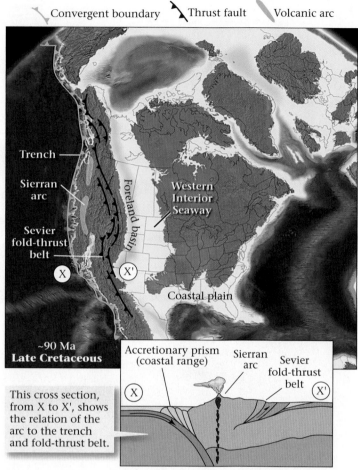

(b) In Late Cretaceous, a continental volcanic arc formed. A fold-thrust belt formed to the east, as did a transcontinental seaway.

California. Compressional stresses along the western North American convergent boundary activated large thrust faults east of the arc, an event geologists refer to as the **Sevier orogeny**. This orogeny produced a thin-skinned fold-thrust belt whose remnants you can see today in the Canadian Rockies and in western Wyoming (Fig. 13.21c). The weight of the orogen helped push the surface of the continent down, forming a wide foreland basin that filled with sediment. Formation of the basin may also reflect warping down of the continental margin in response to the downward pull of the subducting plate beneath it. This foreland basin constituted the western part of the Western Interior Seaway.

At the end of the Cretaceous Period, continued compression along the convergent boundary of western North America caused large faults in the region of Wyoming, Colorado, eastern Utah, and northern Arizona to slip. In contrast to the faults

of fold-thrust belts, these faults penetrated deep into the Precambrian rocks of the continent, and thus movement on them generated basement uplifts (Fig. 13.21d). Overlying layers of Paleozoic strata warped into large monoclines, folds whose shape resembles the drape of a carpet over a step. This event, which geologists call the **Laramide orogeny**, formed the structure of the present Rocky Mountains in the United States (Fig. 13.21c, d). Some geologists have suggested that the contrast in the location of faulting between the Sevier and Laramide orogenies may reflect contrasts in the dip of the subducting plate. During the Laramide orogeny, the subducting plate entered the mantle at a shallower angle, and therefore scraped along, and applied stress to, the base of the continent farther inland. This change in dip angle did not occur in Canada, so during the Laramide orogeny the Canadian Rockies simply continued to grow eastward as a thin-skinned fold-thrust belt (Fig. 13.21c).

FIGURE 13.21 Paleogeography in Late Cretaceous through Eocene time.

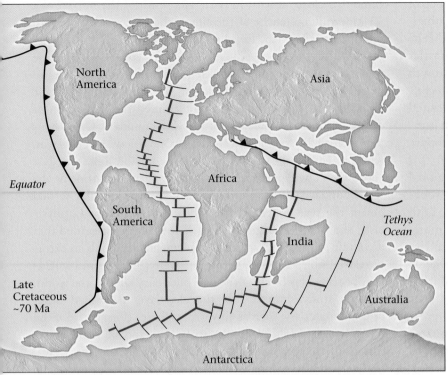

(a) By the Late Cretaceous Period, the Atlantic Ocean had formed, and India was moving rapidly northward to eventually collide with Asia.

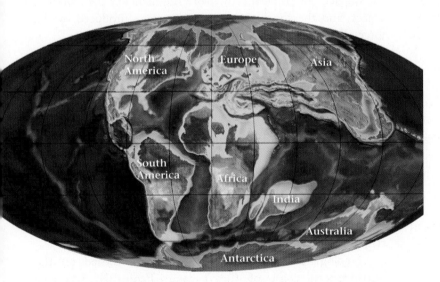

(b) In this Late Cretaceous paleogeographic reconstruction, southern Europe and Asia are beginning to form from a collage of many crustal blocks.

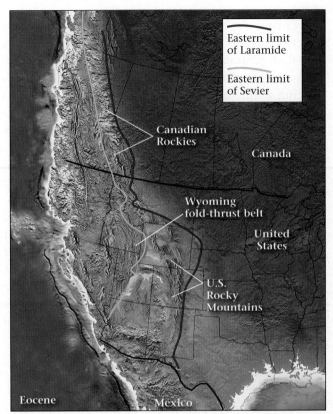

(c) During the Laramide orogeny, deformation shifted eastward in the United States, moving from the Sevier belt to the Rocky Mountains, and the style of deformation changed.

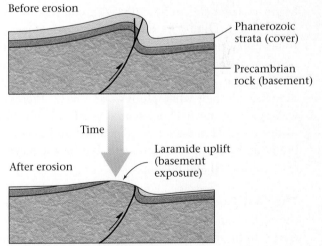

(d) The Laramide orogeny produced "basement-cored uplifts." In these, faults lifted up blocks of basement, causing the overlying strata to bend into a stair-step-like fold.

Geologists have determined that sea-floor spreading rates may have been as much as three times faster during the Cretaceous than they are today. As a result, more of the oceanic crust was younger and warmer than it is today, and since young sea floor lies at a shallower depth than does older sea floor (due to isostasy; see Chapter 11), Cretaceous mid-ocean ridges occupied more volume than they do today. The extra volume of the ridges displaced sea water, causing sea level to rise. (You can simulate this phenomenon by dropping a stone in a glass of water—the water level in the glass rises.) Also during the Cretaceous, huge submarine plateaus formed from basalts erupted at hot-spot volcanoes. The existence of these plateaus implies that particularly active mantle plumes, or **superplumes**, reached the base of the lithosphere. Melting at the top of such plumes produced immense quantities of magma, which erupted and built up the plateaus. Growth of submarine plateaus displaced sea water and thus also contributed to sea-level rise. Notably, the volcanism associated with extra-rapid sea-floor spreading, as well as with submarine plateau growth, likely released CO_2, a greenhouse gas, into the atmosphere. Geologists hypothesize that this increased atmospheric CO_2 concentration led to a global rise in atmospheric temperature. Rising temperatures would cause sea water to expand and polar ice sheets to melt, both phenomena that would make sea level go up even more. Considering all the phenomena that caused sea level to rise during the Cretaceous, it's no surprise that the continents flooded and that large epicontinental seas formed during this era.

Biologic evolution. In the seas of the late Mesozoic world, modern fish appeared and became dominant. In contrast with earlier fish, new fish had short jaws, rounded scales, symmetrical tails, and specialized fins. Huge swimming reptiles and gigantic turtles (with shells up to 4 m across) preyed on the fish. On land, cycads largely vanished, and angiosperms (flowering plants), including hardwood trees, began to compete successfully with conifers for dominance of the forest. Dinosaurs reached their peak of success at this time, inhabiting almost all environments on Earth. Social herds of grazing dinosaurs roamed the plains, preyed on by the fearsome *Tyrannosaurus rex* (a Cretaceous, not a Jurassic, dinosaur, despite what Hollywood says!). Pterosaurs, with wingspans of up to 11 m, soared overhead, and birds began to diversify. Mammals also diversified and developed larger brains and more specialized teeth, but for the most part, they remained small and rat-like.

The "K-T boundary event." Geologists first recognized the K-T boundary (*K* stands for Cretaceous and *T* for Tertiary) from eighteenth-century studies that identified an abrupt global change in fossil assemblages. Until the 1980s, most geologists assumed the faunal turnover took millions of years. But modern dating techniques indicate that this change happened almost instantaneously and that it signaled a sudden mass extinction of most species on Earth. The dinosaurs, which had ruled the planet for over 150 million years, simply vanished, along with 90% of some plankton species in the ocean and up to 75% of plant species. What kind of catastrophe could cause such a sudden and extensive mass extinction? From data collected in the 1970s and 1980s,

most geologists have concluded that the Cretaceous Period came to a close, at least in part, as a result of the impact of a 13-km-wide meteorite at the site of the present-day Yucatán Peninsula in Mexico (**Fig. 13.22a**).

The discoveries leading up to this conclusion provide fascinating insight into how science works. The story began when Walter Alvarez, a geologist studying strata in Italy, noted that a thin layer of clay interrupted the deposition of deep-sea limestone precisely at the K-T boundary. Cretaceous plankton shells constituted the limestone below the clay layer, whereas Tertiary plankton shells made up the limestone just above the clay. Apparently, for a short interval of time at the K-T boundary, all the plankton died, so that only clay settled out of the sea. When Alvarez and his father, Luis (a physicist), and other colleagues analyzed the clay, they learned that it contained iridium, a very heavy element found only in extraterrestrial objects. Soon, geologists were finding similar iridium-bearing clay layers at the K-T boundary all over the world. Further study showed that the clay layer contained other unusual materials, such as tiny glass spheres (formed from the flash freezing of molten rock), wood ash, and shocked quartz (grains of quartz that had been subjected to intense pressure). Only an immense impact could explain all these features. The glass spherules formed when melt sprayed in the air from the impact site, the ash resulted when forests were set ablaze by the impact, the iridium came from fragments of the colliding object, and the shocked quartz grains were produced by the force of the impact.

The impact caused so much destruction because it not only formed a crater, blasting huge quantities of debris into the sky, but probably also generated 2-km-high tsunamis that inundated the shores of continents and generated a blast of hot air that set forests on fire. The blast and the blaze together could have ejected so much debris into the atmosphere that for months there would have been perpetual night and winter-like cold. In addition, chemicals ejected into the air could have combined with water to produce acid rain. These conditions would cause photosynthesis to all but cease, and thus would break the food chain and trigger extinctions.

Geologists suggest that the meteorite landed on the northwestern coast of the Yucatán Peninsula, where a 100-km-wide by 16-km-deep scar called the Chicxulub crater lies buried beneath younger sediment (**Fig. 13.22b**). A layer of glass spherules up to 1 m thick occurs at the K-T boundary in strata near the site. And radiometric dating indicates that igneous melts in the crater formed at 65 ± 0.4 Ma, exactly the time of the K-T boundary event. The discovery of this event has led geologists to speculate that other such collisions may have punctuated the path of life evolution throughout Earth history.

FIGURE 13.22 The Cretaceous-Tertiary impact. The aftermath probably caused extinction of the dinosaurs and other species.

(a) A meteorite with a diameter of ~13 km (8 miles) slammed into the Earth at a point that is now at the north end of the Yucatán Peninsula.

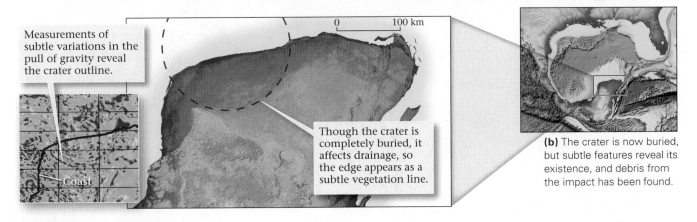

Measurements of subtle variations in the pull of gravity reveal the crater outline.

Coast

Though the crater is completely buried, it affects drainage, so the edge appears as a subtle vegetation line.

0   100 km

(b) The crater is now buried, but subtle features reveal its existence, and debris from the impact has been found.

Take-Home Message

- The Mesozoic is known as the Age of Dinosaurs.
- At the beginning of the Mesozoic, Pangaea broke apart; during this eon, the Atlantic grew, and the western margin of North America was a convergent plate boundary.
- Mesozoic orogenic events built mountain belts in western North America.

THINK: What event may have caused the dinosaurs to go extinct?

13.9 THE CENOZOIC ERA: THE FINAL STRETCH TO THE PRESENT

Paleogeography. During the last 65 million years, the map of the Earth has continued to change, gradually producing the configuration of continents we see today. The final stages of the Pangaea breakup separated Australia from Antarctica and Greenland from North America, and formed the North Sea between Britain and continental Europe. The Atlantic Ocean continued to grow because of sea-floor spreading on the Mid-Atlantic Ridge, and thus the Americas moved relatively westward, away from Europe and Africa. Meanwhile, the continents that once constituted Gondwana drifted northward as the intervening Tethys Ocean was consumed by subduction (see Fig. 13.21). Collisions of the former Gondwana continents with the southern margins of Europe and Asia resulted in the formation of the largest orogenic belt on Earth today, the **Alpine-Himalayan chain** (Fig. 13.23). India and a series of intervening volcanic island arcs and microcontinents collided with Asia to form the Himalayas and the Tibetan Plateau to the north, while Africa along with some volcanic island arcs and microcontinents collided with Europe to produce the Alps in the west and the

Zagros Mountains of Iran in the east. Finally, collision between Australia and New Guinea led to orogeny in Papua New Guinea.

As the Americas moved westward, convergent plate boundaries evolved along their western margins. In South America, convergent-boundary activity built the Andes, which remains an active orogen to the present day. In North America, convergent-boundary activity continued without interruption until about 40 Ma (the Eocene Epoch) yielding, as we have seen, the Laramide orogeny. Then, because of the rearrangement of plates off the western shore of North America, a transform boundary replaced the convergent boundary in the western part of the continent by 25 Ma (Fig. 13.24). When this happened, volcanism and compression ceased in western North America, the San Andreas Fault system formed along the coast of the United States, and the Queen Charlotte Fault system developed off the coast of Canada. Along the San Andreas and Queen Charlotte faults today, the Pacific Plate moves northward with respect to North America at a rate of about 6 cm per year. In the western United States, convergent-boundary tectonics continues only in Washington, Oregon, and northern California, where subduction of the Juan de Fuca Plate generates the volcanism of the Cascade volcanic chain.

FIGURE 13.23 The two main active continental orogenic systems on the Earth today. The Alpine-Himalayan system formed when Africa, India, and Australia collided with Asia (inset). The Cordilleran and Andean systems reflect the consequences of convergent-boundary tectonism along the eastern Pacific Ocean.

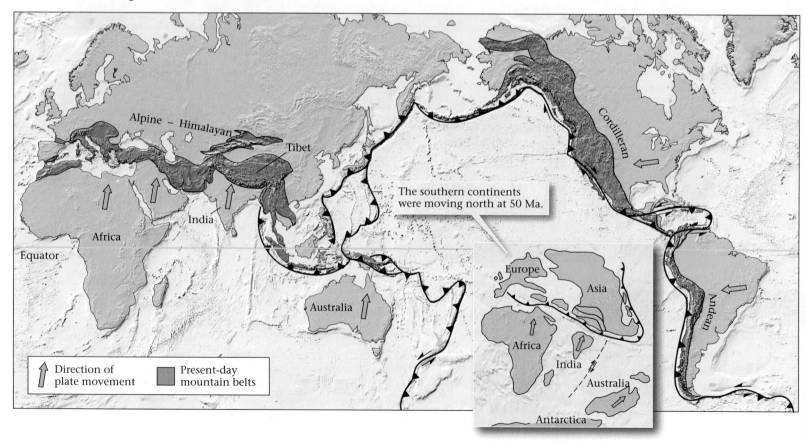

FIGURE 13.24 The western margin changed from a convergent-plate boundary into a transform-plate boundary until subduction of the Farallon Ridge. Then, the San Andreas Fault developed. To the east, the Basin and Range rift was developed.

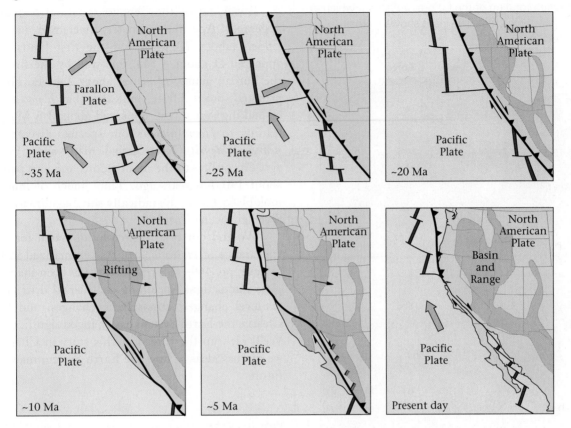

As convergent tectonics ceased in the western United States south of the Cascades, the region began to undergo rifting (extension) in roughly an east-west direction. The result was the formation of the **Basin and Range Province** (See for Yourself M, p. S-24), a broad continental rift that has caused the region to stretch to twice its original width (**Fig. 13.25**). The Basin and Range gained its name from its topography—the province contains long, narrow mountain ranges separated from each other by flat, sediment-filled basins. This topography formed when the crust of the region was broken up by *normal* faults (see Chapter 11). Blocks of crust above these faults slipped down and tilted, producing narrow, wedge-shaped depressions. Crests of the tilted blocks form the ranges, and the depressions, which rapidly filled with sediment eroded from the ranges, became basins.

The Basin and Range Province terminates just north of the Snake River Plain, the track of the hot spot that now lies beneath Yellowstone National Park. As North America drifts westward, volcanic calderas have formed along the Snake River Plain; Yellowstone National Park straddles the most recent caldera (see Chapter 9).

Recall that in the Cretaceous Period, the world experienced greenhouse conditions and sea level rose so that extensive areas of continents were submerged. During the Cenozoic Era, however, the global climate rapidly shifted to icehouse conditions, and by the early Oligocene Epoch, Antarctic glaciers reappeared for the first time since the Triassic. The climate continued to grow colder through the Late Miocene Epoch, leading to the formation of grasslands in temperate climates. About 2.5 Ma, the Isthmus of Panama formed, separating the Atlantic completely from the Pacific, changing the configuration of oceanic currents, perhaps leading the Arctic Ocean to freeze over.

During the overall cold climate of the past 2 million years, the Quaternary Period, continental glaciers have expanded and retreated across northern continents at least 20 times, resulting in the **Pleistocene Ice Age** (**Fig. 13.26**). Each time the glaciers grew, sea level fell so much that the continental shelf became exposed to air. At times, a land bridge formed across the Bering Strait, west of Alaska, providing migration routes for animals and people from Asia into North America. A partial land bridge also formed from southeast Asia to Australia, making human migration to Australia easier. Erosion and deposition by the glaciers created much of the landscape we see today in northern temperate regions. About 11,000 years

FIGURE 13.25 The Basin and Range Province is a rift. Its opening caused rotation of the Sierra Nevada. The opening of the Rio Grande Rift caused rotation of the Colorado Plateau, which is an unrifted block of cratonic crust. The inset shows a cross section along the red line.

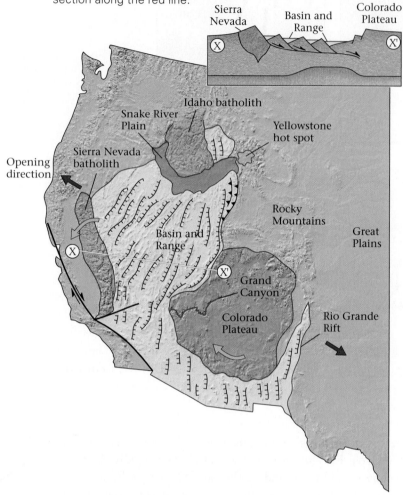

extinct during the past 10,000 years, perhaps because of hunting by humans.

It was during the Cenozoic that our own ancestors first appeared. Ape-like primates diversified in the Miocene Epoch (about 20 Ma), and the first human-like primate appeared at about 4 Ma, followed by the first members of the human genus, *Homo*, at about 2.4 Ma. Fossil evidence, primarily from Africa, indicates that *Homo erectus*, capable of making stone axes, appeared about 1.6 Ma, and the line leading to *Homo sapiens* (our species) diverged from *Homo neanderthalensis* (Neanderthal man) about 500,000 years ago. According to the fossil record, modern people appeared about 190,000 years ago. Thus, much of human evolution took place during the radically shifting climatic conditions of the Pleistocene Epoch.

We end our brief biography of Earth for now with the appearance of *Homo sapiens*. As summarized in **Geology at a Glance**, pp. 460–461, this history has been shaped by complex plate interactions (including continental drift and collisions), sea-level changes, atmospheric changes, and life evolution. Clearly, the Earth System has changed significantly over time. We'll pick up the thread of this story in Chapter 23, where we discuss ideas of how the Earth System may change in the future.

FIGURE 13.26 The maximum advance of the Pleistocene ice sheet in North America.

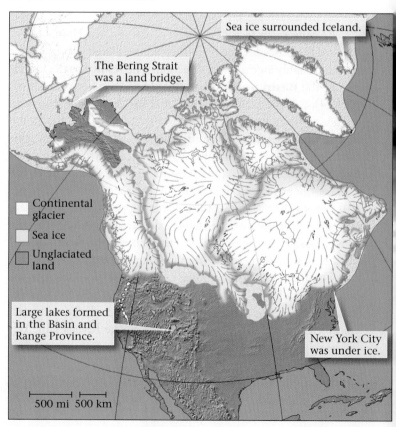

ago, the climate warmed, and we entered the interglacial time interval we are still experiencing today (see Chapter 22). This most recent interval of time is the Holocene.

Biologic evolution. When the skies finally cleared in the wake of the K-T boundary catastrophe, plant life recovered, and soon forests of both angiosperms and gymnosperms reappeared. The grasses, which first appeared in the Cretaceous, spread across the plains in temperate and subtropical climates by the middle of the Cenozoic Era, transforming them into vast grasslands. The dinosaurs, except for their distant relatives, the birds, were gone for good. Mammals rapidly diversified into a variety of forms to take their place. In fact, most of the modern groups of mammals that exist today originated at the beginning of the Cenozoic Era, giving this time the nickname "Age of Mammals." During the latter part of the era, incredible *huge* mammals appeared (such as mammoths, giant beavers, giant bears, and giant sloths), but these became

Take-Home Message

- During the Cenozoic, subduction ceased along most of western North America, as the San Andreas Fault and the Basin and Range rift formed.
- Collision between various continental fragments and Asia led to the formation of the Alpine-Himalayan orogen.
- The last 2 million years of the Cenozoic has been an ice age, during which continental glaciers have advanced and retreated over northern continents.

- After the demise of the dinosaurs, mammals became the dominant animal on land; fossils indicate that humans evolved since 2 Ma.

THINK: According to geologic data, did dinosaurs and humans live at the same time?

Chapter Summary

- Earth formed about 4.57 billion years ago. For part of the first 600 million years, the Hadean Eon, the planet was so hot that its surface was a magma ocean.

- The Archean Eon began about 3.85 Ga, when permanent continental crust that remains formed. The crust was assembled out of volcanic arcs and hot-spot volcanoes that were too buoyant to subduct. The atmosphere contained little oxygen, but the first life-forms—bacteria and archaea—appeared.

- In the Proterozoic Eon, which began at 2.5 Ga, Archean cratons collided and were sutured together along orogenic belts and large Proterozoic cratons. Photosynthesis by organisms added oxygen to the atmosphere. By the end of the Proterozoic, complex but shell-less marine invertebrates populated the planet. Most continental crust accumulated to form a supercontinent called Rodinia at about 1 Ga.

- At the beginning of the Paleozoic Era, rifting yielded several separate continents. Sea level rose and fell a number of times, creating sequences of strata in continental interiors. Continents began to collide and coalesce again, leading to orogenies and, by the end of the era, to another supercontinent, Pangaea. Early Paleozoic evolution produced many invertebrates with shells, and jawless fish. Land plants and insects appeared in the middle Paleozoic. And by the end of the eon, there were land reptiles and gymnosperm trees.

- In the Mesozoic Era, Pangaea broke apart and the Atlantic Ocean formed. Convergent-boundary tectonics dominated along the western margin of North America. Dinosaurs appeared in Late Triassic time and became prominent land animals through the Mesozoic Era. During the Cretaceous Period, sea level was very high, and the continents flooded. Angiosperms appeared at this time, along with modern fish. A huge mass-extinction event, which wiped out the dinosaurs, occurred at the end of the Cretaceous Period, probably because of the impact of a large meteorite.

- In the Cenozoic Era, continental fragments of Pangaea began to collide again. The collision of Africa and India with Asia and Europe formed the Alpine-Himalayan orogen. Convergent tectonics has persisted along the margin of South America, creating the Andes, but ceased in North America when the San Andreas Fault formed. Rifting in the western United States during the Cenozoic Era produced the Basin and Range Province. Various kinds of mammals filled niches left vacant by the dinosaurs, and the human genus, *Homo*, appeared and evolved throughout the radically shifting climate and ice ages of the Pleistocene Epoch.

GEOPUZZLE REVISITED

The Earth has a long and complex history. According to geologic studies, our planet formed at about 4.57 Ga. The first land and water may have appeared as early as 4.04 Ga, but the record of the earliest crust was destroyed by meteorite bombardment between 4.0 and 3.9 Ga. The first long-lived continental crust appeared about 3.85 Ga. Oceans had also appeared by about 3.85 Ga and have lasted ever since. Subtle evidence suggests that life appeared by around 3.5 Ga. Thus, life did not become established until the Earth was a billion years old. Most continental crust had formed by about 2.5 Ga. Some has survived intact, but most has been recycled by tectonic processes. The amount of dry land, however, changes over time due to the rise and fall of sea level. As continents drift, collide, and break apart, mountain belts form and then later erode away. The present ranges on Earth have been around only since the beginning of the Cenozoic.

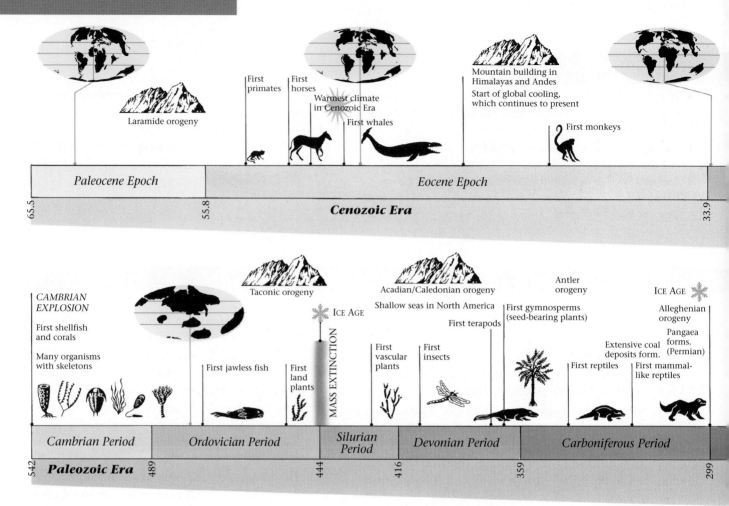

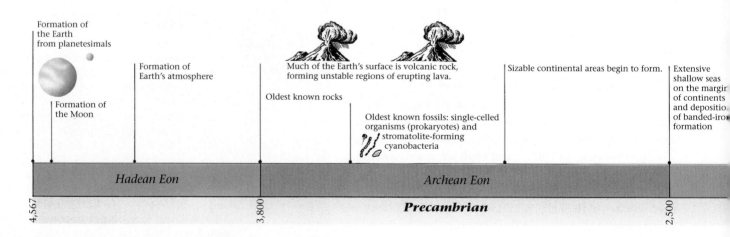

The Evolution of Earth

Earth has not had a static history; because of plate tectonics and its consequences (continental drift, sea-floor spreading, volcanism, and so on), the map of the Earth constantly changes. Distinct mountain-building events, or orogenies, have taken place during this process. The fossil record suggests that life first appeared within the first few hundred million years of our planet's existence, soon after a liquid-water ocean had accumulated; and like the planet itself, life has constantly changed ever since. The progressive change of the assemblage of species of life on Earth is called evolution.

The earliest life-forms were microscopic. By the end of the Precambrian, complex multicellular organisms had formed,

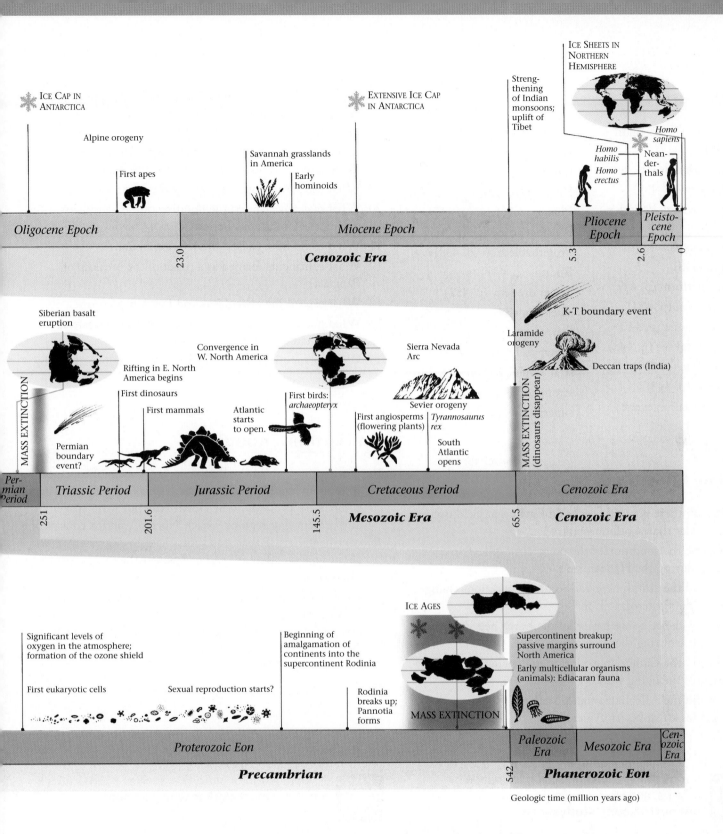

ICE CAP IN ANTARCTICA

Alpine orogeny

First apes

EXTENSIVE ICE CAP IN ANTARCTICA

Savannah grasslands in America

Early hominoids

ICE SHEETS IN NORTHERN HEMISPHERE

Strengthening of Indian monsoons; uplift of Tibet

Homo sapiens

Homo habilis
Homo erectus

Neanderthals

Oligocene Epoch | Miocene Epoch | Pliocene Epoch | Pleistocene Epoch

23.0

Cenozoic Era

5.3

2.6

0

Siberian basalt eruption

Convergence in W. North America

Sierra Nevada Arc

K-T boundary event

Laramide orogeny

Deccan traps (India)

Rifting in E. North America begins

First dinosaurs

First mammals

Atlantic starts to open.

First birds: archaeopteryx

Sevier orogeny

First angiosperms (flowering plants)

Tyrannosaurus rex

South Atlantic opens

MASS EXTINCTION

Permian boundary event?

MASS EXTINCTION (dinosaurs disappear)

Permian Period | Triassic Period | Jurassic Period | Cretaceous Period | Cenozoic Era

251

201.6

145.5

65.5

Mesozoic Era

Cenozoic Era

ICE AGES

Significant levels of oxygen in the atmosphere; formation of the ozone shield

Beginning of amalgamation of continents into the supercontinent Rodinia

Supercontinent breakup; passive margins surround North America

Early multicellular organisms (animals): Ediacaran fauna

First eukaryotic cells

Sexual reproduction starts?

Rodinia breaks up; Pannotia forms

MASS EXTINCTION

Proterozoic Eon | Paleozoic Era | Mesozoic Era | Cenozoic Era

Precambrian

542

Phanerozoic Eon

Geologic time (million years ago)

and a burst of evolution at the Precambrian-Cambrian boundary yielded a diversity of invertebrates with shells. During the past half-billion years, several new major groups of organisms have appeared, and countless species have become extinct. Evidence suggests that evolution is not a continuous, gradual process, but occurs in pulses, separated by intervals of time during which the assemblage of species is fairly stable.

The last few hundred million years of Earth history have seen life leave the ocean and spread across the land. During the Mesozoic Era, dinosaurs roamed the Earth—then vanished abruptly 65 million years ago, perhaps as a result of a meteorite's colliding with the Earth. Since then, mammals have diversified into a great variety of species. And the last 100,000 years or so have witnessed the evolution of our own species, Homo sapiens. Considering the changes that people have brought about to the Earth System, the appearance of our species is clearly a major event in the history of the planet. The time scale is the 2009 time scale (International Committee on Stratigraphy).

Guide Terms

Alpine-Himalayan chain (p. 456)

Ancestral Rockies (p. 446)

Archean Eon (p. 435)

Basin and Range Province (p. 457)

Cambrian explosion (p. 444)

craton (p. 438)

cratonic platform (p. 439)

differentiation (p. 434)

Ediacaran fauna (p. 440)

epicontinental sea (p. 443)

exotic terrane (p. 450)

Gondwana (p. 443)

great oxygenation event (p. 441)

Hadean Eon (p. 434)

Laramide orogeny (p. 452)

Laurentia (p. 443)

Pangaea (p. 446)

Phanerozoic Eon (p. 443)

Pleistocene ice age (p. 457)

Proterozoic Eon (p. 438)

Rodinia (p. 439)

Sevier orogeny (p. 452)

shield (p. 439)

Sierran arc (p. 451)

snowball Earth (p. 443)

stratigraphic sequence (p. 446)

stromatolite (p. 437)

superplume (p. 454)

Review Questions

1. Why are there no whole rocks on Earth that yield isotopic dates older than 4 billion years?

2. Describe the condition of the crust, atmosphere, and oceans during the Hadean Eon.

3. How did the atmosphere and tectonic conditions change during the Proterozoic Eon?

4. What evidence do we have that the Earth nearly froze over twice during the Proterozoic Eon?

5. How did the Cambrian explosion of life change the nature of the living world?

6. How did the Alleghanian and Ancestral Rockies orogenies affect North America?

7. What are the major types of organisms that appeared during the Paleozoic?

8. Describe the plate-tectonic conditions that led to the formation of the Sierran arc and the Sevier thrust belt. What happened during the Laramide orogeny?

9. What life-forms appeared during the Mesozoic?

10. What may have caused the flooding of the continents during the Cretaceous Period?

11. What could have caused the K-T extinctions?

12. What continents formed as a result of the breakup of Pangaea?

13. What caused the Himalayas and the Alps to form?

14. What major tectonic provinces formed in the western United States during the Cenozoic?

15. What major climatic and biologic events happened during the Pleistocene?

On Further Thought

16. During intervals of the Paleozoic, large areas of continents were submerged by shallow seas. Using *Google Earth*™ or a comparable program, tour North America from space. Do any present-day regions *within* North America consist of continental crust that was submerged by seawater? What about regions offshore? (*Hint*: Look at the region just east of Florida.)

17. Geologists have concluded that 80% to 90% of Earth's continental crust had formed by 2.5 Ga. But if you look at a geological map of the world, you find that only about 10% of the Earth's continental crustal surface is labeled "Precambrian." Why?

 For more resources, including animations, quizzes, and Norton's GeoTours, go to **wwnorton.com/studyspace**.

 If your instructor assigns exercises in SmartWork, log in at **smartwork.wwnorton.com**.

ANOTHER VIEW The present-day Bahamas serve as an example of what the interior of the United States might have looked like during intervals of the Paleozoic. Shallow land areas were submerged and became the site of shallow-marine sedimentation.

An airplane view shows us the great pit of the Bingham Mine in Utah. This huge excavation, which can be seen by satellites 2,000 km in space, provides ore that has yielded tons of copper, gold, and silver. Earth materials provide energy, metals, and other resources.

Earth Resources

Many of the products we use in our daily lives come from geologic materials—the Earth itself is, essentially, a natural resource. In Chapter 14, we look at the energy resources that come from the Earth. These include fossil fuels (oil and coal) as well as nuclear fuel and moving water. Chapter 15 focuses on nonenergy resources, particularly the mineral deposits from which we obtain metals.

By the end of Part V, we'll realize that many natural resources are not renewable and thus must be conserved if we are to avoid shortages. Also, we'll see how our use of resources has had and continues to have an impact on the environment and on transnational politics.

CHAPTER 14

Squeezing Power from a Stone: Energy Resources

In 1991, near the end of the Gulf War, the oil wells of Kuwait were set on fire. Flames and smoke filled the sky. The fires, fed by fuels that had been trapped in rocks deep underground for millions of years, are a dramatic display of the energy held in geologic materials. The opaqueness of the sky shows how the use of this energy can have environmental consequences.

GEOPUZZLE

Most of the energy used today comes from the burning of oil, gas, and coal. How much longer will this pattern of energy usage last, and why?

14.1 **INTRODUCTION**

In the extreme chill of a midwinter night, a pan of water freezes almost instantly. But the low temperature doesn't stop a wolf from stalking its prey—the wolf's legs move through the snow, its heart pumps, and its body radiates heat. These life processes require energy. **Energy** provides the capacity to do work, to cause something to happen, or to cause change in a system. The wolf's energy comes from the metabolism of special chemicals such as sugar, protein, and carbohydrates in its body. These chemicals, in turn, came from the food the animal eats. To survive, a wolf must catch and eat mice and rabbits, so to a wolf, these animals are energy resources. In a general sense, we use the term **resource** for any item that can be employed for a useful purpose, and, more specifically, **energy resource** for something that can be used to produce heat, produce electricity, or move vehicles. Matter that stores energy in a readily usable form is also called **fuel**.

The earliest humans needed about the same quantity of fuel per capita as a wolf and thus could maintain themselves simply by hunting and gathering. But when people discovered how to use fire for cooking and heating, their need for energy resources began to exceed that of other animals. Before the dawn of civilization, wood and dried dung provided adequate fuel. As people began to congregate in towns, however, they also required energy for agriculture and transportation, and new resources such as animal power, wind, and flowing water came into use. Energy-resource needs began to outpace the supplies available at the Earth's surface because not only did the population continue to grow, but new industries such as iron smelting became commonplace. In fact, to feed the smelting industry, woodcutters mowed down most European forests during the seventeeth century, so that when the industrial revolution began in the eighteenth century, miners had to extract supplies of underground coal to power steam engines in factories.

Since the industrial revolution, society's hunger for energy has increased almost unabated (**Fig. 14.1a, b**). In the United States today, for example, the average person uses more than 110 times the amount of energy used by a prehistoric hunter-gatherer. Most energy for human consumption in the industrial world now comes from oil and natural gas. To a lesser extent, we also continue to use coal, wind, and flowing water, and in the last half century we've added nuclear energy, geothermal energy, and solar energy to the list of energy resources. In the twenty-first century, we have started to work toward developing new ways of obtaining energy resources, such as extraction of natural gas from coal beds and from gas hydrates (strange, ice-like substances on the sea floor), gasification of coal (turning coal into various burnable gases), production of hydrogen fuel cells, and production of alcohol from plant crops.

Why is a chapter in a geology book devoted to such energy resources? Simply because most of these resources originate in geologic materials or are due to geologic processes. Oil, gas, coal, and the fuel for atomic power plants come from rocks; geothermal energy is a product of Earth's internal heat; and the movement of wind and water involves cycles in the Earth

FIGURE 14.1 Sources of energy used by people. We use a variety of sources.

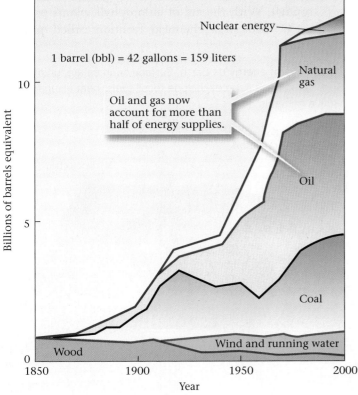

(a) Between 1850 and 2000, the proportions of different types of energy used have changed significantly.

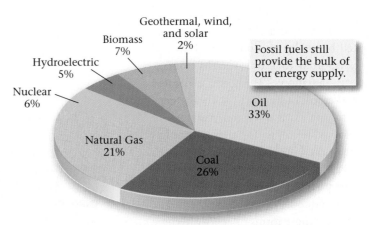

(b) In the twenty-first century, a variety of energy sources have come into use globally. The percentages are approximate.

System. To understand the source and limitations of energy resources and to find new resources, we must understand their geology. That's why the multibillion-dollar-a-year energy industry employs tens of thousands of geologists.

In this chapter, we begin by scanning the various types of energy resources on Earth. Then we focus on fossil fuels (oil, gas, and coal)—combustible materials derived from organisms that lived in the past—in detail and other types of energy sources more briefly. The chapter concludes by outlining the dilemmas we will face as energy resources begin to run out, and as the products of energy consumption enter our environment in dangerous amounts.

Chapter Themes

By the end of this chapter, you should know . . .

- what constitutes an energy resource, and the variety of resources in the Earth System.
- what oil and gas are, how they form, where they come from, and how they are obtained.
- how coal forms, is classified, and is obtained; and how it can be used to produce gas.
- the basic operation of nuclear power plants, and the origin of the fuel they use.
- the variety of alternative energy resources (e.g., geothermal, solar, wind).
- the challenges that society faces as to the sustainability of energy sources in the future.

- *Energy directly from the Sun*: Solar energy, resulting from nuclear fusion reactions in the Sun, bathes the Earth's surface. It may be converted directly into electricity, using solar-energy panels, or it may be used to heat water or warm a house. (No one has yet figured out how to produce sustainable controlled nuclear fusion on Earth, but we *can* produce uncontrolled fusion by exploding a hydrogen bomb.)

- *Energy directly from gravity*: The gravitational attraction of the Moon, and to a lesser extent the Sun, causes ocean tides, the daily up-and-down movement of the sea surface. The flow of water in and out of channels during tidal changes can drive turbines.

- *Energy involving both solar energy and gravity*: Solar radiation heats the air, which becomes buoyant and rises. As this happens, gravity causes cooler air to sink. The resulting air movement, wind, powers sails and windmills. Solar energy also evaporates water, which enters the atmosphere. When the water condenses, it rains and falls on the land, where it accumulates in streams that flow downhill in response to gravity. This moving water powers waterwheels and turbines.

- *Energy via photosynthesis*: Green plants absorb some of the solar energy that reaches the Earth's surface. Their green color comes from a pigment called chlorophyll. With the aid of chlorophyll, plants produce sugar through a chemical reaction called **photo-**

14.2 SOURCES OF ENERGY IN THE EARTH SYSTEM

When you get down to basics, there are only five fundamental sources of energy on the Earth: (1) energy generated by nuclear fusion in the Sun and transported to Earth via electromagnetic radiation; (2) energy generated by the pull of gravity; (3) energy generated by nuclear fission reactions of radioactive atoms; (4) energy that has been stored in the interior of the Earth since the planet's beginning; and (5) energy stored in the chemical bonds of compounds. Let's look at the different ways these forms of energy become resources we can use (Fig. 14.2).

FIGURE 14.2 The diverse sources of energy on Earth. Surface sources are, ultimately, driven by heat from the Sun. Subsurface sources include fossil fuels, heat rising from hot rocks at depth, and radioactive minerals.

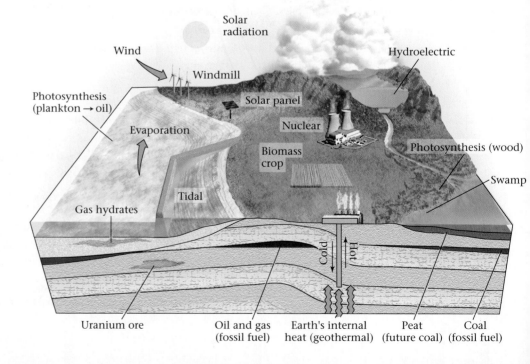

synthesis. In chemist's shorthand, we can write this reaction as:

$$6CO_2 + 12H_2O + light \rightarrow 6O_2 + C_6H_{12}O_6 + 6H_2O$$

carbon   water   oxygen   sugar   water
dioxide

Plants use the sugar produced by photosynthesis to manufacture more complex chemicals, or they metabolize it to provide themselves with energy.

Burning plant matter in a fire releases potential energy stored in the chemical bonds of organic chemicals. During burning, the molecules react with oxygen and break apart to produce carbon dioxide, water, and carbon (soot):

$$plant + O_2 \xrightarrow{\text{burning}} CO_2 + H_2O + C \text{ (soot)}$$
$$+ \text{ other gases} + \text{heat energy}$$

The flames you see in fire consist of glowing gases released and heated by this reaction.

People have burned wood to produce energy for centuries. More recently, plant material (biomass) from corn, sugar cane, switchgrass, and algae has been used to produce ethanol, a flammable alcohol.

- *Energy from chemical reactions*: A number of inorganic chemicals can burn to produce light and energy. The energy results from exothermic (heat-producing) chemical reactions. A dynamite explosion is an extreme example of such energy production. Recently, researchers have been studying electrochemical devices, such as hydrogen fuel cells, that produce electricity directly from chemical reactions.

- *Energy from fossil fuels*: Oil, gas, and coal come from organisms that lived long ago and thus store solar energy that reached the Earth long ago. We refer to these substances as **fossil fuels**, to emphasize that they were derived from ancient organisms and have been preserved in rocks for geologic time.

> **Did you ever wonder...**
> what are the "fossils" in fossil fuel?

- *Energy from nuclear fission*: Atoms of radioactive elements can split into smaller pieces, a process called nuclear fission (see Box 1.4). During fission, a tiny amount of mass transforms into a large amount of energy, called nuclear energy. This type of energy runs nuclear power plants and nuclear submarines.

- *Energy from Earth's internal heat*: Some of Earth's internal energy dates from the birth of the planet, while some is produced by radioactive decay in minerals. This internal energy heats water underground. The resulting hot water, when transformed to steam, provides **geothermal energy** that can drive turbines.

14.3 OIL AND GAS

What Are Oil and Gas?

For reasons of economics and convenience, industrialized societies today rely primarily on oil (petroleum) and natural gas for their energy needs. These materials contain a huge amount of energy per unit of weight; for example, 1 gram of oil provides about 1,000 times more energy than a battery of comparable weight provides. We can say that oil and gas have a relatively high "energy density." Oil and natural gas consist of **hydrocarbons**, chain-like or ring-like molecules made of carbon and hydrogen atoms. For example, bottled gas (propane) has the chemical formula C_3H_9. Chemists consider hydrocarbons to be a type of organic chemical, so named because similar carbon-based chemicals make up living organisms.

Some hydrocarbons are gaseous and invisible, some resemble watery liquids, some appear syrupy, and some are solid (**Fig. 14.3**). The *viscosity* (ability to flow) and the *volatility* (ability to evaporate) of a hydrocarbon product depend on the size of its component molecules. Hydrocarbon products composed of short chains of molecules tend to be less viscous (they can flow more easily) and more volatile (they evaporate more easily) than products composed of long chains, because the long chains tend to tangle with each other. Thus, short-chain molecules occur in gaseous form at room temperature, moderate-length-chain molecules occur in liquid form (gasoline and oil), and long-chain molecules occur in solid form (tar).

Why can we use hydrocarbons as fuel? Simply because hydrocarbons, like wood, *burn*, meaning they react with oxygen to form carbon dioxide, water, and heat. As an example, we can describe the burning of gasoline by the reaction

$$2C_8H_{18} + 25O_2 \rightarrow 16CO_2 + 18H_2O + \text{heat energy}$$

During such reactions, the potential energy stored in the chemical bonds of the hydrocarbon molecules converts into usable heat energy. This energy can be used to run engines or to transform water into the steam that drives the electricity-producing dynamo of a power plant.

FIGURE 14.3 The diversity of hydrocarbon products, in order of increasing viscosity.

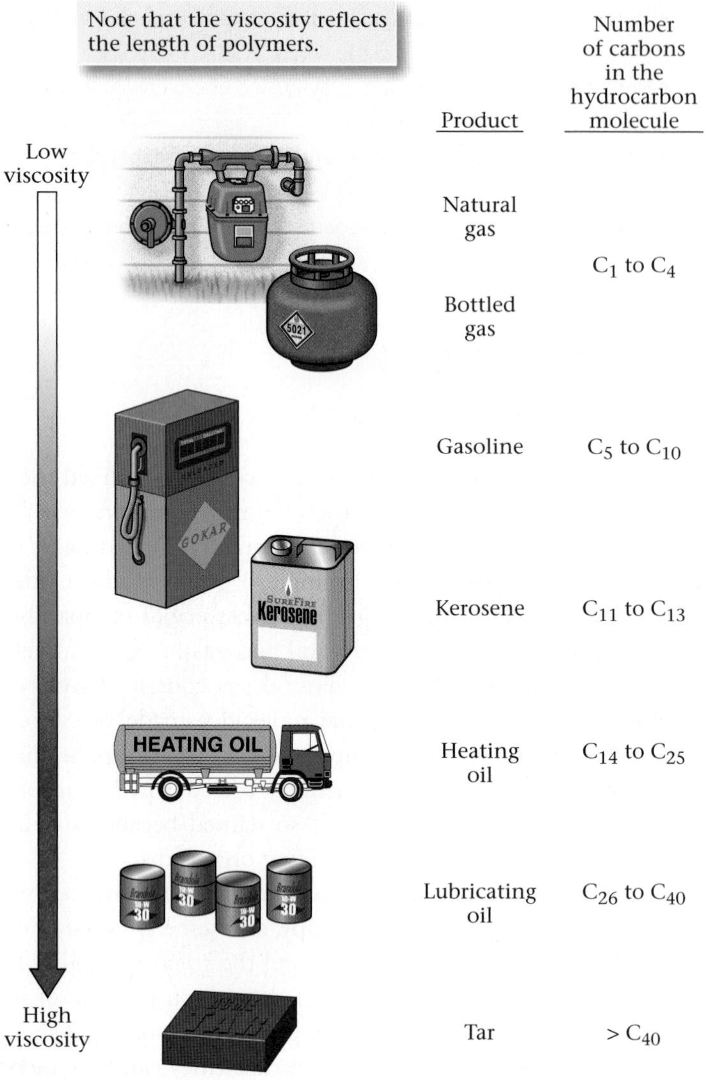

Note that the viscosity reflects the length of polymers.

Product	Number of carbons in the hydrocarbon molecule
Natural gas	
	C_1 to C_4
Bottled gas	
Gasoline	C_5 to C_{10}
Kerosene	C_{11} to C_{13}
Heating oil	C_{14} to C_{25}
Lubricating oil	C_{26} to C_{40}
Tar	$> C_{40}$

Low viscosity

High viscosity

Hydrocarbon Systems

Oil and gas do not occur in all rocks at all locations. That's why the desire to control oil fields, regions that contain significant amounts of accessible oil underground, has sparked bitter wars. A known supply of oil and gas held underground is a **hydrocarbon reserve**; if the reserve consists dominantly of oil, it is usually called an oil reserve.

The development of a reserve requires a specific association of materials, conditions, and time. Geologists refer to this association as a **hydrocarbon system**. We'll now look at the components of a hydrocarbon system, namely the source rock, the thermal conditions of oil formation, the migratory pathway, and the trap.

Source Rocks and Hydrocarbon Generation

Many people incorrectly believe that hydrocarbons come from buried trees or carcasses of dinosaurs. In fact, the primary sources of the hydrocarbons in oil and gas are organic chemicals, such as fatty molecules called lipids, from algae and plankton. (Plankton, as we have seen, are the tiny plants and animals—typically around 0.5 mm in diameter—that float in sea or lake water.) When algae and plankton die, they settle to the bottom of a lake or sea. Because their cells are so tiny, they can be deposited only in quiet-water environments in which clay also settles, so typically the dead cells mix with clay to create an organic-rich, muddy ooze. For this ooze to be preserved, it must be deposited in oxygen-poor water. Otherwise, the organic chemicals in the ooze would react with oxygen or be eaten by bacteria and thus would decompose quickly and disappear. In some quiet-water environments (oceans, lagoons, or lakes), dead algae and plankton get buried by still more sediment before being destroyed. Eventually, the resulting ooze lithifies and becomes black organic shale (in contrast to ordinary shale that consists only of clay and is lighter gray), which contains the raw materials from which hydrocarbons eventually form. Thus, we refer to organic shale as a **source rock**.

If organic shale is buried deeply enough (2 to 4 km), it becomes warmer, since temperature increases with depth in the Earth. Chemical reactions that take place in warm source rocks slowly transform the organic material in the shale into a mass of waxy molecules called **kerogen** (Fig. 14.4 ●)). Shale containing kerogen is called **oil shale**. If oil shale warms to temperatures of greater than about 90°, the kerogen molecules break into smaller oil and natural gas molecules, a process known as **hydrocarbon generation**. At temperatures over about 160°, any remaining oil breaks down to form natural gas; and at temperatures over 225°–250°C, organic matter loses all its hydrogen and transforms into graphite. Thus, oil itself forms only in a relatively narrow range of temperatures, called the **oil window** (Fig. 14.5). For regions with a geothermal gradient of 25°C/km, the oil window lies at depths of about 3.5 to 6.5 km and the gas window extends down to 9 km, so the gas window is larger. If the geothermal gradient is low (15°C/km), oil can survive down to depths of about 11 km and gas down to 15 km. Thus, hydrocarbon reserves can exist only in the topmost 15 to 25% of the crust.

Reservoir Rocks and Hydrocarbon Migration

Wells drilled into source rocks do not yield much oil because kerogen can't flow easily from the rock into the well. Any organic matter in an oil shale remains trapped among the grains and can't move easily. So to obtain oil, companies

FIGURE 14.4 The formation of oil. The process begins when organic debris settles with sediment. As burial depth increases, heat and pressure transform the sediment into black shale in which organic matter becomes kerogen. At appropriate temperatures, kerogen becomes oil, which then seeps upward. ▶❚❚

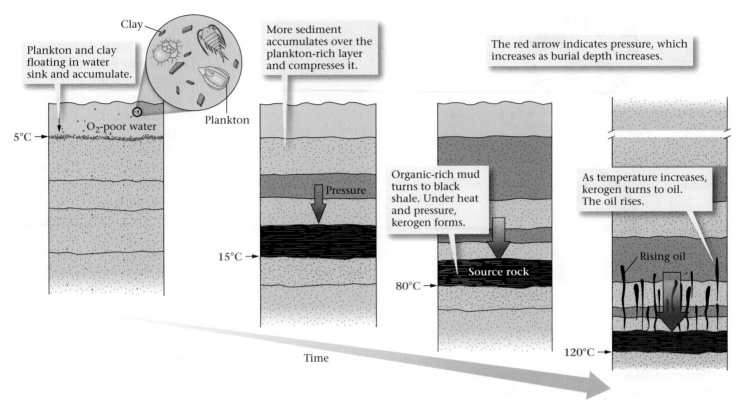

FIGURE 14.5 The "oil window" indicates subsurface conditions in which oil can form and survive. Deeper down, metamorphism transforms organic material to graphite.

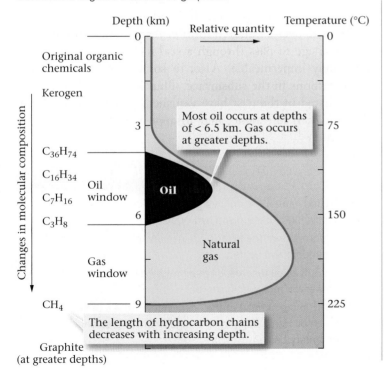

drill instead into **reservoir rocks**, rocks that contain (or could contain) an abundant amount of *easily accessible* oil and gas, meaning hydrocarbons that can be extracted out of the ground.

Did you ever wonder...
are there actually lakes or pools of oil underground?

To be a reservoir rock, a body of rock must have space in which the oil or gas can reside and must have channels through which the oil or gas can move. The space can be in the form of openings, or **pores**, between clastic grains (which exist because the grains didn't fit together tightly and because cement didn't fill all the spaces during cementation) or in the form of cracks and fractures that developed after the rock formed. In some cases, groundwater passing through rock dissolves minerals and creates new space. **Porosity** refers to the amount of open space in a rock (Fig. 14.6). Oil or gas can fill porosity just as water fills the holes in a sponge. Not all rocks have the same porosity. For example, shale typically has a porosity of 10%, while sandstone has a porosity of up to 35%. That means that about a third of a block of sandstone can consist of open space.

FIGURE 14.6 Variations in the porosity and permeability of sedimentary rocks. Each image represents what a rock looks like as viewed through a microscope.

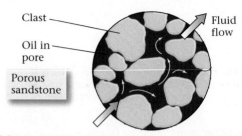

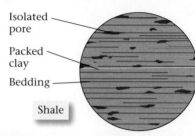

(a) In rocks with high porosity and permeability, a large percentage of space consists of pores, and they are interconnected.

(b) In rocks with low porosity and permeability, there is a small percentage of pore space, and pores are not interconnected.

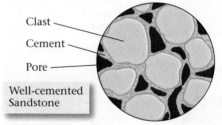

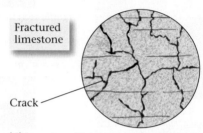

(c) In rocks with high porosity and low permeability, there is lots of pore space, but pores are isolated.

(d) In rocks with low porosity and high permeability, there is not much pore space, but pores are interconnected.

Permeability refers to the degree to which pore spaces connect to each other. Even if a rock has high porosity, it is not necessarily permeable (See Fig. 14.6). In a permeable rock, the pores and cracks are linked, so a fluid is able to flow slowly through the rock, following a tortuous pathway.

Keeping the concepts of porosity and permeability in mind, we can see that a poorly cemented sandstone makes a good reservoir rock because it is both porous and permeable. A highly fractured rock can be porous and permeable, even if there is no pore space between individual grains. The greater the porosity, the greater the capacity of a reservoir rock to hold oil; and the greater the rock's permeability, the easier it is for the oil to be extracted.

To fill the pores of a reservoir rock, oil and gas must first migrate (move) from the source rock into a reservoir rock, takes thousands to millions of years to happen (Fig. 14.7 ▶). Why do hydrocarbons migrate? Oil and gas are less dense than water, so they try to rise toward the Earth's surface to get above groundwater, just as salad oil rises above the vinegar in a bottle of salad dressing. Natural gas, being less dense, ends up floating above oil. In other words, buoyancy drives oil and gas upward. Typically, a hydrocarbon system must have a good **migration pathway**, such as a set of permeable fractures, in order for large volumes of hydrocarbons to move.

Traps and Seals

The existence of reservoir rock alone does not create a reserve, because if hydrocarbons can flow into a reservoir rock, they can also flow out. If oil or gas escapes from the reservoir rock and ultimately reaches the Earth's surface, where it leaks away at an **oil seep**, there will be none left underground to extract. Thus, for an oil reserve to exist, oil and gas must be *trapped* underground in the reservoir rock by means of a geologic configuration called a **trap**.

There are two components to an oil or gas trap. First, a **seal rock**, a relatively impermeable rock such as shale, salt, or unfractured limestone, must lie above the reservoir rock and stop the hydrocarbons from rising further. Second, the seal and reservoir rock bodies must be arranged in a geometry that collects the hydrocarbons in a restricted area. Geologists recognize several types of hydrocarbon trap geometries, four of which are described in **Box 14.1**.

Note that when we talk about trapping hydrocarbons underground, we are talking about a *temporary* process in the context of geologic time. Oil and gas may be trapped for millions to over a hundred million years, but eventually they may manage to pass through a seal rock because no rock is absolutely impermeable. Also, in some cases, microbes eat hydrocarbons in the subsurface. Thus innumerable oil fields that existed in the past have vanished, and the oil fields we find today, if left alone, may disappear millions of years in the future.

Take-Home Message

- Oil and natural gas are hydrocarbons—organic chemicals consisting of carbon and hydrogen.
- Oil forms from the bodies of algae and plankton, buried with sediments and heated.
- To form an oil reserve, oil must migrate from the source rock to a reservoir rock within a trap.

THINK: In the context of the "oil window," does the depth at which oil forms depend on the geothermal gradient in a sedimentary basin?

FIGURE 14.7 Initially, oil resides in the source rock. Because it is buoyant relative to groundwater, the oil migrates into the overlying reservoir rock. The oil accumulates beneath a seal rock in a trap.

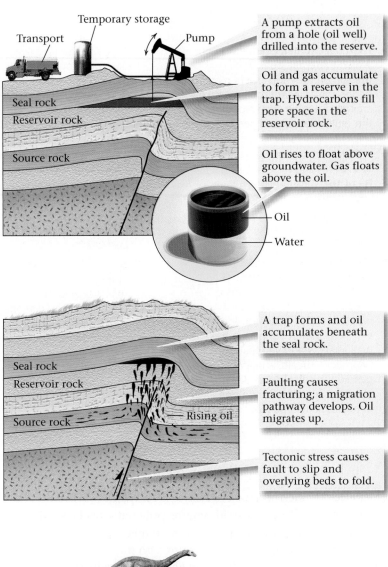

Present

Time

Past

Transport

Temporary storage

Pump

A pump extracts oil from a hole (oil well) drilled into the reserve.

Seal rock

Reservoir rock

Source rock

Oil and gas accumulate to form a reserve in the trap. Hydrocarbons fill pore space in the reservoir rock.

Oil rises to float above groundwater. Gas floats above the oil.

Oil

Water

Seal rock

Reservoir rock

Source rock

Rising oil

A trap forms and oil accumulates beneath the seal rock.

Faulting causes fracturing; a migration pathway develops. Oil migrates up.

Tectonic stress causes fault to slip and overlying beds to fold.

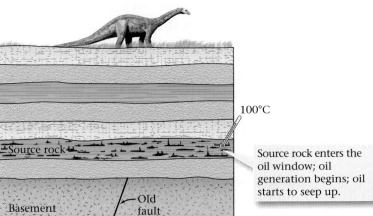

100°C

Source rock

Basement

Old fault

Source rock enters the oil window; oil generation begins; oil starts to seep up.

14.4 OIL EXPLORATION AND PRODUCTION

Birth of the Oil Industry

People have used oil since the dawn of civilization—as a cement, as a waterproof sealant, and even as a preservative to embalm mummies. But originally, the only available oil came from natural seeps, places where oil-filled reservoir rock intersects the Earth's surface or where fractures connect a reservoir rock to the Earth's surface, so that oil flows out on the ground on its own.

In the United States, during the first half of the nineteenth century, people collected "rock oil" (later called petroleum, from the Latin words *petra*, meaning rock, and *oleum*, meaning oil) at seeps and used it to grease wagon axles and to make patent medicines. But such oil was rare and expensive. In 1854, George Bissel, a New York lawyer, came to the realization that oil might have broader uses, particularly as fuel for lamps (to replace increasingly scarce whale oil). Bissel and a group of investors contracted Edwin Drake, a colorful character who had drifted among many professions, to find a way to drill for oil in rocks beneath a hill near Titusville, Pennsylvania, where oily films floated on the water of springs. Using the phony title "Colonel" to add respectability, Drake hired drillers and obtained a steam-powered drill. Work was slow and the investors became discouraged, but the very day that a letter arrived ordering Drake to stop drilling, his drillers found that the hole, which had reached a depth of 21.2 m, had filled with oil. They set up a pump, and on August 27, 1859, for the first time in history, oil was pumped out of the ground. No one had given much thought to the question of how to store the oil, so workers dumped it into empty whisky barrels. This first oil well yielded 10 to 35 barrels a day, which sold for about $20 a barrel (1 barrel equals 42 gallons).

Within a few years, thousands of oil wells had been drilled in many states, and by the turn of the twentieth century, civilization had begun its addiction to oil. Initially, most oil went into the production of kerosene for lamps. Later, when electricity took over from kerosene as the primary source for illumination, gasoline derived from oil became the fuel of choice for

BOX 14.1

Types of Oil and Gas Traps

Geologists who work for oil companies spend much of their time trying to identify underground traps. No two traps are exactly alike, but we can classify most into the following four categories.

- *Anticline trap*: In some places, sedimentary beds are not horizontal, as they are when originally deposited, but have been bent by the forces involved in mountain building. These bends, as we have seen, are called folds. An anticline is a type of fold with an arch-like shape (**Fig. Bx14.1a**; see Chapter 11). If the layers in the anticline include a source rock overlain by a reservoir rock that is overlain by a seal rock, then we have the recipe for an oil reserve. The oil and gas rise from the source rock, enter the reservoir rock, and rise to the crest of the anticline, where they are trapped by a seal.

- *Fault trap*: A fault is a fracture on which there has been sliding. If the slip on the fault crushes and grinds the adjacent rock to make an impermeable layer along the fault, then oil and gas may migrate upward along bedding in the reservoir rock until they stop at the fault surface (**Fig. Bx14.1b**). Alternatively, a fault trap develops if the slip on the fault juxtaposes an impermeable rock layer against a reservoir rock.

- *Salt-dome trap*: In some sedimentary basins, the sequence of strata contains a thick layer of salt, deposited when the basin was first formed and seawater covering the basin was shallow and very salty. Sandstone, shale, and limestone overlie the salt. The salt layer is not as dense as sandstone or shale, so it is buoyant and tends to rise up slowly through the overlying strata. Once the salt starts to rise, the weight of surrounding strata squeezes the salt out of the layer and up into a growing, bulbous **salt dome**. As the dome rises, it bends up the adjacent layers of sedimentary rock. Oil and gas in reservoir rock layers migrate upward until they are trapped against the boundary of the salt dome, for salt is not permeable (**Fig. Bx14.1c**).

- *Stratigraphic trap*: In a stratigraphic trap, a tilted reservoir rock bed "pinches out" (thins and disappears up its dip) between two impermeable layers. Oil and gas migrating upward along the bed accumulate at the pinch-out (**Fig. Bx14.1d**).

the newly invented automobile. Oil was also used to fuel electric power plants. In its early years, the oil industry was in perpetual chaos. When "wildcatters" discovered a new oil field, there would be a short-lived boom during which the price of oil could drop to pennies a barrel. In the midst of this chaos, John D. Rockefeller established the Standard Oil Company, which monopolized the production, transport, and marketing of oil. In 1911, the Supreme Court broke Standard Oil down into several companies (including Exxon, Chevron, Mobil, Sohio, Amoco, Arco, Conoco, and Marathon), some of which have recombined in recent decades. Oil became a global industry governed by the complex interplay of politics, profits, supply, and demand.

The Modern Search for Oil

Wildcatters discovered the earliest oil fields either by blind luck or by searching for surface seeps. But in the twentieth century, when most known seeps had been drilled and blind luck became too risky, oil companies realized that finding new oil fields would require systematic exploration. The modern-day search for oil is a complex, sometimes dangerous, and often exciting procedure with many steps.

Source rocks are always sedimentary, as are most reservoir and seal rocks, so geologists begin their exploration by looking for a region containing appropriate sedimentary rocks. Then they compile a geologic map of the area, showing the distribution of rock units. From this information, it may be possible to construct a preliminary cross section depicting the geometry of the sedimentary layers underground as they would appear on an imaginary vertical slice through the Earth.

To add detail to the cross section, an exploration company makes a **seismic-reflection profile** of the region. To construct a seismic profile, a special vibrating truck or a dynamite explosion sends seismic waves (shock waves that move through the Earth) into the ground (Fig. 14.8a; see Interlude D). The seismic waves reflect off contacts between rock layers, just as

FIGURE Bx14.1 Examples of oil traps. A trap is a configuration of a seal rock over a reservoir rock, in a geometry that keeps the oil underground.

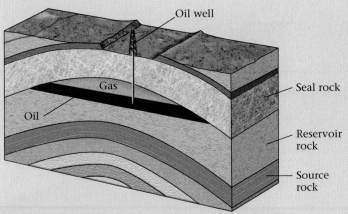

(a) Anticline trap. Oil and gas rise to the crest of the fold.

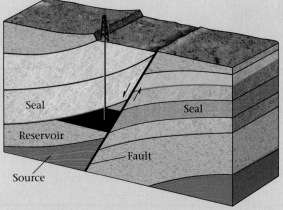

(b) Fault trap. Oil and gas collect in tilted strata adjacent to the fault.

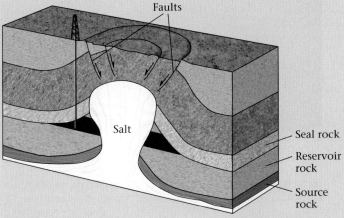

(c) Salt-dome trap. Oil and gas collect in strata on the flanks of the dome, beneath salt.

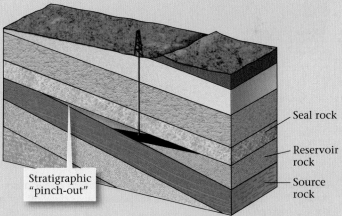

(d) Stratigraphic trap. Oil and gas collect where the reservoir layer pinches it out.

sonar waves sent out by a submarine reflect off the bottom of the sea. Reflected seismic waves then return to the ground surface, where sensitive instruments (geophones) record their arrival. A computer measures the time between the generation of a seismic wave and its return, and from this information defines the depth to the contacts at which the wave reflected. With such information, the computer constructs an image of the configuration of underground rock layers and, in some cases, can "see" reserves of oil. Technological advances now enable geologists to create 3-D seismic-reflection profiles of the subsurface both under land and underwater (**Fig. 14.8b, c**). Seismic-reflection profiles are expensive—it may cost millions of dollars to create just one profile.

Drilling and Refining

If geological studies identify a trap, and if the geologic history of the region indicates the presence of good source rocks and reservoir rocks, geologists make a recommendation to drill. (They do not make such recommendations lightly, as drilling a deep well may cost over $10 million.) Once the decision has been made, drillers go to work. These days, drillers use rotary drills to grind a hole down through rock. A rotary drill consists of a rotating pipe tipped by a bit, a bulb of metal studded with hard metal prongs (**Fig. 14.9a**). As the bit rotates, it scratches and gouges the rock, turning it into powder and chips. Drillers pump "drilling mud," a slurry of water mixed with clay

FIGURE 14.8 Using seismic-reflection profiling to describe the character of underground beds and help locate reservoirs.

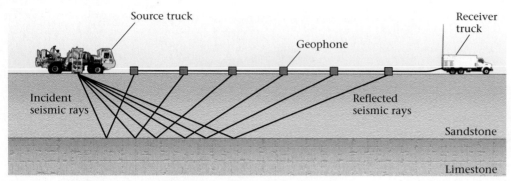

(a) Source truck sends a vibration into the Earth. The vibration reflects off layer boundaries up to geophones on the surface. The time it takes for the vibration to travel indicates the depth to the reflector.

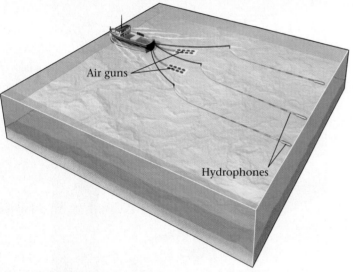

(b) For offshore seismic-reflection studies, a ship sends pulses of sound through the water. These penetrate into the crust and reflect off horizons back to hydrophones.

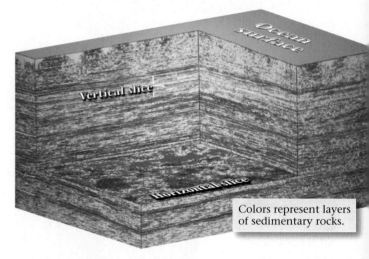

Colors represent layers of sedimentary rocks.

(c) Modern techniques produce 3-D "blocks" of data, so geologists can study both vertical and horizontal slices through the subsurface.

and other materials, down the center of the pipe. The mud squirts out of holes in the drill bit, cooling the bit and flushing rock cuttings up and out of the hole. The weight of the mud also keeps oil down in the hole and prevents "gushers," fountains of oil or gas formed when underground pressure causes the oil to rise out of the hole on its own (**Box 14.2**).

Drillers use derricks (towers) to hoist the heavy drill pipe. To drill in an offshore oil reserve, one that occurs in strata of the continental shelf, the derrick must be constructed on an offshore-drilling facility. These may be built on huge towers rising from the sea floor, or on giant submerged pontoons. Drill holes can be aimed in any direction (not just vertical), so drillers can reach many traps from the same facility. In fact, during such **directional drilling**, the driller sits in a 3-D visualization room and controls the drilling direction with a joystick, as if playing a video game.

On completion of a hole, workers remove the drilling rig and set up a pump. Some pumps resemble a bird pecking for grain; their heads move up and down to pull up oil that has seeped out of pores in the reservoir rock into the drill hole (**Fig. 14.9b**). You may be surprised to learn that simple pumping gets only about 30% of the oil in a reservoir rock out of the ground. Thus oil companies use secondary recovery techniques to coax out more oil (as much as 20% more). For example, a company may drive oil toward a drill hole by forcing steam into the ground nearby. The steam heats the oil in the ground, making it less viscous, and pushes it along. In some cases, drillers create artificial fractures in rock around the hole by pumping a high-pressure water and chemical mixture into a portion of the hole; this process is called **hydrofracturing**. This process creates easy routes for the oil to follow from the rock to the well.

Once extracted directly from the ground, "crude oil" flows first into storage tanks and then into a pipeline or tanker, which transports it to a refinery (**Fig. 14.9c, d**; **See for Yourself N**, p. S-26). At a refinery, workers distill crude oil into several separate components by heating it gently in a vertical pipe called

FIGURE 14.9 The steps involved in searching for, recovering, transporting, and refining oil.

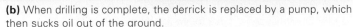

(a) A rotating pipe tipped by a drill bit grinds a hole into the ground. Drilling mud, pumped down through the pipe, comes out through holes in the bit, flushing cuttings out of the hole.

(b) When drilling is complete, the derrick is replaced by a pump, which then sucks oil out of the ground.

(c) The Trans Alaska Pipeline transports oil from fields on the Arctic coast to a tanker port on the southern coast of Alaska.

(d) An oil supertanker. This ship pictured weighs 310,453 tons. The largest ever built weighed almost 650,000 tons when loaded. By comparison, the largest cruise ship weighs 225,000 tons, and the *Titanic* weighed only 46,000 tons.

a **distillation column** (Fig. 14.9e). Lighter molecules rise to the top of the column, while heavier molecules stay at the bottom. The heat may also "crack" larger molecules to make smaller ones. Chemical factories buy the largest molecules left at the bottom and transform them into plastics.

Where Does Oil Occur?

Reserves are not randomly distributed around the Earth (Fig. 14.10a). Currently, countries bordering the Persian Gulf contain the world's largest reserves in 25 supergiant fields (Fig. 14.10b). In fact, this region has almost 60% of the world's reserves, whereas the United States, by far the largest consumer of oil, has only a few percent of the total. Reserves are specified in barrels (bbl); 1 bbl = 42 gallons = 159 liters.

(e) Distilling columns of an oil refinery transform crude oil into gasoline and other hydrocarbon products.

BOX 14.2

Spindletop

As demand for oil increased in the last decade of the nineteenth century, oil became "black gold," worth a fortune to anyone lucky enough to find it. One Texas mechanic, Patillo Higgins, not only dreamed of oil riches, but thought he knew a place to find them. Higgins had noticed a strange gas bubbling slowly up through little water springs on Spindletop Hill, where he took his Sunday school class for walks. When he lit the bubbles with a match, they burned. Did oil lie beneath the hill? Higgins struggled to find investors that would help him find out. Finally, in 1899, his arguments convinced Anthony Lucas, who brought a drilling crew to Spindletop.

Drilling began in late 1900, and soon after, Lucas and his crew reached a small oil pocket, which to their delight yielded 50 barrels a day, more than the average wells of Pennsylvania. But as long as drilling equipment was in place, Lucas decided to keep drilling. The operation proceeded routinely until January 10, 1901, when suddenly, without warning, the drilling platform began to vibrate violently. Workers fled for cover as the drilling pipe shot upward and out of the ground. As tons of equipment rained from the sky, a black fountain of oil erupted from the ground and blasted skyward to a height of almost 80 m (**Fig. Box14.2a**). Oil from the Spindletop "gusher" rained

FIGURE Bx14.2 Oil wells at the Spindletop Oil Field in the early 1900s.

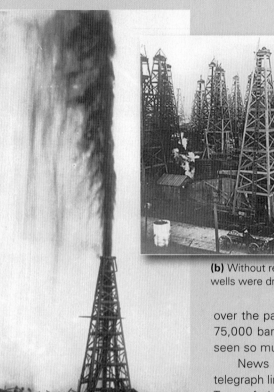

(b) Without regulations to guide well spacing, wells were drilled right next to one another.

(a) A gusher formed when the pressure in the oil underground was great enough to force the oil out of the well.

over the pasture at the astounding rate of 75,000 barrels per day! No one had ever seen so much oil before.

News of the gusher chattered over telegraph lines, and pandemonium ensued. Tens of thousands of people flocked to Spindletop Hill, creating a boomtown of mud streets, plank shacks, and saloons. Land prices exploded from $10 per acre to almost $1 million per acre, and soon the hill looked like a pin cushion, punctured by over 200 wells (**Fig. Bx14.2b**). Oil hysteria had moved to Texas.

Source rock is closely and directly related to the biological activity of a given zone. If the biological activity of an area is high, it will concentrate the organic matter at one place and afterward, if appropriate conditions are met, it will become preserved in that zone. Much of the region that is now the Middle East was situated in tropical areas between latitude 20° south and 20° north between the Jurassic (135 Ma) and the Late Cretaceous (65 Ma). Biological activity and productivity are highest in tropical and equatorial zones (as compared to the areas lying nearer to the poles), making such areas a

more likely location for source rock to form. In addition, thick successions of porous sandstone overlie the source rocks of the Middle East. Crustal compression and resulting shortening later folded the strata to produce excellent traps.

The Middle East is not the only source of oil (Fig. 14.10b). Reserves also occur in sedimentary basins formed along passive continental margins, such as the Gulf Coast of the United States and the Atlantic Coasts of Africa and Brazil, as well as within continents (i.e., in intracratonic and foreland basins, see Chapter 7).

FIGURE 14.10 The distribution of oil reserves around the world.

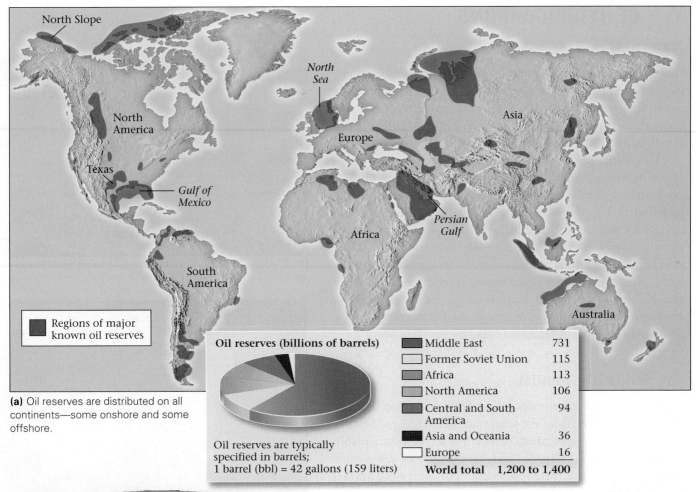

(a) Oil reserves are distributed on all continents—some onshore and some offshore.

Oil reserves (billions of barrels)

Middle East	731
Former Soviet Union	115
Africa	113
North America	106
Central and South America	94
Asia and Oceania	36
Europe	16
World total	**1,200 to 1,400**

Oil reserves are typically specified in barrels; 1 barrel (bbl) = 42 gallons (159 liters)

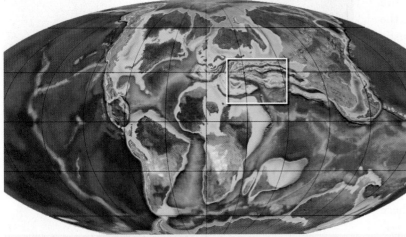

(b) A map of the Late Cretaceous world, during the time when the deposition of sediment from which source rocks in the Mideast are forming. The box shows the region that will become the Middle East oil fields of today. Note that the region lay in the warm subtropics in which marine plankton could thrive.

Take-Home Message

- The oil industry began in the mid-nineteenth century, initially to provide kerosene for lamps.
- J. D. Rockefeller consolidated the industry into the Standard Oil Company, which the Supreme Court broke up into a number of major companies, several of which exist today in some form.
- Geologists search for oil by obtaining seismic-reflection profiles showing the configuration of strata underground.

- Drilling is complex and expensive; once a well is complete, pumps suck oil out of the reservoir.
- Refiners "crack" the oil to produce smaller hydrocarbon molecules.

THINK: Where do most of the oil reserves in the world occur? Explain why.

14.5 ALTERNATIVE RESERVES OF HYDROCARBONS

Natural Gas

Natural gas consists of volatile short-chain hydrocarbons, including methane, ethane, propane, and butane. Some gas occurs in association with oil. In such cases, the gas floats above the oil in a reservoir. But in rocks that were heated above the oil window, gas may occur without oil.

Gas burns more cleanly than oil (burning gas produces primarily carbon dioxide and water, whereas burning oil also produces complex organic pollutants) and thus has become the preferred fuel for home cooking and heating. But gas transportation requires expensive high-pressure pipelines or special ships, so even though gas is much more abundant than oil, the world's population still consumes more oil than gas. However, in recent years many industries and electrical generation plants have switched to natural gas, so demand for gas has increased. New discoveries near population centers in the northeastern United States have led to an increase in drilling for gas (Box 14.3).

Tar Sands (Oil Sands)

So far, we've focused our discussion on hydrocarbon reserves that can be pumped from the subsurface in the form of a liquid or gas. But in several locations around the world, most notably Alberta (in western Canada) and Venezuela, vast reserves of very viscous, tar-like "heavy oil" exist. This heavy oil, known also as bitumen, has the consistency of gooey molasses, and thus cannot be pumped directly from the ground. It fills the pore spaces of sand or of poorly cemented sandstone, constituting up to 12% of the sediment or rock volume. Sand or sandstone containing such high concentrations of bitumen is known as **tar sand** or oil sand (Fig. 14.11a).

The hydrocarbon system that leads to the generation of tar sands begins with the production and burial of a source rock in a large sedimentary basin. When subjected to temperatures of the oil window, the source rock yields oil and gas, which migrate into sandstone layers and then up the dip of tilted layers to the edge of the basin, where they become caught in stratigraphic traps (see Box 14.1). Initially, these hydrocarbons have relatively low viscosity; in the geologic past, they could have been pumped easily. But over time, microbes attacked the oil reserve underground, digested lighter, smaller hydrocarbon molecules, and left behind only the larger molecules, whose presence makes the remaining oil so viscous. Geologists refer to such a transformation process as **biodegradation**. Generation of tar sand by biodegradation is yet another example of the interaction between physical and biological components of the Earth System.

FIGURE 14.11 Examples of alternative hydrocarbon sources.

(a) An open-pit mine in Canada for digging up tar sand. Trucks haul the sand to a plant where it is heated so hydrocarbons can be extracted.

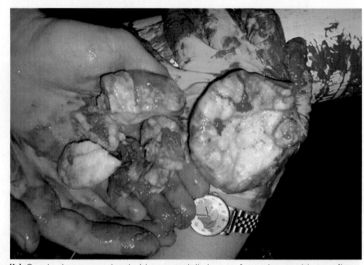

(b) Gas hydrate samples (white material) dug up from the muddy sea floor.

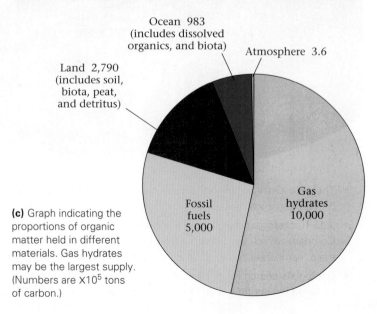

(c) Graph indicating the proportions of organic matter held in different materials. Gas hydrates may be the largest supply. (Numbers are X10⁵ tons of carbon.)

Ocean 983 (includes dissolved organics, and biota)

Atmosphere 3.6

Land 2,790 (includes soil, biota, peat, and detritus)

Fossil fuels 5,000

Gas hydrates 10,000

BOX 14.3

CONSIDER THIS . . .

The Marcellus Gas Play

In 2008, Terry Engelder (Penn State University) and others pointed out that previous reports had greatly underestimated the amount of natural gas that is locked in the organic-rich, black Marcellus Shale of the Devonian age (**Fig. Bx14.3a**). He estimated that this shale (15 to 240 m thick; or 49 to 787 feet), which lies beneath a highly populated part of the United States (New York, Pennsylvania, and Ohio), contained enough gas to serve as a major energy supply for many years.

The Marcellus Shale, like all shale, has relatively low permeability. So even though it contains a huge amount of gas, it would not be a useable resource were it not for the development of two key technologies. The first of these technologies is **directional drilling**. During directional drilling, an oper-

ator can make the pipe bend gradually so that at a specified depth, the drill cuts horizontally through rock (**Fig. Bx14.3b**). Beds of the Marcellus Formation are horizontal. Thus a single drill hole, which would only intersect a maximum of 240 m (787 feet) of shale if drilled vertically, could intersect many kilometers of shale if drilled horizontally. Using directional drilling, many holes can be drilled from the same platform.

The second of these technologies is **hydraulic fracturing** or, informally, "hydrofracking," which gives drillers the capability to greatly increase permeability in the otherwise impermeable shale. To "hydrofrac" a hole, drillers seal off a portion of the hole and then pump in a water solution under enough pressure to rupture

the surrounding rock. In general, a set of vertical cracks are produced.

The combination of increased gas-volume estimates, and of the ability to access the gas by directional drilling and hydrofracturing, has led to a "gas boom" in the northeastern United States called the Marcellus Gas Play. Oil companies have spent billions of dollars to lease the right to obtain gas from large areas of the region, and some local landowners have become millionaires overnight. The boom remains controversial, though, because of lingering questions about the amount of gas that can be recovered and because of environmental concerns. Residents are worried that chemicals in the solutions used during hydraulic fracturing and drilling may contaminate ground water and that increased permeability caused by fracturing may also release gas into groundwater.

FIGURE Bx14.3 Extracting natural gas from the Marcellus Shale

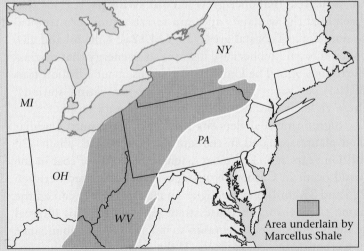

(a) Marcellus Shale underlies regions near population centers; this map is an approximation.

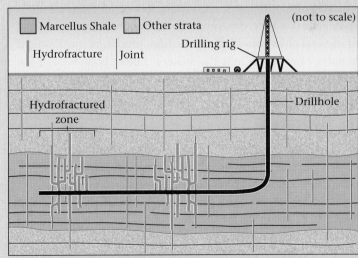

(b) By using directional drilling and hydrofracture, gas can be drawn from broad regions. The hole stays within shale for kilometers.

Production of usable oil from tar sand is difficult and expensive, but not impossible. It takes about 2 tons of tar sand to produce one barrel of oil. Oil companies mine near-surface deposits in vast open-pit mines and then heat the tar sand in a furnace to extract the oil. Producers then crack the heavy oil molecules to produce smaller, more usable molecules. Trucks dump the drained sand back into the mine pit. To extract oil from deeper deposits of tar sand, oil companies drill wells and pump steam or solvents down into the sand to liquefy the oil enough so that it can be pumped out.

Oil Shale

Vast reserves of organic shale have not been subjected to temperatures of the oil window, or if they were, they did not stay within the oil window long enough to complete the transformation to oil. Such rock still contains a high proportion of kerogen. Shale that contains at least 15 to 30% kerogen is called **oil shale**. Oil shale is not the same as coal because the organic matter within it exists in the form of waxy hydrocarbon molecules, not as elemental carbon.

Lumps of oil shale can be burned directly and thus have been used as a fuel since ancient times. In the 1850s, researchers developed techniques to produce liquid oil from oil shale. The process involves heating the oil shale to a temperature of 500°C; at this temperature, the shale decomposes and the kerogen transforms into liquid hydrocarbon and gas. Large supplies of oil shale occur in Estonia, Scotland, China, and Russia, and in the Green River Basin of Wyoming in the United States. As is the case with tar sand, production of oil from oil shale is possible, but very expensive. In addition to the expense of mining and environmental reclamation, producers must pay for the energy needed to heat the shale. It takes about 40% of the energy yielded by a volume of oil shale to produce the oil.

Gas Hydrate

Gas hydrate is a chemical compound consisting of methane (CH_4) molecules surrounded by a cage-like arrangement of water molecules. An accumulation of gas hydrate occurs as a whitish solid that resembles ordinary water ice (Fig. 14.11b). Gas hydrate forms when anaerobic bacteria (bacteria that live in the absence of oxygen) eat organic matter such as dead plankton that have been incorporated into sediments of the sea floor. When the bacteria digest organic matter, they produce methane as a by-product, and the methane bubbles into the cold seawater that fills pore spaces in sediments. Under pressures found at water depths greater than 300 m, the methane dissolves in the pore water and produces gas-hydrate molecules; the reaction can occur at shallower depths at colder, polar latitudes. Exploration tests suggest that gas hydrate occurs as layers interbedded with sediment, and/or as a cement holding together the sediment, at depths of

between 90 and 900 m beneath the sea floor. Geologists estimate that an immense amount of methane lies trapped in gas-hydrate layers. In fact, worldwide there may be more organic carbon stored in gas hydrate than in all other reservoirs combined (Fig. 14.11c)! So far, however, techniques for safely recovering gas hydrate from the sea floor have not been devised.

> ### Take-Home Message
>
> - Natural gas burns relatively cleanly, but is difficult to transport.
> - Abundant hydrocarbons exist in less-accessible or inconvenient sources. In tar sands, hydrocarbons are too viscous to pump, and in oil shale they occur as kerogen within impermeable shale.
> - In gas hydrate, methane molecules are locked in ice crystals beneath the sea floor.
>
> **THINK:** How are usable fuels obtained from organic shales and tar sands?

14.6 COAL: ENERGY FROM THE SWAMPS OF THE PAST

Coal, a black, brittle, sedimentary rock that burns, consists of elemental carbon mixed with minor amounts of organic chemicals, quartz, and clay. Typically, the carbon atoms in coal have bonded to form complicated molecules. Note that coal and oil do not have the same composition or origin. In contrast to oil, coal forms from plant material (wood, stems, leaves) that once grew in **coal swamps**, regions that resembled the wetlands and rain forests of modern tropical to semitropical coastal areas (Fig. 14.12a). Like oil and gas, coal is a fossil fuel because it stores solar energy that reached Earth long ago. The United States burns about 1 billion tons of coal per year, mostly at electricity-generating stations; these yield about 23% of the country's energy supply.

Significant coal deposits could not form until vascular land plants appeared in the late Silurian Period, about 420 million years ago. The most extensive deposits of coal in the world occur in Carboniferous-age strata (deposited between 286 and 354 million years ago). In fact, geologists coined the name *Carboniferous* because strata representing this interval of the geologic column contain so much coal. The abundance of Carboniferous coal reflects (1) the past position of the continents (during the Carboniferous Period, North America, Europe, and northern Asia straddled the equator, and thus had warm climates in which vegetation flourished) and (2) the height of sea level (at this time, shallow seas bordered by coal swamps covered vast parts of continental interiors). Because of their antiquity, Carboniferous coal deposits contain fossils of long-extinct species such as giant tree ferns, primitive conifers, and giant horsetails. Extensive coal

483

FIGURE 14.12 The formation of coal. Coal forms when plant debris becomes deeply buried.

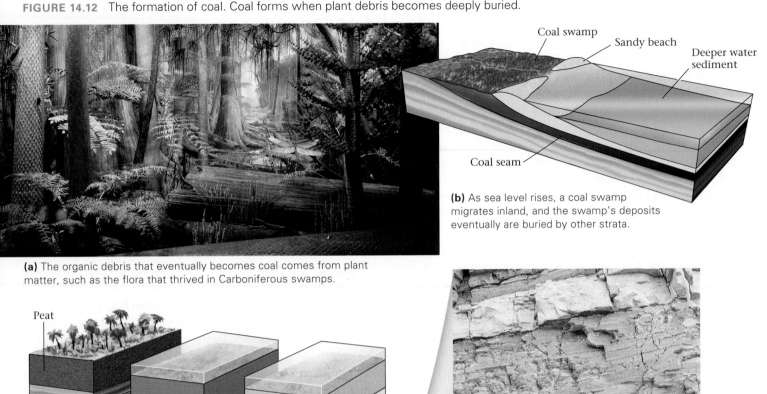

(b) As sea level rises, a coal swamp migrates inland, and the swamp's deposits eventually are buried by other strata.

(a) The organic debris that eventually becomes coal comes from plant matter, such as the flora that thrived in Carboniferous swamps.

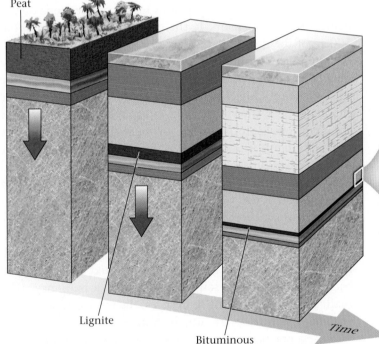

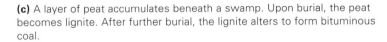

(c) A layer of peat accumulates beneath a swamp. Upon burial, the peat becomes lignite. After further burial, the lignite alters to form bituminous coal.

(d) Coal is found in sedimentary beds interlayered with other strata (sandstone, shale, and limestone).

deposits can also be found in strata of Cretaceous age (about 65 to 144 million years ago). Some of these deposits formed in wetlands bordering lakes.

The Formation of Coal

How do the remains of plants transform into coal? The vegetation of an ancient swamp must fall and be buried in an *oxygen-poor* environment, such as stagnant water, so that it can be incorporated in a sedimentary sequence without first decaying by reacting with oxygen and/or by being eaten by microbes or larger organisms. Compaction and partial decay of the vegetation transforms it into **peat**. Peat, which contains about 50% carbon, itself serves as a fuel in many parts of the world, where thick deposits formed from moss and grasses in bogs during the last several thousand years. It can easily be cut out of the ground, and once dried, it will burn.

To transform peat into coal, the peat must be buried deeply (4–10 km) by overlying sediment. Such deep burial can happen where the surface of the continent gradually sinks, creating a depression, or sedimentary basin, that can collect sediment. As the relative sea level at a location rises, the shoreline migrates inland (transgression takes place), and the location of a coal swamp becomes submerged beneath deeper water. Marine silt and mud then bury the swamp (Fig. 14.12b).

Over time, many kilometers of sediment containing numerous peat layers accumulate. At depth in the pile, the weight of overlying sediment compacts the peat and squeezes out any remaining water. Then, because temperature increases with depth in the Earth, deeply buried peat gradually heats up. Heat accelerates chemical reactions that gradually destroy plant fiber and release elements such as hydrogen, nitrogen, and sulfur in the form of gas. These gases seep out of the reacting peat layer, leaving behind a residue concentrated with carbon. Once the proportion of carbon in the residue exceeds about 70%, the deposit formally becomes coal. With further burial and higher temperatures, chemical reactions allow additional hydrogen, nitrogen, and sulfur to escape, yielding progressively higher concentrations of carbon.

The Classification of Coal

Geologists classify coal according to the concentration of carbon. With increasing burial, peat transforms into a soft, dark-brown coal called **lignite**. At higher temperatures (about 100°–200°C), lignite, in turn, becomes dull, black **bituminous coal**. At still higher temperatures (about 200°–300°C), bituminous coal transforms into shiny, black **anthracite coal** (also called hard coal). The progressive transformation of peat to anthracite coal, which occurs as the coal layer is buried more deeply and becomes warmer, reflects the completeness of chemical reactions that remove water, hydrogen, nitrogen, and sulfur from the organic chemicals of the peat and leave behind carbon (Fig. 14.12c). As the carbon content of coal increases, we say the **coal rank** increases.

Notably, the formation of anthracite coal requires high temperatures that develop only on the borders of mountain belts, where mountain-building processes can push thick sheets of rock up along thrust faults and over the coal-bearing sediment, so the sediment ends up at depths of 8 to 10 km, where temperatures reach 300°C. In the interiors of mountain belts, temperatures become even higher, and rock begins to undergo metamorphism, at which point almost all elements except carbon leave the coal. The remaining carbon atoms rearrange to form graphite, the gray mineral used to make pencils; thus, coal cannot exist in metamorphic rocks.

The burning of coal is a chemical reaction: C (carbon) + O_2 (oxygen) → CO_2 (carbon dioxide). Therefore, the different ranks of coal produce different amounts of energy when burned

(Table 14.1). Anthracite contains more carbon per kilogram than lignite contains, so it produces more energy when burned.

Finding and Mining Coal

Because the vegetation that eventually becomes coal was initially deposited in a sequence of sediment, coal occurs as sedimentary beds ("seams," in mining parlance) interlayered with other sedimentary rocks (Fig. 14.12d). To find coal, geologists search for sequences of strata that were deposited in tropical to semitropical, shallow-marine to terrestrial environments—the environments in which a swamp could exist. The sedimentary strata of continents contain huge quantities of discovered coal, or **coal reserves**. For example, economic seams (beds of coal about 1–3 m thick, thick enough to be worth mining) of Cretaceous age occur in the U.S. and Canadian Rocky Mountain region, while economic seams of Carboniferous age are found throughout the midwestern United States (Fig. 14.13a–c).

The way in which companies mine coal depends on the depth of the coal seam. If the coal seam lies within about 100 m of the ground surface, strip mining proves to be most economical. In strip mines, miners use a giant shovel called a drag line to scrape off soil and layers of sedimentary rock above the coal seam (Fig. 14.14a). Drag lines are so big that the shovel could swallow a two-car garage without a trace. Once the drag line has exposed the seam, it then scrapes out the coal and dumps it into trucks or onto a conveyor belt. Before modern environmental awareness took hold, strip mining left huge scars on the landscape. Without topsoil, the rubble and exposed rock of the mining operation remained barren of vegetation. In many contemporary mines, however, the drag-line operator separates out and preserves soil. Then, when the coal has been scraped out, the operator fills the hole with the rock that had been stripped to expose the coal and covers the rock back up with the saved soil, on which grass or trees may eventually grow. Within years to decades, the former mine site can become a pasture or a forest. In hilly areas, however, miners may use a practice called mountain top removal, during which

TABLE 14.1 **Types of Coal**

Material	% Carbon	Energy Content[1]	Rank
Peat	50	1,500 kcal/kg	
Lignite	70	3,500 kcal/kg	Low-rank coal
Bituminous coal	85	6,500 kcal/kg	Mid-rank coal
Anthracite coal	95	7,500 kcal/kg	High-rank coal

[1] 1 kcal (kilocalorie) = 1,000 calories. A calorie is the heat needed to raise the temperature of 1 gram of water by 1°C. 1 kg (kilogram) = 2.2 pounds.

FIGURE 14.13 The distribution of coal reserves. Vast quantities lie buried in continental sedimentary basins.

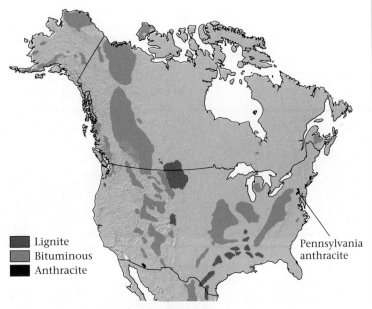

(a) Areas underlain by coal in North America.

Lignite
Bituminous
Anthracite

Pennsylvania anthracite

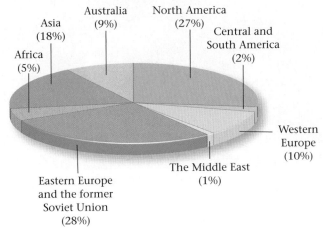

Asia (18%)
Africa (5%)
Australia (9%)
North America (27%)
Central and South America (2%)
Western Europe (10%)
The Middle East (1%)
Eastern Europe and the former Soviet Union (28%)

(b) Global coal reserves.

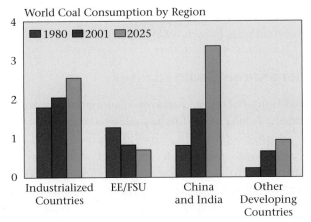

World Coal Consumption by Region

■ 1980 ■ 2001 ■ 2025

Industrialized Countries   EE/FSU   China and India   Other Developing Countries

(c) World coal consumption by region. The vertical axis is billion "short tons," where 1 short ton = 2,000 lbs = 907 kg. EE/FSU = Eastern Europe and the former Soviet Union.

they blast off the top of the mountain and dump the debris into adjacent valleys. This practice can disrupt the landscape permanently.

Deep coal can be obtained only by underground mining. To develop an underground mine, miners dig a shaft down to the depth of the coal seam and then create a maze of tunnels, using huge grinding machines that chew their way into the coal (Fig. 14.14b, c). Depending on circumstances, miners either leave columns of coal behind to hold up the mine roof, or they let the mined area collapse, so the ground level above sinks, once mining is finished.

Underground coal mining can be very dangerous, not only because the sedimentary rocks forming the roof of the mine are weak and can collapse but also because methane gas released by chemical reactions in coal can accumulate in the mine, leading to the danger of a small spark triggering a deadly mine explosion. Unless they breathe through filters, underground miners also risk contracting black lung disease from the inhalation of coal dust. The dust particles wedge into tiny cavities of the lungs and gradually cut off the oxygen supply or cause pneumonia.

Coalbed Methane

As we've noted, the natural process by which coal forms underground yields methane, a type of natural gas. Over time, some of the gas escapes to the atmosphere, but vast amounts remain within the coal in pores or bonded to coal molecules. Such **coalbed methane**, trapped in strata too deep to be reached by mining, is an energy resource that has become a target for exploration in many regions of the world.

Obtaining coalbed methane from deep layers of strata involves drilling, rather than mining. Drillers penetrate a coal bed with a hole and then start pumping out groundwater. As a result of pumping out water, the pressure in the vicinity of the drillhole decreases relative to the surrounding bed. Methane bubbles into the hole and then up to the ground surface, where condensers compress it into tanks for storage. Disposal of the water produced by coalbed methane extraction can be a major problem. If the water is pure, it can be used for irrigation, but in deep coalbeds the water may be saline and thus can't be used for crops. Producers either pump this water back underground or evaporate it in large ponds, so that they can extract and collect the salt.

Coal Gasification

Traditional burning of coal produces clouds of smoke containing fly ash (solid residue left after the carbon in coal has been burned) and noxious gases (including sulfur dioxide, SO_2, which forms because coal contains sulfur-bearing minerals such as pyrite). Today, smoke can be partially cleaned by expensive scrubbers, but pollution remains a major problem. Alternatively, solid coal can be transformed into

FIGURE 14.14 The process of mining coal.

(a) The configuration of a coal strip mine. After removing the overburden, the drag line strips the coal. Then the trench is refilled and the ground surface replanted.

A drag line stripping coal in an Indiana mine.

(b) Special machinery grinds into an underground coal seam and dumps coal on a conveyor. This seam is about 2.5 m thick.

(c) Chunks of freshly mined coal await processing at a mine in Indiana.

various gases, as well as solid by-products, *before* burning. The gases burn quite cleanly. The process of producing relatively clean-burning gases from solid coal is called **coal gasification**; the process was invented in the late eighteenth and early nineteenth centuries and was used extensively to produce fuels during World War II.

Coal gasification involves the following steps. First, pulverized coal is placed in a large container. Then, a mixture of steam and oxygen passes through the coal at high pressure. As a result, the coal heats up to a high temperature but does not ignite; under these conditions, chemical reactions break down and oxidize the molecules in coal to produce hydrogen and other gases such as carbon monoxide (CO), water, and CO_2. Solid ash, as well as sulfur and mercury, concentrate at the bottom of the container and can be removed before the gases are burned, so that the contaminants do not go up the chimney and into the atmosphere. The CO gas and H_2

gas can then be combined to produce hydrocarbons. Alternatively, the CO gas can react with water to produce more H_2 (plus CO_2). Of note, the hydrogen can be concentrated to produce fuel cells (which we will discuss later).

Underground Coalbed Fires

Coal will burn not only in furnaces but also in surface and subsurface mines, as long as the fire has access to oxygen. Coal mining of the past two centuries has exposed much more coal to the air and has provided many more opportunities for fires to begin; once started, a coalbed fire that progresses underground (sucking in oxygen from joints in overlying rock) may be very difficult or impossible to extinguish. Some fires begin as a result of lightning strikes, some from spontaneous combustion (when coal reacts with air, it heats up), some in the aftermath of methane explosions, and some when people intentionally set trash fires in mines.

FIGURE 14.15 Coal mine fires—long-lived disasters.

(a) A burning coal bed, exposed in a mine wall, glows red.

(b) A coalbed fire beneath Centralia, Pennsylvania, produces noxious gas and forced the evacuation of many homes.

Major coalbed fires (Fig. 14.15a, b) can be truly disastrous. The most notorious of these fires in North America began as a result of trash burning in a mine near the town of Centralia, Pennsylvania. For the past 50 years, the fire has progressed underground, eventually burning coal seams beneath the town itself. The fire produces toxic fumes that rise through the ground and make the overlying landscape uninhabitable, and it also causes the land surface to collapse and sink. Eventually, inhabitants had to abandon many neighborhoods of the town. Much longer-lived fires occur elsewhere—one in Australia may have been burning for 2,000 years. And today, over 100 major coalbed fires are burning in northern China; these can be detected by satellite imagery, which shows warm spots on the overlying ground surface. Recent estimates suggest that 200 million tons of coal burn in China every year, an amount equal to approximately 20% of the annual national production of coal in China.

Take-Home Message

- Coal consists of buried plant material; heat and pressure at depth compact the material, and reactions allow volatile components to migrate away, leaving carbon behind.
- Geologists distinguish among grades of coal based on the carbon content; higher-grade coal contains a higher proportion of carbon and yields more energy when burned.
- Coal is a sedimentary rock that occurs in layers called seams; it can be mined in surface or subsurface mines.
- Coal aging naturally produces methane, an explosive hazard during mining but a source of energy if extracted in wells; coal itself can be transformed into burnable gases.

THINK: Why do underground coal fires start, and why are they so hard to stop?

14.7 NUCLEAR POWER

How Does a Nuclear Power Plant Work?

So far, we have looked at fuels (oil, gas, and coal) that release energy when chemically burned. During such burning, a chemical reaction between the fuel and oxygen releases the potential energy stored in the *chemical bonds* of the fuel. The energy that drives a nuclear power plant comes from a totally different process: the fission, or breaking, of the *nuclear bonds* that hold protons and neutrons together in the nucleus. Fission splits an atom into smaller pieces, and during this process, a small amount of mass transforms into thermal and electromagnetic energy.

Nuclear power plants were first built to produce electricity during the 1950s. A **nuclear reactor**, the heart of the plant, commonly lies within a dome-shaped shell (containment building) made of reinforced concrete (Fig. 14.16a, b). The reactor contains nuclear fuel, consisting of pellets of concentrated uranium oxide or a comparable radioactive material that are packed into metal tubes called fuel rods. Fission occurs when a neutron strikes a radioactive atom, causing it to split. For example, radioactive uranium-235 splits into barium-141 and krypton-92, plus three neutrons. The neutrons released during the fission of one atom strike other atoms, thereby triggering more fission in a self-perpetuating process called a **chain reaction** that, overall, produces lots of heat. Pipes carry water close to the heat-generating fuel rods, and the heat transforms the water into high-pressure steam. The pipes then carry this steam to a turbine, where it rotates fan blades; the rotation drives a dynamo that generates electricity. Eventually the steam goes into cooling towers, where it condenses back into water that can be reused in the plant or returned to the environment.

FIGURE 14.16 Producing electricity at a nuclear power plant.

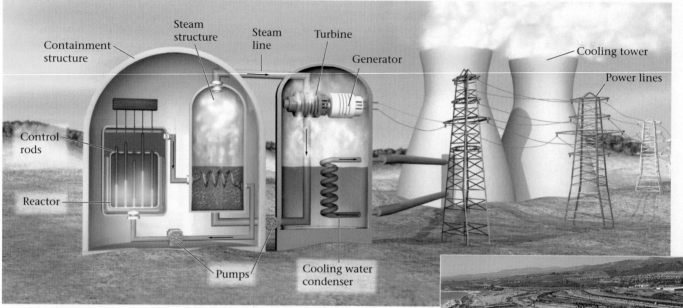

(a) In a nuclear power plant, a reactor heats water, which produces high-pressure steam. The steam drives a turbine that, in turn, drives a generator to produce electricity. A condenser transforms the steam back into water.

(b) This nuclear power plant in California has two reactors, each in its own containment building.

The Geology of Uranium

^{235}U, an isotope of uranium containing 143 neutrons, serves as the most common fuel for conventional nuclear power plants. But ^{235}U accounts for only about 0.7% of naturally occurring uranium; most uranium consists of ^{238}U, an isotope with 145 neutrons. Thus, to make a fuel suitable for use in a power plant, the ^{235}U concentration in a mass of natural uranium must be increased by a factor of 2 or 3, an expensive process called enrichment.

Where does uranium come from? The Earth's radioactive elements, including uranium, probably developed during the explosion of a supernova before the existence of the Solar System. Uranium atoms from this explosion became part of the nebula out of which the Earth formed and thus were incorporated into the planet. They gradually rose into the upper crust, carried in granitic magma.

Even though granite contains uranium, it does not contain very much. But nature has a way of concentrating uranium. Hot water circulating through a pluton after intrusion dissolves the uranium and precipitates it as the mineral pitchblende (UO_2) in veins. Uranium may be further concentrated once plutons, and the associated uranium-rich veins, weather and erode at the ground surface. Sand derived from a weathered pluton washes down a stream, and as it does so, uranium-rich grains stay behind because they are so heavy, relative to quartz and feldspar grains. The world's richest uranium deposits, in fact, occur in ancient streambed gravels. Uranium deposits may also form when groundwater percolates through uranium-rich sedimentary rocks; the uranium dissolves in the water and moves with the water to another location, where it precipitates out of solution and fills the pores of the host sedimentary rock.

Challenges of Using Nuclear Power

Maintaining safety at nuclear power plants requires work. Operators must constantly cool the nuclear fuel with circulating water, and the rate of nuclear fission must be regulated by the insertion of boron-steel control rods, which absorb neutrons and thus decrease the number of collisions between neutrons and radioactive atoms. Without control rods, the number of neutrons dashing around in the nuclear fuel would progressively increase, causing the rate of fission and accompanying heat production to increase. Eventually, the fuel would become so hot that it would melt. Such a **meltdown** might cause a steam explosion that could breach the containment building

Did you ever wonder...
could a nuclear power plant explode like an atomic bomb?

and scatter radioactive debris into the air. In some cases, the water gets so hot that it splits into hydrogen and oxygen gases that can recombine explosively. (Note that a meltdown is *not* the same as an atomic bomb explosion. The fuel of a reactor is not enriched enough for a "critical mass" to exist in which fission reactions happen so quickly that the fuel itself explodes.)

Globally, about 600 nuclear power plants are currently operating. To date, there have been three significant disasters. The first occurred in 1979 at the Three Mile Island plant in Pennsylvania. A stuck valve allowed coolant water to escape, causing the reactor to overheat, and eventually some of that radioactive coolant leaked out into the environment. The next, and most serious nuclear disaster occurred at the power plant in Chernobyl, Ukraine, in April 1986. The fuel pile became too hot, triggering a steam or hydrogen explosion that raised the roof of the containment building and spread fragments of the reactor and its fuel around the plant grounds. Within 6 weeks, 20 people had died from radiation sickness. Some radioactive material entered the atmosphere and dispersed over Eastern Europe and Scandinavia. No one yet knows the full effects of this fallout on the health of exposed populations. These days, there is concern that even a safely operated plant could pose a hazard if it became the target of a terrorist attack. In 2011, a disaster occurred at the multi-reactor Fukushima power plant along the east coast of northern Japan. This disaster, second only to Chernobyl in terms of the amount of radiation released, occurred when the catastrophic tsunami (giant wave) that followed the magnitude 9.0 Tohoku-Oki earthquake knocked out both the power lines providing electricity to pumps that circulated cooling water, and the backup diesel generators. As a result, some of the reactors overheated, and they suffered partial meltdowns and hydrogen gas explosions.

The nuclear industry also needs to manage nuclear waste. **Nuclear waste** is the radioactive material produced in a nuclear plant. It includes spent fuel, which contains radioactive elements, as well as water and equipment that have come in contact with radioactive materials. Radioactive elements emit gamma rays and X-rays that can damage living organisms and cause cancer. Some radioactive material decays quickly (in decades to centuries), but some remains dangerous for thousands of years or more.

Nuclear waste cannot just be stashed in a warehouse or buried in a town landfill. Some of the waste is hot enough that it needs to be cooled with water. Even cooler waste has the potential to leak radioactive elements into municipal water supplies or nearby lakes or streams. Ideally, waste should be sealed in containers that will last for thousands of years (the time needed for the short-lived radioactive atoms to undergo decay) and stored in a place where it will not come in contact with the environment. Finding an appropriate place is not easy. So far, experts disagree about which is the best way to dispose of nuclear waste. The U.S. government favored storing waste at Yucca Mountain in the Nevada desert, for the mountain consists of fairly dry rhyolite and the area is far from a population center. But currently, the idea is on hold.

Take-Home Message

- Nuclear power plants obtain energy by controlled fission reactions of elements such as uranium.
- Heat produced by fission boils water to produce steam that turns turbines of generators.
- Uranium can occur in several different types of deposits.
- Use of nuclear energy presents challenges—for example, how should nuclear waste be stored?

THINK: What is the difference between a meltdown in a reactor and the explosion of an atomic bomb?

14.8 OTHER ENERGY SOURCES

Biofuels

In recent years, farmers have begun to produce rapidly growing crops specifically for the purpose of producing biomass for fuel production. The resulting liquids are called **biofuels**.

The most commonly used biofuel is ethanol, a type of alcohol. Ethanol can substitute for gasoline in car engines. The process of producing ethanol from corn includes the following steps. First, producers grind corn into a fine powder, mix it with water, and cook it to produce a mash of starch. Then, they add an enzyme to the mash, which converts the mash into sugar. The sugar, when mixed with yeast, ferments. Fermentation produces ethanol and CO_2. Finally, the fermented mash is distilled to concentrate the ethanol. While corn is the main source of ethanol in North America, in Brazil, ethanol is produced directly from sugar extracted from sugar cane.

Researchers have also begun to develop processes that yield ethanol from cellulose, permitting perennial grasses to become a source of liquid fuel, or from algae, which naturally produces fatty chemicals from which hydrocarbons can be produced. Such processes could help make ethanol a renewable energy source. And recent technologies have also led to the commercial production of biodiesel, a fuel produced by chemical modification of fats and vegetable oils. Biodiesel can be mixed with petroleum-derived diesel to run trucks and buses.

Geothermal Energy

As the name suggests, **geothermal energy** comes from the internal heat of the Earth. Geothermal energy exists because the crust becomes progressively hotter with increasing depth.

FIGURE 14.17 Geothermal power plants utilize hot groundwater. In some cases, the water is so hot that it becomes steam when it rises and undergoes decompression. Most geothermal plants are in volcanic areas.

Where the geothermal gradient is steep (meaning temperature increases rapidly with depth), we find high temperatures at relatively shallow depths. Groundwater absorbs heat from hot rock and itself becomes hot.

Power companies use geothermal energy to produce heat and electricity in several ways. For example, in some places, they simply pump hot groundwater out of the ground and run it through pipes to heat houses and spas. Elsewhere, the groundwater is so hot that when it rises to the Earth's surface and decompresses, it turns to steam. This steam can drive turbines and generate electricity (Fig. 14.17). In volcanic areas such as Iceland or New Zealand, geothermal energy provides a major portion of energy needs. But on a global basis, it doesn't, because few cities lie near geothermal resources.

Hydroelectric and Wind Power

For millennia, people have used flowing water and air to produce energy. In fact, many towns were founded next to rivers, where streams could rotate the waterwheels that powered mills and factories. And in agricultural areas, farmers used windmills to pump water for irrigation. In the past century, engineers have begun to employ the same basic technology to drive generators that produce electricity. Energy derived from flowing fluids is clean ("green"), in the sense that its production does not release chemical or radioactive pollutants, and is renewable, in the sense that its production does not consume limited resources. But its use does impact the environment.

In a modern **hydroelectric power plant**, the potential energy of water is converted into kinetic energy as the water flows from a higher elevation to a lower elevation. The flowing water turns turbine blades placed in a pipe, and the turbine drives an electrical generator. Most hydroelectric plants rely on water from a reservoir held back by a dam (Fig. 14.18a). Dam construction increases the available potential energy of the water, because the water level in a filled reservoir is higher than the level of the valley floor that the dam spans.

Hydroelectric energy is clean, and reservoirs may have the added benefit of providing flood control, irrigation water, and recreational opportunities. But the construction of dams and reservoirs may bring unwanted changes to a region, so the benefits of their construction must be weighed against potential harm. Damming a river may flood a spectacular canyon, eliminate exciting rapids, or destroy an ecosystem. Further, reservoirs trap sediment and nutrients, thus disrupting the supply of these materials to downstream floodplains or deltas, a process that may adversely affect agriculture.

Not all hydroelectric power generation utilizes flowing river water—engineers have been developing new means to tap the energy in tides (the daily rises and fall of the sea; see Chapter 18). One approach involves building a dam, called a tidal barrage, across the entrance to a bay or estuary (the flooded mouth of a river) in which there are large tides. When the tide rises, water spills into the enclosed area through openings in the dam. When the tide outside drops, water flows back to the sea via a pipe that carries it through a power-generating turbine (Fig. 14.18b). More recently, engineers have developed technologies to place huge fan blades underwater in nearshore regions where tidal currents naturally flow; the blades slowly turn in the current and run generators.

FIGURE 14.18 Alternative energy sources.

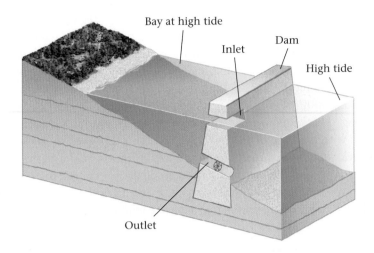

(a) Gravity causes water held back by the Grand Coulee Dam to flow through turbines and generate electricity.

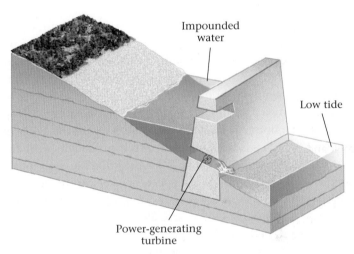

(c) A wind farm in southwestern England. The towers are about 50 m high.

(b) To produce tidal power, a dam traps seawater at high tide; the water can flow through a turbine at low tide.

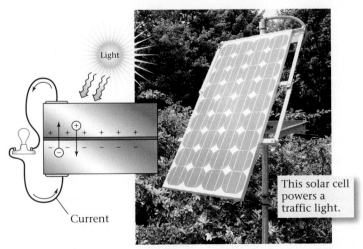

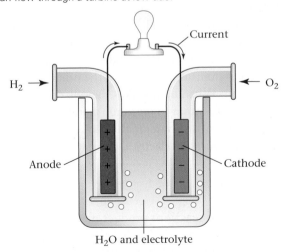

(d) A photovoltaic cell produces electricity directly from solar radiation.

(e) A hydrogen fuel cell.

When you think of wind energy, you may picture a classic Dutch windmill driving a water pump. Modern efforts to harness the wind are on a much larger scale. To produce wind-generated electricity, engineers identify regions, either onshore or just offshore, with steady breezes. In these regions, they build wind farms that consist of numerous towers, each of which holds a wind turbine, a giant fan blade that turns even in a gentle breeze (Fig. 14.18c). Some towers are on the order of 100 m (300 ft) tall, with fan blades that are over 40 m (120 ft) long. The largest wind farm currently has close to 200 towers. As with any type of energy source, wind power has some drawbacks. Cluttering the horizon with towers may spoil a beautiful view, and the constant loud hum of the towers can disturb nearby residents. The towers may also be a hazard to migrating birds, and offshore towers may interfere with marine life.

Solar Energy

The Sun drenches the Earth with energy in quantities that dwarf the amounts stored in fossil fuels. Were it possible to harness this energy directly, humanity would have a reliable and totally clean solution for powering modern technology. But using solar energy is not quite so simple. At present, energy consumers have two options for the direct use of solar energy.

The first option is a **solar collector**, a device that collects energy to produce heat. One class of solar collectors includes mirrors or lenses that focus light striking a broad area into a smaller area. On a small scale, such devices can be used for cooking; and on a large scale, they can produce steam to drive turbines. Another class of solar collectors consists of a black surface placed beneath a glass plate. The black surface absorbs light that has passed through the glass plate and heats up. The glass does not let the heat escape. When a consumer runs water between the glass and the black surface, the water heats up.

Photovoltaic cells (solar cells), the second option, convert light energy directly into electricity (Fig. 14.18d). Most photovoltaic cells consist of two wafers of silicon pressed together. One wafer contains atoms of arsenic, and the other wafer contains atoms of boron. When light strikes the cell, arsenic atoms release electrons that flow over to the boron atoms. If a wire loop connects the back side of one wafer to the back side of the other, this phenomenon produces an electrical current.

Fuel Cells

In a fuel cell, chemical reactions produce electricity directly. Let's consider a hydrogen fuel cell. Hydrogen gas flows through a tube across an anode (a strip of platinum) that has been placed in a water solution containing an electrolyte (a substance that enables the solution to conduct electricity). At the same time, a stream of oxygen gas flows onto a separate platinum cathode that has also been placed into the solution. A wire connects the anode and the cathode to provide an electrical circuit (Fig. 14.18e). In this configuration, hydrogen reacts with oxygen to produce water and electricity. Fuel cells are efficient and clean. Their limitation lies in the need for a design that will protect the cells from damage by impact and that will enable them to store hydrogen, an explosive gas, in a safe way. Also, it takes a significant amount of energy from other sources to produce the hydrogen used in fuel cells.

Take-Home Message

- Society is beginning to explore several alternatives to fossil fuels for producing energy.
- Options include use of renewable biofuels, hydroelectric and wind power, solar cells and solar collectors, and fuel cells.

THINK: Why is there no easy solution, at present, to providing sustainable energy sources for the future?

TABLE 14.2 Statistics of Oil Reserves, Oil Production, and Oil Consumption[1]

Country	Reserves[2] (billion bbl)	Production (millions bbl/day)	Consumption[3] (millions bbl/ day)
Saudia Arabia	265	8.5	1.4
Iraq	115	2.4	0.5
Kuwait	99	1.8	0.3
Iran	96	3.8	1.1
United Arab Emirates	63	2.6	0.3
Russia	54	7.0	2.5
Venezuela	48	3.1	0.5
China	31	3.3	4.9
Libya	30	1.4	0.2
Mexico	27	3.6	1.9
Nigeria	24	2.2	0.3
United States	22	8.1	20.0
Qatar	15	0.6	0.03
Norway	10	3.4	0.2
Algeria	9	1.5	0.2
Brazil	8	1.6	2.1
World	1,032	75.2	76.0

[1]Data comes from various sources. The countries listed are the top 15 in terms of total reserves; [2]Taking population into account: UAE has 39,359 bbl/person, Saudia Arabia has 10,875 bbl/person, Russia has 336 bbl/person, and the U.S. has 77 bbl/person. [3]The six largest consumers worldwide are U.S. (20), Japan (5.4), China (4.8), Germany (2.8), Russia (2.5), South Korea (2.1). Note that the largest consumers are not countries with large reserves.

14.9 ENERGY CHOICES, ENERGY PROBLEMS

The Oil Crunch

Energy usage in industrialized countries grew with dizzying speed through the mid-twentieth century, and during this time people came to rely increasingly on oil (Fig. 14.19a, b). The United States is the largest consumer of oil (at a rate of 7 million barrels per day, about 25% of world consumption), but lost its position as the largest producer in the 1970s. Oil reserves in the United States now account for only about 4% of the world total. Thus, today the United States must import more than half of the oil it uses. The industrialized countries use vastly more oil per capita than do the developing countries. For example, 1,000 people in the United States use 70 barrels per day. In China, 1,000 people use 4 barrels per day; in Ethiopia, the same number of people use only 0.3 barrels per day (see Table 14.2).

Eventually, oil supplies within the borders of industrialized countries could no longer match the demand, and these countries began to import more oil than they produced themselves. Through the 1960s, oil prices remained low (about $1.80 a barrel), so this was not a problem. In 1973, however, a complex tangle of politics and war led the Organization of Petroleum Exporting Countries (OPEC) to limit its oil exports. In the United States, fear of an oil shortage turned to panic, and motorists began lining up at gas stations, in many cases waiting for hours to fill their tanks. The price of oil rose to $18 a barrel, and newspaper headlines proclaimed, "Energy Crisis!" Governments in industrialized countries instituted new rules to encourage oil conservation. During the last two decades of the twentieth century, the oil market stabilized, though political events occasionally led to price jumps and short-term shortages. Since 2004, oil prices rose overall, passing the $147/bbl mark in 2008; but the price collapsed in late 2008 when the Great Recession hit. By 2011, the price was back over $100/bbl. Will

FIGURE 14.19 World energy reserves. Oil is the dominant source of energy, at present, but other sources have more abundant reserves.

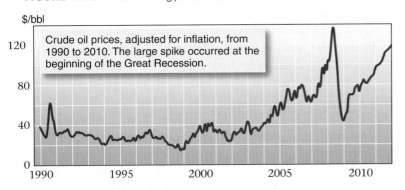

(a) Oil is currently the dominant source of energy worldwide. Recently, prices have been fluctuating dramatically.

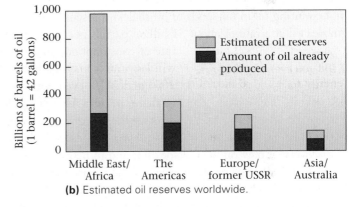

(b) Estimated oil reserves worldwide.

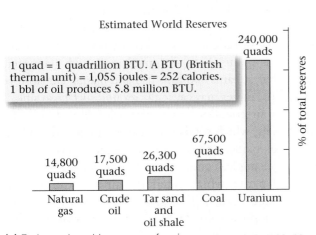

(c) Estimated world reserves of various energy resources, as measured in quads.

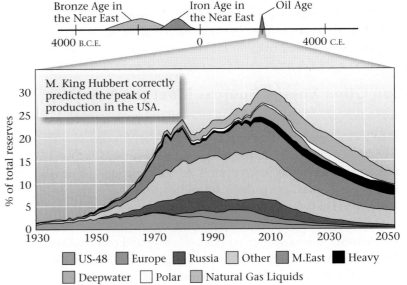

(d) The age of oil will have a relatively short duration. The graph depicts changes in oil production over time. The high point is called "Hubbert's Peak," after M.K. Hubbert, a geologist who first predicted that production would eventually decrease. Numbers after 2010 are predictions.

a day come when shortages arise not because of an embargo or limitations on refining capacity, but because there is no more oil to produce? As highly populous countries such as China and India industrialize, the use of fuels accelerates.

To understand the issues involved in predicting the future of energy supplies, we must first classify energy resources (**Geology at a Glance**, pp. 496–497). As noted earlier, we call a particular resource *renewable* if nature can replace it within a short time relative to a human life span (in months or, at most, decades). A resource is *nonrenewable* if nature takes a very long time (hundreds to perhaps millions of years) to replenish it. Oil is a nonrenewable resource, in that the rate at which humans consume it far exceeds the rate at which nature replenishes it, so we will inevitably run out of oil. The question is, when?

Historians in the future may refer to our time as the **Oil Age** because so much of our economy depends on oil. How long will the Oil Age last? Geologists estimate that as of 2000, there were about 850 to 1,000 billion barrels of proven reserves, meaning oil that had been found (**Fig. 14.19c**). There may be an additional 2,000 billion barrels not yet found. Thus, the world probably holds between 1,000 and 3,000 billion barrels of obtainable oil (not counting oil in tar sands or oil shales). Presently, humanity guzzles oil at a rate of about 31 billion barrels per year. At this

> **Did you ever wonder...**
> how much longer the world's oil supply will last?

rate of consumption, oil supplies will last until sometime between 2050 and 2150, even with more offshore drilling. Already, geologists see the beginning of the end of the Oil Age, for the rate of consumption now exceeds the rate of discovery of new oil by a factor of 3; and in many oil-bearing provinces, the rate of production has already started to decrease. In the end, the Oil Age will probably have lasted less than three centuries (**Fig. 14.19d**). On a time line representing the four thousand years since the construction of the Egyptian pyramids, this looks like a very short blip. We may indeed be living during a unique interval of human history.

Can Other Energy Sources Meet the Need?

Are there alternatives to oil? Perhaps. The world contains vast fossil-fuel supplies in other forms, enough to last for centuries if they can be exploited in an economical and environmentally sound way. These supplies include tar sand, oil shale, natural gas, and coal. All told, there are perhaps 1.5 trillion barrels of hydrocarbons trapped in tar sands and 3 trillion barrels trapped in oil shale. In addition, coal deposits could meet our energy needs for the next few centuries. But extracting tar sand, oil shale, and coal requires the construction of huge mines, which might scar the landscape and may negatively affect the environment as we'll discuss shortly. Natural gas may be our best alternative in the near future.

Can nuclear power or hydroelectric power fill the need? Vast supplies of uranium, the fuel of traditional nuclear plants, remain untapped; and nuclear engineers have designed alternative plants, powered by breeder reactors, that essentially produce new fuel. But many people view nuclear plants with concern because of issues pertaining to radiation, accidents, terrorism, and waste storage, and these concerns have slowed the industry. A substantial increase in hydroelectric power production is not likely, as most appropriate rivers have already been dammed, and industrialized countries have little appetite for taming any more. Similarly, the growth of geothermal-energy output seems limited.

Because of the problems that would result from relying more on coal, hydroelectric, and nuclear energy, researchers have been increasingly exploring clean energy options. One possibility is solar power. Unfortunately, affordable technologies for large-scale solar energy do not yet exist, and the idea of covering the landscape with solar cells is not attractive. Similarly, we can turn to wind power for relatively small-scale energy production, but covering the landscape with windmills is no more appealing. Fusion power may be possible, but physicists and engineers have not yet figured out a way to harness it. Clearly, society will be facing difficult choices in the not-so-distant future about where to obtain energy, and we will need to invest in the research required to discover new alternatives. In the near term, conservation can play an important role by diminishing demand for fuels.

Environmental Issues

Environmental concerns about energy resources begin right at the source. Oil drilling requires substantial equipment, the use of which can damage the land. And as demonstrated by the 2010 Gulf of Mexico offshore well blowout, oil drilling can lead to tragic loss of life and disastrous marine oil spills (**Box 14.4**). Oil spills from pipelines or trucks sink into the subsurface and contaminate groundwater, and oil spills from ships and tankers create slicks that spread over the sea surface and foul the shoreline (**Fig. 14.20a, b**). Coal and uranium mining also scar the land and can lead to the production of **acid mine runoff**, a dilute solution of sulfuric acid that forms when sulfur-bearing minerals such as pyrite (FeS_2) in mines react with rainwater. The runoff enters streams and kills fish and plants. Collapse of underground coal mines may cause the ground surface to sink.

Numerous air-pollution issues also arise from the burning of fossil fuels, which sends soot, carbon monoxide, sulfur dioxide, nitrous oxide, and unburned hydrocarbons into the air. Coal, for example, commonly contains sulfur, primarily in the form of pyrite, which enters the air as sulfur dioxide (SO_2) when coal is burned. This gas combines with rainwater to form dilute sulfuric acid (H_2SO_4), or **acid rain**. For this reason, many countries now regulate the amount of sulfur that coal can contain when it is burned.

But even if pollutants can be decreased, the burning of fossil fuels still releases carbon dioxide (CO_2) into the atmosphere.

FIGURE 14.20 Marine oil spills. These can come from drilling rigs, or from tankers.

(a) An oil tanker leaking oil on the sea surface.

(b) Oil spills can contaminate the shore, and can be very difficult to clean.

CO_2 is important because it traps heat in the Earth's atmosphere much like glass traps heat in a greenhouse—this is the **greenhouse effect**. A global increase of CO_2 will lead to a global increase in atmospheric temperature (global warming), which in turn may alter the distribution of climatic belts. Because of concern about CO_2 production, research efforts are under way to develop techniques to capture CO_2 at power plants, liquefy it, and pump it into reservoir rocks deep underground. This process is called **carbon sequestration**. We'll learn more about this issue in Chapter 23.

> ## Take-Home Message
>
> - Current estimates of reserves suggest that conventional oil reserves may last for only another 50 to 100 years; we live in the "age of oil" that may end up lasting less than 300 years.
> - Coal reserves are much more abundant, and along with natural gas and nuclear power may provide near-term substitutes for oil.
> - Use of fossil fuels produces pollutants, as well as CO_2, a greenhouse gas; new technologies may reduce emissions. CO_2 may be captured and sequestered underground.
>
> **THINK:** What is "Hubbert's peak," and when was it reached in the United States?

Chapter Summary

- Energy resources come in a variety of forms: energy directly from the Sun; energy from tides, flowing water, or wind; energy stored by photosynthesis (either in contemporary plants or in fossil fuels); energy from inorganic chemical reactions; energy from nuclear fission; and geothermal energy (from Earth's internal heat).
- Oil and gas are hydrocarbons, a type of organic chemical. The viscosity and volatility of a hydrocarbon depend on the length of its molecules.

- Oil and gas originally develop from the dead bodies of plankton and algae, which settle out in a quiet-water, oxygen-poor depositional environment and form black organic shale. Later, chemical reactions at elevated temperatures convert the dead plankton into kerogen and then oil.
- For an oil reserve to be usable, oil must migrate from a source rock into a porous and permeable rock called a reservoir rock. Unless the reservoir rock is overlain by an impermeable seal rock, the oil will escape to the ground surface. The subsurface configuration of strata that ensures the entrapment of oil in a good reservoir rock is called an oil trap.

Power from the Earth

Water, Wind, and Tides

The hydrologic cycle carries water over land. Water flows back toward the sea.

Forming and Mining Coal

Plants in coastal swamps and forests die, become buried, and transform into coal.

Coal at shallow depths can be accessed by strip mines.

Forming and Finding Oil

Plankton, algae, and clay settle to the floor of quiet water in a lake or sea. Eventually, the organic sediment becomes buried deeply and becomes a source rock. Chemical reactions yield oil, which percolates upward.

Tectonic processes form oil traps. Oil accumulates in reservoir rock within the trap; a seal rock keeps the oil underground.

Regardless of whether an oil reserve is under land or under sea, modern drilling technology can reach it and pump it.

Exploration for oil utilizes seismic-reflection profiling, which can reveal the configuration of layers underground.

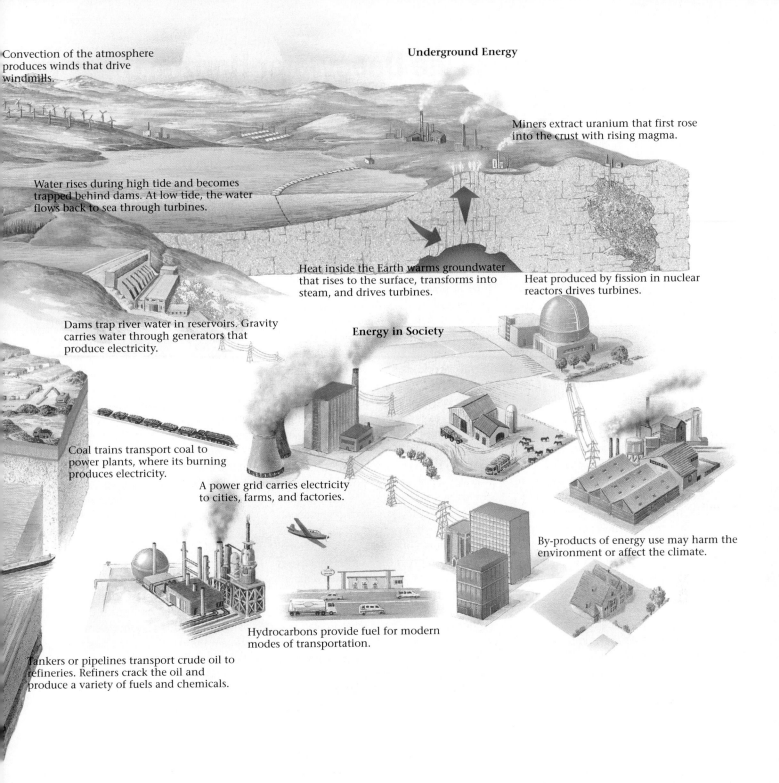

Convection of the atmosphere produces winds that drive windmills.

Miners extract uranium that first rose into the crust with rising magma.

Water rises during high tide and becomes trapped behind dams. At low tide, the water flows back to sea through turbines.

Heat inside the Earth warms groundwater that rises to the surface, transforms into steam, and drives turbines.

Heat produced by fission in nuclear reactors drives turbines.

Dams trap river water in reservoirs. Gravity carries water through generators that produce electricity.

Energy in Society

Coal trains transport coal to power plants, where its burning produces electricity.

A power grid carries electricity to cities, farms, and factories.

By-products of energy use may harm the environment or affect the climate.

Hydrocarbons provide fuel for modern modes of transportation.

Tankers or pipelines transport crude oil to refineries. Refiners crack the oil and produce a variety of fuels and chemicals.

Modern society, for better or worse, uses vast amounts of energy to produce heat, to drive modes of transportation, and to produce electricity (see Appendix). This energy comes either from geologic materials stored in the Earth or from geologic processes happening at our planet's surface. For example, oil and gas fill the pores of reservoir rocks at depth below the surface, coal occurs in sedimentary beds, and uranium concentrates in ore deposits. A hydroelectric power plant taps into the hydrologic cycle, windmills operate because of atmospheric convection, and geothermal energy comes from hot groundwater. Ultimately, the energy in the sources just listed comes from the Sun, from gravity, from Earth's internal heat, and/or from nuclear reactions. Oil, gas, and coal are fossil fuels because the energy they store first came to Earth as sunlight, long ago.

As energy usage grows, easily obtainable energy resources dwindle, the environment can be degraded, and the composition of the atmosphere has changed. The pattern of energy use that forms the backbone of society today may have to change radically in the not-so-distant future, if we wish to avoid a decline in living standards.

BOX 14.4

Offshore Drilling and the Deepwater Horizon Disaster

A substantial proportion of the world's oil reserves reside in the sedimentary basins that underlie the continental shelves of passive continental margins. To access such reserves, oil companies must build offshore drilling platforms. Offshore drilling began in the early twentieth century in water less than 10 m (33 feet) deep, using drills placed on raised platforms or anchored barges. Drilling in deeper water began in the 1960s. Such drilling requires complex technology. In water less than 600 m (2,000 ft) deep, companies position fixed platforms on towers resting on the sea floor. In deeper water, semi-submersible platforms float on huge submerged pontoons (**Fig. Bx14.4a**). With these, oil companies can now access fields lying beneath 3 km (10,000 ft) of water.

North America's largest offshore fields occur in the passive-margin basin that fringes the coast of the Gulf of Mexico (**Fig. Bx14.4b**). This basin started forming in Jurassic time, subsequent to the breakup of Pangaea, and its floor has been slowly sinking ever since. Up to 15 to 20 km (9 to 12 miles) of sediment (mud, salt, sand, marine shells) fills the basin. Some of the mud was rich in organic matter; when buried deeply, this converted to hydrocarbons that now fill pores of reservoir rocks in salt-dome traps and stratigraphic traps (see Box 14.1). More than 3,500 platforms operate in the Gulf at present, together yielding up to 1.7 million bbl/day.

During both onshore and offshore exploration, drillers worry about the possibility of a **blowout**. A blowout happens when the pressure within a hydrocarbon reserve penetrated by a well exceeds the pressure that drillers had planned for, causing the hydrocarbons (oil and/or gas) to rush up the well in an uncontrolled manner and burst out of the well at the surface in an oil gusher or gas plume, which can explode and burn if ignited. Blowouts are rare because, although fluids below the ground are

under great pressure due to the weight of overlying material, engineers always fill the hole with drilling mud (a mixture of water, clay, and other chemicals) with a density greater than that of clear water. The weight of drilling mud can counter the pressure of underground hydrocarbons and hold the fluids underground. But if drillers encounter a bed in which pressures are unexpectedly high, or if they remove the mud before the walls of the well have been sealed with a casing (a pipe, cemented in place by concrete, that lines the hole) a blowout may happen.

A catastrophic blowout occurred on April 20, 2010, when drillers on the Deepwater Horizon, a huge semi-submersible platform leased by BP, were finishing a 5.5-km-long (18,000 ft) hole in 1.5-km-deep (5,000 ft) water south of Louisiana. Due to a series of errors, the casing was not sufficiently strong when workers began to replace the drilling mud with clear water. Thus the high-pressure, gassy oil in the reservoir that the well had punctured rushed up the drill hole. A backup safety device called a blowout preventer failed, so the gassy oil reached the platform and sprayed 100 m (328 ft) into the sky. Sparks from electronic gear triggered an explosion, and the platform became a fountain of flame and smoke that killed 11 workers. An armada of fireboats could not douse the conflagration (**Fig. Bx14.4c**), and after 36 hours, the still-burning platform tipped over and sank.

Robot submersibles sent to the sea floor to investigate found that the twisted mess of bent and ruptured pipes at the well head was billowing oil and gas (**Fig. Bx14.4d**). Estimates as to the amount of hydrocarbons released vary wildly, but most data suggests that on the order of 50,000 to 62,000 bbl/day of hydrocarbons entered the Gulf's water from the well. (By comparison, natural oil seeps at many localities around the Gulf together yield about 1,000 to 4,000 bbl/day.) Stopping

this underwater gusher proved to be an immense challenge, and initial efforts to block the well, or to put a containment dome over the well head, failed. It was not until July 15, almost three months after the blowout, that the flow was finally stopped, and not until September 19 that a new relief well intersected the blown well and provided a conduit to pump concrete down to block the original well permanently.

All told, about 4.2 million bbl of hydrocarbons contaminated the Gulf from the Deepwater Horizon blowout. (By comparison, the 1979 breakup of the *Exxon Valdez* supertanker released about 0.5 million bbl of oil; the largest tanker spill in history released about 2.0 million bbl; the largest on-land gusher released about 9 million bbl in California in 1911; and the destruction of Kuwaiti oil wells at the end of the 1991 war released over 1 billion bbl). The more volatile components of the spill evaporated, but enough liquid hydrocarbons remained to form a slick that, at times, covered an area of over 100,000 sq km (**Fig. Bx14.4e**). In addition, a huge plume of dispersed oil remained suspended in the water column beneath the surface. Thousands of people labored to contain the damage—floating booms were set up to contain the oil, skimmer ships traversed the slick to suck up oil, sandbag barriers were placed along the coast from Louisiana to Florida to keep the oil out of wetlands and beaches. Also, planes spread chemical dispersants on the floating oil to break it into tiny particles. (Of note, the use of dispersants remains controversial because, though successful at breaking up the slick, dispersants consist of chemicals that may in themselves be toxic.) Nevertheless, the spill was devastating to wetlands, wildlife, and the fishing and tourism industries. In the end, only natural processes—microbes that eat the oil droplets—can conquer the spill. But how long this process will take remains unknown.

FIGURE Bx14.4 Offshore drilling and the Deepwater Horizon disaster.

(a) The Deepwater Horizon platform being transported to a drilling site.

(c) Fireboats dousing the burning rig, before the rig sank.

(b) Drilling on the Gulf Coast margin. Most drill sites are on the continental shelf and; more recently, deeper water sites have been explored. The roughness of the slope to the abyssal plain is due to salt domes. Each yellow dot is a drilling platform. The location of the Deepwater Horizon platform is indicated by the larger, white dot.

(d) A plume of oil billows into the water from the well head, as viewed by underwater cameras.

(e) A satellite image showing the oil slick in the Gulf of Mexico, southeast of the Mississippi Delta.

- Substantial volumes of hydrocarbons also exist in tar sand, oil shale, and gas hydrate.

- For coal to form, abundant plant debris must be deposited in an oxygen-poor environment so that it does not completely decompose. Compaction near the ground surface creates peat, which, when buried deeply and heated, transforms into coal. Coal has a high concentration of carbon.

- Geologists rank coal based on the amount of carbon it contains: lignite (low rank), bituminous (medium rank), and anthracite (high rank).

- Coal occurs in beds, interlayered with other sedimentary rocks. Coal beds can be mined by either strip mining or underground mining.

- Coalbed methane and coal gasification provide additional sources of energy.

- Nuclear power plants generate energy by using the heat released from the nuclear fission of radioactive elements. The heat turns water into steam, and the steam drives turbines.

- Some economic uranium deposits occur as veins in igneous rock bodies; some are found in sedimentary beds.

- Nuclear reactors must be carefully controlled to avoid overheating or meltdown. The disposal of radioactive nuclear waste can create environmental problems.

- Geothermal energy uses Earth's internal heat to transform groundwater into steam that drives turbines; hydroelectric power uses the potential energy of water; and solar energy uses solar cells to convert sunlight to electricity.

- We now live in the Oil Age, but oil supplies may last only for another century. Natural gas may become a major energy supply in the near future.

- Most energy resources have environmental consequences. Oil spills pollute the landscape and sea, and the sulfur associated with some fuel deposits causes acid mine runoff. The burning of coal can produce acid rain, and the burning of coal and hydrocarbons produces smog and can contribute to global warming.

GEOPUZZLE REVISITED

Hydrocarbons (oil and gas) and coal are fossil fuels, meaning that they store the energy that was produced by photosynthesis millions to hundreds of millions of years ago. As such, they are *nonrenewable resources*, because geologic processes cannot replace them as fast as humans consume them. Conventional hydrocarbon supplies may last for only another 50 to 150 years. There are other sources of hydrocarbons, but they are expensive to utilize. Coal supplies can last longer, but regardless of supply, the use of fossil fuels has significant environmental consequences ranging from landscape destruction to global warming. To sustain standards of living, humanity will need to switch to alternative resources, which will take time.

Guide Terms

acid mine runoff (p. 494)
acid rain (p. 494)
anthracite coal (p. 484)
biodegradation (p. 480)
biofuel (p. 489)
bituminous coal (p. 484)
blowout (p. 498)
carbon sequestration (p. 495)
chain reaction (p. 487)
coal (p. 482)
coal gasification (p. 486)
coal rank (p. 484)
coal reserves (p. 484)
coal swamps (p. 482)
coalbed methane (p. 485)
directional drilling (pp. 476, 481)
distillation column (p. 477)
energy (p. 467)
energy resource (p. 467)
fossil fuels (p. 469)
fuel (p. 467)
gas hydrate (p. 482)
geothermal energy (pp. 469, 489)
greenhouse effect (p. 495)
hydraulic fracturing (p. 481)
hydrocarbon generation (p. 470)
hydrocarbon reserve (p. 470)
hydrocarbon system (p. 470)

hydrocarbons (p. 469)
hydroelectric power plant (p. 490)
hydrofracturing (p. 476)
kerogen (p. 470)
lignite (p. 484)
meltdown (p. 488)
migration pathway (p. 472)
nuclear reactor (p. 487)
nuclear waste (p. 489)
Oil Age (p. 494)
oil seep (p. 472)
oil shale (pp. 470, 482)
oil window (p. 470)
peat (p. 483)
permeability (p. 472)
photosynthesis (pp. 468–469)
photovoltaic cells (p. 492)
pores (p. 471)
porosity (p. 471)
reservoir rock (p. 471)
resource (p. 467)
salt dome (p. 474)
seal rock (p. 472)
seismic-reflection profile (p. 474)
solar collector (p. 492)
source rock (p. 470)
tar sand (p. 480)
trap (p. 472)

Review Questions

1. What are the fundamental sources of energy?
2. How does the length of a hydrocarbon chain affect its viscosity and volatility?
3. What is the source of the organic material in oil?
4. What is the oil window, and what happens to oil at temperatures higher than the oil window?
5. How is organic matter trapped and transformed to yield an oil reserve?
6. What are the different kinds of oil traps?

7. What are tar sand and oil shale, and how can oil be extracted from them?

8. What are gas hydrates, and where do they occur?

9. How do porosity and permeability affect the oil-bearing potential of a rock?

10. Where is most of the world's oil found? At present rates of consumption, how long will oil supplies last?

11. How is coal formed, and in what class of rocks is coal considered to be?

12. What conditions cause coal to transform in rank from peat to anthracite?

13. What are some of the environmental drawbacks of mining and burning coal?

14. What is coalbed methane, and how is it extracted?

15. Describe the nuclear reactions in a nuclear reactor and the means that engineers use to control reaction rates.

16. Where does uranium form in the Earth's crust? Where does it usually accumulate in minable quantities?

17. What are some of the drawbacks of nuclear energy?

18. Discuss the pros and cons of alternative energy sources.

19. What is geothermal energy? What limits its use?

20. What is the difference between renewable and nonrenewable resources?

21. What are the major environmental consequences of producing and burning fossil fuels?

22. What is the likely future of hydrocarbon production and use in the twenty-first century?

On Further Thought

23. Much of the oil production in the United States takes place at offshore platforms along the coast of the Gulf of Mexico. Consider the geologic setting of the Gulf Coast, in the context of the theory of plate tectonics, and explain why an immensely thick sequence of sediment accumulated in this region, and why so many salt-dome traps formed.

24. Ethanol can potentially be used as an alternative to petroleum as a liquid fuel. Ethanol can be produced by processing corn, sugar cane, or certain perennial grasses (e.g., switch-grass, or *Miscanthus*). What factors should be considered in determining which of these crops would be the most appropriate for use as a source of ethanol in North America?

 For more resources, including animations, quizzes, and Norton's GeoTours, go to **wwnorton.com/studyspace**.

 If your instructor assigns exercises in SmartWork, log in at **smartwork.wwnorton.com**.

ANOTHER VIEW Solar-panel arrays are beginning to carpet the landscape in sunny regions. This example (the Beneixama photovoltaic power plant in Spain) generates 20 megawatts of electricity at peak production, and covers 500,000 square meters (= 0.5 square km, or 123 acres). By comparison, a typical nuclear power plant produces 1,000 megawatts of electricity (enough for a city of 1.2 million). A solar array large enough to produce as much electricity as a nuclear power plant would have to cover an area of about 25 square km.

CHAPTER **15**

Riches in Rock: Mineral Resources

This quarry in northern Illinois produces crushed stone. Truckload by truckload, what began as a living reef hundreds of millions of years ago, and lay buried as bedrock until very recently, becomes the roadbed or building foundation of tomorrow.

GEOPUZZLE

Your lifestyle depends on a great variety of materials—bricks, copper, concrete, steel, gravel, and aluminum are all essential for the construction of buildings, roads, cars, and computers. Where do these materials come from?

Truth, like gold, is to be obtained not by its growth, but by washing away from it all that is not gold.

—Leo Tolstoy (Russian author, 1828–1910)

15.1 INTRODUCTION

In June 1845, over a year after leaving Missouri, James Marshall arrived by horse at Sutter's Fort, in central California, to make a new life. Having just finished a stint as a rancher and a few months as a soldier, Marshall decided to go into the lumber trade and convinced Captain John Sutter to finance the construction of a sawmill in the foothills of the Sierra Nevada mountain range. Marshall's scruffy crew finished the mill by the beginning of 1848. As Marshall stood admiring the new building, he noticed a glimmer of metal in the gravel that littered the bed of the adjacent stream. He picked it up, banged it between two rocks to test its hardness, and shouted, "Boys, by God, I believe I have found a gold mine!" For a short while, Marshall and Sutter managed to keep the discovery secret. But word of the gold soon spread, and within weeks the workers at Marshall's mill had disappeared into the mountains to seek their own fortunes.

All through that year, gold fever spread throughout California and eventually reached the East Coast. As a result, 1849 brought 40,000 prospectors to California. These "forty-niners," as they were called, had abandoned their friends and relatives on the gamble that they could strike it rich (Fig. 15.1). Many traveled for months by sea, sailing through stormy waters around the southern tip of South America to reach San Francisco, a town of mud streets and plank buildings. In many cases, crews abandoned their ships in the harbor to join the scramble to the gold fields. Although $20 million worth of gold was

mined in 1849, few of the forty-niners actually became rich. Most lost whatever wealth they found in the saloons and gambling halls that sprouted around the gold fields.

Gold is but one of many **mineral resources**, meaning minerals extracted from the Earth's crust for use by people. Without these resources, industrialized societies could not function. Geologists divide mineral resources into two categories: metallic mineral resources (materials containing gold, copper, aluminum, iron, etc.) and nonmetallic mineral resources (building stone, gravel, sand, gypsum, phosphate, salt, etc.). In this chapter, we look at the nature of mineral resources, the geologic phenomena responsible for their formation, and the ways people mine them. We conclude by considering the sustainability of mineral reserves.

Chapter Themes

By the end of this chapter, you should understand that . . .

- most materials used in everyday life come, originally, from geologic sources.
- metals are obtained from ore deposits containing ore minerals.
- ore deposits form in many ways and can be found in igneous and sedimentary environments.
- nonmetallic resources include stone, sand, and various salts.
- the sustainability of mineral supplies may be problematic.
- production and use of mineral resources can have environmental consequences.

15.2 METALS AND THEIR DISCOVERY

What Is a Metal?

Metals are opaque, shiny, smooth solids that can conduct electricity and can be bent, drawn into wire, or hammered into thin sheets. They look and behave quite differently from wood, plastic, meat, or rock. This is because, unlike in other substances, the atoms that make up metals are held together by metallic bonds, so electrons can flow from atom to atom fairly easily (see Chapter 5). Despite the mobility of their electrons, metals are solids, so their atoms lie fixed in a regular lattice defining a crystal structure. For example, the atoms in pure copper are arranged in wafer-like layers (Fig. 15.2a).

Not all metals behave the same way. Some are noted for their strength, others for their hardness, and others for their malleability (how easily they can be bent or molded). The behavior of a metal depends on the strength of bonds between atoms and on its crystalline structure. When you bend or stretch

FIGURE 15.1 Forty-niners mining in the Sierra Nevada range.

FIGURE 15.2 What makes a metal, a metal?

(a) A copper crystal consists of wafer-like sheets of copper atoms.

(b) Bending of the crystal can take place because the rows can easily slide past each other. Metals can be rolled into thin sheets without breaking.

Pennies

Sheets of rolled copper are stamped to make coins.

a strip of copper, for example, the wafer-like layers of atoms slip past each other quite readily (Fig. 15.2b). The shape and dimensions of crystals in a piece of metal are determined by how fast the molten metal cools and solidifies when removed from a furnace, by tempering (alternate heating and cooling), and by cold working (manipulating the shape of the metal after it is cooled). Before the development of the modern science of metallurgy, metalworkers learned how to obtain the qualities they wanted in a metal object simply by trial and error. Skilled metalworkers passed trade secrets from master to apprentice for generations. But now engineers with a detailed knowledge of chemistry can control the properties of a metal product.

The Discovery of Metals

Certain metals—namely, copper, silver, gold, and mercury—can occur in rock as **native metals** (Fig. 15.3a, b). Native metals consist only of metal atoms, and thus look

and behave like metal. Prehistoric hunters collected nuggets of native metal from stream beds and pounded them together with stone hammers to make tools. Eventually, because native metals are rare and durable, people began to use them as money. Gold, in particular, attained popularity as a currency because of its unique warm yellow glow and its resistance to tarnish or rust. Though iron does not occur as a native metal in rocks, chunks of iron fall to Earth from space in the form of meteorites. Ancient peoples used meteorite iron for toolmaking when they could find it.

If we had to rely solely on native metals as our source of metal, we would have access to only a tiny fraction of our current metal supply. Most of the metal atoms we use today occurred as ions bonded to nonmetallic elements in a great variety of minerals that themselves look nothing like metal. Only because of the chance discovery by some prehistoric genius that certain rocks, when heated to high temperatures in fire (a process now called **smelting**), decompose to

FIGURE 15.3 Examples of native metals.

Gold bracelets on display in a Kuwaiti marketplace.

(a) Gold occurs as native metal within quartz veins. The quartz breaks up to form sand, leaving nuggets of gold.

(b) Native copper occurs in complex shapes, which remain when the surrounding rock weathers away. Note the penny for scale.

yield metal plus a nonmetallic residue called slag do we now have the ability to produce sufficient metal for the needs of industrialized society.

Of the three principal metals in use today—copper, iron, and aluminum—copper began to be used first, because copper smelting from sulfide minerals is relatively easy. Copper implements appeared as early as 4000 B.C.E. Pure copper has limited value for tool- or weapon-making because it is too soft to retain a sharp edge. By chance, around 2800 B.C.E., Sumerian craftsmen discovered that copper could be mixed with tin to produce bronze, an **alloy** (a compound containing two or more metals) whose strength exceeds that of either metal alone, and warriors came to rely on bronze for their swords.

Iron proves superior to copper or bronze for many purposes, because of its strength, hardness, and abundance. Still, people didn't start using iron widely until 1,500 years after they had begun to use bronze. The delay was due, in part, to the fact that iron has a very high melting temperature that can be difficult to work with. But, in addition, the metal generally occurs in iron-oxide minerals (such as hematite, Fe_2O_3), and the liberation of iron metal from oxide minerals requires a chemical reaction, not just simple heating. The widespread

use of iron became possible only after someone discovered that metallic iron can be produced by heating iron-oxide minerals in the presence of carbon monoxide gas, which comes from burning charcoal (**Fig. 15.4a, b**). Chemists describe this reaction by the formula: $Fe_2O_3 + 3CO \rightarrow 2 Fe + 3CO_2$. More recently, people learned to make steel, an alloy of iron and carbon, and stainless steel, an alloy of iron and chromium, that resists corrosion.

Aluminum is abundant in rocks of the crust and, for many uses, is preferable to iron because it weighs less. But it does not occur in native form, and the extraction of aluminum from minerals requires complex methods using lots of electricity, and have become economically practical only since the 1880s.

These days, in addition to iron, copper, tin, and aluminum, we use a vast array of different metals. Some are known as **precious metals** (gold, silver, and platinum) and others as **base metals** (copper, lead, zinc, and tin) because of the difference in their price. Of the 63 or so metals in use today, people knew of only nine (gold, copper, silver, mercury, lead, tin, antimony, iron, and arsenic) before the year 1700.

> **Did you ever wonder . . .**
> where the iron in the steel bodies of cars comes from?

FIGURE 15.4 The smelting of iron ore in a blast furnace requires temperatures of 1,250°C to 1,400°C. Workers use very hot air to heat iron ore, coke (carbon), and limestone. A chemical reaction ($Fe_2O_3 + 3CO \rightarrow 2Fe + 3CO_2$) produces iron metal.

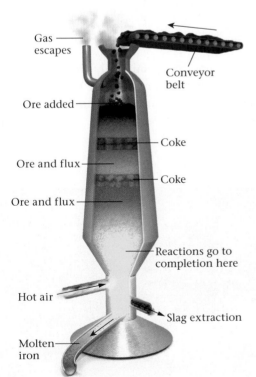

(a) Molten iron sinks to the bottom and "slag" (silicates and oxides) floats. "Flux" decreases the melting point of ore.

(b) In a factory, huge tubs pour the end product, molten iron.

15.3 ORES, ORE MINERALS, AND ORE DEPOSITS

What Is an Ore?

As we have seen, metals occur in two forms—as native elements (copper, silver, or gold) or as ions bonded to nonmetallic elements in a great variety of minerals, many of which can be found in common rocks. For example, if you pick up a chunk of common granite and analyze its mineral content, you will find that it consists mostly of quartz, plagioclase, and potassium feldspar—potassium feldspar ($KAlSi_3O_8$) contains about 8% aluminum. But even though common granite includes metal ions, we don't mine granite to produce these metals. Why? Simply because the minerals in granite contain relatively little metal, so it would cost more to separate the metals from granite than you would get if you were to sell them. To obtain the metals needed for industrialized society, we mine **ore**, rocks containing a significant concentration of native metals or ore minerals.

Ore minerals (or economic minerals) are minerals that contain metal in high concentrations and in a form that can be easily extracted. Galena (PbS), for example, is about 50% lead, so we consider it to be an ore mineral of lead (Fig. 15.5a). We obtain most of our iron from oxide minerals such as hematite and magnetite, so these minerals are iron ores. Geologists have identified many different kinds of ore minerals (Table 15.1). As you can see from the chemical formulas, many ore minerals are sulfides, in which the metal occurs in combination with sulfur (S), or oxides, in which the metal occurs in combination with oxygen (O). Numerous ore minerals are colorful and come in interesting shapes, and some have a metallic luster (Fig. 15.5b).

To be an ore, a rock must contain a sufficient amount of ore minerals to make the rock worth mining. For example, iron constitutes about 6.2% of the continental crust's weight, whereas it makes up 30 to 60% of iron ore (Fig. 15.6). We refer to the concentration of a useful metal in an ore as the **grade** of the ore—the higher the concentration, the higher the grade. Whether or not ore of a certain grade is worth mining depends on the price of metal in the market. For example, in 1880, copper-bearing rocks needed to contain at least 3% copper to be considered "economic ore" (ore worth mining), but as of 1970, rock containing only about 0.3% copper can be considered economic. This change reflects the efficiency of new technologies for mining and processing ore.

How Do Ore Deposits Form?

Ore minerals do not occur uniformly through rocks of the crust. If they did, we would not be able to extract them economically. Fortunately for humanity, geologic processes concentrate these minerals in ore deposits. Simply put, an **ore**

FIGURE 15.5 Examples of ore minerals.

(a) This lead ore, from Missouri, consists of galena (PbS) crystals that grew in dolostone.

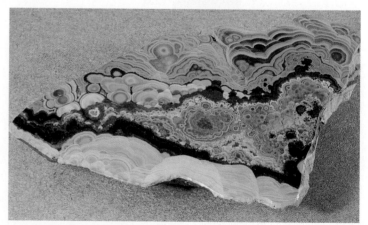

(b) Most copper comes from ore minerals that look nothing like metallic copper. This ore consists of azurite (blue) and malachite (green).

TABLE 15.1 Some Common Ore Minerals

Metal	Mineral Name	Chemical Formula
Copper	Chalcocite	Cu_2S
	Chalcopyrite	$CuFeS_2$
	Bornite	Cu_5FeS_4
	Azurite	$Cu_3(CO_3)_2(OH)_2$
	Malachite	$Cu_2(CO_3)(OH)_2$
Iron	Hematite	Fe_2O_3
	Magnetite	Fe_3O_4
Tin	Cassiterite	SnO_2
Lead	Galena	PbS
Mercury	Cinnabar	HgS
Zinc	Sphalerite	ZnS
Aluminum	Kaolinite	$Al_2Si_2O_5(OH)_4$
	Corundum	Al_2O_3
Chrome	Chromite	$(Fe,Mg)(Cr,Al,Fe)2O_4$
Nickel	Pentlandite	$(Ni,Fe)_9S_8$
Titanium	Rutile	TiO_2
	Ilmenite	$FeTiO_3$
Tungsten	Sheelite	$CaWO_4$
Molybdenum	Molybdenite	MoS_2
Magnesium	Magnesite	$MgCO_3$
	Dolomite	$CaMg(CO_3)_2$
Manganese	Pyrolusite	MnO_2
	Rhodochrosite	$MnCO_3$

FIGURE 15.6 The difference between ore and other rock. Iron in a block of granite is not worth extracting.

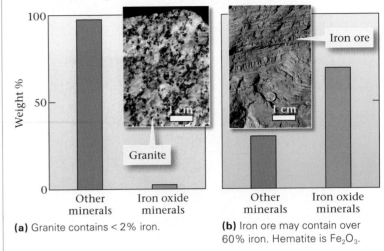

(a) Granite contains < 2% iron.

(b) Iron ore may contain over 60% iron. Hematite is Fe_2O_3.

deposit is an economically significant occurrence of ore. The various kinds of ore deposits differ from each other in terms of which ore minerals they contain and which kind of rock body they occur in. Below, we introduce a few examples.

Magmatic deposits. When a magma cools, sulfide ore minerals crystallize early and then, because sulfides tend to be dense, sink to the bottom of the magma chamber, where they accumulate; this accumulation is a **magmatic deposit**. When the magma freezes solid, the resulting igneous body may contain a concentration of sulfide minerals at its base. Because of their composition, we consider these concentrations to be a type of **massive-sulfide deposit** (Fig. 15.7a).

Hydrothermal deposits. Hydrothermal activity involves the circulation of hot-water solutions through a magma or through the rocks surrounding an igneous intrusion. These fluids dissolve metal ions. When a solution enters a region of lower pressure, lower temperature, different acidity, and/or different availability of oxygen, the metals come out of solution and form ore minerals

that precipitate in fractures and pores, creating a **hydrothermal deposit** (Fig. 15.7b). Such deposits may form within an igneous intrusion or in surrounding country rock. If the resulting ore minerals disperse through the intrusion, we call the deposit a *disseminated* deposit, but if they precipitate to fill cracks in preexisting rock, we call the deposit a *vein* deposit (veins are mineral-filled cracks; Fig. 15.7c). Hydrothermal copper deposits commonly occur in porphyritic igneous intrusions; these examples are known as porphyry copper deposits. Typically, vein deposits include quartz in addition to the ore minerals. For example, native gold commonly appears as flakes in milky-white quartz veins.

In recent years, geologists have discovered that hydrothermal activity at the submarine volcanoes along mid-ocean ridges leads to the eruption of hot water, containing high concentrations of dissolved metal and sulfur, from a vent. When this hot water comes in contact with cold seawater, the dissolved components instantly precipitate as tiny crystals of metal-sulfide minerals (Fig. 15.7d). The erupting water, therefore, looks like a black cloud, so the vents are called black smokers (see Chapter 4). The minerals in the cloud eventually sink and form a pile of ore minerals around the vent. Since the ore minerals typically are sulfides, the resulting hydrothermal deposits constitute another type of massive-sulfide deposit.

Secondary-enrichment deposits. Sometimes groundwater passes through ore-bearing rock long after the rock first formed. This groundwater dissolves some of the ore minerals and carries the dissolved ions away. When the water eventually flows into a different chemical environment (e.g., one with a different amount of oxygen or acid), it precipitates new ore minerals, commonly in concentrations exceeding that of the original deposit. A new ore deposit formed from metals that were dissolved

FIGURE 15.7 Various processes that form ore deposits.

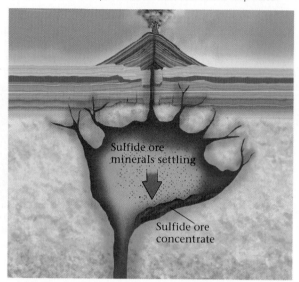

(a) Massive sulfide deposits can form when sulfide ore minerals sink to the bottom of a magma chamber.

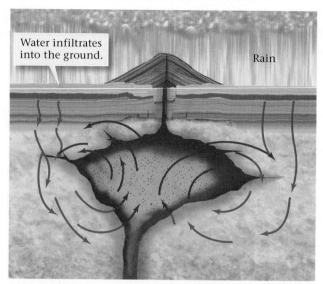

(b) Hydrothermal deposits form when water circulating around and through magma dissolves and redistributes metals (arrows indicate flowing water).

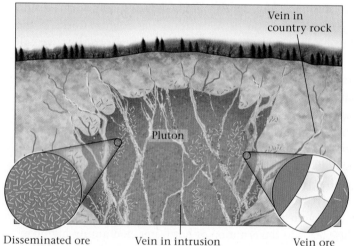

(c) Disseminated ore consists of a rock in which ore minerals are dispersed. Veins consist of minerals precipitated in cracks.

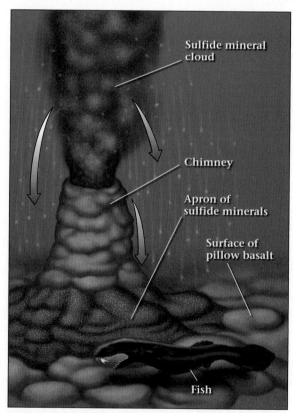

(d) Massive-sulfide deposits also form when ore minerals precipitate around hydrothermal vents (black smokers) along a mid-ocean ridge.

and carried away from a preexisting ore deposit is called a **secondary-enrichment deposit** (Fig. 15.8a, b). Some of these deposits contain spectacularly beautiful copper-bearing carbonate minerals, such as azurite and malachite (Fig. 15.5).

MVT ores. Geologists have discovered that some groundwater beneath mountain belts sinks several kilometers down into the crust, and follows a curving flow path that eventually brings it back to the surface perhaps hundreds of kilometers away from the mountain front. At the depths reached by the groundwater, temperatures become high enough that the water dissolves metals. When the water returns to the surface and enters cooler rock, these metals precipitate in ore minerals.

Ore deposits formed in this way, containing lead- and zinc-bearing minerals, appear in dolomite beds of the Mississippi Valley region, and thus have come to be known as **Mississippi Valley–type (MVT) ores**.

FIGURE 15.8 Secondary enrichment occurs when oxidizing groundwater leaches (extracts) metals from a rock and moves them elsewhere into reducing conditions where they precipitate.

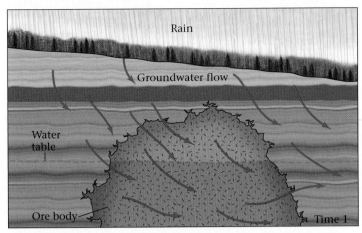

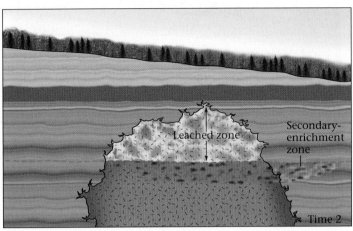

(a) Downward-flowing water dissolves ore minerals. The metal-rich solution moves down below the water table.

(b) Below the water table, the chemical environment changes, and new ore minerals precipitate.

Sedimentary deposits of metals. Some ore minerals accumulate in sedimentary environments under special circumstances. For example, between 2.5 and 1.8 billion years ago, the atmosphere, which previously had contained very little oxygen, gained oxygen because of the evolution of abundant photosynthetic organisms. This change affected the chemistry of seawater so that large quantities of dissolved iron precipitated as iron oxide minerals that settled as sediment on the sea floor. As we learned in Chapter 13, the resulting iron-rich sedimentary deposits are known as **banded iron formations (BIF)** (Fig. 15.9a), because after lithification they consist of alternating beds of gray iron oxide (magnetite or hematite) and red beds of jasper (iron-rich chert). Microbes may have participated in the precipitation process.

The chemistry of seawater in some parts of the ocean today leads to the deposition of manganese-oxide minerals on the sea floor. These minerals grow into lumpy accumulations known as **manganese nodules** (Fig. 15.9b). Mining companies have begun to explore technologies for vacuuming up these nodules; geoscientists estimate that the worldwide supply of nodules contains 720 years' worth of copper and 60,000 years' worth of manganese, at current rates of consumption.

FIGURE 15.9 Examples of ore deposits that originated as sedimentary layers.

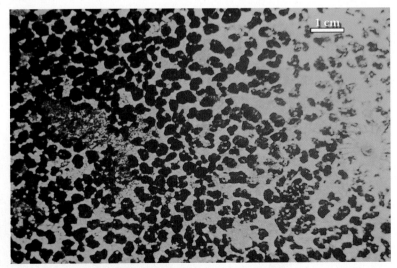

(a) Precambrian banded iron formation from northern Michigan consists of hematite interbedded with jasper.

(b) A view from a submersible of sea floor on which manganese nodules have grown in the sediment.

FIGURE 15.10 The formation of residual mineral deposits, by intense weathering and leaching.

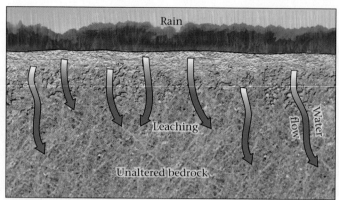

(a) When water sinks through the ground, bedrock weathers and slowly transforms into soil.

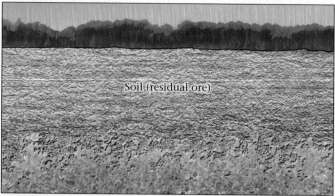

(b) If the amount of water is great, it leaches (dissolves and removes) many elements, leaving behind a residuum rich in iron or aluminum.

Residual mineral deposits. Recall from Interlude B that as rainwater sinks into the Earth, it leaches (dissolves) certain elements and leaves behind others, as part of the process of forming soil. In rainy, tropical environments, the residuum left behind in soils after leaching includes concentrations of iron or aluminum. Locally, these metals become so concentrated that the soil itself becomes an ore deposit (Fig. 15.10a, b). We refer to such deposits as **residual mineral deposits**. Most of the aluminum ore mined today comes from bauxite, a residual mineral deposit created by the extreme leaching of rocks (e.g., granite) containing aluminum-bearing minerals.

Placer deposits. Ore deposits may develop when rocks containing native metals erode, producing a mixture of sand grains and metal flakes or nuggets (pebble-sized fragments).

FIGURE 15.11 Placer deposits form where erosion produces clasts of native metals. Sorting by the stream concentrates the metals.

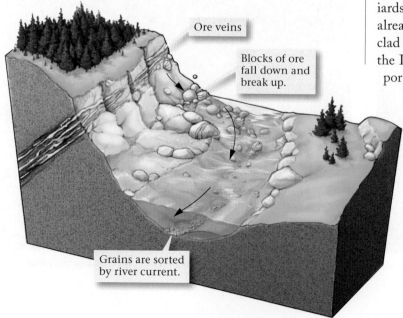

The heavy metal grains (e.g., gold) accumulate in sand or gravel bars along the course of rivers, for the moving water carries away lighter mineral grains but can't move the heavy metal grains so easily. Concentrations of metal grains in stream sediments are a type of **placer deposit** (Fig. 15.11). (The term is also used for concentrations of diamonds.) Panning further concentrates gold flakes or nuggets—swirling water in a pan causes the lighter sand grains to wash away, leaving the gold behind. Placer deposits may eventually be buried and lithify to become part of a new sedimentary rock, which could be mined. By tracking placer deposits upstream, prospectors may find the "mother lode," the bedrock source of the gold.

Where Are Ore Deposits Found?

The Inca Empire of fifteenth-century Peru boasted elaborate cities and temples, decorated with fantastic masks, jewelry, and sculptures made of gold. Then, around 1532, Spanish ships arrived, led by conquistadors who quipped, "We Spaniards suffer from a disease that only gold can cure." The Incas, already weakened by civil war, were no match for the armor-clad Spaniards with their guns and horses. Within six years, the Inca Empire had vanished, and Spanish ships were transporting Inca treasure back to Spain. Why did the Incas possess so much gold? Or to ask the broader question, what geologic factors control the distribution of ore? Once again, we can find the answer by considering the consequences of plate tectonics.

Several of the ore-deposit types mentioned above occur in association with igneous rocks. As we learned in Chapter 6, igneous activity does not happen randomly around the Earth, but rather concentrates along convergent plate boundaries (specifically, in the overriding plate of a subduction zone), divergent plate boundaries (along mid-ocean ridges), continental rifts, or hot spots. Thus, magmatic and hydrothermal deposits (and secondary-enrichment deposits derived from these) occur along plate boundaries, along rifts, or at hot spots. Placer deposits are typically found in the sediments eroded from magmatic or hydrothermal deposits.

Consider the Inca gold. The Inca Empire was situated in the Andes Mountains, which formed as a result of compression and volcanic activity where the Pacific Ocean floor subducts beneath the South American Plate. As the mountains rose, erosion stripped away surface rocks to expose the large granite plutons that had intruded into the continental crust beneath. The magma that froze to make the granite brought gold, copper, and silver atoms with it. Some of the gold precipitated along with quartz to form veins in the plutons. Inca miners quarried these veins and separated the gold, or panned for gold in the streams choked with sediment eroded from the plutons. Plutons that contain similar ore deposits developed in the western United States during the Mesozoic and Cenozoic Eras.

As noted earlier, some massive sulfide deposits accumulate from black smokers along a mid-ocean ridge system and are interlayered with sea-floor basalt. Miners can gain access to such sea-floor deposits only in places where the collision of continents slides a sliver of sea floor up along a fault and onto continental crust.

Some ore deposits are not a direct result of plate tectonics activity, and thus are not directly associated with plate boundaries. For example, banded iron formation (BIF) formed along passive continental margins during the Precambrian. Its occurrence today reflects the present distribution of preserved Precambrian sedimentary rocks, and thus most major exposures occur in shield areas of continents. Bauxite forms where aluminum-rich bedrock occurs, and extreme leaching takes place during soil formation. Thus, some bauxite deposits form on Precambrian granite bedrock in stable continental areas that now lie in tropical regions.

Take-Home Message

- Ore is rock with a concentration of ore minerals; ore minerals contain a high proportion of metal in a form that can be extracted.
- Metals can occur in native form, or in a variety of minerals such as oxides, sulfides, and salts.
- Ore deposits form for a variety of reasons: settling from magmas, precipitation from water, settling with sediment, and extreme weathering of rock.
- Igneous-related ore deposits occur in the remnants of volcanic arcs and in other places where intrusions occur.

THINK: Why don't we use fresh granite as a source for aluminum.

15.4 ORE-MINERAL EXPLORATION AND PRODUCTION

Imagine an old prospector of days past clanking through the desert with a worn-out donkey. To find bedrock ore, such prospectors would eye hillsides for a "show," visible evidence on the ground surface. What does a show look like? It may be an outcropping of milky-white quartz veins, for veins could indicate the presence of hydrothermal ore deposition. It may be the sparkle of minerals with metallic luster disseminated through the outcrop. Or, it may be the presence of orangish, yellowish, or bluish stains in outcrops, for these stains could result either from the presence of brightly colored ore minerals, or from the rusting (oxidation) of oxide or sulfide minerals (Fig. 15.12a). To find placer deposits, prospectors would "pan" a pile of sand or gravel from a stream bed.

On finding a possible ore, a prospector would take samples back to town for an assay, a test to determine how much extractable metal the samples contain. If the assay indicated a significant concentration of metal, the prospector might "stake a claim" by literally marking off an area of ground with stakes. In some locations, a claim gave the prospector rights to all the mineral deposits on or under the land, and the opportunity to develop mines. When one prospector would find ore, word would quickly spread and others would rush to stake neighboring claims.

These days, large mining companies employ professional geologists to survey ore-bearing regions systematically. The geologists focus their studies on rocks that developed in settings appropriate for ore formation. Once such a region has been identified, they may measure the local strength of Earth's magnetic field and the local pull of gravity. These measurements lead them to ore bodies, because ore minerals tend to be denser and more magnetic than average rocks (Fig. 15.12b, c). Geologists also sample rocks and soils to test for metal content, and may even analyze plants in the area to detect traces of metals, for plants absorb metals through their roots. Once geologists have identified a possible ore deposit, they drill holes to sample subsurface rock and to determine the ore deposit's shape and extent. Ore-mineral exploration sometimes takes geologists into jungles, deserts, and tundras worldwide. The largest gold deposit now being mined, at the Grasberg Mine in Papua (Indonesia), was found in 1988 by a geologist who helicoptered from mountaintop to mountaintop looking for shows.

If calculations indicate that mining an ore deposit will yield a profit, and if environmental concerns can be accommodated, a company builds a mine. Mines can be below or above ground, depending on how close the ore deposit is to the surface. To develop an **open-pit mine** (Fig. 15.12d), workers first drill a series of holes into the solid bedrock and then fill the holes with high explosives. They must space the holes carefully and must set off the charges in a precise sequence, so that the bedrock shatters

FIGURE 15.12 Finding and mining ore deposits.

The blue indicates the presence of copper ore.

20 cm

(a) Prospectors look for traces of ore minerals on rock outcrops.

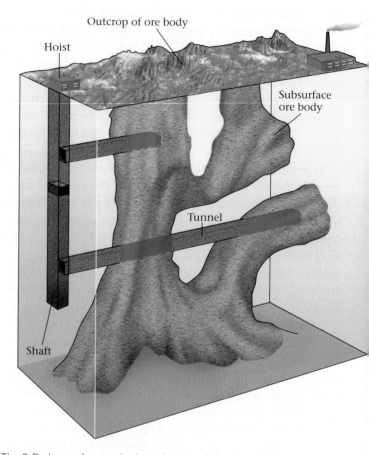

Outcrop of ore body

Hoist

Subsurface ore body

Tunnel

Shaft

(b) The 3-D shape of an ore body underground. Shafts and tunnels access the body.

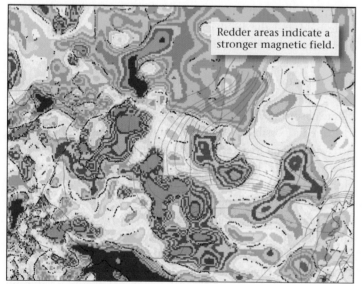

Redder areas indicate a stronger magnetic field.

(c) A map of anomalies in the magnetic field may hint at ore bodies because metal is magnetic.

(d) Open-pit mining utilizes ore that lies fairly close to the ground surface. The steps help wall stability.

into appropriate-sized blocks for handling. When the dust settles, large front-end loaders dump the ore into giant ore trucks, which can carry as much as 200 tons of ore in a single load. (In comparison, a loaded cement mixer weighs about 70 tons.) Tires on these ore trucks are so huge that a tall person comes up only to the base of the hub. The trucks transport waste rock or "tailings" (rock that doesn't contain ore) to a tailings pile and the ore to a crusher, a giant set of moving steel jaws that smash the ore into small fragments. Workers then separate ore minerals from other minerals and send the ore-mineral concentrate to a processing plant, where it undergoes smelting or treatment with acidic solutions to separate metal atoms from other atoms. Eventually, workers melt the metal and then pour it into molds to make ingots (brick-shaped blocks) for transport to a manufacturing facility.

If the ore deposit lies more than about 100 meters below the Earth's surface, miners must make an **underground mine**. To do so, they either dig a tunnel into the side of a mountain (the entrance to the tunnel is called an adit), or they sink a vertical shaft in which they install an elevator. In some cases, they cut a spiral tunnel downward, to provide a gentle ramp to the surface. At the level in the crust where the ore body appears, they build a maze of tunnels into the ore by drilling holes into the rock and then blasting. The rock removed must be carried back to the surface. Rock columns between the tunnels hold up the ceiling of the mine. The deepest mine on the planet, located in South Africa, currently reaches a depth of 3.5 km, where temperatures exceed 55°C, making mining there a very uncomfortable occupation. Miners face danger from mine collapse and rock falls (Box 15.1). Some miners have been killed or injured by "rock bursts," sudden explosions of rock off the ceiling or walls of a tunnel. These explosions happen because the rock surrounding the adit is under such great pressure that it sometimes spontaneously fractures.

Take-Home Message

- Finding ore deposits can be difficult. Some were found by recognizing shows at the surface and others by systematic study using sophisticated methods.
- Some ore deposits are obtained from open-pit mines and some from underground mines.

THINK: How can geologists confirm that an ore deposit lies deep beneath a given location?

15.5 NONMETALLIC MINERAL RESOURCES

So far, this chapter has focused on resources that contain metal. But society uses many other geological materials, commonly known as **industrial minerals**, as well. From the ground we get the stone used to make roadbeds and buildings, the chemicals for fertilizers, the gypsum in drywall, the salt filling salt shakers, and the sand used to make glass—the list is endless. This section looks at a few of these materials and explains where they come from. As was the case with metallic minerals,

all these originated as a result of geologic processes (Geology at a Glance, p. 518–519).

Dimension Stone

The Parthenon, a colossal stone temple rimmed by 46 carved columns, has stood atop a hill overlooking the city of Athens for almost 2,500 years. No wonder—"stone," an architect's word for rock, outlasts nearly all other construction materials. We use stone to make facades, roofs, curbs, steps, countertops, and floors. We value stone for its visual appeal as well as its durability. The names that architects give to various types of stone may differ from the formal rock names that geologists use. For example, architects refer to any polished carbonate rock as marble, whether or not it has been metamorphosed. Likewise, they refer to any rocks containing feldspar and quartz as granite, regardless of whether the rock has an igneous or a metamorphic texture, or a felsic or mafic composition.

To obtain intact slabs and blocks of rock (granite or marble)—known as **dimension stone** in the trade—for architectural purposes, workers must carefully cut rock out of the walls of quarries (Fig. 15.13a). (Note that a quarry provides stone, whereas a mine supplies ore.) To cut stone slabs, quarry operators split rock blocks from bedrock by hammering a series of wedges into the rock. Or they cut it off bedrock by using a wireline saw, a thermal lance, or a water jet. A wireline saw consists of a loop of braided wire moving between two pulleys. In some cases, as the wire moves along the rock surface, the quarry operator spills abrasive (sand or garnet grains) and water onto the wire. The movement of the wire drags the abrasive along the rock and grinds a slice into it. Alternatively, the quarry operator may use a diamond-coated wire, cooled with pure water. A thermal lance looks like a long blowtorch: a flame of burning diesel fuel, stoked by high-pressure air, pulverizes rock and thereby cuts a slot. More recently, quarry operators have begun to use an abrasive water jet, which squirts out water and abrasives at very high pressure, to cut rock.

Crushed Stone and Concrete

Crushed stone forms the substrate of highways and railroads and serves as the raw material for manufacturing cement, concrete, and asphalt. In crushed-stone quarries (Fig. 15.13b), operators use high explosives to break up bedrock into rubble that they then transport by truck to a jaw crusher. This reduces the rubble into usable-size fractions.

Much modern construction utilizes mortar and concrete (see Box 15.2), human-made rock-like materials formed when a slurry composed of aggregate (sand and/or gravel) mixed with cement and water is allowed to harden. The hardening takes place when a complex assemblage of minerals grows by

BOX 15.1

The Amazing Chilean Mine Rescue of 2010

The San José mine, near Copiapó in northern Chile, penetrates an ore-bearing intrusive body of diorite (intermediate igneous rock) that formed due to Cenozoic Andean convergent-margin tectonism. The mine produces copper and gold, like many other mines in the Andes. The ore body extends downward to great depth, and over the years, miners have been extracting ore from progressively deeper levels. The mining operation uses a long, gently sloped ramp that spirals downward to provide access in and out of the mine for miners, supplies, and ore. The ramp intersects the ore body at a depth of between 150 and 800 m (~500 to 2,600 feet) below the ground surface. Off of this ramp, miners have cut numerous horizontal tunnels from which they remove ore. They also have enlarged some areas to produce "rooms" for repair shops and for shelters.

The rocks in the mine are under great stress due to the weight of overlying rock, so when miners cut tunnels, they must set up supports that are strong enough to resist the overlying load and prevent rock from collapsing. In addition, workers may need to bolt, cement, or fence loose rocks to keep them from falling, even in places where the tunnel, overall, has adequate support.

On August 5, 2010, something went wrong. At a depth of around 500 m (~1,600 feet) below the surface, a catastrophic rock fall suddenly filled a portion of the ramp with thousands of tons of debris. The fall isolated a group of 33 miners who were working near the bottom of the mine. Ladders that might have given the miners an escape route through the ventilation shafts hadn't been installed—the miners were trapped! Choking dust from the rock fall blocked visibility and made breathing difficult, but the miners managed to retreat to a refuge room at a depth of over 688 m (2,300 feet) below the surface. Two Empire State Buildings (100-story skyscrapers) stacked one on

top of the other would not have reached the ground surface from the refuge. Immediately, in an amazing example of self-reliance and strength, the miners organized into a functioning group under the leadership of their foreman, carefully rationed emergency food supplies, dug small wells to provide water, and established a daily routine to wait until rescue. Fortunately, there is no methane in copper/gold mines, so the threat of explosion was not a worry.

For 17 days the miners sweltered in the humid, 35°C (95°F) air of their refuge, while unbeknownst to them, rescuers were frantically drilling exploratory holes in hopes of reaching the miners in order to provide ventilation and communication. Finally, a drill broke through the ceiling of the refuge. The miners tapped on the pipe to indicate they were there and taped a paper message to the end of the drill bit announcing to the world, *Estamos bien en el refugio los 33*. ("We are all right in the shelter—all 33.") With contact established, rescuers drilled a wider hole allowing them to send down supplies, letters, and even a video link. But the wait had to continue, for engineers estimated that bringing the men out might be four months away.

Immediately, mining engineers at the surface set to work drilling rescue holes wide enough for miners to fit in. The effort was not easy, for the igneous rocks of the mine are very hard, and drilling would have to avoid earlier workings. Three different plans (A, B, and C), using different drilling technologies, began the race through solid rock to reach the miners. In the end, Plan B reached them first, weeks before the predicted rescue date. Plan B utilized a percussion drilling machine manufactured by a Pennsylvania supplier—four hammers at the end of a device sent down one of the ventilation holes bashed into the surrounding rock and widened the hole. The drilling progressed at a rate of 40 m (130 ft) per day and sent a cascade of rock debris down the

ventilation hole into the workroom. To keep the workroom open, the trapped miners had to haul away the debris—all 700 tons of it.

Once the access hole was complete, rescuers sent down a special rescue cylinder, or capsule, named the Phoenix 2, in which one miner at a time could fit. Before strapping into the cylinder, which came equipped with oxygen and other safety devices, each miner donned sunglasses, so the brightness of light at the Earth's surface would not harm his eyes. One by one, they were pulled to safety by a large winch at a rate of about 1 m/s (~2 mph), on a journey that took up to 18 minutes. As a huge crowd, including the Chilean President, waited at the surface, and an audience of 1 billion people watched on TV worldwide, the miners emerged on October 13, after 69 days underground (**Fig. Bx15.1**).

FIGURE Bx15.1 The rescue capsule emerges at the surface.

FIGURE 15.13 Stone production in quarries.

(a) An active quarrying operation in Missouri that produces large blocks of cut dimension stone.

(b) A crushed limestone quarry in Illinois. The stone is sorted by fragment size.

Did you ever wonder . . .
how concrete differs from rock?

chemical reactions in the slurry; these minerals bind together the grains of sand or gravel in mortar or concrete. (Note that the word *mortar* refers to the substance that holds bricks or stone blocks together, whereas **concrete** refers to the substance that workers shape into roads or walls by spreading it out into a layer or by pouring it into a form.) The cement in mortar or concrete starts out as a powder composed of lime (CaO), quartz (SiO_2), aluminum oxide (Al_2O_3), and iron oxide (Fe_2O_3). Typically, lime accounts for 66% of cement, silica for 25%, and the remaining chemicals for about 9%.

It appears that the ancient Romans were the first to use cement—they made it from a mixture of volcanic glass and limestone. In the 18th and early 19th century, cement was produced by heating specific types of limestone (which happened to contain calcite, clay, and quartz in the correct proportions) in a kiln up to a temperature of about 1450°C; the heating releases CO_2 gas and produces "clinker," chunks consisting of lime and other oxide compounds. Manufacturers crush the clinker into cement powder and pack it in bags for transport. But natural limestone with the exact composition necessary to make good cement is fairly rare, so most cement used today is **Portland cement**, made by mixing limestone, sandstone, and shale in just the right proportions to provide the correct chemical makeup. Isaac Johnson, an English engineer, came up with the recipe for Portland cement in 1844; he named it after the town of Portland, England, because he thought it resembled rock exposed there.

Nonmetallic Minerals for Homes and Farms

We use an astounding variety of nonmetallic geologic resources (Table 15.2) without ever realizing where they come from. Consider the materials in a typical house or apartment. The concrete foundation consists of cement, made from limestone mixed with sand or gravel. The bricks in the exterior walls originated as clay,

TABLE 15.2 Common Nonmetallic Resources

Limestone	Sedimentary rock made of calcite; used for gravel or cement
Crushed stone	Any variety of coherent rock (limestone, quartzite, granite, gneiss)
Siltstone	Beds of sedimentary rock; used to make flagstone
Granite	Coarse igneous rock; used for dimension stone
Marble	Metamorphosed limestone; used for dimension stone
Slate	Metamorphosed shale; used for roofing shingles
Gypsum	A sulfate salt precipitated from salt water; used for wallboard
Phosphate	From the mineral apatite; used for fertilizer
Pumice	Frothy volcanic rock; used to decorate gardens and paths
Clay	Very fine mica-like mineral in sediment; used to make bricks or pottery
Sand	From sandstone, beaches, or riverbeds; quartz sand is used for construction and for making glass
Salt	From the mineral halite, formed by evaporating saltwater; used for food, melting ice on roads
Sulfur	Occurs either as native sulfur, typically above salt domes, or in sulfide minerals; used for fertilizer and chemicals

BOX 15.2

The Sidewalks of New York

Untold tons of concrete have gone into the construction of New York City. In fact, with the exception of a few city parks, most of the walking space in the city consists of concrete (**Fig. Bx15.2**). And concrete skyscrapers tower above the concrete plain. Where does all this concrete come from?

Much of the sand used in New York concrete was deposited during the last ice age. As vast glaciers moved southward over 14,000 years ago, they ground away the igneous and metamorphic rocks that constituted central and eastern Canada. These ancient rocks contained abundant quartz, and since quartz lasts a long time (it does not undergo chemical weathering easily), the sediment transported by the glaciers retained a large amount of quartz. Glaciers deposited this sediment in huge piles called moraines (see Chapter 22). As the glaciers melted, fast-moving rivers of meltwater washed the sediment, sorting sand from mud and pebbles. The sand was deposited in bars in the meltwater rivers, and these relict bars now provide thick lenses of sand that can be economically excavated.

What about the cement? Cement contains a mixture of lime, derived from limestone, and other elements (such as silica) derived from shale and sandstone. The bedrock of New York, though, consists largely of schist and gneiss, not sedimentary rocks. Fortunately, a source of rocks appropriate for making cement lies up the Hudson River. A Silurian-age rock unit, exposed in low hills just west of the river and called the Rosendale Formation, naturally contains exactly the right mixture of lime and silica needed to make durable cement. Beginning in the late 1820s, workers began quarrying the Rosendale Formation for cement. Quarry operators followed the Rosendale beds closely, making horizontal mine tunnels where the beds were horizontal, tilted mine tunnels where the beds tilted, and vertical mine tunnels where the beds were vertical. They then dumped the excavated rock into nearby kilns and roasted it to produce lime mixed with other oxides. The resulting powder was packed into barrels, loaded onto barges, and shipped downriver to New York. As demand for cement increased, operators eventually dug open-pit quarries from which they excavated Devonian limestone and shale units, mixing them together in the correct proportion to make Portland cement.

The rocks making up the strata that provides the source for cement consist of shell fragments and small, reef-like colonies of organisms. In other words, the lime in the concrete of New York sidewalks was originally extracted from seawater by living organisms—brachiopods, crinoids, and bryozoans—over 400 million years ago.

FIGURE Bx15.2 The production of concrete effectively transforms natural rock into human-made stone.

(a) A quarry of limestone. The trucks are carrying rock to a crusher.

(b) The heat of a kiln transforms limestone into lime (CaO).

(c) A concrete mixer pouring wet concrete.

(d) A sidewalk in New York.

formed from the chemical weathering of silicate rocks and perhaps dug from the floodplain of a stream. To make *bricks*, workers mold wet clay into blocks and then bake it. Baking drives out water and causes metamorphic reactions that recrystallize the clay. Clay is also the raw material that pottery, porcelain, and other ceramic materials come from (see Chapter 8).

The glass used to glaze windows consists largely of silica, formed by first melting and then freezing pure quartz sand from a beach deposit or a sandstone formation. Quartz may also be used in the construction of photovoltaic cells for solar panels. Gypsum board (drywall), used to construct interior walls, comes from a slurry of water and the mineral gypsum sandwiched between sheets of paper. Gypsum ($CaSO_4 \cdot 2H_2O$) occurs in evaporite strata precipitated from seawater or saline lake water. Evaporites provide other useful minerals as well, such as halite, and serve as the source for lithium, a key element in computer and camera batteries. Of note, some of the most abundant sources of lithium occur in dry lake beds of Bolivia (Fig. 15.14).

Modern technological innovations have also greatly increased the demand for the **rare earth elements** (REE), a group of 17 elements including the lanthanides (elements that have atomic numbers between 57 and 71 on the periodic table), scandium, and yttrium. While the names of rare earth elements are unfamiliar to most people—and are also quite hard to pronounce—the elements themselves have become essential in the production of lasers, magnets, X-ray tubes, night-vision goggles, camera lenses, high-tech lamps, and chemical catalysts. Rare earth elements aren't actually that rare, in terms of their abundance relative to other elements in the crust. But localities where ores have a high enough concentration to be mined are rare. Most rare earth elements used today come from strip mining certain types of granitic plutons that contain REE-rich veins, or sediments and soils derived from such plutons.

FIGURE 15.14 Tubs of lithium-bearing salt scraped from a dry lake bed in Bolivia.

The asbestos that was once used to make roof shingles is derived from serpentine, a rock created by the reaction of olivine with water. This olivine may come from oceanic lithosphere, thrust onto continental crust during continental collisions. The copper in the house's electrical wiring most likely comes from the ores in a porphyry copper deposit; the iron in the nails probably comes from banded iron formation; and the plastic used in everything from countertops to light fixtures comes from oil formed from the bodies of plankton and algae that died millions of years ago.

Chemicals employed for agricultural purposes also come from the ground. For example, potash (K_2CO_3) comes from the minerals in evaporite deposits. Phosphate (PO_4^{-3}) comes from the mineral apatite, which crystallizes in organic-rich muds that were deposited in shallow, oxygen-free (anoxic) seawater. Much of the phosphate now used in the United States comes from strip mines accessing 10 Ma shale beds (formed from such muds) that lie about 15 m below the land surface of Florida. Smaller quantities come from mines in a Permian stratigraphic unit called the Phosphoria Formation, which underlies portions of the Rocky Mountain states in the western United States.

As you can see, geologic processes acting over thousands to billions of years provide many of the material goods used in homes, farms, and industry. Truly, without the geologic resources of the Earth, modern society would grind to a halt.

> ### Take-Home Message
>
> - Dimension stone, used for everything from curbstones to countertops, must be cut into useful shapes. Crushed stone is used for roadbeds or in concrete.
> - Cement forms when minerals grow in a slurry of lime (from limestone), mixed with other elements.
> - A variety of components from geologic materials comprise key materials in everyday use. For example, gypsum in drywall comes from evaporite deposits.
>
> **THINK:** Why not use sandstone as the source for sand grains to be mixed with cement to make concrete?

15.6 GLOBAL MINERAL NEEDS

How Long Will Resources Last?

The average citizen of an industrialized country uses 25 kilograms (kg) of aluminum, 10 kg of copper, and 550 kg of iron and steel in a year's time (Fig. 15.15; Table 15.3). If you combine these figures with the quantities of energy resources and nonmetallic geologic resources a person uses, you get a total of about 15,000 kg (15 metric tons) of resources used per capita each year. Thus, the population of the United States consumes

Forming and Processing Earth's Mineral Resources

Ore deposits can be obtained either in strip mines or in underground mines.

Circulating groundwater may extract and concentrate metals to form ore deposits.

Clay, when formed into blocks and baked, becomes brick.

Gravel itself may be quarrie construction purposes.

Mud, a mixture of clay minerals and water, accumulates in beds.

Ore minerals may collect on the bottom of a magma chamber.

Hydrothermal vents (black smokers) produce accumulations of massive sulfides.

From Mud to Brick

Miners pan for gold in pl deposits where metal flak and nuggets occur in sand and gravel.

Erosion tears down mountains and produces gravel and sand.

From Magma to Metal

From Stream Channel to Roadbed

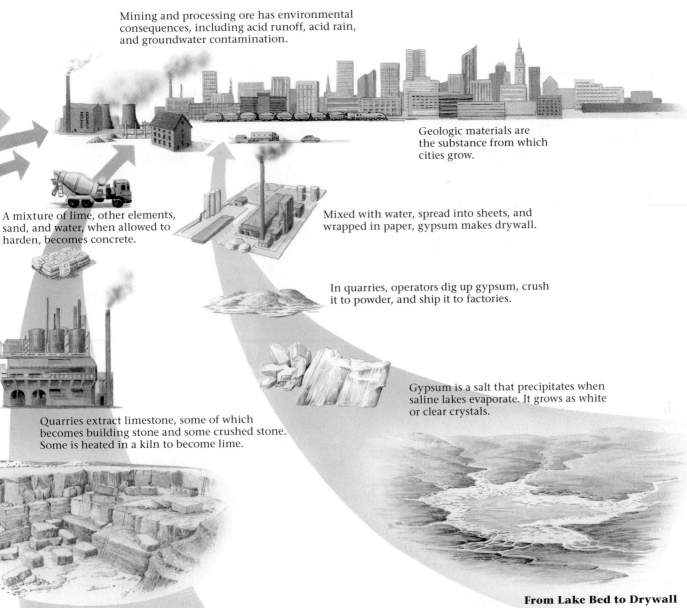

Mining and processing ore has environmental consequences, including acid runoff, acid rain, and groundwater contamination.

Geologic materials are the substance from which cities grow.

A mixture of lime, other elements, sand, and water, when allowed to harden, becomes concrete.

Mixed with water, spread into sheets, and wrapped in paper, gypsum makes drywall.

In quarries, operators dig up gypsum, crush it to powder, and ship it to factories.

Gypsum is a salt that precipitates when saline lakes evaporate. It grows as white or clear crystals.

Quarries extract limestone, some of which becomes building stone and some crushed stone. Some is heated in a kiln to become lime.

From Lake Bed to Drywall

Over millions of years, shells and shell fragments collect and eventually form beds of limestone.

The raw materials from which we manufacture the buildings, roads, wires, and coins of modern society were produced by geologic processes. For example, ore deposits—the concentrations of minerals that are a source of metal—formed during a variety of magmatic or sedimentary processes. Limestone, a rock used for buildings and for making concrete, began as an accumulation of seashells. Brick began as clay, a by-product of chemical weathering. And the gypsum of drywall began as an accumulation of salt along a desert lake. Metal, gravel, lime, and gypsum are all examples of Earth's mineral resources. We can use some mineral resources right from the Earth, simply by digging them out. But most become usable only after expensive processing.

Organisms extract ions from water and construct shells.

From Sea Floor to Sidewalk

519

FIGURE 15.15 Industrialized countries consume vast quantities of mineral resources in a year, as the diagram indicates. The numbers indicate the weight of the material used per person per year.

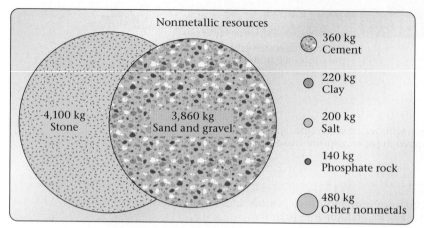
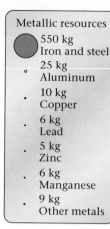

TABLE 15.3 Yearly per Capita Usage of Geologic Materials in the United States

4,100 kg	Stone
3,860 kg	Sand and gravel
3,050 kg	Petroleum
2,650 kg	Coal
1,900 kg	Natural gas
550 kg	Iron and steel
360 kg	Cement
220 kg	Clay
200 kg	Salt
140 kg	Phosphate
25 kg	Aluminum
10 kg	Copper
6 kg	Lead
5 kg	Zinc

1 kg = 2.205 pounds.

about 4 billion metric tons of geologic material per year. To create this supply, workers must mine, quarry, or pump 18 billion metric tons. By comparison, the Mississippi River transports 190 million metric tons of sediment per year.

Mineral resources, like oil and coal, are *nonrenewable* resources. Once mined, an ore deposit or a limestone hill disappears forever. Natural geologic processes do not happen fast enough to replace the deposits as quickly as we use them. Geologists have calculated **reserves** (measured quantities of a commodity) for various mineral deposits just as they have for oil. Based on current definitions of reserves (which depend on today's prices) and rates of consumption, supplies of some metals may run out in only decades to centuries (Table 15.4). But these estimates may change as supplies become depleted and prices rise (making previously uneconomical deposits worth mining).

TABLE 15.4 Expected Lifetimes of Currently Known Ore Resources (in years)

Metal	World Resources	U.S. Resources
Iron	120	40
Aluminum	330	2
Copper	65	40
Lead	20	40
Zinc	30	25
Gold	30	20
Platinum	45	1
Nickel	75	less than 1
Cobalt	50	less than 1
Manganese	70	0
Chromium	75	0

And supplies could increase if geologists discover new reserves or if new ways of mining become available (e.g., providing access to nodules on the sea floor or to deeper parts of the crust). Further, increased efforts at conservation and recycling can cause a dramatic decrease in rates of consumption and thereby stretch the lifetime of existing reserves.

Ore deposits do not occur everywhere because their formation requires special geologic conditions. As a result, some countries possess vast supplies, whereas others have none. In fact, no single country owns all the mineral resources it needs, so nations must trade with each other to maintain supplies, and global politics inevitably affects prices. Many wars have their roots in competition for mineral reserves, and it is no surprise that the outcomes of some wars have hinged on who controls these reserves.

The United States worries in particular about supplies of so-called **strategic minerals**, which include manganese, platinum, chromium, and cobalt—metals alloyed with iron to make the special-purpose steels needed in the aerospace industry. At present, the country must import 100% of the manganese, 95% of the cobalt, 73% of the chromium, and 92% of the platinum it consumes. Principal reserves of these metals lie in the crust of countries that have not always practiced open trade with the United States. As a defense precaution, the United States stockpiles these metals in case supplies are cut off.

We've already noted that resources of lithium, a key component of batteries, occur in nonindustrialized countries, so that industrialized countries are dependent on them. Resources containing rare earth elements became a subject of international tension in 2009. These elements are fairly scarce not because they don't occur in the crust, but because they rarely occur in mineable concentrations. In recent years, they have become increasingly important because of their use in high-tech batteries and magnets. Currently, about 90% of the mined supply occurs in one country (China), which has started to limit exports to the extent that high-tech factories in other countries have had to shut down.

Mining and the Environment

Mining leaves a big footprint in the Earth System. Some of the gaping holes that open-pit mining creates in the landscape have become so big that astronauts can see them from space. Both open-pit (See for Yourself N, p. S-26) and underground mining yield immense quantities of waste rock, which miners dump in tailings piles. Some tailings piles grow into artificial hills 200 meters high and many kilometers long. Lacking soil, tailings piles tend to remain unvegetated for a long time. Mining also exposes ore-bearing rock to the atmosphere, and since many ore minerals are sulfides, they react with rainwater to produce **acid**

mine runoff, which can severely damage vegetation downstream (Fig. 15.16a). In many cases, mining companies douse tailings with acidic solutions to leach out more metals; these acids sometimes escape into the environment.

Ore processing tends to release noxious chemicals that can mix with rain and spread over the countryside, damaging life. Before the installation of modern environmental controls, smoke from ore-processing plants caused severe air pollution; plumes of smoke from the old smelters in Sudbury, Ontario, for example, created a wasteland for many kilometers downwind (Fig. 15.16b).

Recent years have seen efforts to reclaim mining spoils, and new technologies have been developed to extract metals in ways that are less deleterious to the environment and that treat waste more efficiently. Clearly, mining and ore processing can potentially become a scar on the landscape, the size of which depends on the willingness of producers and consumers to minimize damage and the degree to which regulations are successfully designed and enforced.

FIGURE 15.16 Environmental consequences of producing metallic mineral resources.

Much of the color comes from bacteria and archaea living in the water.

Nickel smelter

The "superstack" is 380 m (1,270 ft) high.

Tailings pile

(a) The orange color in this mine runoff is due to iron and sulfide in the water.

(b) Acidic smelter smoke killed off vegetation near Sudbury, Ontario, in the 1970s. A large tailings pile can be seen in the distance.

Take-Home Message

- A mineral reserve is the quantity of given mineral; mineral reserves are not uniformly distributed around the world.
- There are particular concerns over supplies of strategic minerals and of lithium and rare earth elements, used in high-tech applications.

- Production and use of mineral resources is problematic for the environment; sulfide minerals, for example, react with water to produce acid runoff.

THINK: If one country restricts the export of a strategic mineral, would that change the minimum concentration of the mineral necessary to make an ore deposit elsewhere worth mining? Why?

Chapter Summary

- Industrial societies use many types of minerals, all of which must be extracted from the upper crust. We distinguish two general categories: metallic resources and nonmetallic resources.

- Metals are materials in which atoms are held together by metallic bonds. They are malleable and make good conductors.

- Metals come from ore. An ore is a rock containing native metals or ore minerals (sulfide, oxide, or carbonate minerals with a high proportion of metal) in sufficient quantities to be worth mining. An ore deposit is an accumulation of ore.

- Magmatic deposits form when sulfide ore minerals settle to the floor of a magma chamber. In hydrothermal deposits, ore minerals precipitate from hot-water solutions. Secondary-enrichment deposits form when groundwater carries metals away from a preexisting deposit. Mississippi Valley–type deposits precipitate from groundwater that has passed long distances through sedimentary basins. Sedimentary deposits precipitate out of the ocean. Residual mineral deposits in soil are the result of severe leaching in tropical climates. Placer deposits develop when heavy metal grains accumulate in sediment along a stream.

- Many ore deposits are associated with igneous activity in subduction zones, along mid-ocean ridges, along continental rifts, or at hot spots.

- Nonmetallic resources include dimension stone for decorative purposes, crushed stone for cement and asphalt production, clay for brick making, sand for glass production, and many other materials. A large proportion of materials in your home have a geological ancestry.

- Mineral resources are nonrenewable. Many are now or may soon be in short supply.

- The production and processing of mineral resources can harm the environment, if not done carefully.

GEOPUZZLE REVISITED

The stuff of everyday life comes from geologic materials. For example, baking mud makes brick, smelting copper-bearing minerals yields metallic copper, and mixing crushed and baked limestone with sand and water forms concrete. Geologists play a key role in identifying sources of such mineral resources, and in addressing the environmental issues that arise from using such resources.

Guide Terms

acid mine runoff (p. 521)
alloy (p. 505)
banded iron formation (BIF) (p. 509)
base metals (p. 505)
concrete (p. 515)
dimension stone (p. 513)
grade (p. 506)
hydrothermal deposit (p. 507)
industrial minerals (p. 513)
magmatic deposit (p. 507)
manganese nodules (p. 509)
massive-sulfide deposit (p. 507)
metals (p. 503)
mineral resources (p. 503)
Mississippi Valley–type (MVT) ores (p. 508)

native metals (p. 504)
open-pit mine (p. 511)
ore (p. 506)
ore deposit (pp. 506, 507)
ore minerals (economic minerals) (p. 506)
placer deposit (p. 510)
Portland cement (p. 515)
precious metals (p. 505)
rare earth elements (p. 517)
reserves (p. 520)
residual mineral deposit (p. 510)
secondary-enrichment deposit (p. 508)
smelting (p. 504)
strategic minerals (p. 520)
underground mine (p. 513)

Review Questions

1. Describe how people have used copper, bronze, and iron throughout history.

2. Why don't we use an average granite as a source for useful metals?

3. What kinds of concentrations of a metal are required for it to be economically mineable?

4. Describe various kinds of economic mineral deposits.

5. In what geologic setting do massive sulfide deposits form?

6. What procedures are used to locate and mine mineral resources today?

7. How is stone cut from a quarry?

8. What are the ingredients of cement? How is Portland cement made?

9. Name materials in your home that come from mineral resources.

10. How many kilograms of mineral resources does the average person in an industrialized country use in a year?

11. Compare the estimated lifetimes of ore supplies (worldwide and in the United States) of iron, aluminum, copper, gold, and chromium.

12. What are some environmental hazards of large-scale mining?

On Further Thought

13. Imagine that an ore deposit of a certain metal contains 0.6% grade ore. This means that 0.6% by weight of a block of ore consists of the metal. The pure metal, on the open market, sells for $8,000/ton. It costs $15/ton to mine the ore, $15/ton to transport the ore to the processing plant, and $15/ton to process the ore and produce pure metal. Start-up costs (building the mine and building the processing factory) are about $100 million. How much profit does the company make when it sells a ton of metal? How much ore does the operation have to mine to pay back the start-up costs? Considering that a giant dump truck in a mine can carry 200 tons of ore at a time, how many dump-truck loads will have been transported at the break-even point? If the mine has 8 trucks that can each make 6 loads a day, about how many years will it take to break even?

14. An ore deposit at a location in Arizona has the following characteristics: One portion of the ore deposit is an intrusive igneous rock in which tiny grains of copper sulfide minerals are dispersed among the other minerals of the rock. Another nearby portion of the ore deposit consists of limestone in which malachite fills cavities and pores in the rock. What types of ores are these? Describe the geologic history that led to the formation of these deposits.

 For more resources, including animations, quizzes, and Norton's GeoTours, go to **wwnorton.com/studyspace**.

 If your instructor assigns exercises in SmartWork, log in at **smartwork.wwnorton.com**.

ANOTHER VIEW The Bingham open-pit mine near Salt Lake City, Utah is 1.2 km deep and 4 km wide, making it the largest excavation in the world. In this photo, the pit itself is largely hidden by vast tailing piles, the debris left behind once the ore has been extracted. During its century of operation, the mine has yielded about 17 million tons of copper, 700 tons of gold, 6,000 tons of silver, and 400 tons of molybdenum. In 2006 alone, the value of metal produced was $1.8 billion. To obtain this metal, miners have to move up to 450,000 tons of rock a year, for typical ore contains less than 0.3% copper. The ore formed due to hydrothermal fluid circulation that accompanied granitic magmatism about 36 million years ago.

Weathering and erosion of uplifted granite forms dramatic cliffs along the "savage coast" of southern Brittany (France). Over time, rock along the cliffs weathers and collapses into stormy waves, which break and toss the fragments until they disintegrate into sand.

PART VI

Processes and Problems at the Earth's Surface

In the last part of this book, we focus on Earth's surface and near-surface realms. This portion of the Earth System, which encompasses the boundaries between the lithosphere, hydrosphere, and atmosphere, displays great variability, for the dynamic interplay between internal processes (driven by Earth's internal heat) and external processes (driven by the warmth of the Sun), under the influence of Earth's gravitational field, has resulted in a diverse array of landscapes. In Chapters 16 through 22, we examine five of these landscapes, plus groundwater and the atmosphere; and finally, in Chapter 23, we see how forces at work in the Earth System cause the planet to change constantly.

Ever-Changing Landscapes and the Hydrologic Cycle

A **digital elevation map (DEM)** of Oahu, Hawaii. The map was produced using radar data from satellites. This landscape formed when a volcano emerged from the sea and was eroded. Erosion produced sediment, which accumulated to form the flat land of Honolulu.

Talk of mysteries! Think of our life in nature—
daily to be shown matter, to come in contact with it—
rocks, trees, wind on our cheeks! the solid earth!
the actual world! the common sense! Contact!
Contact! Who are we? Where are we?

—Henry David Thoreau (1817–1862)

F.1 INTRODUCTION

The Earth's surface is at once a place of endless variety and intricate detail. Observe the height of its mountains, the expanse of its seas, the desolation of its deserts, and you may be inspired, frightened, or calmed. It's no wonder that artists and writers across the ages have sought inspiration from the **landscape**—the character and shape of the land surface in a region—for landscapes encompass the diversity of human emotion (Fig. F.1a–d). Geologists, like artists and writers, savor the impression of a dramatic landscape; but on seeing one, they can't help but ask, how did it come to be, and how will it change in the future?

The subject of landscape development and evolution, and the **landforms** (individual shapes such as mesas, valleys, cliffs, and dunes) that constitute it, dominate some of the chapters in Part VI. Geologists who study landscape development are known as geomorphologists. You will see that water, both in its liquid and solid forms, plays a key role in the Earth's surface and near-surface processes. The ability of water to move from low areas to high areas, and from sea to land, would not be possible without the circulation of the atmosphere.

This interlude, a general introduction to Earth's surface and near-surface realms, explains the driving forces behind

FIGURE F.1 Examples of the great variety of landscapes on Earth.

(a) A rock and sand seascape along the coast of Brazil.

(b) The peaks of the Grand Tetons in Wyoming.

(c) Buttes of sandstone in Monument Valley, Arizona.

(d) Steep cliffs in Australia's Blue Mountains.

landscape development, identifies factors that control which landscape develops in a given locality, and describes the hydrologic cycle—the pathway water molecules follow as they move from ocean to air to land and back to ocean. We conclude by introducing landscapes on other planets.

F.2 SHAPING THE EARTH'S SURFACE

If the Earth's surface were totally flat, the great diversity of landscapes that embellish our vistas would not exist. But the surface isn't flat, because a variety of geologic processes cause portions of the surface to move up or down relative to adjacent regions. We refer to the upward movement of the land surface as **uplift**, and the sinking or downward movement of the land surface as **subsidence**. Both uplift and subsidence occur for a variety of reasons (Table F.1).

When uplift or subsidence takes place, the elevation difference does not remain the same forever because other components of the Earth System kick into action. Material at higher elevations weathers, fractures, and becomes unstable and thus, is susceptible to **downslope movement** (the tumbling or sliding of rock and sediment from higher elevations to lower ones, driven by gravity); moving water, ice, and air cause **erosion** (the grinding away and removal of the Earth's surface); and where moving fluids slow down, **deposition** of sediment takes place. Downslope movement, erosion, and deposition redistribute rock and sediment, ultimately stripping it from higher areas and collecting it in low areas. As a consequence, a great variety of both **erosional landforms** (those carved by erosion) and **depositional landforms** (those built from an accumulation of sediment) can form.

Because lithosphere "floats" on asthenosphere (see the discussion of isostasy in Chapter 11), the removal of 1 km of rock off the top of a mountain range causes the crust's surface to rise by about 1/3 km, just as the removal of heavy containers from the deck of a cargo ship makes the ship's deck rise. Because the crust rises as erosion takes place, *more than* 5 km of rock must erode from a 5-km-high mountain range to return the land surface to sea level. Similarly, in a depositional setting, the weight of a 1-km-thick layer of sediment depresses the crust by about 1/3 km, producing space for even more sediment.

The energy that drives landscape evolution comes from three sources: **internal energy**, the heat within the Earth, which drives the plate motions and mantle plumes that cause displacement of the crust's surface; **external energy**, energy coming to the Earth from the Sun, which warms the atmosphere and ocean; and **gravitational energy**, which pulls rock down slopes at the surface and, along with external energy, causes convection (i.e., currents and winds). Landscape evolution, in fact, reflects a "battle" between (1) tectonic processes

TABLE F.1 Causes of Uplift and Subsidence

Causes of Uplift

- **Thickening of the crust due to deformation**. At convergent and collisional boundaries, compression causes the crust to shorten horizontally (by development of folds, faults, and foliations) and thicken in the vertical direction. Because of isostasy (see Chapter 11), lithosphere with thickened crust floats relatively higher on the asthenosphere, with the result that the surface of the crust in mountain belts rises.

- **Heating of the lithosphere**. Heating decreases the thickness and density of the lithosphere, so to maintain isostatic equilibrium, lithosphere floats higher. Intrusion or extrusion of igneous rocks thickens the crust or builds volcanoes on top of the surface; all these phenomena cause uplift.

- **Rebound due to unloading**. Removal of a heavy load (such as a glacier or mountain) from the surface causes the Earth's surface to rise, in a manner similar to the way a trampoline's surface rises when you step off of it.

- **Delamination**. If the dense lithospheric mantle separates from the base of the plate, and sinks down into the mantle, the surface of the lithosphere rises. The effect resembles the consequence of unloading ballast from a ship.

Causes of Subsidence

- **Thinning of the crust due to stretching**. In rifts, where the crust undergoes horizontal stretching, the axis of the rift drops down by slip on normal faults.

- **Cooling of the lithosphere**. Cooling thickens the lithosphere and makes it denser, so to maintain isostatic equilibrium, the lithosphere sinks down and its surface lies at a lower elevation.

- **Sinking due to loading**. Where a heavy load (such as a glacier or volcano) forms on the Earth's surface, the lithosphere warps downward, somewhat like the surface of a trampoline warps down when you stand on it.

such as collision, convergence, and rifting, which create **relief** (an elevation difference between two locations; see Box F.1) in an area; and (2) processes such as downslope movement, erosion, and deposition, which destroy relief by removing material from high areas and depositing it in low ones. If, in a particular region, the rate of uplift exceeds the rate of erosion, the land surface rises; if the rate of subsidence exceeds the rate of deposition, the land surface sinks. Without uplift and subsidence, erosion and deposition would have transformed Earth's surface into a flat plain. And without erosion and deposition, high and low areas would have lasted for the entirety of Earth history.

How rapidly do uplift, subsidence, erosion, and deposition take place? The Earth's surface can rise or sink by as much as 3 m during a single major earthquake. But *averaged over time,* the rates of uplift and subsidence range between 0.01 and 10

FIGURE F.2 The processes of uplift, subsidence, erosion, and deposition can be slow or rapid.

Uplifted terrace

New terrance forming

(a) Uplifted beach terraces form where the coast is rising relative to sea level. Present-day wave erosion is forming a new terrace and cutting a cliff on the edge of the old one.

Undermined foundation

(b) So much erosion can take place during a single hurricane that houses built along the beach become undermined.

mm per year (Fig. F.2a). Similarly, erosion can carve out several meters of **substrate**, the material just below the ground surface, during a single flood, storm, or landslide (Fig. F.2b). And deposition during a single event can produce a layer of debris tens of meters thick in a matter of minutes to days. But, averaged over time, erosional and depositional rates also vary between 0.10 and 10 mm per year. Although these rates seem small, a change in surface elevation of just 0.5 mm (the thickness of a fingernail) per year can yield a net change of 5 km in 10 million years. Uplift can build a mountain range, and erosion can whittle one down to near sea level—it just takes time!

In recent years, geologists have used satellites to produce highly detailed digital images of the Earth's surface. The satellite sends a radar beam down to the surface, and the beam bounces back to the satellite. The time it takes for the beam to make its round trip represents land elevation. The resulting maps, called **digital elevation maps** (**DEMs**), provide images that can be analyzed with a computer. The maps can be produced with shading and color that gives the impression of a 3-D shape even on a 2-D sheet of paper (see Interlude opening photo).

F.3 FACTORS CONTROLLING LANDSCAPE DEVELOPMENT

Imagine traveling across a continent. On your journey, you pass plains, swamps, hills, valleys, mesas, and mountains. Some of these features are erosional landforms, in that they result from the breakdown and removal of rock or sediment and develop where **agents of erosion** such as water, ice, or air carve into

the substrate. Of these three agents, water has the greatest effect on a global basis. Other features are depositional landforms, in that they result from the deposition of sediment where the medium carrying the sediment evaporates, slows down, or melts. The specific landforms that develop at a given locality, and that together make up the landscape, reflect six factors.

- *Eroding or transporting agents*: Water, ice, and wind all can cause erosion and transport sediment, but landforms produced by glacial erosion or deposition differ from landforms produced by river erosion or deposition, and both differ from landforms carved by the wind. The differences reflect the varying abilities of water, ice, and wind to carve into the substrate and to carry debris.

- *Elevation (relief)*: In regions where the land surface has risen to a high elevation, rugged mountain ranges with deep valleys and steep cliffs can form; but in places where the land is a wide plain near sea level, valleys and cliffs do not exist. The elevation difference, or relief, between adjacent places in a landscape determines the height and steepness (angle) of slopes.

- *Climate*: The average mean temperature and the volume of precipitation in a region determines whether running water, flowing ice, or wind is the main agent of erosion or deposition and also affects the nature of weathering and soil development.

- *Life activity*: Organisms such as bacteria, worms, trees, and elephants affect landscape development. Although some life activity weakens the substrate, some protects it. The ecology of an area depends on the climate. A

Topographic Maps and Profiles

We can distinguish one landform from another by its shape—for example, as you will see in succeeding chapters, a river-carved valley simply does not look like a glacially carved valley. Landform shapes are manifested by variations in elevation within a region. Geologists use the term **topography** to refer to such variations.

How can we convey information about topography—a three-dimensional feature—on a two-dimensional sheet of paper? Geologists do this by means of a **topographic map**, which uses contour lines to represent variations in elevation (**Fig. BxF.1a**). A **contour line** is an imaginary line along which all points have the same elevation. For example, if you walk along the 200-m contour line on a hill slope, you stay at exactly the same elevation (200 m). As another example, the shoreline on a flat, calm body of water is a contour line. In other words, you can picture the contour line as the intersection between the land surface and an imaginary horizontal plane (**Fig. BxF.1b**).

We can also represent variations in elevation by means of a topographic profile (**Fig. BxF.1c**). A profile is the trace of the ground surface as it would appear on a vertical plane that sliced into the ground—put another way, it's the shape of the ground surface as viewed from the side.

The elevation difference between two adjacent contour lines on a topographical map is called the **contour interval**. For a given topographic map, the contour interval is constant, so the spacing between contour lines represents the steepness of a slope. Specifically, closely spaced contour lines represent a steep slope, whereas widely spaced contour lines represent a gentle slope (**Fig. BxF.1d, e**).

If we add a representation of geologic features under the ground surface, then we have a geologic cross section. In some cases, geologists gain insight into subsurface geology simply by looking at the shape of a landform. For example, a steep cliff in a region of dipping sedimentary strata may indicate the presence of a resistant (difficult to erode) layer; low areas may be underlain with nonresistant (easy to erode) layers (**Fig. BxF.1f**).

FIGURE Bx F.1 Topographic maps and profiles.

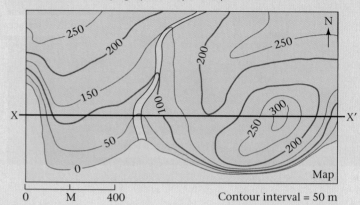

Contour interval = 50 m

(a) A topographic map depicts the shape of the land surface through the use of contour lines. The difference in elevation between two adjacent lines is the contour interval.

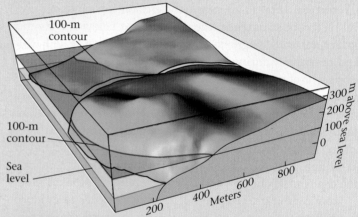

(b) A contour line is the intersection of a horizontal plane with the land surface. This block diagram shows the map area.

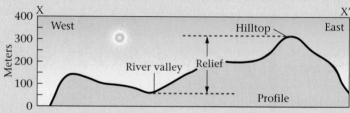

(c) A topographic profile (along section line X′–X′) shows the shape of the land surface as seen in a vertical slice.

landscape that would consist of gently rolling, forested hills in a temperate climate might consist of barren escarpments in a desert climate.

- *Substrate composition*: A substrate's composition determines how it responds to erosion. Regions underlain by hard rock erode less easily than regions underlain by unconsolidated sediment.

- *Time*: Landscapes change in response to continued erosion and/or deposition acting over time. In erosional landscapes, more erosion happens as time passes, whereas

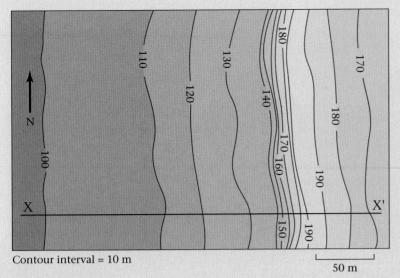

Contour interval = 10 m

50 m

(d) Topographic map showing a north-south trending escarpment, indicated by the closely spaced contour lines.

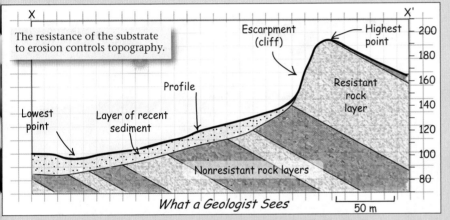

The resistance of the substrate to erosion controls topography.

What a Geologist Sees

50 m

(e) Topographic profile showing a geologic interpretation of the subsurface.

1km

(f) These curving ridges in Wyoming, as seen from an airplane, are underlain by resistant sandstone. The valleys in between are underlain by nonresistant shale.

in depositional landscapes, more deposition takes place as time passes.

Although water, wind, and ice are responsible for the development of most landscapes, human activities have had an increasingly important impact on the Earth's surface. We have dug pits (mines) where once there were mountains, built hills (tailings piles and landfills) where once there were valleys, and made steep slopes gentle and gentle slopes steep (Fig. F.3a–c). By constructing concrete walls, we modify the shapes of coastlines, change the courses of rivers, and fill new lakes (reservoirs). In cities, buildings and pavements completely seal the ground and cause water that might once have seeped down into the ground to spill into streams instead, increasing their flow. The area of land covered by pavement or buildings in the United States now exceeds the area of Ohio! And in the country, agriculture, grazing, water usage, and deforestation substantially alter the rates at which natural erosion and deposition take place. For example, agriculture greatly increases the rate of erosion, because for much of the year farm fields have no vegetation cover.

F.4 THE HYDROLOGIC CYCLE

Nothing that is can pause or stay—
The moon will wax, the moon will wane,
The mist and cloud will turn to rain,
The rain to mist and cloud again,
Tomorrow be today.

—Henry Wadsworth Longfellow (1807–1882)

As is evident from the discussion above, water in its various forms (liquid, gas, and solid) plays a major role in erosion and deposition on Earth's surface. Our planet's water is found in certain distinct reservoirs (containers), namely: the oceans, glacier ice (today, mostly in Antarctica and Greenland), **groundwater**, lakes, soil moisture, living organisms, the atmosphere, and rivers (Table F.2). Together, these constitute the hydrosphere. Water constantly flows from reservoir to reservoir, and this never-ending passage is called the **hydrologic cycle** (see Geology at a Glance, pp. 536–537). Perhaps without realizing it, Longfellow, an American poet fascinated with reincarnation, provided an accurate if somewhat romantic image of the hydrologic cycle. Without this cycle, the erosive forces of running water (rivers and streams) and flowing ice (glaciers) would not exist.

FIGURE F.3   Human influence on a geologic scale.

(a) The pyramids of Egypt are human-made hills that rise above the desert sands. They have lasted for thousands of years.

(b) In the process of making highway cuts, deep valleys are cut through high ridges. This example borders a highway near Denver.

(c) This stone dam holds back a reservoir in Colorado. Think about how long it would take a glacier to pile up so much sediment.

The *average* length of time that water stays in a particular reservoir during the hydrologic cycle is called the **residence time** (Table F.3). Water in different reservoirs has different residence times. For example, a typical molecule of water remains in the oceans for 4,000 years or less, in lakes and ponds for 10 years or less, in rivers for 2 weeks or less, and in the atmosphere for 10 days or less. Groundwater residence times are highly variable and depend on how deep the groundwater flows. Water can stay underground for anywhere from 2 weeks to 10,000 years before it inevitably moves on to another reservoir.

To get a clearer sense of how the hydrologic cycle operates, let's follow the fate of seawater that has just reached the surface of the ocean. Solar radiation heats the water, and the increased thermal energy of the vibrating water molecules allows them to evaporate (break free from the liquid) and drift upward in a gaseous state to become part of the atmosphere. About 417,000 cubic km (102,000 cubic miles), or about 0.03% of the total ocean volume (1.35 billion km^3), evaporates every year. Atmospheric water vapor moves with the wind to higher elevations, where it cools, undergoes condensation (the molecules link together to form a liquid), and rains or snows. About 76% of this water precipitates (falls out of the air) directly back into the ocean. The remainder precipitates onto land; most of this water becomes trapped temporarily in the soil, or in plants and animals, and soon returns directly to the atmosphere by **evapotranspiration**. This is the sum of evaporation from bodies of water, evaporation from the ground surface, and transpiration (release as a metabolic by-product) from plants and animals. Rainwater that did not become trapped in the soil or in living organisms either enters lakes or rivers and ultimately flows back to the sea as surface water, becomes trapped in glaciers, or sinks deeper into the ground to become groundwater. Groundwater also flows and ultimately returns to the Earth's surface

TABLE F.2 Major Water Reservoirs of the Earth

H₂O Reservoir	Volume (km³)	% of Total Water	% of Fresh Water
Oceans and seas	1,338,000,000	96.5	—
Glaciers, ice caps, snow	24,064,000	2.05	68.7
Saline groundwater	12,870,000	0.76	—
Fresh groundwater	10,500,000	0.94	30.1
Permafrost	300,000	0.022	0.86
Freshwater lakes	91,000	0.007	0.26
Salt lakes	85,400	0.006	—
Soil moisture	16,500	0.001	0.05
Atmosphere	12,900	0.001	0.04
Swamps	11,470	0.0008	0.03
Rivers and streams	2,120	0.0002	00.006
Living organisms	1,120	0.0001	00.003

Source: Data from P. H. Gleick, *Encyclopedia of Climate and Weather* (New York: Oxford University Press, 1996).

TABLE F.3 Estimated Residence Time of Water in Earth's Reservoirs

H₂O Reservoir	Average Residence Time
Ice caps	10,000 to 200,000 years
Deep groundwater	3,000 to 10,000 years
Oceans and inland seas	3,000 to 3,500 years
Shallow groundwater	100 to 200 years
Valley glaciers	20 to 100 years
Freshwater lakes	50 to 100 years
Winter snow	2 to 6 months
Rivers and streams	2 to 6 months
Soil moisture	1 to 2 months
Atmosphere	5 to 15 days
Living organisms	hours to days

reservoirs. In sum, during the hydrologic cycle, water moves among the ocean, the atmosphere, reservoirs on or below the land surface, and living organisms.

F.5 LANDSCAPES OF OTHER PLANETS

The dynamic, ever-changing landscape of Earth contrasts markedly with those of other terrestrial planets. Each of the terrestrial planets and moons has its own unique surface landscape features, reflecting the interplay between the object's particular tectonic and erosional processes. Let's look at a few examples: the Moon, Mars, and Venus.

Our Moon has a static, pockmarked landscape generated exclusively by meteorite impacts and volcanic activity. Because no plate tectonics occurs on the Moon, no new mountains form; and because no atmosphere or ocean exists, there is no hydrologic cycle and no erosion from rivers, glaciers, or winds. Therefore, the lunar surface has remained largely unchanged for over billions of years. The landscape can be divided into the Lunar Highlands, the heavily cratered, light-colored regions of the moon exposing rocks over 4.0 billions of years old; and the mare, vast plains of flood basalt possibly formed in response to impacts over 3.8 billion years ago. These impacts were so huge that they caused melting in the Moon's mantle and extrusion of flood basalts (Fig. F.4a, b).

Landscapes on Mars differ from those of the Moon because Mars *does* have an atmosphere (though much less dense than that of Earth) whose winds generate huge dust storms, some of which obscure nearly the entire surface of the planet for months at a time. The landscapes of Mars also differ from the Moon's because Mars probably once had surface water (Box F.2 and Fig. F.4c). Thus, the Martian surface appears to consist of four kinds of materials: volcanic flows and deposits (primarily of basalt), debris from impacts, windblown sediment, and water-laid sediment. There is even evidence that soil-forming processes affected surface materials. Martian winds not only deposit sediment, they also slowly erode impact craters and polish surface rocks.

Landscapes on Mars also differ from those on Earth, because Mars does not have plate tectonics. So, unlike Earth, Mars has no mountain belts or volcanic arcs. In fact, most landscape features on Mars, with the exception of wind-related ones, are over 3 billion years old. Long ago, a huge mantle plume formed. This plume caused the uplift of a 9-km-high bulge (the Tharsis Ridge) that covers an area comparable to that of North America. Thermal activity also led to the eruption of gargantuan hot-spot volcanoes, such as the 22-km-high Olympus Mons, the highest mountain in the Solar System. Mars also boasts the largest known canyon, the Valles Marineris, a gash over 3,000 km long and 8 km deep. No comparable feature exists on Earth. Because Mars has no vegetation and no longer has rain, its surface does not weather and erode like that of Earth, so it still bears the scars of impact by swarms of meteors earlier in the history of the Solar System, scars that have long since disappeared on Earth. Mars does have a

FIGURE F.4 Landscapes of other planets.

(a) The heavily cratered surface of Earth's moon.

(b) A close-up view of the lunar landscape, with the lunar rover and an astronaut for scale.

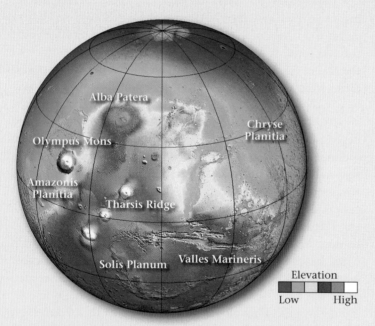

(c) A digital elevation model depicting the surface of Mars. Note the huge bulge of Tharsis Ridge, the giant Olympus Mons volcano, and the deep Valles Marineris canyon.

(d) A radar image of Venus. A thick blanket of clouds obscures this planet's surface, so it can't be seen through a telescope. Red areas are higher, blue areas lower.

hydrologic cycle, of sorts, in that it has ice caps that grow and recede on an annual basis.

Venus is closer to the size of the Earth and may still have operating mantle plumes. Virtually the entire surface of Venus was resurfaced by volcanic eruptions about 300 to 1,600 Ma, making the planet's surface much younger than those of the Moon and Mars. Further, Venus has a dense atmosphere that protects it from impacts by smaller objects. Because there has been relatively little cratering since the resurfacing event, volcanic and tectonic features dominate the landscape of Venus (**Fig. F.4d**). Satellites have used radar to reveal a variety of volcanic constructions (such as shield volcanoes, lava flows, and

Water on Mars?

In 1877, an Italian astronomer named Giovanni Schiaparelli studied the surface of Mars with a telescope and announced that long, straight *canali* crisscrossed the planet's surface. *Canali* should have been translated into the English word *channel*, but perhaps because of the recent construction of the Suez Canal, newspapers of the day translated the word into the English *canal*, with the implication that the features had been constructed by intelligent beings. An eminent American astronomer began to study the "canals" and suggested that they had been built to carry water from polar ice caps to Martian deserts.

Late-twentieth-century satellite mapping of Mars showed that the "canals" do not exist—they were simply optical illusions. There are no lakes, oceans, rainstorms, or flowing rivers on the surface of Mars today. The atmosphere of Mars has such low density, and thus exerts so little pressure on the planet's surface, that any liquid water released at the surface would quickly evaporate. Thus Mars has no hydrologic cycle the way the Earth does. But three crucial questions remain: Does liquid water ever form, even for short periods, on the Martian surface today? Did Mars ever have a significant amount of running water or standing water in the past? If the planet once had significant water, where is the water now? The question of the presence of water lies at the heart of an even more basic question: Given that the simplest life as we know it requires water, is there, or was there, life on Mars?

Many planetary geologists believe that the case for liquid water on Mars is quite strong. Much of the evidence comes from comparing landforms on the planet's surface with landforms of known origin on Earth. High-resolution images of Mars reveal a number of landforms that look as if they formed in response to flowing water. Examples include networks of channels resembling river networks on Earth (**Fig. BxF.2a, b**), scour features, deep gullies, and streamlined deposits of sediment.

Studies by the *Odyssey* satellite in 2003, and by Mars rovers (*Spirit* and *Opportunity*) that landed on the planet in 2004, added intriguing new data to the debate. *Odyssey* detected hints that hydrogen, an element in water, exists beneath the surface of the planet over broad regions, and the Mars rovers have documented the existence of hematite and gypsum, minerals that form in the presence of water. The rovers also have found sedimentary deposits that appear to have been deposited in water. The *Phoenix* lander, in 2008, confirmed the existence of water ice by digging into the surface to expose some. Researchers speculate that Mars was much wetter in its past, perhaps billions of years ago. But since the atmosphere became less dense, the water evaporated and now lies hidden underground or trapped in polar ice caps.

FIGURE BxF.2 Water-carved landscapes on Mars, as photographed by satellites orbiting the planet.

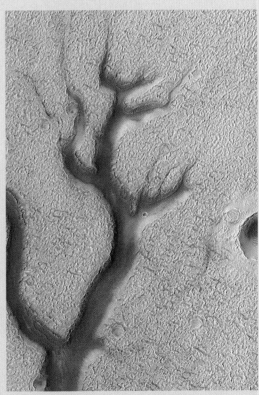

(a) An elongate island in a channel, resembling islands that have been shaped by rivers on Earth.

(b) A network of channels resembling channels cut by rivers on Earth.

The Hydrologic Cycle

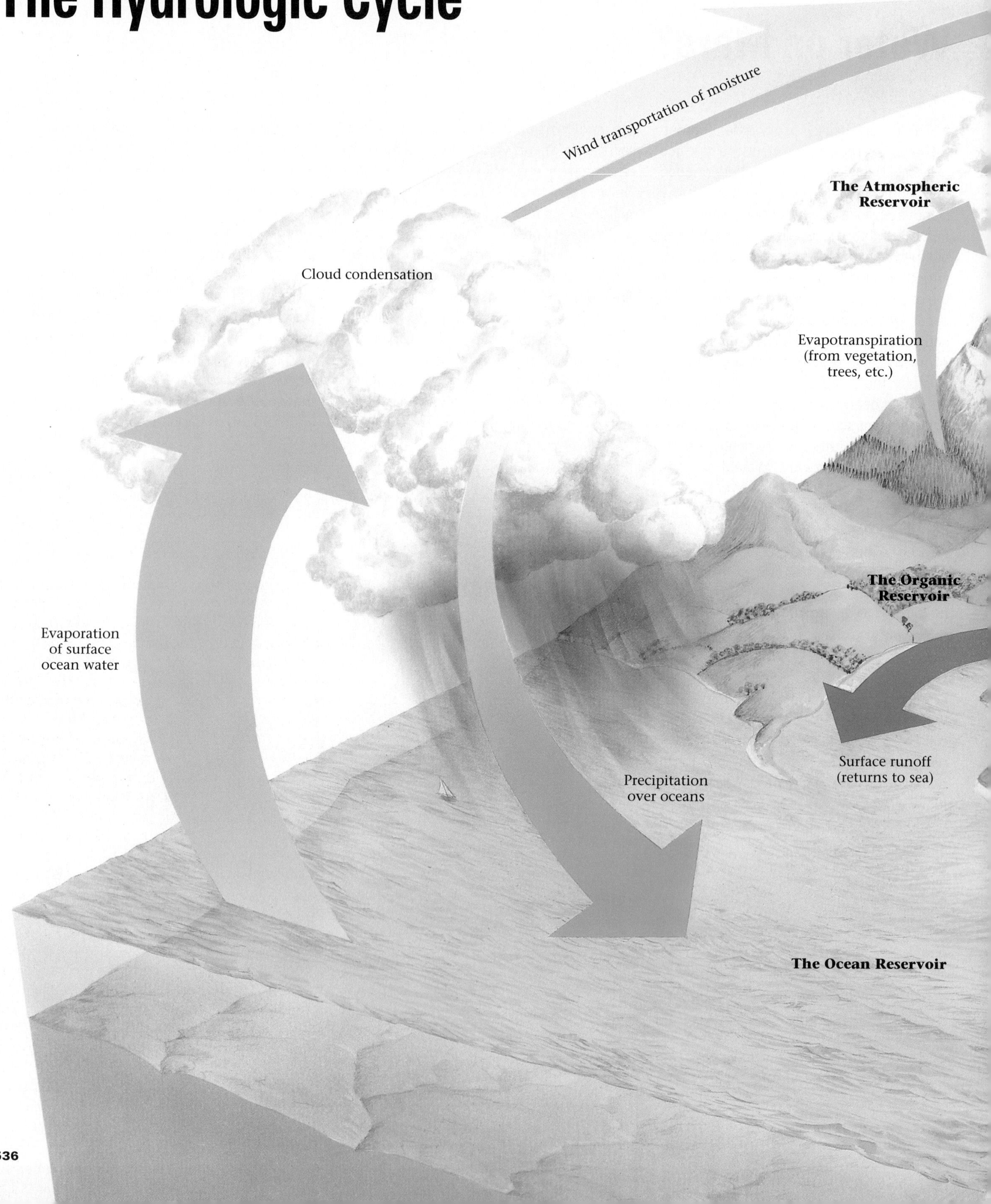

Wind transportation of moisture

The Atmospheric Reservoir

Cloud condensation

Evapotranspiration (from vegetation, trees, etc.)

The Organic Reservoir

Evaporation of surface ocean water

Surface runoff (returns to sea)

Precipitation over oceans

The Ocean Reservoir

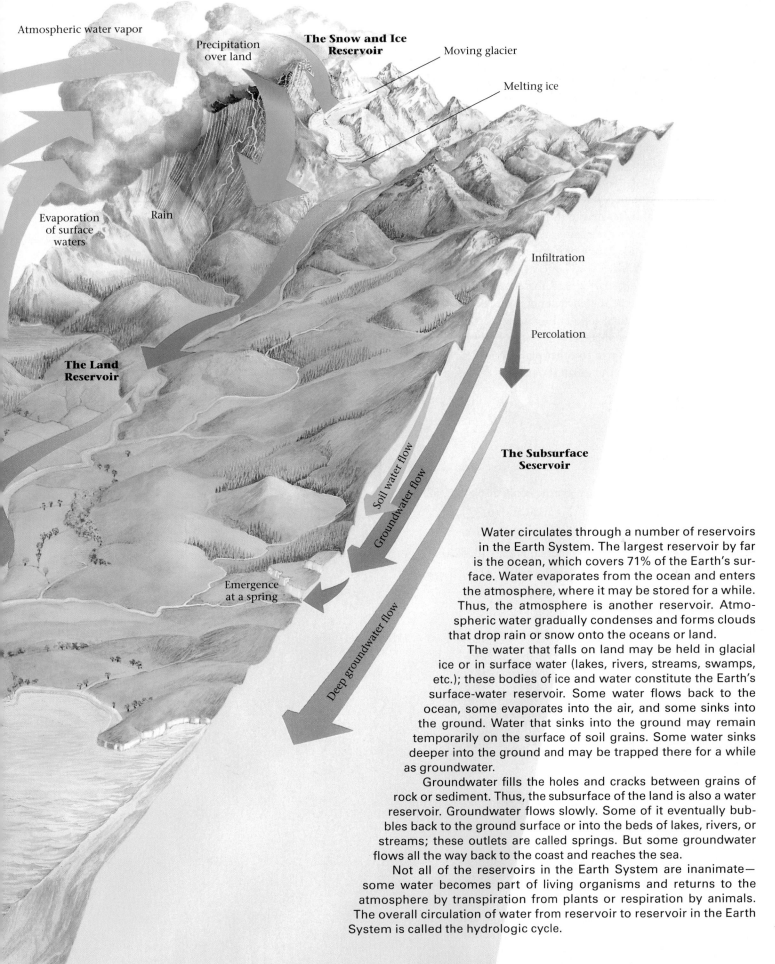

Atmospheric water vapor

Precipitation over land

The Snow and Ice Reservoir

Moving glacier

Melting ice

Evaporation of surface waters

Rain

Infiltration

Percolation

The Land Reservoir

Soil water flow

Groundwater flow

The Subsurface Seservoir

Emergence at a spring

Deep groundwater flow

Water circulates through a number of reservoirs in the Earth System. The largest reservoir by far is the ocean, which covers 71% of the Earth's surface. Water evaporates from the ocean and enters the atmosphere, where it may be stored for a while. Thus, the atmosphere is another reservoir. Atmospheric water gradually condenses and forms clouds that drop rain or snow onto the oceans or land.

The water that falls on land may be held in glacial ice or in surface water (lakes, rivers, streams, swamps, etc.); these bodies of ice and water constitute the Earth's surface-water reservoir. Some water flows back to the ocean, some evaporates into the air, and some sinks into the ground. Water that sinks into the ground may remain temporarily on the surface of soil grains. Some water sinks deeper into the ground and may be trapped there for a while as groundwater.

Groundwater fills the holes and cracks between grains of rock or sediment. Thus, the subsurface of the land is also a water reservoir. Groundwater flows slowly. Some of it eventually bubbles back to the ground surface or into the beds of lakes, rivers, or streams; these outlets are called springs. But some groundwater flows all the way back to the coast and reaches the sea.

Not all of the reservoirs in the Earth System are inanimate—some water becomes part of living organisms and returns to the atmosphere by transpiration from plants or respiration by animals. The overall circulation of water from reservoir to reservoir in the Earth System is called the hydrologic cycle.

calderas). Rifting on Venus produced faults, some of which occur in association with volcanic features. Liquid water cannot survive the scalding temperatures of Venus's surface, so no hydrologic cycle operates there and no life exists. Because of the density of the atmosphere, winds are too slow to cause much erosion or deposition, thus leaving volcanic landforms virtually unchanged.

During the past two decades, spacecraft visiting the moons of the outer planets have sent home amazing images of surface features that differ markedly from any found on Earth. As an example, consider Enceladus, a 500-km-diameter moon of Saturn (Fig. F.5a, b). Much of Enceladus's ice-covered surface is cracked and wrinkled and largely crater free, suggesting that tectonic movements rifted and folded this moon's crust subsequent to the intense meteorite bombardment episodes of early Solar System history. The still-cratered terrains may be older and stabler regions of the crust.

After this brief side trip to other planets, let's now return to Earth. Our planet has the greatest diversity of landscapes in the Solar System.

Guide Terms

agents of erosion (p. 529)	gravitational energy (p. 528)
contour interval (p. 530)	groundwater (p. 531)
contour line (p. 530)	hydrologic cycle (p. 531)
deposition (p. 528)	internal energy (p. 528)
depositional landform (p. 528)	landform (p. 527)
digital elevation map (DEM) (p. 529)	landscape (p. 527)
	relief (p. 528)
downslope movement (p. 528)	residence time (p. 532)
erosion (p. 528)	subsidence (p. 528)
erosional landform (p. 528)	substrate (p. 529)
evapotranspiration (p. 532)	topographic map (p. 530)
external energy (p. 528)	topography (p. 530)
	uplift (p. 528)

FIGURE F.5 The variety of moons surrounding gas-giant planets. The moons come in a variety of compositions and with different amounts of cratering.

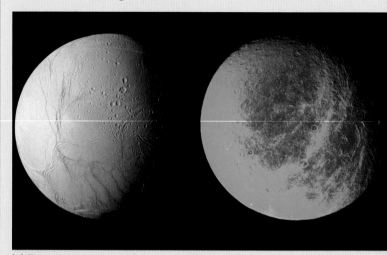

(a) Two of the moons of Saturn—Enceladus (radius = 256 km), on the left, and Rhea (radius = 764 km), on the right.

(b) Four of the moons of Jupiter. From left to right (with radii km): Io (1,821), Europa (1,569), Ganymede (2,634), Callisto (2,410).

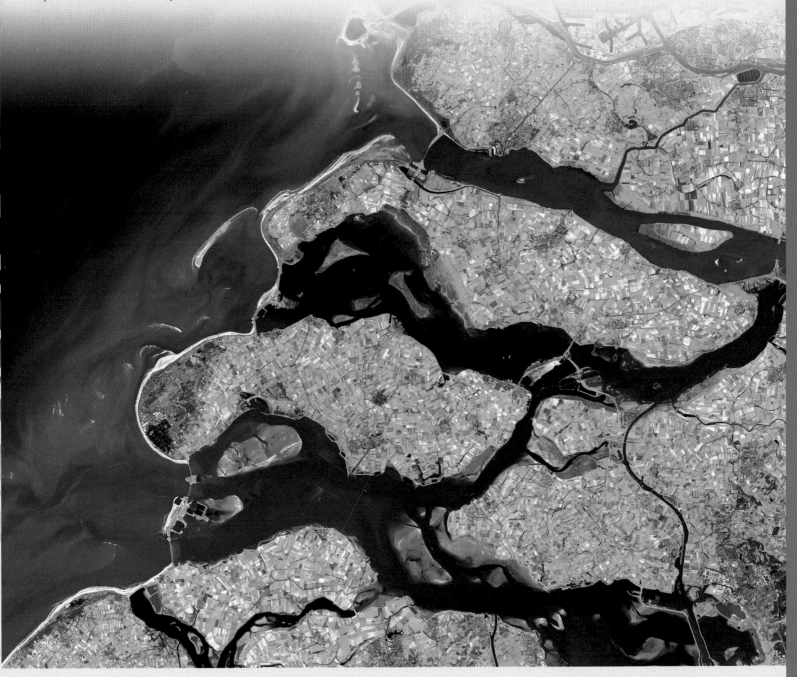

ANOTHER VIEW This ~50-km-wide view of the Netherlands western coast was taken looking straight down from an elevation of 650 km (400 miles) by the Terra satellite. The patchwork pattern represents farm fields. Much of the land in this view was once below sea level but has been isolated by dikes and drained. Thus we see a landscape that has been intensively modifed by humanity. The substrate of the region consists of sediment deposited in a river delta, and the swirls offshore consist of submarine sandbars whose shape has been modified by coastal currents.

CHAPTER 16

Unsafe Ground: Landslides and Other Mass Movements

A mass of rock and sediment, almost 1 km wide and 100 m thick, buried a highway and blocked the Naches River in the Cascade Mountains of Washington. Though destructive, this landslide is actually quite small, compared with some that have happened in historic time. Sloping land can move, sometimes with disastrous consequences.

GEOPUZZLE

Picture a mountain slope covered by an ancient forest. The next day, the forest along with tens of meters of its substrate (soil and bedrock) are gone, having tumbled downhill to form a massive pile of debris in the river valley below. What triggers such catastrophic landslides? Can they be prevented?

16.1 INTRODUCTION

It was Sunday, May 31, 1970, a market day, and thousands of people had crammed into the Andean town of Yungay, Peru, to shop. Suddenly they felt the jolt of an earthquake, strong enough to topple some masonry houses. But worse was yet to come. This earthquake also broke an 800-m-wide ice slab off the end of a glacier at the top of Nevado Huascarán, a nearby 6.6-km-high mountain peak. Gravity instantly pulled the ice slab down the mountain's steep slopes. As it tumbled down over 3.7 km, the ice disintegrated into a chaotic avalanche of chunks traveling at speeds of over 300 km per hour. Near the base of the mountain, most of the avalanche channeled into a valley and thickened into a moving sheet as high as a ten-story building, ripping up rocks and soil along the way. Friction transformed the ice into water, which when mixed with rock and dust created 50 million cubic meters of mud, a slurry viscous enough to carry boulders larger than houses. This mass, sometimes floating on a compressed air cushion that allowed it to pass without disturbing the grass below, traveled over 14.5 km in less than 4 minutes.

At the mouth of the valley, most of the mass overran the village of Ranrahirca before coming to rest and creating a dam that blocked the Santa River. But part of the mass shot up the sides of the valley and became airborne for several seconds, flying over the ridge bordering Yungay. As the town's inhabitants and visitors stumbled out of earthquake-damaged buildings, they heard a deafening roar and looked up to see the churning mud cloud bursting above the nearby ridge. The town was completely buried under several meters of mud and rock. When the dust had settled, only the top of the church and a few palm trees remained visible to show where Yungay once lay (Fig. 16.1a, b)—18,000 people are forever entombed beneath the mass. Today, the site is a grassy meadow with a hummocky (irregular and lumpy) surface, spotted with crosses left by mourning relatives.

Could the Yungay tragedy have been prevented? Perhaps. A few years earlier, climbers had recognized the instability of the glacial ice on Nevado Huascarán, and Peruvian newspapers had published a warning, but alas, no one took notice. In the aftermath of the event, geologists discovered that Yungay had been built on ancient layers of debris, from past avalanches. The government has since prevented new towns from rising in the danger zone.

People often assume that the ground beneath them is *terra firma*, a solid foundation on which they can build their lives. But the catastrophe at Yungay says otherwise. Much of the Earth's surface is unstable and capable of moving downslope in seconds to weeks. Geologists refer to the gravitationally caused downslope transport of rock, regolith (soil, sediment, and debris), snow, and ice as **mass movement**, or mass wasting. Like earthquakes, volcanic eruptions, storms, and floods, mass movements are a type of **natural hazard**, meaning a natural feature of the environment that can cause damage to living organisms and to buildings. Unfortunately, mass movement becomes more of a threat every year, because as the world's population grows, cities expand into areas of unsafe ground. In fact, by some estimates, mass movements may, on average, be *the most costly* type of natural hazard. But mass movement also plays a critical role in the rock cycle, for it's the first step in the transportation of sediment. And it plays a critical role in the evolution of landscapes: it's the most rapid means of modifying the shapes of slopes.

In this chapter, we look at the types, causes, and consequences of mass movement, and the precautions society can take to protect people and property from its dangers. You might want to consider this information when selecting a site for your home or when voting on land-use propositions for your community.

FIGURE 16.1 The May 1970 Yungay landslide disaster in Peru.

(a) Before the landslide, the town of Yungay perched on a hill near the ice-covered mountain Nevado Huascarán.

(b) The landslide completely buried the town beneath debris. A landslide scar is visible on the mountain in the distance.

By the end of this chapter, you should know . . .

- the characteristics and consequences of different types of mass movements.
- factors that determine whether a slope is stable or unstable.
- events that can trigger a mass-movement event.
- why some regions are more susceptible to mass movements than are others.
- how landslide hazards can be evaluated and, in some cases, prevented.

16.2 TYPES OF MASS MOVEMENT

Most people refer to any mass movement of rock and/or regolith (soil and unconsolidated sediment) down a slope as a **landslide**. Geologists and engineers, however, find it useful to distinguish among different kinds of landslides based on four features: (1) the type of material involved (rock or regolith); (2) the velocity of movement (slow, intermediate, or fast); (3) the character of the moving mass (coherent, chaotic, or slurry); and (4) the environment in which the movement takes place (subaerial or submarine). Similarly, most people envision an **avalanche** as a downslope mass movement of snow—geologists and engineers, however, apply the term more broadly to any mass movement that moves like a turbulent cloud. Why bother classifying mass movements? We make these distinctions because different types of mass movements have different consequences and, therefore, represent different kinds of hazard—by characterizing mass movements more completely, we can better prepare for them. Below, we examine mass movements that occur on land roughly in order from slow to very fast (See for Yourself O, p. S-28). We also briefly consider sumbarine mass movements.

Creep, Solifluction, and Rock Glaciers

In temperate climates, the upper few centimeters of ground freeze during the winter, only to thaw again the following spring. Because water increases in volume by 9.2% when it freezes, water-saturated soil and underlying fractured rock expand outward, and particles in the regolith move out perpendicular to the slope during the winter. During the spring thaw, water becomes liquid again, and gravity makes the particles sink vertically and thus migrate downslope slightly. This gradual downslope movement of regolith is called **creep**. You can't see creep by staring at a hill slope, because it occurs too slowly; but over a period of years, creep causes trees, fences, gravestones, walls, and foundations built on a hillside to tilt downslope.

Trees that continue to grow after they have been tilted display a pronounced curvature at their base (Fig. 16.2a–c).

In Arctic or high-elevation regions, regolith freezes solid to great depth during the winter. In the brief summer thaw, only the uppermost 1 to 3 m of the ground thaws. Since meltwater cannot sink into the permafrost (permanently frozen ground below), the melted layer becomes soggy and weak and flows slowly downslope in overlapping sheets. Geologists refer to this kind of creep, characteristic of cold, treeless tundra regions, as **solifluction** (Fig. 16.2d).

Slow mass movement also takes place in a **rock glacier**, a body made of rock fragments embedded in a matrix of ice (Fig. 16.2e). Rock glaciers differ from more familiar ice-dominated glaciers in terms of the proportion of rock fragments to ice—in a rock glacier, most of the volume consists of rock, whereas in an ice-dominated glacier, most of the volume consists of ice. In effect, rock glaciers are breccias cemented by ice. Since ice is very weak, the combined mass of rock and ice in a rock glacier can slowly flow downslope, as does the relatively pure ice of an ice-dominated glacier. Rock glaciers form in two ways. First, they develop where snow or rain percolates down into a pile of rock debris that has accumulated above permafrost at the base of a cliff. This water can't infiltrate the frozen ground below the debris pile, so it freezes to form ice in the pores between clasts within the debris pile. Second, they develop where an ice-dominated glacier already containing abundant rock debris begins to melt. As melting progresses, the proportion of rock to ice in the glacier increases, and if enough melting takes place, the glacier eventually contains more rock than ice.

Slumping

Near Pacific Palisades, along the coast of southern California, Highway 1 runs between the beach and a 120-m-high cliff. On March 31, 1958, a gash developed about 200 m inland of the cliff's edge, and a *semicoherent* mass of sediment and rock began to move downslope. Four days later, when the movement finally stopped, a 1-km-long stretch of the coastal highway had been buried—it took weeks for bulldozers to make this stretch passable again. A similar event happened in northern New York State during the summer of 2011 when, after weeks of drenching rains, a 1.5-km-wide portion of a slope began to move down and out into the floor of Keene Valley. The mass moved at only centimeters to tens of centimeters per day, but even at this slow rate, the accumulated displacement destroyed several expensive homes. In places, the boundary between the moving mass and the unmoving land upslope evolved into a 5-m-high escarpment.

Geologists refer to such mass-movement events, during which moving rock and/or regolith does not disintegrate into a jumble of debris but rather stays somewhat coherent, as a **slump**, and they refer to the moving mass itself as a slump block

FIGURE 16.2 The process and consequences of slow mass movements (creep and solifluction).

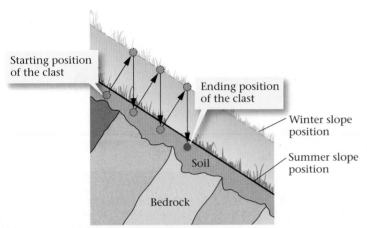

(a) Creep due to freezing and thawing: The clast rises perpendicular to the ground during freezing, and sinks vertically during thawing. After 3 years, it migrates to the position shown.

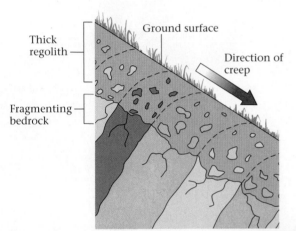

(b) As rock layers weather and break up, the resulting debris creeps downslope.

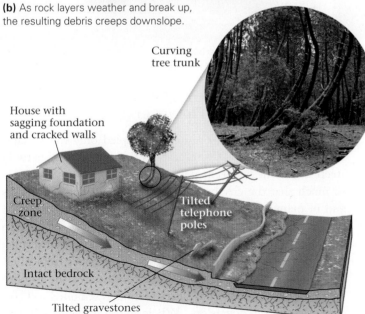

(c) Soil creep causes walls to bend and crack, building foundations to sink, trees to bend, and power poles and gravestones to tilt.

(d) Solifluction on a hill slope in the tundra.

(e) A rock glacier in Alaska. Note how the flow of the ice below wrinkles the rock layer on the glacier's surface.

FIGURE 16.3 The process of slumping on a hillslope. Note the scarps that form at the head of the slump.

(a) A head scarp on the hillslope.

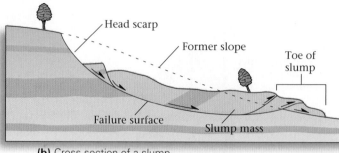

(b) Cross section of a slump.

(c) Slumping dumped sediment into this river in Costa Rica.

(d) A slump beginning to form along a highway in Utah.

(Fig. 16.3a–d). A slump block slides on a **failure surface**. Some failure surfaces are planar, but commonly they curve and resemble a spoon lying concave side up. The exposed upslope edge of a failure surface forms a **head scarp**, a new cliff face. Geologists refer to the downslope end of a slump block as the block's toe. On a hillslope, the toe can move up and over the preexisting land surface to form a curving ridge; but along sea coasts or river banks, the toe ends up in the water where it erodes away. The upslope and downslope ends of a slump block may break into a series of discrete slices, each separated from its neighbor by a small sliding surface. Slumps come in all sizes, from only a few meters across to tens of kilometers across. They move at speeds from millimeters per day to tens of meters per minute. Structures (such as houses, patios, and swimming pools) built on them crack and fall apart.

Mudflows, Debris Flows, and Lahars

Rio de Janeiro, Brazil, originally occupied only the flatlands bordering beautiful crescent beaches that had formed between steep hills. But in recent decades, the population has grown so much that the city has expanded up the sides of the hills, and in many places densely populated communities of makeshift shacks cover the slopes. These communities, which have no storm drains, were built on the thick regolith that resulted from long-term weathering of bedrock in Brazil's tropical climate. In 1988, particularly heavy rains saturated the regolith, which turned into a viscous slurry of mud, resembling wet concrete, that flowed downslope. Whole communities disappeared overnight, replaced by a hummocky muddle of mud and debris. And at the base of the cliffs, the flowing mud knocked over and buried buildings of all sizes. Similar devastation occurred in

FIGURE 16.4 Examples of mudflows and lahars.

(a) Mudslides of 2011 stripped away forests on hillslopes in Brazil.

(c) A recent debris flow in Utah. Note the chaotic mixture of rock chunks and mud.

(b) A 2011 mudslide destroyed a high-rise building at the base of the hill.

(d) The aftermath of a lahar that flowed down the side of Mt. St. Helens following an eruption in 1982.

regions north of Rio de Janeiro in 2011 (Fig. 16.4a, b). Geologists refer to a moving slurry of mud as a **mudflow** or mudslide, and a slurry consisting of a mixture of mud and larger, pebble- to boulder-sized fragments as a **debris flow** or debris slide (Fig. 16.4c and Box 16.1).

The speed at which mud or debris moves depends on the slope angle and on the water content. Flows move faster if they contain more water (i.e., are less viscous), and if they move on steeper slopes. On a gentle slope, drier mud flows move like molasses; but on a steep slope, low-viscosity, very wet mud may move at over 100 km per hour. Because mud and debris flows have greater viscosity than clear water, they can carry large rock chunks as well as houses and cars. They typically follow channels downslope and at the base of the slope they spread out into a broad lobe.

Particularly devastating mudflows spill down the river valleys bordering volcanoes. These mudflows, known as **lahars**, consist of a mixture of volcanic ash from a currently erupting or previously erupted pyroclastic cloud, and water from the snow and ice that melts in a volcano's heat or from heavy rains (Fig. 16.4d; see Chapter 9). One of the most destructive lahars occurred on November 13, 1985, in the Andes Mountains in Colombia. That night, a major eruption melted a volcano's thick snowcap, creating hot water that mixed with ash. A scalding lahar rushed down river valleys and swept over the nearby town of Armero while most inhabitants were asleep. Of the 25,000 residents, 20,000 perished.

Rock and Debris Slides

In the early 1960s, engineers built a huge new dam across a river on the northern side of Monte Toc, in the Italian Alps, to create a reservoir for generating electricity. This dam, the Vaiont Dam, was an engineering marvel, a concrete wall rising

BOX 16.1

CONSIDER THIS . . . . . .

What Goes Up Must Come Down

Along the shore of California, waves slowly erode the land and produce low, flat areas called wave-cut benches (see Chapter 18). While this happens, tectonic motions slowly raise the land surface. When uplifted, these benches form small plateaus, or terraces. One such terrace lies at an elevation of 180 m above sea level, about 500 m east of the present-day beach at La Conchita; the west face of this terrace is a cliff-like bluff (**Fig. Bx16.1a**). Repeated slip along the San Andreas plate boundary has "busted up" the bedrock, so the substrate of the terrace and bluff has weathered into clay and debris.

Relatively little vegetation covers the bluff or the terrace above. Rain that falls on the face of the bluff drains away quickly via a network of small, temporary streams. But the water falling on the terrace infiltrates into the ground, sinks down, and saturates clay and debris meters below the surface of the bluff, turning into very weak mud. When this happens, the weight of surface material causes the bluff to give way, and a mass of mud and debris flows downslope at rates of up to 10 m per second.

If the region of La Conchita were uninhabited, such mass wasting would just be part of the natural process of land-scape evolution—gravity brings down land that had been raised by tectonic activity. But when downslope movements take place in La Conchita, it makes headlines, because, on the modern wave-cut bench between the shore and the base of the bluff, developers built a community housing 350 people. In 1995, a mud and debris flow overwhelmed 9 houses at the base of the bluff. An even more devastating flow happened in 2005, burying 13 houses, damaging 23, and killing 10 people (**Fig. Bx16.1b, c**). Nature has sent a clear message about the importance of reading the landscape before planning construction.

FIGURE Bx16.1 The 2005 La Conchita mudslide along the coast of California.

(a) A housing development was built in a narrow strip between the beach and steep cliffs.

(b) During heavy rains, the slope gave way and heavy mud flowed down, burying houses and taking several lives.

(c) Rescuers at the toe of the mudslide.

FIGURE 16.5 The Vaiont Dam disaster—a catastrophic landslide that displaced the water in a reservoir with rock debris.

(a) Vaiont Dam and the debris now behind it. The exposed failure surface on Monte Toc is visible in the distance.

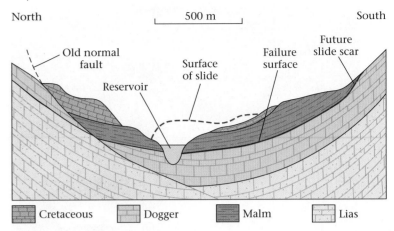

(b) A cross section parallel to the face of Vaiont Dam before the landslide. The failure surface lies at the base of the weak Malm Shale. The dashed line shows the top of the slide debris pile, after the slide had moved. The names in the explanation refer to time intervals in the Mesozoic.

260 m (as high as an 85-story skyscraper) above the valley floor (Fig. 16.5a). Unfortunately, the dam's builders did not recognize the hazard posed by nearby Monte Toc. The side of Monte Toc facing the reservoir was underlain with limestone beds interlayered with weak shale beds. These beds dipped parallel to the surface of the mountain and curved under the reservoir (Fig. 16.5b). As the reservoir filled, the flank of the mountain cracked, shook, and rumbled. Local residents began to call Monte Toc *la montagna che cammina* (the mountain that walks).

After several days of rain, Monte Toc began to rumble so much that on October 9, 1963, engineers lowered the water level in the reservoir. They thought the wet ground might slump a little into the reservoir, with minor consequences, so no one ordered the evacuation of the town of Longarone, a few kilometers down the valley. Unfortunately, the engineers underestimated the problem. At 10:30 that evening, a huge chunk of Monte Toc—600 million tons of rock—detached from the mountain and slid downslope into the reservoir. Some debris rocketed up the opposite wall of the valley to a height of 260 m above the original reservoir level. The displaced water of the reservoir spilled over the top of the dam and rushed down into the valley below. When the flood had passed, nothing of Longarone and its 1,500 inhabitants remained. Though the dam itself still stands, it holds back only debris and has never provided any electricity.

Geologists refer to such a sudden movement of rock and debris down a nonvertical slope as a **rock slide**, if the mass consists only of rock, or a **debris slide**, if it consists mostly of regolith. Once a slide has taken place, it leaves a scar on the slope and forms a debris pile at the base of the slope.

Slides happen when bedrock and/or regolith detaches from a slope, slips rapidly downhill on a failure surface, and breaks up into a chaotic jumble. Slides are more likely to occur where a weak layer of rock or sediment at depth below the ground parallels the land surface. (At the Vaiont Dam, the plane of weakness that would become the failure surface was a weak shale bed.) Slides may move at speeds of up to 300 km per hour; they are particularly fast when a cushion of air gets trapped beneath, so there is virtually no friction between the slide and its substrate, and the mass moves like a hovercraft. Rock and debris slides sometimes have enough momentum to climb the opposite side of the valley into which they fall. Slides, like slumps, come at a variety of scales. Most are small, involving blocks up to a few meters across. Some, such as the Vaiont slide, are large enough to cause a catastrophe.

Avalanches

In the winter of 1999, an unusual weather system passed over the Austrian Alps. First it snowed. Then the temperature warmed and the snow began to melt. But then the weather turned cold again, and the melted snow froze into a hard, icy crust. This cold snap ushered in a blizzard that blanketed the ice crust with tens of centimeters (1–2 ft.) of new snow. With the frozen snow layer underneath acting as a failure surface, 200,000 tons of new snow began to slide down the mountain.

FIGURE 16.6 Examples of avalanches.

(a) Aftermath of a 1999 avalanche in the Austrian Alps. Masses of snow buried several homes.

(c) A dry-snow avalanche in Alaska.

(b) Trees that were flattened by an avalanche are exposed now that the snow has melted.

As it accelerated, the mass transformed into a **snow avalanche**, a chaotic jumble of snow surging downslope. At the bottom of the slope, the avalanche overran a ski resort, crushing and carrying away buildings, cars, and trees, and killing over 30 people. It took searchers and their specially trained dogs many days to find buried survivors and victims under the 5- to 20-m-thick pile of snow that the avalanche deposited (Fig. 16.6a).

Snow avalanches display a variety of behaviors, depending on both the temperature of the snow, and the steepness of the slope down which the snow moves. Specifically, *wet-snow* avalanches, which involve snow that has started to melt and thus contains some liquid water, behave like a viscous slurry, in that they hug the slope as they move and entrain relatively little air. Wet-snow avalanches generally travel at speeds of less than 30 km per hour, and can pick up rock debris and vegetation along their path. *Dry-snow* avalanches contain cold, powdery snow that, on tumbling down a steep slope, disintegrates into a turbulent, air-rich cloud that rushes along at hurricane speeds

FIGURE 16.7 Examples of rockfalls.

(a) Successive rockfalls have littered the base of this sandstone cliff with boulders. Note the talus at the base of the cliff.

(c) A large rock slide buried a portion of the forest bordering a lake in the Uinta Mountains, Utah.

(b) A rockfall covered an Oregon highway in 2005. Note that the rock broke free on joint surfaces.

of up to 250 km per hour (Fig. 16.6b). Both wet-snow and dry-snow avalanches can become powerful enough to flatten forests in their paths (Fig. 16.6c).

What triggers snow avalanches? Some happen when a cornice, a large drift of snow that builds up on the lee side of a windy mountain summit, suddenly gives way and falls onto slopes below where it knocks free additional snow. Others happen when a broad slab of snow on a moderate slope detaches from its substrate along an icy failure surface. Avalanches tend to affect the same localities year after year, because of the characteristics of snow build-up and of slopes at the localities. To protect populated areas downslope of known "avalanche chutes," experts may use special explosives to trigger small, controlled avalanches before the snow piles deep enough to become a dangerous hazard.

Notably, geologists commonly use the word "avalanche" in a broader sense for any mass movement during which the solid fragments are suspended in so much fluid (air *or* water) that the flowing mixture behaves like a turbulent cloud. Thus, a "debris avalanche" contains rock fragments and regolith mixed with air, and a "submarine avalanche" is a cloud composed of sediment suspended in water. Avalanches of all types flow downslope because the mixture of solid and fluid in an avalanche is denser than the surrounding pure fluid.

Rockfalls and Debris Falls

Rockfalls and **debris falls**, as their names suggest, occur when a mass free-falls from a cliff (Fig. 16.7a). Commonly, rockfalls happen when a body of rock separates from a cliff face along a joint. Friction and collision with other rocks may bring some blocks to a halt before they reach the bottom of the slope; these blocks pile up to form a **talus**, a sloping apron of rocks along the base of the cliff. Debris that has fallen a long way can reach speeds of 300 km per hour and may have so much momentum that it keeps moving as an avalanche-like cloud of fragments mixed with air when it reaches the base of a cliff. Large, fast rockfalls push the air in front of them, creating a short blast of hurricane-like wind. For example, the wind in front of a 1996 rockfall in Yosemite National Park flattened over 2,000 trees.

Most rockfalls involve only a few blocks detaching from a cliff face and dropping into the talus. But some falls dislodge immense quantities of rock. In September 1881, a 600-m-high crag of slate, undermined by quarrying, suddenly collapsed onto the town of Elm in a valley of the Swiss Alps. Over 10 million cubic meters of rock fell to the valley floor, burying Elm and its 115 inhabitants to a depth of 10 to 20 m.

Rockfalls happen fairly frequently along steep highway road cuts, leading to the posting of "falling-rock zone" signs (Fig. 16.7b). Such rockfalls take place because frost wedging and/or root wedging pries fragments loose, and weathering weakens bonds holding rock together. Figure 16.7c shows a large rock slide in the Uinta Mountains of Utah. A section of forest that once surrounded the lake is now buried under the debris. Had this slide occurred in a populated area, it could have been devastating.

Did you ever wonder...
why highway engineers erect "falling rock" signs?

Submarine Mass Movements

So far, we've focused on mass movements that occur subaerially, for these are the ones we can see and are affected by most. But mass wasting also happens underwater. The sedimentary record contains abundant evidence of submarine mass movements, because after they take place, they tend to be buried by younger sediments and are preserved.

Geologists distinguish three types of submarine mass movements, according to whether the mass remains coherent or disintegrates as it moves (Fig. 16.8a–c). In **submarine slumps**, semi-coherent blocks slip downslope on weak detachments (Fig. 16.9a). In some cases, the layers constituting the blocks become contorted as they move, like a tablecloth that has slid off a table. In **submarine debris flows**, the moving mass breaks apart to form a slurry containing larger clasts (pebbles to boulders) suspended in a mud matrix. And in **turbidity currents**, sediment disperses in water to create a turbulent cloud of suspended sediment that rushes downslope like a powdery avalanche. When the turbidity current starts to slow, its entrained sediment settles out in a sequence, with coarser grains at the base and finer grains at the top. The resulting deposits, therefore, consist of graded beds (see Chapter 7).

In recent years, geologists have used satellites as well as side-scan sonar (sonar that analyzes a 60 km-wide swath of the sea-floor, instead of just a line, as it moves) to map out the extent of submarine landslides. The shapes of slumps and landslide deposits stand out on the resulting new generation of high-resolution sea-floor maps. Geologists have found that submarine slopes bordering both hot-spot volcanoes and active plate boundaries are scalloped by immense slumps, because tectonic activity frequently jars these areas with earthquakes that set masses of material in motion. For example, slumps up to 200 km long and 100 km wide

FIGURE 16.8 Submarine mass movements.

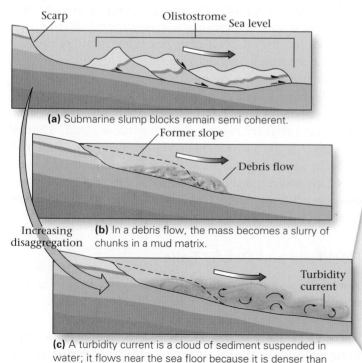

(a) Submarine slump blocks remain semi coherent.

(b) In a debris flow, the mass becomes a slurry of chunks in a mud matrix.

(c) A turbidity current is a cloud of sediment suspended in water; it flows near the sea floor because it is denser than clear water.

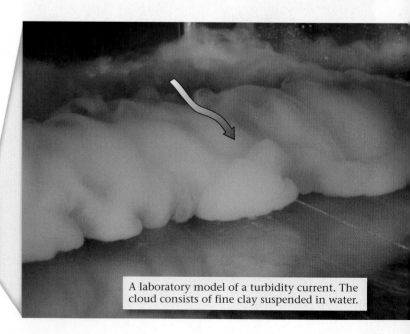

A laboratory model of a turbidity current. The cloud consists of fine clay suspended in water.

FIGURE 16.9 Examples of huge submarine slumps and debris flows.

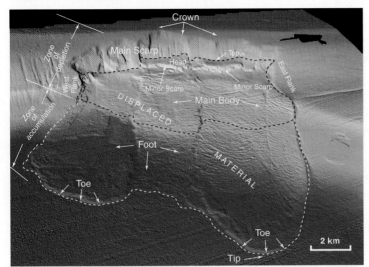

(a) A digital bathymetric map of a slip along the coast of California. The parts of the slump are labeled.

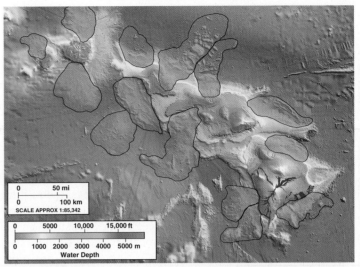

(b) A bathymetric map of the area around Hawaii shows several huge slumps, shaded in tan.

have substantially modified the flanks of the Hawaiian Islands (Fig. 16.9b). Some of the slumping events also carried away large chunks of the subaerial parts of the islands—in fact, the steep portions of the islands' coasts are the head scarps of huge slumps. Studies suggest that huge slumping events off Hawaii happen, on average, about once every 100,000 years. Significantly, passive-margin coasts are not immune to slumping, and immense slumps have been mapped along the coasts of the Atlantic Ocean.

Since a submarine slump can develop fairly quickly, and since its movement can displace a large area of the sea floor, it can trigger a tsunami. A submarine slump set in motion by a 1998 earthquake in Papua New Guinea generated a tsunami that devastated a 40 km-long stretch of coast and killed 2,100 people. A prehistoric tsunami triggered by a slump off the coast of Norway left its trace all around the North Sea (Box 16.2).

Take-Home Message

- Regolith slowly creeps downslope due to freezing and thawing; solifluction occurs where thawed sediments move downslope over permafrost.
- A slump is a semi-coherent mass of material that moves downslope on a failure surface.
- In an avalanche, debris or snow may mix with air to form a chaotic cloud.
- During rock or debris slides, material breaks into a jumble of fragments that rush down a slope. The chunks comprising a rockfall or debris fall drop vertically.

THINK: Can significant mass movements take place under the surface of the sea?

16.3 WHY DO MASS MOVEMENTS OCCUR?

We've seen that mass movements travel at a range of different velocities, from slow (creep) to faster (slumps, mud and debris flows, and rock and debris slides) to fastest (snow avalanches, and rock and debris falls—see Geology at a Glance, pp. 558–559). The velocity depends on the steepness of the slope and the water or air content of the mass. For these movements to take place, the stage must be set by the following phenomena: fracturing and weathering, which weaken materials at Earth's surface so that they cannot hold up against the pull of gravity; and the development of relief, which provides slopes down which masses move.

Weakening the Surface: Fragmentation and Weathering

If the Earth's surface were covered by intact (unbroken) rock, mass movements would be of little concern, for intact rock has great strength and could form stalwart mountain faces that would never tumble. But the rock of the Earth's upper crust has been fractured by jointing and faulting, and in many locations the surface has a cover of regolith resulting from the weathering of rock. Regolith and fractured rock are much weaker than intact rock and can indeed collapse in response to gravitational pull (Fig. 16.10). Thus jointing, faulting, and weathering ultimately make mass movements possible.

Why are regolith and fractured rocks weaker than intact bedrock? The answer comes from looking at the strength of the attachments holding materials together. A mass of intact

BOX 16.2

The Storegga Slide and the North Sea Tsunamis

The Firth of Forth, a long inlet of the North Sea, forms the waterfront of Edinburgh, Scotland. At its western end, it merges with a broad plain in which mud and peat have been accumulating since the last ice age. Around 1865, geologists investigating this sediment discovered an unusual layer of sand containing bashed seashells, marine plankton, and torn-up fragments of substrate. This sand layer lies sandwiched between mud layers at an elevation of up to 4 m above the high-tide limit and 80 km inland from the shore. How could the sand layer have been deposited? The mystery baffled geologists for many decades. During this time, layers of shelly sand, similar to the one found in Scotland, were discovered at many other localities along both sides of the North Sea (**Fig. Bx16.2**). In some cases, the layer occurred 20 m above the high-tide limit. Geologists studying the coast also found locations where coastal cliffs appeared to have been eroded by wave action at elevations well out of reach of normal storm waves.

While land-based geologists puzzled about these unusual sand beds and coastal erosional features, marine geologists investigating the continental shelf off the western coast of Norway discovered a region of very irregular sea floor underlain with a jumble of chaotic blocks, some of which are 10 km by 30 km across and 200 m thick. When mapped out, they indicate that a 290-km-long sector of the continental shelf had collapsed in a series of submarine slides that together involve about 5,580 cubic km of debris—the overall feature is called the Storegga Slide.

Further studies show that the Storegga Slide formed during three movement events: one occurred 30,000 years ago, the second about 7,950 years ago, and the third about 6,000 years ago. Now the pieces of the puzzle were in place. Immense submarine slides displace enough ocean water to create tsunamis. The wave produced by the second

Storegga Slide may coincide with the disappearance of Stone Age tribes along the North Sea coast, suggesting that the tribes were effectively washed away.

If such a calamity happened in the past, could it happen again, with submarine-slide-generated tsunamis inundating coastal cities with large populations? Geologists now realize that submarine landslides can trigger tsunamis every bit as devastating as earthquake- and volcano-generated tsunamis. A slide in 1929 along the coast of Newfoundland, for example, not only created a turbidity current that broke the trans-Atlantic telephone cable but also generated tsunamis that washed away houses and boats along the coast of Newfoundland at elevations of up to 27 m above sea level. With a bit of looking, geologists have found huge boulders flung by tsunamis onto the land, layers of sand and gravel deposited well above the high-tide limit, and erosional features high up on shoreline cliffs along many coastal areas, even in areas (such as the Bahamas and southeastern Australia) far from seismic or volcanic regions. And submarine mapping shows that many large slumps occur all along continental shelves. It's no wonder that Edward Bryant, in his recent book on tsunamis, calls them "the underrated hazard."

FIGURE Bx16.2 Map of the North Sea region showing the location of sites (red dots) where marine sand layers occur significantly above the high-tide limit. These were caused by tsunamis generated by movement of the Storegga Slide. The map shows the estimated position of the tsunamis at 2 hours, 4 hours, and 6 hours.

FIGURE 16.10 The limestone in this outcrop on the coast of western Ireland has broken into blocks bounded by vertical joints and horizontal bedding planes. Storm waves moved some of the blocks.

bedrock is relatively strong because the chemical bonds within its interlocking grains, or within the cements between grains, can't be broken easily. A mass of loose rocks or of regolith, in contrast, is relatively weak because the grains are held together only by friction (the force resisting sliding; friction is caused by roughness along the contact between two bodies), electrostatic attraction (the weak force that develops between clay flakes because the surfaces of the flakes are charged), and surface tension of water (the weak force caused by the attraction of water molecules to one another). All of these forces combined are weaker than chemical bonds holding together the atoms in the minerals of intact rock. To picture this contrast, think how much easier it is to bust up a sand castle (whose strength comes primarily from the surface tension of water films on the sand grains) than it is to bust up a granite sculpture of a castle.

Slope Stability: The Battle between Downslope Force and Resistance Force

Mass movements do not take place on all slopes, and even on slopes where such movements are possible, they occur only occasionally. Geologists distinguish between *stable slopes*, on which sliding is unlikely, and *unstable slopes*, on which sliding will likely happen. When material starts moving on an unstable slope, we say that **slope failure** has occurred. Whether a slope fails or not depends on the balance between two forces—the *downslope force*, caused by gravity, and the *resistance force*, which inhibits sliding. If the downslope force exceeds the resistance force, the slope fails and mass movement results.

Let's examine this phenomenon more closely by imagining a block sitting on a slope. We can represent the gravitational attraction between this block and the Earth by an arrow (a vector) that points straight down, toward the Earth's center of gravity. This arrow can be separated into two components—the downslope force parallel to the slope and the normal force perpendicular to the slope. We can represent the resistance force by an arrow pointing uphill. If the downslope force is larger than the resistance force, then the block moves; otherwise, it stays in place (Fig. 16.11a, b). Note that for a given mass, the magnitude of the downslope force increases as the slope angle increases, so downslope forces are greater on steeper slopes.

What produces a resistance force? As we saw above, chemical bonds in mineral crystals or cement hold intact rock in place, friction holds an unattached block in place, electrical charges and friction hold dry regolith in place, and surface tension holds wet regolith in place.

Because of resistance force, granular debris tends to pile up to produce the steepest slope it can without collapsing. The angle of this slope is called the **angle of repose**, and for most dry, unconsolidated materials (such as dry sand) it typically has a value of between 30° and 37°. The angle depends partly on the shape and size of grains, which determine the amount of friction across grain boundaries. For example, steeper angles of repose (up to 45°) tend to form on slopes composed of large, irregularly shaped grains (Fig. 16.12).

FIGURE 16.11 Forces that trigger downslope movement.

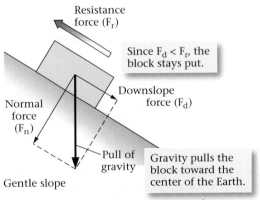

Since $F_d < F_r$, the block stays put.

Resistance force (F_r)

Normal force (F_n)

Downslope force (F_d)

Pull of gravity

Gravity pulls the block toward the center of the Earth.

Gentle slope

(a) Gravity can be divided into a normal force and a downslope force. If the resistance force, caused by friction, is greater than the downslope force, the block does not move.

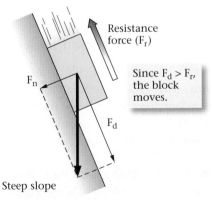

Resistance force (F_r)

Since $F_d > F_r$, the block moves.

F_n

F_d

Steep slope

(b) If the slope angle increases, the downslope force due to gravity increases. If the downslope force becomes greater than the resistance force, the block starts to move.

FIGURE 16.12 The angle of repose is the steepest slope that a pile of unconsolidated sediment can have and remain stable. The angle depends on the shape and size of grains.

Well-rounded sand has a small angle.

Irregularly shaped gravel has a large angle.

In many locations, the resistance force is less than might be expected because a weak surface exists at some depth below ground level. The weak surface separates unstable rock and debris above from the substrate below. If downslope movement begins on the weak surface, we say that the weak surface has become a failure surface. Geologists recognize several different kinds of weak surfaces that are likely to become failure surfaces (Fig. 16.13a–c). These include wet clay layers; wet, unconsolidated sand layers; joints that are parallel to the ground surface; weak bedding planes (shale beds and evaporite beds are particularly weak); and metamorphic foliation planes.

Surfaces that dip parallel to the land surface slope are particularly likely to fail. An example of such failure occurred in Madison Canyon, southwestern Montana, on August 17, 1959. That day, shock waves from a strong earthquake jarred the region. Metamorphic rock with a strong foliation that could serve as planes of weakness formed the bedrock of the canyon's southern wall. When the ground vibrated, rock detached along a foliation plane and tumbled downslope. Unfortunately, 28 campers lay sleeping on the valley floor. They were probably awakened by the hurricane-like winds blasting in front of the moving mass, but seconds later were buried under 45 m of rubble.

Fingers on the Trigger: What Causes Slope Failure?

What triggers an individual mass-wasting event? In other words, what causes the balance of forces to change so that the downslope force exceeds the resistance force, and a slope suddenly fails? Here we look at various phenomena—natural and human-made—that trigger slope failure.

Shocks, vibrations, and liquefaction. Earthquake tremors, storms, the passing of large trucks, or blasting in construction sites may cause a mass that was on the verge of moving actually to start moving. For example, an earthquake-triggered

FIGURE 16.13 Different kinds of weak surfaces can become failure surfaces.

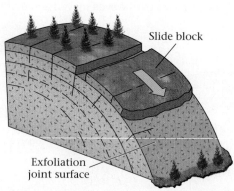

Slide block

Exfoliation joint surface

(a) Exfoliation joints form parallel to slope surfaces in granite and become failure surfaces.

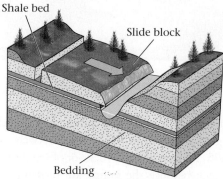

Shale bed

Slide block

Bedding

(b) In sedimentary rock, bedding planes (particularly in weak shale) become failure surfaces.

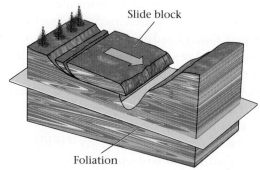

Slide block

Foliation

(c) In metamorphic rock, foliation planes (particularly in mica-rich schist) become failure surfaces.

slide dumped debris into Lituya Bay, in southeastern Alaska, in 1958. The debris displaced the water in the bay, creating a 300-m-high (1,200 feet) splash that washed forests off the slopes bordering the bay and carried fishing boats anchored in the bay many kilometers out to sea. The vibrations of an earthquake break bonds that hold a mass in place and/or cause the mass and the slope to separate slightly, thereby decreasing

FIGURE 16.14 Quick clay mudslides.

(a) Before shaking, clay flakes stick together.

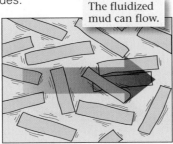

The fluidized mud can flow.

(b) During shaking, clays separate and become suspended in water.

(c) A mudslide near Namsos, Norway, destroyed ten houses. Movement may have been triggered by nearby blasting.

friction. As a consequence, the resistance force decreases, and the downslope force sets the mass in motion.

Shaking produces a unique effect in certain types of sediment. For example, **quick clay**, which consists of damp clay flakes, behaves like a solid when still, for surface tension holds water-coated flakes together. But shaking separates the flakes from one another and suspends them in the water, thereby transforming the clay into wet mud that flows like a fluid (Fig. 16.14a–c). Shaking can also cause **liquefaction** of wet sand, for shaking causes the sand grains to try to compress together more tightly. This compression, in turn, increases the water pressure in the pores between grains, destroying the cohesion between the grains. Without cohesion, the mixture of sand and water turns into a weak slurry capable of flow.

Changing slope loads, steepness, and support. As we have seen, the stability of a slope at a given time depends on the balance between downslope force and resistance force. Factors that change one or the other of these forces can lead

to failure. Examples include changes in slope loads, failure-surface strength, slope steepness, and the support provided by material at the base of the slope.

Slope loads change when the weight of the material above a potential failure plane changes. If the load increases, due to construction of buildings on top of a slope or due to saturation of regolith with water due to heavy rains, the downslope force increases and may exceed the resistance force.

Seepage of water into the ground may also weaken underground failure surfaces, further decreasing resistance force. An example of such failure triggered the largest observed landslide in U.S. history, the Gros Ventre Slide, which took place in 1925 on the flank of Sheep Mountain, near Jackson Hole, Wyoming (Fig. 16.15). The surface of Sheep Mountain is a **dip slope**, meaning that it is parallel to the bedding in underlying bedrock. Heavy rains saturated the permeable sandstone beds of the Tensleep Formation, which lay above the weak Amsden Shale, making the bedrock much heavier. On June 27, the downslope force exceeded the restraining force, and 40 million cubic meters of rock, as well as the overlying soil and forest, detached from the side of the mountain and slid 600 m downslope, filling a valley and forming a 75-m-high natural dam across the Gros Ventre River.

Slope steepness may change when, over time, rivers cut valleys, or construction engineers pile up debris (Fig. 16.16a,b). An increase in steepness causes an increase in the downslope force but does not change the resistance force. If the slope becomes too steep, it becomes unstable and may fail.

Removing support at the base of a slope plays a major role in triggering many slope failures. In effect, the material at the base of a slope acts like a dam holding back the material farther up the slope. When natural erosion, or bulldozer excavations, takes away this "dam," the upslope material can start to move. Such a process contributed to the Gros Ventre Slide, where river erosion had removed support from the base of Sheep Mountain. Cutting terraces into hillslopes for road building can have the same effect, as can wave erosion at the base of a slope along a coast.

In some cases, erosion by a river or by waves eats into the base of a cliff and produces an overhang. When such **undercutting** has occurred, rock making up the overhang eventually breaks away from the slope and falls (Fig. 16.17a, b ▶).

Changing the slope strength. The stability of a slope depends on the strength of the material constituting it. If the material weakens with time, the slope becomes weaker and eventually collapses. Three factors influence the strength of slopes: weathering, vegetation cover, and water.

With time, chemical weathering produces weaker minerals, and physical weathering breaks rocks apart. Thus, a

FIGURE 16.15 Stages leading to the 1925 Gros Ventre Slide in Wyoming.

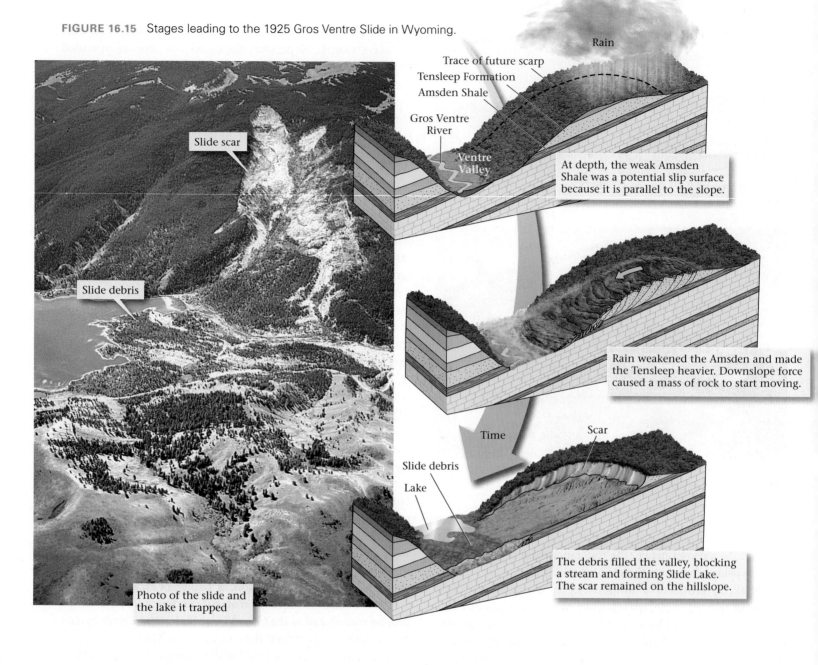

Slide scar

Slide debris

Photo of the slide and the lake it trapped

Rain

Trace of future scarp
Tensleep Formation
Amsden Shale

Gros Ventre River

Ventre Valley

At depth, the weak Amsden Shale was a potential slip surface because it is parallel to the slope.

Rain weakened the Amsden and made the Tensleep heavier. Downslope force caused a mass of rock to start moving.

Time

Scar

Slide debris

Lake

The debris filled the valley, blocking a stream and forming Slide Lake. The scar remained on the hillslope.

FIGURE 16.16 Processes that can steepen slopes and make them unstable.

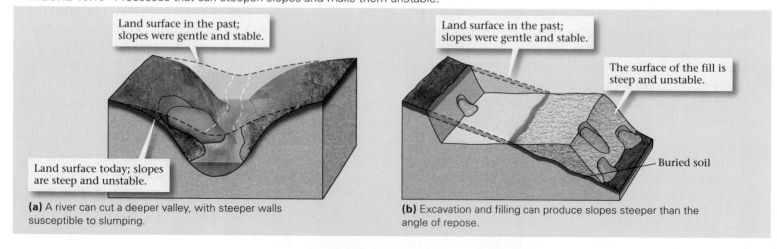

Land surface in the past; slopes were gentle and stable.

Land surface today; slopes are steep and unstable.

Land surface in the past; slopes were gentle and stable.

The surface of the fill is steep and unstable.

Buried soil

(a) A river can cut a deeper valley, with steeper walls susceptible to slumping.

(b) Excavation and filling can produce slopes steeper than the angle of repose.

FIGURE 16.17 Undercutting and collapse of a sea cliff.

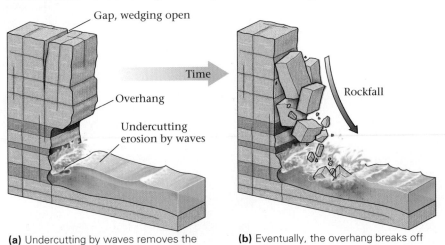

(a) Undercutting by waves removes the support beneath an overhang.

(b) Eventually, the overhang breaks off along joints, and a rockfall takes place.

formerly intact rock composed of strong minerals is transformed into a weaker rock or into regolith.

We've seen that in slightly damp regolith, the surface tension of water binds grains to each other and thus increases the strength of the regolith. You can carve a steep-walled sand castle from damp sand, but not from dry sand. If, however, the amount of water in regolith increases until pores between grains completely fill, then the water can push grains apart and dramatically decrease the overall strength of the regolith—you can't carve a sand castle from a slurry of sand and water. Because of this effect, a mass of regolith can become weak enough to start flowing downslope, as a mud flow or debris flow, after a torrential rain. Addition of water, either by infiltration of rain or by a rise in the water table (the top surface of groundwater) also weakens failure surfaces in bedrock.

Water infiltration has a particularly notable effect in regions underlain by swelling clays. These clays possess a mineral structure that allows them to absorb water. The water molecules form sheets between layers of silica tetrahedra within clay flakes, a process that causes the flakes to swell to several times their original size. Such swelling pushes up the ground surface, making it crack, and weakens the upper layer of the ground, making it susceptible to creep or slip. When the clay dries, it shrinks, and the ground surface subsides. This up-and-down movement can wrinkle road surfaces and crack foundations.

In the case of slopes underlain by regolith, vegetation tends to strengthen the slope because the roots hold otherwise unconsolidated grains together. Also, plants absorb water from the ground, thus keeping it from turning into slippery mud. The removal of vegetation therefore has the net result of making slopes more susceptible to downslope mass movement. In 2003, terrifying wildfires, stoked by strong winds, destroyed the ground-covering vegetation in many areas of California. When heavy rains followed, the barren ground of this hilly region became saturated with water and turned into mud, which then flowed downslope, damaging and destroying many homes and roads. Deforestation in tropical rain forests, similarly, leads to catastrophic mass wasting of the forest's substrate (Fig. 16.18).

FIGURE 16.18 Deforestation decreases slope stability. A large slump has formed on this deforested hill in Brazil.

Slump scar

Take-Home Message

- Mass wasting can occur because fractured rock and regolith are relatively weak.
- The stability of a slope reflects the relative sizes of downslope force and resistance force. If the former is larger, the slope is stable. When resistance force approaches downslope force, the slope becomes unstable.
- Ground shaking can transform wet sediments into liquid-like slurries that flow.
- A number of factors, such as changes in slope steepness, loads, support, and strength, can trigger mass wasting.

THINK: If the slope of a sand pile is less than the angle of repose, will the slope of the pile fail?

Mass Movement

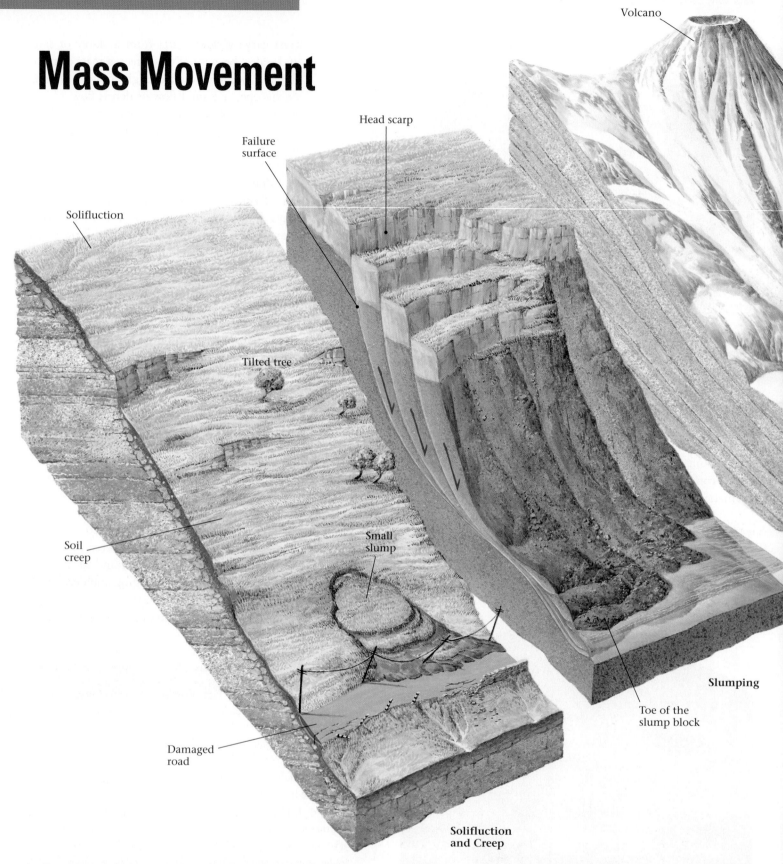

Solifluction

Failure surface

Head scarp

Volcano

Tilted tree

Soil creep

Small slump

Damaged road

Solifluction and Creep

Slumping

Toe of the slump block

In Earth's gravity field, what goes up must come down—sometimes with disastrous consequences. Rock and regolith are not infinitely strong, so every now and then slopes or cliffs give way in response to gravity, and materials slide, tumble, or career downslope. This downslope movement, called mass movement, or mass wasting, is the first step in the process of erosion and sediment formation. The resulting debris may eventually be carried away by water, ice, or wind.

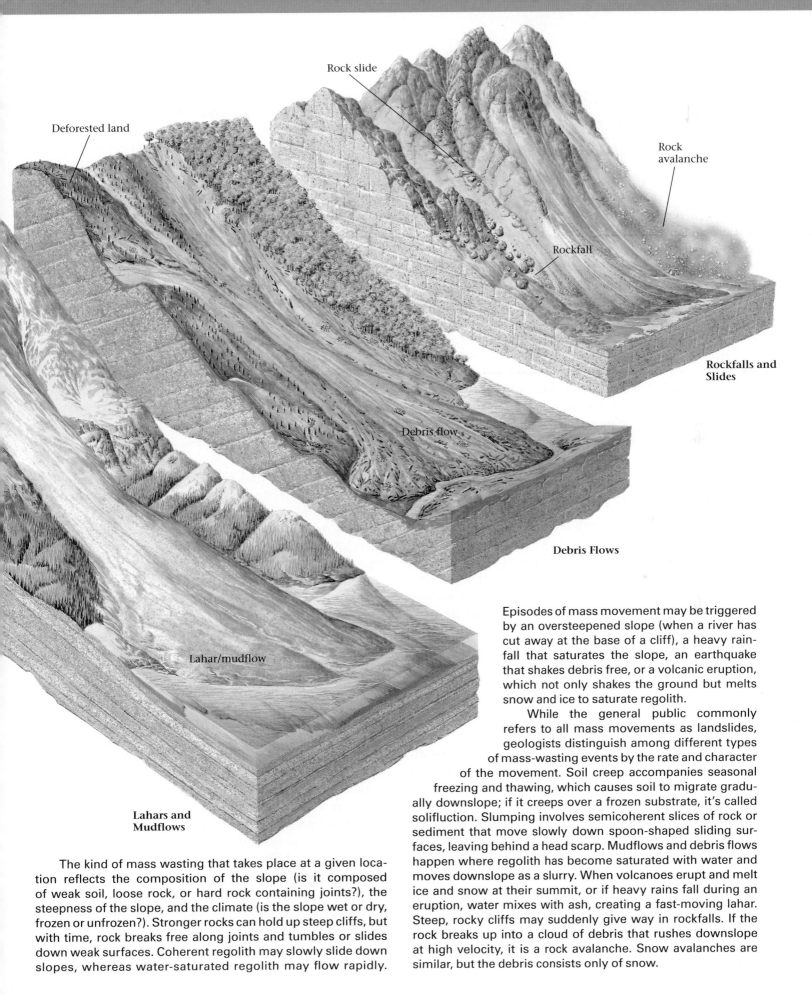

Deforested land

Rock slide

Rock
avalanche

Rockfall

**Rockfalls and
Slides**

Debris flow

Debris Flows

Lahar/mudflow

**Lahars and
Mudflows**

Episodes of mass movement may be triggered by an oversteepened slope (when a river has cut away at the base of a cliff), a heavy rainfall that saturates the slope, an earthquake that shakes debris free, or a volcanic eruption, which not only shakes the ground but melts snow and ice to saturate regolith.

While the general public commonly refers to all mass movements as landslides, geologists distinguish among different types of mass-wasting events by the rate and character of the movement. Soil creep accompanies seasonal freezing and thawing, which causes soil to migrate gradually downslope; if it creeps over a frozen substrate, it's called solifluction. Slumping involves semicoherent slices of rock or sediment that move slowly down spoon-shaped sliding surfaces, leaving behind a head scarp. Mudflows and debris flows happen where regolith has become saturated with water and moves downslope as a slurry. When volcanoes erupt and melt ice and snow at their summit, or if heavy rains fall during an eruption, water mixes with ash, creating a fast-moving lahar. Steep, rocky cliffs may suddenly give way in rockfalls. If the rock breaks up into a cloud of debris that rushes downslope at high velocity, it is a rock avalanche. Snow avalanches are similar, but the debris consists only of snow.

The kind of mass wasting that takes place at a given location reflects the composition of the slope (is it composed of weak soil, loose rock, or hard rock containing joints?), the steepness of the slope, and the climate (is the slope wet or dry, frozen or unfrozen?). Stronger rocks can hold up steep cliffs, but with time, rock breaks free along joints and tumbles or slides down weak surfaces. Coherent regolith may slowly slide down slopes, whereas water-saturated regolith may flow rapidly.

16.4 WHERE DO MASS MOVEMENTS OCCUR?

The Importance of Relief, Climate, and Substrate

The single most important factor in determining whether a locality is susceptible to mass wasting is the relief (elevation difference) of a region, for mass wasting does not occur without slopes. Further, regions with steeper slopes tend to be more susceptible to mass movements than are regions with gentle slopes.

Slope is not the only factor affecting the character and frequency of mass movements, though. Climate also affects mass movements. For example, regions with heavy seasonal rainfalls are subject to mudflows and landslides, especially if fire or deforestation has removed vegetation. In these regions, the climate has led to deep weathering and weakening of the substrate, and rainfalls increase slope loads and weaken failure surfaces. Desert and alpine regions, where slopes tend to be underlain by unweathered bedrock, tend to host rare but dramatic rock falls.

The Importance of the Tectonic Setting

Most unstable ground on Earth ultimately owes its existence to the activity of plate tectonics. As we've seen, plate tectonics causes uplift, generates relief, and causes faulting, which fragments the crust. And, of course, earthquakes on plate boundaries trigger devastating landslides. Spend a day along the steep slopes of the Alpine Fault, a plate boundary that transects New Zealand, and you can *hear* mass movement in progress: during heavy rains, rockfalls and landslides clatter with astounding frequency, as if the mountains were falling down around you.

A Case Study: Southern California

To see the interplay of plate tectonics and other factors, let's consider a case study in southern California. Southern Californians pay immense sums for the privilege of building homes on cliffs overlooking the Pacific. The sunset views from their backyard patios are spectacular. But the landscape is not ideal from the standpoint of stability. Slumps and mudflows on coastal cliffs have consumed many homes over the years, with a cost to their owners (or insurance companies) of untold millions of dollars (see Fig. 10.27a). What is special about southern California that makes it so susceptible to mass wasting?

First, California lies along an active plate boundary, the San Andreas transform fault. Faulting has shattered the rock of California's crust. Fractures not only act as planes of weakness but also provide paths for water to seep into bedrock and cause chemical weathering, producing slippery clay that weakens the rock. Also, the rocks in many areas are weak to begin with because they formed as part of an accretionary prism, a chaotic mass of clay-rich sediment that was scraped off subducting oceanic lithosphere during the Mesozoic Era.

Though most of the movement between the North American and Pacific plates involves strike-slip displacement on the San Andreas fault, there is a component of compression across the fault. This compression leads to uplift and slope formation (see Box 16.1). Since the uplifted region borders the coast, wave erosion steepens and in some places undercuts cliffs. And because it is a plate boundary, numerous earthquakes rock the region, thus shaking regolith loose.

Did you ever wonder... why parts of the California coast slip into the sea?

California is also susceptible to mass movements because of its climate. In general, the region is hot and dry and thus supports only semidesert flora. Brush fires remove much of this cover, leaving large areas with no dense vegetation. But since the region lies on the West Coast, it endures occasional heavy winter rains. The water sinks quickly into the sparsely vegetated ground, adds weight to the mass on the slope, and weakens failure surfaces.

Development of cities and suburbs is the final factor that triggers mass movements in southern California. Development has oversteepened and overloaded slopes and has caused the water content of regolith to change. The consequences can be seen in the Portuguese Bend Slide of the Palos Verdes area near Los Angeles. The Portuguese Bend region is underlain with a thick, seaward-dipping layer of weak volcanic ash (now altered to weak clay) resting on shale. The weak ash has served as a failure plane a few times during the past few thousand years. In 1956, developers deposited a 23-m-thick layer of fill over the ground surface and built homes on top. Residents began to water their lawns and to use septic tanks that were susceptible to leaking. The water seeped into the ground and decreased the strength of the ash layer. Because of the decrease in strength, the added weight, and the erosion of the toe of the hill by the sea, the upper 30 m of land began to move once again. Between 1956 and 1985, the Portuguese Bend Slide moved at rates of up to 2.5 cm per day. Eventually, portions of a 260-acre region slid by over 200 m and, in the process, over 150 homes were destroyed (Fig. 16.19).

FIGURE 16.19 The Portuguese Bend "Slide" viewed from the air.

Head scarp

Approximate edge of slump

Hummocky slump surface

Take-Home Message

- Not all regions are equally susceptible to mass movement. Mass wasting can't occur where there is no relief, and it is more likely in regions with steep slopes.

- Climate and substrate characteristics also affect the frequency and style of mass movement.

- Tectonic setting affects mass wasting because tectonic processes produce slopes and cause ground shaking.

- The California coast is particularly susceptible to mass wasting because it is tectonically active, underlain by weak substrate, and has seasonally heavy rainfalls.

THINK: Are mass movements likely to be frequent on the surface of the Moon? Why or why not?

16.5 HOW CAN WE PROTECT AGAINST MASS-MOVEMENT DISASTERS?

Identifying Regions at Risk

Clearly, landslides, mudflows, and slumps are natural hazards we cannot ignore. Too many of us live in regions where mass wasting has the potential to kill people and destroy property. In many cases, the best solution is avoidance: don't build, live, or work in an area where mass movement will take place. But avoidance is possible only if we know where the hazards are.

To pinpoint dangerous regions, geologists look for landforms known to result from mass movements, for where these movements have happened in the past, they might happen again in the future. Features such as slump head scarps, swaths of forest in which trees have been tilted, piles of loose debris at the base of hills, and hummocky land surfaces all indicate recent mass wasting.

Geologists may also be able to detect regions that are *beginning to move* (Fig. 16.20). For example, roads, buildings, and pipes begin to crack over unstable ground. Power lines may be too tight or too loose because the poles to which they are attached move together or apart. Visible cracks form on the ground at the potential head of a slump, while the ground may bulge up at the toe of the slump. In some cases, subsurface cracks may drain the water from an area and kill off vegetation, whereas in other areas land may sink and form a swamp. Slow movements cause trees to develop pronounced curves at their base.

Even if mass wasting takes place too slowly to be perceptible to people, it can be documented with sensitive surveying techniques that can detect a subtle tilt of the ground or changes in distance between nearby points. Specifically, measurements with satellite data (GPS and LIDAR) permit geologists to identify movements of just a few millimeters that, while small, may indicate the reactivation of a slump. (GPS, which stands for "global positioning system," allows accurate surveying and can indicate whether a point on the surface of the Earth has moved during the time between two measurements. LIDAR, which stands for "light detection and

ranging," uses laser beams to measure the shape of the land and produces very accurate digital elevation maps. Changes in the shape of the land surface shown on such maps provides evidence of movement.)

If various clues indicate that a landmass is beginning to move, and if conditions make accelerating movement likely (e.g., persistent rain, rising floodwaters, or continuing earthquake aftershocks), then officials may order an evacuation. Evacuations have saved lives, and ignored warnings have cost lives. But unfortunately, some mass movements happen without any warning, and some evacuations prove costly but unnecessary.

Even if there is no evidence of recent movement, a danger may still exist: just because a steep slope hasn't collapsed in the recent past doesn't mean it won't in the future. In recent years, geologists have begun to identify such potential hazards by using computer programs that evaluate factors that trigger mass wasting. With this data, they create maps that portray the degree of risk for a certain location. These factors include the following: slope steepness; strength of substrate; degree of water saturation; orientation of bedding, joints, or foliation relative to the slope;

FIGURE 16.20 Surface features warn that a large slump is beginning to develop. Cracks that appear at the head scarp may drain water and kill trees. Power-line poles tilt and the lines become tight. Fences, roads, and houses on the slump begin to crack.

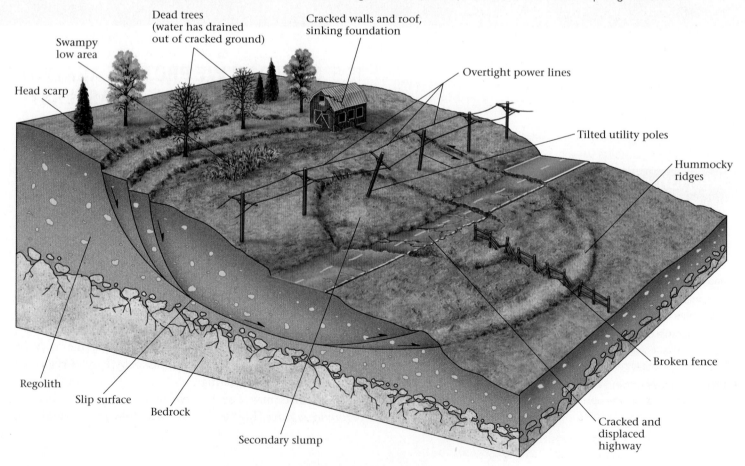

nature of vegetation cover; potential for heavy rains; potential for undercutting to occur; and likelihood of earthquakes. From such hazard-assessment studies, geologists compile **landslide-potential maps**, which rank regions according to the likelihood that a mass movement will occur (Fig. 16.21). In any case, common sense suggests that you should avoid building on or below particularly dangerous slide-prone slopes. In Japan, regulations on where to build in regions susceptible to mass wasting, careful monitoring of ground movements, and well-designed evacuation plans have drastically reduced property damage and the number of fatalities.

Preventing Mass Movements

In areas where a hazard exists, people can take certain steps to remedy the problem and stabilize the slope (Fig. 16.22a–h).

- *Revegetation*: Since bare ground is more vulnerable to downslope movement than is vegetated ground, stability in deforested areas will be greatly enhanced if land owners replant the region with vegetation that sends down deep roots.

- *Regrading*: An oversteepened slope can be regraded or terraced so that it does not exceed the angle of repose.

- *Reducing subsurface water*: Because water weakens material beneath a slope and adds weight to the slope, an unstable situation may be remedied either by improving drainage so that water does not enter the subsurface in the first place, or by removing water from the ground.

- *Preventing undercutting*: In places where a river undercuts a cliff face, engineers can divert the river. Similarly, along coastal regions they may build an offshore breakwater or pile **riprap** (loose boulders or concrete) along the beach to absorb wave energy before it strikes the cliff face.

- *Constructing safety structures*: In some cases, the best way to prevent mass wasting is to build a structure that stabilizes a potentially unstable slope or protects a region downslope from debris if a mass movement does occur. For example, civil engineers can build retaining walls or bolt loose slabs of rock to more coherent masses in the substrate in order to stabilize highway embankments. The danger from rock falls can be decreased by covering a road cut with chain link fencing or by spraying roadcuts with "shotcrete," a cement that coats the wall and prevents water infiltration and consequent freezing and thawing. Highways at the base of an avalanche chute can be covered by an avalanche shed, whose roof keeps debris off the road.

- *Controlled blasting of unstable slopes*: When it is clear that unstable ground or snow threatens a particular region,

FIGURE 16.21 A landslide hazard map of the Seattle area.

the best solution may be to blast the unstable ground or snow loose at a time when its movement can do no harm.

Clearly, the cost of preventing mass-wasting calamities is high, and people might not always be willing to pay the price. In such cases, they have a choice of avoiding the risky area, taking the chance that a calamity will not happen while they are around, buying appropriate insurance, or counting on relief agencies to help if disaster does strike. Once again, geology and society cross paths.

> ## Take-Home Message
>
> - Geologists can map out regions that are more susceptible to mass movements and use the data to produce a landslide-potential map.
> - Sensitive instruments can detect the beginning of mass movements.
> - In regions with high susceptibility to mass movements, people should take precautions to stabilize slopes. Preventive methods include revegetation and regrading as well as changing the water table, preventing undercutting, and changing slope steepness.
>
> **THINK:** Why is mass movement of major concern during production of large roadcuts?

FIGURE 16.22 A variety of remedial steps can stabilize unstable ground.

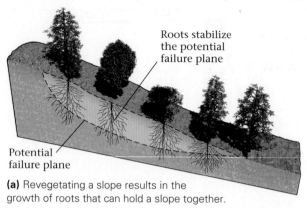

(a) Revegetating a slope results in the growth of roots that can hold a slope together.

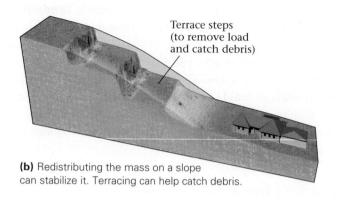

(b) Redistributing the mass on a slope can stabilize it. Terracing can help catch debris.

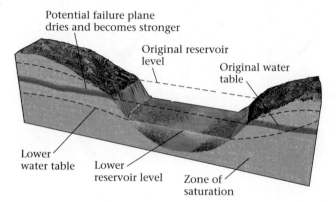

(c) Lowering the level of the water table can strengthen a potential failure surface.

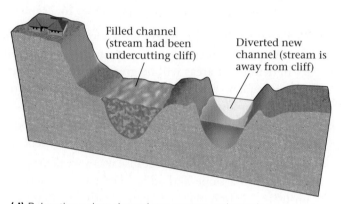

(d) Relocating a river channel can prevent undercutting.

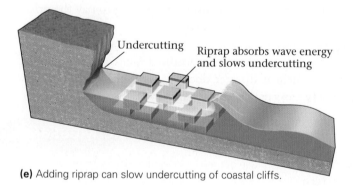

(e) Adding riprap can slow undercutting of coastal cliffs.

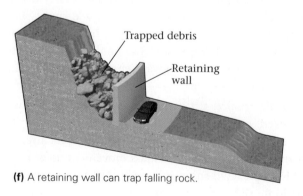

(f) A retaining wall can trap falling rock.

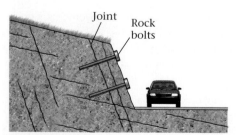

(g) Bolting or screening a cliff face can hold loose rocks in place.

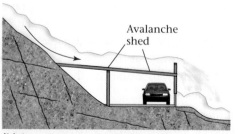

(h) An avalanche shed diverts debris or snow over a roadway.

Chapter Summary

- Rock or regolith on unstable slopes has the potential to move downslope under the influence of gravity. This process, called mass movement, or mass wasting, plays an important role in the erosion of hills and mountains.

- Though in everyday language people refer to most mass movements as landslides, geologists distinguish among different types based on factors such as the composition of the moving materials and the rate of movement.

- Slow mass movement, caused by the freezing and thawing of regolith, is called creep. In places where slopes are underlain with permafrost, solifluction causes a melted layer of regolith to flow down slopes. During slumping, a semicoherent mass of material moves down a spoon-shaped failure surface. Mudflows and debris flows occur where regolith has become saturated with water and moves downslope as a slurry.

- Rock and debris slides move very rapidly down a slope; the rock or debris breaks apart and tumbles. During avalanches, snow or debris mixes with air and moves downslope as a turbulent cloud. And in a debris fall or rockfall, the material free-falls down a vertical cliff.

- Landslides of various types occur undersea, and can trigger tsunamis.

- Intact, fresh rock is too strong to undergo mass movement. Thus, for mass movement to be possible, rock must be weakened by fracturing (joint formation) or weathering.

- Unstable slopes start to move when the downslope force exceeds the resistance force that holds material in place. The steepest angle at which a slope of unconsolidated material can remain without collapsing is the angle of repose.

- Downslope movement can be triggered by shocks and vibrations, a change in the steepness of a slope, removal of support from the base of the slope, a change in the strength of a slope, deforestation, weathering, or heavy rain.

- Geologists produce landslide-potential maps to identify areas susceptible to mass movement. Engineers can help prevent mass movements by using a variety of techniques.

GEOPUZZLE REVISITED

Gravity constantly applies a downslope force. For a time, the strength of material making up the substrate of a slope may be strong enough to resist this relentless pull. But heavy rains, ground shaking, undercutting, deforestation, and/or a change in the water table depth can destabilize a slope until it finally gives way and a landslide takes place. Roots, retaining walls, and other natural or human-built features may delay a landslide, but in the context of geologic time, gravity always wins. Mass-wasting events, such as landslides, contribute to the erosion of uplifted land on islands and continents. They can also take place under the sea.

Guide Terms

angle of repose (p. 553)

avalanche (p. 542)

creep (p. 542)

debris fall (p. 549)

debris flow (debris slide) (pp. 545, 547)

dip slope (p. 555)

failure surface (p. 544)

head scarp (p. 544)

lahar (p. 545)

landslide (p. 542)

landslide-potential maps (p. 563)

liquefaction (p. 555)

mass movement (wasting) (p. 541)

mudflow (mudslide) (p. 545)

natural hazard (p. 541)

quick clay (p. 555)

riprap (p. 563)

rockfall (p. 549)

rock glacier (p. 542)

rock slide (p. 547)

slope failure (p. 553)

slump (p. 542)

snow avalanche (p. 548)

solifluction (p. 542)

submarine debris flow (p. 550)

submarine slump (p. 550)

talus (p. 549)

turbidity current (p. 550)

undercutting (p. 555)

Review Questions

1. What factors do geologists use to distinguish among various types of mass movements?
2. Identify the key differences between a slump, a debris flow, a lahar, an avalanche, a rockslide, and a rockfall.
3. Explain the process of creep, and discuss how it differs from solifluction.
4. Distinguish among different types of submarine mass movements. Which of these types can trigger a major tsunami, and why?
5. Why is intact bedrock stronger than fractured bedrock? Why is it stronger than regolith?
6. Explain the difference between a stable and unstable slope. What factors determine the angle of repose of a material? What features are likely to serve as failure surfaces?
7. Discuss the variety of phenomena that can cause a stable slope to become so unstable that it fails.
8. How can ground shaking cause fairly solid layers or sand or mud to become weak slurries capable of flowing?
9. Discuss the role of vegetation in slope stability. Why can fires and deforestation lead to slope failure?
10. Identify the various factors that make the coast of California susceptible to mass movements.

11. What factors do geologists take into account when producing a landslide-potential map, and how can geologists detect the beginning of mass movement in an area? What steps can people take to avoid landslide disasters?

On Further Thought

12. Imagine that you have been asked by the World Bank to determine whether it makes sense to build a dam in a steep-sided, east-west-trending valley in a small central Asian nation. The local government has lobbied for the dam because the climate of the country has gradually been getting drier, and the farms of the area are running out of water. The World Bank is considering making a loan to finance construction of the dam, a process that would employ thousands of now-jobless people. Initial investigation shows that the rock of the valley floor consists of schist containing a strong foliation that dips south. Outcrop studies reveal that abundant fractures occur in the schist along the valley floor; the surface of most fractures are coated with slickensides. Moderate earthquakes have rattled the region. What would you advise the bank? Explain the hazards and what might happen if the reservoir were filled.

 For more resources, including animations, quizzes, and Norton's GeoTours, go to **wwnorton.com/studyspace**.

 If your instructor assigns exercises in SmartWork, log in at **smartwork.wwnorton.com**.

ANOTHER VIEW The huge Ganges Chasma landslide, which formed along the Valles Marineris, a canyon wall on Mars, has taken part of a meteorite crater with it. The field of view is about 60 km across. The lower photo shows a perspective image of the Ganges Chasma, showing multiple landslides. Note the debris fans of the two largest landslides overlap.

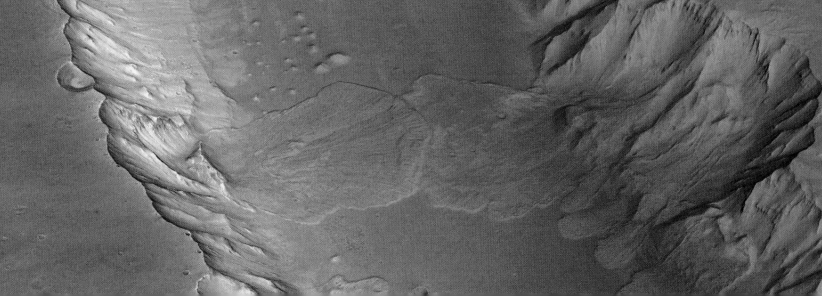

CHAPTER **17**

Streams and Floods: The Geology of Running Water

Some of the water that falls on land drains downslope in stream channels. Where the slope becomes very steep, the water tumbles through the air as a waterfall—a visible display of the power of running water—as seen here in the San Juan Mountains, Colorado.

GEOPUZZLE

Why do stream networks develop, and why do streams sometimes overflow their channels and flood the surrounding landscape?

As many fresh streams meet in one salt sea, as many lines close in the dial's center, so may a thousand actions, once afoot, end in one purpose.

—William Shakespeare (1564–1616)

17.1 INTRODUCTION

By the 1880s, Johnstown, built along the Conemaugh River in scenic western Pennsylvania, had become a significant industrial town. Recognizing the attraction of the surrounding hills as a summer retreat, speculators built a mud-and-gravel dam across the river, upstream of Johnstown, to trap a pleasant reservoir of cool water. A group of industrialists and bankers bought the reservoir and established the exclusive South Fork Hunting and Fishing Club, a cluster of lavish 15-room "cottages" on the shore. Unfortunately, the dam had been poorly designed, and debris blocked its spillway (a passageway for surplus water), setting the stage for a monumental tragedy. On May 31, 1889, torrential rain drenched Pennsylvania, and the reservoir filled until water began to flow over the dam. Despite frantic attempts to strengthen the dam, the soggy structure abruptly collapsed, and the reservoir emptied into the Conemaugh River Valley. A 20-m-high wall of water roared downstream and slammed into Johnstown, transforming bridges and buildings into twisted wreckage (Fig. 17.1). When the water subsided, 2,300 people were dead, and Johnstown became the focus of national sympathy. Clara Barton mobilized the recently founded Red Cross, which set to work building dormitories, and

citizens nationwide donated everything from clothes to beds. Nevertheless, it took years for the town to recover, and many residents simply picked up and left. Despite several lawsuits, no one payed a penny of restitution, but the South Fork Hunting and Fishing Club abandoned its property.

The unlucky inhabitants of Johnstown had experienced the immense power of **running water**, water that flows down the surface of sloping land in response to the pull of gravity. Geologists use the term **stream** for any body of running water that flows along a **channel**, an elongate depression or trough. In everyday English, we also refer to large streams as rivers and medium-sized ones as creeks. Streams drain water from the landscape and carry it into lakes or to the sea, much as culverts drain water from a parking lot. In the process, streams relentlessly erode the landscape and transport sediment and debris to sites of deposition. Generally, a stream stays within the confines of its channel, but when the supply of water entering a stream exceeds the channel's capacity, water spills out and covers the surrounding land, thereby causing a **flood**, such as the one that washed away Johnstown.

Earth is the only planet in the Solar System that currently hosts flowing streams; Mars did at times in the past, but most if not all surface water on the red planet dried up long ago (see Interlude F). Streams are of great importance to human society not only because of how they modify the landscape—during normal flow but especially during floods—but also because they provide avenues for travel and commerce, nutrients and sediments for agriculture, water for irrigation and consumption, and sources for power. In this chapter, we examine how streams operate in the

FIGURE 17.1 During the 1889 Johnstown flood, raging waters could tumble large houses.

Earth System. First, we learn about the origin of running water and about the architecture of streams and stream networks. Then we look at the process of stream erosion and deposition and at the landscapes that form in response to these processes. Finally, we consider the nature and consequences of flooding.

17.2 DRAINING THE LAND

Where Does the Water in Streams Come From?

Stand by a river and watch—you'll see volumes of water traveling past you, from somewhere upstream toward the river's mouth. Geologists refer to all the water that eventually ends up in streams as **runoff**. Where does this water come from? The answer may seem obvious at first: it must originally have fallen from the sky as meteoric water (rain or snow). But on close examination, the story becomes complex, for the water can follow one of several pathways to a stream, passing through one or more of the reservoirs in the hydrologic cycle (**Fig. 17.2**):

- Some water falls directly from the sky onto the surface of a stream. Generally, this water makes only a small contribution to a stream's volume.

- In places where the ground surface is flat or forms a depression, water accumulates in a standing body (puddle, swamp, pond, or lake). When the water level in the standing body becomes higher than the lowest point along the body's bank, an outlet forms, through which water spills into a stream.

- Where the ground surface is permeable, water can infiltrate down and become subsurface water (see Interlude F and Chapter 19). Subsurface water includes soil moisture (water adhering to particles in soil), vadose-zone water (water that partially fills cracks and pores in rock or regolith below the soil but above the water table), and groundwater (water that completely fills cracks and pores in the region below the water table). Subsurface water enters streams either where heavy rains push existing soil water and vadose-zone water back to the surface, or where the stream channel lies below the water table so that groundwater bubbles up through the channel walls in seeps or underwater springs.

- On slopes, water can move as **sheetwash**—a layer up to several millimeters thick—down the ground surface to a lower elevation. This water either enters a stream directly or first enters a standing body and later flows through an outlet into a stream. The volume of sheetwash increases when the soil is saturated with water so that rain cannot soak into the ground.

FIGURE 17.2 Excess surface water (runoff) comes from rain, melting ice or snow, and groundwater springs. On flat ground, water accumulates in puddles or swamps; but on slopes, it flows downslope in streams.

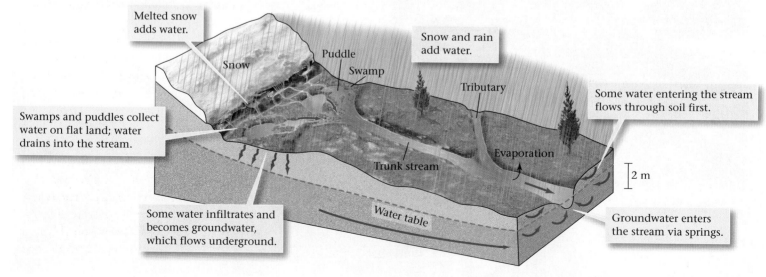

FIGURE 17.3 The formation of stream channels and drainage networks.

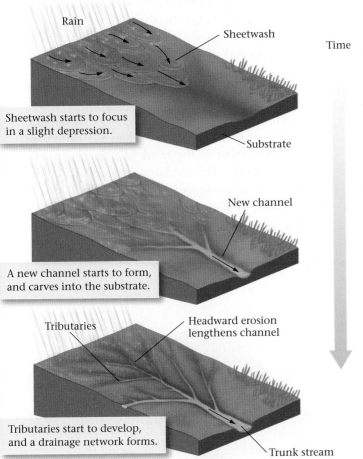

Rain

Sheetwash

Time

Substrate

Sheetwash starts to focus in a slight depression.

New channel

A new channel starts to form, and carves into the substrate.

Tributaries

Headward erosion lengthens channel

Tributaries start to develop, and a drainage network forms.

Trunk stream

(a) Progressive stages during the development of a network of stream channels.

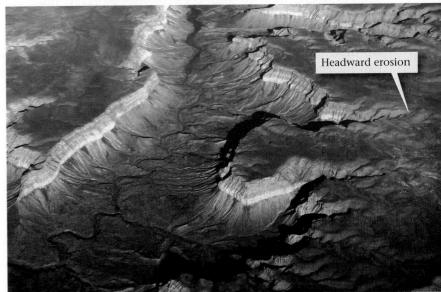

Headward erosion

(b) An example of headward erosion. The main stream flows in a deep valley. Side streams are cutting into the bordering plateau.

Note that if it's cold enough, water remains in solid form (snow or ice) until melting takes place. Meltwater can move to a stream via the surface or subsurface pathways listed above.

Water flowing in streams may pause temporarily in downstream lakes or reservoirs. But all *fresh* standing bodies of water have an outlet through which water escapes and continues to the sea. On a global basis, 36,000 cubic km of water becomes runoff every year, about 10% of the total volume that passes through the hydrologic cycle.

Forming Streams and Drainage Networks

How does a stream channel form in the first place? To find an answer to this question, remember that any flowing fluid can cause **downcutting**, the process of eroding or digging into substrate. The efficiency of downcutting depends on several factors, including (1) the velocity of the flow, for faster flow erodes more rapidly than slower flow; (2) the strength of the

substrate, for weaker substrate can be eroded more rapidly than stronger substrate; and (3) the amount of vegetation cover, for unvegetated ground can be eroded more rapidly than land held together by plant roots. With these controls in mind, we can now complete our answer. Overland flow starts as sheetwash—if you've ever looked down while standing on the smooth concrete of a sidewalk in a heavy rain, you've witnessed sheetwash. But natural ground surfaces, unlike sidewalks, do not have uniform slope, uniform resistance to erosion, or uniform vegetation cover. Thus, the velocity of natural sheetwash and the ease with which it can dig into its substrate varies with location. Where sheetwash is faster, or the substrate weaker and less protected, downcutting proceeds more rapidly and lowers the land relative to its surroundings (Fig. 17.3a). More sheetwash immediately diverts into the newly lowered land surface in which, therefore, flow velocity increases and erosion takes place even more rapidly. If this process continues for enough time, it produces a distinct channel. Note that downcutting is, in effect, a "positive feedback" process. Once downcutting begins to produce a channel, water flow focuses into the channel, so the channel deepening accelerates.

As time passes, a stream channel lengthens up its slope, meaning that the position of the point marking the head of the channel moves in the direction opposite to the direction that water flows down the channel. This process, called **headward erosion** (Fig. 17.3b), can take place for two reasons. First, it occurs because downcutting due to water flowing on the land surface happens faster where water focuses into the head of the channel than where water spreads over a broad, sheet-washed surface. Second, it occurs because groundwater emerging from a spring at the head of a channel erodes adjacent material and, therefore, undermines overlying rock or sediment until eventually the land above the head of the stream channel slumps down into the channel. (This overall process is called "groundwater sapping.") The instant that the slump forms, the scarp of the slump becomes the new head of the channel. Eventually, floodwaters wash the debris out of the now-longer channel, to complete the increment of headward erosion.

Once a channel has formed, new smaller channels form on its slopes, so that an array of linked streams evolves, with the smaller streams, or **tributaries**, flowing into a single larger stream, or **trunk stream**. The array of interconnecting streams together constitute a **drainage network**.

Like transportation networks, drainage networks reach into all corners of a region, providing conduits for the removal of runoff. The configuration of tributaries and trunk streams defines the "map pattern" of a drainage network. This pattern depends on the shape of the landscape and the composition of the substrate. Geologists recognize several types of networks on the basis of their map pattern (Fig. 17.4a):

- *Dendritic*: When rivers flow over a fairly uniform substrate with a fairly uniform initial slope, they develop a dendritic network, which looks like the pattern of branches connecting to the trunk of a deciduous tree (Fig. 17.4b). In fact, the word *dendritic* comes from the Greek *dendros*, meaning tree.
- *Radial*: Drainage networks forming on the surface of a cone-shaped mountain flow outward from the mountain peak, like spokes on a wheel. Such a pattern defines a radial network (Fig. 17.4c).

FIGURE 17.4 Different types of drainage networks.

(a) Block diagrams illustrating five types of drainage networks.

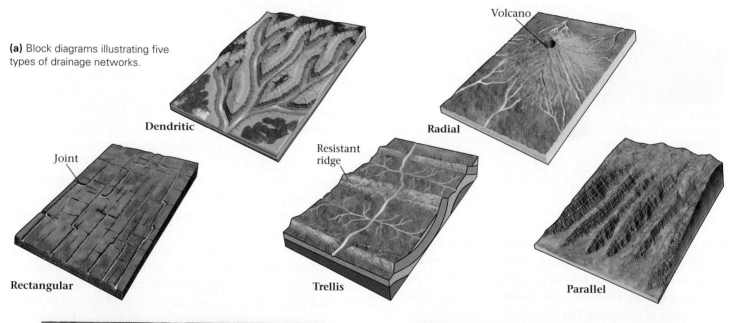

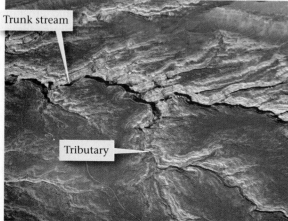

(b) Example of a dendritic drainage network.

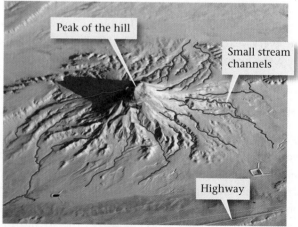

(c) Example of a radial drainage network.

FIGURE 17.5 Drainage divides and basins.

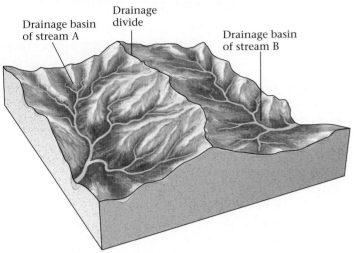

Drainage basin of stream A

Drainage divide

Drainage basin of stream B

(a) A drainage divide is a relatively high ridge that separates two drainage basins.

- - - Mississippi River basin limit
— Drainage divide

Arctic Ocean

Hudson Bay

Atlantic Ocean

Great Basin

Continental divide

Mississippi River

Pacific Ocean

Gulf of Mexico

(c) The major drainage basins of North America.

(b) An air view shows numerous small drainage basins formed during erosion of this ridge in Oregon.

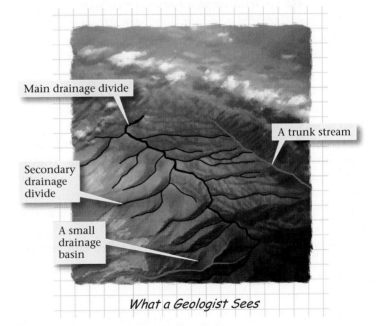

Main drainage divide

A trunk stream

Secondary drainage divide

A small drainage basin

What a Geologist Sees

- *Rectangular*: In places where a rectangular grid of fractures (vertical joints) breaks up the ground, channels form along the preexisting fractures, and streams join each other at right angles, creating a rectangular network.

- *Trellis*: In places where a drainage network develops across a landscape of parallel valleys and ridges, major tributaries flow down a valley and join a trunk stream that cuts across the ridges; the place where a trunk stream cuts across a resistant ridge is a water gap. The resulting map pattern resembles a garden

trellis, so the arrangement of streams constitutes a trellis network.

- *Parallel*: On a uniform slope, several streams with parallel courses develop simultaneously.

Drainage Basins and Divides

A drainage network collects water from a broad region, variously called a drainage basin, catchment, or **watershed**, and feeds it into the trunk stream, which carries the water away. The highland, or ridge, that separates one watershed from another is a

drainage divide (Fig. 17.5a, b). A continental divide separates drainage that flows into one ocean from drainage that flows into another. For example, if you straddle the North American continental divide and pour a cup of water out of each hand, the water in one hand flows to the Atlantic, and the water in the other flows to the Pacific. This continental divide is not, however, the only important divide in North America. A divide runs along the crest of the Appalachians, separating Atlantic Ocean drainage from Gulf of Mexico drainage, and another one runs just south of the border between Canada and the United States, separating Gulf of Mexico drainage from Hudson Bay (Arctic Ocean) drainage. These three divides bound part of the Mississippi drainage basin, which drains the interior of the United States (Fig. 17.5c).

Streams That Last, Streams That Don't: Permanent versus Ephemeral Streams

Some streams flow all year long, whereas others flow for only part of the year; in fact, some flow only for a brief time after a heavy rain. The character of a stream depends on the depth of the water table. If the bed, or floor, of a stream lies below the water table, then the stream flows year-round (Fig. 17.6a). In such **permanent streams**, found in humid or temperate climates, water comes not only from upstream or from surface runoff but also from springs in the stream bed through which groundwater seeps. But if the bed of a stream lies above the water table, then water flows in the channel only when the rate at which water enters the channel exceeds the rate at which water infiltrates the ground below the channel (Fig. 17.6b). Such streams can be permanent only if supplied by abundant water from upstream. In dry climates with intermittent rainfall and high evaporation rates, water entirely sinks into the ground, and the stream dries up when the supply of water stops. Streams that do not flow all year are called **ephemeral streams**. Ephemeral streams flow only during rainstorms

or after spring thaws. A dry ephemeral stream bed (channel floor) is called a dry wash, wadi, or arroyo.

Take-Home Message

- Streams drain runoff from the land as culverts drain water from a parking lot.
- Running water causes downcutting, so streams undergo headward erosion.
- Eventually, drainage networks evolve; their configuration reflects the topography and substrate.
- A drainage network obtains water from a watershed; drainage divides separate watersheds.
- Streams in temperate or tropical regions flow all year, but those in dry regions do not.

THINK: Why do streams develop distinct channels?

17.3 DESCRIBING FLOW IN STREAMS: DISCHARGE AND TURBULENCE

Geologists and engineers use the term **discharge** for the volume of water passing a point on the stream bank in a given time. We can calculate discharge (D) by using a simple equation: $D = A_c \times v_a$. In this equation, A_c is the cross-sectional area of the stream as measured in a vertical plane that lies perpendicular to the stream bank, and v_a is the *average* velocity at which water moves in the downstream direction. For example, if the cross-sectional area is 10 m^2, and the average velocity is 5 m/s, then the discharge is 50 m^3/s. Note that we specify discharge by *volume per unit time*, using units such as cubic meters per second (m^3/s) or cubic feet per second (ft^3/s). Measurements of discharge can be made at a stream-gauging station, with instruments that analyze velocity and depth at various points across the stream (Fig. 17.7a).

FIGURE 17.6 The contrast between permanent and ephemeral streams.

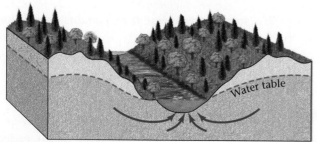

(a) The floor of a permanent stream in a temperate climate lies below the water table. Springs add water from below, so the stream contains water even between rains.

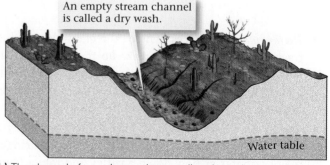

An empty stream channel is called a dry wash.

(b) The channel of an ephemeral stream lies above the water table, so the stream flows only when water enters the stream faster than it can infiltrate into the ground.

FIGURE 17.7 Flow velocity and turbulence in streams.

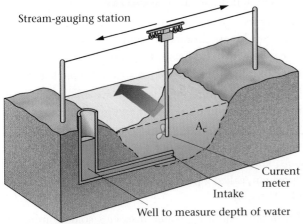

(a) At a stream-gauging station, geologists measure the area (A_c) and depth of the stream. They measure velocity using a current meter at various points in the stream.

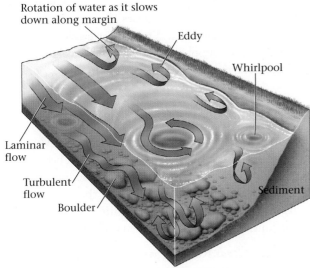

(b) Turbulence can make measuring velocity difficult. In turbulent flow, water swirls in curving paths.

Different streams have different average discharges. Fundamentally, discharge depends on the size of the watershed and on the amount of meteoric water falling in the watershed. The Amazon River has the largest average discharge in the world—about 200,000 m³/s, or 15% of the total amount of runoff on Earth. The next-largest stream, the Congo River, has an average discharge of 40,000 m³/s, whereas the "mighty" Mississippi's is only 17,000 m³/s. Also, the discharge of a given stream varies along its length. For example, the discharge in a temperate region *increases* in the downstream direction, because each tributary that enters the stream adds more water, whereas the discharge in an arid region may *decrease* downstream, as more and more water seeps into the ground or

evaporates. Discharge can also be affected by human activity. If people divert the river's water for irrigation, the river's discharge decreases downstream. Finally, the discharge at a given location can vary with time: in a temperate climate, a stream's discharge during the spring may be double or triple the amount during a hot summer, and a flood may increase the discharge to more than 100 times normal.

The *average* velocity of stream water can be difficult to calculate, because velocity at one point in a stream may differ from that at another point. Such variations happen for two reasons. First, **turbulence**, or turbulent flow, is a twisting, swirling motion that, on a large scale, can create eddies (whirlpools) in which water curves and actually flows upstream or circles in place (**Fig. 17.7b**). Turbulence develops in part because the shearing motion of one volume against its neighbor causes the neighbor to spin, and in part because obstacles such as boulders deflect volumes, forcing them to move in a different direction. Second, friction along the sides and bed of the stream slows the flow. In general, water near the channel walls or the stream bed (the floor of the stream) moves more slowly than water in the middle of the flow, and the fastest-moving part of the stream flow lies near the surface in the center of the channel. In a curved channel, the fastest flow shifts toward the outside curve, somewhat like a car swerves to the outer edge of a curve on a highway. Therefore, the deepest part of a channel, its **thalweg**, lies near the outside curve. In fact, as the water flows toward the outside wall of a curving channel, it begins a spiral motion because as surface water moves toward the outer bank, water deeper down must flow toward the inner bank to replace the surface water.

The amount by which friction slows the flow depends both on the roughness of the walls and bed and on the channel shape. Rougher walls slow the floor more, and a wide, shallow stream channel has a larger wetted perimeter (the area in which water touches the channel walls) than does a semicircular channel, so water flows more slowly in the former than in the latter (**Fig. 17.8a–c**).

Take-Home Message

- The discharge of a stream is the volume of water that passes a point on the stream bank in a given time. Large streams have greater discharges than smaller ones.
- Stream flow tends to be turbulent so that water twists and turns and forms eddies.
- The channel's cross-sectional shape, its roughness, and its curviness affect flow velocity.

THINK: Where is the velocity of flow in a stream fastest? Why?

FIGURE 17.8 Variations in flow velocity in a channel.

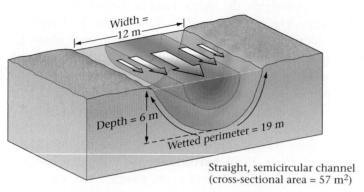

(a) In a straight channel, the fastest velocity occurs near the surface of the stream, equidistant from the banks.

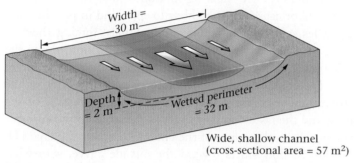

(b) Velocity in a wide, shallow channel is less than that of a semicircular channel with the same cross-sectional area, because the wetted perimeter is greater.

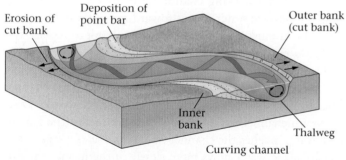

(c) In a curved channel, the fastest flow shifts to the outer edge of the stream, and water follows a spiral-like path.

17.4 **THE WORK OF RUNNING WATER**

How Do Streams Erode?

The energy that makes running water move comes from gravity. As water flows downslope from a higher to a lower elevation, the gravitational potential energy stored in water transforms into kinetic energy. About 3% of this energy goes into the work of eroding the walls and beds of stream channels. Running water causes erosion in four ways.

- *Scouring*: Running water can remove loose fragments of sediment, a process called **scouring**.
- *Breaking and lifting*: In some cases, the push of flowing water can break chunks of solid rock off the channel floor or walls. In addition, the flow of a current over a clast can cause the clast to rise, or lift off the substrate.
- *Abrasion*: Clean water has little erosive effect, but sand-laden water acts like sandpaper and grinds or rasps away at the channel floor and walls, a process called **abrasion**. In places where turbulence produces long-lived whirlpools, abrasion by sand or gravel carves a bowl-shaped depression, called a **pothole**, into the floor of the stream (Fig. 17.9a, b).
- *Dissolution*: Running water dissolves soluble minerals as it passes, and carries the minerals away in solution.

The efficiency of erosion depends on the velocity and volume of water and on its sediment content. A large volume of fast-moving, turbulent, sandy water causes more erosion than does a trickle of quiet, clear water. Thus, most erosion takes place during floods, which supply streams with large volumes of fast-moving, sediment-laden water.

How Do Streams Transport Sediment?

The Mississippi River received the nickname "Big Muddy" for a reason—its water can become chocolate brown because of all the clay and silt it carries. All streams carry sediment, though not the same amount. Geologists refer to the total volume of sediment carried by a stream as its sediment load. The sediment load consists of three components (Fig. 17.9c):

- *Dissolved load*: Running water dissolves soluble minerals in the sediment or rock of its substrate, and groundwater seeping into a stream through the channel walls brings dissolved minerals with it. The ions of these dissolved minerals constitute a stream's **dissolved load**.
- *Suspended load*: The **suspended load** of a stream consists of tiny solid grains (silt or clay size) that swirl along with the water without settling to the floor of the channel; this sediment makes the water brown (Fig. 17.10a–c).
- *Bed load*: The **bed load** of a stream consists of large particles (such as sand, pebbles, or cobbles) that bounce or roll along the stream floor (Fig. 17.10d). Typically, bed-load movement involves **saltation**, a process during which grains on the channel floor get knocked into the water column, follow a curved trajectory downstream, and gradually sink to the bed again, where they knock other grains into the water column.

When describing a stream's ability to carry sediment, geologists specify its competence and capacity. The **competence**

FIGURE 17.9 Erosion and transportation in streams.

(a) Two potholes in the bed of a stream near Ithaca, New York.

(b) This slot canyon in Arizona formed when many potholes linked together.

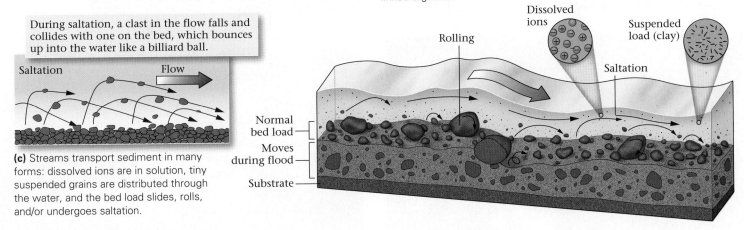

During saltation, a clast in the flow falls and collides with one on the bed, which bounces up into the water like a billiard ball.

(c) Streams transport sediment in many forms: dissolved ions are in solution, tiny suspended grains are distributed through the water, and the bed load slides, rolls, and/or undergoes saltation.

of a stream refers to the maximum particle size it carries; a stream with high competence can carry large particles, whereas one with low competence can carry only small particles. A fast-moving, turbulent stream has greater competence (it can carry bigger particles) than a slow-moving stream, and a stream in flood has greater competence than a stream with normal flow. In fact, the huge boulders that litter the bed of a mountain creek move *only* during floods. The **capacity** of a stream refers to the total quantity of sediment it *can carry*. A stream's capacity depends on its competence and discharge.

How Do Streams Deposit Sediment?

A raging torrent of water can carry coarse and fine sediment—the finer clasts rush along with the water as suspended load, whereas the coarser clasts may bounce and tumble as bed load. If the flow velocity decreases, either because the gradient (downstream slope) of the stream bed becomes shallower or because the channel broadens out and friction between the bed and the water increases, then the competence of the stream decreases and sediment settles out. The size of the clasts that settle at a particular locality depends on the decrease in flow

FIGURE 17.10 Sediment, carried and deposited by streams. The clast size depends on stream velocity.

(a) On a sunny day, this stream in Switzerland is fairly clear; you can see cobbles on the bed.

(b) On a rainy day, the stream's discharge increases and the faster, more turbulent water becomes brown due to its load of sediment.

(c) An air photo shows a muddy river emptying into the Black Sea. Currents carry the sediment away.

(d) Bars of gravel deposited within the channel of a stream in the Wasatch Mountains, Utah.

(e) Gravel in a stream bed of a mountain stream in Denali National Park, Alaska. The large clasts were carried during floods.

(f) Point bars of mud deposited along a gentle, slowly moving stream in Brazil.

velocity. For example, if the stream slows by a small amount, only large clasts settle; if the stream slows by a greater amount, medium-sized clasts settle; and if the stream slows almost to a standstill, the fine grains settle. Thus, coarser sediment tend to settle out farther upstream, where the gradient of the stream is steeper and water flows faster, whereas finer grains settle

out farther downstream, where the water flows more slowly. Because of this process of sediment sorting, stream deposits tend to be segregated by size—gravel accumulates in one location and mud in another.

Geologists refer to sediments transported by a stream as **fluvial** deposits (from the Latin *fluvius*, meaning river) or

alluvium. Fluvial deposits may accumulate along the stream bed in elongate mounds, called **bars** (Fig. 17.10d–f). Some stream channels make broad curves. Water slows along the inner edge of a curve, so crescent-shaped **point bars** bordering the shoreline develop. During floods, a stream may overtop the banks of its channel and spread out over its floodplain, a broad flat area bordering the stream. Friction slows the water on the floodplain, so a sheet of silt and mud settles out to comprise floodplain deposits. (We discuss such deposits further on p. 585.) Where a stream empties at its mouth into a standing body of water, the water slows and a wedge of sediment, called a **delta**, accumulates. We discuss floodplains and deltas in more detail later in this chapter.

Take-Home Message

- Streams can scour and abrade their beds and pick up sediment. They carry sediment as dissolved load, suspended load, and bed load.
- Faster, more turbulent streams have greater competence; they can carry larger clasts.
- The total volume of sediment that can be carried depends on competence and velocity.
- Alluvium collects where flow velocity decreases, and it forms bars and deltas.

THINK: Why do point bars form on the inner arc of a curve in a stream?

17.5 HOW DO STREAMS CHANGE ALONG THEIR LENGTH?

Longitudinal Profiles

In 1803, under President Thomas Jefferson's leadership, the United States bought the Louisiana Territory, a vast tract of land encompassing the western half of the Mississippi drainage basin. At the time, the geography of the territory was a mystery. To fill the blank on the map, Jefferson asked Meriwether Lewis and William Clark to lead a voyage of exploration across the Louisiana Territory to the Pacific.

Lewis and Clark, along with about 40 men, began their expedition at the mouth of the Missouri River, where it joins the Mississippi. At this juncture, the Missouri is a wide, languid stream of muddy water. The group found the Missouri's downstream **reach** (an interval along the length of a stream), where the river's channel is deep and the water smooth, to be easy going. But the farther upstream they went, the more difficult their voyage became, for the **stream gradient** (the slope of the stream channel) became progressively steeper, and the stream's discharge became less. When Lewis and Clark

reached the site of what is now Bismarck, North Dakota, they had to abandon their original boats and haul smaller vessels up rapids where turbulent water plunges over a steep, bouldery bed, and occasionally they had to carry their boats around waterfalls, where water drops over an escarpment. When they reached what is now southwestern Montana, they abandoned these boats as well and trudged along the stream valley on foot or on horseback, struggling up steep gradients until they reached the continental divide.

If Lewis and Clark had been able to plot a graph showing their elevation above sea level relative to their distance along the Missouri, they would have found that the **longitudinal profile** of the Missouri, a cross-sectional image showing the variation in the river's elevation along its length, is roughly a concave-up curve (Fig. 17.11a, b). This curve illustrates that stream gradient is steeper near its headwaters (source) than near its mouth. Real longitudinal profiles are not perfectly smooth curves, but rather display little plateaus and steps, representing interruptions by lakes or waterfalls. Near its headwaters, an idealized stream flows down deep valleys or canyons, whereas near its mouth, it flows over nearly horizontal plains.

The Base Level

Streams progressively deepen their channels by downcutting, but there is a depth below which a stream cannot downcut any further. The lowest elevation a stream channel's floor can reach at a locality is the **base level** of the stream. A *local* base level is one that occurs upstream of a drainage network's mouth, and the *ultimate* base level (i.e., the lowest possible elevation along the stream's longitudinal profile) is sea level. The trunk stream cannot downcut deeper than sea level, for if it did, it would have to flow upslope to enter the sea.

Lakes or reservoirs can act as local base levels along a stream, for where the stream enters such standing bodies of water, it slows almost to a halt and cannot downcut further (Fig. 17.12a). A ledge of resistant rock can also act as a local base level, for the stream level cannot drop below the ledge until the ledge erodes away (Fig. 17.12b). Finally, where a tributary joins a larger stream, the channel of the larger stream acts as the base level for the tributary; thus, the mouth of a tributary lies at the same elevation as the stream that it joins at the point of intersection. Local base levels do not last forever, because running water eventually removes the obstructions that create them.

Streams attain a concave-up longitudinal profile gradually. During this process, steps or ledges defining local base levels along the stream erode away, and low areas fill in, until any point along the stream approaches a condition such that there is no net erosion or deposition—the stream can carry all the sediment that has been supplied to it, and it deposits as

FIGURE 17.11 Drainage basins, and the change in character of a stream along its longitudinal profile.

(b) In general, the longitudinal profile of a stream (elevation change along its length) is concave up.

The cross-sectional profile changes with position along the stream.

(a) A drainage network collects water from a broad drainage basin, or watershed, via numerous tributaries. These carry water to a trunk stream and eventually to a standing body of water. Points 1 to 5 refer to locations along the longitudinal profile (inset).

FIGURE 17.12 The concept of local base levels in stream profiles.

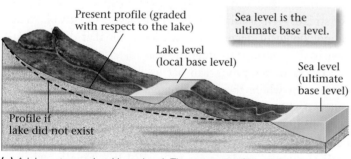

(a) A lake acts as a local base level. The stream profile uphill of the lake lies above the profile that would form if the lake didn't exist.

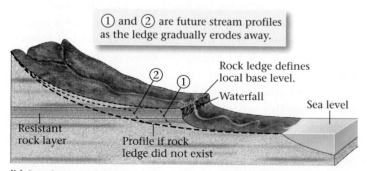

(b) A resistant rock ledge can form a local base level. Headward erosion gradually cuts back into the ledge.

much sediment as it removes. A stream that has reached this condition is called a **graded stream**.

Take-Home Message

- A stream's longitudinal profile tends to be a concave-up curve.
- Streams cannot downcut below their base level.
- Local base levels form steps in the longitudinal profile. The lowest elevation on a profile is sea level, the ultimate base level.

THINK: Why can't a stream downcut more deeply than its base level?

17.6 STREAMS AND THEIR DEPOSITS IN THE LANDSCAPE

Valleys and Canyons

About 17 million years ago, a large block of crust, the region now known as the Colorado Plateau (located in Arizona, Utah, Colorado, and New Mexico), began to rise. Before the rise, the

FIGURE 17.13 Examples of steep-sided canyons. These form where the canyon walls consist of resistant rock.

(a) Erosion has cut deep canyons into the horizontal strata of the Colorado Plateau in northern Arizona.

(b) At Canyon de Chelly, Arizona, vertical walls of Permian sandstone border a flat valley.

Colorado River had been flowing over a plain not far above sea level, causing little erosion. But as the land uplifted, the river began to downcut. Eventually, its channel lay as much as 1.6 km below the surface of the plateau at the base of a steep-walled gash now known as the Grand Canyon (Fig. 17.13a, b). The formation of the Grand Canyon illustrates a general phenomenon. In regions where the land surface lies well above the base level, a stream can carve a deep trough, much deeper than the channel itself. If the walls of the trough slope gently, the landform is a valley. If they slope steeply, the landform is a canyon.

Whether stream erosion produces a valley or a canyon depends on the rate at which downcutting takes place relative to the rate at which mass wasting causes the walls on either side of the stream to collapse. In places where a stream downcuts through its substrate faster than the walls of the stream collapse, erosion creates a slot (steep-walled) canyon. Such canyons typically form in hard rock, which can hold up steep cliffs for a long time (Fig. 17.14a). In places where the walls collapse as fast as the stream downcuts, landslides and slumps gradually cause the slope of the walls to approach the angle of repose. When this happens, the stream channel lies at the floor of a valley whose cross-sectional shape resembles the letter V (Fig. 17.14b); this landform is called a **V-shaped valley**. Where the walls of the stream consist of alternating layers of hard and soft rock, the walls develop a stair-step shape such as that of the Grand Canyon (Fig. 17.14c).

In places where active downcutting occurs, the valley floor remains relatively clear of sediment, for the stream—especially

FIGURE 17.14 The shape of a canyon or valley depends on the resistance of its walls to erosion slumping.

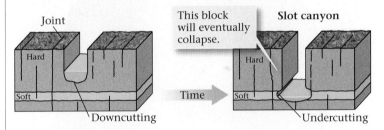

(a) If downcutting by the stream happens faster than mass wasting on the walls, a slot canyon forms. The canyon widens as the stream undercuts the walls.

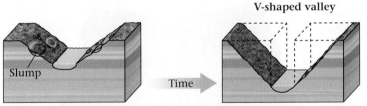

(b) If mass wasting takes place as fast as downcutting occurs, a V-shaped valley develops.

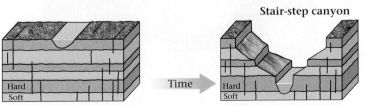

(c) Downcutting through alternating hard and soft layers produces a stair-step canyon.

when it floods—carries away sediment that has fallen or slumped into the channel from the stream walls. But if the stream's base level rises, its discharge decreases, or its sediment load increases, the valley floor fills with sediment, creating an alluvium-filled valley (Fig. 17.15a, b). The surface of the alluvium becomes a broad floodplain. If the stream's base level later drops again and/or the discharge increases, the stream will start to cut down into its own alluvium, a process that generates **stream terraces** bordering the present floodplain (Fig. 17.15c).

Rapids and Waterfalls

When Lewis and Clark forged a path up the Missouri River, they came to reaches that could not be navigated by boat because of **rapids**, particularly turbulent water with a rough surface (Fig. 17.16a). Rapids form where water flows over steps or large clasts in the channel floor, where the channel abruptly narrows, or where its gradient abruptly changes. The turbulence in rapids produces eddies, waves, and whirlpools that roil and churn the water surface, in the process creating whitewater, a mixture of bubbles and water. Modern-day whitewater rafters thrill to the unpredictable movement of rapids (Fig. 17.16b).

A **waterfall** forms where the gradient of a stream becomes so steep that the water literally free-falls down the stream bed (Fig. 17.16c, d). The energy of falling water may scour a depression, called a plunge pool, at the base of the waterfall. Some waterfalls develop where a stream crosses a resistant ledge of rock,

> **Did you ever wonder . . .**
> why does a large waterfall form, and will it always be there?

FIGURE 17.15 The evolution of alluvium-filled stream valleys and the development of terraces.

Time 1

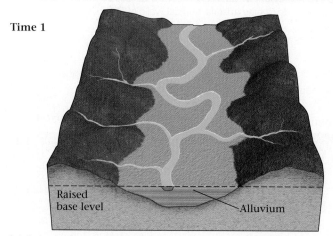

(a) A rise in the base level or a decrease in discharge causes the valley to fill with alluvium.

Time 2

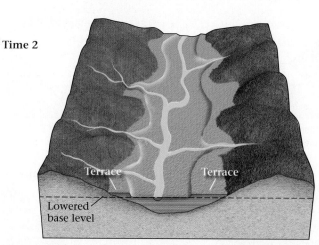

(b) Later, if the base level falls or the discharge increases, the stream downcuts through the alluvium and a new, lower floodplain develops. The remnants of the original alluvial plain remain as a pair of terraces.

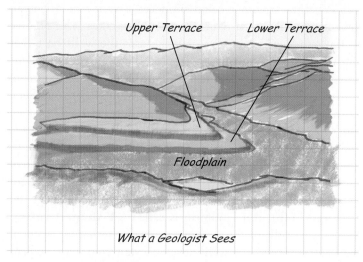

What a Geologist Sees

(c) In this view of a valley in Utah, we can see a floodplain and two terrace levels.

FIGURE 17.16 Examples of rapids and waterfalls.

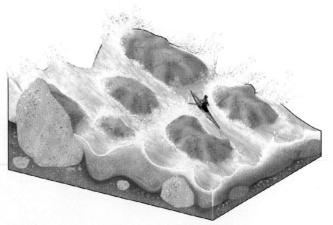

(b) Class V rapids are extremely difficult to navigate.

(a) These rapids in the Grand Canyon formed when a flood from a side canyon dumped debris into the channel of the Colorado River.

(c) Iguaçu Falls, at the Brazil-Argentina border, spills across layers of basalt. The basalt acts as a resistant ledge.

(d) Waterfall spilling into Milford Sound, New Zealand, from a hanging valley.

and some develop as a result of faulting because displacement produces an escarpment. Waterfalls also occur where glacial erosion has deepened a trunk valley relative to tributary valleys to form a "hanging valley" whose mouth is much higher than the floor of the trunk valley (see Chapter 22).

Though a waterfall may appear to be a permanent feature of the landscape, all waterfalls eventually disappear because headward erosion slowly eats back the resistant ledge until the stream reaches grade. We can see a classic example of headward erosion at Niagara Falls. As water flows from Lake Erie to Lake Ontario, it drops over a 55-m-high ledge of resistant Silurian dolostone, which overlies a weak shale. Erosion of the shale leads to undercutting of the dolostone. Gradually, the overhang of dolostone becomes unstable and collapses, with the result that the waterfall migrates upstream. Before the industrial age, the edge of Niagara Falls cut upstream at an average rate of 1 m per year; but since then, the diversion of water from the Niagara River into a hydroelectric power station has decreased the rate of headward erosion to half that. Nevertheless, Niagara Falls will cut all the way back to Lake Erie in about 60,000 years (Fig. 17.17a–d).

FIGURE 17.17 The formation of Niagara Falls, at the border between Ontario, Canada, and New York State. The falls tumble over the Lockport Dolomite, a relatively strong rock layer.

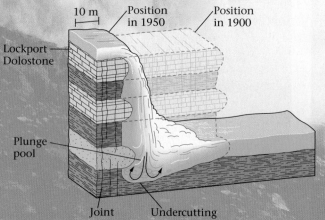

(a) Niagara Falls formed where the outlet of Lake Erie flowed over the Niagara escarpment.

(b) The American Falls, a part of Niagara Falls.

(c) The face of the falls retreats over time. The lower shale erodes, and stronger layers above break off at joints.

(d) In the 1960s, the water was diverted from the falls, making the ledge of dolomite visible.

Alluvial Fans and Braided Streams

Where a fast-moving stream abruptly emerges from a mountain canyon into an open plain at the range front, the water that was once confined to a narrow channel spreads out over a broad surface. As a consequence, the water slows and abruptly drops its sedimentary load, forming a gently sloping apron of sediment (sand, gravel, and cobbles) called an **alluvial fan** (Fig. 17.18a). The stream then divides into a series of small channels that spread out over the fan. During particularly strong floods, debris flows spread over and smooth out the fan's surface.

In some localities, streams carry abundant coarse sediment during floods but cannot carry this sediment during normal flow.

Thus, during normal flow, the sediment settles out and chokes the channel. As a consequence, the stream divides into numerous strands weaving back and forth between elongate bars of gravel and sand. The result is a **braided stream**—the name emphasizes that the streams entwine like strands of hair in a braid (Fig. 17.18b). Strands of the stream branch out at the upstream ends of bars and merge at the downstream end of bars. Because the gravelly sediment of a braided stream can't stick together, the stream cannot cut a deep channel with steep banks—the channel walls simply collapse, so the stream spreads out over a broad area. Braided streams commonly form in landscapes where streams fill with sediment-choked glacier meltwater.

Meandering Streams and Their Floodplains

A riverboat cruising along the lower reaches of the Mississippi River cannot sail in a straight line, for the river channel winds back and forth in a series of snake-like curves called **meanders** (Fig. 17.19a–c ⏵; See for Yourself P, p. S-30). In fact, the boat has to go 500 km along the river channel to travel 100 km as the crow flies. A meandering stream is one that has many meanders. The development of meanders increases the volume of the stream by increasing its length.

How do meanders evolve? Even if a stream starts out with a straight channel, natural variations in the water depth and associated friction (see Fig. 17.7b) cause the fastest-moving current to swing back and forth. The water erodes the side of the stream more effectively where it flows faster, so it begins to cut away faster on the outer arc of the curve. Thus, each curve begins to migrate sideways and grow more pronounced until it becomes a meander. On the outside edge of a meander, erosion continues to eat away at the channel wall, forming a cut bank, whereas on the inside edge, water slows down so that its competence decreases and sediment accumulates, forming a wedge-shaped deposit called a point bar, as noted earlier. (Mark Twain, who worked as a riverboat pilot on the Mississippi River before writing such classic books as *Huckleberry Finn*, took his pen name from the signals the mate of a paddle-wheel steamer called out to the skipper to indicate water depth; "mark twain" means two fathoms, or about 4 m deep.) With continued erosion, a meander may curve through more than 180°, so that the cut bank at the meander's entrance approaches the cut bank at its end, leaving a meander neck, a narrow isthmus of land separating the portions of the meander.

Meandering streams initially form where running water travels over a plain that has only a gentle gradient. For the meander curves to evolve, the substrate of a stream must be strong enough to cut banks to hold up. In unconsolidated substrates, this strength comes from plant roots. Interestingly, recent research suggests that meandering streams could not develop until abundant land plants appeared during the Silurian—before that all streams on plains may have been braided!

People building communities along a riverbank may assume that the shape of a meander remains fixed for a long time. It doesn't. In a natural meandering river system, the river channel migrates back and forth across the floodplain. When erosion eats through a meander neck, a straight reach called a cutoff develops. The meander that has been cut off is called an **oxbow lake** if it remains filled with water, or an abandoned meander if it dries out (Fig. 17.19d).

Most meandering stream channels cover only a relatively small portion of a broad, gently sloping **floodplain** (Fig. 17.19e). Floodplains, as we noted earlier, are so named because during a flood, water overtops the edge of the stream channel and spreads out over the floodplain. In many cases, a floodplain terminates at its sides along a bluff, or escarpment; large floods may cover the entire region from bluff to bluff. As the water leaves the channel, friction between the ground and the thin sheet of water moving over the floodplain slows down the flow. This slowdown decreases the competence of the running water, so sediment settles out along the edge of the channel. Over time, the accumulation of this sediment creates a pair of low ridges, called **natural levees**, on either side of the

FIGURE 17.18 Examples of depositional landforms produced from stream sediment.

(a) An alluvial fan in Death Valley, California, consists of sand, gravel, and debris flows. The curving black line is a road.

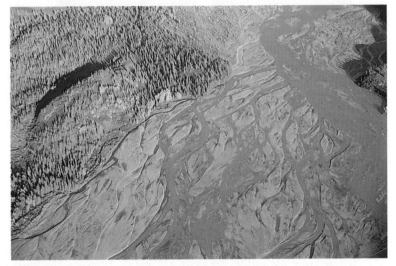

(b) A braided stream, carrying meltwater from a glacier near Denali, Alaska, deposits elongate bars of gravel.

FIGURE 17.19 The character and evolution of meandering streams and floodplains.

(a) A meandering stream wanders across a floodplain.

(b) Availability of water lets farmers use this floodplain in Utah for crops; adjacent slopes are barren.

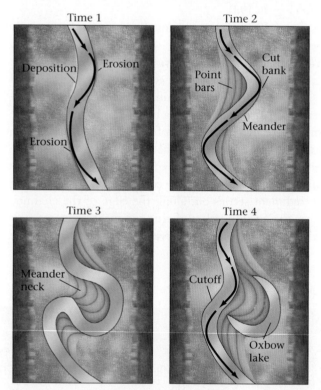

(c) Meanders evolve because erosion occurs faster on the outer bank of a curve, and deposition takes place on the inner curve. Eventually, a cutoff isolates an oxbow lake.

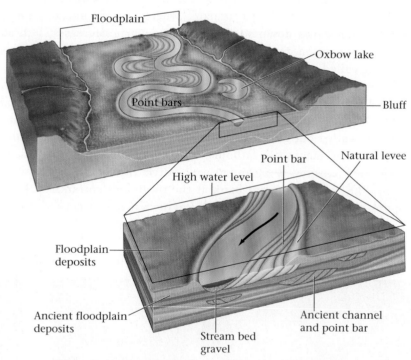

(d) Landforms along meandering streams include natural levees, point bars, and floodplains. Older deposits record the position of ancient channels and floodplains.

stream (Fig. 17.20). Natural levees may grow so large that the floor of the channel may become higher than the surface of the floodplain. In fact, the higher areas of New Orleans, which have remained fairly dry during floods that submerged the rest of the city, are the parts built on the natural levees. In places where large natural levees exist, the region between the bluffs and the levees may become a low, marshy swamp. Also because of the levees, small tributaries may be blocked from joining the trunk stream; the tributaries, called yazoo streams, flow in the floodplain and trend parallel to the main river.

Deltas: Deposition at the Mouth of a Stream

Along most of its length, only a narrow floodplain—covered by green, irrigated farm fields—borders the Nile River in Egypt. But at its mouth, the trunk stream of the Nile divides into a

FIGURE 17.19 (continued)

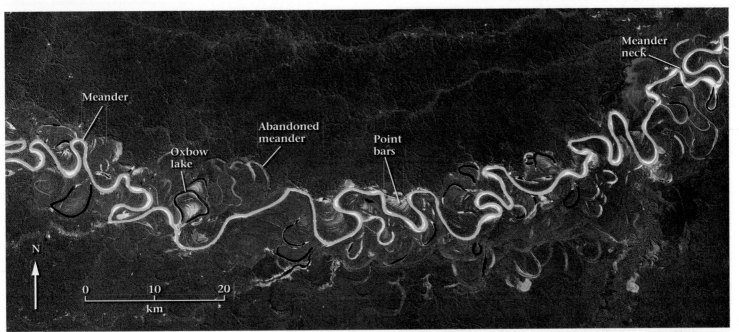

(e) A meandering stream in Brazil, as viewed from space. Note the oxbows and cutoffs. Variations in vegetation color provide visual clues to the position of abandoned meanders.

FIGURE 17.20 On this satellite photo of the southernmost Mississippi River, where it flows into the Gulf of Mexico, the green areas are vegetated levees. Note how natural levees have built up adjacent to every channel. Engineers have built up the levee along the main channel.

FIGURE 17.21 Deltas form where a sediment-laden river enters standing water.

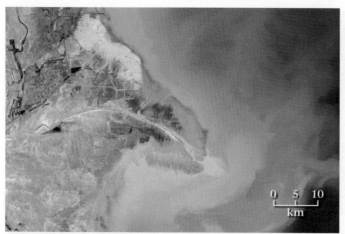

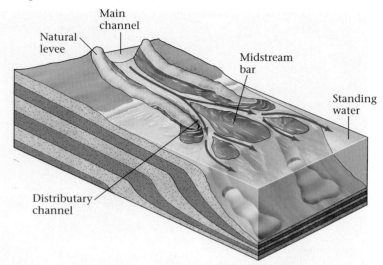

(a) The Yellow River Delta, China, from space. Engineers diverted the river in 1996, so that it would build a new lobe.

(b) Distributaries form because the river deposits sediment in its mouth when it reaches standing water.

(c) A small delta forming in Lake Tekapo, New Zealand, shows several distributaries.

fan of smaller streams, called **distributaries**, and the area of green agricultural lands broadens into a triangular patch. The Greek historian Herodotus noted that this triangular patch resembles the shape of the Greek letter delta (Δ), and so the region became known as the Nile Delta.

Deltas, as we noted earlier, develop where the running water of a stream enters standing water, the current slows, the stream loses competence, and sediment settles out (Fig. 17.21a–c). Geologists refer to any wedge of sediment formed at a river mouth as a delta, even though relatively few have the triangular shape of the Nile Delta (Fig. 17.22a–c). Some deltas curve out into the sea, whereas others consist of many elongate lobes; the latter are called bird's-foot deltas, because they resemble the scrawny toes of a bird. The existence of several toes indicates that the main course of the river in the delta has shifted on several occasions. These shifts occur when a toe builds so far out into the sea that the slope of the stream becomes too gentle to allow the river to

flow. At this point, the river overflows a natural levee upstream and begins to flow in a new direction, an event called an avulsion. The distinct lobes of the Mississippi Delta, a bird's-foot delta, suggest that avulsions have happened several times during the past 9,000 years (Fig. 17.23). New Orleans, built along one of the Mississippi's distributaries, may eventually lose its riverfront, for a break in a levee upstream of the city could divert the Mississippi into the Atchafalaya River channel.

The shape of a delta depends on many factors. Deltas that form where the strength of the river current exceeds that of ocean currents have a bird's-foot shape, since the sediment can be carried far offshore. In contrast, deltas that form where the ocean currents are strong have a Δ shape, for the ocean currents redistribute sediment in bars running parallel to the shore. And in places where waves and currents are strong enough to remove sediment as fast as it arrives, a river has no delta at all.

With time, the sediment of a delta compacts, and the lithosphere beneath the delta subsides (see Chapter 7). As a consequence, the surface of a delta slowly sinks. In a natural delta, distributaries provide sediment that fills the resulting space so that the delta's surface remains at or just above sea level, forming a broad flat area called a delta plain. But if people build up artificial levees to constrain the river to its channel, sediment gets carried directly to the seaward edge of the delta and the delta's interior "starves" (does not receive sediment). When this happens, the land surface drops below sea level. Because of this process, much of New Orleans lies below sea level, so high waters from Hurricane Katrina in 2005 caused the city to flood extensively (see Chapter 18).

FIGURE 17.22 Delta shape varies depending on current activity, waves, and vegetation.

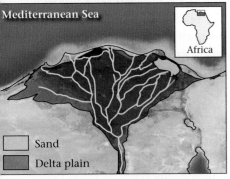

Mediterranean Sea

Africa

Sand

Delta plain

(a) The Nile is a Δ-shaped delta.

Africa

Atlantic Ocean

(b) The Niger is an arc-like delta.

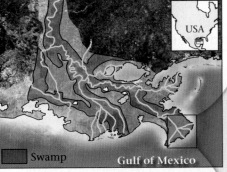

USA

Swamp   Gulf of Mexico

(c) The Mississippi is a bird's-foot delta.

Mouth of the Mississippi seen from space.

Why do rivers divide into distributaries at their mouths? When a river reaches standing water, its velocity slows. The sediment settles out at the mouth to form a midstream bar. The presence of the bar causes the stream to split into two channels. Similar bars created at the mouths of these two subsidiary channels cause each of them to separate in turn, until eventually numerous distributary channels have formed (see Fig. 17.21b).

Take-Home Message

- Streams carve valleys and canyons. V-shaped valleys form where the walls collapse faster than downcutting takes place; canyons form where walls are strong.

- Rapids develop where water becomes very turbulent, and waterfalls develop where streams undergo free fall.

- Streams deposit sediment in alluvial fans at mountain fronts. Sediment-choked streams tend to become braided.

- Streams flowing on gentle gradients meander, commonly within a floodplain.

- Deltas develop where streams enter standing water and deposit sediment.

THINK: What factors affect the map-view character of a delta?

FIGURE 17.23 A map showing ancient lobes of the Mississippi Delta. A major flood could divert water from the Mississippi into the channel of the Atchafalaya.

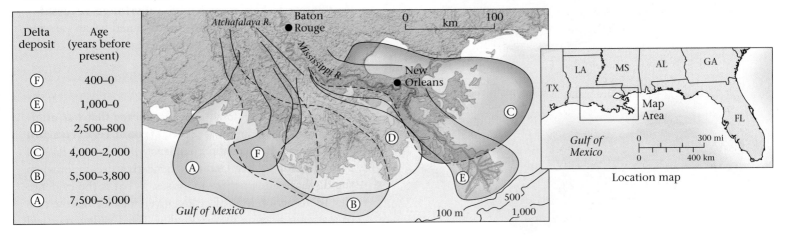

Delta deposit	Age (years before present)
Ⓕ	400–0
Ⓔ	1,000–0
Ⓓ	2,500–800
Ⓒ	4,000–2,000
Ⓑ	5,500–3,800
Ⓐ	7,500–5,000

Atchafalaya R.   Baton Rouge   0   km   100

Mississippi R.   New Orleans

Gulf of Mexico   100 m   1,000   500

LA   MS   AL   GA

TX   Map Area   FL

Gulf of Mexico   0   300 mi   0   400 km

Location map

17.7 THE EVOLUTION OF DRAINAGE

Beveling Topography

Imagine a place where continental collision uplifts a region (Fig. 17.24). At first, rivers have steep gradients, flow over many rapids and waterfalls, and cut deep valleys. But with time, rugged mountains become low, rounded hills; once-deep, narrow valleys broaden into wide floodplains with more gradual gradients. As more time

FIGURE 17.24 Evolution of a fluvial landscape when base level drops.

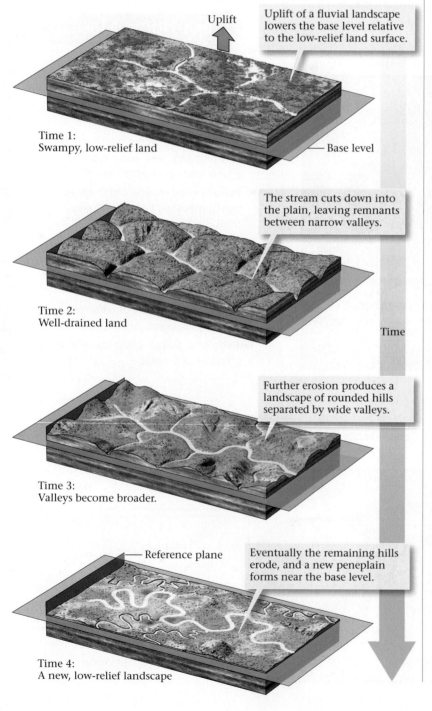

Uplift

Uplift of a fluvial landscape lowers the base level relative to the low-relief land surface.

Time 1:
Swampy, low-relief land

— Base level

The stream cuts down into the plain, leaving remnants between narrow valleys.

Time 2:
Well-drained land

Time

Further erosion produces a landscape of rounded hills separated by wide valleys.

Time 3:
Valleys become broader.

— Reference plane

Eventually the remaining hills erode, and a new peneplain forms near the base level.

Time 4:
A new, low-relief landscape

passes, even the low hills are beveled down, becoming small mounds or even disappearing altogether. (Some geologists have referred to the resulting landscape as a peneplain, from the Latin *paene*, which means almost; it lies at an elevation close to that of a stream's base level.) Through these stages, a fluvial landscape changes or evolves through time, and extensive denudation (the removal of rock and regolith from the Earth's surface) occurs.

Though the above model makes intuitive sense, it is an oversimplification. Plate tectonics can uplift the land again, and/or global sea-level rise or fall can change the base level, so in reality peneplains rarely develop before downcutting begins again. In fact, geologists debate about whether they ever really form at all.

Stream Piracy and Drainage Reversal

Stream piracy sounds like pretty violent stuff. In reality, **stream piracy**, or stream capture, simply refers to a situation in which headward erosion causes one stream to intersect the course of another stream. When this happens, the pirate stream "captures" the water in the stream it intersects, so the captured stream starts flowing into the pirate stream (Fig. 17.25a, b). The piracy of a stream that had been flowing through a water gap transforms the water gap into a wind gap, a dry pathway through a high ridge. In 1775, the pioneer Daniel Boone blazed the "Wilderness Road" through the Cumberland Gap, a wind gap in the Appalachian Mountains at the border of Kentucky and Virginia, to provide other settlers with access to the Kentucky wilderness.

A regional change in the flow direction of a drainage network can also take place, sometimes due to plate tectonics. For example, in the early Mesozoic Era, when South America linked to Africa on its eastern coast, a "proto-Amazon" River flowed westward and drained the interior of Gondwana into the Pacific Ocean. Later, when South America separated from Africa, the Andes rose on South America's western coast. This event caused a **drainage reversal**—flow in the Amazon changed direction and began carrying water eastward to the newly formed Atlantic (Fig. 17.26a, b).

Stream Rejuvenation

Where streams cut down into a landscape that was originally near the stream's base level, **stream rejuvenation** has occurred. Rejuvenation happens when the base level of a stream drops, when land rises beneath a stream, or when the discharge of a stream increases. What is the evidence that rejuvenation took place at a given locality? In the case of a stream flowing in an alluvium-filled valley, renewed

FIGURE 17.25 The concept of stream capture or "piracy."

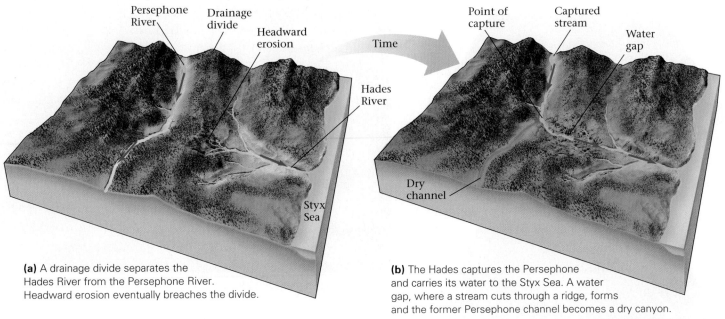

(a) A drainage divide separates the Hades River from the Persephone River. Headward erosion eventually breaches the divide.

(b) The Hades captures the Persephone and carries its water to the Styx Sea. A water gap, where a stream cuts through a ridge, forms and the former Persephone channel becomes a dry canyon.

FIGURE 17.26 An example of continental-scale drainage reversal in South America.

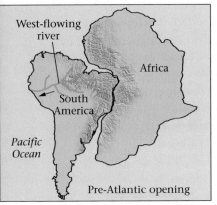

(a) In the early Mesozoic, rivers drained westward, from the interior of Gondwana.

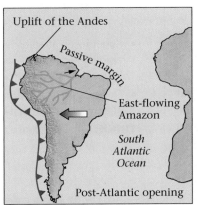

(b) Later, after rifting and the formation of the Andes, the Amazon drainage flowed eastward.

downcutting allows the stream to create a new floodplain at a lower elevation than the original one (see Fig. 17.15b). As we have seen, the younger floodplain tends to be narrower than the older, and the surface of the older floodplain becomes a terrace on either side of the new floodplain. In the case of a stream flowing on bedrock, a drop of the base level causes the stream to incise into the bedrock.

In some locations, rejuvenated streams develop incised meanders that lie at the bottom of a steep-walled canyon; the canyon was carved by the stream when uplift occurred in the region. The "goosenecks" of the San Juan River, in southern Utah, illustrate this geometry (Fig. 17.27a–c).

Superposed and Antecedent Streams

The structure and topography of the landscape do not always appear to control the path, or course, of a stream. For example, imagine a stream that carves a deep canyon straight across a strong mountain ridge—why didn't the stream find a way around the ridge? We distinguish two types of streams that cut across resistant topographic highs:

- *Superposed streams*: Imagine a region in which drainage initially forms on a layer of soft, flat strata that unconformably overlies folded strata. Streams carve channels into the flat strata; when they eventually erode down through the unconformity and start to downcut into the folded strata, they maintain their earlier course, ignoring the structure of the folded strata. Geologists call such streams **superposed streams**, because their preexisting geometry has been laid down on underlying rock structure (Fig. 17.28a, b).

- *Antecedent streams*: In some cases, tectonic activity (such as subduction or collision) causes a mountain range to rise up beneath an already established stream. If the stream downcuts as fast as the range rises, it can maintain its course and will cut right across the range. Geologists call such streams **antecedent streams** (from the Greek *ante*, meaning before) to emphasize

FIGURE 17.27. The formation of incised meanders.

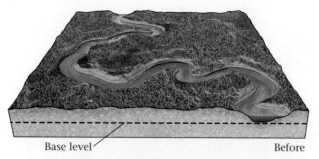

(a) The process begins when the base level drops relative to a stream that is meandering on a plain.

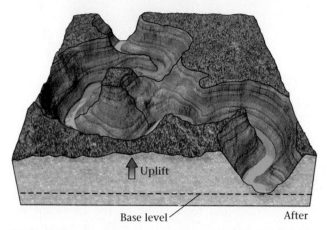

(b) Over time, the stream cuts down into the bedrock. Meanders continue to evolve while this happens.

(c) The "goosenecks" of the San Juan River, Utah, are incised meanders.

that they existed before the range uplifted. Note that if the range rises faster than the stream downcuts, the new highlands divert (change) the stream's course so that it flows parallel to the range face (Fig. 17.29a–c).

Take-Home Message

- When streams start to cut down into elevated land, they can carve deep valleys. Over time, mountains are beveled to low hills, and eventually the land surface may become a peneplain.
- Superposed streams cut across bedrock structure as they erode down into the substrate; erosion by antecedent streams keeps pace with uplift and cuts into rising ridges.
- One stream may capture the flow of another, in an act of stream piracy, when it erodes through a drainage divide.

THINK: What can cause a drainage-reversal event to take place?

FIGURE 17.28 Formation of superposed drainage.

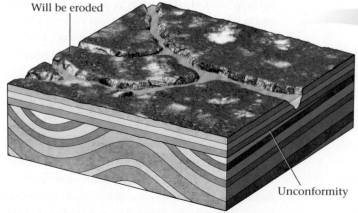

(a) A superposed stream establishes its geometry while flowing over a uniform substrate above an unconformity.

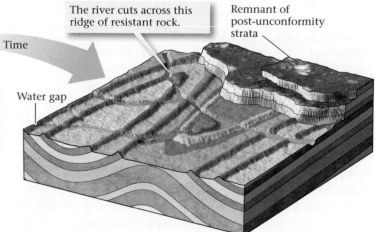

(b) When erosion exposes underlying rock with a different structure, the river is superposed on the structure. As a result, it cuts across resistant ridges instead of flowing around them.

FIGURE 17.29 Development of antecedent and diverted streams.

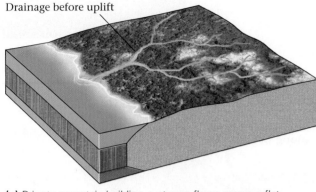

Drainage before uplift

(a) Prior to mountain building, a stream flows across a flat landscape to the sea.

If a mountain range rises across the path of a stream, the stream can either cut across the range or be diverted by the range.

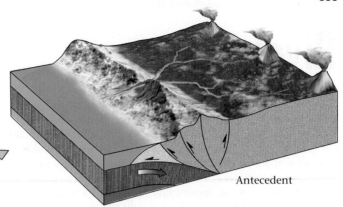

Antecedent

(b) If stream erosion is faster than mountain uplift, the stream cuts across the range and is antecedent.

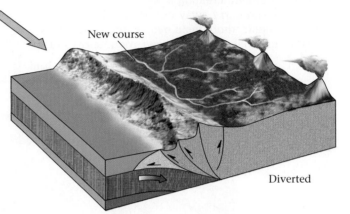

New course

Diverted

(c) If uplift happens faster than erosion, the stream is diverted and flows along the edge of the range.

17.8 **RAGING WATERS**

[And Enhil, the ruler of the gods, said,] "The earth bellows like a herd of wild oxen. The clamor of human beings disturbs my sleep. Therefore, I want Adad [god of the skies] to cause heavy rains to pour down upon the Earth, both day and night. I want a great flood to come like a thief upon the Earth, steal the food of these people and destroy their lives."

—from the *Epic of Gilgamesh*
(written in Sumeria, c. 2100 B.C.E.)

The Inevitable Catastrophe

Up to now, this chapter has focused on the process of drainage formation and evolution and on the variety of landscape features formed by streams (see Geology at a Glance, pp. 594–595). Now we turn our attention to the havoc that a stream can cause when flooding takes place. Floods can be catastrophic—they can strip land of forests and buildings, they can bury land in mud and silt, and they can submerge cities. As noted earlier, a flood occurs when the volume of water flowing down a stream exceeds the volume of the stream channel, so water rises out of the normal channel and spreads out over the floodplain or delta plain (by breaking through levees), or it fills a canyon to a greater depth than normal. The news media may report that a river "crested at 9 feet [3 m] above flood stage at 10 P.M." This means that the water surface in the stream was 3 m

higher than the top of the normal channel at 10 P.M., and after that time it became lower. (The flood crest is the highest level that the stream reaches.) Because of its increased discharge, a stream in flood flows faster than it normally does, so it's more turbulent, has greater competence, and exerts more pressure on structures in its path. Muddy, fast-moving floodwater is denser than clear water, and it pushes harder on objects in its path. It can buoy and transport sediment as well as cars, buildings, and people.

Floods happen (1) during abrupt, heavy rains, when water falls on the ground faster than it can infiltrate and thus immediately becomes surface runoff; (2) after a long period of continuous rain, when the ground has become saturated with water and can hold no more; (3) when heavy snows from the previous winter melt rapidly in response to a sudden hot spell; or (4) when a dam holding back a lake or reservoir, or a levee holding back a river or canal, suddenly collapses and releases the water that it held back.

Geologists find it convenient to divide floods into two general categories. Floods that occur when rainfall is particularly heavy or when winter snows start to melt during part of a year are called **seasonal floods**. Severe floods of this type

River Systems

Rivers, or streams, drain the landscape of surface runoff. Typically, an array of connected streams called a drainage network develops, consisting of a trunk stream into which numerous tributaries flow. The land drained is the network's watershed. A stream starts from a source, or headwaters. Some headwaters are in the mountains, perhaps collecting water from rainfall or from melting ice and snow. In the mountains, streams carve deep, V-shaped valleys and tend to have steep gradients. For part of its course, a river may flow over a steep, bouldery bed, forming rapids, and it may drop off an escarpment, creating a waterfall. Rivers gradually erode landscapes and carry away debris, so after a while, if there is no renewed uplift, mountains evolve into gentle hills. Through time, rivers can bevel once-rugged mountain ranges into nearly flat plains.

Developing drainage networks.

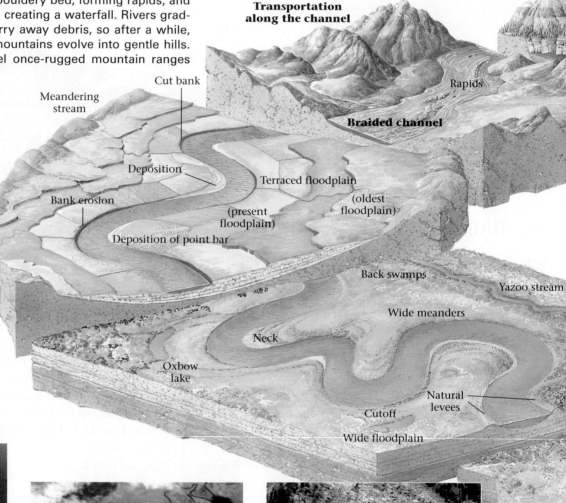

Meandering stream

Cut bank

Transportation along the channel

Rapids

Braided channel

Deposition

Bank erosion

Terraced floodplain

(present floodplain)

(oldest floodplain)

Deposition of point bar

Back swamps

Yazoo stream

Wide meanders

Neck

Oxbow lake

Natural levees

Cutoff

Wide floodplain

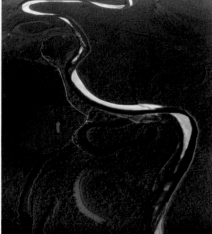

Point bars forming on inner curves.

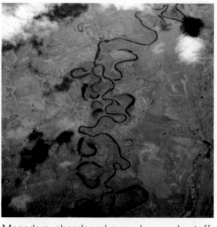

Meanders, abandoned meanders, and cutoffs.

A small delta in a mountain lake.

Headward erosion

Glaciers

Valleys with high relief

Melting ice

Lake

Dendritic drainage

Collection of water in watershed

Waterfall

Streams contribute to carving mountains.

Waterfall in Hawaii spilling over a basalt ledge.

Farther along its length, the river emerges from the mountains. If it is choked with sediment, it may split into numerous entwined channels separated from one another by gravel bars, creating a braided stream. Where a stream that has not been choked by sediment flows over flat ground, it becomes a meandering stream, winding back and forth in snake-like curves called meanders. The current flows faster on the outer arc of a curve, so erosion takes place there, whereas the current flows more slowly on the inner arc, where it drops sediment. Because of erosion and deposition, a meandering stream changes shape over time. Occasionally a meander may be cut off, leaving a curving lake called an oxbow lake. A broad floodplain, covered with water only during floods, may develop on either side of the stream. Natural levees build up between the channel and the floodplain from sediment dropped as a flooding river starts to spill out of its channel. Eventually, a river reaches a standing body of water and slows down, and the sediment it carries gets deposited to form a delta. On a delta, the trunk stream divides into many smaller channels called distributaries.

Deposition at mouth

Delta

Distributaries

Natural levees

Swamps and marsh

Tidal flats

Bar

Banks

FIGURE 17.30 Examples of seasonal floodplain flooding.

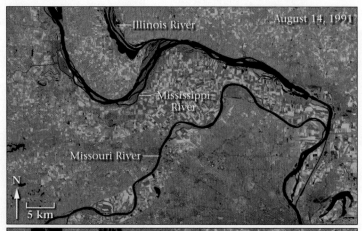

(a) Satellite photos (before and during) show the extent of flooding during the 1993 Mississippi River flood.

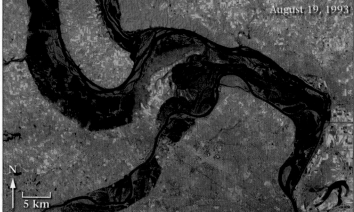

(d) Receding floodwater leaves sediment, here burying an orchard.

(b) Great Falls, Montana, was submerged by floodwaters in 1975.

(c) In 2007, contaminated floodwaters spread disease in Indonesia.

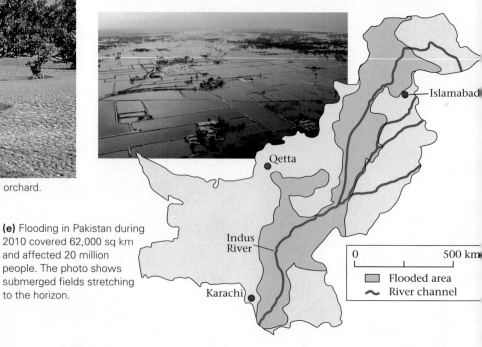

(e) Flooding in Pakistan during 2010 covered 62,000 sq km and affected 20 million people. The photo shows submerged fields stretching to the horizon.

take place in tropical regions that are drenched by monsoons. During the 1990 monsoon season in Bangladesh, for example, rain fell almost continuously for weeks. The delta plain became inundated; the resulting flood killed 100,000 people. The floods in Bangladesh can also be called delta-plain floods because the water submerges the delta plain. Similarly, floodplain floods submerge a floodplain (Fig. 17.30a–d).

Typically, seasonal floods take time—hours or days—to develop. Thus, in many cases, authorities can evacuate potential victims and organize efforts to protect property. Nevertheless, so many people live on deltas and floodplains that these floods can cause a staggering loss of life and property. A 1931 flood of the Yangtze River in China led to a famine that killed 3.7 million people, and an 1887 flood of China's Hwang (Yellow) River, so named because of the yellow silt it carries, killed as many as 2.5 million. More recently, the flooding of Italy's Arno River submerged the art treasures of Florence, and the flooding of North Dakota's Red River submerged the town of Grand Forks and threatened Winnipeg, Manitoba. Flooding of the Souris River in 2011 submerged Minot, North Dakota. Seasonal floods struck Indonesia in 2007, killing dozens of people and displacing almost half a million, nearly half of whom became sick with diarrhea and skin infections from contact with filthy water and mud that submerged 60% of the capital and hundreds of square kilometers of farmland.

One of the most devastating floods of recent times began in July 2010, when a seasonal flood fed by intense monsoonal rains submerged floodplains of the Indus River drainage system in Pakistan (Fig. 17.30e). On some days, it had rained up to 40 cm (1.3 ft) in 24 hours! The floods put almost 70,000 km^2 of the country underwater and severely impacted the lives of over 20 million people (12% of the population)—many people lost all they owned. Crops growing in the fertile floodplain floated away or rotted, and over 200,000 cattle drowned. Due to the destruction of clean-water supplies and of road and rail networks, survivors were stranded for weeks or more, and sadly, disease spread despite relief efforts by organizations from around the world.

Events during which the floodwaters rise so fast that it may be impossible to escape from the path of the water are called **flash floods** (Fig. 17.31a, b). These happen during unusually intense rainfall or as a result of a dam collapse (as in the 1889 Johnstown flood) or levee failure. During a flash flood, a canyon containing a stream may fill meters to tens of meters above normal. In some cases, a wall of water may slam downstream with great force, leaving devastation in its wake, but the floodwaters subside after a short time. Flash floods can be particularly unexpected in arid or semiarid climates, where isolated thundershowers may suddenly fill the channel of an otherwise dry wash, whose unvegetated ground can absorb little water. Such a flood may even affect areas downstream that had not received a drop of rain.

Case Study: A Seasonal Flood (Midwestern United States)

In the spring of 1993, the jet stream, the high-altitude (10–15 km high) wind current that controls weather systems, drifted southward (see Chapter 20). For weeks, the jet stream's cool, dry air formed an invisible wall that trapped warm, moist air

FIGURE 17.31 Flash floods can occur after torrential rains.

(a) A flash flood in a desert region of Israel has washed over a highway forcing the evacuation of truckers.

(b) During the 1976 Big Thompson River flash flood, this house was carried off its foundation and dropped on a bridge.

from the Gulf of Mexico over the central United States. When this air rose to higher elevations, it cooled, and the water it held condensed and fell as rain, rain, and more rain. In fact, almost a whole year's supply of rain fell in just that spring—some regions received 400% more than usual. Because the rain fell over such a short period, the ground became saturated and could no longer absorb additional water, so the excess entered the region's streams, which carried it into the Missouri and Mississippi rivers. Eventually, the water in these rivers rose above the height of levees and spread out over the floodplain. By July, parts of nine states were underwater (see Fig. 17.30a).

The roiling, muddy flood uprooted trees, cars, and even coffins (which floated up from inundated graveyards). All barge traffic along the Mississippi came to a halt, bridges and roads were undermined and washed away, and towns along the river were submerged in muddy water. For example, in Davenport, Iowa, the riverfront district and baseball stadium were covered with 4 m (14 feet) of water. In Des Moines, Iowa, 250,000 residents lost their supply of drinking water when

floodwaters contaminated the municipal water supply with raw sewage and chemical fertilizers. Rowboats replaced cars as the favored mode of transportation in towns where only the rooftops remained visible. In St. Louis, Missouri, the river crested 14 m (47 feet) above flood stage. When the water finally subsided, it left behind a thick layer of silt and mud, filling living rooms and kitchens in floodplain towns and burying crops in floodplain fields. For 79 days, the flooding continued. In the end, more than 40,000 square km of the floodplain had been submerged, 50 people died, at least 55,000 homes were destroyed, and countless acres of crops were buried. Officials estimated that the flood caused over $12 billion in damage.

Comparable flooding happened again in the spring of 2011, in the Mississippi and Missouri drainage basins. These floods came in the wake of multiple, intense storm systems that drenched Midwestern states after the land was already saturated with water from melting winter snows. Floodwaters overtopped levees, despite efforts to reinforce the levees with sandbags, burying fields in silt and submerging whole towns (Fig. 17.32a–d).

FIGURE 17.32 Flooding of the Missouri-Mississippi River network in 2011

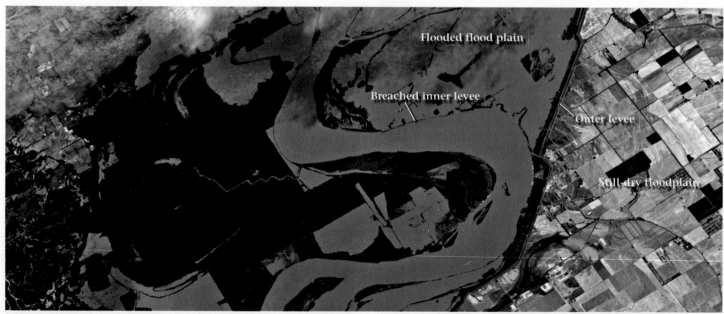

Flooded flood plain

Breached inner levee

Outer levee

Still-dry floodplain

(a) A space station view of the flood. Note that the inner levees have been breached, but the outer levee is still holding.

(b) At a levee breach, water flows from the river to the floodplain.

(c) Strengthening levees with sandbags.

(d) Some homeowners built personal levees.

Case Study: A Flash Flood (Big Thompson Canyon)

On a typical sunny day in the Front Range of the Rocky Mountains, north of Denver, Colorado, the Big Thompson River seems quite harmless. Clear water, dripping from melting ice and snow higher in the mountains, flows down its course through a narrow canyon, frothing over and around boulders. In places, vacation cabins, campgrounds, and motels line the river, for the pleasure of tourists. The landscape seems immutable; but, as is the case with so many geologic features, permanence is an illusion.

On July 31, 1976, easterly winds blew warm, moist air from the Great Plains toward the Rocky Mountain front. As this air rose over the mountains, towering thunderheads built up, and at 7:00 P.M. rain began to fall. It poured, in quantities that even old-timers couldn't recall. In a little over an hour, 19 cm (7.5 inches) of rain drenched the watershed of the Big Thompson River. The river's discharge grew to more than four times the maximum recorded at any time during the previous century. The river rose quickly, in places reaching depths several meters above normal. Turbulent water swirled down the canyon at up to 8 m per second and churned up so much sand and mud that it became a viscous slurry. Slides of rock and soil tumbled down the steep slopes bordering the river and fed the torrent with even more sediment. The water undercut house foundations and washed the houses away, along with their inhabitants (see Fig 17.31b). Roads and bridges disappeared. Boulders that had stood like landmarks for generations bounced along in the torrent like beach balls, striking and shattering other rocks along the way; the largest rock known to be moved by the flood weighed 275 tons. Cars drifted downstream until they finally wrapped like foil around obstacles. When the flood subsided, the canyon had changed forever, and 144 people had lost their lives.

Ice-Age Megafloods

Perhaps the greatest floods chronicled in the geologic record happen when natural ice dams burst. The Great Missoula Floods of about 11,000 years ago illustrate this phenomenon (see Box 22.2). This flood occurred at the end of the last ice age, when a glacier acted like a dam, holding back a large lake called Glacial Lake Missoula. When the glacier melted and the dam suddenly broke, the lake abruptly drained, and water roared over what is now eastern Washington, eventually entering the Columbia River Valley and flowing on out to the Pacific Ocean. The glacier then grew again, and the dam re-formed, trapping a new lake, which drained during a subsequent failure. These floods formed the channeled scablands of eastern Washington, a region where the soil and regolith have been stripped off the land surface, leaving barren, craggy rock.

The hypothesis that the channeled scablands formed as a consequence of catastrophic flooding was first proposed by J. Harlan Bretz, who studied the landscape of the region in the 1920s. Initially, other geologists ridiculed Bretz because his idea seemed to violate the well-accepted principle of uniformitarianism (see Chapter 12). But Bretz steadily fought back, demonstrating that the scablands, an unglaciated region, are littered with boulders too large to have been carried by normal rivers, that hills in the region were giant ripples a thousand times larger than the ripples typically found in a stream bed, that the now-dry Grand Coulee was once a giant waterfall hundreds of times larger than Niagara Falls, and that deep pits in the region are actually huge potholes scoured by whirlpools. Ultimately the geologic community accepted the reality of the Great Missoula Floods and other such catastrophic events during Earth history.

Living with Floods

Mark Twain once wrote of the Mississippi that we "cannot tame that lawless stream, cannot curb it or confine it, cannot say to it, 'go here or go there,' and make it obey." Was Twain right? Since ancient times, people have attempted to confine rivers to set courses so as to prevent undesired flooding. In the twentieth century, flood-control efforts intensified as the population living along rivers increased. For example, since the passage of the 1927 Mississippi River Flood Control Act (drafted after a disastrous flood took place that year), the U.S. Army Corps of Engineers has labored to control the Mississippi. First, engineers built about 300 dams along the river's tributaries so that excess runoff could be stored in the reservoirs and later be released slowly. Second, they built artificial levees of sand and mud, and built concrete flood walls to increase the channel's volume. Artificial levees and flood walls isolate a discrete area of the floodplain (Fig. 17.33a, b).

But although the Corps' strategy worked for floods up to a certain size, it was insufficient to handle the 1993 flood. Because of the volume of water drenching the Midwest during the spring and early summer of that year, the reservoirs filled to capacity, and additional runoff headed downstream. The river rose until it spilled over the tops of some levees and undermined others. Undermining occurs when rising floodwaters increase the water pressure on the river side of the levee, forcing water through sand under the levee. In susceptible areas, water begins to spurt out of the ground on the dry side of the levee, thereby washing away the levee's support. The levee finally becomes so weak that it collapses, and water fills in the area behind it.

Using lessons learned from 1993, the Army Corps of Engineers undertook a number of proactive, but controversial,

FIGURE 17.33   Holding back rivers to prevent floods.

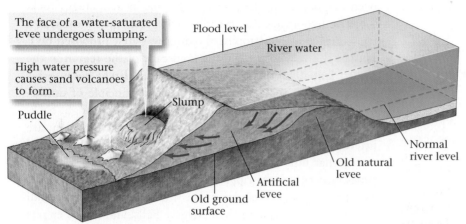

(a) Engineers build artificial levees to keep floodwater in the channel. Levees weaken and may fail when infiltrated by water.

(b) A concrete floodwall at Cape Girardeau, Missouri. When floods threaten, a crane drops a gate into the slot to hold out the river.

Sooner or later a flood comes along that can breach a river's levees, allowing water to spread out over the floodplain. Defensive efforts merely delay the inevitable, for it is infeasible and too expensive to build levees high enough to handle all conceivable floods. And in some cases, building levees may be counterproductive since they constrain water to a smaller area and thus make floodwaters rise to a higher level than they would if they were free to spread over a wide floodplain. Those who build on floodplains must face this reality and consider alternative ways to use the region so that they can accommodate occasional flooding. The cost of flood damage has quadrupled in recent years, despite the billions of dollars that have been spent on flood "control," because more people have settled in floodplains.

There are other ways to prevent floods besides building levees and reservoirs. For example, transforming portions of floodplains back into natural wetlands helps prevent floods, for wetlands absorb water like a sponge. A solution to flooding in some cases may lie in the removal, rather than the construction, of levees, so that there is more room for floodwaters. Property may also be kept safe by defining **floodways**, regions likely to be flooded, and then by moving or abandoning buildings located there. Even the simple act of moving levees farther away from the river and creating natural habitats in the resulting floodways would decrease flooding damage immensely.

When making decisions about investing in flood-control measures, mortgages, or insurance, planners need a basis for defining the hazard or risk posed by flooding. If floodwaters submerge a locality every year, a bank officer would be ill advised to approve a loan that would promote building there. But if floodwaters submerge the locality very rarely, then the loan may be worth the risk. Geologists characterize the risk of flooding in two ways. The **annual probability** (more formally known as the "annual exceedance probability") of flooding indicates the likelihood that a flood of a given size or larger will happen at a specified locality during any given year. For example, if we say that a flood of a given size has an annual probability of 1%, then we mean there is a 1 in 100 chance that a flood of at least this size will happen in any given year. The **recurrence interval** of a flood of a given size is defined as the *average* number of years between successive floods of at least this size. For example, if a flood of a given size happens once in 100 years, *on average*, then it

steps to reduce the impact of the 2011 Mississippi and Missouri River floods. The floodwaters simply didn't fit in the space bounded by existing levees and floodwalls, so at a few places, the Corps had to choose between flooding fields or flooding towns on the floodplain. Generally, they chose the former. For example, to protect Cairo, Illinois, the Corps blasted a 3-km-long gap in a levee upstream of the town. This action diverted enough water into a 530 sq km area of farmed floodplain so that the river level remained lower than the levees at Cairo and water stayed out of the town's streets. Further downstream, the Corps opened floodgates to divert water gradually into the Atchafalaya Basin. Without this action, the river might have broken through natural levees, and if its new course did not reach the port of New Orleans, that could have had huge economic implications.

Did you ever wonder . . .
what do newscasters mean by a "100-year flood"?

is assigned a recurrence interval of 100 years and is called a 100-year-flood. Note that annual probability and recurrence interval are related:

$$\text{annual probability} = \frac{1}{\text{recurrence interval}}$$

For example, the annual probability of a 50-year-flood is 1/50, which can also be written as 0.02 or 2%. To learn how to calculate annual probabilities and recurrence intervals of floods in more detail, see Box 17.1.

As an example of how to think about flood hazards, let's consider the case of Nashville, Tennessee. The "home of country music" endured a "100-year flood" on May 1st and 2nd of 2010 (Fig. 17.34a, b). Put another way, the likelihood of such a flood, based on previous discharge records, is only 0.2% in any given year. But comparable floods had happened in 1927 and 1937. The 2010 disaster began when a storm system stalled and warm, moist air from the Gulf of Mexico channeled over the region. Ferocious storms spawned at least 13 tornadoes, and in a 2-day period, 13.5" (0.34 m) of rain fell over Nashville. (Not far from Nashville, almost 19", half a meter, of rain fell.) The Cumberland River overtopped its banks and rose 3 m (12 ft) above the stage, putting it 15 m (over 52 ft) above its normal height. At least ten people perished, most when they drove into streets submerged by the flood. The churning, muddy water washed out bridges and damaged or destroyed thousands of homes. In the end, much of the downtown, as well as the Grand Ole Opry, lay beneath water, and Nashville's famous music scene was badly affected.

Unfortunately, some people are misled by the meaning of recurrence interval, and think that they do not face future flooding hazard if they buy a home within an area just after a 100-year flood has occurred. Their confidence comes from making the incorrect assumption that because such flooding just happened, it can't happen again until "long after I'm gone." They may regret their decision because two 100-year floods can occur in consecutive years or even in the same year (alternatively, the interval between such floods could be, say, 210 years). Because the term *recurrence interval* can lead to confusion, it may be better to report risk in terms of annual probability.

Knowing the discharge during a flood of a specified annual probability, and knowing the shape of the river channel and the elevation of the land bordering the river, hydrologists can predict the extent of land that will be submerged by such a flood (Fig. 17.35a). Such data, in turn, permit hydrologists to produce **flood-hazard maps**. In the United States, the Federal Emergency Management Agency (FEMA) produces maps that show the 1% annual probability (100-year) flood area and the 0.2% annual probability (500-year) flood risk zones (Fig. 17.35b).

Take-Home Message

- During a flood, water covers areas outside the normal channel.
- Seasonal floods happen annually due to excess meltwater or rain in a given season. Flash floods happen when intense rainfall drenches a region over a short time.
- During flooding, turbulent, sediment-filled water can buoy large clasts.
- The recurrence interval is the average time between floods of a given size; the annual probability indicates the likelihood of such a flood in a given year.
- People try to protect land from flooding by building dams and levees, but these don't always work.

THINK: Why might the flood level (height above flood stage) increase for a given discharge when levees are built along the river?

FIGURE 17.34 The 2010 flooding of Nashville, Tenessee followed intense storms.

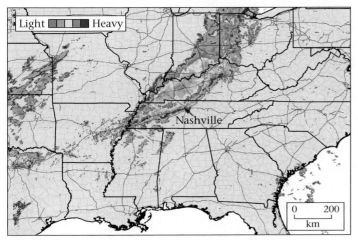

(a) A satellite radar image showing the rain over Nashville.

(b) Flooding in downtown Nashville, May 2010, is being called a 100-year flood. The last time the water rose this high was in 1937.

CONSIDER THIS . . . .

BOX 17.1

Calculating the Threat Posed by Flooding

How do we calculate the probability that a flood of a given size at a locality along a stream will happen in a given year? (Note that "size" in this context is indicated by the stream's discharge, as measured in cubic feet per second or cubic meters per second). First, researchers collect data on the stream's discharge at the locality for at least 10 to 30 years to get a sense of how the discharge varies during a year and from year to year. Then, they pick the largest, or peak, discharge for each year and make a table listing the peak discharges. The largest peak discharge is given a *rank* of 1, the second-largest discharge is given a rank of 2, and so on. Researchers can then calculate the recurrence interval for each different discharge by using a simple equation:

$$R = (n + 1) \div m$$

where R is the recurrence interval in years, n is the total number of years for which there is a record, and m is the rank.

Once the recurrence interval for each peak discharge has been calculated, the researchers plot a graph: the vertical axis represents peak discharge, and the horizontal axis represents recurrence interval. In order for all the data to fit on a reasonable-size graph, the horizontal axis must be logarithmic. Typically, the data for a stream plots roughly along a straight line (**Fig. Bx17.1a**). In the example shown in this figure, a flood with a recurrence interval of 10 years (meaning an annual probability of 10%) has a peak discharge of about 460 cubic feet per second. We can extend the line beyond the data points (the dashed line on Fig. Bx17.1a) to make *predictions*

about the recurrence interval and, therefore, the annual probability (= 1/R), of floods with discharges larger than the ones that have been measured. As more data become available, the graph may need to be modified. In this example, note that the graph predicts that a 1% probability flood (a 100-year flood) will have a discharge of about 650 cubic feet per second.

The peak annual discharge of the Mississippi River at St. Louis has been measured almost continuously since 1850. A bar graph of these data shows that floods characterized as 100-year floods (meaning 1% probability floods) or larger happened in 1844, 1903, and 1993 (**Fig. Bx17.1b**). Note that the time between "100-year floods" is not exactly 100 years.

FIGURE Bx17.1 Flood frequency and peak discharge graphs.

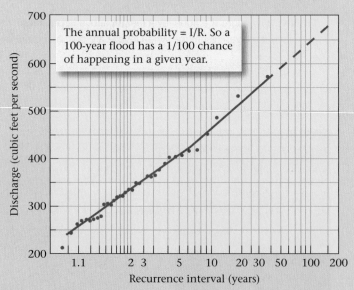

The annual probability = I/R. So a 100-year flood has a 1/100 chance of happening in a given year.

(a) A flood-frequency graph shows the relationship between the recurrence interval and the discharge for an idealized river.

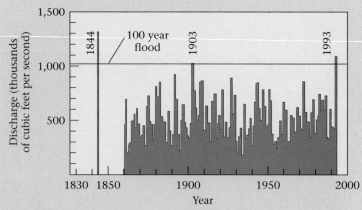

(b) A peak discharge graph for the Mississippi at St. Louis. Each bar represents the largest discharge for a given year.

FIGURE 17.35 The conceptual relationship between flood size and probability.

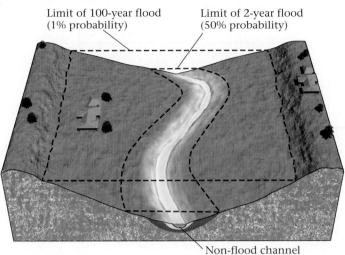

Limit of 100-year flood
(1% probability)

Limit of 2-year flood
(50% probability)

Non-flood channel

(a) A 100-year flood covers a larger area than a 2-year flood and occurs less frequently.

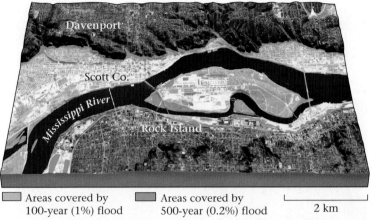

Davenport

Scott Co.

Mississippi River

Rock Island

Areas covered by
100-year (1%) flood

Areas covered by
500-year (0.2%) flood

2 km

(b) A flood-hazard map shows areas likely to be flooded. Here, near Rock Island, Illinois, even large floods are confined to the floodplain.

17.9 RIVERS: A VANISHING RESOURCE?

As *Homo sapiens* evolved from hunter-gatherers into farmers, areas along rivers became attractive places to settle. Rivers serve as avenues for transportation and are sources of food, irrigation water, drinking water, power, recreation, and (unfortunately) waste disposal. Further, their floodplains provide particularly fertile soil for fields, replenished annually by seasonal floods. Considering the multitudinous resources that rivers provide, it's no coincidence that ancient cultures developed in river valleys and on floodplains. The civilization of Mesopotamia arose around the Tigris and Euphrates Rivers, Egypt around the Nile, India in the Indus Valley, and China along the Hwang (Yellow) River. Over the millennia, rivers have killed millions of people in floods, but they have been the lifeblood for hundreds of millions more. Nevertheless, over time, humans have increasingly tended to abuse or overuse the Earth's rivers. Here we note four pressing environmental issues.

Pollution

The capacity of some rivers to carry pollutants has long been exceeded, transforming them into deadly cesspools. Pollutants include raw sewage and storm drainage from urban areas, spilled oil, toxic chemicals from industrial sites, floating garbage, excess fertilizer, and animal waste. Some pollutants directly poison aquatic life, some feed algae blooms that strip water of its oxygen, and some settle out to be buried along with sediments. River pollution has become overwhelming in developing countries, where there are few waste-treatment facilities.

Dam Construction

In 1950, there were about 5,000 large (over 15 m high) dams worldwide, but today there are over 38,000. Damming rivers has both positive and negative results. Reservoirs provide irrigation water and hydroelectric power, and they trap some floodwaters and create popular recreation areas. But in some locations their construction destroys "wild rivers" (the whitewater streams of hilly and mountainous areas) and alters the ecosystem of a drainage network by forming barriers to migrating fish, by decreasing the nutrient supply to organisms downstream, by removing the source of sediment for the delta, and by eliminating seasonal floods that replenish nutrients in the landscape.

Overuse of Water

Because of growing populations, our thirst for river water continues to increase, but the supply of water does not. The use of water has grown especially in response to the "Green Revolution" of the 1960s, during which huge new tracts of land came under irrigation. Today, 65% of the water taken out of rivers is used for agriculture, 25% for industry, and 9% for drinking and sewage transport. Civilization needed three times as much river water in 1995 as it did in 1950.

As a result, in some places human activity consumes the entire volume of a river's water, so that the channel contains little more than a saline trickle, if that, at its mouth. For example, except during unusually wet years, the Colorado River contains almost no water where it crosses the Mexican border, for huge pipes and canals carry the water instead to Phoenix

FIGURE 17.36 The Central Arizona Project canal shunts water from the Colorado River to Phoenix.

FIGURE 17.37 Hydrographs, showing discharge as a function of time, are affected by urbanization.

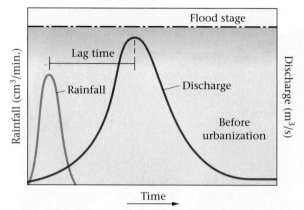

(a) Before urbanization, rain infiltrated the ground so discharge was smaller and peak runoff occurred after a long lag time.

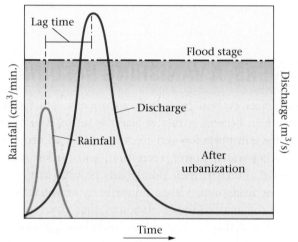

(b) After urbanization, water flows directly into streams, discharge is greater, and lag time is less.

and Los Angeles (Fig. 17.36). In the case of the Colorado River, states along its banks have established legal agreements that divide up the river's water. Unfortunately, the agreements were written during wet years of the early twentieth century, when the river had unusually large discharge. Thus, the amount of water specified in the agreements actually exceeds the amount of water the river carries in most years.

Perhaps the most dramatic consequence of river over-consumption can be seen in central Asia, where almost all the water in the Amu Darya and Syr Darya rivers has been diverted into irrigation. These rivers once fed the Aral Sea. Now the sea has shrunk so much that "coastal" towns lie many kilometers from the coast, fishing trawlers rot in the desert (see Chapter 21), and the catch of fish has dropped from 44,000 to 0 tons per year.

The Effects of Urbanization and Agriculture on Discharge

Although the consumption of water for agricultural and industrial purposes decreases the overall supply of river water, urbanization may actually increase the short-term supply. Cities cover the ground with impermeable concrete or black-top, so rainfall does not soak into the ground but rather runs into storm sewers and then into streams, thereby increasing runoff. Stream discharge during a rainfall thus increases much more rapidly than it would without urbanization. This change can be illustrated by diagrams, called hydrographs, that show how discharge varies with time (Fig. 17.37a, b). Such change can cause flash floods. Similarly, although damming rivers decreases the amount of silt a river carries

downstream, agriculture may increase the sediment supply: agriculture decreases the vegetative cover on the land, so that when it rains, soil washes into streams.

Take-Home Message

- Society depends on streams for water supplies, irrigation, energy, and transport.
- Pollution fouls river water; dam construction changes the character of a drainage network.
- Due to overuse of river waters, some rivers barely flow all the way to their mouth.
- Urbanization may lead to an increase in runoff and a greater likelihood of flash floods.

THINK: How does dam construction affect the sediment load carried by a stream?

Chapter Summary

- Streams are bodies of water that flow down channels and drain the land surface. Channels develop when sheetwash cuts into the substrate and concentrates the water flow; they grow by headward erosion. Streams carry water out of a drainage basin. A drainage divide separates two adjacent catchments.

- Drainage networks consist of many tributaries that flow into a trunk stream.

- Permanent streams exist where the water table lies above the bed of the channel or when large amounts of water enter the channel from upstream. Where the water table lies below the channel bed, streams are ephemeral and dry up between rainfalls to form dry washes.

- The discharge of a stream is the total volume of water passing a point along the bank in a second. Most streams are turbulent, meaning that their water swirls in complex patterns.

- Streams erode the landscape by scouring, lifting, abrading, and dissolving. The resulting sediment provides dissolved loads, suspended loads, and bed loads. The total quantity of sediment carried by a stream is its capacity. Capacity differs from competence, the maximum particle size a stream can carry. When stream water slows, it deposits alluvium.

- The longitudinal profile (the shape of a stream bed in cross section from its source to its mouth) of a stream is concave up. Typically, a stream has a steeper gradient at its headwaters than near its mouth. Streams cannot cut below the base level.

- Streams cut valleys or canyons, depending on the rate of downcutting relative to the rate at which the slopes on either side of the stream undergo mass wasting. Where a stream flows down steep gradients and has a bed littered with large rocks, rapids develop, and where a stream plunges off a vertical face, a waterfall forms.

- A meandering stream wanders back and forth across a floodplain. It erodes its outer bank and builds out sediment into a point bar on the inner bank. Eventually, a meander may be cut off and turn into an oxbow lake. Natural levees form on either side of the river channel. Braided streams consist of many entwined channels.

- Where streams or rivers flow into standing water, they deposit deltas. The shape of a delta depends on the balance between the amount of sediment supplied by the river and the amount of sediment redistributed or carried away by wave activity along the coast.

- With time, fluvial erosion can bevel landscapes to a nearly flat plain. If the base level drops or the land surface rises, stream rejuvenation causes the stream to start downcutting into the peneplain. The headward erosion of one stream may capture the flow of another.

- If an increase in rainfall or spring melting causes more water to enter a stream than the channel can hold, a flood results. Some floods are seasonal in that they accompany monsoonal rains. Some floods submerge broad floodplains or delta plains. Flash floods happen very rapidly. Officials try to prevent floods by building reservoirs and levees.

- Rivers are becoming a vanishing resource because of pollution, damming, and overuse.

GEOPUZZLE REVISITED

As soon as the land surface of a region rises above the ultimate base level (sea level), water starts flowing toward lower elevations. Eventually, the faster flow carves channels, with tributary channels flowing into a trunk channel. Streams eventually cut down to the base level. The channel volume reflects the usual discharge of the stream. If heavy rain, melting, or a dam rupture provides more water than the channel can hold, water spills over the channel walls and floods the surrounding landscape.

Guide Terms

abrasion (p. 576)

alluvial fan (p. 584)

alluvium (p. 579)

annual probability (p. 600)

antecedent stream (p. 591)

bars (p. 579)

base level (p. 579)

bed load (p. 576)

braided stream (p. 584)

capacity (p. 577)

channel (p. 569)

competence (p. 576)

delta (pp. 579, 588)

discharge (p. 574)

dissolved load (p. 576)

distributaries (p. 588)

downcutting (p. 571)

drainage divide (p. 574)

drainage network (p. 572)

drainage reversal (p. 590)

ephemeral stream (p. 574)

flash flood (p. 597)

flood (p. 569)

flood-hazard map (p. 601)

floodplain (p. 585)

floodways (p. 600)

fluvial (p 578)

graded stream (p. 580)

headward erosion (p. 571)

longitudinal profile (p. 579)

meanders (p. 585)

natural levees (p. 585)

oxbow lake (p. 585)

permanent stream (p. 574)

point bar (p. 579)

pothole (p. 576)

rapids (p. 582)

reach (p. 579)

recurrence interval (p. 600)

running water (p. 569)

runoff (p. 570)

saltation (p. 576)

scouring (p. 576)

seasonal floods (p. 593)

sheetwash (p. 570)

stream gradient (p. 579)

stream piracy (p. 590)

stream rejuvenation (p. 590)

stream terrace (p. 582)

stream (p. 569)

superposed stream (p. 591)

suspended load (p. 576)

thalweg (p. 575)

tributaries (p. 572)

trunk stream (p. 572)

turbulence (p. 575)

V-shaped valley (p. 581)

waterfall (p. 582)

watershed (p. 573)

Review Questions

1. What role do streams serve during the hydrologic cycle? Indicate various sources of water in streams.

2. Describe the four different types of drainage networks. What factors are responsible for the formation of each?

3. What factors determine whether a stream is permanent or ephemeral?

4. How does discharge vary according to the stream's length, climate, and position along the stream course?

5. Why is average downstream velocity always less than maximum downstream velocity?

6. How does a turbulent flow differ from a laminar flow?

7. Describe how streams and running water erode the Earth.

8. What are three components of sediment load in a stream?

9. Distinguish between a stream's competence and its capacity.

10. Describe how a drainage network changes, along its length, from headwaters to mouth.

11. What factors determine the position of the base level?

12. What do lakes, rapids, waterfalls, and terraces indicate about the stream gradient and base level? Why do canyons form in some places and valleys in others?

13. How does a braided stream differ from a meandering stream?

14. Describe how meanders form, develop, are cut off, and then are abandoned.

15. Describe how deltas grow and develop. How do they differ from alluvial fans?

16. How does a stream-eroded landscape evolve as time passes?

17. What is stream piracy? What causes a drainage reversal?

18. How are superposed and antecedent drainages similar? How are they different?

19. What human activities tend to increase flood risk and damage?

20. What is the recurrence interval of a flood, and how is it related to the annual probability? Why can't someone say, "The hundred-year flood happened last year, so I'm safe for another hundred years"?

21. How have humans abused and overused the resource of running water?

On Further Thought

22. The northeastern two-thirds of Illinois, in the midwestern United States, was last covered by glaciers only 14,000 years ago. The rest of the state was last covered by glaciers over 100,000 years ago. Until the advent of modern agriculture, the recently glaciated area was a broad, grassy swamp, cut by very few stream channels. In contrast, the area that was glaciated over 100,000 years ago is not swampy and has been cut by numerous stream valleys. Why?

23. Records indicate that flood crests for a given amount of discharge along the Mississippi River have been getting *higher* since 1927, when a system of levees began to block off portions of the floodplain. Why?

24. The Ganges River carries an immense amount of sediment load, which has been building a huge delta in the Bay of Bengal. Look at the region using an atlas or *Google Earth*™, think about the nature of the watershed supplying water to the drainage network that feeds the Ganges, and explain why this river carries so much sediment.

25. Fly to Lat 43°16'20.55"S Long 170°24'8.52"E using *Google Earth*™, zoom to an elevation of 40 km, and look straight down. You will see a portion of the South Island in New Zealand. In this area, the Whataroa River flows northwest from the Southern Alps, a mountain range formed by movement on a plate boundary called the Alpine fault. The fault trace forms the abrupt boundary between the

mountain front and the plains to the northwest. Describe the changes in the nature of the river as it crosses the fault—how is the downstream reach of the stream different from the upstream reach? To help you answer this question, zoom in and out, change the tilt angle of your view, and compare your view to the drawing of Figure 17.11. Find the drainage divide of the Southern Alps, and try to map it along the length of the range. You might want to print an image of your screen to provide a base on which to draw the divide.

26. Look closely at the graph in Box 17.1. What is the recurrence interval of a flood with a discharge of 650 cubic feet per second? In a given year, how much more likely is a flood with a discharge of 200 cubic feet per second than a flood of 400 cubic feet per second? For the sake of discussion, imagine that the floodplain of the river is completely covered when a flood with an annual probability of 1/300 occurs. Would you build a new home in the floodplain? Would it make a difference to you if the last flood with this probability happened 1 year ago? 100 years ago?

For more resources, including animations, quizzes, and Norton's GeoTours, go to **wwnorton.com/studyspace**.

If your instructor assigns exercises in SmartWork, log in at **smartwork.wwnorton.com**.

ANOTHER VIEW   The Ganges River divides into meandering distributaries that cross a delta plain, carrying sediment to the Indian Ocean. Monsoon and typhoon rains cause the low-lying area to flood.

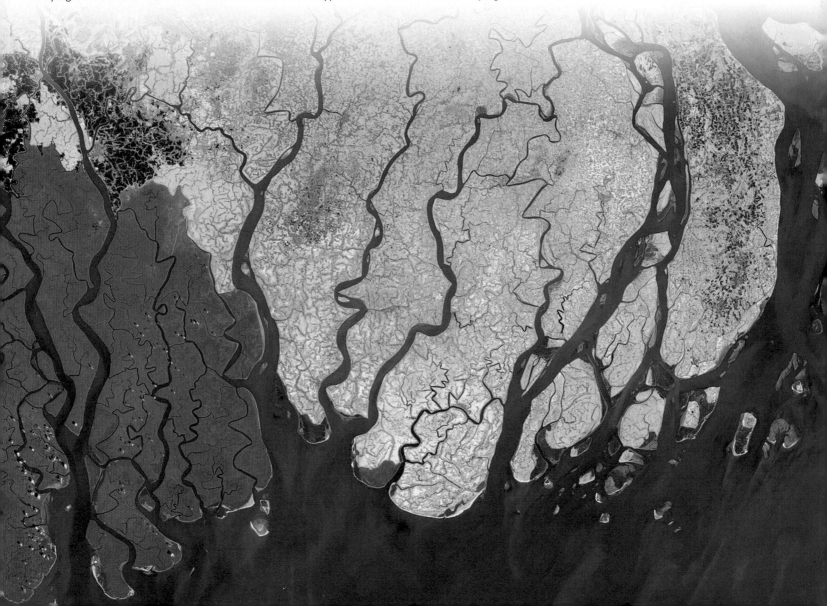

CHAPTER **18**

Restless Realm: Oceans and Coasts

As storm clouds build, frothing waves round and sort sand on a beach near Natal, Brazil. The wind can blow the sand into huge dunes near the shore. The oceans and their fringes are ever changing.

GEOPUZZLE

In 1992, a shipment of 29,000 plastic bath toys fell off a cargo ship and into the middle of the Pacific Ocean. Some of the toys have washed onto the Atlantic Ocean beaches of the UK. Why?

"The three great elemental sounds in nature are the sound of rain, the sound of wind in a primeval wood, and the sound of the outer ocean on a beach."

—Henry Beston (American naturalist, 1888–1968)

18.1 INTRODUCTION

A thousand kilometers from the nearest shore, two scientists and a pilot wriggle through the entry hatch of the research submersible *Alvin*, ready for a cruise to the floor of the ocean and, hopefully, back. *Alvin* consists of a super-strong metal sphere embedded in a cigar-shaped tube (Fig. 18.1a). The sphere protects its crew from the immense water pressures of the deep ocean, and the tube holds motors and oxygen tanks. When the hatch seals, *Alvin* sinks at a rate of 1.8 km per hour. Most of this journey takes place through utter darkness, for light penetrates only the top few hundred meters of ocean water. On reaching the bottom, at a depth of 4.5 km, the cramped explorers turn on outside lights to reveal a stark vista of loose sediment, black rock, and the occasional sea creature. For the next five hours they take photographs and use a robotic arm to collect samples. When finished, they release ballast and rise like a bubble, reaching the surface about two hours later.

Alvin dives began in the 1970s, but humans have explored the ocean for tens of centuries. In fact, Phoenician traders had circumnavigated Africa by 590 B.C.E., and Polynesian sailors used outrigger canoes to travel among South Pacific islands beginning around 700 C.E. Chinese naval ships may have circled the globe in the fifteenth century. European mapmakers have known that the ocean spanned the entire globe since Ferdinand Magellan's round-the-world voyage of 1519–22, but they could not systematically map the ocean until the late eighteenth century, when it became possible to determine longitude accurately. Subsequently, naval officers gathered data on water depths in the ocean (using a plumb line, a lead weight on the end of a cable), and by 1839 had determined that the greatest ocean depths could swallow the highest mountains without a trace.

A converted British navy ship, the HMS *Challenger*, made the first true ocean research cruise (Fig. 18.1b). Beginning in 1872, onboard scientists spent four years dredging rocks from the sea floor, analyzing water composition, collecting specimens of marine organisms, and measuring water depths and currents. But still our knowledge of the ocean remained

FIGURE 18.1 Modern oceanographic research vessels explore the surface and subsurface realms.

(a) *Alvin*, a 3-person submersible explores mid-ocean ridges.

(e) ABE is a robotic submersible.

(b) The HMS *Challenger* (ca. 1876) was the first oceanographic research vessel.

(c) RV *Atlantis* can take 24 scientists to sea for 2 months.

(d) Submersibles use spotlights to see at depth.

spotty. In fact, we knew less about the ocean floor than we did about the surface of the Moon, for at least we could see the Moon with a telescope.

The fields of oceanography (the study of ocean water and its movements), marine geology (the study of the ocean floor and shorelines), and marine biology (the study of lifeforms in the sea) expanded rapidly in the latter half of the twentieth century, as new technology became available and a fleet of oceanographic research ships criss-crossed the seas (Fig. 18.1c–e). It's now commonplace for ships to tow instrument-laden sleds just above the sea floor. The sleds can use sonar to generate detailed bathymetric maps revealing the shape of the floor. Some ships send pulses of sound into the sea floor that reflect off layers in the subsurface and return to the ship to provide an image, called a seismic-reflection profile, of the layering in the oceanic crust (Fig. 18.2; see Interlude D). Other ships, such as the *JOIDES Resolution*, drill holes as deep as 4 km into the sea floor and bring up samples of the oceanic crust. In addition, satellites circling the globe produce high-resolution maps of the ocean and its coasts. And while ships and satellites research the open ocean, land-based geologists continue to study its margins.

When seen from space, Earth glows blue, for the oceans cover 70.8% of its surface. The sea provides the basis for life, tempers Earth's climate, and spawns its storms. It is a vast reservoir for water and chemicals that cycle into the atmosphere and crust, and for sediment washed off the continents. In this chapter, we first learn the fundamental characteristics of ocean basins and seawater, and the role they play in the Earth System. Then we focus on the landforms that develop along the **coast**, the region where the land meets the sea, and where over 60% of the global population lives today. Finally, we consider how to cope with the hazards of living on the coast.

By the end of this chapter, you should know . . .

- how tectonic processes produce bathymetric features of the sea floor.
- the nature and causes of surface currents and deep currents.
- the behavior of tides and the forces that cause tides to happen.
- why waves form and how waves behave as they approach the shore.
- how a great variety of different coastal landforms develop and evolve.
- how changes in sea level, wave erosion, and human activities affect the coast.

18.2 LANDSCAPES BENEATH THE SEA

The oceans exist because oceanic lithosphere and continental lithosphere differ markedly from one another in terms of composition and thickness (Fig. 18.3; see Chapter 2). The surface of the denser and thinner oceanic lithosphere lies deeper than the surface of the relatively buoyant, thicker continental lithosphere, creating oceanic *basins* (low areas) that fill with water. On the present-day map of the world, the ocean encircles the globe. For purposes of reference, however, cartographers divide the ocean into several major parts, with somewhat arbitrary boundaries and significantly different volumes (Fig. 18.4). Of note, most continental crust (81%) lies in the northern hemisphere today; but because of plate tectonics, the map of Earth's surface was different in the past and in the late Paleozoic, most

FIGURE 18.2 A seismic-reflection profile straddling the Aleutian trench, south of Alaska, shows the structure of an accretionary prism and sediment layers of the Pacific Ocean floor. The colored stripes are sediment layers and the black lines are faults.

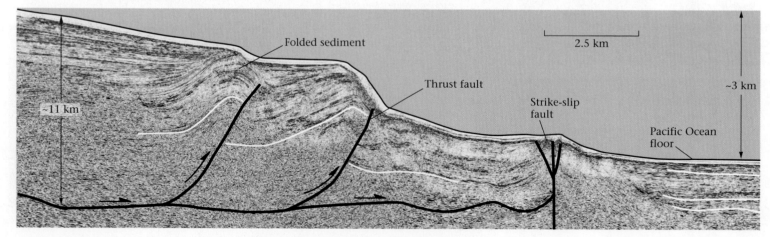

FIGURE 18.3 Contrasts between continental lithosphere and oceanic lithosphere. The crustal portion of continental lithosphere differs markedly from that of oceanic lithosphere.

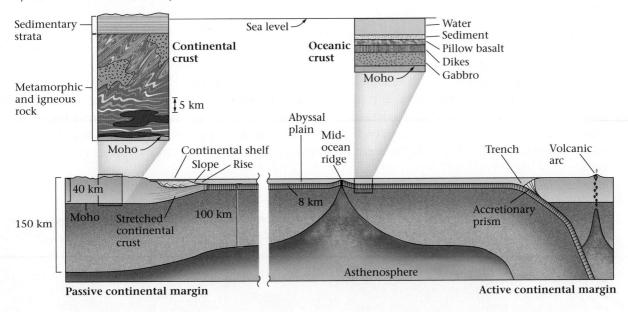

FIGURE 18.4 The oceans of the world. The Pacific is the largest, covering almost half the planet. The Arctic region is an ocean covered by a thin coating of ice, whereas the Antarctic region is a continent.

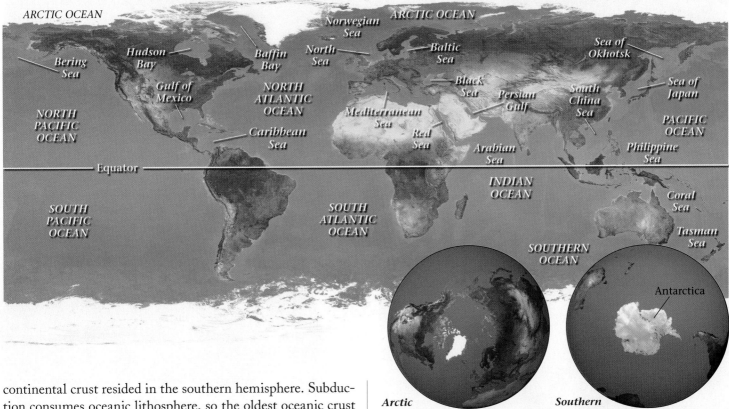

continental crust resided in the southern hemisphere. Subduction consumes oceanic lithosphere, so the oldest oceanic crust visible today is only about 200 million years old.

Have you ever wondered what the ocean floor would look like if all the water evaporated? Marine geologists can now provide a clear image of the ocean's **bathymetry**, or variation in depth, based originally on sonar measurements and more recently on measurements made by satellites. Such studies indicate that the ocean contains broad bathymetric provinces, distinguished from each other by their water depth. Let's now examine each of these provinces.

Continental Shelves, Slopes, and Rises

Imagine you're in a submersible cruising just above the floor of the western half of the North Atlantic. If you start at the shoreline of North America and head east, you will cross the 200- to 500-km-wide **continental shelf**, a relatively shallow portion of the ocean that fringes the continent in which water depth does not exceed 500 m. Across the width of the shelf, the ocean

floor slopes seaward at only about 0.3°, an almost imperceptible angle. At its eastern edge, the continental shelf merges with the continental slope, which descends to depths of nearly 4 km at an angle of about 2°. From about 4 km down to about 4.5 km, a province called the continental rise, the angle decreases until at 4.5 km deep, you find yourself above a vast, nearly horizontal plain: the **abyssal plain**.

Broad continental shelves, like that of eastern North America, form along **passive continental margins**, margins that are not plate boundaries and thus lack seismicity (Fig. 18.5a; see Fig. 18.3 and Chapter 4). Passive margins

FIGURE 18.5 Bathymetric features of the seafloor. The maps are produced by computer using measurements from satellites or submersibles.

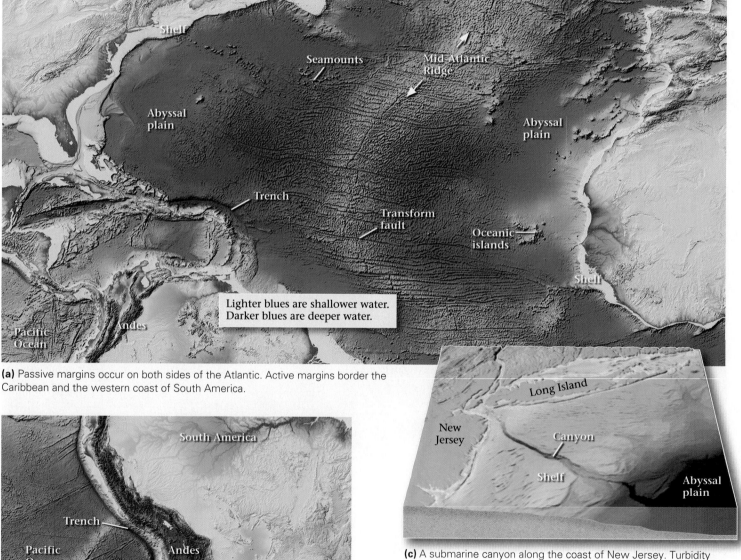

(a) Passive margins occur on both sides of the Atlantic. Active margins border the Caribbean and the western coast of South America.

(b) The western coast of South America is an active continental margin.

(c) A submarine canyon along the coast of New Jersey. Turbidity currents in submarine canyons carry sediment to the abyssal plain.

originate after rifting breaks a continent in two. When rifting stops and sea-floor spreading begins, the stretched lithosphere at the boundary between the ocean and continent gradually cools and sinks. Sand and mud that washed off the continent, along with the shells of marine creatures that grow on the sea floor or settle from the water above, bury the sinking crust, slowly producing a pile of sediment up to 20 km thick. The flat surface of this sedimentary pile constitutes the continental shelf. As discussed in Chapter 16, recent surveys show that large submarine slumps occasionally slip down the continental slope and may have generated tsunamis.

If you were to take your submersible to the western coast of South America and cruise out into the Pacific, you would find a very different continental margin. After crossing a narrow continental shelf, the sea floor falls off at the relatively steep angle of 3.5° down to a depth of over 8 km. South America does not have a broad continental shelf because it is an **active continental margin**, a margin that coincides with a plate boundary and thus hosts many earthquakes (Fig. 18.5b). In the case of South America, the edge of the Pacific Ocean is a convergent plate boundary. The narrow shelf along a convergent plate boundary forms where an apron of sediment spreads out over the top of an accretionary prism, the pile of material scraped off the downgoing subducting plate. Here, the continental slope corresponds to the face of the accretionary prism.

At many locations, relatively narrow and deep valleys called **submarine canyons** dissect continental shelves and slopes (Fig. 18.5c). Some submarine canyons start offshore of major rivers, and for good reason: rivers cut into the continental shelf at times when sea level was low and the shelf was exposed. But river erosion cannot explain the total depth of these canyons—some slice almost 1,000 m down into the continental margin, far deeper than the maximum sea-level change. Submarine exploration demonstrates that much of the erosion of submarine canyons results from erosion by turbidity currents, avalanches of sediment mixed with water (see Chapter 7). When turbidity currents finally reach the base of the continental slope, the velocity of flow decreases, so sediments settle out and turbidites (composed of graded beds) accumulate and build up into a submarine fan.

The Bathymetry of Oceanic Plate Boundaries

You can see all three types of plate boundaries by studying the bathymetry of the ocean floor (see Fig. 18.5a; **See for Yourself M**, p. S-24). Sea-floor spreading at a divergent boundary yields a mid-ocean ridge, a 2-km-high submarine mountain belt. Because crust stretches and breaks as sea-floor spreading continues, the axis of a ridge may be bordered by escarpments, a result of normal faulting. Oceanic transform faults, strike-slip faults along which one plate shears sideways past another, typically link segments of mid-ocean ridges (see Chapter 4). Transforms are delineated by fracture zones, narrow belts of steep escarpments, and broken-up rock. These fracture zones can be traced into the oceanic plate away from the ridge axis where they are not seismically active but still form the boundary between plates of different ages. Subduction at convergent boundaries yields a trench, a deep, elongate trough bordering a volcanic arc. Some trenches reach depths of over 8 km—the deepest point in the ocean, −11,035 m, lies in the Mariana Trench of the western Pacific. Some trenches border continents as we described earlier; others border island arcs, which are curving chains of active volcanic islands.

Abyssal Plains and Seamounts

As oceanic crust ages and moves away from the axis of the mid-ocean ridge, two changes take place. First, the lithosphere cools, and as it does so, its surface sinks (to maintain isostatic compensation; see Chapter 11). Second, a blanket of **pelagic sediment** gradually accumulates and covers the basalt of the oceanic crust (Fig. 18.6a). This blanket consists mostly of microscopic plankton shells and fine flakes of clay, which slowly fall like snow from the ocean water and settle on the sea floor. Because the ocean crust gets progressively older away from the ridge axis, sediment thickness increases away from the ridge axis; sediment has had more time to accumulate on older sea floor (Fig. 18.6b). Eventually, the sediment buries the escarpments that had formed at the mid-ocean ridge, resulting in a flat, featureless surface of the abyssal plain (Fig. 18.6c, d).

Numerous distinct, localized high areas rise above surrounding ocean depths (see Figs. 3.9b and 18.5a). These high areas result from hot-spot volcanic activity. If the product of this activity protrudes above sea level, it forms an oceanic island. Oceanic islands that lie over hot spots host active volcanoes, whereas those that have moved off the hot spot are extinct. In warm climates, coral reefs grow and surround oceanic islands. With time, oceanic islands erode and partially collapse due to slumping. Also, the sea floor beneath them ages and sinks. As a result, each island's peak eventually submerges, and what was once an island becomes a **seamount**. A seamount that submerges after being overgrown by a reef will have a flat top, and can be called a **guyot**. Oceanic islands and seamounts that developed above the same hot spot line up in a chain (a hot-spot track) with the oldest seamount at one end and the youngest seamount or island at the other (see Chapter 4). In places where hot-spot igneous activity was particularly voluminous (forming a LIP; see Chapter 6), a broad **oceanic plateau**, underlain by flood basalt, forms.

FIGURE 18.6 The sedimentary layer of the sea floor.

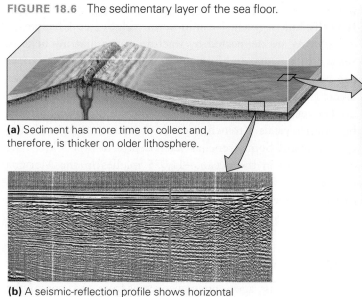

(a) Sediment has more time to collect and, therefore, is thicker on older lithosphere.

0.5 m

(c) A photograph of the abyssal plain shows a muddy surface and a few organisms.

(b) A seismic-reflection profile shows horizontal bedding in abyssal-plain strata.

10 cm

(d) Cores of sea-floor sediments show bedding.

Take-Home Message

- Broad continental shelves form along passive margins, where sediments accumulate above stretched lithosphere. The continental slope and rise descend to the abyssal plain.
- Trenches border active continental margins, and fracture zones define transform faults.
- Abyssal plains are the surface of old oceanic lithosphere. They are interrupted by oceanic islands, seamounts, guyots, and submarine plateaus, the results of hot-spot volcanism.

THINK: How do submarine canyons form?

18.3 OCEAN WATER AND CURRENTS

Composition

If you've ever had a chance to swim in the ocean, you may have noticed that you float much more easily in ocean water than you do in freshwater. That's because ocean water contains an average of 3.5% dissolved salt (**Fig. 18.7**); in contrast, typical freshwater contains less than 0.02% salt. The dissolved ions fit between water molecules without changing the volume of the water, so adding salt to water increases the water's density, and you float higher in a denser liquid.

FIGURE 18.7 The composition of average seawater. The expanded part of the graph shows the proportions of ions in the salt of seawater.

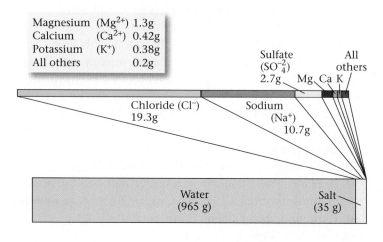

Magnesium	(Mg^{2+})	1.3g
Calcium	(Ca^{2+})	0.42g
Potassium	(K$^+$)	0.38g
All others		0.2g

Sulfate (SO$_4^{-2}$) 2.7g   All others   Mg Ca K

Chloride (Cl$^-$) 19.3g   Sodium (Na$^+$) 10.7g

Water (965 g)   Salt (35 g)

Workers on Cape Verde Island collecting salt produced by evaporation of seawater.

Leonardo da Vinci, the famous Renaissance artist and scientist, speculated that sea salt came from rivers passing through salt mines. But modern studies demonstrate that most cations in sea salt—sodium (Na^+), potassium (K^+), calcium (Ca^{2+}), and magnesium (Mg^{2+})—come from the general chemical weathering of rocks and that the anions, chloride (Cl^-) and sulfate (SO_4^{-2}), come from volcanic gases. Still, da Vinci was right in believing that dissolved ions get carried to the sea by flowing groundwater and river water: rivers deliver over 2.5 billion tons of salt to the sea every year.

There's so much salt in the ocean that if all the water suddenly evaporated, a 60-m-thick layer of salt would coat the ocean floor. This layer would consist of about 75% halite (NaCl) with lesser amounts of gypsum ($CaSO_4 \cdot H_2O$), anhydrite ($CaSO_4$), and other salts. Oceanographers refer to the concentration of salt in water as salinity. Although ocean salinity averages 3.5%, measurements from around the world demonstrate that salinity varies with location, ranging from about 1.0% to about 4.1% (Fig. 18.8a). Salinity reflects the balance between the addition of freshwater by rivers or rain and the removal of freshwater by evaporation, for when seawater evaporates, salt stays behind; salinity may also depend on water temperature, for warmer water can hold more salt in solution than can cold water.

The salinity of the ocean changes with depth. A graph of the variation in salinity with depth (Fig. 18.8b) indicates that such differences in salinity are found in seawater only down to a depth of about 1 km. Deeper water tends to be more homogenous. Oceanographers refer to the gradational boundary between surface-water salinities and deep-water salinities as the halocline.

Temperature

When the *Titanic* sank after striking an iceberg in the North Atlantic, most of the unlucky passengers and crew who

FIGURE 18.8 Physical characteristics of the ocean.

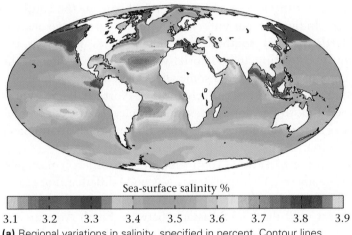

Sea-surface salinity %

3.1　3.2　3.3　3.4　3.5　3.6　3.7　3.8　3.9

(a) Regional variations in salinity, specified in percent. Contour lines separate areas of different salinity.

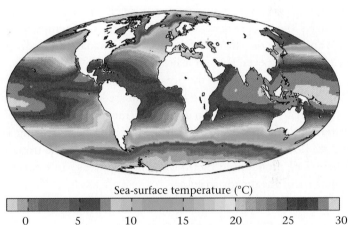

Sea-surface temperature (°C)

0　5　10　15　20　25　30

(c) Regional variations in temperature. Note that the higher temperatures occur in equatorial regions.

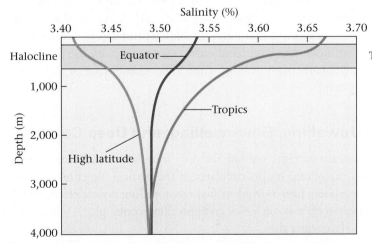

(b) Changes in salinity with depth. Note the halocline (blue band), an interval where salinity changes rapidly.

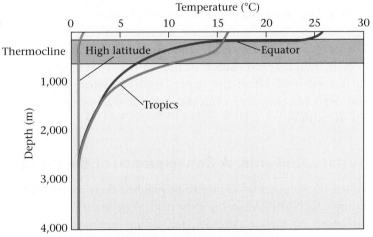

(d) Changes in temperature with depth. Note the thermocline (orange band), an interval where temperatures decrease rapidly.

jumped or fell into the sea died within minutes because the seawater temperature at the site of the tragedy approached freezing, and cold water removes heat from a body very rapidly. Yet swimmers can play for hours in the Caribbean, where sea-surface temperatures reach 28°C (83°F). Though the *average* global sea-surface temperature hovers around 17°C, it ranges between freezing near the poles to almost 35°C in restricted tropical seas (Fig. 18.8c). The *general* correlation of average temperature with latitude exists because the intensity of solar radiation varies with latitude. (As we'll see, temperature at a given latitude may vary because of currents.)

The intensity of solar radiation also varies with the season, so surface seawater temperature varies with the season. But the difference is only around 2° in the tropics, 8° in the temperate latitudes, and 4° near the poles. (By contrast, the seasonal temperature change on land can be much greater—in central Illinois, for example, temperatures may reach 40°C [104°F] in the summer and drop to –32°C [–25°F] in the winter.) Seasonal seawater temperature changes remain in a narrow range because water can absorb or release large amounts of heat without changing temperature very much. Thus, the ocean regulates the temperatures of coastal regions; air temperature in Vancouver, on the Pacific coast of Canada, rarely drops below freezing even though it lies farther north than Illinois.

Water temperature in the ocean varies markedly with depth (Fig. 18.8d). Waters warmed by the Sun are less dense and tend to remain at the surface. An abrupt thermocline—below which water temperatures decrease sharply, reaching near freezing at the sea floor—appears at a depth of about 300 m in the tropics. There is no pronounced thermocline in polar seas, since surface waters there are already so cold.

Currents: Rivers in the Sea

Since first setting sail on the open ocean, people have known that the water of the ocean does not stand still, but rather flows or circulates at velocities of up to several kilometers per hour in fairly well-defined streams called **currents**. Oceanographic studies made since the *Challenger* expedition demonstrate that circulation in the sea occurs at two levels: surface currents affect the upper hundred meters of water, and deep currents keep even the water at the bottom of the sea in motion.

Surface Currents: A Consequence of the Wind

When the skippers of sailing ships planned their routes from Europe to North America, they paid close attention to the directions of surface currents, for sailing against a current slowed down the voyage substantially (Fig. 18.9). If they headed due west at a high latitude, they would find themselves battling an eastward-flowing surface current, the Gulf Stream. Further, they found that the water moving in a surface current does not flow smoothly but displays some turbulence. Isolated swirls or ring-shaped currents of water, called eddies, form along the margins of currents.

Surface currents occur in all the world's oceans. They result from interaction between the sea surface and the wind—as moving air molecules shear across the surface of the water, the friction between air and water drags the water along. But the movement of water resulting from wind shear does not exactly parallel the movement of the wind. This is a consequence of Earth's rotation, which generates the **Coriolis effect** and associated Ekman transport (Box 18.1). This phenomenon causes surface currents in the northern hemisphere to veer toward the right and surface currents in the southern hemisphere to veer toward the left of the average wind direction.

Because of the geometry of ocean basins and the pattern of wind directions, surface currents in the oceans today trace out large circular flow patterns, known as **gyres**, clockwise in the northern seas and counterclockwise in the southern seas. Individual currents within these gyres have names. Northern- and southern-hemisphere gyres merge at the equator, creating an equatorial westward flow. Water in the center of a gyre becomes somewhat isolated from surface currents. Sailors refer to the center of the North Atlantic gyre as the Sargasso Sea, for sargassum, a tropical seaweed, accumulates in this sluggish water. Sadly, floating trash accumulates here too.

In the past, when continents were in different positions, the geometry of ocean currents was quite different. For example, the circum-Antarctic current that exists today did not appear until the Drake Passage, between South America and Antarctica, opened 25 million years ago. Currents that move from the poles to the equator bring cool water toward the equator, whereas currents that move from the equator toward the poles carry warm water poleward, which means that warm water can locally occur at fairly high latitudes. This transport of heat moderates the global climate, so changes in the pattern of currents through geologic time affect the climate of the Earth System.

Upwelling, Downwelling, and Deep Currents

Surface currents are not the only means by which water flows in the ocean; it also circulates in the vertical direction. Oceanographers have now identified downwelling zones, places where near-surface water sinks, and upwelling zones, places where subsurface water rises.

What causes upwelling and downwelling? First, along coastal regions, these two phenomena exist because as the wind

FIGURE 18.9 The major surface currents of the world's oceans.

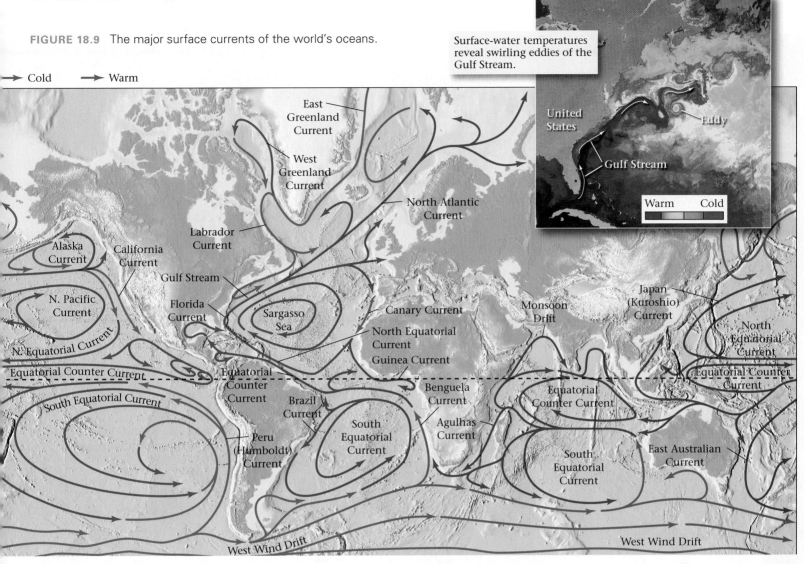

Surface-water temperatures reveal swirling eddies of the Gulf Stream.

blows, it drags surface water along with it. If near-surface water moves toward the coast, then an oversupply of water develops along the shore and excess water must sink—that is, down-welling occurs. Alternatively, if near-surface water moves away from the coast (e.g., due to Ekman transport; see Box 18.1), then a deficit of water develops near the coast and water rises to fill in the gap—upwelling takes place (Fig. 18.10a). Coastal upwelling can be important for the fishing industry, because it brings up nutrients that encourage growth of the plankton on which fish feed. Fishing fleets of South America, for example, pursue their catch in upwelling zones along the Western coast (Fig. 18.10b).

Upwelling of subsurface water also occurs along the equator because the winds blow steadily from east to west. The Coriolis effect causes water to deflect to the right in the northern hemisphere and to the left in the southern hemisphere, resulting in a removal of water along the equator. Upwelling replaces this deficit and causes the surface water at the equator to be cooler and rich in nutrients. The nutrients foster an abundance of life in equatorial water.

Contrasts in water density, caused by *differences in temperature and salinity*, can also drive upwelling and down-welling. We refer to the rising and sinking of water driven by such density contrasts as **thermohaline circulation**. During thermohaline circulation, denser water (cold and/or saltier) sinks, whereas water that is less dense (warm and/or less salty) rises. As a result, the cold water in polar regions sinks and flows back along the bottom of the ocean toward the equator. This process divides the ocean vertically into a number of distinct water masses, which mix only very slowly with one another. In the Atlantic Ocean, for example, the Antarctic Bottom Water sinks along the coast of Antarctica, and the North Atlantic Deep Water sinks in the north polar region (Fig. 18.11a). The combination of surface currents and thermohaline circulation, like a conveyor belt, moves water and heat among the various ocean basins (Fig. 18.11b).

BOX 18.1

The Coriolis Effect

Imagine you are spinning a playground merry-go-round counterclockwise around a vertical axis at a rate of 10 revolutions per minute. The circumference of the outer edge of the merry-go-round is 5 m. Thus Emma, a child sitting at the outer edge, moves at a velocity of 50 m per minute whereas David, a child sitting at the center, moves at zero velocity. If Emma were to try throwing a ball to David by aiming directly along a radius, the ball would veer to the right of the radius and miss David because the ball is moving not only in the direction parallel to a radius line but also in the direction parallel to the edge of the circle. If David were to throw a ball along a radius to Emma, this ball would miss Emma because the revolution of the merry-go-round moves her relative to the ball's trajectory (**Fig. Bx18.1a, b**).

The rotation of the Earth generates the same phenomenon. Earth spins counterclockwise around its axis, so a cannon shell fired along a line of longitude from the North Pole toward the equator veers to the right (west) because the Earth is moving faster to the east at the equator (**Fig. Bx18.1c**). A cannon shell fired parallel to a line of longitude from the equator to the North Pole veers to the right (east) because as it moves north, it is traveling east faster than the land beneath it (**Fig. Bx18.1d**). Similarly, a cannon shell fired from the equator to the South Pole veers to the left (east).

In 1835, a French engineer named Gaspard-Gustave de Coriolis (1792–1843) proposed that a similar effect would cause the deflection of winds and currents on the surface of the Earth. Because of this "Coriolis effect," north-flowing currents in the northern hemisphere deflect to the east, whereas south-flowing currents deflect to the west. The opposite is true in the southern hemisphere.

In 1902, a Swedish oceanographer named V. W. Ekman determined that the surface current would deflect by an angle of about 45° relative to the wind direction.

Because of the Coriolis effect, shear by the uppermost layer would cause the next layer below to deflect even more, and so on. As a result, the mass of water in the upper 100 m of the sea ends up flowing at almost 90° relative to the prevailing wind, a

FIGURE Bx18.1 The Coriolis effect occurs because the velocity of a point at the equator, in the direction of the Earth's spin, is greater than that of a point near the pole.

phenomenon now known as Ekman transport. In the southern hemisphere, Ekman transport is counterclockwise relative to the north-flowing coastal current (the Humbolt Current), causing upwelling along the coast of South America (see Fig. 18.10b).

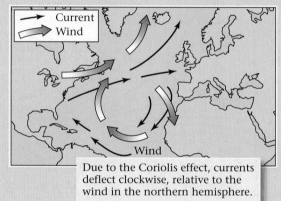

Due to the Coriolis effect, currents deflect clockwise, relative to the wind in the northern hemisphere.

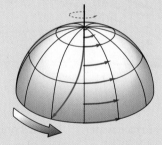

(a) A projectile shot from the pole to the equator deflects to the west.

(b) A projectile shot from the equator to the pole deflects to the east.

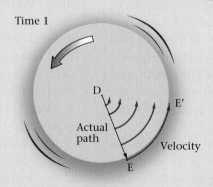

(c) If David throws to Emma on a spinning merry-go-round, the ball misses because Emma has moved to E' by the time the ball reaches her.

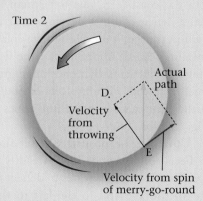

(d) If Emma throws to David, the ball's velocity (yellow arrow) is the sum of the velocity from throwing and the velocity from spin.

FIGURE 18.10 Upwelling and downwelling can happen along coasts.

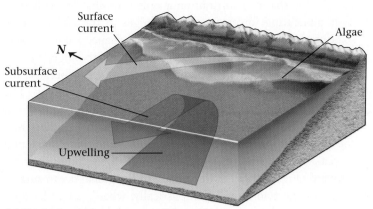

(a) Northerly winds produce an offshore Ekman transport (see Box 18.1) and therefore, upwelling.

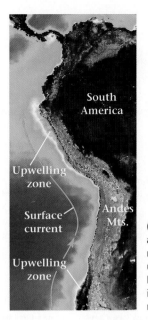

(b) Ocean colors indicate algae concentration, which reflects the amount of upwelling nutrients. Blue is low concentration, green is intermediate, yellow and red are high

FIGURE 18.11 Global-scale upwelling and downwelling of ocean water.

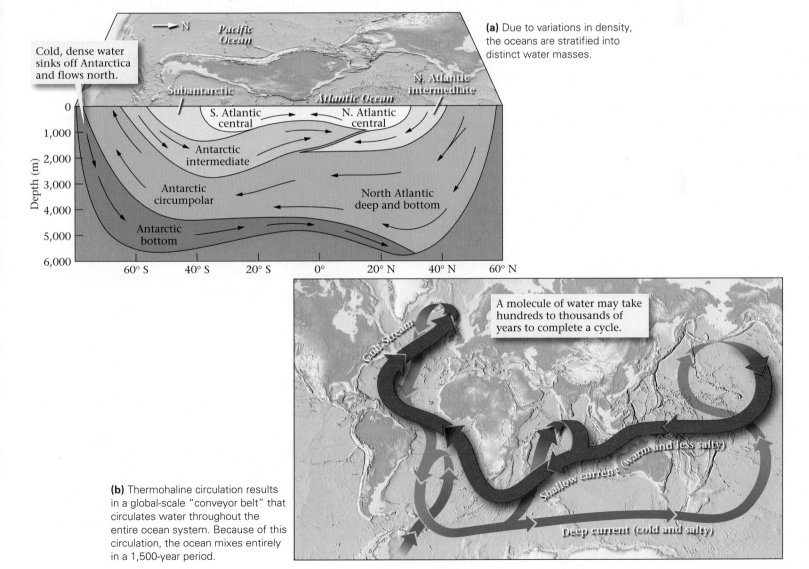

(a) Due to variations in density, the oceans are stratified into distinct water masses.

(b) Thermohaline circulation results in a global-scale "conveyor belt" that circulates water throughout the entire ocean system. Because of this circulation, the ocean mixes entirely in a 1,500-year period.

- Ocean-water salinity averages around 3.5%, but varies significantly. Ions in salt come from rock weathering and volcanic gases.
- Temperature varies with latitude; and locally, temperature can be influenced by currents.
- Variations in temperature and salinity affect water density and drive thermohaline circulation.
- Shallow currents, driven by the wind, form gyres in the ocean.

THINK: Why does downwelling of ocean water occur at polar latitudes?

18.4 THE TIDES GO OUT . . . THE TIDES COME IN. . .

A ship captain seeking to float a ship over reefs, a fisherman hoping to set sail from a shallow port, marines planning to attack a beach from the sea, a tourist eager to harvest shellfish from nearshore mud—all must pay attention to the rise and fall of sea level, a vertical movement called a **tide**, if they are to be successful (Fig. 18.12). If the tide is too low, ships run aground or stay trapped in harbors; and if the tide is too high, a beach may become too narrow to permit access. The **tidal reach**, meaning the difference between sea level at high tide and sea level at low tide, depends on location. The largest tidal reach on Earth is 16.8 m (54.6 feet). The intertidal zone, the region submerged at high tide and exposed at low tide, is a fascinating ecological niche.

During a rising tide, or flood tide, the shoreline (the boundary between water and land) moves inland, whereas during the falling tide, or ebb tide, the shoreline moves seaward. The horizontal distance over which the shoreline migrates between high and low tides depends on both the tidal reach and the slope of the shore surface. Where the tidal reach is large and the slope is gentle, the position of the shore can move a long way during a tidal cycle—at low tide, a broad tidal flat lies exposed to the air in the intertidal zone (Fig. 18.12c–f). Tidal flats can be hazards. For example, in February 2004, fifteen shellfish hunters lost their lives along the coast of northwestern England; they were far offshore, searching for cockles in the mud, when the flood tide came in. Arrival of a flood tide can create a tidal bore, a visible wall of water ranging from a few centimeters to a couple of meters high, that moves at speeds of up to 35 km/h (i.e., faster than a person can run).

Tides are caused by a **tide-generating force**, which is due partly to the gravitational attraction of the Sun and Moon and partly to centrifugal force caused by the revolution of the Earth-Moon system around its center of mass. (To understand the meaning of this complex statement, see Box 18.2.) Gravitational pull by the Moon contributes most of the gravitational part of the tide-generating force. The Sun, even though it is larger, is so far away that its contribution is only 46% that of the Moon. Tide-generating forces create two bulges in the global ocean, making this envelope of water more oval shaped than the solid Earth (Fig. 18.12b). One bulge, the sublunar bulge, lies on the side of the Earth closer to the Moon; it forms because the Moon's gravitational attraction is greatest at this point. The other, the secondary bulge, lies on the opposite (far) side of the Earth (12,000 km—the diameter of the Earth—farther from the Moon); it forms because the Moon's gravitational attraction is weakest at this point. Here, centrifugal force can push water outward (Fig. 18.12a). A depression in the global ocean surface separates the two bulges. Simplistically, when a shore location lies under a tidal bulge, it experiences a high tide, and when it passes under a depression, it experiences low tide.

If the Earth's solid surface were smooth and completely submerged beneath the ocean, so that there were no continents or islands, the timing of tides would be fairly simple to understand. Because the Earth spins on its axis once a day, we would predict two high tides and two low tides at a given point per day. But the story isn't quite that simple—many other factors affect the timing and magnitude of tides. These include the following.

- *Tilt of the Earth's axis*: Because the spin axis of the Earth is not perpendicular to the plane of the Earth-Moon system, a given point passes between a high part of one bulge during one part of the day and through a lower part of the other bulge during another part of the day, so the two high tides at the given point are not the same size (Fig. 18.12b).

- *The Moon's orbit*: The Moon progresses in its 28-day orbit around the Earth in the same direction as the Earth rotates. High tides arrive 50 minutes later each day because of the difference between the time it takes for Earth to spin on its axis and the time it takes for the Moon to orbit the Earth.

- *The Sun's gravity*: When the angle between the direction to the Moon and the direction to the Sun is 90°, we experience extra-low tides (neap tides) because the Sun's gravitational attraction counteracts the Moon's. When the Sun is on the same side as the Moon, we experience extra-high tides (spring tides) because the Sun's attraction *adds* to the Moon's (Fig. 18.12f).

- *Focusing effect of bays*: In the open ocean, the maximum tidal reach is only a few meters. But in the Bay of Fundy, along the eastern coast of Canada, the tidal reach approaches 20 m. In a bay that narrows to a point, such as the Bay of Fundy, the flood tide brings a large volume of water into a small area, so the point experiences an especially large high tide.

FIGURE 18.12 Ocean tides and their manifestation.

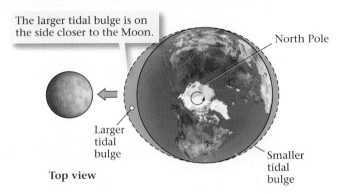

The larger tidal bulge is on the side closer to the Moon.

North Pole

Larger tidal bulge

Smaller tidal bulge

Top view

(a) Tides develop as the Earth spins relative to the two tidal bulges.

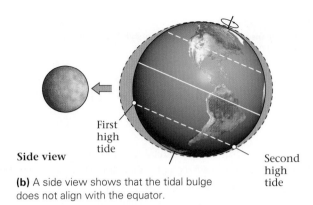

Side view

First high tide

Second high tide

(b) A side view shows that the tidal bulge does not align with the equator.

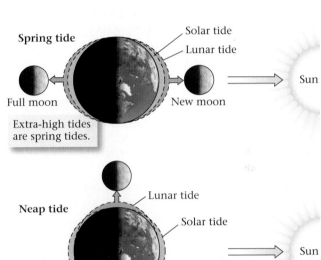

Spring tide

Solar tide

Lunar tide

Full moon

New moon

Sun

Extra-high tides are spring tides.

Neap tide

Lunar tide

Solar tide

Sun

Extra-low tides are neap tides.

(d) Gravitational pull of the Sun can add to that of the Moon to cause extra-high (spring) tides. If the Sun's pull is at right angles to that of the Moon, the tides are particularly low (neap tides).

High tide

Tidal reach

Low tide

(c) The tidal reach is the vertical distance between low tide and high tide. This photo of a French harbor was taken at low tide.

(e) Mont-Saint-Michel, off the western coast of France, is an island at high tide.

(f) At low tide, muddy tidal flats surround Mont-Saint-Michel.

BOX 18.2

The Forces Causing Tides

Fundamentally, tides result from interaction between two forces: gravitational attraction exerted by the Moon and Sun on the Earth, and centrifugal force caused by the revolution of the Earth around the center of mass of the Earth-Moon system. To explain this statement, we must review some key terms from physics.

- **Gravitational pull** is the attractive force that one mass exerts on another. The magnitude of gravitational pull depends on the amount of mass in each object, and on the distance between the two masses.
- **Centrifugal force** is the apparent outward-directed ("center-fleeing") force that material on or in an object feels when the object spins or moves in orbit around a point. Note that centrifugal force differs from *centripetal force*, the "center-seeking" force; this distinction can be confusing. To experience centripetal force, tie a ball to a string and swing it around your head. The string exerts an inward-directed centripetal force on the ball—if the string breaks, the centripetal force ceases to exist and the ball heads off in a straight-line path. To picture centrifugal force, imagine that the ball is hollow and that you've placed a marble inside. As you twirl the ball around your head, the marble moves to the outer edge of the ball. The apparent force pushing the marble outward is the centrifugal force. But as such, centrifugal force is not a real force—it is simply a manifestation of inertia, and it exists *only* from the perspective, or reference frame, of the orbiting object. (A physics book explains this contrast in greater detail.)
- **Earth-Moon system** refers to this pair of objects viewed as a unit, as they move together through space.
- **Center of mass** is the point within an object, or a group of objects, about which mass is evenly distributed; put

another way, it is the location of the average position, or the balance point, of the total mass in a single object or a group of objects. Because the Earth is 81 times more massive than the Moon, the center of mass of the Earth-Moon system actually lies 1,700 km below the surface of the Earth.

With these terms in mind, let's first consider the origin of centrifugal force in the Earth-Moon system. To do this, we must consider the way in which the Earth-Moon system moves. The center of the Earth itself does not follow a simple orbit around the Sun. Rather, it is the *center of mass* of the Earth-Moon system that follows this trajectory; the Earth actually spirals around this trajectory as it speeds around the Sun. To picture this motion, imagine that the Earth-Moon system is a pair of dancers, one of them much heavier than the other. The dancers face each other, hold hands, and whirl in a circle as they drift across the dance floor (**Fig. Bx18.2a**). Each dancer's head orbits the center of mass.

Revolution of the Earth around the Earth-Moon system's center of mass generates centrifugal forces on both the Earth and the Moon that would cause the Earth and the Moon to fly away from one another, were it not for the gravitational attraction holding them together. We can see this by looking again at our dancer analogy (**Fig. Bx18.2b**)—the centrifugal force acting on each dancer points outward, away from his or her partner, and is the *same* for all points on each dancer.

We can represent the direction and magnitude of this centrifugal force by arrows called vectors. (A vector is a number that has magnitude and direction.) In this case, the length of the arrow represents the magnitude of the force, and the orientation of the arrow indicates the direction of the force. If we think of

the dancers as the Earth and the Moon, then centrifugal force vectors at all points on the surface of the Earth point away from the Moon (**Fig. Bx18.2c**). On the Earth, therefore, centrifugal force causes the surface of the ocean to bulge outward, away from the center of mass of the Earth-Moon system, on the far side of the Earth.

Now, let's consider how the force of gravity comes into play in causing tides. To simplify this discussion, we examine only the effect of the Moon's gravity on Earth. Vectors representing the magnitude and direction of the Moon's gravitational pull at any point on the surface of the Earth all point toward the center of the Moon. Because the magnitude of gravity depends on distance, the Moon exerts more attraction on the near side of the Earth than at the Earth's center, and less attraction on the far side of the Earth than at the Earth's center. Gravity, therefore, causes the surface of the ocean on the near side of the Earth to bulge toward the Moon.

In the Earth-Moon system, both centrifugal force and gravitational pull operate at the same time. How do they interact? If we draw vectors representing both centrifugal force and gravitational force at various points on or in the Earth, we see that the vectors representing centrifugal force do not have the same length as those representing gravitational attraction, except at the Earth's center. Moreover, the vectors representing centrifugal force do not point in the same direction as the vectors representing gravitational attraction. The force that the ocean water feels is the sum of the two forces acting on the water. You can determine the sum of two vectors by drawing the vectors so they touch head to tail—the sum is the vector that completes the triangle. This sum is the *tide-generating force*.

The magnitude and direction of the tide-generating forces vary with location on the Earth. For example, on the side of the Earth closer to the Moon, gravitational vectors are larger than centrifugal force vectors, so adding the two gives a net tide-generating force that pulls the sea surface to bulge toward the Moon. On the side of the Earth farther from the Moon, the centrifugal force vectors are larger, so centrifugal force caused by the orbiting of the Earth-Moon system around the center of mass causes the surface of the sea to bulge outward, away from the Moon (**Fig. Bx18.2d**). Thus the ocean has two tidal bulges—one on the side close to the Moon, and one on the opposite side of the Earth. The bulge closer to the Moon is larger.

FIGURE Bx18.2 The concept of the center of mass of the Earth-Moon system, and the tide-generating force.

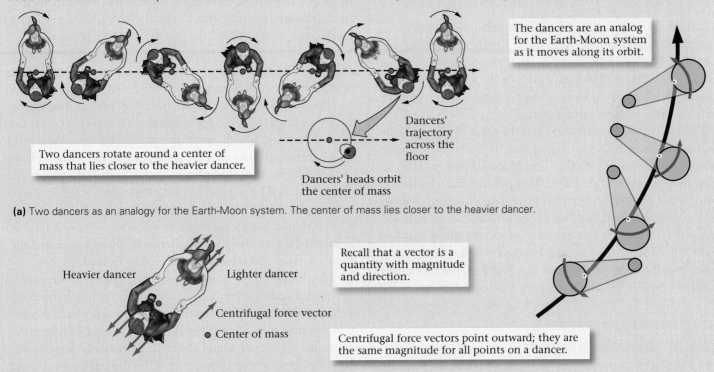

The dancers are an analog for the Earth-Moon system as it moves along its orbit.

Dancers' trajectory across the floor

Two dancers rotate around a center of mass that lies closer to the heavier dancer.

Dancers' heads orbit the center of mass

(a) Two dancers as an analogy for the Earth-Moon system. The center of mass lies closer to the heavier dancer.

Heavier dancer   Lighter dancer

Recall that a vector is a quantity with magnitude and direction.

↗ Centrifugal force vector

● Center of mass

Centrifugal force vectors point outward; they are the same magnitude for all points on a dancer.

(b) Each point on each dancer feels an outward-directed centrifugal force.

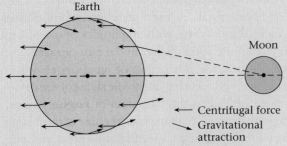

Earth

Moon

← Centrifugal force
↘ Gravitational attraction

(c) Each point on the Earth's surface feels the same centrifugal force, due to the spin of the Earth-Moon system around its center of mass, but feels a different gravitational pull from the Moon because gravitational force depends on distance between objects.

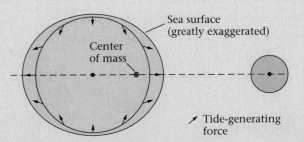

Sea surface (greatly exaggerated)

Center of mass

↗ Tide-generating force

(d) The tide-generating force is the sum of the centrifugal force vector and the gravitational force vectors. On the side closer to the Moon, the gravitational force vector dominates, producing a bulge toward the Moon. On the other side, centrifugal force dominates, producing a smaller bulge.

- *Basin shape*: The shape of the basin containing a portion of the sea influences the sloshing of water back and forth within the basin as tides rise and fall. Depending on the timing and magnitude of this sloshing, this effect can locally add to the global tidal bulge or subtract from it, and thus can affect the rhythm of tides. In some locations, the net effect is to cancel one of the daily tides entirely, so that the locality experiences only one high tide and one low tide in a day.

- *Air pressure*: The effects of air pressure on tides can contribute to disaster. For example, during a hurricane the air pressure drops radically, so the sea surface rises; if the hurricane coincides with a high tide, the storm surge (water driven landward by the wind) can inundate the coast.

Because of the complexity of factors contributing to tides, the timing and magnitude of tides vary significantly along the coast. Nevertheless, at a given location the tides are periodic and can be predicted. Tides gave early civilizations a rudimentary way to tell time. In fact, in some languages the word for tide is the same as the word for time.

Friction between ocean water and the ocean floor causes the movement of the tidal bulge to lag slightly behind the movement of the Moon across the Earth. The Moon, therefore, exerts a slight pull on the side of the bulge. This pull acts like a brake and slows the Earth's spin, so that days grow longer at a rate of about 0.002 seconds per century. Over geologic time, the seconds add up; a day was only 21.9 hours long in the Middle Devonian Period (390 Ma). As the spinning Earth slows, the Moon moves farther away. During the Archean Eon (3.8 Ga), the Moon was 15,000 km closer, so the tidal reach on Earth was larger.

Take-Home Message

- Tides are the twice-daily rise and fall of the sea surface.
- They are caused by the tide-generating force, a consequence of gravitational pull by the Moon and Sun, and of centrifugal force due to revolution of the Earth-Moon system.
- Many factors affect the specific tidal reach at a given locality.

THINK: Why has the magnitude of tidal reach changed over Earth history?

18.5 WAVE ACTION

Wind-driven waves make the ocean surface a restless, ever-changing vista. They develop because of the shear between the molecules of air in the wind and the molecules of water at the surface of the sea. It may seem surprising that so much friction can arise between two fluids, but it can, as Benjamin Franklin demonstrated. Franklin noted that oily waste spilled from ships on a windy day made the water surface smoother. He proposed that the oily coating on the water decreased frictional shear between the air and the water, and therefore prevented waves from forming.

When you watch a wave travel across the open ocean, you may get the impression that the whole mass of water constituting the wave moves with the wave. But drop a cork overboard and it will bob up and down and back and forth as a wave passes; it does not move along with a wave. Within a wave, away from shore, a particle of water moves in a circular motion, as viewed in cross section. The diameter of the circle is greatest at the ocean's surface, where it equals the amplitude (half the height from crest to trough) of the wave. With increasing depth, though, the diameter of the circle decreases until, at a depth equal to about half the wavelength (the horizontal distance between two wave troughs), there is no wave movement at all (Fig. 18.13a). Submarines traveling below this **wave base** cruise through smooth water, while ships toss about above.

The character of waves in the open ocean depends on the strength of the wind (how fast the air moves) and on the fetch of the wind (the distance over which it blows). When the wind first begins to blow, it creates ripples in the water surface, pointed waves whose amplitude and wavelength are small. With continued blowing over a long fetch, swells, larger waves with amplitudes of 2 to 10 m and wavelengths of 40 to 500 m, begin to build. Hurricane wave amplitudes may grow to over 25 m. Swells may travel for thousands of kilometers across the ocean, well beyond the region where they formed.

How large can wind-driven waves in the open ocean get? It's not surprising that huge waves form during hurricanes. Oceanographers calculate that if hurricane-strength winds were to blow across the width of the Pacific for at least 24 hours, 15- to 20-m-high waves would develop. Particularly large waves may also form where two sets of wind-driven waves coming from different directions constructively interfere, so that wave crests add to each other. This happened in 1979, when waves generated by an east-blowing gale collided with waves generated by a west-blowing gale in the waters off Ireland during the Fastnet Yacht Race. Waves with amplitudes of over 15 m developed, and 23 of the 300 sailboats in the race capsized.

Wave interference, the interaction of wind-driven waves with strong currents, and focusing due to the shape of the coastline or sea floor can lead to the formation of **rogue waves**, defined as waves that are more than twice the size of most large waves passing a locality during a specified time interval. Long thought to exist only in the imagination of sailors, rogue waves now have been documented numerous times. For example, wave-measuring instruments on oil platforms in the North Sea recorded almost 500 encounters with rogue waves during a 10-year period. Some of the waves were 3 to 5 times higher

FIGURE 18.13 Ocean waves build in response to the shear of wind blowing over the water surface.

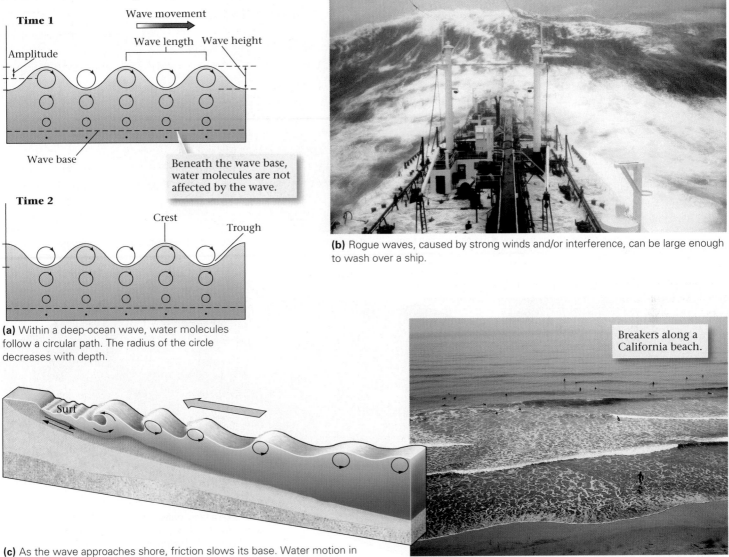

Time 1

Wave movement

Amplitude

Wave length Wave height

Wave base

Beneath the wave base, water molecules are not affected by the wave.

Time 2

Crest

Trough

(a) Within a deep-ocean wave, water molecules follow a circular path. The radius of the circle decreases with depth.

Surf

(b) Rogue waves, caused by strong winds and/or interference, can be large enough to wash over a ship.

Breakers along a California beach.

(c) As the wave approaches shore, friction slows its base. Water motion in the wave becomes more elliptical, and the wave becomes a breaker.

than other large waves. The decks of large ships—including famous cruise ships—have been swamped by immense rogue waves in the open ocean (Fig. 18.13b). In 1995, for example, a 29-m-high wave struck the *Queen Elizabeth II*, and in 1933, a military ship encountered one at least 34 m (112 ft) high. Recent research proves that rogue waves are not so rare as once thought. Radar studies from satellites demonstrate that at any given time, there are about ten rogue waves around the world's oceans. If a rogue wave reaches the shore, it can wash unsuspecting bystanders off shore-side piers or beaches. The mysterious loss of some ships may be a consequence of rogue waves.

Waves have no effect on the ocean floor, as long as the floor lies below the wave base. However, near the shore, where the wave base just touches the floor, it causes a slight back-and-forth motion of sediment. Closer to shore, as the water gets shallower, friction between the wave and the sea floor slows the deeper part of the wave, and the motion in the wave becomes more elliptical. Eventually, water at the top of the wave curves over the base, and the wave becomes a breaker, ready for surfers to ride. Breakers crash onto the shore in the surf zone, sending a surge of water up the beach. This upward surge, or **swash**, continues until friction brings motion to a halt. Then gravity draws the water back down the beach as **backwash** (Fig. 18.13c).

Waves may make a large angle with the shoreline as they're coming in, but they bend as they approach the shore,

Did you ever wonder . . . why the breakers that surfers love form near the shore?

FIGURE 18.14 Wave refraction and its consequences along the shore.

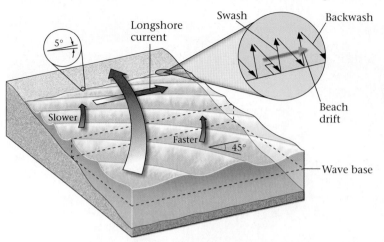

Wave refraction occurs where coastlines are not straight.

(a) Wave refraction occurs when waves approach the shore at an angle. If the wave reaches the beach at an angle, it causes a longshore current and beach drift of sand.

Waves crashing on a rocky headland in Hawaii.

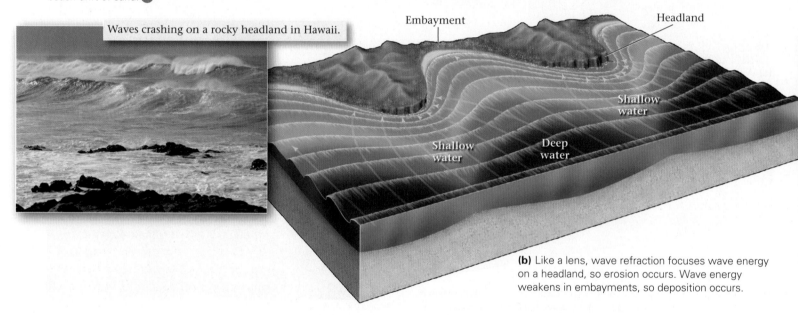

(b) Like a lens, wave refraction focuses wave energy on a headland, so erosion occurs. Wave energy weakens in embayments, so deposition occurs.

a phenomenon called **wave refraction**; right at the shore, their crests make no more than about a 5° angle with the shoreline (Fig. 18.14 ▶️). To understand why this happens, imagine a wave approaching the shore so that its crest makes an angle of 45° with the shoreline. The end of the wave closer to the shore touches bottom first and slows down because of friction, whereas the end farther offshore continues to move at its original velocity, swinging the whole wave around so that it's more parallel with the shoreline.

Although wave refraction decreases the angle at which waves move close to shore, it does not necessarily eliminate the angle. Where waves do arrive at the shore obliquely, water in the nearshore region has a component of motion that trends parallel to the shore. This **longshore current** causes swimmers floating in the water just offshore to drift gradually in

a direction parallel to the beach. And where waves roll onto the shore at an angle, sediment in the surf follows a sawtooth pattern of movement that results in a gradual *net* transport of beach sediment parallel to the beach—such movement is called **longshore drift** (or beach drift). This sawtooth pattern happens because the swash of a wave moves perpendicular to the wave crest, so an oblique wave carries sediment diagonally up the beach, but the backwash must flow straight down the slope of the beach due to gravity.

Waves pile water up on the shore incessantly. As the excess water moves back to the sea, it may localize into a strong seaward flow perpendicular to the beach called a rip current (Fig. 18.15). Rip currents are the cause of many drownings every year along beaches, because they can carry unsuspecting swimmers away from the beach.

FIGURE 18.15 Waves bring water up on shore. The water may return to sea in a narrow rip current perpendicular to the shore.

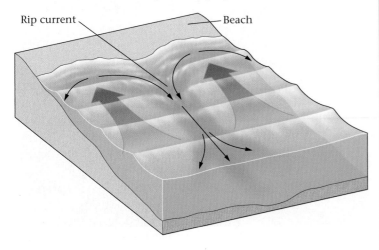

Rip current — — Beach

Take-Home Message

- Friction between wind and the sea surface produces waves. As a wave passes, water moves in a circular motion.
- Below the wave base, a depth equal to half a wave's wavelength, the passing of a wave above does not cause water to move.
- Friction with the sea floor slows the base of waves, so they build into breakers near the shore.
- Interference and other factors can generate huge rogue waves.
- Waves refract along the shore, but if waves reach a beach at an angle, they cause longshore drift and longshore currents.

THINK: What's the difference between swash and backwash?

18.6 WHERE LAND MEETS SEA: COASTAL LANDFORMS

Tourists along the Amalfi coast of Italy thrill to the sound of waves crashing on rocky shores. But in the Virgin Islands sunbathers can find seemingly endless white sand beaches, and along the Mississippi delta, vast swamps border the sea. Large, dome-like mountains rise directly from the sea in Rio de Janeiro, Brazil, but a 100-m-high vertical cliff marks the boundary between the Nullarbor Plain of southern Australia and the Great Southern Ocean (Fig. 18.16a–d). As these examples illustrate, coasts, the belts of land bordering the sea, vary dramatically in terms of topography and associated landforms (Fig. 18.17).

Beaches and Tidal Flats

For millions of vacationers, the ideal holiday includes a trip to a **beach**, a gently sloping fringe of sediment along the shore. Some beaches consist of pebbles or boulders, whereas others consist of sand grains (Fig. 18.18a, b). This is no accident, for waves winnow out finer sediment like silt and mud and carry it to quieter water, where it settles. Storm waves, which can smash cobbles against one another with enough force to shatter them, have little effect on sand, for sand grains can't collide with enough energy to crack. Thus, cobble beaches exist only where nearby cliffs supply large rock fragments.

Did you ever wonder... why beautiful sandy beaches don't form along all coasts?

The composition of sand itself varies from beach to beach, because different sands come from different sources. Sands derived from the weathering and erosion of silicic-to-intermediate rocks consist mainly of quartz; other minerals in these rocks chemically weather to form clay, which washes away in waves. Beaches made from the erosion of limestone or of coral reefs and shells consist of carbonate sand, including masses of sand-sized chips of shells. And beaches derived from the erosion of basalt boast black sand, made of tiny basalt grains.

A beach profile, a cross section drawn perpendicular to the shore, illustrates the shape of a beach (Fig. 18.18c). Starting from the sea and moving landward, a beach consists of a foreshore zone, or intertidal zone, across which the tide rises and falls. The **beach face**, a steeper, concave-up part of the foreshore zone, forms where the swash of the waves actively scours the sand. The backshore zone extends from a small step, cut by high-tide swash to the front of the dunes or cliffs that lie farther inshore. The backshore zone includes one or more **berms**, horizontal to landward-sloping terraces that received sediment during a storm.

Geologists commonly refer to beaches as "rivers of sand," to emphasize that beach sand moves along the coast over time—it is not a permanent substrate. Wave action at the shore moves an active sand layer on the sea floor on a daily basis. Inactive sand, buried below this layer, moves only during severe storms or not at all. Longshore drift, discussed earlier, can transport sand hundreds of kilometers along a coast in a matter of centuries. Where the coastline indents landward, beach drift stretches beaches out into open water to create a **sand spit**. Some sand spits grow across the opening of a bay, to form a baymouth bar (Fig. 18.18d).

The scouring action of waves piles sand up in a narrow ridge away from the shore called an **offshore bar**, which parallels the shoreline. In regions with an abundant sand supply, offshore bars rise above the mean high-water level

FIGURE 18.16 Spectacular coastal scenery from localities around the world.

(a) Rocky cliffs form the shore of eastern Italy.

(c) Rounded "sugar loaf" mountains rise from the bays of Rio de Janeiro.

(b) A sandy beach along the coast of St. John, U.S. Virgin Islands.

(d) The abrupt edge of the Nullarbor Plain forms the south coast of Australia.

and become **barrier islands** (Fig. 18.18e). The water between a barrier island and the mainland becomes a quiet-water **lagoon**, a body of shallow seawater separated from the open ocean.

Though developers have covered some barrier islands with expensive resorts, in the time frame of centuries to millennia, barrier islands are temporary features. Wind and waves pick up sand from the ocean side of the barrier island and drop it on the lagoon side, causing the island to migrate landward. Storms may breach barrier islands and create an inlet (a narrow passage of water). Finally, beach drift gradually transports the sand of barrier islands and modifies their shape.

Tidal flats, regions of mud and silt exposed or nearly exposed at low tide but totally submerged at high tide, develop in regions protected from strong wave action, as we noted earlier

(Fig. 18.12e–f; Fig. 18.18f). They are typically found along the margins of lagoons or on shores protected by barrier islands. Here, mud and silt accumulate to form thick, sticky layers. In tidal flats that provide a home for burrowing organisms such as clams and worms, **bioturbation** ("stirring by life") mixes sediments together.

Because of the movement of sediment, the "sediment budget" (the difference between sand supplied and sand removed) plays an important role in determining the long-term evolution of a beach. Let's look at how the budget works for a small segment of beach (Fig. 18.19). Sand may be supplied to the segment from local rivers or by wind from nearby dune fields; it may also be brought from just offshore by waves or from far away by beach drift. (In fact, the large quantity of sand along beaches of the southeastern United States may have been brought to the region by glaciers during the ice age.) Some

FIGURE 18.17 Examples of different kinds of coasts.

Uplifted terraces

Glacial fjords

Drowned river valleys (estuaries)

Coastal plains and offshore sandbars

Coral reefs off a mangrove swamp

A swampy delta

Coastal sand dunes and a wide beach

of the sand from a stretch of beach may be removed by beach drift, whereas some gets carried offshore by waves, where it either settles locally or tumbles down a submarine canyon into the deep sea. If the lost sand cannot be replaced, the beach segment grows narrower, whereas if the supply of sand exceeds the amount that washes away, the beach becomes wider.

In temperate climates, winter storms tend to be stronger and more frequent than summer ones. The larger, shorter-wavelength waves of winter storms wash beach sand into deeper water and thus make the beach narrower, whereas the smaller, longer-wavelength summer waves bring sand in from offshore and deposit it on the beach (Fig. 18.20a, b).

Rocky Coasts

More than one ship has met its end, smashed and splintered in the spray and thunderous surf of a rocky coast, where bedrock cliffs rise directly from the sea (Fig. 18.16). Lacking the protection of a beach, rocky coasts feel the full impact of ocean breakers. The water pressure generated during the impact of a breaker can pick up boulders and smash them together until they shatter, and it can squeeze air into cracks, creating enough force to widen them. Further, because of its turbulence, the water hitting a cliff face carries suspended sand and thus can abrade the cliff. The combined effects of shattering, wedging, and abrading, together called wave erosion, gradually undercut a

FIGURE 18.18 Characteristics of beaches, barrier islands, and tidal flats.

(a) A gravel beach along the Olympic Peninsula, Washington. The clasts were derived from erosion of adjacent cliffs.

(b) A sand beach on the western coast of Puerto Rico. Wave action has carried away finer sediment.

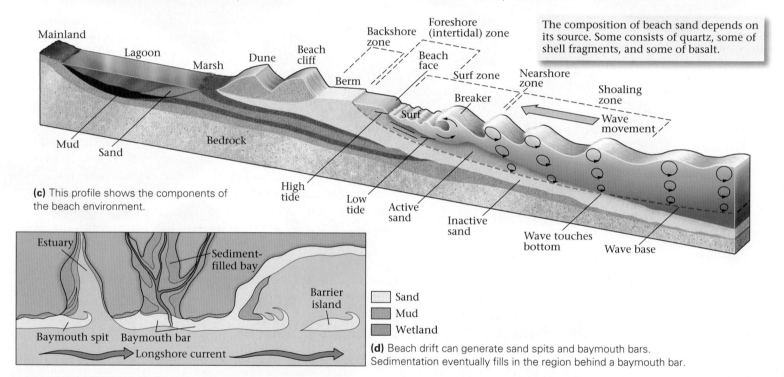

The composition of beach sand depends on its source. Some consists of quartz, some of shell fragments, and some of basalt.

(c) This profile shows the components of the beach environment.

Sand
Mud
Wetland

(d) Beach drift can generate sand spits and baymouth bars. Sedimentation eventually fills in the region behind a baymouth bar.

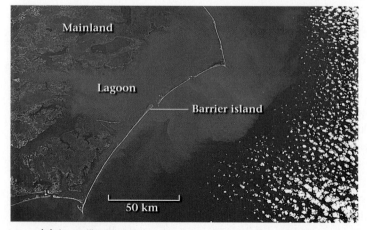

(e) A satellite image showing barrier islands off the coast of North Carolina.

(f) At low tide, boats at anchor rest in the mud of a tidal flat along the coast of Wales.

FIGURE 18.19 The sediment budget along a coast. Sediment is brought into the system by rivers, by the erosion of cliffs and moraines, and by wind. Sediment moves along the coast as a result of beach drift. And sediment leaves the system by being blown off the beach, by sinking into deeper water, or by being carried out by the longshore current.

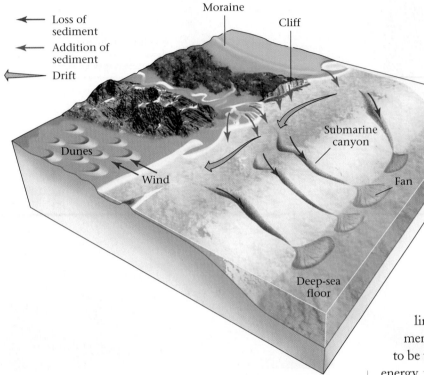

FIGURE 18.20 Contrasts between winter and summer beaches.

(a) In the winter, waters are stormier so sediment moves farther offshore.

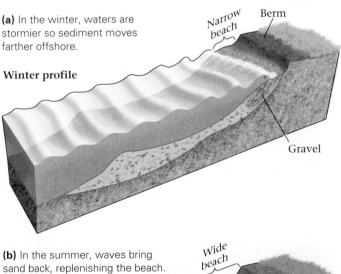

(b) In the summer, waves bring sand back, replenishing the beach.

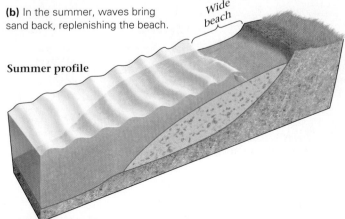

cliff face and make a **wave-cut notch** (Fig. 18.21a). Undercutting continues until the overhang becomes unstable and breaks away at a joint, creating a pile of rubble at the base of the cliff that waves immediately attack and break up. In this process, wave erosion cuts away at a rocky coast, so that the cliff gradually migrates inland. Such cliff retreat leaves behind a **wave-cut bench**, or platform, that becomes visible at low tide (Fig. 18.21b).

Other processes besides wave erosion break up the rocks along coasts. For example, salt spray coats the cliff face above the waves and infiltrates into pores. When the water evaporates, salt crystals grow and push apart the grains, thereby weakening the rock. Biological processes also contribute to erosion, for plants and animals in the intertidal zone bore into the rocks and gradually break them up.

Many rocky coasts start out with an irregular coastline, with headlands protruding into the sea and embayments set back from the sea. Such irregular coastlines tend to be temporary features in the context of geologic time: wave energy focuses on headlands and disperses in embayments, a result of wave refraction. The resulting erosion removes debris at headlands, and sediment accumulates in embayments (Fig. 18.14b and 18.21c); thus, over time the shoreline becomes less irregular.

A headland erodes in stages (Fig. 18.21d). Because of refraction, waves curve and attack the sides of a headland, slowly eating through it to create a sea arch connected to the mainland by a narrow bridge. Eventually the arch collapses,

A berm on a beach.

FIGURE 18.21 Erosion landforms of rocky shorelines.

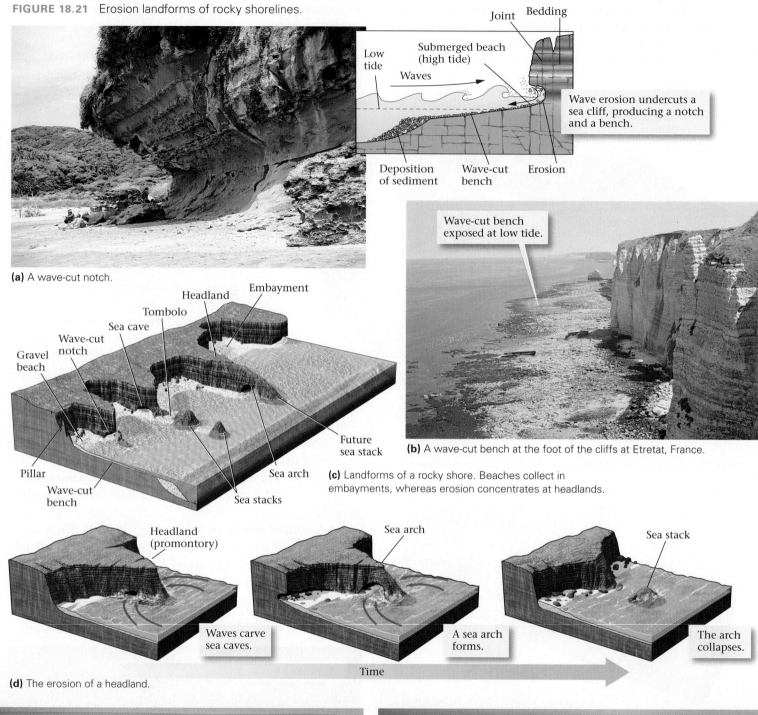

(a) A wave-cut notch.

(b) A wave-cut bench at the foot of the cliffs at Etretat, France.

(c) Landforms of a rocky shore. Beaches collect in embayments, whereas erosion concentrates at headlands.

(d) The erosion of a headland.

(e) Coastal erosion along Australia's southern coast produced a sea arch (left). Eventually, the bridge will collapse, and only sea stacks will remain (right). These two are among several that together are known locally as the Twelve Apostles.

leaving isolated **sea stacks** just offshore (Fig. 18.21e). Once formed, a sea stack protects the adjacent shore from waves. Therefore, sand collects in the lee of the stack, slowly building a tombolo, a narrow ridge of sand that links the sea stack to the mainland.

FIGURE 18.22 The Chesapeake Bay estuary formed when the sea flooded river valleys. The region is sinking relative to other coast areas because it overlies a buried meteor crater.

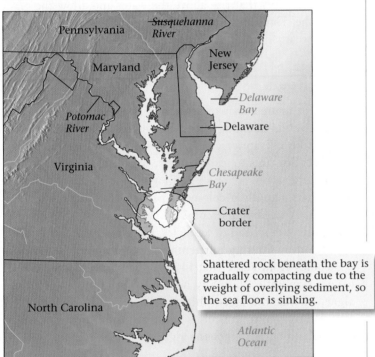

Shattered rock beneath the bay is gradually compacting due to the weight of overlying sediment, so the sea floor is sinking.

Estuaries

Along some coastlines, a relative rise in sea level causes the sea to flood river valleys that merge with the coast, resulting in **estuaries**, where seawater and river water mix. You can recognize an estuary on a map by the dendritic pattern of its river-carved coastline (Fig. 18.22). Oceanic and fluvial waters interact in two ways within an estuary. In quiet estuaries, protected from wave action or river turbulence, the water becomes stratified, with denser oceanic saltwater flowing upstream as a wedge beneath less-dense fluvial freshwater. Such saltwater wedges migrate about 100 km up the Hudson River in New York, and about 40 km up the Columbia River in Oregon. In turbulent estuaries, such as the Chesapeake Bay, oceanic and fluvial water combine to create nutrient-rich brackish water with a salinity between that of oceans and rivers. Estuaries are complex ecosystems inhabited by unique species of shrimp, clams, oysters, worms, and fish that can tolerate large changes in salinity.

Fjords

During the last ice age, glaciers carved deep valleys in coastal mountain ranges. When the ice age came to a close, the glaciers melted away, leaving deep, U-shaped valleys (see Chapter 22). The water stored in the glaciers, along with the water within the vast ice sheets that covered continents during the ice age, flowed back into the sea and caused sea level to rise. The rising sea filled the deep valleys, creating **fjords**, or flooded glacial valleys. Coastal fjords are fingers of the sea surrounded by mountains; because of their deep-blue water and steep walls of polished rock, they are distinctively beautiful (Fig. 18.23a, b). Some of the world's most spectacular fjords decorate the western coasts of Norway, British Columbia, and New Zealand. Smaller examples appear along the coast of Maine and southeastern Canada.

FIGURE 18.23 Fjord landscapes form where relative sea-level rise drowns glacially carved valleys.

Glacial valleys have a U shape, so fjords have steep sides.

Fjord

A fjord in Norway

Take-Home Message

- A beach is a fringe of sediment along a shore and can consist of cobbles or sand.
- Beaches have a distinct profile, which varies with the season.
- Sand can build into sand spits, offshore bars, and barrier islands. In quiet lagoons or other protected areas, large muddy tidal flats develop.
- Rocky coasts are slowly eroded by wave action, producing distinctive landforms such as wave-cut benches and sea stacks.
- Estuaries develop where the sea floods river valleys, whereas fjords form when the sea floods glacially carved valleys.

THINK: Does the sand of a beach stay in the same location for a long time? Why?

18.7 ORGANIC COASTS

Coasts in which living organisms control landforms along the shore are called **organic coasts**. The nature of an organic coast depends on the type of organisms that live there, which, in turn, depends on climate.

Coastal Wetlands

Let's move now from the crashing waves of rocky coasts to the gentlest type of shore, the **coastal wetland**, a vegetated, flat-lying stretch of coast that floods at high tide but does not feel the impact of strong waves. In temperate climates, coastal wetlands include swamps (wetlands dominated by trees), marshes (wetlands dominated by grasses; Fig. 18.24a), and bogs (wetlands dominated by moss and shrubs). So many marine species spawn in wetlands that, despite their relatively small area when compared with the oceans as a whole, wetlands account for 10 to 30% of marine organic productivity.

In tropical or semitropical climates (between 30° north and 30° south of the equator), mangrove swamps thrive in wetlands (Fig. 18.24b). Mangrove tree roots can filter salt out of water, so the trees have evolved to survive in freshwater or saltwater. Some mangrove species form a broad network of roots above the water surface, making the plant look like an octopus standing on its tentacles, and some send up small protrusions from roots that rise above the water and allow the plant to breathe. Dense stands of mangroves counter the effects of stormy weather and thus prevent coastal erosion.

Coral Reefs

Along the azure coasts of Hawaii, visitors swim through colorful growths of living coral. Some corals look like brains, others like elk antlers, still others like delicate fans (Fig. 18.25a). Sea anemones, sponges, and clams grow on and around the coral. Though at first glance coral looks like a plant, it is actually a colony of tiny invertebrates related to jellyfish. An individual coral animal, or polyp, has a tube-like body with a head of tentacles. Corals obtain part of their livelihood by filtering nutrients out of seawater; the remainder comes from algae that live on the corals' tissue. Corals have a symbiotic (mutually beneficial) relationship with the algae, in that the algae photosynthesize and provide nutrients and oxygen to the corals while the corals provide carbon dioxide and nutrients for the algae.

FIGURE 18.24 Examples of coastal wetlands.

(a) A salt marsh along the coast of Cape Cod.

(b) A mangrove swamp in northeastern Brazil.

FIGURE 18.25 The character and evolution of coral reefs.

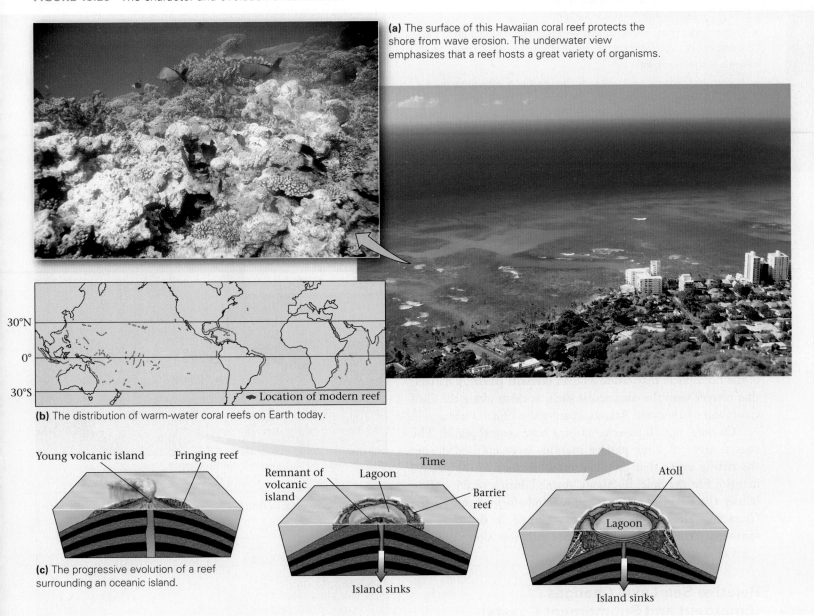

(a) The surface of this Hawaiian coral reef protects the shore from wave erosion. The underwater view emphasizes that a reef hosts a great variety of organisms.

(b) The distribution of warm-water coral reefs on Earth today.

30°N
0°
30°S
← Location of modern reef

Young volcanic island   Fringing reef

Remnant of volcanic island   Lagoon   Time   Atoll
Barrier reef
Lagoon

(c) The progressive evolution of a reef surrounding an oceanic island.

Island sinks

Island sinks

Coral polyps secrete calcite shells, which gradually build into a mound of solid limestone whose top surface lies from just below the low-tide level down to a depth of about 60 m. At any given time, only the surface of the mound lives—the mound's interior consists of shells from previous generations of coral. The realm of shallow water underlain by coral mounds, associated organisms, and debris comprises a **coral reef** (Fig. 18.25a). Reefs absorb wave energy and thus serve as a living buffer zone that protects coasts from erosion. Corals need clear, well-lit, warm (18°–30°C) water with normal oceanic salinity, so coral reefs grow only along clean coasts at latitudes of less than about 30° (Fig. 18.25b).

Marine geologists distinguish three different kinds of coral reef, on the basis of their geometry (Fig. 18.25c). A fringing reef forms directly along the coast, a barrier reef develops offshore (separated from the coast by a lagoon), and an atoll makes a circular ring surrounding a lagoon. As Charles Darwin first recognized back in 1859, coral reefs associated with islands in the Pacific start out as fringing reefs and then later become barrier reefs and finally atolls. Darwin suggested, correctly, that this progression reflects the continued growth of the reef as the island around which it formed gradually sinks. Eventually, the reef itself sinks too far below sea level to remain alive and becomes the cap of a guyot.

18.8 CAUSES OF COASTAL VARIABILITY

Plate Tectonic Setting

The tectonic setting of a coast plays a role in determining whether the coast has steep-sided mountain slopes or a broad plain that borders the sea (see **Geology at a Glance**, pp. 638–639). Along an active margin, compression squeezes the crust and pushes it up, creating mountains like the Andes along the western coast of South America. Along a passive margin, the cooling and sinking of the lithosphere may create a broad **coastal plain**, a flatland that merges with the continental shelf, as exists along the Gulf Coast and southeastern Atlantic coast of the United States.

Of note, not all passive margins have coastal plains. The coastal areas of some passive margins were uplifted during the rifting event that preceded establishment of the passive margin. For example, highlands formed during recent rifting border the Red Sea. Similarly, highlands formed during the Cretaceous rifting persist along portions of the Brazilian and southern African coasts. Highlands also rise along the east coast of Australia.

Relative Sea-Level Changes (Emergent and Submergent Coasts)

Sea level, relative to the land surface, changes during geologic time. Some changes develop due to vertical movement of the land. These may reflect plate-tectonic processes or the addition or removal of a load (such as a glacier) on the crust. Local changes in sea level may reflect human activity—when people pump out groundwater, for example, the pores between grains in the sediment beneath the ground collapse, and the land surface sinks (see Chapter 19). Some relative sea-level changes, however, are due to a *global* rise or fall of the ocean surface. Such **eustatic sea-level changes** may reflect changes in the volume of mid-ocean ridges. An increase in the number or size of ridges, for example, displaces water and causes sea level to rise. Eustatic sea-level changes may also reflect changes in the volume of glaciers, for glaciers store water on land (Fig. 18.26a, b).

FIGURE 18.26 Because of sea-level drop during the ice age, there was more dry land.

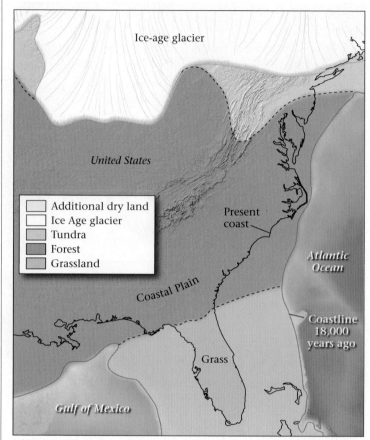

(a) Exposed land along eastern North America.

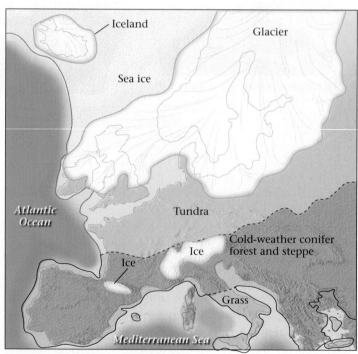

(b) Exposed land along the coast of Europe.

As glaciers grow, sea level falls, and as glaciers shrink, sea level rises.

Geologists refer to coasts where the land is rising or rose relative to sea level as **emergent coasts**. At emergent coasts, steep slopes typically border the shore. A series of step-like terraces form along some emergent coasts (Fig. 18.27a, b). These terraces reflect episodic changes in relative sea level.

Those coasts at which the land sinks relative to sea level become **submergent coasts** (Fig. 18.27c, d). At submergent coasts, landforms include estuaries and fjords that developed when the sea flooded coastal valleys. Many of the coastal landforms of eastern North America are the consequences of submergence.

Sediment Supply and Climate

The quantity and character of sediment supplied to a shore affects its character. That is, coastlines where the sea washes sediment away faster than it can be supplied (erosional coasts) recede landward and may become rocky, whereas coastlines that receive more sediment than erodes away (accretionary coasts) grow seaward and develop broad beaches.

Climate also affects the character of a coast. Shores that enjoy generally calm weather erode less rapidly than those constantly subjected to ravaging storms. A sediment supply large enough to generate an accretionary coast in a calm environment may be insufficient to prevent the development of

FIGURE 18.27 Features of emergent coastlines (relative sea level is falling) and submergent coastlines (relative sea level is rising).

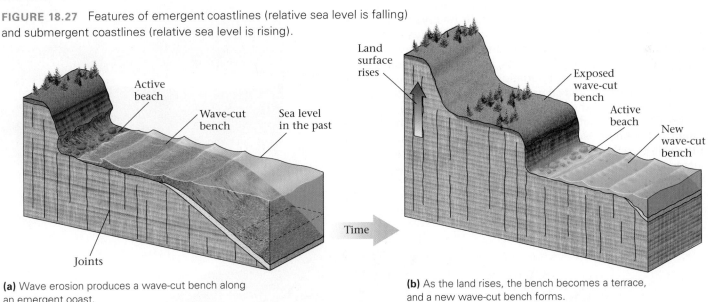

(a) Wave erosion produces a wave-cut bench along an emergent coast.

(b) As the land rises, the bench becomes a terrace, and a new wave-cut bench forms.

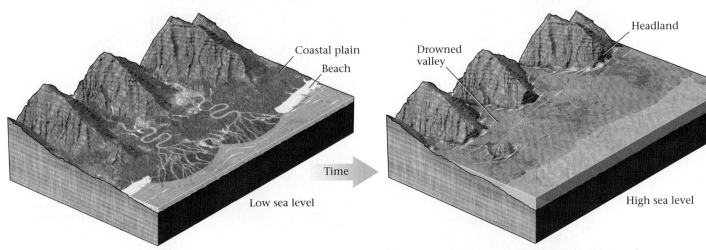

(c) A coast before sea level rises. Rivers drain valleys and deposit sediment on a coastal plain.

(d) As a submergent coast forms, sea level rises and floods the valleys, and waves erode the headlands.

Oceans and Coasts

The oceans of the world host a diverse array of environments and landscapes, illustrating the complexity of the Earth system. Tectonic processes and surface processes working alone or in tandem generate unique features beneath the sea and along its coasts.

The major structures of the ocean floor are the result of plate-tectonic activity. For example, mid-ocean ridges define divergent boundaries, fracture zones form along transform faults, and trenches mark subduction zones. Oceanic islands, seamounts, and plateaus build above hot spots. Along passive continental margins, broad continental shelves develop, locally incised by submarine canyons, as sediment buries stretched continental lithosphere. Along convergent margins, accretionary prisms and volcanic arcs grow.

Within the ocean, currents circulate water due to regional variations in both water salinity and temperature (factors that control density) and to the traction between wind and the water surface. Tides cause sea level to rise and fall, and wind builds waves that churn the sea surface, erode shorelines, and transport sediment.

Coastal landscapes reflect variations in sediment supply, relative sea-level rise or fall, and climate. For example, where the supply of sediment is low and the landscape is rising relative to sea level, rocky shores with dramatic cliffs and sea stacks may evolve. Where sediment is abundant, sandy beaches and bars develop. Regions where glaciers carved deep valleys now feature spectacular fjords. Protected coastal areas, especially those in warm climates, host unique coastal ecosystems inhabited by grasses, mangroves, and/or corals. Corals may contribute to growth of broad reefs along the shore. Human activities can significantly affect coastal landscape.

Wave erosion cuts notches at the base of cliffs and bevels wave-cut benches.

Along sandy shores, sand builds beaches, sand spits, and bars.

Turbidities flowing down submarine canyons produce submarine fans.

In tropical environments, mangroves live along the shore and coral reefs grow offshore.

Along rocky coasts, sea cliffs, sea arches, and sea stacks evolve.

At a passive margin, a broad continental shelf develops. Submarine slumping may occur along the shelf.

The ocean teems with life.

At divergent plate boundaries, a mid-ocean ridge rises. Transform faults, marked by fracture zones, link segments of the ridge.

The Global Conveyor

Surface winds drive surface currents in large gyres.

Cold water sinks at polar regions.

0
1
2
3 km
4
5
6

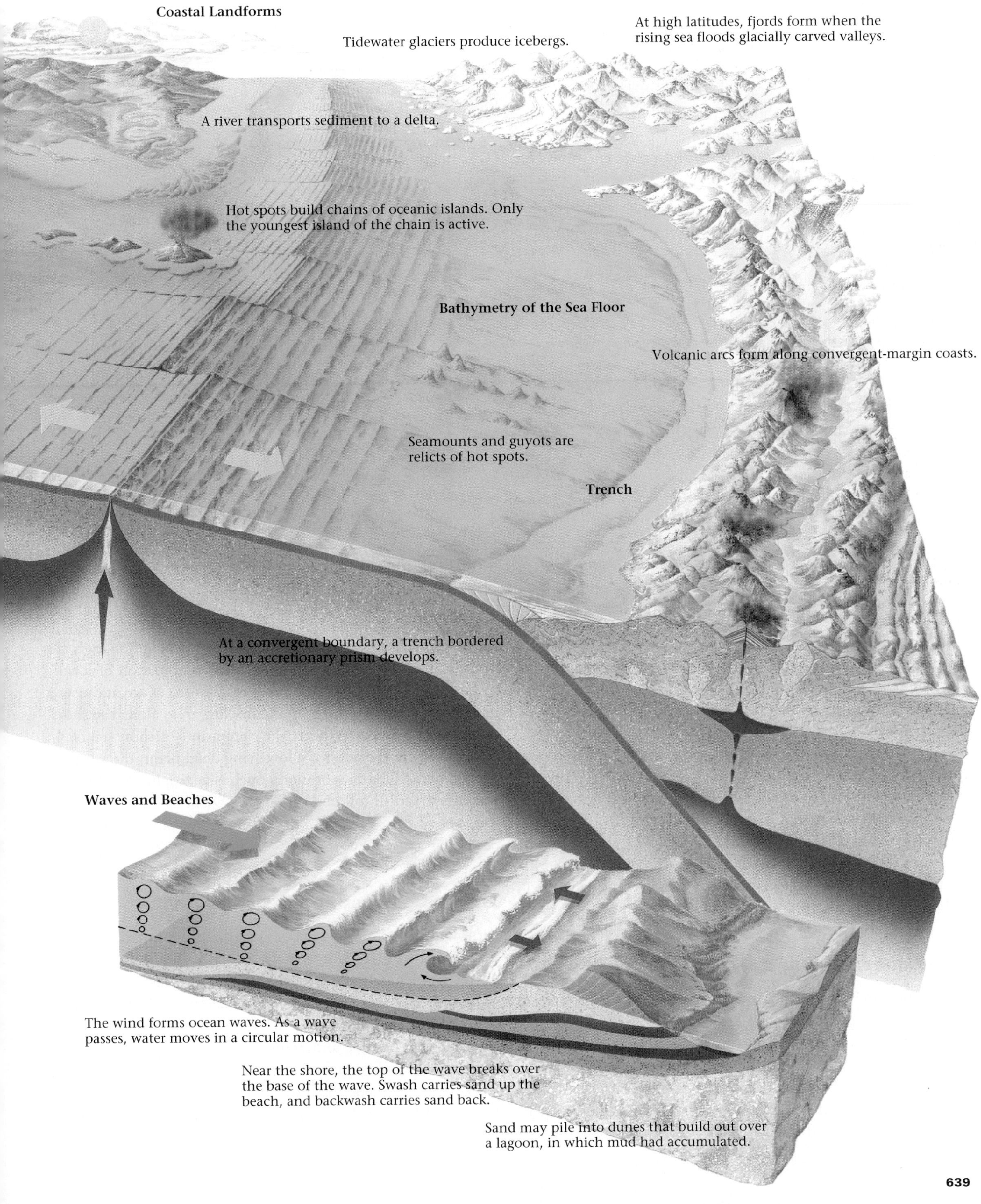

Coastal Landforms

Tidewater glaciers produce icebergs.

At high latitudes, fjords form when the rising sea floods glacially carved valleys.

A river transports sediment to a delta.

Hot spots build chains of oceanic islands. Only the youngest island of the chain is active.

Bathymetry of the Sea Floor

Volcanic arcs form along convergent-margin coasts.

Seamounts and guyots are relics of hot spots.

Trench

At a convergent boundary, a trench bordered by an accretionary prism develops.

Waves and Beaches

The wind forms ocean waves. As a wave passes, water moves in a circular motion.

Near the shore, the top of the wave breaks over the base of the wave. Swash carries sand up the beach, and backwash carries sand back.

Sand may pile into dunes that build out over a lagoon, in which mud had accumulated.

an erosional coast in a stormy environment. The climate also affects biological activity along coasts. For example, in the warm water of tropical climates, mangrove swamps flourish along the shore, and coral reefs form offshore. The reefs may build into a broad carbonate platform such as appears in the Bahamas today. In cooler climates, salt marshes develop, whereas in arctic regions, the coast may be a stark environment of lichen-covered rock and barren sediment.

Take-Home Message

- Tectonic activity can affect the character of coastlines. At emergent coasts, the land is rising relative to sea level, and at submergent coasts, it is sinking.
- Eustatic (global) sea-level change affects the position of the coastline.
- The sediment supply relative to the erosion rate determines whether beaches grow (to form an accretionary coast) or shrink, eventually resulting in a rocky coast.

THINK: What landforms occur at submergent coasts?

18.9 COASTAL PROBLEMS AND SOLUTIONS

Contemporary Sea-Level Changes

People tend to view a shoreline as a permanent entity. But in fact, shorelines are ephemeral geologic features. On a time scale of hundreds to thousands of years, a shoreline moves inland or seaward depending on whether relative sea level rises or falls or whether sediment supply increases or decreases. In places where sea level is rising today, shoreline towns will eventually be submerged. For example, the Persian Gulf now covers about twice the area that it did 4,000 years ago. And if present rates of sea-level rise along the East Coast of the United States continue, major coastal cities such as Washington, New York, Miami, and Philadelphia may be inundated within the next millennium (Fig. 18.28).

Hurricanes and Coastal Floods

Hurricanes are immense storms that grow over the waters of equatorial oceans. Some are born and die at sea, but some move onto land. Winds in hurricanes can exceed 250 km/h (155 mph) and can generate waves in excess of 15 m. Because air rises beneath a hurricane, atmospheric pressure (effectively, the weight of the overlying atmosphere) decreases substantially in the region beneath a hurricane. Without the downward

FIGURE 18.28 Future sea-level rise, due to melting of polar ice, would flood many coastal cities.

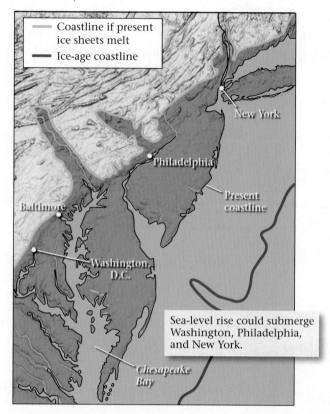

push of the air, the sea surface rises, so a bulge of high sea moves with the hurricane. When this combination of locally high sea level and powerful waves reaches the shore, it causes a calamity. The storm surge inundates low areas along the shore, and the waves batter both the shore and offshore reefs. In regions where the coast is a low-lying delta plain, the land can be submerged for days or more. Such catastrophic flooding has taken a dreadful toll on the Ganges Delta in Bangladesh and on the Gulf Coast of the United States. We discuss hurricanes and their consequences more fully in Chapter 20.

Beach Destruction—Beach Protection?

In a matter of hours, a storm—especially a hurricane—can radically alter a landscape that took centuries or millennia to form. The backwash of storm waves sweeps vast quantities of sand seaward, leaving the beach a skeleton of its former self. The surf submerges barrier islands and shifts them toward the lagoon. Waves and wind together rip out mangrove swamps and salt marshes and fragment coral reefs, thereby destroying the organic buffer that normally protects the coast and leaving it vulnerable to erosion for years to come. Of course, major storms also destroy human constructions:

erosion undermines shoreside buildings, causing them to collapse into the sea; wave impacts smash buildings to bits; and the storm surge—very high water levels created when storm winds push water toward the shore—floats buildings off their foundations (Fig. 18.29a).

But even less-dramatic events, such as the loss of river sediment, a gradual rise in sea level, a change in the shape of a shoreline, or the destruction of coastal vegetation, can alter the balance between sediment accumulation and sediment removal on a beach, leading to beach erosion (Fig. 18.29b). In some places, beaches retreat landward at rates of 1 to 2 m per year, forcing homeowners to pick up and move their houses. Even large lighthouses have been moved to keep them from washing away or tumbling down eroded headlands.

In many parts of the world, beachfront property has great value; but if a hotel loses its beach sand, it probably won't stay in business. Thus property owners often construct artificial barriers to protect their stretch of coastline or to shelter the mouth of a harbor from waves. These barriers alter the natural movement of sand in the beach system and thus change the shape of the beach, sometimes with undesirable results.

For example, people may build groins, concrete or stone walls protruding perpendicular to the shore, to prevent beach drift from removing sand (Fig. 18.30a). Sand accumulates on the updrift side of the groin, forming a long triangular wedge, but sand erodes away on the downdrift side. Needless to say, the property owner on the downdrift side doesn't appreciate this process.

A pair of walls called jetties may protect the entrance to a harbor (Fig. 18.30b, d). But jetties erected at the mouth of a river channel effectively extend the river into deeper water and thus may lead to the deposition of an offshore sandbar. Engineers may also build an offshore wall called a breakwater, parallel or at an angle to the beach, to prevent the full force of waves from reaching a harbor. With time, however, sand builds up in the lee of the breakwater and the beach grows seaward, clogging the harbor (Fig. 18.30c). To protect expensive shoreside homes, people build seawalls out of riprap (large stone or concrete blocks) or reinforced concrete on the landward side of the backshore zone (Fig. 18.30d, e). But seawalls reflect wave energy, which crosses the beach, back to sea, so this process increases the rate of erosion at the foot of the seawall. During a large storm, the seawall may be undermined so that it collapses (Fig. 18.31).

In some places, people have given up trying to decrease the rate of beach erosion and instead have worked to increase the rate of sediment supply. To do this, they pump sand from farther offshore, or truck in sand from elsewhere to replenish a beach. This procedure, called **beach nourishment**, can be hugely expensive and at best provides only a temporary fix, for the backwash and beach drift that removed

FIGURE 18.29 Examples of beach erosion.

(a) When Hurricane Ike hit Galveston, Texas, in September 2008, many beachfront homes were washed away.

(b) Wave erosion has completely removed the beach, and has started to erode a beach cliff, along the coast of Cape Cod.

FIGURE 18.30 Techniques used to preserve beaches.

Before　　　　　　　After

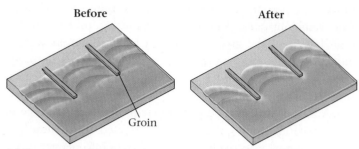

(a) The construction of groins may produce a sawtooth beach.

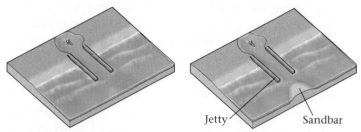

(b) Jetties extend a river farther into the sea, but may cause a sandbar to form at the end of the channel.

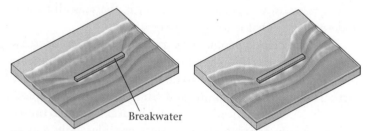

(c) A beach may grow seaward behind a breakwater.

(d) Groins spaced along a beach in southern England are intended to prevent loss of sediment.

(e) Riprap will slow erosion of a parking lot along a California beach.

the sand in the first place continue unabated as long as the wind blows and the waves break. Clearly, beach management remains a controversial issue, for beachfront properties are expensive; but the shore is, geologically speaking, a temporary feature whose shape can change radically with the next storm.

Pollution and the Destruction of Organic Coasts

Bad cases of beach pollution create headlines. Because of beach drift, garbage dumped in the sea in an urban area may drift along the shore and be deposited on a tourist beach far from its point of introduction. For example, hospital waste from New York City has washed up on beaches tens of kilometers to the south. Oil spills, from ships that flush their bilges or from tankers that have run aground or foundered in stormy seas, have contaminated shorelines at several places around the world. Another cause of oil spills are blowouts on offshore oil

rigs, as illustrated by the Deepwater Horizon catastrophe of 2010 (see Chapter 14).

Organic coasts, a manifestation of interaction between the physical and biological components of the Earth System, are particularly susceptible to changes in the environment. The loss of such landforms can increase a coast's vulnerability to erosion and, because they provide spawning grounds for marine organisms, can upset the food chain of the global ocean.

In wetlands and estuaries, sewage, chemical pollutants, and agricultural runoff cause havoc. Toxins settle along with clay and concentrate in the sediments, where they contaminate burrowing marine life and then move up the food chain. Fertilizers and sewage that enter the sea with runoff increase the nutrient content of water, creating algal blooms that absorb oxygen and therefore kill animal and plant life. Coastal wetlands face destruction by development—they have been filled or drained to be converted into farmland or suburbs, and they have been used as garbage dumps. In most parts of the world,

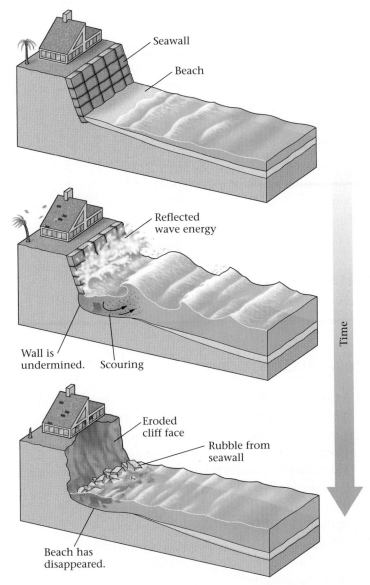

FIGURE 18.31 A seawall protects the sea cliff under most conditions, but during a severe storm the wave energy reflected by the seawall helps scour the beach. As a result, the wall may be undermined and collapse.

Labels in figure: Seawall · Beach · Reflected wave energy · Wall is undermined. · Scouring · Eroded cliff face · Rubble from seawall · Beach has disappeared. · Time

between 20 and 70% of coastal wetlands have been destroyed in the last century.

Reefs, which depend on the health of delicate coral polyps, can be devastated by even slight changes in the environment. Pollutants and hydrocarbons, for example, will poison the polyps. Organic sewage fosters algal blooms that rob water of dissolved oxygen and suffocate the coral. And suspended sediment introduced to coastal water either from agricultural runoff, or due to beach-nourishment projects, reduces the light, killing the algae that live in the coral, and clogs the pores that coral polyps use to filter water. Changes in water temperature or salinity caused by dumping wastewater from power plants into the sea or by global warming of the atmosphere also destroy reefs, for reef-building organisms are very sensitive to temperature changes. People can destroy reefs directly by dragging anchors across reef surfaces, by setting off explosives to kill fish, by touching reef organisms, or by quarrying reefs to obtain construction materials. In the last decade, marine biologists have noticed that reefs around the world have lost their color and died. This process, called **reef bleaching**, may be due to the removal or death of symbiotic algae, in response to the warming of seawater, or it may be a result of the dust carried by winds from desert or agricultural areas.

Take-Home Message

- The character and position of a coastline can change over time due to sea-level change, storms, and other factors.
- Removing sediment from beaches can be a problem for landowners. People try to prevent this process by building groins and other structures.
- Pollution and warming of seawater are affecting wetlands, beaches, and reefs.

THINK: Why do people build seawalls, and how may they be affected by waves?

Chapter Summary

- The landscape of the sea floor depends on the character of the underlying crust. Wide continental shelves form over passive-margin basins, whereas narrow continental shelves form over accretionary prisms. Continental shelves are cut by submarine canyons. Abyssal plains develop on old, cool oceanic lithosphere. Seamounts and guyots form above hot spots.

- The salinity, temperature, and density of seawater vary with location and depth.

- Water in the oceans circulates in currents. Surface currents are driven by the wind and are deflected in their path by the Coriolis effect. The vertical upwelling and downwelling of water create deep currents. Some of this movement is thermohaline circulation, a consequence of variations in temperature and salinity.

- Tides—the daily rise and fall of sea level—are caused by a tide-generating force. The largest contribution to this force comes from the gravitational pull of the Moon.

- Waves are caused by friction where the wind shears across the surface of the ocean. Water particles follow a circular motion in a vertical plane as a wave passes. Waves refract (bend) when they approach the shore because of frictional drag with the sea floor.

- Sand on beaches moves with the swash and backwash of waves. If there is a longshore current, the sand gradually moves along the beach and may extend outward from headlands to form sand spits.
- At rocky coasts, waves grind away at rocks, yielding such features as wave-cut benches and sea stacks. Some shores are wetlands, where marshes or mangrove swamps grow. Coral reefs grow along coasts in warm, clear water.
- The differences in coasts reflect their tectonic setting, whether sea level is rising or falling, sediment supply, and climate.
- To protect beach property, people build groins, jetties, breakwaters, and seawalls.
- Human activities have led to the pollution of coasts. Reef bleaching has become dangerously widespread.

GEOPUZZLE REVISITED

In the past 15 years, ocean currents have carried the armada of yellow duckies and blue turtles nearly around the world. Some of the toys have arrived on the shores of South America and Australia. Others floated through the Bering Strait and were carried by sea ice around the Arctic Ocean to the Atlantic. Their amazing voyage emphasizes how mobile ocean water is. Ocean currents serve as a major conveyor of heat in the Earth System. Also, as the toys have found out, wave action where the sea meets the land not only transports water but also transports sediment and debris, making it a tool that can sculpt coastlines of amazing beauty.

Guide Terms

abyssal plain (p. 612)

active continental margin (p. 613)

backwash (p. 625)

barrier island (p. 628)

bathymetry (p. 611)

beach (p. 627)

beach face (p. 627)

beach nourishment (p. 641)

berm (p. 627)

bioturbation (p. 628)

center of mass (p. 622)

centrifugal force (p. 622)

coast (p. 610)

coastal plain (p. 636)

coastal wetland (p. 634)

continental shelf (p. 612)

coral reef (p. 635)

Coriolis effect (p. 616)

currents (p. 616)

Earth-Moon system (p. 622)

emergent coasts (p. 637)

estuary (p. 633)

eustatic sea-level change (p. 636)

fjord (p. 633)

gravitational pull (p. 622)

guyot (p. 613)

gyre (p. 616)

lagoon (p. 628)

longshore current (p. 626)

longshore drift (p. 626)

oceanic plateau (p. 613)

offshore bar (p. 627)

organic coasts (p. 634)

passive continental margin (p. 612)

pelagic sediment (p. 613)

reef bleaching (p. 643)

rogue wave (p. 624)

sand spit (p. 627)

seamount (p. 613)

sea stack (p. 633)

submarine canyons (p. 613)

submergent coasts (p. 637)

surface currents (p. 616)

swash (p. 625)

thermohaline circulation (p. 617)

tidal reach (p. 620)

tide (p. 620)

tide-generating force (p. 620)

wave base (p. 624)

wave refraction (p. 626)

wave-cut bench (platform) (p. 631)

wave-cut notch (p. 631)

Review Questions

1. How much of the Earth's surface is covered by oceans? What proportion of the world's population lives near a coast?

2. Describe the typical topography of a passive continental margin, from the shoreline to the abyssal plain. How does the lithosphere beneath a passive margin differ from that beneath an abyssal plain?

3. How do the shelf and slope of an *active* continental margin differ from those of a *passive* margin?

4. Where does the salt in the ocean come from? How does the salinity in the ocean vary? How does the temperature of the oceans vary?

5. What factors control the direction of surface currents in the ocean? What is the Coriolis effect, and how does it affect oceanic circulation? Explain thermohaline circulation.

6. What causes the tides? Why does the range and reach of tides vary with location?

7. Describe the motion of water molecules in a wave. How does wave refraction cause longshore currents?

8. Describe the components of a beach profile.

9. How does beach sand migrate as a result of longshore currents? Explain the sediment budget of a coast.

10. Describe how waves affect a rocky coast, and how such coasts evolve.

11. What is an estuary? Why is it such a delicate ecosystem? What is the difference between an estuary and a fjord?

12. Discuss the different types of coastal wetlands. Describe the different kinds of reefs, and how a reef surrounding an oceanic island changes with time.

13. How do plate tectonics, sea-level changes, sediment supply, and climate change affect the shape of a coastline? Explain the difference between emergent and submergent coasts.

14. In what ways do people try to modify or "stabilize" coasts? How do the actions of people threaten the natural systems of coastal areas?

On Further Thought

15. In 1789, the crew of the HMS *Bounty* mutinied. Near Tonga, in the Friendly Islands (approximately 20° S and 175° W), the crew, led by Fletcher Christian, forced the ship's commanding officer, Lieutenant Bligh, along with those crewmen who remained loyal to Bligh, into a rowboat and set them adrift in the Pacific Ocean. The castaways, amazingly, survived; 47 days later, they landed at Timor (near Sumatra), 6,700 km to the west. Why did they end up where they did? Considering how long the journey took them, how fast were they moving?

16. A hotel chain would like to build a new beachfront hotel along a north-south-trending stretch of beach where a strong longshore current flows from south to north. The neighbor to the south has constructed an east-west-trending groin on the property line. Will this groin pose a problem? If so, what solutions could the hotel try?

17. Observations made during the last decade suggest that sea level is rising in response to global warming. This rise happens partly because water expands when it heats up and partly because glacial ice sheets in Greenland and elsewhere are melting. Much of southern Florida lies at elevations of less than 6 m above sea level. What changes will take place to the region as sea level rises? To answer, keep in mind the concepts of emergent and submergent coasts, and assume that southern Florida will continue to lie in the subtropical realm.

 For more resources, including animations, quizzes, and Norton's GeoTours, go to **wwnorton.com/studyspace**.

 If your instructor assigns exercises in SmartWork, log in at **smartwork.wwnorton.com**.

ANOTHER VIEW The Great Barrier Reef forms shoals off the coast of northeastern Australia. It is the largest reef on Earth.

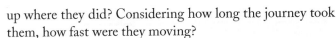

Reef

Mainland

50 km

CHAPTER 19

A Hidden Reserve: Groundwater

The land that is now a checkerboard of fields in this region near Phoenix, Arizona, once looked like the desert on the right. Groundwater, along with water brought from a river by the canal, keeps the fields green. Unfortunately, irrigation is using up the groundwater faster than nature can supply it.

GEOPUZZLE

In centuries past, the social center of a town was the town's well, consisting of a hole dug down several meters into the ground. Usually, townspeople would line the well with stone, and to obtain water they lowered a bucket down the well and then lifted up the water. Where did the water come from?

When the rain falls and enters the earth, when a pearl drops into the depth of the sea, you can dive in the sea and find the pearl, you can dig in the earth and find the water.

—Mei Yao-ch'en (Chinese Poet, 1002–1060)

19.1 INTRODUCTION

Imagine Mae Rose Owens's surprise when, on May 8, 1981, she looked out her window and discovered that a large sycamore tree in the backyard of her Winter Park, Florida, home had suddenly disappeared. It wasn't a particularly windy day, so the tree hadn't blown over—it had just vanished! When Owens went outside to investigate, she found that more than the tree had disappeared. Her whole backyard had become a deep, gaping pit. The pit continued to grow for a few days until finally it swallowed Owens's house and six other buildings, as well as the municipal swimming pool, part of a road, and several expensive Porsches in a car dealer's lot (Fig. 19.1a).

What had happened in Winter Park? The bedrock beneath the town consists of limestone, a fairly soluble rock. **Groundwater**, the liquid water that resides in sediment or rock under the surface of the Earth, had gradually dissolved the limestone, carving open rooms, or caverns, underground. On May 8, the roof of a cavern underneath Owens's backyard began to collapse, forming a circular depression called a **sinkhole** (Fig. 19.1b). The sycamore tree and the rest of the neighborhood simply dropped down into the sinkhole. It would have taken too much effort to fill in the sinkhole with soil, so the community allowed it to fill with water, and now it's a circular lake, the centerpiece of a pleasant municipal park. Similar

FIGURE 19.1 Development of sinkholes in central Florida.

(a) The Winter Park sinkhole, as seen from a helicopter.

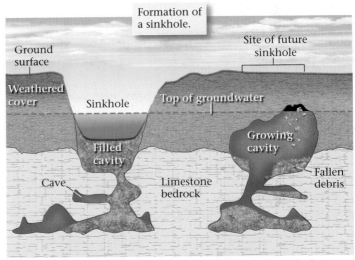

(b) As overburden slowly washes into underlying caves, a cavity forms. When the roof of this cavity collapses, a sinkhole forms (not to scale).

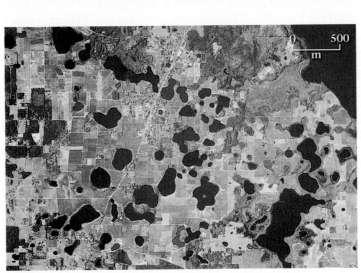

(c) A satellite view of central Florida's sinkholes.

(d) A small sinkhole in Brazil remains forested, while the surrounding land has been cleared.

lakes appear throughout central Florida (Fig. 19.1c); in fact, sinkholes pockmark the ground worldwide (Fig. 19.1d).

The Winter Park sinkhole is one of the more dramatic reminders that the significant quantities of water reside underground. Though we can easily see Earth's *surface water* (in lakes, rivers, streams, marshes, glaciers, and oceans) and *atmospheric water* (in clouds and rain), *groundwater* lies hidden beneath the surface. Nevertheless, groundwater accounts for about two-thirds of the Earth's freshwater resources used by homes, agriculture, and industry.

In this chapter, we first examine where subsurface water resides and discuss how this water flows and interacts with rock and sediment. We then look into how subsurface water has been affected by human activities and conclude with a survey of landscape features that originate in response to interactions between water and rock in the subsurface realm. Because of its value, many scientists specialize in hydrogeology, the study of groundwater, to help find and maintain supplies of clean groundwater.

Chapter Themes

By the end of this chapter, you should know . . .

- what groundwater is, where it comes from, and how it flows.
- the difference between porosity and permeability, and how these features form.
- the difference between aquifers and aquitards, and the nature of the water table.
- the various types of wells and springs that provide access to groundwater.
- how hot springs and geysers originate.
- how groundwater supplies can be damaged or depleted, and how to address these problems.
- how caves and karst landscapes originate and evolve.

19.2 WHERE DOES GROUNDWATER RESIDE?

The Underground Reservoir of the Hydrologic Cycle

As we saw in Interlude F, water moves among various reservoirs (the ocean, the atmosphere, rivers and lakes, groundwater, living organisms, soil, and glaciers) during the hydrologic cycle. Of the water that falls on land, some evaporates directly back into the atmosphere, some gets trapped in glaciers, and some becomes runoff that enters a network of streams and lakes that drains to the sea. The remainder sinks or percolates downward, by a process called infiltration, into the ground. In

effect, the upper part of the crust behaves like a giant sponge that can soak up water.

Of the water that does infiltrate, some descends only into the soil and wets the surfaces of grains and organic material making up the soil. This water, called **soil moisture**, later evaporates back into the atmosphere or gets sucked up by the roots of plants and transpires back into the atmosphere. But some water sinks deeper into sediment or rock, and along with water trapped in rock at the time the rock formed, makes up *groundwater*. Groundwater slowly flows underground for anywhere from a few months to tens of thousands of years before returning to the surface to pass once again into other reservoirs of the hydrologic cycle.

Porosity: Open Space in Rock and Regolith

The existence of liquid water underground requires the existence of open space underground. What is the nature of this space? When asked this question, many people picture networks of caves containing spooky lakes and inky streams. Indeed, as we see later in this chapter, caves locally do provide room for subsurface water to drip and flow. But contrary to popular belief, only a small proportion of underground water actually occurs in caves. Most groundwater resides in small open spaces or voids *within* sediment and within seemingly solid rock. The term **pore** refers to any open space within a volume of sediment, or within a body of rock, and the term **porosity** refers to the total amount of open space within a material, specified as a percentage. For example, if we say that a block of rock has 30% porosity, then 30% of the block consists of pores. Geologists distinguish between two basic kinds of porosity—primary and secondary.

Primary porosity develops during sediment deposition and during rock formation (Fig. 19.2a–c). It includes the pores between clastic grains that exist because the grains don't fit together tightly during deposition. Such pores survive the process of lithification if cementation is incomplete. Primary porosity in clastic sediment and clastic sedimentary rock depends on the size and sorting of the grains and on the amount of compaction. For example, the overall porosity of a poorly sorted sediment is less than that of a well-sorted sediment because smaller clasts can fill spaces between larger grains. Primary porosity decreases with increasing burial depth, because the weight of overburden pushes grains together. In chemical sedimentary rocks or biochemical sedimentary rocks, primary porosity develops because mineral crystals do not all grow snugly against their neighbors during precipitation. In crystalline igneous or metamorphic rock, primary porosity can persist if grains do not interlock perfectly. In fine-grained or glassy igneous rocks, primary porosity may consist of vesicles, relicts of air bubbles that were trapped during cooling. The amount of primary porosity ranges from less than 1% in crystalline igneous rocks and

FIGURE 19.2 Porosity—where water resides underground. Groundwater can fill microscopic pores (holes) and cracks between grains, or it can fill gaps formed along joints.

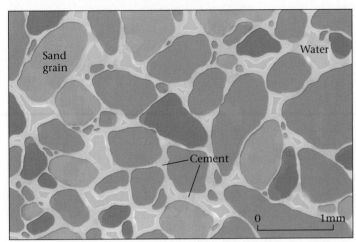

(a) Isolated pores in a sandstone occur in the spaces between grains. Water or air can fill pores.

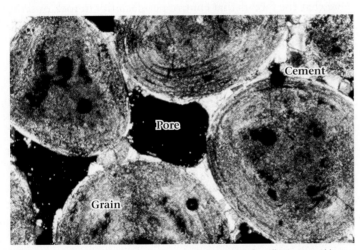

(b) A photograph of a thin section shows tiny pores in a limestone. Here, the grains consist of calcite spheres, and the cement of calcite crystals.

(d) Limestone outcrop on the coast of Ireland contains abundant fractures that provide secondary porosity.

These fractures have been enlarged by dissolution.

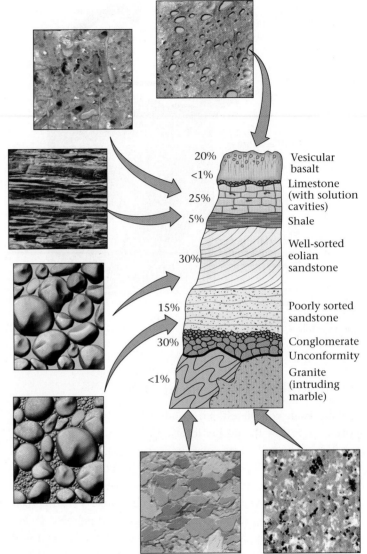

20%	Vesicular basalt
<1%	Limestone (with solution cavities)
25%	
5%	Shale
30%	Well-sorted eolian sandstone
15%	Poorly sorted sandstone
30%	Conglomerate
	Unconformity
<1%	Granite (intruding marble)

(c) Different kinds and amounts of porosity occur in different kinds of rocks. Well-sorted and poorly cemented sandstone contains high porosity, whereas granite contains low porosity.

metamorphic rocks to over 30% in well-sorted sands and poorly cemented sandstone.

Secondary porosity refers to new pore space produced in rocks some time after the rock first formed. For example, when rocks fracture, the opposing walls of the fracture do not fit together tightly, so narrow spaces remain in between. Thus, joints and faults may provide secondary porosity for water (Fig. 19.2d). Faulting may also produce breccia, a jumble of angular fragments—the space between the fragments provides another type of secondary pore space. Finally, as groundwater passes through rock, it may dissolve and remove some minerals, creating solution cavities that can serve as pores.

FIGURE 19.3   Permeability—the ability of water to flow through rock, from pore to pore or along joints.

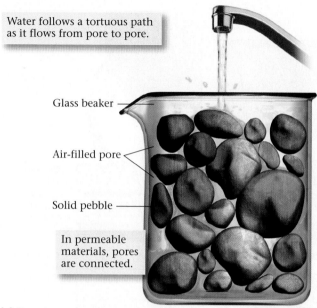

Water follows a tortuous path as it flows from pore to pore.

Glass beaker

Air-filled pore

Solid pebble

In permeable materials, pores are connected.

(a) Gravel contains pore space, because clasts don't fit together tightly. The connection of pores produces permeability, so water can flow through gravel.

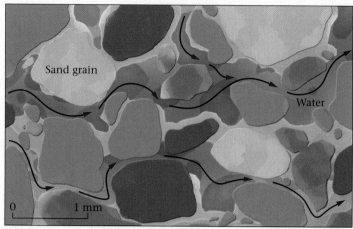

Sand grain

Water

0   1 mm

(b) Even at a microscopic scale, if there are passages connecting pores then water can flow through and the rock has permeability.

Permeability: The Ease of Flow

If solid rock completely surrounds a pore, the water in the pore cannot flow to another location. For groundwater to flow, pores must be linked by conduits (openings). The ability of a material to allow fluids to pass through an interconnected network of pores is a characteristic known as **permeability**. To develop an intuitive image of permeability, take a sturdy glass jar and fill it with gravel composed of rounded pebbles. If you look closely at this gravel through the side of the jar, you will see air-filled pores between the pebbles, because the pebbles don't fit together perfectly. If you pour water into the jar, the water can trickle between grains down to the bottom of the jar, where it displaces air and fills the pores (Fig. 19.3a). And it can move from pore to pore (Fig. 19.3b). Water flows easily through a permeable material. In contrast, water flows slowly or not at all through an impermeable material. The permeability of a material depends on several factors.

- *Number of available conduits*: As the number of conduits increases, permeability increases.
- *Size of the conduits*: More fluids can travel through wider conduits than through narrower ones.
- *Straightness of the conduits*: Water flows more rapidly through straight conduits than it does through crooked ones. In crooked channels, the distance a water molecule actually travels may be many times the straight-line distance between the two end points.

Note that the factors that control permeability in rock or sediment resemble those that control the ease with which traffic moves through a city. Traffic can flow quickly through cities with many straight, multilane boulevards, whereas it flows slowly through cities with only a few narrow, crooked streets.

Porosity and permeability are *not* the same feature. A material whose pores are isolated from each other can have high porosity but low permeability. For example, if you plug the bottom of a water-filled tube with a cork, the water remains in the tube even though cork is very porous; this is because woody cell walls separate adjacent pores in the cork and prevent communication between them, making cork impermeable. Vesicular basalt is an example of a rock that can have high porosity but low permeability.

Aquifers and Aquitards

With the concept of permeability in mind, hydrogeologists distinguish between **aquifers**, sediment or rocks that transmit water easily, and **aquitards**, sediment or rocks that do not transmit water easily and therefore retard the motion of water. The now rarely used term "aquiclude," refers to a geologic material that does not transmit water at all. We can refer to an aquifer that is not overlain by an aquitard as an **unconfined aquifer**. Water can infiltrate down into an unconfined aquifer from the Earth's surface, and groundwater can rise through one to reach the Earth's surface. In contrast, we refer to an aquifer that is overlain by an aquitard as a **confined aquifer**—its water is isolated from the ground surface (Fig. 19.4).

Hydrogeologists distinguish among many diverse types of aquifers on the basis of the nature of the material making up the aquifer and on the configuration of the aquifer. To get a sense of this diversity, let's look at examples of specific aquifers that provide important groundwater supplies in the United States. Each aquifer has a name, based on the region in which it occurs or on the stratigraphic unit that hosts the groundwater.

- *Mahomet aquifer, Illinois*: When glaciers melted during the Pleistocene Ice Age, meltwater streams carried coarse gravel southward. This gravel filled valleys that had previously been cut into bedrock (Fig. 19.5a). Younger glacial sediment later buried the filled valleys. The porous and permeable gravel of the largest buried valley is the Mahomet aquifer, named for a nearby town; this aquifer provides water to over a million people.

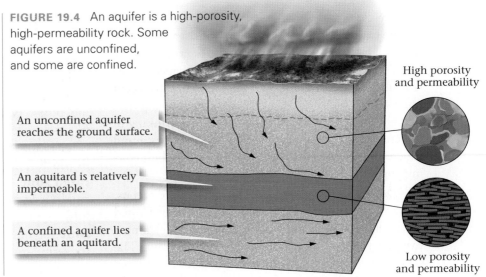

FIGURE 19.4 An aquifer is a high-porosity, high-permeability rock. Some aquifers are unconfined, and some are confined.

An unconfined aquifer reaches the ground surface.

An aquitard is relatively impermeable.

A confined aquifer lies beneath an aquitard.

High porosity and permeability

Low porosity and permeability

FIGURE 19.5 Examples of different types of aquifers. The aquifers are stippled.

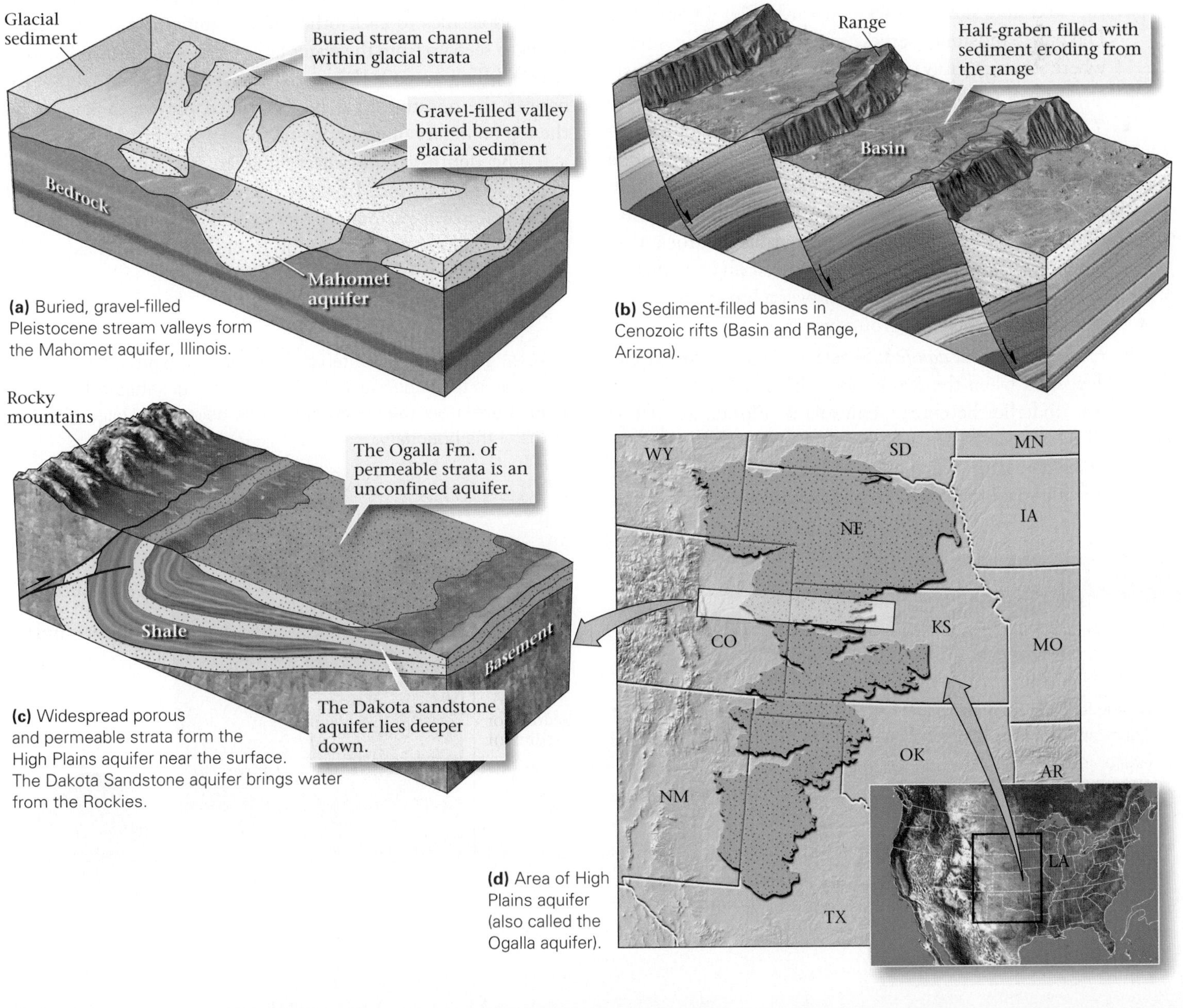

Glacial sediment

Buried stream channel within glacial strata

Gravel-filled valley buried beneath glacial sediment

Bedrock

Mahomet aquifer

(a) Buried, gravel-filled Pleistocene stream valleys form the Mahomet aquifer, Illinois.

Range

Half-graben filled with sediment eroding from the range

Basin

(b) Sediment-filled basins in Cenozoic rifts (Basin and Range, Arizona).

Rocky mountains

The Ogalla Fm. of permeable strata is an unconfined aquifer.

Shale

Basement

The Dakota sandstone aquifer lies deeper down.

(c) Widespread porous and permeable strata form the High Plains aquifer near the surface. The Dakota Sandstone aquifer brings water from the Rockies.

WY SD MN

NE IA

CO KS MO

NM OK AR

TX LA

(d) Area of High Plains aquifer (also called the Ogalla aquifer).

- *Phoenix Basin aquifer, Arizona*: The city of Phoenix resides in the Basin and Range Province, a continental rift. During rifting, downward slip of crustal blocks on normal faults produced deep, wedge-shaped basins that filled with gravels and sands eroded from adjacent ranges (**Fig. 19.5b**). These sediments serve as aquifers that provide groundwater for irrigation and drinking in Phoenix. Similar aquifers elsewhere in the Basin and Range Province provide water for the region's burgeoning population.

- *The Dakota Sandstone aquifer*: During the Cretaceous, rivers flowing from mountains in the western United States dumped sediment into an inland sea, forming a vast sheet of sand that later was buried by mud. The sand layer lithified to become the Dakota Sandstone, an aquifer, and overlying beds became shale aquitards. Continued mountain building warped the strata into a syncline. Water infiltrates into the western limb of the fold and slowly flows eastward (**Fig. 19.5c**). The Dakota Sandstone serves as a groundwater source where it returns toward the ground surface on the eastern limb of the syncline.

- *The High Plains aquifer*: Erosion of the Rocky Mountains 3 million years ago led to the deposition of huge alluvial fans over what is now the High Plains region. The resulting deposits comprise the Ogallala Formation, a sheet of porous and permeable bedrock that remains just below the ground surface (**Fig. 19.5d**). Water can infiltrate this aquifer over a broad area, to be stored underground as groundwater.

- *Floridan limestone aquifers*: Sheets of Cenozoic limestone, deposited when the region was submerged by a shallow sea, underlie the entire peninsula of Florida as well as adjacent areas of Georgia and Alabama. Fractures in the limestone have been widened by dissolution and now provide spaces that hold large quantities of groundwater.

Take-Home Message

- Porosity refers to the open space in rock or sediment, whereas permeability characterizes the degree to which pores are interconnected.
- An aquifer has high porosity and permeability, so it contains significant quantities of groundwater that can flow relatively easily.
- Water can percolate from the ground surface down to unconfined aquifers; confined aquifers are separated from the surface by an aquitard.
- Geologists distinguish among many types of aquifers.

THINK: Why do poorly cemented sandstones make good aquifers?

19.3 GROUNDWATER AND THE WATER TABLE

Infiltrating water can enter permeable sediment and bedrock by percolating along cracks and through conduits connecting pores. Nearer the ground surface, water only partially fills pores, leaving some space that remains filled with air (**Fig. 19.6**). In technical jargon, the region of the subsurface in which water only partially fills pores is called the **unsaturated zone**, or the vadose zone; the water that resides in this region is vadose-zone water. Deeper down, water completely fills, or saturates, the pores. This region is the **saturated zone**, or the phreatic zone. In a strict sense, geologists use the term "groundwater" specifically for subsurface water in the saturated zone, where water completely fills pores.

The term **water table** refers to the horizon that separates the unsaturated zone above from the saturated zone below (Fig. 19.6). Typically, surface tension, the electrostatic attraction of water molecules to each other and to mineral surfaces, causes water to seep up from the water table (just as water rises in a thin straw), filling pores in the **capillary fringe**, a thin layer at the base of the unsaturated zone. Note that the water table forms the top boundary of groundwater in an unconfined aquifer.

The depth of the water table in the subsurface varies greatly with location. In some places, the water table defines the surface of a permanent stream, lake, or marsh, and thus effectively lies above the ground level (**Fig. 19.7a**). Elsewhere,

FIGURE 19.6 The water table is the top of the groundwater reservoir in the subsurface. It separates the unsaturated (vadose) zone above, from the unsaturated zone below. A capillary fringe forms at the boundary.

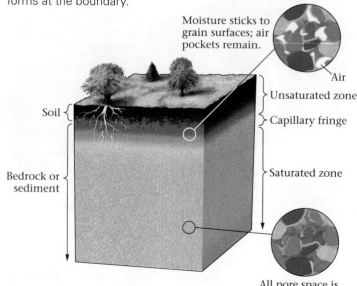

Moisture sticks to grain surfaces; air pockets remain.

Air

Unsaturated zone

Capillary fringe

Soil

Saturated zone

Bedrock or sediment

All pore space is filled with water.

FIGURE 19.7 The depth of the water table depends on the water supply from above.

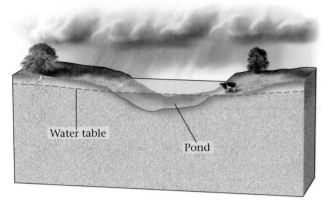

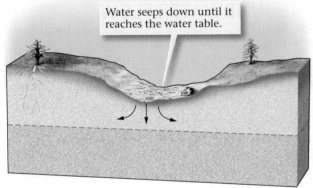

(a) Where the water table lies close to the ground surface, ponds remain filled—the water table is the surface of the pond.

Water seeps down until it reaches the water table.

(b) In dry regions, the water table sinks deep below the surface. Water that collects temporarily in low areas sinks into the subsurface. Temperate plants die, because their roots can't reach the water table.

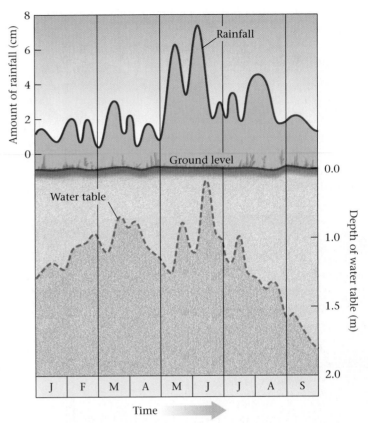

(c) This graph illustrates how the water table rises and falls depending on the amount of rainfall.

the water table lies hidden below the ground surface. In humid regions, it typically lies at depths of only meters to tens of meters. But in arid regions, it may lie hundreds of meters below the surface. Thus permanent lakes or streams cannot exist in arid regions, unless fed by a water supply from elsewhere, because water can infiltrate down through the lake or stream bed to the water table far below (**Fig. 19.7b**). Rainfall rates affect the water table depth in a given locality (**Fig. 19.7c**). Specifically, the water table drops during the dry season and rises during the wet season. Streams or ponds that hold water during the wet season may dry up during the dry season.

We've defined the water table as the *top* of groundwater in the crust. How deep down can we find groundwater? Recent research shows that liquid groundwater does circulate in basement igneous and metamorphic rock, perhaps to depths of over 15 km. Most of this deep circulation takes place along fractures. The ultimate base of groundwater in crust may be taken as the depth where metamorphism and plastic deformation currently take place. At this depth, plastic deformation closes up pore space, and water becomes involved in metamorphic reactions and no longer exists as a free liquid. The depth of this metamorphic horizon at a given location depends on the geothermal gradient.

Topography of the Water Table

In hilly regions, if the subsurface has relatively low permeability, the water table is not a planar surface. Rather, its shape mimics, in a subdued way, the shape of the overlying topography (**Fig. 19.8a**). This means that the water table lies at a higher elevation beneath hills than it does beneath valleys. But the relief (the vertical distance between the highest and lowest elevations) of the water table is not as great as that of the overlying land, so the surface of the water table tends to be smoother than that of the landscape.

At first thought, it may seem surprising that the elevation of the water table varies as a consequence of ground-surface topography. After all, when you pour a bucket of water into a

FIGURE 19.8 Factors that influence the position of the water table.

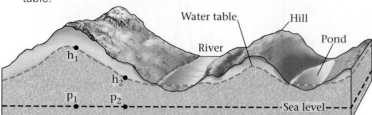

(a) The shape of a water table beneath hilly topography. Point h_1 on the water table is higher than Point h_2, relative to a reference elevation (sea level). The pressure at p_1 is, therefore, more than the pressure at p_2.

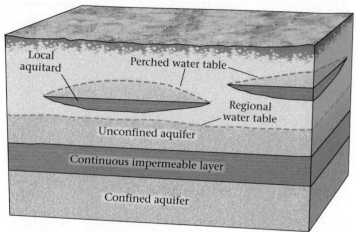

(b) A perched water table occurs where a mound of groundwater becomes trapped above a localized aquitard that lies above the regional water table.

pond, the surface of the pond immediately adjusts to remain horizontal. The elevation of the water table varies because groundwater moves so slowly through rock and sediment that it cannot quickly assume a horizontal surface. When it rains on a hill and water infiltrates down to the water table, the water table rises a little. When it doesn't rain, the water table sinks slowly, but so slowly that when rain falls again, the water table rises before it has had time to sink very far.

Perched Water Tables

In some locations, layers of strata are discontinuous, meaning that they pinch out at their sides. As a result, lens-shaped layers of impermeable rock (such as shale) may lie within a thick aquifer. A mound of groundwater accumulates above such aquitard lenses. The result is a **perched water table**, a groundwater top surface that lies above the regional water table because an underlying lens of impermeable rock or sediment prevents the groundwater from sinking down to the regional water table (Fig. 19.8b).

- Groundwater is water that resides underground in the pores of rock or sediment.
- Below the water table, in the saturated zone, water fills available pore space; above the water table, pores are partially or entirely filled with air.
- The water table is higher in regions with wetter climates than in regions with drier climates; it can rise or fall depending on the amount of rainfall.
- In hilly areas, the water table itself has topography; where the water table intersects valleys or depressions, streams or lakes form.

THINK: Why is the water table higher beneath hills than beneath valleys?

19.4 GROUNDWATER FLOW

Groundwater Flow Paths

What happens to groundwater over time? Does it just sit, unmoving, like the water in a stagnant puddle, or does it flow and eventually find its way back to the surface? Countless measurements confirm that groundwater enjoys the latter fate—groundwater indeed flows, and in some cases it moves great distances underground. In this section we examine factors that drive groundwater flow and determine the path that this flow follows. Then, in the next section, we examine the rate (velocity) at which groundwater moves.

In the unsaturated zone—the region between the ground surface and the water table—water percolates down, like the water passing through a drip coffee maker, for this water moves only in response to the downward pull of gravity. But in the zone of saturation—the region below the water table—water flow is more complex, for in addition to the downward pull of gravity, water responds to differences in pressure. Pressure can cause groundwater to flow sideways, or even upward. (If you've ever watched water spray from a fountain, you've seen pressure pushing water upward.) Thus, to understand the nature of groundwater flow, we must first understand the origin of pressure in groundwater. For simplicity, we'll consider only the case of groundwater in an unconfined aquifer.

Pressure in groundwater at a specific point underground is caused by the weight of all the overlying water from that point up to the water table. (The weight of overlying rock does not contribute to the pressure exerted on groundwater, for the contact points between mineral grains bear the rock's weight.) Thus a point at a greater depth below the water table feels more pressure than does a point at lesser depth. If the water table is horizontal, the pressure acting on an imaginary horizontal reference plane at a specified depth below the water table is the same everywhere. But if the water table is not horizontal, as shown in Figure 19.8a, the pressure at points on a horizontal reference plane at depth changes with location. For example, the pressure acting at point p_1, which lies below the hill in Figure 19.8a, is greater than the pressure acting at point p_2, which lies below the valley, even though both p_1 and p_2 are at the same elevation (sea level, in this case).

Both the elevation of a volume of groundwater and the pressure within the water provide energy that, if given the chance, will cause the water to flow. Physicists refer to such stored energy as potential energy. (To understand why elevation provides potential energy, imagine a bucket of water high on a hill; the water has potential energy due to Earth's gravity, which will cause it to flow downslope if the bucket were suddenly to rupture. To understand why pressure provides potential energy, imagine a water-filled plastic bag sitting on a table; if you puncture the bag and then squeeze the bag to exert pressure, water spurts out.) The potential energy available to drive the flow of a given volume of groundwater at a location is called the **hydraulic head**. To measure the hydraulic head at a point in an aquifer, hydrogeologists drill a vertical hole down to the point and then insert a pipe in the hole. The height above a reference elevation (e.g., sea level) to which water rises in the pipe represents the hydraulic head—water rises higher in the pipe where the head is higher. As a rule, *groundwater flows from regions where it has higher hydraulic head to regions where it has lower hydraulic head*. This statement generally implies that groundwater regionally flows from locations where the water table is higher to locations where the water table is lower.

Hydrogeologists have calculated how hydraulic head changes with location underground, by taking into account both the effect of gravity and the effect of pressure. These calculations reveal that groundwater flows along concave-up curved paths, as illustrated in cross section (**Fig. 19.9a**). (Specialized books on hydrogeology provide the details of why flow paths have such a specific shape.) These curved paths eventually take groundwater from regions where the water table is high (under a hill) to regions where the water table is low (below a valley), but because of flow-path shape, some groundwater may flow deep down into the crust along the first part of its path and then may flow back up, toward the ground surface, along the final part of its path. The location where water enters the ground (i.e., where the flow direction has a downward trajectory) is called the **recharge area**, and the location where groundwater flows back up to the surface is called the **discharge area** (Fig. 19.9a).

Groundwater following short paths close to the Earth's surface travels tens of meters to a few kilometers before returning to the surface. Such movement constitutes *local* flow. Locally flowing groundwater has a residence time (i.e., stays underground) for only hours to weeks. Groundwater following

FIGURE 19.9 The flow of groundwater.

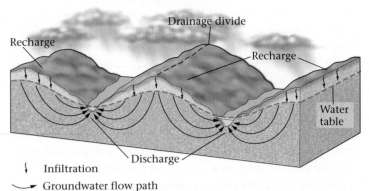

(a) Groundwater flows from recharge areas to discharge areas. Typically, the flow follows curving paths.

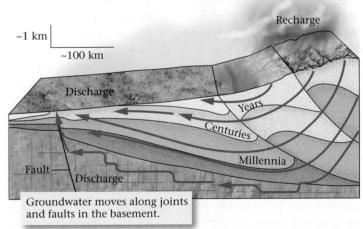

(b) Some groundwater may flow hundreds of kilometers, across regional sedimentary basins. The transit time depends on the flow path. Deeper flow paths take longer.

paths of several kilometers to tens of kilometers constitutes *intermediate* flow and has a residence time of weeks to years. Groundwater following paths that carry it hundreds of kilometers across a sedimentary basin constitutes *regional* flow and stays underground for centuries to millennia (**Fig. 19.9b**).

Rates of Groundwater Flow: Darcy's Law

Flowing water in an ocean current moves at up to 3 km per hour (over 26,000,000 m per year), and water in a steep river channel can reach speeds of up to 30 km per hour (over 260,000,000 m per year). In contrast, groundwater moves at less than a snail's pace. Hydrogeologists can measure the rate (velocity) of groundwater flow in a region by "tagging" water with dye or a radioactive element. They inject the tracer in one well and time how long it takes for the tracer to reach another well. They have found that typical rates range between 0.01 and 1.4 m per day (about 4 to

500 m per year). Groundwater moves much more slowly than surface water, for two reasons. First, groundwater moves by percolating through a complex, crooked network of tiny conduits; it must travel a much greater distance than it would if it could follow a straight path. Second, friction between groundwater and conduit walls slows down the water flow, and the surface tension of water may slow its escape from pores.

The rate at which groundwater flows at a given location depends on the permeability of the material containing the groundwater; groundwater flows faster in a more permeable material than it does in a less permeable material. Rate also depends on the **hydraulic gradient**, the change in hydraulic head per unit of distance between two locations as measured along the flow path. Recall that groundwater flows from a region where the hydraulic head is higher (due to higher elevation and/or greater pressure) to a region where the hydraulic head is lower. If there is a large difference in the hydraulic head over a given distance, then a greater amount of energy drives the flow, so the flow is faster.

To calculate the hydraulic gradient, we simply divide the difference in hydraulic head between two points by the distance between the two points as measured along the flow path. This can be written as a formula:

$$\text{hydraulic gradient} = \Delta h / j$$

where Δh is the difference in head (given in meters or feet, because head can be represented as an elevation), and j is the distance between the two points (Fig. 19.10). A hydraulic gradient exists anywhere that the water table has a slope. Typically, the slope of the water table is so small that the path length is almost the same as the horizontal distance between two points. So *the hydraulic gradient is roughly equivalent to the slope of the water table.*

FIGURE 19.10 The level to which water rises in a drillhole is the hydraulic head (h). The hydraulic gradient (HG) is the difference in head divided by the length of the flow path.

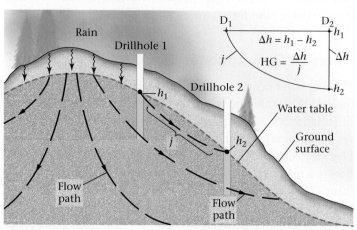

With an understanding of the controls on the rate of groundwater flow, we can consider the practical problem of determining the volume of water that will pass through an area underground during a specified time. In 1856, a French engineer named Henry Darcy addressed this problem because he wanted to find out whether the city of Dijon in central France could supply its water needs from groundwater. Darcy carried out a series of experiments in which he measured the rate of water flow through sediment-filled tubes tilted at various angles. Darcy found that the volume of water that passed through a specified area in a given time, a quantity he called the *discharge* (Q), can be calculated from the equation

$$Q = \text{K}(\Delta h / j)A$$

where $\Delta h / j$ is the hydraulic gradient, A is the area of porous rock or sediment through which the water is passing as measured in a plane perpendicular to flow direction, and K is a number called the hydraulic conductivity. The hydraulic conductivity takes into account permeability as well as gravity, the fluid's viscosity (stickiness), and the fluid's density. Hydrogeologists refer to this equation as **Darcy's law**.

To simplify discussion, geologists sometimes represent Darcy's law as follows:

$$\text{discharge } \alpha \text{ (water table slope)} \times \text{(permeability)}$$

where α means "depends on." In words, Darcy's law states that groundwater flows faster through very permeable rocks than through impermeable rocks and that it flows faster where the water table has a steep slope than where the water table has a shallow slope. It is no surprise, therefore, that groundwater moves very slowly (0.5–3.0 cm per day) through a well-cemented bed of the Dakota Sandstone, but moves relatively quickly (over 15 cm per day) through a steeply sloping layer of unconsolidated gravel on the side of a hill.

Take-Home Message

- Groundwater flows slowly from recharge areas to discharge areas, driven by pressure.
- Geologists represent the pressure in groundwater at a location by specifying the hydraulic head (the elevation to which groundwater would rise in an open pipe penetrating the water table).
- Groundwater flows at rates of generally less than 4 m to 0.5 km/year.
- Darcy's law states that discharge (the amount of groundwater passing through an area in a given time) depends on the hydraulic gradient (~slope of the water table) and the permeability.

THINK: Does steepening the water table increase or decrease the rate of groundwater flow?

19.5 TAPPING GROUNDWATER SUPPLIES

We can obtain groundwater at wells or springs. **Wells** are holes that people dig or drill to obtain water. **Springs** are natural outlets from which groundwater flows. Wells and springs provide welcome sources of water but must be treated with care if they are to last.

Wells

In an **ordinary well**, the base of the well penetrates an aquifer below the water table (Fig. 19.11a,b). Water from the pore space in the aquifer seeps into the well and fills it to the level of the water table. Drilling into an aquitard, or into rock that lies above the water table, will not supply water, and thus yields a *dry* well. Some ordinary wells are seasonal and function only during the rainy season, when the water table rises. During the dry season, the water table lies below the base of the well, so the well is dry.

Ideally, we would like an ordinary well to be as shallow as possible (to decrease the cost of digging). So hydrogeologists search for particularly porous and permeable aquifers in which the water table lies near the surface. Contrary to legend, a dowser cannot find water simply by using a forked stick—when dowsers do strike water, either they have enjoyed dumb luck or they have studied published data about the water table in the area prior to their search.

To obtain water from an ordinary well, you either pull water up in a bucket or pump the water out (Fig. 19.11a, b). As long as the rate at which groundwater fills the well exceeds the rate at which water is removed, the level of the water table near the well remains about the same. However, if users pump water out of the well too fast, then the water table sinks down around the well, in a process called drawdown, so that the water table becomes a downward-pointing, cone-shaped surface called a **cone of depression** (Fig. 19.12). Drawdown may cause shallower wells that have been drilled nearby to run dry.

An **artesian well**, named for the province of Artois in France, penetrates confined aquifers in which water is under enough pressure to rise on its own to a level above the surface of the aquifer. If this level lies below the ground surface, the well is a *nonflowing* artesian well. But if the level lies above the ground surface, the well is a *flowing* artesian well, and water actively fountains out of the ground (Fig. 19.13a). Artesian wells occur in special situations where a confined aquifer lies beneath a sloping aquitard.

We can understand why artesian wells exist if we look first at the configuration of a city water supply (Fig. 19.13b).

FIGURE 19.11 The nature of ordinary wells.

(a) At a traditional ordinary well, villagers obtain water by lifting it with a bucket.

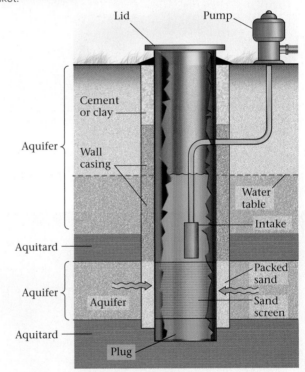

(b) A modern ordinary well sucks up water with an electric pump. The packed sand filters the water.

Water companies pump water into a high tank that has a significant hydraulic head relative to the surrounding areas. If the water were connected by a water main to a series of vertical pipes, pressure caused by the elevation of the water in the high tank would make the water rise in the pipes until it reached an imaginary surface, called a potentiometric surface, that lies above the ground. This pressure drives water through water mains to household water systems without requiring pumps. In an artesian system, water enters a tilted, confined aquifer

FIGURE 19.12 Pumping groundwater at a normal well affects the water table.

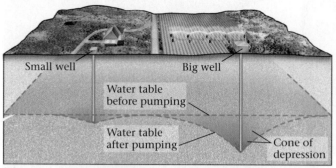

(a) If groundwater is extracted faster than it can be replaced, a cone of depression forms around the well.

(b) Pumping by the big well may lower the water table enough to cause the nearby small well to go dry.

FIGURE 19.13 Artesian wells, where water rises from the aquifer without pumping.

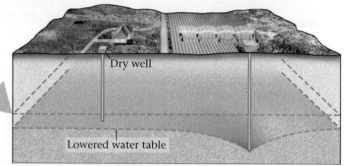

(a) A flowing artesian well in a Missouri field.

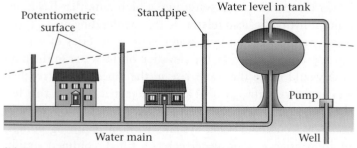

(b) The configuration of a city water supply. Water rises in vertical pipes up to the level of the potentiometric surface.

that intersects the ground in the hills of a high-elevation recharge area (**Fig. 19.13c**). The confined groundwater flows down to the adjacent plains, which lie at a lower elevation. The potentiometric surface to which the water would rise, were it not confined, lies above this aquifer. Pressure in the confined aquifer pushes water up a well. Where the potentiometric surface lies underground, the well will be a nonflowing artesian well, but where the surface lies above the ground, the well will be a flowing artesian well.

Springs

More than one town has grown up around a spring, a place where groundwater naturally flows or seeps onto the Earth's surface, for springs can provide fresh, clear groundwater for drinking or irrigation, without the expense of drilling or digging. Some springs spill water onto dry land. Others bubble up through the bed of a stream or lake. Springs form under a variety of conditions:

- where the ground surface intersects the water table in a discharge area (**Fig. 19.14a**); such springs typically occur in valley floors, where they may add water to lakes or streams.
- where flowing groundwater collides with a steep, impermeable barrier, and pressure pushes it up to the ground along the barrier (**Fig. 19.14b**).
- where a perched water table intersects the surface of a hill (**Fig. 19.14c**).
- where downward-percolating water runs into a relatively impermeable layer and migrates along the top surface of the layer to a hillslope (**Fig. 19.14d**).
- where a network of interconnected fractures channels groundwater to the surface of a hill (**Fig. 19.14e**).
- where the ground surface intersects a natural fracture (joint) that taps a confined aquifer in which the

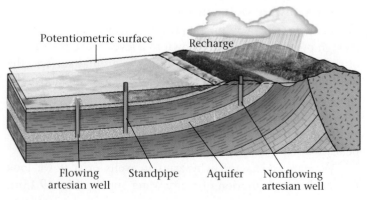

(c) The configuration of a regional artesian system.

FIGURE 19.14 Geological settings in which springs form.

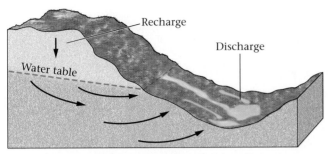

(a) Groundwater reaches the ground surface in a discharge zone.

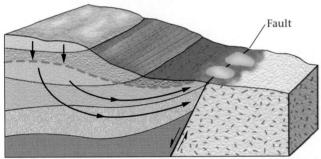

(b) Where groundwater reaches an impermeable barrier, it rises.

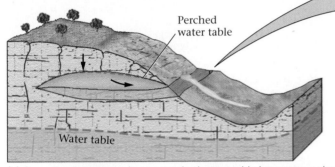

(c) Groundwater seeps where a perched water table intersects a slope.

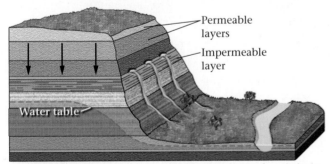

(d) Groundwater seeps out of a cliff face at the top of a relatively impermeable bed.

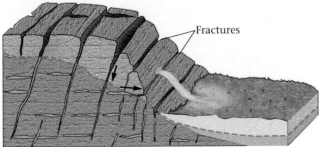

(e) A network of interconnected fractures channels water to the surface of a hill.

A spring on the wall of the Grand Canyon

pressure is sufficient to drive the water to the surface; such an occurrence is an **artesian spring**.

Springs can provide water in regions that would otherwise be uninhabitable. For example, oases in deserts may develop around a spring. An **oasis** is a wet area, where plants can grow, in an otherwise bone-dry region (see Box 19.1).

Take-Home Message

- A well is a human-made conduit that provides access to groundwater. Ordinary wells simply penetrate the water table; in artesian wells, water rises under natural pressure.
- Pumping groundwater too fast produces a cone of depression and can lower the water table.
- Springs are natural conduits through which groundwater emerges at the Earth's surface; they can exist for many reasons.

THINK: Why do some springs emerge along the side of a hill?

BOX 19.1

Oases

The Sahara of northern Africa is now one of the most barren and desolate places on Earth, for it lies in a climatic belt where rain seldom falls. But it wasn't always that way. During the last ice age, when glaciers covered parts of northern Europe on the other side of the Mediterranean, the Sahara enjoyed a more temperate climate, and the water table was high enough that permanent streams dissected the landscape. In fact, using ground-penetrating radar, geologists can detect relict beds of these streams today—they look like ghostly valleys beneath the sand. When the climate warmed after the ice age, rainfall diminished, the water table sank below the floors of most stream beds, and the streams dried up.

Today, the water of the Sahara region lies locked in a vast underground aquifer composed of porous sandstone. Recharge into this aquifer comes from highlands bordering the desert, from occasional downpours, and from the Nile River (particularly Lake Nasser, impounded by the Aswan High Dam). In general, the water of the aquifer can be obtained only by drilling deep wells, but locally, water spills out at the surface—either because folding brings the aquifer particularly close to the ground so that valley floors intersect the water table, or because artesian pressure pushes groundwater up along joints of faults (**Fig. Bx19.1a**). In either case, the aquifer feeds springs that quench the thirst of plants and create an **oasis**, an island of green in the sand sea (**Fig. Bx19.1b**). Oases became important stopping points along caravan routes, allowing both people and camels to replenish water supplies.

In some oases, people settled and used the groundwater to irrigate date palms and other crops. For example, the

Bahariya Oasis, about 400 km southwest of Cairo, Egypt, hosted a town of perhaps 30,000 between 300 B.C.E. and 300 C.E. During that time, the water table lay only 5 m below the ground and could easily be reached by shallow wells. Today, as a result of changing climates and centuries of use, the water table lies 1,500 m below the ground, almost out of reach. Bahariya's glorious past came to light in 1996, quite by accident. A man was riding his donkey in the desert near the oasis when the ground beneath the donkey suddenly caved in. The man had inadvertently opened the roof into a tomb filled with over 150 mummies, along with thousands of well-preserved artifacts. The site has since come to be known as the Valley of the Mummies.

FIGURE Bx19.1   The formation of oases in deserts.

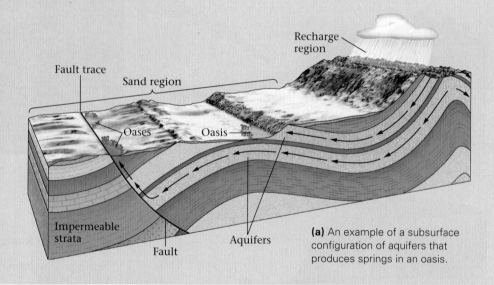

(a) An example of a subsurface configuration of aquifers that produces springs in an oasis.

(b) A small oasis in the Sahara Desert.

19.6 **HOT SPRINGS AND GEYSERS**

Hot springs, springs that emit water ranging in temperature from about 30° to 104°C, are found in two geologic settings. First, they occur where very deep groundwater, heated in warm bedrock at depth, flows up to the ground surface. This water brings heat with it as it rises. Such hot springs form in places where faults or fractures provide a high-permeability conduit for deep water, or where the water emitted in a discharge region followed a trajectory that first carried it deep into the crust. Second, hot springs develop in **geothermal regions**, places where volcanism currently takes place or has occurred recently, so that magma and/or very hot rock resides close to the Earth's surface. Hot groundwater dissolves minerals from rock that it passes through (Fig. 19.15), because water becomes a more effective solvent when hot. So water emitted at hot springs can contain a high concentration of dissolved minerals. People use the water emitted at hot springs to fill relaxing mineral baths (Fig. 19.16a). Natural pools of geothermal water may become brightly colored—the gaudy greens, blues, and oranges of these pools come from

FIGURE 19.15 Geothermal waters can occur in volcanic regions, where magma heats groundwater from below.

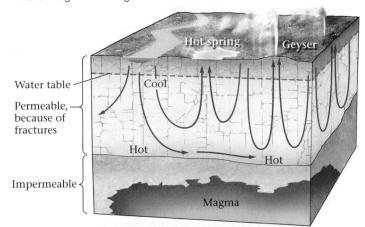

FIGURE 19.16 Examples of hot springs in volcanic provinces.

(a) Hot springs in Iceland, warmed by magma below, attract tourists from around the world.

(b) Colorful bacteria- and archaea-laden pools, Yellowstone National Park, Wyoming.

(c) Mudpots form where boiling water mixes with ash. The ash changes into clay and forms a muddy soup.

(d) Terraces of minerals precipitated at Mammoth Hot Springs, Yellowstone.

thermophyllic (heat-loving) bacteria and archaea that thrive in hot water and metabolize the sulfur-containing minerals dissolved in the groundwater (Fig. 19.16b).

Numerous distinctive geologic features form in geothermal regions as a result of the eruption of hot water (Fig. 19.16c, d). In places where the hot water rises into soils rich in volcanic ash and clay, a viscous slurry forms and fills bubbling mud pots. Bubbles of steam rising through the slurry cause it to splatter about in goopy drops. Where geothermal waters spill out of natural springs and then cool, dissolved minerals in the water precipitate, forming colorful mounds or terraces of travertine and other minerals.

Under special circumstances, geothermal water emerges from the ground in a **geyser** (from the Icelandic spring, Geysir, and the word for *gush*), a fountain of steam and hot water that erupts episodically from a vent in the ground (Fig. 19.17a–e). To understand why a geyser erupts, we first need a picture of its underground plumbing. Beneath a geyser lies a network of irregular fractures in very hot rock; groundwater sinks and fills these fractures. Heat transfers from the rock to the groundwater and makes the water's temperature rise. Since the boiling point of water (the temperature at which water vaporizes) increases with increasing pressure, hot groundwater at depth can remain in liquid form even if its temperature has become greater than the boiling point of water at the Earth's surface. When such "superheated" groundwater begins to rise through a conduit toward the surface, pressure in it decreases until eventually some of the water transforms into steam. The resulting expansion causes water higher up to spill out of the conduit at the ground surface. When this spill happens, pressure in the conduit, from the

FIGURE 19.17 Geysers form where steam erupts from the ground.

(a) A vent in Iceland begins to bubble.

(b) Water spurts out, releasing pressure on super-hot water below.

(c) Water at depth transforms into vapor, which rises and pushes overlying water out of the vent.

Time

(d) Steaming ground, Rotorua, New Zealand.

(e) The Old Faithful geyser, in Yellowstone, erupts predictably.

weight of overlying water, suddenly decreases. A sudden drop in pressure causes the superhot water at depth to turn into steam instantly, and this steam quickly rises, ejecting all the water and steam above it out of the conduit in a geyser eruption. Once the conduit empties, the eruption ceases, and the conduit fills once again with water that gradually heats up, starting the eruptive cycle all over again.

Hot springs are found in many localities around the world: at Hot Springs, Arkansas, where deep groundwater rises to the surface; in Yellowstone National Park, above the magma chamber of a continental hot spot; around the Salton Sea in southern California, where rifting triggers local volcanism; in the Geysers Geothermal Field of California, an area of significant geothermal power generation formed above a felsic magma intrusion; in Iceland, which has grown on top of an oceanic hot spot along the Mid-Atlantic Ridge; and in Rotorua, New Zealand, which lies in an active volcanic field above a subduction zone.

People live in some geothermal regions, though the areas have inherent natural hazards. In Rotorua, signs along the road warn of *STEAM!*, which can obscure visibility, and steam indeed spills out of holes in backyards and parking lots. But all this hot water does offer a benefit—in Rotorua, waters circulate through pipes to provide home heating. In geothermal areas worldwide, steam provides a relatively inexpensive means of generating electricity.

Take-Home Message

- At hot springs, warm or even boiling water emerges at the ground surface.
- Some hot springs form where groundwater rises from warm rock several kilometers down; others form where igneous activity heats water near the surface.
- At a geyser, a fountain of steam and scalding water spurts from the ground episodically.

THINK: Why do geysers generally erupt episodically, instead of continuously?

19.7 GROUNDWATER USAGE PROBLEMS

Since prehistoric times, groundwater has been an important resource that people have relied on for drinking, irrigation, and industry. Groundwater feeds the lushness of desert oases in the Sahara, the amber grain in the North American high plains, and the growing cities of sunny arid regions. Agricultural and industrial usage accounts for about 93% of all water usage, so as once-empty land comes under cultivation and countries become increasingly industrialized, demands on the groundwater supply soar. Groundwater provides only about 20% of the water we use worldwide, but this percentage has increased as surface-water resources decrease. In fact, in many locations groundwater is the only water source.

Though groundwater accounts for about 95% of the liquid freshwater on the planet, *accessible* groundwater cannot necessarily be replenished quickly, and this leads to shortages. Groundwater contamination is also a growing tragedy. Such pollution, caused when toxic wastes and other impurities infiltrate down to the water table, may be invisible to us but may ruin a water supply for generations to come. In this section, we'll take a look at problems associated with the use of groundwater supplies.

Depletion of Groundwater Supplies

Is groundwater a renewable resource? In a time frame of 10,000 years, the answer is yes, for the hydrologic cycle will eventually resupply depleted reserves. But in a time frame of 100 to 1,000 years—the span of a human lifetime or a civilization—groundwater in many regions may be a nonrenewable resource. By pumping water out of the ground at a rate faster than nature replaces it, people are effectively mining the groundwater supply. In fact, in portions of the desert Sunbelt region of the United States, supplies of young groundwater have already been exhausted, and deep wells now extract 10,000-year-old groundwater. Some of this ancient water has been in rock so long that it has become too mineralized to be usable. A number of other problems accompany the depletion of groundwater.

- *Lowering the water table*: When we extract groundwater from wells at a rate faster than it can be resupplied by nature, the water table drops. First, a cone of depression forms locally around the well; then the water table gradually becomes lower in a broad region. As a consequence, existing wells, springs, and rivers dry up (**Fig. 19.18a, b**). To continue tapping into the water supply, we must drill progressively deeper. Pumping has lowered the water table in the High Plains aquifer (see Fig. 19.5d) by 15 m over broad areas, and by over 50 m locally.

  During the past decade, researchers have used data from GRACE satellites to study groundwater depletion. These satellites provide precise measurements of how the magnitude of Earth's gravitational pull in a given region varies over time. In some regions, a very small but detectable decrease in gravitational pull reflects groundwater depletion because when air replaces water in subsurface pores, the mass of the near-surface realm becomes smaller. By calibrating the magnitude of gravity decrease to the amount of groundwater depletion, researchers can produce maps depicting how much the water table has dropped (**Fig. 19.18c, d**)

FIGURE 19.18 Effects of human modification of the water table.

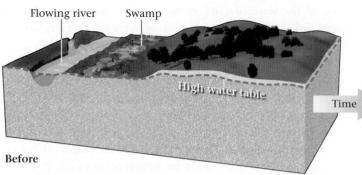

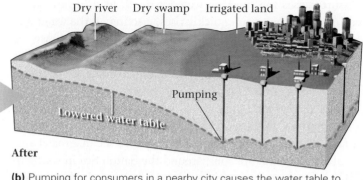

Before

(a) Before humans start pumping groundwater, the water table is high. A swamp and permanent stream exist.

After

(b) Pumping for consumers in a nearby city causes the water table to sink in, so the swamp dries up.

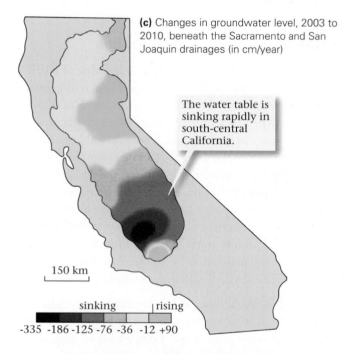

(c) Changes in groundwater level, 2003 to 2010, beneath the Sacramento and San Joaquin drainages (in cm/year)

The water table is sinking rapidly in south-central California.

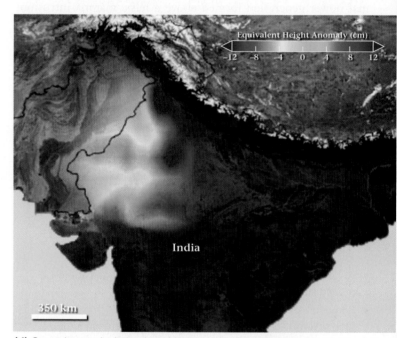

(d) Groundwater depletion in northwest India, based on GRACE satellite data.

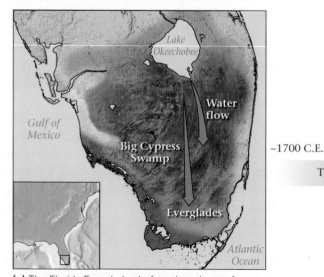

(e) The Florida Everglades before the advent of urban growth and intensive agriculture.

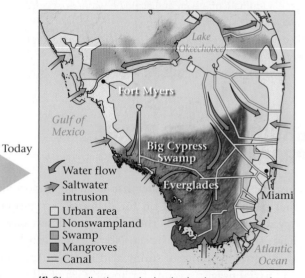

(f) Channelization and urbanization have removed water from recharge areas, disrupting flow paths.

Notably, the water table can also drop when people divert surface water from the recharge area. Such a problem has developed in the Everglades of southern Florida, a huge swamp where, before the expansion of Miami and the development of agriculture, the water table lay at the ground surface (Fig. 19.18e, f). Diversion of water from the Everglades' recharge area into canals has significantly lowered the water table, causing parts of the Everglades to dry up.

- *Reversing the flow direction of groundwater*: The cone of depression that develops around a well creates a local slope to the water table. The resulting hydraulic gradient may be large enough to reverse the flow direction of nearby groundwater (Fig. 19.19a, b). Such reversals can allow pollutants, seeping out of a septic tank, to contaminate a well.

- *Saline intrusion*: In coastal areas, fresh groundwater lies in a layer above saline (salty) water that entered

FIGURE 19.19 Some causes of groundwater problems.

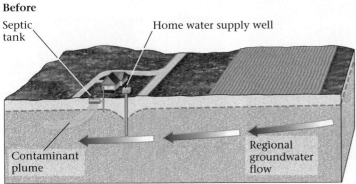

(a) Before pumping, effluent from a septic tank drifts with the regional groundwater flow, and the home well pumps clean water.

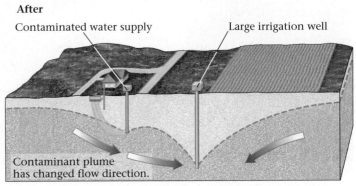

(b) After pumping by a nearby irrigation well, effluent flows into the home well in response to the new local slope of the water table.

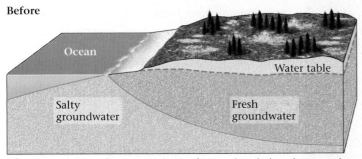

(c) Before pumping, fresh groundwater forms a lens below the ground.

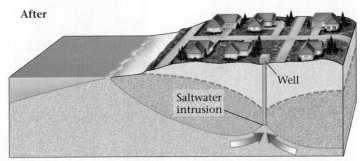

(d) If the freshwater is pumped too fast, saltwater from below is sucked up into the well. This is saltwater intrusion.

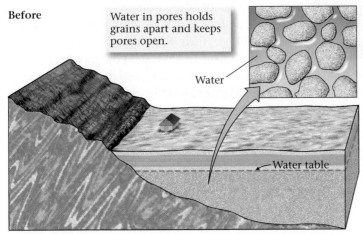

(e) When intensive irrigation removes groundwater, pore space in an aquifer collapses.

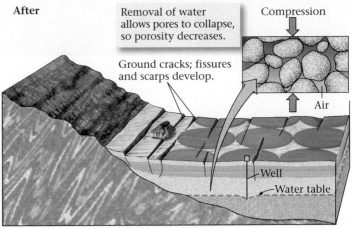

(f) As a result, the land surface sinks, leading to the formation of ground fissures and cracked houses.

FIGURE 19.20 Consequences of groundwater removal, and subsurface collapse.

(a) Ground fissures in Arizona.

(b) The Leaning Tower of Pisa.

(c) Sinking ground surface, California.

(d) Sinking and submergence of Venice, Italy.

the aquifer from the adjacent ocean (Fig. 19.19c, d). Because fresh water is less dense than saline water, it floats above the saline water. If people pump water out of a well too quickly, the boundary between the saline water and the fresh groundwater rises. And if this boundary rises above the base of the well, then the well will start to yield useless saline water.

- *Pore collapse and land subsidence*: When groundwater fills the pore space of a rock, it holds the grains of the rock or

regolith apart, for water cannot be compressed. The extraction of water from a pore eliminates the support holding the grains apart, because the air that replaces the water *can* be compressed. As a result, the grains pack more closely together. Such pore collapse permanently decreases the porosity and permeability of a rock, and thus lessens its value as an aquifer (Fig. 19.19e, f).

Pore collapse also decreases the volume of the aquifer, with the result that the ground above the aquifer sinks. Such land subsidence may cause fissures at the surface to develop and the ground to tilt (Fig. 19.20a). Buildings constructed over regions undergoing land subsidence may themselves tilt, or their foundations may crack. The Leaning Tower of Pisa, in Italy, tilts because the removal of groundwater caused its foundation to subside (Fig. 19.20b). In the San Joaquin Valley of California, the land surface subsided by 9 m between 1925 and 1975, because water was removed to irrigate farm fields (Fig. 19.20c). In coastal areas, land subsidence may even make the land surface sink below sea level. The flooding of Venice, Italy, is due to land subsidence accompanying the withdrawal of groundwater beneath Venice (Fig. 19.20d).

To avoid such problems, communities have sought to prevent groundwater depletion either by directing surface water into recharge areas or by pumping surface water back into the ground. For example, some communities have excavated to lower the land surface of a park and then configured storm sewers so that they drain onto its grassy surface. The park then acts as a catchment for storm water (Fig. 19.21).

Natural Groundwater Quality

As we've seen, groundwater is not distilled water, composed only of H_2O. Rather, it is a solution containing ions obtained by reactions of water with the rock or sediment through which it passes. The reactions between groundwater and rock or sediment are similar to the chemical weathering reactions that take place at the Earth's surface (see Interlude B), and they include dissolution, oxidation, hydrolysis, and hydration. During these reactions, ions go into or drop out of solution, causing existing minerals to disappear and new minerals to grow.

The concentration of dissolved ions in groundwater (i.e., the quantity of ions dissolved per unit volume of water) depends on factors such as the temperature, pressure, and acidity (pH) of the groundwater. For example, warmer groundwater can carry more ions in solution than can cooler groundwater. If groundwater has had a long time to react with the material in which it resides, it becomes "saturated," in that the water contains just as many dissolved ions as possible, under the local environmental conditions.

Because groundwater flows, it eventually enters different environments. If groundwater enters an environment where it has the capacity to contain more ions, it may dissolve surrounding rock or sediment and create secondary porosity. Alternatively, if groundwater enters an environment in which it cannot hold as many dissolved ions, some of the ions in the water bond together and become solid mineral grains that precipitate to form cement or vein fill.

Much of the world's groundwater is crystal clear, and pure enough to drink right out of the ground. Rocks and sediment are natural filters capable of removing suspended solids—these solids get trapped in tiny pores or stick to the surfaces of clay flakes. In fact, the commercial distribution of bottled groundwater ("spring water") has become a major business worldwide. But dissolved chemicals, and in some cases methane, may make some natural groundwater unusable. For example, groundwater that has passed through salt-containing strata may become salty and unsuitable for irrigation or drinking. Groundwater that has passed through limestone or dolomite contains dissolved calcium (Ca^{2+}) and magnesium (Mg^{2+}) ions; this water, called **hard water**, can be a problem because carbonate minerals precipitate from it to form "scale" that clogs pipes. Also, washing with hard water can be difficult because soap won't develop a lather. Groundwater that has passed through iron-bearing rocks may contain dissolved iron oxide that precipitates to form rusty stains. Some groundwater contains dissolved hydrogen sulfide, which comes out of solution when the groundwater rises to the surface; hydrogen sulfide is a poisonous gas that has a rotten-egg smell. In recent years, concern has grown about arsenic in groundwater. Arsenic, a highly toxic chemical, enters groundwater when arsenic-bearing minerals dissolve. In the United States, government standards require drinking water to contain less than 10 micrograms of arsenic per liter. Unfortunately, surveys suggest that about 5% of groundwater supplies contain 20 or more micrograms per liter.

Did you ever wonder... where does bottled "spring water" come from?

Human-Caused Groundwater Contamination

As we've noted, some contaminants in groundwater occur naturally—sulfur, iron, calcium carbonate, methane, and salt can all be introduced to groundwater directly from the rock through which it is flowing. But in recent decades, contaminants have increasingly been introduced into aquifers because of human activity (Fig. 19.22a). These contaminants include agricultural waste (pesticides, fertilizers, and animal sewage), industrial waste (dangerous organic and inorganic chemicals), effluent from "sanitary" landfills and septic tanks (including bacteria and viruses), petroleum products and other chemicals that do not dissolve in water (together referred to as non-aqueous phase liquids or NAPL), radioactive waste (from weapons manufacture, power plants, and hospitals), and acids leached from sulfide minerals in coal and metal mines. Some of these contaminants seep into the ground from subsurface tanks, some infiltrate from the surface, and some are intentionally forced through injection wells (wells in which a liquid is pumped *down* into the ground under pressure so that it passes from the well back into the pore space of the rock or regolith).

FIGURE 19.21 Sketch of an enhanced recharge catchment in a city.

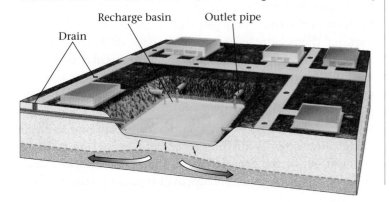

Recharge basin   Outlet pipe

Drain

FIGURE 19.22 Contamination plumes in groundwater.

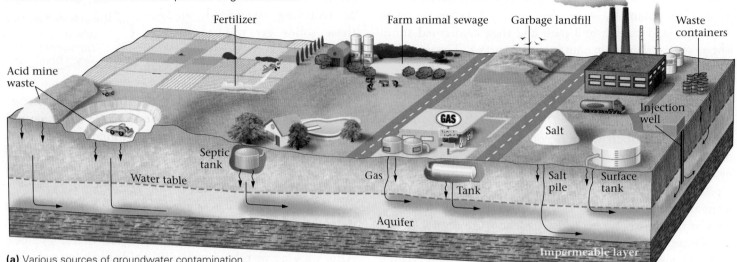

(a) Various sources of groundwater contamination.

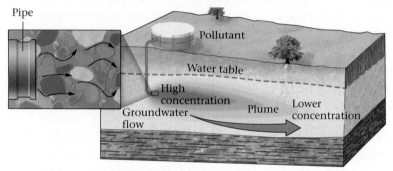

(b) A contaminant plume as seen in cross section. The darker the color, the greater the concentration of contaminant.

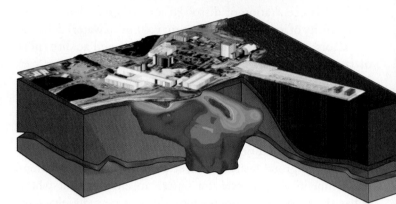

(c) A 3-D image showing a contaminant plume beneath a coastal industrial facility. The red indicates the greatest concentration of pollutant.

The cloud of contaminated groundwater that moves away from the source of contamination is called a **contaminant plume** (Fig. 19.22b–c). Staggering quantities of contaminating liquids (trillions of gallons in the United States alone) enter the groundwater system every year.

The best way to avoid such **groundwater contamination** is to prevent contaminants from entering groundwater in the first place. This can be done by locating potential sources of contamination on impermeable bedrock so that they are isolated from aquifers. If such a site is not available, the storage area should be lined with a thick layer of clay, for the clay not only acts as an aquitard, but it can hold on to contaminants. For this reason, environmental engineers commonly place landfills on top of a liner of packed clay or place waste in durable, sealed containers. Government agencies have studied various options for safely storing the containers. One option involves stockpiling them in tunnels cut into salt domes, for salt is impermeable. Another option involves

hiding waste containers in tunnels above the water table. For example, officials considered a proposal to store American nuclear waste in a network of tunnels 300 m beneath Yucca Mountain, Nevada, for this mountain consists of dry, fairly impermeable tuff high above the water table.

Fortunately, in some cases, natural processes can clean up groundwater contamination. Chemicals may be absorbed by clay, oxygen in the water may oxidize the chemicals, and bacteria in the water may metabolize the chemicals, thereby turning them into harmless substances. Where contaminants do make it into an aquifer, environmental engineers drill test wells to determine which way and how fast the contaminant plume is flowing; once they know the flow path, they can close wells in the path to prevent consumption of contaminated water. Engineers may attempt to clean the groundwater by drilling a series of extraction wells to pump it out of the ground. If the contaminated water does not rise fast enough, engineers drill injection wells to force clean water or steam into the ground beneath the contaminant plume (**Fig. 19.23**). The injected fluids then push the contaminated water up into the extraction wells.

More recently, environmental engineers have begun exploring techniques of **bioremediation**: injecting oxygen and nutrients into a contaminated aquifer to foster growth of bacteria that can react with and break down molecules of contaminants. They have also been experimenting with "permeable reactive barriers," subsurface walls of materials such as iron filings that chemically react with contaminants to transform the contaminants into safer materials. To produce a barrier, engineers dig a trench in the path of the flow, fill it with the reactive material, and bury it. When the plume flows through the barrier, the reactions take place. Needless to say, cleaning techniques are expensive and generally only partially effective.

Unwanted Effects of Rising Water Tables

We've seen the negative consequences of sinking water tables, but what happens when the water table rises? Is that necessarily good? Sometimes, but not always. If the water table rises above the level of a house's basement, water seeps through the foundation and floods the basement floor. As we noted in Chapter 16, catastrophic damage occurs when a rising water table weakens the base of a hillslope or a failure surface underground triggers landslides and slumps (**Fig. 19.24a, b**).

FIGURE 19.23 Steam injected beneath the contamination drives the contaminated water upward in the aquifer, where pumping wells remove it.

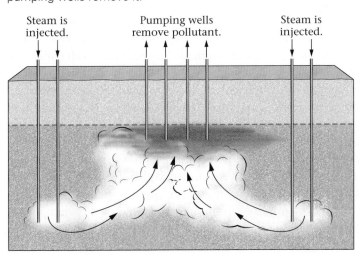

FIGURE 19.24 When the water table rises, material above a weak sliding surface begins to slump, and a landslide may result.

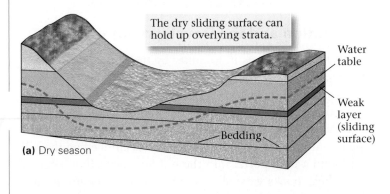

The dry sliding surface can hold up overlying strata.

Water table

Weak layer (sliding surface)

Bedding

(a) Dry season

Time

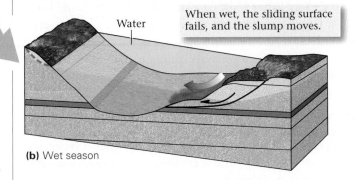

Water

When wet, the sliding surface fails, and the slump moves.

(b) Wet season

Take-Home Message

- Pumping groundwater at rates faster than it can be recharged lowers the water table and can cause wells and springs to go dry and the flow direction to change.

- Near coasts, pumping groundwater may cause underlying saline water to be pulled into wells.

- Removal of groundwater can cause pore space to collapse and overlying land to subside.

- Groundwater reacts with rock and sediment, so it contains ions of many elements, some of which are toxic. Dissolved lime causes water to be "hard."

- Humans contaminate groundwater when toxic waste spills on the ground surface or leaks from buried pipes and tanks. A contaminant plume spreads out from the source.

- A variety of expensive techniques can help remediate contamination.

THINK: How can the introduction of microbes remove contaminants from groundwater?

19.8 CAVES AND KARST: A SPELUNKER'S PARADISE

Dissolution and the Development of Caves

In 1799, as legend has it, a hunter by the name of Houchins was tracking a bear through the woods of Kentucky when the bear suddenly disappeared on a hillslope. Baffled, Houchins plunged through the brambles trying to sight his prey. Suddenly he felt a draft of surprisingly cool air flowing down the slope from uphill. Now curious, Houchins climbed up the hill and found a dark portal into the hillslope beneath a ledge of rocks. Bear tracks were all around—was the creature inside? He returned later with a lantern and cautiously stepped into the passageway. After walking a short distance, he found himself in a large, underground room. Houchins had discovered Mammoth Cave, an immense network of natural tunnels and subterranean chambers—a walk through the entire network would extend for 630 km!

> **Did you ever wonder . . .**
> why do huge underground caverns form?

Large cave networks develop in limestone bedrock because limestone dissolves relatively easily in corrosive groundwater. Generally, the corrosive component groundwater is dilute carbonic acid (H_2CO_3), which forms when water absorbs carbon dioxide (CO_2) from materials it has passed through. The CO_2 in groundwater comes from two sources—a small amount dissolves in rain as it falls through the sky, and a larger amount dissolves in water percolating through organic-rich soil on its way down to the water table. When carbonic acid comes in contact with calcite ($CaCO_3$) in limestone, it reacts to produce HCO_3^{1-} and Ca^{2+} ions.

Geologists debate about the depth at which limestone cave networks form. Some limestone dissolves above the water table, particularly along the joints, which act as conduits through which water flows into the subsurface, and some limestone may dissolve far below the water table. But it appears that *most* cave formation, or speleogenesis, takes place in limestone that lies just below the water table, for in this interval the acidity of the groundwater remains high, the mixture of groundwater and newly added rainwater is undersaturated (meaning it has the capacity to dissolve more ions), and groundwater flow is fastest. The association between cave formation and the water table helps explain why openings in a cave network align along the same horizontal plane.

In recent years, geologists have discovered that not all caves form due to reactions with carbonic acid. An alternative type of cave-forming process, called sulfuric-acid speleogenesis, takes place in regions where limestone overlies strata containing hydrocarbons, and may be responsible for about 5 percent of caves. This process happens because microbes can convert organic sulfur within oil into hydrogen sulfide (H_2S) gas. The gas rises into the limestone, where it oxidizes either by reaction with air or by the action of microbes to produce sulfuric acid. Sulfuric acid is very strong and eats into calcite to produce gypsum and CO_2 gas. This process appears to be responsible for Carlsbad Caverns of New Mexico, an immense network that contains the Big Room, whose floor has an area 14 times that of a football field.

The Character of Cave Networks

As we have noted, caves in limestone usually occur as part of a network. Cave networks include rooms, or chambers, which are large, open spaces sometimes with cathedral-like ceilings, and tunnel-shaped or slot-shaped passages. (See **Geology at a Glance**, pp. 672–673.) Some chambers may host underground lakes, and some passageways may serve as conduits for underground streams. The shape of the cave network reflects variations in permeability and in the composition of the rock from which the caves formed. Larger open spaces developed where the limestone was most soluble and where groundwater flow was fastest. Thus, in a sequence of strata, caves develop preferentially in the more soluble limestone beds. Passages in cave networks typically follow preexisting joints, for the joints provide secondary porosity along which groundwater can flow faster (**Fig. 19.25a**). Because joints commonly occur in orthogonal systems (consisting of two sets of joints oriented at right angles to each other; see Chapter 11), passages form a grid.

Why do extensive cave networks, with large rooms and abundant corridors, develop in some locations but not in others? There are several reasons. First, most caves form in limestone, so without a thick layer of limestone in the subsurface, extensive networks can't form. Second, the dissolution that forms caves occurs primarily in freshwater at the water table, so unless the water table lies above sea level and below the land surface, extensive networks can't form. Third, for caves to develop, sufficient liquid water must be present. Thus extensive networks form preferentially in temperate or tropical regions, drenched by rain—they do not form in polar regions, where ice covers the ground and subsurface water remains permanently frozen, nor do they form in desert regions, where water is scarce. Finally, since percolation through organic matter provides the acidity that groundwater must have in order to dissolve limestone, caves form beneath regions that have organic-rich soil. In fact, caves develop more rapidly in tropical regions, where jungle provides abundant organic litter.

FIGURE 19.25 Development of karst and dripstone.

Joint set 1   Joint set 2

□ More-soluble bed
▨ Less-soluble bed

Bedding

(a) Joints act as conduits for water in cave networks. Caves and passageways follow joints, and preferentially form in more-soluble beds.

Soda straw

Stalactite
Stalagmite

Limestone column

Time

(b) The evolution of a soda straw stalactite into a limestone column.

Squeezing through a joint-controlled passageway, Utah

Stalactites and stalagmites in a cave in Italy

1m

(c) Flowstone on the wall of a cave in Vietnam.

Precipitation and the Formation of Speleothems

When the water table drops below the level of a cave, the cave becomes an open space filled with air. In places where downward-percolating groundwater containing dissolved calcite emerges from the rock above the cave and drips from the ceiling, the surface of the cave gradually changes. As soon as this water reenters the air, it evaporates a little and releases some of its dissolved carbon dioxide. As a result, calcium carbonate (limestone) precipitates out of the water and produces a type of travertine called **dripstone**. The various intricately shaped formations that grow in caves by the accumulation of dripstone are called **speleothems**.

Caves and Karst Landscapes

Limestone pavement, Ireland

Disappearing stream

Sinkhole

Collapsed breccia

Stalagmite

Stalactite

Soda straw

Dissolved joint

Flowstone

Cavern

Stalactite

Limestone column

Underground stream

Underground pool

Corridor

Emerging spring

Sinkholes

Underground pool, Mexico

Natural Bridge, Virginia

Spelunker crawling in a cave

Limestone is soluble in acidic water. Much of the water that falls to the ground as rain, or seeps through the ground as groundwater, tends to be acidic, so in regions of the Earth where bedrock consists of limestone, we find signs of dissolution. Underground openings that develop by dissolution are called caves or caverns. Some of these may be large, open rooms, whereas others are long, narrow passages. Underground lakes and streams may cover the floor. A cave's location depends on the orientation of bedding and joints, for these features localize the flow of groundwater.

Caves originally form at or near the water table. As the water table drops, caves empty of most water and become filled with air. In many locations, groundwater drips from the ceiling of a cave or flows along its walls. As the water evaporates and loses its acidity, new calcite precipitates. Over time, this calcite builds into cave formations, or speleothems, such as stalactites, stalagmites, columns, and flowstone.

Distinctive landscapes, called karst landscapes, develop at the Earth's surface over limestone bedrock. In such regions, the ground may be rough where rock has dissolved along joints; where the roofs of caves collapse, sinkholes develop. If a surface stream flows into a cave network, we say that the stream is "disappearing." The water from such streams may reemerge elsewhere as a spring. In some places, the collapse of subsurface openings leaves behind natural bridges.

Cave explorers (spelunkers) and geologists have developed a detailed nomenclature for different kinds of speleothems (Fig. 19.25b and inset). Where water drips from the ceiling of the cave, the precipitated limestone builds an icicle-like cone called a **stalactite**. Initially, calcite precipitates around the outside of the drip, forming a delicate, hollow stalactite called a soda straw. But eventually, the soda straw fills up, and water migrates down the margin of the cone to form a more massive, solid stalactite. Where the drips hit the floor, the resulting precipitate builds an upward-pointing cone called a **stalagmite**. If the process of dripstone formation in a cave continues long enough, stalagmites merge with overlying stalactites to create travertine columns. In some cases, groundwater flows along the surface of a wall and precipitates to produce cloth-like sheets of travertine called flowstone (Fig. 19.25c). The travertine of caves tends to be translucent and, when lit from behind, glows with an eerie amber light.

The Formation of Karst Landscapes

Limestone bedrock underlies most of the Kras Plateau in Slovenia, along the east coast of the Adriatic Sea. The name *kras*, which means rocky ground, is apt because this region includes abundant rock exposures. Geologists refer to regions such as the Kras Plateau, where surface landforms develop when limestone bedrock dissolves both at the surface and in underlying cave networks, as **karst landscapes**—from the Germanized version of kras (See for Yourself R, p. S-34).

Karst landscapes typically display a number of distinct landforms. Perhaps the most widespread are sinkholes (Fig. 19.26 top), circular depressions that form either when the ground collapses into an underground cave below (as we noted in the introduction to this chapter) or when surface bedrock dissolves in acidic water on the floor of a bog or pond. Not all of the caves or passageways beneath a karst landscape have collapsed, and this situation leads to unusual drainage patterns. Specifically, where surface streams intersect cracks (joints) or holes that link to caverns or passageways below, the water cascades downward into the subsurface and disappears (Fig. 19.26 bottom). Such **disappearing streams** may flow through passageways underground and reemerge from a cave entrance downstream. In cases where the ground collapses over a long, joint-controlled passage, sinkholes may be elongate and canyon-like. Remnants of cave roofs remain as **natural bridges**. Ridges or walls between adjacent sinkholes tend to be steep-sided, for they were originally joint controlled. Over time, the walls erode, leaving only jagged, isolated spires—a karst landscape dominated by such spires is called tower karst. Notably, the surreal collection of pinnacles

constituting the tower karst landscape in the Guilin region of China inspired generations of artists who portray them on scroll paintings (Fig. 19.27a, b).

Karst landscapes form in a series of stages (Fig. 19.28a–c).

- *The establishment of a water table in limestone*: The story of a karst landscape begins after the formation of a thick interval of limestone. Limestone forms in seawater, and thus initially lies below sea level. If rela-

FIGURE 19.26 Features of karst landscapes.

Sinkholes of the Kras Plateau.

A disappearing stream flows into a cave.

FIGURE 19.27 Tower karst forms a spectacular landscape in southern China.

Chinese artists painted scrolls depicting forested towers of karst.

The landscape is treeless today, a consequence of industrialization policies in the 1950s.

FIGURE 19.28 The progressive formation of caves and a karst landscape.

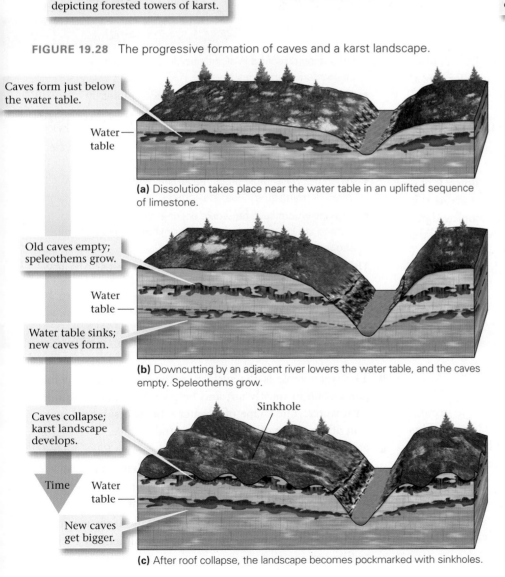

Caves form just below the water table.

Water table

(a) Dissolution takes place near the water table in an uplifted sequence of limestone.

Old caves empty; speleothems grow.

Water table

Water table sinks; new caves form.

(b) Downcutting by an adjacent river lowers the water table, and the caves empty. Speleothems grow.

Sinkhole

Caves collapse; karst landscape develops.

Time Water table

New caves get bigger.

(c) After roof collapse, the landscape becomes pockmarked with sinkholes.

tive sea level drops, a water table can develop in the limestone below the ground surface. If, however, the limestone gets buried deeply, it must first be uplifted and exposed by erosion before it can contain the water table.

- *The formation of a cave network*: Once the water table has been established, dissolution begins and a cave network develops.

- *A drop in the water table*: If the water table later becomes lower, either because of a decrease in rainfall or because nearby rivers cut down through the landscape and drain the region, newly formed caves dry out. Downward-percolating groundwater emerges from the roofs of the caves; dripstone and flowstone precipitate.

- *Roof collapse*: If rocks fall off the roof of a cave for a long time, the roof eventually collapses. Such collapse creates sinkholes and troughs, leaving behind hills, ridges, and natural bridges of limestone.

FIGURE 19.29 Unusual organisms have evolved in the darkness of caves.

(a) A blind fish—there's no need for eyes in a cave.

(b) Gobs of bacteria drip from the ceiling of a cave to form "snotites."

Life in Caves

Despite their lack of light, caves are not sterile, lifeless environments. Caves that are open to the air provide a refuge for bats as well as for various insects and spiders. Similarly, fish and crustaceans enter caves where streams flow in or out. Species living in caves have evolved some unusual characteristics. For example, cave fish lose their pigment and in some cases their eyes (Fig. 19.29a). Recently, explorers discovered caves in Mexico in which warm, mineral-rich groundwater currently flows. Colonies of bacteria metabolize sulfur-containing minerals in this water and create thick mats of living ooze in the complete darkness of the cave. Long gobs of this bacteria slowly drip from the ceiling. Because of the mucus-like texture of these drips, they have come to be known as "snotites" (Fig. 19.29b).

Take-Home Message

- Limestone dissolves in slightly acidic groundwater: this phenomenon produces caves.
- The acidity comes from either carbonic acid, formed by dissolution of CO_2 as water passes through soil above, or sulfuric acid, formed where H_2S bubbles into groundwater from oil below.
- Large cave networks form by dissolution at or near the water table. Passageways follow joints.
- When the water table drops, travertine may precipitate in caves, forming stalactites, stalagmites, and other speleothems.
- Regions where cave networks occur at or near the ground surface are karst landscapes, with sinkholes, towers, natural bridges, and disappearing streams.

THINK: Can organisms live in the pitch black of caves? If so, how?

Chapter Summary

- During the hydrologic cycle, water infiltrates the ground and fills the pores and cracks in rock and sediment. This subsurface water is called groundwater. The amount of open space in rock or sediment is its porosity; and the degree to which pores are interconnected, so that water can flow through, defines its permeability.

- Geologists classify rock and sediment according to their permeability. Aquifers are relatively permeable, and aquitards are relatively impermeable.

- The water table is the surface in the ground above which pores contain mostly air, and below which pores are filled with water. The shape of a water table is a subdued imitation of the shape of the overlying land surface.

- Groundwater flows wherever the water table has a hydraulic gradient, and moves slowly from recharge areas to discharge areas. Darcy's law shows that this rate depends on permeability and on the hydraulic gradient.

- Groundwater contains dissolved ions. These ions may come out of solution to form the cement of sedimentary rocks or to fill veins.

- Groundwater can be extracted in wells. An ordinary well penetrates below the water table, but in an artesian well, water rises on its own. Pumping water out of a well too fast causes drawdown, yielding a cone of depression. At a spring, groundwater exits the ground on its own.

- Hot springs and geysers release hot water to the Earth's surface. This water may have been heated by residing very deep in the crust, or by the proximity of a magma chamber or of recently formed igneous rock.

- Groundwater is a precious resource, used for municipal water supplies, industry, and agriculture. In recent years, some regions have lost their groundwater supply because of overuse or contamination.

- When limestone dissolves just below the water table, underground caves are the result. Soluble beds and joints determine the location and orientation of caves. If the water table drops, caves empty out. Limestone precipitates out of water dripping from cave roofs, and creates speleothems (such as stalagmites and stalactites).

- Regions where abundant caves have collapsed to form sinkholes are called karst landscapes. These terrains can contain sinkholes, natural bridges, and disappearing streams.

GEOPUZZLE REVISITED

Water resides underground in the pores and cracks of rock and regolith. Below the water table, this water completely fills open space. When it rains, some water infiltrates the ground and percolates downward until it reaches the water table—and this water becomes groundwater. The level of water in a well defines the water table.

Guide Terms

aquifer (p. 650)

aquitard (aquiclude) (p. 650)

artesian spring (p. 659)

artesian well (p. 657)

bioremediation (p. 669)

capillary fringe (p. 652)

cone of depression (p. 657)

confined aquifer (p. 650)

contaminant plume (p. 668)

Darcy's law (p. 656)

disappearing stream (p. 674)

discharge area (p. 655)

dripstone (p. 671)

geothermal region (p. 661)

geyser (p. 662)

groundwater (p. 647)

groundwater contamination (p. 668)

hard water (p. 667)

hot springs (p. 661)

hydraulic gradient (p. 656)

hydraulic head (p. 655)

karst landscape (p. 674)

natural bridge (p. 674)

oasis (pp. 659, 660)

ordinary well (p. 657)

perched water table (p. 654)

permeability (p. 650)

pore (p. 648)

porosity (p. 648)

primary porosity (p. 648)

recharge area (p. 655)

saturated zone (phreatic zone) (p. 652)

secondary porosity (p. 649)

sinkhole (p. 647)

soil moisture (p. 648)

speleothem (p. 671)

spring (p. 657)

stalactite (p. 674)

stalagmite (p. 674)

unconfined aquifer (p. 650)

unsaturated zone (vadose zone) (p. 652)

water table (p. 652)

wells (p. 657)

Review Questions

1. How do porosity and permeability differ? Give examples of substances with high porosity but low permeability.

2. What is a water table, and what factors affect the level of the water table? What factors affect the flow direction of the water below the water table?

3. How does the rate of groundwater flow compare with that of moving ocean water or river currents?

4. What does Darcy's law tell us about how the hydraulic gradient and permeability affect discharge?

5. How does the chemical composition of groundwater change with time? Why is "hard water" hard?

6. How does excessive pumping affect the local water table?

7. How is an artesian well different from an ordinary well?

8. Why do natural springs form?

9. Explain why hot springs form and what makes a geyser erupt.

10. Is groundwater a renewable or nonrenewable resource? Explain your answer.

11. Describe some of the ways in which human activities can adversely affect the water table.

12. What are some sources of groundwater contamination? How can it be prevented?

13. Describe the process leading to the formation of caves and the speleothems within caves.

14. Describe the various features of a karst landscape, and explain how they evolve.

On Further Thought

15. The population of Desert Paradise (a fictitious town in the southwestern United States) has been doubling every seven years. Most of the new inhabitants are "snowbirds," people escaping the cold winters of more northerly latitudes. There are no permanent streams or lakes anywhere near DP. In fact, the only standing water in the town occurs in the ponds of the many golf courses that have been built recently. The water in these ponds needs to be replenished almost constantly, for without supplementing it, the water seeps into the ground quickly and dries up. The golf courses and yards of the suburban-style developments of DP all have lawns of green grass. DP has been growing on a flat, sediment-filled basin between two small mountain ranges. Where does the water supply of DP come from? What do you predict will happen to the water table of the area in coming years, and how might the land surface change as a consequence? Is there a policy that you might suggest to the residents of DP that could slow the process of change?

16. You are part of a cave-exploration team that is trying to map a cave network in a temperate region of flat-lying limestone beds that underlie a plateau. A set of NW-SE-trending systematic joints cuts the limestone beds. A river cuts through the region, and the entrance to the cave is along the valley wall. The surface of the river lies about 400 m below the surface of the plateau. What do you predict will be the trend of tunnels in the cave network, and how far below the surface of the plateau do you think the cave network extends?

 For more resources, including animations, quizzes, and Norton's GeoTours, go to **wwnorton.com/studyspace**.

 If your instructor assigns exercises in SmartWork, log in at **smartwork.wwnorton.com**.

ANOTHER VIEW A satellite view of the Arecibo radio telescope in Puerto Rico. The dish was built by smoothing the surface of a 300-m-wide sinkhole in a karst terrain.

CHAPTER 20

An Envelope of Gas: Earth's Atmosphere and Climate

The setting Sun highlights clouds in the atmosphere. Some form naturally, from water evaporating from the sea below, whereas some are human-made, emanating from the contrails of jets. The colors, both the blue up high and the red near the horizon, come from the interaction of sunlight with atmospheric gases.

GEOPUZZLE

Where did the atmosphere come from, and why do different regions of the Earth host different climates? And what's the difference between "climate" and "weather" anyway?

Who has seen the wind?
Neither you nor I:
But when the trees bow down their heads,
The wind is passing by.

— Christina Rossetti (British poet, 1830–1894)

20.1 INTRODUCTION

On March 21, 1999, Bertrand Piccard, a Swiss psychiatrist, and Brian Jones, a British balloon instructor, became the first people to circle the globe nonstop in a balloon (Fig. 20.1). They began their flight on March 1, following more than 20 unsuccessful attempts by various balloonists during the previous two decades. Their airtight gondola, which could float in case they had to ditch in the sea, contained a heater, bottled air, food and water, and instruments for navigation and communication. The nature of their equipment hints at the challenges the balloonists faced, and why it took so many years before anyone finally succeeded in circling the globe.

Balloons can rise from the Earth only because an **atmosphere**, a layer consisting of a mixture of gases called **air**, surrounds our planet. Any object placed in such a fluid feels a buoyancy force, and if the object is less dense than the fluid, then the buoyancy force can lift it off the ground. Balloons rise because the gas (either helium or hot air) in a balloon is less dense than air. Balloonists control their vertical movements by changing either the buoyancy of the balloon or the weight of the payload, but they cannot directly control their horizontal motions—balloons float with the **wind**, the flow

FIGURE 20.1 The balloon and gondola used by Piccard and Jones during their successful attempt to circle the globe in March 1999.

of air from one place to another. In order to reach their destination before running out of supplies, long-distance balloonists must find a fast wind flowing in the correct direction. Thus, round-the-world balloonists study the **weather**, the physical conditions (the temperature, pressure, moisture content, and wind velocity and direction) of the atmosphere at a given time and location, in great detail to decide when and where to take off and how high to fly. Different winds blow at different elevations, so balloonists adjust their elevation to catch the best wind. On leaving their launching point in the Swiss Alps, Piccard and Jones entered a strong wind flowing from west to east. Despite a few problems (with fuel supplies and heaters), the balloonists circled the globe and touched down in Egypt.

In this chapter, we explore the envelope of air—the atmosphere—through which Piccard and Jones traveled. We begin by learning where the gases came from and how the atmosphere evolved in the context of the Earth System. Then we look at the structure of the atmosphere, and the global-scale and local-scale circulation of the lowermost layer; this circulation ultimately controls the weather, and can lead to the growth of storms. We conclude by exploring **climate**, the average weather conditions for a region over many years.

Chapter Themes

By the end of this chapter, you should know . . .

- how the Earth's atmosphere has evolved over time, and where its gases come from.
- that the atmosphere contains layers, and that temperature and pressure vary with elevation.
- why the atmosphere circulates regionally, and what causes prevailing winds.
- what causes fronts and clouds to form, and how they relate to weather.
- why thunderstorms, tornadoes, and hurricanes originate, and how they cause damage.
- the difference between weather and climate, and why climate varies with location.

20.2 THE FORMATION OF THE ATMOSPHERE

When the Earth formed about 4.57 Ga, it was initially surrounded by gas molecules gravitationally attracted to its surface from the protoplanetary disk surrounding the newborn

Sun (see Chapter 1). This *first* atmosphere, which consisted mostly of hydrogen and helium and traces of other gases, survived only a short time. Heat from the Sun caused the light atoms (H and He) in it to move so rapidly that they eventually escaped the attraction of Earth's gravity. Effectively, the primary atmosphere leaked into space and was blown away by the solar wind.

Even as the primary atmosphere was disappearing, volcanic activity on Earth released new gases that could accumulate to form a *second* atmosphere once the Earth had developed a magnetic field to deflect the solar wind. The elements in these volcanic gases had been bonded to minerals inside the Earth, but during melting, the elements separated from the minerals and bubbled out of volcanoes. Volcanic gas consists of about 70 to 90% water (H_2O), with smaller amounts of carbon dioxide (CO_2) and sulfur dioxide (SO_2), along with traces of other gases including nitrogen (N_2) and ammonia (NH_3). The secondary atmosphere consisted of these gases plus, according to some researchers, other gases brought to Earth by comets.

The second atmosphere of the Earth underwent major evolutionary changes over geologic time. Specifically, when the Earth cooled sufficiently for water to condense, definitely by 3.9 to 3.8 Ga (though possibly earlier), most of the water from the atmosphere fell as rain, accumulated on the surface, and either filled oceans, lakes, and streams or sank underground to become groundwater. As a consequence, the proportion of water in the atmosphere decreased. Once liquid water existed at Earth's surface, the concentration of CO_2 in the atmosphere also began to decrease, because CO_2 can dissolve in oceans and then combine with calcium to form solid carbonate minerals that precipitate to the sea floor. CO_2 also reacts with rocks exposed on the surface of continents to produce solid chemical-weathering products (Interlude B). In addition, ultraviolet radiation from the Sun split apart molecules of NH_3 to produce nitrogen and hydrogen atoms. The lightweight hydrogen atoms escaped into space, but the nitrogen atoms combined to form N_2 molecules. Molecular nitrogen is a stable gas that does not chemically react with rocks, so once it forms it remains in the air for a long time. Because of the accumulation of new N_2 molecules and the loss of atmospheric H_2O and CO_2, the proportion of nitrogen in the atmosphere progressively increased.

It is noteworthy that if the Earth's surface had been too hot for liquid water to exist on it, then CO_2 would not have been removed from the atmosphere, and Earth's atmosphere today would resemble the present atmosphere of Venus. Venus's atmosphere currently contains 96.5% CO_2, whereas Earth's contains only 0.033%. CO_2 is a greenhouse gas, meaning that it traps heat in the atmosphere. The high concentration of CO_2 makes Venus's atmosphere so hot that lead can melt at the planet's surface.

If you were suddenly to travel back through time and appear on Earth 3.8 billion years ago, you would instantly suffocate, for the atmosphere back then contained virtually no molecular oxygen (O_2). It took the appearance of life on Earth to add significant quantities of oxygen to the air, for O_2 is produced by photosynthesis. The first photosynthetic organisms, cyanobacteria, appeared on Earth between 3.8 and 3.5 Ga and began to add O_2 to the atmosphere. The first biologic oxygen entered the atmosphere by ~2.7 Ga, and by 2.4 Ga, Earth's atmosphere contained about 1% of its present oxygen level. Oxygen concentrations increased very slowly. At about 1.2 Ga, there may have been a boost in the production of O_2 with the appearance of photosynthetic algae. Only by about 600 Ma did oxygen levels in the air become substantial, ushering Earth's *third* (modern) atmosphere, dominated by N_2 and O_2, into existence. How oxygen levels changed during the Phanerozoic remains the subject of debate, but the occurrence of charcoal in Silurian strata indicates that there must have been at least 13% oxygen in the air by about 425 Ma, when vascular plants appeared, for vegetation cannot burn in air with less than 13 percent O_2. According to one model, oxygen concentration reached a peak of about 35% during the late Paleozoic coal age and has fluctuated since then, declining since the end of the Mesozoic to the present value of 21% (**Fig. 20.2a, b**).

Oxygen is important not only because it allows complex multicellular organisms to breathe but also because it supplies the raw components for the production of **ozone** (O_3), a gas that absorbs harmful ultraviolet (short-wavelength) radiation from the Sun. Ozone, which accumulates primarily at an elevation of about 30 km, forms by a two-step reaction:

(1) O_2 + energy (from the Sun) → 2O
(2) $O_2 + O → O_3$

Only when enough ozone had accumulated in the atmosphere could life leave the protective blanket of seawater, which also absorbs ultraviolet radiation. When this happened, terrestrial plants and animals could evolve. These organisms have themselves interacted with and modified the atmosphere ever since.

In sum, Earth's atmosphere today consists mostly of volcanic gas modified by interactions with sunlight, the land, and life. The proportions of gases reflect these interactions—if all life were to vanish and all volcanic activity to cease, lighter gases would leak into space in only a few million years.

Did you ever wonder...
has our atmosphere always been breathable?

FIGURE 20.2   Stages in the evolution of Earth's atmosphere over time (not to scale).

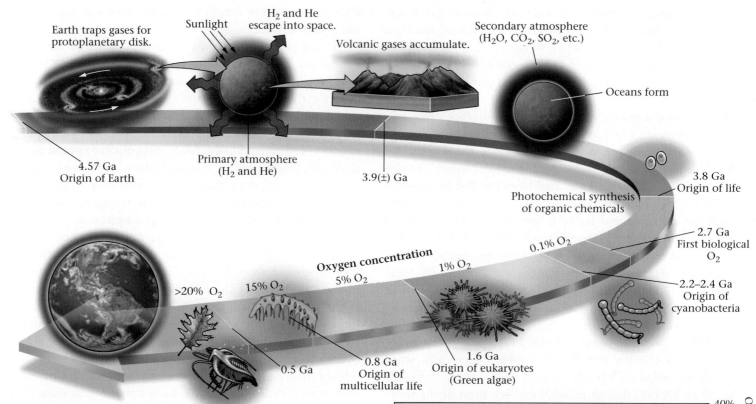

(a) The composition of the atmosphere has changed profoundly during geologic time.

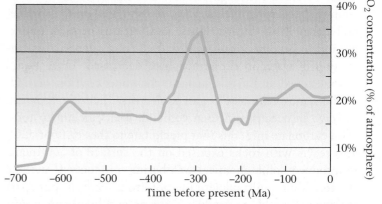

(b) During the Phanerozoic, oxygen composition has varied. It was particularly high during Caboniferous and Permian time.

Take-Home Message

- The composition of the atmosphere has changed radically over Earth's history.
- The early atmosphere consisted of volcanic gases. When the ocean formed, most H_2O became liquid water; CO_2 dissolved in the oceans and eventually precipitated to form limestone.
- The remaining atmosphere became rich in N_2. It does not dissolve or react chemically.
- O_2 in air comes from photosynthesis. Oxygen concentration remained low until about 600 Ma.
- Some oxygen bonds to form ozone (O_3), which protects Earth's surface from ultraviolet rays.

THINK: How does the evolution of life affect oxygen concentrations?

20.3 THE ATMOSPHERE IN PERSPECTIVE

The Components of Air

Completely dry air consists of 78% nitrogen and 21% oxygen. The remaining 1% includes several gases in trace amounts. Trace gases are important in the Earth System. They include carbon dioxide (CO_2) and methane (CH_4), which are greenhouse gases that regulate Earth's atmospheric temperature; greenhouse gases allow solar radiation from the Sun to pass through, but they trap infrared radiation rising from the Earth's surface. Trace gases also include ozone (O_3), which protects the surface from ultraviolet radiation. Not all trace gases are beneficial, though. For example, radon, produced by the natural decay of uranium in rocks, is radioactive and can cause cancer. It can seep from the Earth and collect in dangerous concentrations in the basements of houses.

In addition to gases, the air contains trace amounts of **aerosols**. These tiny particles (mostly 1 to 4 micrometers in diameter) of liquid or solid material are so small that they remain suspended in the air, just like fine mud remains suspended in river water. Aerosols include tiny droplets of water and acid and microscopic particles of sea salt, volcanic ash, clay, soot, and pollen (**Fig. 20.3**).

Atmospheric Pollutants

During the past two centuries, human activity has added substantial amounts of pollutants (both gases and aerosols) to the air, primarily through the burning of fossil fuels and through industrial operations. Pollutants include sulfate (SO_4^{-2}) and nitrate (NO_3^-) molecules, which react with water to make a weak acid that then falls from the sky as **acid rain**. Where acid rain falls, lakes, streams, and the ground become more acidic and thus toxic to fish and vegetation (particularly coniferous trees). The burning of fossil fuels has significantly increased the amount of CO_2 in the atmosphere. As we discuss in Chapter 23, most researchers have concluded that this increase is a cause of global warming, an overall rise in atmospheric temperature. In 1985, researchers discovered that certain pollutants, notably chlorofluorocarbons (CFCs), react with ultraviolet light from the Sun to release chlorine atoms, which, in turn, react with ozone and break it down. These reactions appear to happen mainly in high clouds above polar regions during certain times of the year, thus preferentially removing ozone from these regions to create an ozone hole. We'll examine such human impact on the atmosphere further in Chapter 23.

Pressure and Density Variations

Air is not uniformly distributed in the atmosphere. In the Earth's gravity field, the weight of air at higher elevations presses down on and compresses air at lower elevations (see Chapter 2). **Air pressure**, the push that air can exert on its surroundings, and air density therefore increase toward the surface of the Earth (**Fig. 20.4**). Because the density of a gas reflects the number of gas molecules in a given volume, a gulp of air on the top of

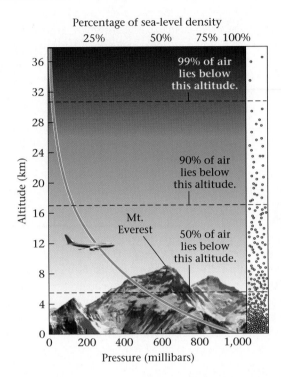

FIGURE 20.4 This graph shows air pressure versus elevation on the Earth. Climbers on top of Mt. Everest breathe an atmosphere that contains only about 33% of the atmospheric gases at sea level.

Mt. Everest, where the air pressure is about one-third that at sea level, contains about a third as many O_2 molecules as air at sea level. Therefore, most climbers seeking to reach Mt. Everest's summit must breathe bottled oxygen. Similarly, the cabin of an airliner must be pressurized to provide adequate oxygen for normal breathing. We measure air pressure in units called atmospheres (atm), where one atm is approximately the pressure exerted by the atmosphere at sea level (about 14.7 pounds per square inch, or 1,035 grams per square centimeter), or in bars, where 1 bar is about 0.986 atm.

Because of the decrease in air density with elevation, 50% of the atmosphere's molecules lie below an elevation of 5.6 km, 90% lie below 16 km, and 99.99997% lie below 100 km. Thus, even though the outer edge of the atmosphere, a vague boundary where the gas density becomes the same as that of interplanetary space, lies as far as 10,000 km from the Earth's surface, most of the atmosphere's molecules lie within a shell only 0.5% as wide as the solid Earth. Though thin, the atmospheric shell contains sufficient gas to turn the sky blue (**Box 20.1**).

Heat and Temperature in Air

The molecules that constitute the atmosphere, or any gas, do not stand still but are constantly moving. We refer to the *total* kinetic energy (energy of motion) resulting from the movement

FIGURE 20.3 A recent forest fire produces smoke that adds aerosols (e.g., soot), as well as CO_2 gas, to the atmosphere.

BOX 20.1

Why Is the Sky Blue?

When astronauts standing on the Moon looked up, even during the day, they saw a black sky filled with stars and a nearly white Sun. On Earth, when we look up during the day, we see a blue sky when there are no clouds and a white sky when there are. The color we see in the sky is the result of the dispersal of energy that occurs when light interacts with air molecules and other particles in the atmosphere, a process called scattering. When light is scattered, a single ray divides into countless beams, each heading off in a different direction. It's similar to what happens when you shine a spotlight on a mirror ball over a dance floor.

Sunlight consists of a broad spectrum of electromagnetic radiation. Different components of the atmosphere scatter different wavelengths of this light, because the ability of a particle to scatter light depends on its size relative to the wavelength. Aerosols (soot, water droplets, dust, etc.) are larger than all wavelengths of light, so all wavelengths reflect off these particles. As a result, light scattered from these particles appears white—that's why clouds are white.

The interaction of light with tiny gas molecules in the air is different. The molecules absorb some wavelengths and shortly afterward, re-emit them. Blue light (with shorter wavelengths) is absorbed and re-emitted more than red light (with longer wavelengths). Put another way, blue light is scattered more effectively by air molecules, while red light passes through more easily. Since scattered light heads in all directions, some of it returns back to space; this light is called backscattered light. Because of backscattering, the intensity of light received on Earth's surface is less than it would be if Earth had no atmosphere. It's also because of scattering that shadows are not completely dark. Light is able to enter regions (beneath trees, for example) that are blocked from the Sun.

When you look at an object, the color you see is the color of light either emitted from the object or reflected off it. A plum appears violet because it reflects violet light and absorbs other wavelengths, and a traffic light appears green because it emits green light. On a clear day, with the Sun high in the sky, the gas in the atmosphere scatters primarily blue light. When we look up, we are seeing the blue light scattered off the gas molecules (**Fig. Bx20.1a**). Because blue light is scattered, not all of it reaches the Earth (some has been backscattered to space), so the Sun appears yellow rather than white—if you subtract a little blue light from white light, you get yellow light. On a cloudy or hazy day, there are more aerosols (such as water droplets) in the atmosphere, and these scatter all wavelengths of light. Therefore, the sky appears white unless the clouds are so dense that they do not let light through, in which case they appear gray.

At the beginning or end of the day, when the Sun lies close to the horizon, light passes through a thicker amount of the atmosphere, and so much of the blue light scatters back to space that the sunlight reaching Earth contains mostly red wavelengths—if you subtract a lot of blue light from white light, you are left with red light. Thus, the Sun appears red, as does the light that reflects off clouds (**Fig. Bx20.1b**).

FIGURE Bx20.1 Atmospheric color depends on the thickness of the atmosphere that light passes through.

(a) Scattering of light by gas molecules leads to the brilliant blue of a clear sky at noon in Hawaii.

(b) So much scattering happens when sunlight enters the atmosphere at a low angle, during sunrise, that all that's left is red light.

of molecules in a gas as its thermal energy, or heat. Note that heat and temperature are not the same—a gas's temperature is a measure of the *average* kinetic energy of its molecules (see Chapter 2). A volume of gas with a small number of rapidly moving molecules has a higher temperature but may contain less heat than a volume with a large number of slowly moving molecules. If we add heat to a gas, its molecules move faster and its temperature rises, and the gas will try to expand to occupy a larger volume.

Relations between Pressure and Temperature

When air moves from a region of greater pressure to a region of less pressure, without adding or subtracting heat, or mixing with its surroundings, it expands. When this happens, the air temperature decreases. Such a process is called **adiabatic cooling** (from the Greek *adiabatos*, meaning impassable). Air cools at 6° to 10°C per kilometer that it rises, depending on its moisture content. Dry air cools faster. The reverse is also the case: if air moves from a region of less pressure to a region of greater pressure, without adding or subtracting heat, it contracts, and the air temperature increases. Such a process is called **adiabatic heating**. Adiabatic cooling and heating are important processes in the atmosphere, because pressure changes with elevation. When air near the ground surface (where pressure is greater) flows up to higher elevations (where pressure is less), it undergoes adiabatic cooling; but when air from high elevations flows down and compresses, it undergoes adiabatic heating.

Water in the Air

Earlier, we examined the percentages of gases in completely dry (water-free) air. In reality, however, air contains variable amounts of water vapor—from 0.3% above a hot desert to 4% in a rain forest before a heavy downpour.

Meteorologists, scientists who study the weather, specify the water content of air by a number called the **relative humidity**, the ratio between the measured water vapor content and the maximum possible amount of water vapor that the air could hold, expressed as a percentage.* The maximum possible amount varies with temperature. Warmer air can hold more water vapor than can colder air. When air contains as much water vapor as possible, it is saturated, whereas air that doesn't is unsaturated. If we say that air at a given temperature has a relative humidity of 20%, we mean the air contains only 20% of the water that it could hold at that temperature when saturated. Such air feels dry. Air with a relative humidity of 100% is saturated and feels very humid, or damp.

---

* "Relative humidity" differs from "absolute humidity." The latter term refers to the mass of water in a volume of air. Absolute humidity is given in grams per cubic meter.

Because cold air can't hold as much water vapor as warm air, air that is unsaturated when warm may become saturated when cooled, without the addition of any new water. The temperature at which the air becomes saturated is called the **dewpoint temperature**; dew forms when unsaturated air cools at night and becomes saturated, so that liquid water condenses on surfaces. When the temperature is below freezing, frost develops. And when saturated air rises and adiabatically cools, its moisture condenses to form a **cloud**, a mist of tiny water droplets or tiny ice crystals (Fig. 20.5). Clouds contain about 50% vapor and 50% liquid or solid water in droplets.

When water in the air changes from liquid to gas or vice versa (a process called a change of state), the temperature of the air also changes. That is, when water evaporates, it absorbs heat, so that molecules can break free from the liquid. The condensation of water reverses this process and therefore releases heat. The heat released during condensation was "hidden," in the sense that it comes only from the change of state and does not require an external energy source. Thus, it is called the *latent* heat of condensation.

Atmospheric Layers

Temperature, in contrast to pressure, does not decrease continuously from the surface of the Earth to the outer edge of the atmosphere. In fact, starting from the surface and going

FIGURE 20.5 As air rises and enters regions of lower pressure, it expands and adiabatically cools. Here, we see moist air rising and becoming less dense; its moisture condenses at elevations above 3 km and produces a cloud.

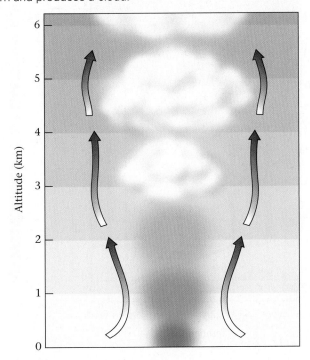

up, temperature decreases, then increases, then decreases, then increases. Elevations where temperatures stop decreasing and start increasing or vice versa divide the Earth's atmosphere into four layers, separated by **pauses** (Fig. 20.6).

- *Troposphere*: This layer starts at the surface of the Earth and rises to an elevation of 5 km at the poles and 18 km at the equator. Within this layer, the temperature decreases gradually from an average of 18°C at the surface to about –55°C at the top, a boundary called the tropopause. The name **troposphere** comes from the Greek *tropos*, which means turning. It is an appropriate name, because air in the troposphere constantly undergoes convection (see Chapter 2). The heat that initiates movement in the troposphere comes primarily from infrared radiation rising from the Earth's surface. The radiation heats air at the base of the troposphere—in effect, the Earth bakes air from below. This warm air then rises, and cold air sinks to take its place. As we will see, this movement causes most weather phenomena, so the troposphere can also be thought of as the "weather layer."

- *Stratosphere*: Beginning at the tropopause and continuing up for about 10 km, the temperature stays about the same. Then it slowly rises, reaching a maximum of about

0°C at an elevation of about 47 km, a boundary called the stratopause. The layer between the tropopause and the stratopause is the **stratosphere**, so named because it doesn't convect much and thus remains relatively stable and stratified. The stratosphere generally doesn't mix much with the underlying troposphere, because at the tropopause hotter (less dense) air already lies on top of cooler (denser) air. Most of the ozone in Earth's atmosphere resides in the stratosphere. Heating in the stratosphere happens because ozone absorbs ultraviolet radiation from the Sun.

- *Mesosphere*: The temperature decreases in the interval, called the **mesosphere**, from about 47 to 82 km. At the mesopause, the top of the mesosphere, the temperature has dropped to –85°C. The mesosphere does not absorb much solar energy and thus cools with increasing distance from the hotter stratosphere below. Most meteors (shooting stars) begin burning here and have vaporized by the time they reach an altitude of 25 km.

- *Thermosphere*: The outermost layer of the atmosphere, the **thermosphere**, contains very little of the atmosphere's gas (less than 1%). The temperature increases with elevation in this layer because gases of the thermosphere absorb short-wavelength solar energy. (The Sun broils this layer from above.) Because the thermosphere has so little gas, it contains very little heat, even though it registers a high temperature. Thus, an astronaut walking in space at an elevation of 200 km doesn't feel hot.

The densities of gases in the lower three layers of the atmosphere are great enough that moving atoms and molecules frequently collide. Like billiard balls, they bounce off each other and shoot off in different directions. This constant, chaotic motion stirs the gases sufficiently to make a homogeneous mixture, so that the air in the lower three layers has essentially the same proportion of different gases regardless of location. For this reason, atmospheric scientists refer to the troposphere, stratosphere, and mesosphere together as the homosphere. In contrast, atoms and molecules in the low-density thermosphere collide so infrequently that this layer does not homogenize. Rather, gases separate into distinct layers based on composition, with the heaviest (nitrogen) on the bottom, followed in succession by oxygen, helium, and at the top, hydrogen, the lightest atom. To emphasize this composition, atmospheric scientists refer to the thermosphere as the heterosphere.

So far, we've distinguished atmospheric layers according to their thermal structure (troposphere, stratosphere, mesosphere, and thermosphere) and according to the degree their gases mix (homosphere and heterosphere). We need to add

FIGURE 20.6 The principal layers of the atmosphere.

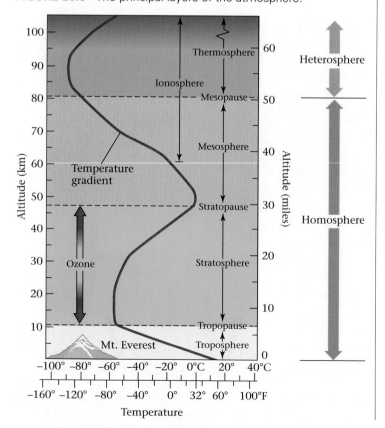

one more "sphere" to our discussion. The **ionosphere** is the interval between 60 and 400 km, and thus includes most of the mesosphere and the lower part of the thermosphere. It was given its name because in this layer, short-wavelength solar energy strips nitrogen molecules and oxygen molecules of their electrons and transforms them into positive ions. The ionosphere plays an important role in modern communication in that, like a mirror, it reflects radio transmissions from Earth back down so that they can be received over great distances.

The ionosphere also hosts a spectacular atmospheric phenomenon, the auroras (*aurora borealis* in the northern hemisphere and *aurora australis* in the southern), which look like undulating, ghostly curtains of varicolored light in the night sky (Fig. 20.7). They appear when charged particles (protons and electrons) ejected from the Sun, especially when solar flares erupt, reach the Earth and interact with the ions in the ionosphere, making them release energy. Auroras occur primarily at high latitudes because Earth's magnetic field traps solar particles and carries them to the poles.

FIGURE 20.7 The splendor of an aurora borealis lights up the night sky in Arctic Canada.

Take-Home Message

- Modern air consists of 78% N_2 and 21% O_2; the remaining components, including CO_2, are trace gases.
- Air also contains aerosols (consisting of tiny liquid or solid particles) and pollutants.
- Air pressure decreases upward, so 90% of air lies beneath an elevation of only 16 km.
- As air rises, pressure decreases so it expands and cools; as it sinks, it compresses and warms.
- Air contains 0.3% to 4% water; the amount of water it can contain depends on temperature. We represent the concentration of water as relative humidity.
- Temperature decreases in elevation up to about 5 to 18 km, then increases up to about 47 km, then decreases up to 82 km, then increases to the top. These changes delimit layers (in succession: troposphere, stratosphere, mesosphere, and thermosphere).

THINK: What causes the ionosphere to form?

20.4 WIND AND GLOBAL CIRCULATION IN THE ATMOSPHERE

A gusty breeze on a summer day, the steady currents of air that once blew clipper ships across the oceans, and a fierce hurricane all are examples of wind, the movement of air from one place to another. We can feel wind because of the impacts of air molecules as they strike us. The existence of wind emphasizes that the lower part of the atmosphere is in constant motion, swirling and overturning at rates between a fraction of a kilometer and a few hundred kilometers per hour. This circulation happens on two scales, local and global. Local circulation refers to the movement of air over a distance of tens to a thousand kilometers. Global circulation refers to the movement of volumes of air in paths that ultimately carry it around the entire planet. (We can picture local circulation as eddies in global-scale "rivers" of air.) To understand both kinds of circulation, we must first see what drives air from one place to another, and examine energy inputs into the atmosphere.

Lateral Pressure Changes and the Cause of Wind

The air pressure of the atmosphere not only changes vertically, it also changes horizontally at a given elevation. The rate of pressure change over a given horizontal distance, called a pressure gradient, can be represented by the slope of a line on a graph plotting pressure on the vertical axis and map distance on the horizontal axis (Fig. 20.8a). We can use a map to illustrate how air pressure at a given elevation varies with location. A line on a map along which the air has a specified pressure is called an **isobar** (Fig. 20.8b); the pressure is the same at all points on an isobar. Isobars can never touch, because they represent different values of pressure.

Winds form wherever a pressure gradient exists. In a general sense, air accelerates from a high-pressure region to a low-pressure region; in other words, it starts flowing down a pressure gradient. To see why, step on one end of a balloon filled with air; you momentarily increase the pressure at that end, so the air flows toward the other end (Fig. 20.8c). A difference in pressure exists between one isobar and the next, so the air starts flowing perpendicular to these lines. As we will see, however, the Coriolis effect modifies wind direction so in general, flow is not perpendicular to isobars.

FIGURE 20.8 The concept of a pressure gradient.

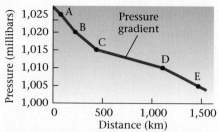

(a) A graph of pressure changes from A to E. The gradient is greater where the slope is steeper.

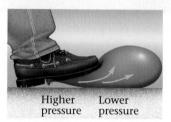

(c) Air flows from a region of high pressure to a region of low pressure.

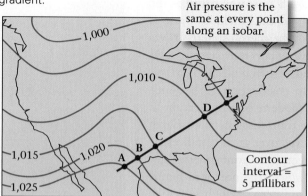

Air pressure is the same at every point along an isobar.

Contour interval = 5 millibars

(b) The air pressure decreases between point A and point E. Note that the rate of change, the pressure gradient, is greater between A and C than between C and D. Spacing of isobars reflects the gradient.

Regional Circulation in the Atmosphere

Solar energy constantly bathes the Earth. Of this energy, 30% reflects back to space (off clouds, water, and land); air or clouds absorb 19%; and land and water absorb 51%. The energy absorbed by land and water later reradiates as infrared radiation and thus bakes the atmosphere from below. As noted earlier, greenhouse gases absorb part of this reradiated energy before it can return to space. Because the Earth is a sphere, not all latitudes receive the same amount of incoming solar energy, or **insolation**: portions of the Earth's surface hit by direct rays of the Sun receive more energy per square meter than portions hit by oblique rays. We can simulate this contrast with a flashlight. If you point a flashlight beam straight down, you get a small but bright spot of light on the ground; but if you aim the beam so that it hits the ground at an angle of 45°, the spot covers a broader area but does not appear as bright (**Fig. 20.9**). Higher latitudes thus receive less energy than lower latitudes. Because of the tilt of Earth's axis, the amount of solar radiation that any point on the surface receives changes during the year, which is why we have seasons (**Box 20.2**).

The contrast in the amount of solar radiation received by different latitudes means that polar regions are cooler at the surface than are equatorial regions. In 1735, George Hadley, a British mathematician, realized that this contrast could cause global air to circulate over broad areas, because warmer air is less dense and tends to rise, whereas cooler air is denser and tends to sink. Specifically, he suggested that warm air at the equator would rise and flow toward the pole, to be replaced by cool polar air, which would flow to

the equator at lower elevations (**Fig. 20.10a**). Hadley's proposal of hemispheric-scale circulation, however, did not take into account an important factor, namely the Earth's rotation and the resulting Coriolis effect. The Coriolis effect, as we learned in Chapter 18, refers to the deflection that happens to an object as it moves from the circumference to the center of a rotating disk or from the equator to the pole of a rotating sphere (and vice versa) (**Fig. 20.10b**).

Because of the Coriolis effect, northward-moving high-altitude air in the northern hemisphere deflects to the east, so by the latitude of 30° N, it basically moves due east and cannot

FIGURE 20.9 The amount of insolation depends on the angle at which sunbeams strike the Earth, so the poles are colder.

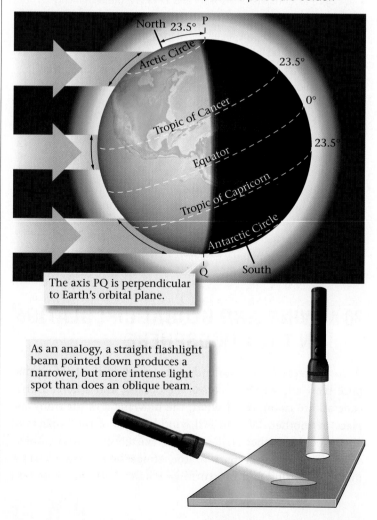

The axis PQ is perpendicular to Earth's orbital plane.

As an analogy, a straight flashlight beam pointed down produces a narrower, but more intense light spot than does an oblique beam.

FIGURE 20.10 Global circulation patterns in the atmosphere.

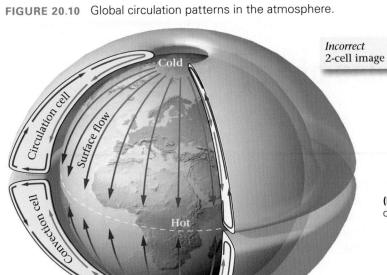

(a) If the Earth did not rotate, two simple circulation cells would exist, each stretching from the equator to the pole.

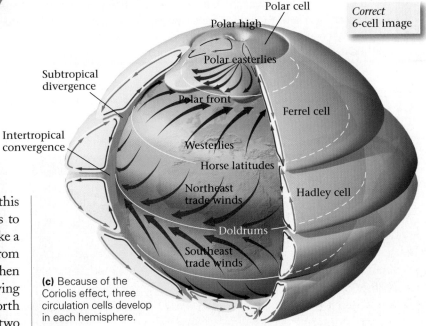

(b) But, because the Earth spins, the Coriolis effect influences the motion of moving objects and materials.

(c) Because of the Coriolis effect, three circulation cells develop in each hemisphere.

make it the rest of the way to the pole. By the time it reaches this latitude, the air has also cooled significantly and thus starts to sink. This means that a circulation cell, a current that looks like a loop in cross section, develops that conveys warm air north from the equator to the subtropics (latitude 30°), where it sinks. When the sinking air reaches low elevations, it divides, some moving back toward the equator near the surface and some moving north near the surface. A place where sinking air separates into two flows moving in opposite directions is a **divergence zone**.

Meanwhile, cool air from the polar region moves south near the surface and deflects to the west. At latitudes of 40° to 60°, the near-surface polar air collides, or converges, with the northward-moving, near-surface mid-latitude air. This air must rise, because there is nowhere for the extra air to go but up. A place where two surface air flows meet so that air has to rise is called a **convergence zone**. The convergence zone between 40° and 60° is called the **polar front**. The air that rises along the polar front divides at the top of the troposphere, with some air heading toward the equator at high altitude and some heading toward the pole, so a circulation cell develops at mid-latitudes, and another develops at polar latitudes. Because of seasonal changes on Earth, the latitude of the polar front changes during the year.

In sum, because of the Coriolis effect, circulating air in the troposphere splits into three circulation cells in each hemisphere. The low-latitude cells, extending from the equator to a latitude of about 30°, are called **Hadley cells**, in honor of George Hadley. The mid-latitude cells are called **Ferrel**

cells, in honor of the American meteorologist William Ferrel, who proposed them. The high-latitude cells are simply called **polar cells** (Fig. 20.10c).

In this three-cell-per-hemisphere model of atmospheric circulation, several belts of high and low pressure form around the Earth. The equatorial regions are marked by a belt of convergence called the intertropical convergence zone. In these regions, intense insolation makes air hot, so that it rises. The upward flow of hot, rising air that diverges higher up produces an area of low pressure; thus, the convergence zone at the equator is also called the equatorial low. The low-pressure area in equatorial regions fills with cooler air flowing in from higher latitudes at the base of the Hadley cells.

At high elevations, cool air at the top of the Hadley cell flows toward cool air at the top of the Ferrel cell. On reaching a latitude of about 30°, this air converges, sinks, and becomes

BOX 20.2

The Earth's Tilt: The Cause of Seasons

A satellite view emphasizes that snow cover distribution changes over the course of the year (**Fig. Bx20.2a**). This change occurs because the Earth has seasons. Seasons exist because the Earth's axis tilts (currently at 23.5°) relative to the plane of its orbit. As our planet moves around the Sun, the direction of tilt relative to the Sun varies, so the insolation at a given latitude and, therefore, the average monthly temperature, changes during the course of a year (**Fig. Bx20.2b**). For example, when the northern hemisphere tilts toward the Sun, it receives more

radiation, warms up, and enjoys summer, but when the northern hemisphere tilts away from the Sun it receives less radiation, cools down, and endures winter.

At any given time, half the Earth has daylight, and half experiences night. The boundary between these two hemispheres is called the terminator. Because of the Earth's tilt, the terminator does *not* always pass through the North and South Poles (Fig. Bx20.2b). On June 21, a special day called a solstice, the terminator lies 23.5° away from the poles, as measured along

the Earth's surface. The line of latitude at this position is called the Arctic Circle in the northern hemisphere and the Antarctic Circle in the southern hemisphere. On June 21, a person standing on the Arctic Circle sees the midnight sun (the Sun is visible for an entire day), whereas anyone south of the Antarctic Circle has night for a full 24 hours. During the northern summer, regions north of the Arctic Circle see the midnight sun for more than a day, and the North Pole itself experiences the midnight sun for a full 6 months. Similarly, regions south of the Ant-

FIGURE Bx20.2 Because of the tilt of Earth's axis, we have seasons.

January

July

(a) Satellite photos emphasize the contrast in snow cover of the northern hemisphere as seasons change.

denser. More air, from still higher elevation, constantly fills in the space left by the sinking air. The presence of sinking air causes a high-pressure zone, called the subtropical high, to form at the surface of the Earth. At a latitude of about 60°, surface air of the polar cell converges with surface air of the Ferrel cell. The converging air has nowhere to go but up, so it rises and expands, resulting in the development of a low-pressure zone at the surface. At high elevation, this rising air diverges. Some of it moves northward toward the pole, where it sinks to create a polar high at the surface.

Recall that where warm, moist air rises, it cools adiabatically. Cooler air can hold less moisture, so this air becomes supersaturated. The excess moisture condenses and forms

clouds, which produce rain. Therefore, the equatorial lows are regions of heavy rainfall, which leads to the growth of tropical rain forests. In contrast, in high-pressure belts where cool, dry air sinks, air contracts and heats adiabatically. The resulting hot air can absorb moisture, so that it rains only rarely. Thus, as we will see in Chapter 21, the subtropical regions at latitude 30° include some of the Earth's major deserts.

Prevailing Surface Winds

Global-scale atmospheric flow in the six major circulation cells on Earth creates belts in which surface air generally moves in a consistent direction. Such airflows are called **prevailing**

arctic Circle experience 24-hour nights for more than a day, and the South Pole itself sees nothing but night for a full 6 months. December 21 is the other solstice, but on this day all regions south of the Antarctic Circle see the midnight sun, whereas all regions north of the Arctic Circle have per-petual night. During the southern summer, the South Pole sees daylight for 6 months.

On the June 21 solstice, the Sun's rays are exactly perpendicular to the Earth (the Sun is directly overhead) at latitude 23.5° N, the Tropic of Cancer, and on December 21 the Sun's rays are exactly perpendicular to the Earth at 23.5° S, the Tropic of Capricorn. On September 22 and March 20, each known as an "equinox," the Sun is directly overhead at the equator. The two solstices and two equinoxes divide the year into four astronomical seasons.

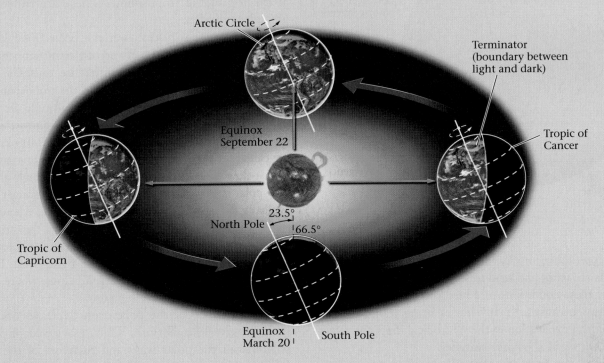

(b) In the northern hemisphere's summer solstice, the noonday Sun is directly overhead at the Tropic of Cancer, and the region above the Arctic Circle sees the midnight sun, whereas at the winter solstice, the noonday Sun is directly overhead at the Tropic of Capricorn and the region south of the Antarctic Circle sees the midnight sun. At the equinoxes, the noonday sun is directly overhead at the equator.

winds. (Note that here we're examining *surface* winds, the base of the circulation cell.) These simple patterns tend to be disrupted by local-scale winds caused by storms or affected by local topography. When describing winds, meteorologists label them according to the direction the air comes from. Thus, a westerly wind blows from west to east.

Let's start our tour of prevailing winds at the base of the Hadley cell in the northern hemisphere. Near-surface winds start to flow from 30° N to the south and are deflected west. Thus, between the equator and 30° N, surface winds come out of the northeast and are called the northeast **trade winds**, so named because they once carried trading ships westward from Europe to the Americas. Trade winds in the southern hemisphere, which start flowing northward and then deflect to the west, end up flowing from southeast to northwest and are called the southeast trade winds. Where the southeast and northeast trade winds merge at the equator, they are flowing almost due west (**Fig. 20.11**). But winds along the equator are very slow, because the air is mostly rising. Ships tended to be becalmed in this belt, which came to be called the **doldrums**.

At the base of a Ferrel cell, at mid-latitudes, surface air starts to move toward the north, but because of the Coriolis effect it curves to the east. Thus, throughout much of North America and Europe, the prevailing surface winds come out of the west or southwest and are known as the surface westerlies.

FIGURE 20.11 If the Earth had a uniform surface, distinct high- and low-pressure zones would form on its surface. But surface winds flowing from high-pressure zones to low-pressure zones are deflected by the Coriolis effect, so that for much of their course they flow almost parallel to pressure zones.

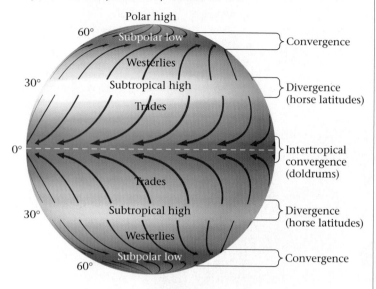

In the subtropical high itself, where airflow is primarily downward, winds are weak and tend to shift in different directions. In the past, these conditions inhibited the progress of sailing ships. Perhaps because so many horses being transported by ship died of heat exhaustion in the subtropical high, the region came to be known as the horse latitudes.

Finally, at the base of polar cells, surface air starts by flowing from the pole southward, but deflects to the west. The resulting prevailing winds, which are known as the polar easterlies, flow from the polar high to the subpolar low, and converge with the westerlies of mid-latitudes at the polar front.

Note that prevailing winds, for much of their course, follow paths almost parallel to the high- and low-pressure belts that surround the Earth, and thus flow nearly parallel to isobars. This situation seems to contradict the statement made earlier that wind flows from regions of high pressure to regions of low pressure, and thus should flow at almost right angles to isobars. This apparent contradiction is a consequence of the Coriolis effect.

High-Altitude Winds in the Troposphere: The Jet Streams

In the upper atmosphere, a global-scale pressure gradient exists because of temperature differences between the equator and the pole. At the equator, air is warmer and thus expands. This causes the top of the troposphere there to rise relative to the top of the troposphere in polar regions, so, as

noted earlier, the troposphere is thicker over the equator than over the poles. As a consequence, the air pressure at a given elevation above the equator is greater than at the same elevation above the poles. This pressure gradient causes overall high-altitude air to flow north; in fact, some air from the top of the Hadley cell spills over and moves north over the top of the Ferrel cell (**Fig. 20.12a**). Once again, the Coriolis effect comes into play and makes this air deflect to the east, and so in the northern hemisphere we have generally westerly winds at the top of the troposphere. These are called the high-altitude westerlies.

In two special places, over the polar front and over the horse latitudes, air masses of very different temperatures come in contact. This temperature difference causes a particularly steep pressure gradient at the top of the troposphere. Because of the steepness of the gradient, high-altitude westerlies flow particularly fast. These zones of rapid movement, where winds typically flow at speeds of between 200 and 400 km per hour, are called **jet streams** (**Fig. 20.12b**). Jet streams can be viewed as fast-flow zones within an overall westerly high-altitude flow. The polar jet stream, which tends to be the stronger of the two, typically flows at 7–11 km above the surface, whereas the subtropical jet stream flows at 9–14 km above the surface. Thus, airliners commonly encounter the jet streams. Planes flying east have shorter flying times than those flying west, because the former are helped along by strong tailwinds, whereas the latter battle headwinds that slow them down.

The latitude of the polar front tends to undulate over time in wave-like motions. Thus, as viewed on a map, the positions of the polar front and therefore of the jet stream follow large, curving trajectories that sometimes bring the jet stream down to southern latitudes of North America and sometimes to northern latitudes of Canada. Further, the average position varies with the season (**Fig. 20.12c**).

Take-Home Message

- Air pressure changes horizontally, over a distance, producing a pressure gradient.
- Air starts to move from high-pressure to low-pressure regions; the Coriolis force modifies the flow direction.
- Due to latitudinal variation in insolation, three regional-scale air circulation cells (Hadley, Ferrel, and polar) develop in each hemisphere.
- The surface flow of air in the circulation cells causes convergence and low pressure, both at the equator and at about 60° N or S.
- Regional circulation cells drive prevailing surface winds and the jet streams.

THINK: What are doldrums, and why do they form?

FIGURE 20.12 Jet streams form just above the tropopause and follow wavy paths that go around the Earth.

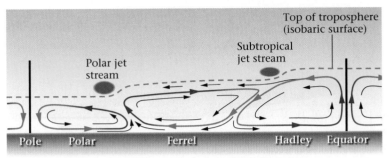

(a) The tropopause is an isobaric surface. At steps in this surface, high-altitude winds are stronger and jet streams form.

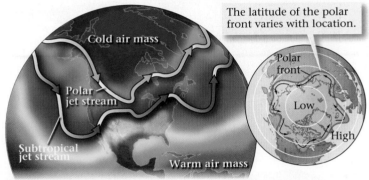

The latitude of the polar front varies with location.

b) Jet streams form at the boundary between regions of the troposphere with different temperatures. The boundaries are wavy, so the jet streams are wavy.

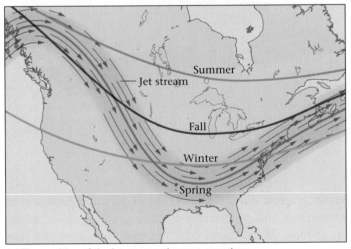

(c) The position of the jet stream changes over time.

20.5 **WEATHER AND ITS CAUSES**

If the surface of the Earth were a perfectly uniform sphere, airflow might follow the global circulation patterns just described. But, in fact, the Earth's surface is heterogeneous—parts are covered with water, parts with land. In the oceans, water currents affect the temperature of the sea surface, and on land, there are variations in altitude and vegetation cover. All these factors ensure that the heat that rises into the base of the atmosphere from the surface of the Earth varies with time and location in a complex way, and that where moving air interacts with obstacles, it must be forced to rise or to deflect sideways. Thus, it's not surprising that on a *local scale*, atmospheric flow can be quite turbulent.

The term *weather* refers to local-scale conditions as defined by temperature, air pressure, relative humidity, rainfall or snow, cloud cover, and wind speed. Weather conditions can vary dramatically (see Table 20.1). A specific set of weather conditions, reflecting the configuration or system of air movement in the atmosphere, may affect a region for a period of time. Such weather systems can move across the surface of the Earth, carried by prevailing winds. In this section, we'll briefly review some of the more general aspects of weather and see how they result from the interaction of air masses along fronts.

Air Masses and Fronts

Air that remains or passes over a certain region for a length of time takes on characteristics that reflect both its interaction with the Earth's surface and the amount of insolation it receives. For example, air that has hovered over a warm sea tends to become warm and moist, whereas air stuck over cold land becomes cold and dry. A body of air, on the order of 1,500 km across, that has recognizable physical characteristics (temperature and moisture content) is called an **air mass**. Air masses move within the overall global circulation of the atmosphere, and their paths are controlled by prevailing winds. Meteorologists name

TABLE 20.1 Weather Extremes

Lowest recorded air pressure at sea level	870 millibars (25.7 inches); during Typhoon Tip (1979), Pacific Ocean
Highest air pressure at sea level	1,084 millibars (32.02 inches); Agata, Siberia
Lowest air temperature (world)	−89°C (−129°F); Vostok, Antarctica
Lowest air temperature (North America)	−63°C (−81°F); Yukon, Canada
Highest air temperature (world)	58°C (136°F); Libya
Highest air temperature (North America)	57°C (134°F); Death Valley, California
Highest recorded wind speed (world), not in a tornado	372 km per hour (231 mph); Mt. Washington, New Hampshire

the different air masses (Fig. 20.13). The weather of a region can change drastically when one air mass replaces another. For example, a summer day in a midwestern state will be cool and dry under a continental polar air mass but hot and humid under a maritime tropical air mass.

The boundary between two air masses is called a **front**. Meteorologists recognize several kinds of fronts of which we introduce three. At a *cold* front, a cold air mass pushes underneath a warm air mass (Fig. 20.14a). As a consequence, the warm, moist air flows up to higher elevations, where it expands and cools adiabatically. The moisture it contains then condenses to form large clouds from which heavy rains may fall. As a *warm* front moves into a region, the warm air slowly rises over the cool air (Fig. 20.14b). Again, the air expands and cools, and the moisture condenses, so clouds develop over the boundaries between the two air masses. It's important to note that even though a front can be portrayed on a map as a line, it is in fact a sloping surface. Cold fronts typically have steeper surfaces than warm fronts.

Not all air masses move at the same velocity. Typically, cold fronts move faster than warm fronts and overtake them. Where this happens, the cold front lifts up the base of the warm front, so that the warm front no longer intersects the ground surface. Meteorologists refer to the geometry that results when a cold front pushes underneath a warm front as an *occluded* front (Fig. 20.14c).

Cyclonic and Anticyclonic Flow

Divergent flow means that air moves outward, away from a region, resulting in a net *loss* of air molecules within the region. Where divergent flow takes place aloft, near the top of the troposphere, a low-pressure system develops at the surface. A

FIGURE 20.13 Air masses that form in different places have been assigned different names. The arrows indicate the average directions in which the air masses move.

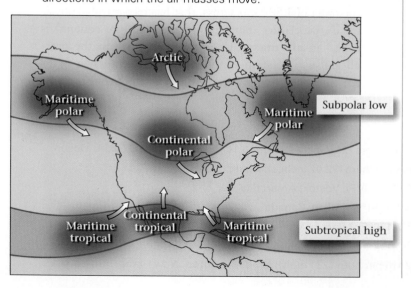

FIGURE 20.14 The formation of fronts between air masses.

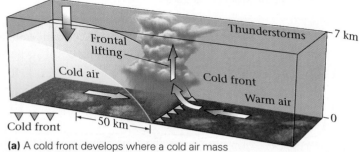

(a) A cold front develops where a cold air mass moves under a warm air mass.

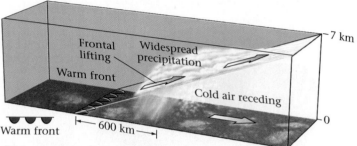

(b) A warm front develops where a warm air mass moves over a cold air mass.

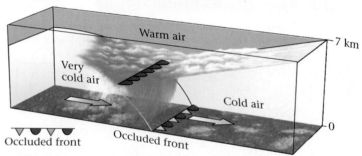

(c) An occluded front develops where a fast-moving cold front overtakes a warm front, and lifts the base of the warm front off the ground. Note the symbols used to represent fronts.

pressure gradient forms between the low-pressure center and its surroundings and, wherever a pressure gradient exists, winds start to blow. If the Earth were not spinning, air would flow from high pressure to low pressure, roughly perpendicular to the isobars surrounding the low-pressure region. But because of the apparent force caused by the Coriolis effect, winds deflect to follow paths around the center of the low pressure. If there were no friction between the air and the Earth's surface, a balance would develop between the pressure-gradient force and the Coriolis force, and the path of air flowing around the low-pressure center would be nearly circular. Friction, however, slows the wind and decreases the Coriolis force, so the pressure-gradient force dominates. As a consequence, winds follow a spiral path toward the interior of the low-pressure center. The rotational flow is called **cyclonic flow** (Fig. 20.15). In the northern hemisphere, this flow is counterclockwise. Because of the inward spiral of winds, convergence takes place in the center of a cyclonic low-pressure system at lower elevations. The excess

FIGURE 20.15 Air spirals downward and clockwise (creating an anticyclone) at a high-pressure mass and upward and counterclockwise (creating a cyclone) at a low-pressure mass.

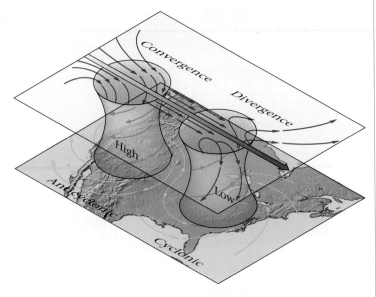

FIGURE 20.16 Formation of a mid-latitude (wave) cyclone.

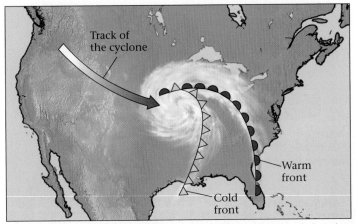

(a) An extratropical cyclone treks from west to east across North America.

air must move up and rises until it reaches the divergent zone near the top of the troposphere. This rising air expands and cools, so the water vapor it contains condenses to form clouds that may produce rain. As a result, a low-pressure system tends to be associated with cloudy, rainy weather.

If there is convergence aloft, air moves inward and down, resulting in a net *gain* of molecules with the region and the production of a high-pressure system. The high-pressure system must also be surrounded by a pressure gradient and, therefore, winds. These winds spiral outward and follow clockwise paths around the high-pressure mass in the northern hemisphere, producing an **anticyclonic flow** (see Fig. 20.15). As a consequence, divergence takes place near the ground and air from above sinks down to fill the deficit. The sinking air compresses, warms, and undergoes a reduction in relative humidity. Thus, a high-pressure system tends to be associated with fair (clear and dry) weather.

In the mid-latitudes that encompass the United States and much of Canada as well as much of Europe, the weather commonly reflects the movement of a large low-pressure center that moves from west to east. Air circulates counterclockwise around the low, creating a system known as an **extratropical cyclone**, a mid-latitude cyclone, or a wave cyclone (Fig. 20.16a). Simplistically, wave cyclones develop when air on one side of a cold front shears sideways past air on the other side (Fig. 20.16b). The shear of air along the cold front warps the face of the front into the shape of a wave. When this happens, warm air starts to move north, up and over the cold air mass, whereas cold air circles around and starts to move south and downward, pushing the cold front forward. The two fronts meet at a V, the point of which lies near the center of the low-pressure mass. In a satellite image, a mid-latitude cyclone is a huge spiral mass of clouds, rotating counterclockwise; the clouds develop along both fronts and are centered on the low-pressure mass. The cold front of a wave cyclone tends to move faster than the warm front, so eventually the warm front becomes occluded, and the cyclone dies out.

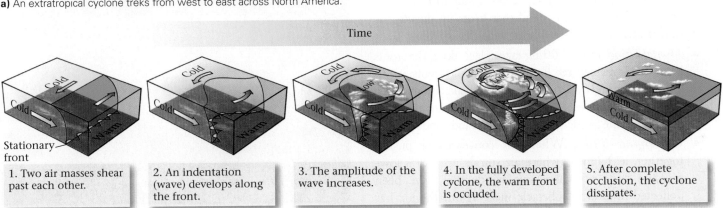

Time

1. Two air masses shear past each other.

2. An indentation (wave) develops along the front.

3. The amplitude of the wave increases.

4. In the fully developed cyclone, the warm front is occluded.

5. After complete occlusion, the cyclone dissipates.

(b) Researchers can define many stages during the evolution of an extratropical (mid-latitude or wave) cyclone.

FIGURE 20.17 Clouds can create a dramatic spectacle. Here, a towering anvil cloud forms, as viewed from an airplane.

Clouds and Precipitation

Let's now look at clouds a little more closely (Fig. 20.17). As noted earlier, a cloud is a region of the atmosphere where about half of the atmospheric moisture exists in the form of tiny water droplets (about 20 micrometers across, less than a third of the diameter of a human hair; 1 micrometer = 0.001 mm), or in the form of tiny ice crystals. Clouds that form at ground level make up **fog**. Because clouds reflect and scatter incoming sunlight, they keep the ground cooler during the day, but at night they prevent infrared radiation from escaping and thus keep the ground warmer.

Droplets or ice grains in clouds form by condensation or deposition, respectively, when the air becomes saturated with water vapor. During cloud formation, water condenses on **condensation nuclei**, preexisting solid aerosols. Air can become saturated (so that clouds develop) when evaporation from the ground surface provides additional water or when the air cools so that its capacity to hold water decreases. Cooling may take place at night, simply because of the loss of sunlight, or at any time that air rises and adiabatically cools. Meteorologists recognize several conditions that cause air to rise, and they refer to these as **lifting mechanisms**.

- *Convective lifting*: This process occurs where air warms, becomes buoyant, and rises. Clouds may form by convective lifting over an island that heats up during the day, relative to the surrounding sea (Fig. 20.18a).
- *Frontal lifting*: Frontal lifting takes place along the fronts between air masses (see Fig. 20.14). At cold fronts, warm air pushes up and over a steep wall of cold air, rising rapidly to form large clouds. At warm fronts, the advancing warm air rides up the gentle slope of the front and condenses.
- *Convergence lifting*: Where air converges or pushes together, as happens where air spirals up into a low-pressure zone or where two winds that have been deflected around an obstacle meet again, the air has

FIGURE 20.18 Causes of cloud formation.

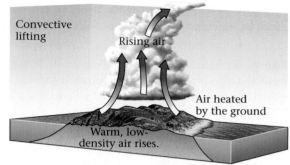

(a) Convective lifting occurs where warm air starts to rise.

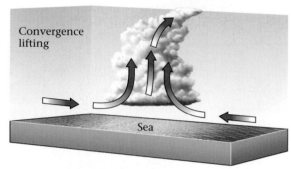

(b) Convergence lifting takes place where winds merge—the air has nowhere to go but up.

(c) Orographic lifting happens where moist winds run into a mountain range and are forced up.

Lifting forms clouds over a ridge in Utah.

nowhere to go but up, and thus rises and cools, forming clouds (Fig. 20.18b).

- *Orographic lifting*: This type of lifting happens where moisture-laden wind blows toward a mountain range, and on meeting the mountain range can go no farther and must rise. As a result, clouds form above the mountain range (Fig. 20.18c).

Rain, snow, hail, and sleet fall from clouds in two ways, depending on the temperature of the cloud. In warm clouds, rain develops by **collision and coalescence**, during which the tiny droplets that compose the cloud collide and stick together to create a larger drop (Fig. 20.19a). Eventually, water drops that are too big to be held in suspension by circulating air start to fall, incorporating more droplets as they descend. Depending on how far the drops have fallen and on how much moisture is available, drops reach different sizes before they hit ground. Typical raindrops have a diameter of 2 mm and fall at a velocity of about 20 km per hour. Any drops larger than 5 mm tend to break into smaller ones upon collision. If rain falls through colder air near the ground, it freezes to become sleet.

In cold clouds, the mist contains a mixture of very cold water droplets and tiny ice crystals. The water droplets evaporate faster than the ice (because water molecules are less tightly bound to liquid than to solid) and provide moisture that condenses onto preexisting ice crystals, leading to the growth of hexagonal snowflakes. If the air below the cloud is very cold, the snow falls as powder-like flakes; if the air is close to the melting temperature, large, wet clumps of snowflakes fall; and if the air is warmer than 0°C, the snow transforms to rain before it hits the ground. This kind of formation, involving the growth of ice crystals in a cloud at the expense of water droplets, is called the **Bergeron process**, after Tor Bergeron, the Swedish meteorologist who discovered it (Fig. 20.19b).

Many kinds of clouds form in the troposphere. It wasn't until 1803, however, that Luke Howard, a British pharmacist, proposed a simple terminology for describing clouds (Fig. 20.20). First, we divide clouds into types based on their shape: puffy, cotton-ball- or cauliflower-shaped clouds are **cumulus** (from the Latin word for "stacking"). Clouds that occur in relatively thin, stable layers and thus have a sheet-like or layered shape are called **stratus**. High clouds that have a wispy shape and taper into delicate, feather-like curls are called **cirrus**. We can then add a prefix to the name of a cloud to indicate its elevation: high-altitude clouds (above about 7 km) take the prefix *cirro*, mid-altitude clouds take the prefix *alto*, and low-altitude clouds (below 2 km) do not have a prefix. Finally, we add the suffix *nimbus* or the prefix *nimbo* if the cloud produces rain.

Applying this cloud terminology, we see that a nimbostratus is a layered, sheet-like rain cloud, and a cumulonimbus is a rain-producing puffy cloud. Cumulonimbus clouds can be truly immense, with their bases lying at less than 1 km high and their tops butting up against the tropopause at an elevation of over 14 km. Large cumulonimbus clouds spread laterally at the tropopause to form broad, flat-topped clouds called anvil clouds.

The differences in cloud types depend on whether the clouds develop in stable or unstable air. *Stable* air does not have a tendency to rise, because it is colder than its surroundings. *Unstable* air has a tendency to rise, because it is warmer than its surroundings. Cumulus clouds, which in time-lapse photography look as if they're boiling, form in unstable air. They billow because of updrafts (upward-moving air) and downdrafts (downward-moving air). Plane flights through these clouds will be rather bumpy. In contrast, stratus clouds indicate stable air.

FIGURE 20.19 Mechanisms of raindrop formation.

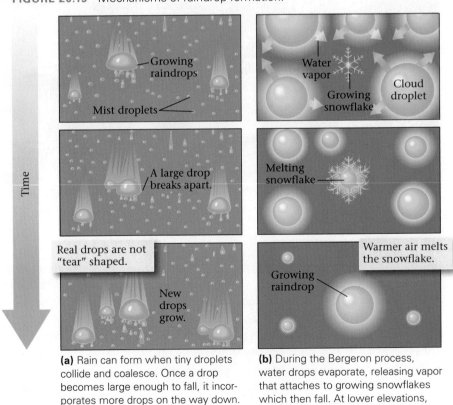

(a) Rain can form when tiny droplets collide and coalesce. Once a drop becomes large enough to fall, it incorporates more drops on the way down. Really large drops can break apart.

(b) During the Bergeron process, water drops evaporate, releasing vapor that attaches to growing snowflakes which then fall. At lower elevations, where air is warmer, the flakes melt.

FIGURE 20.20 The various types of clouds on Earth.

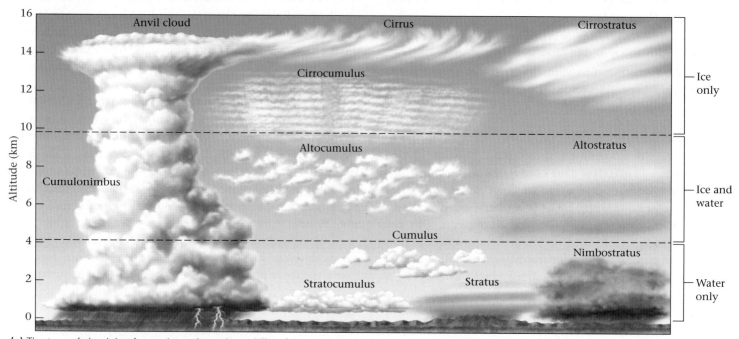

(a) The type of cloud that forms depends on the stability of the air, the temperature at which moisture condenses, and the wind speed.

(b) A satellite photo of the Earth displays the distribution of clouds. At times more than half of the surface has cloud cover.

Take-Home Message

- An air mass is a volume of air, about 1,500 km across, whose temperatures and pressures distinguish it from adjacent masses.

- Air masses interact along a front. At a cold front, a colder air mass pushes under a warmer one, whereas at a warm front, warmer air rises over cooler air. Clouds typically form along fronts.

- Air spirals toward the center of a low-pressure mass to cause cyclonic flow and away from the center of a high-pressure mass to cause anticyclonic flow.

- In mid-latitudes, large extratropical cyclones form, resulting in huge, spiral masses of clouds and strong storms.

- Air rises and cools for a variety of reasons; as a result, water in air condenses to form clouds.

- Water droplets in clouds can collide and coalesce to form raindrops, or the water may crystallize onto tiny ice particles which can melt as they fall. Falling water (as liquid or ice) is precipitation.

- Atmospheric scientists distinguish among kinds of clouds based on shape and elevation, and on whether they produce rain.

THINK: What determines whether precipitation falls as rain or snow?

20.6 STORMS: NATURE'S FURY

The rain from an overcast sky may be inconvenient for a picnic, but it won't threaten life or property and generally will be appreciated by farmers. In contrast, a storm poses a threat. A **storm** is an episode of severe weather, when winds, rainfall, snowfall, and, in some cases, lightning become strong enough to be bothersome and even dangerous (**Fig. 20.21**). Storms are commonly associated with large pressure gradients (as may exist across a front), for pressure gradients produce strong winds. Storms also form where local conditions cause warm, moist air to rise. Rising moist air can trigger a storm because when the air reaches higher elevations, it condenses and, as we have seen, releases latent heat. This heat warms the air, makes it more buoyant, and thus causes it to rise still further, until it becomes cool enough to produce clouds. Meanwhile, at ground level, new moist air flows in beneath the clouds to replace the air that has already risen. This new air then immediately starts to rise and causes the clouds to build upward. Warm, moist air effectively feeds the storm. Once the clouds become thick enough to start producing heavy rain, and/or the wind becomes strong enough to be troublesome, we can say that a storm has been born. We'll now look at various types of storms.

Thunderstorms

An individual thunderstorm may occur in isolation, but along strong fronts, a row of thunderstorms may develop to comprise a squall line. At any given time, over 2,000 thunderstorms are active around the globe, and over 100,000 take place in the United States every year. Thunderstorms form where a cold front moves into a region of particularly warm, moist air (as happens, for example, in North America's mid-latitudes during the summer where cold polar air masses collide with warm Gulf air masses); where convection driven by solar radiation takes place in a region with an immense supply of moisture (as occurs over tropical rain forests); or where orographic lifting causes clouds to form over mountains. A particularly large, slowly rotating thunderstorm is called a **supercell**. Supercells develop where there is a large source of warm, moist air, and where high-elevation air is flowing rapidly relative to lower-elevation air. The resulting shear between the air at different elevations causes rotation to begin, initially with a horizontal axis. Downdrafts push the rotation axis down so it becomes steep, and the clouds of the supercell swirl into a spiral with an updraft in the middle.

A typical thunderstorm has a relatively short lifespan, lasting from under an hour to a few hours (**Fig. 20.22**). The storm begins when a cumulonimbus cloud, fed by a steady supply of warm, moist air, grows large. The rising hot air, kept warm by the addition of energy from the latent heat of condensation, creates updrafts that cause the cloud to stack, or billow upward, toward the tropopause. When this air adiabatically cools, precipitation begins.

If updrafts in the cloud are strong enough, ice crystallizes in the higher levels of the cloud, where temperatures are below freezing, building into ice balls known as **hail** (or hailstones). As a hailstone tosses about high in the cloud, more ice accretes (attaches) to its surface. A discrete shower of hail may fall from a cloud over a few minutes to form a hail streak on the ground, typically 2 km by 10 km and elongated in the direction that the storm moves.

FIGURE 20.21 A thunderstorm can drench an area with rain and attack it with lightning.

FIGURE 20.22 Evolution of a thunderstorm takes place in three stages.

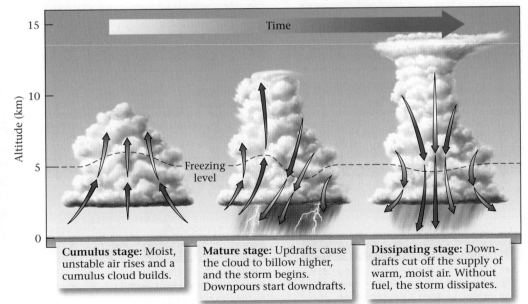

Time

Altitude (km)

Freezing level

Cumulus stage: Moist, unstable air rises and a cumulus cloud builds.

Mature stage: Updrafts cause the cloud to billow higher, and the storm begins. Downpours start downdrafts.

Dissipating stage: Downdrafts cut off the supply of warm, moist air. Without fuel, the storm dissipates.

Though most hailstones are pea-sized, the largest recorded hailstone reached a diameter of 14 cm and weighed 0.7 kilograms.

Once precipitation begins, a thunderstorm has reached its mature stage. Falling rain pulls air down with it, and evaporative cooling takes place, creating strong downdrafts. By this stage, the top of the cloud has reached the top of the troposphere and begins to spread laterally to form an anvil cloud. Because of the simultaneous occurrence of updrafts and downdrafts, a mature thunderstorm produces gusty winds and the greatest propensity for lightning. Eventually, downdrafts become the overwhelming wind; their cool air cuts off the supply of warm, moist air, so the thunderstorm dissipates.

Thunderstorms are so named because of the crashes and rumbles of thunder that typically accompany them. The thunder, as we will see, is a consequence of lightning. A **lightning bolt** (or lightning stroke) is a giant spark that forms between two parts of a cloud, between two clouds, or between a cloud and the ground. It resembles the spark that shocks you when you shuffle along a rug and then touch a door handle. But the spark from the door handle is only about a millimeter long and a fraction of a millimeter in diameter, whereas a single lightning bolt may be up to 10 km long, up to 3 cm in diameter, and can contain enough energy to power a house for many months.

Lightning develops, in part, because of movement of various kinds of ice particles (hail and ice crystals) within storm clouds. Simplistically, the particles transfer electrons to each other when they come in contact, and different types of particles with different charges move to different parts of the storm cloud. Overall, positive charges accumulate toward the top of the cloud, and negative charges accumulate toward the bottom (**Fig. 20.23a**).

The negative charges at the base of the cloud repel negative charges on the ground below, creating a zone of positive charge on the ground. Air is a good insulator, so the charge separation can become very large until a giant spark or pulse of current, the lightning stroke, jumps across the gap. Essentially, lightning is like a giant short circuit across which a huge (30-million-volt) pulse of electricity flows.

As we noted, lightning strokes can jump from one part of a cloud to another, or from a cloud to the ground. In the case of cloud-to-ground lightning, the stroke begins when electrons leak from the negatively charged base of the cloud incrementally downward across the insulating air gap, creating a conductive path, or leader. While this happens, positive charges flow upward to the cloud through conducting materials, such as trees or buildings (**Fig. 20.23b**). The instant that the charge flows connect, a strong current develops, allowing electrons to flow groundward; this current is called the "return stroke" (**Fig. 20.23c**).

We hear **thunder**, the cracking or rumbling noise that accompanies lightning, because the immense energy of a flash almost instantaneously heats the surrounding air to a temperature of 8,000° to 33,000°C, and this abrupt expansion, like an explosion, creates sound waves that travel through the air to our ears. But because sound travels so much more slowly than light, we hear thunder after we see lightning. A 5-second time delay between the two means that the lightning flashed about 1.6 km (1 mile) away.

> **Did you ever wonder . . .**
> why do you hear claps of thunder during a storm?

Over 80 people a year die from lightning strikes in the United States alone, and many more are seriously burned or shocked. Lightning that strikes trees heats the sap so quickly that the trees literally explode. Lightning can spark devastating forest fires and set buildings on fire. You can reduce the hazard to buildings by installing lightning rods, upward-pointing iron spikes that conduct electricity directly to the ground so that it doesn't pass through the building.

FIGURE 20.23 Formation of a lightning bolt happens because a charge separation forms in a cloud as rain fails.

(a) The negative charge at the cloud's base repels negative charges on the ground, so the ground develops a positive charge.

(b) As a leader descends from the cloud, positive charges flow upward from an object on the ground.

(c) When connection is complete, a return stroke from the ground to the cloud produces the main part of the flash.

Tornadoes

Thunderstorms that grow to be intense supercells may spawn one or more tornadoes. A **tornado** is a near-vertical, funnel-shaped cloud in which air rotates extremely rapidly around the axis (center line) of the funnel (Fig. 20.24a, b). In other words, a tornado is an intense vortex beneath a severe thunderstorm. The word probably comes either from the Spanish *tonar*, meaning to

FIGURE 20.24 Tornadoes and the damage they cause.

(a) A moderate-sized tornado touches down near Mulvane, Kansas.

(b) A huge tornado near Eureka, Illinois.

(c) The swath of destruction left by a 1999 tornado in Oklahoma.

(d) Even downtowns are not immune—this tornado struck Miami in 1997.

(e) Close-up of the damage due to a 2007 tornado in Kansas.

(f) A helicopter view of the devastation track left by one of the April, 2011 tornadoes that hit Tuscaloosa, Alabama.

turn, or *tronar*, meaning thunder (perhaps referring to the loud noise generated by a tornado).

In the mid-latitudes of the northern hemisphere, where most tornadoes form, air in the funnel normally rotates counterclockwise around the center and spirals upward. Air in the fiercest tornadoes may move at speeds of up to 500 km per hour (about 300 mph). The diameter of the base of the funnel in a small tornado may be only 5 m across, while in the largest tornadoes it may be as wide as 1,500 m across. Because of the centrifugal force caused by rotating air, air diverges from a tornado, so the tornado itself is an intense low-pressure center.

In North America, tornadoes drift with a thunderstorm from southwest to northeast, because of the prevailing wind direction, traveling at speeds of up to 100 km per hour. They tend to hopscotch across the landscape, touching down for a stretch, then rising up into the air for a while before touching down again. This characteristic leads to a bizarre incidence of damage—one house may be blasted off its foundation while its next-door neighbor remains virtually unscathed. Winds on one side of a tornado move in the same direction as the overall funnel drift whereas those on the other side move in the opposite direction. Thus, for a northeast-moving counterclockwise funnel, damage is greater on the southeast side of the funnel.

Small tornadoes may cut a swath less than a kilometer long, but large tornadoes raze the ground for tens of kilometers, and the largest have left a path of destruction up to 500 km long (Fig. 20.24c–f). In 1925, the Tristate tornado, one of the most enormous tornadoes on record, ripped across Missouri, Illinois, and Indiana, killing 689 people before it dissipated. In some cases, two or three tornadoes may erupt from a single thunderstorm. Massive thunderstorm fronts may produce a tornado swarm or "outbreak," dozens of tornadoes out of the same storm. In April 1974, a single thunderstorm system generated a swarm of at least 148 individual tornadoes, which killed 307 people over 11 states. On November 11, 2002, a chain of thunderstorms covering a belt from Ohio to Alabama spawned 66 tornadoes that left 66 people dead. The death toll would have been higher were it not for warnings broadcast by the U.S. National Weather Service, which sent people scrambling for safety. Between April 25 and April 28, 2011, a "super outbreak" of tornadoes occurred in midwestern and southeastern states. At least 336 tornados were recorded.

Tornadoes cause damage because of the force of their rapidly moving wind. The wind lifts trucks and tumbles them for hundreds of meters, uproots trees, and flattens buildings (Fig. 20.24e). Particularly large tornadoes can even rip asphalt off a highway. In some cases, tornadoes cause strange kinds of damage: they have been known to drive straw through wood, lift cows and carry them unharmed for hundreds of meters, and raise railroad cars right off the ground.

Because of the range of damage a tornado can cause, T. T. Fujita of the University of Chicago proposed a scale that distinguishes among tornadoes on the basis of damage (Table 20.2). The wind speeds in the **Enhanced Fujita scale** are estimates, based on the damage assessment.

Most tornadoes in North America evolve from supercells or develop in squall lines where strong westerlies exist at high altitudes while strong southeast winds develop near the ground surface (Fig. 20.25). These opposing winds shear the air between them, so that the air begins to rotate in a horizontal cylinder. Tornado watchers search for the resulting funnel cloud, the precursor to a true tornado. As the associated thunderstorm matures, updrafts tilt one end of the cylinder up and downdrafts push the other end down, until the air in the cylinder starts spiraling inward and upward, gaining speed as do spinning figure skaters who pull their arms inward. At this stage, the funnel consists only of rotating moisture and may look grayish white. But if the process continues, the low end of

TABLE 20.2 Enhanced Fujita Scale for Tornadoes

Scale	Category	Wind Speed km/h (mph)	Average Path Length; Average Path Width	Typical Damage
EF0	Weak	104–137 (65–85)	0–1.6 km; 0–17 m	Branches and windows broken.
EF1	Moderate	138–177 (86–110)	1.6–5.0 km; 18–55 m	Trees broken; shingles peeled off; mobile homes moved off their foundations.
EF2	Strong	178–217 (111–135)	5–16 km; 56–175 m	Large trees broken; mobile homes destroyed; roofs torn off.
EF3	Severe	218–266 (136–165)	16–50 km; 176–556 m	Trees uprooted; cars overturned; well-constructed roofs and walls removed.
EF4	Devastating	267–322 (166–200)	50–160 km; 0.56–1.5 km	Strong houses destroyed; buildings torn off foundations; cars thrown; trees carried away.
EF5	Incredible	over 322 (over 200)	160–500 km; 1.5–5.0 km	Cars and trucks carried more than 90 m; strong houses disintegrated; bark stripped off trees; asphalt peeled off roads.

FIGURE 20.25 Progressive development of a tornado.

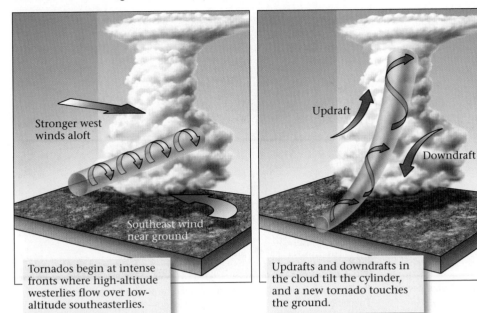

Stronger west winds aloft

Southeast wind near ground

Tornados begin at intense fronts where high-altitude westerlies flow over low-altitude southeasterlies.

Updraft

Downdraft

Updrafts and downdrafts in the cloud tilt the cylinder, and a new tornado touches the ground.

Storm movement

The spiraling wind of the funnel also circulates in the main cloud. The tornado we see is only part of the funnel.

the funnel touches ground, and at the instant of contact, dirt and debris get sucked into the tornado, giving it a dark color.

The special weather conditions that spawn tornadoes in the midwestern United States and Florida develop when cold polar air from Canada collides with warm tropical air from the Gulf of Mexico. These conditions happen most frequently during the months of March through September. So many tornadoes occur during the summer in a belt from Texas to Indiana that this region has the unwelcome nickname "tornado alley" (Fig. 20.26a). During a 30-year span, the number of reported tornadoes per year ranged between about 420 and 1,100 in the United States (with an average of 770 per year; fewer than 20 strike Canada annually). On average, about 80 people a year die in tornadoes but in some years, the toll may be in the hundreds. The 2011 super outbreak alone killed about 350 people because some of its tornadoes rampaged through cities (Fig. 20.26b).

Because of the threat tornadoes pose to life and property, meteorologists have worked hard to be able to forecast them. First they search for appropriate weather conditions. If these conditions exist, meteorologists issue a "tornado watch." If observers spot an actual tornado forming, they issue a "tornado warning" for the region in its general path (the exact path can't be predicted). If you hear a warning, it's best to take cover immediately in a basement, or at least in an interior room away from windows. With the invention of Doppler radar, which uses the Doppler effect (see Chapter 1) to identify rain moving in strong winds, meteorologists may detect tornadoes without even going outside. Probable tornadoes appear in Doppler images as a distinct hook-like shape at the edge of

a thunderstorm (Fig. 20.26c). The ball of debris carried by a tornado causes intense reflectivity on a Doppler image.

Extratropical Cyclones and Nor'easters

Extratropical (mid-latitude or wave) cyclones can produce incredible storms. In the Midwest, they can spawn intense thunderstorms and associated flash floods. In colder weather they can produce blizzards (snowstorms of immense proportions). A blizzard in 1888 dumped up to 1.5 m (60 in.) of snow, and a blizzard in 1996 buried New York under 1.2 m (48 in.) of snow. Large extratropical cyclones of North America that affect the Atlantic coast are called **nor'easters** because the cold, counterclockwise winds of these cyclones come out of the northeast. Some nor'easters are truly phenomenal storms. During some storms, winds were not so strong as those of a hurricane, but they covered such a large area that waves in the open ocean built to a height of over 11 m, a disaster for ships. When the waves reached shore, they eroded huge stretches of beach.

Hurricanes

During the summer and early fall, cyclonic wind systems called *tropical disturbances* develop off the western coast of Africa, near the Cape Verde Islands (latitude 20° N) or in the Caribbean Sea. In these low-pressure regions, air converges and rises and, because of the Coriolis effect, begins to circulate counterclockwise. But because these storms form over the warm tropical waters of the central Atlantic, the air that

FIGURE 20.26 North American tornadoes are most common in "tornado alley," a band extending from Texas to Indiana, where the polar air mass collides with the Gulf Coast maritime tropical air mass. The storm systems are move eastward.

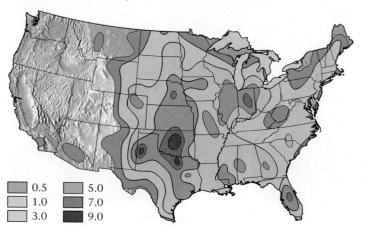

0.5		5.0	
1.0		7.0	
3.0		9.0	

(a) Number of tornadoes per year (per 26,000 sq. km, for a 27-year period)

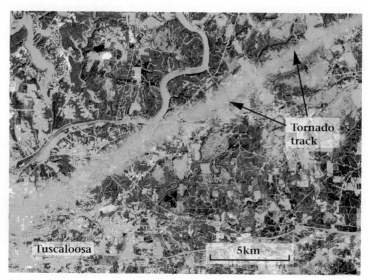

(b) This false-color satellite image shows the track of a tornado near Tuscaloosa, Alabama. Red areas are forested—the winds ripped out the trees.

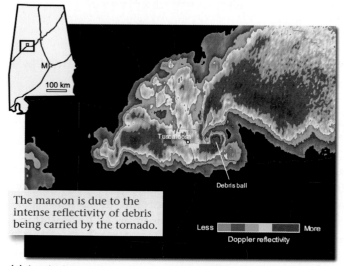

The maroon is due to the intense reflectivity of debris being carried by the tornado.

Less [ ] More
Doppler reflectivity

(c) A radar image shows the characteristic "hook" of a tornado, as viewed by satellite, as it passes near Vilonia, Arkansas, in 2011.

rises within them is particularly warm and moist. As it rises, it cools, and moisture condenses, releasing latent heat of condensation. This latent heat is a fuel that provides energy to the storm—it causes the air to rise still higher, creating even lower pressure at the Earth's surface, which in turn sucks up even more warm, moist air. As winds blow, waves form and spray water into the air, allowing even more water to enter the storm. Thus, the storm grows in strength until it becomes a *tropical depression* whose winds may reach 61 km per hour.

As long as there continues to be a supply of warm, moist air, the storm continues to grow broader, and air within begins to rotate still faster. When the sustained wind speed exceeds 119 km per hour, a hurricane has been born. In other words, a **hurricane** (named for the Carib god of evil) is a huge, rotating

storm, resembling a giant spiral in map view, in which sustained winds are greater than 119 km (74 miles) per hour (Fig. 20.27a, b). Within a hurricane, air pressure becomes much lower because centrifugal force causes divergence. Before the days of satellite forecasting, mariners watched their barometers closely, knowing that an extreme drop signaled the approach of a hurricane.

Once a storm reaches tropical-depression status, it is assigned a name. Each year, the first Atlantic depression has a name beginning with *A*, the second with *B*, and so on in alphabetical order, with alternating male and female names. Names for less-significant hurricanes may be reused, but the names of particularly notorious hurricanes are retired. Meteorologists classify hurricanes according to the **Saffir-Simpson scale** (Table 20.3). The most intense hurricanes have sustained winds of over 230 km per hour.

Traditionally, the designation "hurricane" applies to storms born in the east-central Atlantic that drift first westward and then northward with the prevailing winds. The path, or hurricane track, that they follow allows them to inflict damage to land regions in the Caribbean and the Gulf Coast and east coast of North America (Fig. 20.27d). Occasionally, hurricanes make it across Central America and then thrash the Pacific coast of Mexico and the western United States; some may even drift northwestward to Hawaii. Because hurricanes require warm water (warmer than 27°C), they develop only at latitudes south of about 20°. They do not form close to the equator because there is not enough Coriolis effect at that latitude. Similar storms that form between latitudes 20° N and 20° S in the western Pacific are called *typhoons*; those at latitude 20° N in the Indian Ocean are called *cyclones* (a second use of the word; Fig. 20.27c).

FIGURE 20.27 The structure and distribution of hurricanes.

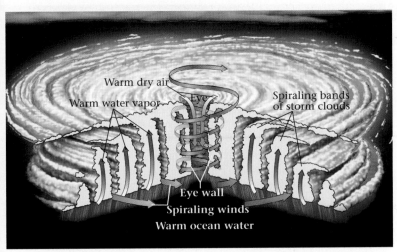

(a) A cutaway shows the spirals of clouds, and the eye of a hurricane. Dry air descends in the eye, producing a small region of calm.

(b) Hurricane Hugo, approaching the Carolinas in 1989. This is an oblique view from a satellite.

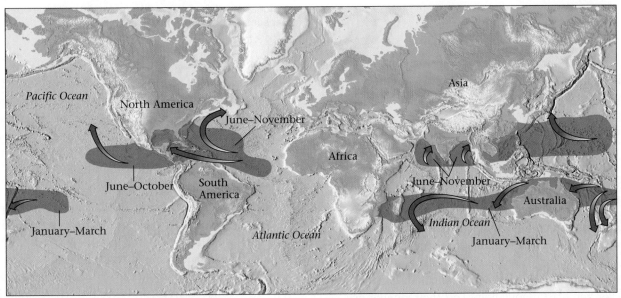

(c) Hurricanes form only in certain regions of the ocean, where water temperatures are high. They generally follow regional tracks and occur during certain times of the year.

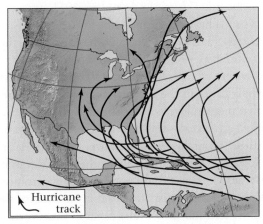

(d) Not all tracks are the same, as these examples show.

A typical hurricane consists of several spiral arms, called rain bands, extending inward to a central zone of relative calm known as the hurricane's eye. A rotating vertical cylinder of clouds, called the eye wall, surrounds the eye (Fig. 20.27a). The entire width of a hurricane ranges from 100 to 1,500 km, with the average around 600 km. Winds spiral toward the eye, and thus their angular momentum, and therefore speed, increases toward the center; as a result, the greatest rotary wind velocity occurs in the eye wall.

Hurricanes, in general, move along their track because of the prevailing motion of the atmosphere. Typically, a hurricane's velocity along its track, the storm-center velocity, ranges from 0 to 60 km per hour (on rare occasions, as fast as 100 km per hour). Because of the rotary motion of wind around the

TABLE 20.3 **Saffir-Simpson Scale for Hurricanes**

Scale	Category	Wind Speed km/h (mph)	Air Pressure in Eye (millibars)	Damage
1	Minimal	119–153 (74–95)	980 or more	Branches broken; unanchored mobile homes damaged; some flooding of coastal areas; no damage to buildings; storm surge of 1.2 to 1.5 m.
2	Moderate	154–177 (96–110)	965–979	Some roofs, doors, and windows damaged; mobile homes seriously damaged; some trees blown down; small-boat moorings broken; storm surge of 1.6 to 2.4 m.
3	Extensive	178–209 (111–130)	945–964	Some structural damage to small buildings; large trees blown down; mobile homes destroyed; structures along coastal areas destroyed by flooding and battering; storm surge of 2.5 to 3.6 m.
4	Extreme	210–250 (131–155)	920–944	Some roofs completely destroyed; extensive window and door damage; major damage and flooding along coast; storm surge of 3.7–5.4 m. Widespread evacuation of regions within up to 10 km of the coast may be necessary.
5	Catastrophic	over 250 (over 155)	less than 920	Many roofs and buildings completely destroyed; extensive flooding; storm surge greater than 5.4 m. Widespread evacuation of regions within up to 16 km of the coast may be necessary.

eye, winds on one side of the eye move in the opposite direction of those on the other side. On the side of the hurricane where winds move in the same direction as the storm's center, the storm-center velocity adds to the rotary motion, making surface winds particularly fast. On the other side of the hurricane, the storm-center velocity subtracts from the rotary velocity, so the winds are slower.

Hurricanes cause damage in several ways (Fig. 20.28).

- *Wind*: Hurricane-force winds may reach such intensity that buildings cannot stand up to them. Hurricanes can tear off branches, uproot trees, rip off roofs, collapse walls, force vehicles off the road, knock trains from their tracks, and blast mobile homes off their foundations.

- *Waves*: The force of hurricane winds shearing across the sea surface generates huge waves, sometimes tens of meters high. Out in the ocean, these waves can swamp or capsize even large ships. Near shore, they batter beachside property, rip up anchored boats and carry them inland, and strip beaches of their sand.

- *Storm surge*: Coastal areas may be severely flooded by a **storm surge**, excess water that is carried landward by the hurricane. In the portion of a hurricane where winds blow onshore, water piles up and becomes deeper over a region of 60 to 80 km, allowing waves to break at a higher elevation than they otherwise would. Storm surges during hurricanes are exacerbated by the very low air pressure in a hurricane, which allows the sea surface to rise still higher—if a hurricane hits at high tide, the

damage may be even worse. A storm surge during the 1900 hurricane that hit Galveston, Texas, flooded the city and killed 6,000 people. Typhoon surges in Bangladesh cause horrendous loss of life because during a typhoon, water of the Indian Ocean submerges the low-lying delta plain of the Ganges. During one 1970 cyclone, the death toll reached 500,000.

- *Rainfall*: The intense rainfall during a hurricane (in some cases, half a meter of rain has fallen in a single day) causes streams far inland to flood. The flooding itself can submerge towns and cause death. Intense rains can also weaken the soil of steep slopes, especially in deforested areas, and thus can trigger mudslides.

On average, five Atlantic hurricanes happen every year. Many Atlantic hurricanes veer northward and move up the eastern seaboard. (Hurrican Irene, in 2011, was an example of this type.) These hurricanes can impact many large population centers. Hurricanes die out when they lose their source of warm, moist air—when they cross onto land or move over the cold waters of high latitudes. 2005 was a particularly bad year for hurricanes in the North Atlantic. There were so many storms that meteorologists went through the entire alphabet of names and had to start over. At one time, a deadly procession of four hurricanes marched westward over the Atlantic at once. Of these, Hurricane Katrina was by far the most devastating (Box 20.3). Because hurricanes are nourished by warm ocean water, some atmospheric scientists fear that a warming of the climate may lead to more and fiercer hurricanes in the future.

FIGURE 20.28 In 1992, Hurricane Andrew caused immense damage in southern Florida before crossing the Gulf of Mexico and slamming into Louisiana.

Take-Home Message

- Thunderstorms develop where convection causes warm, moist air to rise. Latent heat released as air rises causes updrafts to carry air even higher, resulting in billowing clouds and rain.
- Charge separation develops in clouds, resulting in giant sparks (lightning bolts).
- Tornadoes develop because of extreme shear between different levels of air. The spiral of air in a tornado produces wind speeds of up to 500 km/h, resulting in severe damage.
- Large extratropical (wave) cyclones can produce severe storms, including nor'easters.
- Hurricanes begin over warm seawater. Energy from latent heat of condensation enlarges the storm—the largest are up to 1,500 km across. At the eye, winds can exceed 119 km/h.

THINK: Why do hurricanes tend to die after they cross onto land or head to higher latitudes?

20.7 GLOBAL CLIMATE

When we talk about the weather, we're referring to the atmospheric conditions at a certain location at a specified time. But when we speak of the characteristic weather conditions, the typical range of conditions, and the nature of seasons of that region over a long time (say, 30 years), we use the term **climate**. For example, on a given summer day in Winnipeg, central Canada, it may be sunny, hot, and humid, and on a given winter day in southern Florida, it may drop below freezing. But averaged over a year, the weather in Winnipeg is more likely to be cooler and drier than the weather in southern Florida. The variables that characterize the climate of a region include its temperature (both the yearly average and the yearly range), its humidity, its precipitation (both the yearly amount and the distribution during the year), its wind conditions, and the character of its storms.

Climate Controls and Belts

Climatologists, scientists who study the Earth's climate, suggest that several distinct factors control the climate of a region.

- *Latitude*: This is perhaps the most significant factor, for the latitude determines the amount of solar energy a region receives as well as the contrasts between seasons. Polar regions, which receive much less solar radiation over the year, have colder climates than equatorial regions. And the contrast between winter and summer is greater in mid-latitudes than at the poles or the equator. We can easily see the influence of latitude by examining the global distribution of temperature, represented on a map by **isotherms**, lines along which the temperature is exactly the same (Fig. 20.29). Because land and sea do not heat up at the same rate, because the distribution of clouds is not uniform, and because ocean currents transfer heat across latitudes, isotherms are not perfect circles.

- *Altitude*: Because temperature decreases with elevation, cold climates exist at high elevations, even at the equator. Hiking from the base of a high mountain in the Andes to its summit takes you through the same range of climate belts you would pass through on a hike from the equator to the pole.

- *Proximity of water*: Land and water have very different heat capacities (ability to absorb and hold heat). Land absorbs or loses heat quickly, whereas water absorbs or loses heat slowly. Also, water can absorb and hold on to more heat than land can because water is semitransparent; sunlight heats water down to a depth of up to 100 m, whereas sunlight heats land down to a depth of only a few centimeters. Thus, the proximity of the sea tempers the climate of a region: as a rule, locations in the interior of a continent experience a much greater range of weather conditions than regions along the coast.

- *Proximity to ocean currents*: Where a warm current flows, it may warm the overlying air, and where a cold current flows, it may cool down the overlying air. For example, the Gulf Stream brings warm water north from the Gulf of Mexico and keeps Ireland, the United Kingdom, and Scandinavia much warmer than they would be otherwise.

BOX 20.3

Hurricane Katrina!

Hurricane Katrina, in 2005, stands as the most expensive hurricane to strike the United States in historic time. Because of where Katrina passed, worst-case damage-prediction scenarios came to be reality, particularly in the historic city of New Orleans. Let's look at this storm's history.

Tropical storm Katrina came into existence over the Bahamas and headed west. Just before landfall in southeastern Florida, winds strengthened and the storm became Hurricane Katrina (**Fig. Bx20.3a**). This hurricane sliced across the southern tip of Florida, causing several deaths and millions of dollars in damage. It then entered the Gulf of Mexico and passed directly over the Loop Current, an eddy (circular flow) of summer-heated water from the Caribbean that had entered the Gulf of Mexico. Water in the Loop Current reaches temperatures of 32°C (90°F), and thus stoked the storm, injecting it with a burst of energy sufficient for the storm to morph into a Category 5 monster whose swath of hurricane-force winds reached a width of 325 km. When it entered the central Gulf of Mexico, Katrina turned north and began to bear down on the Louisiana-Mississippi coast. The eye of the storm passed just east of New Orleans and then across the coast of Mississippi (**Fig. Bx20.3b**). Storm surges broke records, in places rising 7.5 m above sea level, and then washed coastal communities off the map along a broad stretch of the Gulf Coast (**Fig. Bx20.3c, d**). In addition to the devastating wind and surge damage, Katrina caused the drowning of New Orleans.

To understand what happened to New Orleans, we must consider the city's environmental history. New Orleans grew on the Mississippi Delta between the banks of the Mississippi River on the south and Lake Pontchartrain (actually a bay in the Gulf of Mexico) on the north (see Fig. Bx20.3b). The oldest part of the town, a neighborhood called the French Quarter, was built on the relatively high land of the Mississippi's natural levee. Newer parts of the city, however, spread out over the lower delta plain. As the decades passed, people modified the surrounding delta landscape by draining wetlands, by constructing artificial levees that confined the Mississippi River, and by

extracting groundwater. Sediment beneath the delta compacted and the delta's surface has been starved of new sediment, so large areas of the delta sank below sea level. Today, most of New Orleans lies in a bowl-shaped depression as much as 2 m below sea level—the hazard implicit in this situation had been recognized for years (see Fig. 17.5b).

The winds of Hurricane Katrina ripped off roofs, toppled trees, smashed windows, and triggered the collapse of weaker buildings, but their direct consequences were not catastrophic. However, when the winds blew storm surge into Lake Pontchartrain, its water level rose beyond most expectations and the weight of the high water pressed against the system of artificial levees and flood walls that had been built to protect New Orleans (**Fig. Bx20.3e**). Upkeep of the levee system had been neglected for many years. Thus, hours after the hurricane eye had passed, the high water of Lake Pontchartrain finally found a weakness along the flood walls bordering the 17th Street Canal and pushed out a 100-m-long section. The canal was one of four that had been built to carry water pumped out of New Orleans away to Lake Pontchartrain. There was no gate at its mouth to keep water out when the storm surge caused the lake to rise, and the flood walls were too weak to resist the push of the rising water. Preliminary studies suggest that the foundations of the walls along the canal had not been driven deeply enough, so that the bases of the walls were anchored in a very weak bed of plant debris, remnants of an ancient swamp, rather than in the firmer sand below—under pressure, the bases of the walls just gave way. Breaks eventually formed in other locations as well. A day after the hurricane was over, New Orleans began to flood. As the water line climbed the walls of houses, brick by brick, residents fled first upstairs, then to their attics, then to their roofs. Water

spread across the city until finally the bowl of New Orleans filled to the same level as Lake Pontchartrain and 80% of the city was submerged (**Fig. Bx20.3f**). The disaster grew to be of national significance as the trapped population sweltered without food, drinking water, or adequate shelter. With no communications, no hospitals, and few police, the city almost descended into anarchy. It took days for outside relief to reach the city, and by then, many had died and much of New Orleans, a cultural landmark and major port, had become uninhabitable. Even years later, many neighborhoods remain wastelands (**Fig. Bx20.3g**).

FIGURE Bx20.3 Hurricane Katrina, in 2005, overwhelmed regions of the Gulf Coast.

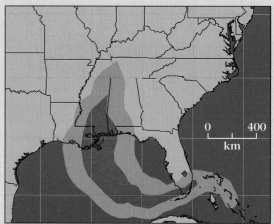

(a) A wind-swath map of Hurricane Katrina. Red areas represent hurricane winds; orange areas represent tropical-storm winds.

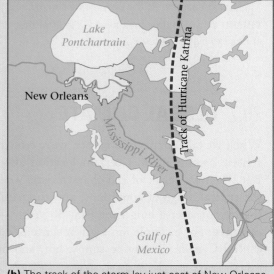

(b) The track of the storm lay just east of New Orleans.

(c) Surge from the storm destroyed homes along the Alabama coast.

(d) Officials survey the storm damage.

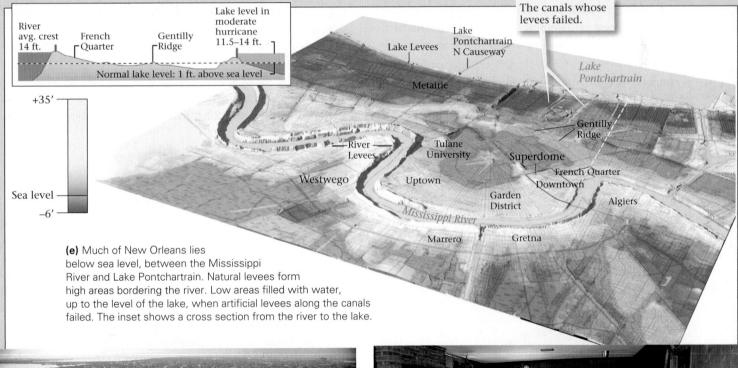

River avg. crest 14 ft.

French Quarter

Gentilly Ridge

Lake level in moderate hurricane 11.5–14 ft.

Normal lake level: 1 ft. above sea level

+35'

Sea level

–6'

The canals whose levees failed.

Lake Levees

Lake Pontchartrain N Causeway

Metairie

Lake Pontchartrain

Gentilly Ridge

River Levees

Tulane University

Superdome

French Quarter

Westwego

Uptown

Downtown

Garden District

Algiers

Mississippi River

Marrero

Gretna

(e) Much of New Orleans lies below sea level, between the Mississippi River and Lake Pontchartrain. Natural levees form high areas bordering the river. Low areas filled with water, up to the level of the lake, when artificial levees along the canals failed. The inset shows a cross section from the river to the lake.

(f) Water flowing across the levees bordering the 17th Street Canal after the hurricane had passed.

(g) The interior of a New Orleans home after the flood waters receded.

FIGURE 20.29   The temperature of the atmosphere changes with the seasons, as depicted by isotherms. The Gulf Stream deflects the isotherms northward, so the United Kingdom and Ireland are warmer than lands at equivalent latitudes in North America.

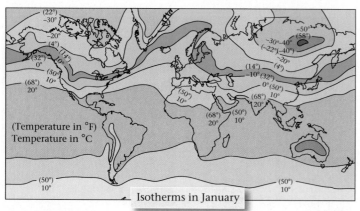

Isotherms in January

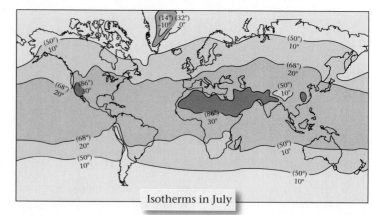

Isotherms in July

■ *Proximity to orographic barriers*: An **orographic barrier** is a landform (such as a mountain range) that diverts airflow upward or laterally. This diversion affects the amount of precipitation and wind a region receives.

■ *Proximity to high- or low-pressure zones*: Zones of high and low pressure, roughly parallel to the equator, encircle the planet (see Fig. 20.11). Because land and sea have different heat capacities, they modify the zones, so that high-pressure zones tend to be narrower over land. Meteorologists refer to the resulting somewhat elliptical regions of high or low pressure as semipermanent pressure cells (Fig. 20.30). These influence prevailing wind direction and relative humidity.

FIGURE 20.30   Because of landmasses, atmospheric pressure belts vary in width to create lens-shaped, semipermanent high- and low-pressure cells.

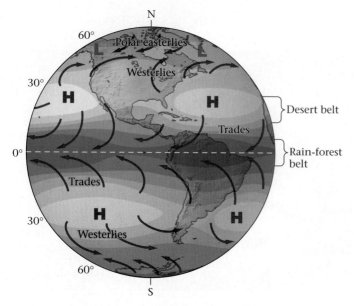

Climatologists, who have studied the distribution of climatic conditions around the globe, have developed a classification scheme for climates, based on such factors as the average monthly and annual temperatures and the total monthly and yearly amounts of precipitation. The vegetation of a region proves to be an excellent indicator of climate, because plants are sensitive to temperature and to the amount and distribution of rainfall. Table 20.4 lists the principal types of climate belts (Fig. 20.31a, b).

Climate Variability: Monsoons and El Niño

The climate at a certain location may change during the course of one or more years. Here, we look at two important examples of climate variability that affect human populations in notoriously significant ways.

A **monsoon** is a major reversal in the wind direction that causes a shift from a very dry season to a very rainy season. In southern Asia, home to about half the world's population, people depend on monsoonal rains to bring moisture for their crops.

The Asian monsoon develops primarily because Asia is so large that it includes vast tracts of land far from the sea. Further, a substantial part of this land, the Tibet Plateau, lies at a high elevation. During the winter, central Asia becomes very cold, much colder than coastal regions to the south. This coldness creates a stable high-pressure cell over central Asia. Dry air sinks and spreads outward from this cell and flows southward over southern Asia, pushing the intertropical convergence zone (ITCZ) out over the Indian Ocean, south of Asia (Fig. 20.32a). Thus, during the winter, southern Asia experiences a dry season. During the summer, central Asia warms up dramatically. As warm air rises and expands over central Asia, a pronounced low-pressure cell develops, and the intertropical convergence zone moves north. When this happens, warm air flows northward from the Indian Ocean, bringing with it

TABLE 20.4 Climate Types of the Earth

Climate Type	Regions and Characteristics
Tropical rainy	Tropical rain forests lie at equatorial latitudes and experience rain throughout the year. Rain commonly falls during afternoon thunderstorms. Tropical savanna (grasslands with brush and drought-resistant trees), which lie on either side of a rain forest, have a rainy season and a dry season. Rain forests and savannas may receive tropical monsoons.
Dry	Dry regions include deserts (regions with very little moisture or vegetation cover; vegetation that does exist has adapted to long periods without moisture) and steppes. Steppe regions (vast, grassy plains with no forest) border the desert and have somewhat more precipitation. Some steppe regions occur at high elevations, in latitudes where the climates would otherwise be more humid.
Humid mesothermal (temperate)	This category includes humid subtropical climates, with moist air and warm temperatures for much of the year, in which mixed deciduous-coniferous forest thrives; Mediterranean climates, coastal regions with most rainfall in the winter, very hot summers, and scrub forests; and marine west-coast climates, where the sea tempers the climate and may create a coastal temperate rain forest.
Humid microthermal (cold)	These higher-latitude temperate climates, which occur only in the northern hemisphere, include humid continental regions with long summers (as in the U.S. Midwest and mid-Atlantic states), in which deciduous forest thrives; humid continental regions with short summers, characterized by mixed deciduous-coniferous forest or coniferous-only forest; and subarctic climates with very short, cool summers and coniferous forest that becomes lower and scrubbier at higher latitudes.
Polar	These cold climates include tundra and ice caps. Tundra are regions with no summer and an extremely cold winter, in which only low, cold-resistant plants (moss, lichen, and grass) can survive. Much of the ground in tundra is permafrost (permanently frozen ground). In ice-cap regions, near the poles, the climate is subfreezing year-round, and any land not covered by ice has essentially no vegetation cover. Highlands are regions that lie at lower (nonpolar) latitudes but have such a high elevation that they have polar-like climates. When you enter a region above the tree line, you have entered a highland polar climate.

FIGURE 20.31 Climate belts, and their effect on vegetation.

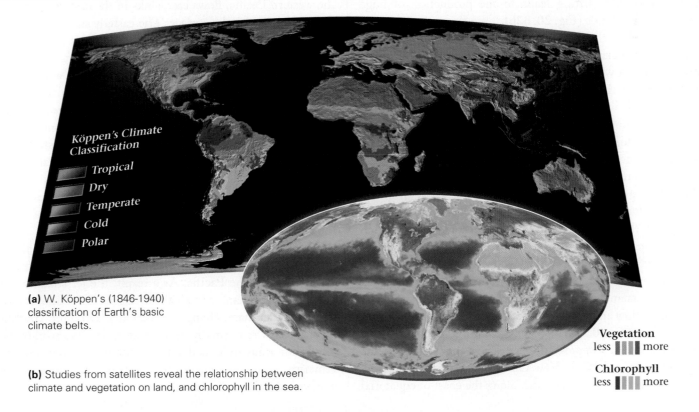

Köppen's Climate Classification

Tropical
Dry
Temperate
Cold
Polar

(a) W. Köppen's (1846-1940) classification of Earth's basic climate belts.

(b) Studies from satellites reveal the relationship between climate and vegetation on land, and chlorophyll in the sea.

Vegetation
less ▌▌▌ more

Chlorophyll
less ▌▌▌ more

FIGURE 20.32   The causes of monsoons. Because of monsoons, Asia has a wet season and a dry season every year.

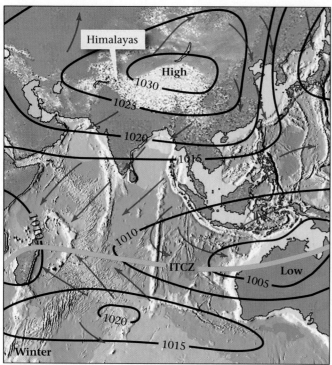

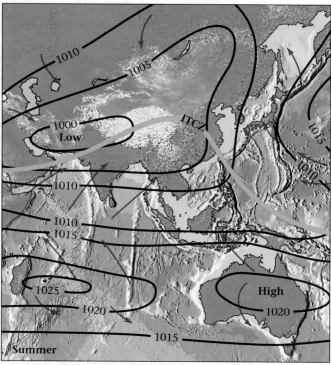

(a) In winter, the dry season, a high-pressure cell develops over central Asia and the intertropical convergence zone lies south of Asia.

(b) In summer, the wet season, a low-pressure cell develops over Asia, so warm moisture-laden air flows landward. Orographic lifting over the Himalayas causes clouds to form, and intense rain begins.

substantial moisture, and the summer rains begin. Rainfalls are especially heavy on the southern slope of the Himalayas, because orographic lifting leads to the production of huge cumulonimbus clouds (Fig. 20. 32b).

Long before the modern science of meteorology became established, fishermen from Peru and Ecuador who ventured into the coastal waters west of South America knew that in late December, the fish population that provided their livelihood diminished. Because of the timing of this event, it came to be known as **El Niño**, Spanish for "little boy" or "the Christ child." Why did the fish vanish? Fish are near the top of a food chain that begins with plankton, which live off nutrients in the water. These nutrients increase when cold water upwells from the deep along the coast of South America. During El Niño, warm water currents flow eastward from the central Pacific, and the cold, nutrient-rich water that supports the marine food chain remains at depth. With fewer nutrients, there are fewer plankton, and without the plankton, the fish migrate elsewhere.

To understand why El Niño occurs, we need to look at atmospheric flow and related surface ocean currents in the equatorial Pacific (Fig. 20.33a, b). When El Niño is *not* in progress, a major equatorial low-pressure cell exists in the western Pacific over Indonesia and Papua New Guinea, while a high-pressure cell forms over the eastern Pacific, along the coast of equatorial

South America. This geometry (called La Niña, Spanish for "little girl," when particularly intense) means that air rises in the western Pacific, flows east, sinks in the eastern Pacific, and then flows west at the surface. The easterly surface winds blow warm surface water westward, so that it pools in the western Pacific. Cold water from the deep ocean rises along South America to replace the warm water that moved west. It is this rising cold water that brings nutrients to the surface. During El Niño, the low-pressure cell moves eastward over the central Pacific, and a high-pressure cell develops over Indonesia; so two circulation cells develop. As a result, surface winds start to blow east in the western Pacific, driving warm surface water back to South America. This warm surface water prevents deep cold water from rising, sending the fish away. In effect, pressure cells oscillate back and forth across the Pacific, an event now called the **southern oscillation**.

El Niño gained world notoriety in late 1982 and 1998 when a particularly large low-pressure cell developed in the eastern part of the Pacific. As a result, the jet streams stayed farther north than is typical. In effect, El Niño caused a temporary climate change worldwide. Drought conditions persisted in the normally rainy western Pacific, while unusually heavy rains drenched western South America. In North America, rains swamped the southern United States and storms battered California, winters were warmer than usual

FIGURE 20.33 El Niño exists because of a change in winds and currents in the central Pacific.

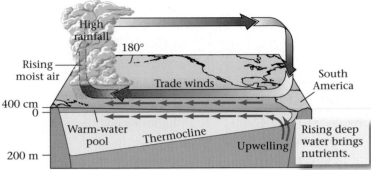

(a) During times between El Niño, low pressure lies over the western Pacific and surface winds blow west. These winds drive warm surface water away so cold water rises along the coast of South America.

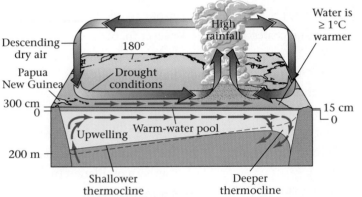

(b) During El Niño, the low-pressure cells move eastward and the westward flow stops, so upwelling ceases.

in Canada, and snowfalls were heavier in the Sierra Nevada, leading to spring floods.

Climatologists have been working intensely to understand the periodicity of El Niño. It is clear that strong El Niños take place around once every four years, with even stronger ones possibly happening at other intervals.

Take-Home Message

- Climate refers to the long-term overall weather conditions in a region, as characterized by the range of weather conditions and the nature of seasons.
- In general, climate varies with latitude because insolation varies with latitude, and regional atmospheric circulation cells are delimited by latitude.
- In detail, factors such as latitude, proximity to the ocean or to orographic barriers, the nature of ocean currents, and location relative to high- or low-pressure zones affect local climate.
- Monsoons, seasonal reversals in the wind direction that trigger a shift from a dry to a rainy season, form because of changes in the position of the intertropical convergence zone.
- El Niño develops because of shifts in the position of air circulation at low latitudes that in turn influence surface wind directions.

THINK: How does El Niño affect the food supply for fish near the coast of South America?

Chapter Summary

- The early atmosphere of the Earth contained high concentrations of water, carbon dioxide, and sulfur dioxide; these are gases erupted by volcanoes.
- After the oceans formed, much of the carbon dioxide was removed from the atmosphere. When photosynthetic organisms evolved, they produced oxygen, and the concentration of this gas gradually increased.
- Air now consists mostly of nitrogen (78%) and oxygen (21%). Several other gases occur in trace amounts. The atmosphere also contains aerosols. Air pressure decreases with elevation. Thus, 90% of the air in the atmosphere occurs below an elevation of 16 km.
- When air rises, it expands and cools, a process called adiabatic cooling. If air sinks and undergoes compression, it heats up, a process called adiabatic heating.

- Air generally contains water. The ratio between the measured water content and the maximum possible amount of water that the air can hold is the air's relative humidity.
- The atmosphere is divided into layers. In the lowest layer, the troposphere, temperature decreases with elevation. The troposphere convects. The other layers are the stratosphere, the mesosphere, and, at the top, the thermosphere.
- Air circulates on two scales, local and global. Winds blow because of pressure gradients: air begins to move from regions of higher pressure to regions of lower pressure. It's path will be affected by the Coriolis effect.
- High latitudes receive less solar energy than low latitudes. This contrast initiates convection in the atmosphere. Because of the Coriolis effect, air moving north from the equator to the pole deflects to the east, and in each hemisphere, three convection cells develop (the Hadley, Ferrel, and polar cells).

- Prevailing surface winds—the northeast trade winds, the surface westerlies, and the polar easterlies—develop because of circulation in global convection cells.

- Air pressure at a given latitude decreases from the equator to the pole, causing a poleward flow of air. In the northern hemisphere, the Coriolis effect deflects this flow to generate high-altitude westerlies. These winds, where particularly strong, are known as jet streams.

- *Weather* refers to the temperature, air pressure, wind speed, and relative humidity at a given time in a given location. Weather reflects the interaction of air masses. The boundary between two air masses is a front.

- Air sinks in high-pressure air masses and rises in low-pressure air masses. Because of the Coriolis effect, the air begins to rotate around the center of the mass, generating cyclones or anticyclones.

- Clouds, which consist of tiny droplets of water or tiny crystals of ice, form when the air is saturated with water and contains condensation nuclei on which water condenses.

- Thunderstorms begin when cumulonimbus clouds grow large. Friction between air and water molecules separates positive and negative charges. Lightning flashes when a giant spark jumps across charge separations.

- Tornadoes, rapidly rotating funnel-shaped clouds, develop in violent thunderstorms. Nor'easters are large storms associated with wave cyclones. Hurricanes, huge rotating storms, originate over oceans where the water temperature exceeds 27°C.

- *Climate* refers to the typical range of conditions, the nature of seasons, and the possible weather extremes of a region averaged over a long time (say 30 years). Climate is controlled by latitude, altitude, proximity to water, ocean currents, orographic barriers, and high- or low-pressure zones. Climate classes can be recognized by the vegetation they support.

- Monsoonal climates occur where there is a seasonal shift in the wind direction, causing a wet season to alternate with a dry season. El Niño is a temporary shift in weather conditions triggered by shifts in the position of high- and low-pressure cells in the Pacific.

GEOPUZZLE REVISITED

The atmosphere initially formed from gas emitted by volcanoes. The composition of the atmosphere evolved over time, due to the formation of the oceans and later, due to the evolution of photosynthesis. Climate is the overall condition of temperature and humidity, as well as their variability, averaged over time. Climate largely reflects variation in solar radiation, but also topography and distance from the ocean. Weather refers only to the condition in the atmosphere at a given time, and may change rapidly at a given location.

Guide Terms

acid rain (p. 683)

adiabatic cooling, heating (p. 685)

aerosols (p. 682)

air (p. 680)

air mass (p. 693)

air pressure (p. 683)

anticyclonic flow (p. 695)

atmosphere (p. 680)

Bergeron process (p. 697)

cirrus (p. 697)

climate (pp. 680, 707)

cloud (p. 685)

collision and coalescence (p. 697)

condensation nuclei (p. 696)

convergence zone (p. 689)

cumulus (p. 697)

cyclonic flow (p. 694)

dewpoint temperature (p. 685)

divergence zone (p. 689)

doldrums (p. 691)

El Niño (p. 712)

Enhanced Fujita scale (p. 702)

extratropical cyclone (p. 695)

Ferrel cell (p. 689)

fog (p. 696)

front (p. 694)

Hadley cell (p. 689)

hail (hailstones) (p. 699)

hurricane (p. 704)

insolation (p. 688)

ionosphere (p. 687)

isobar (p. 687)

isotherm (p. 707)

jet stream (p. 692)

lifting mechanism (p. 696)

lightning bolt (lightning stroke) (p. 700)

mesosphere (p. 686)

monsoon (p. 710)

nor'easter (p. 703)

orographic barrier (p. 710)

ozone (p. 681)

pauses (p. 686)

polar cell (p. 689)

polar front (p. 689)

prevailing winds (pp. 691–692)

relative humidity (p. 685)

Saffir-Simpson scale (p. 704)

southern oscillation (p. 712)

storm (p. 699)

storm surge (p. 706)

stratosphere (p. 686)

stratus (p. 697)

supercell (p. 699)

thermosphere (p. 686)

thunder (p. 700)

tornado (p. 701)

trade winds (p. 691)

troposphere (p. 686)

weather (p. 680)

wind (p. 680)

Review Questions

1. Describe the stages in the formation and evolution of Earth's atmosphere. Where does the ozone in the atmosphere come from, and why is it important?

2. Describe the composition of air (considering both its gases and its aerosols). Why are trace gases important?

3. How does air pressure change with elevation? Does the density of the atmosphere also change with elevation? Explain why or why not.

4. Describe the atmosphere's structure from base to top. What characteristics define the boundaries between layers?

5. What is the relative humidity of the atmosphere? What is the latent heat of condensation, and what is its relevance to the evolution of a thunderstorm or hurricane?

6. Explain the relation between the wind and variations in air pressure.

7. Why do changes in atmospheric temperature depend on latitude and the seasons? Why does global circulation break into three distinct convection belts in each hemisphere?

8. Why do prevailing winds develop at the Earth's surface? Why do the jet streams form?

9. Explain the origin of cyclones and anticyclones, and note their relationship to high-pressure and low-pressure air masses. What is a mid-latitude cyclone?

10. How does a cold front differ from a warm front and from an occluded front?

11. Why do clouds form? (Include a discussion of lifting mechanisms.) What are the basic categories of clouds?

12. Under what conditions do thunderstorms develop? What provides the energy that drives clouds to the top of the troposphere? How do meteorologists explain lightning?

13. What conditions lead to the formation of a tornado? Where do most tornadoes appear?

14. Describe the stages in the development of a hurricane. Describe a hurricane's basic geometry.

15. What factors control the climate of a region? What special conditions cause monsoons? El Niño?

On Further Thought

16. When Columbus set sail from Spain, his route first took him southward to the Canary Islands and then westward. His landfall in the new world, on an island southeast of Florida, was much farther to the south than his point of departure. Why?

17. Explain why *large* rain forests occur in equatorial Africa (the Congo) and in equatorial South America (the Amazon). Why do *small* rain forests occur along the coastal regions of Washington State (northwestern North America) and along the coast of southern Chile (southwestern South America)?

18. Much more snow falls on the eastern and southern shores of Lake Michigan, one of the Great Lakes, than on the western shore. Why does this phenomenon, which weather reporters refer to as "lake-effect snow," occur?

19. A typhoon is approaching the east coast of Asia. Weather forecasters have predicted where the eye of the storm will make landfall on a narrow, north-south-trending island. People in the path of the storm have been told to evacuate but cannot head inland because they are on a narrow island. To escape the highest winds, should they move to the north or to the south of the eye?

 For more resources, including animations, quizzes, and Norton's GeoTours, go to **wwnorton.com/studyspace**.

 If your instructor assigns exercises in SmartWork, log in at **smartwork.wwnorton.com**.

ANOTHER VIEW A 3-D rendering showing clouds in the Earth's atmosphere, and water temperature of the sea surface; red is warm, blue is cool.

CHAPTER **21**

Dry Regions:
The Geology of Deserts

At Zabriskie Point, near Death Valley, it's too dry for vegetation to grow. So, during infrequent storms, rushing water can carve quickly into the soft, tan rock.

GEOPUZZLE

Hollywood movies typically portray deserts as endless seas of hot sand, in which nothing at all grows. Is this portrayal accurate?

The bare hills are cut out with sharp gorges, and over their stone skeletons scanty earth clings . . . A white light beat down, dispelling the last tract of shadow, and above hung the burnished shield of hard, pitiless sky.

—Clarence King (1842–1901; first director of the U.S. Geological Society, describing a desert)

21.1 INTRODUCTION

For generations, nomadic traders have saddled camels to traverse the Sahara Desert in northern Africa (Fig. 21.1a). The Sahara, the world's largest desert, receives so little rainfall that it has hardly any surface water or vegetation. So camels must be able to walk for up to 3 weeks without drinking or eating. They can survive these journeys because they sweat relatively little, thereby conserving their internal water supply; they have the ability to metabolize their own body fat (up to 20 kg of fat make up the animal's hump alone) to produce new water; and they can withstand severe dehydration (loss of water). Most mammals die after losing only 10 to 15% of their body fluid, but camels can survive 30% dehydration with no ill effects. Camels do get thirsty, though. After a marathon trek across the desert, a camel may guzzle up to 100 liters of water in less than 10 minutes.

The survival challenges faced by a camel emphasize that deserts are lands of extremes—extreme dryness, heat, cold, and, in some places, beauty. Desert vistas include everything from sand seas to sagebrush plains, cactus-covered hills to endless stony pavements. Although less populated than other regions on Earth, deserts cover a significant percentage (about 25%) of the land surface, and thus constitute an important component of the Earth System (Fig. 21.1b). In this chapter, we take a look at the desert landscape. We learn why deserts occur where they do, and how erosion and deposition shape their surface. We conclude by exploring life in the desert and by examining the problem of desertification, the gradual transformation of temperate lands into desert.

Chapter Themes

By the end of this chapter, you should know . . .

- why certain regions of the land have been classified as deserts, and what factors cause these regions to have arid climates.
- how weathering and erosional processes in deserts differ from those in temperate lands.
- that distinctive landforms and landscapes form in deserts.
- how certain species of plants and animals survive desert climates, and how human activity may transform vegetated regions into deserts.

21.2 THE NATURE AND LOCATION OF DESERTS

What Is a Desert?

Formally defined, a **desert** is a region that is so arid (dry) that it contains no permanent streams, except for rivers that bring water in from temperate regions elsewhere, and supports vegetation on no more than 15% of its surface. In general, desert

FIGURE 21.1 Deserts and their hardy inhabitants.

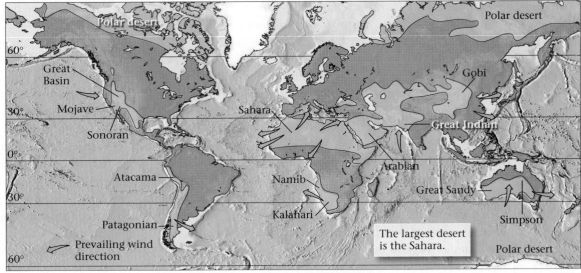

(a) Camels can survive the harsh desert conditions of the Arabian Desert.

(b) The global distribution of deserts. Arid regions cover 25% of the land surface.

conditions exist where less than 25 cm of rain falls per year, on average. But rainfall alone does not determine the aridity of a region. Aridity also depends on rates of evaporation and on whether rainfall occurs only sporadically or more continuously during the year. If all the rain in a region drenches the land during isolated downpours only once every few years, the region becomes a desert, because the intervals of drought last so long that plants and permanent streams cannot survive. Similarly, if high temperatures and dry air cause evaporation rates from the ground to exceed the rate at which rainfall wets the ground, then the region becomes a desert even if it receives more than 25 cm per year of rain.

Note that the definition of a desert depends on a region's aridity, *not* on its temperature. Geologists distinguish between cold deserts, where temperatures generally stay below about 20°C for the year, and hot deserts, where summer daytime temperatures exceed 35°C. Cold deserts exist at high latitudes where the Sun's rays strike the Earth obliquely and thus don't provide much energy, at high elevations where the air is too thin to hold much heat, or in lands adjacent to cold oceans, where the cold water absorbs heat from the air above. Hot deserts develop at low latitudes where the Sun's rays strike the desert at a high angle, at low elevations where dense air can hold a lot of heat, and in regions distant from the cooling effect of cold ocean currents. The hottest recorded temperatures on Earth occur in low-latitude, low-elevation deserts—58°C (136°F) in Libya and 57°C (133°F) in Death Valley, California.

Did you ever wonder . . .
how hot can it become in a desert?

Heat contributes to aridity by increasing the rate of evaporation. In fact, evaporation rates in hot deserts may be so great that even when it rains, the ground stays dry because raindrops evaporate in midair. But even the hottest of hot deserts become cold at night: because of the dryness of the air, the lack of cloud cover, and the lack of foliage, deserts radiate their heat back into space at night. As a consequence, the air temperature at the ground surface in a desert may change by as much as 80°C in a single day. Of note, the ground surface absorbs so much heat in hot deserts that a layer of very hot air (up to 77°C, or 170°F) forms just above the ground. This layer refracts sunlight, creating a mirage, a wavering pool of light, on the ground. Mirages make the dry sand of a desert wasteland look like a shimmering lake and distant mountains look like islands (**Fig. 21.2**).

Aridity causes weathering, erosion, and depositional processes in deserts to be different from those in temperate or tropical regions. Without plant cover, rain and wind batter and scour the ground, and during particularly heavy rains water accumulates into flash floods of immense power. In deserts, rocks and sediment do not undergo rapid chemical weathering, and humus (organic matter) does not collect on the ground surface. Thus, the desert land surface consists of any of the following: exposed bedrock, accumulations of clasts, relatively

FIGURE 21.2 The shimmering in this desert mirage may look like water, but it's not. Mirages result from the interaction of light with a thin layer of hot air just above the ground surface.

unweathered sediment, precipitated salt, or windblown sand. Overall, therefore, desert landscapes tend to be harsher and more rugged than temperate or tropical ones. If eastern North America were a desert, the Appalachians would not be gentle, forested hills but rather would consist of stark, rocky ridges.

Types of Deserts

Each desert on Earth has unique characteristics of landscape and vegetation that distinguish it from others. Geologists group deserts into five different classes, based on the environment in which the desert forms (**Fig. 21.3**).

- *Subtropical deserts*: Subtropical deserts (e.g., the Sahara, Arabian, Kalahari, and Australian) form because of the pattern of air circulation in the atmosphere (see Chapter 20). At the equator, the air becomes warm and humid, for sunlight is intense and water rapidly evaporates from the ocean. The hot, moisture-laden air rises to great heights above the equator. As this air rises, it expands and cools, and can no longer hold so much moisture. Water condenses and falls in downpours that feed the lushness of the equatorial rain forest. The now-dry air high in the troposphere spreads laterally north or south. When this air reaches latitudes of 20° to 30°, a region called the subtropics, it has become cold and dense enough to sink. Because the air is dry, no clouds form, and intense solar radiation strikes the Earth's surface. The sinking, dry air condenses and heats up, soaking up any moisture present. In the regions swept by this hot

FIGURE 21.3 Subtropical deserts form because the air that convectively flows downward in the subtropics warms and absorbs water as it sinks.

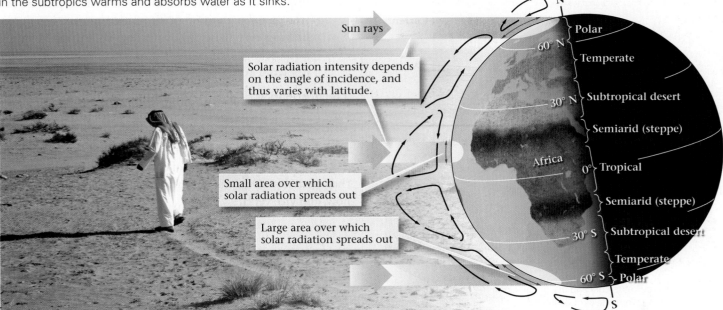

air on its journey back to the equator, evaporation rates greatly exceed rainfall rates.

In subtropical deserts that border the sea, episodes of high sea level flood coastal areas. The eventual evaporation of stranded seawater leaves broad regions, called sabkhas, where a salt crust covers a mire of organic-rich mud.

FIGURE 21.4 The formation of a rain-shadow desert.

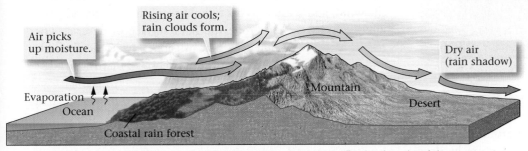

(a) Moist air rises and drops rain on the coastal side of the range. By the time the air has crossed the mountains, it is dry.

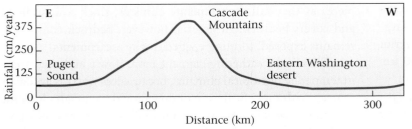

(b) A graph shows that measured rainfall is much greater on the western side of the Cascade Mountains in Washington than it is on the eastern side.

■ *Deserts formed in rain shadows*: As air flows over the sea toward a coastal mountain range, the air must rise (Fig. 21.4a, b). As the air rises, it expands and cools. The water it contains condenses and falls as rain on the seaward flank of the mountains, nourishing a coastal rain forest. When the air finally reaches the inland side of the mountains, it has lost all its moisture and can no longer provide rain. As a consequence, a **rain shadow** forms, and the land beneath the rain shadow becomes a desert. A rain-shadow desert can be found east of the Cascade Mountains in Washington.

■ *Coastal deserts formed along cold ocean currents*: Cold ocean water cools the overlying air by absorbing heat, thereby decreasing the capacity of the air to hold moisture. For example, the cold Humboldt Current, which carries water northward from Antarctica to the western coast of South America, cools the air that blows east, over the coast. Thus, the air is so dry when it reaches the coast that rain rarely falls on the coastal areas of Chile and Peru. As a result, this region hosts a desert landscape, including one of the driest deserts in the world, the Atacama (Fig. 21.5a, b). Portions of this narrow (less than 200-km-wide) desert, which lies between the Pacific coast on the west and the Andes on the east, received no rain at all between 1570 and 1971.

FIGURE 21.5 The formation of a coastal desert.

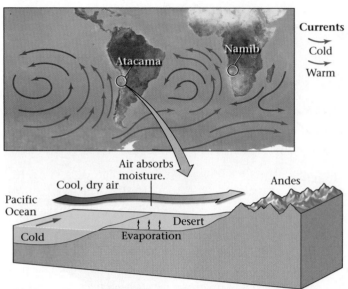

(a) Cold ocean currents cool and dry the air along the coast. This air absorbs moisture from adjacent coastal land.

(b) The Atacama Desert is the driest place on Earth.

that these areas are, in fact, arid. Polar regions are dry, in part, for the same reason that the subtropics are dry (the global pattern of air circulation means that the air flowing over these regions is dry) and, in part, for the same reason that coastal areas along cold currents are dry (cold air holds little moisture).

The distribution of deserts around the world through geologic time reflects the process of plate tectonics, for plate movements determine the latitude of landmasses, the position of landmasses relative to the coast, and the proximity of landmasses to a mountain range. Because of continental drift, some regions that were deserts in the past are temperate or tropical regions now, and vice versa.

Take-Home Message

- Deserts are arid regions that receive less than 25 cm of rain per year and support vegetation on less than 15% of their surface.
- Not all deserts are hot, but temperature extremes can develop in deserts. The highest temperature recorded is almost 58°C.
- Deserts form in a variety of settings, including subtropical latitudes, rain shadows, coasts adjacent to cold currents, the interior of continents, and polar regions.

THINK: Why does the largest desert, the Sahara, exist?

21.3 WEATHERING AND EROSIONAL PROCESSES IN DESERTS

Without the protection of foliage to catch rainfall and slow the wind, and without roots to hold regolith in place, rain and wind can attack and erode the land surface of deserts. The result, as we have noted, is that hillslopes are typically bare, and plains can be covered with stony debris or drifting sand.

Arid Weathering and Desert Soil Formation

In the desert, as in temperate climates, physical weathering happens primarily when joints (natural fractures) split rock into pieces. Joint-bounded blocks eventually break free of bedrock and tumble down slopes, fragmenting into smaller pieces as they fall. In temperate climates, thick soil develops and covers bedrock. In deserts, however, bedrock commonly remains exposed, forming rugged, rocky escarpments.

Chemical weathering happens more slowly in deserts than in temperate or tropical climates, because less water is available to react with rock. Still, rain or dew provides enough moisture for some weathering to occur. This water seeps into rock and leaches (dissolves and carries away) calcite, quartz, and various

- *Deserts formed in the interiors of continents*: As air masses move across a continent, they lose moisture by dropping rain, even in the absence of a coastal mountain range. Thus, when an air mass reaches the interior of a particularly large continent such as Asia, it has grown quite dry, so the land beneath becomes arid. The largest present-day example of such a continental-interior desert, the Gobi, lies in central Asia, over 2,000 km away from the nearest ocean.

- *Deserts of the polar regions*: So little precipitation falls in Earth's polar regions (north of the Arctic Circle, at 66°30′ N, and south of the Antarctic Circle, at 66°30′ S)

salts. Leaching effectively rots the rock by transforming it into a poorly cemented aggregate. Over time, the rock will crumble and form a pile of unconsolidated sediment, susceptible to transport by water or wind. If water seeping into the rock contains dissolved salts, salt crystals may grow in the rock when the water dries out. The growth of such crystals pushes neighboring crystals apart and can also weaken the rock.

Although enough rain falls in deserts to leach chemicals out of rock, there is insufficient water to flush the dissolved minerals entirely away. Thus, when the water percolates down and dries up, minerals precipitate in regolith below the surface. If calcite precipitates, it cements loose grains together, forming solid rock-like material called calcrete (Fig. 21.6a). The growth of calcrete can occur rapidly enough to incorporate tools abandoned by prospectors.

Shiny **desert varnish**, a dark, rusty brown coating of iron oxide, manganese oxide, and clay, covers the surface of many rock varieties in deserts (Fig. 21.6b). Desert varnish was once thought to form when water from rain or dew seeped into a rock, dissolved iron and magnesium ions, and carried the ions back to the surface of the rock by capillary action. More recent studies, however, suggest that desert varnish is not necessarily derived from the rock it coats. It may form when windborne dust settles on the surface of the rock, for in the presence of moisture, microorganisms (bacteria and archaea) can extract elements from the dust and transform it into iron or manganese oxide. Such varnish won't form in humid climates, because rain washes away dust and associated ions.

Desert varnish takes a long time to form. In fact, the thickness of a desert varnish layer provides an approximate estimate of how long a rock has been exposed at the ground surface. In past centuries, Native Americans used desert-varnished rock as a medium for art: by chipping away the varnish to reveal the underlying lighter-colored rock, they were able to create figures or symbols on a dark back-ground. The resulting drawings are called **petroglyphs** (Fig. 21.6b).

Because of the lack of plant cover in deserts, variations in bedrock and soil color stand out. Slight variations in the concentration of iron, or in the degree of iron oxidation, in adjacent beds result in spectacular color bands in rock layers and the thin soils derived from them. The Painted Desert of northern Arizona earned its name from the brilliant and varied hues of oxidized iron in the region's shale bedrock (Fig. 21.6c).

Water Erosion

Although rain rarely falls in deserts, when it does come, it can radically alter a landscape in a matter of minutes. Since deserts lack plant cover, rainfall, sheetwash, and stream flow are all extremely effective agents of erosion. It may seem surprising, but water generally causes more erosion than does wind in most deserts (Fig. 21.7a).

Water erosion begins with the impact of raindrops, which eject sediment from the ground into the air. On a hill, the ejected sediment lands downslope; and thus, during a rain, sediment gradually migrates to lower elevations. The ground quickly becomes saturated with water during a heavy rain, so water starts flowing across the surface, carrying the loose sediment with it. Within minutes after a heavy downpour begins, dry stream channels fill with a turbulent mixture of water and sediment, which rushes downstream as a flash flood. When the rain stops, the water sinks into the stream bed's gravel and disappears—such streams are called intermittent, or ephemeral streams (see Chapter 17). Because of the relatively high viscosity of the water (owing to its load of suspended sediment) and the velocity and turbulence of the flow, flash floods in deserts cause intense erosion—they undercut cliffs and transport huge boulders downstream. As rocks roll and tumble along,

FIGURE 21.6 Soil formation and chemical weathering in deserts.

Stream erosion exposed this calcrete layer.

(a) Calcrete forms when $CaCO_3$ precipitates and binds together rock fragments.

(b) Native Americans chiseled this petroglyph of a snake into desert varnish on a Utah boulder.

(c) The red hues of the Painted Desert in Arizona are due to the oxidation of iron in the rock.

they strike each other and shatter, creating smaller pieces that can be carried still farther. Between floods, the stream floor consists of gravel littered with boulders.

Flash floods can carve steep-sided channels into the ground; and scouring of bedrock walls by sand-laden water may polish the walls and create grooves. Dry stream channels in desert regions of southwestern North America are called **dry washes**, or **arroyos**, and in the Middle East and North Africa they are called wadis (Fig. 21.7b).

Wind Erosion

In temperate and humid regions, plant cover protects the ground surface from the wind, but in deserts, the wind has direct access to the ground. In hot deserts, gusts of hot air feel like blasts from a furnace. Wind, just like flowing water, can carry sediment both as suspended load and as bed load. **Suspended load** (fine-grained sediment such as dust and silt held in suspension) stays in the air for a long time and moves with it. The suspended sediment can be carried so high into the atmosphere (up to several kilometers above the Earth's surface) and so far downwind (tens to thousands of kilometers) that it may move completely out of its source region. In some cases, tiny vortices can churn up dust. In general, these vortices are very small (centimeters to meters high). But, in some cases, they become "dust devils" up to 100 m high, and look like miniature tornadoes. Cars driving down dirt roads in deserts have the same effect; they break dust free from the ground and generate plumes of suspended sediment. Particularly strong winds, such as the downdrafts in front of a thunderstorm, can generate a dramatic **dust storm**, or haboob, that can be 100 km long and up to 1.5 km

high (Fig. 21.8a, b). A particularly large one rolled over Phoenix, Arizona in July, 2011, shutting down the airport.

Moderate to strong winds can roll and bounce sand grains along the ground, a process called **saltation**. Saltating sand constitutes the wind's **surface load**. Saltation begins when turbulence caused by wind shearing along the ground surface lifts sand grains. The grains move downwind, following an asymmetric, arch-like trajectory. Eventually, they return to the ground, where they strike other sand grains, causing the new grains to bounce up and drift or roll downwind. The collisions between sand grains make the grains rounded and frosted. Saltating grains generally rise no more than 0.5 m. But where sand bounces on bedrock during a desert sandstorm, the grains may rise 2 m, and can strip the paint off a car.

The size of clasts that wind can carry depends on the wind velocity. Wind, therefore, does an effective job of sorting sediment, sending dust-sized particles skyward and sand-sized particles bouncing along the ground, while pebbles and larger grains remain behind. In some cases, wind carries away so much fine sediment that pebbles and cobbles become concentrated at the ground surface (Fig. 21.9a, b). An accumulation of coarser sediment left behind when fine-grained sediment blows away is called a **lag deposit**.

In many locations, the desert surface resembles a tile mosaic in that it consists of separate stones that fit together tightly, forming a fairly smooth surface layer above a soil composed of silt and clay. Such natural mosaics constitute **desert pavement** (Fig. 21.10a, b). Typically, desert varnish coats the top surfaces of the stones forming desert pavement. Geologists have come up with several explanations for the origin of desert pavements. Traditionally, pavements were thought to

FIGURE 21.7 Evidence of erosion by running water in deserts.

(a) These hills in the desert near Las Vegas, Nevada, are bone dry, but their shape indicates erosion by water. Note the numerous stream channels.

(b) Gravel and sand are left behind on the floor of a dry wash after a flash flood in Death Valley. Erosion by the water is cutting a channel.

FIGURE 21.8 Dust storms in the desert develop when turbulent winds pick up sediment and are suspended in the air.

(a) A 2005 dust storm (haboob) approaches an army base in Iraq. The surging layer of dust-laden air is hundreds of meters high.

(b) Dust storms can move huge volumes of sediment and cause intense erosion.

be lag deposits, formed when wind blows away the fine sediment between clasts, so that the clasts can settle down and fit together. More recently, researchers suggest that pavements form when windblown dust slowly sifts down onto the stones and then washes down between the stones. In this model, the pavement is "born at the surface," meaning that the stones forming the pavement were never buried, but have been progressively lifted up as sediment collects and builds up beneath (Fig. 21.10c). Over time, the rocks at the surface crack, perhaps due to extreme heating by the desert sun. Sheetwash, during downpours, may wash away fine sediment between fragments, and when soils dry and shrink between storms, the clasts settle together, locking into a stable, jigsaw-like arrangement.

Desert pavements are remarkably durable and can last for hundreds of thousands of years if they are left alone. But like many features of the desert, they can be disrupted by human activity. For example, people driving vehicles across the pavement indent and crack its surface, making it susceptible to erosion. In parts of Arizona, vast desert pavements have become parking lots for campers who migrate to the desert in motor homes for the winter season.

Just as sandblasting cleans the grime off the surface of a building, windblown sand and dust grind away at surfaces in the desert. Over long periods of time, such wind abrasion creates smooth faces, or facets, on pebbles, cobbles, and boulders. If a rock rolls or tips relative to the prevailing wind direction after it has been faceted on one side, or if the wind shifts direction, a new facet with a different orientation forms, and the two facets join at a sharp edge. Rocks whose surface has been faceted by the wind are faceted rocks, or **ventifacts** (Fig. 21.11). Wind abrasion also gradually polishes and bevels down irregularities

FIGURE 21.9 Sediment transport by wind in deserts.

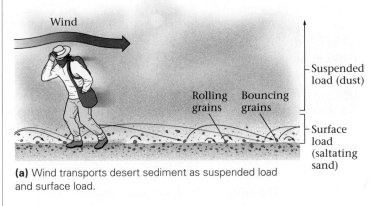

(a) Wind transports desert sediment as suspended load and surface load.

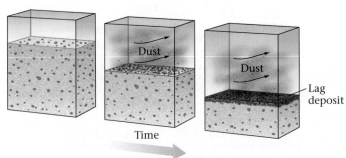

(b) A lag deposit develops when wind blows away finer sediment, leaving behind a layer of coarser grains.

on a desert pavement and polishes the surfaces of desert-varnished outcrops, giving them a reflective sheen.

In places where a resistant layer of rock overlies a softer layer of rock, wind abrasion may create a formation consisting

FIGURE 21.10 Desert pavement, and a hypothesis for how it forms by building up a soil from below.

(a) A well-developed desert pavement in the Sonoran Desert, Arizona.

(b) Students standing at the edge of a trench cut into desert pavement. Note the soil between the pavement and the underlying alluvium.

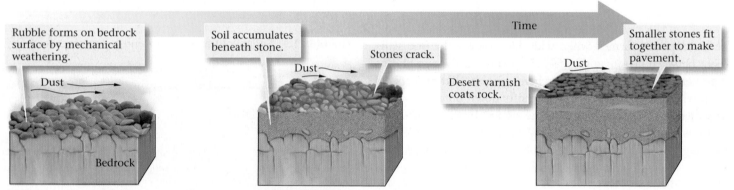

(c) Desert pavement forms in stages. First, loose pebbles and cobbles collect at the surface. Dust settles among the stones and builds up a soil layer below. The stones eventually crack into smaller pieces and settle to form a mosaic-like pavement.

FIGURE 21.11 The progressive development of a ventifact.

Time 1

Windblown sand abrades the face of a rock, forming a facet.

Time 2

The wind shifts direction, and a new facet forms.

A ventifact from the Dry Valleys of Antarctica.

of a resistant block perched on an eroding mushroom-like column of softer rock. These unusual features are called **yardangs** (Fig. 21.12). If a strong wind blows in only one direction, the yardangs become elongate, aligned with the wind direction.

Over time, in regions where the substrate consists of soft sediment, wind picks up and removes so much sediment that the land surface becomes lower. The process of lowering the land surface by wind erosion is called **deflation**. Shrubs can stabilize a small patch of sediment with their roots, so after deflation a forlorn shrub with its residual pedestal of soil stands isolated above a lowered ground surface (Fig. 21.13). In some places, the shape of the land surface twists the wind into a turbulent vortex that causes enough deflation to scour a deep, bowl-like depression called a blowout.

Satellite exploration in the last two decades shows that wind affects the desert-like surface of Mars just as it does the

FIGURE 21.12 Yardangs are small landforms sculpted by the wind, where a resistant rock layer overlies a softer rock layer. These yardangs formed in the Sahara of Egypt.

FIGURE 21.14 This NASA photo shows the wind abrasion of clasts and small sand dunes on the surface of Mars.

deserts on Earth. In fact, during the Martian fall and spring, when differences in temperature between the ice-covered poles and the ice-free equator are greatest, strong winds blow from the poles to the equator, transporting so much dust that the planet's surface, as seen from Earth, visibly changes. At times, the entire planet becomes enveloped in a cloud of dust. Close-up images taken by spacecraft that have landed on Mars show that rocks have been abraded by saltating sand and that sand has accumulated on the lee side of the rocks (Fig. 21.14).

FIGURE 21.13 Wind erosion in Death Valley has left bushes perched on mounds; roots keep soil from blowing away.

Take-Home Message

- Though water is scarce, rocks still undergo chemical and physical weathering in deserts.
- Desert varnish, a thin rusty-brown layer of iron oxide, manganese oxide, and clay, coats rock in desert regions.
- Water is the dominant agent of erosion in desert; it does its work during infrequent but intense downpours that can cause flash floods in dry washes.
- Unprotected by vegetation, rock may also be abraded by sediment blowing in the wind. Wind carries sediment both as suspended load and as a saltating surface load.
- Removal of finer sediment may leave behind lag deposits. Buildup of soil beneath fractured surface rocks may eventually produce a desert pavement.

THINK: Why do the unusual shapes of ventifacts and yardangs develop?

21.4 DEPOSITIONAL ENVIRONMENTS IN DESERTS

We've seen that erosion relentlessly eats away at bedrock and sediment in deserts. Where does the debris go? Below, we examine the various desert settings in which sediment accumulates.

Talus Aprons

Over time, joint-bounded blocks of rock break off ledges and cliffs on the sides of hills. Under the influence of gravity, the resulting debris tumbles downslope and accumulates as **talus**, a pile of debris at the base of a hill. Talus can survive for a long time in desert climates, so we typically see them fringing the bases of cliffs in deserts (Fig. 21.15a). The angular clasts constituting talus gradually become coated with desert varnish.

Alluvial Fans

Flash floods can carry sediment downstream in a steep-walled canyon. When the turbulent water flows out into a plain at the mouth of a canyon, it spreads out over a broader surface and slows. As a consequence, sediment in the water settles out. In some cases, debris flows also emerge from the canyon and spread out. The resulting lenses of sediment cause the channel that has emerged from the mountains to subdivide into a number of subchannels (distributaries) that diverge outward in a broad fan. The fan of distributaries spreads the sediment, or alluvium, out into a broad **alluvial fan**, a wedge- or apron-shaped pile of sediment (Fig. 21.15b). Alluvial fans emerging from adjacent valleys may merge and overlap along the front of a mountain range, producing an elongate wedge of sediment called a **bajada**. Over long periods, the sediment of bajadas can fill in adjacent valleys to depths of several kilometers.

Playas and Salt Lakes

Water from a flash flood may make it out to the center of an alluvium-filled basin, but if the supply of water is relatively small, it quickly sinks into the permeable alluvium without accumulating as a standing body of water. During a particularly large storm or an unusually wet spring, however, a temporary lake may develop over the low part of a basin. During drier times, such desert lakes evaporate entirely, leaving behind a dry, flat, exposed lake bed known as a **playa** (Fig. 21.16a, b). Over time, a smooth crust of clay and various salts (halite, gypsum, borax, and other minerals) accumulates on the surface of playas. Some of these minerals have industrial uses and thus have been mined. When it rains lightly, clay-covered playa surfaces become very slippery. Racetrack Playa, in California, becomes so slippery, in fact, that the wind sends stones sliding out across the surface, leaving grooves behind them that mark their path (Fig. 21.16c).

Where sufficient water flows into a desert basin, it creates a permanent lake. If the basin is an **interior basin**, with no outlet to allow water to flow out, the lake becomes very salty, because although its water escapes by evaporation in the desert sun, its salt cannot. The Great Salt Lake, in Utah, exemplifies this process. Even though the streams feeding the lake are fresh enough to drink, their water contains trace amounts of dissolved salt ions. Because the lake has no outlet, these ions have become concentrated in the lake over time, making it even saltier than the ocean.

Deposition from the Wind

As mentioned earlier, wind carries two kinds of sediment loads—a suspended load of dust-sized particles and a surface load of sand. Much of the dust is carried out of the desert and accumulates elsewhere. Locally, it builds into layers of fine-grained sediment called **loess**. Sand, however, cannot travel far, and accumulates

FIGURE 21.15 Production and transportation of debris and sediment in deserts.

(a) This talus apron along the base of a desert cliff formed from rocks that broke off and tumbled down the cliff.

(b) Sediment in this alluvial fan in Death Valley was carried by flash floods. The fan forms at the mountain front where water slows.

FIGURE 21.16 Playas form where a shallow, salty lake dries up.

An oblique air photo

Playa

Bajada

Range

(a) This playa in California formed at the base of a bajada.

(b) White salt crystals encrust the floor of a playa in Death Valley.

within the desert in piles called **dunes**, ranging in size from less than a meter to over 300 m high. In favorable locations, dunes accumulate to form vast sand seas hundreds of meters thick. We'll look at dunes in more detail later in this chapter.

Take-Home Message

- Erosional debris accumulates at the bases of cliffs to form talus; sediment transported by flash floods accumulates in alluvial fans at the mouths of canyons.

- In basins between desert ranges, shallow lakes may form after rainstorms. When these dry up, they leave behind dry lake beds called playas.

- Desert lakes that do not have an outlet become salty when water evaporates and salt stays behind.

THINK: How do bajadas develop?

(c) Winds can push rocks (that leave tracks) along the slippery clay surface of Racetrack Playa in Death Valley when the surface is wet. Drying later produces polygonal mudcracks.

21.5 DESERT LANDSCAPES AND LIFE

The popular media commonly portray deserts as endless seas of sand, piled into dunes that hide the occasional palm-studded oasis. In reality, immense sand seas are merely one type of desert landscape. Some deserts are vast, rocky plains, others sport a stubble of cacti and other hardy desert plants, and still others contain intricate rock formations that look like medieval castles (See for Yourself S, p. S-36). Explorers of the Sahara, for example,

Did you ever wonder . . .
are all deserts completely covered by sand?

traditionally distinguished among hamada (barren, rocky highlands), reg (vast, stony plains), and erg (sand seas in which large dunes form). In this section, we'll see how the erosional and depositional processes described above lead to the formation of such contrasting landscapes.

Rocky Cliffs and Mesas

In hilly desert regions, the lack of soil exposes rocky ridges and cliffs. As noted earlier, cliffs erode when rocks split away along joints. When this happens, the cliff face retreats but retains roughly the same form. The process, commonly referred to as **cliff retreat**, or scarp retreat, occurs in fits and starts. A cliff may remain unchanged for decades or centuries, and then suddenly a block of rock falls off and crumbles into rubble at the foot of the cliff (Fig. 21.17a). Cliff height depends on bed thickness: in places where particularly thick, resistant beds crop out, tall cliffs develop. This is because large, widely spaced joints form in thick beds, so the collapse of a portion of the wall generates huge blocks. In thinly bedded shales, joints are small and closely spaced, so shale beds erode to make an overall gradual slope consisting of many tiny stair steps. Thus, cliffs formed from strata of contrasting strength develop a step-like shape; strong layers (sandstone or limestone) become vertical cliffs, and weak layers (shale) become rubble-covered slopes (Fig. 21.17b).

With continued erosion and cliff retreat, a plateau of rock slowly evolves into a cluster of isolated hills, ridges, or columns (Fig. 21.18a, b). Flat-lying strata or flat-lying layers of volcanic rocks erode to make flat-topped hills. These go by different names, depending on their size. Large examples (with a top surface area of several square km) are **mesas**, from the Spanish word for table. Medium-sized examples are **buttes** (Fig. 21.19a, b). Small examples, whose height greatly exceeds their top surface area, are **chimneys**. Erosion of strata has resulted in the skyscraper-like buttes of Monument Valley, Arizona, and the stark cliffs of Canyonlands National Park. Bryce Canyon National Park in Utah contains countless chimneys of brightly colored shale and sandstone—locally, these chimneys are known as hoodoos (Fig. 21.19c). Natural arches, such as those of Arches National Monument, form when erosion along joints leaves narrow walls of rock. When the lower part of the wall erodes while the upper part remains, an arch results. (see Geology at a Glance, pp. 732–733.)

In places where bedding dips at an angle to horizontal, flat-topped mesas and buttes don't form; rather, asymmetric ridges called **cuestas** develop. A joint-controlled cliff forms the steep front side of a cuesta, and the tilted top surface of a resistant bed forms the gradual slope on the backside (Fig. 21.20a). Because the angle of the gradual slope is the same as the dip angle of the bed (the angle the bed surface makes with respect to horizontal), it is known as a dip slope. If the bedding dip is steep to near vertical, a narrow symmetrical ridge, called a hogback, forms. If desert hills consist of homogeneous rock such as granite rather than stratified rock, they typically erode to make a pile of rounded blocks (see Fig. B.8b).

FIGURE 21.17 Cliff retreat in a desert environment.

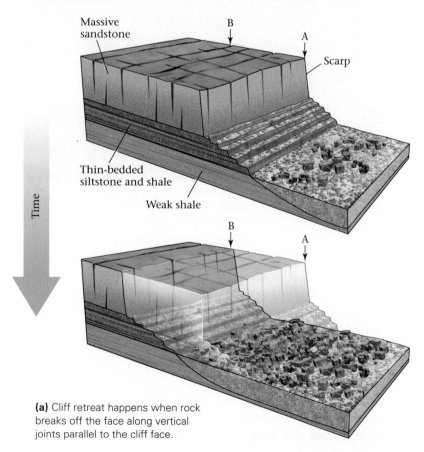

Time

Massive sandstone

B

A

Scarp

Thin-bedded siltstone and shale

Weak shale

B

A

(a) Cliff retreat happens when rock breaks off the face along vertical joints parallel to the cliff face.

Stronger sandstone

Weak shale

Debris from a recent rock fall

(b) On a hill in Utah, the strong sandstone holds up a cliff face, whereas weak shale forms a slope.

FIGURE 21.18 Contrasts between desert hill slopes and temperate landscapes.

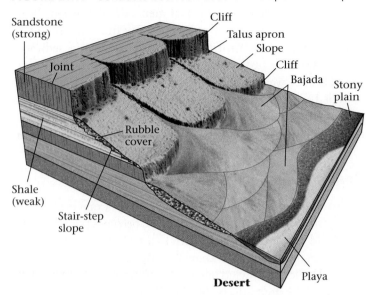

Sandstone (strong)
Cliff
Talus apron
Slope
Cliff
Bajada
Stony plain
Joint
Rubble cover
Shale (weak)
Stair-step slope
Desert
Playa

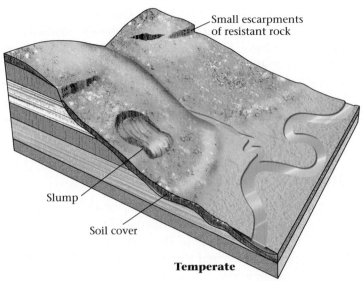

Small escarpments of resistant rock
Slump
Soil cover
Temperate

(a) In a desert, steep cliffs form out of resistant rock, fans of debris form a bajada at the base of the cliffs, and playas form on the plain.

(b) In a temperate region, thick soils form, and slumping prevents steep slopes from developing. Vegetation covers the surface.

FIGURE 21.19 Mesas and buttes form in deserts as cliffs retreat over time.

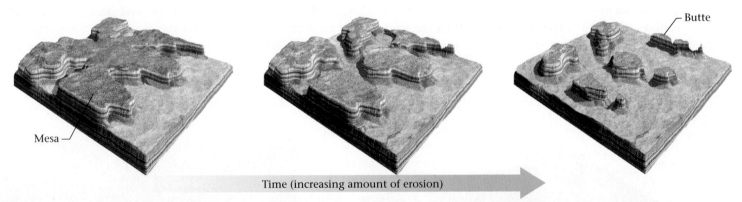

Butte

Mesa

Time (increasing amount of erosion)

(a) Because of cliff retreat, a once-continuous layer of rock evolves into a series of isolated remnants. If the bedding is horizontal, the resulting landforms have flat tops.

(b) Buttes and mesas tower above the floor of Monument Valley, Arizona.

(c) Erosion produced "hoodoos," chimney-like columns of rock, in Bryce Canyon, Utah.

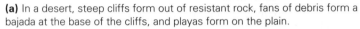

With progressive cliff retreat on all sides of a hill, finally all that remains of the hill is a relatively small island of rock, surrounded by alluvium-filled basins. Geologists refer to such islands of rock by the German word **inselberg** (island mountain; Fig. 21.20b). Depending on the rock type or the orientation of stratification in the rock, and on rates of erosion, inselbergs may be sharp-crested, plateau-like, or loaf-shaped (steep sides and a rounded crest). Inselbergs with a loaf geometry, as exemplified by Uluru (Ayers Rock) in central Australia (Box 21.1), are also known as bornhardts.

Stony Plains and Pediments

The coarse sediment eroded from desert mountains and ridges washes into the lowlands and builds out to form gently sloping alluvial fans. The surfaces of these gravelly piles are strewn with pebbles, cobbles, and boulders, and are dissected by dry washes (wadis or arroyos). Portions of these stony plains evolve into desert pavements.

When travelers began trudging through the desert of the southwestern United States during the nineteenth century, they found that in many locations the wheels of their wagons were rolling over flat or gently sloping bedrock surfaces. These bedrock surfaces extended outward like ramps from the steep cliffs of a mountain range on one side, to alluvium-filled valleys on the other (see Fig. 21.20b). Geologists now refer to such surfaces as **pediments**. Pediments develop when sheetwash during floods carries sediment away from the mountain front, during mountain-front retreat. The moving sediment grinds away the bedrock that it tumbles over. Between erosional events, weathering weakens the surface of the pediment so that the next flood has more material to move. Alluvium that is washed off pediments accumulates farther downslope, and may eventually build up sufficiently to bury the pediment.

Seas of Sand: The Nature of Dunes

A **sand dune** is a pile of sand deposited by a moving current. Dunes in deserts start to form where sand becomes trapped on the windward side of an obstacle, such as a rock or a shrub. (Dunes formed around shrubs are known as coppice dunes.) Gradually, the sand builds downwind into the lee of the obstacle. Once initiated, the dune itself affects the wind flow, and sand accumulates on the lee (downwind) side of the dune. Here, the sand eventually slides down the lee surface of the dune (Fig. 21.21).

In places where abundant sand accumulates, sand seas (ergs) bury the landscape. The wind builds the sand in these ergs into dunes that display a variety of shapes and sizes, depending on the character of the wind and the sand supply (Fig. 21.22a). Where the sand is relatively scarce and the wind blows steadily

FIGURE 21.20 The formation of cuestas and inselbergs, due to erosion and deposition in deserts.

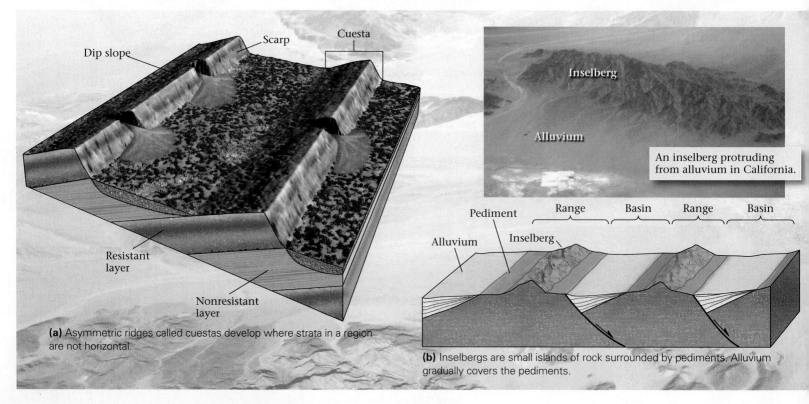

(a) Asymmetric ridges called cuestas develop where strata in a region are not horizontal.

An inselberg protruding from alluvium in California.

(b) Inselbergs are small islands of rock surrounded by pediments. Alluvium gradually covers the pediments.

BOX 21.1

Uluru (Ayers Rock)

Much of central Australia is an immense desert. Parts are barren and sandy, but much of the land is covered with scrub brush, which provides meager grazing for cattle and kangaroo. Uluru, also known by its English name, Ayers Rock, is a huge bornhardt that towers 360 m above the desert plain (**Fig. Bx21.1**). This rock mass, 3.6 km long and 2 km wide, consists of nearly vertical dipping sandstone beds. It makes up one limb of a huge regional syncline. The other limb is also a bornhardt, known locally as The Olgas, and the entire area in between is buried in alluvium. Uluru formed because its sandstone resisted erosion, whereas adjacent rock formations did not. Thus, over geologic time, alluvium buried the surrounding landscape, but Uluru remained high. Because its strata dip vertically, it has not developed the stair-step shape of mesas and buttes.

Because of its grandeur, Uluru plays a sacred role in Australian Aboriginal traditions. In the dreamtime (time of creation) legends of the Aboriginal people, erosional features on the surface of the rock are scars from a fierce battle between ancient clans. In recent years, the rock has attracted tourists from around the world, who risk life and limb to climb to its top. Plaques at the base of the rock record the names of those who slipped and fell from its steep side.

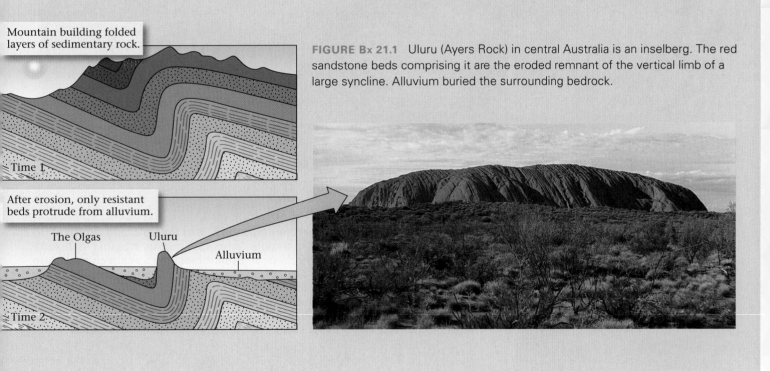

Mountain building folded layers of sedimentary rock.

Time 1

After erosion, only resistant beds protrude from alluvium.

The Olgas   Uluru

Alluvium

Time 2

FIGURE Bx 21.1 Uluru (Ayers Rock) in central Australia is an inselberg. The red sandstone beds comprising it are the eroded remnant of the vertical limb of a large syncline. Alluvium buried the surrounding bedrock.

FIGURE 21.21 Progressive stages in the growth of a small sand dune.

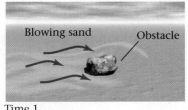

Blowing sand   Obstacle

Time 1

Trapped sand

Time 2

Sand grows around obstacle.

Time 3

Obstacle is buried and dune grows.

Time 4

The Desert Realm

The desert of the Basin and Range Province in Utah, Nevada, and Arizona consists of alternating basins (grabens or half-grabens) separated by narrow ranges (tilted fault blocks). The Sierra Nevada, underlain largely by granite, borders the western edge of the province, while the Colorado Plateau, underlain by flat-lying sedimentary strata, borders the eastern edge. The overall climate of the region is dry. Because of the great variety of elevations and rock types, the region hosts different desert landscapes.

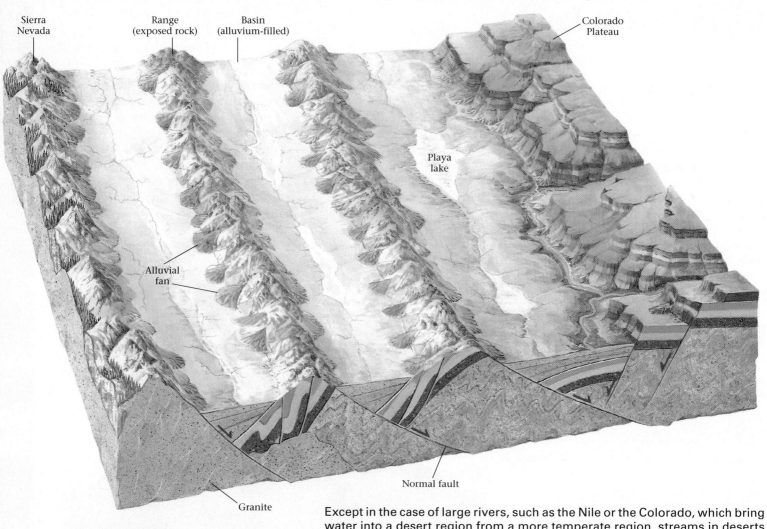

Sierra Nevada

Range (exposed rock)

Basin (alluvium-filled)

Colorado Plateau

Playa lake

Alluvial fan

Normal fault

Granite

Barchan dune

Cross beds

Flash flood

Except in the case of large rivers, such as the Nile or the Colorado, which bring water into a desert region from a more temperate region, streams in deserts fill with water only after heavy rains. At other times, the stream channels are dry. These channels are called dry washes, arroyos, or wadis. When there is a heavy rain, water cannot be absorbed into the ground fast enough, so runoff enters dry washes and fills them very quickly, creating a flash flood. The turbulent, muddy water of a flash flood can transport even large boulders. This flash flood is rushing down a stream in the Sonoran Desert of Arizona.

Where there is a large supply of sand, a variety of sand dunes develop. The geometry of a particular sand dune (such as barchan, longitudinal, or star) depends on the sand supply and the wind. Inside sand dunes, we find cross beds.

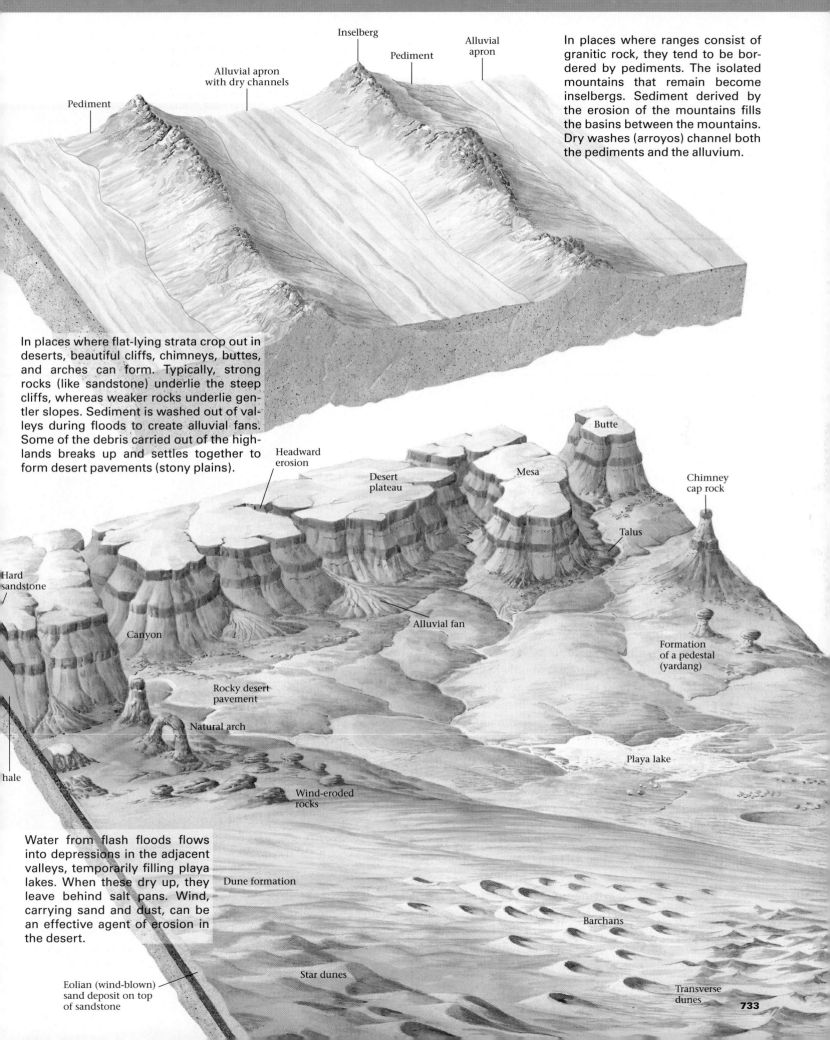

Inselberg

Alluvial apron

Pediment

Alluvial apron
with dry channels

Pediment

Pediment

In places where ranges consist of granitic rock, they tend to be bordered by pediments. The isolated mountains that remain become inselbergs. Sediment derived by the erosion of the mountains fills the basins between the mountains. Dry washes (arroyos) channel both the pediments and the alluvium.

In places where flat-lying strata crop out in deserts, beautiful cliffs, chimneys, buttes, and arches can form. Typically, strong rocks (like sandstone) underlie the steep cliffs, whereas weaker rocks underlie gentler slopes. Sediment is washed out of valleys during floods to create alluvial fans. Some of the debris carried out of the highlands breaks up and settles together to form desert pavements (stony plains).

Butte

Headward
erosion

Mesa

Desert
plateau

Chimney
cap rock

Talus

Hard
sandstone

Formation
of a pedestal
(yardang)

Canyon

Alluvial fan

Rocky desert
pavement

Natural arch

Playa lake

hale

Wind-eroded
rocks

Water from flash floods flows into depressions in the adjacent valleys, temporarily filling playa lakes. When these dry up, they leave behind salt pans. Wind, carrying sand and dust, can be an effective agent of erosion in the desert.

Dune formation

Barchans

Eolian (wind-blown)
sand deposit on top
of sandstone

Star dunes

Transverse
dunes

733

FIGURE 21.22 The types of sand dunes and the cross beds within them.

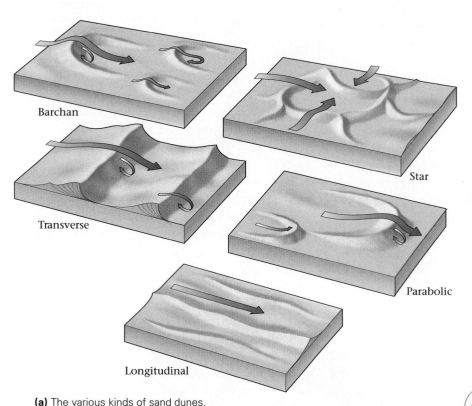

Barchan

Transverse

Star

Parabolic

Longitudinal

(a) The various kinds of sand dunes.

(b) A sand dune with surface ripples.

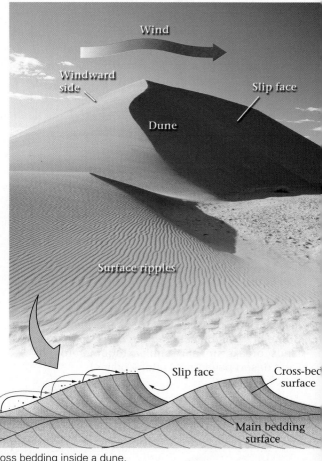

Wind

Windward side

Slip face

Dune

Surface ripples

Slip face

Cross-bed surface

Main bedding surface

(c) Cross bedding inside a dune.

Main bed

Cross bed

(d) Cross beds preserved in Mesozoic sandstone beds of Zion National Park.

in one direction, beautiful crescents called barchan dunes develop, with the tips of the crescents pointing downwind. If the wind shifts direction frequently, a group of crescents pointing in different directions overlap one another, creating a constantly changing star dune. Where enough sand accumulates to bury the ground surface completely, and only moderate winds blow, sand piles into simple, wave-like shapes called transverse dunes. The crests of transverse dunes lie perpendicular to the wind direction. Strong winds may break through transverse dunes and change them into parabolic dunes whose ends point in the upwind direction. Finally, if there is abundant sand and a strong, steady wind, the sand streams into longitudinal dunes (also called seif dunes after the Arabic word for sword) whose axis lies parallel to the wind direction. In the southern third of the Arabian Peninsula, a region called the Empty Quarter because of its lack of population, a vast erg called the Rub al Khali contains seif dunes that stretch for almost 200 km and reach heights of over 300 m.

In a sand dune, sand saltates up the windward side of the dune, blows over the crest of the dune, and then settles on the steeper, lee face of the dune. The slope of this face attains the angle of repose, the slope angle of a freestanding pile of sand. As sand collects on this surface, it may become unstable and

slide down the slope, so geologists refer to the lee side of a dune as the **slip face**. As more and more sand accumulates on the slip face, the crest of the dune migrates downwind, and former slip faces become preserved inside the dune. In cross section, these slip faces appear as cross beds (**Fig. 21.22b–d**).

The surfaces of dunes are not, in general, smooth, but rather are covered with delicate ripples.

With the exception of star and longitudinal dunes, sand dunes migrate downwind as the wind continuously picks up sand from the gently dipping windward slope and drops it onto the leeward side, or slip face. Rates of migration can exceed 25 m per year. Because of moving sand on an active dune, vegetation can't grow there. If a change in climate brings more rain or less wind, plant cover may grow and stabilize the dunes. At the end of the last ice age, for example, the Sand Hills region of western Nebraska was a vast dune field, but in the past 11,000 years it has been covered by grasslands, and the dunes have become stabilized.

Life in the Desert

In the midst of a large erg, there seems to be nothing growing or moving at all. But most desert landscapes do include plants and animals. These organisms must possess special characteristics to enable them to survive in the desert: they must be able to withstand extremes of temperature—oppressive heat during the day and chilling cold at night—and to survive without abundant water.

Plants have evolved a number of different means to survive desert conditions. Some produce thick-skinned seeds that last until a heavy rainfall, then quickly germinate, grow, and generate new seeds only while water is available. The new generation of seeds then waits until the next rainfall to start the cycle over again. Other plants have evolved the ability to send long roots down to find deep groundwater. Still others have shallow root systems that spread over a broad area so they can efficiently soak up water when it does rain.

Many desert plants have thick, fleshy stems and leaves. These plants, known as **succulents**, can store water for long periods of time (Fig. 21.23a–c). Because succulents may be the only source of water during a time of drought, they have developed threatening thorns or needles to keep away thirsty animals. Plant life is much more diverse in desert oases, the verdant islands that crop up where natural springs spill groundwater onto the surface (see Chapter 19). The nearly year-round supply of water in an oasis nourishes a variety of palms and other nonsucculent plants.

Animal life in the desert includes scavengers, hunters, and plant eaters. Animals face the same challenge as plants—they must be able to retain water and survive extreme temperatures. To accomplish these goals, desert animals have also evolved numerous strategies. Frogs, for example, burrow beneath the ground and remain dormant for months, waiting until the next rain. Reptiles escape the midday heat by crawling into dark cracks between rocks. Rodents forage for food only during the cool night. And kit foxes, jack rabbits, and mule deer have disproportionately large ears through which they efficiently lose body heat. Many desert mammals, such as camels, retain body water by not sweating, as we noted earlier.

Humans are not meant to live in the desert. The loss of body moisture in extreme heat can be so rapid that a person will die in less than 24 hours unless shaded from the Sun and supplied with at least 8 liters of water per day. Nevertheless,

FIGURE 21.23 Plants of the Sonoran Desert, Arizona, are well adapted for dry conditions.

(a) Saguaro cactus can become huge. Don't try to hug one.

(b) "Teddy bear" cholla look soft, but the spines are extremely sharp and hard to remove.

(c) Cactus flowers stand out against brown sand and rock.

FIGURE 21.24 Living in the desert is a challenge because of the need for water.

(a) A west African village struggles with blowing sand and the lack of vegetation.

(b) Suburbs of Las Vegas, Nevada, push into the desert. Residents use groundwater, and water brought from the Colorado River, to keep grass green.

all but the most barren deserts are inhabited, though sparsely. Before technology provided water wells, pipelines, and mechanized transportation, desert peoples lived in small nomadic groups, spaced far enough apart that they could live off the land; deserts have too little water to sustain agriculture or husbandry (Fig. 21.24a). Australian Aboriginal groups, for example, rarely included more than a dozen individuals. In the past, nomadic desert dwellers either built temporary shelters out of local materials or traveled with tents. Locally, people carved underground dwellings in sediment or soft rock, for rock is such a good insulator that a few meters below the ground surface, it stays close to the region's mean temperature year round.

Take-Home Message

- Portions of some deserts are covered by extensive seas of sand; the sand builds into dunes of various shapes. But not all areas of deserts are covered by sand; some are stone-covered plains.
- Cliff retreat may eventually isolate buttes, mesas, and chimneys.
- After a while, sediment may completely surround island-like erosional remnants (inselbergs) of bedrock. In some cases, erosion produces a gently sloped pediment around an inselberg.
- In the harsh climate of deserts, only very hardy organisms, such as succulents, can survive.

THINK: What factors control the shape, dimensions, and orientation of sand dunes?

21.6 DESERT PROBLEMS

More and more people are moving into desert regions. In fact, the population of the desert in the southwestern United States is growing faster than in any other part of the country, and as desert cities grow, environmental problems soon follow (Fig. 21.24b). As noted in Chapter 19, growing cities must either suck water out of the ground or bring in water via canals from rivers or reservoirs to meet their needs. So water tables in deserts are dropping, rivers are drying up, and the land surface is cracking. People also have imported exotic plants and animals that invade the countryside and upset the ecological balance.

The modern era has seen a remarkable change in desert margins. Natural droughts (periods of unusually low rainfall), aggravated by overpopulation, overgrazing, careless agriculture, and diversion of water supplies, have transformed semi-arid grasslands into true deserts, leading to tragic famines that have killed millions of people. **Desertification**, the process of transforming nondesert areas to desert, has accelerated.

The consequences of desertification have devastated the Sahel, the belt of semi-arid land that fringes the southern margin of the Sahara. In the past, the Sahel provided sufficient vegetation to support a small population of nomadic people and animals. But during the second half of the twentieth century, large numbers of people migrated into the Sahel to escape overcrowding in central Africa. The immigrants began farming and maintained large herds of cattle and goats. Plowing and overgrazing removed soil-preserving grass and caused the soil to dry out. In addition, the trampling of animal hooves compacted the ground so it could no longer soak up water. In the 1960s and again in the 1980s, a series of natural droughts hit the region, bringing catastrophe (Fig. 21.25a–c). Wind erosion stripped off the remaining topsoil. Without vegetation, the air grew drier, and the semiarid grassland of the Sahel became desert, with mass starvation as the result.

Other regions on Earth are developing the same problem. The Aral Sea in Kazakhstan, for example, has almost entirely dried up. Diversion of the rivers that once fed the sea has so diminished its water supply that the area of the sea has shrunk dramatically, and boats that once plied its waters now lie as rusting hulks in a sea of dust (Fig. 21.26).

Desertification does not happen only in less industrialized nations. People in the western Great Plains of the United States and Canada suffered from the problem beginning in 1933, the fourth year of the Great Depression. Banks had failed, workers had lost their jobs, the stock market had crashed, and hardship burdened all. No one needed yet another disaster—but that year, even nature turned hostile. All through the fall, so little rain fell in the plains of Texas and Oklahoma that the region's grasslands and croplands browned and withered, and the topsoil turned to powdery dust. Then, on November 12 and 13, strong storms blew eastward across the plains. Without vegetation to protect the ground, the wind lapped at it, stripped off the topsoil, and sent it skyward to form rolling black clouds that literally blotted out the sun (Fig. 21.27). People caught in the resulting dust storm found themselves choking and gasping for breath. When the dust finally settled, it had buried houses and roads under huge drifts, and dirtied every nook and cranny. The dust blew east as far as New England, where it turned the snow brown. What had once been a

FIGURE 21.25 Desertification is happening in parts of Africa.

(a) The Sahel is the semi-arid land along the southern edge of the Sahara. Large parts have undergone desertification.

(b) Drought in the Sahel has brought deadly consequences. Here, residents seek water from a dwindling pond.

(c) In the Dead Vlei of Namibia (in southwest Africa) streams have dried up, causing trees to die. Large sand dunes tower in the background.

FIGURE 21.26 Desertification of the Aral Sea in central Asia occurred when the Soviet Union diverted the rivers that flowed into the sea for irrigation projects. Once home to a fishing fleet, all that remains of the sea are a few ultra-salty ponds.

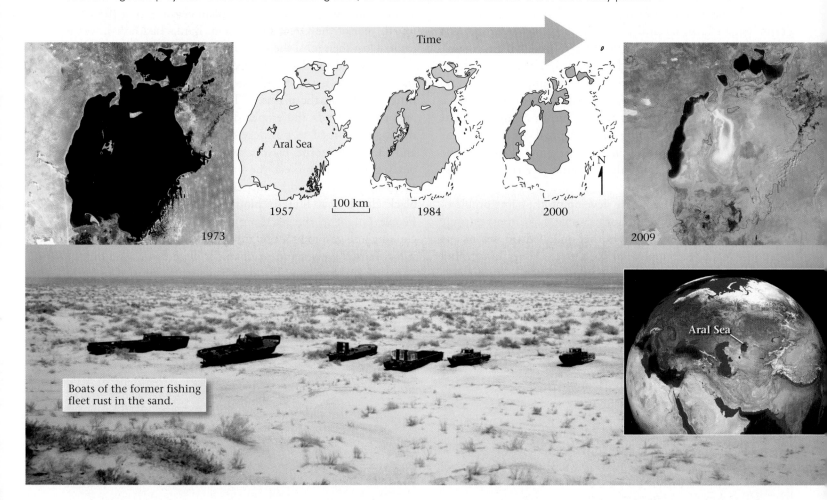

Boats of the former fishing fleet rust in the sand.

rich farmland in the southwestern plains turned into a wasteland that soon acquired a nickname, the Dust Bowl.

For several more years the drought persisted, leading to starvation and bankruptcy. Many of the region's residents were forced to move to wherever they could find work. People from Texas and Oklahoma piled into jalopies and drove on old Route 66 out to California, looking for jobs in the state's still-green agricultural regions. Many people were subjected to exploitation and persecution once they arrived in California. John Steinbeck dramatized this staggering human tragedy, which came to symbolize the Depression, in his novel *The Grapes of Wrath*.

Why did the fertile soils of the southern Great Plains suddenly dry up? The causes were complex; some were natural and some were human-induced. Typically, the region has a semi-arid climate in which only thin soil develops. But the plains were settled in the 1880s and 1890s, which were unusually wet years. Not realizing its true character, far more people moved into the region than it could sustain, and the land was

farmed too intensively. When farmers used steel plows, they destroyed the fragile grassland root systems that held the thin soil in place. And when the drought of the 1930s came, it brought catastrophe.

Desertification can be reversed, but at a price. Planting and irrigation may transform desert into farm fields, orchards, forests, or lawns. But water to nourish the plants has to come from somewhere, and people obtain it by diverting rivers or by pumping groundwater—activities that create their own set of problems. River diversion robs regions downstream of their water supply, and pumping out too much groundwater lowers the water table so substantially that the pore space in aquifers collapses and the ground surface subsides. In short, people will need to rethink land-use policies in semi-arid lands to avoid catastrophe.

The Dust Bowl of the 1930s and the fate of the Sahel in Africa remind us of how fragile the Earth's green blanket of vegetation really is. Global climate changes can shift climatic belts sufficiently to transform agricultural regions into

FIGURE 21.27 Iconic photograph of a farmer walking through a 1930s dust storm in Oklahoma. Much of the topsoil of the region was blown away.

FIGURE 21.28 In this satellite image, a huge dust cloud that originated in the Sahara blows across the Atlantic.

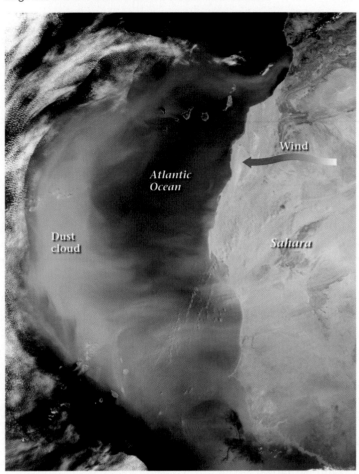

deserts. Some 5,000 years ago, the swath of land between the Nile Valley in Egypt and the Tigris-Euphrates Valley of Mesopotamia was known as the fertile crescent; here, people first abandoned their nomadic ways and settled in agricultural communities. Yet the original "land of milk and honey" has become a barren desert, in need of intensive irrigation to maintain any agriculture. The change in landscape reflects a change in climate—the beginning of Western civilization occurred during the warmest and wettest period of global climate since before the last ice age. So much water drenched the Middle East and North Africa that rivers flowed where Saharan sands now blow. Unfortunately, if the current trend in climate change continues, our present agricultural belts could someday become new Saharas.

Desertification has an additional dangerous side effect—global transmission of chemicals and pathogens by blowing dust. As desert areas expand in response to desertification, and as desert pavement gets disrupted, windblown dust becomes more of a problem. Not only do winds have larger areas of dry, dusty land to churn, but the dust generated from lands that were once agricultural and are now desert may contain harmful chemicals (e.g., residues of herbicides and pesticides) that can themselves become windborne. In recent years, satellite images have revealed that windblown dust from deserts can

travel across oceans and affect regions on the other side. For example, dust blown off the Sahara can traverse the Atlantic and settle over the Caribbean (Fig. 21.28). Geologists are concerned that this dust, along with the fungi, toxic chemicals, and microbes that it carries, may infect corals with disease or in some other way inhibit their life processes. Thus, windblown dust can contribute to the destruction of coral reefs.

Take-Home Message

- Desertification, the transformation of non-desert regions into desert, may happen if climate changes, regions are overgrazed, or water sources are diverted.
- When vegetation disappears, wind may pick up sediment and produce dust storms.

THINK: What factors transformed Oklahoma and Texas into a dust bowl during the 1930s?

Chapter Summary

- Deserts generally receive less than 25 cm of rain per year. Vegetation covers no more than 15% of their surface.

- Subtropical deserts form between latitudes of 20° and 30°, rain-shadow deserts are found on the inland side of mountain ranges, coastal deserts are located on the land adjacent to cold ocean currents, continental-interior deserts exist in landlocked regions far from the ocean, and polar deserts form at high latitudes.

- In deserts, chemical weathering happens slowly, so rock bodies tend to erode primarily by physical weathering. Desert varnish forms on rock surfaces, and soils tend to accumulate soluble minerals.

- Water causes significant erosion in deserts, mostly during heavy downpours. Flash floods carry large quantities of sediments down ephemeral streams. When the rain stops, these streams dry up, leaving steep-sided dry washes.

- Wind causes significant erosion in deserts because it picks up dust and silt as suspended load, and it causes sand to saltate. Where wind blows away finer sediment, a lag deposit remains. Windblown sediment abrades the ground, creating a variety of features such as ventifacts and yardangs.

- Desert pavements are mosaics of varnished stones armoring the surface of the ground.

- Talus piles form when rock fragments accumulate at the base of a slope. Alluvial fans form at a mountain front where water in ephemeral streams deposits sediment. When temporary desert lakes dry up, they leave playas.

- In some desert landscapes, erosion causes cliff retreat, eventually resulting in the formation of mesas, buttes, and inselbergs. Pediments of nearly flat or gently sloping bedrock surround some inselbergs.

- Where sand is abundant, the wind builds it into dunes. Common types include barchan, star, transverse, parabolic, and longitudinal (seif) dunes.

- Deserts contain a great variety of plant and animal life. All are adapted to survive extremes in temperature and without abundant water.

- Changing climates and land abuse may cause desertification, the transformation of semi-arid land into deserts. Windblown dust, sometimes carrying microbes and toxins, may waft from deserts across oceans.

GEOPUZZLE REVISITED

Deserts are very arid regions that can sustain very little vegetation. They host only hardy plants and animals adapted to living with little water. Vast areas of sand dunes do cover some portions of some deserts, but desert landscapes can also host stony plains and rocky ridges. And, although many deserts are very hot, some occur at high elevations or at polar latitudes and can be very cold.

Guide Terms

alluvial fan (p. 726)	lag deposit (p. 722)
arroyo (p. 722)	loess (p. 726)
bajada (p. 726)	mesa (p. 728)
butte (p. 728)	pediment (p. 730)
chimney (p. 728)	petroglyph (p. 721)
cliff retreat (p. 728)	playa (p. 726)
cuesta (p. 728)	rain shadow (p. 719)
deflation (p. 724)	saltation (p. 722)
desert (p. 717)	sand dune (p. 730)
desert pavement (p. 722)	slip face (p. 734)
desert varnish (p. 721)	succulents (p. 735)
desertification (p. 736)	surface load (p. 722)
dry wash (p. 722)	suspended load (p. 722)
dune (p. 727)	talus (p. 726)
dust storm (p. 722)	ventifact (p. 723)
inselberg (p. 730)	yardang (p. 724)
interior basin (p. 726)	

Review Questions

1. What factors determine whether a region can be classified as a desert?

2. Explain why deserts form.

3. Have today's deserts always been deserts? (Hint: Keep in mind the consequences of plate tectonics.)

4. How do weathering processes in deserts differ from those in temperate or humid climates?

5. Describe how water modifies the landscape of a desert. Be sure to discuss both erosional and depositional landforms.

6. Explain the ways in which desert winds transport sediment.

7. Explain how the following features form: (a) desert varnish; (b) desert pavement; (c) ventifacts; (d) yardangs.

8. Describe the process of formation of alluvial fans, bajadas, and playas.

9. Describe the process of cliff (scarp) retreat and the landforms that result from it.

10. What are the various types of sand dunes? What factors determine which type of dune develops at a particular location?

11. Discuss various adaptations that lifeforms have evolved in order to survive in desert climates.

12. What is the process of desertification, and what causes it? How can desertification in Africa affect the Caribbean?

On Further Thought

13. Death Valley, California, lies to the east of a high mountain range, and its floor lies below sea level. During the summer, Death Valley is very hot and dry. Explain why it has such weather.

14. You are working for an international nongovernmental organization (NGO) and have been charged with the task of providing recommendations to an African nation that wishes to slow or halt the process of desertification within its borders. What are your recommendations?

15. The Namib Desert lies to the north and west of the Kalahari Desert, in southern Africa. The reason that the former region is a desert is not the same as the reason that the latter is a desert. Explain this statement.

16. Lake Havasu, Arizona, is a resort town along the California–Arizona border, in the western part of the Sonoran Desert (find in on *Google Earth*™). Water sports are one of the attractions of the town. Where does the standing water of this location come from, and what factors can affect the supply in the future?

 For more resources, including animations, quizzes, and Norton's GeoTours, go to **wwnorton.com/studyspace**.

 If your instructor assigns exercises in SmartWork, log in at **smartwork.wwnorton.com**.

ANOTHER VIEW Under appropriate conditions, sand can build into large dunes. The dunes in this satellite image are up to 300 m high. They developed in the Namib Desert of Namibia, in southwestern Africa.

CHAPTER 22

Amazing Ice: Glaciers and Ice Ages

The blue ice of a glacier in the French Alps has played a role in carving the surrounding mountains. Ice like this covered vast areas of land during the ice age.

GEOPUZZLE

If modern society had existed about 12,000 years ago, would it have been possible to build New York City (USA) or Edinburgh (Scotland) at their present locations? Why?

I seemed to vow to myself that some day I would go to the region of ice and snow and go on and on till I came to one of the poles of the earth.

—Ernest Shackleton
(British polar explorer, 1874–1922)

22.1 INTRODUCTION

There's nothing like a good mystery, and one of the most puzzling in the annals of geology came to light in northern Europe early in the nineteenth century. When farmers of the region prepared their land for spring planting, they occasionally broke their plows by running them into large boulders buried randomly through otherwise fine-grained sediment. Many of these boulders did not consist of local bedrock, but rather came from outcrops hundreds of kilometers away. Because the boulders had apparently traveled so far, they came to be known as **erratics** (from the Latin *errare*, to

FIGURE 22.1 Agassiz's thoughts about the Ice Age.

(a) Agassiz found boulders protruding from the ground in places that are not currently glaciated. He proposed that the boulders are erratics left by now-vanished glaciers.

(b) Agassiz envisoned that vast areas of the northern hemisphere were once covered by vast ice sheets, comparable to the one covering Antarctica today.

wander). Similar occurrences were discovered in northern regions of North America (**Fig. 22.1a**).

The mystery of the wandering boulders became a subject of great interest to early nineteenth-century geologists, who realized that such deposits of extremely *unsorted* sediment (which contain a variety of clast sizes) could not be examples of typical stream alluvium, for running water sorts sediment by size. Most attributed the deposits to a vast flood that they imagined had been powerful enough to spread a slurry of boulders, sand, and mud across the continent. In 1837, however, a young Swiss geologist named Louis Agassiz proposed a radically different interpretation. Agassiz often hiked among **glaciers** (slowly flowing masses of ice that survive the summer melt) in the Alps near his home. He observed that glacial ice could carry enormous boulders as well as sand and mud, because ice is solid and has enough strength to support the weight of rock. Agassiz realized that because solid ice does not sort sediment as it flows, glaciers leave behind unsorted sediment when they melt. On the basis of these observations, he proposed that the mysterious sediment and erratics of Europe were deposits left by **ice sheets**, vast glaciers that had once covered much of the continent (**Fig. 22.1b**). In Agassiz's mind, Europe had once been in the grip of an **ice age**, a time when the climate was significantly colder and glaciers grew.

Agassiz's radical proposal faced intense criticism for the next two decades. But he didn't back down, and instead challenged his opponents to visit the Alps and examine the sedimentary deposits that alpine glaciers had left behind. By the late 1850s, most doubters had changed their minds, and the geological community concluded that the notion that Europe once had Arctic-like climates was correct. Later in life, Agassiz traveled to the United States and documented many glacier-related features in North America's landscape, proving that an ice age had affected vast areas of the planet.

Glaciers, which have many forms, cover only about 10% of the land on Earth today, but during the most recent ice age, which ended only about 11,000 years ago, as much as 30% of continental land surface had a coating of ice. New York City, Montréal, and many of the great cities of Europe now occupy land that once lay beneath hundreds of meters to a few kilometers of ice. The work of Louis Agassiz brought the subject of glaciers and ice ages into the realm of geologic study and led people to recognize that major climate changes happen in Earth history. In this chapter, after considering the nature of ice, we see how glaciers form, why they move, and how they modify landscapes by erosion and deposition. A substantial portion of the chapter concerns the last ice age, known as the **Pleistocene Ice Age**, for its impact on the landscape can still be seen today, but we briefly introduce ice ages that happened earlier in Earth history too. We conclude by considering hypotheses to explain why ice ages happen.

Chapter Themes ➤

By the end of this chapter, you should know . . .

- how glacial ice forms and flows, and how to categorize various kinds of glaciers.
- how glaciers advance and retreat, and how their flow modifies the landscape.
- how to recognize sedimentary deposits and associated landforms left by glaciers.
- that glaciers covered large areas of continents and that sea level dropped during ice ages.
- why ice ages happen, and why glaciations during an ice age happen periodically.

22.2 ICE AND THE NATURE OF GLACIERS

What Is Ice?

Ice consists of solid water, formed when liquid water cools below its freezing point. We can consider a single ice crystal to be a mineral specimen: it is a naturally occurring, inorganic solid, with a definite chemical composition (H_2O) and a regular crystal structure. Ice crystals have a hexagonal form, so snowflakes grow into six-pointed stars (Fig. 22.2a). We can think of a layer of fresh snow as a layer of sediment, and a layer of snow that has been compacted so that the grains stick together as a layer of sedimentary rock (Fig. 22.2b). We can also think of the ice that appears on the surface of a pond as an igneous rock, for it forms when molten ice (liquid water) solidifies. Glacier ice, in effect, is a metamorphic rock. It develops when preexisting ice *recrystallizes* in the solid state, meaning that the molecules in solid water rearrange to form new crystals (Fig. 22.2c).

Pure ice has the transparency of glass, but if ice contains tiny air bubbles or cracks that disperse light, it becomes milky white. Like glass, ice has a high **albedo**, meaning that it reflects light well—so well, in fact, that if you walk on ice without eye protection, you risk blindness from the glare. Ice differs from most other familiar materials in that its solid form is *less dense* than its liquid form—the architecture of an ice crystal holds water molecules apart. Ice, therefore, floats on water. This unusual characteristic has benefits; if ice didn't float, ice in oceans would sink, leaving room for new ice to form until the oceans froze solid.

Ice also has the unusual property of being slippery—that's why skaters can skate! Surprisingly, researchers still don't completely understand why ice is slippery. An older explanation—that skaters can glide on ice because a film of liquid water forms on the surface of the ice in response to frictional heating or to pressure-induced melting—can't explain how ice remains slippery even when it's so cold that water can't exist as liquid.

Modern studies suggest that ice can remain slippery even at very low temperatures because the surface of ice consists of a layer of water molecules that are not completely fixed within a crystal lattice. Specifically, the molecules are chemically bonded to the solid ice below, but not to the air above. The existence of unattached bonds permits the surface molecules to behave somewhat like a liquid, even while attached to the solid.

How a Glacier Forms

In order for a glacier to form, three conditions must be met. First, the local climate must be sufficiently cold that winter snow does not melt entirely away during the summer; second, there must be, or must have been, sufficient snowfall for a large amount of snow to accumulate; and third, the slope of the surface on which the snow accumulates must be gentle enough that the snow does not slide away in avalanches, and must be protected enough that the snow doesn't blow away.

Glaciers develop in polar regions because even though relatively little snow falls today, temperatures remain so low that most ice and snow survive all year. Glaciers develop in mountains, even at low latitudes, because temperature decreases with elevation; at high elevations, the mean temperature stays low enough for ice and snow to survive all year. Since the temperature of a region depends on latitude, the specific elevation at which mountain glaciers form also depends on latitude. In Earth's present-day climate, glaciers form only at elevations above 5 km between 0° and 30° latitude and down to sea level at 60° to 90° latitude. Thus, you can see high-latitude glaciers from a cruise ship, but at the equator, you have to climb high mountains to find glaciers. Mountain glaciers tend to develop on the side of mountains that receives less wind, and on the side that receives less sunlight. Glaciers do not exist on slopes greater than about 30°, because avalanches clear such slopes.

The transformation of snow to glacier ice takes place slowly, as younger snow progressively buries older snow. Freshly fallen snow consists of delicate hexagonal crystals with sharp points. The crystals do not fit together tightly, so this snow contains about 90% air. With time, the points of the snowflakes become blunt because they either **sublimate** (evaporate directly into vapor) or melt, and the snow packs more tightly. As snow becomes buried, the weight of the overlying snow increases pressure, which causes remaining points of contact between snowflakes to melt. This process of melting at points of contact where the pressure is greatest is another example of "pressure solution" (see Chapter 8). Gradually, the snow transforms into a packed granular material called **firn**, which contains only about 25% air (Fig. 22.2d). Melting of firn grains at contact points produces water that crystallizes in the spaces between grains until eventually the firn transforms into a solid mass of glacial ice composed of interlocking ice crystals. Such glacial ice, which may still contain up to 20% air trapped in bubbles, tends to

FIGURE 22.2 The nature of ice and the formation of glaciers. Snow falls like sediment and metamorphoses to ice when buried.

A boundary between layers

The layers in the photo at left are part of this glacier in the Alps.

(a) The hexagonal shape of snowflakes. No two are alike.

(b) Layers of snow accumulate. They recrystallize to become ice.

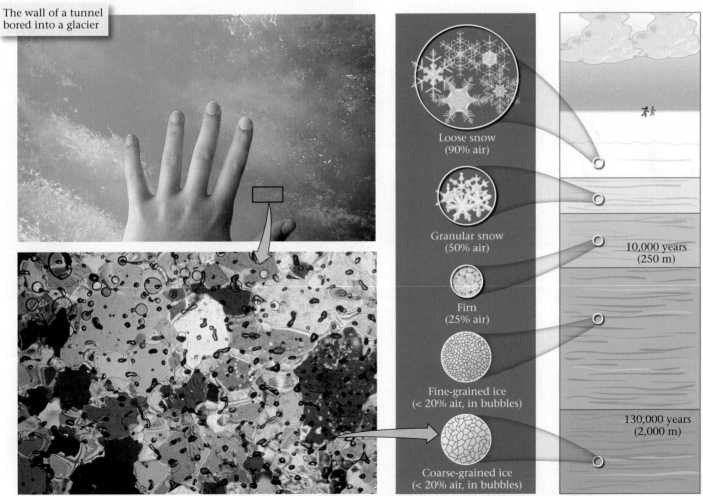

The wall of a tunnel bored into a glacier

Loose snow (90% air)

Granular snow (50% air)

10,000 years (250 m)

Firn (25% air)

Fine-grained ice (< 20% air, in bubbles)

130,000 years (2,000 m)

Coarse-grained ice (< 20% air, in bubbles)

(c) Glacial ice is blue. As revealed by a microscope, the ice has coarse grains and contains air bubbles.

(d) Snow compacts and melts to form firn, which recrystallizes into ice. Crystal size increases with depth.

absorb red light and thus has a bluish color. The transformation of fresh snow to glacier ice can take as little as tens of years in regions with abundant snowfall, or as long as thousands of years in regions with little snowfall.

Categories of Glaciers

Glaciers are streams or sheets of recrystallized ice that last all year long and flow under the influence of gravity. Today, they highlight coastal and mountain scenery in Alaska, the Cordillera of western North America, the Alps of Europe, the Southern Alps of New Zealand, the Himalayas of Asia, and the Andes of South America, and they cover most of Greenland and Antarctica (**See for Yourself T**, p. S-38). Geologists distinguish between two main categories, mountain glaciers and continental glaciers.

Mountain glaciers (also called alpine glaciers) exist in or adjacent to mountainous regions (**Fig. 22.3a**). Topographical

FIGURE 22.3 A great variety of glaciers form in mountainous areas.

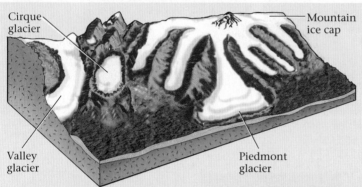

(a) Mountain glaciers are classified based on shape and position.

(d) A valley glacier and cirque glaciers in Switzerland.

(b) An ice cap in Alaska.

(e) A large trunk valley glacier, and tributary glaciers, in Pakistan.

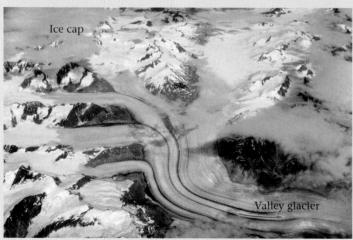

(c) Valley glaciers draining a mountain ice cap in Alaska.

(f) A piedmont glacier near the coast of Greenland.

FIGURE 22.4 Two major continental glaciers exist today—one on Antarctica and one on Greenland.

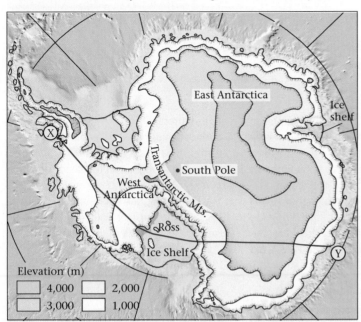

(a) A contour map of the Antarctic ice sheet. Valley glaciers carry ice from the ice sheet of East Antarctica down to the Ross Ice Shelf.

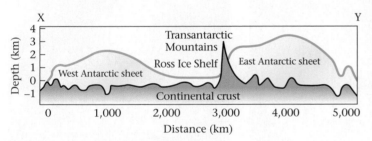

(b) A cross section X to Y of the Antarctic ice sheet. The Transantarctic Mountains separate East Antarctica from West Antarctica.

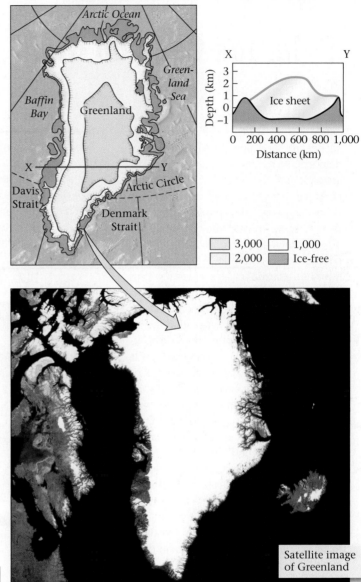

(c) Greenland is also covered by an ice sheet that at its thickest, is 1 km thinner than Antarctica's.

features of the mountains control their shape; overall, mountain glaciers flow from higher elevations to lower elevations. Mountain glaciers include cirque glaciers, which fill bowl-shaped depressions, or cirques, on the flank of a mountain; valley glaciers, rivers of ice that flow down valleys; mountain ice caps, mounds of ice that submerge peaks and ridges at the crest of a mountain range; and piedmont glaciers, fans or lobes of ice that form where a valley glacier emerges from a valley and spreads out into the adjacent plain (Fig. 22.3b–f). Mountain glaciers range in size from a few hundred meters to a few hundred kilometers long.

Continental glaciers are vast ice sheets that spread over thousands of square kilometers of continental crust. Continental glaciers now exist only on Antarctica and Greenland (Fig. 22.4a–c). Remember that Antarctica is a continent, so the ice beneath the South Pole rests *mostly* on solid ground—new research reveals that at least three lakes underlie the glacier. (The largest, Lake Vostok, is 5,400 square km in area.)

In contrast, the ice beneath the North Pole forms part of a thin sheet of sea ice floating on the Arctic Ocean. Continental glaciers flow outward from their thickest point (up to 3.5 km thick) and thin toward their margins, where they may be only a few hundred meters thick. The front edge of the glacier may divide into several tongue-shaped lobes, because not all of the glacier flows at the same speed. Of note, Earth is not alone in hosting polar ice sheets—Mars has them too (Box 22.1).

Geologists also find it valuable to distinguish between types of glaciers on the basis of the thermal conditions in which the glaciers exist. **Temperate glaciers** occur in regions where atmospheric temperatures become high enough for the

CONSIDER THIS . . . .

BOX 22.1

Polar Ice Caps on Mars

The discovery that Mars has polar ice caps dates back to 1666, when the first telescopes allowed astronomers to resolve details of the red planet's surface. By 1719, astronomers had detected that Martian polar caps change in area with the season, suggesting that they partially melt and then refreeze. You can see these changes on modern images (**Fig. Bx22.1a, b**). The question of what the ice caps consist of remained a puzzle until fairly recently. Early studies revealed that the atmosphere of Mars consists mostly of carbon dioxide, so researchers first assumed that the ice caps consisted of frozen carbon dioxide. But data from modern spacecraft led to the conclusion that this initial assumption is false. It now appears that the Martian ice caps consist mostly of water ice (mixed with dust) in layers from 1 to 3 km thick. During the winter, atmospheric carbon dioxide freezes and covers the north polar cap with a 1-m-thick layer of dry ice. During the summer this layer melts away. The south polar cap is different, for its dry ice blanket is 8 m thick and doesn't melt away entirely in the summer. The difference between the north and south poles

may reflect elevation, for the south pole is 6 km higher and therefore remains colder. The amount of melting appears to have increased as the years pass, hinting that Mars is now undergoing climate change.

High-resolution photographs reveal that distinctive canyons, up to 10 km wide and 1 km deep, spiral outward from the center of the north polar ice cap. Why did

this pattern form? Recent calculations suggest that if the ice sublimates (transforms into gas) on the sunny side of a crack and refreezes on the shady side, the crack will migrate sideways over time. If the cracks migrate more slowly closer to the pole, where it's colder, than they do farther away, they will naturally evolve into spirals.

FIGURE Bx22.1 The ice caps of Mars.

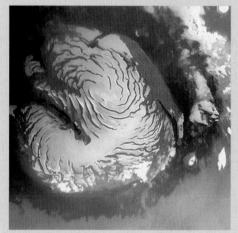

(a) During the winter, the ice caps expand to lower latitudes.

(b) A close-up of the northern polar cap in summer.

glacial ice to be at or near its melting temperature during part or all of the year. **Polar glaciers** occur in regions where atmospheric temperatures stay so low all year long that the glacial ice remains below melting temperature throughout the year.

The Movement of Glacial Ice

When Louis Agassiz became fascinated by glaciers, he decided to find out how fast the ice in them moved, so he hammered stakes into an Alpine glacier and watched the stakes change position during the year. More recently, researchers have observed glacial movement with the aid of time-lapse photography, which shows the evolution of a glacier over several

years in a movie that lasts a few minutes. In such movies, the glacial ice seems to flow across the screen. Now, let's consider the two mechanisms that allow glaciers to move—plastic deformation and basal sliding.

At conditions found below depths of about 60 m in a glacier, ice deforms by **plastic deformation**, meaning that the grains within it change shape very slowly, and new grains grow while old ones disappear (Fig. 22.5a, b). Simplistically, we can picture such changes to be a consequence of the rearrangement of water molecules within ice grains. If ice is warm enough for thin water films to form along grain boundaries, plastic

Did you ever wonder . . .
how a glacier moves?

deformation may also involve the microscopic slip of ice grains past their neighbors along the water films.

In cases where significant quantities of meltwater accumulate at the base of a glacier, to form a layer either of liquid or of slurry-like wet sediment, glaciers can move by **basal sliding**. During this process, the liquid water or water-saturated slurry layer holds the glacial ice above bedrock and thereby decreases friction; effectively, the glacier glides along on a wet cushion (Fig. 22.5c). Where does the liquid water at the base of glaciers come from? Some forms when sunlight and atmospheric warming heats the glacier sufficiently to produce meltwater, either on the surface of the glacier or within the glacier. Surface meltwater ponds may drain abruptly (in a matter of minutes to hours) into cracks or tunnels that provide a conduit between the surface and the base of the glacier. Melting may also occur due to the trapping of heat rising from the ground beneath the glacier (for ice is an insulator), or due to the weight of overlying ice (for at elevated pressures, ice can melt even if its temperature remains below 0°C).

In the case of polar glaciers, which are so cold that they have no internal water films and have a dry base, flow takes place generally by plastic deformation alone. In the case of temperate glaciers, which have some intergranular water and/or a wet base, flow can involve both plastic deformation and basal sliding. Note that not all parts of a given glacier necessarily flow in the same way. For example, imagine a continental glacier that forms in a very cold polar realm, but eventually flows into warmer realms at lower latitudes. Near its cold origin, this "dry-based" glacier moves by plastic deformation, but near its warmer margin, it becomes "wet-based" and moves by both plastic deformation and basal sliding. Similarly, a long valley glacier may be dry-based and flow only plastically at higher, colder elevations, but may become wet-based and flow by both plastic deformation and basal sliding at lower, warmer elevations.

As we noted earlier, plastic deformation takes place only at depths of greater than about 60 meters in a glacier—above this depth, known as the brittle–plastic transition, ice is too brittle to flow. (Note that, by comparison, plastic deformation in silicate rocks of the Earth's crust occurs primarily under metamorphic temperatures greater than about 300°C;

thus, the brittle–plastic transition in continental crust occurs at depths of about 10 to 15 km; see Chapters 8 and 11.) As a glacier overall undergoes movement, its upper 60 meters of ice deforms predominantly by cracking. A crack that develops by brittle deformation of a glacier is called a **crevasse** (Fig. 22.6). In large glaciers, crevasses can be hundreds of meters long and tens of meters deep, and they may open into gashes that are many meters across. Tragically, explorers, hikers, and skiers have died by falling into crevasses, whose openings may be hidden by a bridge of weak, windblown snow. Crevasse formation tends to localize in regions where a glacier flows over steps or hills in the underlying bedrock surface, for the ice of the glacier must bend to accommodate the surface shape of the substrate.

Why do glaciers move? Ultimately, because the pull of gravity is strong enough to make ice flow (Fig. 22.7a). A glacier flows in the direction in which its top surface slopes. Thus, valley glaciers flow down their valleys, and continental ice sheets spread outward from their thickest point. To picture the movement of a continental ice sheet, imagine that a thick pile of ice builds up. Gravity causes the top of the pile to push down on the ice at the base. Eventually, the basal ice can no longer support the weight of the overlying ice and begins to deform plastically and/or slide on its substrate. When this happens, the basal ice starts squeezing out to the side, carrying the overlying ice with it. The greater the volume of ice that builds up, the wider the sheet of ice can become. You've seen a similar process of gravitational spreading if you've ever poured honey onto a plate. The honey can't build up into a narrow

FIGURE 22.5 Mechanisms of glacial movement.

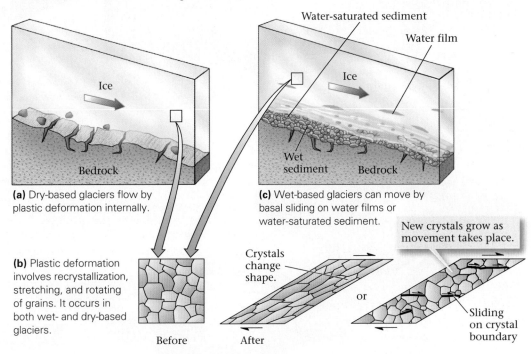

(a) Dry-based glaciers flow by plastic deformation internally.

(b) Plastic deformation involves recrystallization, stretching, and rotating of grains. It occurs in both wet- and dry-based glaciers.

(c) Wet-based glaciers can move by basal sliding on water films or water-saturated sediment.

New crystals grow as movement takes place.

Crystals change shape.

Sliding on crystal boundary

Before   After   or

FIGURE 22.6 Crevasses form in the upper layer of a glacier, in which the ice is brittle. Commonly, cracking takes place where the glacier bends while flowing over steps or ridges in its substrate.

Crevasse

Crevasses up to 10 m wide in an Antarctic glacier

Brittle–plastic transition

Ice cannot crack at depths below 60 m.

Crevasses formed in an Alpine glacier

Step in the substrate

Meters: 0, 50, 100, 150, 200, 250

FIGURE 22.7 Forces that drive the movement of glaciers.

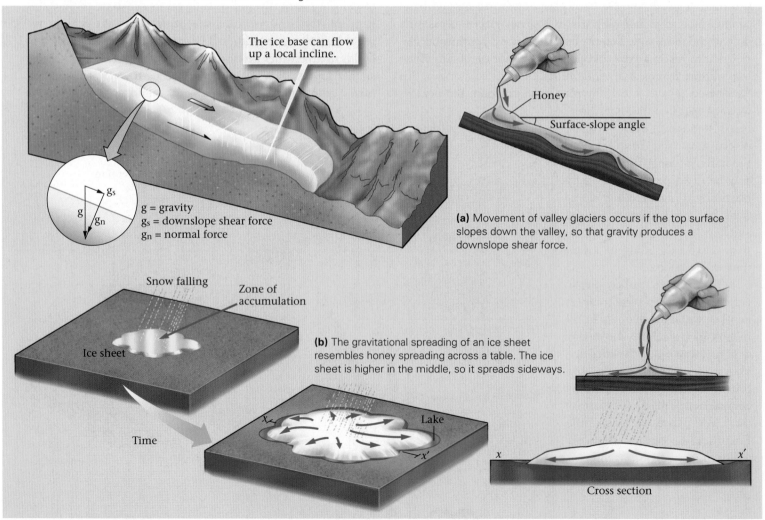

The ice base can flow up a local incline.

g = gravity
g_s = downslope shear force
g_n = normal force

Honey

Surface-slope angle

(a) Movement of valley glaciers occurs if the top surface slopes down the valley, so that gravity produces a downslope shear force.

Snow falling

Zone of accumulation

Ice sheet

Time

(b) The gravitational spreading of an ice sheet resembles honey spreading across a table. The ice sheet is higher in the middle, so it spreads sideways.

Lake

Cross section

FIGURE 22.8 Flow velocities vary with location in a glacier. Overall, ice flows from the zone of accumulation to the toe.

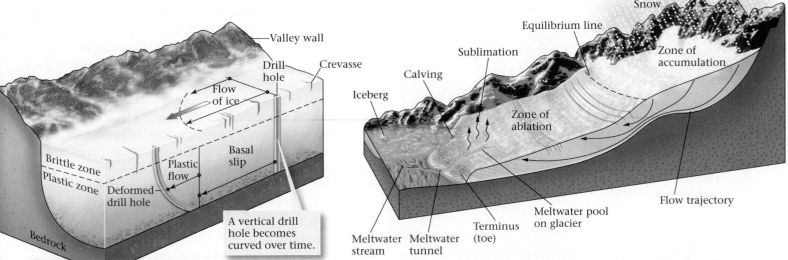

(a) Different parts of a glacier flow at different velocities, due to friction with the substrate. The center flows fastest.

(b) The equilibrium line separates the zone of accumulation from the zone of ablation. As indicated by arrows, ice flows down in the zone of accumulation and up in the zone of ablation.

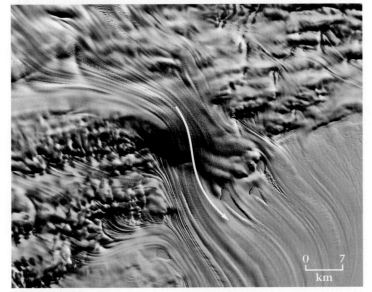

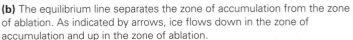

(c) A satellite image of ice flowing from the Polar Plateau of Antarctica, down a 400-m-high ice fall to the Lambert Glacier. Curving lines indicate the flow directions; the roughness is due to crevasses.

Not all parts of a glacier move at the same rate. For example, friction between rock and ice slows a glacier, so the center of a valley glacier moves faster than its margins, and the top of a glacier moves faster than its base (Fig. 22.8a–c). And because water at the base of a glacier allows it to travel more rapidly, portions of a continental glacier that flow over water or wet sediment become "ice streams" that travel 10 to 100 times faster than adjacent dry-based portions of the glacier. Similarly, if water builds up beneath a valley glacier to the point where it lifts the glacier off its substrate, basal sliding starts and the glacier undergoes a **surge**. During a surge, the glacier flows much faster for a limited time (rarely more than a few months), until the water escapes and basal sliding ceases. During surges, glaciers have been clocked at speeds of 10 to 110 m per day! Sudden surges may generate **ice quakes**, whose seismic vibrations travel through the glacier and through the rock below.

Glacial Advance and Retreat

Glaciers resemble bank accounts: snowfall adds to the account, while **ablation**—the removal of ice by sublimation (the evaporation of ice into water vapor), melting (the transformation of ice into liquid water, which flows away), and calving (the breaking off of chunks of ice at the edge of the glacier)—subtracts from the account. Snowfall adds to the glacier in the **zone of accumulation**, whereas ablation subtracts in the **zone of ablation**; the boundary between these two zones is the **equilibrium line**. The zone of accumulation occurs where the temperature remains cold enough year round so that winter snow does not melt or sublimate away

column because it's too weak; rather, it flows laterally away from the point where it lands to form a wide, thin layer (Fig. 22.7b).

Glaciers generally flow at rates of between 10 and 300 m per year—far slower than a river, but far faster than a silicate rock even under high-grade metamorphic conditions. The velocity of a particular glacier depends, in part, on the magnitude of the force driving its motion. For example, a glacier whose surface slopes steeply moves faster than one with a gently sloping surface. Flow velocity also depends on whether liquid water exists at places along the base of the glacier; wet-based glaciers tend to move faster than dry-based glaciers.

FIGURE 22.9   Glacial advance and retreat. 🔊

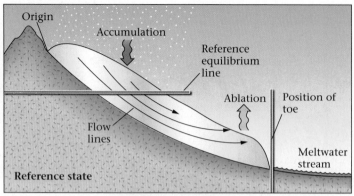

(a) The position of the toe represents a balance between addition by accumulation and loss by ablation.

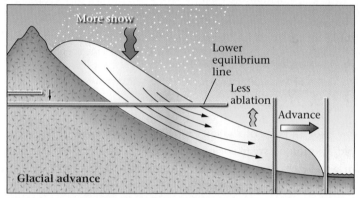

(b) If accumulation exceeds ablation, the glacier advances, the toe moves farther from the origin, and the ice thickens.

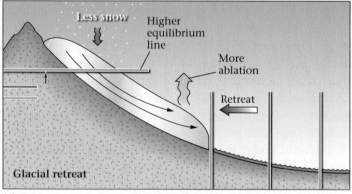

(c) If ablation exceeds accumulation, the glacier retreats and thins. The toe moves back, even though ice continues to flow toward the toe.

entirely during the summer. Therefore, elevation and latitude control the position of the equilibrium line.

The leading edge or margin of a glacier is called its **toe**, or terminus (Fig. 22.9a 🔊). If the rate at which ice builds up in the zone of accumulation exceeds the rate at which ablation occurs below the equilibrium line, then the toe moves forward into previously unglaciated regions. Such a change is called a **glacial advance** (Fig. 22.9b). In mountain glaciers,

the position of a toe moves downslope during an advance, and in continental glaciers, the toe moves outward, away from the glacier's origin. If the rate of ablation below the equilibrium line equals the rate of accumulation, then the position of the toe remains fixed. But if the rate of ablation exceeds the rate of accumulation, then the position of the toe moves back toward the origin of the glacier; such a change is called a **glacial retreat** (Fig. 22.9c). During a mountain glacier's retreat, the position of the toe moves upslope. It's important to realize that when a glacier retreats, it's only the *position* of the toe that moves back toward the origin, for ice continues to flow toward the toe. Glacial ice cannot flow back toward the glacier's origin.

One final point before we leave the subject of glacial flow: beneath the zone of accumulation, a volume of ice gradually moves down toward the base of the glacier as new ice accumulates above it, whereas beneath the zone of ablation, a volume of ice gradually moves up toward the surface of the glacier, as overlying ice ablates. Thus, as a glacier flows, ice volumes follow curved trajectories (see Fig. 22.9a–c). For this reason, rocks picked up by ice at the base of the glacier may slowly move to the surface. The upward flow of ice where the Antarctic ice sheet collides with the Transantarctic Mountains, for example, brings up meteorites long buried in the ice (Fig. 22.10a, b).

Ice in the Sea

On the moonless night of April 14, 1912, the great ocean liner *Titanic* plowed through the calm but frigid waters of the North Atlantic on her maiden voyage from Southampton, England, to New York. Although radio broadcasts from other ships warned that **icebergs**, large (>6 m high and 15 m across) blocks of ice floating in the water, had been sighted in the area and might pose a hazard, the ship sailed on, its crew convinced that they could see and avoid the biggest bergs, and that smaller ones would not be a problem for the steel hull of this "unsinkable" vessel. But in a story now retold countless times, their confidence was fatally wrong. At 11:40 P.M., while the first-class passengers danced, the *Titanic* struck a large iceberg. Lookouts had seen the ghostly mass of frozen water only minutes earlier and had alerted the ship's pilot, but the ship had been unable to turn fast enough to avoid disaster. The force of the blow split the steel hull spanning 5 of the ship's 16 water-tight compartments. The ship could stay afloat if four compartments flooded, but the flooding of five meant it would sink. At about 2:15 A.M., the bow disappeared below the water, and the stern rose until the ship protruded nearly vertically from the water. Without water to support its weight, the hull buckled and split in two. The stern section fell back down onto the water and momentarily bobbed horizontally before following the bow, settling downward through over 3.5 km of water to the silent sea floor below. Because of an inadequate number of lifeboats, only 705 passengers survived; 1,500

FIGURE 22.10 Meteorites accumulate along the Transantarctic Mountains.

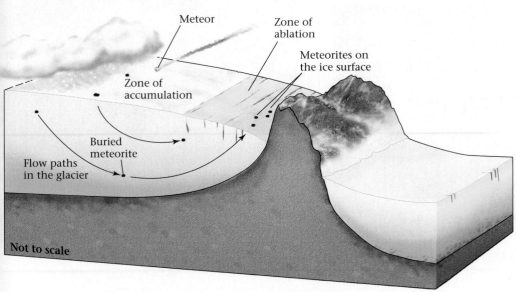

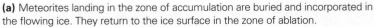

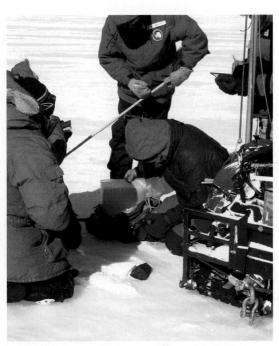

(a) Meteorites landing in the zone of accumulation are buried and incorporated in the flowing ice. They return to the ice surface in the zone of ablation.

(b) Researchers document a new meteorite discovery.

expired in the frigid waters of the Atlantic. The *Titanic* remained lost until 1985, when a team of oceanographers led by Robert Ballard located the sunken hull and photographed its eerie form.

Where do icebergs, such as the one responsible for the *Titanic*'s demise, originate? In high latitudes, mountain glaciers and continental ice sheets flow down to the sea, and they either stop at the shore or flow into the sea. Glaciers whose terminus lies in the water are called **tidewater glaciers**. Valley glaciers may protrude further into the ocean to become ice tongues (Fig. 22.11a). Continental glaciers entering the sea become broad, flat sheets called **ice shelves** (Fig. 22.11b). In shallow water, glacial ice remains grounded (Fig. 22.11c). But where the water becomes deep enough, the ice floats with four-fifths of the ice below the water's surface. At the boundary between glacier and ocean, blocks of ice calve off and tumble into the water with an impressive splash. As we've noted, if a free-floating chunk rises 6 m above the water and is at least 15 m long, it is formally called an iceberg. Smaller pieces, formed when ice blocks fragment before entering the water or after icebergs have had time to melt, include "bergy bits," rising 1 to 5 m above the water and covering an area of 100 to 300 square m, and "growlers," rising less than 1 m above the water and covering an area of about 20 square m—still big enough to damage a ship. Growlers get their name because of the sound they make as they bob in the sea and grind together.

Most large icebergs form along the western coast of Greenland or along the coast of Antarctica. Icebergs that calve off valley glaciers tend to be irregularly shaped with pointed peaks rising upward. Such glaciers are called castle bergs or pinnacle bergs. One of the largest on record protruded about 180 m above

the sea. Since four-fifths of the ice lies below the surface of the sea, the base of a large iceberg may actually be a few hundred meters below the surface (Fig. 22.11d). Icebergs that originate in Greenland float into the "iceberg alley" region of the North Atlantic. These are the bergs that threaten ships, although the danger has diminished in modern times because of less ice and because of ice patrols that report the locations of floating ice. Blocks that calve off the vast ice shelves of Antarctica tend to have flat tops and nearly vertical sides—such glaciers are called tabular bergs. Some of the tabular bergs in the Antarctic are truly immense; air photos have revealed bergs over 160 km across.

Not all ice floating in the sea originates as glaciers on land. In polar climates, the surface of the sea itself freezes, forming **sea ice** (Fig. 22.11e). Some sea ice, such as that covering the interior of the Arctic Ocean, floats freely; but some protrudes outward from the shore (Fig. 22.11f). Icebreakers can crunch through sea ice that is up to 2.5 m thick; the icebreaker rides up on the ice, then crushes it. Vast areas of ice shelves and of sea ice have been disintegrating in recent years, perhaps because of global warming. For example, open regions develop in the Arctic Ocean during the summers, and the ice shelf in Antarctica has been decreasing rapidly in area. In some locations, large openings known as polynyas have developed in the sea ice of Antarctica. Some sea ice forms in winter and melts away in summer, but at high latitudes, sea ice may last for several years. For example, in the Arctic Ocean, sea ice may last long enough to make a 7- to 10-year voyage around the Arctic Ocean in response to currents.

The existence of icebergs leaves a record in the stratigraphy of the sea floor, for icebergs carry ice-rafted sediment (Fig. 22.11g). Larger rocks that drop from the ice to the sea floor are called

FIGURE 22.11 Ice along the edge of continents—shelves, tongues, and bergs.

(a) An ice tongue protruding into the Ross Sea along the coast of Antarctica.

(b) The Larsen Ice Shelf, along the coast of Antarctica, as viewed from a satellite in 2002.

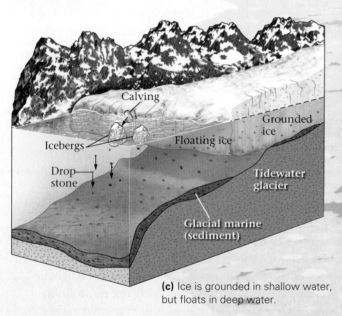

(c) Ice is grounded in shallow water, but floats in deep water.

(d) This artist's rendition of an iceberg emphasizes that most of the ice is underwater.

(e) In summer, some of the sea ice of Antarctica breaks up to form tabular icebergs.

(f) A "growler" of floating ice off Alaska. Note the layering in the ice.

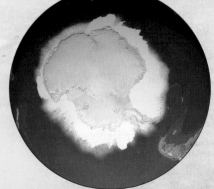

(g) Sea ice covers most of the Arctic ocean (left) and surrounds Antarctica (right).

drop stones (Fig. 22.11c). In ancient glacial deposits, drop stones appear as isolated blocks surrounded by mud. Icebergs and smaller fragments also drop sand and gravel, derived by the erosion of continents, onto the sea floor. In cores extracted by drilling into sea-floor sediment, horizons of such land-derived sediment, sandwiched between layers of sediment formed from marine plankton shells, indicate times in Earth history when glaciers were breaking up and icebergs became particularly abundant.

Take-Home Message

- Glacial ice consists of interlocking ice crystals formed when deeply buried snow recrystallizes.
- Mountain glaciers build local ice caps, carve cirques, or flow down valleys. Continental glaciers are thick sheets of ice that cover broad areas of continents.
- Glacial ice is capable of flowing very slowly, due to plastic deformation of ice crystals and/or to basal sliding of the glacier on an underlying wet layer.
- The upper, brittle, portion of a glacier cracks to form crevasses.
- Glaciers build in the zone of accumulation and diminish in the zone of ablation; if the rate of accumulation increases, the toe of the glacier advances; if the rate decreases, the toe retreats.
- Some glaciers flow into the sea, where blocks of ice calve off to form icebergs; in polar regions, the surface of the ocean freezes to form sea ice.

THINK: Does ice actually flow uphill during a glacial retreat?

22.3 CARVING AND CARRYING BY ICE

Glacial Erosion and Its Products

The Sierra Nevada mountains of California consist largely of granite that formed during the Mesozoic Era in the crust beneath a volcanic arc. During the past 10 to 20 million years, the land surface slowly rose, and erosion stripped away overlying rock to expose the granite. The style of erosion formed rounded domes. Then, during the last ice age, valley glaciers cut deep, steep-sided valleys into the range. In the process, some of the domes were cut in half, leaving a rounded surface on one side and a steep cliff on the other. Half Dome, in Yosemite National Park, formed in this way (Fig. 22.12a); its steep cliff has challenged many rock climbers. Such glacial erosion also produces the knife-edge ridges and pointed spires of high mountains (Fig. 22.12b) and broad expanses where rock outcrops have been stripped of overlying sediment and polished smooth (Fig. 22.12c). Glacial erosion in the mountains can lower a valley floor by over 100 m, and continental glaciation during the last ice age stripped up to 30 m of material off the land in northern Canada.

As glaciers flow, clasts embedded in the ice act like the teeth of a giant rasp and grind away the substrate. This process,

glacial abrasion, produces very fine sediment called rock flour, just as sanding wood produces sawdust. Rasping by large clasts produces long gouges, grooves, or scratches [1 cm to 1 m (0.4 in to 3 ft) across] called **glacial striations** (Fig. 22.12d). As you might expect, striations run parallel to the flow direction of the ice. When boulders entrained in the base of the ice strike bedrock below, as the ice moves, the impact may break off asymmetric wedges of bedrock, leaving behind indentations called chatter marks (Fig. 22.12e). Rasping by embedded sand yields shiny **glacially polished surfaces** (Fig. 22.12f, g). In regions of wet-based glaciers, sediment-laden water rushing through tunnels at the base of the glaciers can carve substantial channels.

Glaciers pick up fragments of their substrate in several ways (Fig. 22.13a, b). During glacial incorporation, ice surrounds debris so the debris starts to move with the ice. During glacial plucking (or glacial quarrying), a glacier breaks off fragments of bedrock. Plucking occurs when ice freezes around rock that has just started to separate from its substrate; movement of the ice lifts off pieces of the rock (Fig. 22.14a, b). At the toe of a glacier, ice may actually bulldoze sediment and trees slightly before flowing over them.

Let's now look more closely at the erosional features associated with a mountain glacier (Fig. 22.15a). Freezing and thawing during the fall and spring help fracture the rock bordering the head of the glacier (the ice edge high in the mountains). This rock falls on the ice or gets picked up at the base of the ice, and moves downslope with the glacier. As a consequence, a bowl-shaped depression, or **cirque**, develops on the side of the mountain. If the ice later melts, a lake called a **tarn** may form at the base of the cirque. An **arête** (French for ridge), a residual knife-edge ridge of rock, separates two adjacent cirques. A pointed mountain peak surrounded by at least three cirques is called a **horn**. The Matterhorn, a peak in Switzerland, is a particularly beautiful example of a horn (Fig. 22.15b).

Glacial erosion severely modifies the shape of a valley. To see how, compare a river-eroded valley with a glacially eroded valley. If you look along the length of a river in unglaciated mountains, you'll see that it flows down a V-shaped valley, with the river channel forming the point of the V. The V develops because river erosion occurs only in the channel, and mass wasting causes the valley slopes to approach the angle of repose. But if you look down the length of a glacially eroded valley, you'll see that it resembles a U, with steep walls. A **U-shaped valley** (Fig. 22.15c) forms because the combined processes of glacial abrasion and plucking not only lower the floor of the valley but also bevel its sides.

Glacial erosion in mountains also modifies the intersections between tributaries and the trunk valley. In a river system, the trunk stream serves as the local base level for tributaries (see Chapter 17), so the mouths of the tributary valleys lie at the *same* elevation as the trunk valley. The ridges (spurs) between valleys taper to a point when they join the trunk valley floor. During

FIGURE 22.12 Products of glacial erosion. Ice is a very aggressive agent of erosion.

(a) Half Dome, in Yosemite National Park, California.

(b) The spectacular sharp peaks of the Alps were carved by the action of ice.

(c) Glacially polished outcrop in Central Park, New York City.

(d) Glacial striations in Victoria, British Columbia.

(e) Close-up of striations and chatter marks, Switzerland.

(f) Glacial polish shining in the sun, along Lake Superior.

(g) Glacially polished mountains in the Wind River Range of Wyoming. Ice once completely covered the field of view.

FIGURE 22.13 The processes of incorporation and plowing.

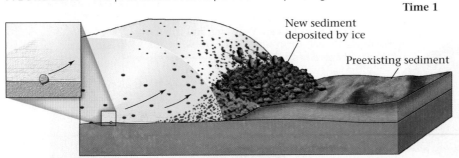

Time 1

New sediment deposited by ice

Preexisting sediment

(a) Glacial ice can pick up and incorporate chunks of rock that it flows over. The chunks then move with the ice, following the ice's flow trajectories, until deposition at the toe of the glacier.

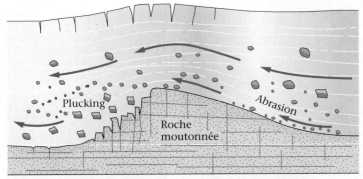

→ Advance Time 2

Ice picks up rock at the base of the glacier.

Ice plows up older sediment and flows over it.

(b) At the toe of the glacier, ice can flow up and over preexisting sediment. Locally, it "plows" sediment, pushing it up into a ridge.

FIGURE 22.14 A roche moutonnée is an asymmetric bedrock hill shaped by the flow of glacial ice.

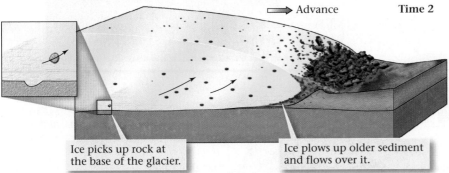

Plucking

Abrasion

Roche moutonnée

(a) Abrasion rasps the upstream side, and plucking carries away fracture-bounded blocks on the downstream side.

(b) An example of a roche moutonnée in the Sierra Nevada. The glacier flowed from right to left.

glaciation, tributary glaciers flow down side valleys into a trunk glacier. But the trunk glacier cuts the floor of its valley down to a depth that far exceeds the depth cut by the tributary glaciers. Thus, when the glaciers melt away, the mouths of the tributary valleys perch at a higher elevation than the floor of the trunk valley. Such side valleys are called **hanging valleys**. The water in post-glacial streams that flow down a hanging valley cascades over a spectacular waterfall to reach the post-glacial trunk stream (**Fig. 22.15d**). As they erode, trunk glaciers also chop off the ends of spurs (ridges) between valleys, to produce truncated spurs.

Now let's look at the erosional features produced by continental ice sheets. To a large extent, these depend on the nature of the preglacial landscape. Where an ice sheet spreads over a region of low relief, such as the Canadian Shield, glacial erosion creates a vast region of polished, flat, striated surfaces. Where an ice sheet spreads over a hilly area, it deepens valleys and smooths hills. In central New York, for example, continental glaciers carved the deep valleys that now cradle the Finger Lakes, and in Maine, glaciers smoothed and streamlined the granite and metamorphic rock hills of Acadia National Park. Glacially eroded hills end up being elongate in the direction of flow and are asymmetric; glacial rasping smoothes and bevels the upstream part of the hill, creating a gentle slope, whereas glacial plucking eats away at the downstream part, making a steep slope. Ultimately, the hill's profile resembles that of a sheep lying in a meadow—such a hill is called a **roche moutonnée**, from the French for sheep rock (Fig. 22.14b).

Fjords: Submerged Glacial Valleys

As noted earlier, where a valley glacier meets the sea, the glacier's base remains in contact with the ground until the water depth exceeds about four-fifths of the glacier's thickness, at which point the glacier floats. Thus, glaciers can continue carving U-shaped valleys even below sea level. In addition, during an ice age, water extracted from the sea becomes locked in the ice sheets on land, so sea level drops significantly. Therefore, the floors of valleys cut by coastal glaciers during the Pleistocene Ice Age were cut much deeper than present sea level. Today, the sea has flooded these deep valleys, producing **fjords** (see Chapter 18). In the spectacular fjord-land regions along the coasts of Norway, New Zealand, Chile, and Alaska, the walls of submerged U-shaped valleys rise straight from the sea as vertical cliffs up to 1,000 m high (**Fig. 22.16**). Fjords also develop where an inland glacial valley fills to become a lake.

FIGURE 22.15 Landscape features formed by the glacial erosion of a mountainous landscape.

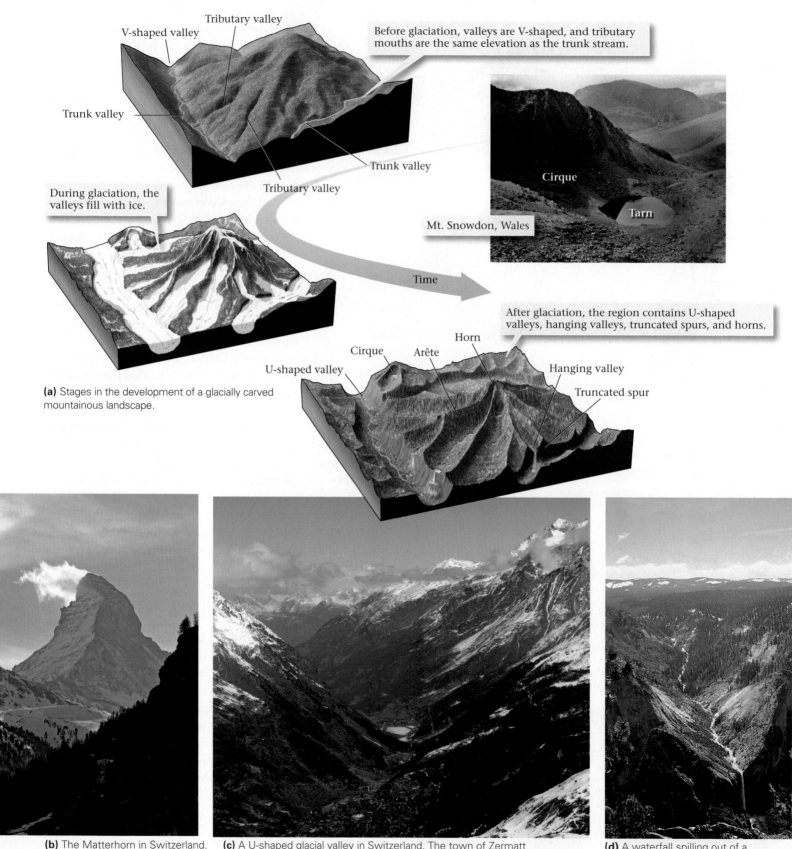

(a) Stages in the development of a glacially carved mountainous landscape.

Before glaciation, valleys are V-shaped, and tributary mouths are the same elevation as the trunk stream.

During glaciation, the valleys fill with ice.

Mt. Snowdon, Wales

After glaciation, the region contains U-shaped valleys, hanging valleys, truncated spurs, and horns.

(b) The Matterhorn in Switzerland. The first ascent was in 1865.

(c) A U-shaped glacial valley in Switzerland. The town of Zermatt now occupies the valley floor.

(d) A waterfall spilling out of a U-shaped hanging valley.

FIGURE 22.16 One of the many spectacular fjords of Norway. The water is an arm of the sea that fills a glacially carved valley. Tourists are standing on Pulpit Rock (Prekestolen).

Take-Home Message

- Sediment carried by glaciers abrades underlying rock, yielding polished surfaces and striations.
- In mountains, glacial erosion carves cirques, separated from each other by arêtes; valley glaciers carve U-shaped valleys.
- Fjords are elongate bodies of water formed where the sea fills glacially carved valleys.

THINK: Why do we find hanging valleys spilling waterfalls into trunk valleys, in regions that have been eroded by mountain glaciers?

22.4 DEPOSITION ASSOCIATED WITH GLACIATION

The Glacial Conveyor: The Transport of Sediment by Ice

Glaciers can carry sediment of any size and, like a conveyor belt, transport it in the direction of flow (i.e., toward the toe; Fig. 22.17a). The sediment load either falls onto the surface of the glacier from bordering cliffs or gets plucked and lifted from the substrate and incorporated into the moving ice.

The word **moraine** was a local term used by Alpine farmers and shepherds for piles of rock and dirt. The term now applies exclusively to debris piles carried by or left by glaciers. Sediment dropped on the glacier's surface moves with the ice and becomes a stripe of debris. Stripes formed along the side edges of the glacier are **lateral moraines**. When a glacier melts, lateral moraines lie stranded along the side of the glacially carved valley, like bathtub rings. If flowing water runs along the edge of the glacier and sorts the sediment of a lateral moraine, a stratified sequence of sediment, called a **kame**, forms. Where two valley glaciers merge, the debris constituting two lateral moraines merges to become a **medial moraine**, running as a stripe down the interior of the composite glacier (Fig. 22.17b, c). Trunk glaciers created by the merging of many tributary glaciers contain several medial moraines. Sediment transported to a glacier's toe by the glacial conveyor accumulates in a pile at the toe and builds up to form an **end moraine**.

So far, we've emphasized sediment moved by ice. But in temperate glacial environments, the flowing water at the base of the glacier, moving through channels, transports much of the sediment load, eventually depositing it beyond the terminus of the glacier.

Types of Glacial Sedimentary Deposits

If you drill through the soil throughout much of the upper midwestern and northeastern United States and adjacent parts of Canada, the drill penetrates a layer of sediment deposited during the Pleistocene Ice Age. A similar story holds true for much of northern Europe. Thus, many of the world's richest agricultural regions rely on soil derived from sediment deposited by glaciers during the Ice Age. This sediment buries a pre–Ice Age landscape, like frosting fills the irregularities on a cake. Pre-glacial valleys may be completely filled with sediment.

Several different types of sediment can be deposited in glacial environments; all of these types together constitute **glacial drift**. (The term dates from pre-Agassiz studies of glacial deposits, when geologists thought that the sediment had "drifted" into place during an immense flood.) Specifically, glacial drift includes the following:

- *Till*: Sediment transported by ice and deposited beneath, at the side, or at the toe of a glacier is called **glacial till**. Glacial till is unsorted (it is a type of diamicton), because the solid ice of glaciers can carry clasts of all sizes (Fig. 22.18a).
- *Erratics*: Glacial erratics (Fig. 22.18b) are cobbles and boulders that have been dropped by a glacier. Some

FIGURE 22.17   The glacial conveyor and the formation of lateral and medial moraines on glaciers.

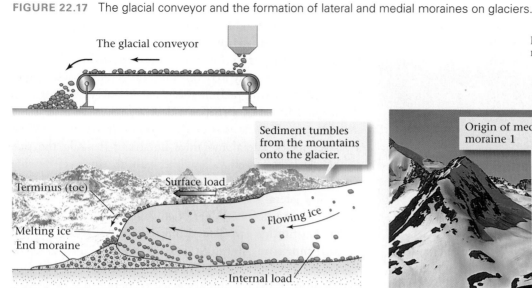

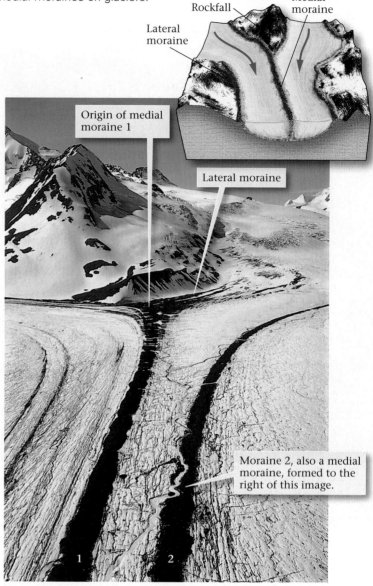

(a) Sediment falls on a glacier from bordering mountains and gets plucked up from below. Glaciers are like conveyor belts, moving sediment toward the toe of the glacier.

(b) This glacier in the French Alps carries lots of sediment.

(c) A medial moraine forms where lateral moraines of two valley glaciers merge.

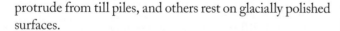

protrude from till piles, and others rest on glacially polished surfaces.

- *Glacial marine*: Where a sediment-laden glacier flows into the sea, icebergs calve off the toe and raft clasts out to sea. As the icebergs melt, they drop the clasts, which settle into the muddy sediment on the sea floor. Pebble- and larger-size clasts deposited in this way, as we have seen, are called dropstones. Sediment consisting of ice-rafted clasts mixed with marine sediment makes up glacial marine. Glacial marine can also consist of sediment carried into the sea by water flowing at the base of a glacier.

- *Glacial outwash*: Till deposited by a glacier at its toe may be picked up and transported by meltwater streams that sort the sediment. The clasts are deposited by a braided

stream network in a broad area of gravel and sandbars called an outwash plain. This sediment is known as **glacial outwash** (Fig. 22.18c).

- *Loess*: When the warmer air above ice-free land beyond the toe of a glacier rises, the cold, denser air from above the glacier rushes in to take its place; a strong wind, called katabatic wind, therefore blows at the margin of a glacier. This wind picks up fine clay and silt and transports it away from the glacier's toe. Where the winds die down, the sediment settles and forms a thick layer. This sediment, called **loess**, sticks together because of the electrical charges on clay flakes; thus, steep escarpments develop by erosion of loess deposits (Fig. 22.18d).

- *Glacial lake-bed sediment*: Streams transport fine clasts, including rock flour, away from the glacial front. This

FIGURE 22.18 Sedimentation processes and products associated with glaciation. Glacial sediment is distinctive.

(a) This glacial till in Ireland is unsorted, because ice can carry sediment of all sizes.

(b) Glacial erratics resting on a glacially polished surface in Wyoming.

(c) Braided streams choked with glacial outwash in Alaska. The streams carry away finer sediment and leave the gravel behind.

(d) Thick loess deposits underlie parts of the prairie in Illinois.

(e) In the quiet water of an Alaskan glacial lake, fine-grained sediments accumulate. Alternating layers in the sediment (varves), now exposed in an outcrop near Puget Sound, Washington, reflect seasonal changes.

sediment eventually settles in meltwater lakes, forming a thick layer of glacial lake-bed sediment. This sediment commonly contains varves. A **varve** is a pair of thin layers deposited during a single year. One layer consists of silt brought in during spring floods and the other of clay deposited in winter when the lake's surface freezes over and the water is still (**Fig. 22.18e**).

Till, which contains no layering, is sometimes called unstratified drift; glacial sediments that have been redistributed by flowing water are called stratified drift.

Depositional Landforms of Glacial Environments

Picture a hunter, dressed in deerskin, standing at the toe of a continental glacier in what is now southern Canada, waiting for an unwary woolly mammoth to wander by. It's a sunny summer day 12,000 years ago, and milky, sediment-laden streams gush

from tunnels and channels at the base of the glacier and pour off the top as the ice melts. No mammoths venture by today, so the bored hunter climbs to the top of the glacier for a view. The climb isn't easy, partly because of the incessant katabatic wind and partly because deep crevasses interrupt his path. Reaching the top of the ice sheet, the hunter looks northward, and the glare almost blinds him. Squinting, he sees the white of snow, and where the snow has blown away, he sees the rippled, glassy surface of bluish ice (see Fig. 22.1b). Here and there, a rock protrudes from the ice. Now looking southward, he surveys a stark landscape of low, sinuous ridges separated by hummocky (bumpy) plains (see **Geology at a Glance**, pp. 764–765). Braided streams, which carry meltwater out across this landscape, flow through the hummocky plains and supply a number of lakes. Dust fills the air because of the wind.

All of the landscape features that the hunter observes as he looks southward were formed by deposition in glacial environments (**Fig. 22.19a, b**). The low, sinuous ridges, called end

FIGURE 22.19 The formation of depositional landforms associated with continental glaciation.

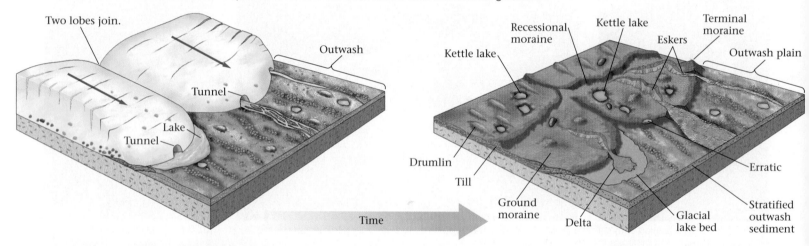

(a) The ice in continental glaciers flows toward the toe; sediment accumulates at the base and at the toe of the ice sheet.

(b) Several distinct depositional landforms form during glaciation; some developed under the ice and some at the toe.

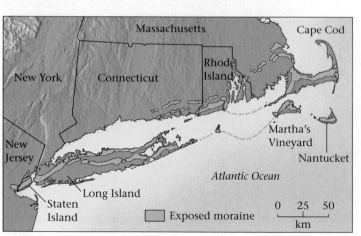

(c) Cape Cod, Long Island, and other landforms in the northeastern United States formed at the end of the continental ice sheet.

(d) This glacial moraine, in Wyoming, formed when glaciers covered the region between the moraine and the mountains in the distance.

FIGURE 22.20 Drumlins and knob-and-kettle topography characterize some areas that were once glaciated.

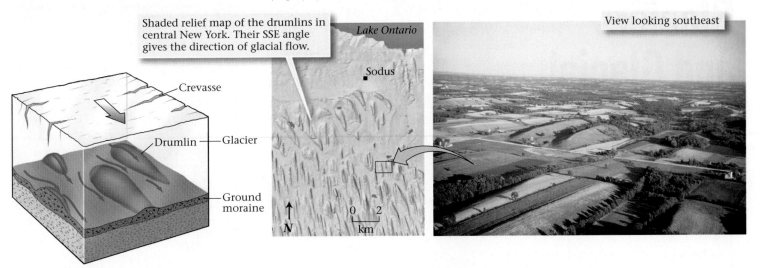

Shaded relief map of the drumlins in central New York. Their SSE angle gives the direction of glacial flow.

Lake Ontario

Sodus

Crevasse

Drumlin — Glacier

Ground moraine

0 2
km
N

View looking southeast

(a) The formation of a drumlin beneath a glacier.

(b) Drumlins dominate this landscape near Rochester, New York.

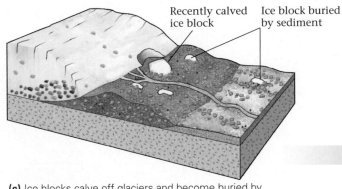

Recently calved ice block

Ice block buried by sediment

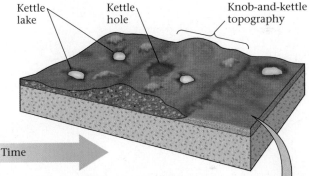

Kettle lake

Kettle hole

Knob-and-kettle topography

Time

(c) Ice blocks calve off glaciers and become buried by sediment. When the ice melts, a kettle forms.

(d) If the water table is high, kettles fill with water and turn into roughly circular lakes.

moraines, develop when the toe of a glacier stalls in one position for a while; the ice keeps flowing to the toe and, like a giant conveyor belt, transports sediment with it, so the sediment accumulates in a pile at the toe. The end moraine at the farthest limit of glaciation is called the **terminal moraine**. (The ridge of sediment that makes up Long Island, New York, and continues east-northeast into Cape Cod, Massachusetts, is part of the terminal moraine of the ice sheet that covered New England and eastern Canada during the last ice age; Fig. 22.19c.) The end moraines that form when a glacier stalls for a while as it recedes are **recessional moraines**.

Till that has been released at the base of a flowing glacier and remains after the glacier has melted away is called lodgment till (Fig. 22.19d). Clasts in lodgment till may be aligned and scratched during their movement or as ice flows over them. The flow of the glacier may mold till and other subglacial sediment into streamlined, elongate hills called **drumlins** (from the Gaelic word for hills). Drumlins tend to be asymmetric along their length, with a gentle downstream slope, tapered in the

(e) Knob-and-kettle topography make the surface of this moraine in Yellowstone Park, Wyoming, very hummocky.

direction of flow, and a steeper upstream slope (Fig. 22.20a, b). The till left behind during rapid recession forms a thin, hummocky layer on the land surface; this till, together with

Glaciers and Glacial Landforms

Continental ice sheet

Crevasses

Ice shelf

Higher sea level

Lower sea level

Drop stones

Iceberg

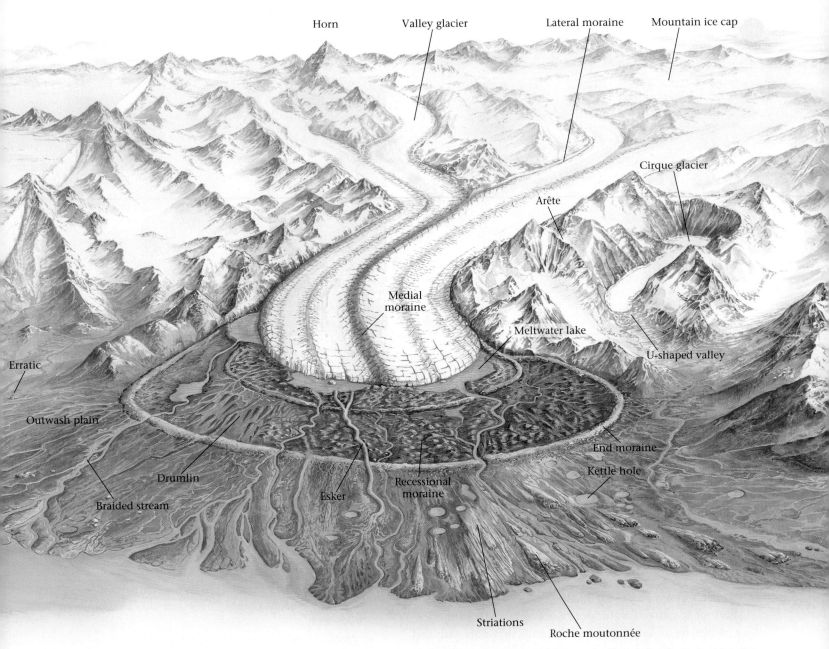

Horn   Valley glacier   Lateral moraine   Mountain ice cap

Cirque glacier

Arête

Medial moraine

Meltwater lake

U-shaped valley

Erratic

Outwash plain

End moraine

Kettle hole

Drumlin

Esker   Recessional moraine

Braided stream

Striations

Roche moutonnée

Glaciers are rivers or sheets of ice that last all year and slowly flow. Continental glaciers, vast sheets of ice up to a few kilometers thick, covered extensive areas of land during times when Earth had a colder climate. Continental glaciers form when snow accumulates at high latitudes, then, when buried deeply enough, packs together and recrystallizes to make glacial ice. Ice, though solid, is weak, and thus ice sheets spread over the landscape like syrup over a pancake. At the peak of the last ice age, ice sheets covered almost all of Canada, much of the United States, northern Europe, and parts of Russia.

The upper part of a sheet is brittle and may crack to form crevasses. Because ice sheets store so much of the Earth's water, sea level becomes lower during an ice age. When a glacier reaches the sea, it becomes an ice shelf. Rock that the glacier has plucked up along the way is carried out to sea with the ice; when the ice melts, the rocks fall to the sea floor as dropstones. At the edge of the shelf, icebergs calve off and float away.

A second class of glaciers, called mountain or alpine glaciers, exist in mountainous areas because snow can last all year at high elevations. During an ice age, mountain glaciers grow and flow out onto the land surface beyond the moun-

tain front. The glacier at the right has started to recede after formerly advancing and covering more of the land. Glacial recession may happen when the climate warms, so ice melts away faster at the toe (terminus) of the glacier than it can be added at the source. In front of the glacier, you can find consequences of glacial erosion such as striations on bedrock and roches moutonnées.

When the glacier pauses, till (unsorted glacial sediment) accumulates to form an end moraine. Meltwater lakes gather at the toe, and streams carry sediment and deposit it as glacial outwash. Sediment that accumulates in ice tunnels, exposed when the glacier melts, make up sinuous ridges called eskers. Even when the toe remains fixed in position for a while, the ice continues to flow, and thus molds underlying sediment into drumlins. Ice blocks buried in till melt to form kettle holes. (Though the examples shown here were left after the melting away of a piedmont glacier, most actually form during continental glaciation.)

In the mountains, glaciers fill valleys or form ice caps. Sediment falling from the mountains creates lateral and medial moraines. Glaciers carve distinct landforms in the mountains, such as cirques, arêtes, horns, and U-shaped valleys.

FIGURE 22.21 Eskers are snake-like ridges of sand and gravel that form when sediment fills meltwater tunnels at the base of a glacier.

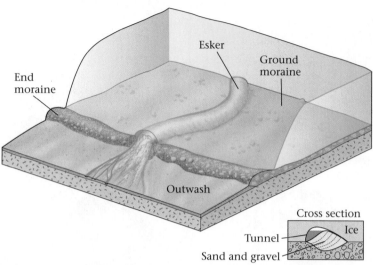

(a) At the time of formation, an esker develops beneath an ice sheet. In cross section (inset), cross-bedded sand accumulates in the tunnel.

(b) An example of an esker in an area once glaciated, but now farmed.

lodgment till, forms a landscape feature known as **ground moraine**.

The hummocky surface of moraines reflects partly the variations in the amount of sediment supplied by the glacier and partly the occurrence of **kettle holes**, circular depressions made when blocks of ice calve off the toe of the glacier, become buried by till, and then melt to leave a depression (Fig. 22.20c, d). A land surface with many kettle holes separated by round hills of till displays knob-and-kettle topography (Fig. 22.20e).

Sediment-choked water pours out of tunnels in the ice. This water feeds large braided streams that sort till and redeposit it as glacial outwash, stratified layers, and lenses of gravel and sand. Thus, outwash plains form between recessional moraines and beyond the terminal moraine. Some of the meltwater forms lakes in lowlands between recessional moraines, or in ice-margin lakes between the glacier's toe and the nearest recessional moraine. Even long after the glacier has melted away, the lowlands between recessional moraines persist as lakes or swamps. Glacial lake beds provide particularly fertile soil for agriculture. When a glacier eventually melts away, ridges of sorted sand and gravel, deposited in subglacial meltwater tunnels, snake across the ground moraine. These ridges are called **eskers** (Fig. 22.21a, b). Sediments in glacial outwash plains and in eskers are important sources of sand and gravel for road building and construction.

Take-Home Message

- Glacial ice carries clasts of all sizes, so deposits left by glaciers consist of unsorted till.

- Till accumulates in lateral moraines along the sides of glaciers. Medial moraines form where two valley glaciers merge.

- End moraines form at the toe of a glacier; the ice acts like a conveyor belt, bringing sediment; the end moraine at the farthest limit of glaciation is the terminal moraine.

- Beyond the toe of a glacier, meltwater may deposit glacial outwash, wind may deposit loess, and lakes accumulate varved sediment.

- When a glacier retreats, end moraines remain as distinctive ridges; moraines may include kettle holes, formed where ice blocks calved, were buried, and then later melted.

THINK: What is the difference between a drumlin and an esker?

22.5 OTHER CONSEQUENCES OF CONTINENTAL GLACIATION

Ice Loading and Glacial Rebound

When a large ice sheet (more than 50 km in diameter) grows on a continent, its weight causes the surface of the lithosphere to sink. In other words, ice loading causes **glacial subsidence**. Lithosphere, the relatively rigid outer shell of the Earth, can sink because the underlying asthenosphere is soft enough to flow slowly out of the way (Fig. 22.22a). As an analogy, imagine the following simple experiment. Fill a bowl with honey and then place a thin rubber sheet over the honey. The rubber represents the lithosphere, and the honey represents the asthenosphere. If you place an ice cube on the rubber sheet, the sheet sinks because the weight of the ice pushes it down; the honey flows out of the way to make room. Because of ice loading, much of Antarctica and Greenland now lie below sea level (see Fig. 22.4), so if their ice were *instantly* to melt away, these continents would be flooded by a shallow sea.

What happens when continental ice sheets do melt away? Gradually, the surface of the underlying continent rises back up, by a process called **glacial rebound**, and the asthenosphere flows back underneath to fill the space (Fig. 22.22b). Where rebound affects coastal areas, beaches along the shoreline rise several meters above sea level and become terraces. In the honey and rubber analogy, when you remove the ice cube, the rubber sheet slowly returns to its original shape. This process doesn't take place instantly, because the honey can only flow slowly. Similarly, because the asthenosphere flows *so* slowly (at rates of a few millimeters per year), it takes thousands of years for ice-depressed continents to rebound. Thus, glacial rebound is still taking place in some regions that were burdened by ice during the Pleistocene Ice Age. Recently, researchers in North America have documented this movement by using GPS measurements (Fig. 22.22c). Regions north of a line passing through the Great Lakes are rising, relative to sea level. Similar measurements have revealed glacial rebound of mountainous areas in Alaska as mountain glaciers and ice caps melt away.

Sea-Level Changes: The Glacial Reservoir in the Hydrologic Cycle

More of the Earth's surface and near-surface freshwater is stored in glacial ice than in any other reservoir. In fact, glacial ice accounts for 2.15% of Earth's total water supply, while lakes, rivers, soil, and the atmosphere together contain only 0.03%. During the last ice age, when glaciers covered almost three times as much land area as they do today, they held significantly more water (70 million cubic km, as opposed to 25 million cubic km today). In effect, water from the ocean reservoir transferred to the glacial reservoir and remained trapped on land. As a consequence, sea level dropped by as much as 100 m, and extensive areas of continental shelves became exposed as the coastline migrated seaward, in places

FIGURE 22.22 The concept of subsidence and rebound, due to continental glaciation and deglaciation. (Not to scale)

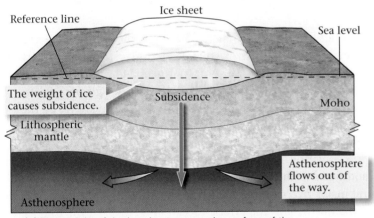

(a) The weight of the ice sheet causes the surface of the lithosphere to sink (subside).

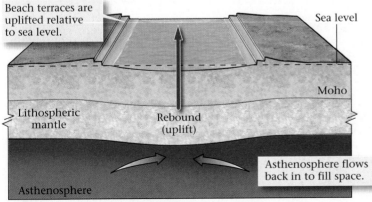

(b) After the glacier melts, the land surface rebounds and rises. This process uplifts beaches relative to sea level.

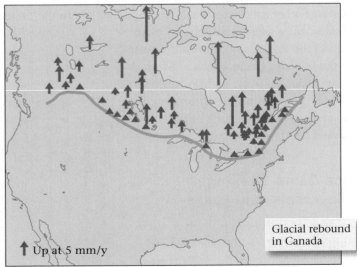

(c) GPS measurements show that the region north of the green line is rebounding. Different rates of uplift occur at different locations.

FIGURE 22.23 The link between sea level and global glaciation: glaciers store water on land, so when glaciers grow, sea level falls, and when glaciers melt, sea level rises.

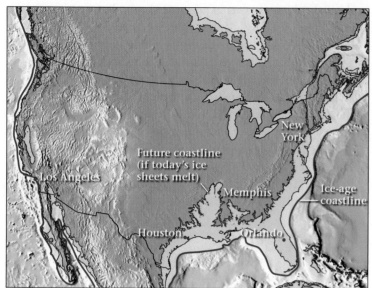

(a) The red line shows the coastline during the last ice age; much of the continental shelf was dry. If present-day ice sheets melt, coastal lands will flood.

(b) Prehistoric people migrated across the Bering Strait land bridge.

(c) Sea-level rise between 17,000 and 7,000 B.C.E. was due to the melting of ice-age glaciers.

by more than 100 km (Fig. 22.23a–c). People and animals migrated into the newly exposed coastal plains; in fact, fishermen dragging their nets along the Atlantic Ocean floor off New England today occasionally recover artifacts. The drop in sea level also created land bridges across the Bering Strait between North America and northeastern Asia and between Australia and Indonesia, providing convenient migration routes.

If today's ice sheets in Antarctica and Greenland were to melt, the crust of these continents would undergo glacial rebound, global sea level would rise substantially, and low-lying areas of other continents would undergo flooding. In the United States, large areas of the coastal plain along the East Coast and Gulf Coast would flood, and cities such as Miami, Houston, New York, and Philadelphia would disappear beneath the waves. In Canada, substantial parts of the shield would flood.

Ice Dams, Drainage Reversals, and Lakes

When ice freezes over a sewer opening in a street, neither meltwater nor rain can enter the drain, and the street floods. Ice sheets play a similar role in glaciated environments. The ice may block the course of a river, leading to the formation of a lake. In addition,

the weight of a glacier changes the tilt of the land surface and therefore the gradients of streams, and glacial sediment may fill preexisting valleys. In sum, continental glaciation destroys preexisting drainage networks. While the glacier exists, streams find different routes and carve out new valleys; by the time the glacier melts away, these new streams have become so well established that old river courses may remain abandoned.

Glaciation during the Pleistocene Ice Age profoundly modified North America's drainage. Before this Ice Age, several major rivers drained much of the interior of the continent to the north, into the Arctic Ocean (Fig. 22.24a, b). The ice

FIGURE 22.24 Ice-age glaciation changed the position of the divide between north-draining and south-draining river networks.

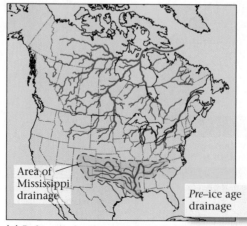

(a) Before the last ice age, more rivers flowed north; the Mississippi network was smaller.

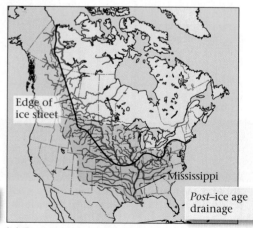

(b) Glaciation blocked northward drainage; the Mississippi network grew larger.

FIGURE 22.25 Ice age lakes in North America.

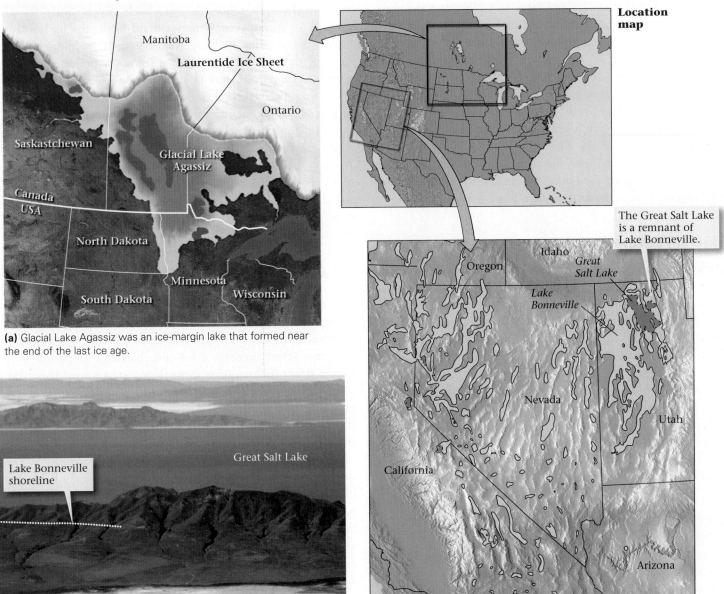

(a) Glacial Lake Agassiz was an ice-margin lake that formed near the end of the last ice age.

(b) Pluvial lakes occurred throughout the Basin and Range Province during the last ice age, due to the wetter climate. The largest of these was Lake Bonneville. Subtle horizontal terraces define the remnants of beaches, now over 100 m above the present level of the Great Salt Lake.

sheet buried this drainage network and diverted the flow into the Mississippi–Missouri network, which became larger. Then after the ice receded, regions covered by knob-and-kettle topography became spotted with thousands of small lakes, as now occur in central and southern Minnesota, and regions of the midwest covered by a smooth frosting of till became swampland, with little or no drainage at all. In the Canadian Shield, scouring left innumerable depressions that have now become lakes.

Inevitably, the ice dams that held back large ice-margin lakes melted and broke. In a matter of hours to days, the contents of the lakes drained, creating immense floodwaters that stripped the land of soil and left behind huge ripple marks. For example, glacial Lake Missoula, in Montana, filled when glaciers advanced and blocked the outlet of a large valley. When the glaciers retreated, the ice dam broke, releasing immense torrents that scoured eastern Washington, creating a barren, soil-free landscape called the channeled scablands (see Chapter 17). Recent evidence suggests that this process occurred many times.

The largest known ice-margin lake covered portions of Manitoba and Ontario, in south-central Canada, and North Dakota and Minnesota in the United States (Fig. 22.25a). This body of water, Glacial Lake Agassiz, existed between

11,700 and 9,000 years ago, a time during which the most recent phase of the last ice age came to a close and the continental glacier retreated north. At its largest, the lake covered over 250,000 square km (100,000 square miles), an area greater than that of all the present Great Lakes combined. Eventually, the ice sheet receded from the north shore of Glacial Lake Agassiz, so near the end of its life, the lake was surrounded by ice-free land. Field evidence suggests that the lake's demise came when it drained catastrophically. How did geologists reconstruct the history of Glacial Lake Agassiz? The present landscape holds the clues. Broad plains define the region that was once the lake floor, and beach terraces (now high and dry) define its former shoreline.

Pluvial Features

During ice ages, regions to the south of continental glaciers were wetter than they are today. Fed by enhanced rainfall, lakes accumulated in low-lying land even at a great distance from the ice front. The largest of these **pluvial lakes** (from the Latin *pluvia*, rain) in North America flooded interior basins of the Basin and Range Province in Utah and Nevada (Fig. 22.25b). These basins received drainage from the adjacent ranges but had no outlet to the sea, so they filled with water. The largest pluvial lake, Lake Bonneville, covered almost a third of western Utah. When this lake suddenly drained after a natural dam holding it back broke, it left a bathtub ring of shoreline rimming the mountains near Salt Lake City. Today's Great Salt Lake itself is but a small remnant of Lake Bonneville.

> ## Take-Home Message
>
> - The weight of glacial ice, especially in a continental glacier, can push the surface of a continent down below sea level.
> - When a large glacier melts away, the surface of the continent rises by a process called rebound.
> - The buildup of glacial ice causes sea level to drop; in fact, during ice ages, much of the continental shelf was dry land.
> - During the Pleistocene Ice Age, large pluvial lakes formed in desert areas of Utah and adjacent states.
>
> **THINK:** Why does the process of glacial rebound take thousands of years?

22.6 PERIGLACIAL ENVIRONMENTS

In polar latitudes today, and in regions adjacent to the fronts of continental glaciers during the last ice age, the mean annual temperature stays low enough (below −5°C) that soil moisture and groundwater freeze and, except in the upper few meters, stay solid all year. Such permanently frozen ground, or **permafrost**, may extend to depths of 1,500 m below the ground surface. Regions with widespread permafrost that do not have a cover of snow or ice are called periglacial environments (the Greek *peri* means around, or encircling; periglacial environments appear around the edges of glacial environments; Fig. 22.26a).

The upper few meters of permafrost may melt during the summer months, only to refreeze again when winter comes. As a consequence of the freeze-thaw process, the ground of some permafrost areas splits into pentagonal or hexagonal shapes, creating a landscape called **patterned ground** (Fig. 22.26b). Water fills the gaps between the cracks and freezes to create wedge-shaped walls of ice. In some places, freeze-and-thaw cycles in permafrost gradually push cobbles and pebbles up from the subsurface. Because the expansion of the ground is not even, the stones gradually collect between adjacent bulges to form stone rings (Fig. 22.26c). Some stone rings may also form when mud at depth pushes up from beneath a permafrost layer and forces stones aside.

Permafrost presents a unique challenge to people who live in polar regions or who work to extract resources from these regions. For example, heat from a building may warm and melt underlying permafrost, creating a mire into which the building settles. For this reason, buildings in permafrost regions must be placed on stilts, so that cold air can circulate beneath them to keep the ground frozen. When geologists discovered oil on the northern coast of Alaska, oil companies faced the challenge of shipping the oil to markets outside of Alaska. After much debate over the environmental impact, the Trans-Alaska Pipeline was built, and now it carries oil for 1,000 km to a seaport in southern Alaska (see Chapter 14). The oil must be warm during transport, or it would be too viscous to flow; thus, to prevent the warm pipeline from melting underlying permafrost, it had to be built on a frame that holds it above the ground for most of its length.

> ## Take-Home Message
>
> - In high latitudes today, and in regions just south of the ice sheets during the ice age, the ground remains frozen all year, or at least most of the year; such ground is called permafrost.
> - Freeze-and-thaw cycles may produce patterned ground or stone rings in permafrost.
>
> **THINK:** Can buildings be constructed in permafrost regions without melting the ground?

FIGURE 22.26 Periglacial regions are not ice covered but do include substantial areas of permafrost.

(a) The present-day distribution of periglacial environments in North America.

(b) An example of patterned ground near a pond in Manitoba, Canada.

22.7 THE PLEISTOCENE ICE AGE

The Pleistocene Glaciers

Today, most of the land surface in New York City lies hidden beneath concrete and steel, but in Central Park it's still possible to see land in a seminatural state. If you stroll through the park and study the rock outcrops, you'll find that their top surfaces are smooth and polished, and in places have been grooved and scratched—you can also find erratics. You are seeing evidence that an ice sheet once scraped along this now-urban ground. Geologists estimate that the ice sheet that overrode the New York City area may have been 250 m thick, enough to bury the Empire State Building up to the 75th floor.

Glacial features such as those on display in Central Park first led Louis Agassiz to propose the idea that vast continental glaciers advanced over substantial portions of North America, Europe, and Asia during a great ice age. Since Agassiz's day, thousands of geologists, by mapping out the distribution of glacial deposits and landforms, have gradually defined the extent of ice-age glaciers and a history of their movement (**Box 22.2**).

The fact that these glacial features decorate the surface of the Earth today means that the most recent ice age occurred fairly recently during Earth history. This ice age, responsible for the glacial landforms of North America and Eurasia, happened mostly during the Pleistocene Epoch, which began 1.8 million years ago (see Chapter 13), so as we've noted earlier, it is commonly known as the Pleistocene Ice Age. This traditional title is a bit of a misnomer. Recent studies demonstrate that the glaciations of this ice age actually began between 3.0 and 2.5 million years ago, during the Pliocene

(c) Stone circles of Spitzbergen form due to repeated freeze and thaw that separates gravel from silt.

Epoch, and continued through the Pleistocene. Further, there was not just a single ice advance, but, as we will see, there were many—probably over 20. (Thus, all of these events together might better be called the "Plio-Pleistocene ice ages", but the traditional name is too well established to be changed.) Geologists use the name Holocene to refer to the last 11,000 years, the time since the last glaciation.

Based on their mapping of glacial striations and deposits, geologists have determined where the great Pleistocene ice sheets originated and flowed. In North America, the Laurentide ice sheet started to grow over northeastern Canada, then merged with the Keewatin ice sheet, which originated in northwestern Canada. Together, these ice sheets eventually covered all of Canada east of the Rocky Mountains and

BOX 22.2

So You Want to See Glaciation?

Though the last of the Pleistocene continental ice sheet that once covered much of North America vanished about 6,000 years ago, you can find evidence of its power quite easily. The Great Lakes, along the U.S.–Canada border, the Finger Lakes and drumlins of New York, the low-lying moraines and outwash plains of Illinois, and the polished outcrops of southern Canada all formed in response to the existence of this glacier. But if you want to see continental glaciers in action today, you must trek to Greenland or Antarctica.

Mountain glaciers are easier to reach. A trip to the mountains of western North America (including Alaska), the Alps of France or Switzerland, the Andes of South America, or the mountains of southern New Zealand will bring you in contact with active glaciers. You can even spot glaciers from the comfort of a cruise ship.

Some of the most spectacular glacial landscapes in North America formed during the Pleistocene Epoch, when mountain glaciers were more widespread. These are now on display in national parks.

- *Glacier National Park (Montana)*: This park, which borders Waterton Lakes National Park in Canada, displays giant cirques, U-shaped valleys, hanging valleys, and terminal moraines. In 1850, there were about 150 glaciers in the park, and some of these were quite large. Now there are only 25 active relics of formerly larger glaciers, all in a mountainous terrain that reaches elevations of over 3 km. Unfortunately, these glaciers are melting away quickly and may vanish entirely before 2030.
- *Yosemite National Park (California)*: A huge U-shaped valley, carved into the Sierra Nevada granite batholith, makes up the centerpiece of this park. Waterfalls spill out of hanging valleys bordering the valley.
- *Voyageurs National Park (Minnesota)*: This park lacks the high peaks of mountainous parks, but shows the dramatic consequences of glacial scouring and deposition on the Canadian Shield. The low-lying landscape, dotted with lakes, contains abundant polished surfaces, glacial striations, and erratics, along with moraines, glacial lake beds, and outwash plains.
- *Acadia National Park (Maine)*: During the last ice age, the continental ice sheet overrode low bedrock hills and flowed into the sea along the coast of Maine. This park provides some of the best examples of the consequences. Its hills were scoured and shaped into large roches moutonnées by glacial flow. Some of the deeper valleys have now become small fjords.
- *Glacier Bay National Park (Alaska)*: In Glacier Bay, huge tidewater glaciers fringe the sea, creating immense ice cliffs from which icebergs calve off. Cruise ships bring tourists up to the toes of these glaciers. More adventurous visitors can climb the coastal peaks and observe lateral and medial moraines, crevasses, and the erosional and depositional consequences of glaciers that have already retreated up the valley.

---

extended southward across the border as far as southern Illinois (Fig. 22.27a). At their maximum, the ice sheets attained a thickness of 2 to 3 km; each thinned toward its toe. In northeastern Canada, the ice sheet eroded the land surface. Farther south and west, it deposited sediment (Fig. 22.27b, c). These ice sheets also eventually merged with the Greenland ice sheet to the northeast and the Cordilleran ice sheet to the west; the Cordilleran covered the mountains of western Canada as well as the southern third of Alaska.

During the Pleistocene Ice Age, mountain ice caps and valley glaciers also grew in the Rocky Mountains, the Sierra Nevada, and the Cascade Mountains. In Eurasia, a large ice sheet formed in northernmost Europe and adjacent Asia, and gradually covered all of Scandinavia and northern Russia. This ice sheet flowed southward across France until it reached the Alps and merged with Alpine mountain glaciers;

it also covered all of Ireland and almost all of the United Kingdom. A smaller ice sheet grew in eastern Siberia and expanded in the mountains of central Asia. In the southern hemisphere, Antarctica remained ice covered, and mountain ice caps expanded in the Andes, but there were no continental glaciers in South America, Africa, or Australia.

In addition to continental ice sheets, sea ice in the northern hemisphere expanded to cover all of the Arctic Ocean and parts of the North Atlantic during the Pleistocene. Sea ice surrounded Iceland and approached Scotland and also fringed most of western Canada and southeastern Alaska.

Life and Climate in the Pleistocene World

During the Pleistocene Ice Age, all climatic belts shifted southward (Fig. 22.28a, b). Geologists can document this shift

FIGURE 22.27 Pleistocene ice sheets and their consequences.

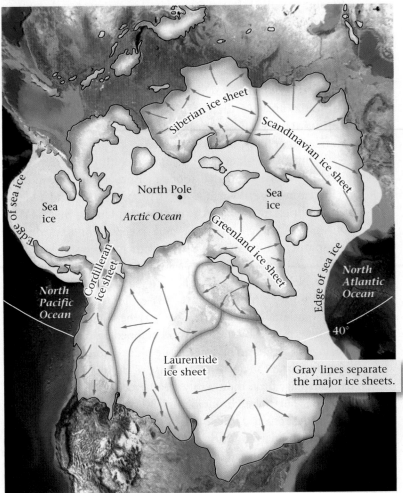

(a) During the Pleistocene, several distinct ice sheets formed. In several places, neighboring sheets came in contact.

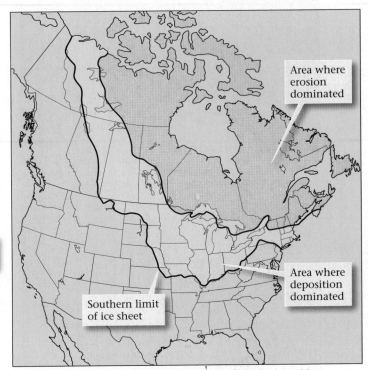

(b) Erosion dominates beneath the interior of the glacier, and deposition dominates along its margins.

(c) Erosion dominated in northern and eastern Canada; deposition dominated in the Great Plains.

by examining fossil pollen, which can survive for thousands of years if preserved in the sediment of bogs. Presently, the southern boundary of North America's **tundra**, a treeless region supporting only low shrubs, moss, and lichen capable of living on permafrost, lies at a latitude of 68° N; during the Pleistocene Ice Age, it moved down to 48° N. Much of the interior of the United States, which now has temperate, deciduous forest, harbored cold-weather spruce and pine forest. Ice-age climates also changed the distribution of rainfall on the planet: increased rainfall in North America led to the filling of pluvial lakes in Utah and Nevada, whereas decreased rainfall in equatorial regions led to shrinkage of the rain forest. Overall, the contrast between colder, glaciated regions and warmer, unglaciated regions created windier conditions worldwide. These winds sent glacial rock flour skyward, creating a dusty atmosphere (and, presumably, spectacular sunsets). The dust settled to create extensive deposits of loess. And because glaciers trapped so much water, as we have seen, sea level dropped.

Numerous species of now-extinct large mammals inhabited the Pleistocene world (**Fig. 22.28c**). Giant mammoths

and mastodons, relatives of the elephant, along with woolly rhinos, musk oxen, reindeer, giant ground sloths, bison, lions, saber-toothed cats, giant cave bears, and hyenas wandered forests and tundra in North America. Early human-like species were already foraging in the woods by the beginning of the Pleistocene Epoch, and by the end modern *Homo sapiens* lived on every continent except Antarctica, and had discovered fire and invented tools. Rapidly changing climates may have triggered a global migration of early humans, who gained access to the Americas, Indonesia, and Australia via land bridges that became exposed when sea level dropped.

Timing of the Pleistocene Ice Ages

Louis Agassiz assumed that only one ice age had affected the planet. But close examination of the stratigraphy of glacial deposits on land revealed that paleosol (ancient soil preserved in the stratigraphic record), as well as beds containing fossils of warmer-weather animals and plants, separated distinct layers of glacial sediment. This observation suggested that between

FIGURE 22.28 Climate belts during the Pleistocene.

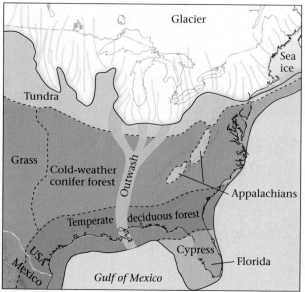

(a) Tundra covered parts of the United States, and southern states had forests like those of New England's today.

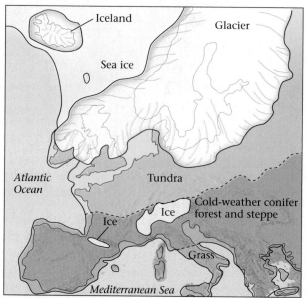

(b) Regions of Europe that support large populations today would have been barren tundra during the Pleistocene.

(c) Cold-adapted, now-extinct, large mammals roamed regions that are now temperate.

episodes of glacial deposition, glaciers receded and temperate climates prevailed. In the second half of the twentieth century, when modern methods for dating geological materials became available, the difference in ages between the different layers of glacial sediment could be confirmed. Clearly, glaciers had advanced and then retreated more than once during the Pleistocene. Times during which the glaciers grew and covered substantial areas of the continents are called glacial periods, or **glaciations**, and times between glacial periods are called interglacial periods, or **interglacials**.

Using the on-land sedimentary record, geologists recognized five Pleistocene glaciations in Europe (named, in order of increasing age: Würm, Riss, Mindel, Gunz, and Donau) and, traditionally, four in the midwestern United States (Wisconsinan, Illinoian, Kansan, and Nebraskan, named after the southernmost states in which their till was deposited; Fig. 22.29). Since the mid-1980s, geologists no longer recognize Nebraskan and Kansan; they are lumped together as "pre-Illinoian." With the advent of radiometric dating in the mid-twentieth century, the ages of the younger glaciations were determined by dating wood trapped in glacial deposits. Geologists estimate the ages of the older glaciations by identifying fossils in the deposits. Because of their greater age, these deposits have been thoroughly weathered.

The four-stage chronology of North American glaciation was turned on its head in the 1960s, when geologists began to study submarine sediment containing the fossilized shells of microscopic marine plankton. Because the assemblage of plankton species living in warm water is not the same as the assemblage living in cold water, geologists can track changes

FIGURE 22.29 Pleistocene glacial deposits in the north-central United States. Curving moraines reflect the shape of glacial lobes.

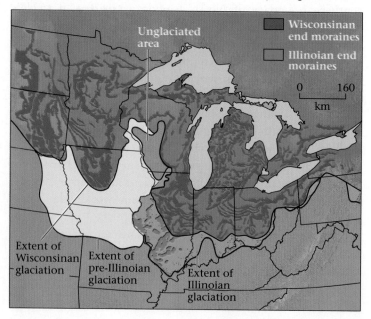

in the temperature of the ocean by studying plankton fossils. Researchers found that in sediment of the last 2 million years, assuming that cold water indicates a glacial period and warm water an interglacial period, there were 20 to 30 different glacial advances during the Pleistocene Epoch. The four traditionally recognized glaciations probably represent only the largest of these. Sediments deposited on land by other glaciations were eroded and redistributed during subsequent glaciations, or were eroded by streams and wind during interglacials.

Geologists refined their conclusions about the frequency of Pleistocene glaciations by examining the *isotopic composition* of fossil shells. Shells of many plankton species consist of calcite ($CaCO_3$). The oxygen in the shells includes two isotopes, a heavier one (^{18}O) and a lighter one (^{16}O). The ratio of these isotopes tells us about the water temperature in which the

plankton grew; this is because as water gets colder, plankton incorporate a higher proportion of ^{18}O into their shells (see Chapter 23). Thus, intervals in the stratigraphic record during which plankton shells have a large ratio of ^{18}O to ^{16}O define times when Earth had a colder, glacial climate. The record also indicates that 20 to 30 of these events occurred during the last 2 million years (Fig. 22.30a).

Older Ice Ages during Earth History

So far, we've focused on the Pleistocene Ice Age because of its importance in developing Earth's present landscape. Was this the only ice age during Earth history, or do ice ages

Did you ever wonder...
how many ice ages have happened during Earth history?

FIGURE 22.30 The timing of glaciations. Ice ages have occurred at several times in the geologic past.

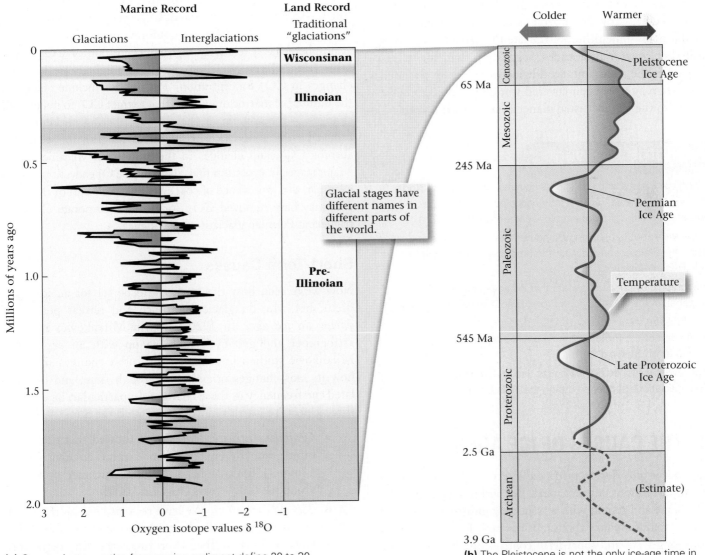

(a) Oxygen–isotope ratios from marine sediment define 20 to 30 glaciations in the Pleistocene. Tan bands represent traditional glacial stages of the midwestern United States.

(b) The Pleistocene is not the only ice-age time in Earth history. Glacial events happened in colder intervals of earlier eras, too.

happen frequently? To answer such questions, geologists study the stratigraphic record and search for ancient glacial deposits that have hardened into rock. These deposits, called **tillites**, consist of larger clasts distributed throughout a matrix of sandstone and mudstone. In many cases, tillites are deposited on glacially polished surfaces.

By using the stratigraphic principles described in Chapter 12, geologists have determined that tillites were deposited about 280 million years ago, in Permian time; these are the deposits Alfred Wegener studied when he argued in favor of continental drift (**Fig. 22.30b**). Tillites were also deposited about 600 to 700 million years ago (at the end of the Proterozoic Eon), about 2.2 billion years ago (near the beginning of the Proterozoic), and perhaps about 2.7 billion years ago (in the Archean Eon). Strata deposited at other times in Earth history do not contain tillites. Thus, it appears that glacial advances and retreats have not occurred steadily throughout Earth history, but rather are restricted to specific time intervals, or ice ages, of which there were four or five: Pleistocene, Permian, late Proterozoic, early Proterozoic, and perhaps Archean. Of particular note, some tillites of the late Proterozoic event were deposited at equatorial latitudes, suggesting that, for at least a short time, the continents worldwide were largely glaciated, and the sea may have been covered worldwide by ice. Geologists refer to the ice-encrusted planet as **snowball Earth**.

Take-Home Message

- During the Pleistocene, ice sheets covered large areas of North America and Europe; regions now occupied by temperate forests were either tundra or cold-weather forest.

- There were about 20 advances and retreats of the glaciers during the Pleistocene Ice Age; the times between glaciations are called interglacials.

- Stratigraphic evidence suggests that ice ages have also happened in the Paleozoic and Proterozoic, and possibly during the Archean.

- During part of the Proterozoic, the entire surface of the planet froze to form "snowball Earth."

THINK: What evidence do geologists use to determine that multiple glaciations happened during the Pleistocene?

22.8 THE CAUSES OF ICE AGES

Ice ages occur only during restricted intervals of Earth history, hundreds of millions of years apart. But within an ice age, glaciers advance and retreat with a frequency measured in tens of thousands to hundreds of thousands of years. Thus, there must be both long-term and short-term controls on glaciation. The nature of these controls emphasizes the complexity of interactions among components of the Earth System.

Long-Term Causes

Plate tectonics exercises some long-term control over glaciation for several reasons. First, continental drift due to plate tectonics determines the distribution of continents relative to the equator. If all continents straddled the equator, none could become cold enough to host continental glaciations. Second, the distribution of continents relative to upwelling and downwelling zones of the mantle may influence overall land elevation, and the global volume of mid-ocean ridges (which reflects sea-floor spreading rates) influences sea level. At times when continents are relatively low and sea level is relatively high, large areas of continents flood and cannot host glaciers. Finally, global climate can be affected by heat redistributed by oceanic currents. Growth of island arcs and drift of continents can influence the configuration of currents and determine whether high-latitude regions can become cold enough to host ice sheet formation.

The concentration of carbon dioxide in the atmosphere may also determine whether an ice age can occur. Carbon dioxide is a greenhouse gas—it traps infrared radiation rising from the Earth—so if the concentration of CO_2 increases, the atmosphere becomes warmer. Ice sheets cannot form during periods when the atmosphere has a relatively high concentration of CO_2, even if other factors favor glaciation. But what might cause long-term changes in CO_2 concentration? Possibilities include changes in the number of marine organisms that extract CO_2 to make shells; changes in the amount of chemical weathering on land (determined by the abundance of mountain ranges), for weathering absorbs CO_2; and changes in the amount of volcanic activity. Major stages in evolution may also affect CO_2 concentration. For example, the appearance of coal swamps at the end of the Paleozoic may have removed CO_2, for plants incorporate CO_2, thus triggering Permian glaciations of Pangaea.

Short-Term Causes

Now we've seen how the stage could be set for an ice age to occur, but why do glaciers advance and retreat periodically *during* an ice age? In 1920, Milutin Milanković, a Serbian astronomer and geophysicist, came up with an explanation. Milanković studied how the Earth's orbit changes shape and how its axis changes orientation through time, and he calculated the frequency of these changes. In particular, he evaluated three aspects of Earth's movement around the Sun.

- *Orbital eccentricity*: Milanković showed that the Earth's orbit gradually changes from a more circular shape to a more elliptical shape. This eccentricity cycle takes around 100,000 years (**Fig. 22.31a** 🔊).
- *Tilt of Earth's axis*: We have seasons because the Earth's axis is not perpendicular to the plane of its orbit. Milanković calculated that over time, the tilt angle varies between 22.5° and 24.5°, with a frequency of 41,000 years (**Fig. 22.31b**).

■ *Precession of Earth's axis*: If you've ever set a top spinning, you've probably noticed that its axis gradually traces a conical path. This motion, or wobble, is called precession (**Fig. 22.31c**). Milanković determined that the Earth's axis wobbles over the course of about 23,000 years. Right now, the Earth's axis points toward Polaris, making Polaris the north star, but 12,000 years ago the axis pointed to Vega. Precession determines the relationship between the timing of the seasons and the position of Earth along its orbit around the Sun.

Milanković showed that precession, along with variations in orbital eccentricity and tilt, combine to affect the total annual amount of insolation (exposure to the Sun's rays) and the seasonal distribution of insolation that the Earth receives at the mid- to high-latitudes (such as 65° N) by as much as 25%. For example, such regions receive more insolation when the Earth's axis is almost perpendicular to its orbital plane than when its axis is greatly tilted. According to Milanković, glaciers tend to advance during times of cool summers at 65° N, which occur periodically (**Fig. 22.31d**). When geologists began to study the climate record, they found climate cycles with the frequency predicted by Milanković. These climate cycles, controlled by "orbital forcing," are now called **Milanković cycles**.

The discovery of Milanković cycles in the geologic record strongly supports the contention that changes in the Earth's orbit and tilt help trigger short-term advances and retreats during an ice age. But orbit and tilt changes cannot be the whole story, because they could cause only about a 4°C temperature decrease (relative to today's temperature), and during glaciations the temperature decreased 5° to 7°C along coasts and 10° to 13°C inland. Geologists suggest that several other factors may come into play in order to trigger a glacial advance.

■ *Solar variability*: Changes in the radiation output of the Sun could affect the amount of energy the Earth receives.

■ *A changing albedo*: When snow remains on land throughout the year, or clouds form in the sky, the albedo (reflectivity) of the Earth increases, so Earth's surface reflects incoming sunlight and thus becomes even cooler.

■ *Interrupting the global heat conveyor*: As the climate cools, evaporation rates from the sea decrease, so seawater does not become as salty. Decreasing salinity might stop the system of thermohaline currents that brings warm water to high latitudes (see Chapter 18). Thus, the high latitudes become even colder than they would otherwise.

■ *Biological processes that change CO_2 concentration*: Several kinds of biological processes may have amplified climate changes by altering the concentration of carbon dioxide in the atmosphere. For example, a greater amount of plankton growing in the oceans could absorb more carbon dioxide and thus remove it from the atmosphere.

FIGURE 22.31 Milanković cycles influence the amount of insolation received at high latitudes. 🔊

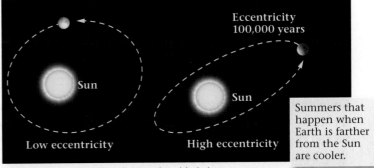

(a) Variations caused by changes in orbital shape.

Summers that happen when Earth is farther from the Sun are cooler.

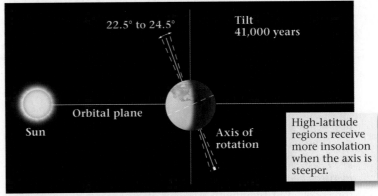

(b) Variations caused by changes in axis tilt.

High-latitude regions receive more insolation when the axis is steeper.

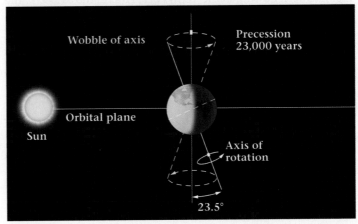

(c) Variations caused by the precession of Earth's axis.

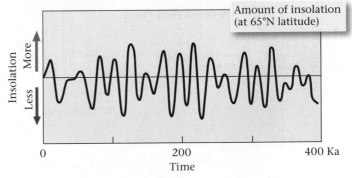

(d) Combining the effects of eccentricity, tilt, and precession produces distinct periods of more or less insolation.

The last three of the above processes are called positive-feedback mechanisms: they enhance the process that causes them. Because of positive feedback, the Earth could cool more than it would otherwise during the cooler stage of a Milanković cycle, and this could trigger a glacial advance.

A Model for Pleistocene-Ice-Age History

Long-term cooling in the Cenozoic Era. Taking all of the above causes into account, we can now propose a scenario for the events that led to the Pleistocene glacial advances. Our story begins in the Eocene Epoch, about 55 million years ago (Fig. 22.32). At that time, climates were warm and balmy not only in the tropics, but even above the Arctic Circle. At the end of the Middle Eocene (37 million years ago), the climate began to cool, and by Early Oligocene time (33 million years ago), Antarctica became glaciated. The Antarctic ice sheet came and went until the middle of the Miocene Epoch (15 million years ago), when an ice sheet formed that has lasted ever since. Ice sheets did not appear in the Arctic, however, until 2 to 3 million years ago, when the Pleistocene Ice Age began.

These long-term climate changes may have been caused, in part, by changes in the pattern of oceanic currents that happened, in turn, because of plate tectonics. For example, in the Eocene, the collision of India with Asia cut off warm equatorial currents that had been flowing in the Tethys Sea. And in the Miocene and Oligocene, Australia and South America drifted away from Antarctica, allowing the cold circum-Antarctic current to develop. This new current prevented warm, southward-flowing currents from reaching Antarctica, allowing ice to form and survive in the region. With the loss of the warm currents, the climate of Antarctica overall underwent cooling. Changes to atmospheric circulation and temperature may also have happened at this time. Models suggest that the uplift of the Himalayas and Tibet diverted winds in a way that cooled the climate. Further, this

uplift exposed more rock to chemical weathering, perhaps leading to extraction of CO_2 from the atmosphere (for chemical weathering reactions absorb CO_2); a decrease in the concentration of this greenhouse gas would contribute to atmospheric cooling.

So far, we've examined hypotheses that explain long-term cooling since 55 million years ago, but what caused the sudden appearance of the Laurentide ice sheet about 2–3 million years ago? This event coincides with another plate-tectonic event, the closing of the gap between North and South America by the growth of the Isthmus of Panama. When this land bridge formed, it separated the waters of the Caribbean from those of the tropical Pacific for the first time, and when this happened, warm currents that previously flowed out of the Caribbean into the Pacific were blocked and diverted northward to merge with the Gulf Stream. This current transfers warm water from the Caribbean, up the Atlantic Coast of North America, and ultimately to the British Isles. As the warm water moves up the Atlantic Coast, it generates warm, moisture-laden air that provides a source for the snow that falls over New England, eastern Canada, and Greenland. In other words, the Arctic has long been cold enough for ice caps, but until the Gulf Stream was diverted northward by the growth of Panama, there was no source of moisture to make abundant snow and ice.

Short-term advances and retreats in the Pleistocene Epoch. Once the Earth's climate had cooled overall, short-term processes such as the Milanković cycles led to periodic advances and retreats of the glaciers. To understand how, let's look at a possible case history of a single advance and retreat of the Laurentide ice sheet. (Note that such models remain the subject of vigorous debate.)

- *Stage 1*: During the overall cooler climates of the late Cenozoic Era, the Earth reaches a point in the Milanković cycle when the average mean temperature in temperate latitudes drops. Because of glacial rebound, the ice-free surface of northern Canada has risen to an altitude of several hundred meters above sea level. With lower temperatures and higher elevations, not all of winter's snow melts away during the summer. Eventually, snow covers the entire region of northern Canada, even during the summer. Because of the snow's high albedo, it reflects sunlight, so the region grows still colder (a positive-feedback effect) and even more snow accumulates. Precipitation rates are high, because evaporation off the Gulf Stream provides moisture. Finally, the snow at the base of the pile turns to ice, and the ice begins to spread outward under its own weight. A new continental glacier has been born.

- *Stage 2*: The ice sheet continues to grow as more snow piles up in the zone of accumulation. And as the ice sheet grows, the atmosphere continues to cool because of the albedo effect. But now, the weight of the ice loads the continent and makes it sink, so the elevation

FIGURE 22.32 Until recently, Earth's atmosphere has been gradually cooling, overall, since the Cretaceous.

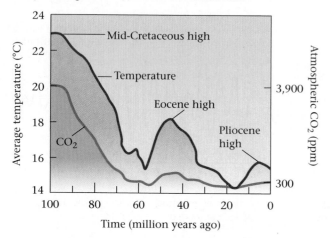

FIGURE 22.33 The Little Ice Age and its demise. Glaciers that advanced between 1550 and 1850 have since retreated.

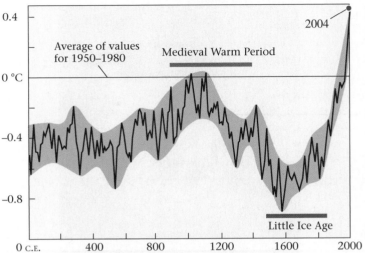

(a) A model of global temperature for the past 2,000 years. Overall trends display the Medieval Warm Period followed by the Little Ice Age. Since 1850, temperatures have warmed.

(b) Skaters (c. 1600) on the frozen canals of the Netherlands during the Little Ice Age.

of the glacier decreases, and its surface approaches the equilibrium line. Also, the temperature becomes cold enough that in high latitudes the Atlantic Ocean begins to freeze. As the sea ice covers the ocean, the amount of evaporation decreases, so the source of snow is cut off and the amount of snowfall diminishes. The glacial advance pretty much chokes on its own success. The decrease in the glacier's elevation (leading to warmer summer temperatures) on the ice surface, as well as the decrease in snowfall, causes ablation to occur faster than accumulation, and the glacier begins to retreat.

■ *Stage 3*: As the glacier retreats, temperatures gradually increase, and the sea ice begins to melt. The supply of water to the atmosphere from evaporation increases once again, but with the warmer temperatures and lower elevations, this water precipitates as rain during the summer. The rain drastically accelerates the rate of ice melting, and the retreat progresses quite rapidly.

Will There Be Another Glacial Advance?

What does the future hold? Considering the periodicity of glacial advances and retreats during the Pleistocene Epoch, we may be living in an interglacial period. Pleistocene interglacials lasted about 10,000 years, and since the present interglacial began about 11,000 years ago, the time seems ripe for a new glaciation. If a glacier on the scale of the Laurentide ice sheet were to develop, major cities and agricultural belts would be overrun by ice, and their populations would have to migrate southward. Long before the ice front arrived, though, the climate would become so hostile that the cities would already be abandoned.

(c) During the Little Ice Age, a glacier filled this valley. In this 2003 photo, most of the glacier has vanished. Most of the retreat has happened in the last century.

The Earth actually had a brush with ice-age conditions between the 1300s and the mid-1800s, when average annual temperatures in the northern hemisphere fell sufficiently for mountain glaciers to advance significantly. During this period, now known as the **Little Ice Age**, sea ice surrounded Iceland and canals froze in the Netherlands, leading to that country's tradition of skating (**Fig. 22.33a, b**). Some researchers

FIGURE 22.34 Greenland's melting glaciers. Melting has accelerated in the last few decades.

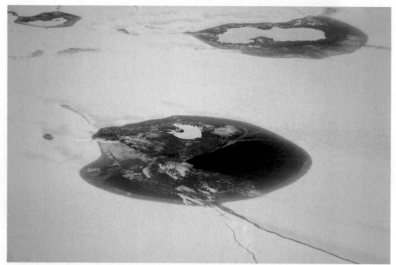

(a) Lakes of meltwater accumulate on the surface of the ice sheet during the summer.

(b) Lakes suddenly drain through cracks that carry the water to the base of the glacier, 1 km down. Addition of liquid water to the base allows the glacier to move faster, causing a "surge."

(c) Where glaciers meet the sea, huge masses calve off and crash into the water. This is happening so fast that the ice front is retreating.

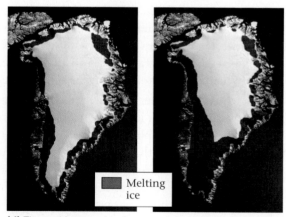

Melting ice

(d) The melting area has increased dramatically in recent years.

speculate that the depopulation of the western hemisphere, in the wake of European conquest, caused temporary reforestation, for without inhabitants, farmlands went untended. The new forests absorbed CO_2 and caused atmospheric concentrations of CO_2 to decrease, leading to the cooler conditions that triggered the Little Ice Age. Others speculate that the change reflects increased cloud cover, not a change in CO_2 concentration. Researchers will likely propose additional ideas as work on this problem continues.

During the past 150 years, temperatures have warmed, and most mountain glaciers have retreated significantly (Fig. 22.33c). We no longer see the icebergs that once threatened Atlantic shipping lanes and sank the *Titanic*, and large slabs frequently calve off the Antarctic ice sheet. In fact, the Larsen B Ice Shelf of Antarctica, an area larger than Rhode Island, disintegrated in 2002. Greenland's glaciers, in particular, are showing signs of accelerating retreat (Fig. 22.34a–d). Large meltwater ponds are forming on the surface of the ice sheet—some of these drain abruptly through cracks to the base of the glacier a kilometer below. Most researchers suggest that this global-warming trend is due to the addition of CO_2 to the atmosphere from the burning of fossil fuels (see Chapter 23). Global warming could conceivably cause a "super-interglacial."

If the climate were to become significantly warmer than it is today, the ice sheet of West Antarctica might begin to float and then break up rapidly. If all of today's ice caps melted, global sea level would rise by 70 m (230 feet), extensive areas of coastal plains would be flooded, and major coastal cities such as New York, Miami, and London would be submerged (see Fig. 22.23). Instead of protruding from ice, the tip of the Empire State Building would protrude from the sea. Icehouse or greenhouse? We may not know which scenario will play out in the future until it happens. However, researchers have voiced concern that, at least in the near term, glacial melting will be the order of the day, as global temperatures seem to be rising. The next chapter addresses such change.

Take-Home Message

- Special conditions must occur on Earth for an ice age to occur; substantial continental areas must be at high latitudes, and CO_2 concentration in the atmosphere must be low.
- Changes in ocean currents may also play a role in triggering ice ages.
- During an ice age, the timing of glaciations and interglacials may be controlled by the Milanković cycle of changes in the Earth's orbit and in the tilt of its axis.
- The Earth might have been heading into another glaciation now, were it not for the effects of recent global warming.

THINK: How does positive feedback contribute to a glaciation during an ice age, and why does the glaciation eventually cease?

Chapter Summary

- Glaciers are streams or sheets of recrystallized ice that survive for the entire year and flow in response to gravity. Mountain glaciers exist in high regions and fill cirques and valleys. Continental glaciers (ice sheets) spread over substantial areas of the continents.

- Glaciers form when snow accumulates over a long period of time. With progressive burial, the snow first turns to firn and then to ice.

- Ice in temperate glaciers is melting at least during part of the year. Polar glaciers are frozen solid. Glaciers move by basal sliding over water or wet sediment, and/or by plastic deformation of ice grains. In general, glaciers move tens of meters per year.

- Glaciers move because of gravitational pull; they flow in the direction of their surface slope.

- Whether the toe of a glacier stays fixed in position, advances farther from the glacier's origin, or retreats back toward the origin depends on the balance between the rate at which snow builds up in the zone of accumulation and the rate at which glaciers melt or sublimate in the zone of ablation.

- Icebergs break off glaciers that flow into the sea. Continental glaciers that flow out into the sea along a coast make ice shelves. Sea ice forms where the ocean's surface freezes.

- As glacial ice flows over sediment, it incorporates clasts. The clasts embedded in glacial ice act like a rasp that abrades the substrate.

- Mountain glaciers carve numerous landforms, including cirques, arêtes, horns, U-shaped valleys, hanging valleys, and truncated spurs. Fjords are glacially carved valleys that filled with water when sea level rose after an ice age.

- Glaciers can transport sediment of all sizes. Glacial drift includes till, glacial marine, glacial outwash, lake-bed mud, and loess. Lateral moraines accumulate along the sides of valley glaciers, and medial moraines form down the middle of a glacier. End moraines accumulate at a glacier's toe.

- Glacial depositional landforms include moraines, knob-and-kettle topography, drumlins, kames, eskers, meltwater lakes, and outwash plains.

- Continental crust subsides as a result of ice loading. When the glacier melts away, the crust rebounds.

- When water is stored in continental glaciers, sea level drops. When glaciers melt, sea level rises.

- During past ice ages, the climate in regions south of the continental glaciers was wetter, and pluvial lakes formed. Permafrost (permanently frozen ground) exists in periglacial environments.

- During the Pleistocene Ice Age, large continental glaciers covered much of North America, Europe, and Asia.

- The stratigraphy of Pleistocene glacial deposits preserved on land records five European and four North American glaciations, times during which ice sheets advanced. The record preserved in marine sediments records 20 to 30 such events. The land record, therefore, is incomplete.

- Long-term causes of ice ages include plate tectonics and changes in the concentration of CO_2 in the atmosphere. Short-term causes include the Milanković cycles (caused by periodic changes in Earth's orbit and tilt).

GEOPUZZLE REVISITED

Even if people had the ability to build large cities 12,000 years ago, they couldn't have built at the localities that are now New York or Edinburgh, for the land surface at these locations lay beneath hundreds of meters of ice. The last glacial advance, when huge ice sheets covered substantial areas of North America and Eurasia, ended 12,000 years ago.

Guide Terms

ablation (p. 751)
albedo (p. 744)
arête (p. 755)
basal sliding (p. 749)
cirque (p. 755)
continental glacier (ice sheet) (p. 747)
crevasse (p. 749)
drop stone (p. 755)
drumlin (p. 763)
equilibrium line (p. 751)
end moraine (p. 759)
erratic (p. 743)
esker (p. 766)
firn (p. 744)
fjord (p. 757)
glacial advance (p. 752)
glacial drift (p. 759)
glacial outwash (p. 760)
glacial rebound (p. 767)
glacial retreat (p. 752)
glacial striation (p. 755)
glacial subsidence (p. 766)
glacial till (p. 759)
glacially polished surface (p. 755)
glaciation (p. 774)
glacier (p. 743)
ground moraine (p. 766)
hanging valley (p. 757)

horn (p. 755)
ice age (p. 743)
iceberg (p. 752)
ice quake (p. 751)
ice sheet (p. 743)
ice shelf (p. 753)
interglacial (p. 774)
kame (p. 759)
kettle hole (p. 766)
lateral moraine (p. 759)
Little Ice Age (p. 779)
loess (p. 760)
medial moraine (p. 759)
Milanković cycles (p. 777)
moraine (p. 759)
mountain (alpine) glacier (p. 746)
patterned ground (p. 770)
permafrost (p. 770)
plastic deformation (p. 748)
Pleistocene Ice Age (p. 743)
pluvial lake (p. 770)
polar glacier (p. 748)
recessional moraine (p. 763)
roche moutonnée (p. 757)
sea ice (p. 753)
snowball Earth (p. 776)
sublimation (p. 744)
surge (p. 751)
tarn (p. 755)

temperate glacier (p. 747)
terminal moraine (p. 763)
tidewater glacier (p. 753)
tillite (p. 776)
toe (p. 752)
tundra (p. 773)

U-shaped valley (p. 755)
varve (p. 762)
zone of ablation (p. 751)
zone of accumulation (p. 751)

Review Questions

1. What evidence did Louis Agassiz offer to support the idea of an ice age?

2. How do mountain glaciers and continental glaciers differ in terms of dimensions, thickness, and patterns of movement?

3. Describe the transformation from snow to glacial ice.

4. Explain how arêtes, cirques, and horns form.

5. Describe the mechanisms that enable glaciers to move, and explain why they move.

6. How fast do glaciers normally move? How fast can they move during a surge?

7. Explain how the balance between ablation and accumulation determines whether a glacier advances or retreats.

8. How can a glacier continue to flow toward its toe even though its toe is retreating?

9. How does a glacier transform a V-shaped river valley into a U-shaped valley? Discuss how hanging valleys develop.

10. Describe the various kinds of glacial deposits. Be sure to note the materials from which the deposits are made and the landforms that result from deposition.

11. How do the crust and mantle respond to the weight of glacial ice?

12. How was the world different during the glacial advances of the Pleistocene Ice Age? Be sure to mention the relation between glaciations and sea level.

13. How was the standard four-stage chronology of North American glaciations developed? Why was it so incomplete? How was it modified with the study of marine sediment?

14. Were there ice ages before the Pleistocene? If so, when?

15. What are some of the long-term causes that lead to ice ages? What are the short-term causes that trigger glaciations and interglacials?

On Further Thought

16. If you fly over the barren cornfields of central Illinois during the early spring, you will see slight differences in soil color due to variations in moisture content—wetter soil is darker. These variations outline the shapes of polygons that are tens of meters across. What do these patterns represent, and how might they have formed? What do they tell us about the climate of central Illinois at the end of the last ice age?

17. Recent observations suggest that glaciers of southern Greenland have started to flow much faster in the past 15 years. Researchers suggest that this change might be a manifestation of the warming of the region's climate.

What mechanism could account for the acceleration of the glaciers?

18. An unusual late Precambrian rock unit crops out in the Flinders Range, a small mountain belt in South Australia, near Adelaide. Structures in the belt formed at the beginning of the Paleozoic. This unit consists of clasts of granite and gneiss, in a wide range of sizes, suspended through a matrix of slate. The clasts are now elliptical, with their long axes parallel to the plane of slaty cleavage. The rock unit lies unconformably above a basement of granite and gneiss, and if you dig out the unconformity surface, you will find that it is polished and striated. What is the "unusual rock," and why does it have cleavage and elliptical clasts? (Hint: Look back at Chapter 11.)

 For more resources, including animations, quizzes, and Norton's GeoTours, go to **wwnorton.com/studyspace**.

 If your instructor assigns exercises in SmartWork, log in at **smartwork.wwnorton.com**.

ANOTHER VIEW This digital elevation model (DEM), produced from Space Shuttle radar measurements, illustrates how the Malaspina Glacier and its neighbors spill out of mountain valleys in southeastern Alaska and spread out onto the coastal plain. They carry huge amounts of debris, which build into moraines.

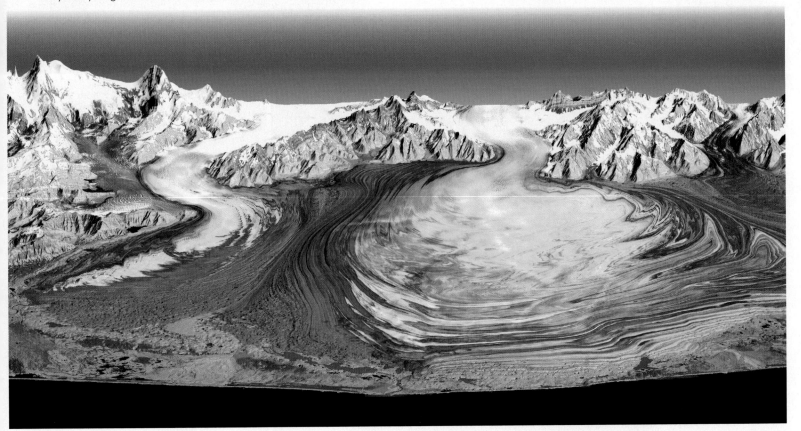

CHAPTER **23**

Global Change in the Earth System

This patchwork of farm fields in central France replaces what was once dense forest. The view emphasizes that in about 0.00002% of Earth's history, humanity has modified much of our planet's surface. We have become a major agent of global change.

GEOPUZZLE

Climate change has been in the news almost daily during the past few years. Why has there been increasing interest in this issue? Is climate the only aspect of the Earth System that changes through time?

All we in one long caravan
are journeying since the world began,
we know not whither, we know . . . all must go.

—Bhartrihari
(Indian poet, c. 500 C.E.)

23.1 INTRODUCTION

Did the Earth's surface look the same in the Jurassic Period as it does today? Definitely not—the Earth of 200 million years ago differed from that of today in many ways. During the Jurassic, the North Atlantic Ocean was a narrow sea and the South Atlantic Ocean didn't exist at all, so most dry land connected to form a single vast continent (Fig. 23.1). Today, both parts of the Atlantic are wide oceans, and the Earth has seven separate continents. Moreover, during the Jurassic, the call of the wild rumbled from the throats of dinosaurs, whereas today, the largest land animals are mammals. In essence, what we see of the Earth today is just a snapshot, an instant in the life story of a *constantly* changing planet. This idea arguably stands as geology's greatest philosophical contribution to humanity's understanding of our Universe.

Why has the Earth changed so much over geologic time, and why can it continue to change? Ultimately, change happens both because the Earth's internal heat makes the asthenosphere weak enough to flow, and because the Sun's heat keeps most of the Earth's surface at temperatures above the freezing point of water. Flow in the asthenosphere permits plate tectonics (sea-floor spreading, subduction, and transform motion), which in turn leads to continental drift, volcanism, and mountain building. These phenomena produce and modify rocks and generate topography. In addition, volcanism provides raw materials of the gases and liquids from which the atmosphere and oceans form. The presence of liquid water and an atmosphere permits a variety of phenomena ranging from the evolution of life to weathering and erosion. Further, the interaction between internally driven processes (plate tectonics) and externally driven processes (atmospheric circulation) leads to the hydrologic cycle and the rock cycle, among many other phenomena. Biological and physical phenomena also interact. For example, photosynthetic organisms affect the composition of the atmosphere by producing oxygen, and atmospheric composition, in turn, determines the nature of chemical weathering in rocks.

No other object in our solar system has a mobile asthenosphere and a surface whose temperature straddles the freezing point of water, so no other object of our solar system undergoes the kinds of changes that Earth does. The Moon, for example, changes so little through time that it would have looked essentially the same to a Jurassic dinosaur as it does to you. For purposes of discussion, we refer to the global interconnecting web of physical and biological phenomena on Earth as the **Earth System**, and we define **global change** as the transformations or modifications of physical and biological components of the Earth System over time (see **Geology at a Glance**, pp. 786–787).

Geologists distinguish among different types of global change, on the basis of the rate or way in which change progresses with time. *Gradual* change takes place over long periods of geologic time (millions to billions of years). *Catastrophic* change takes place relatively rapidly in the context of geologic time (seconds to millennia). *Unidirectional* change involves transformations that never repeat. *Cyclic* change repeats the same steps over and over, though not necessarily with the same results. Some types of cyclic change are periodic, in that the cycles happen with a definable frequency; others are not.

In this chapter, we begin by reviewing examples of global change involving phenomena discussed earlier in the book. Then we introduce the concept of a biogeochemical cycle, the exchange of chemicals among living and nonliving reservoirs, for some kinds of global change reflect changes in the proportions of chemicals held in different reservoirs through time. Finally, we focus on global climate change, transformations or modifications in Earth's climate over time, some of which appear to be anthropogenic (human-caused). We conclude this chapter, and this book, by considering hypotheses that describe the ultimate global change—the end of the Earth in the very distant future.

FIGURE 23.1 The map of Earth's surface changes over time because of plate motions.

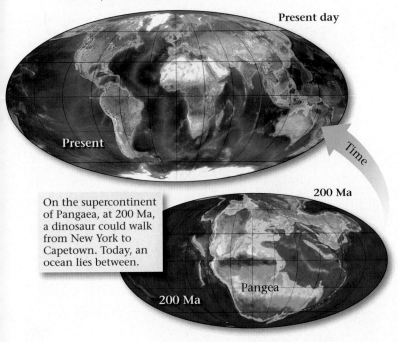

Present day

Present

Time

200 Ma

On the supercontinent of Pangaea, at 200 Ma, a dinosaur could walk from New York to Capetown. Today, an ocean lies between.

Pangea

200 Ma

The Earth System

External energy

Sun

Thunderhead

Lightning

Mountain uplift

Rain and snow

Continental glacier

City

Ocean

Rocky coastline

Desert

Valley

Arid mountains

Mining

Field pattern

Lakes

Deciduous forest

Beach

Forested mountains

The Earth's surface is the interface among the solid Earth (lithosphere); the ice and liquid water of oceans, lakes, streams, groundwater, and glaciers (the hydrosphere); and the planet's gaseous envelope (the atmosphere). Countless species of life, ranging from nearly invisible bacteria to giant whales and trees, make up the complex ecosystems of Earth's biosphere. All of these components— the lithosphere, hydrosphere, atmosphere, and biosphere—interact with each other. These components and the interactions among them consti- tute the Earth System.

Various materials cycle among living and nonliving components of the Earth System. In the hydrologic cycle, for example, water evapo- rates from the sea, rains on the land, and eventually flows back to the sea. During this process, water may be trapped temporarily in living organ- isms, clouds, subsurface pores, or ice sheets. Carbon dioxide can be stored in the air, dissolved in water, or

Tropical rain forest

Shark

Coral reef

Internal energy

786

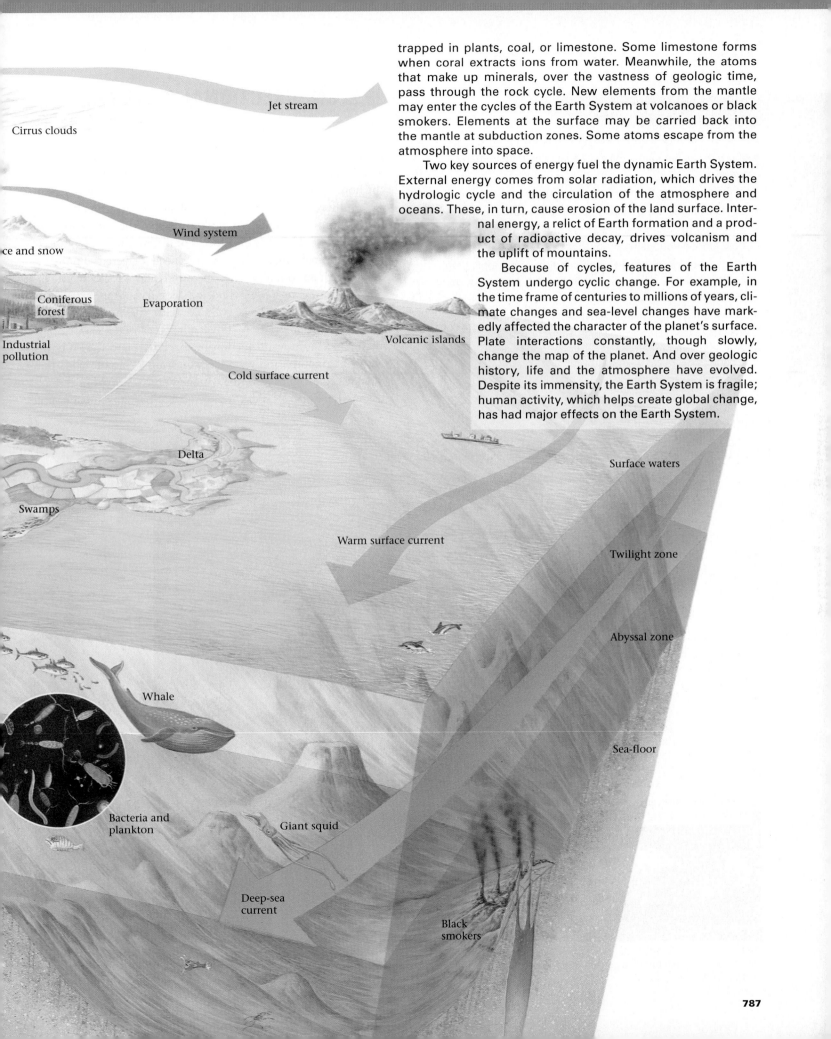

Cirrus clouds

Jet stream

Wind system

ce and snow

Coniferous forest

Evaporation

Industrial pollution

Volcanic islands

Cold surface current

Delta

Swamps

Warm surface current

Surface waters

Twilight zone

Abyssal zone

Whale

Sea-floor

Bacteria and plankton

Giant squid

Deep-sea current

Black smokers

trapped in plants, coal, or limestone. Some limestone forms when coral extracts ions from water. Meanwhile, the atoms that make up minerals, over the vastness of geologic time, pass through the rock cycle. New elements from the mantle may enter the cycles of the Earth System at volcanoes or black smokers. Elements at the surface may be carried back into the mantle at subduction zones. Some atoms escape from the atmosphere into space.

Two key sources of energy fuel the dynamic Earth System. External energy comes from solar radiation, which drives the hydrologic cycle and the circulation of the atmosphere and oceans. These, in turn, cause erosion of the land surface. Internal energy, a relict of Earth formation and a product of radioactive decay, drives volcanism and the uplift of mountains.

Because of cycles, features of the Earth System undergo cyclic change. For example, in the time frame of centuries to millions of years, climate changes and sea-level changes have markedly affected the character of the planet's surface. Plate interactions constantly, though slowly, change the map of the planet. And over geologic history, life and the atmosphere have evolved. Despite its immensity, the Earth System is fragile; human activity, which helps create global change, has had major effects on the Earth System.

Chapter Themes

By the end of this chapter, you should know . . .

- the Earth is a dynamic planet and can change in many ways.
- some changes are unidirectional, whereas others are cyclic.
- some changes are gradual and some are catastrophic.
- during biogeochemical cycles, elements or compounds flow among various reservoirs.
- the climate has changed significantly over geologic time, and greenhouse gases play an important role in regulating the climate.
- humans have significantly changed aspects of the Earth's system.
- addition of greenhouse gases during the past two centuries appears to be causing warming.

23.2 UNIDIRECTIONAL CHANGES

The Evolution of the Solid Earth

Recall from Chapter 1 that Earth began as a fairly homogenous mass, formed by the coalescence of planetesimals. But the homogeneous proto-Earth did not last long—within about 100 million years of its birth the planet began to melt, yielding a liquid iron alloy that sank rapidly to the center to form the core (Fig. 23.2a). This process of differentiation represents major unidirectional change: it produced a layered, onion-like planet with an iron alloy core surrounded by a rocky mantle.

According to a well-suported model, a Mars-sized protoplanet appears to have collided with the newborn Earth soon after differentiation. This collision caused a catastrophic change—a significant portion of the Earth and the colliding object fragmented and vaporized, creating a ring of debris that quickly coalesced to form the Moon (Fig. 23.2b). After the collision, the Earth's mantle was probably partially molten, and

FIGURE 23.2 Examples of major unidirectional change in Earth history.

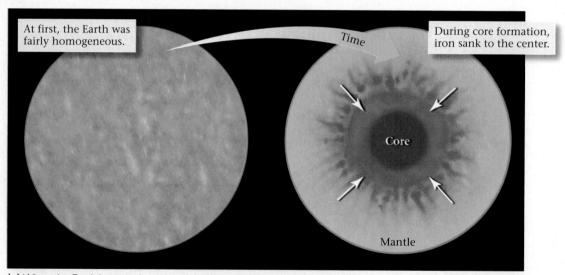

(a) When the Earth became hot enough inside, rock comprising it started to melt, and droplets of iron sank to the center, accumulating to form a core. Once core formation happened, it could not happen again.

(b) Moon formation happened early during Earth history, but probably after core formation. According to a popular theory, the Moon coalesced from debris resulting from the collision of a Mars-sized protoplanet with the Earth.

the planet's surface became a sea of magma. The Earth endured intense bombardment by asteroids and comets between about 4.0 and 3.9 Ga, so any crust that had formed prior to 3.9 Ga was largely pulverized or melted. Eventually, however, bombardment ceased and our planet gradually cooled, permitting a crust to form at its surface and plate tectonics to begin operating. Subduction, and/or the upwelling of mantle plumes, produced relatively low-density rocks such as granite. These rocks could not be subducted and thus remained buoyant blocks at the Earth's surface. Plate motion eventually caused these buoyant blocks to collide and suture together, forming the first continents. The amount of continental crust has increased over time; most appeared by the end of the Archean.

Overall, therefore, the transition from the Hadean Eon to the Archean Eon saw remarkable unidirectional change in the nature of the Earth. By early Archean time, our planet had distinct continents and ocean basins, and thus looked radically different from the other terrestrial planets (see Chapter 13).

The Evolution of the Atmosphere and Oceans

Like its surface, the Earth's atmosphere has also changed over time. Partial melting in the mantle produced magma and also released large quantities of gases that belched from volcanoes. More gases may have arrived when comets collided with our new planet. Eventually, Earth accumulated an early atmosphere composed dominantly of carbon dioxide (CO_2) and water (H_2O). Other gases, such as nitrogen (N_2), composed only a minor proportion of the early atmosphere. When the Earth's surface cooled, however, water condensed and fell as rain, collecting in low areas to form oceans. This may have happened before 4.0 Ga, but it had certainly happened by 3.8 Ga. Gradually, CO_2 dissolved in the oceans and was absorbed by chemical-weathering reactions on land, so its concentration in the atmosphere decreased. Nitrogen, which doesn't react with other chemicals, was left behind. Thus, the atmosphere's composition changed to become dominated by nitrogen. Photosynthetic organisms appeared early in the Archean. But it probably wasn't until between 2.5 and 2.0 Ga, the early Proterozoic, that oxygen (O_2) became a significant proportion of the atmosphere. Present concentrations of oxygen may have existed only for the past 400 million years.

Did you ever wonder ... has the Earth's atmosphere always been breathable?

The Evolution of Life

During most of the Hadean Eon, Earth's surface was probably lifeless, for carbon-based organisms could not survive the high temperatures of the time. The fossil record indicates that life had appeared at least by 3.8 billion years ago and has undergone unidirectional change (evolution) in fits and starts ever since (see Interlude E). Though simple organisms such as archaea and bacteria still exist, life evolution during the late Proterozoic and early Phanerozoic yielded multicellular plants and animals (Fig. 23.3). Life now inhabits regions from a few kilometers below the surface to a few kilometers above, yielding a diverse biosphere.

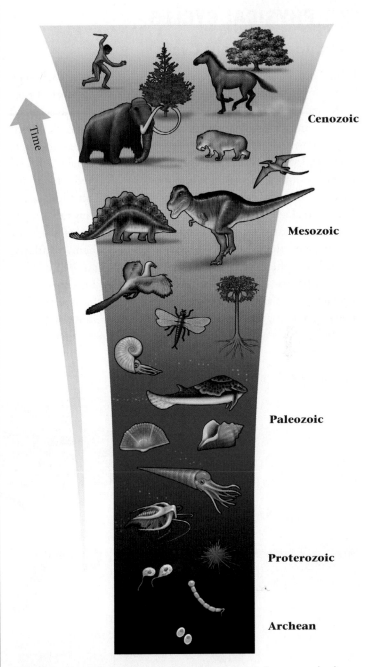

FIGURE 23.3 New species of life have evolved over geologic time. Though some of the simplest still exist, more complex organisms have appeared more recently.

- A unidirectional change is a transformation that doesn't repeat.
- Examples of unidirectional changes include formation of the core, accumulation of water in the oceans, atmospheric evolution, and life evolution.

THINK: How are life evolution and atmosphere evolution linked?

23.3 PHYSICAL CYCLES

The Supercontinent Cycle

Today, our planet's surface hosts seven continents, each separated from its neighbors by ocean basins. At discrete times in the past, however, almost all continental crust was sutured together into a single supercontinent. Indeed, the rock record suggests that Earth's history has been punctuated multiple times (0.3 Ga; 0.6 Ga; 1.1 Ga; 2.1 Ga; and perhaps 2.7 Ga) by the assembly of supercontinents. Once formed, supercontinents don't last forever. After less than 200 million years, rifting breaks a supercontinent apart, yielding several smaller continents that then drift away from each other as sea-floor spreading produces new ocean basins between them. Similarly, ocean lithosphere can't last forever. In less than 200 million years, it begins to sink back into the mantle. This process of subduction consumes ocean basins and brings the smaller continents back together until they eventually collide with one another to form a new supercontinent. Geologists refer to such

alternating episodes of continental dispersal and reassembly as the **supercontinent cycle** (Fig. 23.4). Because plates move at only 1 to 15 cm per year, the changes to Earth's surface that result from the supercontinent cycle take a very long time.

Notably, during the supercontinent cycle, ocean basins do not simply open and close like accordions; plate motions tend to be more complex. Thus, the relative positions of individual crustal blocks comprising one supercontinent won't be the same in another. For example, South American crust lay adjacent to African crust in Pangaea (0.3 Ga), whereas it lay to the east of North American crust in Rodinia (1.1 Ga), as illustrated in Chapter 13.

The Sea-Level Cycle

Global sea level rose and fell by as much as 300 m during the Phanerozoic, and likely did the same in the Precambrian. When sea level rises, the shoreline migrates inland, and low-lying plains in the continents become submerged. In fact, during periods of particularly high sea level, more than half of Earth's continental area can be covered by shallow seas; at such times, sediment buries continental regions (Fig. 23.5a, b). When sea level falls, the continents become dry again, and regional unconformities develop. For example, the sedimentary strata of the midwestern United States record at least six continent-wide advances and retreats of the sea, each of which left behind a blanket of sediment called a **sedimentary sequence**; unconformities define the boundaries between the sequences (Fig. 23.5c). Of note, the sequence deposited during the Pennsylvanian contains at least 30 shorter repeated

FIGURE 23.4 The stages of the supercontinent cycle.

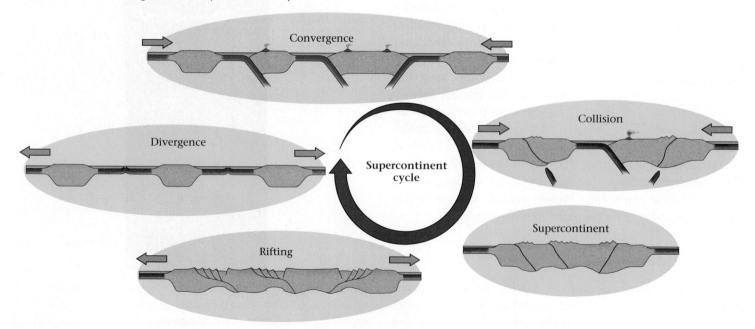

FIGURE 23.5 Sea-level change, and its manifestations, over geologic time.

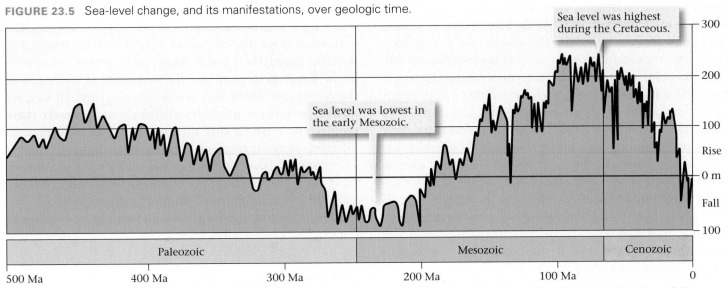

Sea level was highest during the Cretaceous.

Sea level was lowest in the early Mesozoic.

Rise

0 m

Fall

Paleozoic | Mesozoic | Cenozoic

500 Ma — 400 Ma — 300 Ma — 200 Ma — 100 Ma — 0

(a) This chart provides one interpretation of sea-level change during the past half-billion years, based on the stratigraphic record. There is not full agreement about this interpretation; it remains the subject of research.

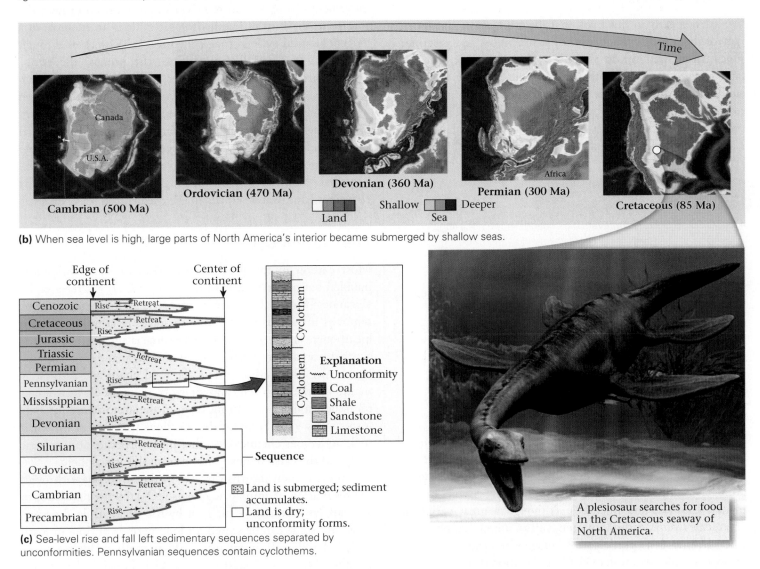

Cambrian (500 Ma)

Ordovician (470 Ma)

Devonian (360 Ma)

Permian (300 Ma)

Cretaceous (85 Ma)

Shallow | Deeper
Land | Sea

(b) When sea level is high, large parts of North America's interior became submerged by shallow seas.

A plesiosaur searches for food in the Cretaceous seaway of North America.

(c) Sea-level rise and fall left sedimentary sequences separated by unconformities. Pennsylvanian sequences contain cyclothems.

intervals, called cyclothems, each of which contains a specific succession of sedimentary beds. At the base of each cyclothem you'll find sandstone, and in the middle you'll find coal. Cyclothems represent short-term cycles of sea-level rise and fall.

After studying sedimentary sequences around the world, geologists at Exxon Corporation pieced together a chart defining the succession of global transgressions and regressions during the Phanerozoic Eon. The global sedimentary cycle chart may largely reflect the cycles of eustatic (worldwide) sea-level change. However, the chart probably does not give us an exact image of sea-level change, because the sedimentary record reflects other factors as well, such as changes in sediment supply. Eustatic sea-level changes may be due to a variety of factors, including advances and retreats of continental glaciers, changes in the volume of mid-ocean ridge systems, and changes in continental elevation and area.

The Rock Cycle

We learned early in this book that the crust of the Earth consists of three rock types: igneous, sedimentary, and metamorphic. Atoms making up the minerals of one rock type may later become part of another rock type. In effect, rocks are simply reservoirs of atoms, and the atoms move from reservoir to reservoir over time. As we learned in Interlude C, this process is the rock cycle. Each stage in the rock cycle changes the Earth by redistributing and modifying material.

Take-Home Message

- Physical cycles refer to reversible changes in the nonliving aspects of the Earth.

- Examples include the rise and fall of sea level (the sea-level cycle); the assembly, breakup, and reassembly of continents (the supercontinent cycle); and the movement of atoms among different rock types (the rock cycle).

THINK: What evidence indicates that sea level rises and falls over time?

23.4 BIOGEOCHEMICAL CYCLES

A **biogeochemical cycle** involves the passage of a chemical among nonliving and living reservoirs in the Earth System, mostly on or near the surface. Nonliving reservoirs include the atmosphere, the crust, and the ocean; living reservoirs include plants, animals, and microbes. Although a great variety of chemicals (water, carbon, oxygen, sulfur, ammonia, phosphorus, and nitrogen) participate in biogeochemical cycles, here we look at only two: water (H_2O) and carbon (C).

Some stages in a biogeochemical cycle may take only hours, some may take thousands of years, and others may take millions of years. The transfer of a chemical from reservoir to reservoir during these cycles doesn't really seem like a change in the Earth in the way that the movement of continents or the metamorphism of rock seems like a change. In fact, for intervals of time, biogeochemical cycles attain a **steady-state condition**, meaning that the proportions of a chemical in different reservoirs remain fairly constant even though there is a constant flux (flow) of the chemical among reservoirs. When we speak of global change in a biogeochemical cycle, we mean a change in the relative proportions of a chemical held in different reservoirs at a given time—in other words, a change in the steady-state condition.

The Hydrologic Cycle

As we learned in Interlude F, the hydrologic cycle involves the movement of water from reservoir to reservoir on or near the surface of the Earth. The hydrologic cycle is an example of a biogeochemical cycle in that a chemical (H_2O) passes through both nonliving and living entities—the oceans, the atmosphere, surface water, groundwater, glaciers, soil, and living organisms. Global change in the hydrologic cycle occurs when a change in global climate alters the ratio between the amount of water held in the ocean and the amount held in continental ice sheets. For example, during an ice age, water that had been stored in oceans moves into glacial reservoirs. Thus, the continents become covered with ice, and sea level drops. When the climate warms, water returns to the oceans, and sea level rises.

The Carbon Cycle

Most carbon in the near-surface realm of Earth originally bubbled out of the mantle in the form of CO_2 gas released by volcanoes (Fig. 23.6). Once it enters the atmosphere, it can be removed in various ways. Some dissolves in seawater to form bicarbonate (HCO_3^-) ions, whereas some is absorbed by photosynthetic organisms that convert it into sugar and other organic chemicals. This carbon enters the food chain and ultimately makes up the flesh, fat, and sinew of animals. In fact, about 63 billion tons of carbon move from the atmosphere into life forms every year. Some of the reactions that take place when rock undergoes chemical weathering incorporate atmospheric CO_2 and thus also remove carbon from the atmosphere.

Some carbon returns directly to the atmosphere through the respiration of animals (again as CO_2), by the flatulence of animals (as methane [CH_4]), or by the decay of dead organisms. But some can be stored for long periods of time in fossil fuels (oil and coal), in organic shale, in methane hydrates (see Chapter 14), or in limestone ($CaCO_3$). Fossil fuel deposits,

FIGURE 23.6 In the carbon cycle, carbon transfers among various reservoirs at or near the Earth's surface. Red arrows indicate release to the air, and green arrows indicate absorption from air.

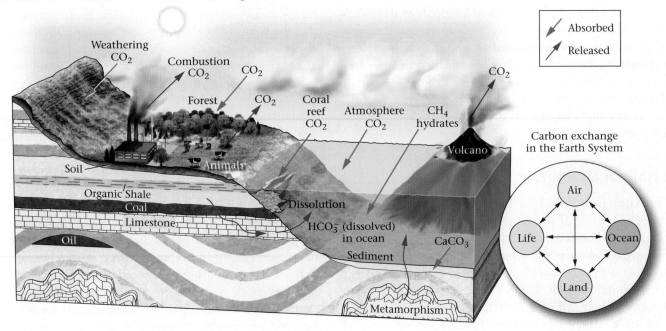

limestone, methane hydrates, and the organic portion of shale contain most of the carbon in the near-surface realm of Earth and can hold on to it for long periods of time. But this carbon either returns to the atmosphere in the form of CO_2, as a result of the burning of fossil fuels and the metamorphism of rocks containing carbonate, or returns to the sea after undergoing chemical weathering followed by dissolution as HCO_3^- in river water or groundwater. Melting of methane hydrates releases methane to the atmosphere.

Take-Home Message

- During a biogeochemical cycle, atoms transfer among different living and nonliving reservoirs.
- The hydrologic cycle involves movement of water from ocean to air to land and back to ocean.
- During the carbon cycle, carbon moves from underground reservoirs (such as fossil fuels), into the air, then into the oceans or living organisms, and eventually back underground.

THINK: What do we mean by a "reservoir" in the context of biogeochemical cycles?

23.5 GLOBAL CLIMATE CHANGE

How often have you seen a newspaper proclaim, "Record High Temperatures!"? In the summer of 2011, such a claim became reality for much of the United States, where thermometers registered weeks of temperatures several degrees above "normal." Places that rarely endured "heat waves" saw many days during which temperatures exceeded 38°C (100°F). Does this mean that the climate—the average range of weather conditions for a given region—is changing?

As discussed in Chapter 20, the atmospheric conditions during a specific time interval in a given region define the region's *weather* for the time interval. The term *climate* indicates overall weather conditions as well as daily to seasonal variability of weather conditions over a period of many decades. So a newspaper's headline about a single hot spell or cold snap does not mean that the climate is changing. But if a new set of conditions—say, increased *average* temperature, rising snow line, a longer growing season, or a change in storm frequency or intensity—becomes the new norm for a region, then climate change has occurred. The stratigraphic record clearly shows that **global climate change** (the transformation of Earth's climate over time) has happened repeatedly throughout Earth's history for natural reasons. Recent measurements indicate that global climate change is now occurring, and that human activities may be contributing to this change.

In this section, we begin by reviewing the fundamental role that greenhouse gases play in regulating atmospheric temperature. Then, we describe how researchers study past climates, how the climate has changed over geologic time, and why most researchers have concluded that global climate change over the last two centuries reflects human activities. We finish by discussing possible implications of current climate change. For purposes of discussion, we distinguish between *long-term* climate change, which takes place over millions to tens of millions of years, and *short-term* climate change, which takes place over tens to hundreds of thousands of years. If the average atmospheric and sea-surface temperature rises, we have **global warming**, and if it falls, we have **global cooling**. Some changes are great enough to cause oceanic islands and large regions of continents to be submerged by shallow seas or to be covered by ice, whereas others are subtle, creating only a slight latitudinal shift in vegetation belts and a sea-level change measured in meters or less.

The Role of Greenhouse Gases

The Sun constantly bathes the Earth in visible light. Some of this energy reflects off the atmosphere and the Earth's surface, so that our planet shines when viewed from space. The Earth's surface absorbs the remainder of the incoming visible light, and then releases it in the form of thermal energy (infrared radiation) that heads back upward. If the Earth had no atmosphere, all of this thermal energy would escape back into space. But our planet does have an atmosphere, and certain gases (H_2O, CO_2, CH_4, NO_2, and O_3) in air absorb thermal radiation and *re-radiate* it. Some of the re-radiated energy continues up into space, but some heads downward and warms the lower atmosphere (Fig. 23.7; see Chapter 20). In effect, these gases trap infrared radiation and keep the lower atmosphere warm, somewhat like the way glass traps heat in a greenhouse. Thus, the overall trapping process is called the **greenhouse effect**, and the gases that cause it are called **greenhouse gases**.

Researchers estimate that were it not for the greenhouse effect, global average surface temperature of the Earth would be about 33°C lower than it is today. Measurements indicate that the average global temperature between 1901 and 2000 was about 14°C, so an Earth without greenhouse gases would have an average global temperature of about –19°C, and our planet's surface would be a frozen wasteland. Because greenhouse gases trap heat, any process that transfers these gases from underground, oceanic, or biomass reservoirs into the atmospheric reservoir will cause the climate to warm. Similarly, any process that removes greenhouse gases from the atmospheric reservoir and transfers them in biomass, or into oceanic or underground reservoirs, will cause the climate to cool.

Of the various greenhouse gases in the atmosphere, H_2O occurs in the greatest quantity, and plays the biggest role in the greenhouse effect (30 to 70%). While H_2O is the dominant cause of the greenhouse effect, researchers emphasize that changes in its overall concentration *depend* on changes in atmospheric temperature; an increase in H_2O concentration cannot independently cause atmospheric warming. Why? Because atmospheric temperature determines the maximum amount of H_2O that can be held in the atmosphere. If the relative humidity gets too high at a location, H_2O simply rains out of the atmosphere and falls on the Earth's surface (see Chapter 20). H_2O molecules, on average, remain in the atmosphere for only nine days (i.e., the "residence time" of water is nine days), such a short time that they can't mix thoroughly with other gases in the atmosphere. For this reason, humidity in a desert region can differ radically from that of a temperate region a few hundred kilometers away.

FIGURE 23.7 The greenhouse effect shows how thermal energy can be trapped in the atmosphere.

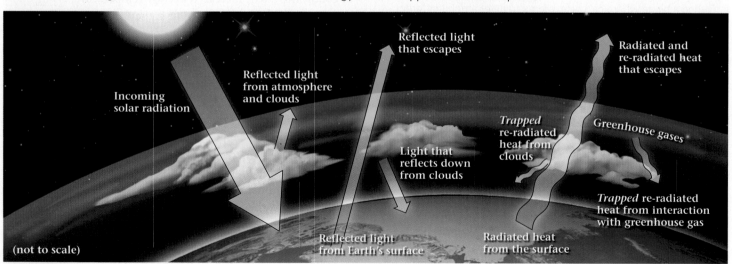

Of the other greenhouse gases, CO_2 and CH_4 are the most significant in influencing temperature. A CO_2 molecule has about 20 times the efficiency of an H_2O molecule, and a CH_4 molecule has about 70 times the efficiency of a CO_2 molecule, in absorbing and re-radiating infrared radiation. So even though these gases occur in very low concentrations, they contribute significantly to the greenhouse effect (9 to 30%, and 4 to 9%, respectively). Also, these gases have relatively long residence times, so they mix thoroughly with other gases in the lower atmosphere. As a consequence, addition of CO_2 and CH_4 to the atmosphere from another biogeochemical reservoir can increase the concentration of these gases in the atmosphere long enough to cause overall average atmospheric temperature to increase.

Warming or cooling due to changes in the concentration of greenhouse gases can be amplified by positive or negative **feedback mechanisms**. *Negative* feedback slows a process down or even reverses it, whereas *positive* feedback enhances a process and amplifies its consequences. Let's consider an example of a feedback mechanism in the greenhouse effect. Imagine that global average atmospheric temperature has increased due to an increase in CO_2. This increase will, in turn, cause the oceans to warm and evaporate more, so more water transfers into the atmospheric reservoir globally. The increase also causes some of the CO_2 dissolved in the oceans to come out of solution and return to the atmosphere, and it causes CH_4 to enter the atmosphere from decay of organic matter in melting permafrost. Thus, the warming due to the initial addition of CO_2 can cause the concentration of other greenhouse gases to increase overall, forcing the atmosphere to warm even more than it would have in the first place—this is a positive feedback.

Methods of Study

Geologists and climatologists are working hard to define the nature of climate change, the rates at which change can take place, and the effects that change may have on our planet. There are two basic approaches to studying global climate change: (1) researchers measure past climate change, as indicated by the stratigraphic record, to document the magnitude of changes that are possible and the rate at which such changes occurred; (2) researchers develop computer programs to calculate how factors such as atmospheric composition, topography, ocean currents, and Earth's orbit affect the circulation of the atmosphere and, therefore, the distribution of climate belts. They then use such **general circulation models (GCMs)** to develop **climate-change models** that seek to provide insight into when and why changes took place in the past and whether they will happen in the future. Climate-change models try to

predict changes in rainfall, sea level, ice cover, and other physical features that may be influenced by the warming or cooling of the atmosphere.

Let's look first at how geologists study the **paleoclimate** (past climate), so as to document climate changes throughout Earth history. Any feature whose character depends on the climate and whose age can be determined can be a clue to defining paleoclimate.

> **Did you ever wonder . . .**
> how do researchers study past climates on Earth?

- *The stratigraphic record*: The nature of sedimentary strata deposited at a certain location reflects the climate at that location. For example, an outcrop exposing cross-bedded sandstone, overlain successively by coal and glacial till, indicates that the site of the outcrop has endured different climates (desert, then tropical, then glacial) over time.

- *Paleontological evidence*: Different assemblages of species survive in different climatic belts. Thus, the succession of fossils in a sedimentary sequence provides clues to the changes in climate at that site. For example, a record of short-term climate change can be obtained by studying the succession of plankton fossils in sea-floor sediments, for cold-water species of plankton are different from warm-water species. Fossil pollen also yields clues to the paleoclimate. Pollen, tiny grains involved in plant reproduction, looks like dust to the unaided eye. But under a microscope, each grain has a distinctive structure, and grains of one species look different from grains of another species (Fig. 23.8a). Further, pollen grains have a tough coating and can survive burial. By studying pollen in sediment, palynologists (scientists who study pollen) can determine whether the sediment accumulated in a cold-climate coniferous forest or in a warm-climate deciduous forest. And by recording changes in the pollen assemblages found in successive layers of sediment, palynologists can track the movement of climate belts over the landscape (Fig. 23.8b). Such studies show that spruce forests, indicative of cool climates, have slowly migrated north since the last ice age.

- *Oxygen–isotope ratios*: Two isotopes of an element have the same atomic number but different atomic weights. For instance, oxygen occurs as ^{16}O (8 protons and 8 neutrons) and ^{18}O (8 protons and 10 neutrons). Geologists have found that the ratio of ^{18}O to ^{16}O in glacial ice indicates the atmospheric temperature in which the snow that made up the ice formed: simplistically, the ratio is larger in snow that forms in warmer air,

FIGURE 23.8 Changes in the assemblage of pollen in sediment indicates a shift in climate belts.

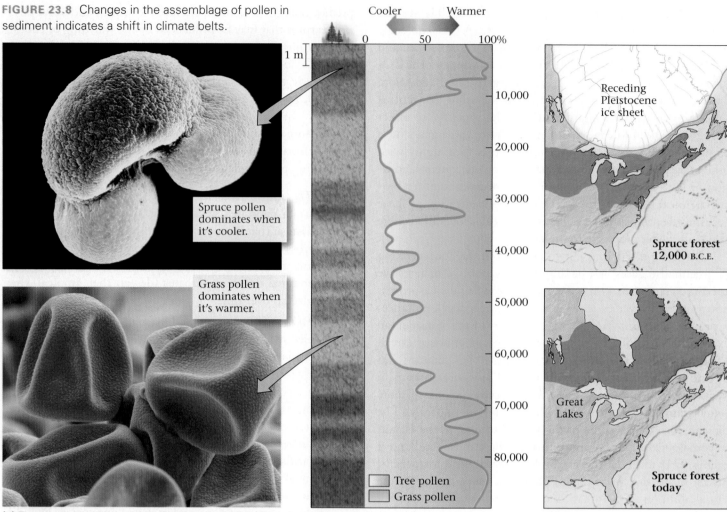

(a) The proportion of tree pollen relative to grass pollen can change in a sedimentary sequence through time. Researchers plot changes in the proportion of pollen types over time by examining samples from a column of sediment.

(b) Spruce forests (green) grew farther south 12,000 years ago than they do today.

but smaller in snow that forms in colder air. Because of this relationship, the isotope ratio of the oxygen in H_2O measured in a succession of ice layers in a glacier indicates temperature change over time. Researchers have now obtained ice cores down to a depth of almost 3 km in Antarctica and in Greenland; this record spans up to 720,000 years (Fig. 23.9a). For similar reasons, the $^{18}O/^{16}O$ ratio in the $CaCO_3$ making up plankton shells also gives geologists an indication of past temperatures. Thus, measurement of oxygen–isotope ratios in drill cores of marine sediment extends the record of temperature change back over millions of years (Fig. 23.9b, c).

- *Bubbles in ice*: Bubbles in ice trap the air present at the time the ice forms. By analyzing these bubbles, geologists can measure the concentration of CO_2 in the atmosphere back through time. This information can be used to correlate CO_2 concentration with past

atmospheric temperature. The CO_2 record has been extended back through 240,000 years.

- *Growth rings*: If you've ever looked at a tree stump, you'll have noticed the concentric rings visible in the wood. Each ring represents one year of growth, and the thickness of the ring indicates the rate of growth in a given year. Trees grow faster during warmer, wetter years and more slowly during cold, dry years (Fig. 23.10a). Thus, the succession of ring widths provides an easily calibrated record of climate during the lifetime of the tree. Bristlecone pines supply a record back through 4,000 years. To go further into the past, dendrochronologists (scientists who study tree rings) look at the record of rings in logs dated by the radiocarbon technique or in logs whose ages overlap with that of the oldest living tree. Growth rings in corals and shells can provide similar information.

FIGURE 23.9 The proportion of isotopes transferred between reservoirs during evaporation or precipitation depends on temperature. The $^{18}O/^{16}O$ ratio can be studied in glacial ice (H_2O) and fossil shells ($CaCO_3$).

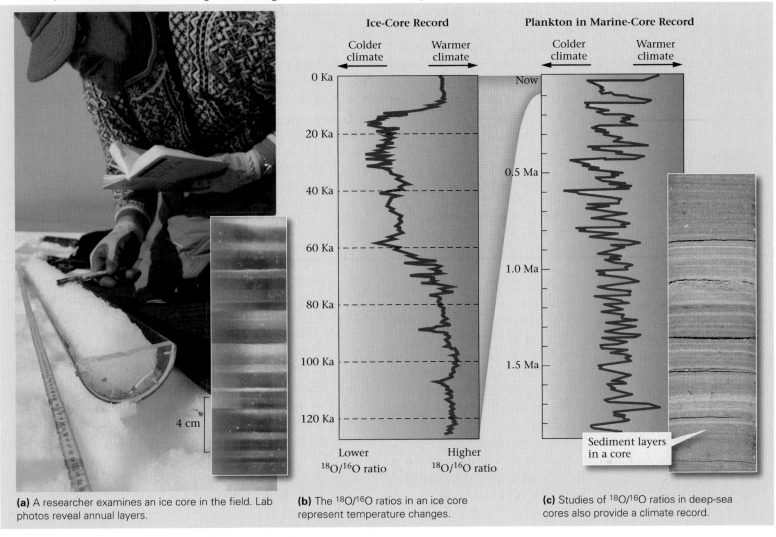

(a) A researcher examines an ice core in the field. Lab photos reveal annual layers.

(b) The $^{18}O/^{16}O$ ratios in an ice core represent temperature changes.

(c) Studies of $^{18}O/^{16}O$ ratios in deep-sea cores also provide a climate record.

FIGURE 23.10 Records of recent climate change.

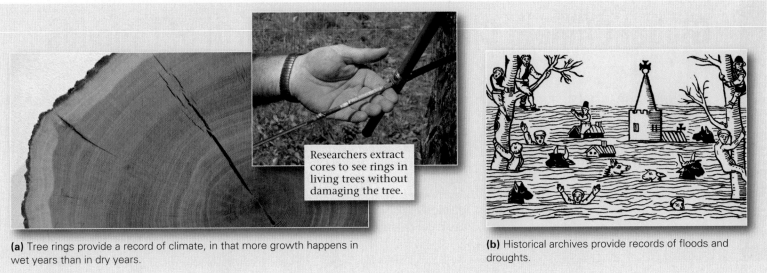

(a) Tree rings provide a record of climate, in that more growth happens in wet years than in dry years.

(b) Historical archives provide records of floods and droughts.

- *Human history*: Researchers have been able to make careful, direct measurements of climate changes only for the past few decades. This record is not long enough to document long-term climate change. But history, both written and archaeological, contains important clues to climates at times in the past. Periods of unusual cold or drought leave an impression on people, who record them in paintings, stories, and records of crop success or failure (Fig. 23.10b; Box 23.1).

Long-Term Climate Change

Using the variety of techniques described above, geologists have reconstructed an approximate record of global climate, represented by mean temperature, for geologic time. The record shows that Earth's surface temperature has stayed well within the freezing point of water and the boiling point of water since the beginning of the Archean. But at times in the past, the Earth's atmosphere was significantly warmer than it is today; whereas, at other times, it was significantly cooler. The warmer periods have come to be known as **greenhouse** (or hothouse) **periods** and the colder as **icehouse periods**. (The more familiar term, *ice age*, refers to the times during an icehouse period when the Earth was cold enough for ice sheets to advance and cover substantial areas of the continents.) As the chart in Figure 23.11a shows, there have been at least five major icehouse periods during geologic time.

Let's look a little more closely at the climate record of the last 100 million years, for this time interval includes the transition between a greenhouse and an icehouse period. Paleontological and other data suggest that the climate of the Mesozoic Era, the

Age of Dinosaurs, was much warmer than the climate of today. At the equator, average annual temperatures may have been 2° to 6°C warmer, while at the poles, temperatures may have been 20° to 60°C warmer. In fact, during the Cretaceous Period, dinosaurs were able to live at high latitudes, and there were no polar ice caps on Earth. But starting about 80 million years ago, the Earth's atmosphere began to cool. It warmed up for about 10 m.y., during the "Eocene climatic optimum," and then about 33 million years ago, we entered an icehouse period and the Antarctic ice sheet formed (Fig. 23.11b). The climate reached its coldest condition about 2 million years ago, during the Pleistocene Ice Age.

What caused long-term global climate changes? The answer may lie in the complex relationships among the various solar, geologic, and biogeochemical cycles of the Earth System, as described earlier (See also Box 23.2). For example:

- *Positions of continents*: Continental drift influences the climate by controlling the pattern of oceanic currents, which redistribute heat around the planet's surface (Fig. 23.12a). Drift also determines whether the land is at high or low latitudes (and thus how much solar radiation strikes it), and whether or not there are large continental interior regions where extremely cold winter temperatures can develop.
- *Volcanic activity*: A long-term global increase in volcanic activity may contribute to long-term global warming, because it increases the concentration of CO_2 in the atmosphere. For example, when Pangaea broke up during the Cretaceous Period, numerous rifts formed, and sea-floor-spreading rates were particularly high, so volcanoes were more abundant than they are today.

BOX 23.1

Global Climate Change and the Birth of Legends

Some geologists suggest that myths passed down from the early days of civilization may have their roots in global climate change. For example, recent evidence suggests that before 7,600 years ago, the region that is now the Black Sea contained a much smaller freshwater lake surrounded by settlements. When the most recent ice-age glaciers retreated, sea level rose, and the Mediterranean eventually broke through a natural dam at the site of the present Bosporus Strait. Researchers suggest that seawater from the Mediterranean spilled into the Black Sea basin via a waterfall 200 times larger than Niagara Falls. This influx of water caused the lake level to rise by as much as 10 cm per day, and within a year 155,000 square km (60,000 square miles) of populated land had become submerged beneath hundreds of meters of water. This traumatic flooding presumably forced many people to migrate, and its timing has led some researchers to speculate that it may have inspired the Babylonian *Epic of Gilgamesh* (2000 B.C.E.) and, later, the biblical epic of Noah's Ark.

FIGURE 23.11 Estimates of global temperature change over geologic time, relative to a reference value.

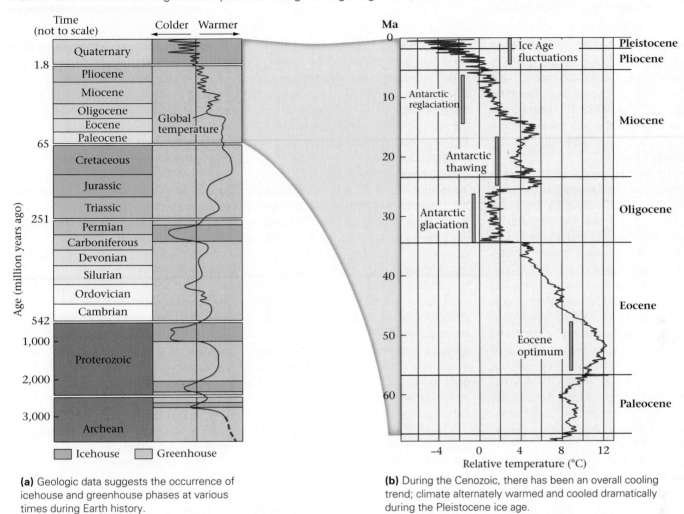

(a) Geologic data suggests the occurrence of icehouse and greenhouse phases at various times during Earth history.

(b) During the Cenozoic, there has been an overall cooling trend; climate alternately warmed and cooled dramatically during the Pleistocene ice age.

Thus, volcanic activity may have triggered Cretaceous greenhouse conditions.

- *The uplift of land surfaces*: Tectonic events that lead to the long-term uplift of the land affect atmospheric CO_2 concentration, because such events expose land to weathering, and chemical-weathering reactions absorb CO_2. Thus, uplift decreases the greenhouse effect and causes global cooling. For example, uplift of the Himalayas and Tibet may have triggered Cenozoic icehouse conditions (Fig. 23.12b). Such uplift will also affect atmospheric circulation and rainfall rates (see Chapter 20).

- *The formation of coal and oil:* At various times during Earth history, environments suitable for the growth, accumulation, and burial of abundant organic material become particularly widespread. Once buried, the organic material transforms into coal or oil, and can remain trapped underground. This overall process removes CO_2 from the atmosphere and may result in

global cooling. The cooling that occurred in the late Paleozoic, a time when coal swamps were abundant, may be a manifestation of the process.

- *Life evolution*: The appearance or extinction of certain life forms may have affected climate significantly. For example, some researchers speculate that the appearance of lichens in the late Proterozoic may have decreased atmospheric CO_2 concentration and thus could have triggered icehouse conditions. Similarly, the appearance of grass about 30 to 35 Ma may have triggered Cenozoic icehouse conditions.

It is important to note the feedback among some of the above phenomena. For example, as global temperature rises because of an increase in CO_2, rates of evaporation and therefore amounts of rainfall increase. As a consequence, weathering rates (which absorb CO_2) increase, so the concentration of CO_2 may then go down.

BOX 23.2

CONSIDER THIS . . .

The Goldilocks Effect, through Time

Like Baby Bear's porridge in the tale of *Goldilocks and the Three Bears*, Earth is not too hot, and it's not too cold . . . it's just right for liquid water and, therefore, life, to exist (**Fig. Bx23.2**). Researchers informally refer to the two related factors that make the Earth "just right" as the **Goldilocks effect**.

What factors keep Earth's surface temperature in the habitable range? The first is our planet's distance from the Sun, which determines the intensity of radiation that reaches the surface, and the second is the concentration of CO_2 and other greenhouse gases, which governs how much of that radiation remains trapped in the atmosphere. If the Earth orbited too close to the Sun, solar radiation would be so intense that water could not exist in liquid form, regardless of atmospheric composition. And without seas of liquid water, most CO_2 emitted by volcanoes would not have dissolved in the sea, so the atmosphere would contain so much water and CO_2 that surface temperatures on Earth would become hot enough to melt lead. Without liquid water, life could not have evolved, so CO_2 would not eventually become trapped in limestone or organic-rock reservoirs underground. If the Earth orbited too far from the Sun, temperatures would have remained so cold, despite atmospheric composition, that any water present would freeze solid, so life could not have evolved.

Astronomers define the distance from the Sun at which the Goldilocks effect is possible as the **habitable zone** of the Solar System. The habitable zone *currently* extends roughly from 0.8 AU to 1.3 AU. (An AU, or astronomical unit, is the mean distance between the Earth and the Sun.) Over the course of Solar System history, however, the position of the habitable zone has changed. That's because the intensity of radiation emitted by the Sun has increased over time. This change has taken place because the Sun's energy comes from the fusion of four hydrogen atoms to form one helium atom, and one helium atom takes up less space than four hydrogen atoms. Thus, over time, production of helium has caused the Sun to contract. The resulting increase in internal pressure and temperature within the Sun, in turn, has caused the rate of fusion reactions to increase. Indeed, the Sun may be 30% brighter today than it was when the Earth first formed.

If the Sun's intensity were the only factor controlling Earth's temperature, our planet should have been over 20°C cooler during the Archean than it is today, and all water should have been frozen. But this wasn't the case. Stratigraphic and fossil records indicate that water has existed in liquid form on our planet's surface at least since the beginning of the Archean (~ 3.8 Ga). Researchers refer to this apparent contradiction between the calculated temperature and the observed temperature of the early Earth as the **faint young Sun paradox**. Most researchers agree that the paradox can be resolved by keeping in mind that earlier in Earth history, before the widespread appearance of photosynthetic life, the atmosphere contained *more* CO_2 than it does today. The greenhouse effect caused by the additional CO_2 increased the temperature of the Earth's atmosphere enough to counteract the lack of radiation from the faint young Sun, and this kept surface temperatures above freezing.

With the faint young Sun paradox in mind, astronomers speculate that when the Solar System was younger, Venus may also have orbited within the habitable zone and may have hosted liquid water. But as the Sun became brighter, Venus warmed up until its surface water evaporated. Addition of water to Venus's atmosphere caused the planet's surface to warm even more, leading to a drastic positive feedback that could not be stopped. Such a situation is called the **runaway greenhouse effect**. As a consequence, Venus's atmosphere eventually became so hot that water molecules broke apart, forming hydrogen atoms that escaped to space and oxygen atoms that reacted with rocks on the planet's surface to produce iron oxide minerals (rusty rocks). Volcanic CO_2 became the dominant gas of Venus's atmosphere, forming a dense blanket that now keeps the surface temperature of the planet at about 460°C.

FIGURE Bx23.2 The "Goldilocks effect" as applied to the Earth and its neighbors.

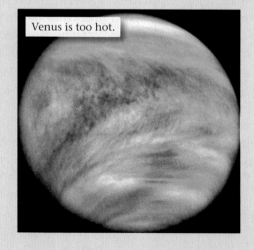
Venus is too hot.

Mars is too cold.

Earth is just right.

FIGURE 23.12 Changes in the distribution and elevation of landmasses can affect climate.

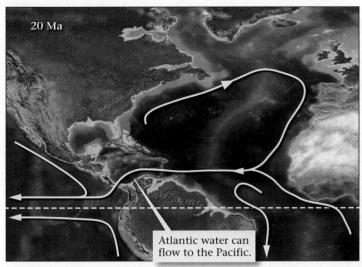
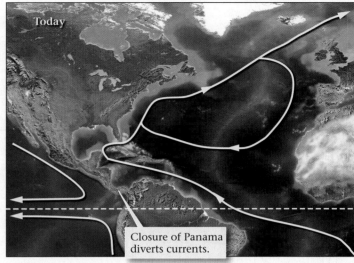

Atlantic water can
flow to the Pacific.

Closure of Panama
diverts currents.

(a) When the Isthmus of Panama, a volcanic arc, formed during the Miocene, the patterns of oceanic currents in the North Atlantic changed. The white dashed line is the equator.

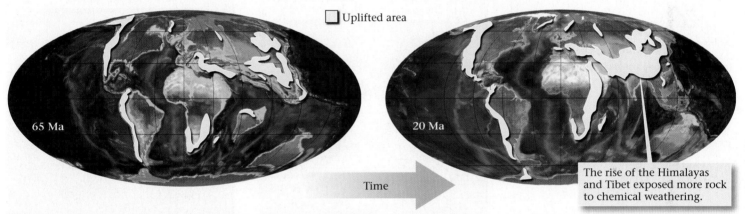

☐ Uplifted area

65 Ma

20 Ma

Time

The rise of the Himalayas
and Tibet exposed more rock
to chemical weathering.

(b) The percentage of elevated land increased between 65 Ma and 20 Ma because of orogeny. The exposure of more rock leads to more weathering, potentially affecting the concentration of greenhouse gases in the atmosphere.

Natural Short-Term Climate Change

So far, we've emphasized how climate has changed over periods of tens to hundreds of millions of years. The geologic record of the past few million years provides enough detail to allow us to detect cyles of climate change that have durations of centuries to hundreds of thousands of years. Such short-term climate change events must be a consequence of factors that can operate quickly in the context of geologic time.

We can get a sense of short-term climate change events by considering the Pleistocene stratigraphic record. Changes in fossil assemblages, sediment composition, and isotopic ratios of minerals indicate that continental glaciers advanced and retreated about 20 to 30 times in the northern hemisphere. Each advance (glaciation) represents an interval of global cooling, and each retreat (interglacial) represents an interval of global warming. If we focus on the last 15,000 years, a period

that includes all of the Holocene, we see trends of cooling or warming that last thousands of years, within which there are events whose duration lasts centuries or less (Fig. 23.13a). Recent evidence suggests that the transition between warming and cooling phases of a cycle can take as little as ten years.

Specifically, the time between about 15,000 and 10,500 B.C.E. was a warming period, during which the last ice-age glaciers retreated. This warming trend was followed by the Younger Dryas, an interval of cooler temperatures named for an Arctic flower that became widespread during the interval. Then, climate warmed again, reaching a peak at 5,000 to 6,000 years ago, a period called the Holocene maximum, when average temperatures were about 2°C above temperatures of today. This warming peak led to increased evaporation and therefore precipitation, making the Middle East unusually wet and fertile—conditions that may partially account for the rise of civilization in this part of the world.

FIGURE 23.13 Climate during the Holocene. Measurements suggest that temperature has varied significantly.

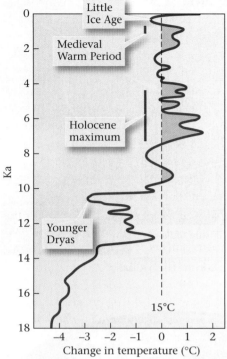

(a) There have been several temperature highs and lows during the Holocene.

(b) During the Medieval Warm Period, Vikings settled in Greenland, where the climate was warm enough for agriculture.

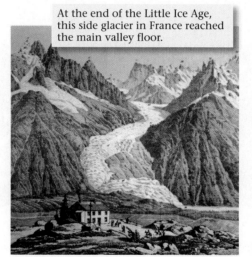

At the end of the Little Ice Age, this side glacier in France reached the main valley floor.

Today, glaciers extend only part-way down the side valleys.

(c) Glaciers advanced in Europe during the Little Ice Age.

The temperature dipped to a low about 3,000 years ago, before returning to a high during the Middle Ages, a time called the Medieval Warm Period. During this time, Vikings established self-supporting agricultural settlements along the coast of Greenland (Fig. 23.13b). The temperature dropped again from 1500 C.E. to about 1800 C.E., a period known as the Little Ice Age, when Alpine glaciers advanced and the canals of the Netherlands froze over in winter (Fig. 23.13c; see Fig. 22.33). Overall, the climate has warmed since the end of the Little Ice Age, and today it is as warm or warmer than it was during the Medieval Warm Period.

Geologists have focused on four factors to explain short-term climate change.

- *Fluctuations in solar radiation and cosmic rays*: The amount of energy produced by the Sun varies with the **sunspot cycle**. This cycle involves the appearance of large numbers of sunspots (black spots thought to be magnetic storms on the Sun's surface) about every 9 to 11.5 years (Fig. 23.14a–c). There may be longer-term cycles that have not yet been identified.

  Some researchers have speculated that changes in the rate of influx of cosmic rays may affect climate, perhaps by generating clouds. Specifically, recent research suggests that cosmic rays striking the atmosphere produce clusters of ions that become condensation nuclei around which water molecules congregate, thus forming the mist droplets making up clouds. But how cloud formation changes climate remains uncertain. High-elevation clouds could reflect incoming solar radiaton and would cool the planet, whereas low-elevation clouds could absorb infrared rays rising from the Earth's surface and would warm the planet.

- *Changes in Earth's orbit and tilt*: As Milanković first recognized in 1920, the change in the tilt of Earth's axis over a period of 41,000 years, the Earth's 23,000-year precession cycle, and changes in the eccentricity of its orbit over

FIGURE 23.14 The abundance of sunspots varies with time and affects the radiation emitted by the Sun.

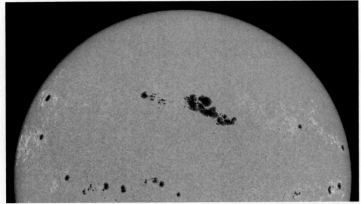

(a) Sunspots are magnetic storms that slow convection at the Sun's surface, producing a cooler area that appears as a dark patch.

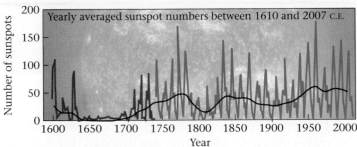

(b) The abundance of sunspots varies cyclically. When there are more sunspots, the Sun radiates less heat.

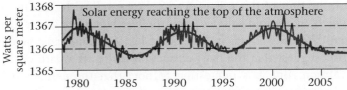

(c) Measurements since 1978 suggest that the solar energy reaching the atmosphere fluctuates periodically over time.

a period of 100,000 years together cause the amount of summer heat in high latitudes to vary. They also cause the overall amount of heat reaching Earth to vary (see Chapter 22). These changes correlate with observed ups and downs in atmospheric and oceanic temperature.

■ *Changes in volcanic emissions*: Not all of the sunlight that reaches the Earth penetrates its atmosphere and warms the ground. Some is reflected by the atmosphere. The degree of reflectivity, or **albedo**, of the atmosphere increases not only if cloud cover increases, as we have seen, but also if the concentration of volcanic aerosols in the atmosphere increases.

The short-term effect of volcanism on global temperature is abundantly clear. For example, the year following the 1815 eruption of Mt. Tambora in the western Pacific became known as the "year without a summer,"

for sulfur aerosols that erupted encircled the Earth and blocked the Sun. Snow fell in Europe throughout the spring, and the entire summer was cold. Recent studies suggest that the immense eruption of Toba (in Indonesia), 70,000 years ago, disrupted climate for so many years that it caused worldwide extinctions.

■ *Changes in ocean currents*: Recent studies suggest that the configuration of currents can change quite quickly, and that this configuration affects the climate. The Younger Dryas may have resulted when a layer of freshwater from melting glaciers spread out over the North Atlantic and prevented thermohaline circulation in the ocean, thereby shutting down the Gulf Stream (see Chapter 18).

■ *Changes in surface albedo*: Regional-scale changes in the nature of continental vegetation cover, and/or the proportion of snow and ice on the Earth's surface, and/or the sudden deposition of reflective volcanic ash would affect our planet's albedo. Increasing albedo causes cooling, whereas decreasing albedo causes warming. The Toba eruption covered vast areas with reflective white ash, which may have contributed to cooling.

■ *Abrupt changes in concentrations of greenhouse gases*: A sudden change in greenhouse gas concentration in the atmosphere could affect climate. One such change might happen if sea temperature warmed or sea level dropped, causing some of the methane hydrate that crystallized in sediment on the sea floor to melt suddenly. Such melting would release CH_4 to the atmosphere. Algal blooms and reforestation (or deforestation) conceivably could change CO_2 concentrations.

Catastrophic Climate Change and Mass-Extinction Events

Changes that happen on Earth almost instantaneously are called *catastrophic changes*. For example, a volcanic explosion, an earthquake, a tsunami, or a landslide can change a local landscape in seconds or minutes. But such events affect only relatively small areas. Can such catastrophes happen on a global scale? In the past decades, geoscientists have come to the conclusion that the answer is yes. The stratigraphic record shows that Earth history includes several **mass-extinction events**, during which large numbers of species abruptly vanished (**Fig. 23.15a**). Some of these events define boundaries between geologic periods. A mass-extinction event decreases the biodiversity (the number of different species that exist at a given time) of life on Earth. It takes millions of years after a mass-extinction event for biodiversity to increase again, and the new species that appear differ from those that vanished, for evolution is unidirectional.

Geologists speculate that some mass-extinction events reflect a catastrophic change in the planet's climate, brought about by

FIGURE 23.15 Life evolution has proceeded in fits and starts. During geologic time, there have been several catastrophic extinction events, when a larger percentage of the genera on Earth goes extinct and biodiversity abruptly decreases.

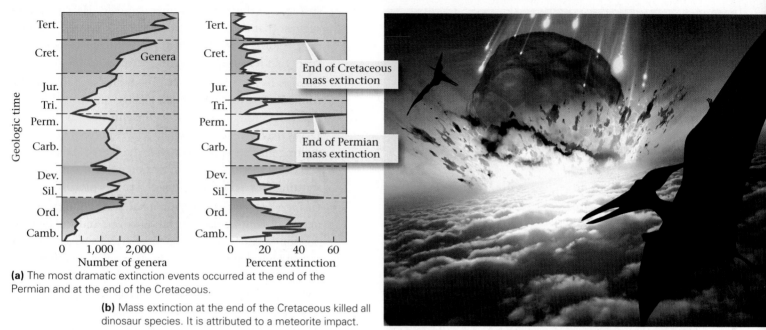

(a) The most dramatic extinction events occurred at the end of the Permian and at the end of the Cretaceous.

(b) Mass extinction at the end of the Cretaceous killed all dinosaur species. It is attributed to a meteorite impact.

unusually voluminous volcanic eruptions or by the impact of a comet or an asteroid with the Earth (Fig. 23.15b). Either of these events could eject enough debris into the atmosphere to block sunlight. Without the warmth of the Sun, winter-like or night-like conditions would last for weeks to years, long enough to disrupt the food chain. In addition, either event could eject aerosols that would turn into global acid rain, scatter hot debris that would ignite forest fires, or give off chemicals that, when dissolved in the ocean, would make the ocean either toxic, killing marine life, or so nutritious that oxygen-consuming algae could thrive.

Let's examine possible causes for two of the more profound mass-extinction events in Earth history. The first event marks the boundary between the Permian and Triassic periods (i.e., between the Paleozoic and Mesozoic eras). During the Permian-Triassic extinction event, over two-thirds of the species on Earth became extinct. In fact, this boundary was first defined in the nineteenth century, precisely because the assemblage of fossils from rocks below the boundary differs so markedly from the assemblage in rocks above. Isotopic dating suggests that the extinction event roughly coincided with the eruption of vast quantities of basalt in Siberia. So much basalt erupted that geologists attribute its source to a superplume, a mantle plume many times larger than the one currently beneath Hawaii. Because of the correlation between the time of the basalt eruptions and the time of the mass extinction, geologists suggest that the former caused the latter. Recent, and still controversial, evidence suggests that a large asteroid collided with the Earth at the Permian-Triassic boundary. A minority of researchers suggest that the mass extinction resulted, instead, from this collision.

The second event, called the K-T boundary event, caused the mass extinction that marks the boundary between the Cretaceous and Tertiary Periods (i.e., the boundary between the Mesozoic and Cenozoic Eras). As discussed in Chapter 13, the time of this event correlates very well with the time at which an asteroid collided with the Earth at a site now called the Chicxulub crater in Yucatán, Mexico. Thus, most geologists suggest that the mass extinction is the aftermath of this collision. But it is interesting to note that the extinction is comparable in age to the eruption of extensive basalt flows in India, so the possibility remains that volcanic activity played a role.

Take-Home Message

- Global climate change (warming or cooling) has happened through geologic time.
- Geologists study many sources of data (e.g., the stratigraphic record, isotope changes in ice or shells, growth rings) to constrain the timing and character of climate change.
- Causes of long-term change include continental drift, mountain building, and life evolution.
- Short-term changes may be due to changes in solar output, changes in Earth's orbit and tilt, volcanic eruptions, and changes in greenhouse gas concentrations.
- Some changes may cause mass extinction.

THINK: How might continental drift cause climate change?

23.6 HUMAN IMPACT ON THE EARTH SYSTEM

During the Stone Age, the human population worldwide had not yet topped 10 million. By the dawn of civilization, 4000 B.C.E., it was still at most a few tens of millions. But by the beginning of the nineteenth century, revolutions in industrial methods, agriculture, medicine, and hygiene had substantially lowered death rates and raised living standards, so that the population began to grow at accelerating rates. It took tens of thousands of years to grow from Stone Age populations to 1 billion people worldwide in 1850, but it took only 80 years for the population to double again, reaching 2 billion in 1930. The growth rate increased during the twentieth century, with the population reaching 4 billion in 1975. Now, the doubling time is only 44 years, and the population passed the 6 billion mark just before the year 2000 (Fig. 23.16). Global population will surpass 7 billion during 2012.

As the population grows and the standard of living improves, per capita usage of resources increases. We use land for agriculture and grazing, forests for wood, rock and dirt for construction, oil and coal for energy or plastics, and ores for metals. Without a doubt, our usage of resources has affected the Earth System profoundly, and thus humanity has become a major agent of global change (See for Yourself U, p. S-40). Here, we examine some of these anthropogenic impacts.

The Modification of Landscapes

Every time we pick up a shovel and move a pile of rock or soil, we redistribute a portion of the Earth's crust, an activity that prior to humanity was accomplished only by rivers, the wind, rodents,

FIGURE 23.16 Population now doubles about every 44 years. The Black Death pandemic caused an abrupt drop that lasted for a few decades.

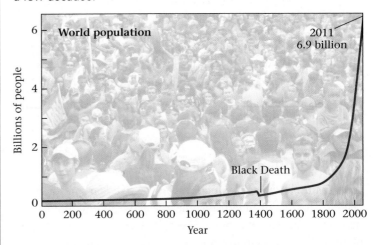

and worms. In the last century, the pace of Earth movement has accelerated, for now we have shovels in coal mines that can move 300 cubic meters of coal in a single scoop, trucks that can carry 200 tons of ore in a single load, and tankers that can transport 500 million liters (~3 million barrels) of oil during a single journey. In North America, human activity now moves more sediment each year than rivers do. The extraction of rock during mining, the building of levees and dams along rivers or of sea walls along the coast, and the construction of highways and cities all involve the redistribution of Earth materials (Fig. 23.17). In addition, people clear and plow fields, drain and fill wetlands, and pave over the land surface (see also chapter opening photo). All these activities change the landscape.

FIGURE 23.17 Excavation, agriculture, and construction modify topography, drainage, infiltration, and ecology.

Humans move vast amounts of rock.

Agriculture eliminates diverse ecosystems.

Urbanization changes the water table.

Landscape modification has side effects. For example, it may make the ground unstable and susceptible to landslides. And it may expose the land to erosion, thereby changing the volume of sediment transported by natural agents (such as running water and wind). Locally, flood-control projects may diminish the sediment supply downstream, also with unfortunate consequences; for example, the damming of the Nile by the Aswan High Dam has cut off the sediment supply to the Nile Delta, so ocean waves along the Mediterranean coast of the delta have begun to erode the coastline by more than 1 m per year.

The Modification of Ecosystems

In undisturbed areas, the **ecosystem** of a region (an interconnected network of organisms and the physical environment in which they live) is the product of evolution for an extended period of time. The ecosystem's flora (plant life) include species that

have adapted to living together in that particular climate and on the substrate available, whereas its fauna (animal life) can survive local climate conditions and utilize local food supplies. Human-caused deforestation, overgrazing, agriculture, and urbanization disrupt ecosystems and lead to a decrease in biodiversity.

Archaeological studies have found that the earliest example of human modification of an ecosystem occurred in the Stone Age, when hunters played a major role in causing the mass extinction of many species of large mammals (mammoths, giant sloths, giant bears). Today, less than 5% of Europe retains its original habitats. The same number can be applied to the eastern United States, which lost its original forest and prairie. Tropical rain forests cover less than about half the area worldwide that they covered before the dawn of civilization, and they are disappearing at a rate of about 1.8% per year (**Fig. 23.18a, b**). Much of this loss comes from slash-and-burn agriculture, in which farmers and ranchers destroy forest to make open land for farming and grazing (**Fig. 23.18c**). Unfortunately, the heavy rainfall of tropical regions removes nutrients from the soil, making the soil useless in just a few years. Overgrazing by domesticated animals can remove vegetation so completely that some grasslands have undergone desertification. And urbanization replaces the natural land surface with concrete or asphalt, a process that completely destroys an ecosystem and radically changes the amount of rain that infiltrates the land surface to become groundwater.

Human-caused changes to ecosystems affect the broader Earth System because they modify biogeochemical cycles and Earth's albedo (surface reflectivity). For example, deforestation increases the CO_2 concentration in the atmosphere, for much of the carbon that was stored in trees is burned and rises. And the replacement of forest cover with concrete or fields increases Earth's albedo.

Pollution

The environment has always contained contaminants such as soot, dust, and the by-products of organisms. But when human populations grew, urbanization, industrial and agricultural activity, the production of electricity, and modern modes of transportation greatly increased both the quantity and diversity of contaminants that entered the air, surface water, and groundwater. These contaminants, or **pollution**, include both natural and synthetic materials (in liquid, solid, or

FIGURE 23.18 The area of the Earth covered by forests has been shrinking.

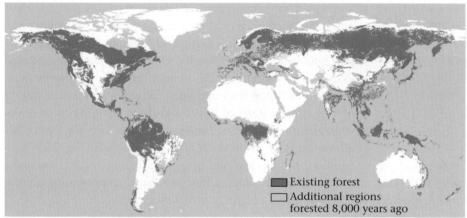

Existing forest
Additional regions forested 8,000 years ago

(a) Comparison of exisiting forest area (green) to precivilization forest area (gold) emphasizes changes in Earth's forest cover. Tropical rain-forest areas are shrinking rapidly.

(b) Satellite images of a region in the Amazon show the increase in cleared areas (light).

(c) Slash-and-burn agriculture.

FIGURE 23.19 Acid rain forms when the sulfur dioxide in industrial smoke dissolves in water and produces sulfuric acid. Acid rain is a problem in all industrialized countries.

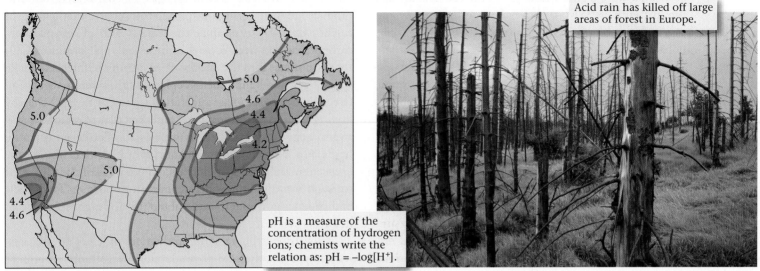

Acid rain has killed off large areas of forest in Europe.

pH is a measure of the concentration of hydrogen ions; chemists write the relation as: pH = –log[H$^+$].

(a) Acid rain in North America occurs downwind of major industrial cities. We can specify the acidity of rainwater by stating its pH.

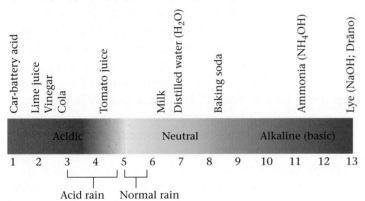

(b) Acid rain has a pH of between 3 and 5.

gaseous form). They have become a problem because they are produced too fast for the Earth System to naturally absorb or modify. For example, small quantities of sewage can be absorbed by clay minerals in the soil or destroyed by bacterial metabolism, but large quantities overwhelm natural controls and can accumulate into destructive concentrations. Further, because many contaminants are not produced in nature, they are not easily removed by natural processes. Pollution of the Earth System is a type of global change because it is a redistribution and reformulation of materials. Some key problems associated with this change include the following:

- *Smog*: The term was originally coined to refer to the dank, dark air that resulted when smoke from the burning of coal mixed with fog in London and other industrial cities. Another kind of smog, called photochemical smog, is the ozone-rich brown haze that blankets cities when exhaust from cars and trucks reacts with air in the presence of sunlight.

- *Water contamination*: We dump a great variety of chemicals into surface water and groundwater. Examples include gasoline, other organic chemicals, radioactive waste, acids, fertilizers—the list could go on for pages. These chemicals affect biodiversity.

- *Acid runoff*: Dissolution of sulfide-containing minerals in ores or coal by groundwater or stream water makes the water acidic (it increases the concentration of hydrogen ions in the water) and toxic to life.

- *Acid rain*: When rain passes through air that contains sulfur-containing aerosols (emitted from power plants), the water dissolves the sulfur and creates sulfuric acid, or **acid rain**. Wind can carry aerosols far from a power plant, so acid rain can damage a broad region (Fig. 23.19a, b).

- *Radioactive materials*: Nuclear weapons, nuclear energy, and medical waste transfer radioactive materials from rock to Earth's surface environment. Also, human-caused nuclear reactions produce new, nonnatural radioactive isotopes, some of which have relatively short half-lives. Thus, society has changed the distribution and composition of radioactive material worldwide.

- *Ozone depletion*: When emitted into the atmosphere, human-produced chemicals, most notably chlorofluorocarbons (CFCs), react with ozone in the stratosphere. This reaction, which happens most rapidly on the surfaces of tiny ice crystals in polar stratospheric clouds, destroys ozone molecules, thus creating an **ozone hole** over high-latitude regions, particularly during the spring (Fig. 23.20a, b). Note that the "hole" is not really an area where no ozone is present, but rather is a region where atmospheric ozone has been reduced substantially. The ozone hole is more prominent in the

Antarctic than in the Arctic because a current of air circulates around the landmass of Antarctica and traps the air, with its CFCs, above the continent, preventing it from mixing with air from elsewhere. Ozone holes have dangerous consequences, for they affect the ability of the atmosphere to shield the Earth's surface from harmful ultraviolet radiation. In 1987, a summit conference in Montréal proposed a global reduction of (ozone-destroying) CFC emissions. Reduction of such emissions may substantially reduce ozone depletion.

Recent Global Warming

We've seen that greenhouse gases, most notably CO_2 and CH_4, play a major role in the regulation of the Earth's surface temperatures—without these gases, the Earth could not be a home for life. Both CO_2 and CH_4 cycle through various biogeochemical reservoirs of the Earth System, and the fluxes between reservoirs determine the amount in any given reservoir at any given time. For most of geologic time, the concentration of these gases in the atmosphere was governed by natural processes—volcanic eruptions, life evolution events (appearance of new genera or extinction of existing ones), forest fires, orogenic uplift and related weathering, warming and cooling due to the Milanković cycle, changes in solar activity or cosmic-ray flux, and even meteor impact. But beginning around 8,000 years ago, human society began to significantly modify the environment, first with the invention of agriculture and then, during the past two centuries, with the advent of major industrialization.

Both industry and agriculture produce greenhouse gases that transfer carbon from underground reservoirs or biomass reservoirs into the atmospheric reservoir. For example: the burning of fossil fuels oxidizes vast quantities of carbon that had previously been held in oil, gas, or coal underground to produce CO_2 that mixes into the atmosphere; the heating of calcite ($CaCO_3$) to produce the lime (CaO) of cement takes carbon that had been locked in limestone and produces CO_2 that also mixes into the air; the clear-cutting of forests to make way for grazing land or fields replaces high-biomass vegetation (trees) with low-biomass vegetation (grasses or crops) thereby leaving CO_2 in the atmosphere; and the decay of organic material in soggy rice paddies, as well as the flatulence of cattle herds, produces significant quantities of CH_4 that would otherwise remain locked in biomass. About 85% of the CO_2 that we send into the atmosphere comes from burning fossil fuels and producing cement, while about 15% is a consequence of deforestation.

How does the rate of production of CO_2 by humanity compare to the rate of production by volcanic eruptions? Researchers estimate that in a year all volcanic eruptions together (including both submarine and subaerial eruptions) emit about 0.15 to 0.26 gigatons of CO_2. By comparison, activities of humanity emit about 35 gigatons of CO_2 every year—i.e., at least 135 times as much. Put another way, human activities now produce more CO_2 in three days than do all of the volcanoes on Earth in a typical year. A large volcanic explosion, such as that of Mt. Pinatubo in 1991, emits only about as much CO_2 as people do in one day, and a supervolcano explosion (see Chapter 9), which only happens once every 100,000 to 300,000 years, produces about as much CO_2 as people do in one year.

Significantly, not all of the greenhouse gases that society sends into the atmosphere stay there. In the case of CO_2, some dissolves in the ocean, some reacts with minerals in rocks during chemical weathering, and some is utilized by plants during

FIGURE 23.20 The ozone hole over Antartica.

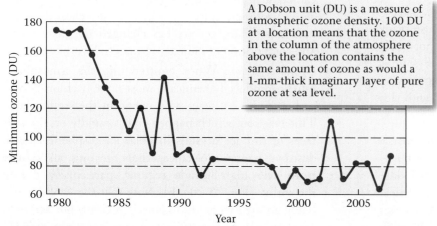

A Dobson unit (DU) is a measure of atmospheric ozone density. 100 DU at a location means that the ozone in the column of the atmosphere above the location contains the same amount of ozone as would a 1-mm-thick imaginary layer of pure ozone at sea level.

(a) The concentration of ozone over Antarctica diminished significantly between 1980 and 1990. Regulation of several chemicals has slowed and perhaps reversed this trend.

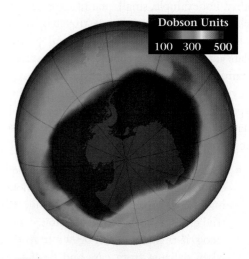

(b) A map showing the dimensions of the ozone hole in 2006—the largest hole ever recorded.

FIGURE 23.21 Antropogenic CO_2 enters the atmosphere from many sources. Since about 1900, natural sinks can no longer absorb it all.

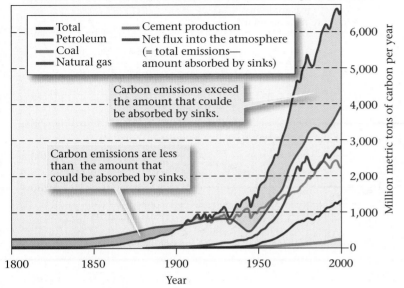

Late Pleistocene, CO_2 concentration varied between about 180 and 300 ppm (Fig. 23.22c). Thus, the increases that have happened since the beginning of the industrial revolution are beyond the range of natural fluctuations that occurred during the last 800,000 years. Have CO_2 concentrations ever been higher than they are today? Probably. For example, a recent study suggests that CO_2 concentration during the Eocene climate optimum (49 to 56 Ma) was over 1,000 ppm, about three times what it is today. CH_4 concentrations have also demonstrably increased over the past two centuries (Fig. 23.22d).

The fundamental principles of the greenhouse effect require that an increase in CO_2 concentration cause warming. But has warming taken place during the past 200 years? During the past 30 years, researchers have published thousands of observations suggesting that it has. For example:

- Large ice shelves, such as the Larsen B Ice Shelf, along the Antarctic Peninsula, and the Ayles Ice Shelf, along Ellesmere Island in northernmost Canada, are breaking up (Fig. 23.23a).

- The Greenland ice sheet is melting at an accelerating pace. Studies suggest that the rate of ice loss has increased from 90 to 220 cubic km per year in the last 10 years and that, in places, the sheet is thinning by about 1 m per year. In addition, the annual melt zone along the margins of the ice sheet has widened dramatically because the elevation of the equilibrium line (see Chapter 22) has risen (Fig. 23.23b). Valley glaciers draining the ice sheet flowed 50% faster in 2003 than they did in 1992.

- The area covered by sea ice in the Arctic Ocean has decreased substantially. In fact, this ice covered 14% less area in 2005 than in 2004. This observation leads to the prediction that it may be possible to sail across the Arctic Ocean within decades (Fig. 23.23c).

- Valley glaciers worldwide have been retreating rapidly, so that areas that were once ice covered are now bare. The change is truly dramatic in many locations (Fig. 23.23d).

- Worldwide, glacial volumes have been diminishing. About 400 km^3 of ice disappears every year (Fig. 22.23e).

- Biological phenomena that are sensitive to climate are being disrupted. For example, the time at which sap in the maple trees of the northeastern United States starts to flow has changed, and the mosquito line (the elevation at which mosquitoes can survive) has risen substantially. Also, the average weight of polar bears has been decreasing, because the bears can no longer walk over pack ice to reach their hunting grounds in the sea.

photosynthesis. But calculations and isotopic studies suggest that only 43% of the CO_2 that society produces (what researchers refer to as *anthropogenic* CO_2) returns into the Earth in biogeochemical cycles—the remaining 57% stays in the atmosphere. Before 1900, natural "sinks" (the ocean, rock weathering, plants) could absorb all anthropogenic CO_2. But since then, the amount of CO_2 has exceeded the ability of natural sinks to absorb it (Fig. 23.21).

In the early 1960s, a chemist named Charles Keeling wondered whether changes in atmospheric CO_2 concentration could be detected, and he set out to measure the concentration of CO_2 in the atmosphere using the most accurate methods available. To avoid areas with urban pollution, he decided to analyze air samples collected every month at the summit of Mauna Kea volcano in Hawaii. After completing many years of measurements, Keeling showed that not only that distinct seasonal variations occurred (CO_2 concentration goes down in the warm summer when rates of photosynthesis increase, and it goes up in the winter when organic matter dies and decays), but that the *average annual concentration* of CO_2 was steadily rising (Fig. 23.22a)! The average yearly CO_2 concentration was 320 ppm in 1965, and it reached 360 ppm in 1995. Charles Keeling died in 2005, but measurements at Mauna Kea have continued, and in 2010, the average yearly CO_2 concentration reached 390 ppm.

Using records in ice cores, researchers extended the record of CO_2 concentration further back in time and have found that in 1750, CO_2 concentration was only 280 ppm (Fig. 23.22b). Thus, atmospheric CO_2 concentration has increased by about 40% since the beginning of the industrial revolution. Ice cores from the Antarctic ice sheet now provide a CO_2 record back through the last 800,000 years, and they demonstrate that during the alternating glacial and interglacial periods of the

FIGURE 23.22   Changes in carbon dioxide (CO₂) and methane (CH₄) concentrations over time.

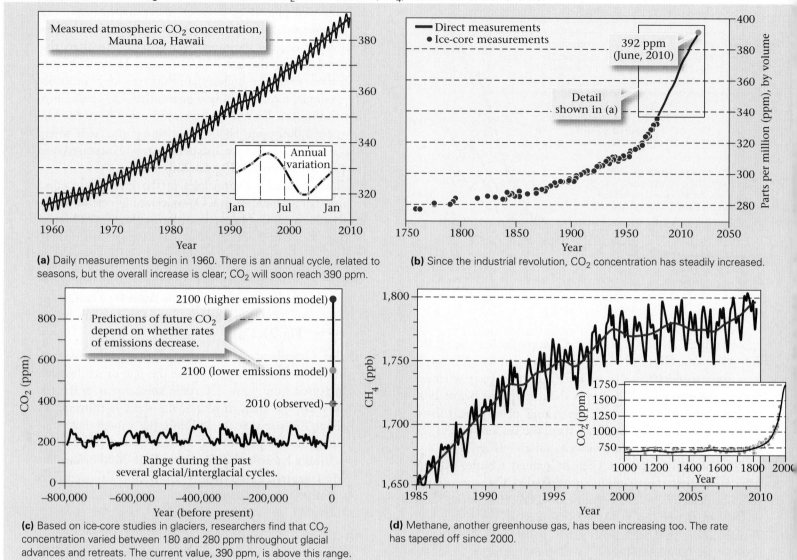

(a) Daily measurements begin in 1960. There is an annual cycle, related to seasons, but the overall increase is clear; CO₂ will soon reach 390 ppm.

(b) Since the industrial revolution, CO₂ concentration has steadily increased.

(c) Based on ice-core studies in glaciers, researchers find that CO₂ concentration varied between 180 and 280 ppm throughout glacial advances and retreats. The current value, 390 ppm, is above this range.

(d) Methane, another greenhouse gas, has been increasing too. The rate has tapered off since 2000.

- The area of permafrost in high latitutdes has substantially decreased, and melt ponds have replaced frozen land. Large regions that stayed frozen all year are now melting in the summer, so trees are tipping over.

- Water vapor in the atmosphere has been increasing due to evaporation of warmer seas, possibly leading to more severe hurricanes.

- Measured values of average global surface atmospheric temperature are rising. Coming up with a value that represents average global surface temperature in a given year is not an easy task, because weather at individual measurement localities can change rapidly. But statistical analyses of surface and satellite measurements have led researchers to conclude that, overall, the mean temperature of this planet's climate has been increasing, a phenomenon widely referred to as **global warming** (Fig. 23.24a, b).

Notably, the rate of warming has been faster in the northern hemisphere than in the southern, probably because there is more ocean area in the southern hemisphere and water absorbs and releases heat very slowly.

Global warming does not necessarily mean that the temperature at all locations is warmer all the time. GCMs suggest that global warming may cause climate to change in different ways in different places, and some places will cool (Fig. 23.24c).

- Measured values of near-surface ocean-water temperatures are rising. Studies of global averages of temperatures in the sea indicate that overall, the oceans have been warming over the past century and a half (Fig. 22.25).

Climate data are not easy to obtain and are not easy to interpret, for individual measurements may have large

FIGURE 23.23 Visible examples of large-scale changes in Earth's ice cover.

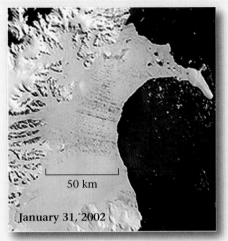

50 km

January 31, 2002   March 7, 2002

(a) The Larsen B Ice Shelf consisted of glacial ice flowing off the Antarctic Peninsula. In 2002, the shelf disintegrated.

June 14, 2001

The gray area, spotted with puddles, is the melt zone.

June 13, 2002

Summer 1979   Summer 2005

(c) The northern polar ice cap shrank significantly between 1979 and 2005.

June 17, 2003

(b) The summer melt line has risen along the edge of Greenland's ice sheet. Each photo shows the same area.

1941

2004

(d) The Muir Glacier, Alaska, retreated 12 km between 1941 and 2004.

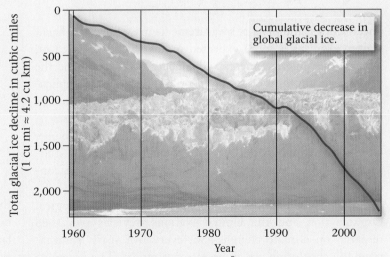

Cumulative decrease in global glacial ice.

Total glacial ice decline in cubic miles (1 cu mi ≈ 4.2 cu km)

0

500

1,000

1,500

2,000

1960   1970   1980   1990   2000

Year

(e) On a global basis, about 100 cu mi (~400 km³) of glacial ice melts every year.

FIGURE 23.24 Measurements of global warming and global temperature anomalies.

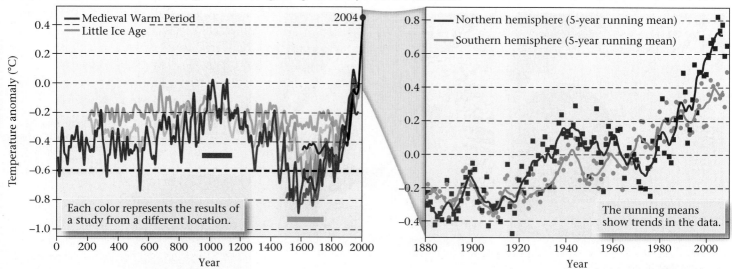

(a) Reconstructions of temperature during the past 2,000 years. Each color is a published estimate from a research team. The black line is the average of direct measurements.

(b) Graphs showing the change in global average temperature since 1880 relative to a reference value. Note that both the northern and southern hemispheres show a temperature increase.

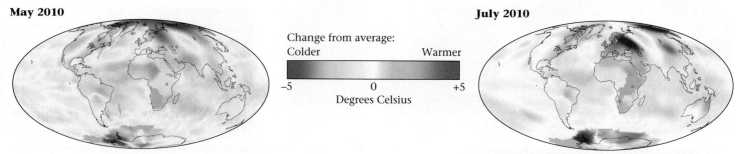

(c) Colors represent anomalies (differences) relative to the average temperature of the period from 1951 to 1980. Redder areas are warmer than the 1951–1980 average, and blue areas are cooler. The position of anomalies changes over time.

uncertainties, and plots of measurements may show significant scatter. And, as is to be expected in any scientific endeavor, in some cases measurements and conclusions do not stand the test of time and turn out to be incorrect. Because the implications of climate studies potentially can have major implications for global policy decisions, in 1988, a group of leading scientists founded the Intergovernmental Panel on Climate Change (IPCC), whose purpose is to evaluate published climate studies from a broad perspective and provide assessments of conclusions from the studies.

This panel, sponsored by the World Meteorological Organization and the United Nations, reviews published research on climate change and, every five years, summarizes the conclusions in an assessment report. The language describing the likelihood that global warming is happening, and that humans have contributed significantly to causing it, has become gradually less equivocal in successive versions of the report. *The Fourth Assessment Report*, published in 2007, states:

Warming of the climate system is unequivocal, as is now evident from observations of increases in global average air and ocean temperatures, widespread melting of snow and ice, and rising global average sea level. . . . The understanding of anthropogenic [human-caused] warming and cooling influences on climate has improved since the Third Assessment Report, leading to very high confidence that the globally averaged net effect of human activities since 1750 has been one of warming.

In other words, most climate researchers have concluded that global warming is real, and that the actions of people—burning fossil fuels, cutting down forests, paving over wetlands—have played a significant role in causing it. Other possible causes, such as changes in solar radiation or cosmic-ray flux, do not appear to be of sufficient magnitude to cause all observed warming. Specifically, global mean atmospheric temperature has risen by almost 1°C during the last century, and the rate of increase during the past 50 years appears to be greater than the rate for the previous 50 years. Although such a change may

FIGURE 23.25 Measurements of average global temperature in shallow waters of the ocean.

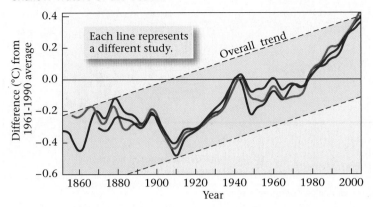

seem small, the magnitude of temperature change between the last ice age and now, by comparison, was only 3° to 5°C. A small change in average temperature may have major consequences.

Some researchers suggest that human impact on climate became noticeable as far back as 8,000 years ago, and that climate has been trending toward warmer conditions ever since. Deviations from the warming trend have been attributed, speculatively, to times when human population abruptly decreased (due to pandemics) so that production of greenhouse gas slowed and forests returned, to long-term sunspot cycles, or to volcanic events. It has been suggested that without the advent of agriculture, deforestation, and the burning of fossil fuels, Earth's climate would be in a cooling trend and we might be heading toward another ice age.

If the fourth assessment of the IPCC is correct, then society faces the challenge of either slowing climate change or dealing with the consequences. The effects of global warming over the coming decades to centuries remains the subject of intense debate, because predictions depend on computer models and not all researchers agree on how to construct or interpret these models. The role of clouds in climate change, for example, remains poorly understood and inadequately addressed by such models. In the worst-case scenario, global warming will continue into the future at the present rate, so that by 2050—within the lifetime of many readers of this book—the average annual temperature will have increased in some parts of the world by 1.5° to 2.0°C. At these rates, by the end of the century, temperatures could be almost 4°C warmer depending on the model used (**Fig. 23.26a**), and by 2150, global temperatures may be 5° to 11°C warmer than at present—the warmest since the Eocene Epoch, 40 million years ago. Models predict that warming will not be the same everywhere—the greatest impact will be in the Arctic (**Fig. 23.26b**).

The effects of such a change are controversial, but according to some climate models, the following events might happen:

- *A shift in climate belts:* As the climate overall warms, temperate and desert regions would occur at higher latitudes. Thus, regions that are now agricultural areas may dry out and become unfarmable (**Fig. 23.27a**). The change in climate would also affect the amount and distribution of precipitation (rain and snow) that falls. For example, in North America, summers would become drier and winters would become wetter (**Fig. 23.27b**). One way to picture this change is to think of how the "climatic latitude" of a state in the USA might change over time. Models suggest that if current warming rates

FIGURE 23.26 Model predictions of future global warming

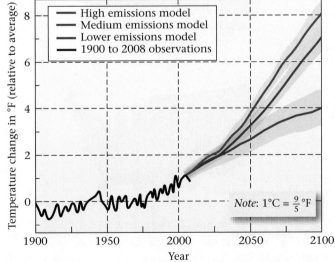

(a) Model calculations of global warming. Different models assume different rates of CO_2 emissions. All suggest significant global temperature increase by 2100.

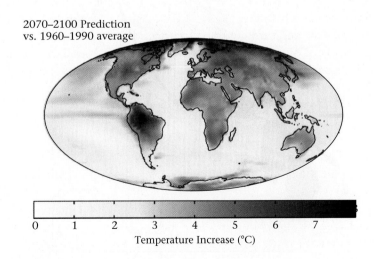

2070–2100 Prediction vs. 1960–1990 average

Temperature Increase (°C)

(b) Global warming does not mean that all locations warm by the same amount. This map shows a model prediction of how temperature may vary with location for the time period 2070 to 2100.

FIGURE 23.27 Model predictions of changes that may happen, if current global warming continues.

☐ Tundra ■ Deciduous forest ■ Evergreen forest ■ Boreal forest ☐ Shrub and grassland ■ Sparse vegetation

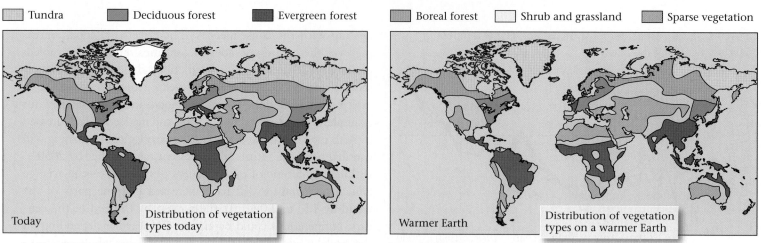

Today — Distribution of vegetation types today

Warmer Earth — Distribution of vegetation types on a warmer Earth

(a) The distribution of climate belts, indicated by vegetation type, will change if the global temperature rises.

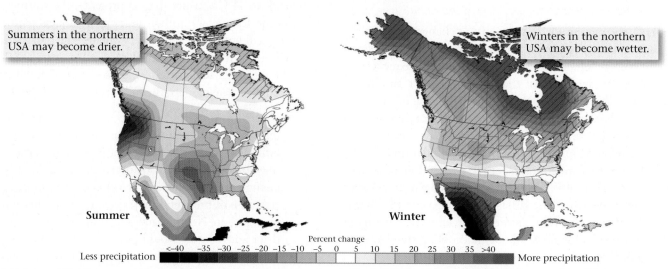

Summers in the northern USA may become drier.

Winters in the northern USA may become wetter.

Summer

Winter

Percent change

Less precipitation   <−40 −35 −30 −25 −20 −15 −10 −5 0 5 10 15 20 25 30 35 >40   More precipitation

(b) Models suggest that the amount of precipitation (rain and snow) at locations in North America about 100 years from now will be different than it is today.

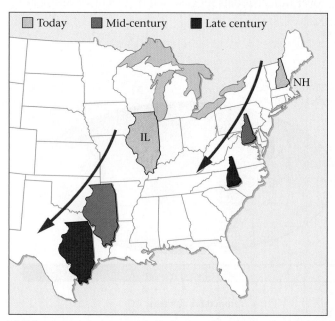

☐ Today ■ Mid-century ■ Late century

NH

IL

(c) This map represents, symbolically, how the climate of a given state may change if the climate warms, according to models. At the end of the century, northern states (such as Illinois) may have climates that are like those of southern states today.

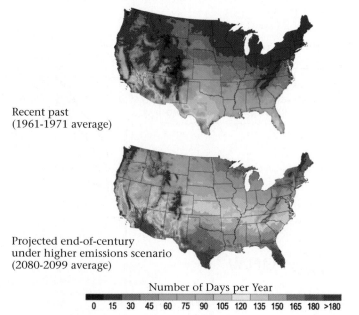

Recent past (1961-1971 average)

Projected end-of-century under higher emissions scenario (2080-2099 average)

Number of Days per Year

0 15 30 45 60 75 90 105 120 135 150 165 180 >180

(d) A prediction of the increase in the number of days above 32°C (90°F) at the end of the century, in comparison to today, in the USA.

continue, the future climate of Illinois would be like the present climate of Texas, and the future climate of New Hampshire might be like the current climate of North Carolina (Fig. 23.27c). Models also suggest that global warming would dramatically increase the number of health-threatening heat waves (loosely defined as days above 32°C, or 90°F) that would affect a region (Fig. 23.27d).

- *Glacial retreat, a rise in the snow line, and melting permafrost:* We've already seen evidence that the volume of glacial ice worldwide is decreasing. This is manifested by the shrinkage and retreat of most glaciers, exposing land that was once buried by ice. Of note, some glaciers are advancing; advance appears to be a consequence either of surging, due to addition of liquid water to the base of the glacier (see Chapter 22), or of local increases in snowfall due to the addition of moisture to the atmosphere as the oceans warm. As temperatures increase, the snow line in mountains rises to higher latitudes and higher elevations. This change affects ecosystems, and even tourism, by making it challenging for winter resorts to host snow-based activities. As climate warms, permafrost in tundra is beginning to melt. Already, the southern limit of permafrost appears to have moved 100 km north of where it was a century ago. Permafrost melting may provide a positive feedback to global warming, for as the permafrost melts, organic matter it contains begins to decay and release CH_4, a greenhouse gas. Similarly, as oceans warm, frozen methane hydrate trapped in sediment of the sea floor may start to melt, adding even more CH_4 to the atmosphere.

- *A rise in sea level:* Water expands when heated, so warming of the global oceans will cause sea level to rise. Glacial melting due to global warming will add to the rise. These two processes together have already changed sea level notably. Since the last ice age, sea level has risen by about 120 m (about 400 feet). The rise due to melting of the ice sheets that once covered portions of North America, Europe, and Asia tapered off about 8,000 years ago (Fig. 23.28a). But a closer look at sea level for the past 130 years shows that it is continuing to rise (Fig. 23.28b). Measurements indicate that there has been a rise of almost 12 cm in the past century. Sea-level rise is already causing flooding of coastal wetlands and has submerged some islands. The amount of sea-level rise in the future depends on the rate of global warming. Models suggest that by 2100, it may rise by an additional 20 to 60 cm (Fig. 23.28c). A map showing the elevations of coastal areas emphasizes that a meter or two of sea-level rise could inundate regions of the world where 20% of the human population currently lives (Fig. 23.28d).

- *Stronger storms*: An increase in average ocean temperatures would mean that more of the ocean could evaporate when a tropical depression passed over. This evaporation might nourish stronger hurricanes. In nondesert areas, there might be more precipitation and, therefore, flooding.

- *An increase in wildfires*: Warmer temperatures may lead to an increase in the frequency of wildfires because the moisture content of plants is lower.

- *An interruption of the oceanic heat conveyor*: Oceanic currents play a major role in transferring heat across latitudes. If global warming melts polar ice, the resulting freshwater would dilute surface ocean water at high latitudes. This water could not sink, and thus thermohaline circulation would be shut off (see Chapter 18), preventing the water from conveying heat.

The potential changes described above, along with other studies that estimate the large economic cost of global warming, imply that the issue needs to be addressed seriously and soon. But what can be done? The 160 nations that signed the 1997 Kyoto Accord, at a summit meeting held in Japan, propose that the first step would be to slow the input of greenhouse gases into the atmosphere by decreasing the burning of fossil fuels and/or by forcing the CO_2 produced at power plants down wells into pore space underground by a process called carbon capture and sequestration (CCS). Motivating such actions, needless to say, involves controversial political decisions. Some researchers suggest that more aggressive solutions may be possible. Speculations include the intentional addition of sulfur dioxide to the atmosphere in order to increase the amount of sunlight reflected back to space, or the intentional addition of iron to the sea to encourage algal blooms that would absorb CO_2. But the feasibility of such approaches remains very far from certain.

Take-Home Message

- Humans have become significant causes of change in the Earth System.
- Much of the Earth's surface has been modified by human excavations and agriculture.
- Human activities change ecosystems, cause extinctions, and produce pollution.
- Growing evidence indicates that industrialization and deforestation have caused the CO_2 concentration in air to increase enough to cause global warming.
- Evidence of warming comes from measurements of glacial ice loss, atmospheric temperature, and sea-surface temperature.
- Continued warming could shift climate belts and cause sea-level rise.

THINK: How do researchers develop predictions of future climate change?

FIGURE 23.28 Sea-level rise of the recent past, and possible sea-level rise of the future.

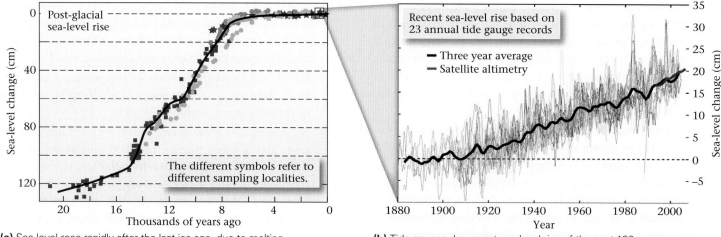

(a) Sea level rose rapidly after the last ice age, due to melting of continental glaciers.

(b) Tide gauges document sea-level rise of the past 130 years.

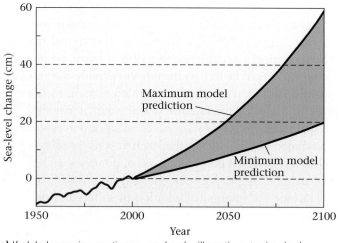

(c) If global warming continues, sea level will continue to rise, both because of the addition of glacial meltwater and the expansion of water that happens when water becomes warmer.

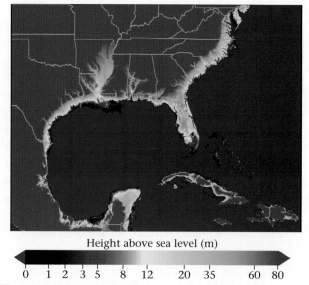

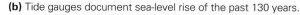

Height above sea level (m)

(d) The red and black areas of this map are so low that they could be flooded if sea level rose a few meters.

23.7 THE FUTURE OF THE EARTH: A SCENARIO

Most of the discussion in this book has focused on the past, for the geologic record preserved in rocks tells us of earlier times. Let's now bring this book to a close by facing in the opposite direction and speculating what the world might look like in geologic time to come.

In the geologic near term, the future of the world depends largely on human activities. Whether the Earth System undergoes a major disruption and shifts to a new equilibrium, whether a catastrophic mass-extinction event takes place, or whether society achieves **sustainable growth** (an ability to prosper within the constraints of the Earth System) will depend on our own foresight and ingenuity. Projecting

thousands of years into the future, we might well wonder if the Earth will return to ice-age conditions, with glaciers growing over major cities and the continental shelf becoming dry land, or if the ice age is over for good because of global warming. No one really knows for sure.

If we project millions of years into the future, it is clear that the map of the planet will change significantly because of the continuing activity of plate tectonics. For example, during the next 50 million years or so, the Atlantic Ocean will probably become bigger, the Pacific Ocean will shrink, and the western part of California will migrate northward. Eventually, Australia will crush against the southern margin of Asia, and the islands of Indonesia will be flattened in between.

Predicting the map of the Earth beyond that is hard, because we don't know for sure where new subduction zones will develop. Most likely, subduction of the Pacific Ocean will lead to the collision of the Americas with Asia, to produce a supercontinent ("Amasia"). A subduction zone eventually will form on one side of the Atlantic Ocean, and the ocean will be consumed. As a consequence, the eastern margin of the Americas will collide with the western margin of Europe and Africa. The sites of major cities—New York, Miami, Rio de Janeiro, Buenos Aires, and London—will be incorporated in a collisional mountain belt, and likely will be subjected to metamorphism and igneous intrusion before being uplifted and eroded. Shallow seas may once again cover the interiors of continents and then later retreat, and glaciers may once again cover the continents—it happened in the past, so it could happen again. And if the past is the key to the future, we *Homo sapiens* might not be around to watch our cities enter the rock cycle, for biological evolution may have introduced new species to the biosphere, and there is no way to predict what these species will be like.

And what of the end of the Earth? Geologic catastrophes resulting from asteroid and comet collisions will undoubtedly occur in the future as they have in the past. We can't predict when the next strike will come, but unless the object can be diverted, Earth is in for another radical readjustment of surface conditions. But it's not likely that such collisions will destroy our planet. Rather, astronomers predict that the end of the Earth will occur some 5 billion years from now, when the Sun begins to run out of nuclear fuel. When this happens, thermal pressure caused by fusion reactions will no longer be able to prevent the Sun from collapsing inward, because of the immense gravitational pull of its mass. Were the Sun a few times larger than it is, the collapse would trigger a supernova explosion that would blast matter of the Universe out into space to form a new nebula, perhaps surrounding a black hole. But since the Sun is not that large, the thermal energy generated when its interior collapses inward will heat the gases of its outer layers sufficiently to cause them to expand. As a result, the Sun will become a red giant, a huge star whose radius would grow beyond the orbit of Earth (Fig. 23.29a, b). Our planet will then vaporize, and its atoms will join an expanding ring of gas—the ultimate global change. If this happens, the atoms that once formed Earth and all its inhabitants through geologic time may eventually be incorporated in a future solar system, where the cycle of planetary formation and evolution will begin anew.

> ## Take-Home Message
>
> - The Earth will undergo major changes in the future. Climate will warm and cool, plates will move and change the map of the surface, and life will evolve.
> - The future of humanity may depend on whether society devises the means for sustainable growth.
> - Catastrophic collisions with extraterrestrial objects will happen in the future, but the end of the Earth probably will wait until the Sun becomes a red giant, 5 billion years from now.
>
> **THINK:** What might happen to the atoms comprising the Earth long after the red-giant stage of Solar System evolution?

FIGURE 23.29 In about 5 billion years, the Sun will become a red giant. When that happens, the Earth will first dry out, then vaporize; initially, it may look like a giant comet.

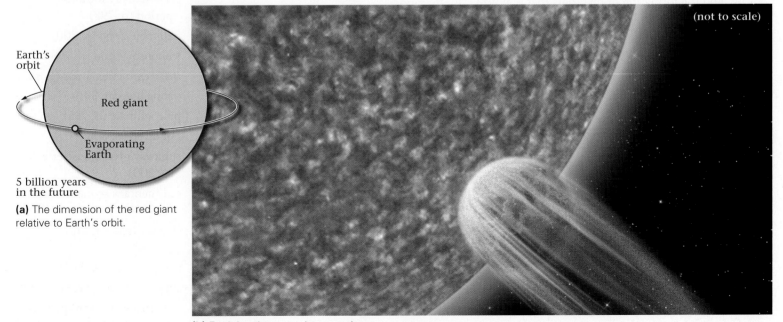

(a) The dimension of the red giant relative to Earth's orbit.

(b) Earth heating up and evaporating.

Chapter Summary

- We refer to the global interconnecting web of physical and biological phenomena on Earth as the Earth System. Global change involves the transformations or modifications of physical and biological components of the Earth System through time. Unidirectional change results in transformations that never repeat, whereas cyclic change involves repetition of the same steps over and over.

- Examples of unidirectional change include the gradual evolution of the solid Earth from a homogeneous collection of planetesimals to a layered planet, the formation of the oceans and the gradual change in the composition of the atmosphere, and the evolution of life.

- Examples of physical cycles that take place on Earth include the supercontinent cycle, the sea-level cycle, and the rock cycle.

- A biogeochemical cycle involves the passage of a chemical among nonliving and living reservoirs. Examples include the hydrologic cycle and the carbon cycle. Global change occurs when factors change the relative proportions of the chemical in different reservoirs.

- Tools for documenting global climate change include the stratigraphic record, paleontology, oxygen–isotope ratios, bubbles in ice, growth rings in trees, and human history.

- Studies of long-term climate change show that, at times in the past, the Earth experienced greenhouse (warmer) periods; at other times, there were icehouse (cooler) periods. Factors leading to long-term climate change include the positions of continents, volcanic activity, the uplift of land, and the formation of materials that remove CO_2, an important greenhouse gas.

- Short-term climate change can be seen in the record of the last million years. In fact, during only the past 15,000 years, we see that the climate has warmed and cooled a few times. Causes of short-term climate change include fluctuations in solar radiation and cosmic rays, changes in Earth's orbit and tilt, changes in reflectivity, and changes in ocean currents.

- Mass extinction, a catastrophic change in biodiversity, may be caused by the impact of a comet or asteroid or by intense volcanic activity associated with a superplume.

- During the last two centuries, humans have changed landscapes; modified ecosystems; and added pollutants to the land, air, and water at rates faster than the Earth System can process.

- The addition of CO_2 and CH_4 to the atmosphere appears to be causing global warming, which could shift climate belts and lead to a rise in sea level. Sources for the added CO_2 include fossil fuel burning, cement production, and deforestation.

- In the future, in addition to climate change, the Earth will witness a continued rearrangement of continents resulting from plate tectonics and will likely suffer the impact of asteroids and comets. The end of the Earth may come in about 5 billion years when the Sun runs out of fuel and becomes a red giant.

GEOPUZZLE REVISITED

Climate change has become a hot topic because evidence that it is taking place and will affect human society has been growing. Phenomena such as glacial melting are occurring at rates faster than anticipated. But climate isn't the only aspect of the Earth System undergoing change. Natural geologic phenomena such as erosion, continental drift, and sea-level rise and fall have changed our planet slowly but surely over geologic time.

Guide Terms

acid rain (p. 807)

albedo (p. 803)

biogeochemical cycle (p. 792)

climate-change model (p. 795)

Earth System (p. 785)

ecosystem (p. 806)

faint young Sun paradox (p. 800)

feedback mechanism (p. 795)

general circulation model (p. 795)

global change (p. 785)

global climate change (p. 793)

global cooling (p. 794)

global warming (pp. 794, 810)

Goldilocks effect (p. 800)

greenhouse effect (p. 794)

greenhouse gas (p. 794)

greenhouse (hothouse) period (p. 798)

habitable zone (p. 800)

icehouse period (p. 798)

mass-extinction event (p. 803)

ozone hole (p. 807)

paleoclimate (p. 795)

pollution (p. 806)

runaway greenhouse effect (p. 800)

sedimentary sequence (p. 790)

steady-state condition (p. 792)

sunspot cycle (p. 802)

supercontinent cycle (p. 790)

sustainable growth (p. 816)

Review Questions

1. Why do we use the term *Earth System* to describe the processes operating on this planet?

2. How have the Earth's crust and atmosphere changed since they first formed?

3. What processes control the rise and fall of sea level on Earth?

4. How does carbon cycle through the various Earth systems?

5. How do paleoclimatologists study ancient climate change?

6. Contrast icehouse and greenhouse conditions.

7. What are the possible causes of long-term climatic change?

8. What factors explain short-term climatic change?

9. Give some examples of events that cause catastrophic change.

10. Give some examples of how humans have changed the Earth.

11. What is the ozone hole, and how does it affect us?

12. Describe how carbon-dioxide-induced global warming takes place, and how humans may be responsible. What effects might global warming have on the Earth System?

13. What are some likely scenarios for the long-term future of the Earth?

On Further Thought

14. If global warming continues, how will the distribution of grain crops change? Might this affect national economies? Why? How will the distribution of spruce forests change?

15. Currently, tropical rain forests are being cut down at a rate of 1.8% per year. At this rate, how many more years will the forests survive? In the eastern United States, the proportion of land with forest cover today has increased over the past century. In fact, most of the farmland that existed in New York State in 1850 is forestland today. Why? How might this change affect erosion rates in the region?

16. Using the library or the Web, examine the change in the nature of world fisheries that has taken place in the last 50 years. Is the world's fish biomass sustainable if these patterns continue? What has happened to whale populations during the past 50 years?

 For more resources, including animations, quizzes, and Norton's GeoTours, go to **wwnorton.com/studyspace**.

 If your instructor assigns exercises in SmartWork, log in at **smartwork.wwnorton.com**.

ANOTHER VIEW   Numerous fjords make western Iceland's coast resemble the shape of a leaf. The glaciers that carved these fjords vanished over 10,000 years ago. Their growth and demise is one manifestation of climate change on Earth.

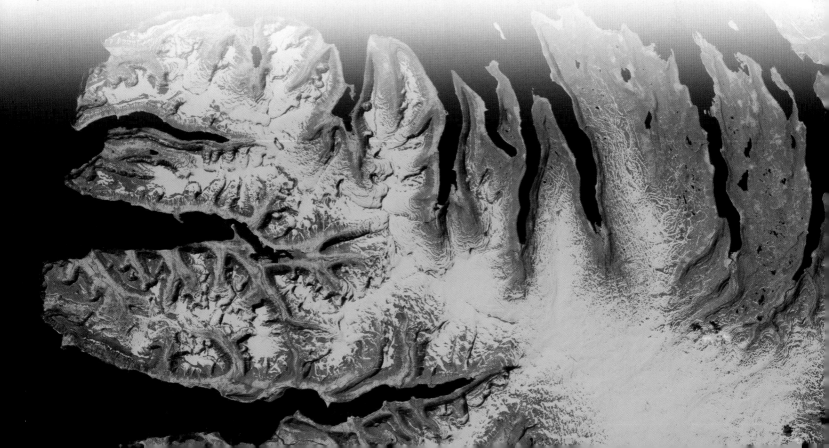

Additional Maps and Charts

This appendix contains several maps and charts for general reference. We list the purpose of each below.

The *Periodic Table of Elements* (Fig. a.1) has been provided as a reference since certain key topics of chemistry are an essential background to this text.

Mineral Identification Flowcharts: Geologists use these charts to identify unknown mineral specimens (Fig. a.2a, b). A mineral flowchart is simply an organized series of questions concerning the mineral's physical properties. The questions are arranged in a sequence such that appropriate answers ultimately lead you on a path to a specific mineral. To understand this concept, let's imagine that we are trying to identify a shiny, bronze- or gold-colored, metallic-looking mineral specimen. We start by observing the specimen's luster. It is metallic, so we follow the path on the chart for metallic-luster minerals (see Fig. a.2a). Next, we determine if the mineral is magnetic or nonmagnetic. If it is nonmagnetic, we follow the path for nonmagnetic minerals. Then, we look at the mineral's color. Since it is bronze- or gold-colored, our path ends at pyrite.

Notice that one of the flowchart questions in Fig. a.2b asks about the reaction of the specimen with hydrochloric acid (HCl). Only calcite and dolomite react, so the question allows definitive identification of these minerals. Another question pertains to striations, faint parallel lines on cleavage planes. Only plagioclase has striations.

World Magnetic Declination Map: This map shows the variation of magnetic declination with location on the surface of the Earth (Fig. a.3a). Declination exists because the position of the Earth's magnetic pole does not coincide exactly with that of the geographic pole. In fact, the magnetic pole location constantly moves, currently at a rate of about 20 km per year. It now lies off the north coast of Canada, and in the not too distant future, it may lie along the coast of Siberia. For a compass to give an accurate indication of direction, it must be adjusted to accommodate for the declination at the location of measurement.

US Magnetic Declination Map: This map shows the magnetic declination for the United States (Fig. a.3b).

Volcanoes of the Past Few Million Years: This map shows the location of volcanoes that have erupted in relatively recent time (Fig. a.4). The map also shows the locations of earthquakes.

Earthquake Epicenter Map: This map shows the positions of epicenters for earthquakes that were large enough to be detected at seismic stations over a broad region (Fig. a.5). An epicenter is the point on the surface of the Earth above the earthquake's hypocenter (focus). The hypocenter is the place where the earthquake energy is generated. Different colors indicate the different depths of the earthquake hypocenters.

World Soils Map: This map displays areas of different types of soils, using one of the standard classification schemes for soils (Fig. a.6).

A Satellite Image of the Earth at Night: This image shows light sources on the surface of the planet (Fig. a.7). Light sources indicate both the density of human habitation and the degree of industrialization. Note, for example, that most of the lights in Australia occur in coastal cities.

A Geologic Province Map: This United States Geological Survey (USGS) map shows the variety of distinct crustal provinces on Earth (Fig. a.8).

The North America Tapestry of Time and Terrain: Another map provided by USGS superimposes a digital elevation map of North America over a geologic map of the same to get a full geologic picture of our continent (Fig. a.9).

The Principal Aquifers of the United States: This map shows aquifers in the U.S. in relation to the six types of rocks and deposits in which they occur (Fig. a.10).

Elevation Map of the World: A modern elevation map of the world has been provided as an overall reference point for many topics throughout this book (Fig. a.11).

Metric Conversion Chart: This chart shows the correlation between U.S. standard units and metric units for length, area, volume, mass, pressure, and temperature (including formulas for converting temperatures between Fahrenheit, Celsius, and Kelvin).

Legend:

	Example
Symbol	He
Atomic number	2
Name	Helium
Atomic weight	4.002

Alkali metals · Transition elements (metals) · Nonmetals · Inert gases

Group	1	2	3	4	5	6	7	8	9	10	11	12	13	14	15	16	17	18
1	H 1 Hydrogen 1.007																	He 2 Helium 4.002
2	Li 3 Lithium 6.941	Be 4 Beryllium 9.0121											B 5 Boron 10.811	C 6 Carbon 12.011	N 7 Nitrogen 14.006	O 8 Oxygen 15.999	F 9 Fluorine 18.998	Ne 10 Neon 20.179
3	Na 11 Sodium 22.989	Mg 12 Magnesium 24.305											Al 13 Aluminum 26.981	Si 14 Silicon 28.085	P 15 Phosphorus 30.973	S 16 Sulfur 32.066	Cl 17 Chlorine 35.452	Ar 18 Argon 39.948
4	K 19 Potassium 39.098	Ca 20 Calcium 40.078	Sc 21 Scandium 44.955	Ti 22 Titanium 47.88	V 23 Vanadium 50.941	Cr 24 Chromium 51.996	Mn 25 Manganese 54.938	Fe 26 Iron 55.847	Co 27 Cobalt 58.933	Ni 28 Nickel 58.693	Cu 29 Copper 63.546	Zn 30 Zinc 65.39	Ga 31 Gallium 69.723	Ge 32 Germanium 72.61	As 33 Arsenic 74.921	Se 34 Selenium 78.96	Br 35 Bromine 79.904	Kr 36 Krypton 83.80
5	Rb 37 Rubidium 85.467	Sr 38 Strontium 87.62	Y 39 Yttrium 88.905	Zr 40 Zirconium 91.224	Nb 41 Niobium 92.906	Mo 42 Molybdenum 95.94	Tc 43 Technetium 98.907	Ru 44 Ruthenium 101.07	Rh 45 Rhodium 102.905	Pd 46 Palladium 106.42	Ag 47 Silver 107.868	Cd 48 Cadmium 112.411	In 49 Indium 114.82	Sn 50 Tin 118.710	Sb 51 Antimony 121.757	Te 52 Tellurium 127.60	I 53 Iodine 126.904	Xe 54 Xenon 131.29
6	Cs 55 Cesium 132.905	Ba 56 Barium 137.327	La 57 Lanthanum 138.905	Hf 72 Hafnium 178.49	Ta 73 Tantalum 180.947	W 74 Tungsten 183.85	Re 75 Rhenium 186.207	Os 76 Osmium 190.2	Ir 77 Iridium 192.22	Pt 78 Platinum 195.08	Au 79 Gold 196.966	Hg 80 Mercury 200.59	Tl 81 Thallium 204.383	Pb 82 Lead 207.2	Bi 83 Bismuth 208.980	Po 84 Polonium 208.982	At 85 Astatine 209.987	Rn 86 Radon 222.017
7	Fr 87 Francium 223.019	Ra 88 Radium 226.025	Ac 89 Actinium 227.027															

Lanthanides:

Ce 58 Cerium 140.115	Pr 59 Praseodymium 140.907	Nd 60 Neodymium 144.24	Pm 61 Promethium 144.912	Sm 62 Samarium 150.36	Eu 63 Europium 151.965	Gd 64 Gadolinium 157.25	Tb 65 Terbium 158.925	Dy 66 Dysprosium 162.50	Ho 67 Holmium 164.930	Er 68 Erbium 167.26	Tm 69 Thulium 168.934	Yb 70 Ytterbium 173.04	Lu 71 Lutetium 174.967

Actinides:

Th 90 Thorium 232.038	Pa 91 Protactinium 231.035	U 92 Uranium 238.028	Np 93 Neptunium 237.048	Pu 94 Plutonium 244.064	Am 95 Americium 243.061	Cm 96 Curium 247.070	Bk 97 Berkelium 247.070	Cf 98 Californium 251.079	Es 99 Einsteinium 252.083	Fm 100 Fermium 257.095	Md 101 Mendelevium 258.10	No 102 Nobelium 259.100	Lr 103 Lawrencium 262.11

FIGURE a.1 The modern periodic table of the elements. Each column groups elements with related properties. For example, inert gases are listed in the column on the right. Metals are found in the central and left parts of the chart.

FIGURE a.2 Simplified mineral identification flowcharts.

Mineral-Identification Flow Chart (metallic or dark-colored nonmetallic)

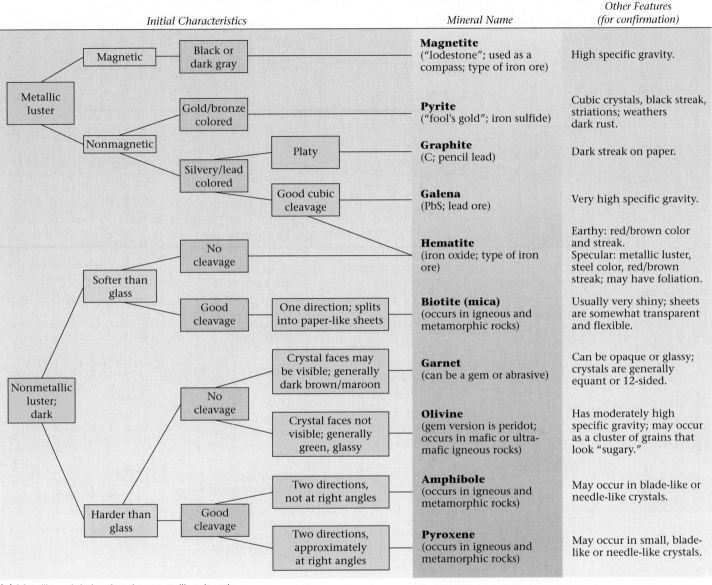

(a) Metallic and dark-colored nonmetallic minerals.

(continued)

FIGURE a.2　Simplified mineral identification flow charts. *(continued)*

Mineral-Identification Flow Chart (light-colored nonmetallic)

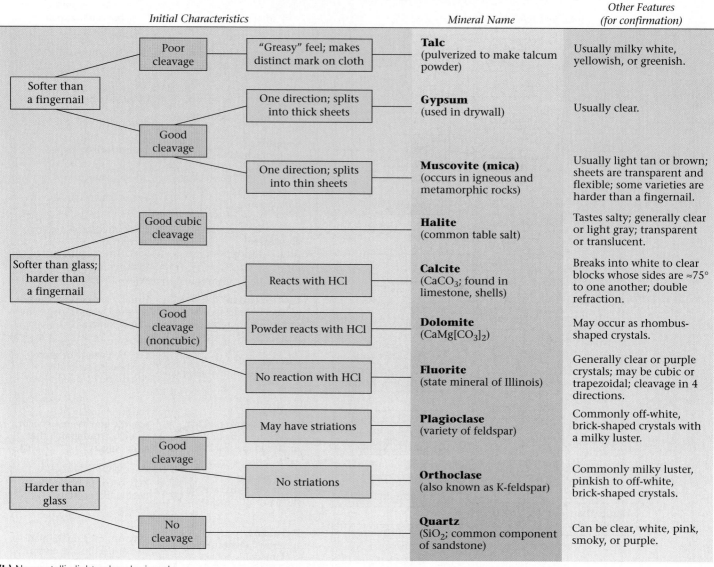

(b) Nonmetallic light-colored minerals.

FIGURE a.3 Magnetic declination charts.

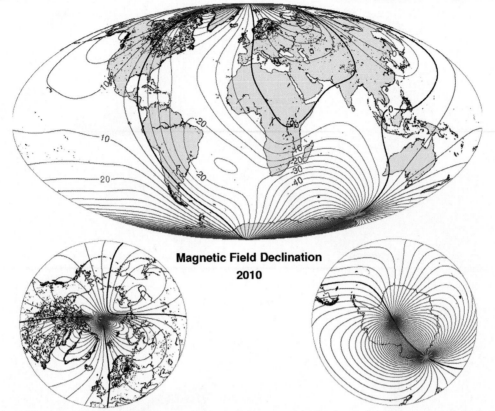

**Magnetic Field Declination
2010**

(a) A simplified declination map for the world. Blue areas are west declination and red areas are east declination.

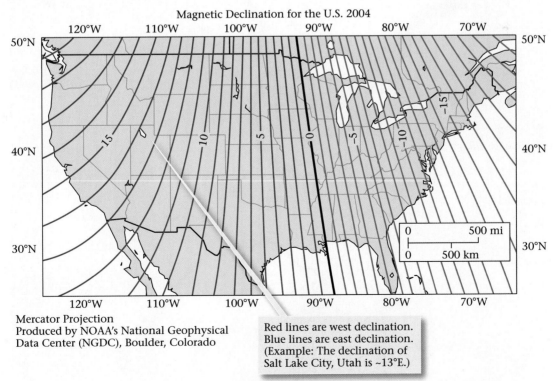

Mercator Projection
Produced by NOAA's National Geophysical
Data Center (NGDC), Boulder, Colorado

Red lines are west declination.
Blue lines are east declination.
(Example: The declination of
Salt Lake City, Utah is ~13°E.)

(b) A simplified declination map of the United States for 2004. Because of changes in the field, most of the lines are drifting westward at about 5′ (minutes) per year. (Note: 1° = 60′)

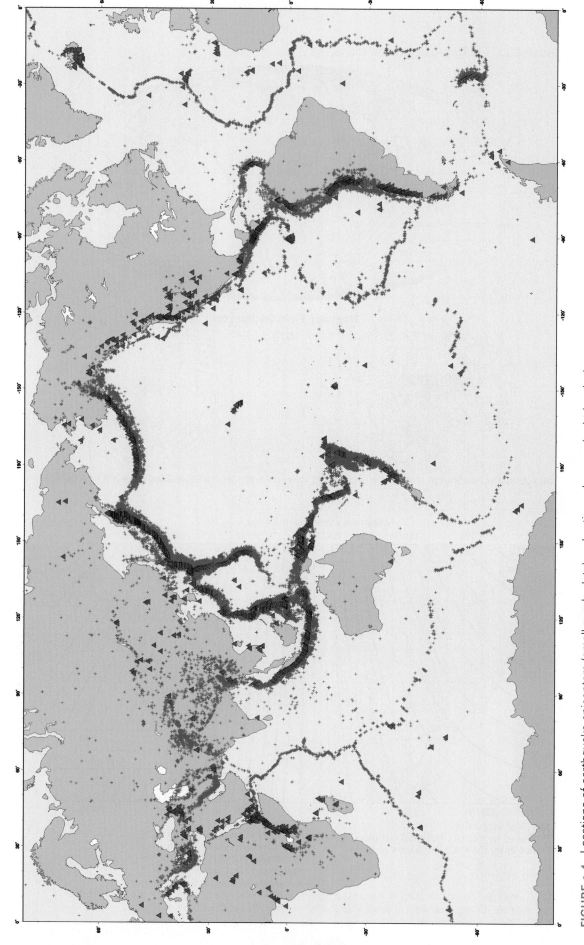

FIGURE a.4 Locations of earthquake epicenters (small purple dots) and active volcanoes (red triangles). Note that most volcanoes occur along plate boundaries, but some occur at hot spots in plate interiors.

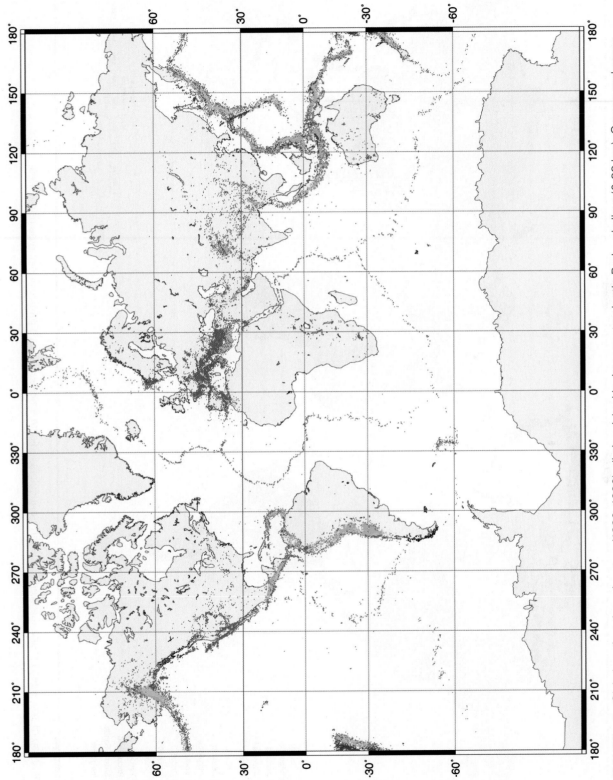

FIGURE a.5 Global earthquake epicenters (1990–1996), distinguished by hypocenter depth. Red = shallow (0-33 km); Orange and green = intermediate (33-300 km); Blue = deep (300-700 km). Note that deep earthquakes occur only at convergent margins.

Global Soil Regions

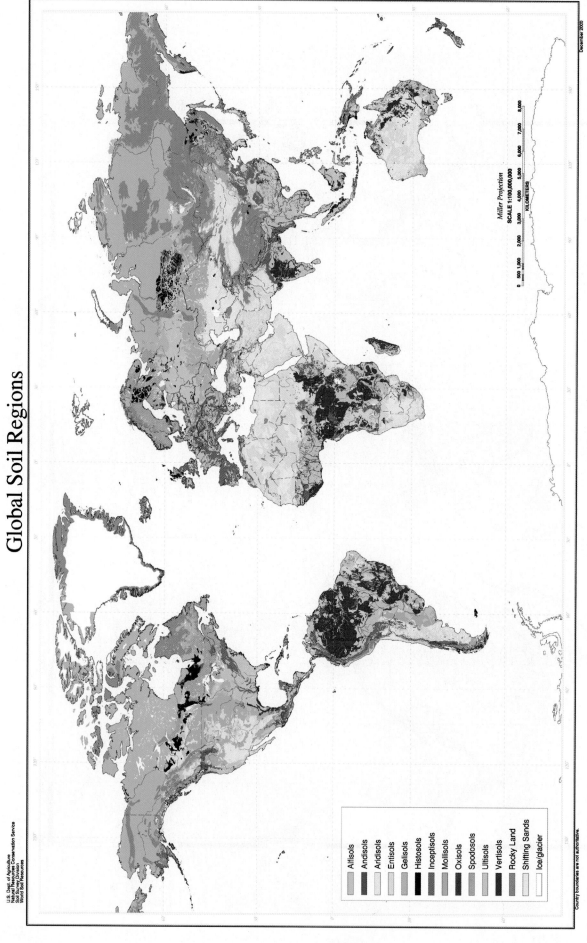

U.S. Dept. of Agriculture
Natural Resources Conservation Service
Soil Survey Division
World Soil Resources

December 2000

| Alfisols |
| Andisols |
| Aridisols |
| Entisols |
| Gelisols |
| Histosols |
| Inceptisols |
| Mollisols |
| Oxisols |
| Spodosols |
| Ultisols |
| Vertisols |
| Rocky Land |
| Shifting Sands |
| Ice/glacier |

Miller Projection
SCALE 1:100,000,000

KILOMETERS
0 500 1,000 2,000 3,000 4,000 5,000 6,000 7,000 8,000

Country boundaries are not authoritative.

FIGURE a.6 Map of soil types around the world. The different colors indicate areas in which a particular soil type dominates.

FIGURE a.7 A composite satellite image of the Earth at night. The white areas are electrically lit. The map gives a sense of the distribution of both population and technology. Dark areas have either small populations or relatively small amounts of electrical power.

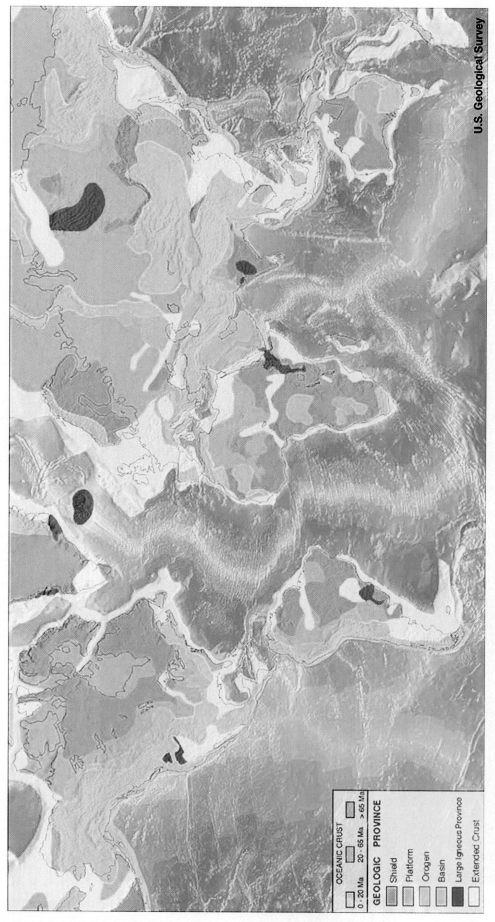

U.S. Geological Survey

OCEANIC CRUST

| 0 - 20 Ma | 20 - 65 Ma | > 65 Ma |

GEOLOGIC PROVINCE

Shield
Platform
Orogen
Basin
Large Igneous Province
Extended Crust

FIGURE a.8 Map showing the variety of distinct crustal provinces on Earth.

**The North American
Tapestry of Time and Terrain**

Quaternary -
Neogene
Paleogene

Cretaceous
Jurassic
Triassic

Permian
Pennsylvanian
Mississippian
Devonian
Silurian
Ordovician
Cambrian

Later Proterozoic
Middle Proterozoic
Early Proterozoic
Later Archean
Middle Archean
Early Archean

Glacial ice
Age unknown

| 0 | 200 | 400 | 600 | 800 | 1,000 mi |

| 0 | 300 | 600 | 900 | 1,200 | 1,500 km |

≋**USGS**
science for a changing world

U.S. DEPARTMENT OF THE INTERIOR
U.S. GEOLOGICAL SURVEY

FIGURE a.9 Geologic map of North America, superimposed on a digital elevation map.

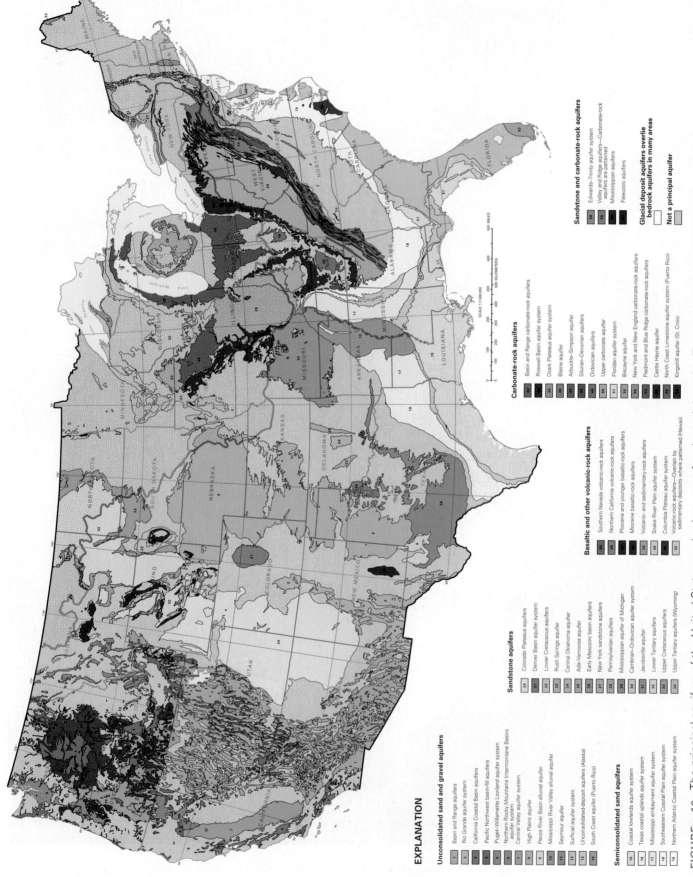

EXPLANATION

Unconsolidated sand and gravel aquifers

1 Basin and Range aquifers
2 Rio Grande aquifer system
3 California Coastal Basin aquifers
4 Pacific Northwest basin-fill aquifers
5 Puget–Willamette Lowland aquifer system
6 Northern Rocky Mountains Intermontane Basins aquifer system
7 Central Valley aquifer system
8 High Plains aquifer
9 Pecos River Basin alluvial aquifer
10 Mississippi River Valley alluvial aquifer
11 Seymour aquifer
12 Surficial aquifer system
13 Unconsolidated-deposit aquifers (Alaska)
14 South Coast aquifer (Puerto Rico)

Semiconsolidated sand aquifers

15 Coastal lowlands aquifer system
16 Texas coastal uplands aquifer system
17 Mississippi embayment aquifer system
18 Southeastern Coastal Plain aquifer system
19 Northern Atlantic Coastal Plain aquifer system

Sandstone aquifers

20 Colorado Plateaus aquifers
21 Denver Basin aquifer system
22 Lower Cretaceous aquifers
23 Rush Springs aquifer
24 Central Oklahoma aquifer
25 Ada–Vamoosa aquifer
26 Early Mesozoic basin aquifers
27 New York sandstone aquifers
28 Pennsylvanian aquifers
29 Mississippian aquifer of Michigan
30 Cambrian–Ordovician aquifer system
31 Jacobsville aquifer
32 Lower Tertiary aquifers
33 Upper Cretaceous aquifers
34 Upper Tertiary aquifers (Wyoming)

Basaltic and other volcanic-rock aquifers

35 Southern Nevada volcanic-rock aquifers
36 Northern California volcanic-rock aquifers
37 Pliocene and younger basaltic-rock aquifers
38 Miocene basaltic-rock aquifers
39 Volcanic- and sedimentary-rock aquifers
40 Snake River Plain aquifer system
41 Columbia Plateau aquifer system
42 Volcanic-rock aquifers—Overlain by sedimentary deposits where patterned (Hawaii)

Carbonate-rock aquifers

43 Basin and Range carbonate-rock aquifers
44 Roswell Basin aquifer system
45 Ozark Plateaus aquifer system
46 Blaine aquifer
47 Arbuckle–Simpson aquifer
48 Silurian–Devonian aquifers
49 Ordovician aquifers
50 Upper carbonate aquifer
51 Floridan aquifer system
52 Biscayne aquifer
53 New York and New England carbonate-rock aquifers
54 Piedmont and Blue Ridge carbonate-rock aquifers
55 Castle Hayne aquifer
56 North Coast Limestone aquifer system (Puerto Rico)
57 Kingshill aquifer (St. Croix)

Sandstone and carbonate-rock aquifers

58 Edwards–Trinity aquifer system
59 Valley and Ridge aquifers—Carbonate-rock aquifers are patterned
60 Mississippian aquifers
61 Paleozoic aquifers

Glacial deposit aquifers overlie bedrock aquifers in many areas

Not a principal aquifer

SCALE 1:7,500,000

FIGURE a.10 The principal aquifers of the United States are in six types of rocks and deposits. The colored areas show the extent of each principal aquifer at or near the land surface.

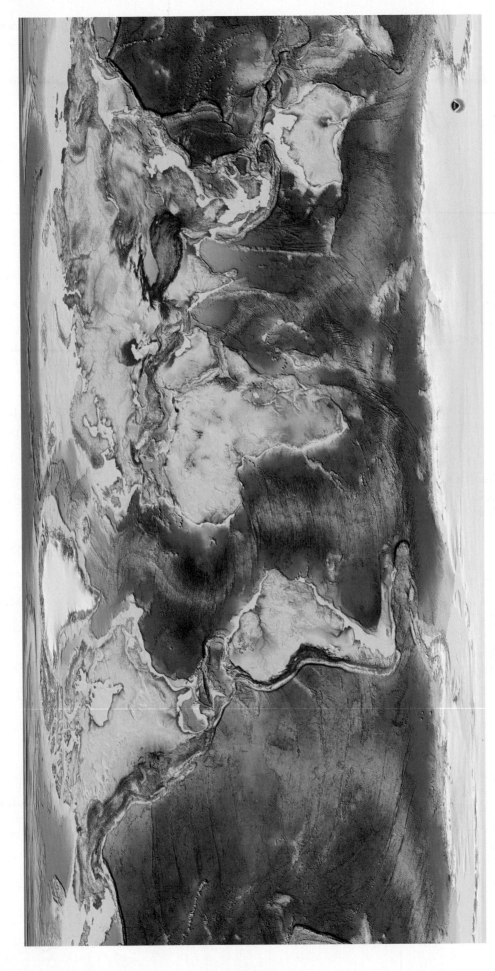

FIGURE a.11 Elevation map of the world. On land, brown is higher elevation and green is lower. In the sea, light blue is shallower water and dark blue is deeper water.

Metric Conversion Chart

Length

1 kilometer (km) = 0.6214 mile (mi)

1 meter (m) = 1.094 yards = 3.281 feet

1 centimeter (cm) = 0.3937 inch

1 millimeter (mm) = 0.0394 inch

1 mile (mi) = 1.609 kilometers (km)

1 yard = 0.9144 meter (m)

1 foot = 0.3048 meter (m)

1 inch = 2.54 centimeters (cm)

Area

1 square kilometer (km^2) = 0.386 square mile (mi^2)

1 square meter (m^2) = 1.196 square yards (yd^2)

= 10.764 square feet (ft^2)

1 square centimeter (cm^2) = 0.155 square inch (in^2)

1 square mile (mi^2) = 2.59 square kilometers (km^2)

1 square yard (yd^2) = 0.836 square meter (m^2)

1 square foot (ft^2) = 0.0929 square meter (m^2)

1 square inch (in^2) = 6.4516 square centimeters (cm^2)

Volume

1 cubic kilometer (km^3) = 0.24 cubic mile (mi^3)

1 cubic meter (m^3) = 264.2 gallons

= 35.314 cubic feet (ft^3)

1 liter (1) = 1.057 quarts

= 33.815 fluid ounces

1 cubic centimeter (cm^3) = 0.0610 cubic inch (in^3)

1 cubic mile (mi^3) = 4.168 cubic kilometers (km^3)

1 cubic yard (yd^3) = 0.7646 cubic meter (m^3)

1 cubic foot (ft^3) = 0.0283 cubic meter (m^3)

1 cubic inch (in^3) = 16.39 cubic centimeters (cm^3)

Mass

1 metric ton = 2,205 pounds

1 kilogram (kg) = 2.205 pounds

1 gram (g) = 0.03527 ounce

1 pound (lb) = 0.4536 kilogram (kg)

1 ounce (oz) = 28.35 grams (g)

Pressure

1 kilogram per square
centimeter (kg/cm^2)* = 0.96784 atmosphere (atm)

= 0.98066 bar

= 9.8067×10^4 pascals (Pa)

1 bar = 10 megapascals (Mpa)

= 1.0×10^5 pascals (Pa)

= 29.53 inches of mercury (in a barometer)

= 0.98692 atmosphere (atm)

= 1.02 kilograms per square centimeter (kg/cm^2)

1 pascal (Pa) = 1 $kg/m/s^2$

1 pound per square inch = 0.06895 bars

= 6.895×10^3 pascals (Pa)

= 0.0703 kilogram per square
centimeter

Temperature

To change from Fahrenheit (F) to Celsius (C):

$$°C = \frac{(°F - 32°)}{1.8}$$

To change from Celsius (C) to Fahrenheit (F):

$$°F = (°C \times 1.8) + 32°$$

To change from Celsius (C) to Kelvin (K):

$$K = °C + 273.15$$

To change from Fahrenheit (F) to Kelvin (K):

$$K = \frac{(°F - 32°)}{1.8} + 273.15$$

*Note: Because kilograms are a measure of mass whereas pounds are a unit of weight, pressure units incorporating kilograms assume a given gravitational constant (g) for Earth. In reality, the gravitational for Earth varies slightly with location.

See for Yourself A

How to Use *Google Earth*™ to See Geologic Features

A static photo or map cannot convey the full three-dimensional character of the Earth's surface features. Fortunately, a free, web-based computer tool, *Google Earth*™, solves this problem by providing a global compilation of satellite imagery in a format that permits you to look at the land surface from any angle and from any elevation. With *Google Earth*™, you can zip anywhere on our planet in a matter of seconds, can fly like a plane over the land surface, and can see features of the ocean floor as if the water weren't there. Also, you can descend low enough to see individual boulders on a mountain, or rocket high enough to see the context of a mountain relative to a continent.

The following pages help you use *Google Earth*™ to learn geology by seeing geologic features for yourself. In each of the following *See for Yourself* spreads, we provide locations (latitude and longitude), thumbnail photographs, and descriptions of localities at which you can study a feature or process. To find a site, you can type in its latitude and longitude, or you can simply click on the site number provided by a downloadable computer file. Given the location information, your computer will fly you right to the site.

For many additional opportunities to learn geology using *Google Earth*™, we also offer a separate *Google Earth Geotour Workbook* (by Scott Wilkerson, Beth Wilkerson, and Stephen Marshak). This *Workbook* provides questions that utilize the *See for Yourself* sites, as well as many other sites, to help you understand how to "read" the landscape and understand the clues it provides about our planet's past and present. The sites provided by the *Workbook* include many additional features, such as map overlays and photographs.

In *See for Yourself A*, we introduce you to the basic tools of *Google Earth*™. Since the program is constantly being updated by its producer, some details of our descriptions may become superseded. But the overall architecture of *Google Earth*™ remains the same across versions, and the changes are generally self explanatory. Subsequent *See for Yourself* spreads introduce you to sites associated with specific geologic themes.

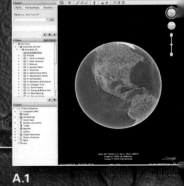

A.1

Opening *Google Earth*™

To use a See for Yourself, start by opening *Google Earth*™. As the program initializes, an image of the globe set in a background of stars appears in a window. The toolbar across the top provides a window icon. Click on this icon and a left sidebar appears (**Image A.1**)—click on the icon again and the sidebar disappears so that the image becomes larger. The toolbar also provides a pushpin tool for marking locations, and a ruler tool for measuring distances. The sidebar contains information about locations and provides options for adding information to the screen image. For example, when you click on "borders," "roads," and "Populated Places" in the Layers panel, political boundaries, highways, and city names appear to provide a visual reference frame (**Image A.2**). Any location you have marked with a pushpin appears in the sidebar within the Places panel. Double-clicking on the location in the sidebar will make the pushpin appear and will take you to the location. If you type a location in the Search panel at the top, a click on the magnifying glass icon will fly you to the location.

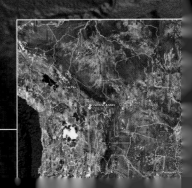

The Basic Tools of *Google Earth*™

In *Google Earth 4.3*™, three navigation tools appear in the upper right-hand corner of the screen (**Image A.3**): (1) The **Look tool** (circle with an eye in the center) allows you to rotate the view. You can do this either by clicking and dragging the mouse on or within the outer ring, or by rotating the view clockwise or counterclockwise by clicking on the left or right arrows, respectively. Double-clicking the N button reorients the view with north at the top of the screen. Clicking on the top arrow of the Look tool tilts the image to show an oblique view, whereas clicking on the bottom arrow rotates the perspective back towards vertical. You can see this contrast by comparing the "Basic Tools-Vertical" and the "Basic Tools-Oblique" images of a site in the Andes Mountains of Bolivia (Lat 18°21'36.08"S, Long 66°0'13.95"W) from an altitude of 10 km (**Images A.4** and **A.5**). Note that, because of the nature of the imagery used by *Google Earth*™, steep cliffs are distorted. Clicking and dragging the mouse in the inner circle of the Look tool performs both actions simultaneously. (2) The **Move tool** (circle with a hand in the center) pans across the image. By clicking the arrows, you move a finite distance in the specified direction. Clicking and dragging the mouse in the inner circle acts as a joystick and translates the view in any direction. (3) The vertical **Zoom slider** zooms the view up or down. You can either move the slider or click on the end of the bar. Clicking on the + end takes you to a lower elevation, whereas clicking on the – end takes you to a higher elevation. The slider offers better control. Using the "View" menu bar, you can click to display a bar scale in the lower left portion of the screen. Note that the default setting in *Google Earth 4.3*™ is to swoop (tilt) into a perspective view as you zoom closer. This auto-tilt behavior may be turned off in the Preferences (Options) menu.

On the Earth image, you will see a hand-shaped cursor. By dragging the cursor across the screen, you can move the image. If you quickly drag the hand cursor, while holding down the mouse, and then let go, the movement will continue. For additional information, see the *Google Earth*™ User Guide.

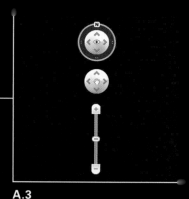

A.3

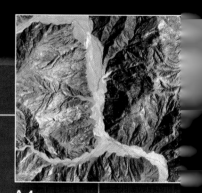

A.4

A.5

Working with Location and Elevation (Lat 43°8'18.84"N, Long 77°34'18.36"W)

On the bottom rule of the window, you will see three information items. The location of the point just beneath the hand-shaped cursor on the screen is specified on the left in terms of latitude (degrees, minutes, and seconds north or south of the equator) and longitude (degrees, minutes, and seconds east or west of the of the prime meridian). Just to the right of the Lat/Long information, a number indicates the elevation of the ground surface just below the hand-shaped cursor. The next number to the right tells you how much of the image has streamed onto your computer—an image starts out blurry, and then as streaming approaches 100%, it becomes clearer. The number on the far right ("Eye alt") indicates your viewing elevation.

To practice using these tools, enter the latitude and longitude provided above and zoom to a viewpoint 25,000 km (15,500 miles) out in space. You will see all of North America (**Image A.6**). If you zoom down to 1800 m (5900 ft), you can see the details of Cobb's Hill Park in Rochester, New York (**Image A.7**). Tilt your image so you just see the horizon, and rotate the image so you are looking SW (**Image A.8**). With this perspective, you realize that the bean-shaped reservoir sits on top of an elongate ridge.

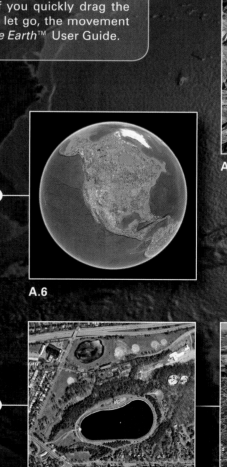

A.6

Image Boundaries
(Lat 16°6'3.12"N, Long 75°48'25.84"E)

The globe that you see on your screen is a compilation of many separate satellite images merged by computer; some regions, therefore, look like a patchwork. To see this effect, fly to the coordinates provided in India and look down from an elevation of 600 km (**Image A.9**). Image boundaries may reflect one or more of the following factors. (1) *Resolution*: The sharpness or clarity of images on *Google Earth*™ depend on the resolution of the image that is available in the database. High-resolution images are clear at high magnification, whereas low-resolution images appear grainy or pixelated at high magnification. Depending on the speed of your internet connection, it may take time for the full available resolution of an image to show up. The image is fully loaded when the blue circle in the lower right-hand corner has become solid. (2) *Season*: Some images show regions in the winter, when leaves are down and the ground is visible, whereas some show regions in summer, when the canopy of trees hides the ground. (3) *Photo Tint*: Not all satellite images are taken in the same way, so not all have the same basic color tint. *Google Earth*™ constantly updates images, as new ones become available, so a region that looks like a patchwork one day might not the next.

A.9

Finding Locations Using Latitude and Longitude
(Lat 48°52'25.26"N, Long 2°17'42.18"E)

You can find towns, parks, and landmarks on the Earth using *Google Earth*™ by entering the name in the search panel in the upper left corner. For example, enter Arc de Triomphe, hit the return button, and you fly to this well known landmark in the center of Paris, France (**Image A.10**). Many of the places you will visit in Geotours of this book are not, however, near a well-known landmark. Thus, to get you to a locality, we provide a latitude and longitude in the form:

   Lat 48°52'25.44"N, Long 2°17'42.31"E

The first number is the latitude in degrees, minutes (60' = 1°), and seconds (60.00" = 1'). Simply enter the latitude and longitude on your screen. When typing numbers into the space provided on the program, you can abbreviate as:

   48 52 25.44N, 2 17 42.31E

Note that instead of typing in latitude and longitude, you can simply copy the ".kmz" files provided in this book's website onto your computer. When you double-click on a file, it will open *Google Earth*™ and add all the locations referred to in the book to the "Temporary Places" folder. Then you simply need to click on the particular image referred to in a Geotour and the program will fly you straight there.

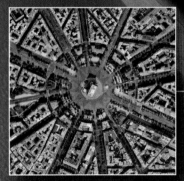

A.10

Enjoy!

Please take advantage of the See for Yourself sites provided in this book. They provide you with an understanding of geology that simply can't appear on a printed page, and they are, arguably, as fun as a video game. Take control of the *Google Earth*™ tools, change your elevation and perspective, and fly around the landscape. Geology will come alive for you, and exploration of the Earth may become your hobby.

The Variety of Earth's Surface

Google Earth™ allows you to sense what a space traveler orbiting the Earth would see if there were no clouds and almost no sea ice, and if features of the sea floor were visible. We can even see relicts of meteorite impact craters. Let's orbit the Earth, and then look more closely at examples of its surface. The thumbnail images provided on this page are only to help identify tour sites. Go to wwnorton. com/studyspace to experience each flyover tour.

Distant View of the Earth

Zoom to view the planet from an elevation of about 12,000 km (about 8,000 miles). Turn on the "Lat/Long Grid" to see lines of latitude and lines of longitude (**Image B.1**). Orient the image so that the equator is horizontal and examine the numbering scheme of these lines relative to the equator and the prime meridian. Click and drag from right to left and then let go to simulate the Earth's rotation and see the distribution of land and sea. Note that different shades of blue distinguish between deeper and shallower water, as you can see in the western Caribbean and Bahamas region, from an elevation of 3,000 km (1,864 miles) (**Image B.2**).

B.1

B.2

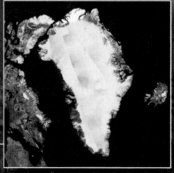

B.3

Eastern Greenland
(Lat 74°55'21.75"N, Long 22°5'32.62"W)

Ice covers the land surface of Greenland. Fly to the designated locality, then zoom to 3,200 km (2,000 miles) above sea level (**Image B.3**). You can see the vast sheet of ice.

Zoom to 160 km (100 miles) above sea level. Now you can see that the ice sheet drains to the sea via slowly moving "rivers of ice" (**Image B.4**). These valley glaciers were once longer and they carved deep valleys. Rising sea level filled the valleys to form fingers of the ocean called fjords. Note that the water in these fjords has frozen to form sea ice. Next, zoom down to 16 km (10 miles) and tilt the image so you just see the Earth's horizon (**Image B.5**). Fly along the coast to see spectacular valleys and cliffs.

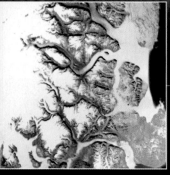

B.4

B.5

B.6

B.7

Southern Alps, New Zealand (Lat 43°30'45.61"S, Long 170°50'41.97"E)

Zoom to an elevation of 20 km (about 12 miles) and you can see an example of rugged, mountainous topography and sediment-choked streams. Next, zoom to 16 km (10 miles), tilt so that you can see the horizon, and fly northwest up the river valley (**Image B.6**). Turn 180° and fly down the river. Eventually, the river leaves the mountains and crosses the plains. Here, farmers have divided the land into fields (**Image B.7**). The river eventually reaches the coast and drains into the Pacific Ocean.

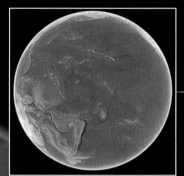

B.8

Pacific Ocean (Lat 3°43'48.84"N, Long 163°51'15.32"W)

Zoom out to an elevation of 17,000 km (10,500 miles) above sea level. The Pacific Ocean almost fills your field of view (**Image B.8**). You can see a few islands and can note that the floor of the sea is not perfectly flat—it has seamounts, trenches, and oceanic ridges.

B.9

Midwestern United States (Lat 39°51'02.69"N, Long 87°24'30.44"W)

Zoom to 9,000 m (30,000 feet) to see the view from a jet plane (**Image B.9**). The image displays the Wabash River, farm fields, wooded areas along small streams, and a town (Newport, Indiana). What percentage of this landscape has been changed by human hands?

B.10

A Sand Sea in Saudi Arabia (Lat 28°55'2.02"N, Long 39°36'59.07"E)

Zoom to 35 km (22 miles). You see golden sand, blown by the wind into ridges called dunes. Tilt the image so that the horizon just appears, and then fly slowly north (**Image B.10**). You'll cross the occasional barren rocky hill without a tree or shrub in sight.

Manicouagan Crater, Quebec, Canada (Lat 51°23'58.50"N, Long 68°41'44.11"W)

Fly to the coordinates of the crater and zoom to an elevation of 800 km (500 miles). Note that the crater is really obvious even from this elevation. Now zoom to a lower elevation. At what elevation does the crater fill the field of view? Tilt the field of view to see the horizon and, using the look tool, fly around the crater to see its context (**Image B.11**). If you descend to 4,570 m (15,000 feet), do you even realize that you are within a crater? Notice that the interior of the crater has risen, relative to the rim. This is due to the crust rebounding after the crater was excavated by impact. Manicouagan Crater formed between 206 and 214 million years ago. The preserved portion of the crater is about 70 km (43.5 miles) in diameter, but before erosion, it was probably 100 km (62 miles) in diameter. Due to the construction of a dam, the depressed outer edge of the crater filled with a lake.

B.11

Plate Boundaries

Where do you go if you want to see a plate boundary for yourself? First, it's important to realize that a plate boundary is not a simple line on the surface of the Earth, but is a zone perhaps 40 to 200 km wide. Still, there are places where you can go to study plate-boundary characteristics in person. Visit the following localities and you'll see for yourself. The thumbnail images provided on this page are only to help identify tour sites. Go to wwnorton.com/studyspace to experience each fly-over tour.

Divergent Plate Boundaries

Most mid-ocean ridges that define the trace of a divergent plate boundary lie submerged below 2 km (1.2 miles) of water. Thus, to get close to see the faults and volcanic vents of a ridge, geologists descend in research submersibles. *Google Earth™*, however, can help you see the shape of a mid-ocean ridge without the submersible, by taking you to Iceland, one of the few places on this planet where a ridge rises above sea level.

The Mid-Atlantic Ridge (Overall View)

Zoom out to about 8,000 km (5,000 miles), and orient your view to show the north Atlantic Ocean (**Image C.1**). See how the coastline of northwest Africa matches that of eastern North America? Now focus on the ocean floor between these coastlines and you'll see an image of the Mid-Atlantic Ridge—its trace follows the curve of the coastlines. Though this image has only low resolution, you can see the segmentation of the ridge axis and can recognize the oceanic fracture zones that link the end of one segment to the end of the next. Fracture zones appear to extend beyond the junctions with ridge segments. But only the portion of a fracture zone between two ridge segments is an active transform fault.

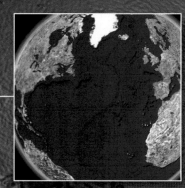

C.1

The Mid-Atlantic Ridge in Iceland (Lat 64°15'11.95"N, Long 20°56'12.66"W)

Trace the ridge northward to Iceland, an island that straddles the ridge. Zoom in to an elevation of 35 km (22 miles) at the above coordinates, tilt, and look northeast (**Image C.2**). Along the south coast, east of Reykjavik, you can see a set of northeast trending ridges whose faces are fault scarps along the ridge axis, and if you look around, you'll see some volcanoes. A glacier covers part of the ridge—after all, it is Iceland!

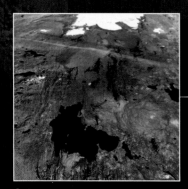

C.2

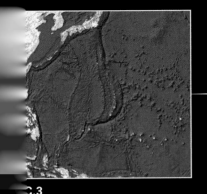

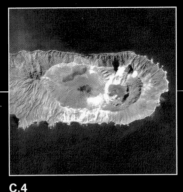

C.3   C.4

The Mariana Trench
(Lat 16°20'49.97"N, Long 145°41'48.16"E)

Fly to the coordinates given above, and you'll find yourself over the western Pacific Ocean. Zoom to an elevation of 6,000 km (3,700 miles), and you will be looking at a curving band of dark blue, south of Japan (**Image C.3**). This is the Mariana Trench, whose floor—the deepest point in the ocean—marks the boundary where the Pacific Plate subducts beneath the Philippine Plate. The curving chain of islands (the Marianas) to the west of the trench is the volcanic island arc on the edge of the Philippine Plate.

Zoom down to an elevation of 12 km (7.5 miles). You will be looking down on the volcano of Anatahan (**Image C.4**). The central portion of the island has collapsed to form a depression called a caldera.

C.5

Cascade Volcanic Arc
(Lat 46°12'25.47"N, Long 121°29'25.80"W)

Fly to the coordinates given (Mt. Adams volcano), and zoom to an elevation of 45 km (28 miles). You'll be looking down on the peak of Mt. Adams, a volcano in the Cascade volcanic arc, a continental arc. Tilt your image so you just see the northern horizon, and use the compass to reorient your image to look along the volcanic chain (**Image C.5**). You can see the spacing between the volcanoes.

Transform Plate Boundaries

Many major earthquakes happen on the San Andreas Fault. All along its length, slip on this continental transform plate boundary has affected the landscape—valleys, elongate ponds, and narrow ridges follow the fault. Also, the fault has offset stream channels.

C.6

San Andreas Transform Fault
(Lat 34°30'43.93"N, Long 118°01'06.19"W)

Fly to these coordinates and zoom to an elevation of 12 km (7.5 miles). You are looking down on the San Andreas Fault, southeast of Palmdale, California. This is the transform boundary between the Pacific Plate and the North American Plate. The trace of the fault is the boundary between flat land to the northeast and the hilly area to the southwest (**Image C.6**). At this locality, you can see a gravelly river channel that has been offset at the fault. Zoom closer, tilt the image, and rotate the view so you are looking along the fault. Follow its trace and see all the buildings, canals, and roads that cross it.

Diamond Mines

Diamonds fetch such a high price that prospectors seek them even in very remote areas. In the 1860s, attention focused on Kimberley, South Africa—at the time, a very remote area. In recent years, new deposits have been found in Arctic Canada. The thumbnail images provided on this page are only to help identify tour sites. Go to wwnorton.com/studyspace to experience this flyover tour.

Kimberley, South Africa
(Lat 28°44'17.06"S, Long 24°46'30.77"E)

Fly to the coordinates provided and zoom out to an elevation of about 15 km (9 miles). You will see the town of Kimberley in the dry interior of South Africa. Zoom down to an elevation of 5 km (3 miles). You will be hovering over a large circular pit (**Image D.1**). This is one of the diamond mines for which Kimberley is famous.

D.1

Mine near Yellowknife, Canada
(Lat 64°43'14.74"N, Long 110°37'32.76"W)

At these coordinates, you will see a remote region of Arctic Canada in which prospectors found diamond pipes in the early 1990s, after a 20-year search. If you zoom to an altitude of 200 km (125 miles), you can get a feel for the tundra landscape of scrubby vegetation and frigid lakes (**Image D.2**). From an altitude of 10 km (6 miles), you can see one of the mines where a pipe is being excavated (**Image D.3**). For the story of this discovery, please see Krajick, K., 2001, *Barren Lands*; the complete reference is at the end of the chapter.

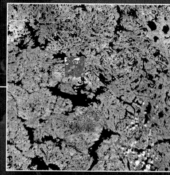

D.2

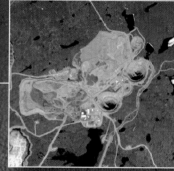

D.3

Diamond Prospect, Diamantina, Brazil
(Lat 18°15'4.35"S, Long 43°34'57.21"W)

Fly to the coordinates, and hover at 15 km (9 miles). You are looking at the town of Diamantina, nestled in the Espinhaço Range of eastern Brazil (**Image D.4**). The town's name means "diamond," and for a good reason. Prospectors (known as *Bandeirantes*) discovered diamonds at the locale in the 1720s, and the stones have been mined ever since. These diamonds are detrital—in the Precambrian, they weathered out of kimberlites and then were transported as sediment. Eventually, they accumulated with quartz grains and pebbles to form a thick unit of quartzite and conglomerate. These rocks form the ridges of the Espinhaço. Weathering of this rock frees the diamonds once again. Zoom to 3 km (2 miles), tilt, and look north, and you see a small prospect pit (**Image D.5**)

D.4

D.5

Wegener's Evidence

Alfred Wegener could not measure plate motions directly, but he did gain insight simply from looking at the map of Earth's surface. Specifically, he recognized the "fit of the continents" and found landforms whose shape seemed to be the result of continental movement. The thumbnail images provided on this page are only to help identify tour sites. Go to wwnorton.com/studyspace to experience this fly-over tour.

Matching Coastlines

View the planet from 12,000 km (roughly 8,000 miles) and orient the image so that you see South America on the left and Africa on the right (**Image E.1**). Do you notice the similarity of the coastlines on opposite sides of the ocean? To see other examples, rotate the globe and compare the coast of eastern United States with the coast of northwest Africa (**Image E.2**). Finally, compare the south coast of Australia with the nearest coast of Antarctica (**Image E.3**).

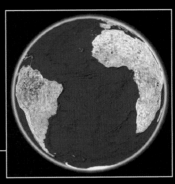

E.1

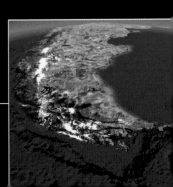

E.2

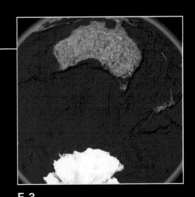

E.3

The Scotia Arc
(Lat 57°44'06.79"S, Long 46°25'43.11"W)

Go to the Scotia Sea and zoom to an elevation of about 5,000 km (3,100 miles). Wegener was impressed that the southern tip of South America and the northern tip of the Antarctic Peninsula both curve to the east where they border the Scotia Sea, and he wondered whether the curves meant that the land masses bent as they drifted westward (**Image E.4**). Modern studies suggest that such bending did indeed happen. Move to Lat 53° 56'8.91"S Long 70° 34'12.99"W, zoom to 1,950 km (1.2 miles), and tilt the image so you are

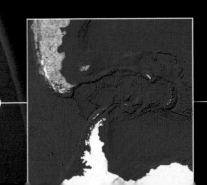

Exposures of Igneous Rocks

What do igneous rocks look like in the field? They are exposed in many places around the world. Here, we take you on tour to see a few of the better examples. The thumbnail images provided on this page are only to help identify tour sites. Go to wwnorton.com/studyspace to experience this flyover tour.

F.1

Yosemite National Park, California
(Lat 37°47'25.70"N, Long 119°29'16.74"W)

Fly to the coordinates provided and zoom up to an elevation of 7 km (4 miles). You can see a vast expanse of whitish-gray outcrop in Yosemite National Park in the Sierra Nevada (Image F.1). This outcrop consists of granite and similar rocks formed by slow cooling of felsic magma many kilometers beneath the surface. The magma that became these rocks formed when Pacific Ocean floor subducted beneath North America, along a convergent plate boundary, about 80 to 100 million years ago. Subsequent uplift and erosion stripped away volcanic rocks that once lay above the granite. During the last twenty thousand years, glacial erosion polished the outcrops.

Now tilt the image so you just see the horizon, and rotate so that you are looking southwest. This points you downstream along Merced Canyon (Image F.2). If you fly slowly down the canyon, you will pass famous mountains—Half Dome (on your left) and eventually El Capitan on the right—whose hard surfaces appeal to mountain climbers.

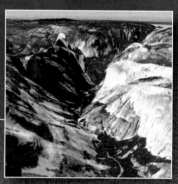

F.3

F.2

Shiprock, New Mexico
(Lat 36°41'17.03"N, Long 108°50'09.31"W)

Fly to the coordinates provided and zoom out to an elevation of 10 km (6 miles). You are looking down on Shiprock. This is the eroded remnant of an explosive volcano that last erupted about 30 Ma (Image F.3). Erosion stripped away layers of lava and ash that once formed the volcano, leaving behind dark intrusive rock that froze inside and just below the volcano. Shiprock consists of a rock type similar to basalt; this rock shattered into fragments as it approached the ground surface. Three dikes cut into the countryside, emanating from Shiprock like spokes from a wheel. These dikes are the remnants of wall-like intrusions.

Zoom down to 5 km (3 miles), tilt the image so you just see the horizon, and rotate so you're looking NW (Image F.4). You can see the steep-sided mountain and two of the dikes. The mountain is 500 m (1,600 feet) in diameter and 600 m (2,000 feet) high.

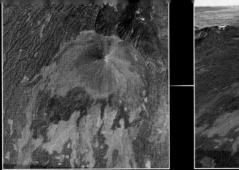

F.5

F.6

Izalco Volcano, El Salvador
(Lat 13°48'50.40"N, Long 89°37'57.74"W)

This volcano was active almost continuously from 1770 to 1958. Zoom to an elevation of 10 km (6 miles). Basalt lava flows flooded down Izalco's flanks and are still clearly visible (**Image F.5**). Younger flows are darker colored than older flows, because the younger flows have had less time to react with the atmosphere and water to undergo weathering (see Chapter 7). The gray, evenly distributed material closer to the summit consists of tiny pellets of cooled lava. Zoom down to 7 km (4 miles), and tilt the image so that you are looking north and just see the horizon (**Image F.6**). You can see nearby volcanoes. The summit of one, Santa Ana, has collapsed to form a circular depression called a caldera.

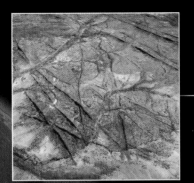

F.7

Dikes, Western Australia
(Lat 22°50'13.69"S, Long 117°23'59.74"E)

Fly to this locality and zoom down to an elevation of 3.5 km (2 miles). You are hovering over the desert of northwestern Australia, an area where extensive areas of Precambrian rocks are exposed. Here, a number of dikes stand out in relief, because they are harder than the surrounding rock, which has eroded away. Tilt the image, look north, and you can see the dikes more clearly (**Image F.7**).

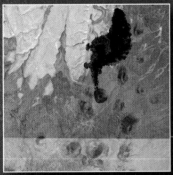

F.8

F.9

Cinder Cones, Arizona
(Lat 35°34'56.58"N, Long 111°37'55.10"W)

Fly to these coordinates and zoom to 19 km (12 miles) (**Image F.8**). You can see many of the cinder cones that spot the landscape north of San Francisco Peak, a stratovolcano near Flagstaff, Arizona. The darkest spot is SP Crater, which erupted about 71 Ka. The black apron to the north of SP Crater is a 30 m-thick basalt flow. Descend to 2.5 km (1.5 miles), tilt and rotate so you are looking south for a better view (**Image F.9**). At an elevation of 10 km (6 miles), fly 16.5 km (10 miles) to the SSE, to find Sunset Crater, which erupted an apron of bright orange tephra at 1 Ka. At this site (Lat 35°21'51.82"N, Long 111°30'10.87"W), zoom to 4 km, tilt the view, and look east (**Image F.10**).

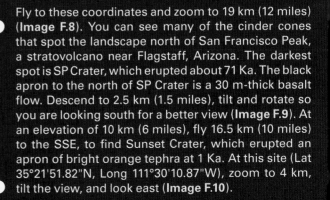

Volcanic Features

There are more than 1,500 active volcanoes on Earth and thousands more volcanic landscapes. The thumbnail images provided on this page are only to help identify tour sites. Go to wwnorton.com/studyspace to experience this flyover tour.

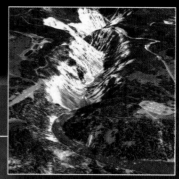

G.1

Yellowstone Falls (Lat 44°43'4.96"N, Long 110°29'45.44"W)

At these coordinates, hover above the Grand Canyon of the Yellowstone from an elevation of 5 km (3 miles). The canyon walls expose bright yellow tuffs deposited during cataclysmic eruptions that occurred during the past 2 million years. Drop to an elevation of 3 km (2 miles), look downstream, and tilt to get a better view (**Image G.1**).

**Mt. Saint Helens, Washington
(Lat 46°12'1.24"N, Long 122°11'20.73"W)**

Box 5.1 discussed the explosion of Mt. Saint Helens. To see the damage for yourself, fly to the coordinates and zoom to 30 km (18.5 miles). The breached crater, the blowdown zone, the slumps, and the lahars are all visible (**Image G.2**). To get a better sense of the devastation, descend to 13 km (8 miles), tilt your image, and fly around the mountain (**Image G.3**).

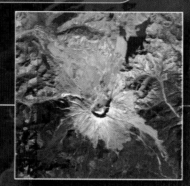

G.2

G.3

**Smoking Volcano, Ecuador
(Lat 1°27'58.96"S, Long 78°26'58.24"W)**

Fly to this locality and zoom to 75 km (46.5 miles). You are looking at the Andean Volcanic Arc in Ecuador, a consequence of subduction of Pacific Ocean floor beneath South America. Here, you see four volcanoes (**Image G.4**). When this image was taken, Tungurahua was erupting, producing a plume of ash that winds blew to the southwest. The largest volcano in view is Chimborazo, the snow-covered peak at the western edge of the image.

G.4

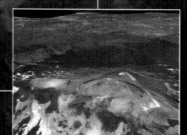

G.5

Mt. Etna, Sicily (Lat 37°45'5.49"N, Long 14°59'38.41"E)

At these coordinates, from an elevation of 50 km (31 miles), you can see Mt. Etna, which dominates the landscape of eastern Sicily (**Image G.5**). This volcano erupts fairly frequently—in this image small clouds of volcanic smoke rise from the 3200 m-high summit. Note the numerous lava flows on its flanks. Zoom to 5 km (3 miles), rotate the image, and tilt it so you are looking east.

G.7

Mt. Vesuvius, Italy (Lat 40°49'18.24"N, Long 14°25'32.04"E)

Eruption of Mt. Vesuvius in 79 C.E. destroyed Herculaneum and Pompeii. Fly to the coordinates provided and you'll be hovering over the volcano's crater. Zoom to 50 km (31 miles), and you can see the entire bay of Naples—several volcanic calderas lie on the west side of Naples. In fact, the entire bay is a caldera. Zoom to 15 km (9 miles), tilt the view, and fly around Vesuvius (**Image G.7**). The central peak of the volcano lies within a larger caldera (Somma) that formed 17,000 years ago.

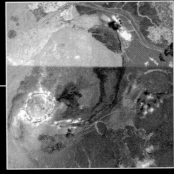

G.8

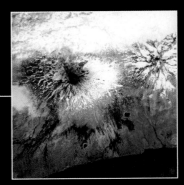

G.9

Hawaiian Volcanoes (Lat 19°28'20.47"N, Long 155°35'32.82"W)

At these coordinates, you'll be above the caldera of Mauna Loa, one of the shield volcanoes on Hawaii. From 25 km (15.5 miles) , you can see that the caldera is elliptical, and that other, smaller calderas occur to the SW. Tilt the image, and you can see a large fissure cutting the frozen lava lake in the Caldera. Distinctive lava flows spilled out of the ends of the caldera (**Image G.8**). Fly about 33 km (20.5 miles) ESE to find the caldera of Kilauea (**Image G.9**). From here, fly 41 km (25.5 miles) to the NNE to cross Mauna Kea, home to an observatory (white buildings).

Mt. Shishildan, Alaska (Lat 54°45'36.59"N, Long 163°58'28.65"W)

Unimak Island, in the Aleutian chain, hosts a large stratovolcano, Mt. Shishildan. From an elevation of 35 km (22 miles), you'll note that the snowline starts halfway up the mountain, but that the peak itself is black (**Image G.10**). That's because the volcano is active, and recent eruptions have buried and melted snow at the peak. From an elevation of 10 km (6 miles), you can see a red glow at the peak. Note that another, smaller volcano lies to the east.

G.10

G.11

East African Rift (Two Locations)

To understand volcanism in the East African rift, it is necessary to visit several locations. Here we provide two: (1) Mt. Kilimanjaro (Lat 3°3'53.63"S, Long 37°21'31.02"E): From a height of 10 km, you can see the caldera at the top of Africa's highest volcano (**Image G.11**). The glaciers on the summit have been shrinking rapidly and may be gone in 20 years. Slumping has produced steep cliffs. Fly 45 km NE to find a long chain of cinder cones marking eruptions along a fault in the rift. (2) Goma region (Lat 1°39'27.40"S Long 29°14'15.27"E): At these coordinates, zoom to 80 km (50 miles), and tilt your view to look north (**Image G.12**). The chain of lakes marks the western arm of the East African Rift. Active volcanoes lie just north of the city of Goma. Zoom to 3 km (2 miles) and tilt the image to look north—note the lava flow covering a portion of the airport runway and the city (**Image G.13**).

G.12

G.13

Images provided by *Google Earth*™ mapping service/DigitalGlobe

can see dramatic exposures of sedimentary rocks at many localities across the
net. Stratification in these exposures tends to be better where climate is drier
vegetation sparse. With a little searching, you can also find many examples of
es where sediment is now accumulating. The thumbnail images provided on
page are only to help identify tour sites. Go to wwnorton.com/studyspace to
erience this flyover tour.

Grand Canyon, Arizona
Lat 36°08'00.34"N, Long 112°13'38.95"W)

Fly to these coordinates and zoom out to an elevation
of 50 km (31 miles). You can see the entire width of the
Grand Canyon, in northern Arizona (**Image H.1**). Here,
the Canyon is about 20 km (12 miles) wide. It cuts into
the Colorado Plateau, a region of dominantly flat-lying
Paleozoic and Mesozoic strata. Zoom down to about
6 km (4 miles), and tilt the image to see the horizon
Image H.2). You now get a sense of the rugged topography of the canyon, and can see how the different
layers of sedimentary rock stand out. More resistant
rock units hold up cliffs. Fly down the Colorado River to
develop a sense of the Canyon's majesty.

H.1

H.2

Lewis Range, Montana (Lat 47°48'02.31"N, Long 112°45'30.97"W)

Fly to these coordinates and zoom out to an elevation of 25 km (15 miles). You will
be looking at a series of north–south-trending cuestas underlain by tilted layers
of late Paleozoic and Cretaceous strata (**Image H.3**). A cuesta is an asymmetric
ridge—one side is a cliff cutting across bedding, and the other side is a slope parallel to the tilted bedding. Faulting has caused units to be repeated. Thus, some of
the adjacent cuestas contain the same strata. Tilt your field of view so the horizon
just appears, and you get a clear sense of how bedding can control topography.
Zoom down to 6 km (4 miles) and you'll be able to pick out individual beds.

H.3

Death Valley, California (Lat 36°20'49.15"N, Long 116°50'8.14"W)

Death Valley includes the lowest point in North America. It's a hot, dry desert
area. When rain does fall, brief but intense floods carry sediment out of the bordering mountains and deposit it in alluvial fans at the foot of the mountains.
Water that makes it out into the valley eventually evaporates and leaves behind
salt. Fly to the coordinates given, zoom to an elevation of 10 km (6 miles), and
tilt the image (**Image H.4**). You can see contemporary sites of alluvial-fan and
evaporite deposition.

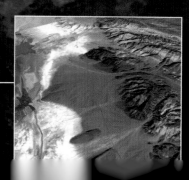

H.5

Great Exhuma, Bahamas (Lat 23°27'03.08"N, Long 75°38'15.68"W)

Here, in the warm waters of the Bahamas, you can see numerous examples of carbonate depositional environments. If you fly to the coordinates given, zoom to an elevation of 12 km (7 miles), and then tilt the image, you'll see reefs, beaches, lagoons, and deeper-water settings (**Image H.5**). The white sand consists of broken shell fragments.

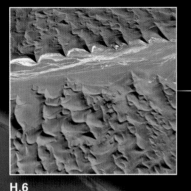

H.6

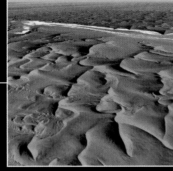

H.7

Sand Dunes, Namibia (Lat 24°44'40.42"S, Long 15°30'5.34"E)

Fly to these coordinates and zoom to an elevation of about 18 km (11 miles). Tilt your image so you gain some 3-D perspective (**Image H.6**). You are looking at a sea of sand piled in huge dunes by the wind. In the field of view, you can also see the deposits of a stream that occasionally washes through the area. Were the sand dunes to be buried and transformed into rock, they would become thick layers of cross-bedded sandstone. Zoom to 5.5 km (3 miles), tilt, and look NW (**Image H.7**). You can determine the wind direction.

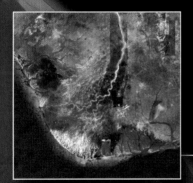

H.8

Niger Delta, Nigeria (Lat 5°24'54.48"N, Long 6°31'26.25"E)

Head to the west coast of Africa, near the equator. The Niger River drains into the Atlantic and has built a 400 km (249 mile)-wide delta. Enter the coordinates given and zoom up to an elevation of 450 km (280 miles) to see the entire delta at once (**Image H.8**). The present surface of the delta sits on top of a sedimentary accumulation that is many kilometers thick.

Now zoom down to 35 km (22 miles). You can see sandbars accumulating at bends in the river (**Image H.9**). If the abundant vegetation of the dense surrounding jungle was preserved and buried, it would transform into coal.

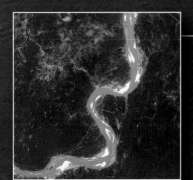

Precambrian Metamorphic Terranes

A metamorphic terrane is a region of crust composed of metamorphic rock. In unvegetated exposures you can see regional grain (the map trend of foliations), nonconformities between basement (metamorphic rock) and cover (overlying sedimentary strata), and contacts between different rock types. Explore these examples! The thumbnail images provided on this page are only to help identify tour sites. Go to wwnorton.com/studyspace to experience this flyover tour.

I.1

Wind River Mountains, Wyoming
(Lat 42°51'30.49"N, Long 109°02'40.81"W)

At these coordinates, zoom to 750 km (466 miles), and you can see all of Wyoming. The NW-SE-trending Wind River Mountains lie in west-central Wyoming (**Image I.1**). Zoom to 250 km (155 miles). A large fault bounds the SW side of the range. During mountain building, between 80 and 40 Ma, movement on the fault pushed up Precambrian metamorphic basement and warped overlying beds of Paleozoic and Mesozoic sedimentary rock. Erosion of sedimentary rock exposed the basement. Zoom to 40 km (25 miles) to see the contrast between tilted bedding of the sedimentary strata and knobby Precambrian gneiss. Tilt your field of view and look NW; the contrast becomes clearer (**Image I.2**).

I.2

Canadian Shield, East of Hudson Bay
(Lat 61°09'50.15"N, Long 76°42'43.88"W)

At these coordinates, zoom to 230 kilometers. You can see the east coast of Hudson Bay and a 150 km-wide swath of the Canadian shield, here covered only by sparse tundra vegetation (**Image I.3**). The land surface displays the trend of metamorphic foliation. Note that a belt of E-W trending grain truncates a region of N-S-trending grain—the contact between these two provinces represents the contact between Archean rocks (to the south) and Proterozoic rocks (which form the E-W-trending band). This contact originally formed deep below the Earth's surface and was subsequently exhumed.

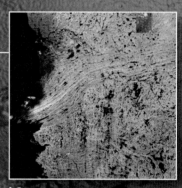

I.3

Pilbara Craton, Western Australia
(Lat 21°14'10.00"S, Long 119°09'39.15"E)

Archean rocks underlie much of northwestern Australia, a crustal province called the Pilbara craton. Here, felsic intrusives and high-grade gneisses occur in light-colored dome-shaped bodies (circular to elliptical regions in map view). Low- to intermediate-grade "greenstone" (mafic and ultramafic metavolcanic rocks, interlayered locally with metasedimentary rocks) comprise curving belts that surround the domes. Zoom to an altitude of about 65 km (40 miles) at the coordinates given, and you can see domes surrounded by greenstone belts (**Image I.4**). Zoom closer, and you can detect folded layering within the greenstone belts.

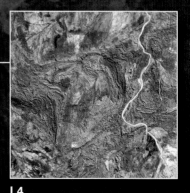

I.4

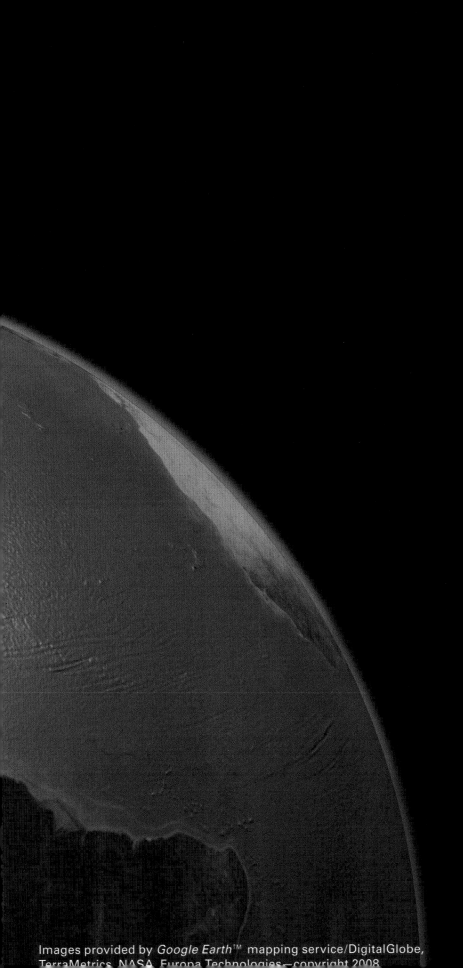

In many locations, faulting during an earthquake ruptures the ground. Below, we visit examples of fault zones that have left a mark on the land surface. The journey also illustrates potential hazards associated with seismic activity. The thumbnail images provided on this page are only to help identify tour sites. Go to wwnorton.com/studyspace to experience this flyover tour.

J.1

San Andreas Fault near Palmdale, California
(Lat 34°30'35.20"N, Long 118°1'8.12"W)

If you fly to the above coordinates and zoom up to an elevation of 350 km (220 miles), you will see the coast of California, the San Gabriel Mountains, and a triangular area of high, flatter land that comprises the western Mojave Desert (**Image J.1**). The San Andreas fault is the NW-trending line at the bottom of the triangle and the Garlock fault is the NE-trending line at the top. Zoom to an elevation of 9 km (5 miles), and you'll see a stream flowing north from the San Gabriels, near Palmdale. Where the stream crosses the fault, the channel steps to the right by over 1 km (0.6 miles), due to the displacement on the fault (**Image J.2**). Narrow pressure ridges and sag ponds delineate the trace of the fault. Tilt the image and reorient so you are looking northwest, along the trace of the fault. You can see canals, roads, and reservoirs that cross the fault (**Image J.3**). Fly along the fault and you will get a sense of how the fault disrupts the landscape.

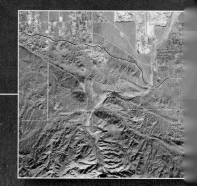

J.2

J.3

San Andreas Fault, San Francisco
(Lat 37°34'41.53"N, Long 122°24'41.88"W)

Here, from an altitude of 20 km (13 miles), the trace of the San Andreas fault runs parallel to Highway 280, along the peninsula between San Francisco Bay on the east and the Pacific Ocean on the west. Erosion along the fault trace has produced a narrow NW-trending valley (**Image J.4**). The dam straddling the fault holds back the San Andreas Reservoir. Zoom down to an elevation of 2 km (1.3 miles), tilt your view, and orient the image so that you are looking along the fault. Now, slowly fly NW, along the trace of the fault. You'll reach the end of the San Andreas Reservoir, cross housing developments, and eventually reach the ocean (**Image J.5**). Fly a short distance up the coast to Lat 37°40'35.70"N, Long 122°29'35.51"W, head slightly out to sea, turn your view so you are looking east, and look back at the coast. You'll see a cliff, with waves lapping at its base (**Image J.6**). This cliff formed because of tectonic uplift associated with faulting. Along the cliff, there are several spoon-shaped indentations—these are slump scarps, places where the cliff face slid down to the sea relatively recently. Places along the cliff face that are devoid of vegetation represent the most recent slumps. Slumping may be triggered by earthquakes. Note that developers have built homes up to the cliff edge, and onto the protrusions between existing slumps.

J.4

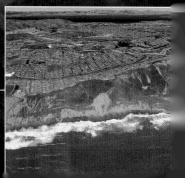

J.7

J.8

J.9

Salt Lake City, Utah
(Lat 40°55'22.81"N, Long 111°51'52.48"W)

Fly to these coordinates just north of Salt Lake City and rise to an elevation of about 1500 km (930 miles). You can see the Basin and Range Province, an active rift that has stretched the crust between Salt Lake City and Reno (**Image J.7**). The Colorado Plateau lies to the east, and the Sierra Nevada to the west. The Wasatch Mountains form a high ridge of uplifted land along the western edge of the Colorado Plateau. Zoom down to an elevation of 8 km (5 miles) and tilt the image (**Image J.8**). You are looking at the Wasatch Front, an escarpment that delineates the western face of the Wasatch Mountains, just east of I-15. The Wasatch Front is a somewhat eroded fault scarp, formed when normal faulting dropped the Great Salt Lake basin down, relative to the Wasatch Mountains. Note how the scarp truncates the ridges between the streams flowing out of the mountains. Fly south along the range front to Lat 40°35'36.44"N, Long 111°47'27.01"W, and you will get another view of the Wasatch Front. Zoom to 7 km (4 miles), tilt the image to see the horizon, and look east (**Image J.9**). This image also emphasizes how mountain building can be a long-term consequence of continued faulting.

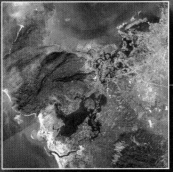

J.10

J.11

J.12

Tsunami Damage, Banda Aceh
(Lat 5°30'46.75"N, Long 95°16'14.68"E)

Fly to these coordinates and zoom up to an elevation of 16 km. You will see an isthmus near the city of Banda Aceh, at the north end of Sumatra, that was attacked on both sides by the December 2004 tsunami (**Image J.10**). The land that was submerged has turned brown, because the saltwater waves washed away most vegetation and killed the remainder. Now, zoom down to the northern part of the image (Lat 5°33'25.50"N, Long 95°17'12.52"E), drop to an elevation of 1 km (0.6 miles), and tilt the image to look north (**Image J.11**). The seaside part of the city was totally destroyed. Rotate the image so you are looking south, and fly to about Lat 5°31'54.31"N, Long 95°17'38.25"E. Here you see the boundary between the land that was submerged and the land that was not—the unscathed land is much greener (**Image J.12**). Note that the highway in the center of the image nearly disappears as it crosses the boundary.

Images provided by *Google Earth*™ mapping service/DigitalGlobe, TerraMetrics, NASA, Europa Technologies—copyright 2008.

Mountains and Structures

Some of the best examples of geologic structures, and some of the most spectacular examples of mountain ranges, are in fairly remote areas. Your computer can now take you there. Note that structures control erosion, and that you can trace them out more easily in dry regions. The thumbnail images provided on this page are only to help identify tour sites. Go to wwnorton.com/studyspace to experience this flyover tour.

Erosion in the Alps
(Lat 46°12'56.50"N, Long 10°33'59.43"E)

Fly to the coordinates provided and zoom to an elevation of about 1,100 km (680 miles). You are looking down on the northern end of the Italian Peninsula. The Alps, a Cenozoic collisional mountain belt, is the snowy arc of peaks that encompasses northern Italy and parts of France and Switzerland (**Image K.1**). Drop to an elevation of 47 km (29 miles), tilt the image, and look north. You'll see a typical mountain landscape—its ruggedness a consequence of erosion by glaciers and rivers (**Image K.2**).

K.1

K.2

Mt. Everest—the Top of the World
(Lat 27°59'17.84"N, Long 86°55'18.01"E)

Ever wonder what the view looks like from the top of Mt. Everest, the highest point on Earth? Fly to the coordinates provided, zoom to 1,300 km (800 miles) and you'll be looking down on the summit of Mt. Everest, in the Himalayas, a range that formed due to the collision of India with Asia. Tilt your view so you are looking north (**Image K.3**). In the foreground is the Ganges Plain, and in the distance, the dry Tibet Plateau. Zoom to 15 km (9 miles) and rotate to look southeast. You can see the surrounding peaks and glaciers and the plains in the far distance (**Image K.4**). Fly around the summit to see the view in all directions.

K.3

K.4

Joints in Arches National Park
(Lat 38°47'52.13"N, Long 109°35'26.37"W)

Zoom to an elevation of 5 km (3 miles) at the specified coordinates in southeastern Utah and you'll see a prominent set of NW-SE-trending joints cutting horizontal white sandstone beds (**Image K.5**). Weathering along the joints made them into narrow troughs. A second set of smaller NE-SW-trending joints also cut the rocks. The major joints are about 20–50 m (65–165 feet) apart. Now, fly SSE for about 5 km (3 miles) and you'll find a set of joints cutting red sandstone. Zoom to 3.3 km (2 miles), tilt the image and look NNW (**Image K.6**). Arches of the park develop where holes weather through the walls between joint-controlled troughs.

K.5

K.6

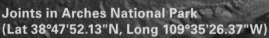

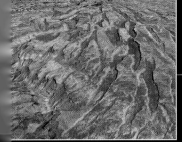

K.7

Horsts and Grabens, Canyonlands (Lat 38°6'1.40"N, Long 109°54'40.14"W)

From an elevation of 20 km (12 miles) at these coordinates in Canyonlands Park, SE Utah, you can see a set of NE-trending grabens—troughs bordered on each side by a normal fault. Fly down to 7 km (4 miles), tilt, and look NE (**Image K.7**). You'll be gazing down the axis of grabens which formed when the region southeast of the river valley stretched in a NW-SE direction. This stretching occurred because river erosion removed lateral support from a sequence of strata that lie above a weak salt layer.

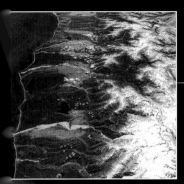

K.8

Alpine Fault, New Zealand (Lat 43°21'15.86"S, Long 170°15'23.18"E)

At these coordinates near the west coast of South Island, New Zealand, zoom to 50 km (30 miles). You can see a high, ice-capped mountain range, called the Southern Alps, formed by motion on the Alpine Fault. Though a significant component of displacement on the fault is strike-slip, its motion also accommodates thrusting—eastern Zealand is being pushed up to the west. Tilt your image and look NE and you'll be looking along the fault (**Image K.8**). The fault runs along the abrupt break in slope between the mountains to the SE, and the plains to the NW.

Faults and Folds, Makran Range (Lat 25°54'0.31"N, Long 64°14'8.63"E)

Fly to the coordinates provided and zoom to 10 km (6 miles) (**Image K.9**). You can see a series of ENE-trending ridges that represent beds of strata tilted by folding and faulting in the Makran Range of southern Pakistan, about 63 km (40 miles) north of the coast. Note that ridges are offset where they are cut by a series of NE-SW-trending valleys. The valleys are eroded left-lateral strike-slip faults. Zoom to 2 km (1 mile), and tilt the view so you are looking up the valley of the most prominent fault. A dry stream follows the valley—faulting broke up the rock, making it easier to erode.

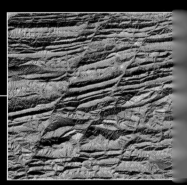

K.9

K.10

Appalachian Fold Belt, Pennsylvania (Lat 40°53'24.28"N, Long 77°4'28.11"W)

From an elevation of 114 km (70 miles), at these coordinates, you can see the curving traces of large folds in the Appalachian Mountains. These folds formed about 280 million years ago when compression wrinkled layers of Paleozoic sedimentary rock. The hinge zones of anticlines have been eroded; resistant ledges of sandstone define the fold limbs. Zoom to 20 km (12 miles), tilt the view, and rotate so you are looking west. You will be peering down the hinge of the folds (**Image K.10**). This region is the Pennsylvania Valley and Ridge Province. Note that ridges remained forested.

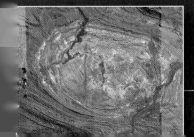

Dome and Syncline, Western Australia (Lat 22°52'9.85"S, Long 117°18'1.58"E)

Here we see an exposure of Precambrian crust, part of the Pilbara craton. Zoom to 50 km (30 miles) and you'll see a 30 km (18 miles)-wide circular region of light rock—a dome of granite and gneiss—surrounded by folded metasedimentary and metavolcanic rock (**Image K.11**). The layering within the sedimentary and volcanic rocks stands out—resistant layers form ridges, and outlines the shape of folds.

The Strata of the Colorado Plateau

The Colorado Plateau encompasses portions of Arizona, Utah, New Mexico, and Colorado. Uplift of the region during the past 20 million years raised the land surface to elevations of 2 to 3 km. Erosion then cut down into the plateau to reveal portions of a 2 km (1.2 miles)-thick sequence of strata. These rocks, which are well exposed because of the plateau's present dryness, tell a story of changing climates and changing sea levels over the past half billion years. *Google Earth*™ allows you to fly through the canyons and along the cliffs of this magical landscape to see the record in the rocks for yourself. The thumbnail images provided on this page are only to help identify tour sites. Go to wwnorton.com/studyspace to experience this flyover tour.

Grand Canyon National Park, Arizona
(Lat 36°5'36.05"N, Long 112°7'0.88"W)

These coordinates take you to Plateau Point, on the edge of the inner gorge. You can reach the Point by trail from Park Headquarters on the south rim of the canyon. Zoom to an elevation of 4 km (2.5 miles), then tilt your view so you are looking due north (**Image L.1**). You can see a stratigraphic section from Precambrian basement, through the Paleozoic to the Kaibab Limestone at the top. Try to spot the formations shown on Figure 10.11. Now, fly to the river, pivot your view, and proceed downstream (**Image L.2**). You'll be able to see continuous exposure of the unconformity at the base of the Tapeats Sandstone. Note that the narrow inner gorge cuts down into hard Precambrian gneiss—the wider outer gorge cuts into sedimentary rock.

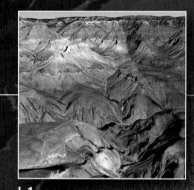

L.1

L.2

Vermillion Cliffs, Arizona
(Lat 36°49'4.81"N, Long 111°37'56.59"W)

Highway 89 crosses Marble Canyon at the coordinates provided. Just upstream of the bridge, rafters enter the Colorado River for a journey downstream into the Grand Canyon. Zoom to 5 km (3 miles), tilt your view, and look SSW. You can see the gash of Marble Canyon (**Image L.3**) bounded by cliffs of Permian Kaibab Limestone. In the right foreground, you can see small hills surrounded by subtle rings that trace out beds of horizontal strata from the Lower Triassic Moenkopi Formation. Rotate your view to look NW and you'll see the steep escarpment of the Vermillion Cliffs, which expose a complete section of

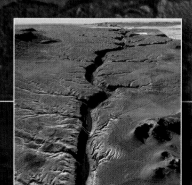

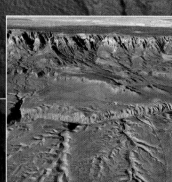

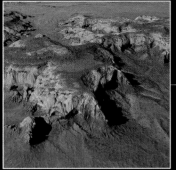

L.5

Monument Valley, Arizona
(Lat 36°55'25.00"N, Long 110°7'8.98"W)

The towering cliffs at these coordinates, along the northern border of Arizona, expose brilliant red strata deposited at the end of the Permian and beginning of the Triassic. Tilt your view and pivot to look due west (**Image L.5**). Weak beds (the Organ Rock Shale) comprise a moderate slope of stairstep-like ledges at the base of the cliff. The vertical cliff consists of massive de Chelly Sandstone, and the thin-bedded unit at top is Moenkopi Shale, capped locally by beds of the Chinle Formation. Note that the Permian strata here differ from the same-age strata of Marble Canyon—this difference demonstrates that the environment of deposition can change with location at a given time.

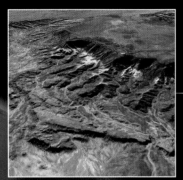

L.6

Painted Desert, Arizona
(Lat 35°5'8.13"N, 109°47'14.78"W)

This portion of the Colorado Plateau exposes a colorful sequence of Late Triassic fluvial deposits, including a variety of varicolored sandstones and shales. Fly to the coordinates given, zoom to an elevation of 3 km (1.9 miles), tilt your view, and pivot to see the gentle escarpment (**Image L.6**). The sequence of strata was deposited in a terrestrial setting. Numerous soil horizons formed as the strata were exposed prior to each new phase of deposition. The colors represent the paleosols—ancient soil horizons—preserved in the stratigraphic record.

L.7   **L.8**

Upper Zion Canyon Park, Utah
(Lat 37°13'34.59"N, Long 112°53'12.03"W)

Fly to these coordinates and zoom to an elevation of 10 km (6 miles). You're looking down on Route 9, traversing the upper part of Zion Park (**Image L.7**). Note that the erosional pattern is strongly controlled by N-S-trending joints. Now zoom down to 3.5 km (2 miles), then tilt and pivot your view so you are looking west along Route 9 (**Image L.8**). From this vantage, you see a thick sequence of cross-bedded Jurassic sandstone recording a time when much of the Colorado Plateau region was a Sahara-like desert covered by large sand dunes.

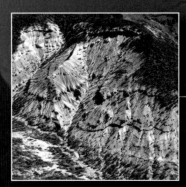

L.9

Cedar Breaks, Utah
(Lat 37°39'12.14"N, Long 112°50'33.93"W)

At Cedar Breaks National Monument, you're at the west edge of the Colorado Plateau. A view to the west looks out over the Basin and Range Rift of western Utah. For our purpose, fly to an elevation of 2 km (1.2 miles) at the coordinates given. Tilt and pivot your view so you are gazing east, and you will see the multicolored Tertiary strata eroded along an escarpment (**Image L.9**). Note that the section includes beds of resistant sandstone, standing out as ridges. But much of the sequence consists of weak shale, the deposits of lakes and floodplains. You would see similar rocks at Bryce Canyon, about 40 km (25 miles) to the east, but as of this writing, the imagery available for Bryce is very poor.

Earth Has a History

The Earth displays the record of a long and complex geologic history, marked by the collision of continents, the rifting of crust, the building and eventual erosion of mountains, and the deposition of strata. Geologists can unravel the character and sequence of events in this history by studying accumulations of sedimentary rocks, the age and character of metamorphic and igneous rocks, and the geometry of geologic structures. Different geologic provinces display the consequences of different events. Thus, you can see the evidence of Earth's geologic history in its landscapes. The following sites provide examples from North America. The thumbnail images provided on this page are only to help identify tour sites. Go to wwnorton.com/studyspace to experience this flyover tour.

Folds of the Canadian Shield
(Lat 58°4'44.52"N, Long 69°36'31.14"W)

Let's start by looking at some of the oldest crust of this continent, which is exposed in the Canadian Shield. This region is part of the craton, a tectonically inactive region of continental crust that formed in the Precambrian. Though it had a complex geologic past, it has not endured mountain building during the last billion years—instead, it has been subjected to erosion. As a result, rocks that were formed at depths of over 10 km (6 miles) now crop out. Fly to the coordinates provided and you'll find yourself in northeastern Canada, west of Ungava Bay (Quebec). Zoom to an elevation of 70 km (45 miles) (**Image M.1**). You can see a subdued landscape—none of the hills are very high, and none of the valleys are very deep. Nevertheless, the ups and downs define swirling patterns, some emphasized by dark lakes. These patterns are controlled by the folded foliation of Precambrian gneiss.

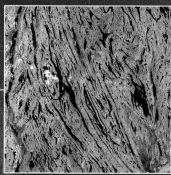

M.1

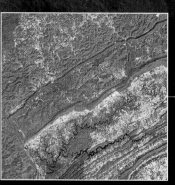

M.2

M.3

Appalachian Mountains, Tennessee
(Lat 36°28'33.93"N, Long 83°45'28.63"W)

Fly to this locality and hover at an elevation of 100 km (62 miles). You can see a portion of the fold belt (known geographically as the Appalachian Valley and Ridge province) in northern Tennessee and southern Kentucky (**Image M.2**). If you zoom down to an elevation of 20 km (12 miles) in the lower right corner of the image, tilt to see the horizon, and pivot the image so you are looking northeast, the Valley and Ridge topography is clear (**Image M.3**). Resistant stratigraphic formations hold up ridges, and weaker ones erode away to form valleys. The deformation here involves Cambrian through Pennsylvania strata, so geologists deduce that the deformation is a consequence of a late Paleozoic event. This event is the Alleghanian orogeny, due to the collision of Africa with North America. This collision compressed the crust and caused it to wrinkle along its eastern margin, somewhat like a rug on a slippery floor.

M.4

Fly to the coordinates given and you'll find a set of peaks known as the Maroon Bells, south of Aspen, Colorado. Zoom to an elevation of 7.5 km (4.5 miles), tilt so you just see the horizon, and rotate your view so you are looking due east (**Image M.4**). You can now see layering of the reddish strata that make up these mountains—fly around the mountains to get a broader perspective. Here, we have the record of two major orogenic events: (1) The late Paleozoic Ancestral Rockies event, which formed deep basins and uplifts. The reddish strata of the mountains were deposited in one of the basins. Because the event happened at roughly the same time as the Alleghanian orogeny, it might also be due to compression caused by the collision of Africa and South America; and (2) The Mesozoic/Cenozoic Laramide orogeny, during which the region was uplifted by several kilometers. Erosion subsequent to the Laramide orogeny produced the present landscape. Some valleys were carved by glaciers during the last ice age.

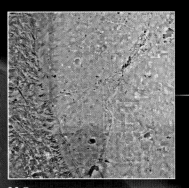

M.5

M.6

The Rocky Mountain Front, Colorado
(Lat 39°44'55.71"N, Long 104°59'44.74"W)

Fly to this location and zoom to an elevation of 70 km (45 miles) and you are hovering over Denver, Colorado. The field of view (**Image M.5**) encompasses the Rocky Mountain front which marks the boundary between the Rocky Mountains to the west and the plains to the east. Now, zoom down to an elevation of 14 km (8.7 miles), tilt your view so you just see the horizon, and rotate so you are looking southwest. You can now see the abrupt face of the mountain front (**Image M.6**). The snow-capped peak on the horizon is Mt. Evans, one of several 4.3 km (14,000 feet)-high peaks in Colorado. Fly about 12 km (7.5 miles) to the mountain front, just where highway US-285 crosses the front (Lat 39° 38'3.75"N Long 105° 10'14.28"W), drop down to an elevation of 3.5 km (2 miles), and tilt your view so you are looking north. Here you see the hogbacks at the front of the range (**Image M.7**). These are asymmetric ridges composed of strata dipping toward the east at about 35°. We see two components of orogeny—uplift and deformation. These developed during the Laramide orogeny.

M.8

M.7

Basin and Range Rift
(Lat 38°50'30.06"N, Long 114°46'41.99"W)

Fly to these coordinates and zoom to an elevation of 150 km (93 miles). Here, you see an approximately 120 km (75 miles)-wide block of the Basin and Range Province, in Nevada, just west of the Utah border (**Image M.8**). This region is a continental rift, formed during the Cenozoic, after the Laramide orogeny ceased. The tilted fault blocks now form N-S-trending narrow ranges. Because of their higher elevation, they are lightly forested. The basins between these blocks have filled with sediment. Their flat surfaces are at lower elevation and have desert climates. Streams occasionally flow along the axes of the basins. If you zoom down to 14 km (8.7 miles), tilt the image to just see the horizon, and look due east at the easternmost range in the previous view, then you can see how a range rises between two basins (**Image M.9**). The floor of the basin in this view occasionally fills with water, forming a desert feature called a playa lake.

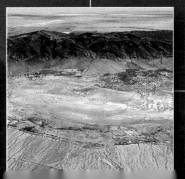

Sources of Energy and Mineral Resources

The geologic features associated with resources are not all visible at the ground surface, but you can see evidence of where resources are being extracted, and of how they are being transported. The thumbnail images provided on this page are only to help identify tour sites. Go to wwnorton.com/studyspace to experience this flyover tour.

Ackerly Oil Field near Lamesa, Texas
(Lat 32°33'18.42"N, Long 101°46'55.39"W)

Fly to this locality and zoom to 8 km (5 miles). You can see a grid of farm fields (1.6 km, or 1 mile, wide) within which small roads lead to patches of dirt (**Image N.1**). Each patch hosts or hosted a single pump for extracting oil from underground. In effect, an oil field viewed from the air looks like a pattern of spots connected by a web of roads. These wells are in the Ackerly Oil field of the Permian Basin in west Texas. Drilling here began in the 1940s to tap oil reserves in Permian sandstone beds that now lie at depths of between 1,500 and 2,500 m (5,000 and 8,200 feet). The beds tilt slightly and grade laterally into impermeable shales, so producers are exploiting stratigraphic traps. To date, 50 million barrels of oil have been extracted from this field.

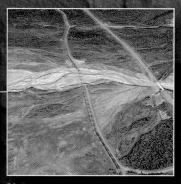

N.1

Ghawar Oil Field, Saudi Arabia
(Lat 24°47'48.02"N, Long 49°10'42.04"E)

Fly to the coordinates provided, zoom to 4 km, and you'll find an active portion of the field where wells feed a processing station linked to pipelines (**Image N.2**). Oil of the Ghawar field resides in an anticline trap whose reservoir rock consists of sandstone, 2,100 m (7,000 feet) below the surface. Oil companies have drilled 3,500 wells into the anticline. So far, the field has yielded 55 *billion* barrels of oil.

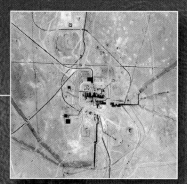

N.2

Trans-Alaska Pipeline
(Lat 64°55'38.64"N, Long 147°37'23.71"W)

The oil of the isolated North Slope oil fields near Prudhoe Bay moves through the Alaska pipeline across the entire state, crossing permafrost (permanently frozen ground) and seismic belts, to reach ports on the south coast where it is loaded into supertankers. Fly to the coordinates provided, zoom to 650 m (0.4 miles), tilt your image, and look northwest. You can see the Trans-Alaska pipeline just north of Fairbanks (**Image N.3**). At Lat 63°24'16.92"N, Long 145°44'31.72"W, zoom to 1.5 km (0.9 miles) and tilt the image so you are looking north. Here, you can see a section of the pipeline near the active Denali fault (**Image N.4**). The pipeline sits on sliding pedestals that allow it to move without breaking during an earthquake.

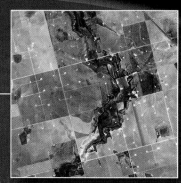

N.3

N.4

N.5

(Lat 39°14'57.59"N, Long 87°20'53.20"W)

Fly to this locality and zoom to 8 km (5 miles). You are looking at an active coal strip mine exploiting Pennsylvanian-age coal seams. Note several points about this mine. First, the active part of the mine consists of a 3.4 km (2 mile)-long trench. To produce the coal, giant bulldozers first scrape off the soil layers to expose bedrock. Then workers drill a series of holes, fill them with explosives, and blast the bedrock into fragments. The dragline scrapes out the rock and sets it aside. The dragline then piles the coal into giant trucks, which can carry as much as 150 tons in a load. Once a dragline completes a trench, bulldozers fill it back in and resurface it with topsoil. The reclaimed land can be cultivated or reforested. Zoom down to 900 m (2,950 feet) and, in the image online at the time of this writing, you can see the giant drag haul, with a tiny pickup truck nearby (**Image N.5**).

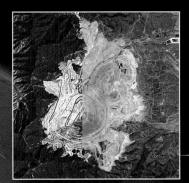

N.6

N.7

Bingham Copper Mine, Utah
(Lat 40°31'14.66"N, Long 112°9'1.97"W)

Fly to the coordinates provided and zoom to 15 km (9 miles). You can see the huge open pit of the Bingham Copper mine, dug in a Cenozoic porphyritic igneous rock of the Oquirrh Mountains (**Image N.6**). This is the largest open-pit mine in the world, and can be seen from at least 2,000 km (1,200 miles) out in space. Zoom down to 3 km (2 miles) and tilt the image to look north. You can see terraces built to stabilize the walls of the pit, and ramps that provide access for giant trucks that carry 240 tons of ore. The mine has produced 17 million tons of copper, as well as vast quantities of gold and silver. Zoom down to 3 km and tilt the image (**Image N.7**) to get a sense of the pit's size.

Iron Mine near Ouro Preto, Brazil
(Lat 20°9'54.45"S, Long 43°30'7.86"W)

Fly to these coordinates and zoom to 3 km (1.8 miles), tilt the image, and look north (**Image N.8**). You are looking into an iron mine in eastern Brazil. The ore consists of banded iron formation, deposited 2.2 billion years ago. Subsequently, these rocks were deformed and metamorphosed, and now include bands of magnetite and/or hematite. The pools of iron-colored water are settling ponds. The ore from this mine is transported to factories in Brazil, China, and Japan.

Examples of Landslides

Landslides cause distinctive scars on the Earth's surface, such as head scarps, hummocky ground, and/or disrupted vegetation. Examples of these features occur worldwide. The thumbnail images provided on this page are only to help identify our sites. Go to wwnorton.com/studyspace to experience this flyover tour.

O.1

Slumping Hawaii (Lat 19°18'21.37"N, Long 155°5'2.55"W)

These coordinates take you to Volcanoes National Park, along the south coast of the Big Island of Hawaii, where active lava flows reach the sea. From an altitude of 5 km (3 miles), you can see black lava flows that descended over the Holei Pali cliff (**Image O.1**). The 250 m (820 foot)-high cliff is the head scarp of the Hilina slump. Here, we see only the upper 2.5 km (1.5 miles) of the slump on land. The rest is submerged and extends about 40 km (25 miles) offshore. Tilt your view to look east along the scarp (**Image O.2**).

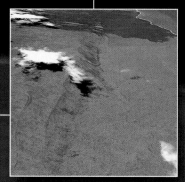

O.2

1964 Slumps, Anchorage, Alaska (Lat 61°12'52.06"N, Long 149°54'31.39"W)

During the 1964 earthquake, several coastal slumps moved. Fly to the coordinates given and zoom to 2 km (1.2 miles)—you are hovering over downtown Anchorage (**Image O.3**). Note the curving head scarps near the shore marking slumps that sank by about 7 m (21 feet) during the 1964 quake. Fly SW along the coast until you see the airport. At Lat 61°11'51.51"N, Long 149°58'25.11"W you'll find a crescent-shaped band of woods along the shore (**Image O.4**). This is Earthquake Park, what's left of the Turnagain Heights neighborhood, destroyed by 1964 slumping.

O.3

O.4

Roadside Landslide, Pacific Coast Highway, California (Lat 34°3'31.44"N, Long 118°58'15.85"W)

Portions of the Pacific Coast Highway lie at the base of unstable slopes—landslides down these slopes block the road. At the coordinates provided, from an elevation of 800 m (2,600 feet), you can see the subtle scar left by one of these landslides, and the remnants of debris cleared by bulldozers along the base of the cliff (**Image O.5**).

O.5

Portuguese Bend Landslide, California (Lat 33°44'46.94"N, Long 118°22'7.83"W)

The land here started to slump about 37,000 years ago. At the coordinates given, from an altitude of 5 km (3 miles), you can see the head scarp of the 3 km (1.8 mile)-wide slump (**Image O.6**). In the 1950s, developers built a housing project on the hummocky land of the slump. The southeastern portion of the slump began moving again, ultimately destroying 150 homes.

O.7

La Conchita Mudslide, California (Lat 34°21'50.29"N, Long 119°26'46.85"W)

At La Conchita, developers built a housing project in the narrow strip of land between the cliffs and the shore. When heavy rains drench this landscape, the slope becomes unstable. In 1995 and again in 2005, mudflows buried homes. From an elevation of 2 km (1.2 miles), you can see the development (**Image O.7**). Note that the cliff bounds a broad, uplifted terrace. This terrace was a wave-cut platform formed at sea level. Tectonic movements caused the terrace to rise. In addition to the mudslide, which also destroyed a road traversing the hill, you can see gullies and canyons that have been incised by successive floods.

O.8

Mt. Saint Helens Lahars, Washington (Lat 46°10'36.32"N, Long 122°10'4.44"W)

Fly to these coordinates, and you are over the southeast flank of Mt. Saint Helens. Zoom to an elevation of 12 km (7.5 miles), tilt your view, and pivot to look NW (**Image O.8**). The southeast flank was not destroyed by the 1980 explosion, but rather was the site of numerous lahars. The gray lahars look like spills of gravy down the side of the mountain. In the stream valleys, the lahars traveled much farther.

Gros Ventre Slide, Wyoming (Lat 43°37'29.78"N, Long 110°33'3.49"W)

Fly to the coordinates provided. From an elevation of 10 km (6 miles), you can see Lower Slide Lake, northeast of Jackson, Wyoming (**Image O.9**). This lake formed in 1925, when the Gros Ventre slide tumbled down Sheep Mountain and blocked the Gros Ventre River. Zoom to 6 km (3.7 miles), tilt your view, and look southeast (**Image O.10**). The scarp left by the landslide is obvious, as is the hummocky ground on top of the debris.

O.9

O.10

Debris Fall, Yungay, Peru (Lat 9°8'53.88"S, Long 77°43'20.35"W)

Remnants of the debris fall of 1970 that buried parts of Yungay, Peru, can still be seen decades later. Fly to the latitude and longitude provided, zoom to 15 km (9 miles), tilt to see the horizon, and look NE (**Image O.11**). Note the towering, ice-covered peak of Nevado Huascarán in the far distance, and the valley down which the debris flow traveled in the middle distance. In the foreground, the broad apron of debris is now farmed, but it does not host many homes.

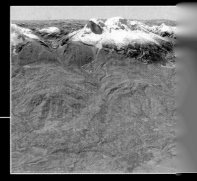

O.11

O.12

O.13

Rockfalls, Canyonlands National Park, Utah (Lat 38°29'52.65"N, Long 110°0'50.20"W)

In southeastern Utah, at the coordinates provided, you can see evidence of rockfalls along the banks of the Green River. Zoom to an elevation of 3 km (1.8 miles) and look down (**Image O.12**). Note that NW-trending joints cut the whitish, resistant Permian sandstone layer that forms the top of the mesa. When the cliff breaks away along a joint, blocks topple down the slope and break up. Zoom down to 2 km (1.2 miles), tilt to see the horizon, and pivot so you are looking east (**Image O.13**). You can see a large rockfall that sent debris almost down to the bottom of the slope.

Fluvial Landscapes

Streams stand out in the landscape, for they carve intricate shapes as their waters flow from high areas to low. In this Geotour, we visit a variety of landscapes whose features are a consequence of deposition and erosion in a fluvial setting. The thumbnail images provided on this page are only to help identify tour sites. Go to wwnorton.com/studyspace to experience this flyover tour.

P.1

Deep Gorge in the Himalayas (Lat 28°9'40.37"N, Long 85°25'53.21"E)

Fly to these coordinates, zoom to 10 km (6 miles), and tilt your image so you are looking up the valley to the east (**Image P.1**). You are 52 km (42 miles) NNE of Katmandu, Nepal. This image illustrates how the upper reaches of a mountain stream have steep gradients and flow down deep gorges. If the rock walls of the gorge are relatively weak, the stream valley attains a V-shape.

Headward Erosion, Canyonlands, Utah (Lat 38°17'58.16"N, Long 109°50'18.76"W)

At these coordinates, in Canyonlands National Park, fly to an altitude of 8 km (5 miles). You see a side canyon carved into horizontal strata by intermittent tributaries that flow into the Colorado River (**Image P.2**). An abrupt scarp marks the upstream limit of each tributary. As the streams erode, they cut into the land by headward erosion, so the scarp migrates upstream. Zoom down to 4 km (2.5 miles), tilt the image, and look southwest to better visualize the concept of headward erosion (**Image P.3**).

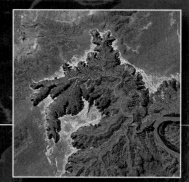

P.2

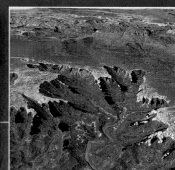

P.3

Meanders Along Rio Ucayali, Peru (Lat 7°27'1.80"S, Long 75°2'48.92"W)

At this locality, 560 km (347 miles) NNE of Lima, Peru, you will find a nearly horizontal landscape on the east side of the Andes. Because of the low gradient here, the Rio Ucayali has become a meandering stream. This view, from an elevation of 100 km (62 miles), shows numerous meanders within a flood plain (**Image P.4**). You also see abandoned meanders, point bars, and oxbow lakes. You can find very similar meanders along the Mississippi River (USA) at Lat 31°29'29.67"N, Long 91°38'0.23"W.

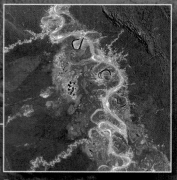

P.4

Incised Meanders, Canyonlands, Utah (Lat 38°13'56.26"N, Long 109°49'27.64"W)

Return to Canyonlands National Park, Utah, zoom to 6.5 km (4 miles), and tilt the image to look northeast. You can see a meander of the Colorado River that has almost cut through the meander neck (**Image P.5**). The meanders incised down through bedrock, and now lie at the floor of a steep-walled canyon.

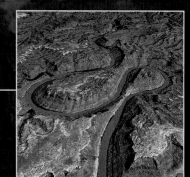

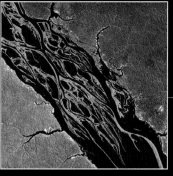

P.6

Mid-Stream Bars, Rio Negro, Brazil (Lat 2°43'31.82"S, Long 60°42'50.50"W)

At this locality, along the Rio Negro (a major tributary of the Amazon), 105 km (65 miles) upstream of Manaus, the river is about 15 km (9 miles) wide (**Image P.6**). From an altitude of 60 km (37 miles), you can see numerous mid-stream bars of sediment that emerge from the river. These bars will be submerged during a flood, and may be shifted by the force of flowing water.

P.7

P.8

Point Bars, Trinity River, Texas (Lat 30°10'2.73"N, Long 94°48'58.24"W)

Fly to this locality and zoom to an elevation of 14 km (8.7 miles). You are looking at a reach of the Trinity River in the vicinity of Dayton Lakes, 70 km (43 miles) to the northeast of Houston (**Image P.7**). Along the inner curve of each meander, a bright tan point bar of sand has formed. In the green landscape surrounding the river, you will also be able to spot a number of abandoned meanders and oxbow lakes. Zoom down to 3 km (1.9 miles) and tilt so you see the horizon (**Image P.8**). The bars and abandoned meanders stand out in the landscape.

P.9

Trellis Drainage Pattern, Pennsylvania (Lat 40°45'31.75"N, Long 76°50'45.41"W)

Fly to these coordinates and zoom to 150 km (93 miles). You can see the Valley and Ridge Province of Pennsylvania, produced by the erosion of folded strata. Here, we see a trellis drainage pattern—tributaries flow down valleys and intersect the Susquehanna at right angles (**Image P.9**). The Susquehanna itself cuts across ridges.

P.10

Radial Drainage, Mt. Shasta (Lat 41°24'17.85"N, Long 122°11'42.87"W)

Looking straight down from an elevation of 25 km (15.5 miles), you can see the peak of Mt. Shasta, a volcano in California. The streams that flow down its slopes, away from the peak, define a radial network (**Image P.10**). This pattern resembles the spokes of a wheel.

Dendritic Drainage, Pennsylvania (Lat 41°30'29.73"N, Long 78°14'18.04"W)

Fly to these coordinates, zoom to an elevation of 30 km (18.6 miles), and you are looking down on a region of the Appalachian Plateau near the town of Emporium, in north-central Pennsylvania. The strata here consist of flat-lying beds of shale. Streams cutting down into this fairly homogeneous and soft substrate carve a dendritic network (**Image P.11**). This means that the pattern of streams and tributaries resembles the pattern of veins on a leaf.

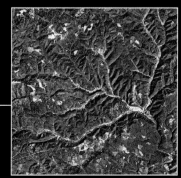

P.11

Landscapes of Oceans and Coasts

You can find a huge variety of different coastal and bathymetric features with *Google Earth*™ or comparable programs. In addition to visiting the stops identified in this Geotour, simply fly to a coast, tilt your view, and set the image in motion—you'll be amazed at what you can see! The thumbnail images provided on this page are only to help identify tour sites. Go to wwnorton.com/studyspace to experience this flyover tour.

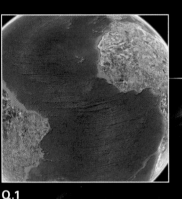

Q.1

Mid-Atlantic Ridge Bathymetry
(Lat 0°3'51.75"N, Long 24°47'35.66"W)

Fly to the coordinates provided and zoom to 8000 km (5000 miles) (**Image Q.1**). You can see the entire South Atlantic Ocean. Examine the bathymetry of the sea floor. Ridge segments and transforms are obvious.

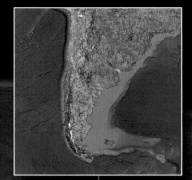

Q.2

Southern South America Bathymetry
(Lat 39°22'15.03"S, Long 65°9'1.31"W)

Zoom to 5000 km (3100 miles) at the coordinates provided. You are looking down on southern South America (**Image Q.2**). Compare the bathymetry of the west coast (an active convergent margin) to that of the east coast (a passive margin). Where does a broad continental shelf occur? Now, fly south to Lat 54°29'29.28"S, Long 52°1'4.39"W, zoom to 4500 km (2800 miles), and tilt to look north. You can see the Scotia Sea, between South America and Antarctica, and the Scotia volcanic arc (**Image Q.3**). Can you identify the pattern of plate boundaries?

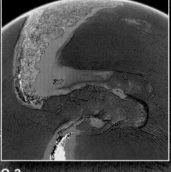

Q.3

Coral Reefs, Pacific (Lat 16°47'26.92"S, Long 150°58'1.27"W)

Fly to the coordinates given and zoom to 6 km (3.7 miles). You are looking down on the coral reef and lagoon of Huahine, one of the Society Islands in the South Pacific. Zoom down to 1.5 km (1 mile), tilt, and look north (**Image Q.4**). Note how the reef absorbs wave energy and protects the shore. Now, fly west to Lat 16°37'25.65"S, Long 151°29'32.80"W, and zoom to 25 km (15 miles). You are looking down on the island of Tahaa, an atoll surrounded by an offshore reef (**Image Q.5**).

Q.4

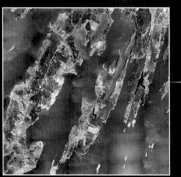

Rocky Coast, Maine
(Lat 43°46'34.77"N, Long 69°58'28.58"W)

Fly to these coordinates and zoom to 10 km (6 miles). From this viewpoint, you can see the rocky coast of Maine (**Image Q.6**). Islands here were carved by glaciers during the last ice age. Post-ice age sea-level rise submerged the landscape.

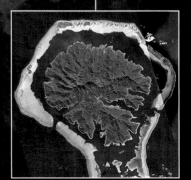

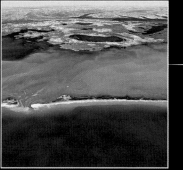

Q.7

Offshore Bar, near Cape Hatteras, North Carolina (Lat 35°8'24.05"N, Long 75°53'34.62"W)

At the coordinates provided, zoom to 28 km (17 miles) and tilt your view so that you are looking NW. You can see an offshore sand bar (**Image Q.7**). The bar formed due to redistribution of sand by waves and currents. Contrast this coast with the one you saw in Maine.

Q.8

Chicago Shoreline (Lat 41°54'59.76"N, Long 87°37'36.41"W)

This locality, on the west shore of Lake Michigan, provides an excellent example of how the moving sand of a beach interacts with groins. Zoom to 2.5 km (1.5 miles) and look down on Chicago's beach front (**Image Q.8**). Note that sand has accumulated in asymmetric wedges due to a south-flowing current.

Fjords of Norway (Lat 60°53'56.09"N, Long 5°12'31.84"E)

At the coordinates provided, zoom to 80 km (50 miles) and you see the intricate coast of Norway (**Image Q.9**). During the last ice age, glaciers carved deep valleys which filled with sea water when the ice melted and sea level rose. Zoom to 13 km (8 miles), tilt, and look north to see the steep cliffs bordering fjords. Now fly to Lat 61°12'34.22"N, Long 5°7'18.20"E. Here, the image resolution is better, and if you zoom to 5 km (3 miles), tilt, and look east, you can look inland, along the axis of a fjord (**Image Q.10**).

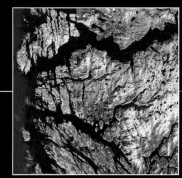

Q.9

Q.10

Organic Coast, Florida (Lat 25°7'48.94"N, Long 80°59'3.18"W)

Much of southern Florida, the Everglades, is a vast swamp, through which fresh water flows slowly south. Here, the transition from land to sea is gradual. Fly to the coordinates provided and look down from 40 km (25 miles) (**Image Q.11**). You see lagoons, swamps and bars, all colored by vegetation. Darker green areas are mangrove thickets. Zoom to 700 m (2300 feet) and tilt to look north (**Image Q.12**). You can see how the thickets stabilize the shore.

Q.11

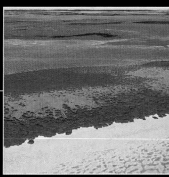

Q.12

Sandspit, Cape Cod (Lat 42°2'40.10"N, Long 70°11'31.46"W)

Cape Cod formed when a glacier deposited a 200 m (600 foot)-thick layer of sand and gravel to form a ridge called a moraine about 18,000 years ago. Eventually, sea level rose and currents began to transport sand along the shore. Fly to the coordinates provided and look down from an elevation of 20 km (12 miles) (**Image Q.13**). Note the large sand spit that protects Provincetown. On the north shore, you can see a beach and dunes.

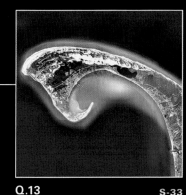

Q.13

Images provided by *Google Earth*™ mapping service/DigitalGlobe, TerraMetrics, NASA, Europa Technologies—copyright 2008.

water at springs or due to pumping, or from underground erosion by groundwater. The thumbnail images provided on this page are only to help identify tour sites. Go to wwnorton.com/studyspace to experience this flyover tour.

Irrigation in the Saudi Desert (Lat 25°18'18.36"N, Long 44°28'5.72"E)

Fly to the coordinates provided and zoom to 70 km (43 miles). You can see the desert of Saudi Arabia. Note the sand dunes in the upper right corner (**Image R.1**). Porous rock holds large reserves of groundwater under the surface. Each circle in the image is a field, irrigated by groundwater pumped from a well at the center. Move to Lat 25° 17'53.71"N, Long 44°28'27.17"E and zoom to an elevation of 2 km (1.2 miles) to see a field close up (**Image R.2**).

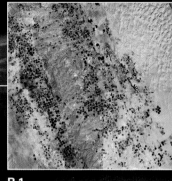

R.1

R.2

Water Table, Minnesota (Lat 46°34'49.28"N, Long 95°43'55.13"W)

Fly to this locality and look down from 12 km (7.5 miles) (**Image R.3**). A large number of ponds fill kettles, depressions that formed at the end of the ice age when blocks of ice buried in glacial sediment melted away. The surface of the lakes represents the surface of the water table. Dry land exists where the ground surface lies slightly above the water table.

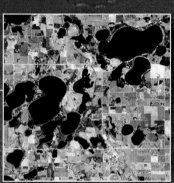

R.3

Everglades, Florida (Lat 25°40'12.62"N, Long 80°42'1.11"W)

Zoom to 15 km (9 miles) at the coordinates provided, and you are hovering over the Everglades (**Image R.4**). This swamp is actually a broad, slowly moving stream that flows from Lake Okeechobee to the tip of Florida. During the wet season, the water table effectively lies above ground, so only vegetated hammocks rise above the water table. But during the dry season, the water table lies below the surface. Why are the hammocks elongated? Some of the water has been diverted into canals, one of which occurs on the right side of this view.

R.4

R.5

Hot Springs, Yellowstone (Lat 44°27'44.65"N, Long 110°51'17.98"W)

Zoom to an elevation of 3 km (2 miles) at the coordinates provided and you're looking at Black Sand Basin, near Old Faithful geyser in Yellowstone National Park (**Image R.5**). Magma heats the groundwater of Yellowstone. The hot water rises and either fills pools or erupts at geysers. This view shows examples of the pools. Bacteria grow in the pools and add color.

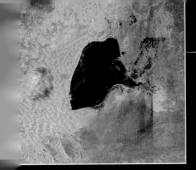

Desert Oasis, Egypt (Lat 30°14'41.27"N, Long 28°55'58.95"E)

The Sahara Desert contains vast seas of sand. Yet here and there, ponds or vegetated areas fill depressions where springs bring groundwater to the surface. You can see such an oasis from 10 km (6 miles) at the locality specified above (**Image R.6**).

R.6

Sinking Venice, Italy (Lat 45°26'9.01"N, Long 12°20'0.02"E)

The city of Venice covers an island in the Venice Lagoon. From 20 km (12 miles) red roofs in the city stand out (**Image R.7**). Zoom to 2.5 km (1.5 miles), and you can see that the main thoroughfares of Venice are canals; the city has subsided in part due to groundwater removal (**Image R.8**).

R.7

R.8

Sinkholes in Florida (Lat 28°38'36.16"N, Long 81°22'18.69"W)

Fly to the coordinates provided and zoom to 6 km (4 miles). You are looking at the landscape around Interstate Highway 4. Here, the circular ponds fill sinkholes (**Image R.9**). Superficially, they look like the kettles of Minnesota, but they form in a totally different manner—sinkholes develop when an underground cavern collapses. The water surface represents the water table.

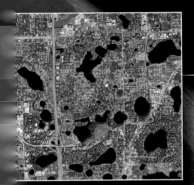

R.9

R.10

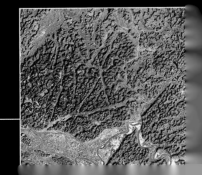

R.11

Karst Landscape, Puerto Rico (Lat 18°24'40.75"N, Long 66°25'7.08"W)

At this locality, a view looking down from 10 km (6 miles) reveals part of an extensive karst region (**Image R.10**). 36 km (22.3 miles) WSW of this area (Lat 18°20'39.41"N, Long 66°45'10.10"W), a sinkhole was converted into the Arecibo Observatory radio telescope (**Image R.11**).

Tower Karst of China (Lat 22°32'29.97"N, Long 107°26'18.17"E)

In southeastern China, at these coordinates, a view from 20 km (12 miles) illustrates a tower karst landscape (**Image R.12**). The thick layer of limestone now at the surface was a cave network. When the roof of the network collapsed, rooms became sinkholes, and tunnels became valleys. Joints localized solution.

Desert Landscapes

In the desert, the unvegetated land surface lies exposed, so landform shapes are clear and dramatic. High-resolution images allow you to see cliffs, dunes, fans, and even individual boulders. Here, we tour several deserts to see the variety of landscapes. Zoom out into space to remind yourself of where you are on the globe. The thumbnail images provided on this page are only to help identify tour sites. Go to wwnorton.com/studyspace to experience this flyover tour.

S.1

Sand Dunes, Namib Desert, Africa (Lat 24°43'57.60"S, Long 15°26'59.90"E)

Fly to the coordinates provided, zoom to 20 km (12 miles), and you will see a reddish sand sea in the Namib Desert on the west coast of southern Africa. The color is due to oxidized iron. A dry wash cut through the sand and washed it away. Zoom to 8 km (5 miles), and tilt to look NNE to see star dunes, formed by shifting winds (**Image S.1**).

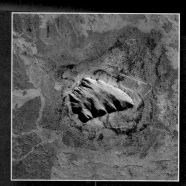

S.2

Uluru (Ayers Rock), Australia (Lat 25°20'52.42"S, Long 131°2'24.98"E)

At these coordinates, from 10 km (6 miles), you can see the famous inselberg, Uluru (Ayers Rock) in the "red center" of Australia (**Image S.2**). Zoom to an elevation of 4 km (2.5 miles), tilt your view, and look southeast. Note that the beds of sandstone comprising Uluru have a vertical dip—they are one limb of a syncline. If you fly to Lat 24°10'33.30"S, Long 132°55'19.57"E, and zoom to 20 km (12 miles), you can see another fold whose shape, without vegetation cover, stands out in the landscape (**Image S.3**).

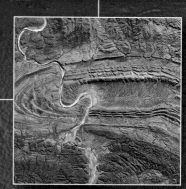

S.3

S.4

Atacama Desert near Chala, Peru (Lat 15°50'40.47"S, Long 74°16'8.10"W)

The Atacama Desert, on the west coast of South America, exists because a cold current flows northwards just offshore. At the coordinates provided, from an elevation of 15 km (9 miles), you can see the stark dryness of the land, right up to the shore of the blue Pacific Ocean (**Image S.4**).

Tarim Basin of Western China (Lat 37°39'56.20"N, Long 82°19'7.20"E)

Fly to this location, zoom to an elevation of 60 km (37 miles), and tilt the view to look north (**Image S.5**). The desert stretches as far as the eye can see. Sand dunes cover the land surface except in the green, irrigated oasis of the middle distance.

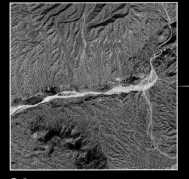

S.6

Dry Wash in Arizona (Lat 33°39'49.26"N, Long 111°35'10.25"W)

Stream channels in deserts fill only intermittently. At other times, the water table lies below the surface of the ground so the channels are dry. At these coordinates, from 9 km (5.6 miles), you see the floor of a dry wash (Image S.6). Adjacent slopes display a dendritic drainage pattern, because even though water flows only intermittently, water is still the dominant agent of erosion.

S.7

Urbanizing a Desert, Tucson, Arizona (Lat 32°15'53.09"N, Long 110°51'19.87"W)

Much of southern Arizona lies in the Sonoran Desert. There's hardly any surface water, but the warm climate attracts a growing population, and the region's cities have ballooned. Zoom to 5 km (3 miles) and tilt your view to look north (Image S.7). Suburbs of Tucson sprawl to the foot of the Catalina Mountains; intense watering keeps the golf courses green.

Playa in Death Valley, California (Lat 36°13'40.64"N, Long 116°45'55.28"W)

Here at Badwater, in Death Valley, you see the lowest point in North America (85 m or 280 feet below sea level). Zoom to an elevation of 3 km (2 miles), tilt your view, and look east (Image S.8). The white surface is the salt crust of a dry playa lake. Note an alluvial fan building out onto the playa.

S.8

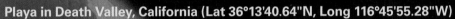

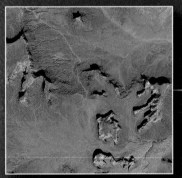

S.9

Buttes of Monument Valley, Arizona (Lat 36°57'6.66"N, Long 110°5'19.66"W)

Fly to this locality and look down from an elevation of 10 km (6 miles) (Image S.9). You are seeing the buttes and chimneys of Monument Valley. These landforms consist of massive sandstone eroding by cliff retreat when joint-bounded walls of rocks collapse.

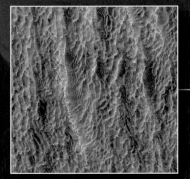

S.10

Sand Sea in the Sahara, Egypt (Lat 26°53'42.61"N, Long 26°5'5.02"E)

The Sahara is the largest continuous desert on Earth. At this locality, from an elevation of 10 km (6 miles), you see two scales of dunes (Image S.10). Very large longitudinal dunes trend from NNW to SSE. On top of these, smaller dunes developed.

On *Google Earth*™ imagery, you can easily spot glacial erosional and depositional features. And at high latitudes and/or elevations, you can still see glacial ice. The thumbnail images provided on this page are only to help identify tour sites. Go to wwnorton.com/studyspace to experience this flyover tour.

Continental Glacier, Antarctica
(Lat 89°59'52.49"S, Long 144°50'31.19"E)

At the southern end of the Earth, zoom 8,000 km (5,000 miles) and look down to see the continental glacier that covers Antarctica (**Image T.1**). Rock crops out only along the Transantarctic Mountains and along the coast. The Antarctic Peninsula protrudes toward the southern tip of South America.

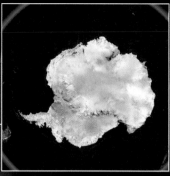

T.1

Greenland and the Arctic Ocean
(Lat 85°4'55.25"N, Long 35°8'59.46"W)

At the northern end of the Earth, from 11,000 km (6,800 miles) out in space, you can see the Arctic Ocean. *Google Earth*™ does not generally show sea ice, so it's clear that, unlike the southern end, the northern end of the Earth is an ocean. A continental glacier covers the large island of Greenland (**Image T.2**).

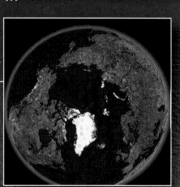

T.2

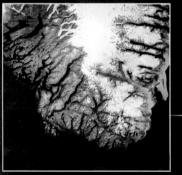

Southern Tip of Greenland (Lat 60°18'48.05"N, Long 44°28'1.45"W)

From an elevation of 250 km (155 miles), you can see the ice cap, valley glaciers, and fjords of southernmost Greenland (**Image T.3**). Zoom to 15 km (9 miles), at Lat 60°9'20.86"N, Long 44°3'34.30"W, tilt your view to look north, and you can see the effects of glacial erosion in detail.

T.3

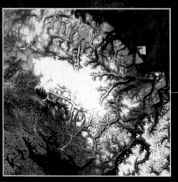

Baffin Island, Canada
(Lat 67°3'38.31"N, Long 65°33'2.88"W)

Zoom to 400 km (250 miles) and your view encompasses a portion of Baffin Island. You can see a small ice cap, and several valley glaciers ending in fjords (**Image T.4**). Move to Lat 67°8'18.30"N, Long 65°0'56.99"W and zoom to 60 km (37 miles) to see valley glaciers "draining" the ice cap. Zoom down to 15 km (9 miles), tilt, and look upstream along a glacier (**Image T.5**). Note how medial moraines form from the lateral moraines of valley glaciers that merge.

T.4

T.5

T.6

Matterhorn, Switzerland (Lat 46°5'15.04"N, Long 7°38'2.72"E)

Ice-age glaciers sculpted the peaks that we see today. At these coordinates, zoom to 10 km (6 miles), tilt, and look south. You can see a glacially carved valley with the Matterhorn at its end. Multiple cirque formation produced the peak (**Image T.6**).

Malaspina Glacier Area, Alaska (Lat 59°57'5.59"N, Long 140°32'54.35"W)

At the coordinates provided, zoom to 65 km (40 miles) and tilt your view to look NNE. You are gazing at the Malaspina Glacier, a piedmont glacier spreading out on a coastal plain (**Image T.7**). A terminal moraine outlines the toe of the glacier. Note that the flow of the ice produced complex folds traced out by the pattern of debris on the glacier's surface. Zoom down to 8 km (5 miles) at Lat 59°50'35.76"N, Long 140°1'27.22"W. You can see knob-and-kettle topography and a braided outwash stream (**Image T.8**).

T.7

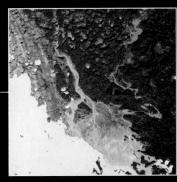
T.8

Glaciated Peaks, Montana (Lat 48°54'53.37"N, Long 113°50'51.04"W)

At the coordinates provided, zoom to 7 km (4 miles), tilt your view, and look east. You can see the peak of Mt. Cleveland at the head of two cirques separated by a sharp arête (**Image T.9**). Move a little south, to Lat 48°45'35.56"N, Long 113°37'47.02"W, and you are in Glacier National Park. Zoom to 8.5 km (5.3 miles) and tilt your view to look southwest, and you can see a U-shaped valley with a tarn at its head (**Image T.10**).

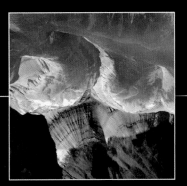
T.9

T.10

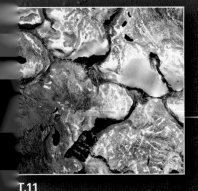
T.11

T.12

Sierra Nevada, California (Lat 37°44'33.69"N, Long 119°16'21.06"W)

Fly to the coordinates provided and zoom to 7 km (4 miles). You can see several cirques and arêtes in the high peaks of the Sierra Nevada (**Image T.11**). Fly to Lat 37°46'23.82"N, Long 119°30'10.68"W, zoom to 5 km (3 miles), and tilt to look SSW. You are looking down the Yosemite Valley, a U-shaped valley at the heart of Yosemite National Park. Waterfalls spill out of hanging valleys into the main valley (**Image T.12**).

Pluvial Lake Shore, Salt Lake City, Utah (Lat 40°31'18.45"N, Long 111°49'54.14"W)

Zoom to 3.5 km (2 miles), tilt, and look south to see the terrace of sediment that marks the beach of a pluvial lake, Lake Bonneville (**Image T.13**). This lake covered the entire area of what is now Salt Lake City during the Pleistocene.

See for Yourself U

Aspects of Global Change

Human activity has become a major agent of change on planet Earth. The consequences of this activity are clearly evident in the character of landscapes on the surface. The thumbnail images provided on this page are only to help identify tour sites. Go to wwnorton.com/studyspace to experience this flyover tour.

Clear Cutting the Amazon, Brazil
(Lat 4°3'51.60"S, Long 54°50'47.96"W)

At these coordinates, and an elevation of 100 km (62 miles), you can see swaths of clear cutting (**Image U.1**). The pattern reflects access roads. Move east to Lat 3°51'7.92"S, Long 54°11'12.93"W, where the resolution is better, and from 5 km (3 miles), you can see clear-cut areas converted into rangeland (**Image U.2**).

Long-Term Deforestation, Brazil
(Lat 16°41'18.46"S, Long 43°58'0.76"W)

At the coordinates provided, the view from 7 km (4 miles) shows hilly land (**Image U.3**). Deforestation here took place decades ago. Forest remains only in valleys. Fly east to the Brazilian highlands at Lat 19°20'0.43"S, Long 41°14'25.05"W. From 25 km (15 miles), you can see that only the highest hills remain forested (**Image U.4**). Yearly burning keeps the forest from regrowing.

Edge of the Everglades, Florida
(Lat 25°33'41.34"N, Long 80°35'2.75"W)

Looking down from 45 km (28 miles) at these coordinates, you see the transition between the natural Everglades on the west and intensively cultivated farmland on the east (**Image U.5**). This radical change in landscape affects many aspects of the Earth System. Fly to Lat 25°53'27.04"N, Long 80°8'25.19"W, zoom to 6 km (4 miles) and look down at what was once a sandbar, but is now a forest of concrete (**Image U.6**).

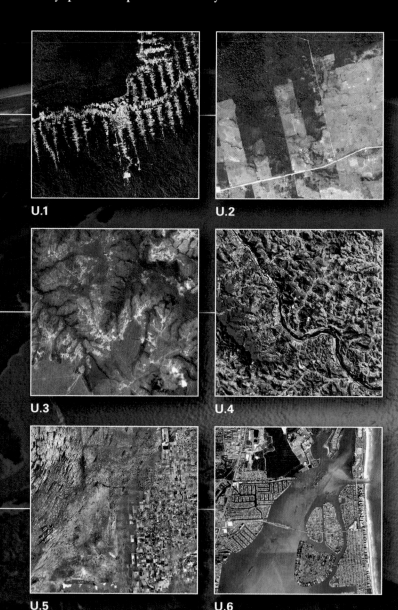

U.1

U.2

U.3

U.4

U.5

U.6

Urbanizing the Desert, Arizona
(Lat 33°33'48.47"N, Long 111°48'33.37"W)

At the coordinates provided, the view from 5 km (3 miles) shows the edge of a Phoenix suburb (**Image U.7**). Note that construction has nearly eliminated natural drainage. The transformation of the Sonoran Desert of Arizona into cityscapes and farmland requires huge amounts of water. Some of the water comes from pumping underground aquifers, and some from canals. A canal traverses this image.

U.8

Receding Glacier, Switzerland
(Lat 46°34'21.13"N, Long 8°22'40.61"E)

Most glaciers have been retreating at accelerating paces in recent decades. Here, near the town of Gletsch, zoom to 7 km (4 miles), tilt, and look northeast (**Image U.8**). You can see the empty U-shaped valley that was occupied by ice a century ago.

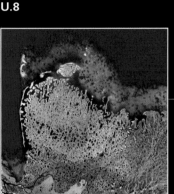

U.9

Melting Permafrost, Siberia
(Lat 73°20'58.61"N, Long 124°43'59.76"E)

Here, on the Arctic coast of Siberia, a landscape that had been permafrost appears to be melting. Zoom to 200 km (125 miles), and you can see the pockmarks on land and beneath shallow coastal waters of the Arctic (**Image U.9**). This image gives the impression that recent sea-level rise has submerged the landscape.

Intense Urbanization, Tokyo, Japan
(Lat 35°41'3.94"N, Long 139°48'35.69"E)

The view of Tokyo from an elevation of 1.5 km (1 mile) shows the results of complete transformation of a natural coastal area into a human-controlled environment (**Image U.10**). Vehicles generate pollutants, the concrete absorbs and retains heat, and geologic materials have been transformed into buildings. Zoom out to 20 km (12 miles) to see how human activity has affected the coastline.

U.10

U.11

Village and Fields, China
(Lat 35°7'3.20"N, Long 114°27'16.86"E)

From an elevation of 5 km (3 miles), at these coordinates, you see a landscape in central China that has been transformed from forest to farmland (**Image U.11**). Every square meter that does not lie beneath a house or road has been turned into a farm field.

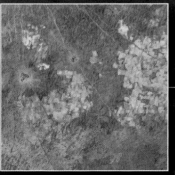

U.12

Desertification of the Sahel, Africa
(Lat 13°1'46.67"N, Long 26°3'49.89"E)

Here, the view from 12 km (7 miles) reveals a dry landscape (**Image U.12**). The radial pattern of paths on the left side of the image comes from cattle walking to a tiny watering hole. Cattle compact the land, preventing water infiltration, farming removes nutrients, and grazing causes devegetation. As a result, the landscape has become more desert-like.

S-42

a'a A lava flow with a rubbly surface.

abandoned meander A meander that dries out after it was cut off.

ablation The removal of ice at the toe of a glacier by melting, sublimation (the evaporation of ice into water vapor), and/or calving.

abrasion The process in which one material (such as sand-laden water) grinds away at another (such as a stream channel's floor and walls).

absolute age Numerical age (the age specified in years).

absolute plate velocity The movement of a plate relative to a fixed point in the mantle.

absolute zero The lowest temperature possible (−273.15°C); at absolute zero, vibrations and movements of atoms in a material cease.

abyssal plain A broad, relatively flat region of the ocean that lies at least 4.5 km below sea level.

Acadian orogeny A convergent mountain-building event, that occurred around 400 million years ago, during which continental slivers accreted to the eastern edge of the North American continent.

accreted terrane A block of crust that collided with a continent at a convergent margin and stayed attached to the continent.

accretionary coast A coastline that receives more sediment than erodes away.

accretionary lapilli Hailstone-like clumps of wet ash that fall from a volcanic eruptive cloud.

accretionary orogen An orogen formed by the attachment of numerous buoyant slivers of crust to an older, larger continental block.

accretionary prism A wedge-shaped mass of sediment and rock scraped off the top of a downgoing plate and accreted onto the overriding plate at a convergent plate margin.

acid mine runoff A dilute solution of sulfuric acid, produced when sulfur-bearing minerals in mines react with rainwater, that flows out of a mine.

acid rain Precipitation in which air pollutants react with water to make a weak acid that then falls from the sky.

active continental margin A continental margin that coincides with a plate boundary.

active fault A fault that has moved recently or is likely to move in the future.

active sand The top layer of beach sand, which moves daily because of wave action.

active volcano A volcano that has erupted within the past few centuries and will likely erupt again.

adiabatic cooling The cooling of a body of air or matter without the addition or subtraction of thermal energy (heat).

adiabatic heating The warming of a body of air or matter without the addition or subtraction of heat.

advection A process of heat transfer in which heat is carried into a solid by a liquid or gas moving through fractures or pores in the solid.

aerosols Tiny solid particles or liquid droplets that remain suspended in the atmosphere for a long time.

aftershocks The series of smaller earthquakes that follow a major earthquake.

air The mixture of gases that make up the Earth's atmosphere.

air-fall tuff Tuff formed when ash settles gently from the air.

air mass A body of air, about 1,500 km across, that has recognizable physical characteristics.

air pressure The push that air exerts on its surroundings.

albedo The reflectivity of a surface.

Alleghenian orogeny The orogenic event that occurred about 270 million years ago when Africa collided with North America.

alloy A metal containing more than one type of metal atom.

alluvial fan A gently sloping apron of sediment dropped by an ephemeral stream at the base of a mountain in arid or semiarid regions.

alluvium Sorted sediment deposited by a stream.

alluvium-filled valley A valley whose floor fills with sediment.

amber Hardened (fossilized) ancient sap or resin.

amphibolite facies A set of metamorphic mineral assemblages formed under intermediate pressures and temperatures.

amplitude The height of a wave from crest to trough.

Ancestral Rockies The late Paleozoic uplifts of the Rocky Mountain region; they eroded away long before the present Rocky Mountains formed.

angiosperm A flowering plant.

angle of repose The angle of the steepest slope that a pile of uncemented material can attain without collapsing from the pull of gravity.

angularity The degree to which grains have sharp or rounded edges or corners.

angular unconformity An unconformity in which the strata below were tilted or folded before the unconformity developed; strata below the unconformity therefore have a different tilt than strata above.

anhedral grains Crystalline mineral grains without well-formed crystal faces.

anion A negatively charged ion.

Antarctic bottom water mass The mass of cold, dense water that sinks along the coast of Antarctica.

antecedent stream A stream that cuts across an uplifted mountain range; the stream must have existed before the range uplifted and must then have been able to downcut as fast as the land was rising.

anthracite coal Shiny black coal formed at temperatures between 200° and 300°C. A high-rank coal.

anticline A fold with an arch-like shape in which the limbs dip away from the hinge.

anticyclone The clockwise flow of air around a high-pressure mass.

anticyclonic flow A circulation of air around a high-pressure region in the atmosphere; it rotates clockwise in the northern hemisphere.

Antler orogeny The Late Devonian mountain-building event in which slices of deep-marine strata were pushed eastward, up and over the shallow-water strata on the western coast of North America.

anvil cloud A large cumulonimbus cloud that spreads laterally at the tropopause to form a broad, flat top.

aphanitic A textural term for fine-grained igneous rock.

apparent polar-wander path A path on the globe along which a magnetic pole appears to have wandered over time; in fact, the continents drift, while the magnetic pole stays fairly fixed.

aquiclude Sediment or rock that transmits no water.

aquifer Sediment or rock that transmits water easily.

aquitard   Sediment or rock that does not transmit water easily and therefore retards the motion of the water.

archaea   A kingdom of "old bacteria," now commonly found in extreme environments like hot springs. (Also called archaeobacteria.)

Archean   The middle Precambrian Eon.

Archimedes' principle   The mass of the water displaced by a block of material equals the mass of the whole block of material.

arête   A residual knife-edge ridge of rock that separates two adjacent cirques.

argillaceous sedimentary rock   Sedimentary rock that contains abundant clay.

arroyo   The channel of an ephemeral stream; dry wash; wadi.

artesian well   A well in which water rises on its own.

ash   *See* Volcanic ash.

ash fall   Ash that falls to the ground out of an ash cloud.

ash flow   An avalanche of ash that tumbles down the side of an explosively erupting volcano.

assimilation   The process of magma contamination in which blocks of wall rock fall into a magma chamber and dissolve.

asteroid   One of the fragments of solid material, left over from planet formation or produced by collision of planetesimals, that resides between the orbits of Mars and Jupiter.

asthenosphere   The layer of the mantle that lies between 100–150 km and 350 km deep; the asthenosphere is relatively soft and can flow when acted on by force.

atm   A unit of air pressure that approximates the pressure exerted by the atmosphere at sea level.

atmosphere   A layer of gases that surrounds a planet.

atoll   A coral reef that develops around a circular reef surrounding a lagoon.

atom   The smallest piece of an element that has the properties of the element; it consists of a nucleus surrounded by an electron cloud.

atomic mass   The amount of matter in an atom; roughly, it is the sum of the number of protons plus the number of neutrons in the nucleus.

atomic number   The number of protons in the nucleus of a given element.

atomic weight   The number of protons plus the number of neutrons in the nucleus of a given element. (Also known as atomic mass.)

aurora australis   The same phenomenon as the aurora borealis, but in the southern hemisphere.

aurora borealis   A ghostly curtain of varicolored light that appears across the night sky in the northern hemisphere when charged particles from the Sun interact with the ions in the ionosphere.

avalanche   A turbulent cloud of debris mixed with air that rushes down a steep hill slope at high velocity; the debris can be rock and/or snow.

avalanche chute   A downslope hillside pathway along which avalanches repeatedly fall, consequently clearing the pathway of mature trees.

avulsion   The process in which a river overflows a natural levee and begins to flow in a new direction.

axial plane   The imaginary surface that encompasses the hinges of successive layers of a fold.

axial trough   A narrow depression that runs along a mid-ocean ridge axis.

axis   An imaginary line around which an object spins.

backscattered light   Atmospheric scattered sunlight that returns to space.

backshore zone   The zone of beach that extends from a small step cut by high-tide swash to the front of the dunes or cliffs that lie farther inshore.

backswamp   The low marshy region between the bluffs and the natural levees of a floodplain.

backwash   The gravity-driven flow of water back down the slope of a beach.

bajada   An elongate wedge of sediment formed by the overlap of several alluvial fans emerging from adjacent valleys.

Baltica   A Paleozoic continent that included crust that is now part of today's Europe.

banded-iron formation (BIF)   Iron-rich sedimentary layers consisting of alternating gray beds of iron oxide and red beds of iron-rich chert.

bar   (1) A sheet or elongate lens or mound of alluvium; (2) a unit of air-pressure measurement approximately equal to 1 atm.

barchan dune   A crescent-shaped dune whose tips point downwind.

barrier island   An offshore sand bar that rises above the mean high-water level, forming an island.

barrier reef   A coral reef that develops offshore, separated from the coast by a lagoon.

basal sliding   The phenomenon in which meltwater accumulates at the base of a glacier, so that the mass of the glacier slides on a layer of water or on a slurry of water and sediment.

basalt   A fine-grained, mafic, igneous rock.

base level   The lowest elevation a stream channel's floor can reach at a given locality.

basement   Older igneous and metamorphic rocks making up the Earth's crust beneath sedimentary cover.

basement uplift   Uplift of basement rock by faults that penetrate deep into the continental crust.

base metals   Metals that are mined but not considered precious. Examples include copper, lead, zinc, and tin.

basin   A fold or depression shaped like a right-side-up bowl.

Basin and Range Province   A broad, Cenozoic continental rift that has affected a portion of the western United States in Nevada, Utah, and Arizona; in this province, tilted fault blocks form ranges, and alluvium-filled valleys are basins.

batholith   A vast composite, intrusive, igneous rock body up to several hundred km long and 100 km wide, formed by the intrusion of numerous plutons in the same region.

bathymetric map   A map illustrating the shape of the ocean floor.

bathymetric profile   A cross section showing ocean depth plotted against location.

bathymetry   Variation in depth.

bauxite   A residual mineral deposit rich in aluminum.

baymouth bar   A sandspit that grows across the opening of a bay.

beach drift   The gradual migration of sand along a beach.

beach erosion   The removal of beach sand caused by wave action and longshore currents.

beach face   A steeply concave part of the foreshore zone formed where the swash of the waves actively scours the sand.

bedding   Layering or stratification in sedimentary rocks.

bed load   Large particles, such as sand, pebbles, or cobbles, that bounce or roll along a stream bed.

bedrock   Rock still attached to the Earth's crust.

Bergeron process   Precipitation involving the growth of ice crystals in a cloud at the expense of water droplets.

berm   A horizontal or landward-sloping terrace in the backshore zone of a beach that receives sediment during a storm.

Big Bang   A cataclysmic explosion that scientists suggest represents the formation of the Universe; before this event, all matter and all energy were packed into one volumeless point.

biochemical sedimentary rock   Sedimentary rock formed from material (such as shells) produced by living organisms.

biodiversity   The number of different species that exist at a given time.

biofuel   Gas or liquid fuel made from plant material (biomass). Examples of biofuel include alcohol (from fermented sugar), bio diesel from vegetable oil, and wood.

biogenic minerals   Substances that meet the definition of a mineral and are produced naturally by organisms (e.g., calcite in shells).

biogeochemical cycle   The exchange of chemicals between living and nonliving reservoirs in the Earth System.

bioremediation The injection of oxygen and nutrients into a contaminated aquifer to foster the growth of bacteria that will ingest or break down contaminants.

biosphere The region of the Earth and atmosphere inhabited by life; this region stretches from a few km below the Earth's surface to a few km above.

bioturbation The mixing of sediment by burrowing animals such as clams and worms.

bituminous coal Dull, black intermediate-rank coal formed at temperatures between 100° and 200°C.

black-lung disease Lung disease contracted by miners from the inhalation of too much coal dust.

black smoker The cloud of suspended minerals formed where hot water spews out of a vent along a mid-ocean ridge; the dissolved sulfide components of the hot water instantly precipitate when the water mixes with seawater and cools.

blind fault A fault that does not intersect the ground surface.

blocking temperature The temperature below which isotopes in a mineral are no longer free to move, so the radiometric clock starts.

blocky lava Lava that is so viscous that it breaks into boulder-like blocks as it moves; typically, such lavas are andesitic or rhyolitic.

blowout A deep, bowl-like depression scoured out of desert terrain by a turbulent vortex of wind.

blue shift The phenomenon in which a source of light moving toward you appears to have a higher frequency.

body fossil A relict of an organism's body, preserved in rock.

body waves Seismic waves that pass through the interior of the Earth.

bog A wetland dominated by moss and shrubs.

bolide A solid extraterrestrial object such as a meteorite, comet, or asteroid that explodes in the atmosphere.

bornhardt An inselberg with a loaf geometry, like that of Uluru (Ayers Rock) in central Australia.

Bowen's reaction series The sequence in which different silicate minerals crystallize during the progressive cooling of a melt.

braided stream A sediment-choked stream consisting of entwined subchannels.

breaker A water wave in which water at the top of the wave curves over the base of the wave.

breakwater An offshore wall, built parallel or at an angle to the beach, that prevents the full force of waves from reaching a harbor.

breccia Coarse sedimentary rock consisting of angular fragments; or rock broken into angular fragments by faulting.

breeder reactor A nuclear reactor that produces its own fuel.

brine Water that is not fresh but is less salty than seawater; brine may be found in estuaries.

brittle deformation The cracking and fracturing of a material subjected to stress.

brittle-ductile transition (brittle-plastic transition) The depth above which materials behave brittlely and below which materials behave ductilely (plastically); this transition typically lies between a depth of 10 and 15 km in continental crustal rock, and 60 m deep in glacial ice.

buoyancy The upward force acting on an object immersed or floating in fluid; the tendency of an object to float when placed in a fluid.

butte A medium-size, flat-topped hill in an arid region.

caldera A large circular depression with steep walls and a fairly flat floor, formed after an eruption as the center of the volcano collapses into the drained magma chamber below.

caliche A solid mass created where calcite cements the soil together (also called calcrete).

calorie A unit of energy approximately equal to 4.2 joules; 1 calorie can raise the temperature of 1 gm of water by 1°C.

calving The breaking off of chunks of ice at the edge of a glacier.

Cambrian explosion of life The remarkable diversification of life, indicated by the fossil record, that occurred at the beginning of the Cambrian Period.

Canadian Shield A broad, low-lying region of exposed Precambrian rock in the Canadian interior.

canyon A trough or valley with steeply sloping walls, cut into the land by a stream.

capillary fringe The thin subsurface layer in which water molecules seep up from the water table by capillary action to fill pores.

carbonate rocks Rocks containing calcite and/or dolomite.

carbon-14 A radioactive isotope of the element carbon; the ratio of C^{14} to C^{12} can provide an isotopic date of organic carbon.

carbon-14 dating A radiometric dating process that can tell us the age of organic material containing carbon originally extracted from the atmosphere.

carbon sequestration The process of extracting carbon dioxide from sources (e.g., power plants) and sending it back underground to keep it out of the atmosphere and diminish the greenhouse effect.

cast Sediment that preserves the shape of a shell it once filled before the shell dissolved or mechanically weathered away.

catabatic winds Strong winds that form at the margin of a glacier where the warmer air above ice-free land rises and the cold, denser air from above the glaciers rushes in to take its place.

catastrophic change Change that takes place either instantaneously or rapidly in geologic time.

catchment *See* Drainage network.

cation A positively charged ion.

Celsius scale A metric-system measure of temperature in which the difference between the freezing point (0°C) and boiling point (100°C) of water is divided into 100 units; equivalent to centigrade scale.

cement Mineral material that precipitates from water and fills the spaces between grains, holding the grains together.

cementation The phase of lithification in which cement, consisting of minerals that precipitate from groundwater, partially or completely fills the spaces between clasts and attaches each grain to its neighbor.

Cenozoic The most recent era of the Phanerozoic Eon, lasting from 65 Ma up until the present.

chalk Very fine-grained limestone consisting of weakly cemented plankton shells.

change of state The process in which a material changes from one phase (liquid, gas, or solid) to another.

channel A trough dug into the ground surface by flowing water.

channeled scablands A barren, soil-free landscape in eastern Washington, scoured clean by a flood unleashed when a large glacial lake drained.

chatter marks Wedge-shaped indentations left on rock surfaces by glacial plucking.

chemical A material consisting of a distinct element or compound.

chemical bond The invisible link that holds together atoms in a molecule and/or in a crystal.

chemical formula The "recipe" that specifies the elements and their proportions in a compound.

chemical fossil Distinctive molecules or molecular fragments, formed from the remains of living organisms, that can be preserved in rock.

chemical reaction Interactions among atoms and/or molecules involving breaking or forming chemical bonds.

chemical sedimentary rocks Sedimentary rocks made up of minerals that precipitate directly from water solution.

chemical weathering The process in which chemical reactions alter or destroy minerals when rock comes in contact with water solutions and/or air.

chert A sedimentary rock composed of very fine-grained silica (cryptocrystalline quartz).

Chicxulub crater A circular excavation buried beneath younger sediment on the Yucatán Peninsula; geologists suggest that a meteorite landed there 65 Ma.

chimney (1) A conduit in a magma chamber in the shape of a long vertical pipe through which magma rises and erupts at the surface; (2) an isolated column of strata in an arid region.

cinder cone A subaerial volcano consisting of a cone-shaped pile of tephra whose slope approaches the angle of repose for tephra.

cinders Fragments of glassy rock ejected from a volcano.

cirque A bowl-shaped depression carved by a glacier on the side of a mountain.

cirrus cloud A wispy cloud that tapers into delicate, feather-like curls.

clastic (detrital) sedimentary rock Sedimentary rock consisting of cemented-together detritus derived from the weathering of preexisting rock.

cleavage (1) The tendency of a mineral to break along preferred planes; (2) a type of foliation in low-grade metamorphic rock.

cleavage planes A series of surfaces on a crystal that form parallel to the weakest bonds holding the atoms of the crystal together.

cliff (or scarp) retreat The change in the position of a cliff face caused by erosion.

climate The average weather conditions, along with the range of conditions, of a region over a year.

cloud A mist of tiny water droplets in the sky.

coal A black, organic rock consisting of greater than 50% carbon; it forms from the buried and altered remains of plant material.

coal rank A measurement of the carbon content of coal; higher-rank coal forms at higher temperatures.

coal reserve The quantities of discovered, but not yet mined, coal in sedimentary rock of the continents.

coal swamp A swamp whose oxygen-poor water allows thick piles of woody debris to accumulate; this debris transforms into coal upon deep burial.

coastal plain Low-relief regions of land adjacent to the coast.

cold front The boundary at which a cold air mass pushes underneath a warm air mass.

collision The process of two buoyant pieces of lithosphere converging and squashing together.

color The characteristic of a material due to the spectrum of light emitted or reflected by the material, as perceived by eyes or instruments.

columnar jointing A type of fracturing that yields roughly hexagonal columns of basalt; columnar joints form when a dike, sill, or lava flow cools.

comet A ball of ice and dust, probably remaining from the formation of the solar system, that orbits the Sun.

compaction The phase of lithification in which the pressure of the overburden on the buried rock squeezes out water and air that was trapped between clasts, and the clasts press tightly together.

composite volcano *See* Stratovolcano.

compositional banding A type of metamorphic foliation, found in gneiss, defined by alternating bands of light and dark minerals.

compound A material composed of two or more elements that cannot be separated mechanically; the smallest piece is a molecule.

compressibility The degree to which a material's volume changes in response to squashing.

compression A push or squeezing felt by a body.

compressional waves Waves in which particles of material move back and forth parallel to the direction in which the wave itself moves.

concentration The proportion of one substance (the solute) dissolved within another (the solvent).

conchoidal fractures Smoothly curving, clamshell-shaped surfaces along which materials with no cleavage planes tend to break.

condensation The process of gas molecules linking together to form a liquid.

condensation nuclei Preexisting solid or liquid particles, such as aerosols, onto which water condenses during cloud formation.

conduction A process of heat transfer involving progressive migration of thermal energy from cooler to warmer regions in a material, without the physical flow of the material itself.

cone of depression The downward-pointing, cone-shaped surface of the water table in a location where the water table is experiencing drawdown because of pumping at a well.

confined aquifer An aquifer that is separated from the Earth's surface by an overlying aquitard.

conglomerate Very coarse-grained sedimentary rock consisting of rounded clasts.

consuming boundary *See* Convergent plate boundary.

contact The boundary surface between two rock bodies (as between two stratigraphic formations, between an igneous intrusion and adjacent rock, between two igneous rock bodies, or between rocks juxtaposed by a fault).

contact metamorphism *See* Thermal metamorphism.

contaminant plume A cloud of contaminated groundwater that moves away from the source of the contamination.

continental crust The crust beneath the continents.

continental divide A highland separating drainage that flows into one ocean from drainage that flows into another.

continental-drift hypothesis The idea that continents have moved and are still moving slowly across the Earth's surface.

continental glacier A vast sheet of ice that spreads over thousands of square km of continental crust.

continental-interior desert An inland desert that develops because by the time air masses reach the continental interior, they have lost all of their moisture.

continental lithosphere Lithosphere topped by continental crust; this lithosphere reaches a thickness of 150 km.

continental margin A continent's coastline.

continental rift A linear belt along which continental lithosphere stretches and pulls apart.

continental rifting The process by which a continent stretches and splits along a belt; if it is successful, rifting separates a larger continent into two smaller continents separated by a divergent boundary.

continental rise The sloping sea floor that extends from the lower part of the continental slope to the abyssal plain.

continental shelf A broad, shallowly submerged fringe of a continent; ocean-water depth over the continental shelf is generally less than 200 meters; the widest continental shelves occur over passive margins.

continental slope The slope at the edge of a continental shelf, leading down to the deep sea floor.

continental volcanic arc A long, curving chain of subaerial volcanoes on the margin of a continent adjacent to a convergent plate boundary.

contour lines Lines on a map along which a parameter has a constant value; for example, all points along a contour line on a topographic map are at the same elevation.

control rod Rods that absorb neutrons in a nuclear reactor and thus decrease the number of collisions between neutrons and radioactive atoms.

convection Heat transfer that results when warmer, less dense material rises while cooler, denser material sinks.

convective cell A distinct flow configuration for a volume of material that is moving during convective heat transport; simplistically, the material rises when warm and sinks when cool, and thus follows a loop-like path.

convergence zone A place where two surface air flows meet so that air has to rise.

convergent margin *See* Convergent plate boundary.

convergent plate boundary A boundary at which two plates move toward each other so that one plate sinks (subducts) beneath the other; only oceanic lithosphere can subduct.

coral reef A mound of coral and coral debris forming a region of shallow water.

core The dense, iron-rich center of the Earth.

core-mantle boundary An interface 2,900 km below the Earth's surface separating the mantle and core.

Coriolis effect The deflection of objects, winds, and currents on the surface of the Earth owing to the planet's rotation.

cornice A huge, overhanging drift of snow built up by strong winds at the crest of a mountain ridge.

correlation The process of defining the age relations between the strata at one locality and the strata at another.

cosmic rays Nuclei of hydrogen and other elements that bombard the Earth from deep space.

cosmology The study of the overall structure of the Universe.

country rock (wall rock) The preexisting rock into which magma intrudes.

covalent bonding The attachment of one atom to another that develops when the atoms share electrons; one type of chemical bond.

crater (1) A circular depression at the top of a volcanic mound; (2) a depression formed by the impact of a meteorite.

craton A long-lived block of durable continental crust commonly found in the stable interior of a continent.

cratonic platform A province in the interior of a continent in which Phanerozoic strata bury most of the underlying Precambrian rock.

creep The gradual downslope movement of regolith.

crevasse A large crack that develops by brittle deformation in the top 60 m of a glacier.

critical mass A sufficiently dense and large mass of radioactive atoms in which a chain reaction happens so quickly that the mass explodes.

cross section A diagram depicting the geometry of materials underground as they would appear on an imaginary vertical slice through the Earth.

crude oil Oil extracted directly from the ground.

crust The rock that makes up the outermost layer of the Earth.

crustal root Low-density crustal rock that protrudes downward beneath a mountain range.

crustal thickening The process by which the continental crust increases in thickness, becoming up to 70 km thick (vs. normal thickness of about 35–40 km); it can occur during continental collision.

crystal A single, continuous piece of a mineral bounded by flat surfaces that formed naturally as the mineral grew.

crystal form The geometric shape of a crystal, defined by the arrangement of crystal faces.

crystal habit The general shape of a crystal or cluster of crystals that grew unimpeded.

crystal lattice The orderly framework within which the atoms or ions of a mineral are fixed.

crystalline Containing a crystal lattice.

cuesta An asymmetric ridge formed by tilted layers of rock, with a steep cliff on one side cutting across the layers and a gentle slope on the other side; the gentle slope is parallel to the layering.

cumulonimbus cloud A rain-producing, puffy cloud.

cumulus cloud A puffy, cotton-ball-shaped cloud.

current (1) A well-defined stream of ocean water; (2) the moving flow of water in a stream.

cut bank The outside bank of the channel wall of a meander, which is continually undergoing erosion.

cutoff A straight reach in a stream that develops when erosion eats through a meander neck.

cyanobacteria Blue-green algae; a type of archaea.

cycle A series of interrelated events or steps that occur in succession and can be repeated, perhaps indefinitely.

cyclone (1) The counterclockwise flow of air around a low-pressure mass; (2) the equivalent of a hurricane in the Indian Ocean.

cyclonic flow A circulation of air around a low-pressure region in the atmosphere; it rotates counterclockwise in the northern hemisphere.

cyclothem A repeated interval within a sedimentary sequence that contains a specific succession of sedimentary beds.

Darcy's law A mathematical equation stating that a volume of water, passing through a specified area of material at a given time, depends on the material's permeability and hydraulic gradient.

daughter isotope The decay product of radioactive decay.

day The time it takes for the Earth to spin once on its axis.

debris avalanche An avalanche in which the falling debris consists of rock fragments and dust.

debris flow A downslope movement of mud mixed with larger rock fragments.

debris slide A sudden downslope movement of material consisting only of regolith.

decompression melting The kind of melting that occurs when hot mantle rock rises to shallower depths in the Earth so that pressure decreases while the temperature remains unchanged.

deep current An ocean current at a depth greater than 100 m.

deep-focus earthquake An earthquake that occurs at a depth between 300 and 670 km; below 670 km, earthquakes do not happen.

deflation The process of lowering the land surface by wind abrasion.

deformation A change in the shape, position, or orientation of a material, by bending, breaking, or flowing.

dehydration Loss of water.

delamination (plate tectonics) The process by which dense lithospheric mantle separates from the base of a plate and sinks into the mantle.

delta A wedge of sediment formed at a river mouth when the running water of the stream enters standing water, the current slows, the stream loses competence, and sediment settles out.

delta plain The low, swampy land on the surface of a delta.

delta-plain flood A flood in which water submerges a delta plain.

dendritic network A drainage network whose interconnecting streams resemble the pattern of branches connecting to a deciduous tree.

dendrochronologist A scientist who analyzes tree rings to determine the geologic age of features.

density Mass per unit volume.

denudation The removal of rock and regolith from the Earth's surface.

deposition The process by which sediment settles out of a transporting medium.

depositional environment A setting in which sediments accumulate; its character (fluvial, deltaic, reef, glacial, etc.) reflects local conditions.

depositional landform A landform resulting from the deposition of sediment where the medium carrying the sediment evaporates, slows down, or melts.

desert A region so arid that it contains no permanent streams, except for those that bring water in from elsewhere, and has very sparse vegetation cover.

desertification The process of transforming nondesert areas into desert.

desert pavement A mosaic-like stone surface forming the ground in a desert.

desert varnish A dark, rusty-brown coating of iron oxide and magnesium oxide that accumulates on the surface of the rock.

detachment fault A nearly horizontal fault at the base of a fault system.

detritus The chunks and smaller grains of rock broken off outcrops by physical weathering.

dewpoint temperature   The temperature at which air becomes saturated so that dew can form.

differential stress   A condition causing a material to experience a push or pull in one direction of a greater magnitude than the push or pull in another direction; in some cases, differential stress can result in shearing.

differential weathering   What happens when different rocks in an outcrop undergo weathering at different rates.

diffraction   The splitting of light into many tiny beams that interfere with one another.

digital elevation map (DEM)   A computer-produced portrayal of elevation differences commonly using shading to simulate shadows; the data used to produce the map assigns elevations to each point on the map.

dike   A tabular (wall-shaped) intrusion of rock that cuts across the layering of country rock.

dimension stone   An intact block of granite or marble to be used for architectural purposes.

dipole   A magnetic field with a north and south pole, like that of a bar magnet.

dipole field (for Earth)   The part of the Earth's magnetic field, caused by the flow of liquid iron alloy in the outer core, that can be represented by an imaginary bar magnet with a north and south pole.

dip-slip fault   A fault in which sliding occurs up or down the slope (dip) of the fault.

dip slope   A hill slope underlain by bedding parallel to the slope.

directional drilling   The process of controlling the trajectory of a drill bit to make sure that the drill hole goes exactly where desired.

disappearing stream   A stream that intersects a crack or sinkhole leading to an underground cavern, so that the water disappears into the subsurface and becomes an underground stream.

discharge   The volume of water in a conduit or channel passing a point in 1 second.

discharge area   A location where groundwater flows back up to the surface and may emerge at springs.

disconformity   An unconformity parallel to the two sedimentary sequences it separates.

displacement (or offset)   The amount of movement or slip across a fault plane.

disseminated deposit   A hydrothermal ore deposit in which ore minerals are dispersed throughout a body of rock.

dissolution   A process during which materials dissolve in water.

dissolved load   Ions dissolved in a stream's water.

distillation column   A vertical pipe in which crude oil is separated into several components.

distributaries   The fan of small streams formed where a river spreads out over its delta.

divergence zone   A place where sinking air separates into two flows that move in opposite directions.

divergent plate boundary   A boundary at which two lithosphere plates move apart from each other; they are marked by mid-ocean ridges.

diversification   The development of many different species.

DNA (deoxyribonucleic acid)   The complex molecule, shaped like a double helix, containing the code that guides the growth and development of an organism.

doldrums   A belt with very slow winds along the equator.

dome   Folded or arched layers with the shape of an overturned bowl.

Doppler effect   The phenomenon in which the frequency of wave energy appears to change when a moving source of wave energy passes an observer.

dormant volcano   A volcano that has not erupted for hundreds to thousands of years but does have the potential to erupt again in the future.

downcutting   The process in which water flowing through a channel cuts into the substrate and deepens the channel relative to its surroundings.

downdraft   Downward-moving air.

downgoing plate (or slab)   A lithosphere plate that has been subducted at a convergent margin.

downslope force   The component of the force of gravity acting in the downslope direction.

downslope movement   The tumbling or sliding of rock and sediment from higher elevations to lower ones.

downwelling zone   A place where near-surface water sinks.

drag fold   A fold that develops in layers of rock adjacent to a fault during or just before slip.

drainage divide   A highland or ridge that separates one watershed from another.

drainage network (or basin)   An array of interconnecting streams that together drain an area.

drawdown   The phenomenon in which the water table around a well drops because the users are pumping water out of the well faster than it flows in from the surrounding aquifer.

drilling mud   A slurry of water mixed with clay that oil drillers use to cool a drill bit and flush rock cuttings up and out of the hole.

dripstone   Limestone (travertine in a cave) formed by the precipitation of calcium carbonate out of groundwater.

drop stone   A rock that drops to the sea floor once the iceberg that was carrying the rock melts.

drumlin   A streamlined, elongate hill formed when a glacier overrides glacial till.

dry-bottom (polar) glacier   A glacier so cold that its base remains frozen to the substrate.

dry wash   The channel of an ephemeral stream when empty of water.

dry well   (1) A well that does not supply water because the well has been drilled into an aquitard or into rock that lies above the water table; (2) a well that does not yield oil, even though it has been drilled into an anticipated reservoir.

ductile (plastic) deformation   The bending and flowing of a material (without cracking and breaking) subjected to stress.

dune   A pile of sand generally formed by deposition from the wind.

dust storm   An event in which strong winds hit unvegetated land, strip off the topsoil, and send it skyward to form rolling dark clouds that block out the Sun.

dynamic metamorphism   Metamorphism that occurs as a consequence of shearing alone, with no change in temperature or pressure.

dynamo   A power plant generator in which water or wind power spins an electrical conductor around a permanent magnet.

dynamothermal metamorphism   Metamorphism that involves heat, pressure, and shearing.

earthquake   A vibration caused by the sudden breaking or frictional sliding of rock in the Earth.

earthquake belt   A relatively narrow, distinct belt of earthquakes that defines the position of a plate boundary.

earthquake engineering   The design of buildings that can withstand shaking.

earthquake warning system   A communications network that provides an alert within microseconds after the first earthquakes waves arrive at a seismograph near the epicenter, but before damaging vibrations reach population centers.

earthquake zoning   The determination of where land is relatively stable and where it might collapse because of seismicity.

Earth System   The global interconnecting web of physical and biological phenomena involving the solid Earth, the hydro sphere, and the atmosphere.

ebb tide   The falling tide.

eccentricity cycle   The cycle of the gradual change of the Earth's orbit from a more circular to a more elliptical shape; the cycle takes around 100,000 years.

ecliptic   The plane defined by a planet's orbit.

ecosystem   An environment and its inhabitants.

eddy   An isolated, ring-shaped current of water.

effusive eruption   An eruption that yields mostly lava, not ash.

Ekman spiral   The change in flow direction of water with depth, caused by the Coriolis effect.

Ekman transport   The overall movement of a mass of water, resulting from the Eckman spiral, in a direction 90° to the wind direction.

elastic strain   A change in shape of a material; the change disappears instantly when stress is removed.

electromagnet   An electrical device that produces a magnetic field.

electron   A negatively charged subatomic particle that orbits the nucleus of an atom; electrons are about 0.0005× the size of a proton.

electron microprobe   A laboratory instrument that can focus a beam of electrons on a small part of a mineral grain in order to create a signal that defines its chemical composition.

element   A material consisting entirely of one kind of atom; elements cannot be subdivided or changed by chemical reactions.

El Niño   The flow of warm water eastward from the Pacific Ocean that reverses the upwelling of cold water along the western coast of South America and causes significant global changes in weather patterns.

embayment   A low area of coastal land.

emergent coast   A coast where the land is rising relative to sea level or sea level is falling relative to the land.

end moraine (terminal moraine)   A low, sinuous ridge of till that develops when the terminus (toe) of a glacier stalls in one position for a while.

energy   The capacity to do work.

energy resource   Something that can be used to produce work; in a geologic context, a material (such as oil, coal, wind, flowing water) that can be used to produce energy.

eon   The largest subdivision of geologic time.

epeirogenic movement   The gradual uplift or subsidence of a broad region of the Earth's surface.

epeirogeny   An event of epeirogenic movement; the term is usually used in reference to the formation of broad mid-continent domes and basins.

ephemeral (intermittent) stream   A stream whose bed lies above the water table, so that the stream flows only when the rate at which water enters the stream from rainfall or melt water exceeds the rate at which water infiltrates the ground below.

epicenter   The point on the surface of the Earth directly above the focus of an earthquake.

epicontinental sea   A shallow sea overlying a continent.

epoch   An interval of geologic time representing the largest subdivision of a period.

equant   A term for a grain that has the same dimensions in all directions.

equatorial low   The area of low pressure that develops over the equator because of the intertropical convergence zone.

equinox   One of two days out of the year (September 22 and March 21) in which the Sun is directly overhead at noon at the equator.

era   An interval of geologic time representing the largest subdivision of the Phanerozoic Eon.

erg   Sand seas formed by the accumulation of dunes in a desert.

erosion   The grinding away and removal of Earth's surface materials by moving water, air, or ice.

erosional coast   A coastline where sediment is not accumulating and wave action grinds away at the shore.

erosional landform   A landform that results from the breakdown and removal of rock or sediment.

erratic   A boulder or cobble that was picked up by a glacier and deposited hundreds of kilometers away from the outcrop from which it detached.

eruptive style   The character of a particular volcanic eruption; geologists name styles based on typical examples (e.g., Hawaiian; Strombolian).

esker   A ridge of sorted sand and gravel that snakes across a ground moraine; the sediment of an esker was deposited in subglacial meltwater tunnels.

estuary   An inlet in which seawater and river water mix; created when a coastal valley is flooded because of either rising sea level or land subsidence.

Eubacteria   The kingdom of "true bacteria."

euhedral crystal   A crystal whose faces are well formed and whose shape reflects crystal form.

eukaryote   An organism whose cells contain a nucleus; all plants and animals consist of eukaryotic cells.

eukaryotic cell   A cell with a complex internal structure, capable of building multicellular organisms.

eustatic sea-level change   A global rising or falling of the ocean surface.

evaporate   To change from liquid to vapor.

evapotranspiration   The sum of evaporation from bodies of water and the ground surface and transpiration from plants and animals.

exfoliation   The process by which an outcrop of rock splits apart into onion-like sheets along joints that lie parallel to the ground surface.

exhumation   The process (involving uplift and erosion) that returns deeply buried rocks to the surface.

exotic terrane   A block of land that collided with a continent along a convergent margin and attached to the continent; the term exotic implies that the land was not originally part of the continent to which it is now attached.

expanding Universe theory   The theory that the whole Universe must be expanding because galaxies in every direction seem to be moving away from us.

explosive eruptions   Violent volcanic eruptions that produce clouds and avalanches of pyroclastic debris.

external process   A geomorphologic process—such as down slope movement, erosion, or deposition—that is the consequence of gravity or of the interaction between the solid Earth and its fluid envelope (air and water). Energy for these processes comes from gravity and sunlight.

extinction   The death of the last members of a species so that there are no parents to pass on their genetic traits to offspring.

extinct volcano   A volcano that was active in the past but has now shut off entirely and will not erupt in the future.

extraordinary fossil   A rare fossilized relict, or trace, of the soft part of an organism.

extratropical cyclone (syn. wave cycolone, mid-latitude cyclone)   A large, rotating storm system, in mid-latitudes, associated with a regional-scale low-pressure zone.

extrusive igneous rock   Rock that forms by the freezing of lava above ground, after it flows or explodes out (extrudes) onto the surface and comes into contact with the atmosphere or ocean.

eye   The relative calm in the center of a hurricane.

eye wall   A rotating vertical cylinder of clouds surrounding the eye of a hurricane.

facies   (1) Sedimentary: a group of rocks and primary structures indicative of a given depositional environment; (2) Metamorphic:a set of metamorphic mineral assemblages formed under a given range of pressures and temperatures.

Fahrenheit scale   An English-system measure of temperature in which the difference between the freezing point (32°F) and the boiling point (212°F) is divided into 180 units.

faint young Sun paradox   The apparent contradiction implied by the fact that much of the Earth's surface temperature has remained above the melting point of water during the past 4 Ga, even though calculations indicate that the Sun produced much less energy when young.

fault A fracture on which one body of rock slides past another.

fault-block mountains An outdated term for a narrow, elongate range of mountains that develops in a continental-rift setting as normal faulting drops down blocks of crust or tilts blocks.

fault breccia Fragmented rock in which angular fragments were formed by brittle fault movement; fault breccia occurs along a fault.

fault creep Gradual movement along a fault that occurs in the absence of an earthquake.

fault gouge Pulverized rock consisting of fine powder that lies along fault surfaces; gouge forms by crushing and grinding.

faulting Slip events along a fault.

fault scarp A small step on the ground surface where one side of a fault has moved vertically with respect to the other.

fault system A grouping of numerous related faults.

fault trace (or line) The intersection between a fault and the ground surface.

feedback mechanism A condition that arises when the consequence of a phenomenon influences the phenomenon itself.

felsic An adjective used in reference to igneous rocks that are rich in elements forming feldspar and quartz.

Ferrel cells The name given to the middle-latitude convection cells in the atmosphere.

fetch The distance across a body of water along which a wind blows to build waves.

field force A push or pull that applies across a distance (i.e., without contact between objects); examples are gravity and magnetism.

fine-grained A textural term for rock consisting of many fine grains or clasts.

firn Compacted granular ice (derived from snow) that forms where snow is deeply buried; if buried more deeply, firn turns into glacial ice.

fission A nuclear reaction during which the nucleus of a large atom splits to form two nuclei of smaller atoms; the process also releases neutrons and energy.

fission track A line of damage formed in the crystal lattice of a mineral by the impact of an atomic particle ejected during the decay of a radioactive isotope.

fissure A conduit in a magma chamber in the shape of a long crack through which magma rises and erupts at the surface.

fjord A deep, glacially carved, U-shaped valley flooded by rising sea level.

flank eruption An eruption that occurs when a secondary chimney, or fissure, breaks through the flank of a volcano.

flash flood A flood that occurs during unusually intense rainfall or as the result of a dam collapse, during which the floodwaters rise very fast.

flexing The process of folding in which a succession of rock layers bends and slip occurs between the layers.

flocculation The clumping together of clay suspended in river water into bunches that are large enough to settle out.

flood An event during which the volume of water in a stream becomes so great that it covers areas outside the stream's normal channel.

flood basalt Vast sheets of basalt that spread from a volcanic vent over an extensive surface of land; they may form where a rift develops above a continental hot spot, and where lava is particularly hot and has low viscosity.

flood-hazard map A representation of a portion of the Earth's surface that is designed to show how the danger of flooding varies with location.

floodplain The flat land on either side of a stream that becomes covered with water during a flood.

floodplain flood A flood during which a floodplain is submerged.

flood stage The stage when water reaches the top of a stream channel.

flood tide The rising tide.

floodway A mapped region likely to be flooded, in which people avoid constructing buildings.

flow fold A fold that forms when the rock is so soft that it behaves like weak plastic.

flowstone A sheet of limestone that forms along the wall of a cave when groundwater flows along the surface of the wall.

fluvial deposit Sediment deposited in a stream channel, along a stream bank, or on a floodplain.

flux Flow.

flux melting The transformation of hot solid to liquid that occurs when a volatile material injects into the solid.

focus The location where a fault slips during an earthquake (hypocenter).

fog A cloud that forms at ground level.

fold A bend or wrinkle of rock layers or foliation; folds form as a consequence of ductile deformation.

fold axis An imaginary line that, when moved parallel to itself, can trace out the shape of a folded surface.

fold-thrust belt An assemblage of folds and related thrust faults that develop above a detachment fault.

foliation Layering formed as a consequence of the alignment of mineral grains, or of compositional banding in a metamorphic rock.

foraminifera Microscopic plankton with calcitic shells, components of some limestones.

foreland sedimentary basin A basin located under the plains adjacent to a mountain front, which develops as the weight of the mountains pushes the crust down, creating a depression that traps sediment.

foreshocks The series of smaller earthquakes that precede a major earthquake.

foreshore zone The zone of beach regularly covered and uncovered by rising and falling tides.

formation *See* Stratigraphic formation.

fossil The remnant, or trace, of an ancient living organism that has been preserved in rock or sediment.

fossil assemblage A group of fossil species found in a specific sequence of sedimentary rock.

fossil correlation A determination of the stratigraphic relation between two sedimentary rock units, reached by studying fossils.

fossil fuel An energy resource such as oil or coal that comes from organisms that lived long ago and thus stores solar energy that reached the Earth then.

fossiliferous limestone Limestone consisting of abundant fossil shells and shell fragments.

fossilization The process of forming a fossil.

fractional crystallization The process by which a magma becomes progressively more silicic as it cools, because early-formed crystals settle out.

fracture zone A narrow band of vertical fractures in the ocean floor; fracture zones lie roughly at right angles to a mid-ocean ridge, and the actively slipping part of a fracture zone is a transform fault.

frequency The number of waves that pass a point in a given time interval.

fresh rock Rock whose mineral grains have their original composition and shape.

friction Resistance to sliding on a surface.

fringing reef A coral reef that forms directly along the coast.

front The boundary between two air masses.

frost wedging The process in which water trapped in a joint freezes, forces the joint open, and may cause the joint to grow.

fuel rod A metal tube that holds the nuclear fuel in a nuclear reactor.

Fujita scale A scale that distinguishes among tornadoes on the basis of wind speed, path dimensions, and possible damage.

fusion A type of nuclear reaction during which nuclei collide and bond; fusion occurs in stars and hydrogen bombs.

Ga Billions of years ago (abbreviation).

gabbro A coarse-grained, intrusive, mafic igneous rock.

Gaia The term used for the Earth System, with the implication that it resembles a complex living entity.

galaxy An immense system of hundreds of billions of stars.

gene An individual component of the DNA code that guides the growth and development of an organism.

general circulation model A numerical calculation that simulates the flow of the atmosphere and resulting phenomena, due to changes in atmospheric temperature and other parameters.

genetics The study of genes and how they transmit information.

geocentric Universe concept An ancient Greek idea suggesting that the Earth sat motionless in the center of the Universe while stars and other planets and the Sun orbited around it.

geochronology The science of dating geologic events in years.

geode A cavity in which euhedral crystals precipitate out of water solutions passing through a rock.

geographical pole The locations (north and south) where the Earth's rotational axis intersects the planet's surface.

geologic column A composite stratigraphic chart that represents the entirety of the Earth's history.

geologic history The sequence of geologic events that has taken place in a region.

geologic map A map showing the distribution of rock units and structures across a region.

geologic time The span of time since the formation of the Earth.

geologic time scale A scale that describes the intervals of geologic time.

geology The study of the Earth, including our planet's composition, behavior, and history.

geotherm The change in temperature with depth in the Earth.

geothermal energy Heat and electricity produced by using the internal heat of the Earth.

geothermal gradient The rate of change in temperature with depth.

geothermal region A region of current or recent volcanism in which magma or very hot rock heats up groundwater, which may discharge at the surface in the form of hot springs and/or geysers.

geyser A fountain of steam and hot water that erupts periodically from a vent in the ground in a geothermal region.

glacial abrasion The process by which clasts embedded in the base of a glacier grind away at the substrate as the glacier flows.

glacial advance The forward movement of a glacier's toe when the supply of snow exceeds the rate of ablation.

glacial drift Sediment deposited in glacial environments.

glacial incorporation The process by which flowing ice surrounds and incorporates debris.

glacial marine Sediment consisting of ice-rafted clasts mixed with marine sediment.

glacial outwash Coarse sediment deposited on a glacial outwash plain by meltwater streams.

glacially polished surface A polished rock surface created by the glacial abrasion of the underlying substrate.

glacial plucking (or quarrying) The process by which a glacier breaks off and carries away fragments of bedrock.

glacial rebound The process by which the surface of a continent rises back up after an overlying continental ice sheet melts away and the weight of the ice is removed.

glacial retreat The movement of a glacier's toe back toward the glacier's origin; glacial retreat occurs if the rate of ablation exceeds the rate of supply.

glacial subsidence The sinking of the surface of a continent caused by the weight of an overlying glacial ice sheet.

glacial till Sediment transported by flowing ice and deposited beneath a glacier or at its toe.

glaciation A period of time during which glaciers grew and covered substantial areas of the continents.

glacier A river or sheet of ice that slowly flows across the land surface and lasts all year long.

glass A solid in which atoms are not arranged in an orderly pattern.

glassy igneous rock Igneous rock consisting entirely of glass, or of tiny crystals surrounded by a glass matrix.

glide horizon The surface along which a slump slips.

global change The transformations or modifications of both physical and biological components of the Earth System through time.

global circulation The movement of volumes of air in paths that ultimately take it around the planet.

global climate change Transformations or modifications in Earth's climate over time.

global cooling A fall in the average atmospheric temperature.

global positioning system (GPS) A satellite system people can use to measure rates of movement of the Earth's crust relative to one another, or simply to locate their position on the Earth's surface.

global warming A rise in the average atmospheric temperature.

gneiss A compositionally banded metamorphic rock typically composed of alternating dark- and light-colored layers.

Gondwana A supercontinent that consisted of today's South America, Africa, Antarctica, India, and Australia. Also called Gondwanaland.

graben A down-dropped crustal block bounded on either side by a normal fault dipping toward the basin.

graded stream A stream that has attained an equilibrium longitudinal profile in which the sediment input into an area equals sediment removal.

gradualism The theory that evolution happens at a constant, slow rate.

grain A fragment of a mineral crystal or of a rock.

grain rotation The process by which rigid, inequant mineral grains distributed through a soft matrix may rotate into parallelism as the rock changes shape owing to differential stress.

granite A coarse-grained, intrusive, silicic igneous rock.

granulite facies A set of metamorphic mineral assemblages formed at very high pressures and temperatures.

gravitational spreading A process of lateral spreading that occurs in a material because of the weakness of the material; gravitational spreading causes continental glaciers to grow and mountain belts to undergo orogenic collapse.

gravity The attractive force that one mass exerts on another; the magnitude depends on the size of the objects and the distance between them.

graywacke An informal term used for sedimentary rock consisting of sand-size up to small-pebble-size grains of quartz and rock fragments all mixed together in a muddy matrix; typically, graywacke occurs at the base of a graded bed.

greenhouse conditions (greenhouse period) Relatively warm global climate leading to the rising of sea level for an interval of geologic time.

greenhouse effect The trapping of heat in the Earth's atmosphere by carbon dioxide and other greenhouse gases, which absorb infrared radiation; somewhat analogous to the effect of glass in a greenhouse.

greenhouse gases Atmospheric gases, such as carbon dioxide and methane, that regulate the Earth's atmospheric temperature by absorbing infrared radiation.

greenschist facies A set of metamorphic mineral assemblages formed under relatively low pressures and temperatures.

greenstone A low-grade metamorphic rock formed from basalt; if foliated, the rock is called greenschist.

Greenwich mean time (GMT) The time at the astronomical observatory in Greenwich, England; time in all other time zones is set in relation to GMT.

Grenville orogeny The orogeny that occurred about 1 billion years ago and yielded the belt of deformed and metamorphosed rocks that underlie the eastern fifth of the North American continent.

groin A concrete or stone wall built perpendicular to a shoreline in order to prevent beach drift from removing sand.

ground moraine A thin, hummocky layer of till left behind on the land surface during a rapid glacial recession.

groundwater Water that resides under the surface of the Earth, mostly in pores or cracks of rock or sediment.

group A succession of stratigraphic formations that have been lumped together, making a single, thicker stratigraphic entity.

growth ring A rhythmic layering that develops in trees, travertine deposits, and shelly organisms as a consequence of seasonal changes.

gusher A fountain of oil formed when underground pressure causes the oil to rise on its own out of a drilled hole.

guyot A seamount that had a coral reef growing on top of it, so that it is now flat-crested.

gymnosperm A plant whose seeds are "naked," not surrounded by a fruit.

gyre A large, circular flow pattern of ocean surface currents.

habitable zone (astronomy) The region in the Solar System where the intensity of radiation is sufficient to allow water to exist in liquid form on the surface of a planet.

Hadean The oldest of the Precambrian eons; the time between Earth's origin and the formation of the first rocks that have been preserved.

Hadley cells The name given to the low-latitude convection cells in the atmosphere.

hail Falling ice balls from the sky, formed when ice crystallizes in turbulent storm clouds.

hail streak An approximately 2-by-10-km stretch of ground, elongate in the direction of a storm, onto which hail has fallen.

half-graben A wedge-shaped basin in cross section that develops as the hanging-wall block above a normal fault slides down and rotates; the basin develops between the fault surface and the top surface of the rotated block.

half-life The time it takes for half of a group of a radioactive element's isotopes to decay.

halocline The boundary in the ocean between surface-water and deep-water salinities.

hamada Barren, rocky highlands in a desert.

hanging valley A glacially carved tributary valley whose floor lies at a higher elevation than the floor of the trunk valley.

hanging wall The rock or sediment above an inclined fault plane.

hard water Groundwater that contains dissolved calcium and magnesium, usually after passing through limestone or dolomite.

head (1) The elevation of the water table above a reference horizon; (2) the edge of ice at the origin of a glacier.

headland A place where a hill or cliff protrudes into the sea.

head scarp The distinct step along the upslope edge of a slump where the regolith detached.

headward erosion The process by which a stream channel lengthens up its slope as the flow of water increases.

headwaters The beginning point of a stream.

heat Thermal energy resulting from the movement of molecules.

heat capacity A measure of the amount of heat that must be added to a material to change its temperature.

heat flow The rate at which heat rises from the Earth's interior up to the surface.

heat-transfer melting Melting that results from the transfer of heat from a hotter magma to a cooler rock.

heliocentric Universe concept An idea proposed by Greek philosophers around 250 B.C.E. suggesting that all heavenly objects including the Earth orbited the Sun.

heliosphere A bubble-like region in space in which solar wind has blown away most interstellar atoms.

Hercynian orogen The late Paleozoic orogen that affected parts of Europe; a continuation of the Alleghenian orogen.

heterosphere A term for the upper portion of the atmosphere, in which gases separate into distinct layers on the basis of composition.

hiatus The interval of time between deposition of the youngest rock below an unconformity and deposition of the oldest rock above the unconformity.

high-altitude westerlies Westerly winds at the top of the troposphere.

high-grade metamorphic rocks Rocks that metamorphose under relatively high temperatures.

high-level waste Nuclear waste containing greater than 1 million times the safe level of radioactivity.

hinge The portion of a fold where curvature is greatest.

hogback A steep-sided ridge of steeply dipping strata.

Holocene The period of geologic time since the last glaciation.

Holocene climatic maximum The period from 5,000 to 6,000 years ago, when Holocene temperatures reached a peak.

homosphere The lower part of the atmosphere, in which the gases have stirred into a homogenous mixture.

hoodoo The local name for the brightly colored shale and sandstone chimneys found in Bryce Canyon National Park in Utah.

horn A pointed mountain peak surrounded by at least three cirques.

hornfels Rock that undergoes metamorphism simply because of a change in temperature, without being subjected to differential stress.

horse latitudes The region of the subtropical high in which winds are weak.

horst The high block between two grabens.

hot spot A location at the base of the lithosphere, at the top of a mantle plume, where temperatures can cause melting.

hot-spot track A chain of now-dead volcanoes transported off the hot spot by the movement of a lithosphere plate.

hot-spot volcano An isolated volcano not caused by movement at a plate boundary, but rather by the melting of a mantle plume.

hot spring A spring that emits water ranging in temperature from about 30° to 104°C.

hummocky surface An irregular and lumpy ground surface.

hurricane A huge rotating storm, resembling a giant spiral in map view, in which sustained winds blow over 119 km per hour.

hurricane track The path a hurricane follows.

hyaloclastite A rubbly extrusive rock consisting of glassy debris formed in a submarine or sub-ice eruption.

hydration The absorption of water into the crystal structure of minerals; a type of chemical weathering.

hydraulic conductivity The coefficient K in Darcy's law; hydraulic conductivity takes into account the permeability of the sediment or rock as well as the fluid's viscosity.

hydraulic gradient The slope of the water table.

hydrocarbon A chain-like or ring-like molecule made of hydrogen and carbon atoms; petroleum and natural gas are hydrocarbons.

hydrocarbon system The association of source rock, migration pathway, reservoir rock, seal, and trap geometry that leads to the occurrence of a hydrocarbon reserve.

hydrofracturing The process of injecting high-pressure water and other chemicals into a drill hole to generate cracks in surrounding rock.

hydrogen bond The attraction of a hydrogen atom to a negatively charged atom or molecule (e.g., hydrogen bonds attach water molecules to each other).

hydrologic cycle The continual passage of water from reservoir to reservoir in the Earth System.

hydrolysis The process in which water chemically reacts with minerals and breaks them down.

hydrosphere The Earth's water, including surface water (lakes, rivers, and oceans), groundwater, and liquid water in the atmosphere.

hydrothermal deposit An accumulation of ore minerals precipitated from hot-water solutions circulating through a magma or through the rocks surrounding an igneous intrusion.

hypsometric curve A graph that plots surface elevation on the vertical axis and the percentage of the Earth's surface on the horizontal axis.

ice age An interval of time in which the climate was colder than it is today, glaciers occasionally advanced to cover large areas of the continents, and mountain glaciers grew; an ice age can include many glacials and interglacials.

iceberg A large block of ice that calves off the front of a glacier and drops into the sea.

icehouse period A period of time when the Earth's temperature was cooler than it is today and ice ages could occur.

ice-margin lake A meltwater lake formed along the edge of a glacier.

ice-rafted sediment Sediment carried out to sea by icebergs.

ice sheet A vast glacier that covers the landscape.

ice shelf A broad, flat region of ice along the edge of a continent formed where a continental glacier flowed into the sea.

ice stream A portion of a glacier that travels much more quickly than adjacent portions of the glacier.

ice tongue The portion of a valley glacier that has flowed out into the sea.

igneous rock Rock that forms when hot molten rock (magma or lava) cools and freezes solid.

ignimbrite Rock formed when deposits of pyroclastic flows solidify.

inactive fault A fault that last moved in the distant past and probably won't move again in the near future, yet is still recognizable because of displacement across the fault plane.

inactive sand The sand along a coast that is buried beneath a layer of active sand and moves only during severe storms or not at all.

incised meander A meander that lies at the bottom of a steep-walled canyon.

index minerals Minerals that serve as good indicators of metamorphic grade.

induced seismicity Seismic events caused by the actions of people (e.g., filling a reservoir, that lies over a fault, with water).

industrial minerals Minerals that serve as the raw materials for manufacturing chemicals, concrete, and wallboard, among other products.

inequant A term for a mineral grain whose length and width are not the same.

inertia The tendency of an object at rest to remain at rest.

infiltrate Seep down into.

injection well A well in which a liquid is pumped down into the ground under pressure so that it passes from the well back into the pore space of the rock or regolith.

inner core The inner section of the core, extending from 5,155 km deep to the Earth's center at 6,371 km and consisting of solid iron alloy.

inselberg An isolated mountain or hill in a desert landscape created by progressive cliff retreat, so that the hill is surrounded by a pediment or an alluvial fan.

insolation Exposure to the Sun's rays.

intensity (seismology) A measure of the relative size of an earthquake (the severity of ground shaking) at a location, as determined by examining the amount of damage caused.

interglacial A period of time between two glaciations.

interior basin A basin with no outlet to the sea.

interlocking texture The texture of crystalline rocks in which mineral grains fit together like pieces of a jigsaw puzzle.

internal process A process in the Earth System, such as plate motion, mountain building, or volcanism, ultimately caused by Earth's internal heat.

intertidal zone The area of coastal land across which the tide rises and falls.

intertropical convergence zone The equatorial convergence zone in the atmosphere.

intraplate earthquakes Earthquakes that occur away from plate boundaries.

intrusive contact The boundary between country rock and an intrusive igneous rock.

intrusive igneous rock Rock formed by the freezing of magma underground.

ion A version of an atom that has lost or gained electrons, relative to an electrically neutral version, so that it has a net electrical charge.

ionic bond The attachment of one atom to another that happens when one atom transfers electrons to another; one type of chemical bond.

ionosphere The interval of Earth's atmosphere, at an elevation between 50 and 400 km, containing abundant positive ions.

iron catastrophe The proposed event very early in Earth history when the Earth partly melted and molten iron sank to the center to form the core.

isobar A line on a map along which the air has a specified pressure.

isograd (1) A line on a pressure-temperature graph along which all points are taken to be at the same metamorphic grade; (2) A line on a map making the first appearance of a metamorphic index mineral.

isostasy (or isostatic equilibrium) The condition that exists when the buoyancy force pushing litho sphere up equals the gravitational force pulling litho sphere down.

isostatic compensation The process in which the surface of the crust slowly rises or falls to reestablish isostatic equilibrium after a geologic event changes the density or thickness of the lithosphere.

isotherm Lines on a map or cross section along which the temperature is constant.

isotopes Different versions of a given element that have the same atomic number but different atomic weights.

jet stream A fast-moving current of air that flows at high elevations.

jetty A man-made wall that protects the entrance to a harbor.

joints Naturally formed cracks in rocks.

joint set A group of systematic joints.

Jovian A term used to describe the outer gassy, Jupiter-like planets (gas-giant planets).

kame A stratified sequence of lateral-moraine sediment that's sorted by water flowing along the edge of a glacier.

karst landscape A region underlain by caves in limestone bedrock; the collapse of the caves creates a landscape of sinkholes separated by higher topography, or of limestone spires separated by low areas.

Kelvin (K) scale A measure of temperature in which 0 K is absolute zero and the freezing point of water is 273.15 K; divisions in the Kelvin scale have the same value as those in the Celsius scale.

kerogen The waxy molecules into which the organic material in shale transforms on reaching about 100°C. At higher temperatures, kerogen transforms into oil.

kettle hole A circular depression in the ground made when a block of ice calves off the toe of a glacier, becomes buried by till, and later melts.

knob-and-kettle topography A land surface with many kettle holes separated by round hills of glacial till.

K-T boundary event The mass extinction that happened at the end of the Cretaceous Period, 65 million years ago, possibly due to the collision of an asteroid with the Earth.

Kuiper Belt A diffuse ring of icy objects, remnants of Solar System formation, that orbit our Sun outside the orbit of Neptune.

lag deposit The coarse sediment left behind in a desert after wind erosion removes the finer sediment.

lagoon A body of shallow seawater separated from the open ocean by a barrier island.

lahar A thick slurry formed when volcanic ash and debris mix with water, either in rivers or from rain or melting snow and ice on the flank of a volcano.

landslide A sudden movement of rock and debris down a nonvertical slope.

landslide-potential map A map on which regions are ranked according to the likelihood that a mass movement will occur.

land subsidence Sinking elevation of the ground surface; the process may occur over an aquifer that is slowly draining and decreasing in volume because of pore collapse.

La Niña Years in which the El Niño event is not strong.

lapilli Any pyroclastic particle that is 2 to 64 mm in diameter (i.e., marble-sized); the particles can consist of frozen lava clots, pumice fragments, or ash clumps.

Laramide orogeny The mountain-building event that lasted from about 80 Ma to 40 Ma, in western North America; in the United States, it formed the Rocky Mountains as a result of basement uplift and the warping of the younger overlying strata into large monoclines.

large igneous province (LIP) A region in which huge volumes of lava and/or ash erupted over a relatively short interval of geologic time.

latent heat of condensation The heat released during condensation, which comes only from a change in state.

lateral moraine A strip of debris along the side margins of a glacier.

laterite soil A hard, brick-red, soil formed from iron-rich rock in a tropical environment; it consists primarily of insoluble iron and aluminum oxide and hydroxide and forms due to extreme leaching.

Laurentia A continent in the early Paleozoic Era composed of today's North America and Greenland.

Laurentide ice sheet An ice sheet that spread over northeastern Canada during the Pleistocene ice age(s).

lava Molten rock that has flowed out onto the Earth's surface.

lava dome A dome-like mass of rhyolitic lava that accumulates above the eruption vent.

lava flows Sheets or mounds of lava that flow onto the ground surface or sea floor in molten form and then solidify.

lava lake A large pool of lava produced around a vent when lava fountains spew forth large amounts of lava in a short period of time.

lava tube The empty space left when a lava tunnel drains; this happens when the surface of a lava flow solidifies while the inner part of the flow continues to stream downslope.

leach To dissolve and carry away.

leader A conductive path stretching from a cloud toward the ground, along which electrons leak from the base of the cloud, and which provides the start for a lightning flash to the ground.

lightning bolt (syn. lightning flash, lightning stroke) A giant spark or pulse of current that jumps across a gap of charge separation.

light year The distance that light travels in one Earth year (about 6 trillion miles or 9.5 trillion km).

lignite Low-rank coal that consists of 50% carbon.

limb The side of a fold, showing less curvature than at the hinge.

limestone Sedimentary rock composed of calcite.

liquidus The lowest temperature at which all the components of a material have melted and transformed into liquid.

liquification The process by which wet sediment becomes a slurry; liquification may be triggered by earthquake vibrations.

lithification The transformation of loose sediment into solid rock through compaction and cementation.

lithologic correlation A correlation based on similarities in rock type.

lithosphere The relatively rigid, nonflowable, outer 100- to 150-km-thick layer of the Earth, constituting the crust and the top part of the mantle.

little ice age A period of cooler temperatures, between 1500 and 1800 C.E., during which many glaciers advanced.

loam A type of soil consisting of roughly equal parts of sand, silt, and clay; it tends to be good for growth of crops.

local base level A base level upstream from a drainage network's mouth.

lodgment till A flat layer of till smeared out over the ground when a glacier overrides an end moraine as it advances.

loess Layers of fine-grained sediments deposited from the wind; large deposits of loess formed from fine-grained glacial sediment blown off outwash plains.

longitudinal (seif) dune A dune formed when there is abundant sand and a strong, steady wind, and whose axis lies parallel to the wind direction.

longitudinal profile A cross-sectional image showing the variation in elevation along the length of a river.

longshore current The flow of water parallel to the shore just off a coast, because of the diagonal movement of waves toward the shore.

longshore drift The movement of sediment laterally along a beach; it occurs when waves wash up a beach diagonally.

lower mantle The deepest section of the mantle, stretching from 670 km down to the core-mantle boundary.

low-grade metamorphic rocks Rocks that underwent metamorphism at relatively low temperatures.

low-velocity zone The asthenosphere underlying oceanic lithosphere in which seismic waves travel more slowly, probably because rock has partially melted.

luster The way a mineral surface scatters light.

L-waves (Love waves) Surface seismic waves that cause the ground to ripple back and forth, creating a snake-like movement.

Ma Millions of years ago (abbreviation).

macrofossil A fossil large enough to be seen with the naked eye.

mafic A term used in reference to magmas or igneous rocks that are relatively poor in silica and rich in iron and magnesium.

magma Molten rock beneath the Earth's surface.

magma chamber A space below ground filled with magma.

magma contamination The process in which flowing magma incorporates components of the country rock through which it passes.

magmatic deposit An ore deposit formed when sulfide ore minerals accumulate at the bottom of a magma chamber.

magnetic anomaly The difference between the expected strength of the Earth's magnetic field at a certain location and the actual measured strength of the field at that location.

magnetic declination The angle between the direction a compass needle points at a given location and the direction of true north.

magnetic dipole An imaginary vector that points from the north magnetic pole to the south magnetic pole of a magnetic field.

magnetic field The region affected by the force emanating from a magnet.

magnetic field lines The trajectories along which magnetic particles would align, or charged particles would flow, if placed in a magnetic field.

magnetic force The push or pull exerted by a magnet.

magnetic inclination The angle between a magnetic needle free to pivot on a horizontal axis and a horizontal plane parallel to the Earth's surface.

magnetic poles The ends of a magnetic dipole; all magnetic dipoles have a north pole and a south pole.

magnetic reversal The change of the Earth's magnetic polarity; when a reversal occurs, the field flips from normal to reversed polarity, or vice versa.

magnetic-reversal chronology The history of magnetic reversals through geologic time.

magnetism An attractive or repulsive field force generated by permanent magnets or by an electrical current.

magnetization The degree to which a material can exert a magnetic force.

magnetometer An instrument that measures the strength of the Earth's magnetic field.

magnetosphere The region protected from the electrically charged particles of the solar winds by Earth's magnetic field.

magnetostratigraphy The comparison of the pattern of magnetic reversals in a sequence of strata, with a reference column showing the succession of reversals through time.

manganese nodules Lumpy accumulations of manganese-oxide minerals precipitated onto the sea floor.

mantle The thick layer of rock below the Earth's crust and above the core.

mantle plume A column of very hot rock rising up through the mantle.

marble A metamorphic rock composed of calcite and transformed from a protolith of limestone.

mare The broad, darker areas on the Moon's surface; they consist of flood basalts that erupted over 3 billion years ago and spread out across the Moon's lowlands.

marginal sea A small ocean basin created when sea-floor spreading occurs behind an island arc.

maritime tropical air mass A mass of air that originates over tropical or subtropical oceanic regions.

marsh A wetland dominated by grasses.

mass The amount of matter in an object; mass differs from weight in that its value does not depend on the strength of gravity.

mass-extinction event A time when vast numbers of species abruptly vanish.

mass movement (or mass wasting) The gravitationally caused downslope transport of rock, regolith, snow, or ice.

matter The material substance of the universe; it consists of atoms and has mass.

matrix Finer-grained material surrounding larger grains in a rock.

meander A snake-like curve along a stream's course.

meandering stream A reach of stream containing many meanders (snake-like curves).

meander neck A narrow isthmus of land separating two adjacent meanders.

mean sea level The average level between the high and low tide over a year at a given point.

mechanical force A push, pull, or shear applied by one object on another; it can be applied only if the objects are in contact.

mechanical weathering *See* Physical weathering.

medial moraine A strip of sediment in the interior of a glacier, parallel to the flow direction of the glacier, formed by the lateral moraines of two merging glaciers.

Medieval warm period A period of high temperatures in the Middle Ages.

melt Molten (liquid) rock.

meltdown The melting of the fuel rods in a nuclear reactor that occurs if the rate of fission becomes too fast and the fuel rods become too hot.

melting curve The line defining the range of temperatures and pressures at which a rock melts.

melting temperature The temperature at which the thermal vibration of the atoms or ions in the lattice of a mineral is sufficient to break the chemical bonds holding them to the lattice, so a material transforms into a liquid.

meltwater lake A lake fed by glacial meltwater.

mesa A large, flat-topped hill (with a surface area of several square km) in an arid region.

mesopause The boundary that marks the top of the mesosphere of Earth's atmosphere.

mesosphere The cooler layer of atmosphere overlying the stratosphere.

Mesozoic The middle of the three Phanerozoic eras; it lasted from 245 Ma to 65 Ma.

metal A solid composed almost entirely of atoms of metallic elements; it is generally opaque, shiny, smooth, malleable, and can conduct electricity.

metallic bond A chemical bond in which the outer atoms are attached to each other in such a way that electrons flow easily from atom to atom.

metamorphic aureole The region around a pluton, stretching tens to hundreds of meters out, in which heat transferred into the country rock and metamorphosed the country rock.

metamorphic facies A set of metamorphic mineral assemblages indicative of metamorphism under a specific range of pressures and temperatures.

metamorphic foliation A fabric defined by parallel surfaces or layers that develop in a rock as a result of metamorphism; schistocity and gneissic layering are examples.

metamorphic mineral assemblage A group of minerals that form in a rock as a result of metamorphism.

metamorphic rock Rock that forms when preexisting rock changes into new rock as a result of an increase in pressure and temperature and/or shearing under elevated temperatures; metamorphism occurs without the rock first becoming a melt or a sediment.

metamorphic zone The region between two metamorphic isograds, typically named after an index mineral found within the region.

metamorphism The process by which one kind of rock transforms into a different kind of rock.

metasomatism The process by which a rock's overall chemical composition changes during metamorphism because of reactions with hot water that bring in or remove elements.

meteoric water Water that falls to Earth from the atmosphere as either rain or snow.

meteorite A piece of rock or metal alloy that fell from space and landed on Earth.

micrite Limestone consisting of lime mud (i.e., very fine-grained limestone).

microfossil A fossil that can be seen only with a microscope or an electron microscope.

mid-ocean ridge A 2-km-high submarine mountain belt that forms along a divergent oceanic plate boundary.

migmatite A rock formed when gneiss is heated high enough so that it begins to partially melt, creating layers, or lenses, of new igneous rock that mix with layers of the relict gneiss.

Milanković cycles Climate cycles that occur over tens to hundreds of thousands of years because of changes in Earth's orbit and tilt.

mine A site at which ore is extracted from the ground.

mineral A homogenous, naturally occurring, solid inorganic substance with a definable chemical composition and an internal structure characterized by an orderly arrangement of atoms, ions, or molecules in a lattice. Most minerals are inorganic.

mineral classes Groups of minerals distinguished from each other on the basis of chemical composition.

mineralogist A geoscientist specializing in the study of minerals.

mineral resources The minerals extracted from the Earth's upper crust for practical purposes.

Mississippi Valley–type (MVT) ore An ore deposit, typically in dolostone, containing lead- and zinc-bearing minerals that precipitated from groundwater that had moved up from several km depth in the upper crust; such deposits occur in the upper Mississippi Valley.

mixture A material consisting of two or more substances that can be separated mechanically (i.e., without chemical reactions).

Modified Mercalli scale An earthquake characterization scale based on the amount of damage that the earthquake causes.

Moho The seismic-velocity discontinuity that defines the boundary between the Earth's crust and mantle.

Mohs hardness scale   A list of ten minerals in a sequence of relative hardness, with which other minerals can be compared.

mold   A cavity in sedimentary rock left behind when a shell that once filled the space weathers out.

molecule   The smallest piece of a compound that has the properties of the compound; it consists of two or more atoms attached by chemical bonds.

monocline   A fold in the land surface whose shape resembles that of a carpet draped over a stair step.

monsoon   A seasonal reversal in wind direction that causes a shift from a very dry season to a very rainy season in some regions of the world.

moraine   A sediment pile composed of till deposited by a glacier.

mountain front   The boundary between a mountain range and adjacent plains.

mountain (alpine) glacier   A glacier that exists in or adjacent to a mountainous region.

mountain ice cap   A mound of ice that submerges peaks and ridges at the crest of a mountain range.

mouth   The outlet of a stream where it discharges into another stream, a lake, or a sea.

mudflow   A downslope movement of mud at slow to moderate speed.

mud pot   A viscous slurry that forms in a geothermal region when hot water or steam rises into soils rich in volcanic ash and clay.

mudstone   Very fine-grained sedimentary rock that will not easily split into sheets.

mylonite   Rock formed during dynamic metamorphism and characterized by foliation that lies roughly parallel to the fault (shear zone) involved in the shearing process; mylonites have very fine grains formed by the nonbrittle subdivision of larger grains.

native metal   A naturally occurring pure mass of a single metal in an ore deposit.

natural arch   An arch that forms when erosion along joints leaves narrow walls of rock; when the lower part of the wall erodes while the upper part remains, an arch results.

natural levees   A pair of low ridges that appear on either side of a stream and develop as a result of the accumulation of sediment deposited naturally during flooding.

natural selection   The process by which the fittest organisms survive to pass on their characteristics to the next generation.

neap tide   An especially low tide that occurs when the angle between the direction of the Moon and the direction of the Sun is 90°.

nebula   A cloud of gas or dust in space.

nebula theory of planet formation   The concept that planets grow out of rings of gas, dust, and ice surrounding a newborn star.

negative anomaly   An area where the magnetic field strength is less than expected.

negative feedback   Feedback that slows a process down or reverses it.

neocrystallization   The growth of new crystals, not in the protolith, during metamorphism.

neutron   A subatomic particle, in the nucleus of an atom, that has a neutral charge.

Nevadan orogeny   A convergent-margin mountain-building event that took place in western North America during the Late Jurassic Period.

nonconformity   A type of unconformity at which sedimentary rocks overlie basement (older intrusive igneous rocks and/or metamorphic rocks).

nonflowing artesian well   An artesian well in which water rises on its own up to a level that lies below the ground surface.

nonfoliated metamorphic rock   Rock containing minerals that recrystallized during metamorphism but has no foliation.

nonmetallic mineral resources   Mineral resources that do not contain metals; examples include building stone, gravel, sand, gypsum, phosphate, and salt.

nonplunging fold   A fold with a horizontal hinge.

nonrenewable resource   A resource that nature will take a long time (hundreds to millions of years) to replenish or may never replenish.

nonsystematic joints   Short cracks in rocks that occur in a range of orientations and are randomly placed and oriented.

nor'easter   A large, midlatitude North American cyclone; when it reaches the east coast, it produces strong winds that come out of the northeast.

normal fault   A fault in which the hanging-wall block moves down the slope of the fault.

normal force   The component of the gravitational force acting perpendicular to a slope.

normal polarity   Polarity in which the paleomagnetic dipole has the same orientation as it does today.

normal stress   The push or pull that is perpendicular to a surface.

North Atlantic deep-water mass   The mass of cold, dense water that sinks in the north polar regions.

northeast tradewinds   Surface winds that come out of the northeast and occur in the region between the equator and 30°N.

nuclear bond   The force that attaches subatomic particles to each other within the nucleus of an atom.

nuclear fuel   Pellets of concentrated uranium oxide or a comparable radioactive material that can provide energy in a nuclear reactor.

nuclear fusion   The process by which the nuclei of atoms fuse together, thereby creating new, larger atoms.

nuclear reaction   A process that results in changing the nucleus of an atom by breaking or forming nuclear bonds.

nuclear reactor   The part of a nuclear power plant where the fission reactions occur.

nucleus   The central ball of an atom that consists of protons and neutrons (except for hydrogen, whose nuclei contains only a proton).

nuée ardente   *See* Pyroclastic flow.

oasis   A verdant region surrounded by desert, occurring at a place where natural springs provide water at the surface.

oblique-slip fault   A fault in which sliding occurs diagonally along the fault plane.

obsidian   An igneous rock consisting of a solid mass of volcanic glass.

occluded front   A front that no longer intersects the ground surface.

oceanic crust   The crust beneath the oceans; composed of gabbro and basalt, overlain by sediment.

oceanic plateau   A region of oceanic floor that is higher than surrounding areas; such regions have particularly thick oceanic crust and are relicts of submarine large igneous provinces.

oceanic lithosphere   Lithosphere topped by oceanic crust; it reaches a thickness of 100 km.

offshore bar   A narrow ridge of sand that forms off the shore of a beach; some offshore bars rise above sea level, and separate a lagoon on one side from the open ocean on the other.

oil   (geology) A liquid hydrocarbon that can be burned as a fuel.

Oil Age   The period of human history, including our own, so named because the economy depends on oil.

oil field   A region containing a significant amount of accessible oil underground.

oil reserve   The known supply of oil held underground.

oil shale   Shale containing kerogen.

oil trap   A geologic configuration that keeps oil underground in the reservoir rock and prevents it from rising to the surface.

oil window   The narrow range of temperatures under which oil can form in a source rock.

olistotrome   A large, submarine slump block, buried and preserved.

Oort Cloud A cloud of icy objects, left over from Solar System formation, that orbit the Sun in a region outside of the heliosphere.

ophiolite A slice of oceanic crust that has been thrust onto continental crust.

ordinary well A well whose base penetrates below the water table and can thus provide water.

ore Rock containing native metals or a concentrated accumulation of ore minerals.

ore deposit An economically significant accumulation of ore.

ore minerals Minerals that have metal in high concentrations and in a form that can be easily extracted.

organic carbon Carbon that has been incorporated in an organism.

organic chemical A carbon-containing compound that occurs in living organisms, or that resembles such compounds; it consists of carbon atoms bonded to hydrogen atoms along with varying amounts of oxygen, nitrogen, and other chemicals.

organic coast A coast along which living organisms control landforms along the shore.

organic sedimentary rock Sedimentary rock (such as coal) formed from carbon-rich relicts of organisms.

organic shale Lithified, muddy, organic-rich ooze that contains the raw materials from which hydrocarbons eventually form.

orogen (or orogenic belt) A linear range of mountains.

orogenic collapse The process in which mountains begin to collapse under their own weight and spread out laterally.

orogeny A mountain-building event.

orographic barrier A landform that diverts air flow upward or laterally.

outcrop An exposure of bedrock.

outer core The section of the core, between 2,900 and 5,150 km deep, that consists of liquid iron alloy.

outwash plain A broad area of gravel and sandbars deposited by a braided stream network, fed by the melt water of a glacier.

overburden The weight of overlying rock on rock buried deeper in the Earth's crust.

overriding plate (or slab) The plate at a subduction zone that overrides the downgoing plate.

oversaturated solution A solution that contains so much solute (dissolved ions) that precipitation begins.

oversized stream valley A large valley with a small stream running through it; the valley formed earlier when the flow was greater.

oxbow lake A meander that has been cut off yet remains filled with water.

oxidation reaction A reaction in which an element loses electrons; an example is the reaction of iron with air to form rust.

ozone O_3, an atmospheric gas that absorbs harmful ultraviolet radiation from the Sun.

ozone hole An area of the atmosphere, over polar regions, from which ozone has been depleted.

pahoehoe A lava flow with a surface texture of smooth, glassy, rope-like ridges.

paleoclimate The past climate of the Earth.

paleomagnetism The record of ancient magnetism preserved in rock.

paleopole The supposed position of the Earth's magnetic pole in the past, with respect to a particular continent.

paleosol Ancient soil preserved in the stratigraphic record.

Paleozoic The oldest era of the Phanerozoic Eon.

Pangaea A supercontinent that assembled at the end of the Paleozoic Era.

Pannotia A supercontinent that may have existed sometime between 800 Ma and 600 Ma.

parabolic dunes Dunes formed when strong winds break through transverse dunes to make new dunes whose ends point upwind.

parallax The apparent movement of an object seen from two different points not on a straight line from the object (e.g., from your two different eyes).

parallax method A trigonometric method used to determine the distance from the Earth to a nearby star.

parent isotope A radioactive isotope that undergoes decay.

partial melting The melting in a rock of the minerals with the lowest melting temperatures, while other minerals remain solid.

passive margin A continental margin that is not a plate boundary.

passive-margin basin A thick accumulation of sediment along a tectonically inactive coast, formed over crust that stretched and thinned when the margin first began.

patterned ground A polar landscape in which the ground splits into pentagonal or hexagonal shapes.

pause An elevation in the atmosphere where temperature stops decreasing and starts increasing, or vice versa.

peat Compacted and partially decayed vegetation accumulating beneath a swamp.

pedalfer soil A temperate-climate soil characterized by well-defined soil horizons and an organic A-horizon.

pediment The broad, nearly horizontal bedrock surface at the base of a retreating desert cliff.

pedocal soil Thin soil, formed in arid climates. It contains very little organic matter, but significant precipitated calcite.

pegmatite A coarse-grained igneous rock containing crystals of up to tens of centimeters across and occurring in dike-shaped intrusions.

pelagic sediment Microscopic plankton shells and fine flakes of clay that settle out and accumulate on the deep-ocean floor.

Pelé's hair Droplets of basaltic lava that mold into long, glassy strands as they fall.

Pelé's tears Droplets of basaltic lava that mold into tear-shaped, glassy beads as they fall.

peneplain A nearly flat surface that lies at an elevation close to sea level; thought to be the product of long-term erosion.

perched water table A quantity of groundwater that lies above the regional water table because an underlying lens of impermeable rock or sediment prevents the water from sinking down to the regional water table.

percolation The process by which groundwater meanders through tiny, crooked channels in the surrounding material.

peridotite A coarse-grained ultramafic rock.

periglacial environment A region with widespread perma frost but without a blanket of snow or ice.

period An interval of geologic time representing a subdivision of a geologic era.

permafrost Permanently frozen ground.

permanent magnet A special material that behaves magnetically for a long time all by itself.

permanent stream A stream that flows year-round because its bed lies below the water table, or because more water is supplied from upstream than can infiltrate the ground.

permeability The degree to which a material allows fluids to pass through it via an interconnected network of pores and cracks.

permineralization The fossilization process in which plant material becomes transformed into rock by the precipitation of silica from groundwater.

petrified A term used by geologists to describe plant material that has transformed into rock by permineralization.

petroglyph Drawings formed by chipping into the desert varnish of rocks to reveal the lighter rock beneath.

petroleum *See* Oil.

phaneritic A textural term used to describe coarse-grained igneous rock.

Phanerozoic Eon   The most recent eon, an interval of time from 542 Ma to the present.

phenocryst   A large crystal surrounded by a finer-grained matrix in an igneous rock.

photochemical smog   Brown haze that blankets a city when exhaust from cars and trucks reacts in the presence of sunlight.

photosynthesis   The process during which chlorophyll- containing plants remove carbon dioxide from the atmo sphere, form tissues, and expel oxygen back to the atmo sphere.

phreatomagmatic eruption   An explosive eruption that occurs when water enters the magma chamber and turns into steam.

phyllite   A fine-grained metamorphic rock with a foliation caused by the preferred orientation of very fine-grained mica.

phyllitic luster   A silk-like sheen characteristic of phyllite, a result of the rock's fine-grained mica.

phylogenetic tree   A chart representing the ideas of paleontologists show- ing which groups of organisms radiated from which ancestors.

physical weathering   The process in which intact rock breaks into smaller grains or chunks.

piedmont glacier   A fan or lobe of ice that forms where a valley glacier emerges from a valley and spreads out into the adjacent plain.

pillow basalt   Glass-encrusted basalt blobs that form when magma extrudes on the sea floor and cools very quickly.

placer deposit   Concentrations of metal grains in stream sediment that develop when rocks containing native metals erode and create a mixture of sand grains and metal fragments; the moving water of the stream carries away lighter mineral grains.

planetesimal   Tiny, solid pieces of rock and metal that collect in a planetary nebula and eventually accumulate to form a planet.

plankton   Tiny plants and animals that float in sea or lake water.

plastic deformation   The deformational process in which mineral grains behave like plastic and, when compressed or sheared, become flattened or elongate without cracking or breaking.

plate   One of about 20 distinct pieces of the relatively rigid lithosphere.

plate boundary   The border between two adjacent litho sphere plates.

plate-boundary earthquakes   The earthquakes that occur along and define plate boundaries.

plate-boundary volcano   A volcanic arc or mid-ocean ridge volcano, formed as a consequence of movement along a plate boundary.

plate interior   A region away from the plate boundaries that consequently experiences few earthquakes.

plate tectonics   *See* Theory of plate tectonics.

playa   The flat, typically salty lake bed that remains when all the water evaporates in drier times; forms in desert regions.

Pleistocene ice age(s)   The period of time from about 2 Ma to 14,000 years ago, during which the Earth experienced an ice age.

plunge pool   A depression at the base of a waterfall scoured by the energy of the falling water.

plunging fold   A fold with a tilted hinge.

pluton   An irregular or blob-shaped intrusion; can range in size from tens of m across to tens of km across.

pluvial lake   A lake formed to the south of a continental glacier as a result of enhanced rainfall during an ice age.

point bar   A wedge-shaped deposit of sediment on the inside bank of a meander.

polar cell   A high-latitude convection cell in the atmo sphere.

polar easterlies   Prevailing winds that come from the east and flow from the polar high to the subpolar low.

polar front   The convergence zone in the atmosphere at latitude 60°.

polar glacier   *See* Dry-bottom glacier.

polar high   The zone of high pressure in polar regions created by the sink- ing of air in the polar cells.

polarity   The orientation of a magnetic dipole.

polarity chron   The time interval between polarity reversals of Earth's magnetic field.

polarity subchron   The time interval between magnetic reversals if the interval is of short duration (less than 200,000 years long).

polarized light   A beam of filtered light waves that all vibrate in the same plane.

polar wander   The phenomenon of the progressive changing through time of the position of the Earth's magnetic poles relative to a location on a conti- nent; significant polar wander probably doesn't occur—in fact, poles seem to remain fairly fixed, while continents move.

polar-wander path   The curving line representing the apparent progres- sive change in the position of the Earth's magnetic pole, relative to a locality X, assuming that the position of X on Earth has been fixed through time (in fact, poles stay fixed while continents move).

pollen   Tiny grains involved in plant reproduction.

polymorphs   Two minerals that have the same chemical composition but a different crystal lattice structure.

pore   A small, open space within sediment or rock.

pore collapse   The closer packing of grains that occurs when groundwa- ter is extracted from pores, thus eliminating the support holding the grains apart.

porosity   The total volume of empty space (pore space) in a material, usually expressed as a percentage.

porphyritic   A textural term for igneous rock that has phenocrysts distrib- uted throughout a finer matrix.

positive anomaly   An area where the magnetic field strength is stronger than expected.

positive-feedback mechanism   A mechanism that enhances the process that causes the mechanism in the first place.

potentiometric surface   The elevation to which water in an artesian system would rise if unimpeded; where there are flowing artesian wells, the potentiometric surface lies above ground.

pothole   A bowl-shaped depression carved into the floor of a stream by a long-lived whirlpool carrying sand or gravel.

Precambrian   The interval of geologic time between Earth's formation about 4.57 Ga and the beginning of the Phanerozoic Eon 542 Ma.

precession   The gradual conical path traced out by Earth's spinning axis; simply put, it is the "wobble" of the axis.

precious metals   Metals (like gold, silver, and platinum) that have high value.

precipitate   (chem, n) A solid substance formed when atoms attach and settle out of a solution, or attach to the walls of the container holding the solution; (chem, v) the action of forming a solid substance from a solution; (meteorology, v) the dropping of snow or rain from the sky.

precipitation   (1) The process by which atoms dissolved in a solution come together and form a solid; (2) rainfall or snow.

preferred mineral orientation   The metamorphic texture that exists where platy grains lie parallel to one another and/or elongate grains align in the same direction.

pressure   Force per unit area, or the "push" acting on a material in cases where the push (compressional stretch) is the same in all directions.

pressure gradient   The rate of pressure change over a given horizontal distance.

pressure solution   The process of dissolution at points of contact, between grains, where compression is greatest, producing ions that then precipitate elsewhere, where compression is less.

prevailing winds   Surface winds that generally flow in the same direction for long time periods.

primary porosity   The space that remains between solid grains or crystals immediately after sediment accumulates or rock forms.

principal aquifer   The geologic unit that serves as the primary source of groundwater in a region.

principle of baked contacts   When an igneous intrusion "bakes" (metamorphoses) surrounding rock, the rock that has been baked must be older than the intrusion.

principle of cross-cutting relations   If one geologic feature cuts across another, the feature that has been cut is older.

principle of fossil succession   In a stratigraphic sequence, different species of fossil organisms appear in a definite order; once a fossil species disappears in a sequence of strata, it never reappears higher in the sequence.

principle of inclusions   If a rock contains fragments of another rock, the fragments must be older than the rock containing them.

principle of original continuity   Sedimentary layers, before erosion, formed fairly continuous sheets over a region.

principle of original horizontality   Layers of sediment, when originally deposited, are fairly horizontal.

principle of superposition   In a sequence of sedimentary rock layers, each layer must be younger than the one below, for a layer of sediment cannot accumulate unless there is already a substrate on which it can collect.

principle of uniformitariansim   The physical pro cesses we observe today also operated in the past in the same way, and at comparable rates.

product   (chem) Materials produced or formed by a chemical reaction.

prograde metamorphism   Metamorphism that occurs as temperatures and pressures are increasing.

prokaryote   An organism whose cells do not contain a nucleus; archaea and bacteria consist of prokaryotic cells.

Proterozoic   The most recent of the Precambrian eons.

protocontinent   A block of crust composed of volcanic arcs and hot-spot volcanoes sutured together.

protolith   The original rock from which a metamorphic rock formed.

proton   A positively charged subatomic particle in the nucleus of an atom.

protoplanet   A body that grows by the accumulation of planetesimals but has not yet become big enough to be called a planet.

protoplanetary nebula   A ring of gas and dust that surrounded the newborn Sun, from which the planets were formed.

protostar   A dense body of gas that is collapsing inward because of gravitational forces and that may eventually become a star.

pumice   A glassy igneous rock that forms from felsic frothy lava and contains abundant (over 50%) pore space.

pumice lapilli   Marble-sized chunks consisting of frothy, siliceous igneous rock that fall from a volcanic eruptive cloud.

punctuated equilibrium   The hypothesis that evolution takes place in fits and starts; evolution occurs very slowly for quite a while and then, during a relatively short period, takes place very rapidly.

P-waves   Compressional seismic waves that move through the body of the Earth.

P-wave shadow zone   A band between 103° and 143° from an earthquake epicenter, as measured along the circumference of the Earth, inside which P-waves do not arrive at seismograph stations.

pycnocline   The boundary between layers of water of different densities.

pyroclastic debris   Fragmented material that sprayed out of a volcano and landed on the ground or sea floor in solid form.

pyroclastic flow   A fast-moving avalanche that occurs when hot volcanic ash and debris mix with air and flow down the side of a volcano.

pyroclastic rock   Rock made from fragments that were blown out of a volcano during an explosion and were then packed or welded together.

quarry   A site at which stone is extracted from the ground.

quartzite   A metamorphic rock composed of quartz and transformed from a protolith of quartz sandstone.

quenching   A sudden cooling of molten material to form a solid.

quick clay   Clay that behaves like a solid when still (because of surface tension holding the water-coated clay flakes together) but that flows like a liquid when shaken.

radial network   A drainage network in which the streams flow outward from a cone-shaped mountain and define a pattern resembling spokes on a wheel.

radiation   (physics) Electromagnetic energy traveling away from a source through a medium or space.

radioactive decay   The process by which a radioactive atom undergoes fission or releases particles, thereby being transformed into a new element.

radioactive isotope   An unstable isotope of a given element.

radiometric dating   The science of dating geologic events in years by measuring the ratio of parent radioactive atoms to daughter product atoms.

rain band   A spiraling arm of a hurricane radiating outward from the eye.

rain shadow   The inland side of a mountain range, which is arid because the mountains block rain clouds from reaching the area.

range (for fossils)   The interval of a sequence of strata in which a specific fossil species appears.

rapids   A reach of a stream in which water becomes particularly turbulent; as a consequence, waves develop on the surface of the stream.

reach   A specified segment of a stream's path.

reactant   (chem) The starting materials of a chemical reaction.

recessional moraine   The end moraine that forms when a glacier stalls for a while as it recedes.

recharge area   A location where water enters the ground and infiltrates down to the water table.

recrystallization   The process in which ions or atoms in minerals rearrange to form new minerals.

rectangular network   A drainage network in which the streams join each other at right angles because of a rectangular grid of fractures that breaks up the ground and localizes channels.

recurrence interval   The average time between successive geologic events.

red giant   A huge red star that forms when Sun-sized stars start to die and expand.

red shift   The phenomenon in which a source of light moving away from you very rapidly shifts to a lower frequency; that is, toward the red end of the spectrum.

reef bleaching   The death and loss of color of a coral reef.

reflected ray   A ray that bounces off a boundary between two different materials.

refracted ray   A ray that bends as it passes through a boundary between two different materials.

refraction   The bending of a ray as it passes through a boundary between two different materials.

refractory materials   Substances that have a relatively high melting point and tend to exist in solid form.

reg   A vast stony plain in a desert.

regional metamorphism   *See also* Dynamothermal metamorphism; metamorphism of a broad region, usually the result of deep burial during an orogeny.

regolith   Any kind of unconsolidated debris that covers bedrock.

regression   The seaward migration of a shoreline caused by a lowering of sea level.

relative age   The age of one geologic feature with respect to another.

relative humidity   The ratio between the measured water content of air and the maximum possible amount of water the air can hold at a given condition.

relative plate velocity   The movement of one litho sphere plate with respect to another.

relief   The difference in elevation between adjacent high and low regions on the land surface.

renewable resource   A resource that can be replaced by nature within a short time span relative to a human life span.

reservoir rock   Rock with high porosity and permeability, so it can contain an abundant amount of easily accessible oil.

residence time   The average length of time that a substance stays in a particular reservoir.

residual mineral deposit   Soils in which the residuum left behind after leaching by rainwater is so concentrated in metals that the soil itself becomes an ore deposit.

resonance   (seismology) A situation that arises when earthquake waves of a particular frequency cause particularly large-amplitude movements because energy input happens at just the right time.

resurgent dome   The new mound, or cone, of igneous rock that grows within a caldera as an eruption begins anew.

retrograde metamorphism   Metamorphism that occurs as pressures and temperatures are decreasing; for retrograde metamorphism to occur, water must be added.

return stroke   An upward-flowing electric current from the ground that carries positive charges up to a cloud during a lightning flash.

reversed polarity   Polarity in which the paleomagnetic dipole points north.

reverse fault   A steeply dipping fault on which the hanging-wall block slides up.

rhythmic layering   Banding in sediments, shells, trees, corals, or ice that repeats periodically; it may be correlated to annual cycles.

Richter magnitude scale   A scale that defines earthquakes on the basis of the amplitude of the largest ground motion recorded on a seismogram.

ridge axis   The crest of a mid-ocean ridge; the ridge axis defines the position of a divergent plate boundary.

right-lateral strike-slip fault   A strike-slip fault in which the block on the opposite fault plane from a fixed spot moves to the right of that spot.

rip current   A strong, localized seaward flow of water perpendicular to a beach.

riprap   Loose boulders or concrete piled together along a beach to absorb wave energy before it strikes a cliff face.

roche moutonnée   A glacially eroded hill that becomes elongate in the direction of flow and asymmetric; glacial rasping smoothes the upstream part of the hill into a gentle slope, while glacial plucking erodes the downstream edge into a steep slope.

rock   A coherent, naturally occurring solid, consisting of an aggregate of minerals or a mass of glass.

rock burst   A sudden explosion of rock off the ceiling or wall of an underground mine.

rock cycle   The succession of events that results in the transformation of Earth materials from one rock type to another, then another, and so on.

rock flour   Fine-grained sediment produced by glacial abrasion of the substrate over which a glacier flows.

rock glacier   A slow-moving mixture of rock fragments and ice.

rock slide   A sudden downslope movement of rock.

rocky coast   An area of coast where bedrock rises directly from the sea, so beaches are absent.

Rodinia   A proposed Precambrian supercontinent that existed around 1 billion years ago.

rotational axis   The imaginary line through the center of the Earth around which the Earth spins.

R-waves (Rayleigh waves)   Surface seismic waves that cause the ground to ripple up and down, like water waves in a pond.

sabkah   A region of formerly flooded coastal desert in which stranded seawater has left a salt crust over a mire of mud that is rich in organic material.

salinity   The degree of concentration of salt in water.

saltation   The movement of a sediment in which grains bounce along their substrate, knocking other grains into the water column (or air) in the process.

salt dome   A rising bulbous dome of salt that bends up the adjacent layers of sedimentary rock.

salt wedging   The process in arid climates by which dissolved salt in groundwater crystallizes and grows in open pore spaces in rocks and pushes apart the surrounding grains.

sand dune   A relatively large ridge of sand built up by a current of wind (or water); cross bedding typically occurs within the dune.

sandspit   An area where the beach stretches out into open water across the mouth of a bay or estuary.

sandstone   Coarse-grained sedimentary rock consisting almost entirely of quartz.

sand volcano (or sand blow)   A small mound of sand produced when sand layers below the ground surface liquify as a result of seismic shaking, causing the sand to erupt onto the Earth's surface through cracks or holes in overlying clay layers.

saprolite   A layer of rotten rock created by chemical weathering in warm, wet climates.

Sargasso Sea   The center of North Atlantic Gyre, named for the tropical seaweed sargassum, which accumulates in its relatively noncirculating waters.

saturated solution   Water that carries as many dissolved ions as possible under given environmental conditions.

saturated zone   The region below the water table where pore space is filled with water.

scattering   The dispersal of energy that occurs when light interacts with particles in the atmosphere.

schist   A medium-to-coarse-grained metamorphic rock that possesses schistosity.

schistosity   Foliation caused by the preferred orientation of large mica flakes.

scientific method   A sequence of steps for systematically analyzing scientific problems in a way that leads to verifiable results.

scoria   A glassy, mafic, igneous rock containing abundant air-filled holes.

scoria cone   (also "cinder cone") An accumulation of lapilli-sized or larger fragments formed from a volcanic eruption that spatters clots of basaltic lava.

scouring   A process by which running water removes loose fragments of sediment from a streambed.

sea arch   An arch of land protruding into the sea and connected to the mainland by a narrow bridge.

sea-floor spreading   The gradual widening of an ocean basin as new oceanic crust forms at a mid-ocean ridge axis and then moves away from the axis.

sea ice   Ice formed by the freezing of the surface of the sea.

seal   A relatively impermeable rock, such as shale, salt, or unfractured limestone, that lies above a reservoir rock and stops the oil from rising further.

seam   A sedimentary bed of coal interlayered with other sedimentary rocks.

seamount   An isolated submarine mountain.

seasonal floods   Floods that appear almost every year during seasons when rainfall is heavy or when winter snows start to melt.

seasonal well   A well that provides water only during the rainy season when the water table rises below the base of the well.

sea stack   An isolated tower of land just offshore, disconnected from the mainland by the collapse of a sea arch.

seawall A wall of riprap built on the landward side of a backshore zone in order to protect shore cliffs from erosion.

second The basic unit of time measurement, now defined as the time it takes for the magnetic field of a cesium atom to flip polarity 9,192,631,770 times, as measured by an atomic clock.

secondary enrichment The process by which a new ore deposit forms from metals that were dissolved and carried away from preexisting ore minerals.

secondary porosity New pore space in rocks, created some time after a rock first forms.

secondary recovery technique A process used to extract the quantities of oil that will not come out of a reservoir rock with just simple pumping.

sediment An accumulation of loose mineral grains, such as boulders, pebbles, sand, silt, or mud, that are not cemented together.

sedimentary basin A depression, created as a consequence of subsidence, that fills with sediment.

sedimentary rock Rock that forms either by the cementing together of fragments broken off preexisting rock or by the precipitation of mineral crystals out of water solutions at or near the Earth's surface.

sedimentary sequence A grouping of sedimentary units bounded on top and bottom by regional unconformities.

sediment budget The proportion of sand supplied to sand removed from a depositional setting.

sediment load The total volume of sediment carried by a stream.

sediment maturity The degree to which a sediment has evolved from a crushed-up version of the original rock into a sediment that has lost its easily weathered minerals and become well sorted and rounded.

sediment sorting The segregation of sediment by size.

seep A place where oil-filled reservoir rock intersects the ground surface, or where fractures connect a reservoir to the ground surface, so that oil flows out onto the ground on its own.

seiche Rhythmic movement in a body of water caused by ground motion.

seismic belts (or zones) The relatively narrow strips of crust on Earth under which most earthquakes occur.

seismicity Earthquake activity.

seismic-moment magnitude scale A scale that defines earthquake size on the basis of calculations involving the amount of slip, length of rupture, depth of rupture, and rock strength.

seismic ray The changing position of an imaginary point on a wave front as the front moves through rock.

seismic-reflection profile A cross-sectional view of the crust made by measuring the reflection of artificial seismic waves off boundaries between different layers of rock in the crust.

seismic retrofitting The strengthening of an already existing structure (building, bridge, etc.) so that it can withstand earthquake vibrations.

seismic tomography Analysis by sophisticated computers of global seismic data in order to create a three-dimensional image of variations in seismic-wave velocities within the Earth.

seismic velocity The speed at which seismic waves travel.

seismic-velocity discontinuity A boundary in the Earth at which seismic velocity changes abruptly.

seismic (earthquake) waves Waves of energy emitted at the focus of an earthquake.

seismogram The record of an earthquake produced by a seismograph.

seismograph (seismometer) An instrument that can record the ground motion from an earthquake.

semipermanent pressure cell A somewhat elliptical zone of high or low atmospheric pressure that lasts much of the year; it forms because high-pressure zones tend to be narrower over land than over sea.

Sevier orogeny A mountain-building event that affected western North America between about 150 Ma and 80 Ma, a result of convergent margin tectonism; a fold-thrust belt formed during this event.

shale Very fine-grained sedimentary rock that breaks into thin sheets.

shatter cones Small, cone-shaped fractures formed by the shock of a meteorite impact.

shear strain A change in shape of an object that involves the movement of one part of a rock body sideways past another part so that angular relationships within the body change.

shear stress A stress that moves one part of a material sideways past another part.

shear waves Seismic waves in which particles of material move back and forth perpendicular to the direction in which the wave itself moves.

shear zone A fault in which movement has occurred ductilely.

sheetwash A film of water less than a few mm thick that covers the ground surface during heavy rains.

shell (biol) A relatively hard, protective structure formed of minerals and surrounding the soft part of an invertebrate organism.

shield An older, interior region of a continent.

shield volcano A subaerial volcano with a broad, gentle dome, formed either from low-viscosity basaltic lava or from large pyroclastic sheets.

shocked quartz Grains of quartz that have been subjected to intense pressure such as occurs during a meteorite impact.

shoreline The boundary between the water and land.

shortening The process during which a body of rock or a region of crust becomes shorter.

short-term climate change Climate change that takes place over hundreds to thousands of years.

Sierran arc A large continental volcanic arc along western North America that was initiated at the end of the Jurassic Period and lasted until about 80 million years ago.

silica SiO_2.

silicate minerals Minerals composed of silicon-oxygen tetrahedra linked in various arrangements; most contain other elements too.

silicate rock Rock composed of silicate minerals.

siliceous sedimentary rock Sedimentary rock that contains abundant quartz.

silicic Rich in silica with relatively little iron and magnesium.

sill A nearly horizontal tabletop-shaped tabular intrusion that occurs between the layers of country rock.

siltstone Fine-grained sedimentary rock generally composed of very small quartz grains.

sinkhole A circular depression in the land that forms when an underground cavern collapses.

slab-pull force The force that downgoing plates (or slabs) apply to oceanic lithosphere at a convergent margin.

slate Fine-grained, low-grade metamorphic rock, formed by the metamorphism of shale.

slaty cleavage The foliation typical of slate, and reflective of the preferred orientation of slate's clay minerals, that allows slate to be split into thin sheets.

slickensides The polished surface of a fault caused by slip on the fault; lineated slickensides also have groves that indicate the direction of fault movement.

slip face The leeward slope of a dune; sand that builds up at the crest of the dune slides down this face; slip faces are preserved as cross beds within sandstone layers.

slip lineations Linear marks on a fault surface created during movement on the fault; some slip lineations are defined by grooves, some by aligned mineral fibers.

slope failure The downslope movement of material on an unstable slope.

slumping Downslope movement in which a mass of regolith detaches from its substrate along a spoon-shaped, sliding surface and slips downward semicoherently.

smelting   The heating of a metal-containing rock to high temperatures in a fire so that the rock will decompose to yield metal plus a nonmetallic residue (slag).

snotite   A long gob of bacteria that slowly drips from the ceiling of a cave.

snow line   The boundary above which snow remains all year.

soda straw   A hollow stalactite in which calcite precipitates around the outside of a drip.

soil   Sediment that has undergone changes at the surface of the Earth, including reaction with rainwater and the addition of organic material.

soil erosion   The removal of soil by wind and runoff.

soil horizon   Distinct zones within a soil, distinguished from each other by factors such as chemical composition and organic content.

soil moisture   Underground water that wets the surface of the mineral grains and organic material making up soil, but lies above the water table.

soil profile   A vertical sequence of distinct zones of soil.

Solar System   Our Sun and all the materials that orbit it (including planets, moons, asteroids, Kuiper Belt objects, and Oort Cloud objects).

solar wind   A stream of particles with enough energy to escape from the Sun's gravity and flow outward into space.

solid-state diffusion   The slow movement of atoms or ions through a solid.

solidus   The highest temperature at which all the components of a material are solid; at the solidus temperature, the material begins to melt.

solifluction   The type of creep characteristic of tundra regions; during the summer, the uppermost layer of permafrost melts, and the soggy, weak layer of ground then flows slowly down slope in overlapping sheets.

solstice   A day on which the polar ends of the terminator (the boundary between the day hemisphere and the night hemisphere) lie 23.5° away from the associated geographic poles.

solution   A material containing dissolved ions.

Sonoma orogeny   A convergent-margin mountain-building event that took place on the western coast of North America in the Late Permian and Early Triassic periods.

sorting   (1) The range of clast sizes in a collection of sediment; (2) the degree to which sediment has been separated by flowing currents into different-size fractions.

source rock   A rock (organic-rich shale) containing the raw materials from which hydrocarbons eventually form.

southeast tradewinds   Tradewinds in the southern hemisphere, which start flowing northward, deflect to the west, and end up flowing from southeast to northwest.

southern oscillation   The movement of atmospheric pressure cells back and forth across the Pacific Ocean, in association with El Niño.

specific gravity   A number representing the density of a mineral, as specified by the ratio between the weight of a volume of the mineral and the weight of an equal volume of water.

speleothem   A formation that grows in a limestone cave by the accumulation of travertine precipitated from water solutions dripping in a cave or flowing down the wall of a cave.

sphericity   The measure of the degree to which a clast approaches the shape of a sphere.

spreading boundary   *See* Divergent plate boundary.

spreading rate   The rate at which sea floor moves away from a mid-ocean ridge axis, as measured with respect to the sea floor on the opposite side of the axis.

spring   A natural outlet from which groundwater flows up onto the ground surface.

spring tide   An especially high tide that occurs when the Sun is on the same side of the Earth as the Moon.

stable air   Air that does not have a tendency to rise rapidly.

stable slope   A slope on which downward sliding is unlikely.

stalactite   An icicle-like cone that grows from the ceiling of a cave as dripping water precipitates limestone.

stalagmite   An upward-pointing cone of limestone that grows when drips of water hit the floor of a cave.

standing wave   A wave whose crest and trough remain in place as water moves through the wave.

star dune   A constantly changing dune formed by frequent shifts in wind direction; it consists of overlapping crescent dunes pointing in many different directions.

star   An object in the Universe in which fusion reactions occur pervasively, producing vast amounts of energy; our Sun is a star.

stellar nucleosynthesis   The production of new, larger atoms by fusion reactions in stars; the process generates more massive elements that were not produced by the Big Bang.

stick-slip behavior   Stop-start movement along a fault plane caused by friction, which prevents movement until stress builds up sufficiently.

stone rings   Ridges of cobbles between adjacent bulges of permafrost ground.

stoping   A process by which magma intrudes; blocks of wall rock break off and then sink into the magma.

storm   An episode of severe weather in which winds, precipitation, and in some cases lightning become strong enough to be bothersome and even dangerous.

storm-center velocity   A storm's (hurricane's) velocity along its track.

storm surge   Excess seawater driven landward by wind during a storm; the low atmospheric pressure beneath the storm allows sea level to rise locally, increasing the surge.

strain   The change in shape of an object in response to deformation (i.e., as a result of the application of a stress).

stratified drift   Glacial sediment that has been redistributed and stratified by flowing water.

stratigraphic column   A cross-section diagram of a sequence of strata summarizing information about the sequence.

stratigraphic formation   A recognizable layer of a specific sedimentary rock type or set of rock types, deposited during a certain time interval, that can be traced over a broad region.

stratigraphic sequence   An interval of strata deposited during periods of relatively high sea level, and bounded above and below by regional unconformities.

stratopause   The temperature pause that marks the top of the stratosphere.

stratosphere   The stable, stratified layer of atmosphere directly above the troposphere.

stratovolcano   A large, cone-shaped subaerial volcano consisting of alternating layers of lava and tephra.

stratus cloud   A thin, sheet-like, stable cloud.

streak   The color of the powder produced by pulverizing a mineral on an unglazed ceramic plate.

stream   A ribbon of water that flows in a channel.

streambed   The floor of a stream.

stream capacity   The total quantity of sediment a stream carries.

stream capture (or piracy)   The situation in which headward erosion causes one stream to intersect the course of another, previously independent stream, so that the intersected stream starts to flow down the channel of the first stream.

stream competence   The maximum particle size that a stream can carry.

stream gradient   The slope of a stream's channel in the downstream direction.

stream rejuvenation   The renewed downcutting of a stream into a floodplain or peneplain, caused by a relative drop of the base level.

stress   The push, pull, or shear that a material feels when subjected to a force; formally, the force applied per unit area over which the force acts.

stretching   The process during which a layer of rock or a region of crust becomes longer.

striations   Linear scratches in rock.

strike-slip fault   A fault in which one block slides horizontally past another (and therefore parallel to the strike line), so there is no relative vertical motion.

strip mining   The scraping off of all soil and sedimentary rock above a coal seam in order to gain access to the seam.

stromatolite   Layered mounds of sediment formed by cyanobacteria; cyanobacteria secrete a mucous-like substance to which sediment sticks, and as each layer of cyanobacteria gets buried by sediment, it colonizes the surface of the new sediment, building a mound upward.

structural control   The condition in which geologic structures, such as faults, affect the distribution and drainage of water or the shape of the land surface.

subaerial   Pertaining to land regions above sea level (i.e., under air).

subduction   The process by which one oceanic plate bends and sinks down into the asthenosphere beneath another plate.

subduction zone   The region along a convergent boundary where one plate sinks beneath another.

sublimation   The evaporation of ice directly into vapor without first forming a liquid.

submarine canyon   A narrow, steep canyon that dissects a continental shelf and slope.

submarine fan   A wedge-shaped accumulation of sediment at the base of a submarine slope; fans usually accumulate at the mouth of a submarine canyon.

submarine slump   The underwater downslope movement of a semicoherent block of sediment along a weak mud detachment.

submergent coast   A coast at which the land is sinking relative to sea level.

subpolar low   The rise of air where the surface flow of a polar cell converges with the surface flow of a Ferrel cell, creating a low-pressure zone in the atmosphere.

subsidence   The vertical sinking of the Earth's surface in a region, relative to a reference plane.

substrate   A general term for material just below the ground surface.

subtropical high (subtropical divergence zone)   A belt of high pressure in the atmosphere at 30° latitude formed where the Hadley cell converges with the Ferrel cell, causing cool, dense air to sink.

subtropics   Desert climate regions that lie on either side of the equatorial tropics between the lines of 20° and 30° north or south of the equator.

summit eruption   An eruption that occurs in the summit crater of a volcano.

sunspot cycle   The cyclic appearance of large numbers of sunspots (black spots thought to be magnetic storms on the Sun's surface) every 9 to 11.5 years.

supercontinent cycle   The process of change during which supercontinents develop and later break apart, forming pieces that may merge once again in geologic time to make yet another supercontinent.

supernova   A short-lived, very bright object in space that results from the cataclysmic explosion marking the death of a very large star; the explosion ejects large quantities of matter into space to form new nebulae.

superplume   A huge mantle plume.

superposed stream   A stream whose geometry has been laid down on a rock structure and is not controlled by the structure.

supervolcano   A volcano that erupts a vast amount (more than 1,000 cubic km) of volcanic material during a single event; none have erupted during recorded human history.

surface current   An ocean current in the top 100 m of water.

surface water   Liquid or seasonally frozen water that resides at the surface of the Earth in oceans, lakes, streams, and marshes.

surface waves   Seismic waves that travel along the Earth's surface.

surface westerlies   The prevailing surface winds in North America and Europe, which come out of the west or southwest.

surf zone   A region of the shore in which breakers crash onto the shore.

surge (glacial)   A pulse of rapid flow in a glacier.

suspended load   Tiny solid grains carried along by a stream without settling to the floor of the channel.

swamp   A wetland dominated by trees.

swash   The upward surge of water that flows up a beach slope when breakers crash onto the shore.

S-waves   Seismic shear waves that pass through the body of the Earth.

S-wave shadow zone   A band between 103° and 180° from the epicenter of an earthquake inside of which S-waves do not arrive at seismograph stations.

swelling clay   Clay possessing a mineral structure that allows it to absorb water between its layers and thus swell to several times its original size.

symmetry   The condition in which the shape of one part of an object is a mirror image of the other part.

syncline   A trough-shaped fold whose limbs dip toward the hinge.

systematic joints   Long planar cracks that occur fairly regularly throughout a rock body.

tabular intrusions   Sheet intrusions that are planar and of roughly uniform thickness.

Taconic orogeny   A convergent mountain-building event that took place around 400 million years ago, in which a volcanic island arc collided with eastern North America.

tailings pile   A pile of waste rock from a mine.

talus   A sloping apron of fallen rock along the base of a cliff.

tar   Hydrocarbons that exist in solid form at room temperature.

tarn   A lake that forms at the base of a cirque on a glacially eroded mountain.

tar sand   Sandstone reservoir rock in which less viscous oil and gas molecules have either escaped or been eaten by microbes, so that only tar remains.

taxonomy   The study and classification of the relationships among different forms of life.

tectonic foliation   A planar fabric, such as cleavage, schistocity, or gneissic banding, that develops in rocks; caused by compression or shearing during deformation (e.g., during mountain building).

temperature   A measure of the hotness or coldness of a material.

tension   A stress that pulls on a material and could lead to stretching.

tephra   Unconsolidated accumulations of pyroclastic grains.

terminal moraine   The end moraine at the farthest limit of glaciation.

terminator   The boundary between the half of the Earth that has daylight and the half experiencing night.

terrace   The elevated surface of an older floodplain into which a younger floodplain had cut down.

terrestrial planets   Planets that are of comparable size and character to the Earth and consist of a metallic core surrounded by a rock mantle.

thalweg   The deepest part of a stream's channel.

theory   A scientific idea supported by an abundance of evidence that has passed many tests and failed none.

theory of plate tectonics The theory that the outer layer of the Earth (the lithosphere) consists of separate plates that move with respect to one another.

thermal energy The total kinetic energy in a material due to the vibration and movement of atoms in the material.

thermal metamorphism Metamorphism caused by heat conducted into country rock from an igneous intrusion.

thermocline A boundary between layers of water with differing temperatures.

thermohaline circulation The rising and sinking of water driven by contrasts in water density, which is due in turn to differences in temperature and salinity; this circulation involves both surface and deep-water currents in the ocean.

thermosphere The outermost layer of the atmosphere containing very little gas.

thin section A 3/100-mm-thick slice of rock that can be examined with a petrographic microscope.

thin-skinned deformation A distinctive style of deformation characterized by displacement on faults that terminate at depth along a subhorizontal detachment fault.

thrust fault A gently dipping reverse fault; the hanging-wall block moves up the slope of the fault.

tidal bore A visible wall of water that moves toward shore with the rising tide in quiet waters.

tidal flat A broad, nearly horizontal plain of mud and silt, exposed or nearly exposed at low tide but totally submerged at high tide.

tidal reach The difference in sea level between high tide and low tide at a given point.

tide The daily rising or falling of sea level at a given point on the Earth.

tide-generating force The force, caused in part by the gravitational attraction of the Sun and Moon and in part by the centrifugal force created by the Earth's spin, that generates tides.

tidewater glacier A glacier that has entered the sea along a coast.

till A mixture of unsorted mud, sand, pebbles, and larger rocks deposited by glaciers.

tillite A rock formed from hardened ancient glacial deposits and consisting of larger clasts distributed through a matrix of sandstone and mudstone.

toe (terminus) The leading edge or margin of a glacier.

tombolo A narrow ridge of sand that links a sea stack to the mainland.

topographical map A map that uses contour lines to represent variations in elevation.

topography Variations in elevation.

topsoil The top soil horizons, which are typically dark and nutrient-rich.

tornado A near-vertical, funnel-shaped cloud in which air rotates extremely rapidly around the axis of the funnel.

tornado swarm Dozens of tornadoes produced by the same storm.

tower karst A karst landscape in which steep-sided residual bedrock towers remain between sinkholes.

trace fossil Fossilized imprints or debris that an organism leaves behind while moving on or through sediment; examples include footprints, burrows, and fecal matter.

transform fault A fault marking a transform plate boundary; along mid-ocean ridges, transform faults are the actively slipping segment of a fracture zone between two ridge segments.

transform plate boundary A boundary at which one litho sphere plate slips laterally past another.

transgression The inland migration of shoreline resulting from a rise in sea level.

transition zone The middle portion of the mantle, from 400 to 670 km deep, in which there are several jumps in seismic velocity.

transpiration The release of moisture as a metabolic by-product.

transverse dune A simple, wave-like dune that appears when enough sand accumulates for the ground surface to be completely buried, but only moderate winds blow.

travel-time curve A graph that plots the time since an earthquake began on the vertical axis, and the distance to the epicenter on the horizontal axis.

trellis network A drainage system that develops across a landscape of parallel valleys and ridges so that major tributaries flow down the valleys and join a trunk stream that cuts through the ridge; the resulting map pattern resembles a garden trellis.

trench A deep, elongate trough bordering a volcanic arc; a trench defines the trace of a convergent plate boundary.

triangulation The method for determining the map location of a point from knowing the distance between that point and three other points; this method is used to locate earthquake epicenters.

tributary A smaller stream that flows into a larger stream.

triple junction A point where three lithosphere plate boundaries intersect.

tropical depression A tropical storm with winds reaching up to 61 km per hour; such storms develop from tropical disturbances, and may grow to become hurricanes.

tropical disturbance Cyclonic winds that develop in the tropics.

tropopause The temperature pause marking the top of the troposphere.

troposphere The lowest layer of the atmosphere, where air undergoes convection and where most wind and clouds develop.

truncated spur A spur (elongate ridge between two valleys) whose end was eroded off by a glacier.

trunk stream The single larger stream into which an array of tributaries flow.

tsunami A large wave along the sea surface triggered by an earthquake or large submarine slump.

tuff A pyroclastic igneous rock composed of volcanic ash and fragmented pumice, formed when accumulations of the debris cement together.

tundra A cold, treeless region of land at high latitudes, supporting only species of shrubs, moss, and lichen capable of living on permafrost.

turbidite A graded bed of sediment built up at the base of a submarine slope and deposited by turbidity currents.

turbidity current A submarine avalanche of sediment and water that speeds down a submarine slope.

turbulence The chaotic twisting, swirling motion in flowing fluid.

typhoon The equivalent of a hurricane in the western Pacific Ocean.

ultimate base level Sea level; the level below which a trunk stream cannot cut.

ultramafic A term used to describe igneous rocks or magmas that are rich in iron and magnesium and very poor in silica.

unconfined aquifer An aquifer that intersects the surface of the Earth.

unconformity A boundary between two different rock sequences representing an interval of time during which new strata were not deposited and/or were eroded.

unconsolidated Consisting of unattached grains.

undercutting Excavation at the base of a slope that results in the formation of an overhang.

undersaturated A term used to describe a solution capable of holding more dissolved ions.

unsaturated zone The region of the subsurface above the water table.

unstable air Air that is significantly warmer than air above and has a tendency to rise quickly.

unstable ground   Land capable of slumping or slipping downslope in the near future.

unstable slope   A slope on which sliding will likely happen.

updraft   Upward-moving air.

upper mantle   The uppermost section of the mantle, reaching down to a depth of 400 km.

upwelling zone   A place where deep water rises in the ocean, or where hot magma rises in the asthenosphere.

U-shaped valley   A steep-walled valley shaped by glacial erosion into the form of a U.

vacuum   Space that contains very little matter in a given volume (e.g., a region in which air has been removed).

valley   A trough with sloping walls, cut into the land by a stream.

valley glacier   A river of ice that flows down a mountain valley.

Van Allen radiation belts   Belts of solar wind particles and cosmic rays that surround the Earth, trapped by Earth's magnetic field.

van der Waals bonding   The relatively weak attachment of two elements or molecules due to their polarity and not due to covalent or ionic bonding.

varve   A pair of thin layers of glacial lake-bed sediment, one consisting of silt brought in during the spring floods and the other of clay deposited during the winter when the lake's surface freezes over and the water is still.

vascular plant   A plant with woody tissue and seeds and veins for transporting water and food.

vein   A seam of minerals that forms when dissolved ions carried by water solutions precipitate in cracks.

vein deposit   A hydrothermal deposit in which the ore minerals occur in veins that fill cracks in preexisting rocks.

velocity-versus-depth curve   A graph that shows the variation in the velocity of seismic waves with increasing depth in the Earth.

ventifact (faceted rock)   A desert rock whose surface has been faceted by the wind.

vesicles   Open holes in igneous rock formed by the preservation of bubbles in magma as the magma cools into solid rock.

viscosity   The resistance of material to flow.

volatiles (syn. volatile materials)   Elements or compounds such as H_2O and CO_2 that evaporate at relatively low temperatures and can exist in gaseous forms at the Earth's surface.

volatility   A specification of the ease with which a material evaporates.

volcanic agglomerate   An accumulation consisting dominantly of volcanic bombs and other relatively large chunks of igneous material.

volcanic arc   A curving chain of active volcanoes formed adjacent to a convergent plate boundary.

volcanic ash   Tiny glass shards formed when a fine spray of exploded lava freezes instantly upon contact with the atmosphere.

volcanic bomb   A large piece of pyroclastic debris thrown into the atmosphere during a volcanic eruption.

volcanic danger-assessment map   A map delineating areas that lie in the path of potential lava flows, lahars, debris flows, or pyroclastic flows of an active volcano.

volcanic gas   Elements or compounds that bubble out of magma or lava in gaseous form.

volcanic island arc   The volcanic island chain that forms on the edge of the overriding plate where one oceanic plate subducts beneath another oceanic plate.

volcaniclastic rock   A material composed of cemented-together grains of volcanic material; it includes both pyroclastic rocks and rocks formed from accumulations of water-transported volcanic debris.

volcano   (1) A vent from which melt from inside the Earth spews out onto the planet's surface; (2) a mountain formed by the accumulation of extrusive volcanic rock.

V-shaped valley   A valley whose cross-sectional shape resembles the shape of a V; the valley probably has a river running down the point of the V.

Wadati-Benioff zone   A sloping band of seismicity defined by intermediate- and deep-focus earthquakes that occur in the downgoing slab of a convergent plate boundary.

wadi   The name used in the Middle East and North Africa for a dry wash.

warm front   A front in which warm air rises slowly over cooler air in the atmosphere.

waste rock   Rock dislodged by mining activity yet containing no ore minerals.

waterfall   A place where water drops over an escarpment.

water gap   An opening in a resistant ridge where a trunk river has cut through the ridge.

watershed   The region that collects water that feeds into a given drainage network.

water table   The boundary, approximately parallel to the Earth's surface, that separates substrate in which groundwater fills the pores from substrate in which air fills the pores.

wave base   The depth, approximately equal in distance to half a wavelength in a body of water, beneath which there is no wave movement.

wave-cut bench   A platform of rock, cut by wave erosion, at the low-tide line that was left behind a retreating cliff.

wave-cut notch   A notch in a coastal cliff cut out by wave erosion.

wave erosion   The combined effects of the shattering, wedging, and abrading of a cliff face by waves and the sediment they carry.

wave front   The boundary between the region through which a wave has passed and the region through which it has not yet passed.

wavelength   The horizontal difference between two adjacent wave troughs or two adjacent crests.

wave refraction (ocean)   The bending of waves as they approach a shore so that their crests make no more than a 5° angle with the shoreline.

weather   Local-scale conditions as defined by temperature, air pressure, relative humidity, and wind speed.

weathered rock   Rock that has reacted with air and/or water at or near the Earth's surface.

weathering   The processes that break up and corrode solid rock, eventually transforming it into sediment.

weather system   A specific set of weather conditions, reflecting the configuration of air movement in the atmosphere, that affects a region for a period of time.

welded tuff   Tuff formed by the welding together of hot volcanic glass shards at the base of pyroclastic flows.

well   A hole in the ground dug or drilled in order to obtain water.

Western Interior Seaway   A north-south-trending seaway that ran down the middle of North America during the Late Cretaceous Period.

wet-bottom (temperate) glacier   A glacier with a thin layer of water at its base, over which the glacier slides.

wetted perimeter   The area in which water touches a stream channel's walls.

wind abrasion   The grinding away at surfaces in a desert by windblown sand and dust.

wind gap   An opening through a high ridge that developed earlier in geologic history by stream erosion, but that is now dry.

xenolith   A relict of wall rock surrounded by intrusive rock when the intrusive rock freezes.

yardang A mushroom-like column with a resistant rock perched on an eroding column of softer rock; created by wind abrasion in deserts where a resistant rock overlies softer layers of rock.

yazoo stream A small tributary that runs parallel to the main river in a floodplain because the tributary is blocked from entering the main river by levees.

Younger Dryas An interval of cooler temperatures that took place 4,500 years ago during a general warming/glacier-retreat period.

zeolite facies The metamorphic facies just above diagenetic conditions, under which zeolite minerals form.

zone of ablation The area of a glacier in which ablation (melting, sublimation, calving) subtracts from the glacier.

zone of accumulation (1) The layer of regolith in which new minerals precipitate out of water passing through, thus leaving behind a load of fine clay; (2) the area of a glacier in which snowfall adds to the glacier.

zone of aeration *See* Unsaturated zone.

zone of leaching The layer of regolith in which water dissolves ions and picks up very fine clay; these materials are then carried downward by infiltrating water.

Credits

Opening photograph: Stephen Marshak

Table of contents

ix top Stephen Marshak; **bottom** NASA and STScI; **x** Stephen Marshak; **top center** Courtesy of Thomas N. Taylor; **bottom** Elliot Lim/NOAA; **xi top** Stephen Marshak **bottom** Hal Lott/Corbis; **xii top** Stephen Marshak; **center** Stephen Marshak; **bottom** Stephen Marshak; **xiii top** Stephen Marshak; **center** Stephen Marshak; **bottom** National Geographic/Getty Images; **xiv top** REUTERS/Eduardo Munoz; **bottom** © Gary Hincks; **xv** USGS; **center** Stephen Marshak; **bottom** Stephen Marshak; **xvi** Stephen Marshak; **xvii top** Department of Defense; **bottom** Stephen Marshak; **xviii top** JPL/NASA; **bottom** permission from DNR, Washington/Flickr; **xix top** Stephen Marshak; **bottom** Stephen Marshak; **xx top** Stephen Marshak; **bottom** Stephen Marshak; **xxi top** Stephen Marshak **bottom** Emma Marshak; **xxii** Stephen Marshak.

Prelude

1 Stephen Marshak; **2 top** Stephen Marshak; **2 bottom left** Stephen Marshak; **2 bottom right** Stephen Marshak; **3 (a)** Stephen Marshak; **3 (b)** Stephen Marshak; **4** AP Photo/Hurriyet; **5** Stephen Marshak; **9** Stephen Alvarez/Getty Images; **9 inset** Stephen Marshak; **11** NASA.

Part opener 1

12, 13 NASA Archive/Alamy.

Chapter 1

14 NASA and STScI; **16 left** Richard E. Hill/Visuals unlimited; **16 right** Stephen Marshak; **17 top** Rare Book Division, The New York Public Library, Astor, Lenox and Tilden Foundation; **17 bottom** Mary Evans; Picture Library/Alamy; **18 left** Stephen Marshak; **18 right** Stephen Marshak; **18 middle** Stephen Marshak; **20** NASA/JPL-Caltech; **20 insert left** NASA; **insert right** Miloslav Druckmuller; **23** Image courtesy of Arthur Smith and Randall Feenstra, Carnegie Mellon University; **25** Moonrunner design; **26** NASA; **27 top** J. Hester and P. Scowen/NASA; **27 bottom** SOHO (ESA & NASA); **28 left** NASA **28 right** Photograph by Pelisson, SaharaMet; **33 left** NASA/HST; **33 right** Robertgendlerastropix.com.

Chapter 2

34 Stephen Marshak; **37 left** NASA/JPL/Cal Tech; **37 right** Photo Researchers; **37 bottom** NASA/NSSDC/GSFC; **38 (a–c)** JPL/NASA; **38 (d)** NASA, James Bell (Cornell Univ.), Michael Wolff (Space Science Inst.), and The Hubble Heritage Team (STScI/AURA); **39** NASA/Photo Researchers; **40** NASA; **41** NOAA; **43 left** Tate Gallery, London/Art Resource,

N.Y.; **43 right** Stephen Marshak; **47 left** Fred Espenak, Photo Researchers, Inc.; **47 top** Tom Bean; **47 bottom** Marli Miller/Visuals Unlimited, Inc.; **49** NASA/JPL-Caltech.

Chapter 3

55 Courtesy of Thomas N. Taylor; **56 left** Alfred Wegener Institute for Polar and Marine Research; **56 right** Ron Blakey, Northern Arizona University; **59** Ron Blakey, Northern Arizona University; **65** NOAA; **73** Images Courtesy of Gary A.Glatzmaier (University of California, Santa Cruz) And Paul H. Roberts (University of California, Los Angeles), Taken from their computer simulation; **76** Christoph Hormann.

Chapter 4

77 Elliot Lim/NOAA; **83** University of Texas Institute of Geophysics; **84 left** NOAA/University of Washington; **84 right** NOAA/University of Washington; **85** J.R. Delaney and D.S. Kelley, University of Washington; **89 top** Elliot Lim CIRES, University of Colorado at Boulder, and NOAA/National Geophysical Data Center; Source for data: "Age, Spreading Rates and Spreading Symmetry of the World's Ocean Crust," By R. Dietmar Muller et al.; **89 bottom** Kevin Schafer/Alamy; **93 top** Johnson Space Center, NASA; **93 bottom** Cindy Ebinger, University of Rochester.

Part opener 2

102, 103 Stephen Marshak.

Chapter 5

104 Stephen Marshak; **105** Tim Graham/Getty Images; **106 top** Ken Lucas/ Visuals Unlimited; **106 bottom** Stephen Marshak; **107** Nicholas Vasilakes/Alamy; **108 left** Jay Schomer; **108 right** Wally McNamee/Corbis; **109 top** Charles O'Rear/Corbis; **109 bottom** Courtesy of John A. Jaszczak Michigan Technological University; **114** Javier Trueba/MSF/Photo Researchers, Inc.; **115** 1996 Jeff Scovil; **117 left** Erich Schrempp/Photo Researchers Inc.; **117 right (top and bottom)** Courtesy of Prof. Huifang Xu, Dept. of Geology and Geophysics, Uni. Of Wisconsin-Madison; **118 (a)** Richard P. Jacobs /JLM Visuals; **118 (b)** Richard P. Jacobs/JLM Visuals; **118 (c)** Breck P. Kent/JLM Visuals; **118 (d)** Richard P. Jacobs/ JLM Visuals; **118 (e) inset** Stephen Marshak; **118 (e)** Andrew Silver/USGS; **118 (f)** Charles D. Winters/Photo Researchers, Inc.; **118 (g)** Scientifica/Visuals Unlimited/Alamy; **121 (a)** Richard P. Jacobs/JLM Visuals; **121 (b)** 1992 Jeff Scovil; **121 (c)** Marli Miller/Visuals Unlimited, Inc.; **121 (d)** Richard P. Jacobs/JLM Visuals; **121 (e)** Richard P. Jacobs/JLM Visuals; **121 (f) left** Arco Images GmbH/Alamy; **121 (f) right** Ann Bryant/Geology.com; **121 (h)** Stephen Marshak; **124** 1996 Smithsonian Institution; **125 top** 1985 Darrel Plowes1985 Darrel Plowes; **125**

bottom Ken Lucas/Visuals Unlimited; **127 top** A.J. Copley/ Visuals Unlimited; **127 top inset** A.J. Copley/Visuals Unlimited; **127 bottom** Chip Clark; **129** © 1998 Jeff Scovil.

Interlude A

130 Hal Lott/Corbis; **132 top left** Richard P. Jacobs/JLM Visuals; **132 top right** Courtesy David W. Houseknecht, USGS; **132 bottom left** sciencephotos/Alamy; **132 bottom right** Courtesy of Kent; Ratajeski, Dept. of Geology and Geophysics, U of Wisconsin, Madison; **133 (a–d)** Stephen Marshak; **134 (a–c)** Stephen Marshak; **136 (a)** Tom Bean; **136 (b)** H. Johnson; **137 (c)** Stephen Marshak; **137 (d)** Scenics & Science/Alamy; **138 top** Product photo courtesy of JEOL, USA; **138 bottom** Stephen Marshak.

Chapter 6

139 Stephen Marshak; **140 (a)** Jim Sugar Photography/Corbis; **140 (b)** J.D. Griggs/ U.S. Geological Survey; **140 (c)** Roger Ressmeyer/Corbis; **140 (d)** Stephen Marshak; **141 (b)** Uri Ten Brink (USGS); **141 (c, d)** Stephen Marshak; **149 (a–c)** Stephen Marshak; **149 (d)** U.S. Geological Survey; **149 (e)** Stephen Marshak; **150 (a) inset** Stephen Marshak; **150 (b)** Stephen Marshak; **150 (c)** 1998 Tom Bean; **151 all** Stephen Marshak; **156 (a)** Courtesy of Dr. Kent Ratajeski Department of Geology and Geophysics, University of Wisconsin, Madison; **156 (a) right** Courtesy of Dr. Kent Ratajeski Department of Geology and Geophysics, University of Wisconsin, Madison; **156 (b)** Doug Sokell/Visuals Unlimited; **156 (c) top left** Dirk Wiersma/ Photo Researchers, Inc.; **156 (c) middle left** Stephen Marshak; **156 (c) bottom left** John S. Shelton; **156 (c) top right** Stephen Marshak; **156 (c) middle right** Dirk Wiersma/ Photo Researchers, Inc; **156 (c) bottom right** Stephen Marshak; **158 (a–d)** Stephen Marshak; **162** Stephen Marshak; **164 top** Stephen Marshak; **164 bottom** Peter Kresan; **167 top** Tony Linck/ SuperStock; **167 bottom** Jacques Descloitres, MODIS team, NASA Visible Earth.

Interlude B

168 Stephen Marshak; **169** Stephen Marshak; **170 (a, b)** Stephen Marshak; **171 insets** Stephen Marshak; **171 top** Marli Miller/ Visuals Unlimited, Inc.; **171 (b) bottom left** Stephen Marshak; **171 (b) right** Stephen Marshak; **172 (b) left** Stephen Marshak; **172 (b) right** Stephen Marshak; **172 (c)** Stephen Marshak; **173 (a)** Stephen Marshak; **173 (b)** Courtesy Carlo Giovanella; **173 (c)** PR/91-05C British Geology Survey. NERC.; **176 left** Stephen Marshak; **176 center** Stephen Marshak; **177 left** Stephen Marshak; **177 center** Stephen Marshak; **177 right** Stephen Marshak; **178 (b)** Stephen Marshak; **178 (c)** Stephen Marshak; **178 (d)** Stephen Marshak; **179** Stephen Marshak; **181** US Dept. of Agriculture/Natural Resources Conservation Services; **182** Stephen Marshak; **183** Jim Richardson/Corbis.

Chapter 7

184 Stephen Marshak; **185** Stephen Marshak; **190 (a) left** Stephen Marshak; **190 (a) right** Stephen Marshak; **190 (b) left** Stephen Marshak; **190 (b) right** Stephen Marshak; **190 (c) left** Stephen Marshak; **190 (c) right** Scottsdale Community College;

191 (d) left Stephen Marshak; **191 (d) right** E.R. Degginger/ Color-Pic, Inc.; **191 (e) left** Stephen Marshak; **191 (e) right** Stephen Marshak; **192 (a–d)** Stephen Marshak; **193 top left** Stephen Marshak; **193 top right** Stephen Marshak; **193 middle** Gerald and Buff Corsi/Visuals Unlimited; **193 middle inset** Stephen Marshak; **193 bottom** Markus Matzel/Peter Arnold Inc.; **194 (a)** Marli Miller/Visuals Unlimited, Inc.; **194 (b)** 1998 M.W. Schmidt; **195 (a, b)** Stephen Marshak; **196 (a)** Stephen Marshak; **197 left** Stephen Marshak; **197 right** Grand Canyon Natural History Assocation; **198 (c, d)** Stephen Marshak; **199 (a)** IMAGINA Photography/Alamy; **199 (c)** Stephen Marshak; **199 (d)** Stephen Marshak; **200 all** Stephen Marshak; **202** Stephen Marshak; **204 (a)** Emma Marshak; **204 (b)** Stephen Marshak; **204 (c)** Marli Miller/Visuals Unlimited, Inc.; **204 (d)** Stephen Marshak; **204 (e)** Stephen Marshak; **204 (f)** John S. Shelton; **206 top** Stephen Marshak; **206 bottom** Yann Arthus-Bertrand/ Corbis; **207 (a)** Belgian Federal Science Policy Office; **207 (b)** G.R. Roberts NSIL; **211** William Manning/Corbis.

Chapter 8

212 Stephen Marshak; **214 (a) left** Stephen Marshak; **214 (a) right** L.S. Stephanowicz/Visuals Unlimited; **214 (b) left** Stephen Marshak; **214 (b) right** Courtesy Kurt Freihauf, Kutztown University of Pennsylvannia; **214 (b) inset** Stephen Marshak; **214 (c) left** Stephen Marshak; **214 (c) right** Stephen Marshak; **219** Stephen Marshak; **220 (a)** Ludovic; Maisant/Corbis; **220 (a) inset** Emma Marshak; **220 (b)** Stephen Marshak **221 (a)** Stephen Marshak; **221 (b)** Stephen Marshak; **221 (c)** Photograph of Muscovite Schist by Ann Bryant of Geology.com; **222** Stephen Marshak; **223 top** Stephen Marshak; **223 bottom** Stephen Marshak **224 all** Stephen Marshak; **229** Stephen Marshak; **234 top** Stephen Marshak; **234 bottom** Dr. Terry Wright; **236 (a)** David Muench/Corbis; **236 (b)** Stephen Marshak; **238** C.E. Brown, 1989, USGS Map I-1725.

Interlude C

239 top Stephen Marshak; **239 left** Stephen Marshak; **239 right** Stephen Marshak; **243** Stephen Marshak.

Part opener 3

246, 247 Stephen Marshak.

Chapter 9

248 National Geographic/Getty Images; **250 (a)** Mount Vesuvius in Eruption, 1817 (w/c on paper), Turner, Joseph Mallord William (1775–1851)/Yale Center for British Art, Paul Mellon Collection, USA/The Bridgeman Art Library; **250 (b)** Stephen Marshak; **250 (c)** Stephen Marshak; **250 (d)** Jack Repchek; **250 (e) left** Stephen Marshak; **250 (e) middle** Jack Repchek; **250 (e) right** Jack Repchek; **251 right** Images provided by Google Earth mapping services/DigitalGlobe, Terra Metrics, NASA, Europa Technologies, © 2011; **251 left** Photograph by C. Nye, Alaska Division of Geological and Geophysical Surveys, August, 1991, USGS; **252 (a)** Roger Ressmeyer/Corbis; **252 (b)** Stephen Marshak; **252 (c) left** USGS, Hawaiian Volcano Observatory; **252 (c) right** Stephen Marshak; **252 (d, e)** Stephen Marshak; **253 (a)** Thomas Hallstein/Alamy; **253 (b)** Stephen Marshak;

254 (a) AP Images; 254 (b) Stephen Weaver; 254 (b) inset Stephen Marshak; 255 (b) A.M. Sarna-Wojcicki/U.S. Geological Survey; 255 (c–d) Stephen Marshak; 255 (e) NASA; 255 (f) Tarko SUDIARNO/AFP/Getty Images; 255 (g) Stephen and Donna O'Meara, Volcano Watch International; 257 (a) USGS; 257 (b–d) Stephen Marshak; 258 (a, b) Stephen Marshak; 258 bottom right N. Banks/U.S. Geographical Survey; 259 Marli Miller/Visuals Unlimited, Inc.; 260 (a) Marli Miller/Visuals Unlimited, Inc.; 260 (b) 1985 Tom Bean; 260 (c) inset Roger Ressmeyer/Corbis; 261 J.D. Griggs, USGS; 262 (a) Westend 61 GmbH/Alamy; 262 (b) Alvaro Vidal/AFP/Getty Images; 262 (c) Photograph by Dave Harlow, taken on June 12, 1991, USGS; 262 (d) Lothar Slabon/AFP/Getty Images; 264 Images provided by Google Earth mapping services/DigitalGlobe, Terra Metrics, NASA, Europa Technologies, Copyright 2011; 265 top Stephen Marshak; 265 bottom Gary Braasch/Corbis; 271 Stephen Marshak; 272 (a) Stephen Marshak; 272 (c) NOAA; 273 (b) David Sandwell/Scripps Institute of Oceanography; 273 (c) Alan Tuchman; 274 (a) J.D. Griggs/U.S. Geographical Survey; 274 (b) Vittoriano Rastelli/Corbis; 274 (c) AP Photo/Sayyid Azim; 274 (d) AP Images; 274 (e) Alberto Garcia/Corbis; 274 (f) US Geological Survey; 274 (g) Philippe Bourseiller; 276 (a) J. Marso, U.S. Geographical Survey; 276 (b) Peter Turnley/Corbis; 277 Stephen Marshak; 278 Charles Wicks, USGS; 279 (a) Vittoriano Rastelli/Corbis; 279 (b) Sigurgeir Jonasson/Frank Lane Picture Agency/Corbis; 282 (a) Gail Mooney/Corbis; 282 (b) Birke Schreiber; 282 (c) Image courtesy of Earth Sciences and Image Analysis Laboratory, NASA Johnson Space Center; 283 (a) Julian Baum/ Photo Researchers, Inc.; 283 (b–e) NASA/JPL; 283 (e) inset NASA/JPL; 284 Images provided by Google Earth mapping services/DigitalGlobe, Terra Metrics, NASA, Europa Technologies, © 2011; 285 top left Images provided by Google Earth mapping services/DigitalGlobe, Terra Metrics, NASA, Europa Technologies, © 2011; 285 top right Images provided by Google Earth mapping services/DigitalGlobe, Terra Metrics, NASA, Europa Technologies, © 2011; 285 bottom left Tom Pfeiffer/www.volcanodiscovery.com; 285 bottom center Image courtesy Jacques Descloitres, MODIS Rapid Response Team at NASA GSFC; 285 bottom right Image courtesy Jacques Descloitres, MODIS Rapid Response Team at NASA GSFC.

Chapter 10

286 REUTERS/Eduardo Munoz ; 288 (a) New Zealand Herald, Mark Mitchell/AP; 288 (b) Kyodo News/AP; 290 (a) Photo Courtesy of Paul "Kip" Otis-Diehl, USMC, 29 Palms CA; 290 (b) Photo Courtesy of Paul "Kip" Otis-Diehl, USMC, 29 Palms CA; 293 Jacques Descloitres, MODIS Rapid Response Team, NASA/ GSFC/Visible Earth; 295 Peltzer, G., Crampe F., and King G., 1999, Evidence of nolinear elasticity of the Crust. Science, v. 286, p. 272–276; 307 Patrick Robert/Corbis Sygma; 308 (b) Corbis; 308 (c) George Hall/Corbis; 310 (a) REUTERS/Eduardo Munoz; 310 (b) USGS; 310 (c) Anthony Sladen, Tectonics Observatory, California Institute of Technology; 311 (d, e) USGS; 311 (f) Images provided by Google Earth mapping services/ DigitalGlobe, Terra Metrics, NASA, Europa Technologies, Copyright 2011; 315 (a) AP Photo/Hurriyet; 315 (b) Pacific Press Service/

Almay; 315 (c) M. Celebi, USGS 315 (d) Reuters; 316 (a) Barry Lewis/Alamy; 316 (b) Bettmann/Corbis; 316 (c) NGDC; 316 (d) Karl V. Steinbrugge Collection Earthquake Engineering Research Center; 317 (c) Nisee-PEER, University of California, Berkeley; 317 (d) James Mori, Research Center for Earthquake Prediction, Disaster Prevention Institute Kyoto University; 317 (e) Martin Luff/Flickr; 318 NGDC; 319 U.S. Department of the Interior/NOAA; 320 (a) Vasily V. Titov, Associate Director, Tsunami Inundation Mapping Efforts (TIME), NOAA/ PMEL-UW/JISAO, USA; 320 (b) AFP/Getty images; 320 (c) David Rydevik; 320 (d) Space Imaging; 321 (b) top Air Photo Service/Reuters /Landov; 321 (a) AP Photo/Kyodo News; 321 (b) bottom EPA/Landov; 324 (a, b) Produced by the Global Seismic Hazard Assessment Program; 326 NOAA/NOA Center for Tsunami Research; 329 NOAA Center for Tsunami Research.

Interlude D

330 © Gary Hincks; 333 George Resch/Fundamental Photographs, NYC; 336 Jenifer Jackson and Jay Bass, University of Illinois.; 339 (d) Matthew Fouch, Arizona State University; 340 Adapted/Reproduced from Naliboff and Kellogg, "Dynamic effects of a step-wise increase in thermal conductivity and viscosity in the lowermost mantle", Geophysical Research Letters, 33, L12S09, 2006; 341 (a) John Q. Thompson, courtesy of Dawson Geophysical Co.; 341 (b) Shell Oil Company; 341 (c) John Q. Thompson, courtesy of Dawson Geophysical Co.; 341 (d) Courtesy Sercel and CGG Veritas; 342 Stephen Marshak; 343 top left ESA; 343 top right Figure provided courtesy F. Lemoine & J. Frawley, NASA Goddard Space Flight Center; 343 bottom USGS; 347 USGS.

Chapter 11

348 Stephen Marshak; 349 National Geophysical Data Center/ NOAA; 350 Stephen Marshak; 351 (c) Stephen Marshak; 351 (d) Stephen Marshak; 352 all Stephen Marshak; 354 (b) Stephen Marshak; 354 (d) Stephen Marshak; 357 (c) Stephen Marshak; 357 (d) Stephen Marshak; 358 (a) Galen Rowell/Corbis; 358 (b) Stephen Marshak; 358 (c) Stephen Marshak; 360 (a) John S. Shelton; 360 (b) Stephen Marshak; 360 (c) U.S. Geological Survey; 361 (a) Stephen Marshak; 361 (b) Stephen Marshak; 361c Stephen Marshak; 361 bottom Stephen Marshak; 362 Stephen Marshak; 364 (a) Stephen Marshak; 364 (b) Stephen Marshak; 364 (c) Stephen Marshak; 364 (d) Stephen Marshak; 364 (e) John S. Shelton; 365 top Stephen Marshak; 365 bottom Stephen Marshak; 367 (b) Stephen Marshak; 367 bottom Marli Miller/Visuals Unlimited, Inc.; 369 Stephen Marshak; 370 (b) Based on a photo by J. Beckett of a model by J. Malavielle © American Museum of Natural History; 370 (c) USGS; 371 top Stephen Marshak; 371 bottom Stephen Marshak; 374 Stephen Marshak; 376 (a) Stephen Marshak; 376 (b) Stephen Marshak; 377 USGS; 380 (a) USGS; 380 (b) Modified from the original: Miles, C.E., compiler (2008). Geologic shaded-relief map of Pennsylvania. Pennsylvania Geological Survey, 4th ser., map 67; 380 bottom Courtesy E. Norabuena, et al. (1998). Science 279:359.; 383 Image created by J. Varner and E. Lim, CIRES, Univeristy of Colorado at Boulder.

Part opener 4
384, 385 Stephen Marshak.

Interlude E
386 Stephen Marshak; 387 (a) Stephen Marshak; 387 (b) Richard T. Nowitz/Corbis; 387 (c) Photo by William L. Jones from the Stones & Bones Collection, http://www.stones-bones.com; 387 right John Reader/Photo Researchers Inc.; 390 (a) Zoological Institute of Russian Academy of Sciences and Zoological Museum; 390 (b) Doug Lundeberg (www.ambericawest.com); 390 (c) Stephen Marshak; 390 (d) Stephen Marshak; 390 (e) Stephen Marshak; 390 (f) Depatment of Earth and Space Sciences, UCLA; 390 (g) Kevin Schafer/Corbis; 390 (h) Stephen Marshak; 390 (i) Stephen Marshak; 391 UCL Micropalaeontology Collections, UCL Museums & Collections; 392 (a) Courtesy of Senckenberg, Messel Research Department; 392 (b) Humboldt-Universität zu Berlin Museum für Naturkunde. Photo by W.Harre.; 392 (c) O. Louis Mazzatenta, National Geographic/Getty Images; 392 (d) Illustration by Karen Carr and Karen Carr Studio, Inc. © Smithsonian Institution; 398 Stephen Marshak.

Chapter 12
399 Stephen Marshak 400 The Granger Collection, NYC 401 Stephen Marshak 403 (a) left Stephen Marshak; 403 (a) right Stephen Marshak; 403 (b) left Tom Pfeiffer/www.volcanodiscovery.com; 403 (b) right Stephen Marshak; 403 (c) Stephen Marshak; 403 (d) Stephen Marshak; 404 Stephen Marshak; 406 Layne Kennedy/ Corbis; 408 (a, b) Stephen Marshak; 409 Stephen Marshak; 411 Stephen Marshak; 413 Jenning, C.W., 1997, California Dept. of Mines & Geology/ USGS; 416 all Stephen Marshak; 421 A.J. Copley/Visuals Unlimited; 422 (a) Damon Runberg, USGS; 422 (b) Stephen Marshak; 422 (c) Stephen Marshak; 423 Courtesy of Yong Il Lee, School of Earth and Environment Sciences, Seoul National University; 426 (a) With Permission of the Royal Ontario Museum © ROM; 426 (b) NASA; 429 USGS National Cventer of EROS and NASA, Landsat Project Science office.

Chapter 13
430 Stephen Marshak; 431 Stephen Marshak; 431 inset Stephen Marshak; 435 Bonestell Space Art; 437 (a) Courtesy of Dr. J. William Schopf/UCLA; 437 (b) Stephen Marshak; 437 (c) Frans Lanting/Corbis; 438 USGS; 441 (a) Lisa-Ann Gershwin/ U.C. Museum of Paleontology; 441 (b) Stephen Marshak; 441 (c) Courtesy of Dr. Paul Hoffman, Harvard University; 444 (b) Ronald C. Blakey, Colorado Plateau Geosystems, Inc.; 444 (c) Ronald C. Blakey, Colorado Plateau Geosystems, Inc.; 445 Jonathan J. Havens, Irving Materials, Inc.; 448 (a) Ronald C. Blakey, Colorado Plateau Geosystems, Inc.; 448 (c) AFP/Getty Images; 448 (d) Stephen Marshak; 449 top right Ronald C. Blakey, Colorado Platea Geosystems, Inc.; 449 bottom left NOAA; 449 bottom right Images provided by Google Earth mapping services/DigitalGlobe, Terra Metrics, NASA, Europa Technologies, Copyright 2011; 450 E.R. Degginger/Color-Pic, Inc.; 450 inset Dan Osipov; 451 (a) inset Ronald C. Blakey, Colorado Plateau Geosystems, Inc.; 451 (b) Stephen

Marshak; 451 (d) Richard Bizley FIAAA; 452 (a) Ronald C. Blakey, Colorado Plateau Geosystems, Inc.; 452 (b) Ronald C. Blakey, Colorado Plateau Geosystems, Inc.; 453 (b, c) Ronald C. Blakey, Colorado Plateau Geosystems, Inc.; 455 (a, inset) Ronald C. Blakey, Colorado Plateau Geosystems; 455 (a) NASA, JPL; 455 (b) center NASA, JPL; 455 (b) left Reproduced with permission of Natural Resources Canada 2009, courtesy of Geological Survey of Canada; 455 (b) right NOAA; 463 Jacques Descloitres, MODIS Team, NASA Visible Earth.

Part opener 5
464, 465 Stephen Marshak.

Chapter 14
466 U.S. Department of Defense; 476 (c) Science Photo Library/ Photo Researchers, Inc.; 477 (b) Stephen Marshak; 477 (c) Robert Holden/Photophile; 477 (d) Courtesy Auke Visser, www.aukevisser.nl; 477 (e) Stephen Marshak; 478 (a) American Petrolum Institute; 478 (b) American Petrolum Institute; 479 (b) Ronald C. Blakey, Colorado Plateau Geosystems, Inc.; 480 (a) AP Photo/Jeff McIntosh; 480 (b) William J.Winters, USGS; 483 (a) Ron Testa/The Field Museum, Chicago; 483 (d) Courtesy University of Melbourne; 486 (a) Stephen Marshak; 486 (b) Roger Ressmeyer/CORBIS; 486 (c) Stephen Marshak; 487 (a) Philadelphia Inquirer; 487 (b) Courtesy of Anupma Prakahs, Geophysical Institute; 488 James Blank/Photophile; 490 G.R. Roberts NSIL; 491 (c) Stephen Marshak; 491 (d) Stephen Marshak; 491 (c) Stephen Marshak; 495 (a) AP Photo/ Stapleton; 495 (b) AFP/Getty Images; 499 (b) NASA 499 (c) USCG/Landov; 499 (d) Reuters/BP/Landov; 499 (e) NASA.

Chapter 15
502 Stephen Marshak; 503 Courtesy of the Oakland Museum of California; 504 top Courtesy of Nightflyer; 504 top inset Stephen Marshak; 504 (a) Stephen Marshak; 504 bottom inset Layne Kennedy/Corbis; 504 (b) Stephen Marshak; 505 Science VU-ASIS/Visual Unlimited; 506 (a) Richard P. Jacobs/ JLM Visuals; 506 (b) Richard P. Jacobs/JLM Visuals; 507 left inset Stephen Marshak; 507 right inset Stephen Marshak; 509 (a) Stephen Marshak; 509 (b) Science VU/Visuals Unlimited; 512 (a) R.W. Gerling/Visual Unlimited; 512 (c) Courtesy of John McBride, Illinois State Geological Survey and Thomes G.Hildenbrand, US Geological Survey; 512 (d) Stephen Marshak; 514 Rodrigo Arangua/AFP/Getty Images; 515 (a) Richard P. Jacobs/JLM Visuals; 515 (b) Connie Toops; 516 (a) Stephen Marshak; 516 (b) Stephen Marshak; 516 (c) Stephen Marshak; 516 (d) Stephen Marshak; 517 Bloomberg/Getty Images; 521 (a) Doug Sokell; 521 (b) A.J. Copley/Visual Unlimited; 523 Stephen Marshak.

Part opener 6
524, 525 Stephen Marshak.

Interlude F
526 JLP/NASA; 527 (a–d) Stephen Marshak; 529 (a) G.R. Roberts NSIL; 529 (b) Photo by Davie Pierson/ St. Petersburg Times; 531 Stephen Marshak; 532 (a–c) Stephen Marshak; 534

seismologists, 289
self-exciting dynamo, **346**
semiprecious stones, *126*
septic tanks, *665*, 667
serpentine, 517
Sevier fold-thrust belt, 451–52, *453*
Sevier orogeny, **452**, *461*
Shackleton, Earnest, 743
Shakespeare, William, 569
shale, 187, **188**, *188*, *191*, *203*, 231, *358*, *531*,
    S-22, S-23, S-31
    at Niagara Falls, 583, *584*
    black organic (oil and gas from), 470, *470*
    by lithification, 188
    and cement, 515, 516
    and cliff retreat, *728*
    and coal seam, *483*
    from river sediments, 204
    in geologic history illustration, *405*, 406
    in Great Plains, 351
    and Gros Ventre slide, *559*
    as impermeable, 654
    lake-bed, 205
    metamorphism of, 213, *214*, 223, *225*
    and Monte Toc landslide, 547, *547*
    organic material in, 191
    prograde metamorphism of, 224
    as prone to become failure surfaces, 554
    in rock cycle, 241
    and sandstone, *191*, *193*
    and slate, *231*, 351
    undeformed, 366
    unmetamorphosed, *231*
shallow earthquakes, 305, *305*, 306
shallow-marine clastic deposits, 205
shallow-water carbonate environment, 205
shark tooth, fossilized, 401, *401*
Shasta, Mt., *261*, S-31
shatter cones, 8, *9*
shear, *218*, 361
    between wind and water, *625*
shearing, *230*, *231*, 233–34, *233*, 366, *367*
    under metamorphic conditions, *218*
shear strain, 352, *353*
shear stress, 217, *217*, **354**, 355, 366
shear waves, **296**, *297*, 332, *332*
shear zone, **361**, *361*
Sheep Mountain, *364*, 555
sheet silicates, 122, *123*
sheetwash, **570**, *570*
    and desert pavements, 723
    pediments from, 730
shell, with growth rings, *422*
Shelley, Mary, 280
Shell Oil Company, *341*

shell-secreting organisms, 443
    and numerical age determinants, 421
shells (orbitals), **110**, *111*
shield, **235**, **377**, *438*, **439**
shield volcanoes, **259**, 261, *261*, 277, 761
    oceanic hot-spot volcanoes as, 267, *267*
    Olympus Mons as, 282
shingles, slate, *220*
Shiprock, N.Mex., *150*, 277, S-10
ships for research, *See* research vessels
Shishildan, Mt., S-13
shock metamorphism, 234–35, **235**
shortening, 352, *353*
short-term climate change, 795, 801–3
short-term predictions of earthquakes, 322,
    324–25
shows of ore, 511
Siberia, 270, *272*, A-1, S-41
    basalt in, 804
    in Cambrian Period, 443
    and creation of Pannotia, *440*
    folklore of, on earthquakes, 289
    mammoth found in, *390*
    in Paleozoic Era, 445, 446
    in Pangaea, 445
    in Pleistocene ice ages, 772
Siberian basalt eruption, *461*
Siberian Shield, *236*
Siberian Traps, 270
Siccar Point, Scotland, Hutton's observations
    at, 407, *408*
Sicily, Mt. Etna on, 268, *274*, *279*, S-12
Sierran arc, 450, **451–52**
Sierra Nevada, Calif., *458*, S-10, S-39
    roche moutonnée in, *757*
Sierra Nevada arc, *461*
Sierra Nevada Batholith, *151*
Sierra Nevada Mountains, 451, *732*, 755, S-18
    glaciers in, 772
    gold rush in, 131, 503, *503*
        and railroad building through, 131
    Yosemite in, 772
silber, *523*
silica, **42**, 144–45, *145*, 146
    in basic metamorphic rocks, 223
    in cement, 516
    and chert, 191
    in glass, 517
    and lava viscosity, 251
    and magma type, 152
    and viscosity, 146
silicate minerals, 42, **122**, 131, **135**
silicate rocks, **42**
silicic (felsic) magma, 164, S-10
silicic lava, 251
silicic melt, *145*

silicic rocks, *157*
silicon, *113*
    in crust, *48*, 131
    in quartz, 106
silicon-oxygen tetrahedron, **135**, *135*, 146
Silliman, Benjamin, 105
sillimanite, 105, 216, *216*, 218, *225*, *227*, *232*
sills, **149**, *150*, 154, *167*, *268*, *269*
    cooling of, *157*
    in geologic history illustration, *405*, 406
    and inclusions, *405*
silt, *171*, *176*, 179, 186
    and bedding, *196*
    and deltas, 205
    in desert, 722
    in floodplains, 188
    lithification of, 188
    in river, *206*
siltstone, **188**, *188*, *203*, 204, 205, S-22
    fossils in, 388
    worm burrows on, *390*
Silurian Period, 445, *460*, *448*
    and coal formation, 482
    life forms in, 414
silver, S-27
    of Andes, 511
    as native metal, 503
    as precious metal, 504, 505
Sinai Peninsula, *93*
single-chain silicates, 122, *123*
sinkholes, *317*, **647**, *647*, 672, 673, 674, *674*,
    S-35
    in Florida, 647–48, *647*
sky, blueness of, 683, *684*
slab-pull force, **95**, **98**
slag, 505
slash-and-burn agriculture, 806, *806*
slate, **219**, 224, 231, *231*, *232*
    in Alps, 351, 358
    and shale, 224, *225*
slaty cleavage, 219, *220*, 231, 366, *367*
sleet, 697
slickensides, 361
slip, 84
    in faults, 289–91, *290*, *291*, 295, 357–62
    flexural, *365*
    in Turkish earthquakes, 324, *325*
slip face, 198, *734*, **734**
slip lineations, 361
slope failure, **553**
slope failure, factors in causing of, 554–55,
    557, 563
slope stability, 553–54, *553*, 555, *556*, 557
slope steepness, and soil characteristics, 180, *180*
Sloss, Larry, 446–47
Slovenia, Kras Plateau in, 674, *674*